Igor N. Toptygin

Electromagnetic Phenomena in Matter

Related Titles

Zelevinsky, V.

Quantum Physics
Volume 1: From Basics to Symmetries and Perturbations

616 pages with 125 figures
2011
Softcover
ISBN: 978-3-527-40979-2

Zelevinsky, V.

Quantum Physics
Volume 2: From Time-Dependent Dynamics to Many-Body Physics and Quantum Chaos

580 pages with 85 figures
2011
Softcover
ISBN: 978-3-527-40984-6

Mandl, F., Shaw, G.

Quantum Field Theory

492 pages
2010
Softcover
ISBN: 978-0-471-49684-7

Paar, H.

An Introduction to Advanced Quantum Physics

224 pages
2010
Hardcover
ISBN: 978-0-470-68676-8

Geyi, W.

Foundations of Applied Electrodynamics

522 pages
2009
Hardcover
ISBN: 978-0-470-68862-5

Fayngold, M.

Special Relativity and How it Works

668 pages with 190 figures
2008
Softcover
ISBN: 978-3-527-40607-4

Van Bladel, J. G.

Electromagnetic Fields

1100 pages
2007
E-Book
ISBN: 978-0-470-12457-4

Sengupta, D. L., Liepa, V. V.

Applied Electromagnetics and Electromagnetic Compatibility

486 pages
2005
Hardcover
ISBN: 978-0-471-16549-1

Igor N. Toptygin

Electromagnetic Phenomena in Matter

Statistical and Quantum Approaches

WILEY-VCH
Verlag GmbH & Co. KGaA

The Author

Prof. Dr. Igor N. Toptygin
State Polytechnic University
Dept. of Theoretical Physics
St. Petersburg, Russia

Library of Congress Card No.: applied for

British Library Cataloguing-in-Publication Data
A catalogue record for this book is available from the British Library.

Bibliographic information published by the Deutsche Nationalbibliothek
The Deutsche Nationalbibliothek lists this publication in the Deutsche Nationalbibliografie; detailed bibliographic data are available on the Internet at <http://dnb.d-nb.de>.

Print ISBN: 978-3-527-41178-8
ePDF ISBN: 978-3-527-41316-4
ePub ISBN: 978-3-527-41318-8
Mobi ISBN: 978-3-527-41317-1

Cover Design Schulz Grafik-Design, Fußgönheim, Germany
Typesetting Laserwords Private Limited, Chennai, India
Printing and Binding Markono Print Media Pte Ltd., Singapore

Printed on acid-free paper

Contents

Preface *IX*
Basis Notations *XIII*
Fundamental Constants and Frequently Used Numbers *XVII*

1 Equations of Steady Electric and Magnetic Fields in Media *1*
1.1 Averaging Microscopic Maxwell Equations. Vectors of Electromagnetic Fields in Media *2*
1.2 Equations of Electrostatics and Magnetostatics in Medium *4*
1.3 Polarization of Media in a Constant Field *7*
Problems *12*
1.4 Answers and Solutions *17*

2 Electrostatics of Conductors and Dielectrics *37*
2.1 Basic Concepts and Methods of Electrostatics *37*
Problems *41*
2.2 Special Methods of Electrostatics *45*
Problems *54*
2.3 Energy, Forces, and Thermodynamic Relations for Conductors and Dielectrics *59*
Problems *71*
2.4 Answers and Solutions *76*

3 Stationary Currents and Magnetic Fields in Media *115*
3.1 Stationary Current *115*
Problems *123*
3.2 Magnetic Field in Magnetic Media *129*
Problems *131*
3.3 Energy, Forces, and Thermodynamic Relations for Magnetics *133*
Problems *145*
3.4 Electric and Magnetic Properties of Superconductors *149*
Problems *153*
Problems *155*

Problems *160*
3.5 Answers and Solutions *164*

4 Quasi-Stationary Electromagnetic Field *193*
4.1 Quasi-Stationary Phenomena in Linear Conductors *193*
Problems *197*
4.2 Eddy Currents and Skin-Effect *201*
Problems *205*
4.3 Magnetic Hydrodynamics *207*
Problems *222*
4.4 Answers and Solutions *228*

5 Maxwell Equations for Alternating and Inhomogeneous Fields *275*
5.1 Different Forms of Maxwell Equations in Media. Coupling Equations and Electromagnetic Response Functions *275*
Problems *287*
5.2 Causality Principle and Dispersion Relations *291*
Problems *296*
5.3 Energy Relations for Alternating Electromagnetic Field in Media. Longitudinal Electric Oscillations *297*
Problems *302*
5.4 Magnetic Oscillations and Magnetic Resonance *304*
Problems *306*
5.5 Electrodynamics of Moving Media *308*
Problems *311*
Problems *321*
5.6 Energy–Momentum Tensor in Dispersive Media *322*
Problems *327*
5.7 Answers and Solutions *327*

6 Propagation of Electromagnetic Waves *363*
6.1 Transverse Waves in Isotropic Media. Reflection and Refraction of Waves *363*
Problems *377*
6.2 Plane Waves in Anisotropic and Gyrotropic Media *382*
Problems *387*
6.3 Scattering of Electromagnetic Waves by Macroscopic Bodies. Diffraction *390*
Problem *393*
Problems *401*
6.4 Diffraction of X-Rays *405*
Problems *408*
6.5 Answers and Solutions *410*

7 Coherence and Nonlinear Waves *463*
7.1 Coherence and Interference *463*
Problems *472*
7.2 Random Waves and Waves in Randomly Inhomogeneous Media *477*
Problems *489*
7.3 Waves in Nonlinear and Active Media *490*
Problems *503*
7.4 Answers and Solutions *504*

8 Electromagnetic Oscillations in Finite Bodies *521*
8.1 Electromagnetic Waves in Waveguides *521*
Problems *524*
8.2 Electromagnetic Oscillations in Resonators *530*
Problems *531*
8.3 Answers and Solutions *536*

9 Interaction of Charged Particles with Equilibrium and Nonequilibrium Media *565*
9.1 Ionization and Radiation Energy Losses of Fast Particles in Media *565*
Problems *590*
9.2 Macroscopic Mechanisms of Radiation of Fast Particles in Media *591*
Problems *605*
9.3 Channeling and Radiation Emitted by Fast Particles in Crystals *609*
Problems *624*
9.4 Acceleration of Particles in Turbulent Plasma Media *624*
Problems *647*
9.5 Answers and Solutions *649*
Appendix: Turbulence and Its Description with the Aid of Correlation Tensors *681*

Bibliography *689*

Index *697*

Preface

This book presents the foundations of macroscopic electrodynamics, namely, the electromagnetic processes in matter and, particularly, the electric and magnetic properties of natural and artificial materials. Due to this feature, it differs essentially from the preceding manual of the same author: "Foundations of Classical and Quantum Electrodynamics," Wiley-VCH, 2014, which is devoted to electrodynamics in vacuum. Both books are closely connected by the common concept and style of presentation. The general peculiarities of the style of the first book are preserved: these are a combination of the brief textbook and detailed collection of problems; a wide range of examples that are offered to a reader for comprehensive study to stimulate the creative activity; a consideration of the material of various complexity for both the bachelor and master level preparations and also for the extension of knowledge and skills of research engineers and young scientific fellows.

In this edition, like in the first one, the more complicated problems are indicated by asterisks and the most difficult problems by double asterisks. In addition, the bullet is used to mark the problems concerning general and important concepts. It is recommended that these problems are solved first. The number of problems in each chapter significantly exceeds the necessary minimum for adoption of the presented subject, and the reader may solve only part of them and/or study the methods of solution considered in the book, depending on the scope of interests or professional level.

Macroscopic electrodynamics finds its numerous and versatile applications in statistical physics, solid state physics, astrophysics and geophysics, theory of electromagnetic wave propagation, theory of the interaction of charged particles with various media and in many other fields of science. As for applications of macroscopic electrodynamics in the areas of industry and technology, they are almost unlimited. Along with the presentation of general regularities, the author has attempted to give attention to the issues of application of electrodynamics in the neighboring fields to the extent permissible for text books of general purpose. All the examples and many problems are analyzed in great detail. This will allow the user to make use of the knowledge and skills obtained through the work on the book for solving similar problems.

Though the main laws of electromagnetism were formulated in its contemporary form even as early as the in the nineteenth century (the Maxwell equations), most of their applications in the interdisciplinary sciences, especially those concerning the structure of matter, were carried out sometime later, in the twentieth century. These applications were based not only on the laws of electromagnetism but also substantially on the two fundamental theories, developed early in the twentieth century, namely, theory of relativity and quantum mechanics, which proved their fruitfulness in the past decades. The author has tried to present the main results of these profound achievements, using his experience in scientific research and in lecturing the course of theoretical physics at St. Petersburg State Polytechnic University (SPbSPU, the former Leningrad Polytechnic Institute). The founders of theoretical studies at SPbSPU are well-known scientists: A.A. Friedman, V.R. Bursian, V.K. Frederiks, Ya.I. Frenkel and others. A significant role in the development of the incipient theoretical physics in St. Petersburg has been played by Paul Erenfest.

The specific character of the book is adequately reflected in its contents. The reader should keep in mind that the most fundamental laws are considered in Chapters 1 and 5, which by no means could be omitted in studying the book. The mathematical formalism of electrodynamics is briefly presented in Chapter 1 of the book "Foundations of Classical and Quantum Electrodynamics," which is numerously cited in the present edition (specifically in the context of mathematical basis). These references are indicated by a Roman "I," for example, (I.2.142) denotes the reference to Equation 142 from Chapter 2 of "Foundations..." Other chapters of the present book contain the most important, and in many cases, nontrivial applications of electrodynamics in the contiguous fields such as ferroelectrics (Chapter 2), ferromagnetics and superconductivity (Chapter 3), magnetic hydrodynamics including turbulent state (Chapter 4), nonlinear waves (Chapter 7), and others. They could be read at random, depending on the interest of the reader.

Initially, the proposed book has been published by "Regular and Chaotic Dynamics" Press in 2005 in Russian as a second volume of the two-volume course under the general name "Contemporary Electrodynamics." As compared with the Russian edition, this book contains more up-to-date material. Chapter 5 is supplemented by the section "Energy-momentum tensor in dispersive media." Chapter 6 offers the information on meta-materials, on cloaking of material bodies by electromagnetic waves, on waves in the gyrotropic media.

The English translation of this book became possible thanks to the grant of the Russian Ministry of Education and Science (project 11. G34.31.0001) and the project 5–100–2020 of Saint-Petersburg State Polytechnical University in Russia. I greatly appreciate the assistance of my colleagues from the A.F. Ioffe Physical Technical Institute and from Saint-Petersburg State Polytechnical University: A.M. Bykov, A.M. Krassilchtchikov, D.A. Varshalovich, D.G. Yakovlev, D.V. Kupriyanov, V.V. Dubov, A.I. Borovkov (vice-rector of SPbSPU), S.B. Makarov (director of Institute of Physics of SPbSPU). I am grateful to the translators of the book – Arseny and Elena Berezins, and also to D.G. Yakovlev, scientific editor of English translation of this book. Special thanks are due to Master A. Egorov for drawing a large number of figures. Undergraduate student K. Levina has checked up answers to some problems and also has helped with proof-reading.

Saint Petersburg
June 2014

Igor N. Toptygin

Basic Notations

A	work
$\boldsymbol{A}$, $\mathbf{A}$	vector potential.
a_B	Bohr radius.
a, b	constants of crystal lattice.
$\boldsymbol{b}$	vector of a reciprocal lattice.
$\widehat{a}_s^\dagger$, $\widehat{a}_s$	creation and annihilation operators for Dirac particles.
$\widehat{b}_s^\dagger$, $\widehat{b}_s$	creation and annihilation operators for Dirac antiparticles.
$\boldsymbol{B}$, $\boldsymbol{\mathcal{B}}$, $\boldsymbol{b}$	magnetic induction.
c	light velocity in vacuum (limiting velocity).
c_s	sound velocity.
C	electric capacity; thermal capacity.
c_{ik}	capacity coefficients.
$\widehat{c}_s^\dagger$, $\widehat{c}_s$	creation and annihilation operators for photons.
$D(p)$	diffusion coefficient in momentum space.
$\boldsymbol{D}$, $\boldsymbol{\mathcal{D}}$	electric induction.
$\boldsymbol{d}$, $\boldsymbol{p}$	electric dipole moment.
∂_k, $k = 0, 1, 2, 3$	covariant components of 4-gradient.
e, q, Q	charge of particle or macroscopic body.
$\boldsymbol{e}$	unit vector.
$\boldsymbol{E}$, $\mathbf{E}$	strength of the electric field.
E, W	energy.
$\mathcal{E}$	electromotive force (EMF); total energy of relativistic particle.
F	Helmholtz free energy per unit volume.
$\mathcal{F}$	Helmholtz free energy of all body.
$\boldsymbol{F}$, $\boldsymbol{\mathcal{F}}$	force.
$\boldsymbol{f}$	density of force.
F^{ik}	4-tensor of electromagnetic field.
$f(\boldsymbol{p}, \boldsymbol{r}, t)$, $F(\boldsymbol{p}, \boldsymbol{r}, t)$	distribution functions.
$\boldsymbol{g}$	density of field's momentum.
G	Green's function.
$H(p, q)$	classical Hamiltonian function.
$\widehat{H}$, $\widehat{\mathcal{H}}$	quantum operator Hamiltonian.

$\boldsymbol{H}$, $\mathbf{H}$	strength of the magnetic field.
$\hbar$	reduced Planck constant.
H^{ik}	4-tensor of electromagnetic induction.
p, P	pressure.
$\boldsymbol{p}$	momentum of particle.
$Q_{\alpha\beta}$	electric quadrupole moment's tensor.
$\boldsymbol{i}$	density of surface current.
I	intensity of emission.
$\boldsymbol{j}$	current density (volumetric).
J	total current passing through the surface.
K, T	kinetic energy.
$K_{\lambda\mu}$	tensor of electromagnetic fluctuations in equilibrium system.
$\boldsymbol{k}$	wave vector.
L	inductance; length (characteristic scale); Langevin function.
L_{ik}	inductance coefficients.
l	length (characteristic scale).
M, m	mass.
$\boldsymbol{m}$	magnetic moment.
$\boldsymbol{M}$	vectors of magnetic polarization.
$\boldsymbol{\mathcal{M}}$	magnetic moment of a macroscopic body.
n, N	density of particle's number.
$\boldsymbol{n}$	unit normal vector.
N	total number of particles.
$N(p, \boldsymbol{r}, t)$	isotropic part of distribution function.
P, p	pressure.
$\boldsymbol{P}$	vectors of electric polarization.
$\boldsymbol{\mathcal{P}}$	electric moment of a macroscopic body.
q, Q	charge of a single particle or macroscopic body.
Q	heat.
$\boldsymbol{q}$	density of a heat current; momentum change of scattered particle.
R	electric resistance; Reynolds number; reflection coefficient.
$\boldsymbol{r}, \boldsymbol{R}$	radius-vector.
S	surface; entropy (per unit volume).
s	entropy (per unit mass).
$\mathcal{S}$	entropy of a macroscopic body.
$\hat{S}$	scattering matrix.
t	time.
T	temperature in energetic units of the Kelvin scale; kinetic energy.
$\hat{T}$	chronological operator.
Tr	symbol of trace (sum of diagonal elements).
T_{ik}	energy-momentum tensor of the electromagnetic field.
U	potential energy; internal energy of a body per unit volume.
$\mathcal{U}$	internal energy of a macroscopic body.
v, V, u	velocity.

V	volume; interaction energy.
$\hat{V}$	quantum interaction operator.
$\mathcal{V}$	volume.
w	energy density (volumetric).
$dw_{i\to f}$	probability of transition for unit time.
W	energy.
Z	charge number of atomic nucleus; impedance (complex resistance).
$\mathbf{Z}$	Hertz vector.
α	angle; dielectric susceptibility.
$\hat{\boldsymbol{\alpha}}$	Dirac's matrixes.
$\beta = v/c$	ratio of particle speed to light speed.
$\hat{\beta}$	Dirac's matrix.
β_{ik}	polarizability tensor (electric or magnetic).
γ	relativistic (Lorentz) factor, damping constant, spectral line width, the complex degree of coherence, index of power-law spectrum.
$\boldsymbol{\gamma}$	Poynting vector, generalized Poynting vector (in medium)
$\hat{\gamma}^k$, $k = 0, 1, 2, 3, 5$	Dirac's matrixes.
Γ	mutual coherence function; spectral line width.
δ	δ-function, symbol of increment, thickness of a shock front, thickness of a skin-layer.
Δ	symbol of increment; Laplace operator.
ϵ	internal energy per unit mass.
ϵ_F	Fermi energy.
ϵ, ε	dielectric permittivity.
ζ	chemical potential per unit mass, surface impedance.
η	gyromagnetic ratio, dynamical viscosity.
ϑ, θ	angle.
Θ	step function.
κ	electric conductivity, diffusion coefficient.
λ	wave length, constant of molecular Weiss field, depth of penetration of a field in superconductor.
$\Lambda_C = \hbar/m_e c$	Compton's length of wave for electron.
Λ	transport (free) path of the particle.
μ	magnetic permeability; chemical potential per one particle.
ν	kinematic viscosity; magnetic viscosity; relaxation frequency.
$d\nu$	number of quantum states in continuous spectrum.
$\Pi_\pm$	projection operators.
ρ	charge density (volumetric).
$\hat{\rho}$	quantum density operator.
σ	density of surface charge; efficient cross-section.

$\sigma_{\alpha\beta}$	Maxwell tension tensor of electromagnetic field.
$\widehat{\sigma}_\alpha$ ($\alpha = 1, 2, 3$)	Pauli matrixes.
Σ	efficient cross-section; surface.
τ	relaxation time; correlation time; mass density (per unit volume).
ϕ	angle.
φ	scalar potential of the electromagnetic field; angle.
Φ	magnetic stream.
ψ, Ψ	wave function; pseudoscalar potential of magnetic field.
ω, Ω	frequency.

Fundamental Constants and Frequently Used Numbers

Ratio of the circumference of a circle to its diameter $\pi = 3.142$.
Base of natural logarithms $e = 2.718282$.
Speed of light in vacuum (maximum speed) $c = 299\,792\,458$ m/s $\approx 3 \times 10^{10}$ cm/s.
Reduced Planck constant $\hbar = 1.055 \times 10^{-27}$ erg·s.
Electron rest mass $m_e = 0.9110 \times 10^{-27}$ g.
Proton rest mass $m_p = 1.673 \times 10^{-24}$ g.
Elementary charge $e_0 = 4.803 \times 10^{-10}$ CGS.
Fine structure constant $\alpha = 0.007297$; $\alpha^{-1} = 137.036$.
Bohr radius $a_B = 0.5292 \times 10^{-8}$ cm.
Bohr magneton $\mu_B = \frac{e_0 \hbar}{2m_e c} = 0.9274 \times 10^{-20}$ erg/G.
Nuclear magneton $\mu_N = \frac{e_0 \hbar}{2m_p c} = 0.5051 \times 10^{-23}$ erg/G.
Neutron magnetic moment $\mu_n = 1.913\mu_N$.
Proton magnetic moment $\mu_p = 2.793\mu_N$.
Classical electron radius $r_0 = \frac{e_0^2}{m_e c^2} = 2.818 \times 10^{-13}$ cm.
Compton wavelength of an electron $\Lambda_C = \frac{\hbar}{m_e c} = 3.862 \times 10^{-11}$ cm.
Compton wavelength of a proton $\Lambda_C = \frac{\hbar}{m_p c} = 2.103 \times 10^{-14}$ cm.
1 erg = 10^{-7} J = 6.24×10^{11} eV.
1 eV = 1.602×10^{-12} erg.
1 dyn = 10^{-5} N.
1 year = 3.156×10^{7} s.
1 pc (parsec, unit of length used in astronomy) = 3.086×10^{18} cm.
1 AU (astronomical unit – the mean distance between the Earth and the Sun).
= 1.496×10^{13} cm.
Mass of the Earth $M_{\oplus} = 5.976 \times 10^{27}$ g.
Mass of the Sun $M_{\odot} = 1.989 \times 10^{33}$ g.

1
Equations of Steady Electric and Magnetic Fields in Media

Maxwell equations (for instance, Equations (I.2.82)–(I.2.85)[1]) also hold in the presence of matter, viz., dielectrics, conductors, magnetized media, and so on. However, the matter while as a whole electrically neutral in most cases, consists of the majority of charged particles, electrons, and atomic nuclei. The resulting electromagnetic field in matter is due to both **external charges and currents**, which do not belong to the matter, and particles of the matter itself. The field produced by external charges causes redistribution of charges and currents and leads to the occurrence of an additional field. For this reason, in matter we are generally dealing with a self-consistent electromagnetic field due to both external and intrinsic charges.

It is a priori clear that the presence of a great variety of natural and artificial materials differing in magnetic and electric properties implies many specific approaches to their description. Currently, there is no unified general method for studying electromagnetic phenomena in the macroscopic electrodynamics as in the microscopic vacuum theory. Therefore, along with consistent microscopic approaches considering the specific atomic structure of matter, one has to use phenomenological laws that generalize the data obtained in macroscopic experiments.

In this book, we will first consider the most general laws that hold for any matter, and thereafter turn our attention to more specific (though relatively simple) models of media exposed to polarization and magnetization, such as plasma, ferromagnetics, conductors, superconductors, dielectrics and modern artificial media, that is, metamaterials. Whenever necessary, we will use quantum mechanics, thermodynamics, statistical physics, and physical kinetics, which are the most significant for the consistent analysis of electromagnetic phenomena in media.

1) Recall that labels (of equations, figures, chapters, examples, problems, appendices, and sections) which start with "I" refer to the monograph by Toptygin (2014). For instance, Equation (I.2.82) means Equation (2.82) from Toptygin (2014).

Electromagnetic Phenomena in Matter: Statistical and Quantum Approaches, First Edition. Igor N. Toptygin.

1.1
Averaging Microscopic Maxwell Equations. Vectors of Electromagnetic Fields in Media

In this section, the microscopic (exact) values of the electric and magnetic field strengths are denoted by calligraphic capital letters $\boldsymbol{\mathcal{E}}$ and $\boldsymbol{\mathcal{H}}$, respectively. The Maxwell equations (I.2.82)–(I.2.85) for the microscopic fields can be written as

$$rot\, \boldsymbol{\mathcal{E}}(\boldsymbol{r}, t) = -\frac{1}{c}\frac{\partial \boldsymbol{\mathcal{H}}(\boldsymbol{r}, t)}{\partial t}, \tag{1.1}$$

$$rot\, \boldsymbol{\mathcal{H}}(\boldsymbol{r}, t) = \frac{1}{c}\frac{\partial \boldsymbol{\mathcal{E}}(\boldsymbol{r}, t)}{\partial t} + \frac{4\pi}{c}(\boldsymbol{j}_{int}(\boldsymbol{r}, t) + \boldsymbol{j}_{ext}(\boldsymbol{r}, t)), \tag{1.2}$$

$$div\, \boldsymbol{\mathcal{E}}(\boldsymbol{r}, t) = 4\pi(\rho_{int}(\boldsymbol{r}, t) + \rho_{ext}(\boldsymbol{r}, t)), \tag{1.3}$$

$$div\, \boldsymbol{\mathcal{H}}(\boldsymbol{r}, t) = 0. \tag{1.4}$$

The charge and current densities in the right-hand sides of these equations consist of two parts associated with the particles of matter and external sources; they are labeled by subscripts "int" and "ext" respectively. We will assume that the external sources are given. Particles of the matter are affected by the fields $\boldsymbol{\mathcal{E}}$ and $\boldsymbol{\mathcal{H}}$, and the quantities ρ_{int} and $\boldsymbol{j}_{int}$ are generally complicated functionals of these fields. In order to determine the motion of particles in the medium and calculate the charge and current densities, one has to use the equations of classical or (in most cases) quantum mechanics. Simultaneous solutions of the Maxwell equations and the equations of the particle motion in medium ensure, in principle, the determination of microscopic values of the fields $\boldsymbol{\mathcal{E}}$ and $\boldsymbol{\mathcal{H}}$.

However, such a detailed description of the field is impossible due to the presence of a large amount of particles in the medium, and it is actually not needed in most cases. The quantities that are measured in macroscopic experiments are fields averaged over a statistical ensemble of medium states (i.e., over regular and random motions of the particles). In accordance with the general principles of statistical physics (Landau and Lifshitz (1980)), such averaging is equivalent to the averaging over a certain time interval Δt. Moreover, when measuring the fields by means of macroscopic devices, an additional averaging is performed over macroscopically small volumes ΔV containing a large number of elementary charges (this issue has already been outlined in the beginning of Section I.2.1). The necessity of such an averaging stems from the fact that a microscopic field in medium undergoes very large and irregular changes in space and time, for example, on length-scales of the order of 3×10^{-8} cm in a condensed medium (Problems I.2.15 and I.2.16).

A quantity averaged over space and time will be denoted by a bar and defined by

$$\bar{f}(\boldsymbol{r}, t) = \frac{1}{\Delta V \Delta t}\int_{\Delta V} dV \int_{-\Delta t/2}^{\Delta t/2} d\tau f(\boldsymbol{r} + \boldsymbol{\rho}, t + \tau), \tag{1.5}$$

where the integration over coordinates is performed within the volume ΔV. The macroscopic field defined in this way remains a function of coordinates and time. Differentiation of both sides of Equation (1.5) with respect to any coordinate or time yields the following relation:

$$\frac{\partial \bar{f}}{\partial t} = \overline{\frac{\partial f}{\partial t}}, \tag{1.6}$$

that is, a derivative of an average value is equal to an average value of a derivative.

Since the Maxwell equations (1.1)–(1.4) contain derivatives with respect to coordinates and time, from Equation (1.6) we have

$$\overline{rot\,\mathcal{E}} = rot\,\overline{\mathcal{E}}, \qquad \overline{\frac{\partial \mathcal{H}}{\partial t}} = \frac{\partial \overline{\mathcal{H}}}{\partial t},$$

and so on. Let us introduce the following notations for the strengths of macroscopic fields:

$$\overline{\mathcal{E}}(\boldsymbol{r}, t) = \boldsymbol{E}(\boldsymbol{r}, t), \qquad \overline{\mathcal{H}}(\boldsymbol{r}, t) = \boldsymbol{B}(\boldsymbol{r}, t). \tag{1.7}$$

The first quantity is referred to as an **electric field vector** and the second as a **magnetic induction vector**. In these notations, the Maxwell equations take the form

$$rot\,\boldsymbol{E}(\boldsymbol{r}, t) = -\frac{1}{c}\frac{\partial \boldsymbol{B}(\boldsymbol{r}, t)}{\partial t}, \tag{1.8}$$

$$rot\,\boldsymbol{B}(\boldsymbol{r}, t) = \frac{1}{c}\frac{\partial \boldsymbol{E}(\boldsymbol{r}, t)}{\partial t} + \frac{4\pi}{c}(\bar{\boldsymbol{j}}_{int}(\boldsymbol{r}, t) + \boldsymbol{j}_{ext}(\boldsymbol{r}, t)), \tag{1.9}$$

$$div\,\boldsymbol{E}(\boldsymbol{r}, t) = 4\pi(\overline{\rho}_{int}(\boldsymbol{r}, t) + \rho_{ext}(\boldsymbol{r}, t)), \tag{1.10}$$

$$div\,\boldsymbol{B}(\boldsymbol{r}, t) = 0. \tag{1.11}$$

Here, the external charges and currents must be macroscopic quantities. The macroscopic vectors $\boldsymbol{E}$ and $\boldsymbol{B}$ are the analogs of microscopic field strengths $\mathcal{E}$ and $\mathcal{H}$ (even though the name of the vector $\boldsymbol{B}$ was changed for historical reasons). They are just these quantities that are used to express the force acting on a small macroscopic body with a charge q moving in a medium with velocity $\boldsymbol{u}$:

$$\boldsymbol{F} = q\left(\boldsymbol{E} + \frac{1}{c}\boldsymbol{u} \times \boldsymbol{B}\right). \tag{1.12}$$

The system of Equations (1.8)–(1.11) is incomplete since the quantities $\bar{\boldsymbol{j}}_{int}$ and $\overline{\rho}_{int}$ are not known in advance. They should be expressed in terms of macroscopic vectors $\boldsymbol{E}$ and $\boldsymbol{B}$. The most consistent approach to this problem is based on the use of the distribution functions (in the classical case) or the density matrices (in the quantum case) for describing the particle motion in matter. This requires invoking corresponding kinetic equations and rather detailed information concerning microscopic parameters, which characterize the state of particles in matter. Such an approach can be consistently implemented only for the simplest models of medium. In most cases, one has to use various phenomenological models and experimental data.

1.2 Equations of Electrostatics and Magnetostatics in Medium

Vectors of electric and magnetic polarization. In a static case, the electric and magnetic fields may exist separately: from the system of Equations (1.8)–(1.11) at $\partial \boldsymbol{B}/\partial t = \partial \boldsymbol{E}/\partial t = 0$ we obtain

$$rot\, \boldsymbol{E} = 0, \qquad div\, \boldsymbol{E} = 4\pi(\overline{\rho}_{int} + \rho_{ext}), \tag{1.13}$$

$$rot\, \boldsymbol{B} = \frac{4\pi}{c}(\bar{\boldsymbol{j}}_{int} + \boldsymbol{j}_{ext}), \qquad div\, \boldsymbol{B} = 0. \tag{1.14}$$

The macroscopic densities $\overline{\rho}_{int}$ and $\bar{\boldsymbol{j}}_{int}$ can be conveniently expressed in terms of the vectors of the electric $\boldsymbol{P}(\boldsymbol{r})$ and magnetic $\boldsymbol{M}(\boldsymbol{r})$ polarization of matter, which are by definition the electric and magnetic dipole moments per unit volume:

$$\boldsymbol{P} = \frac{\sum_i \boldsymbol{p}_i}{\Delta V}, \qquad \boldsymbol{M} = \frac{\sum_i \boldsymbol{m}_i}{\Delta V}. \tag{1.15}$$

Here $\boldsymbol{p}_i$ and $\boldsymbol{m}_i$ are, respectively, the electric and magnetic moments of individual structural units of the medium (atoms or molecules); summation is over all particles in a macroscopically small volume ΔV.

Example 1.1

Show that the density of the induced volume charge inside medium is related to the electric polarization vector by the expression

$$\overline{\rho}_{int}(\boldsymbol{r}) = -div\, \boldsymbol{P}(\boldsymbol{r}). \tag{1.16}$$

What significance does this relation acquire on the boundary of a body?

Solution. Consider an electrically neutral body placed in vacuum in the presence of external electric charges ρ_{ext}. The total electric dipole moment of the body $\mathcal{P}$ can be written as the integral of the electric polarization vector $\mathcal{P} = \int \boldsymbol{P}(\boldsymbol{r})\, dV$ over the body's volume. On the other hand, the dipole moment can be written in terms of the charge macroscopic density $\mathcal{P} = \int \boldsymbol{r}\overline{\rho}_{int}(\boldsymbol{r})\, dV$. The latter integral is independent of the choice of the coordinate origin provided the condition for electric neutrality $\int \overline{\rho}_{int}(\boldsymbol{r})\, dV = 0$ is satisfied. We equate these two expressions for $\mathcal{P}$ and multiply them by a constant vector $\boldsymbol{a}$:

(1) $$\int (\boldsymbol{a}\cdot\boldsymbol{P})\, dV = \int \overline{\rho}_{int}(\boldsymbol{a}\cdot\boldsymbol{r})\, dV.$$

We use the identities $\boldsymbol{a}\cdot\boldsymbol{P} = (\boldsymbol{P}\cdot\nabla)(\boldsymbol{a}\cdot\boldsymbol{r}) = \nabla[\boldsymbol{P}(\boldsymbol{a}\cdot\boldsymbol{r})] - (\boldsymbol{a}\cdot\boldsymbol{r})(\nabla\cdot\boldsymbol{P})$ and apply the Gauss–Ostrogradskii theorem to the integral in Equation (1):

(2) $$\int \overline{\rho}_{int}(\boldsymbol{a}\cdot\boldsymbol{r})\, dV = \oint_S [\boldsymbol{P}(\boldsymbol{a}\cdot\boldsymbol{r})]\cdot d\boldsymbol{S} - \int (\boldsymbol{a}\cdot\boldsymbol{r})(\nabla\cdot\boldsymbol{P})\, dV.$$

The surface that encloses the integration volume in Equation (2) can be chosen outside the body, where $\boldsymbol{P} = 0$. Omitting the vector $\boldsymbol{a}$ in the remaining equation, we have

(3) $$\int r\overline{\rho}_{int}(r)\,dV = -\int r\,div\,\boldsymbol{P}\,dV.$$

It follows from this equation that the density of induced charges may be identified with the divergence of the electric polarization vector according to Equation (1.16). On the boundary of the body, the vector $\boldsymbol{P}$ jumps to zero. In this case, we have to perform the limiting transition in Equation (1.16), by analogy with that which led to Equation (I.2.18), and take into account that $\boldsymbol{P} = 0$ outside the body. As a result, we find the density of the surface macroscopic charges induced on the surface of the polarized body:

$$\sigma_{int} = P_n. \quad \blacksquare \tag{1.17}$$

The obtained charge densities (1.16) and (1.17) are restricted to dielectrics, that is, to media whose internal charges can be displaced only by microscopic distances. For this reason, these quantities are also referred to as densities of **bound charges**. Inside conductors, the charges move freely. Therefore, in these cases $\overline{\rho}_{int} = 0$ and $\boldsymbol{P} = 0$. However, a surface charge, which is expressed in terms of the external field, may be present (Chapter 2).

Example 1.2

Show that in the absence of charges, which can freely propagate through the body ***(free charges)****, the volume current density induced inside medium* ***(magnetization current)*** *is related to the magnetic polarization vector by the expression*

$$\overline{\boldsymbol{j}}_{int}(\boldsymbol{r}) = c\,rot\,\boldsymbol{M}(\boldsymbol{r}). \tag{1.18}$$

Write down the limiting form of this relation on the boundary of the body.

Solution. We use the same approach as in Example 7.1, that is, equate two integrals for the total magnetic moment of the body, $\mathcal{M}$, and multiply them by a constant vector $\boldsymbol{a}$. This gives

(1) $$\boldsymbol{a}\cdot\mathcal{M} = \frac{1}{2c}\int \boldsymbol{a}\cdot[\boldsymbol{r}\times\overline{\boldsymbol{j}}_{int}]\,dV = \int \boldsymbol{a}\cdot\boldsymbol{M}\,dV.$$

Then we use the identity

(2) $$2\boldsymbol{a}\cdot\boldsymbol{M} = \boldsymbol{M}\cdot rot[\boldsymbol{a}\times\boldsymbol{r}] = \boldsymbol{a}\cdot[\boldsymbol{r}\times\ rot\,\boldsymbol{M}] - \nabla\cdot[\boldsymbol{M}\times(\boldsymbol{a}\times\boldsymbol{r})]$$

and reduce Equation (1) to the form

(3) $$\frac{1}{2c}\int \boldsymbol{a}\cdot[\boldsymbol{r}\times\overline{\boldsymbol{j}}_{int}]\,dV = \frac{1}{2}\int \boldsymbol{a}\cdot[\boldsymbol{r}\times rot\,\boldsymbol{M}]\,dV,$$

from which Equation (7.18) follows. The limiting form of this formula on the body boundary, where $\boldsymbol{M}$ undergoes a discontinuity, can be obtained by analogy with Equation (I.2.58). A jump of the magnetization vector determines the surface current density according to the relation

$$\boldsymbol{i}_{int} = c\,\boldsymbol{n}\times\boldsymbol{M}, \tag{1.19}$$

where $\boldsymbol{n}$ is the unit vector of the normal to the surface. $\blacksquare$

With the aid of Equations (1.16) and (1.18), Equations (1.13) and (1.14) for static fields take the form

$$rot\,\boldsymbol{E} = 0, \qquad div\,\boldsymbol{D} = 4\pi\rho_{ext}, \tag{1.20}$$

$$rot\,\boldsymbol{H} = \frac{4\pi}{c}\boldsymbol{j}_{ext}, \qquad div\,\boldsymbol{B} = 0. \tag{1.21}$$

Here, two new field vectors are introduced: the **electric induction vector**

$$\boldsymbol{D} = \boldsymbol{E} + 4\pi\boldsymbol{P}, \tag{1.22}$$

and the **magnetic field strength vector**

$$\boldsymbol{H} = \boldsymbol{B} - 4\pi\boldsymbol{M}. \tag{1.23}$$

Coupling equations. The sets of Equations (1.20) and (1.21) are not closed until the coupling between the vectors $\boldsymbol{D}$ and $\boldsymbol{E}$, and also between $\boldsymbol{H}$ and $\boldsymbol{B}$, is specified. It is possible to find the coupling equations for a broad class of media based on general physical considerations and experimental data of general character, without invoking accurate data on the internal structure of the medium. This holds true for media where in the absence of external fields the electric and magnetic polarizations are also absent. If external fields are weak compared with the interatomic fields, the electric polarization vector $\boldsymbol{P}$ and the magnetization vector $\boldsymbol{M}$ are linear functions of the components of the corresponding external fields. In this case, for isotropic bodies we have

$$\boldsymbol{P} = \alpha\boldsymbol{E}, \qquad \boldsymbol{M} = \chi\boldsymbol{H}, \tag{1.24}$$

where the coefficients α and χ are independent of external fields; they are referred to as **dielectric and magnetic susceptibilities**, respectively. Certainly, these quantities are not the same for different media, and depend on density and, in the general case, on temperature. In a constant electric field, we always have $\alpha > 0$, whereas the magnetic susceptibility may be either positive or negative. Substances with $\chi > 0$ are called **paramagnetics** and those with $\chi < 0$ **diamagnetics**.

If a conducting body, that is, a body containing free charges, presents a closed contour, then with the aid of a suitable source of energy (external electromotive force) a nonzero electric field and current can be maintained inside it. In the case of sufficiently weak fields, the relationship between the current density $\boldsymbol{j} \equiv \bar{\boldsymbol{j}}_{int}$ and the electric field $\boldsymbol{E}$ is given by the **Ohm law**

$$\boldsymbol{j} = \kappa\boldsymbol{E}, \tag{1.25}$$

where the **electric conductivity** κ is a macroscopic characteristic of medium and is independent of $\boldsymbol{E}$.

Using Equations (1.22)–(1.24) we find the coupling equations

$$\boldsymbol{D} = \varepsilon\boldsymbol{E}, \qquad \boldsymbol{B} = \mu\boldsymbol{H}, \tag{1.26}$$

where the coefficients of proportionality

$$\varepsilon = 1 + 4\pi\alpha, \quad \text{and} \quad \mu = 1 + 4\pi\chi \tag{1.27}$$

are called dielectric **permittivity** and magnetic **permeability**, respectively. In anisotropic media the susceptibility, permittivity, permeability and electric conductivity are second-rank tensors representing the apparent generalization of Equation (1.27):

$$\varepsilon_{\sigma\nu} = \delta_{\sigma\nu} + 4\pi\alpha_{\sigma\nu}, \qquad \mu_{\sigma\nu} = \delta_{\sigma\nu} + 4\pi\chi_{\sigma\nu}. \tag{1.28}$$

In this case, the coupling Equation (1.26) takes the form

$$D_\sigma = \varepsilon_{\sigma\nu} E_\nu, \qquad B_\sigma = \mu_{\sigma\nu} H_\nu. \tag{1.29}$$

In very strong external fields, linear relations (1.24)–(1.26) do not hold, and the coupling equations become nonlinear. Moreover, there exist the media that exhibit spontaneous polarization at certain temperatures (ferroelectrics, ferromagnetics). In all these cases, the coupling equations are very complicated.

1.3 Polarization of Media in a Constant Field

Electric polarization. The electric permittivity or magnetic permeability of a certain medium, as a rule, can be calculated based on some simplified models. For rarefied molecular gases, the permeabilities are connected with polarizabilities of individual molecules by simple relations. In the general case, the polarization of a certain system of particles is described by a second-rank tensor, which relates the vector of the induced dipole moment $\boldsymbol{p}$ to the vector of the field $\boldsymbol{E}$ acting on the system:

$$p_\kappa = \beta_{\kappa\nu} E_\nu, \tag{1.30}$$

where $\beta_{\kappa\nu}$ is the polarizability tensor. Polarizabilities of atoms and molecules are calculated by quantum mechanical methods (Problems I.6.62, 1.5, 1.17, 1.19). If the molecules possess electric dipole moments in the absence of an external field, then in order to calculate the polarization vector in a given field, one should use the Boltzmann distribution

$$dN(q_i) = C \exp\left[-\frac{U(q_i)}{T}\right] d\Gamma. \tag{1.31}$$

Here dN is the number of particles in an elementary volume $d\Gamma$ in the space of generalized coordinates, $U(q_i)$ is the potential energy of one particle in an external field, q_i is a set of generalized coordinates characterizing position and orientation of the particle, T is the temperature in energy units, and C is the normalization constant. The distribution of particles over momenta in the equilibrium state is the Maxwell distribution:

$$dN = f_0(p) d^3p = \frac{n}{(2\pi m T)^{3/2}} \exp\left(-\frac{p^2}{2mT}\right) d^3p, \tag{1.32}$$

where n is the number density of the particles with all energies, which may depend on coordinates in a nonhomogeneous system, and m is a molecular mass. The Maxwell distribution assumes that particles move according to the classical laws.

Example 1.3

*A rarefied, statistically equilibrated gas consists of identical dipole molecules of number density N, having an individual dipole moment p. Calculate the dependence of the polarization vector **P** on an applied electric field **E**, neglecting the inter-molecule interaction and deformation of molecular electron shells. Determine also the dielectric permittivity and find the applicability criterion of the linear dependence given by Equation (1.24).*

Solution. We calculate the projection P of the polarization vector on the direction of an external field with the aid of the Boltzmann distribution (1.31):

$$P = Np\frac{\int_0^\pi \cos\vartheta \exp\left[(pE/T)\cos\vartheta\right]\sin\vartheta\, d\vartheta}{\int_0^\pi \exp\left[(pE/T)\cos\vartheta\right]\sin\vartheta\, d\vartheta} = \frac{\partial}{\partial a}\ln\int_{-1}^{1} e^{ax}\,dx, \tag{1}$$

where $U = -pE\cos\vartheta$ is the interaction energy between the dipole and the external field, and $a = pE/T$, $x = \cos\vartheta$. Calculations yield

$$P = Np\,\frac{d}{da}\ln\left(\frac{2}{a}\sinh a\right) = NpL(pE/T), \tag{2}$$

where $L(a)$ is called the **Langevin**[2] function:

$$L(a) = \coth a - \frac{1}{a}. \tag{1.33}$$

The Langevin function is close to unity when $a \gg 1$, while for small a it can be expanded into a series

$$L(a) = \frac{a}{3} - \frac{a^3}{45} + \cdots, \qquad a \ll 1. \tag{3}$$

For this reason, at $pE \gg T$ the saturation takes place, and all dipoles are oriented along the field. When $pE \ll T$, the polarization depends linearly on the field with the proportionality coefficient $\alpha = Np^2/3T$. This leads to the dielectric permittivity

$$\varepsilon = 1 + \frac{4\pi Np^2}{3T}. \tag{1.34}$$

The dipole moments of simple molecules are of the order of the product of the elementary charge and a linear molecular size (the Bohr radius), that is, $p \approx 10^{-18}$ CGS units. The transition from the linear dependence of polarization on the field strength and the saturation range takes place at the field value $E_c \approx T/p$, or $E_c \approx 10^4$ CGSE $\approx 3 \times 10^6$ V/cm for $T \approx 300$ K ≈ 0.03 eV. For a gas under normal conditions ($N \approx 3 \times 10^{19}$ cm^{-3}), one has $\varepsilon - 1 \approx 4 \times 10^{-3}$. ■

Example 1.4

The molecules of a dielectric are spherically symmetric and have no dipole moments in the absence of the field. The number density N and polarizability β of

2) Paul Langevin (1872–1946), outstanding French physicist, founder of numerous physical schools (L. de Broglie, M. de Broglie, F. Joliot-Curie, F. Perren, etc.)

the molecules are known. Find the dependence of the dielectric permittivity ε on N and polarizability. Take into account a possible difference between the electric field $\boldsymbol{\mathcal{E}}$ acting on a molecule and the average (macroscopic) field **E** *due to other molecules.*

Solution. If the field acting on the molecule is equal to the average field, that is, $\boldsymbol{\mathcal{E}} = \boldsymbol{E}$, the induced dipole moment of a single molecule is $\boldsymbol{p} = \beta \boldsymbol{E}$, and the polarization vector is $\boldsymbol{P} = N\boldsymbol{p} = N\beta\boldsymbol{E} = \alpha\boldsymbol{E}$,

$$\varepsilon = 1 + 4\pi\alpha = 1 + 4\pi N\beta. \tag{1.35}$$

In order to include the effect of surrounding molecules, we enclose a given molecule by a sphere of radius $a \gg N^{-1/3}$ (Figure 1.1) and present the acting field as a sum of two fields $\boldsymbol{\mathcal{E}} = \boldsymbol{\mathcal{E}}_1 + \boldsymbol{\mathcal{E}}_2$, where $\boldsymbol{\mathcal{E}}_1$ is the field which is produced by the external charges and all molecules localized outside the sphere of radius a.

The field $\boldsymbol{\mathcal{E}}_1$ is macroscopic and can be calculated as the field in the center of the sphere carved in a uniformly polarized dielectric. In accordance with the superposition principle, $\boldsymbol{\mathcal{E}}_1 = \boldsymbol{E} - \boldsymbol{E}_i$, where $\boldsymbol{E}_i = -4\pi\boldsymbol{P}/3$ is the field inside the uniformly polarized sphere (Problem 1.3). As a result, $\boldsymbol{\mathcal{E}}_1 = \boldsymbol{E} + 4\pi\boldsymbol{P}/3$.

The field $\boldsymbol{\mathcal{E}}_2$ can be easily calculated in two limiting cases: (a) molecules inside the sphere are distributed quite randomly; the field of a single dipole, averaged over the sphere volume, is zero, that is, $\boldsymbol{\mathcal{E}}_2 = 0$; (b) molecules are localized at the sites of a cubic lattice, and, hence, $\boldsymbol{\mathcal{E}}_2 = 0$ from symmetry considerations (see the field of a single dipole in Example I.2.5). For these cases

$$\boldsymbol{\mathcal{E}} = \boldsymbol{E} + \frac{4\pi}{3}\boldsymbol{P}, \tag{1.36}$$

and $\boldsymbol{P} = N\beta\boldsymbol{\mathcal{E}} = N\beta(\boldsymbol{E} + 4\pi\boldsymbol{P}/3)$. Hence

$$\boldsymbol{P} = \frac{N\beta\boldsymbol{E}}{1 - 4\pi N\beta/3} = \alpha\boldsymbol{E}, \qquad \varepsilon = 1 + 4\pi\alpha = 1 + \frac{N\beta}{1 - 4\pi N\beta/3}$$

or

$$\frac{\varepsilon - 1}{\varepsilon + 2} = \frac{4\pi}{3}N\beta. \tag{1.37}$$

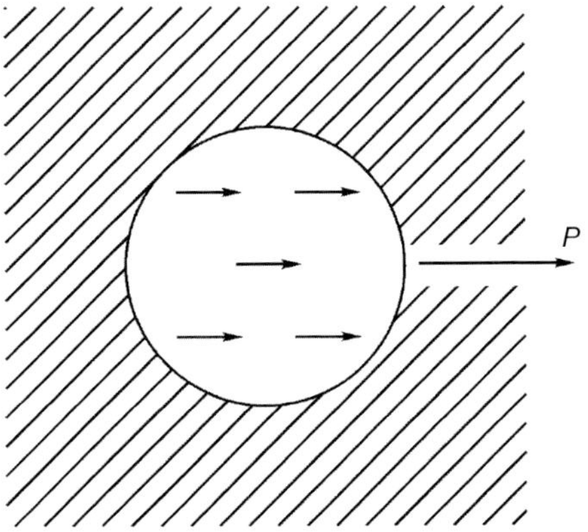

Figure 1.1 To calculation of the field acting on an individual molecule.

Equations (1.36) and (1.37) are called the **Clausius–Mossotti formulas.**[3] For the optical range of spectrum they are expressed in terms of the refraction coefficient $n = \sqrt{\varepsilon}$ (Chapter 6) and referred to as the **Lorenz–Lorentz.**[4] The validity of these relations is confirmed by experimental data on polarization of liquids consisting of molecules with quasi-elastic dipoles; however, they do not describe a matter with solid dipoles. ■

Magnetic polarization of matter is a purely quantum phenomenon. When a system of charged particles is in a magnetic field and is in a state of statistic equilibrium, its magnetic moment is zero only if the motion of particles is governed by classical laws (Problems 1.15 and 1.21). The magnetic susceptibility is therefore calculated on the basis of quantum mechanics (Problems 1.17, 1.19, 1.22, and 1.23). Even the models that suggest the existence of classical electron orbits in atoms (Problems 1.16 and 1.18) are essentially of quantum character since in classical physics an atom is unstable and the stationary orbits cannot exist (see, for instance, Problem I.5.119).

Electric conductivity. In order to calculate the electric conductivity, it is necessary to find an electric current produced by a weak electric field. These problems are solved by using either the classical distribution function or the quantum mechanical density matrix, depending on the character of particle motion. The equation for the density matrix is presented in (I.C42). The distribution function $f(\boldsymbol{r}, \boldsymbol{p}, t)$ of charged particles exposed to an electromagnetic field satisfies the **Boltzmann equation**

$$\frac{\partial f}{\partial t} + \boldsymbol{v}\cdot\frac{\partial f}{\partial \boldsymbol{r}} + e\left(\boldsymbol{E} + \frac{1}{c}\boldsymbol{v}\times\boldsymbol{B}\right)\cdot\frac{\partial f}{\partial \boldsymbol{p}} = I[f]. \tag{1.38}$$

The distribution function is normalized by the condition

$$\int f(\boldsymbol{r}, \boldsymbol{p}, t)\, d^3p = n(\boldsymbol{r}, t), \tag{1.39}$$

where $n(\boldsymbol{r}, t)$ is the particle number density. In a multi-component system, each component should be described by its own distribution function.

The right-hand side of Equation (1.38) is called the **collision Integral**. The collision integral describes processes of mutual scattering of particles and must contain detailed information their interaction. As a rule, it has a specific and rather complicated form for each system. In most (may be in all) cases, Equation (1.38) is an integro-differential equation with respect to the distribution function. The collision integral is frequently written in the **relaxation time approximation** to obtain semi-qualitative results:

$$I[f] = -\frac{f - f_0}{\tau(p)}, \tag{1.40}$$

3) Clausius Rudolf (1822–1888), German physicist theorist. His basic works are in thermodynamics and kinetic theory of gases. He formulated the second law of thermodynamics and introduced the concept of entropy. Mosotti Ottaviano Fabricio (1791–1863), Italian physicist.

4) Lorentz Ludvig Valentin (1829–1891), Danish physicist, constructed, independently of Maxwell, the electromagnetic theory of light.

where f_0 is the equilibrium distribution function and $\tau(p)$ is the relaxation time that may depend on the energy of particles. This quantity should be considered a phenomenological adjustable parameter, which must show the best correlation with experimental data or with a more rigorous theory. The physical meaning of this parameter becomes apparent from a consideration of a spatially homogeneous system in the absence of external fields:

$$\frac{\partial f(p,t)}{\partial t} = -\frac{1}{\tau}[f(p,t) - f_0(p)].$$

It gives the exponential relaxation of the nonequilibrium distribution:

$$f(p,t) = f_0(p) + \delta f(p,0)\, e^{-t/\tau(p)}. \tag{1.41}$$

Here $f_0(p) + \delta f(p,0)$ is the initial nonequilibrium distribution function. One should bear in mind that this relaxation may become more complicated as the equilibrium is approached (e.g., several different relaxation times are possible); therefore, in each specific case it is necessary to verify that approximation (1.40) can be used.

Example 1.5

Calculate the electric conductivity of a semiconductor, in the relaxation time approximation $\tau = const$. The number density n of free charge carriers is sufficiently low. Therefore, the equilibrium distribution function can be treated as classical (Maxwell distribution), see Equation (1.31).

Solution. The stationary kinetic equation for a homogeneous system of charged particles in a uniform electric field is

(1) $$e\boldsymbol{E}\cdot\frac{\partial f}{\partial \boldsymbol{p}} = -\frac{\delta f}{\tau}.$$

Let $\boldsymbol{E}$ be small ($eE\tau \ll p$) and linearize Equation (1) remembering that the nonequilibrium additive correction to the distribution function is also of the order of E:

(2) $$e\boldsymbol{E}\cdot\frac{\partial f_0}{\partial \boldsymbol{p}} = -\frac{\delta f}{\tau}.$$

Substituting the Maxwell distribution (1.31) for f_0 in Equation (2), we find the nonequilibrium component of the distribution function:

(3) $$\delta f = \frac{\tau e}{T}\boldsymbol{E}\cdot\boldsymbol{v} f_0(p).$$

The electric current is calculated using the formula

(4) $$\boldsymbol{j} = e\int \boldsymbol{v}\,\delta f(\boldsymbol{p})\, d^3p.$$

Then for the electric conductivity tensor we have

(5) $$\kappa_{\alpha\beta} = \frac{e^2\tau}{T}\int v_\alpha v_\beta f_0(p)\, d^3p.$$

If the distribution of particles over momenta is isotropic, the electric conductivity is also isotropic: $\kappa_{\alpha\beta} = \kappa\delta_{\alpha\beta}$, where

$$\kappa = \frac{ne^2\tau}{m} \tag{1.42}$$

is the **Drude formula.**[5] ■

Suggested literature: Maxwell (1989); Frenkel (1926a,b); Landau *et al.* (1984); Landau and Lifshitz (1980); Landau and Lifshitz (1977); Tamm (1976); Bredov *et al.* (2003); Sivukhin (1977); Peierls (1979); Pitaevskii and Lifshitz (1980); Frohlich (1958); Toptygin (2014)

Problems

1.1 Derive Equation (1.16) by a physically transparent method: assume that the electric polarization $\boldsymbol{P}$ is produced in matter by identical elementary dipoles $\boldsymbol{p} = e\boldsymbol{l}$, and calculate the charge inside an arbitrary closed surface.

1.2 Derive Equation (1.18) by a physically transparent method: assume that the magnetic polarization $\boldsymbol{M}$ is due to the closed circular microscopic currents circulating in matter, and calculate the current through an arbitrary surface inside the matter.

1.3 A dielectric sphere of radius a is uniformly polarized (polarization vector $\boldsymbol{P} = const$) and is in vacuum. Calculate the electric field inside and outside the sphere using the model that considers a small relative displacement of positive and negative charges.

1.4 Calculate the polarizability β of a hydrogen atom in a weak electric field with the aid of the classical model that suggests that the electron cloud density is described by the function $\rho(r) = -(e_0/\pi a_B^3)\exp(-2r/a_B)$, where e_0 is the elementary charge and a_B is the constant (Bohr radius). Neglect the deformation of the electron cloud. Find the change in the polarizability assuming that the electron cloud has a constant density inside a sphere of radius a_B.

1.5* Calculate the polarizability of a hydrogen atom in the ground state by the quantum mechanical method (see the general formula in Problem I.6.62).

1.6* A molecule consists of two atoms that are at a distance a from each other. The atoms are spherically symmetric and have the polarizabilities β' and β''. Find the polarizability tensor of the molecule assuming the atomic radii are small in comparison with a. Consider, in particular, the case $\beta' = \beta''$.

1.7 On the basis of energy conservation, prove that the polarizability tensor of a molecule in a constant field is symmetric.

5) Drude Paul (1863–1906), German physicist, who laid the foundation for the electronic theory of metals.

1.8 A dielectric consists of identical molecules whose dipole moment is zero in the absence of an external field. The polarizability tensor β_{ik} of an individual molecule is known. Find the dielectric polarization coefficient α for two cases: (i) all molecules are equally oriented and (ii) all molecules are randomly oriented.[6] Take account of the difference between the field acting on the molecule and the average field with the aid of Clausius–Mossotti formula.

1.9* If the polarizability of a molecule is different in different directions, then the energy of interaction between the molecule and an external field will depend on the orientation of the molecule. For this reason, along with the deformation mechanism of polarization, the orientation mechanism will operate even though the molecule does not possess any constant electric moment. This results in the temperature dependence of the dielectric constant of the matter consisting of randomly oriented nonpolar molecules. Investigate this effect by considering a diatomic gas in a weak constant electric field. Calculate the dielectric p olarizability coefficient α. The longitudinal polarizability of a molecule of the gas is β_1, and the transverse polarizability is β_2.

1.10 Two molecules in a gas have the dipole moments p_1 and p_2 and are at a distance R from each other. Their orientation changes due to collisions with other molecules. The probability of a specified mutual orientation is determined by the Boltzmann formula (1.31), where U is the energy of interaction between the two dipoles. Assume that the condition $U \ll kT$ is satisfied and show that the quantity U averaged over the Boltzmann distribution[7] has the form

$$U(R) = -\frac{2p_1^2 p_2^2}{3kTR^6}.$$

1.11 A molecule with an electric dipole moment p interacts with a nonpolar molecule of polarizability β. Show that the interaction energy averaged over orientations of the dipole moment has the form

$$U(R) = -\frac{\beta p^2}{R^6},$$

where R is the distance between the molecules.

1.12* In a dielectric placed in a constant electric field, apart from the dipole moment (the polarization vector $\boldsymbol{P}$), there exist higher-order moments. Find the densities of the volume and surface charges that are equivalent to the quadrupole polarization Q_{ik} (Q_{ik} are the components of the quadrupole moment per unit volume of the dielectric).

6) Case (i) may take place in solid bodies, either crystalline or amorphous, and case (ii) can happen in gases, liquids, and solids. Note, however, that solid body unlike a gas is a complicated system of strongly interacting particles. Therefore, the idea of isolated molecules in the solid body may be questionable.

7) When averaging over the directions of dipole moments in Problems 1.10 and 1.11, use the formulas obtained in Problem I.1.33.

1.13 For polar substances, the Clausius–Mossotti relation is inapplicable and their dielectric permittivity may be calculated by the following approximate method proposed by L. Onsager.

Consider a sphere as small as to contain a single molecule. Assume that outside the sphere there is a dielectric with permittivity ε, and inside the sphere there is vacuum, while the field is equal to the effective field acting on the molecule. This field is determined by solving the macroscopic electrostatic equations. Find the relation between the dielectric permittivity ε of the substance and the polarizability of its molecules β.

1.14* A homogeneous isotropic dielectric of permittivity ε does not exhibit any spontaneous polarization in the absence of an external field. Due to this, in any given macroscopic volume V, in the absence of the external field, the dipole moment averaged over the equilibrium configurations of the charge distribution, is $\langle \boldsymbol{\mathcal{P}} \rangle_0 = 0$. However, an instantaneous value of the moment is fluctuating and $\boldsymbol{\mathcal{P}} \neq 0$. Hence, generally, $\langle \mathcal{P}^2 \rangle_0 \neq 0$. Show that $\langle \mathcal{P}^2 \rangle_0$ is expressed through the dielectric permittivity as

$$\langle \mathcal{P}^2 \rangle_0 = \frac{VT(1+2\varepsilon)(\varepsilon-1)}{4\pi\varepsilon},$$

where T is the temperature, V is the volume of the macroscopic sphere inside the dielectric, and $\langle \mathcal{P}^2 \rangle_0$ is the mean square of the fluctuating dipole moment of this sphere (in the absence of any external field).

1.15* Show that the magnetic moment of a system of charged particles, which are moving in a magnetic field according to the **laws of classical mechanics**, is zero in the stationary state (the **Bohr–Van Leeuween theorem**). For this purpose, write down the energy of the system, averaged over Gibbs ensemble, in the presence and in the absence of the magnetic field and show that the energy is independent of the external field.

1.16 Atoms (molecules) of a statistically equilibrated rarefied gas (of number density N and temperature T) have the intrinsic magnetic moment $\boldsymbol{\mu}$. Assuming that atomic electrons move in stationary classical orbits, show that in the presence of a magnetic field each atom will acquire an additional kinetic energy $\Delta K = -\boldsymbol{\mu}\cdot\boldsymbol{H}$. Calculate the magnetic polarization vector $\boldsymbol{M}$ and the paramagnetic susceptibility of the gas due to the orientation of atomic magnetic moments. Are the results obtained in agreement with the theorem proved in Problem 1.15?

1.17* Carry out a quantum mechanical calculation of the magnetization and the **paramagnetic susceptibility** of a statistically equilibrated rarefied atomic gas placed in a weak magnetic field. The quantum mechanical operator of the total magnetic moment of an individual atom has the form (cf. Equation (I.6.86))

$$\hat{\boldsymbol{\mu}} = \mu_B(\hat{\boldsymbol{J}} + \hat{\boldsymbol{S}}), \tag{1.43}$$

where μ_B is the Bohr magneton, while $\hat{\boldsymbol{J}} = \hat{\boldsymbol{L}} + \hat{\boldsymbol{S}}$ and $\hat{\boldsymbol{S}}$ are, respectively, the dimensionless operators of the total and spin mechanical moment of an electron shell of the atom. The atom is in the ground state with fixed quantum numbers J, L, and S. The magnetic field is weak and unable to break the LS-binding.

1.18 Let atoms in the quasi-classical model (Problem 1.16) be spherically symmetric and have no intrinsic magnetic moments. Calculate the **diamagnetic susceptibility** of the atoms due to the Larmor precession of the electron shells in an external magnetic field.

1.19⋆ Carry out a quantum mechanical calculation of the diamagnetic susceptibility of an atomic gas. The electron shells of the atoms have the quantum numbers $L = S = 0$.

1.20 An atom with spherically symmetric charge distribution is placed in an external uniform magnetic field $\boldsymbol{H}$. Show that the additional field near the nucleus due to a diamagnetic current (Larmor precession of electrons) is equal to

$$\Delta H = -\frac{eH}{3mc^2}\varphi(0),$$

where $\varphi(0)$ is the electrostatic potential near the nucleus due to atomic electrons; e and m are the electron charge and mass respectively.

1.21⋆ Consider a system consisting of particles with charge e and mass m, each moving at a fixed distance a from a certain center (classic rotators). The system is in a magnetic field in the state of statistical equilibrium. Show that the total magnetic susceptibility of this system is zero.

1.22 In the simplest model, free electrons in metals can be considered an ideal Fermi-gas at temperature T close to absolute zero. Calculate the paramagnetic susceptibility of the electron gas due to orientation of the spin magnetic moments of the electrons in a weak magnetic field. The number density of electrons is N and the temperature $T = 0$. Ignore the effect of the magnetic field on the motion of the electrons in space.

Hint. The Fermi energy ϵ_F (i.e., the energy of the highest occupied level) in the absence of the magnetic field is $\epsilon_F = \hbar^2(3\pi^2 N)^{2/3}/2m$, where m is the electron mass.

1.23 A rarefied electron gas at temperature T is in a weak uniform magnetic field and obeys the Maxwell–Boltzmann statistics. Calculate the magnetic susceptibility of the electron gas and separate its part, which is due to orientation of spin magnetic moments and the contribution associated with the effect of the magnetic field on the orbital motion of particles. Make use of the quantum mechanical expression for the electron energy in the magnetic field.

Hint. An electron in a homogeneous magnetic field has the energy (Landau and Lifshitz (1977)) (cf. also Problem I.6.76)

$$\mathcal{E}_n = \left(n + \frac{1}{2}\right)\hbar\omega_B + \frac{p_z^2}{2m} - \mu_B m_s B. \tag{1.44}$$

Here $\omega_c = \omega_B = |e|B/mc$ is the cyclotron frequency, $n = 0, 1 \ldots,$ $m_s = \pm 1/2$. The values of energy degenerate due to the position uncertainty of

Larmor circle. Operators of coordinates

$$\hat{x}_0 = \hat{x} + \frac{c\hat{p}_y}{eB}, \quad \hat{y}_0 \frac{c\hat{p}_x}{eB} \tag{1.45}$$

of Larmor circle centre mutually do not permutable. The number of quantum states in volume V per dp_z interval is

$$dQ = \frac{eBV}{(2\pi\hbar)^2 c} dp_z. \tag{1.46}$$

1.24★ An ionized gas consists of ions (charge Ze, average number density N_0) and electrons (charge $-e$, average number density n_0). The gas as a whole is electrically neutral, that is, $ZN_0 = n_0$, and is in statistical equilibrium at temperature T. Find the charge density distribution near an individual ion on the assumption that the gas is described by classical statistics and that the energy of particle-particle interaction is low compared with the thermal energy T.

1.25 An infinite conducting plate, bounded by planes $x = h$ and $x = -h$, is placed in a constant and uniform transverse electric field E_0. The plate as a whole is electrically neutral, the average number density of "free charges" is N_0, and the dielectric permittivity is ε. Assuming that the change in the number density under the action of the applied field is small ($|N - N_0| \ll N_0$), find the field distribution inside the plate and the thickness of the layer in which the "surface" charge is concentrated. The charge-carrying particles obey the Boltzmann distribution.

1.26★ A layer of electrolyte is placed between two infinite plane electrodes, $x = h$ and $x = -h$, at a potential difference $2\varphi_0$. The electrolyte consists of ions of two types with charges $+e$ and $-e$. Their average number density is N_0 in the absence of an external field, and the dielectric permittivity of the electrolyte is ε. Find the potential distribution between the electrodes. The particles obey the Boltzmann distribution.

Hint. Use the same method as in Problem 1.24.

1.27★ Find the charge and potential distribution around an impurity ion with the charge Ze in metal. Use the simplest model of the metal as degenerate electron gas ($T \to 0$) with an average number density n_0, whose charge is neutralized by motionless positive ions.

Hint. Make use of the Thomas–Fermi quasi-classical model (Landau and Lifshitz (1977)).

1.28 Calculate an electric conductivity of degenerate electron gas of number density n in the relaxation time approximation $\tau(\epsilon)$.

1.29 An equilibrium plasma with electron number density n and motionless ions is in a weak uniform magnetic field $\boldsymbol{B} = const$. In the relaxation time approximation $\tau = const$, calculate the current in the plasma induced by a weak electric field, and also the electric conductivity tensor. In the expression for the current, take into account the terms not higher than the first-order terms with respect to B.

1.30* Solve Problem 1.29 without limitations on the magnetic field strength. Calculate the anisotropic electric conductivity tensor and analyze specific cases of weak and strong magnetic fields.

1.4
Answers and Solutions

1.1 Consider an arbitrary volume V inside a dielectric, which is enclosed by surface S, and calculate the electric charge $q_{int} = \int_V \rho_{int}\, dV$ inside this volume. This charge is produced only by the dipoles, which are intersected by the surface S (Figure 1.2).

All remaining dipoles are either wholly inside or outside the volume V and make no contribution to the total charge. A surface element $d\boldsymbol{S}$ meets $N\boldsymbol{l}{\cdot}d\boldsymbol{S}$ dipoles on average. Their charge enclosed by this surface element is $dq_{int} = -eN\boldsymbol{l}{\cdot}d\boldsymbol{S} = -\boldsymbol{P}{\cdot}d\boldsymbol{S}$. Hence, $q_{int} = -\oint_S \boldsymbol{P}{\cdot}d\boldsymbol{S} = -\int_V div\,\boldsymbol{P}\, dV$, which leads to Equation (1.16).

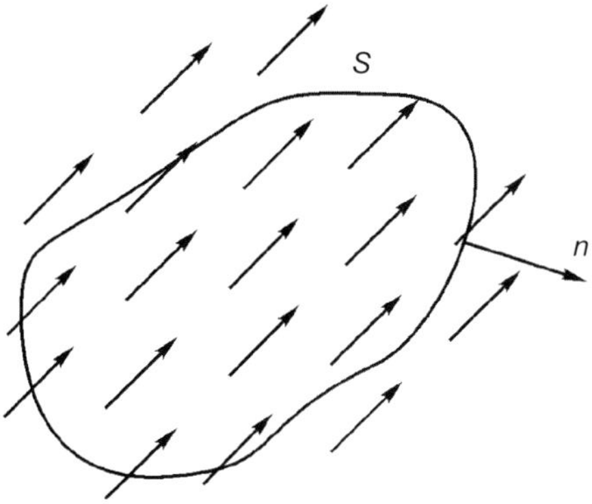

Figure 1.2 To calculation of the density of bound charges in a dielectric.

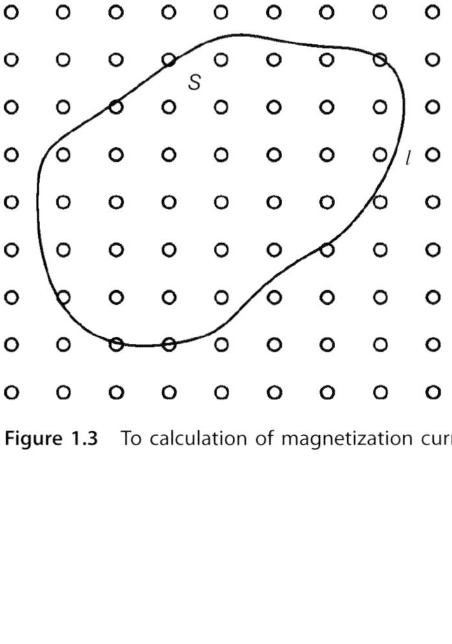

Figure 1.3 To calculation of magnetization current.

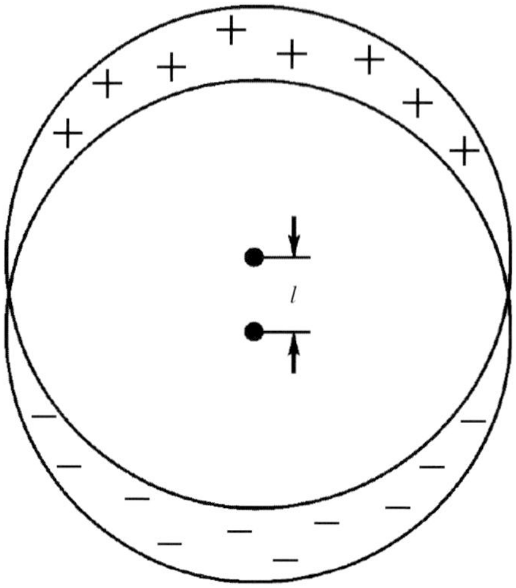

Figure 1.4 To calculation of an electric field of polarized sphere.

1.2 The magnetic moment of an individual elementary current can be written as $\boldsymbol{m} = is\boldsymbol{n}/c$, where i is the elementary molecular current, s is the circle area, and $\boldsymbol{n}$ is a unit normal to the plane of the circle (Equation (I.2.60)). Let us plot an arbitrary closed contour inside matter (Figure 1.3).

The magnetization current J_{int} through the surface S bounded by contour l is due to those closed elementary currents that are pierced by this contour. The remaining currents either cross the surface S twice or do not cross it at all, and, hence, do not contribute to J_{int}. The segment $d\boldsymbol{l}$ of the contour crosses $s(\boldsymbol{n}\cdot d\boldsymbol{l})N$ molecular currents on average and makes the contribution $dJ_{int} = is(\boldsymbol{n}\cdot d\boldsymbol{l})N = c\boldsymbol{M}\cdot d\boldsymbol{l}$ to the total current. Hence, we have $J_{int} = c\oint_l \boldsymbol{M}\cdot d\boldsymbol{l} = c\int_S rot\,\boldsymbol{M}\cdot d\boldsymbol{S}$. Since the contour has been chosen arbitraryly, from the latter relation follows we obtain Equation (1.18).

1.3 If the number density of elementary dipoles $\boldsymbol{p} = e\boldsymbol{l}$ in a polarized sphere is N, then the total dipole moment of the sphere is $\boldsymbol{\mathcal{P}} = 4\pi a^3\boldsymbol{P}/3$, where $\boldsymbol{P} = N\boldsymbol{p}$ is the polarization vector. At $l \ll a$, the polarized sphere can be treated as a system of two spheres with the charges $q = \pm 4\pi a^3 Ne/3$, whose centers are separated by the distance l (Figure 1.4). In the outer region, each sphere produces a field similar to that of a point charge localized at the corresponding center; in other words, the two spheres produce the field of the dipole with the moment $q\boldsymbol{l} = \boldsymbol{\mathcal{P}}$, whose potential is given by

$$\varphi_e(\boldsymbol{r}) = \frac{\boldsymbol{\mathcal{P}}\cdot\boldsymbol{r}}{r^3} \tag{1}$$

(Equation (I.2.21)). Inside the polarized sphere at a distance $r < a$ from its center the field is only due to internal charges localized at distances smaller than r. The external charges produce no field in the inner region, and therefore formula (1) may be applied provided $\boldsymbol{\mathcal{P}}$ is replaced by the dipole moment of the inner region, $\mathcal{P}r^3/a^3$:

(2) $$\varphi_i(\boldsymbol{r}) = \frac{\mathcal{P}\cdot\boldsymbol{r}}{a^3}.$$

The field strength in the outer region was calculated in Problem I.2.21. For the inner region we have

(3) $$\boldsymbol{E}_i = \nabla\varphi_i(\boldsymbol{r}) = -\frac{\mathcal{P}}{a^3} = -\frac{4\pi}{3}\boldsymbol{P}.$$

1.4 $\beta = 3a_B^3/4$. In the case of uniform charge distribution in an electron cloud, $\beta = a_B^3$.

1.5 A hydrogen atom in the ground state is spherically symmetric and its polarizability tensor is diagonal. The general formula (see Equation (3) from the solution of Problem I.6.62) takes the form

(1) $$\beta = 2\sum_n{}' \frac{e^2|\langle n|z|0\rangle|^2}{\mathcal{E}_n - \mathcal{E}_0}.$$

The most difficult procedure in this expression is summation over intermediate states n. Following (Landau and Lifshitz (1977)), we replace the z-coordinate by the auxiliary operator

(2) $$z = \frac{m}{\hbar}\frac{d\hat{\zeta}}{dt},$$

where m is the electron mass. The matrix element of the time derivative operator with the wave functions of stationary states is calculated according to the rule $\langle n|\dot{\hat{\zeta}}|0\rangle = i(\mathcal{E}_0 - \mathcal{E}_n)\langle n|\hat{\zeta}|0\rangle/\hbar$, which allows one to eliminate the energy denominator in Equation (1) and carry out the summation:

(3) $$\beta = \frac{2ime^2}{\hbar^2}\langle 0|z\hat{\zeta}|0\rangle.$$

We will now find the action of $\hat{\zeta}$ on the wave function of the ground state of an hydrogen atom by writing it in the form

(4) $$\hat{\zeta}|0\rangle = q(\boldsymbol{r})|0\rangle,$$

where $q(\boldsymbol{r})$ is a new unknown function. With the aid of equations (2), (4), and (I.C34), we have

(5) $$z|0\rangle = \frac{m}{\hbar}\frac{d\hat{\zeta}}{dt}|0\rangle = \frac{im}{\hbar^2}(\hat{\mathcal{H}}\hat{\zeta} - \hat{\zeta}\hat{\mathcal{H}})|0\rangle = \frac{im}{\hbar^2}(\hat{\mathcal{H}} - \mathcal{E}_0)q(\boldsymbol{r})|0\rangle.$$

Substituting Equation (5) into the Hamiltonian of a hydrogen atom, given by

(6) $$\hat{\mathcal{H}} = -\frac{\hbar^2}{2m}\Delta + U(r),$$

we obtain the equation for determining $q(\boldsymbol{r})$:

(7) $$\frac{1}{2}|0\rangle\Delta q + \nabla q\cdot\nabla|0\rangle = iz|0\rangle.$$

We need only the specific solution of this equation which is expressed in terms of quantities entering its right-hand side.

Substituting the wave function of the ground state $|0\rangle = \exp(-r/a_B)/\sqrt{\pi a_B^3}$, where $a_B = \hbar^2/me^2$ is the Bohr radius, and using the dependence on the polar angle ϑ defined by the right-hand side of $z = r\cos\vartheta$, we seek the solution in the form $q(\boldsymbol{r}) = f(r)\cos\vartheta$ and obtain the equation

(8) $$\frac{1}{2}f'' + \left(\frac{1}{r} - \frac{1}{a_B}\right)f' - \frac{1}{r^2}f = ir,$$

whose particular solution is given by

(9) $$f(r) = -ia_B r\left(a_B + \frac{r}{2}\right).$$

Finally, with the aid of Equations (2), (3), and (9) we find

(10) $$\beta = \frac{2i}{a_B}\langle 0|rf(r)\cos^2\vartheta|0\rangle = \frac{9}{2}a_B^3.$$

A comparison with the result of Problem 1.4 shows that the classical models give the correct order of magnitude (provided the value of atomic radius is borrowed from quantum mechanics) but do not allow us to obtain correct numerical factor.

1.6 It is clear from the symmetry of the molecule that one of the main polarization tensor axes lies along the axis of the molecule while two others can be chosen arbitrarily in the plane perpendicular to the molecular axis. Hence, only two of the three main values of the polarizability tensor are different, namely, $\beta^{(1)}$ and $\beta^{(2)} = \beta^{(3)}$. The following cases should be considered separately to find these values:

(a) The external field is directed along the axis of the molecule. It is clear that the induced dipole moment of each atom is directed along the external field. Denote these moments by $\boldsymbol{p}'$ and $\boldsymbol{p}''$, respectively, and obtain two equations for them:

(1) $$\boldsymbol{p}' = \beta'(\boldsymbol{E} + \boldsymbol{E}'), \qquad \boldsymbol{p}'' = \beta''(\boldsymbol{E} + \boldsymbol{E}''),$$

where $\boldsymbol{E}$ is the external field, and $\boldsymbol{E}'$ and $\boldsymbol{E}''$ are the additional fields in the center of each atom due to the presence of another atom. The fields $\boldsymbol{E}'$ and $\boldsymbol{E}''$ can be expressed in terms of dipole moments of the respective atoms with the aid of the formula for the strength of the field due to the dipole with moment $\boldsymbol{p}$, taking into account that all vectors are directed along the axis of the molecule. Determining p' and p'' from Equation (1), and using the formula $p = p' + p'' = \beta^{(1)}E$, we find

(2) $$\beta^{(1)} = \left[\frac{1}{\beta'} - \frac{2(a^3 + 2\beta')}{a^3(a^3 + 2\beta'')}\right]^{-1} + \left[\frac{1}{\beta'} - \frac{2(a^3 + 2\beta'')}{a^3(a^3 + 2\beta')}\right]^{-1}.$$

(b) The external field is perpendicular to the axis of the molecule. In a similar way, we arrive at

(3) $$\beta^{(2)} = \beta^{(3)} = \left[\frac{1}{\beta'} + \frac{a^3 - \beta'}{a^3(a^3 - \beta'')}\right]^{-1} + \left[\frac{1}{\beta''} + \frac{a^3 - \beta''}{a^3(a^3\beta')}\right]^{-1}.$$

At $\beta' = \beta''$ the expressions $\beta^{(1)}$ and $\beta^{(2)}$ are simplified:

$$(4) \qquad \beta^{(1)} = \frac{2\beta'}{(1 - 2\beta'/a^3)}, \qquad \beta^{(2)} = \frac{2\beta'}{(1 + 2\beta'/a^3)}.$$

The average polarizability is given by

$$(5) \qquad \overline{\beta} = \frac{1}{3}(\beta^{(1)} + \beta^{(2)}) = \frac{2}{3}\beta'\left(\frac{1}{1 - 2\beta'/a^3} + \frac{2}{1 + \beta'/a^3}\right).$$

1.8 (a) The dielectric is anisotropic as a whole. The principal values of the dielectric polarizability tensor (cf. Equation (1.37)) are given by

$$\alpha^{(i)} = \frac{N\beta^{(i)}}{1 - 4\pi N\beta^{(i)}/3}.$$

(b) In the case of a random orientation of molecules, there are no physically selected directions in macroscopic volumes of dielectric, except for the direction of external field. Hence, the average dipole moment of the molecule $\overline{\boldsymbol{p}}$ is proportional to the field $\boldsymbol{\mathcal{E}}$ acting on the molecule:

$$\overline{\boldsymbol{p}} = \beta\boldsymbol{\mathcal{E}}.$$

On the other hand, it is clear that

$$\overline{p}_i = \overline{\beta_{ik}\mathcal{E}_k} = \overline{\beta}_{ik}\mathcal{E}_k,$$

where the averaging is carried out over a macroscopically small volume. It follows from comparison of the two last formulas that

$$\beta = \overline{\beta}_{11} = \overline{\beta}_{22} = \overline{\beta}_{33}, \qquad \overline{\beta}_{ik} = 0 \quad (\text{at } i \neq k).$$

Thus,

$$\beta = \frac{1}{3}\overline{(\beta_{11} + \beta_{22} + \beta_{33})}.$$

However, the sum of the diagonal components of the tensor is Invariant, which is equal to the sum of the principal values $\beta^{(1)} + \beta^{(2)} + \beta^{(3)}$ (Equation (I.1.263)). Hence,

$$\beta = \frac{1}{3}(\beta^{(1)} + \beta^{(2)} + \beta^{(3)}).$$

The polarization coefficient of dielectric α relates to β by a usual formula (Example 1.4).

1.9 If the axis of the molecule is oriented at an angle θ to the direction of an external field $\boldsymbol{E}_0$, the energy of the molecule can be written down as

$$W = -\frac{1}{2}\boldsymbol{p}\cdot\boldsymbol{E}_0 = -\frac{1}{2}(\beta_1\cos^2\theta + \beta_2\sin^2\theta)E_0^2.$$

The number of particles per unit volume with axes directed at an angle θ to the field $\boldsymbol{E}_0$ is given by the Boltzmann formula (1.31). The polarization vector is defined by the formula $\boldsymbol{P} = N\overline{\boldsymbol{p}}$, where $\overline{\boldsymbol{p}}$ is the dipole moment of an individual molecule, averaged over the Boltzmann distribution, and N is the number density of particles. Since in the absence of the field, the molecules are chaotically oriented, the vector $\overline{\boldsymbol{p}}$ has the same direction as the external field.

In line with this, we calculate the quantity $\overline{p}$ by the formula

$$\overline{p} = \frac{1}{N}\int p_{\|}\,dN = \frac{E_0\int_0^{\pi}\exp\left(-\frac{W(\theta)}{kT}\right)(\beta_1\cos^2\theta + \beta_2\sin^2\theta)\sin\theta\,d\theta}{\int_0^{\pi}\exp\left(-\frac{W(\theta)}{kT}\right)\sin\theta\,d\theta},$$

where $p_{\|}$ denotes the component of the molecule dipole moment parallel to the field. According to the condition of the problem, the field is weak. Hence, it is sufficient to consider the terms linear in $a = (\beta_1 - \beta_2)E_0^2/2kT \ll 1$ only. Finally, with the use of formulas $P = N\overline{p} = \alpha E_0$ we obtain

$$\alpha = N\beta_2 + \frac{1}{3}N(\beta_1 - \beta_2)\left[1 + \frac{2}{15}\frac{(\beta_1 - \beta_2)E_0^2}{kT}\right].$$

It is clear from this formula that the dependence of P on E_0 appears to be nonlinear, and that α is not a proportionality coefficient, which is independent of E_0. Let us estimate the value of the correction term at room temperatures (T = 300 K). If $\beta_1 - \beta_2$ is of the order of 10^{-24} cm^3, then $T/(\beta_1 - \beta_2) \approx 10^6$. Thus, this term is small at $E_0 \ll 10^3$ V cm^{-1}. Neglecting the correction term, we obtain the previous expression for α:

$$\alpha = \frac{1}{3}N(\beta_1 + 2\beta_2)$$

(Problem 1.8).

1.12 The additional potential due to the quadrupole polarization of the dielectric can be written in the form

(1) $$\varphi = \frac{1}{2}\int \frac{\partial^2(1/R)}{\partial x_i\partial x_k}Q_{ik}\,dV,$$

where R is a distance between the volume element dV and point of observation; the integration is carried out over the volume of the dielectric. On the other hand, the potential of the volume and surface charges is generally given by

(2) $$\varphi = \int \frac{\rho'}{R}\,dV + \int \frac{\sigma'}{R}\,dS + \int \boldsymbol{\tau}'\cdot\nabla\left(\frac{1}{R}\right)dS,$$

where ρ' is the volume charge density, σ' is the surface charge density, and τ' is the thickness of the double layer. Reducing Equation (1) to the form of Equation (2), we get

(3) $$\rho' = \frac{1}{2}\frac{\partial^2 Q_{ik}}{\partial x_i\partial x_k},\quad \sigma' = -\frac{1}{2}\frac{\partial Q_{in}}{\partial x_i},\quad \tau_k' = \frac{1}{2}Q_{ki}n_i.$$

Thus, the quadrupole polarization is determined by the volume charges ρ' inside the dielectric, the surface charges σ', and the double electric layer with the power τ' on the surface of the dielectric. Since the densities of the volume and surface charges are related to the polarization vector by formulas $\rho' = -div\, \boldsymbol{P}'$, $\sigma' = P'_n$, it follows from Equation (3) that the quadrupole polarization is determined by the additional dipole polarization

$$P'_k = -\frac{1}{2}\frac{\partial Q_{ik}}{\partial x_i}$$

and the double layer with the power τ'. Equation (3) can also be obtained from the energy of dielectric in the presence of quadrupole polarization.

1.13 $\varepsilon = \frac{1}{4}\left[1 + 3x + 3\left(1 + \frac{2}{3}x + x^2\right)^{\frac{1}{2}}\right]$,

where $x = 4\pi N\beta$. The polarizability β of polar materials in weak fields is given by the formula

$$\beta = \frac{p^2}{3T},$$

where p is the dipole moment of the molecule and T is the temperature in energy units.

For $x \ll 1$, when the difference between the field acting on the molecule and the average field becomes very small, we have

$$\varepsilon = 1 + x = 1 + 4\pi N\beta.$$

1.14 We will describe the charges inside the sphere using the microscopic approach, on the base of the classical Boltzmann distribution (1.31), while outside the sphere we consider the dielectric as a continuous medium of permittivity ε. Let an ath charge inside the sphere be displaced by a vector $\boldsymbol{u}_a$ with respect to the equilibrium position. A set of such displacements is denoted by $Q = Q(\boldsymbol{u}_1, \ldots \boldsymbol{u}_a, \ldots)$. In the absence of an external field, these displacements are of pure fluctuation character. The interaction between the charges is described by the potential energy $U_0(Q)$, which includes also the interaction between the charges inside the sphere. The interaction with the charges outside the sphere occurs on the surface and is negligibly small by virtue of the macroscopic sizes of the sphere. The dipole moment of the sphere is given by

$$\mathcal{P} = \sum_a e_a \boldsymbol{u}_a. \tag{1}$$

Since there is no spontaneous polarization, the statistical average of the dipole moment is equal to zero:

$$\langle \mathcal{P} \rangle_0 = \int \mathcal{P} \exp\left(-\frac{U_0(Q)}{T}\right) dQ = 0. \tag{2}$$

The presence of an external field gives rise to a uniform electric field inside the sphere whose strength is given by (Problem 2.11)

$$\mathcal{E} = \frac{3\varepsilon}{2\varepsilon + 1}\boldsymbol{E}. \tag{3}$$

This field is produced by the external sources and external (with respect to the sphere) charges of the dielectric. As a result, the potential energy of the charges inside the sphere acquires the additional term

(4) $$U(Q,\boldsymbol{E}) = U_0(Q) - \sum_a e_a \boldsymbol{u}_a \cdot \boldsymbol{\mathcal{E}} = U_0(Q) - \frac{3\varepsilon}{2\varepsilon+1} \boldsymbol{\mathcal{P}} \cdot \boldsymbol{E}.$$

This potential energy should be used now in the Boltzmann distribution to calculate the dipole moment induced by the external field:

(5) $$\langle \boldsymbol{\mathcal{P}}(\boldsymbol{E}) \rangle = \int \boldsymbol{\mathcal{P}} \exp\left(-\frac{U}{T}\right) dQ \Big/ \int \exp\left(-\frac{U}{T}\right) dQ.$$

Assuming the field $\boldsymbol{E}$ to be weak, we expand the exponent into a series to obtain

(6) $$\exp\left(-\frac{U}{T}\right) = \left[1 + \frac{3\varepsilon}{2\varepsilon+1} \frac{\boldsymbol{\mathcal{P}} \cdot \boldsymbol{E}}{T}\right] \exp\left(-\frac{U_0}{T}\right).$$

Using Equation (2) we find

(7) $$\langle \mathcal{P}_\mu(\boldsymbol{E}) \rangle = \frac{3\varepsilon E_\nu}{(2\varepsilon+1)T} \frac{\int \mathcal{P}_\mu \mathcal{P}_\nu \exp\left(-\frac{U_0}{T}\right) dQ}{\int \exp\left(-\frac{U_0}{T}\right) dQ} = \frac{3\varepsilon E_\nu}{(2\varepsilon+1)T} \langle \mathcal{P}_\mu \mathcal{P}_\nu \rangle_0.$$

We transform the average over components of the dipole moment using the symmetry considerations:

(8) $$\langle \mathcal{P}_\mu \mathcal{P}_\nu \rangle_0 = \frac{1}{3} \langle \mathcal{P}^2 \rangle_0 \delta_{\mu\nu}.$$

The left-hand side of Equation (7) can be written in terms of the projection P_μ of the electric polarizability vector:

(9) $$\langle \mathcal{P}_\mu(\boldsymbol{E}) \rangle = VP_\mu = \frac{V(\varepsilon-1)}{4\pi} E_\mu.$$

Substituting Equations (8) and (9) into Equation (7), we obtain the formula given in the statement of the problem.

We stress that this formula relates the response of the medium, characterized by permittivity ε, to an external perturbation with the fluctuation of the internal parameter of the medium $\langle \mathcal{P}^2 \rangle_0$ corresponding to the statistical equilibrium in the absence of any external perturbation. This relation is actually a specific case of the fluctuation-dissipative theorem (FDT). The general relation of such kind for the time-dependent disturbance can be found in Section 7.2.

1.15 Consider a system of particles whose Hamiltonian function in the absence of any external magnetic field is given by

(1) $$\mathcal{H}_0 = \sum_a \frac{p_a^2}{2m_a} + U,$$

where the potential energy U is the function of coordinates. In the presence of a magnetic field, the Hamiltonian function takes the form [(I.4.65′)]

(2) $$\mathcal{H} = \sum_a \frac{1}{2m_a}\left(\boldsymbol{P}_a - \frac{e_a}{c}\boldsymbol{A}_a\right)^2 + U,$$

where $\boldsymbol{P}_a$ is the generalized momentum of an ath particle, and $\boldsymbol{A}_a$ is the vector potential of the external field at the point where the particle is localized. The energy of the system averaged over the Gibbs ensemble (its internal energy in the thermodynamic approach) is expressed as an integral over the phase space

(3) $$\overline{\mathcal{E}} = \frac{1}{Z}\int \mathcal{H}\exp\left(-\frac{\mathcal{H}}{T}\right)d\Gamma,$$

where $d\Gamma = \prod_i dP_i\,dx_i$ is an element of phase space. Let us change momentum variables in Equation (3)

(4) $$\boldsymbol{P}_a - \frac{e_a}{c}\boldsymbol{A}_a \to \boldsymbol{p}_a,$$

keeping the same coordinates. With such a replacement, $\mathcal{H}$ becomes $\mathcal{H}_0$, and $d\Gamma = \prod_i dp_i\,dx_i$ because the Jacobian of transition to the new variables equals 1. As a result, the internal energy of the system in a magnetic field is expressed just as it is expressed in the absence of the field, that is, the energy is field-independent. A body that does not possess the magnetic moment in the absence of the field, will not acquire it in the presence of the field. However, such a result holds only in the classical case and fails when particles move according to quantum mechanical laws.

1.16 Let electrons in atoms move with velocity $\boldsymbol{v}_a$. When a field $\boldsymbol{H}$ is applied, each electron undergoes Larmor precession with the angular velocity $\boldsymbol{\Omega}_L = -e\boldsymbol{H}/2mc$ (Problem I.4.102) and acquires an additional velocity

(1) $$\Delta\boldsymbol{v}_a = \boldsymbol{\Omega}_L \times \boldsymbol{r}_a.$$

A change in the kinetic energy of the electrons is given by

(2) $$\Delta K = \frac{m}{2}\sum_a [(\boldsymbol{v}_a + \Delta\boldsymbol{v}_a)^2 - \boldsymbol{v}_a^2] \approx m\sum_a \boldsymbol{v}_a\cdot\Delta\boldsymbol{v}_a = m\boldsymbol{\Omega}_L\cdot\sum_a \boldsymbol{r}_a \times \boldsymbol{v}_a,$$

where the second-order term in small velocity $\Delta\boldsymbol{v}$ is omitted.

Using the evident expression $\boldsymbol{\mu} = \sum_a \boldsymbol{\mu}_a$ for the total magnetic moment of an atom, and the coupling Equation (I.4.76) for the mechanical and magnetic orbital moments, $\boldsymbol{\mu}_a = e\boldsymbol{r}_a \times \boldsymbol{p}_a/2mc$, we obtain the relation given in the statement of the problem. Note that, conceptually, it would be more reasonable to write $\boldsymbol{B}$ instead of $\boldsymbol{H}$ in the expression for $\boldsymbol{\Omega}_L$. However, since the magnetic susceptibility of gases is small, the error introduced in this case is negligible.

Further calculations of the magnetization and magnetic susceptibility can be carried out in the same manner as for the electric polarization in Example 1.3, that is, with the aid of the Boltzmann distribution. In the general case, the magnetization is expressed in terms of the Langevin function

(3) $$M = N\mu L\left(\frac{\mu H}{T}\right).$$

At $\mu H \ll T$, the dependence between M and H is linear, and the magnetic susceptibility is given by

(4) $$\chi = \frac{N\mu^2}{3T} \propto \frac{1}{T}$$

(the **Curie**[8] **law for paramagnetics**). The results obtained are not at variance with the Bohr–Van Leeuween theorem since the model considered in the present problem is not purely classical: the assumption on the existence of stationary orbits of electrons in atoms is inconsistent with classical mechanics and electrodynamics.

1.17 The operator of interaction between an atom and a field in the approximation, which is linear in the field has the form

(1) $$\hat{V} = -\hat{\boldsymbol{\mu}}\cdot\boldsymbol{H}.$$

In the absence of the external field, the energy levels at given L, S, J, and M_J degenerate in M_J with multiplicity $2J+1$, but the degeneracy is removed in the magnetic field. In order to solve the problem, one needs to find corrections to the energy levels based on the perturbation theory and thereafter calculate the average magnetic moment of the atom in a magnetic field making use of the density matrix in the energy representation (Equations (I.C43) and (I.C44)).

Let the direction of the field $\boldsymbol{H}$ be the quantization axis Oz. Hence, in accordance with the stationary perturbation theory, the corrections to the energy levels must be calculated as

(2) $$\Delta\mathcal{E}_{LSJM_J} = -\langle LSJM_J|\hat{\mu}_z|LSJM_J\rangle H.$$

In order to calculate this matrix element, it is convenient to present the magnetic moment operator in the form

(3) $$\hat{\boldsymbol{\mu}} = \hat{G}\hat{\boldsymbol{J}},$$

where $\hat{G}$ is a certain scalar operator. Using the explicit form of $\hat{\boldsymbol{\mu}}$, given in the statement of the problem, we have

(4) $$\hat{G}\hat{\boldsymbol{J}} = \mu_B(\hat{\boldsymbol{J}} + \hat{\boldsymbol{S}}),$$

and consider the scalar product of this equation and the operator $\hat{\boldsymbol{J}}$. Then, we calculate the diagonal matrix elements of both sides of the obtained equation:

(5) $$\langle LSJM_J|\hat{G}\hat{\boldsymbol{J}}^2|LSJM_J\rangle = \mu_B\langle LSJM_J|\hat{\boldsymbol{J}}^2 + \hat{\boldsymbol{S}}\cdot\hat{\boldsymbol{J}}|LSJM_J\rangle.$$

8) Curie Pierre (1859–1906), outstanding French physicist, studied magnetism, physics of crystals, radioactivity; Nobel Prize laureate.

It is clear that

(6) $$\hat{\boldsymbol{J}}^2|LSJM_J\rangle = J(J+1)|LSJM_J\rangle, \quad \hat{\boldsymbol{L}}^2|LSJM_J\rangle = L(L+1)|LSJM_J\rangle,$$
$$\hat{\boldsymbol{S}}^2|LSJM_J\rangle = S(S+1)|LSJM_J\rangle.$$

Finally, we square both sides of the equality $\hat{\boldsymbol{J}} - \hat{\boldsymbol{S}} = \hat{\boldsymbol{L}}$ and use Equation (6) to find

(7) $$2\hat{\boldsymbol{S}}\cdot\hat{\boldsymbol{J}}|LSJM_J\rangle = [J(J+1) - L(L+1) + S(S+1)]|LSJM_J\rangle.$$

With the aid of Equations (6) and (7) we obtain the matrix element

(8) $$\langle LSJM_J|\hat{G}|LSJM_J\rangle = g\mu_B,$$

and calculate the energy levels

(9) $$\Delta\mathcal{E}_{LSJM_J} = -g\mu_B H M_J,$$

where

(10) $$g = 1 + \frac{J(J+1) - L(L+1) + S(S+1)}{2J(J+1)}$$

is the **gyromagnetic factor** or Landé g-factor.

The magnetization can be calculated with the aid of the probability (I.C44) by substituting the values of the energy levels given by Equation (9):

(11) $$M = Ng\mu_B \frac{\sum_{m=-J}^{m=J} me^{am}}{\sum_{m=-J}^{m=J} e^{am}} = Ng\mu_B \frac{d}{da}\ln\left(\sum_{m=-J}^{m=J} e^{am}\right),$$

where $a = g\mu_B H/T$, and the sum over m under the logarithm is a statistical sum for an individual atom. Its calculation with the aid of the formula for a finite geometric progression yields

(12) $$M = M_0 L_J(aJ), \quad \text{where}$$

$$L_J(x) = \left(1 + \frac{1}{2J}\right)\coth\left[\left(1 + \frac{1}{2J}\right)x\right] - \frac{1}{2J}\coth\left(\frac{x}{2J}\right)$$

is the quantum Langevin function or **Brillouin**[9] **function**). The quantity $M_0 = Ng\mu_B J$ is the saturation magnetization which is reached at low temperatures ($aJ \gg 1$). If the opposite inequality is satisfied, then the magnetization is proportional to the magnetic field, and the paramagnetic susceptibility is expressed in the form

(13) $$\chi = \frac{Ng^2\mu_B^2 J(J+1)}{3T}.$$

At $J \gg 1$ we pass to the semi-classical model considered in Problem 1.16. In this case, the atomic magnetic moment is expressed in the form $\mu \approx g\mu_B J$,

9) Brillouin Léon (1889–1969), French physicist, worked mainly in quantum mechanics, radiophysics, physics of solids.

and the quantum Langevin function becomes the classical one, $L_J(x) \to L(x)$ as $J \to \infty$.

The paramagnetic susceptibility of gases under normal conditions is very low. Substituting $N \approx 3 \times 10^{19}$ cm^{-3} and $T \approx 10^{-14}$ erg in Equation (13), we obtain $\chi \approx 10^{-7}$. For comparison, the dielectric susceptibility under the same conditions is $\alpha \approx 10^{-3}$. The difference arises from the fact that the magnetization is related to the motion of particles, owing to which the magnetic susceptibility possesses an additional small factor $(v/c)^2 \approx 10^{-4}$, where v is the velocity of atomic electrons.

1.18 Due to Larmor precession, each electron acquires an additional velocity $\Delta \boldsymbol{v}$ which is given by Equation (1) of the solution of Problem 1.16. Introducing the volume charge density of the electron shell $\rho(r)$, we calculate the atomic magnetic moment induced by the Larmor precession:

(1) $$\boldsymbol{\mu} = \frac{1}{2c}\int \boldsymbol{r} \times \rho \Delta \boldsymbol{v}\, dV = \frac{1}{2c}\int \rho \boldsymbol{r} \times [\boldsymbol{\Omega}_L \times \boldsymbol{r}]\, dV.$$

In view of spherical symmetry of $\rho(r)$, the vector $\boldsymbol{\mu}$ is parallel to $\boldsymbol{\Omega}_L$:

(2) $$\boldsymbol{\mu} = \frac{1}{2c}\boldsymbol{\Omega}_L \int \rho(x^2 + y^2)\, dV,$$

where the Ox- and Oy-axes are perpendicular to the vector $\boldsymbol{H}$. Evidently, it is possible to write down

(3) $$\int \rho(x^2 + y^2)\, dV = \frac{2}{3} Zea^2,$$

where Ze is the total charge of the electron shell and a^2 is the squared distance of electrons from the nucleus averaged over all electrons. As a result, from Equations (2) and (3) we get

(4) $$\boldsymbol{M} = N\boldsymbol{\mu} = \chi \boldsymbol{H}, \quad \text{where} \quad \chi = -\frac{NZe^2a^2}{6mc^2} < 0$$

is the diamagnetic susceptibility. The diamagnetic atomic moment is opposite to the magnetic field, and the diamagnetic susceptibility is negative.

1.19 If the orbital and spin moments of an electron shell of an atom are zero, then the operator (1.43) will give the zero value of the magnetic moment in all orders of the perturbation theory while the nonzero value will be associated with the term quadratic in vector-potential in the interaction operator (I.6.57). Since we deal with the constant but non-quantized external field, we can substitute $\boldsymbol{A} = \boldsymbol{H} \times \boldsymbol{r}/2$ in Equation (I.6.57) and write the operator of interaction of the electron shell with the magnetic field in the form

(1) $$\hat{V} = \frac{e^2}{8mc^2}\sum_a [\boldsymbol{H} \times \boldsymbol{r}_a]^2.$$

The correction to the atomic energy in the first order of the perturbation theory is expressed by the integral

(2) $$\Delta \mathcal{E} = \int \psi^*(q)\hat{V}\psi(q)\, dq,$$

where q is a set of coordinates of all atomic electrons. Owing to spherical symmetry of the electron shell, in integrating over the angles, the quantity $\sin^2 \vartheta_a$ entering the vector product may be substituted by its average value which is equal to 2/3. The final result is

(3) $$\Delta\mathcal{E} = \frac{e^2}{12mc^2} H^2 \sum_a \langle\psi|r_a^2|\psi\rangle.$$

The atomic magnetic moment can be calculated as $\mu = -\partial\Delta\mathcal{E}/\partial H$ (see Section 3.3 below). For the number density of atoms N, the calculation of the magnetization yields

(4) $$\boldsymbol{M} = N\boldsymbol{\mu} = \chi\boldsymbol{H}, \quad \text{where} \quad \chi = -\frac{Ne^2}{6mc^2} \sum_a \langle\psi|r_a^2|\psi\rangle.$$

The sum of r_a^2 for atomic electrons in Equation (4), averaged over the quantum state of the atom, may be replaced by the product Za^2, where a^2 is the mean squared distance for all electrons. Thereafter the quantum formula for the diamagnetic susceptibility will coincide with the semi-classical formula obtained in Problem 1.18. The numerical values of the diamagnetic susceptibility of gases are very small. An estimate can be made by substituting the squared Bohr radius (4). Under normal pressure, we will have $\chi \approx 10^{-10}$. For condensed matter, where number density of particles is four orders of magnitude higher, we will obtain $\chi \approx 10^{-6}$.

1.21 The total magnetic susceptibility is equal to the sum of the paramagnetic and diamagnetic susceptibilities (Problems 1.16, 1.18, 1.19*):

(1) $$\chi = \frac{N\mu^2}{3kT} - \frac{Ne^2}{6mc^2}\overline{r^2}.$$

The magnetic moment μ of an individual rotator entering this formula can be calculated in the following way. Based on the well-known theorem (Equation (I.2.61)), we have

(2) $$\boldsymbol{\mu} = \frac{e}{2mc}\boldsymbol{K},$$

where $\boldsymbol{K}$ is the angular momentum of the particle. In the case of a rotator, the $\boldsymbol{K}$ is related to the kinetic energy through the formula

(3) $$W_k = \frac{K^2}{2ma^2}.$$

The statistical average $\overline{K^2}$ is therefore expressed in terms of the average kinetic energy

(4) $$\overline{K^2} = 2ma^3\overline{W}_k.$$

The average kinetic energy $\overline{W}_k$ can be found from the theorem on uniform energy distribution over degrees of freedom. Since the rotator has two degrees of freedom, $\overline{W}_k = T$. Substituting Equations (4) and (2) in Equation (1), we find $\chi = 0$. This result is consistent with the general

Bohr–Van Leeuween theorem (Problem 1.15) according to which the total magnetic moment of a body obeying the classical statistics is zero. A nonzero magnetic moment will only be obtained assuming discrete electron orbits in atoms. However, such an assumption goes beyond the scope of the classical theory.

1.22 In the absence of a field, each pair of electrons of equal energy has, according to the Pauli principle, the opposite spin projections, and the microscopic magnetic moment of the system equals zero. In the presence of the field, each of the electrons acquires an additional energy $\pm\mu_B B$ depending on the direction of its spin projection. In other words, the Fermi levels of electrons with different spin projections shift by $2\mu_B B$. However, in a statistically equilibrated system there must be a single Fermi level for all particles. It means that some electrons will change the spin projection. The number density N' of spin-flip electrons can be determined from the condition $\epsilon_{F+} = \epsilon_{F-}$, or

$$\frac{\hbar^2}{2m}(6\pi)^{2/3}\left(\frac{N}{2} - N'\right)^{2/3} + \mu_B B = \frac{\hbar^2}{2m}(6\pi)^{2/3}\left(\frac{N}{2} + N'\right)^{2/3} - \mu_B B. \quad (1)$$

The same condition ensures minimum of internal and free energies at $T = 0$.

For a weak field ($\mu_B B \ll \epsilon_F$) we will have $N' \ll N$, and from Equation (1) we obtain

$$N' = \frac{3\mu_B B}{4\epsilon_F} N. \quad (2)$$

The magnetization is calculated from the formula $M = 2\mu_B N'$, and the paramagnetic susceptibility of a completely degenerate electron gas ($T = 0$) is given by

$$\chi_{para} = \frac{3\mu_B^2 N}{2\epsilon_F}. \quad (3)$$

The temperature corrections to the quantity (Equation (3)) are of the order of $(T/\epsilon_F)^2 \approx 10^{-4}$ for the majority of metals at room temperature. For this reason, the "Pauli paramagnetism" (Equation (3)) unlike the Curie law [see Equation (4) in Problem 1.16 and Equation (13) in Problem 1.17] is almost independent of temperature.

A degenerate electron gas exhibits also the diamagnetic susceptibility (Landau diamagnetism):

$$\chi_{dia} = -\frac{1}{3}\chi_{para}, \qquad \chi = \chi_{para} + \chi_{dia} = \frac{2}{3}\chi_{para} \quad (4)$$

(see Landau and Lifshitz (1980) and Problem 1.23).

1.23 In order to calculate the magnetization, we use the formulas of statistical physics for the free energy: $F = -T \ln Z$ ((I.C46)) and $M = -\partial F/\partial B$ (Section 3.3). Since the gas is rarefied and homogeneous, we ignore the electron-electron interaction and calculate the free energy per unit volume from the formula $F = -NT \ln z$, where z is the statistical sum of an individual electron regarded as a quasi-independent equilibrium subsystem. With the aid of the data given in the statement of the problem, we find

$$(1) \qquad z = z_s z_{orb}, \quad z_s = \sum_{m_s=-1/2}^{m_s=1/2} \exp\left(-\frac{2\mu_B B m_s}{T}\right) = 2\cosh\left(\frac{\mu_B B}{T}\right),$$

$$z_{orb} = \sum_{n=0}^{\infty} \exp\left(-\frac{\hbar\omega_B}{2T}(2n+1)\right) \int \exp\left(-\frac{p_z^2}{2mT}\right) \frac{eBV dp_z}{(2\pi\hbar)^2 c}$$

$$(2) \qquad = \sqrt{2\pi m T}\frac{eBV}{(2\pi\hbar)^2 c} \frac{1}{2\sinh(\hbar\omega_B/2T)},$$

where z_s relates to the spin states and z_{orb} to the states of orbital motion. Further on, we take into consideration that $\hbar\omega_B/2 = \mu_B B$ and $\mu_B B \ll T$. Carrying out the small-argument expansion of the hyperbolic functions, we find

$$(3) \qquad M = NT\left(\frac{\partial z_s}{\partial B} + \frac{\partial z_{orb}}{\partial B}\right) = \frac{N\mu_B^2 B}{T} - \frac{N\mu_B^2 B}{3T}.$$

The first and second terms in the right-hand side describe the paramagnetic and diamagnetic effects, respectively. Now the corresponding susceptibilities depend on temperature; however the relation between them remains the same as for degenerate gas:

$$(4) \qquad \chi_{para} = \frac{N\mu_B^2}{T}, \qquad \chi_{dia} = -\frac{1}{3}\chi_{para}.$$

1.24 The number densities of ions (N) and electrons (n) can be determined from the Boltzmann formula (1.31)[10)]

$$(1) \qquad N = N_0 \exp\left(-\frac{Ze\varphi}{T}\right), \quad n = n_0 \exp\left(\frac{e\varphi}{T}\right),$$

where $k_B \approx 1.38 \times 10^{-16}$ erg/K is the Boltzmann constant and $\varphi(x, y, z)$ is the electric potential. The pre-exponents are chosen so that as $T \to \infty$ (when the interaction between the particles dies out) N and n become N_0 and n_0. With Equation (1), the charge density becomes

$$(2) \qquad \rho = ZeN - en = eZN_0 \exp\left(-\frac{Ze\varphi}{k_B T}\right) - en_0 \exp\left(\frac{e\varphi}{k_b T}\right).$$

The potential φ should be determined by solving the Poisson equation:

$$(3) \qquad \triangle\varphi = -4\pi\rho = -4\pi e Z N_0 \exp\left(-\frac{Ze\varphi}{k_B T}\right) + 4\pi e n_0 \exp\left(\frac{e\varphi}{k_B T}\right).$$

10) In this and next three problems, the quantities being considered are averaged over statistical ensemble but not over volume.

To do this, we take into account that the interaction energy is small compared with the thermal energy:

$$\left|\frac{Ze\varphi}{k_B T}\right| \ll 1, \qquad \left|\frac{e\varphi}{k_B T}\right| \ll 1.$$

Expanding the exponents into a series to within the terms linear in φ, and using the condition of electrical neutrality of the gas, $ZN_0 = n_0$, we obtain

(4) $$\rho = -\frac{\kappa^2}{4\pi}\varphi, \qquad \kappa^2 = \frac{4\pi e^2(Z^2 N_0 + n_0)}{k_B T}.$$

Equation (3) can therefore be written in the form

(5) $$\triangle\varphi = \kappa^2\varphi.$$

The potential φ can depend only on distance r to the ion under consideration. The spherically symmetric solution of Equation (5) is

$$\varphi = C_1\frac{e^{-\kappa r}}{r} + C_2\frac{e^{\kappa r}}{r}.$$

The potential cannot grow up at infinity, and hence, $C_2 = 0$. The constant C_1 is determined from the condition that at $r \ll 1/\kappa$ it must become the purely Coulomb potential of the considered ion:

$$\varphi\big|_{r\ll 1/\kappa} = \frac{Ze}{r} = \frac{C_1}{r}, \qquad C_1 = Ze.$$

Thus, the ion is surrounded by a "cloud" of electrons and other ions. The density of the cloud exponentially decreases. The lower the temperature, the less the average radius $1/\kappa = r_D$.

The presented method for calculating the potential was proposed by Debye[11] and Hückel[12] and was employed by them in the theory of strong electrolytes. The constant $1/\kappa = r_D$ is referred to as the *Debye–Hückel radius*.

1.25 The electric induction inside the plate is described by the formula

$$D(x) = E_0\frac{\cosh\kappa x}{\cosh\kappa h},$$

where $\kappa = \sqrt{4\pi e^2 n_0/\varepsilon T}$. At $\kappa h \gg 1$, for the region near the surfaces $x = \pm h$, we have

$$D(x) = E_0 e^{-\kappa(h-|x|)};$$

hence, as $|x - h| \gg 1/\kappa$ we have $D(x) \to 0$, that is, the field penetrates into the conductor at a depth of about $1/\kappa$. The charge is concentrated in the

11) Debye Peter (1884–1966), outstanding European scientist who worked in Switzerland, Holland, Germany, and United States, had a wide range of scientific interests: theory of crystal lattice, diffraction of X-rays, quantum theory of atoms, Compton effect, theory of strong electrolytes, and so on. Nobel Prize laureate in chemistry.

12) Hückel Erich (1896–1980), German physicist-theorist. Worked mainly in the field of quantum mechanics, quantum chemistry and electrochemistry.

layer of the same thickness,

$$\rho = \frac{1}{4\pi}\frac{\partial D}{\partial x} = \pm\frac{\kappa E_0}{4\pi} e^{-\kappa(h-|x|)}.$$

The "surface" charge density, which is considered in the macroscopic theory, is determined by integrating ρ. On the boundary $x = h$, we obtain

$$\sigma = \int \rho \, dx = -\frac{\kappa E_0}{4\pi}\int_0^{\infty} e^{-\kappa x'}\, dx' = \frac{E_0}{4\pi},$$

which is the same as the usual boundary condition on the surface of a conductor.

1.26

$$\varphi = \varphi_0 \frac{\sinh \kappa x}{\sinh \kappa h}, \qquad \kappa = \sqrt{\frac{8\pi e^2 n_0}{\varepsilon T}}.$$

In the present case the value of κ^2 is twice larger than in Problem 1.24 since there are mobile ions of two species.

1.27 Unlike Problem 1.24, this one deals with the electrons obeying the Fermi distribution. We start from the assumption that the states of the electrons in the vicinity of the ion are semi-classical. The origin is chosen at the point of the ion location and the electron number density is denoted by $n(r)$. Far from the ion the boundary conditions are given by $n(r)|_{r\to 0} \to n_0$, $\varphi(r)|_{r\to 0} \to 0$, where $\varphi(r)$ is the electrostatic potential. At temperature $T \to 0$ all electron energy levels from the zero-level to the Fermi level ϵ_F are occupied. Far from the ion the Fermi level is determined by the conditions

$$(1) \qquad \frac{2}{(2\pi\hbar)^3}\cdot\frac{4}{3}\pi p_F^3 = n_0, \qquad \epsilon_F = \frac{p_F^2}{2m} = \frac{(3\pi^2 n_0)^{2/3}\hbar^2}{2m}.$$

Here, the Fermi momentum p_F is calculated from the fact that the number of occupied states $p_F^3/3\pi^2\hbar^3$ is equal to the number of the electrons n_0 per unit volume.

The Fermi energy retains the value given by Equation (1) throughout space, since, otherwise, the electrons will move to the sites with lower ϵ_F. However, at finite r the Fermi energy ϵ_F should be regarded as a sum of the kinetic and potential energies,

$$(2) \qquad \epsilon_F = \frac{(3\pi^2)^{2/3}\hbar^2}{2m} n^{2/3}(r) - e\varphi(r).$$

Expressing $n(r)$ from Equations (1) and (2), we have

$$(3) \qquad n(r) = n_0\left[1 + \frac{e\varphi(r)}{\epsilon_F}\right]^{3/2} \approx n_0\left[1 + 3\frac{e\varphi(r)}{2\epsilon_F}\right].$$

Using the last equation, we find the charge density distribution in the vicinity of the ion, $\rho(r) = -(3e^2 n_0/2\epsilon_F)\varphi(r)$. For the potential we obtain

Equation (5) from the solution of Problem 1.24, where the screening constant is given by

(4) $$\kappa^2 = \frac{4me^2}{\hbar^2}\left(\frac{3n_0}{pi}\right)^{1/3} = \frac{8}{a_B \lambda_F}.$$

Here, $a_B = \hbar^2/me^2$ is the Bohr radius, and $\lambda_F = 2\pi\hbar/p_F$ is the De Broglie wavelength of the electron at the Fermi level.

1.28 Let us use the kinetic equation as in Example 1.5. For the equilibrium distribution function, we substitute the Fermi step-function $f_0(\epsilon)$ ($f_0(\epsilon) = 1$, $\epsilon \leq \epsilon_F$; $f_0(\epsilon) = 0$, $\epsilon > \epsilon_F$, where ϵ_F is the Fermi energy expressed through particle number density – see the statement of Problem 1.22). Using the relation $\boldsymbol{v} = \partial\epsilon/\partial\boldsymbol{p}$, we write

(1) $$\delta f = -e(\boldsymbol{E}\cdot\boldsymbol{v})\tau(\epsilon)\frac{\partial f_0}{\partial \epsilon},$$

and obtain the expression for the current density

(2) $$\boldsymbol{j} = -e^2 \int \tau(\epsilon)\boldsymbol{v}(\boldsymbol{E}\cdot\boldsymbol{v})\frac{\partial f_0}{\partial \epsilon}\frac{2\,d^3p}{(2\pi\hbar)^3},$$

where the last fraction under the integral presents the number of the quantum states including two spin projections.

For a degenerate electron gas, we have $\partial f_0/\partial\epsilon = -\delta(\epsilon - \epsilon_F)$. Substituting the required quantities into Equation (2) and integrating over the energy with the aid of the delta-function, we obtain the electric conductivity, which is the coefficient of proportionality between the current and the electric field strength:

(3) $$\kappa = \frac{ne^2\tau(\epsilon_F)}{m}.$$

This result coincides with the Drude formula (1.42).

In real metals, an electron gas interacts with a crystal lattice. As a result, the energy dependence of the particle on its momentum, $\epsilon(\boldsymbol{p})$, becomes more complicated, the effective mass of charge carriers may be anisotropic, and the Fermi surface in momentum space non-spherical. In this case, the simple model considered above is inapplicable.

1.29 A nonequilibrium additive correction to the electron distribution function, produced by weak electric and magnetic fields in the first order, is calculated from the kinetic equation and has the form

(1) $$\delta f = \left(-e(\boldsymbol{E}\cdot\boldsymbol{v})\tau + \frac{e^2\tau^2}{mc}[\boldsymbol{E}\times\boldsymbol{B}]\cdot\boldsymbol{p}\right)\frac{\partial f_0}{\partial \epsilon}.$$

Using this distribution function, we find the current density

(2) $$\boldsymbol{j} = \kappa\boldsymbol{E} + \boldsymbol{E}\times\boldsymbol{a},$$

where

(3) $$\kappa = \frac{ne^2\tau}{m}, \qquad \boldsymbol{a} = \frac{ne^3\tau^2}{m^2c}\boldsymbol{B}.$$

The electric conductivity, under the action of the magnetic field, becomes anisotropic,

(4) $$\kappa_{\alpha\beta} = \kappa\delta_{\alpha\beta} - e_{\alpha\beta\gamma}a_\gamma,$$

and the electric current perpendicular to the magnetic field is induced (**the Hall[13] current**). The inverse dependence between the current and the electric field in the same approximation is of the form

(5) $$\boldsymbol{E} = \frac{1}{\kappa}\boldsymbol{j} - R[\boldsymbol{j} \times \boldsymbol{B}],$$

where $R = 1/cen$ is the Hall constant.

1.30 We write down the stationary kinetic equation with allowance for the first-order terms in E, but without restriction on values of B:

(1) $$e\boldsymbol{E}\cdot\frac{\partial f_0}{\partial \boldsymbol{p}} + \frac{e}{c}[\boldsymbol{v} \times \boldsymbol{B}]\cdot\frac{\partial \delta f}{\partial \boldsymbol{p}} = -\frac{\delta f}{\tau}.$$

Multiply both parts of Equation (1) by $e\boldsymbol{v}$, integrate over momenta, and transform this equation into the algebraic equation for the current $\boldsymbol{j} = e\int \boldsymbol{v}\delta f\, d^3p$. Multiplying the result by τ, we obtain

(2) $$\kappa\boldsymbol{E} = \boldsymbol{j} - \tau\boldsymbol{j} \times \boldsymbol{\omega}_B,$$

where κ is given by Equation (3) of Problem 1.29, and $\boldsymbol{\omega}_B = e\boldsymbol{B}/mc$ is the cyclotron frequency. Resolving Equation (2) with respect to the components of $\boldsymbol{j}$, we find $j_\alpha = \kappa_{\alpha\beta}E_\beta$, where the electric conductivity tensor is

(3) $$\kappa_{\alpha\beta} = \begin{pmatrix} \kappa_\perp & \kappa_H & 0 \\ -\kappa_H & \kappa_\perp & 0 \\ 0 & 0 & \kappa_\parallel \end{pmatrix}, \text{ where } \kappa_\perp = \frac{\kappa}{1+(\omega_B\tau)^2},$$

$$\kappa_H = \frac{\kappa(\omega_B\tau)}{1+(\omega_B\tau)^2}, \quad \kappa_\parallel = \kappa.$$

The subscripts $\perp$ and $\parallel$ denote the directions perpendicular and parallel to the magnetic field, respectively. The component κ_H is responsible for the Hall current.

The quantity $\omega_B\tau$ is the rotation angle of the transverse momentum of the particle during relaxation time, when it is moving along a spiral trajectory. If $\omega_B\tau \ll 1$, the effect of the magnetic field is insignificant and the electric conductivity is almost isotropic. In the opposite case, when $\omega_B\tau \gg 1$, the conductivity is strongly anisotropic: $\kappa_\parallel \gg \kappa_H \gg \kappa_\perp$.

13) Hall Adven Herbert (1855–1938), American physicist.

2
Electrostatics of Conductors and Dielectrics

2.1
Basic Concepts and Methods of Electrostatics

Equations and boundary conditions. The electrostatic field in a dielectric is characterized by the electric field vector $\boldsymbol{E}$ and the electric induction vector $\boldsymbol{D}$ which satisfy, according to Equation (1.20), the following equations:

$$\left.\begin{array}{lllll} rot\,\boldsymbol{E} & = & 0, & \oint_l \boldsymbol{E}\cdot d\boldsymbol{l} & = 0, \\ div\,\boldsymbol{D} & = & 4\pi\rho_{ext}, & \oint_S \boldsymbol{D}\cdot d\boldsymbol{S} & = 4\pi q, \end{array}\right\} \tag{2.1}$$

and the coupling equation (1.26)

$$\boldsymbol{D} = \varepsilon \boldsymbol{E} \quad \text{or} \quad D_\mu = \varepsilon_{\mu\nu} E_\nu \tag{2.2}$$

(the latter form is used for an anisotropic dielectric). Here ρ_{ext} is the density of charges external with respect to the dielectric and q is the total external charge enclosed by the surface S. The density of bound macroscopic volume and surface charges of the dielectric (denoted in the present chapter by ρ and σ without any index) are given by Equations (1.16) and (1.17), respectively. The field vectors at the boundary separating two different dielectrics satisfy the boundary conditions, which can be derived from the integral equations (2.1). They are of the form (see the derivation of Equations (I.2.17) and (I.2.18))

$$(\boldsymbol{E}_2 - \boldsymbol{E}_1) \times \boldsymbol{n} = 0, \qquad (\boldsymbol{D}_2 - \boldsymbol{D}_1)\cdot\boldsymbol{n} = 4\pi\sigma_{ext}. \tag{2.3}$$

For the description of the field, it is convenient to use a scalar quantity, namely, the electrostatic potential $\varphi(\boldsymbol{r})$:

$$\boldsymbol{E} = -\nabla\varphi, \qquad \varphi(\boldsymbol{r}) = \int_{\boldsymbol{r}}^{\boldsymbol{r}_0} \boldsymbol{E}\cdot d\boldsymbol{r}, \tag{2.4}$$

where $\varphi(\boldsymbol{r}_0) = 0$. The potential satisfies the equation, which follows from Equations (2.1) and (2.2):

$$div\,(\varepsilon grad\,\varphi) = -4\pi\rho_{ext}. \tag{2.5}$$

In those regions, where the dielectric is uniform, the latter equation reduces to the Poisson equation

$$\triangle\varphi = -\frac{4\pi\rho_{ext}}{\varepsilon}. \tag{2.6}$$

Electromagnetic Phenomena in Matter: Statistical and Quantum Approaches, First Edition. Igor N. Toptygin.

The boundary conditions for the electrostatic potential at the boundary between two media with different dielectric properties follow from Equations (2.1) and (2.2). For isotropic dielectrics they read (cf. Equations (I.2.19) and (I.2.20)):

$$\varphi_1 = \varphi_2, \qquad \varepsilon_1 \frac{\partial \varphi_1}{\partial n} - \varepsilon_2 \frac{\partial \varphi_2}{\partial n} = 4\pi\sigma_{ext}. \tag{2.7}$$

The unit vector of the normal $\boldsymbol{n}$ is directed from the first medium into the second.

Inside a conductor the distribution of free charges is such that the resulting electric field is zero. In this case, according to the Gauss electrostatic theorem, $\rho = 0$ inside the conductor, and free charges are localized in a thin surface layer (Problems 1.25 and 1.26). The boundary conditions on the surface of a conductor, S, are

$$\boldsymbol{E}_\tau|_S = 0, \qquad \varphi|_S = const. \tag{2.8}$$

The surface charge density on the conductor is given by

$$\sigma = \frac{1}{4\pi} D_n = -\frac{\varepsilon}{4\pi}\frac{\partial \varphi}{\partial n}. \tag{2.9}$$

In the general case, the distribution of the surface charge in the conductor is nonuniform and cannot be specified in advance. However, the total charge of the conductor, q, can be given, and in this case, the boundary conditions should be supplemented by

$$\int_S \nabla_\mu \varphi \, dS_\mu = -\frac{4\pi q}{\varepsilon}, \tag{2.10}$$

where the integral is taken over the surface of the conductor.

Example 2.1

Consider a set of conductors and generally anisotropic dielectrics, which occupies a finite region of space. Prove that the electrostatic problem has a unique solution if the distribution of external charges, ρ_{ext}, in dielectrics is given, and also if either the electrostatic potential of each conductor, φ_i, or the total charge, q_i, is given. Use the symmetry property of the dielectric tensor $\varepsilon_{\mu\nu} = \varepsilon_{\nu\mu}$, and the fact that its principal values are positive, $\varepsilon^{(\mu)} > 1$ (Section 2.3).

Solution. For the proof, let us assume the opposite (that the statement is incorrect). Let φ_1 and φ_2 be two different potentials, which satisfy the equation

(1) $$\nabla_\mu \varepsilon_{\mu\nu} \nabla_\nu \varphi_{1,2} = -4\pi\rho_{ext}$$

and the boundary conditions

(2) $$\varphi_{1,2}|_{r\to\infty} \to 0, \qquad \varphi_{1,2}|_{S_i} = \varphi_i,$$

or

(3) $$\varphi_{1,2}|_{r\to\infty} \to 0, \qquad \varphi_{1,2}|_{S_i} = const, \qquad \oint_{S_i} \varepsilon_{\mu\nu}\nabla_\nu \varphi \, dS_\mu = -4\pi q_i.$$

We apply the Gauss–Ostrogradskii theorem to the vector $a_\mu = \psi\varepsilon_{\mu\nu}\nabla_\nu\psi$, where $\psi = \varphi_1 - \varphi_2$:

$$(4) \qquad \int \psi\nabla_\mu\varepsilon_{\mu\nu}\nabla_\nu\psi\, dV + \int \varepsilon_{\mu\nu}\nabla_\mu\psi\nabla_\nu\psi\, dV = \oint_S \psi\varepsilon_{\mu\nu}\nabla_\nu\psi\, dS_\mu.$$

Here, the integration is carried out over the entire space occupied by dielectrics, while S includes an infinitely distant surface and surfaces of all conductors. The first integral in the left-hand side vanishes by virtue of equality (Equation (1)), and the integral in the right-hand side turns to zero due to boundary conditions (2) and (3). Thus,

$$(5) \qquad \int \varepsilon_{\mu\nu}(\nabla_\mu\psi)(\nabla_\nu\psi)\, dV = 0.$$

By writing the integrand in the principal axes of the dielectric tensor, we verify that the potentials φ_1 and φ_2 differ by no more than a constant and describe the same electrostatic field. ■

Potential and capacitance coefficients. The capacitance C of a capacitor is defined as the ratio of a charge q on one of its plates (the first one) to the potential difference $V = \varphi_1 - \varphi_2$ between the plates:

$$C = \frac{q}{V}. \tag{2.11}$$

The capacitance of an isolated conductor is defined as the ratio of the charge on the conductor to its potential (the potential must vanish at infinity).

If n conductors are placed in a dielectric medium having the linear relation between the electric induction and the field strength, then the dependence of the potentials of the conductors, V_i, on their charges q_i will also be linear:

$$V_i = \sum_{k=1}^{n} s_{ik} q_k \qquad (i = 1, 2, \ldots, n). \tag{2.12}$$

The quantities s_{ik} are called **potential coefficients**. They depend on relative position, shape, and geometric dimensions of conductors as well as on the dielectric permittivity of the surrounding medium. The matrix $\hat{s}$ is symmetric:

$$s_{ik} = s_{ki} \tag{2.13}$$

(Problem 2.27). The quantity s_{ik} is the potential acquired by the ith conductor if the charge $q_k = 1$ is imparted to the kth conductor, while the other conductors remain uncharged. All $s_{ik} > 0$.

It follows from Equation (2.12) that the charges of the conductors are linear functions of the potentials

$$q_i = \sum_{k=1}^{n} c_{ik} V_k \qquad (i = 1, 2, \ldots, n). \tag{2.14}$$

The quantities c_{ik} are called **capacitance coefficients**, with $c_{ii} > 0$ (self-capacitances); $c_{ik} = c_{ki} < 0$, $i \neq k$ (mutual capacitances). The quantities c_{ik} represent the charge acquired by the ith conductor when all conductors, except for kth conductor, are grounded, and kth conductor has the potential $V_k = 1$. The matrices s_{ik} and c_{ik} are mutually inverse.

In the case of an isolated conductor, there is a single capacitance coefficient, c_{11}, which is called **simply capacitance**. The capacitance of the capacitor (Equation (2.11)) may be expressed in terms of the capacitance coefficients of its electrodes (Problem 2.31).

In many cases, it is necessary to know the polarizability of a conducting or dielectric body in the external uniform electric field. It is described by the second-rank tensor $\beta_{\mu\nu}$, which relates the projections of a dipole moment $\mathcal{P}_\mu$ of the body with the field strength $\mathcal{E}_\nu$ at a large distance from the body:

$$\mathcal{P}_\mu = V\beta_{\mu\nu}\mathcal{E}_\nu. \tag{2.15}$$

Here, the volume of the body is separated as an individual factor, and the tensor $\beta_{\mu\nu}$ is therefore the dimensionless quantity, which characterizes the shape anisotropy and, in the case of a dielectric, the dielectric susceptibility of its material as well (cf. Equation (1.29)). Sometimes, the factor V in Equation (2.15) is not separated, and the polarizability tensor acquires the dimension of volume (see, for instance, Problems 1.4 and 1.5).

Example 2.2

Prove the electrostatic Green's reciprocity theorem: if the potentials of n conductors are equal to $V_1, V_2, \dots, V_n$ when their charges are $q_1, q_2, \dots, q_n$, and are equal to $V_1', V_2', \dots, V_n'$ when their charges are $q_1', q_2', \dots, q_n'$, then the following relation holds:

$$\sum_{i=1}^{n} q_i V_i' = \sum_{i=1}^{n} q_i' V_i. \tag{2.16}$$

At the same time, the dielectric medium between the conductors may be inhomogeneous and anisotropic. However, the external charges in the dielectric must be absent.

Solution. Proceed from the identity representing the generalization of Equation (4) in Example (2.1):

$$\int (\varphi'\nabla_\mu\varepsilon_{\mu\nu}\nabla_\nu\varphi - \varphi\nabla_\mu\varepsilon_{\mu\nu}\nabla_\nu\varphi')\,dV = \oint_S (\varphi'\varepsilon_{\mu\nu}\nabla_\nu\varphi dS_\mu - \varphi\varepsilon_{\mu\nu}\nabla_\nu\varphi')\,dS_\mu.$$

Due to the absence of external charges in the dielectric, the differential terms in the integrand in the left-hand side of this equation are zero. The right-hand side retains the sum of integrals over the surfaces of the conductors, which leads to Equation (2.16) by virtue of the boundary conditions on the conductors. ■

Suggested literature: Tamm (1976); Landau *et al.* (1984); Frenkel (1928); Smythe (1950); Bredov *et al.* (2003); Sommerfeld (1952); Vlasov (1955); Toptygin (2014)

Problems

2.1 A point charge q is located on the plane between two homogeneous infinite dielectrics with permittivities ε_1 and ε_2. Find the potential φ, the electric field strength $\boldsymbol{E}$, and the induction $\boldsymbol{D}$.

2.2 Three half-planes radiate outward from a straight line on which a point charge q is located. The half-planes form three dihedral angles α_1, α_2, and α_3 ($\alpha_1 + \alpha_2 + \alpha_3 = 2\pi$). The space between each pair of angles is filled with homogeneous dielectrics with permittivities ε_1, ε_2, and ε_3. Find the potential φ, the electric field strength $\boldsymbol{E}$, and the induction $\boldsymbol{D}$.

2.3 The center of a conducting sphere carrying a charge q lies on the plane boundary between two infinite homogeneous dielectrics with permittivities ε_1 and ε_2. Find the potential φ of the electric field and the charge distribution σ on the sphere. The radius of the sphere is a.

2.4 The space between the electrodes of a spherical capacitor is partly filled with a dielectric, which occupies a region inside a solid angle Ω with a vertex in the center of the spheres. The radii of the spherical electrodes are a and b, and the dielectric permittivity is ε. Find the capacitance C of the capacitor.

2.5 Inside a spherical capacitor with the electrodes of radii a and b the dielectric permittivity varies as

$$\varepsilon(r) = \begin{cases} \varepsilon_1 & = const \quad \text{at} \quad a \le r < c, \\ \varepsilon_2 & = const \quad \text{at} \quad c \le r \le b, \end{cases}$$

where $a < c < b$. Find the capacitance C of the capacitor, the distribution of bound charges σ_{int}, and the total bound charge in the dielectric.

2.6 A spherical capacitor, whose electrodes have radii a and b, is filled with a dielectric, whose permittivity is given by $\varepsilon(r) = \varepsilon_0 a^2/r^2$, where r is the distance from the center. Show that the capacitance of this capacitor is equal to the capacitance of a plane capacitor filled with a homogeneous dielectric with permittivity ε_0, electrode area $4\pi a^2$, and distance between the electrodes equal to $b - a$ (neglect edge effects).

2.7 A plane parallel capacitor is filled with a dielectric with the permittivity given by $\varepsilon = \varepsilon_0(x + a)/a$, $0 \le x \le a$, where a is the distance between the plates, and the x-axis is perpendicular to the plates having the area S. Neglecting the edge effects, find the capacitance C of this capacitor and the distribution of bound charges if a potential difference V is applied between the plates.

2.8 A point charge q is placed at a point A lying at a distance a from the plane boundary separating two homogeneous infinite dielectrics with permittivities ε_1 and ε_2 (Figure 2.1). Find the electric field potential φ by the method of images.

Hint. Look for the solution in the form

$$\varphi_1 = \frac{q}{\varepsilon_1 r_1} - \frac{q'}{\varepsilon_1 r_2} \quad \text{at} \quad z \ge 0, \qquad \varphi_2 = \frac{q''}{\varepsilon_2 r_1} \quad \text{at} \quad z < 0,$$

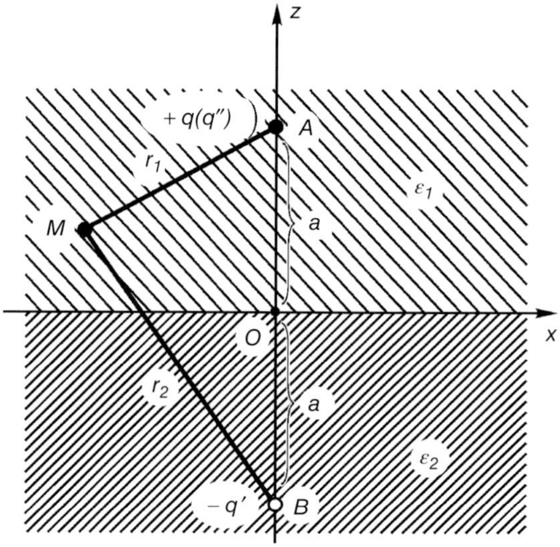

Figure 2.1 The arrangement of images-charges.

where q' and q'' are the required effective charges placed at points B and A, respectively; r_1 and r_2 are shown in the figure.

2.9 Find the density σ_{int} of bound surface charges on the plane boundary between two homogeneous dielectrics with permittivities ε_1 and ε_2, which is due to the point charge q (Problem 2.8). What result can be expected when $\varepsilon_2 \to \infty$? What is its physical meaning?

2.10 The dihedral angle between two grounded conducting planes is α_0. There is a point charge q inside the angle. Find the electric field by the method of electric images. Consider the cases $\alpha_0 = 90°$, $\alpha_0 = 60°$, and $\alpha_0 = 45°$.

2.11* A homogeneous sphere of radius a and dielectric permittivity ε_1 is immersed in an unbounded homogeneous dielectric of permittivity ε_2. At a large distance from the sphere an electric field in the dielectric is uniform and has strength $\boldsymbol{E}_0$. Find the field φ in the whole of space. Sketch the lines of force for the two cases: $\varepsilon_1 > \varepsilon_2$ and $\varepsilon_1 < \varepsilon_2$; find the distribution of bound charges.

2.12 An infinite dielectric was at first homogeneous and uniformly polarized (polarization vector $\boldsymbol{P} = const$). A spherical cavity was then made inside it. Determine the change in the electric field $\Delta\boldsymbol{E}$ in the cavity for the two cases: (i) the presence of the cavity does not affect the polarization in the surrounding dielectric;[1] (ii) the change in the field induces the change in the polarization, that is, $\boldsymbol{P} = (\varepsilon - 1)\boldsymbol{E}/4\pi$.

1) Such a situation takes place when the dielectric (electret) consists of polar molecules of fixed orientation.

2.13 An uncharged metal sphere of radius R is introduced into an electric field which was uniform and equal to $\boldsymbol{E}_0$ in the absence of the sphere. The dielectric permittivity of the surrounding medium is $\varepsilon_0 = const$. Find the resulting field φ and density of the surface charges σ on the sphere.

2.14⋆ A conducting sphere of radius R is placed in the field of a point charge q located at a distance $a > R$ from the center of the sphere. The system is immersed in a homogeneous dielectric of permittivity ε. Find the field potential φ and the distribution of charges induced on the sphere σ for the two cases; (i) the potential of the sphere is V ($\varphi = 0$ at infinity) and (ii) the charge on the sphere is Q. Represent the potential in the form of the sum of potentials due to a number of point-charges images.

Hint. Use the solution of the Laplace equation as a series in terms of spherical harmonics and the expansion of the point-charge field which was obtained in Problem I.2.30.

2.15 A conductor having the potential V contains a spherical cavity of radius R, which is filled with a dielectric of permittivity ε. A point charge q is placed at a distance a from the center of the cavity ($a < R$). Determine the field in the cavity and the equivalent system of image-charges.

2.16 A grounded conducting plane has a hemispherical boss of radius a. The center of the sphere lies at the plane. The point charge q is placed on the axis of the symmetry of the system at a distance $b > a$ from the plane. Use the method of images to find the field φ and the charge q' induced on the boss.

2.17 A conducting sphere of radius R_1 is placed in a homogeneous dielectric of permittivity ε_1. The sphere contains a spherical cavity of radius R_2 which is filled with a homogeneous dielectric of permittivity ε_2. The point charge q is located in the cavity at a distance a from its center ($a < R_2$). Find the field φ in the entire space.

2.18 The point charge q is placed at a distance a from the center of a dielectric sphere of radius R and permittivity ε_1. The dielectric permittivity of the medium outside the sphere is ε_1. Find the field φ in the whole of space. Consider the special case $a = 0$ (charge at the center of the sphere).

2.19⋆ An insulated metal sphere of radius a is placed inside a hollow metal sphere of radius b. The distance between the centers of the two spheres is c, where $c \ll a$ and $c \ll b$. These spheres form the spherical capacitor. Use the first nonvanishing approximation to evaluate the correction ΔC to the capacitance, which is caused by the departure from the exact concentricity.

2.20 An electrostatic field is produced by two conducting cylinders with parallel axes, radii R_1 and R_2, and charges $\pm\kappa$ per unit length. A distance between the axes is $a > R_1 + R_2$. Find the capacitance C per unit length of the capacitor whose electrodes are these cylinders ($C = \kappa/(\varphi_1 - \varphi_2)$), where φ_1 and φ_2 are the potentials of the cylinders.

Hint. Use the result of Problem I.2.51.

2.21 The axes of two identical conducting cylinders with radii R are placed at a distance a from each other. The cylinders carry a charge $\pm\kappa$ per unit length. Find the distribution of charge σ on the surfaces of the two cylinders.

2.22 A capacitor consists of two cylindrical conducting surfaces of radii R_1 and $R_2 > R_1$. A distance between the axes of the cylinders is $a < R_2 - R_1$. Find the capacitance C of this capacitor.

2.23 Determine the field φ due to a point charge in a homogeneous anisotropic medium characterized by a permittivity tensor ε_{ik}.

2.24 A plane parallel plate made of an anisotropic homogeneous dielectric having a permittivity tensor ε_{ik} is placed in vacuum. A uniform electric field outside the plate is $\boldsymbol{E}_0$. Using the boundary conditions for the field vector, find the field $\boldsymbol{E}$ inside the plate.

2.25 Find the capacitance C of a plane parallel capacitor having the plate area S, and a distance a between the plates. The space between the plates is filled with an anisotropic dielectric of permittivity ε_{ik}. The edge effect should be neglected.

2.26 Find the change in the direction of the lines of a vector $\boldsymbol{E}$ on passing from vacuum to an anisotropic dielectric. Use the results of Problem 2.24.

2.27 Prove the symmetry property of the potential coefficients $s_{ik} = s_{ki}$ with the aid of the Green's reciprocity theorem (2.16).

2.28 A system consists of two conductors at a large distance from all other conductors. The conductor 1 is placed inside the hollow conductor 2. Express the capacitances C and C' of the capacitor and the isolated conductor forming this system in terms of its capacitance coefficients. Prove that mutual capacitance of conductor 1 and any other conductor located outside conductor 2 is equal to zero.

2.29 Express the potential coefficients s_{ik} in terms of the capacitance coefficients c_{ik} in the case of a system of two conductors.

2.30 The capacitances of two isolated conductors are C_1 and C_2. These conductors are placed in a homogeneous dielectric of permittivity ε at a distance r from each other, where r is large compared with the linear dimensions of the conductors. Calculate the capacitance coefficients of the system.

Hint. First, determine the potential coefficients to within terms of the order of $1/r$.

2.31 The capacitance coefficients of a system of two conductors equal c_{11}, c_{22}, and $c_{12} = c_{21}$. Find the capacitance C of a capacitor formed by the two conductors.

2.32 Four identical conducting spheres are placed at the corners of a square. Sphere 1 carries a charge q. It is then connected by a thin wire in turn to spheres 2, 3, and 4 (cyclic numeration) until equilibrium is established in each case. Find the charge on each of the conductors at the end of all operations. The potential coefficients of the system are given.

2.33 Three identical conducting spheres of radii a are placed at the corners of an equilateral triangle with sides $b \geq a$. All the spheres carry equal charges q.

They are then grounded in turn until the equilibrium is established. What is the charge of each sphere at the end of all operations?

2.34 A closed conducting surface at a potential V_1 contains a conductor at a potential V_0. At the same time, the potential at a point P in the space between the two conducting surfaces is equal to V_P. The conductors are then grounded and a charge q is placed at the point P. Find the charges induced on the conductors.

2.35 Show that the geometric locus of points from which a unit charge induces in a grounded conductor a charge of equal magnitude is an equipotential surface in the field of this conductor in the absence of the point charge.

2.36 Two conductors with self-capacitances c_{11} and c_{22} and mutual capacitance c_{12} constitute a part of a system of insulated conductors connected by a thin wire. Determine the self-capacitance of the joint conductor and the coefficients of mutual capacitance between this conductor and all other conductors of the system.

2.37⋆ A conductor is charged by means of successive contacts with the discharge sphere of an electrophorus. The sphere is recharged to a charge Q after each contact. At the first contact a charge given to the conductor from the discharge sphere is q. What is the charge received by the conductor after a very large number of contacts?

2.2
Special Methods of Electrostatics

This section contains the problems whose solution requires special mathematical methods and approaches. The various methods that are used to solve electrostatic problems are discussed in several textbooks and monographs, such as Grinberg (1948); Landau *et al.* (1984); Panofsky and Phillips (1963); Smythe (1950); Stratton (1948); Simonyi (1956), and Jackson (1999). The most frequently used methods are listed below. They are illustrated by some classical examples.

The method of separation of variables in the curvilinear coordinates, see Problems 2.38*–2.42, 2.56*, and so on.

The method of integral transformations, see Problems 2.50*, 2.54*, and so on.

The method of inversion is based on inverse transformations. Inversion is defined as the transformation of space in which any point of space is transformed into a conjugate point with respect to some properly chosen sphere of inversion with radius R. If the spherical coordinates (with the origin at the center of the sphere of inversion) of an original point are r, ϑ, and α, then the spherical coordinates of the inverted point are $r' = R^2/r$, ϑ, and α. In vector form,

$$\boldsymbol{r}' = \frac{R^2\boldsymbol{r}}{r^2}, \quad \boldsymbol{r} = \frac{R^2\boldsymbol{r}'}{r'^2}. \tag{2.17}$$

The inversion transformation acquires the property of **conformity**. The inversion transforms a sphere into a sphere. If, in particular, the center of inversion lies on the sphere to be transformed, then the sphere becomes a plane (and vice versa).

The Laplace equation is invariant with respect to inversion. If a function $\varphi(\boldsymbol{r})$ is a solution of the Laplace equation in the original space, then

$$\varphi'(\boldsymbol{r}') = \frac{r}{R}\varphi(\boldsymbol{r}) = \frac{R}{r'}\varphi\left(\frac{R^2}{r'^2}\boldsymbol{r}'\right) \tag{2.18}$$

is a solution of the Laplace equation in the inverted space.

The basic problem that can be solved by the method of inversion is to find the field produced by a system of grounded conductors and point charges q_i placed at points $\boldsymbol{r}_i$. The potential at infinity is $V = const$.

In order to solve the problem, we use the method of inversion to transform the surfaces of conductors into a simpler form. The point charges q_i are then replaced by the charges

$$q_i' = \frac{R}{r_i}q_i, \tag{2.19}$$

located at the points $\boldsymbol{r}_i' = R^2\boldsymbol{r}_i/r_i^2$. Moreover, a point charge

$$q_0 = -RV \tag{2.20}$$

appears at the point $r' = 0$.

Now we can solve the electrostatic problem in the inverted system, that is, find the potential $\varphi'(\boldsymbol{r}')$. The potential $\varphi(\boldsymbol{r})$ is then obtained by means of the reciprocal transformation (Problems 2.61 and 2.62).

Solving boundary problems with Green's functions. Let us have a region of volume V bounded by a surface S, which contains a homogeneous dielectric of permittivity ε. The external charges with volume density ρ_{ext} may also be placed inside the dielectric, and in either of the electrostatic potential (Dirichlet boundary conditions)

$$\varphi|_S = f(\boldsymbol{r}) \tag{2.21}$$

or its normal derivative (Neumann boundary conditions)

$$\left.\frac{\partial\varphi}{\partial n}\right|_S = F(\boldsymbol{r}) \tag{2.22}$$

is given on the surface S. Find the potential $\varphi(\boldsymbol{r})$ inside the volume bounded by the surface S, which satisfies the Poisson equation (2.6).

In order to solve this problem, we use the Green's identity (I.1.98)

$$\int_V (\varphi\triangle\psi - \psi\triangle\varphi)dV = \oint_S \left(\varphi\frac{\partial\psi}{\partial n} - \psi\frac{\partial\varphi}{\partial n}\right)dS. \tag{2.23}$$

Let φ be the potential to be found, and ψ be the Green's function $G(\boldsymbol{r},\boldsymbol{r}')$ for point charge, satisfying the Poisson equation,

$$\triangle' G(\boldsymbol{r},\boldsymbol{r}') = -\frac{4\pi}{\varepsilon}\delta(\boldsymbol{r}-\boldsymbol{r}'). \tag{2.24}$$

If the whole space contains a single point charge, the Green's function has the form $G(\boldsymbol{r},\boldsymbol{r}') = 1/(\varepsilon|\boldsymbol{r}-\boldsymbol{r}'|)$ (cf. Equation (I.1.225)). However, when other charges, which are given implicitly by means of the corresponding boundary condition S, are present outside the volume, the Green's function becomes more complicated:

$$G(\boldsymbol{r}, \boldsymbol{r}') = \frac{1}{\varepsilon|\boldsymbol{r} - \boldsymbol{r}'|} + \mathcal{G}(\boldsymbol{r}, \boldsymbol{r}'). \tag{2.25}$$

Here, $\mathcal{G}(\boldsymbol{r}, \boldsymbol{r}')$ satisfies the Laplace equation $\triangle'\mathcal{G}(\boldsymbol{r}, \boldsymbol{r}') = 0$ and hence includes the effect of the charges located outside the given volume.

When the Dirichlet boundary conditions (2.21) are given, we should choose the function $\mathcal{G}(\boldsymbol{r}, \boldsymbol{r}')$ so that at any $\boldsymbol{r}$ the following condition be satisfied:

$$G(\boldsymbol{r}, \boldsymbol{r}')|_S = 0, \tag{2.26}$$

for the position vector $\boldsymbol{r}'$ lying on the surface S. Substituting the right-hand sides of Equations (2.6) and (2.24) in Equation (2.23) and using the boundary condition (2.26), we obtain

$$\varphi(\boldsymbol{r}) = \int_V G(\boldsymbol{r}, \boldsymbol{r}')\rho_{ext}(\boldsymbol{r}')\, dV' - \frac{\varepsilon}{4\pi} \oint_S \varphi(\boldsymbol{r}') \frac{\partial G(\boldsymbol{r}, \boldsymbol{r}')}{\partial n'}\, dS'. \tag{2.27}$$

Since the potential φ on the surface S is given, the right-hand side of the last equation is also known provided the Green's function is constructed, that is, this equation provides the solution of the problem in the given region.

In the case of the Neumann boundary conditions (2.22) for the potential, it is necessary to impose a certain condition on the derivative $\partial G/\partial n'$. This derivative satisfies the integral condition

$$\oint_S \frac{\partial G(\boldsymbol{r}, \boldsymbol{r}')}{\partial n'}\, dS' = -\frac{4\pi}{\varepsilon}, \tag{2.28}$$

which follows from Equation (2.24). The simplest condition that is consistent with Equation (2.28) is Jackson (1999)

$$\left.\frac{\partial G(\boldsymbol{r}, \boldsymbol{r}')}{\partial n'}\right|_S = -\frac{4\pi}{\varepsilon S}, \tag{2.29}$$

for the position vector $\boldsymbol{r}'$ lying on the surface S, while $\boldsymbol{r}$ is arbitrary. In this case, Equation (2.23) yields

$$\varphi(\boldsymbol{r}) = \int_V G(\boldsymbol{r}, \boldsymbol{r}')\rho_{ext}(\boldsymbol{r}')\, dV' + \frac{\varepsilon}{4\pi} \oint_S G(\boldsymbol{r}, \boldsymbol{r}') \frac{\partial \varphi(\boldsymbol{r}')}{\partial n'}\, dS' + \overline{\varphi}, \tag{2.30}$$

where $\overline{\varphi} = \oint_S \varphi\, dS/S$ is the average value of the potential on the surface S, which may be chosen arbitrarily. Equation (2.30) gives the solution of the electrostatic problem inside the region when the Neumann boundary conditions (2.22) are given.

Example 2.3

A point charge q is placed at a distance $b < a$ from the center of a metallic spherical shell of radius a, which is filled with a homogeneous dielectric. The potential of the shell is maintained to be zero. Find the field inside the sphere.

Solution. We construct the Green's function, which satisfies Equation (2.24) inside the sphere and the condition (2.21) on its surface. This can be accomplished by choosing the function $\mathcal{G}(\boldsymbol{r}, \boldsymbol{r}')$ in the form of the point charge potential

(1) $$\mathcal{G}(\boldsymbol{r}, \boldsymbol{r}') = \frac{q_*}{|\boldsymbol{r}_* - \boldsymbol{r}'|},$$

whose value q_* and position $r_*(r)$ are taken so that Equation (2.21) is satisfied. It is clear from symmetry arguments that r and r_*, which are measured from the center of the sphere, must be directed along the same straight line. Considering the spherical system of coordinates with the Oz-axis along this straight line, we get (from Equation (2.21)) the condition for determining the parameters q_* and r_*:

(2) $$\frac{1}{\sqrt{r^2 + a^2 - 2ar\cos\theta}} + \frac{q_*}{\sqrt{r_*^2 + a^2 - 2ar_*\cos\theta}} = 0.$$

This equation has two solutions: $q_* = -1$, $r_* = r$ (a trivial solution, with the zero potential throughout space) and $q_* = -a/r$, $r_* = a^2/r$. Using the second solution, we can write the following Green's function in the spherical coordinates:

(3) $$G(\boldsymbol{r}, \boldsymbol{r}') = \frac{1}{\sqrt{r^2 + r'^2 - 2rr'\cos\vartheta'}} - \frac{a/r}{\sqrt{r_*^2 + r'^2 - 2r_*r'\cos\vartheta'}}.$$

Here θ is the angle between the vectors $\boldsymbol{r}$ and $\boldsymbol{r}'$. Calculating the potential with the aid of Equation (2.27), where we should set $\rho_{ext}(\boldsymbol{r}') = q\,\delta(\boldsymbol{b} - \boldsymbol{r}')$ and $\varphi = 0$ on the surface S, we obtain

(4) $$\varphi(\boldsymbol{r}) = \frac{q}{\varepsilon\sqrt{r^2 + b^2 - 2br\cos\vartheta}} - \frac{q}{\varepsilon\sqrt{r^2b^2/a^2 + a^2 - 2br\cos\vartheta}}.$$

Here θ is the angle between the line, on which the charges lie, and the position vector $\boldsymbol{r}$. ■

The method of conformal mapping by functions of complex variable $z = x + iy$ is extensively used for solving two-dimensional electrostatic problems. In this method, the plane (x, y) plays the role of a complex plane of variable z. The function $w(z) = u(x, y) + iv(x, y)$ maps the points of the z- surface into the corresponding points on the complex surface w, on whose axes the quantities (u, v) are plotted. Of special importance for physical applications are those functions $w(z)$ which are single-valued and differentiable (analytical) within a certain region of the complex plane (x, y) except for a finite number of singular points. The differentiability of functions of complex variable imposes on them more rigid restrictions than the differentiability of functions of real variable. Namely, the derivative $w'(z_0)$ at point z_0, which is defined as the limit

$$w'(z_0) = \lim_{\triangle z \to 0} \frac{w(z_0 + \triangle z) - w(z_0)}{\triangle z}$$

must be independent of the way by which $\triangle z$ tends to zero (in particular, of the direction along which the points $z_0 + \triangle z$ and z_0 come closer together in the complex plane). It is clear that the arbitrary **differentiable** function of two real

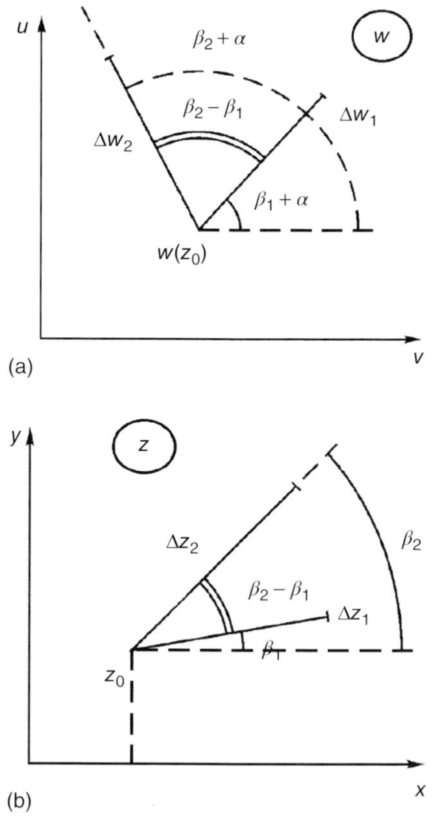

Figure 2.2 Conservation of angles between intersecting lines (conformity).

variables, $f(x, y)$, does not possess such a property since in the general case, for example, $\partial f/\partial x \neq \partial f/\partial y$.

The differentiability of the function of complex variable $w(z)$ mapping the z-plane into the w-plane leads to two important consequences.

(1) **Conformity** is conservation of angles between intersecting lines in the case of mapping by an analytic function. This property becomes clear if we express at the intersection point z_0 on the z-plane small increments along each curve in terms of the modulus and phase: $\triangle z_{1,2} = \varrho_{1,2}e^{i\beta_{1,2}}$. Clearly, the angle between the curves is equal to the difference between the arguments $\beta_2 - \beta_1$ (Figure 2.2). On the other hand, on the w-plane we have the increments $\triangle w_{1,2} \approx w'(z_0)\triangle z_{1,2}$ along the corresponding curves, and if $w'(z_0) = \varrho e^{i\alpha} \neq 0$, the difference between these increments is also equal to $\beta_2 - \beta_1$, since the derivative of analytical function is independent of the direction of increment. Both curves at the intersection point on the w-plane proved to be rotated through the same angle α, whereas the angle between them remains unchanged.

(2) The real $u(x, y)$ and imaginary $v(x, y)$ parts of the analytical function are **harmonic** functions, that is, they satisfy the two-dimensional Laplace equation. To

be certain, calculate the first derivative dw/dz in two ways, by pulling together the points along the $0x$ axis and then along the $0y$ axis. In the first case, we have $\triangle z = \triangle x$ and

$$\frac{dw}{dz} = \frac{\partial u}{\partial x} + i\frac{\partial v}{\partial x};$$

in the second case $\triangle z = i\triangle y$ and

$$\frac{dw}{dz} = -i\frac{\partial u}{\partial x} + \frac{\partial v}{\partial y}.$$

Equating the two expressions for dw/dz, we find the **Cauchy-Riemann**[2] **conditions**

$$\frac{\partial u}{\partial x} = \frac{\partial v}{\partial y}, \qquad \frac{\partial u}{\partial y} = -\frac{\partial v}{\partial x}, \tag{2.31}$$

which were originally obtained by d'Alembert and Euler. The Cauchy-Riemann conditions and the fact that the cross derivatives are equal, $\partial^2 f/\partial x\,\partial y = \partial^2 f/\partial y\,\partial x$, give the Laplace equations

$$\frac{\partial^2 u}{\partial x^2} + \frac{\partial^2 u}{\partial y^2} = 0, \qquad \frac{\partial^2 v}{\partial x^2} + \frac{\partial^2 v}{\partial y^2} = 0, \tag{2.32}$$

which are in turn satisfied by the real and imaginary parts of the analytical function.

Owing to the properties considered above, the analytical functions of complex variable can be used to solve two-dimensional electrostatic problems of determining the fields of charged metallic bodies immersed in a homogeneous dielectric medium. If a real part $u(x, y)$ of a complex function w is chosen so that the equation $u(x, y) = const$ describes the surface of a two-dimensional metal electrode, then $u(x, y)$ in the surrounding dielectric will represent the electrostatic potential satisfying the Laplace equation. Let us calculate the electrostatic induction:

$$D_x = -\varepsilon\frac{\partial u}{\partial x} = -\varepsilon\frac{\partial v}{\partial y}, \qquad D_y = -\varepsilon\frac{\partial u}{\partial y} = \varepsilon\frac{\partial v}{\partial x}.$$

Here, we used the Cauchy-Riemann relations (2.31). By writing down the equation for the lines of force of the induction vector

$$\frac{dx}{D_x} = \frac{dy}{D_y},$$

we get

$$\frac{\partial v}{\partial x}dx + \frac{\partial v}{\partial y}dy = 0 \quad \text{or} \quad v(x, y) = const.$$

Hence, the imaginary part $v(x, y) = const$ of the analytic function w at different $const$ describes a family of the force lines of an electrostatic field, which are orthogonal to the equipotential surfaces $u(x, y) = const$. It is equally possible to use the imaginary part $v(x, y)$ as the electrostatic potential. In this case, a family of the lines of force will be described by the equation $u(x, y) = const$.

2) Riemann Bernhard (1826–1866), outstanding German mathematician.

Let us calculate the capacitance per unit length measured along the normal to the (x, y) plane, of the capacitor formed by two charged metallic surfaces, $u(x, y) = c_1$ and $u(x, y) = c_2$ (Figure 2.3), maintained at the potentials V_1 and V_2. For the charge on the inner surface we have

$$q = \frac{1}{4\pi}\oint D_n dS = -\frac{\varepsilon}{4\pi}\oint\left(\frac{\partial u}{\partial x}(\boldsymbol{e}_x\cdot\boldsymbol{n}) + \frac{\partial u}{\partial y}(\boldsymbol{e}_y\cdot\boldsymbol{n})\right)$$

$$dS = \frac{\varepsilon}{4\pi}\oint\left(\frac{\partial v}{\partial x}(\boldsymbol{e}_x\cdot\boldsymbol{\tau}) + \frac{\partial v}{\partial y}(\boldsymbol{e}_y\cdot\boldsymbol{\tau})\right)dS = \frac{\varepsilon}{4\pi}\oint\nabla v\cdot\boldsymbol{\tau}dS. \tag{2.33}$$

Here, $\boldsymbol{\tau}$ is the unit vector tangential to the equipotential surface (Figure 2.3), and the Cauchy-Riemann conditions have been used. Note that the integral over the closed circle is not zero, if the surface is charged! The capacitance is given by Equation (2.11).

If the charge on a part of the surface should be found, then, according to Equation (2.33), it can be calculated as

$$q = \frac{\varepsilon}{4\pi}(v_2 - v_1), \tag{2.34}$$

where v_1 and v_2 refer to the initial and final points of the integration path.

Example 2.4

By using the analytical function $w(z) = \ln z$, $z = x + iy = re^{i\alpha}$, calculate the capacitance of a capacitor formed by two co-axial round cylindrical shells, and investigate the structure of the electric field due to the metallic planes with different potentials, which diverge like a fan from the Oz-axis.

Solution. We have $w = u + iv = \ln r + i\alpha$, that is,

(1) $$u(x, y) = \ln r = \ln\sqrt{x^2 + y^2}, \quad v(x, y) = \alpha = \arctan\frac{y}{x}.$$

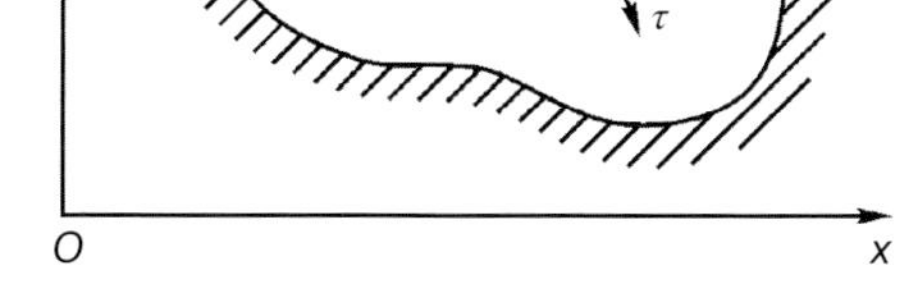

Figure 2.3 To calculation of the capacitance of the cylindrical capacitor of arbitrary form.

Taking u for the potential function, we find the equipotential lines on the (x, y) plane: $r^2 = x^2 + y^2 = const$, that is, circles. The lines of force are determined by the condition $v = \alpha = \arctan(y/x) = const$, that is, they are straight lines directed outward from the origin of coordinates. In the w-plane, the system under consideration is a plane capacitor (Figure 2.4). If the internal cylinder of radius a is at potential V, while the outer cylinder of radius b is grounded, then the corresponding constants $w(z) = -(V/\ln(b/a))\ln z + V\ln b/\ln(b/a)$ should be introduced into the analytical function. Accordingly, the imaginary part $v(x, y)$ takes the form $v(x, y) = -V\arctan(y/x)/\ln(b/a) = -V\alpha/\ln(b/a)$. On calculation, the charge from Equation (2.34) and assuming that $\alpha = 2\pi$ at point 1 and $\alpha = 0$ at point 2, we find $q = \varepsilon V/2\ln(b/a)$ and obtain the capacitance $C = \varepsilon/2\ln(b/a)$.

Taking the function v for the potential, we obtain the surfaces of equal potential in the form of the planes diverging from the Oz-axis (Figure 2.5). The potential in the regions 1 and 2 is described, respectively, by the following functions,

(2)

$$\varphi_1 = \frac{V}{\alpha_0}\alpha = \frac{V}{\alpha_0}\arctan\frac{y}{x}, \varphi_2 = -\frac{V}{2\pi - \alpha_0}(\alpha - 2\pi) = -\frac{V}{2\pi - \alpha_0}\left(\arctan\frac{y}{x} - 2\pi\right).$$

The lines of force of the electric field are circular arcs. The nonzero components are described in the cylindrical coordinates as

(3) $$E_{\alpha 1} = -\frac{V}{\alpha_0 r}, \qquad E_{\alpha 2} = \frac{V}{(2\pi - \alpha_0)r}.$$

When $\alpha_0 \ll 1$ the field is largely concentrated between the plates (Figure 2.6); at a given r we have $|E_{\alpha 2}/E_{\alpha 1}| \approx \alpha_0/2\pi \ll 1$. When $\alpha_0 = \pi$, the plates having different potentials are butted (Figure 2.7), and the field structure is symmetric with respect to the $y = 0$ plane. ■

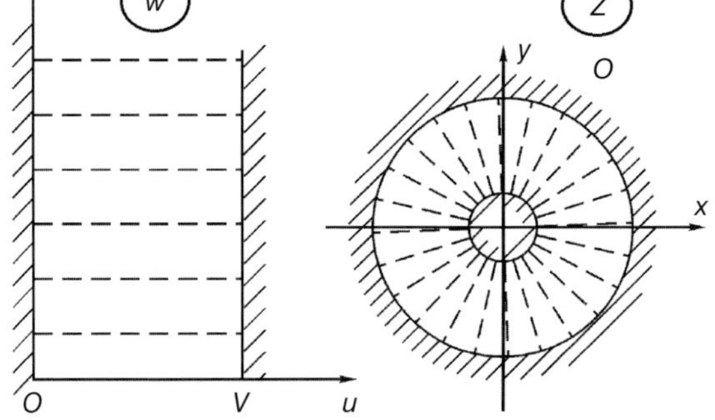

Figure 2.4 In plane w, the system under consideration is a plane capacitor, in plane (x, y) it is cylindrical.

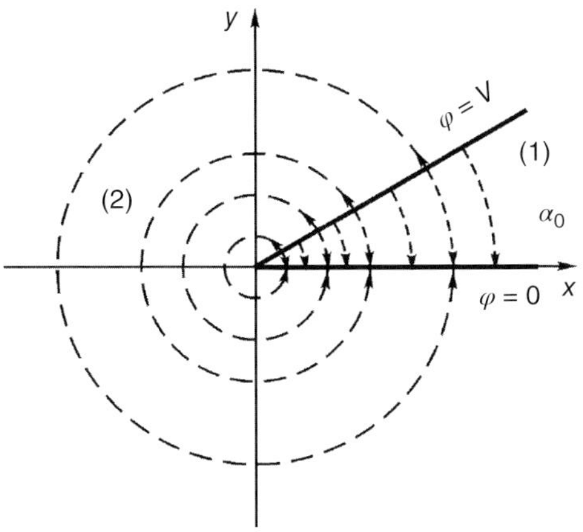

Figure 2.5 The surfaces of equal potential are the planes emanating from the *Oz*-axis.

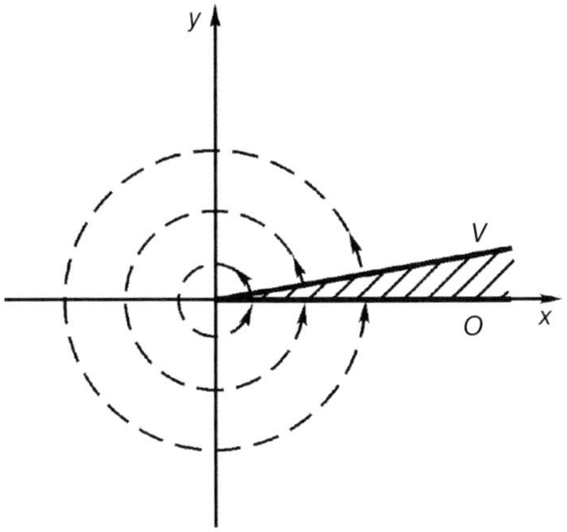

Figure 2.6 At a small angle α the field is concentrated between the plates.

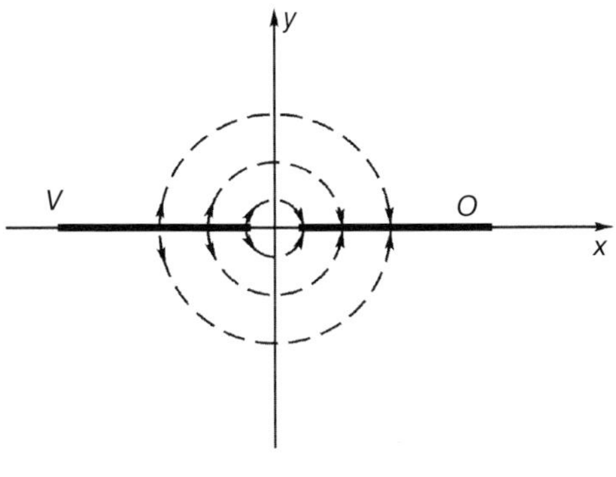

Figure 2.7 The plates at different potentials are butted.

Problems

2.38⋆ A conducting ellipsoid having the charge q and semiaxes a, b, and c is placed in a homogeneous dielectric of permittivity ε. Find the potential φ and the capacitance of the ellipsoid C, and also its surface charge density σ.

Hint. Use the ellipsoid coordinates (Problem I.1.94). Look for the potential in the form $\varphi(\xi)$.

2.39 On the basis of the results of Problem 2.38, find the potentials and capacitances of a prolate and an oblate ellipsoid of revolution. Consider the special cases of a long, thin rod and a thin disk. Find also the capacitance C and the potential φ of a prolate ellipsoid of revolution using the results of Problem 2.8.

2.40⋆ A conducting ellipsoid carrying a charge q is placed in vacuum in a uniform external field whose strength $\boldsymbol{E}_0$ is parallel to one of the ellipsoid axes. Find the potential φ of the resultant electric field.

Hint. Use the ellipsoid coordinates of Problem I.1.94. The boundary conditions on the surface of the ellipsoid ($\xi = 0$) can be satisfied only if the dependence of the potential φ' due to the induced charges, on η, ζ is the same as for the external field:

$$\varphi' = \varphi_0(\xi, \eta, \zeta)F(\xi).$$

2.41 The field strength in a plane capacitor is E_0. A grounded plate has a boss in the form of one half of a prolate ellipsoid of revolution whose axis of symmetry is perpendicular to the plates. The distance between the plates is large compared with the linear dimensions of the boss. Find the electric potential φ in the capacitor and the ratio of the maximum value E_{max} of the field to the field E_0.[3)]

2.42 An uncharged conducting ellipsoid is placed in a uniform field $\boldsymbol{E}_0$ oriented arbitrarily with respect to the ellipsoid axes. Find the total electric field φ. Consider the field at a large distance from the ellipsoid by expressing it in terms of the depolarization coefficients:

$$n^{(x)} = \frac{abc}{2}\int_0^\infty \frac{ds}{(s+a^2)R_s}, \qquad n^{(y)} = \frac{abc}{2}\int_0^\infty \frac{ds}{(s+b^2)R_s},$$

$$n^{(z)} = \frac{abc}{2}\int_0^\infty \frac{ds}{(s+c^2)R_s} \qquad \left(R_s = \sqrt{(s+a^2)(s+b^2)(s+c^2)}\right).$$

2.43 Find the expressions for the depolarization coefficients introduced in Problem 2.42 for a prolate ellipsoid of revolution ($a > b = c$). Consider the special cases of a very elongated ellipsoid (rod) and an almost spherical ellipsoid.

2.44 Find the depolarization coefficients for an oblate conducting ellipsoid ($a = b > c$). Consider, in particular, the case of a disk.

3) The result of the problem explains the principle of operation of a lightning conductor.

2.45★ A dielectric ellipsoid with semiaxes a, b and c is placed in a uniform external field $\boldsymbol{E}_0$. The dielectric permittivity of the ellipsoid and of the homogeneous dielectric surrounding it are ε_1 and ε_2, respectively. Find the potential φ of the resultant electric field (use the hint given in Problem 2.40*). Find the strength of the electric field $\boldsymbol{E}$ inside the ellipsoid and the potential φ_2 at large distances from it by expressing the potential in terms of the components of the polarizability of the ellipsoid along its principal axes.

2.46 An ellipsoid of revolution with a dielectric permittivity ε_1 is placed in an external uniform field $\boldsymbol{E}_0$ in a homogeneous dielectric medium with permittivity ε_2. Find the energy U of the ellipsoid in this field and the torque N acting on it. Consider also a conducting ellipsoid of revolution.

2.47★ Show that when a spherical conducting liquid drop has a sufficiently high charge it becomes unstable. Find the critical value of charge q_c. The radius of the drop is R and the surface tension coefficient is α.

Hint. Compare the energy of the spherical drop with the energy of a drop in the form of a prolate ellipsoid of revolution. The surface area of this ellipsoid is

$$S = 2\pi b^2 + \frac{2\pi b a^2}{\sqrt{a^2 - b^2}} \arccos\frac{b}{a} \qquad (a > b = c).$$

2.48★ A uniform electric field $\boldsymbol{E}_0 \parallel Oz$ in the half-space $z > 0$ is bounded by a grounded conducting plane $z = 0$, which contains a circular aperture of radius a. Find the field φ in the entire space. Consider, in particular, the field at a large distance from the aperture (in the half-space $z > 0$).

Hint. Use the oblate spheroidal coordinates (Problem I.1.95) with $c=0$. Look for the solution in the form $\varphi = -E_0 z F(\xi)$.

2.49 Find the distribution of charges σ on the conducting plane of Problem 2.48.

2.50★ A charge q is placed at point $N(\boldsymbol{r}_0)$ (Figure 2.8) between two grounded conducting half-planes OA and OB, which intersect at an angle β. The cylindrical coordinates of the charge are $(r_0, \gamma, 0)$; the Oz-axis lies along the line of intersection of the two planes, and the azimuthal angle α is measured from the plane OA. Show that the potential $\varphi(r, \alpha, z)$ can be written in the form

$$\varphi(r, \alpha, z) = \int_0^\infty \varphi_k(r, \alpha) \cos kz \, dk,$$

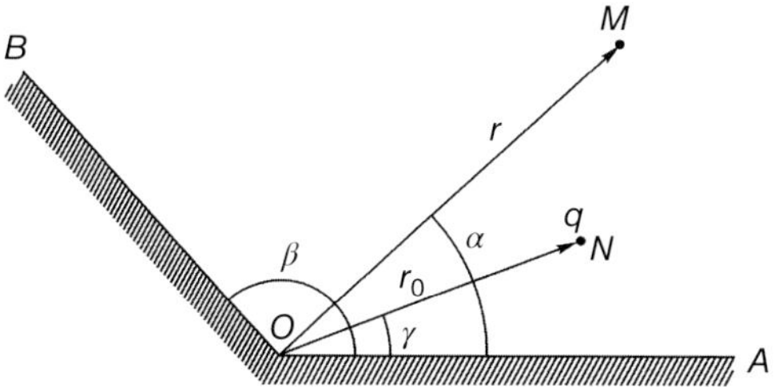

Figure 2.8 A point charge between two conducting half-planes.

where

$$\varphi_k(r,\alpha) = \frac{8q}{\beta}\begin{cases}\sum_{n=1}^{\infty} K_{n\pi/\beta}(kr_0) I_{n\pi/\beta}(kr)\sin(n\pi\gamma/\beta)\sin(n\pi\alpha/\beta) & \text{at } r < r_0,\\ \sum_{n=1}^{\infty} I_{n\pi/\beta}(kr_0) K_{n\pi/\beta}(kr)\sin(n\pi\gamma/\beta)\sin(n\pi\alpha/\beta) & \text{at } r > r_0,\end{cases}$$

with $I_{n\pi/\beta}$ and $K_{n\pi/\beta}$ being the cylinder functions.

2.51 Show that the potential produced by the point-charge between the two planes of Problem 2.50 may be presented in the form

$$\varphi(r,\alpha,z) = \frac{q}{\beta\sqrt{2rr_0}}\int_{\eta}^{\infty}\left[\frac{\sinh(\pi\zeta/\beta)}{\cosh(\pi\zeta/\beta) - \cos(\pi(\alpha-\gamma)/\beta)} - \frac{\sinh(\pi\zeta/\beta)}{\cosh(\pi\zeta/\beta) - \cos(\pi(\alpha+\gamma)/\beta)}\right]\frac{d\zeta}{\sqrt{\cosh\zeta - \cosh\eta}},$$

where

$$\cosh\eta = \frac{r_0^2 + r^2 + z^2}{2rr_0}, \qquad \eta > 0.$$

Hint. Use the formulas:

$$\int_0^{\infty} K_\nu(kr) I_\nu(kr_0)\cos kz\, dk = \frac{1}{2\sqrt{2rr_0}}\int_{\eta}^{\infty}\frac{e^{-\xi\nu}d\xi}{\sqrt{\cosh\xi - \cosh\eta}}$$

and

$$\sum_{n=1}^{\infty} p^n \cos nx = \frac{1}{2}\left(\frac{1-p^2}{1-2p\cos x + p^2} - 1\right).$$

2.52 Find the field φ produced by a charge q located at the point r_0 with the cylindrical coordinates $(r_0, \gamma, z = 0)$, which is close to the conducting half-plane $\alpha = 0$.

Hint. Use the results of Problem 2.51. In order to calculate the integral, use the substitution $\cosh(\zeta/2) = \cosh(\eta/2)\cosh u$, where $0 < u < \infty$.

2.53 Find the distribution of surface charge σ near the edge of a conducting wedge with a dihedral angle β (which is measured outside the conductor). The wedge is in the field of an arbitrary charge distribution.

Hint. First, consider the case of a single point charge near the wedge; use the results of Problem 2.50 and the formula

$$\int_0^{\infty} K_\nu(k\rho)k^\nu \cos kz\, dk = 2^{\nu-1}\sqrt{\pi}\,\Gamma\left(\nu + \frac{1}{2}\right)\frac{\rho^\nu}{(\rho^2 + z^2)^{\nu+1/2}}.$$

2.54* A point charge q is placed at a distance a from a homogeneous plane parallel dielectric plate of thickness c. Find the electric field using the fact that both the product $J_0(kr_1)e^{\pm kz}$ and $\int_0^{\infty} A(k)J_0(kr_1)e^{\pm kz}dk$ satisfy the Laplace equation (r_1, z are cylindrical coordinates of the point, J_0 is the Bessel function and $A(k)$ is an arbitrary function of k).

Hint. Use the expansion:

$$\frac{1}{\sqrt{r_1^2 + z^2}} = \int_0^{\infty} e^{-k|z|}J_0(kr_1)\, dk.$$

2.55 A plane parallel dielectric plate of thickness $a/2$ and permittivity ε is inserted between the electrodes of a plane capacitor, which are at a distance a from each other. The dielectric plate is in contact with one of the grounded electrodes. The dielectric surface is then given a charge q, which can be considered as a point charge. Find the field potential φ in the capacitor and determine, in particular, its form near the charge. Represent this field as the superposition of images.

2.56* The radii of the electrodes of a nonconcentric spherical capacitor are a_1 and a_2, and the distance between their centers is b $(a_1 + b < a_2)$. The outer sphere is grounded, while the inner one is maintained at a potential V. Find the field φ inside the capacitor. Find also the capacitance C of the spherical capacitor.

Hint. Use bispherical coordinates (Problem I.1.97) and the substitution $\varphi = \sqrt{2\cosh\xi - 2\cos\eta}\,\psi$; separate variables in the equation for ψ, and use the properties of the Legendre polynomials, especially the expansion

$$\frac{1}{\sqrt{\cosh\xi - \cos\eta}} = \sqrt{2}\sum_{l=0}^{\infty} e^{-(l+1/2)|\xi|} P_l(\cos\eta).$$

2.57 Find the capacitance of a slightly nonconcentric spherical capacitor $(b \leq a_1, a_2)$ to within terms of the order of b^2 by using the results of Problem 2.56 (cf. Problem 2.19*).

2.58 The distance between the centers of two conducting spheres of radii a_1 and a_2 is equal to b $(b > a_1 + a_2)$. Find the capacitance coefficients c_{ik} of the system using bispherical coordinates.

2.59 The two conducting spheres considered in Problem 2.58 are at a large distance from each other $(b \gg a_1, a_2)$. Find the capacitance coefficients c_{ik} to within terms of the order $1/b^4$.

2.60 Show that function (2.18) satisfies the Laplace equation $\triangle'\varphi'(\boldsymbol{r}') = 0$, if $\triangle\varphi(\boldsymbol{r}) = 0$.

2.61* Two conducting spheres of equal radii a touch each other. Find the capacitance C of the system by the inversion method. Find also the electric field φ of the system when a charge q is given to the spheres.

Hint. Use the results of Problem 2.55.

2.62 Use the inversion method to determine the field of a grounded sphere of radius R with a point charge q placed at a distance a $(a > R)$ from its center (Problem 2.14).

Hint. The potential of a uniformly charged sphere in the absence of a point charge may be regarded as known.

2.63* The surface of a conductor is formed by two spheres of radii R_1 and R_2 which intersect on a circle of radius a. Find the capacitance C of this conductor using the results of the problem on a conducting wedge in the field of a point charge (Problem 2.51) and the inversion method.

Hint. The surface of the considered conductor can be described in terms of toroidal coordinates (Problem I.1.98) by the equations

$$\xi = \xi_1 = const, \qquad \xi = \xi_2 = const \qquad (\sin\xi_1 = \pm a/R_1,\ \sin\xi_2 = \pm a/R_2).$$

It is sufficient to consider the transformation of the coordinates in the plane perpendicular to the edge of wedge and passing through the center of inversion, which must be taken on the curve of intersection of the two spheres. In order to determine the charge q on the conductor at a given potential, it is convenient to use the fact that the field at large distances from the conductor has the form $\varphi = q/r - V$, where $-V$ is the potential at infinity.

2.64 Find the capacitance C of the following conductors:

(a) a hollow spherical segment of radius R, which subtends an angle 2θ at the center;

(b) a hemisphere of radius R.

2.65 A conductor is formed by two spheres of equal radii a, whose surfaces intersect at an angle $\pi/3$. Find the capacitance C of the conductor.

2.66 Use the Green's function (3) from Example 2.3 to calculate the potential due to a uniformly charged ring of radius b with charge q located inside the grounded spherical shell of radius $a > b$. The center of the ring is at the center of the sphere.

2.67 Solve Problem 2.66 for the case where the field inside the spherical shell is produced by a uniformly charged segment, which lies on one of the diameters and carries a charge q.

Hint. In the spherical coordinates with the polar axis lying along the segment, the volume distribution of charge is of the form

$$\rho_{ext}(r', \vartheta') = \frac{q}{2a}\frac{1}{2\pi r'^2}[\delta(\cos\vartheta' - 1) + \delta(\cos\vartheta' + 1)].$$

2.68 Examine the field whose potential and lines of force are described by the analytical function $z = l\cosh w$, where $l = const$:

(a) determine the form of conductors producing the field that can be represented by this function;

(b) determine the lines of force and the equipotential surfaces due to an infinitely long and thin charged metallic strip of finite width;

(c) determine the lines of force and the equipotential surfaces due to the charged metallic plane and half-plane, which are mutually perpendicular. The edge of the half-plane lies at a finite distance from the plane.

2.69⋆ Two infinitely long circular cylinders having the radii R_1 and R_2, with a distance between the parallel axes L, are immersed in a homogeneous dielectric medium of permittivity ε. Using the analytical function

$$w = \ln\frac{z + ia}{z - ia},$$

calculate the capacitance of the system per unit length. Consider the cases $L < R_2 - R_1$ and $L > R_2 + R_1$.

2.3
Energy, Forces, and Thermodynamic Relations for Conductors and Dielectrics

Thermodynamic functions of a dielectric. In order to calculate the thermodynamic functions of a system consisting of charged dielectrics and conductors, it is necessary to find an elementary work associated with a change in the electric field. Let a charge of *i*th conductor be increased by δq_i, while the density of external charges in a dielectric be increased by $\delta\rho_{ext}(\boldsymbol{r})$. Since the work required to transport a small charge δq from infinity (where the potential equals zero) to a given point, is equal to $\varphi(\boldsymbol{r})\,\delta q$, the total work to be done can be represented in the form

$$\delta A = \sum_i \varphi_i\,\delta q_i + \int \varphi\delta\rho_{ext}\,dV, \tag{2.35}$$

where the sum is taken over all conductors, and the integration is carried out over the entire space outside them including both the regions filled with dielectric and vacuum.

The increase of charge on the conductor can be expressed in terms of the increase of induction near its surface, with the aid of Equation (2.9):

$$\delta q_i = -\frac{1}{4\pi}\oint_{S_i} \delta D_n\,dS_i,$$

where the normal is directed inward the conductor and is outer with respect to the dielectric. The increase of the external charge density in the dielectric can be written, with the aid of Equation (1.20), in the form $div\,\delta\boldsymbol{D}/4\pi$, and the volume integral can be transformed using the identity $div\,\varphi\delta\boldsymbol{D} = \varphi\,div\,\delta\boldsymbol{D} + \delta\boldsymbol{D}\cdot\nabla\varphi$:

$$\int \varphi\delta\rho_{ext}\,dV = \frac{1}{4\pi}\sum_i \varphi_i \oint_{S_i} D_n\,dS_i + \frac{1}{4\pi}\oint_{S\to\infty} \varphi D_n\,dS + \frac{1}{4\pi}\int \boldsymbol{E}\cdot\delta\boldsymbol{D}\,dV.$$

Here, the normal is external with respect to the region occupied by the dielectric or vacuum. Substituting this expression into Equation (2.35), we get

$$\delta A = \frac{1}{4\pi}\int \boldsymbol{E}\cdot\delta\boldsymbol{D}\,dV. \tag{2.36}$$

The integral here is extended over the entire space, the relationship between $\boldsymbol{E}$ and $\boldsymbol{D}$ is not specified.

Since the thermodynamic state of a dielectric depends on its density, temperature, electric field, and possibly on other parameters, the elementary work (Equation (2.36)) may characterize changes in different thermodynamic potentials, depending on the surrounding conditions. In the case of a thermally insulated, statistically equilibrated body, the work of electric forces is performed at constant entropy and represents a change in the internal energy $\mathcal{U}$. The total change in the internal energy, according to the first law of thermodynamics, is a combination of heat and work,

$$\delta\mathcal{U} = T\delta S + \frac{1}{4\pi}\int \boldsymbol{E}\cdot\delta\boldsymbol{D}\,dV. \tag{2.37}$$

Here, contrary to (I.C46), the internal energy of a body is denoted by $\mathcal{U}$, $\mathcal{S}$ is its entropy, and T its temperature; a symbol δ is used for small changes of thermodynamic quantities in order to distinguish them from the integration volume element dV. A mechanical work due to a change in sizes of the body is not included in Equation (2.37).

If the body under consideration is in a thermal contact with surrounding equilibrated bodies and its temperature does not change, then the work represents a change in the Helmholtz free energy $\mathcal{F}$, and its total change can be written in the form

$$\delta\mathcal{F} = \mathcal{S}dT + \frac{1}{4\pi}\int \boldsymbol{E}\cdot\delta\boldsymbol{D}\,dV. \tag{2.38}$$

The changes in thermodynamic potentials discussed above can also be written for the specific quantities, that is, values per unit volume of the dielectric:

$$dU = TdS + \zeta d\tau + \frac{1}{4\pi}\boldsymbol{E}\cdot d\boldsymbol{D}, \tag{2.39}$$

$$dF = -SdT + \zeta d\tau + \frac{1}{4\pi}\boldsymbol{E}\cdot d\boldsymbol{D}. \tag{2.40}$$

Since the volume of study can exchange particles with the surrounding dielectric, Equations (2.38) and (2.39) include the corresponding term describing the change in the mass density $d\tau$ and chemical potential per unit mass, $\zeta = \mu/m$, where μ is the chemical potential that enters (I.C47) and is related to a single particle of mass m.

In Equations (2.38) and (2.39), an induction $\boldsymbol{D}$, which is determined by charges on the conductors (and by external charges in the dielectric), enters as an independent electric quantity. It is easy to pass to an independent variable $\boldsymbol{E}$ which is determined by potentials of the conductors. To do this, the new thermodynamic potentials are introduced:

$$\tilde{U}(S,\tau,\boldsymbol{E}) = U - \frac{1}{4\pi}\boldsymbol{E}\cdot\boldsymbol{D}, \qquad \tilde{F}(T,\tau,\boldsymbol{E}) = F - \frac{1}{4\pi}\boldsymbol{E}\cdot\boldsymbol{D}, \tag{2.41}$$

whose differentials are written in the form

$$d\tilde{U} = T\,dS + \zeta\,d\tau - \frac{1}{4\pi}\boldsymbol{D}\cdot d\boldsymbol{E}, \qquad d\tilde{F} = -S\,dT + \zeta\,d\tau - \frac{1}{4\pi}\boldsymbol{D}\cdot d\boldsymbol{E}. \tag{2.42}$$

The electric field and induction in the dielectric can be expressed in terms of thermodynamic functions:

$$\boldsymbol{E} = 4\pi\left(\frac{\partial U}{\partial \boldsymbol{D}}\right)_{S,\tau} = 4\pi\left(\frac{\partial F}{\partial \boldsymbol{D}}\right)_{T,\tau}, \qquad \boldsymbol{D} = -4\pi\left(\frac{\partial \tilde{U}}{\partial \boldsymbol{E}}\right)_{S,\tau} = -4\pi\left(\frac{\partial \tilde{F}}{\partial \boldsymbol{E}}\right)_{T,\tau}. \tag{2.43}$$

Example 2.5

Calculate the changes of thermodynamic functions U, $\tilde{U}$, F, $\tilde{F}$, S, ζ of a dielectric, which are due to the presence of the electric field. The dielectric is isotropic, its coupling equation is $\boldsymbol{D} = \varepsilon(T,\tau)\boldsymbol{E}$.

Solution. We proceed from relation (2.43)

(1) $$\left(\frac{\partial F}{\partial \boldsymbol{D}}\right)_{T,\tau} = \frac{\boldsymbol{E}}{4\pi} = \frac{\boldsymbol{D}}{4\pi\varepsilon(T,\tau)}.$$

Integration of both sides of this equation yields

$$F(T,\tau,D) - F_0(T,\tau) = \frac{D^2}{8\pi\varepsilon}. \tag{2.44}$$

Dielectric permittivity ε is independent of the field and it can, therefore, be expressed only in terms of the entropy and density of the dielectric: $\varepsilon(S,\tau)$. Using the thermodynamic relation $U = F + TS$ we obtain

$$U(S,\tau,D) - U_0(S,\tau) = \frac{D^2}{8\pi\varepsilon}. \tag{2.45}$$

Further, we find

$$\tilde{U} - U_0(S,\tau) = -\frac{\varepsilon E^2}{8\pi}, \qquad \tilde{F} - F_0(T,\tau) = -\frac{\varepsilon E^2}{8\pi}, \tag{2.46}$$

$$S = -\left(\frac{\partial F}{\partial T}\right)_{\tau,\mathbf{D}} = S_0(T,\tau) + \frac{D^2}{8\pi\varepsilon^2}\left(\frac{\partial \varepsilon}{\partial T}\right)_\tau, \tag{2.47}$$

$$\zeta = \left(\frac{\partial F}{\partial \tau}\right)_{T,\mathbf{D}} = \zeta_0(T,\tau) - \frac{D^2}{8\pi\varepsilon^2}\left(\frac{\partial \varepsilon}{\partial \tau}\right)_T, \tag{2.48}$$

where it is certainly possible to write down $D^2/\varepsilon^2 = E^2$.

We have now obtained the densities of thermodynamic quantities for the system consisting of charged conductors and dielectrics, where the external charges may exist. Hence, the electric energy density is not zero outside the dielectric (in vacuum), whereas S and ζ in vacuum are zero. Conducting bodies contribute to the thermodynamic functions S_0, U_0 and others in the absence of the field. However the field change is independent of the intrinsic properties of conductors because the field does not penetrate in them. ■

Example 2.6

Write down the energy of an electrostatic field (i.e., the field-dependent part of the internal energy of a system consisting of conductors and dielectrics) $W = \mathcal{U} - \mathcal{U}_0$ in terms of the charges of the conductors q_i, and their potentials V_i, provided external charges are absent in the dielectric. Express this energy in terms of the potential (Equation (2.12)) and capacitance (Equation (2.14)) coefficients.

Solution. We write down the energy (Equation (2.44)) in the form

(1) $$W = \int \frac{\boldsymbol{E}\cdot\boldsymbol{D}}{8\pi}\, dV,$$

where the conductors with $\boldsymbol{E} = 0$ inside them are excluded from the volume integration. Further we use the equalities $\boldsymbol{E} = -\nabla\varphi$, $\nabla(\varphi\boldsymbol{D}) = \varphi\nabla\cdot\boldsymbol{D} + \boldsymbol{D}\cdot\nabla\varphi$, the Maxwell equation $\nabla\cdot\boldsymbol{D} = 0$ and the Gauss–Ostrogradskii theorem. We have

$$W = -\sum_i V_i \oint_{S_i} \frac{D_n}{8\pi}\, dS_i = \frac{1}{2}\sum_i q_i V_i. \tag{2.49}$$

The sum is taken over all conductors. With the aid of Equations (2.12) and (2.14), we find

$$W = \frac{1}{2}\sum_i s_{ik}q_iq_k = \frac{1}{2}\sum_i c_{ik}V_iV_k. \tag{2.50}$$

The formulas obtained are applicable for an inhomogeneous dielectric as well.

Example 2.7

A dielectric or a conducting uncharged body is introduced into a uniform external field $\mathcal{E}$ in vacuum. Express a change in the internal energy in terms of the polarizability of the body assuming the body to be thermally insulated, and neglecting the change in its volume.

Solution. For a dielectric, we have

$$\mathcal{U} - \mathcal{U}_0 = \int \frac{\boldsymbol{E}\cdot\boldsymbol{D} - \mathcal{E}^2}{8\pi}\,dV = \frac{1}{8\pi}\int (\boldsymbol{E} + \boldsymbol{\mathcal{E}})\cdot(\boldsymbol{D} - \boldsymbol{\mathcal{E}})\,dV - \frac{1}{8\pi}\int \boldsymbol{\mathcal{E}}\cdot(\boldsymbol{D} - \boldsymbol{E})\,dV.$$

The integrand of the first integral in the right-hand side of this expression can be written in the form

$$(\boldsymbol{E} + \boldsymbol{\mathcal{E}})\cdot(\boldsymbol{D} - \boldsymbol{\mathcal{E}}) = -[\nabla(\varphi + \varphi_0)]\cdot(\boldsymbol{D} - \boldsymbol{\mathcal{E}}) = -\nabla\cdot[(\varphi + \varphi_0)(\boldsymbol{D} - \boldsymbol{\mathcal{E}})],$$

where φ_0 and φ are the potentials of the field distorted and undistorted by the dielectric, respectively. Here, we take into account that $\nabla\cdot\boldsymbol{D} = \nabla\cdot\boldsymbol{\mathcal{E}} = 0$. The integral of the total divergence is zero because the difference $\boldsymbol{D} - \boldsymbol{\mathcal{E}}$ tends to zero at large distances from the body. As a result, substituting $\boldsymbol{D} - \boldsymbol{E} = 4\pi\boldsymbol{P}$, we have

$$\mathcal{U} - \mathcal{U}_0 = -\frac{1}{2}\int \mathbf{P}\cdot\boldsymbol{\mathcal{E}}dV = -\frac{1}{2}\boldsymbol{\mathcal{P}}\cdot\boldsymbol{\mathcal{E}} = -\frac{1}{2}V\beta_{\mu\nu}\mathcal{E}_\mu\mathcal{E}_\nu. \tag{2.51}$$

The last equation is the consequence of the external field uniformity,

$$\boldsymbol{\mathcal{P}} = \int \boldsymbol{P}\,dV, \qquad \mathcal{P}_\mu = V\beta_{\mu\nu}\mathcal{E}_\nu$$

is the dipole moment of the dielectric body and $\beta_{\mu\nu}$ is its polarizability (Equation (2.15)). On the other hand, the field and polarizability inside the body can be inhomogeneous. Homogeneity of the field is retained only for the bodies having a form of ellipsoid (Problem 2.45).

For a conductor,

$$\mathcal{U} - \mathcal{U}_0 = \int \frac{E^2 - \mathcal{E}^2}{8\pi}\,dV = \frac{1}{8\pi}\int_{out} (\boldsymbol{E} + \boldsymbol{\mathcal{E}})\cdot(\boldsymbol{E} - \boldsymbol{\mathcal{E}})\,dV - \frac{1}{8\pi}\int_V \mathcal{E}^2\,dV,$$

where the first integral is evaluated over the evacuated region and the second integral over the volume V of the conductor inside which $\boldsymbol{E} = 0$. Performing the transformations analogous to those made above, we reduce the required quantity to the form

$$\mathcal{U} - \mathcal{U}_0 = -\frac{1}{8\pi}\oint_S \varphi_0\boldsymbol{E}\cdot d\boldsymbol{S},$$

where the integral is taken over the surface of the conductor. We write $\varphi_0 = -\boldsymbol{\mathcal{E}}\cdot\mathbf{r}$ for the uniform field, where the origin of the position vector is arbitrary, and obtain the integrand $\boldsymbol{r}E_n\, dS/8\pi = -\boldsymbol{r}\sigma\, dS/2$, with σ being the surface charge density on the conductor. Finally, we get the electric dipole moment of the conductor in an external uniform field:

$$\boldsymbol{\mathcal{P}} = \oint_S \boldsymbol{r}\sigma\, dS, \qquad \mathcal{P}_\mu = \beta_{\mu\nu}\mathcal{E}_\nu. \tag{2.52}$$

The energy of the conductor is written in terms of its dipole moment in the same form (Equation (2.51)) as for a dielectric. ■

Forces acting on conductors and dielectrics in an electric field. Let a conductor be surrounded by a dielectric of permittivity ε. In accordance with Equation (2.9), its surface charge density is $\sigma = D_n/4\pi$. The force applied to the element $dq = \sigma\, dS$ of the surface charge of the conductor can be expressed as $d\boldsymbol{F} = \boldsymbol{E}_{ef}\, dq$, where $\boldsymbol{E}_{ef}$ is the field strength due to all charges except for dq. The elementary area dS, at distances that are small compared to its linear sizes, produces the electrostatic induction $\boldsymbol{D}' = \pm 2\pi\sigma\boldsymbol{n}$ of opposite directions on either side of this elementary area, similar to the case of an infinite charged plane. All other charges located beyond the chosen surface element produce the homogeneous induction $\boldsymbol{D}_{ef}$ in its neighborhood. The total induction inside the conductor is zero: $\boldsymbol{D}_{ef} - 2\pi\sigma\boldsymbol{n} = 0$ (the normal is external relative to the conductor). Hence, we find $\boldsymbol{D}_{ef} = 2\pi\sigma\boldsymbol{n}$, and the force per unit area of the conductor surface is

$$\boldsymbol{f}_s = \frac{d\boldsymbol{F}}{dS} = \frac{2\pi\sigma^2\boldsymbol{n}}{\varepsilon} = \frac{\varepsilon E^2}{8\pi}\boldsymbol{n}. \tag{2.53}$$

The force is directed along the normal to the surface and tends to increase the volume of the conductor.

Now we calculate the density of the force $\boldsymbol{f}$ acting on the elements of dielectric medium in the case of their small displacement and deformation. If $\boldsymbol{u}(\boldsymbol{r})$ denotes the vector of a small displacement of an element of the medium, which is located at the point $\boldsymbol{r}$, then the work done can be written in the form

$$\delta A = \int \boldsymbol{u}(\boldsymbol{r})\cdot\boldsymbol{f}(\boldsymbol{r})\, dV,$$

where the integration is carried out over the volume of the dielectric. Since the work of the electric forces is accompanied by the decrease in the energy, the magnitude of the elementary work, taken with the opposite sign, is equal to the increment of the electric energy, that is,

$$\delta W = -\int \boldsymbol{u}(\boldsymbol{r})\cdot\boldsymbol{f}(\boldsymbol{r})\, dV. \tag{2.54}$$

In a thermodynamical sense, this energy should be understood as the internal energy, provided the dielectric is insulated, or as the free energy if the deformation process is isothermal. The use of Equation (2.44) or (2.45) yields

$$W = \int \frac{\varepsilon E^2}{8\pi}\, dV. \tag{2.55}$$

If the field in the dielectric is due to the external charges, then the same energy (Equation 2.55)) can be written as

$$\tilde{W} = \frac{1}{2}\int \rho_{ext}\varphi\, dV \tag{2.56}$$

(Section I.2.1).

The energy change at small displacements of elements of the medium can be presented in the form

$$\delta\tilde{W} = \frac{1}{2}\int \varphi\delta\rho_{ext}\, dV + \frac{1}{2}\int \delta\varphi\rho_{ext}\, dV$$

and in the form

$$\delta W = \frac{1}{8\pi}\int E^2\delta\varepsilon\, dV + \frac{1}{4\pi}\int \boldsymbol{D}\cdot\delta\boldsymbol{E}\, dV = \frac{1}{8\pi}\int E^2\delta\varepsilon\, dV + \frac{1}{4\pi}\int \nabla\cdot\boldsymbol{D}\delta\varphi$$

$$\times\, dV - \frac{1}{4\pi}\int_{S\to\infty} \delta\varphi\boldsymbol{D}\cdot d\boldsymbol{S},$$

where symbol δ denotes a small change in the corresponding quantity at a given point of space (local change). The integral over infinitely distant surface in the last equality becomes zero in the case of the finite system. As a result, we have

$$\delta W = 2\delta\tilde{W} - \delta W = -\frac{1}{8\pi}\int E^2\delta\varepsilon\, dV + \int \varphi\delta\rho_{ext}\, dV. \tag{2.57}$$

We consider the quantity $\varepsilon(\tau)$ as a function of a mass density (the second argument, either the temperature or entropy, is constant). In this case,

$$\delta\varepsilon = (\partial\varepsilon/\partial\tau)\,\delta\tau. \tag{2.58}$$

A local change in the density consists of two parts: one part is due to the displacement of the element of the medium with another density to a given point, $\tau(\boldsymbol{r}-\boldsymbol{u}) \approx \tau(\boldsymbol{r}) - \boldsymbol{u}\cdot\nabla\tau$, and the second part is due to the contraction or expansion of the element during its displacement. Small changes are additive. The second part of the density change is expressed in terms of divergence of the deformation vector:

$$\delta'\tau = -\tau\frac{\delta' V}{V} = -\tau\frac{\oint u_n\, dS}{V} = -\tau\, div\,\boldsymbol{u}.$$

Combining the two results, we get

$$\delta\tau = -\tau\nabla\cdot\boldsymbol{u} - \boldsymbol{u}\cdot\nabla\tau = -\nabla\cdot(\tau\boldsymbol{u}). \tag{2.59}$$

The external charge density is subject to precisely the same change,

$$\delta\rho_{ext} = -\nabla\cdot(\rho_{ext}\boldsymbol{u}). \tag{2.60}$$

Substituting Equations (2.58)–(2.60) into Equation (2.57), we have

$$\delta W = -\frac{1}{8\pi}\int E^2\frac{\partial\varepsilon}{\partial\tau} div\,(\tau\boldsymbol{u})\, dV + \int \varphi div\,(\rho_{exs}\boldsymbol{u})\, dV.$$

The use of identity (I.1.85) and the Gauss–Ostrogradskii theorem allows us to present the right-hand side in the form similar to Equation (2.54), where

$$\boldsymbol{f} = \rho_{ext}\boldsymbol{E} - \frac{1}{8\pi}E^2\nabla\varepsilon + \frac{1}{8\pi}\nabla\left(E^2\frac{\partial\varepsilon}{\partial\tau}\tau\right). \tag{2.61}$$

The first term is the force on the external charges in the dielectric. The second term is associated with inhomogeneity of the dielectric itself. The third term does not contribute to the total force acting on the dielectric body. However, it affects the distribution of internal stresses in it **(electrostriction)**.

Equation (2.61) is applicable only for liquid and gaseous dielectrics where changes in dielectric properties are due to changes in density. Solid dielectrics may experience a shear deformation that does not change the density but change the internal stresses and the electric properties (the distribution of the bound charges) which is disregarded here. Moreover, the above formula only gives the force due to the electric field. In the stationary liquid, there is also the force of the hydrostatic pressure $-\nabla p_0(\tau, T)$, which should be introduced in Equation (2.61) in order to obtain the total force. The quantity p_0 is the pressure in the liquid.

The stress tensor. Volume forces acting on external and bound charges in a given volume V can be substituted by the equivalent system of the surface stresses applied to the surface S of this volume:

$$\boldsymbol{F} = \int_V \boldsymbol{f}\,dV = \oint_S \boldsymbol{\sigma}_{\mathbf{n}}\,dS, \tag{2.62}$$

where $\boldsymbol{\sigma}_{\mathbf{n}}$ is the surface force applied to the unit area with the outer normal $\mathbf{n}$. The surface stresses are described by the stress tensor $\sigma_{\mu\nu}$, and the quantity $\boldsymbol{\sigma}_{\mathbf{n}}$ is the projection of $\sigma_{\mu\nu}$ onto the direction of the outer normal $\mathbf{n}$ to the element dS: $(\boldsymbol{\sigma}_{\mathbf{n}})_\mu = \sigma_{\mu\nu}n_\nu$.

The transformation from $\boldsymbol{f}$ to $\sigma_{\mu\nu}$ is easy to perform by using the identity (I.1.275) and Equation (2.62):

$$\int_V f_\mu\,dV = \oint_S \sigma_{\mu\nu}n_\nu\,dS = \int_V \frac{\partial\sigma_{\mu\nu}}{\partial x_\nu}\,dV,$$

hence

$$f_\mu = \frac{\partial\sigma_{\mu\nu}}{\partial x_\nu}. \tag{2.63}$$

The stress tensor may be constructed based on Equations (2.63) and (2.61). It has the form

$$\sigma_{\mu\nu} = \frac{\varepsilon}{4\pi}E_\mu E_\nu - \frac{1}{8\pi}E^2\left(\varepsilon - \frac{\partial\varepsilon}{\partial\tau}\tau\right)\delta_{\mu\nu}. \tag{2.64}$$

It generalizes the case of dielectric with the Maxwell stress tensor defined by Equation (I.4.131) at $H = 0$. In order to include the hydrostatic pressure in Equation (2.64), one must add the term $p_0\delta_{\mu\nu}$. The term containing the derivative of density with respect to permittivity (the striction term) is generally not small. However, this term does not contribute to the total forces applied to a dielectric body and can be omitted (see Tamm (1976), & 34 for details). In this case, it is possible to use a simpler (Maxwellian) tensor

$$\sigma'_{\mathrm{n}} = \frac{\varepsilon}{4\pi}\left(E_n\boldsymbol{E} - \frac{1}{2}E^2\boldsymbol{n}\right). \tag{2.65}$$

Hence, on the surface of the conductor $\boldsymbol{E} \parallel \boldsymbol{n}$ and

$$\boldsymbol{f} = \sigma_{\mathrm{n}} = \boldsymbol{n}\frac{\varepsilon E^2}{8\pi}. \tag{2.66}$$

Example 2.8

*Write down the equilibrium condition for a plane boundary between the atmosphere ($\varepsilon = 1$) and a liquid dielectric with an electrostatic field **E** inside it.*

Solution. The forces applied to the surface on both sides must be equal and have opposite directions: $(\sigma_{\mu\nu} + p_0\delta_{\mu\nu})n_\nu = -(\sigma'_{\mu\nu} + p_{atm}\delta_{\mu\nu})n'_\nu$, where $\boldsymbol{n}' = -\boldsymbol{n}$. Moreover, we have the boundary conditions for the field components: $\boldsymbol{E}'_t = \boldsymbol{E}_t$, $E'_n = \varepsilon E_n$. From these equalities, we find

$$p_0(\tau, T) - p_{atm} = -\frac{\varepsilon - 1}{8\pi}(\varepsilon E_n^2 + E_t^2) + \frac{\tau E^2}{8\pi}\frac{\partial\varepsilon}{\partial\tau}. \quad \blacksquare \tag{2.67}$$

Ferroelectrics. Previously we have considered the occurrence of electric polarization in a dielectric under the action of external charges or an external electric field. In some ionic crystals, the polarization arises due to their deformation (either rarefaction or compression in the definite direction). This phenomenon is referred to as **piezoelectric effect**. Examples of piezoelectrics are tourmaline, tartaric acid, Seignette's salt, and barium titanate. Certain piezoelectrics exhibit polarization even in the absence of external stresses or external electric field. The materials that exhibit the **spontaneous polarization** are called **pyroelectrics** (tourmaline is one of the best known pyroelectrics). The most important pyroelectrics are the **ferroelectrics** exhibiting the following basic properties.

1) They have spontaneous polarization within a certain temperature range where they exist in a crystalline modification (ordered phase) characterized by some spontaneous polarization. The transition temperatures from the ordered phase to the disordered phase are referred to as **Curie dielectric temperatures,** T_C. For example, barium titanate, $BaTiO_3$, displays the spontaneous polarization at $T < T_C = 393$ K= 120°C; Seignette's salt, $NaKC_4H_4O_6 \cdot 4H_2O$, has two Curie points: $T_{C1} = 255$ K= −18°C, and $T_{C2} = 297$ K= 24°C, and shows spontaneous polarization in the range $T_{C1} < T < T_{C2}$.
2) The transition from the ordered phase to the disordered phase may be either second-order or first-order phase transition. In the former case, the spontaneous polarization increases gradually as the temperature becomes farther from the Curie point; in the latter case, the polarization arises steeply.
3) The straight line parallel to the direction of spontaneous polarization in a ferroelectric is termed the polar axis. Ferroelectrics may have either a single

polar axis (Seignette's salt) or several polar axes (barium titanate). By applying an external field along the polar axis, the direction of polarization can be changed to the opposite.

4) Near the Curie temperature, the dielectric permittivity is abnormally high (up to 10^4). The dependence of induction D on an electric field is nonlinear and ambiguous.
5) Ferroelectricity is a widespread phenomenon with an increasing number of applications. A great variety of ferroelectric materials are currently available.

The basic properties of ferroelectrics can be explained in terms of phenomenological thermodynamics. We begin from the consideration of spontaneous polarization, based on the Landau theory of second-order phase transitions (Landau and Lifshitz (1980)). V.L. Ginzburg applied this theory to the analysis of ferroelectric properties[4] (Ginzburg (1949), also see Ginzburg (2001)). The analysis is based on the assumption that the specific Gibbs thermodynamic potential Φ in the vicinity of Curie point and in the absence of an external electric field can be expanded in even powers of the polarization vector:

$$\Phi(p, T, P) = \Phi_0(p, T) + \frac{1}{2}\alpha(p, T)P^2 + \frac{1}{4}\beta(p, T)P^4, \tag{2.68}$$

where the higher powers are neglected and the dielectric is assumed to be isotropic. The quantity P is the projection of the polarization vector on the polar axis; it must enter in even powers since the energy of isotropic body cannot depend of the direction of $\boldsymbol{P}$. In this case, the polarization plays the role of the "order parameter" which is nonzero in the ordered phase. Its equilibrium value at a given pressure p and temperature T can be determined from the condition of thermodynamic potential minimum,

$$\frac{\partial \Phi}{\partial P} = 0, \qquad \frac{\partial^2 \Phi}{\partial P^2} > 0. \tag{2.69}$$

By applying the latter to Equation (2.68), from the conditions (2.69) we have

$$P(\alpha + \beta P^2) = 0, \qquad \alpha + 3\beta P^2 > 0. \tag{2.70}$$

We consider now two temperature regions. When $T > T_C$ (the disordered phase), the equilibrium polarization is $P = 0$, and from Equation (2.70) we have $\alpha > 0$. When $T < T_C$ (the ordered phase), from Equation (2.70), we find the equilibrium polarization

$$P_0^2 = -\frac{\alpha}{\beta} > 0, \qquad \alpha < 0, \qquad \beta > 0. \tag{2.71}$$

For the equilibrium polarization to be absent in the disordered phase, one should assume that $\beta > 0$ at $T > T_C$ as well. Hence, near the Curie temperature

4) Vitaly Lazarevich Ginzburg (1916–2009), outstanding Soviet physicist-theorist, obtained the most important results in studies of ferroelectricity, superfluidity and superconductivity, classical and quantum electrodynamics, physics of cosmic rays and plasma physics, radioastronomy, astrophysics, and so on. Nobel Prize laureate (2003) for the works in the field of superconductivity and superfluidity (Section 3.4).

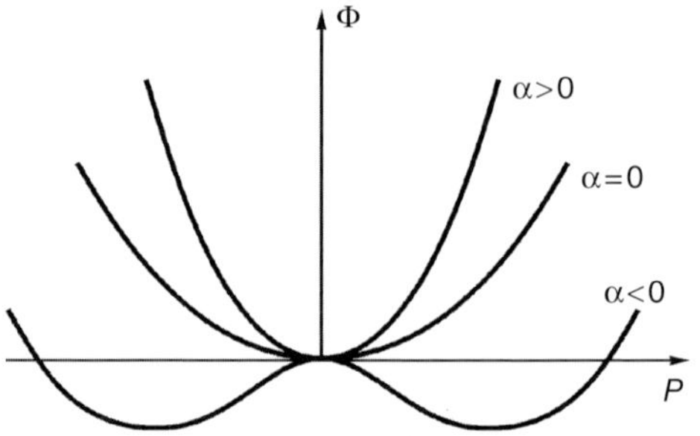

Figure 2.9 The Gibbs thermodynamic potential as a function of polarization, and occurrence of the spontaneous polarization.

($|T - T_C| \ll T_C$) the temperature dependence for the expansion coefficients α and β is

$$\alpha(p, T) \approx a(p)(T - T_C), \qquad \beta(p, T) \approx \beta(p, T_C) = const. \tag{2.72}$$

With the aid of these relations, we can find the temperature dependence for the spontaneous polarization near the Curie temperature:

$$P_0^2 = \frac{a}{\beta}(T_C - T), \qquad T < T_C. \tag{2.73}$$

The dependence of the Gibbs thermodynamic potential on the polarization, which illustrates an occurrence of spontaneous polarization is shown in Figure 2.9.

In order to calculate the dielectric susceptibility, we include an external electric field to obtain the additional term $-PE$ in Equation (2.68) for the thermodynamic potential density:

$$\Phi(p, T, P) = \Phi_0(p, T) + \frac{1}{2}\alpha(p, T)P^2 + \frac{1}{4}\beta(p, T)P^4 - PE. \tag{2.74}$$

The equilibrium conditions (2.69) for a ferroelectric in an external field take the form

$$P(\alpha + \beta P^2) - E = 0, \qquad \alpha + 3\beta P^2 > 0. \tag{2.75}$$

At $T > T_C$ the spontaneous polarization is absent and the cubic term can be neglected in a weak field: $P = E/\alpha = (\varepsilon - 1)E/4\pi$; hence,

$$\varepsilon \approx \frac{C}{T - T_C}. \tag{2.76}$$

Here $C = 4\pi/a$ is the Curie constant, and the temperature dependence (2.76) is called the **Curie–Weiss**[5] **law**. The Curie constant takes values $10^3 - 10^5$, and the dielectric permittivity near the Curie temperature is rather high.

The total polarization in the ordered phase ($T < T_C$). In the presence of an external field, it depends on the field according to the first equation (2.75):

$$E = \alpha P + \beta P^3. \tag{2.77}$$

This dependence is shown in Figure 2.10.

5) Weiss Pierre Ernest (1865–1940), French physicist, known for his studies of magnetic phenomena.

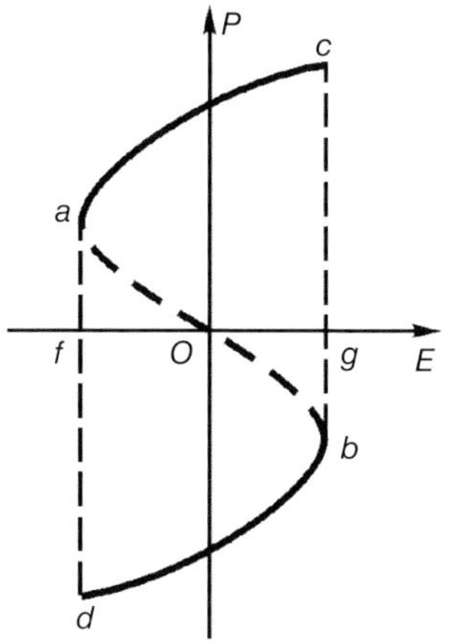

Figure 2.10 The polarization as a function of the field for the ordered phase.

The section *ab* of the curve $P(E)$ corresponds to unstable state. It is clear that on this section $dE/dP < 0$, whereas the thermodynamic potential minimum takes place at $dE/dP > 0$, in agreement with the second condition (2.75). Also, as seen from the figure, when the field E decreases and changes its sign, the polarization P does not change the sign immediately but when the field reaches a certain finite value determined by points f and g. The polarization direction changes steeply (the transition from the point a to the point d and from the point b to the point c).

In the case of weak field E, it is easy to write down separately the polarization P_{ind} induced by the field. We write $P = P_0 + P_{ind}$ and $P_{ind} \ll P_0$, where $P_0 = -\alpha/\beta$, and find from Equation (2.76) that $P_{ind} = -E/2\alpha$. A comparison with the analogous formula for the disordered phase shows that the dielectric permittivity at the Curie point jumps and steeply decreases by a factor of two at transition to the ordered phase.

As mentioned above, there exist ferroelectrics, where the spontaneous polarization occurs steeply at a certain temperature T_0, while the temperature T_C, at which the dielectric permittivity has singularity, is a few degrees lower than T_0. Such cases are interpreted as first-order phase transitions and can be explained by some modifications of the theory given above.

Let the parameter $\beta < 0$ in the expansion (2.74). A medium with such a Gibbs free energy becomes unstable with respect to spontaneous growth of P at all temperatures at which this inequality holds. The expansion presented above becomes incorrect and should be extended as

$$\Phi(p, T, P) = \Phi_0(p, T) + \frac{1}{2}\alpha(p, T)P^2 + \frac{1}{4}\beta(p, T)P^4 + \frac{1}{6}\gamma(p, T)P^6 - PE, \quad (2.78)$$

on condition that the inequality $\gamma > 0$ is valid (Ginzburg-Devonshire expansion). For the spontaneous polarization to appear steeply at a certain temperature T_0, the thermodynamic potential should depend on temperature and polarization, as shown schematically in Figure 2.11.

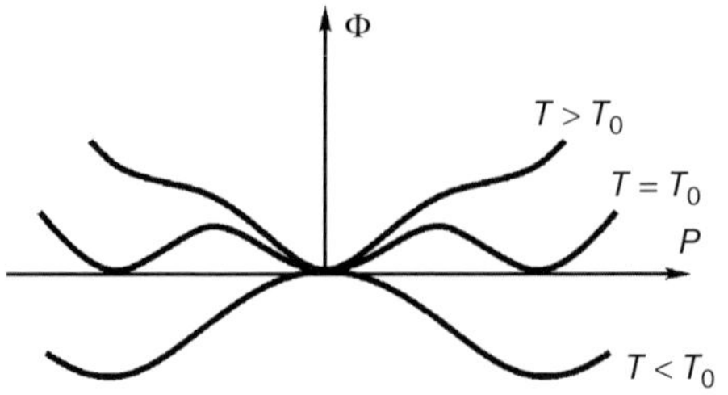

Figure 2.11 Onset of spontaneous polarization.

At $T = T_0$ the curve has three minima of the same deepness. Accordingly, three equilibrium polarizations, $P_0 = 0$ and $\pm P_0 \neq 0$, are possible. Spontaneous polarization at the phase transition point must satisfy the condition

$$\alpha + \frac{1}{2}\beta P_0^2 + \frac{1}{3}\gamma P_0^4 = 0, \tag{2.79}$$

which follows from Equation (2.78), and the extremum condition

$$\frac{\partial \Phi}{\partial P} = P_0(\alpha + \beta P_0^2 + \gamma P_0^4) = 0. \tag{2.80}$$

On excluding γP_0^4 from the two last equations, we find

$$P_0^2 = -\frac{4\alpha}{\beta} > 0, \qquad \gamma = \frac{3\beta^2}{16\alpha}. \tag{2.81}$$

Since $\beta < 0$, the inequality $\alpha > 0$ must hold at the phase transition. Hence, on passing from the region $T > T_0$ to the region $T < T_0$, the polarization jumps from zero to $\sqrt{3|\beta|/4\gamma}$. A jump of the dielectric permittivity at phase transition is calculated in Problem 2.85.

Let us outline the drawbacks and limitations of the theory of segnetoelectrics given above. This theory is phenomenological and cannot be used to determine the expansion coefficients α, β, and γ. The theory establishing the relationship between these coefficients and microscopic parameters of the matter can be found in monographs Blinc and Zeks (1974). The theory given above is applicable near the phase transition point but not too close to it. The last limitation is due to polarization fluctuations, which should not exceed the average polarization value $\overline{P_0}$. It is just this average value that has been analyzed in the above thermodynamic theory. Finally, a ferroelectric of large sizes in the absence of an external field is broken in regions with opposite polarization, so-called domains. We have studied only one-domain segnetoelectrics.

Suggested literature: Landau and Lifshitz (1980); Kubo (1965b); Landau *et al.* (1984); Bredov *et al.* (2003); Sivukhin (1977); Toda *et al.* (1983); Burfoot (1967); Lines and Glass (1980); Feynman (1972); Kittel (1978)

Problems

2.70 Equation (2.45) gives the change in the internal energy density of a thermally insulated body when an electric field is applied. The body temperature may also change. Calculate the change in the internal energy density assuming the process to be isothermal ($T = const$). Ignore the changes in volume and mass density.

2.71 Calculate the input or output heat Q per unit volume of dielectric when the field changes isothermally from 0 to E.

2.72 Write down the density of thermodynamic functions for an anisotropic dielectric having the linear coupling equation $D_\mu = \varepsilon_{\mu\nu} E_\nu$. On the basis of thermodynamic relations, prove that the permittivity tensor of an anisotropic dielectric is symmetric, $\varepsilon_{\mu\nu} = \varepsilon_{\nu\mu}$.

2.73 Calculate the derivative dT/dD^2, which characterizes a change in the temperature of a thermally insulated dielectric in the process of its polarization (electrocaloric effect). Neglect changes in density.

2.74 In a gaseous dielectric consisting of rigid dipole molecules, the electric field increases adiabatically from zero up to E. Calculate the change in the temperature $T_2 - T_1$ (Kelvin temperature scale) using Equation (1.34) for the dielectric permittivity. Obtain an order-of-magnitude estimation of the cooling effect $\Delta T/T$ for a gas under normal conditions.

2.75 A field $\boldsymbol{E}_1$ was first produced in an isotropic dielectric medium having permittivity ε_1. An uncharged dielectric body of volume V with dielectric permittivity ε_2 was then introduced into the medium. Show that the internal energy of the system will change by

$$U = \frac{1}{8\pi} \int_V (\varepsilon_1 - \varepsilon_2) \boldsymbol{E}_1 \cdot \boldsymbol{E}_2 \, dV, \tag{2.82}$$

where $\boldsymbol{E}_2$ is the electric field after the the body was introduced in the medium. The quantity U can be regarded as the energy of interaction between the dielectric body and the external field.

2.76 Calculate the change in the heat capacity of a dielectric $C_{VD} - C_{V0}$ at constant volume and electric induction due to an electric field. Compare this difference with the difference $C_{VE} - C_{V0}$, assuming that the coupling equation between D and E is linear.

2.77 Calculate the difference between the specific heat capacities of dielectric $C_E - C_D$ without an assumption on the linear dependence of D on E. In calculating, assume that either volume V or pressure p is fixed.

2.78 Prove that specific heat capacities of dielectric can be written as

$$C_D = \left(\frac{\partial U}{\partial T}\right)_D, \quad C_E = \left(\frac{\partial U}{\partial T}\right)_E - E\left(\frac{\partial P}{\partial T}\right)_D,$$

where P is the specific dipole moment of the dielectric, and the subscripts indicate the parameter which is kept constant.

2.79 In the nonlinear case, the dielectric susceptibility $\alpha = \partial P/\partial E$ depends on the conditions of polarization process, that is, whether the dielectric remains thermally insulated or it is in contact with the thermostat (maintaining constant temperature). Find the relation between the adiabatic and isothermal susceptibilities:

$$\alpha_S = \frac{C_P}{C_E}\alpha_T,$$

where the heat capacities C_P and C_E correspond to the cases when the polarization and the electric field are constant, respectively.

Hint. Apply the Jacobian method.

2.80* A dielectric body is polarized by an external electric field $\mathcal{E}$ in air. The shape of the body is such that the electric field E inside it remains uniform. Assuming the polarization process to be isothermal and isobaric, calculate the change in the volume $\Delta V \ll V$ of the body (electrostriction) and express it in terms of polarizability, compressibility and external field.

2.81 Under the conditions of Problem 2.80, calculate the heat Q absorbed by the body in the course of polarization.

2.82 A field in a liquid dielectric increases isothermally from zero up to a given value E. Neglecting changes in the density and using formula (2.61), show that the pressure increases by

$$p - p_0 = \frac{\tau E^2}{8\pi}\frac{\partial \varepsilon}{\partial \tau}. \tag{2.83}$$

2.83 Calculate the density of the electric force $\boldsymbol{f}$ acting in a homogeneous uncharged gaseous dielectric. Use Equation (1.34) for the dielectric permittivity of a gas.

2.84 Perform the same procedure using Equation (1.37) for the dielectric permittivity of a gas.

2.85 Use the expansion (2.78) to calculate a jump of the dielectric permittivity of ferroelectric due to first-order phase transition. Find the relation between the temperatures T_0 and T_C.

2.86 (a) Find the force of attraction f_0 per unit area between the plates of a plane parallel capacitor in vacuum when the distance between the plates is a, and the potential difference is V. (b) Find the new value f of this force when the charged capacitor is separated from the battery and then is filled with a liquid dielectric of permittivity ε, or when a solid dielectric of the same permittivity ε and thickness just smaller than a is inserted between the plates (not touching them). (c) Find the force of attraction f between the plates when the capacitor is first filled with a liquid dielectric, or when a solid dielectric plate is inserted into it, and the capacitor is then charged from a battery.

2.87 The electrodes of a plane capacitor are at a distance h_1 from each other and have the form of rectangles with sides a and b. A dielectric plate of permittivity ε is placed between the electrodes in such a way that it is parallel to them. The plate is in the form of a parallelepiped of thickness h_2 and base area $a \times b$. The plate is not completely inserted into the capacitor so that only a section x of the side a is inside. Find the force F with which the plate is drawn inside the capacitor in the following two cases: (a) a constant potential difference V is maintained across the capacitor and (b) the charge q on each of the plates is constant. The edge effects should be neglected.

2.88⋆ A plane capacitor is immersed into incompressible liquid of permittivity ε and density τ so that its plates are vertical. The distance between the plates is d and the potential difference between them is V. Determine the height h to which the liquid will rise in the capacitor.

Hint. Use Equation (2.67).

2.88 Find the direction of the Maxwell stress σ'_{n} acting on the area element dS whose normal $\boldsymbol{n}$ makes an angle ϑ with the direction of the field $\boldsymbol{E}$. Find also the magnitude of σ'_{n} and determine the direction of the electrostrictionl tension σ''_{n}.

2.90 Two identical point charges q are placed in a homogeneous liquid dielectric of permittivity ε at a distance a from each other. Find the force $\boldsymbol{F}$ acting on each of the charges with the aid of the Maxwell or the total stress tensor. Find also the components of the force of interaction $q^2/a^2\varepsilon$ between the two charges. For comparison, calculate the forces applied to (a) the plane of symmetry perpendicular to the line joining the charges and to (b) the surface of a small sphere centered on one of the charges.

2.91 An uncharged conducting sphere of radius R and mass m floats in a liquid of dielectric permittivity ε and density τ so that one quarter of its volume is immersed. Find the potential φ at which the sphere must be charged in order to ensure that half of its volume will be immersed. Solve this problem using (a) the Maxwell stress tensor and (b) the total stress tensor including the electrostriction term.

2.92 Find the force F on the point charge from Problem 2.8 (electric image force). Solve the problem by a number of methods, in particular, with the use of the Maxwell stress tensor. Describe qualitatively the motion of the charge for the case if it is free to move through the dielectric.

2.93⋆ Two identical dielectrics of permittivities ε_1 and ε_2 occupy the entire space and are in contact along an infinite plane. Two charges q_1 and q_2 are located at equal distances a from this plane on a straight line perpendicular to it. Find the forces F_1 and F_2 acting on each of these charge. Why are these forces not equal?

2.94 A point charge q is placed in a homogeneous dielectric at a distance a from the plane boundary of an infinitely extended conductor. Find the electric potential φ in the dielectric, the distribution of charges σ induced on metal surfaces, and the force F acting on the charge q.

2.95 An electric dipole of moment $\boldsymbol{p}$ is placed in a homogeneous dielectric near the plane boundary of an infinitely extended conductor. Find the potential energy U of interaction between the dipole and the induced charges, and also the force $\boldsymbol{F}$ and the torque $\boldsymbol{N}$ acting on the dipole.

2.96⋆ Two identical point charges $q_1 = q_2 = q$ are placed at a distance a from each other in a solid dielectric of permittivity ε_1. The charges are at the centers of small spherical cavities of radius R. Find the forces acting on these charges. Compare them with the electric tensions in the plane of symmetry perpendicular to the line joining the charges.

2.97★ A dielectric sphere of radius R and permittivity ε_1 is placed in a homogeneous dielectric of permittivity ε_1. A point charge q is located at a distance $a > R$ from the center of the sphere. Find the field φ in entire space. As a limiting case, determine the field for the conducting sphere. Also find the force acting on the charge q due to the polarization of the sphere produced by the charge. How does this force change if the other identical point charge is placed at a point symmetric relative to the center of the dielectric sphere?

2.98★ An insulated metal sphere of radius a is placed inside a hollow metal sphere of radius b. The distance between their centers is c, where $c \ll a$ and $c \ll b$. The total charge on the inner sphere is q. Find the charge distribution σ on the inner sphere and the force F acting on it up to the terms linear in c.

2.99 Find the energy U of interaction between a point charge q and a grounded conducting sphere of radius R. Also find the force F on the point charge. The charge is at a distance a from the center of the sphere and the system is placed in a homogeneous dielectric medium of permittivity ε.

2.100 A point charge q is placed in a dielectric at a distance a from the center of an insulated sphere of radius R. The sphere charge is Q. Find the energy U of interaction between the point charge and the sphere and also the force F on the point charge.

2.101 What are the conditions that must be satisfied by a test charge q (in the sense of its magnitude and position in space) in order that it can be used to investigate the field produced by a system of charges located on conducting and dielectric bodies, and in particular, the field produced by a charged sphere in a homogeneous dielectric?

2.102★ An electric dipole $\boldsymbol{p}$ is placed in a homogeneous dielectric at a distance r from the center of a grounded conducting sphere of radius R. Find the system of images which is equivalent to the induced charges, the energy of interaction U between the dipole and the sphere, the force $\boldsymbol{F}$ and the torque $\boldsymbol{N}$ on the dipole. Consider the limiting case $r \to R$ $(r > R)$.

2.103 A spherical cavity of radius R is formed in a conductor. An electric dipole of moment p is placed in the center of the cavity. Find the distribution of the charges induced on the surface of the cavity. Determine the field $\boldsymbol{E}'$ produced by these charges inside the cavity.

2.104★ An electric field is produced in a homogeneous dielectric of permittivity ε. Its potential in the neighborhood of a point O can be presented in the form

$$\varphi_1 = \sum_{l,m} \sqrt{\frac{4\pi}{2l+1}}\, a_{lm} r^l Y_{lm}(\vartheta, \alpha).$$

The homogeneity and neutrality of the dielectric in the neighborhood of the point O is then disturbed (e.g., by an inclusion of a conductor, which in general may carry a charge, or a dielectric of permittivity $\varepsilon_1 \neq \varepsilon$). As a consequence, the electric field potential outside the disturbed region becomes $\varphi = \varphi_1 + \varphi_2$, where

$$\varphi_2 = \sum_{l,m} \sqrt{\frac{4\pi}{2l+1}}\,\varepsilon^{-1} r^{-(l+1)} Q_{lm} Y_{lm}(\vartheta, \alpha)$$

is the potential of the field due to free and bound charges in the region of the inhomogeneity (the factor ε is introduced for convenience). Find the potential energy U of interaction between the inhomogeneity and the external field φ_1.

Hint. Consider the electric stresses acting on the surface enclosing the inhomogeneity. Use the result of Problem I.2.66.

2.105 Find the energy of interaction U_0 of a small inhomogeneous region in a dielectric considered in Problem 2.104 with a slowly varying external field. Owing to the rapid convergence of the series, it is possible to take into consideration only the terms with $l = 0$ and 1. Give the result in the vector form. Find the force $\boldsymbol{F}$ and the torque $\boldsymbol{N}$ on the inhomogeneous region in this approximation.

2.106 Show that an uncharged dielectric body of permittivity ε_0, which is placed in a dielectric with permittivity ε, is attracted toward regions of higher electric fields when $\varepsilon_0 > \varepsilon$, and repelled from such regions when $\varepsilon_0 < \varepsilon$.

Hint. Use Equation (2.68).

2.107 In general, the components of the dipole moment $\boldsymbol{p}$ which are induced in a dielectric body placed in an external uniform electric field $\boldsymbol{E}$ can be given by $p_i = \beta_{ik} E_k$, where β_{ik} is the symmetric polarizability tensor. What orientation will the body tend to take in the external uniform field? The body is uncharged, $\beta_{ik} x_i x_k > 0$, where x_i ($i = 1, 2, 3$) is an arbitrary vector.

2.108 A rod made of dielectric with permittivity ε_1 is immersed into a homogeneous liquid dielectric with permittivity ε_2. What will be its orientation if the system is placed in a uniform external field? What would be the orientation of a thin disk placed in liquid dielectric?

2.109 Find the force $\boldsymbol{F}$ acting on a dielectric sphere due to a point charge q (see the conditions of Problem 2.97). Consider the limiting case of a conducting sphere. Solve the problem by two methods: the method of Problem 2.104 and using Equation (2.68).

2.110 The self-capacitances of two conductors that are placed in a homogeneous dielectric are C_1 and C_2, respectively; the distance between the conductors, r, is much larger than their linear sizes. Find the force $\boldsymbol{F}$ between the conductors if either their potentials V_1 and V_2 or charges q_1 and q_2 are maintained constant.

2.111 Two identical spherical capacitors with inner and outer radii a and b are insulated and placed at a large distance r from each other. The inner spheres are charged (with charges q_1 and q_2), after which the outer spheres are joined by a wire. Find (approximately) the change ΔW in the energy of the system.

2.112 The grounded outer electrode (sphere) of a spherical capacitor is thin and a small aperture is cut in it. An insulated wire passes through the aperture and connects the inner electrode (sphere) to the third conductor placed at a large distance r from the capacitor. The self-capacitance of the third conductor is C, and the total charge carried by this conductor and the inner sphere is q. The radii of the outer and inner spheres is b and a, respectively. Find the force $\boldsymbol{F}$ on the third conductor.

2.113⋆ In order to remove a single charged particle from a conductor, it is necessary to do a certain work $A > 0$ which is called the **work function** and depends on the type of the conductor, its thermodynamic state, and the nature of the particle (electron, ion).

(a) Express the work function in terms of the surface density κ of the dipole moment which forms the double electrical layer on the surface of the conductor (the double electrical layer is considered in Example I.2.6 and Problems I.2.42 and I.2.43).

(b) Express the potential difference $\varphi_b - \varphi_a$ which is established between two conductors as they come in contact, in terms of their work functions A_a and A_b.

(c) Express the electric potential φ and the electric field inside the dihedral angle γ formed by two conductors in terms of their work functions A_a and A_b. The conductors are placed in vacuum and are in contact along the infinite half-plane (Figure 2.32).

2.114 A plane capacitor has the electrodes in the form of disks of radius a, which are at a distance $h \ll a$ from each other. The capacitor is filled with a dielectric of permittivity ε. The charge $q(t)$ on the disks is slowly varying. Ignoring the edge effects, calculate the change dW/dt in the electrostatic energy of the capacitor and show that this change is equal to the Poynting vector flux through the side surface.

Hint. Calculate the magnetic field in the capacitor with the aid of the Maxwell equation (5.12) and symmetry considerations.

2.4 Answers and Solutions

2.1

$$\varphi_1 = \varphi_2 = \frac{2}{\varepsilon_1 + \varepsilon_2}\frac{q}{r}, \qquad \boldsymbol{D}_1 = \frac{2\varepsilon_1}{\varepsilon_1 + \varepsilon_2}\frac{q\boldsymbol{r}}{r^3}, \qquad \boldsymbol{D}_2 = \frac{2\varepsilon_2}{\varepsilon_1 + \varepsilon_2}\frac{q\boldsymbol{r}}{r^3}.$$

2.2

$$\varphi_1 = \varphi_2 = \varphi_3 = \frac{2\pi}{\varepsilon_1\alpha_1 + \varepsilon_2\alpha_2 + \varepsilon_3\alpha_3}\frac{q}{r}, \qquad \boldsymbol{D}_i = \frac{2\pi\varepsilon_i}{\varepsilon_1\alpha_1 + \varepsilon_2\alpha_2 + \varepsilon_3\alpha_3}\frac{q\boldsymbol{r}}{r^3}.$$

2.3 The boundary conditions ($\varphi = const$ on the surface of the conductor, and $\varphi = 0$ as $r \to \infty$) can be satisfied by a potential of the form $\varphi = C/r$; the constant C is determined from the conditions $\oint_S D_n\, dS = 4\pi q$, $C = 2/(\varepsilon_1 + \varepsilon_2)$. Hence, we have the potential $\varphi = 2q/[(\varepsilon_1 + \varepsilon_2)r]$ and the surface charge distributions:

$$\sigma_1 = \frac{q\varepsilon_1}{2\pi a^2(\varepsilon_1 + \varepsilon_2)}, \qquad \sigma_2 = \frac{q\varepsilon_2}{2\pi a^2(\varepsilon_1 + \varepsilon_2)},$$

$$\sigma_{1int} = \frac{q(\varepsilon_1 - 1)}{2\pi a^2(\varepsilon_1 + \varepsilon_2)}, \qquad \sigma_{2int} = \frac{q(\varepsilon_2 - 1)}{2\pi a^2(\varepsilon_1 + \varepsilon_2)}.$$

2.4

$$C = \left[\frac{(\varepsilon - 1)\Omega}{4\pi} + 1\right]\frac{ab}{b - a}.$$

2.5

$$C = \left[\frac{1}{\varepsilon_1}\left(\frac{1}{a} - \frac{1}{c}\right) + \frac{1}{\varepsilon_2}\left(\frac{1}{c} - \frac{1}{b}\right)\right]^{-1}.$$

The bound charges are located where the dielectric is nonuniform, that is, on the spheres with radii a, b, and c:

$$\sigma_{a\,int} = -\frac{q}{4\pi a^2}\cdot\frac{\varepsilon_1 - 1}{\varepsilon_1}, \qquad \sigma_{b\,int} = -\frac{q}{4\pi b^2}\cdot\frac{\varepsilon_2 - 1}{\varepsilon_2},$$

$$\sigma_{c\,int} = -\frac{q}{4\pi c^2}\left(\frac{1}{\varepsilon_2} - \frac{1}{\varepsilon_1}\right),$$

where q is the charge of the inner electrode of the capacitor. The total bound charge in the capacitor is zero.

2.7 The capacitance of the capacitor is

$$C = \frac{\varepsilon_0 S}{4\pi a \ln 2}.$$

The surface density of the bound charges is given by

$$\sigma_{int} = -\sigma\left(1 - 1/\varepsilon_0\right) \qquad \text{at} \qquad x = 0,$$

$$\sigma_{int} = \sigma\left(1 - 1/2\varepsilon_0\right) \qquad \text{at} \qquad x = a.$$

The volume density is $\rho_{int} = -\sigma a/\varepsilon_0(x + a)^2$ (σ is the charge of the plate at $x = 0$).

2.8 At $z \geq 0$:

$$\varphi = \varphi_1 = \frac{q}{\varepsilon_1 r_1} + \frac{(\varepsilon_1 - \varepsilon_2)}{\varepsilon_1(\varepsilon_1 + \varepsilon_2)}\,\frac{q}{r_2}.$$

At $z \leq 0$:

$$\varphi = \varphi_2 = \frac{2}{\varepsilon_1 + \varepsilon_2}\,\frac{q}{r_1}$$

2.9

$$\sigma_{int} = \frac{1}{4\pi}\left[(\varepsilon_2 - 1)\frac{\partial\varphi_2}{\partial z} - (\varepsilon_1 - 1)\frac{\partial\varphi_1}{\partial z}\right]\Bigg|_{z=0} = \frac{qa}{2\pi r^3}\,\frac{\varepsilon_1 - \varepsilon_2}{\varepsilon_1(\varepsilon_1 + \varepsilon_2)},$$

where $r = \sqrt{x^2 + y^2 + a^2} = r_1\big|_{z=0} = r_2\big|_{z=0}$.

When $\varepsilon_2 \to \infty$ we have the case of a point charge q placed in the dielectric ε_1 near the boundary with the plane conductor. In this case, the surface density is $\sigma_{int} \to -qa/2\pi\varepsilon_1 r^3$. This limiting density is in fact the sum of the density of bound charges on the surface of the dielectric and the density of free charges on the surface of the conductor.

2.10 The field between the conducting planes forming the dihedral angle is produced by the charge systems shown in Figure 2.12.

2.11 Consider polar coordinates with the origin at the center of the sphere and the Oz-axis parallel to $\boldsymbol{E}_0$. The potential may be sought as a series in Legendre polynomials (cf. the solution of Problem 2.14). The final result is

$$\varphi_1 = -\frac{3\varepsilon_2}{\varepsilon_1 + 2\varepsilon_2} E_0 r \cos\vartheta \quad \text{at} \quad r < a,$$

$$\varphi_2 = -E_0 r \cos\vartheta + \frac{(\varepsilon_1 - \varepsilon_2) E_0 a^3 \cos\vartheta}{(\varepsilon_1 + 2\varepsilon_2) r^2} \quad \text{at} \quad r > a.$$

The electric field inside the sphere is uniform and its strength is given by

$$E_1 = \frac{3\varepsilon_2}{\varepsilon_1 + 2\varepsilon_2} E_0 \begin{cases} > E_0 & \text{at} \quad \varepsilon_2 > \varepsilon_1, \\ < E_0 & \text{at} \quad \varepsilon_2 < \varepsilon_1. \end{cases}$$

Outside the sphere the field of the electric dipole of moment

$$p = E_0 a^3 \frac{\varepsilon_1 - \varepsilon_2}{\varepsilon_1 + 2\varepsilon_2}$$

is imposed on the external uniform field E_0. This secondary field is produced by bound charges on the surface of the dielectric sphere:

$$\sigma_{int} = \frac{3}{4\pi} \frac{\varepsilon_1 - \varepsilon_2}{\varepsilon_1 + 2\varepsilon_2} E_0 \cos\vartheta, \qquad \rho_{int} = 0.$$

It is easy to understand the reason for such a charge distribution by considering any small element of the polarized dielectric as an elementary dipole.

2.12 For the dielectric with unchanged polarization, $\Delta\boldsymbol{E} = 4\pi\boldsymbol{P}/3$ (Problem I.2.38).

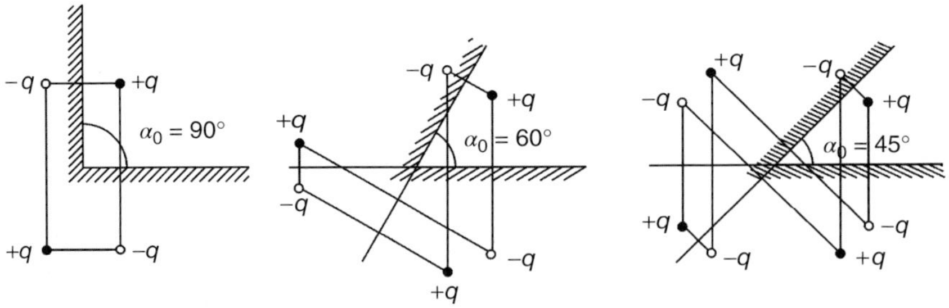

Figure 2.12 Real charge is located between the planes of the dihedral angle, other charges are images.

For a normal dielectric,

$$\Delta \boldsymbol{E} = -\frac{12\pi\varepsilon}{(2\varepsilon + 1)(\varepsilon - 1)}\boldsymbol{P}.$$

2.13

$$\varphi = -\boldsymbol{E}_0\cdot\boldsymbol{r} + \frac{\boldsymbol{p}\cdot\boldsymbol{r}}{r^3} \ (r \geq R),$$

where $\boldsymbol{p} = R^3\boldsymbol{E}_0$ is the polarizability of the sphere and $\sigma = 3\varepsilon_0 E_0 \cos\vartheta/4\pi$.

2.14 Let the origin of spherical polar coordinates be at the center of the sphere (Figure 2.13) and the polar axis pass through the point charge. The potential will be sought in the form

(1) $$\varphi(r, \vartheta, \alpha) = \frac{q}{\varepsilon r_1} + \sum_{l,m}\left(a_{lm}r^l + \frac{b_{lm}}{r^{l+1}}\right)P_{lm}(\cos\vartheta)e^{im\alpha},$$

where r_1 is the distance between q_1 and the point of observation. Evidently, the series in Equation (1) describes the field due to charges induced on the sphere. This field must vanish at infinity and, therefore, $a_{lm} = 0$. Because of the symmetry, the potential is independent of the angle α, and hence the terms with $m \neq 0$ will also vanish. The remaining constants $b_l \equiv b_{l0}$ will be determined from the boundary conditions.

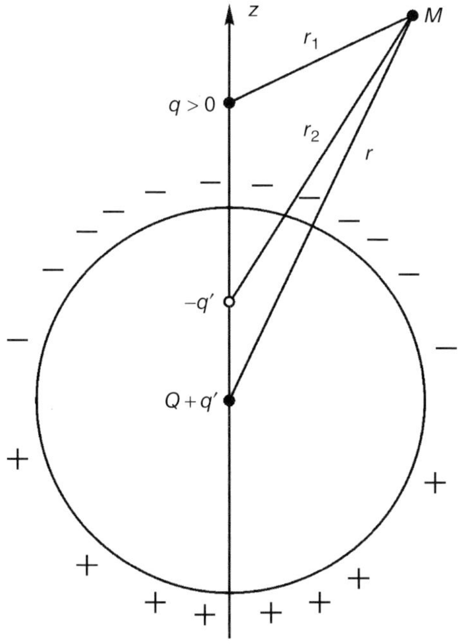

Figure 2.13 A point charge near a conducting sphere.

In case (a) the potential of the sphere is $\varphi(R, \vartheta) = V = const$. Using the expansion for q/r_1 from Problem I.2.30, we have:

$$\varphi(R, \vartheta) = \sum_{l=0}^{\infty} \left(\frac{qR^l}{\varepsilon a^{l+1}} + \frac{b^l}{R^{l+1}} \right) P_l(\cos\vartheta) = V.$$

Hence, $b_l = -qR^{2l+1}/\varepsilon a^{l+1}$ when $l \neq 0$, $b_0 = VR - Rq/\varepsilon a$, so that the potential outside the sphere is

$$\varphi(r, \vartheta) = \frac{q}{\varepsilon r_1} + \frac{VR}{r} - \frac{qR}{\varepsilon a} \sum_{l=0}^{\infty} \left(\frac{R^2}{a} \right)^l \frac{P_l(\cos\vartheta)}{r^{l+1}}. \tag{2}$$

The density of charges induced on the surface of the sphere can now be found:

$$\sigma(R, \vartheta) = -\frac{\varepsilon}{4\pi} \frac{\partial \varphi}{\partial r}\bigg|_{r=R} = \frac{\varepsilon V}{4\pi R} - \frac{q}{4\pi} \sum_{l=0}^{\infty} (2l+1) \frac{R^{l+1}}{a^{l+1}} P_l(\cos\vartheta). \tag{3}$$

In case (b) the potential V is unknown and should be expressed in terms of the charge Q of the sphere. Clearly,

$$Q = 2\pi \int \sigma(R, \vartheta) R^2 \sin\vartheta \, d\vartheta = \varepsilon VR - \frac{qR}{a},$$

and hence $V = Q/\varepsilon R + q/\varepsilon a$. Using the solution of Problem I.2.30, Equation (2) can be written in the form

$$\varphi = \frac{q}{\varepsilon r_1} + \frac{Q + q'}{\varepsilon r} - \frac{q'}{\varepsilon r_2}, \tag{4}$$

where

$$q' = q\frac{R}{a}, \qquad r_2 = \sqrt{r^2 + a'^2 - 2a'r\cos\vartheta}, \qquad a' = \frac{R^2}{a}.$$

Thus, in the region $r > a$ the potential produced by the point charge and the charged sphere is equivalent to the potential produced by four point charges lying along the axis of symmetry: a charge q at a distance a from the origin and its three images, that is, charges Q and $q' = qR/a$ at the origin and a charge $-q'$ at the point $a' = R^2/a$ harmonically conjugated with respect to the surface of the sphere. The charge $-q'$ represents the effect of charges induced on the surface of the sphere which is nearest to q. Evidently, the sign of these charges is opposite to the sign of q. The charge $+q'$ represents the effect of charges induced on the distant side of the sphere and is of the same sign as q.

When the sphere is neutral, the term containing Q is absent. When the sphere is grounded ($V = 0$), the potential assumes the form

$$\varphi = \frac{q}{\varepsilon r_1} - \frac{q'}{\varepsilon r_2}. \tag{5}$$

2.15

$$\varphi(M) = \frac{q}{\varepsilon r_1} - \frac{q'}{\varepsilon r_2} + V$$

(Figure 2.14), where

$$q' = q\frac{R}{a}, \qquad a' = \frac{R^2}{a}.$$

2.16

$$\varphi(M) = \frac{q}{r_1} - \frac{q'}{r_2} + \frac{q'}{r_3} - \frac{q}{r_4}V$$

(Figure 2.15), where

$$q' = \frac{qa}{b}, \qquad b' = \frac{a^2}{b}.$$

The charge on the boss is

$$Q = -q\left[1 - \frac{b^2 - a^2}{b\sqrt{a^2 + b^2}}\right].$$

2.17 Outside the sphere $\varphi \equiv \varphi_1 = q/\varepsilon r_1$; in the conductor $\varphi \equiv \varphi_3 = q/\varepsilon_1 R_1$; in the cavity $\varphi \equiv \varphi_2 = q/\varepsilon_2 r_1 - q'/\varepsilon_2 r_2 + q/\varepsilon_1 R_1$ (Figure 2.16), where $q' = qR_2/a$, and $a' = R_2^2/a$.

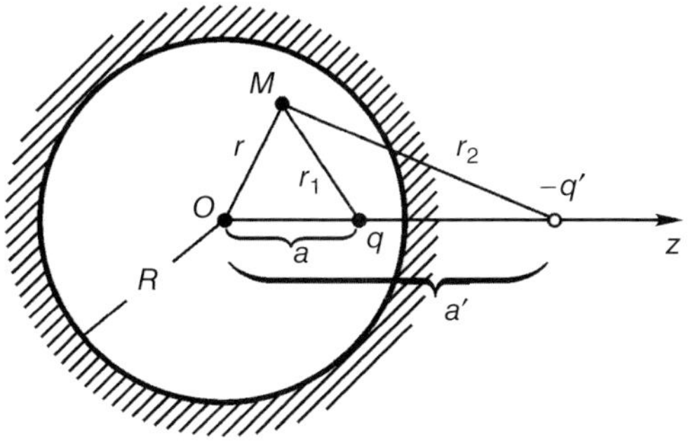

Figure 2.14 A point charge in a cavity filled with dielectric.

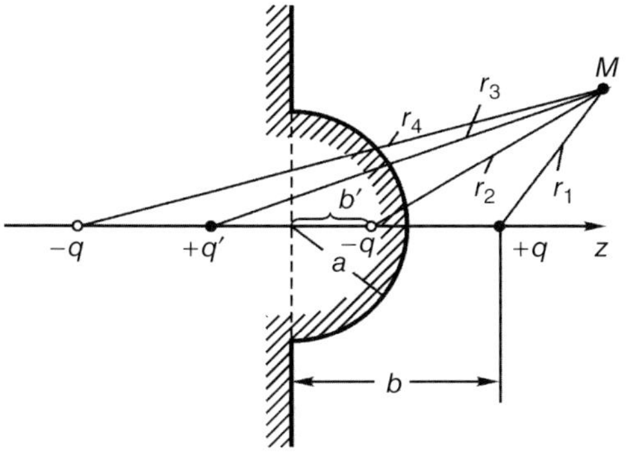

Figure 2.15 A point charge near a hemispherical boss on a conducting plane.

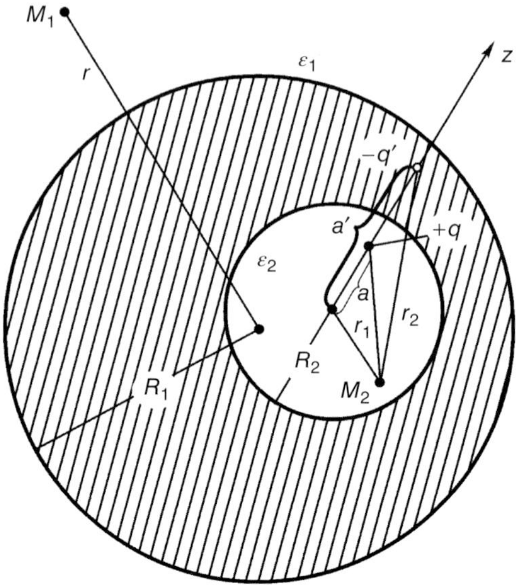

Figure 2.16 A point charge near the cavity filled with dielectric in a conducting sphere immersed in another dielectric.

2.18

$$\varphi_1 = \frac{q}{\varepsilon_1 r_1} + q\frac{\varepsilon_1 - \varepsilon_2}{\varepsilon_1}\sum_{l=0}^{\infty}\frac{l+1}{\varepsilon_1 l + \varepsilon_2(l+1)}\frac{a^l r^l}{R^{2l+1}}P_l(\cos\vartheta) \text{ at } r \le R;$$

$$\varphi_2 = q\sum_{l=0}^{\infty}\frac{2l+1}{\varepsilon_1 l + \varepsilon_2(l+1)}\frac{a^l}{r^{l+1}}P_l(\cos\vartheta) \text{ at } r \ge R,$$

where r_1 is the distance between the charge and the point of observation q.

At $a = 0$,

$$\varphi_1 = \frac{q}{\varepsilon_1 r} + \left(1 - \frac{\varepsilon_2}{\varepsilon_1}\right)\frac{q}{\varepsilon_2 R}, \qquad \varphi_2 = \frac{q}{\varepsilon_2 r}.$$

2.19

$$\Delta C = \frac{a^2 b^2 c^2}{(b-a)^2(b^3 - a^3)}.$$

2.20

$$C = \frac{\kappa}{\varphi_1 - \varphi_2} = \frac{1}{2}\left[\operatorname{arccosh}\frac{a^2 - R_1^2 - R_2^2}{2R_1R_2}\right]^{-1}.$$

2.21

$$\sigma = \pm\frac{|E|}{4\pi} = \pm\frac{4b^2k^2}{\pi\left[(x^2 - y^2 - b^2)^2 + 4x^2y^2\right]}, \quad \text{where} \quad b = \frac{2R^4}{a^2\sqrt{a^2 - 4R^2}}.$$

The origin is at the center of the segment that joins the axes of the cylinders and is the Ox-axis in the problem.

2.22

$$C = \frac{1}{2}\left[\operatorname{acrcosh}\frac{R_1^2 + R_2^2 - a^2}{2R_1R_2}\right]^{-1}.$$

2.23 If the axes Ox, Oy, and Oz are parallel to the main axes of the tensor ε_{ik}, then

(1) $$\varphi(x, y, z) = \frac{e'}{r'} = \frac{e}{\sqrt{\varepsilon^{(x)}\varepsilon^{(y)}\varepsilon^{(z)}}}\left[\frac{x^2}{\varepsilon^{(x)}} + \frac{y^2}{\varepsilon^{(y)}} + \frac{z^2}{\varepsilon^{(z)}}\right]^{-\frac{1}{2}}.$$

For an arbitrary orientation of the coordinate system, Equation (1) can be written in the form

(2) $$\varphi(\boldsymbol{r}) = \frac{e}{\sqrt{|\varepsilon_{ik}|\varepsilon_{ik}^{-1}x_ix_k}},$$

where $|\varepsilon_{ik}|$ is the determinant of the tensor ε_{ik}.

2.24

$$\boldsymbol{E} = \boldsymbol{E}_0 - \frac{(\varepsilon_{ik} - \delta_{ik})n_iE_{0k}}{\varepsilon_{lm}n_ln_m}\boldsymbol{n}.$$

2.25 $C = S\varepsilon^{(z)}/4\pi a$, where z is measured along the normal to the plates of the capacitor.

2.26 If the Ox- and Oz-axes lie in the $(\boldsymbol{E}_0, \boldsymbol{n})$ plane, and $Oz \parallel \boldsymbol{n}$, then

$$\tan\vartheta = \frac{E_x}{E_z} = \frac{\varepsilon_{zz}\tan\vartheta_0}{1 - \varepsilon_{zx}\tan\vartheta_0},$$

where $\tan\vartheta_0 = E_{0x}/E_{0z}$. Then the line of force in the dielectric will remain in the $(\boldsymbol{E}_0, \boldsymbol{n})$ plane.

2.27 From the Green's reciprocity theorem and Equation (2.12) we find $\sum_{i,k}(s_{ik} - s_{ki})q_iq'_k = 0$. Since the charges of the conductors, q_i and q'_k, are independent, the required relation follows from this equation.

2.28 Let the charge of the first conductor be q_1 and the charge on the outer surface of the second conductor be q' (the charge on the inner surface of the second conductor is $-q_1$ according to the electrostatic Gauss theorem). Then the system (2.14) takes the form

(1) $$\left.\begin{aligned} q_1 &= c_{11}V_1 + c_{12}V_2, \\ -q_1 + q' &= c_{12}V_1 + c_{22}V_2. \end{aligned}\right\}$$

Combining these equations, we obtain

(2) $$q' = (c_{11} + c_{12})V_1 + (c_{12} + c_{22})V_2.$$

Once q' is specified, the field in the external space is defined, including the potential V_2 of the second conductor. Equation (2) must therefore be valid for all values V_1 at given q' and V_2, which can only be true if

(3) $$c_{11} + c_{12} = 0.$$

The first equation (1) then takes the form

(4) $$q_1 = c_{11}(V_1 - V_2).$$

It follows from Equations (2)–(4) that

$$C = c_{11} = -c_{12} = -c_{21}, \quad C' = c_{12} + c_{22}.$$

2.29

$$s_{11} = \frac{c_{22}}{c_{11}c_{22} - c_{12}^2}, \quad s_{22} = \frac{c_{11}}{c_{11}c_{22} - c_{12}^2}, \quad s_{12} = s_{21} = -\frac{c_{12}}{c_{11}c_{22} - c_{12}^2}.$$

2.30

$$c_{11} \approx C_1, \quad c_{22} \approx C_2, \quad c_{12} = c_{21} \approx -\frac{C_1C_2}{\varepsilon r}.$$

2.31

$$C = \frac{c_{11}c_{22} - c_{12}^2}{c_{11} + c_{22} + 2c_{12}}.$$

2.32

$$q_1 = \frac{s_{11} - 2s_{12} + s_{13}}{s_{11} - s_{12}}\frac{q}{8}, \quad q_2 = \frac{q}{2}, \quad q_3 = \frac{q}{4}, \quad q_4 = \frac{s_{11} - s_{13}}{s_{11} - s_{12}}\frac{q}{8}.$$

2.33

$$q_1 = -\frac{2a}{b}q, \quad q_2 = -\frac{a}{b}, \quad q_3 = \frac{3a^2}{b^2}q.$$

2.34

$$q_1 = q\frac{V_0 - V_P}{V_1 - V_0}, \quad q_0 = q\frac{V_1 - V_P}{V_0 - V_1}.$$

2.36 The self-capacitance of the joined conductors is

$$c_{00} = c_{11} + c_{22} + 2c_{12}.$$

The mutual capacity between the joined conductors and the ith conductor of the system is

$$c_{0i} = c_{1i} + c_{2i}.$$

2.37 When the sphere and the conductor touch, they have the same potential

$$V_1 = qs_{11} + (Q - q)s_{12} = qs_{12} + (Q - q)s_{22} = V_2,$$

and hence

(1) $$\frac{s_{11} - s_{12}}{s_{22} - s_{12}} = \frac{Q}{q} - 1,$$

where s_{ik} are the potential coefficients (subscripts 1 and 2 refer to the sphere and the conductor, respectively). Let q_k be the charge on the conductor after the kth contact. Since the potentials of the sphere and the conductor are equal on contact, then

$$q_k s_{11} + (Q + q_{k-1} - q_k)s_{12} = q_k s_{12} + (Q - q + q_{k-1})s_{22}.$$

Hence using Equation (1) we obtain the recurrence relation for q_{k-1} and q_k:

(2) $$q_k = q + \frac{q}{Q} q_{k-1}.$$

Successive application of formula (2) followed by passing to the limit $k \to \infty$ yields

$$q = \lim_{k\to\infty} q_k = q\left[1 + \frac{q}{Q} + \left(\frac{q}{Q}\right)^2 + \left(\frac{q}{Q}\right)^3 + \ldots\right] = \frac{qQ}{Q-q}.$$

2.38 The Laplace equation takes the form:

$$\frac{d}{d\xi}\left(R_\xi \frac{d\varphi}{d\xi}\right) = 0, \qquad R_\xi = \sqrt{(\xi + a^2)(\xi + b^2)(\xi + c^2)}.$$

This equation must be integrated subject to boundary conditions $\varphi = const$ at $\xi = 0$ (the surface of the ellipsoid), and $\varphi \to 0$ as $\xi \to \infty$. We can find the constant of integration by taking into account that $r = \sqrt{x^2 + y^2 + z^2} \to \infty$ at $\xi \to r^2$. The integration yields

$$\varphi(\xi) = \frac{q}{2\varepsilon} \int_\xi^\infty \frac{d\xi}{R_\xi}, \qquad \frac{1}{C} = \frac{1}{2\varepsilon} \int_0^\infty \frac{d\xi}{R_\xi}.$$

Hence,

$$\sigma = -\frac{\varepsilon}{4\pi} \frac{\partial\varphi}{\partial n}\bigg|_{\xi=0} = -\frac{\varepsilon}{4\pi}\left(\frac{1}{h_1}\frac{\partial\varphi}{\partial\xi}\right)_{\xi=0} = \frac{q}{4\pi abc}\left(\frac{x^2}{a^4} + \frac{y^2}{b^4} + \frac{z^2}{c^4}\right)^{-1/2}.$$

The charge densities at the ends of the semiaxes are proportional to the lengths of the axes, so that $\sigma_a : \sigma_b : \sigma_c = a : b : c$.

2.39 At $a = b > c$ (oblate ellipsoid):

$$\varphi = \frac{q}{\sqrt{a^2 - c^2}} \arctan \sqrt{\frac{\xi + c^2}{a^2 - c^2}}, \qquad C = \frac{\sqrt{a^2 - c^2}}{\arccos(c/a)}.$$

In particular, at $c = 0$ (a disk) we have $C = 2a/\pi$.

At $a > b = c$ (prolate ellipsoid):

$$\varphi = \frac{q}{2\varepsilon\sqrt{a^2 - b^2}} \ln \frac{\sqrt{\xi + a^2} + \sqrt{a^2 - b^2}}{\sqrt{\xi + a^2} - \sqrt{a^2 - b^2}}, \qquad C = \frac{\varepsilon\sqrt{a^2 - b^2}}{\ln[(a + \sqrt{a^2 - b^2})/b]}.$$

In particular, at $b \ll a$ (a rod):

$$C = \frac{\varepsilon a}{\ln(2a/b)}.$$

2.40 Let us suppose that the ellipsoid is initially uncharged. that is, $q = 0$. If the external uniform field $\boldsymbol{E}_0$ is parallel to the Ox-axis, then

$$\varphi_0 = -E_0 x = \mp E_0 \sqrt{\frac{(\xi + a^2)(\eta + a^2)(\xi + a^2)}{(b^2 - a^2)(c^2 - a^2)}}.$$

The negative sign corresponds to $x > 0$ and the positive sign to $x < 0$. Both the function φ_0 and the potential φ' due to charges induced on the ellipsoid satisfy the Laplace equation. Substituting $\varphi' = \varphi_0 F(\xi)$ into the Laplace equation we obtain the equation for the unknown function $F(\xi)$:

$$\frac{d^2F}{d\xi^2} + \frac{dF}{d\xi}\frac{d}{d\xi}\ln[R_\xi(\xi + a^2)] = 0.$$

This equation is easily integrated. The solution satisfying the boundary conditions is

$$\varphi\big|_{q=0} = \varphi_0\left\{1 - \int_\xi^\infty \frac{d\xi}{(\xi + a^2)R_\xi} \Bigg/ \int_0^\infty \frac{d\xi}{(\xi + a^2)R_\xi}\right\}.$$

If the ellipsoid has an inherent charge q, then the solution satisfying the conditions $\varphi\big|_{\xi=0} = const$ and $-\oint_S \partial\varphi/\partial n\, dS = 4\pi q$ (S is a closed surface surrounding the ellipsoid) can be obtained from the superposition principle (Problem 2.42):

$$\varphi\big|_q = \varphi_{q=0} + \frac{1}{2}q\int_\xi^\infty \frac{d\xi}{R_\xi}.$$

2.41 The potential has the same form as in the preceding problem. The integrals entering the expression for the potential can be expressed in terms of elementary functions whenever the ellipsoid has rotational symmetry. As a result, we obtain

$$\varphi = -E_0x + E_0x\left[\ln\frac{\sqrt{1+\xi/a^2}+e}{\sqrt{1+\xi/a^2}-e} - \frac{2e}{\sqrt{1+\xi/a^2}}\right]\left[\ln\frac{1+e}{1-e} - 2e\right]^{-1},$$

where a and b are the large and small semiaxes, $e = \sqrt{1 - b^2/a^2}$ is the eccentricity of the ellipsoid, and the Ox-axis is perpendicular to the plane,

$$x = \frac{a}{e}\sqrt{\left(1 + \frac{\xi}{a^2}\right)\left(1 + \frac{\zeta}{a^2}\right)}$$

(Problem I.1.96). The field strength reaches maximum at the apex of the ellipsoid:

$$\frac{E_{max}}{E_0} = -\frac{1}{E_0 h_\xi}\frac{\partial\varphi}{\partial\xi}\Big|_{\xi,0,\zeta=-b^2} = \frac{2e^3(1-e^2)^{-1}}{\ln[(1+e)/(1-e)] - 2e} = \frac{1}{n^{(x)}},$$

where $n^{(x)}$ is a depolarization coefficient (Problem 2.43). For a sphere, we have $e = 0$ and $E_{max}/E_0 = 3$. In the case of a very long rod (lightning conductor):

$$\frac{E_{max}}{E_0} = \frac{a^2}{b^2}\left(\ln\frac{2a}{b} - 1\right)^{-1}, \qquad a \gg b,$$

and, therefore, a spark discharge in air is more likely to occur at the end of such a lightning conductor than in other places.

2.42 The field at an arbitrary distance from the ellipsoid can be obtained by superimposing three fields of the form established in Problem 2.40 (the field $\boldsymbol{E}_0$ should be separated into the components parallel to the principal axes of the ellipsoid). At large distances from the ellipsoid

$$\varphi = \varphi_0 + \frac{\boldsymbol{p}\cdot\boldsymbol{r}}{r^3}, \qquad p_x = \beta^{(x)}E_x, \qquad p_y = \beta^{(y)}E_y, \qquad p_z = \beta^{(z)}E_z.$$

The principal values of the polarizability tensor of the ellipsoid are

$$\beta^{(x)} = \frac{abc}{3n^{(x)}}, \qquad \beta^{(y)} = \frac{abc}{3n^{(y)}}, \qquad \beta^{(z)} = \frac{abc}{3n^{(z)}}.$$

2.43

$$n^{(x)} = \frac{1-e^2}{2e^2}\left(\ln\frac{1+e}{1-e} - 2e\right) \le \frac{1}{3}, \qquad n^{(y)} = n^{(z)} = \frac{1-n^{(x)}}{2},$$

where $e = \sqrt{1 - b^2/a^2}$ is the eccentricity of the ellipsoid.

At $e \to 1$ (a rod):

$$n^{(x)} = 0, \qquad n^{(y)} = n^{(z)} = \frac{1}{2}.$$

At $e \ll 1$ (nearly spherical shape),

$$n^{(x)} = \frac{1}{3} - \frac{2}{15}e^2, \qquad n^{(y)} = n^{(z)} = \frac{1}{3} + \frac{1}{15}e^2.$$

2.44

$$n^{(z)} = \frac{1+e^2}{e^3}(e - \arctan e) \ge \frac{1}{3}, \quad n^{(x)} = n^{(y)} = \frac{1-n^{(z)}}{2}, \quad e = \sqrt{a^2/c^2 - 1}.$$

In the specific case of a disk $n^{(z)} = 1$, $n^{(x)} = n^{(y)} = 0$.

2.45 $\varphi = \varphi_x = \varphi_y = \varphi_z$.

Inside the ellipsoid,

$$\varphi_x = \varphi_{1x} = -\frac{E_0 x}{1 + (\varepsilon_1 - \varepsilon_2)n^{(x)}/\varepsilon_2}.$$

Outside the ellipsoid,

$$\varphi_x = \varphi_{2x} = -E_0 x + E_0 x \frac{abc(\varepsilon_1 - \varepsilon_2)}{2[\varepsilon_2 + (\varepsilon_1 - \varepsilon_2)n^{(x)}]}\int_\xi^\infty \frac{d\xi}{(\xi + a^2)R_\xi},$$

where

$$n^{(x)} = \frac{1}{2}abc\int_0^\infty \frac{d\xi}{(\xi + a^2)R_\xi}.$$

The potentials φ_y and φ_z are given by analogous expressions, where x should be replaced by y and z, and a by b and c, respectively. Inside the ellipsoid the field is uniform and is given by

$$\boldsymbol{E}_1 = \frac{E_{0x}\boldsymbol{e}_x}{1 + (\varepsilon_1 - \varepsilon_2)n^{(x)}/\varepsilon_2} + \frac{E_{0y}\boldsymbol{e}_y}{1 + (\varepsilon_1 - \varepsilon_2)n^{(y)}/\varepsilon_2} + \frac{E_{0z}\boldsymbol{e}_z}{1 + (\varepsilon_1 - \varepsilon_2)n^{(z)}/\varepsilon_2}.$$

At large distances from the ellipsoid

$$\varphi_2 = -\boldsymbol{E}_0\cdot\boldsymbol{r} + \frac{\boldsymbol{p}\cdot\boldsymbol{r}}{r^3},$$

where $p_x = \beta^{(x)} E_x$,

$$\beta^{(x)} = \frac{abc}{3(\varepsilon_2/(\varepsilon_1 - \varepsilon_2) + n^{(x)})},$$

and so on.

2.46 Using Equation (2.82), we obtain

$$U = E_0^2 \frac{abc(\varepsilon_2-\varepsilon_1)\{2[\varepsilon_2+(\varepsilon_1-\varepsilon_2)n]\sin^2\vartheta + [\varepsilon_1+\varepsilon_2+(\varepsilon_2-\varepsilon_1)n]\cos^2\vartheta\}}{6[\varepsilon_2+\varepsilon_1+n(\varepsilon_2-\varepsilon_1)][\varepsilon_2+(\varepsilon_1-\varepsilon_2)n]},$$

$$N = -\frac{\partial U}{\partial \vartheta} = E_0^2 \frac{abc(\varepsilon_2-\varepsilon_1)^2(3n-1)\sin 2\vartheta}{6[\varepsilon_2+\varepsilon_1+n(\varepsilon_2-\varepsilon_1)][\varepsilon_2+(\varepsilon_1-\varepsilon_2)n]},$$

where ϑ is the angle between the axis of symmetry and the field $\boldsymbol{E}_0$, and n is the depolarization coefficient with respect to the symmetry axis of the ellipsoid (see, for instance, the solution of the preceding problem).

It follows from the latter formula that the external field tends to turn over the axis of symmetry of prolate ($n < 1/3$) and oblate ($n > 1/3$) ellipsoids into positions, respectively, parallel and perpendicular to the field.

For a conducting ellipsoid, $\varepsilon_1 \to \infty$ and

$$N = E_0^2 \frac{abc(3n-1)\sin 2\vartheta}{6n(1-n)}.$$

2.47 The potential energy of a charged liquid drop in the form of an ellipsoid of revolution with an eccentricity $e = \sqrt{1 - b^2/a^2}$ and volume equal to the volume of a sphere of radius R (charge q) can be expressed by the formula (1)

$$U(e) = \frac{q^2}{2C} + \alpha S = \frac{q^2\sqrt[3]{1-e^2}}{4Re}\ln\frac{1+e}{1-e} + 2\pi R^2\alpha\left(\sqrt[3]{1-e^2} + \frac{\arcsin e}{e\sqrt[3]{1-e^2}}\right),$$

where one should use the expression for the capacitance C of the prolate ellipsoid of revolution given in the solution of Problem 2.39.

In order to investigate the stability of a charged spherical drop, it is necessary to determine the dependence of the energy (Equation (1)) on eccentricity e for small e. We expand the function U into powers of e up to terms of order e^4:

$$U(e) = \frac{q^2}{2R} + 4\pi R^2\alpha + \frac{e^4}{45}\left(8\pi R^2\alpha - \frac{q^2}{2R}\right).$$

It is clear from this formula that if the charge on the drop is $q < q_c = \sqrt{16\pi R^3\alpha}$, then for a small deformation the drop tends to return to the spherical state, that is, the drop is stable. When $q > q_c$, the drop is unstable because the deformation, once initiated, continues to increase. The process ends with fragmentation of the drop into two or more stable smaller drops.[6] The process ends with the formation of stable drops. This is apparent from the expression for the critical charge q_c. As the sizes of the drop are reduced, the critical charge q_c decreases in proportion to the square root of the volume, whereas the total charge q on the drop decreases, on the average, in proportion to the volume. Hence, the stability conditions will be satisfied for sufficiently small sizes of the drop.

2.48

$$\varphi = -\frac{E_0 z}{\pi}\left(\arctan\frac{a}{\sqrt{\xi}} - \frac{a}{\sqrt{\xi}}\right) = -\frac{E_0}{\pi}\sqrt{-\eta}\left(\frac{\sqrt{\xi}}{a}\arctan\frac{a}{\sqrt{\xi}} - 1\right),$$

where $\sqrt{\xi}$ should be taken with positive sign for $z > 0$ and with negative sign for $z < 0$. At large distances from the aperture one has $\xi \approx r^2$, and the field is

$$\varphi \approx \frac{E_0 a^3 z}{3\pi r^3} \qquad \text{at } z > 0.$$

This is the field of an electric dipole whose axis is parallel the Oz-axis and whose dipole moment is $p = E_0 a^3/3\pi$. Hence, the lines of force passing through the aperture close on the other side of the metal screen.

2.49

$$\sigma = -\frac{E_0}{4\pi^2}\left(\pi - \arcsin\frac{a}{r_1} + \frac{a}{\sqrt{r_1^2 - a^2}}\right) \qquad \text{at } z = -0,$$

$$\sigma = -\frac{E_0}{4\pi^2}\left(\frac{a}{\sqrt{r_1^2 - a^2}} - \arcsin\frac{a}{r_1}\right) \qquad \text{at } z = +0,$$

where $r_1 = \sqrt{\xi + a^2}$ is the distance between the center and observation point on the plane.

2.50 The equation to be solved is $\Delta\varphi = -4\pi q\delta(\boldsymbol{r} - \boldsymbol{r}_0)$. The delta-function should be written in cylindrical coordinates:

$$\delta(\boldsymbol{r} - \boldsymbol{r}_0) = \frac{1}{r_0}\,\delta(r - r_0)\,\delta(\alpha - \gamma)\,\delta(z).$$

The Fourier component

$$\varphi_k(r,\alpha) = \frac{1}{\pi}\int_{\infty}^{+\infty}\varphi(r,\alpha,z)\cos kz\,dk \tag{1}$$

of the potential $\varphi(r,\alpha,z)$ satisfies the equation

$$\frac{1}{r}\frac{\partial}{\partial r}\left(r\frac{\partial\varphi_k}{\partial r}\right) + \frac{1}{r^2}\frac{\partial^2\varphi_k}{\partial\alpha^2} - k^2\varphi_k = -\frac{4q}{r_0}\,\delta(r - r_0)\,\delta(\alpha - \gamma) \tag{2}$$

and the boundary conditions (Figure 2.8):

$$\varphi_k(r,0) = \varphi_k(r,\beta) = 0, \tag{3}$$

6) It is easy to verify, for example, that when a charged drop separates into two identical spherical drops the energy decreases by a factor $2^{2/3}$.

(4) $$\varphi_k(\infty, \alpha) = 0.$$

Consider the homogeneous equation corresponding to Equation (2). The particular solutions satisfying Equation (3) are the products $R_n(r)$ $\sin(n\pi\alpha/\beta)$ $(n = 1, 2, 3, \ldots)$, where the quantity $R_n(r)$ is equal either to $I_{n\pi/\beta}(kr)$ or to $K_{n\pi/\beta}(kr)$ (up to a constant factor). The solution of the nonhomogeneous Equation (2) will be sought in the form of superposition of these particular solutions:

(5) $$\varphi_k = \begin{cases} \sum_{n=1}^{\infty} A_n I_{n\pi/\beta}(kr)\sin(n\pi\alpha/\beta) & \text{at } r < a, \\ \sum_{n=1}^{\infty} B_n K_{n\pi/\beta}(kr)\sin(n\pi\alpha/\beta) & \text{at } r > a. \end{cases}$$

While writing Equation (5), we have kept in mind that the potential φ_k must satisfy Equation (4) and be bounded at $r = 0$.

In order to determine the constants A_n and B_n, we will first use the fact that the potential at $r = r_0$ is continuous. This gives

(6) $$\frac{B_n}{A_n} = \frac{I_{n\pi/\beta}(kr_0)}{K_{n\pi/\beta}(kr_0)}.$$

Second, we will require that the potential given by Equation (5) should satisfy Equation (2). Substitute Equation (5) into Equation (2), multiply both sides by $\sin(m\pi\alpha/\beta)$ $(m = 1, 2, \ldots)$ and integrate over α from 0 to β. Since the functions $\sin(n\pi\alpha/\beta)$ are orthogonal in the above interval, we have

(7) $$\frac{1}{r}\frac{d}{dr}\left(r\frac{dR_m}{dr}\right) - \left(k^2 + \frac{m^2\pi^2}{\beta^2 r^2}\right)R_m = -\frac{8q}{\beta r_0}\delta(r - r_0)\sin\frac{m\pi\gamma}{\beta},$$

where

$$R_m(r) = \begin{cases} A_m I_{m\pi/\beta}(kr) & \text{at } r < a, \\ B_m K_{m\pi/\beta}(kr) & \text{at } r > a. \end{cases}$$

The function $R_m(r)$ is continuous at $r = r_0$, but its first derivative with respect to r undergoes a jump at that point:

$$b \equiv R'_m(r_0 + 0) - R'_m(r_0 - 0) = kB_m K'_{m\pi/\beta}(kr_0) - kA_m I'_{m\pi/\beta}(kr_0).$$

The second derivative of $R_m(r)$ is therefore equal to $R''_m(r) = b\delta(r - r_0)$.

Substituting this expression into Equation (7) and omitting the terms that are bounded at $r = r_0$, we obtain the second equation for A_n and B_n:

(8) $$kB_n K'_{n\pi/\beta}(kr_0) - kA_n I'_{n\pi/\beta}(kr_0) = -\frac{8q}{\beta r_0}\sin\frac{n\pi\gamma}{\beta}.$$

In simplifying the expressions for A_n and B_n it is useful to employ the formula

$$K_\nu(x)I'_\nu(x) - K'_\nu(x)I_\nu(x) = \frac{1}{x}.$$

2.52

$$\varphi(r, \alpha, z) = \frac{2q}{\pi}\left(\frac{1}{R_0}\arctan\sqrt{\frac{\cosh(\eta/2) + \cos[(\alpha - \gamma)/2]}{\cosh(\eta/2) - \cos[(\alpha - \gamma)/2]}}\right.$$

$$-\frac{1}{R_0'}\arctan\sqrt{\frac{\cosh(\eta/2)+\cos[(\alpha+\gamma)/2]}{\cosh(\eta/2)-\cos[(\alpha+\gamma)/2]}}\Bigg),$$

where

$$R_0 = \sqrt{r_0^2 + r^2 + z^2 - 2rr_0\cos(\gamma-\alpha)} = \sqrt{2rr_0(\cosh\eta - \cos(\gamma-\alpha))},$$

$$R_0' = \sqrt{r_0^2 + r^2 + z^2 - 2rr_0\cos(\gamma+\alpha)} = \sqrt{2rr_0(\cosh\eta - \cos(\gamma+\alpha))}.$$

2.53 $\sigma = const \cdot r^{(\pi/\beta-1)}$,
where r is the distance from the edge of wedge. In the special case where the wedge is placed in the field of a point charge (Problem 2.50), we have

$$const = -\frac{q\sqrt{\pi}r_0^{\pi/\beta}\sin(\pi\gamma/\beta)}{\beta^2(r_0^2+z^2)^{\pi/\beta+1/2}}\frac{\Gamma(\pi/\beta+1/2)}{\Gamma(\pi/\beta+1)}.$$

It is clear that $\sigma \to 0$ as $r \to 0$ and $\beta < \pi$; and $\sigma \to \infty$ as $r \to 0$ and $\beta > \pi$. In the special case, where a charge is located near the edge of a plane, we have

$$\sigma \propto \frac{1}{\sqrt{r}}.$$

2.54 Suppose that the charge q is at the origin and the Oz-axis is perpendicular to the surface of the plate. The equations for the front and rear surfaces will then take the form $z = a$ and $z = a + c$, respectively. The potential will be sought in the form

(1)

$$\left.\begin{aligned}
\varphi_1 &= q\int_0^\infty J_0(kr_1)e^{-k|z|}\,dk + \int_0^\infty A_1(k)J_0(kr_1)e^{kz}\,dk \qquad (-\infty < z < a),\\
\varphi_2 &= \int_0^\infty B_1(k)J_0(kr_1)e^{-kz}\,dk + \int_0^\infty B_2(k)J_0(kr_1)e^{kz}\,dk \qquad (a < z < b),\\
\varphi_3 &= \int_0^\infty A_2(k)J_0(kr_1)e^{-kz}\,dk \qquad (b = a + c < z < \infty).
\end{aligned}\right\}$$

The boundary conditions on the surfaces of the plate give the system of four algebraic equations for the coefficients A_1, A_2, B_1, and B_2. Solving this system, we have

$$\text{(2)}\qquad A_1 = q\beta\frac{e^{-2kb}-e^{-2ka}}{1-\beta^2e^{-2kc}}, \quad A_2 = q\frac{1-\beta^2}{1-\beta^2e^{-2kc}},$$

$$B_1 = \frac{q(1-\beta)}{1-\beta^2e^{-2kc}}, \qquad B_2 = \frac{q\beta(1-\beta)e^{-2kb}}{1-\beta^2e^{-2kc}},$$

where $\beta = (\varepsilon-1)/(\varepsilon+1)$ and $b = a + c$.

Equations (2) and (1) together give the solution of our problem. At large distances from the plate ($z > 0$), the field is of the form:

$$\varphi(r_1, z) \approx \frac{q}{\sqrt{r_1^2+z^2}} + \frac{pz}{(r_1^2+z^2)^{3/2}},$$

where $r_1 = \sqrt{x^2+y^2}$, $p = -(\varepsilon-1)^2cq/2\varepsilon$.

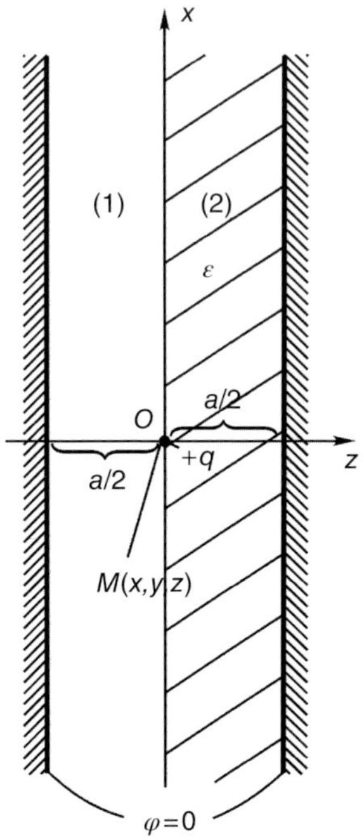

Figure 2.17 Plane parallel capacitor with a dielectric plate and point charge.

2.55

$$\varphi(M) = \frac{2q}{\varepsilon + 1} \int_0^{\infty} \frac{\sinh k(a - |z|)}{\cosh ka} J_0(kr_1)\, dk,$$

where $r_1 = \sqrt{x^2 + y^2}$ (Figure 2.17).

As $\sqrt{z^2 + r_1^2} \to 0$ (near the charge),

$$\varphi \to \frac{2q}{(\varepsilon + 1)\sqrt{r_1^2 + z^2}}$$

(cf. Problem 2.1).

The potential φ can be written in the form

$$\varphi = \frac{2q}{\varepsilon + 1} \sum_{n=-\infty}^{\infty} \frac{(-1)^n}{\sqrt{r_1^2 + (z - 2an)^2}}.$$

The corresponding system of images is shown in Figure 2.19a.

2.56 We use the bispherical coordinates so that the surfaces of the inner and outer electrodes will be the coordinate surfaces $\xi = \xi_1$ and $\xi = \xi_2$, respectively. To achieve this the Oz-axis must be drawn through the centers of the spheres of the capacitor as shown in Figure 2.18. The coordinates of the centers will then be $z_1 = a \coth \xi_1$, and $z_2 = a \coth \xi_2$ (a is the parameter of the bispherical coordinates). The relations between the radii of

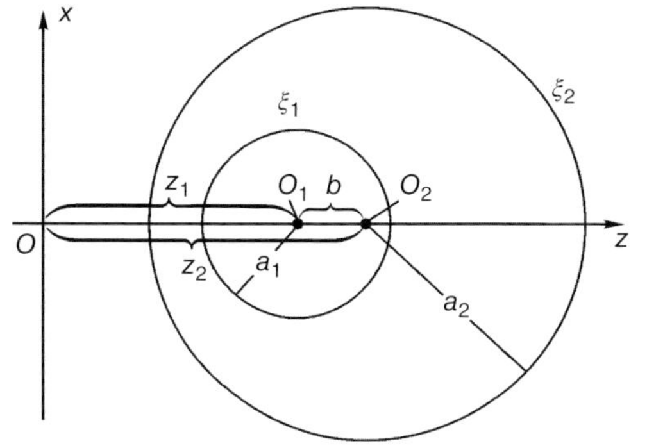

Figure 2.18 Nonconcentric spherical capacitor.

the spherical electrodes and the quantities a, ξ_1, ξ_2 are $a = a_1 \sinh \xi_1$, $a = a_2 \sinh \xi_2$, $b = z_2 - z_1 = a(\coth \xi_2 - \coth \xi_1)$; and hence,

$$(1) \qquad \cosh \xi_1 = \frac{a_2^2 - a_1^2 - b^2}{2a_1 b}, \qquad \cosh \xi_2 = \frac{a_2^2 + b^2 - a_1^2}{2a_2 b}.$$

In the space between the electrodes of the capacitor, the function ψ satisfies the equation

$$(2) \qquad \frac{\partial^2 \psi}{\partial \xi^2} + \frac{1}{\sin \eta} \frac{\partial}{\partial \eta}\left(\sin \eta \frac{\partial \psi}{\partial \eta}\right) + \frac{1}{\sin^2 \eta} \frac{\partial^2 \psi}{\partial \alpha^2} - \frac{1}{4}\psi = 0.$$

Separating the variables in Equation (2) and taking into account that ψ is independent of the azimuthal angle α, we find the particular solutions that are bounded at $\eta = 0$ and π:

$$(3) \qquad \psi_l(\xi, \eta) = \left[A_l \cosh\left(l + \frac{1}{2}\right)\xi + B_l \sinh\left(l + \frac{1}{2}\right)\xi\right] P_l(\cos \eta),$$

where $l = 0, 1, 2, 3, \ldots$

The function ψ will be sought in the form of the series $\psi(\xi, \eta) = \sum_{l=0}^{\infty} \psi_l(\xi, \eta)$. The coefficients A_i and B_i are determined from the boundary conditions $\psi(\xi_2, \eta) = 0$,

$$\psi(\xi_1, \eta) = V(2\cosh \xi_1 - 2\cos \eta)^{-1/2} = V \sum_{l=0}^{\infty} \exp\left[-\left(l + \frac{1}{2}\right)\xi_1\right] P_l(\cos \eta).$$

Finally, we obtain

$$(4) \quad \varphi(\xi, \eta) = V\sqrt{2\cosh \xi - 2\cos \eta} \times \sum_{l=0}^{\infty} \frac{\exp[-(l + 1/2)\xi_1] \sinh(l+1/2)(\xi - \xi_2)}{\sinh(l + 1/2)(\xi_1 - \xi_2)} P_l(\cos \eta).$$

The capacitance of the capacitor is given by

$$C = \frac{q_1}{V} = \frac{1}{4\pi V} \int_0^{\pi} \int_0^{2\pi} \frac{1}{h_\xi} \frac{\partial \varphi}{\partial \xi} h_\eta h_\alpha \bigg|_{\xi=\xi_1} d\eta \, d\alpha.$$

The sign "+" in the last formula is chosen because the ξ-coordinate decreases along the outer normal to the inner electrode. Substituting Equation (4) into this expression and recalling that Legendre polynomials form an orthogonal set, we have:

$$C = \frac{a_1}{2} + a_1 \sinh \xi_1 \sum_{l=0}^{\infty} e^{-(2l+1)\xi_1} \coth\left(l + \frac{1}{2}\right)(\xi_1 - \xi_2).$$

2.57

$$C = \frac{a_1 a_2}{a_2 - a_1} + \frac{a_1^2 a_2^2 b^2}{(a_2 - a_1)^2 (a_2^3 - a_1^3)}.$$

2.58

$$c_{11} = \frac{a_1}{2} + a_1 \sinh \xi_1 \sum_{l=0}^{\infty} \exp\left[-\left(l + \frac{1}{2}\right)\xi_1\right] \coth\left(l + \frac{1}{2}\right)(\xi_1 + \xi_2),$$

$$c_{22} = \frac{a_2}{2} + a_2 \sinh \xi_2 \sum_{l=0}^{\infty} \exp\left[-\left(l + \frac{1}{2}\right)\xi_2\right] \coth\left(l + \frac{1}{2}\right)(\xi_1 + \xi_2),$$

$$c_{12} = -a_1 \sinh \xi_1 \sum_{l=0}^{\infty} \frac{\exp[-(l + 1/2)(\xi_1 + \xi_2)]}{\sinh(l + 1/2)(\xi_1 + \xi_2)},$$

where $\cosh \xi_1 = (b^2 + a_1^2 - a_2^2)/2ba_1$, $\cosh \xi_2 = (b^2 - a_1^2 + a_2^2)/2ba_2$.

The surfaces of the first and second conductor are given by the equations $\xi = -\xi_1$ and $\xi = \xi_2$, respectively, where $a_1 \sinh \xi_1 = a_2 \sinh \xi_2$.

2.59

$$c_{11} = a_1(1 + mn + mn^3 + m^2n^2),$$
$$c_{12} = -a_1 n(1 + mn),$$
$$c_{22} = a_2(1 + mn + m^3 n + m^2 n^2),$$

where $m = a_1/b$, $n = a_2/b$.

2.61 Suppose that the potential of the spheres is zero and that the potential at infinity is $-V$. We carry out the inversion of the system with respect to the sphere of radius $R = 2a$ whose center lies at the point of contact of the conducting spheres (Figure 2.19a). After inversion, the system takes the

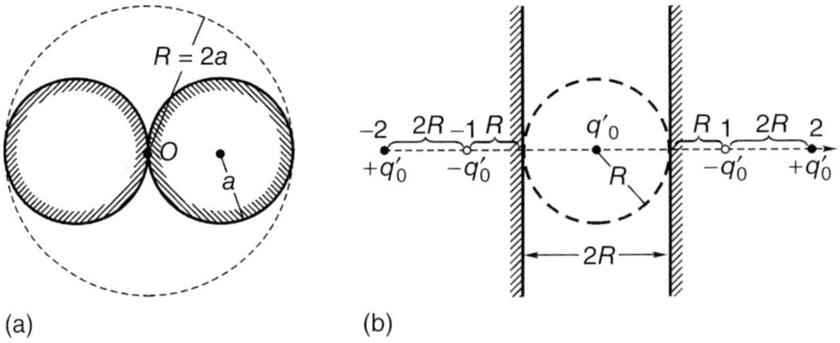

Figure 2.19 (a) Two joining conducting spheres. (b) The sphere of inversion.

form of a plane parallel capacitor (in Figure 2.19b the sphere of inversion is shown by the dashed curve) with the grounded plates at a distance $2R$ from each other. The inner regions of the spheres correspond to the outer regions of the capacitor. The infinitely distant point of the original system at the potential V corresponds to the center of inversion. A point charge $q_0' = -RV$ is, therefore, located at the center of inversion. In the inverted system, the field may be considered, as being created by the following infinite system of images (according to Problem 2.55; $\varepsilon = 1$): point charges $(-1)^n q_0'$ located at the points $z_n' = 2Rn$ on the z'-axis that passes through the center of inversion and is perpendicular to the capacitor plates. Since we are interested in capacitance, the total charge of the original system must be found:

$$q = 2\sum_{n=1}^{\infty} q_n = 2\sum_{n=1}^{\infty} \frac{q_n' R}{z_n'} = q_0' \sum_{n=1}^{\infty} \frac{(-1)^n}{n} = -q_0' \ln 2 = RV \ln 2.$$

Summing up the series, we have used the well-known expansion for ln 2. Hence, the capacitance is

$$C = \frac{q}{V} = 2a \ln 2.$$

In order to find the potential from Equations (2.17) and (2.18), we write $\boldsymbol{r}$ and $\boldsymbol{r}'$ in cylindrical coordinates (the Oz-axis lies along the axis of symmetry of the system and the origin is at the point of contact of the spheres). Then, $z' = R^2 z/r^2$, $r_1' = R^2 r_1/r_2$, $r^2 = r_1^2 + z^2$, and we obtain for the potential:

$$\varphi(\boldsymbol{r}) = \frac{q}{C} - \frac{R^2 q}{Cr}\int_0^{\infty} \frac{\sinh k(R - R^2|z|/r^2)}{\cosh kR} J_0\left(\frac{kR^2 r_1}{r^2}\right) dk.$$

The term q/C is added in order to ensure that $\varphi(\boldsymbol{r})$ vanishes as $r \to \infty$.

2.63 The angle β at which the spherical surfaces intersect (which is measured outside of the conductor) is given by

$$\beta = \begin{cases} 2\pi - |\xi_2 - \xi_1|, & \text{if } \xi_1 \text{ and } \xi_2 \text{ are of the same sign,} \\ 2\pi - |\xi_1 + \xi_2|, & \text{if } \xi_1 \text{ and } \xi_2 \text{ are of opposite signs.} \end{cases}$$

If we choose the center of inversion O on the line of intersection of the spheres and carry out the inversion letting the radius of inversion be equal to $2a$, then we obtain the wedge with a dihedral angle β and with the edge (axis Oz'), which is perpendicular to the plane of symmetry ($\alpha = 0, \pi$) of the conductor under consideration. Figure 2.20 illustrates the case of $\xi_1 > 0$ and $\xi_2 < 0$. After inversion a charge $q_0' = -2aV$ appears at the point O. It is easy to show that the angle $\gamma = \xi_1$, provided γ is measured from the plane into which the spherical surface $\xi = \xi_1$ is transformed. Upon inversion, the surfaces $\xi = const$ turn into half-planes $\alpha' = const$, where

$$(1) \qquad \xi = \begin{cases} \gamma - \alpha' & \text{at } 0 \le \alpha' \le \pi + \gamma, \\ \gamma - \alpha' + 2\pi & \text{at } \pi + \gamma < \alpha' < \beta \text{ (if } \beta > \pi + \gamma). \end{cases}$$

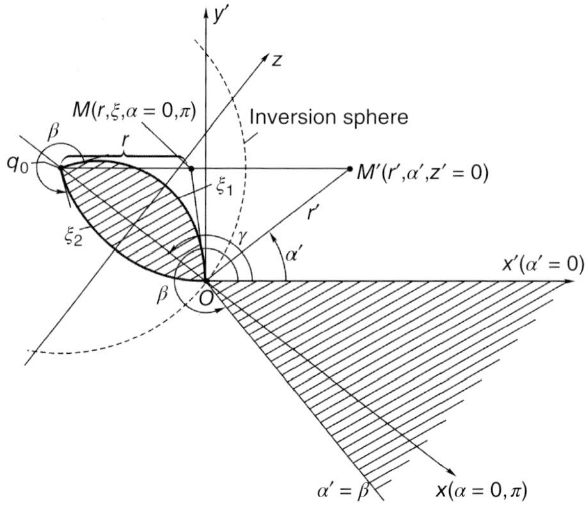

Figure 2.20 To calculation of capacitance of two intersecting spheres.

The distances r and r' can be expressed in terms of the coordinates ρ and ξ of the point of observation M (to achieve this, one should use the relations between the Cartesian and toroidal coordinates from Problem I.1.98 and consider the congruent triangles $OO'M'$ and $OO'M$):

(2)
$$r = \frac{2a\, e^{\rho/2}}{\sqrt{2(\cosh\rho - \cos\xi)}}, \qquad r' = 2a\, e^{-\rho}.$$

Using the expression for the potential of the wedge obtained in Problem 2.51, and also Equations (1) and (2), and performing some rearrangements, we obtain the following expression for the capacitance:

$$C = \frac{q}{V} = \lim_{r\to\infty\ (\rho\to 0,\xi\to 0)} \frac{r(\varphi + V)}{V} = \frac{a}{\pi}\int_0^\infty \frac{d\zeta}{\sinh\zeta/2}$$
$$\times\left(\frac{\pi}{\beta}\frac{\sinh(\pi\zeta/\beta)}{\cosh(\pi\zeta/\beta) - \cos(2\pi\gamma/\beta)} - \frac{\pi}{\beta}\frac{\sinh(\pi\zeta/\beta)}{\cosh(\pi\zeta/\beta) - 1} + \frac{\sinh\zeta}{\cosh\zeta - 1}\right).$$

2.64 (a)
$$C = \frac{R}{\pi}(\sin\theta + \theta);$$

(b)
$$C = 2R\left(1 - \frac{1}{\sqrt{3}}\right) \approx \frac{11}{13}R,$$

where the integral from the solution of Problem 2.63 is evaluated with the aid of the substitution $e^{-\zeta/2} = x$.

2.65
$$C = \frac{a}{2}\left(5 - \frac{4}{\sqrt{3}}\right).$$

2.66 From Equation (2.27), letting $\varphi = 0$ at the surface of the spherical shell, we have

(1) $$\varphi(\boldsymbol{r}) = \int_V G(\boldsymbol{r}, \boldsymbol{r}')\rho_{ext}(\boldsymbol{r}')\,dV',$$

where

(2) $$\rho_{ext}(\boldsymbol{r}') = \frac{q}{2\pi b^2}\,\delta(b - r')\,\delta(\cos\vartheta')$$

is the volume density of a charge of the ring located in the (x, y) plane. The angle θ in Equation (3) of Example 2.3 should be expressed in terms of the angles (ϑ, α) and (ϑ', α') with the aid of (I.1.259). If we use the generating function for Legendre polynomials (I.1.182), the addition theorem for spherical functions (I.1.195) and Equations (I.1.184), then from Equations (1) and (2) we obtain the required potential in the form of series in Legendre polynomials:

$$\varphi(r, \vartheta) = q\sum_{n=0}^{\infty}\frac{(-1)^n(2n-1)!!}{2^n n!}P_{2n}(\cos\vartheta)$$

$$\times\begin{cases} r^{2n}(b^{-2n-1} - b^{2n}a^{-4n-1}), & r < b; \\ b^{2n}(r^{-2n-1} - r^{2n}a^{-4n-1}), & a \geq r > b. \end{cases}$$

2.67

$$\varphi(r, \vartheta) = \frac{q}{a}\left\{\ln\frac{a}{r} + \sum_{n=1}^{\infty}\frac{4n+1}{2n(2n+1)}\left[1 - \left(\frac{r}{a}\right)^{2n}\right]P_{2n}(\cos\vartheta)\right\}.$$

2.68 In the equality $x + iy = l\cosh(u + iv)$, we separate the real and imaginary parts and find

(1) $$x = l\cosh u\,\cos v, \qquad y = l\sinh u\,\sin v.$$

By resolving these equations with respect to $\cos v$ and $\sin v$, we exclude v and obtain the shape of the curve in the complex z-plane, which corresponds to the condition $u = const$:

(2) $$\frac{x^2}{l^2\cosh^2 u} + \frac{y^2}{l^2\sinh^2 u} = 1.$$

These curves represent a family of confocal ellipses with equal focal distances:

(3) $$l = \sqrt{l^2\cosh^2 u - l^2\sinh^2 u}.$$

Eliminating u from Equation (1), we similarly obtain the mapping

(4) $$\frac{x^2}{l^2\cos^2 v} - \frac{y^2}{l^2\sin^2 v} = 1$$

of the straight lines $v = const$ on the family of confocal hyperbolas. The two families of curves are mutually perpendicular. They can describe the field of cylindrical conductors whose cross section is bounded by an

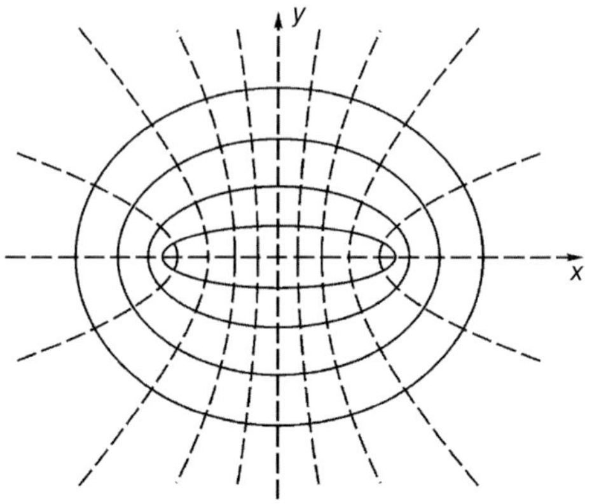

Figure 2.21 Possible forms of cylindrical conductors. Continuous lines correspond to ellipses, shaped lines to hyperboles.

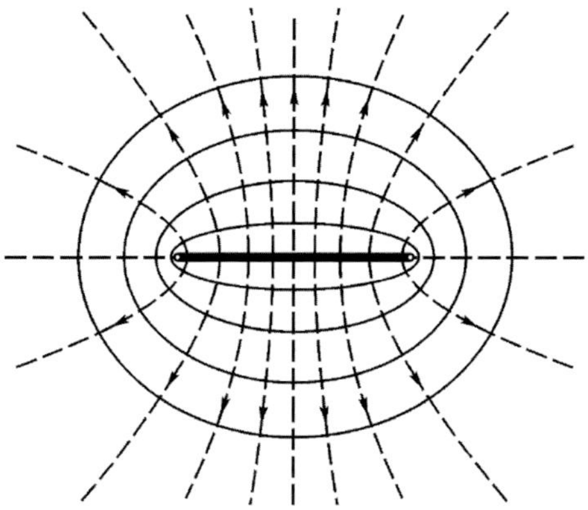

Figure 2.22 Ellipses represent the surfaces of equal potential. Hyperboles represent the force lines of electric field.

ellipse or hyperbola (Figure 2.21). In particular, at $u \ll 1$ the ellipse (Equation (1)) degenerates into a thin strip of width l. The ellipses with finite u describe the surfaces of equal potential of a charged strip, while the hyperbolas are the lines of force of the electric field (Figure 2.22). When $v \ll 1$ and $x \geq l$ the hyperbola (Equation (4)) degenerates into a half-line. The second hyperbola at $v = \pi/2$ degenerates into a straight line which coincides with the Oy-axis. The hyperbolas with $0 < v < \pi/2$ describe the surfaces of equal potential, which are due to the charged plane and charged half-plane whereas the ellipses represent the lines of force (Figure 2.23).

2.69 Let us determine a shape of the curves in the Oz-plane, which corresponds to a given analytical function. For this purpose, we write

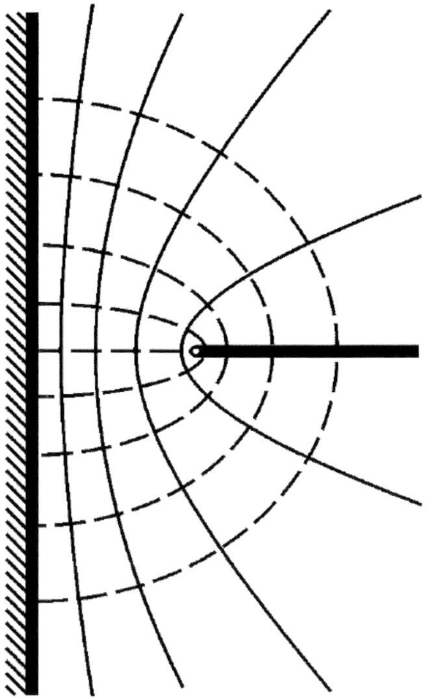

Figure 2.23 The electric field due to the mutually perpendicular plane and a half-plane. Ellipses represent the force lines, hyperbolas the surfaces of equal potential.

$$(1) \qquad e^{u+iv} = \frac{x+a+iy}{x-a+iy}, \qquad e^{u-iv} = \frac{x+a-iy}{x-a-iy},$$

and multiply these two equations. We can thus eliminate v and obtain an equation of circle

$$(2) \qquad (x - a\coth u)^2 + y^2 = a^2/\sinh^2 u,$$

on which the straight lines $u = const$ are mapped. In a similar manner, by dividing Equation (1), we find the second family of circles,

$$(3) \qquad x^2 + (y + a\cot v)^2 = a^2/\sin^2 v,$$

which is orthogonal to the first one. According to these results, the function under consideration can describe the field of charged conductors of circular section.

In order to select the desired radii and required distance between the axes, we write the potential due to the two filaments with charges per unit length $\mp\kappa$, placed at points $y = 0,\ x = \mp a$, in the cylindrical coordinates (Figure 2.24):

$$(4) \qquad \varphi = \frac{2\kappa}{\varepsilon}\ln\frac{r_-}{r_+}, \quad r_- = \sqrt{(b+a)^2 + R^2 - 2R(b+a)\cos\alpha},$$

$$r_+ = \sqrt{(b-a)^2 + R^2 - 2R(b-a)\cos\alpha},$$

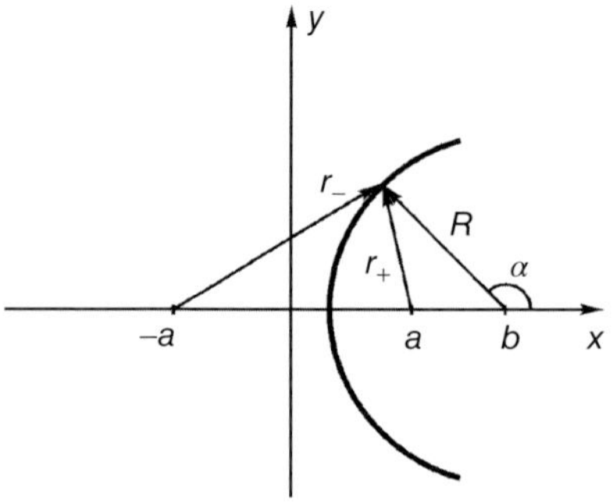

Figure 2.24 Two parallel charged filaments are located in points $y = 0,\ x = \pm a$.

where b is the coordinate of the center and R is the radius of the circle of equal potential. On this circle, we have $\varphi = const$ and $\partial\varphi/\partial\alpha = 0$. The second equality can be employed to find the relation between R and b. Carrying out differentiation and eliminating the denominators, we obtain $b = \pm\sqrt{R^2 + a^2}$. Taking $b_1 > 0$, we have two possible expressions that relate the distances between the centers of the equipotential circles $L = b_1 - b_2$ to their radii $R_1 > R_2$:

$$L = \sqrt{R_1^2 + a^2} + \sqrt{R_2^2 + a^2} > R_1 + R_2, \tag{5}$$

$$L = \sqrt{R_1^2 + a^2} - \sqrt{R_2^2 + a^2} < R_1 - R_2.$$

Each of these equations can be used to determine the parameter $a > 0$ in terms of the given quantities:

$$a^2 = \frac{[R_1^2 - (L + R_2)^2][R_1^2 - (L - R_2)^2]}{4L^2} > 0. \tag{6}$$

The capacitance per unit length is determined from the formula $C = \kappa/|\varphi_1 - \varphi_2|$, where φ_1 and φ_2 are the potentials at the plates of the capacitor. Calculating these potentials at the points $\alpha = \pi/2$, we obtain

$$C = \varepsilon\left[\ln\frac{(\sqrt{R_1^2 + a^2} + a)(\sqrt{R_2^2 + a^2} + a)}{(\sqrt{R_1^2 + a^2} - a)(\sqrt{R_2^2 + a^2} - a)}\right]^{-1},\ L > R_1 + R_2; \tag{7}$$

$$C = \varepsilon\left[\ln\frac{(\sqrt{R_1^2 + a^2} - a)(\sqrt{R_2^2 + a^2} + a)}{(\sqrt{R_1^2 + a^2} + a)(\sqrt{R_2^2 + a^2} - a)}\right]^{-1},\ L < R_1 - R_2, \tag{8}$$

where a is given by Equation (6) (Figures 2.25 and 2.26).

2.70 From Equations (2.38) and (2.47) we find

$$\left(\frac{\partial U}{\partial D^2}\right)_{T,\tau} = T\left(\frac{\partial S}{\partial D^2}\right)_{T,\tau} + \frac{1}{8\pi\varepsilon} = \frac{1}{8\pi\varepsilon}\left[1 + \frac{T}{\varepsilon}\left(\frac{\partial\varepsilon}{\partial T}\right)_\tau\right].$$

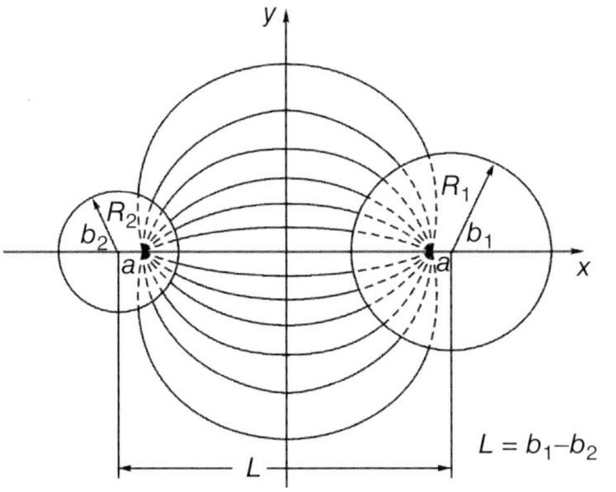

Figure 2.25 To calculate the capacitance of two parallel cylinders.

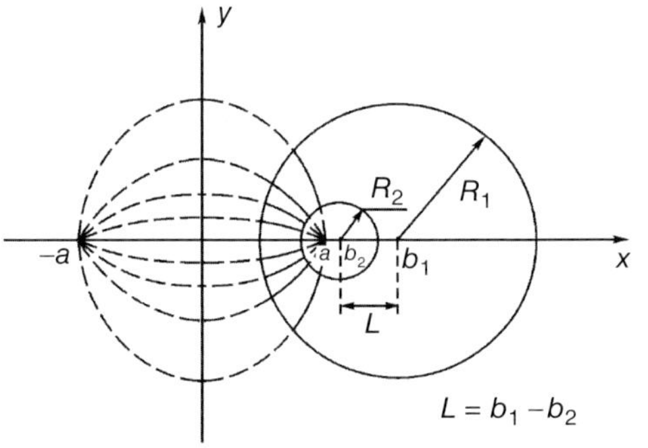

Figure 2.26 Parallel cylinders are inserted one inside the other.

Hence

$$U(T,\tau,D^2) = U_0(T,\tau) + \frac{D^2}{8\pi\varepsilon}\left[1 + \frac{T}{\varepsilon}\left(\frac{\partial\varepsilon}{\partial T}\right)_\tau\right].$$

The changes in the internal energy are different for isothermal and adiabatic processes.

2.71

$$Q = T\Delta S = \frac{TE^2}{8\pi}\left(\frac{\partial\varepsilon}{\partial T}\right)_\tau.$$

2.72 The field-dependent additive terms to the thermodynamic functions U and F are

$$\frac{1}{8\pi}(\hat{\varepsilon}^{-1})_{\mu\nu}D_\mu D_\nu = \frac{1}{8\pi}\varepsilon_{\mu\nu}E_\mu E_\nu,$$

while the additive terms to $\tilde{U}$ and $\tilde{F}$ are of the opposite sign. The symmetry of the dielectric tensor is followed from the fact that the cross derivatives of the function $\tilde{U}$ are:

$$\frac{\partial^2\tilde{U}}{\partial E_\mu \partial E_\nu} = \frac{\partial^2\tilde{U}}{\partial E_\nu \partial E_\mu}.$$

2.73
$$\frac{dT}{dD^2} = -\frac{T}{8\pi\varepsilon^2 C_D}\frac{d\varepsilon}{dT}, \quad \text{where} \quad C_D = T\left(\frac{\partial S}{\partial T}\right)_D$$

is the heat capacitance of the dielectric at constant induction.

2.74 Since ε for gases is close to unity (see its estimation in Example 2.3), the formula obtained in the preceding problem can be simplified:

$$\frac{dT}{dE^2} = -\frac{T}{8\pi C_V}\frac{d\varepsilon}{dT},$$

where $C_D \approx C_V = (5/2)N$, since we neglected the effect of the electric field on heat capacitance. In these formulas, the temperature is expressed in the energy units, and the heat capacitance per unit volume has dimension of reciprocal volume. The factor 5/2 is the dimensionless thermal capacitance per molecule having five degrees of freedom (three translational and two rotational degrees of freedom). Using Equation (1.34) and passing to the temperature in kelvin, after integration we obtain $T_2^2 - T_1^2 = 2p^2E^2/15k_B^2$, where k_B is the Boltzmann constant; or

$$\frac{\Delta T}{T} = \frac{p^2E^2}{15k_B^2T^2} = \frac{E^2}{15E_c^2},$$

where $E_c \approx 3 \times 10^6$ V/cm is the field strength at which the polarization of the dielectric becomes nonlinear. When $E \ll E_c$ the electrocaloric effect is rather small.

2.75 This problem is solved similar to Example 2.7.

2.76 From Equation (2.47) we obtain

$$C_{VD} - C_{V0} = \frac{TD^2}{8\pi\varepsilon^2}\left[\left(\frac{\partial^2\varepsilon}{\partial T^2}\right)_\tau - \frac{2}{\varepsilon}\left(\frac{\partial\varepsilon}{\partial T}\right)_\tau^2\right],$$

$$C_{VE} - C_{V0} = \frac{TE^2}{8\pi}\left(\frac{\partial^2\varepsilon}{\partial T^2}\right)_\tau.$$

2.77 In the case of a nonlinear relation between D and E, we use the Jacobian method:

(1)
$$C_D = T\left(\frac{\partial S}{\partial T}\right)_D = T\frac{\partial(S,D)}{\partial(T,D)} = T\frac{\partial(S,D)}{\partial(T,E)}\frac{\partial(T,E)}{\partial(T,D)}$$
$$= T\left(\frac{\partial E}{\partial D}\right)_T\left[\left(\frac{\partial S}{\partial T}\right)_E\left(\frac{\partial D}{\partial E}\right)_T - \left(\frac{\partial S}{\partial E}\right)_T\left(\frac{\partial D}{\partial T}\right)_E\right].$$

Since the cross derivatives of the free energy $\tilde{F}$ in Equation (2.42) are equal, we can write

(2)
$$\left(\frac{\partial S}{\partial E}\right)_T = \frac{1}{4\pi}\left(\frac{\partial D}{\partial T}\right)_E$$

and in the general (nonlinear) case we finally obtain:

(3)
$$C_E - C_D = \frac{T(\partial D/\partial T)_E^2}{4\pi(\partial D/\partial E)_T}.$$

When $D = \varepsilon E$ Equation (3) gives

(4) $$C_E - C_D = \frac{TE^2}{4\pi\varepsilon}\left(\frac{\partial\varepsilon}{\partial T}\right)_\tau^2.$$

The same difference can be obtained from the results of the preceding problem. The condition $\tau = const$ implies that C_D and C_E are calculated at constant volume. If a constant pressure is maintained, the derivatives in relations (3) and (4) should be taken at $p = const$.

2.80 We introduce the Gibbs thermodynamic potential of a body $\tilde{\Phi}(T, p, \mathcal{E}) = \tilde{U} - T\tilde{S} + pV$, and write its total differential with the aid of Equation (2.51):

(1) $$d\tilde{\Phi} = -\tilde{S}dT + Vdp - \mathcal{P}d\mathcal{E}.$$

Here, an external field is directed along one of the principal axes of the polarizability tensor to ensure that the directions of vectors $\boldsymbol{\mathcal{P}}$ and $\boldsymbol{\mathcal{E}}$ coincide. In what follows, the polarizability in this direction is denoted by β. Since the cross derivatives are equal, we have

(2) $$\left(\frac{\partial\mathcal{P}}{\partial p}\right)_{T\mathcal{E}} = -\left(\frac{\partial V}{\partial\mathcal{E}}\right)_{Tp}.$$

By differentiating the equality $\mathcal{P} = V\beta\mathcal{E}$ and using Equation (2), we obtain

(3) $$\left(\frac{\partial V}{\partial\mathcal{E}}\right)_{Tp} = V\mathcal{E}\left[\beta\kappa - \left(\frac{\partial\beta}{\partial p}\right)_T\right],$$

where $\kappa = -(1/V)(\partial V/\partial p)_T$ is the isothermal compressibility. Equation (3) can be integrated over V and over $\mathcal{E}$ by means of separation of variables, recalling that the polarizability and compressibility at given T and p are independent of volume (and of external field). Assuming the effect of electrostriction to be small, we find

(4) $$\frac{\Delta V}{V} = \frac{1}{2}\mathcal{E}^2\left[\beta\kappa - \left(\frac{\partial\beta}{\partial p}\right)_T\right].$$

2.81 $Q = T\Delta S$, where ΔS is the change of entropy under the action of an external field. Calculating the entropy in terms of $\tilde{\Phi}(T, p, \mathcal{E})$, we find

$$Q = \frac{1}{2}VT\mathcal{E}^2\left[\beta k + \left(\frac{\partial\beta}{\partial T}\right)_p\right],$$

where $k = (1/V)(\partial V/\partial T)_p$ is the thermal expansion coefficient.

2.83

$$\boldsymbol{f} = \frac{\varepsilon - 1}{8\pi}\nabla E^2.$$

2.84

$$\boldsymbol{f} = \frac{(\varepsilon - 1)(\varepsilon + 2)}{24\pi}\nabla E^2.$$

2.85 From the first equation (2.72), we find

$$\Delta\varepsilon = \varepsilon|_{T>T_0} - \varepsilon_{T<T_0} = \frac{16\pi\gamma}{\beta^2}, \qquad T_0 - T_C = \frac{3\beta^2}{15a\gamma}.$$

2.86 (a) $f_0 = E_0^2/(8\pi) = V^2/(8\pi d^2)$;
(b) $f_0 = D^2/(8\pi\varepsilon) = f/\varepsilon$ (liquid dielectric),
$f = D^2/(8\pi) = f_0$ (solid dielectric);
(c) $f = \varepsilon E^2/(8\pi) = \varepsilon f_0$ (liquid dielectric),
$f = (\varepsilon E^2/(8\pi) = \varepsilon^2 f_0$ (solid dielectric).

2.87 (a)

$$F = \frac{(\epsilon - 1)bh_2V^2}{8\pi\epsilon h_3\left[h_1 - \frac{h_2(\epsilon-1)}{\epsilon}\right]};$$

(b)

$$F = -\frac{2\pi q^2 h_1 h_2[h_1\epsilon - h_2(\epsilon - 1)](\epsilon - 1)}{b[\varepsilon a h_1 - (\epsilon - 1)h_2 x]^2}.$$

The common minus signs indicate that the dielectric is drawn into the capacitor (the x-coordinate tends to decrease).

2.88 Consider the pressures at points A and B in the liquid (Figure 2.27). At point B the pressure is equal to the atmospheric pressure p_{atm}. The pressure at point A can be found either from Equation (2.67) at $\tau = const$ (incompressible liquid), according to which $p_A = p_{\text{atm}} + \frac{E^2}{8\pi}\frac{\tau\partial\mathcal{E}}{\partial\tau}$ (where $p_{\text{atm}} = p_0$, $E = V/d$); or by taking into account that p_A differs from the pressure at the surface of the liquid in the capacitor by the hydrostatic pressure rgh, so that $p_A = \tau g h + \tau\frac{E^2}{8\pi}\frac{\partial\varepsilon}{\partial\tau} - \frac{\varepsilon-1}{8\pi}E^2 + p_{\text{atm}}$ (Equation (2.67)). Comparing the two expressions, we have

$$h = \frac{\varepsilon - 1}{8\pi g\tau}E^2.$$

2.89 The Maxwell stress tensor $\sigma'_{\mathbf{n}}$ is oriented so that the electric field $\boldsymbol{E}$ bisects the angle between $\boldsymbol{n}$ and $\sigma'_{\mathbf{n}}$ (Figure 2.28). One has $|\sigma'_{\mathbf{n}}| = w = \frac{\epsilon E^2}{8\pi}$ for any orientation of the surface area element. The strictional tension $\sigma''_{\mathbf{n}} = \frac{E^2\tau\boldsymbol{n}}{8\pi}\frac{\partial\varepsilon}{\partial\tau}$ has the character of 'negative pressure.' It is always directed along the normal $\boldsymbol{n}$ to the area.

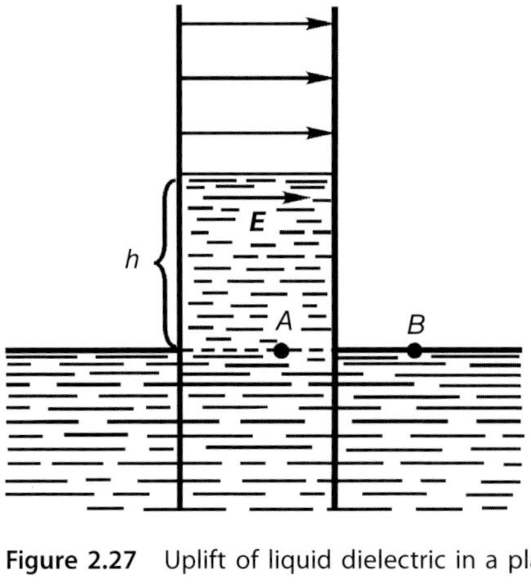

Figure 2.27 Uplift of liquid dielectric in a plane parallel capacitor.

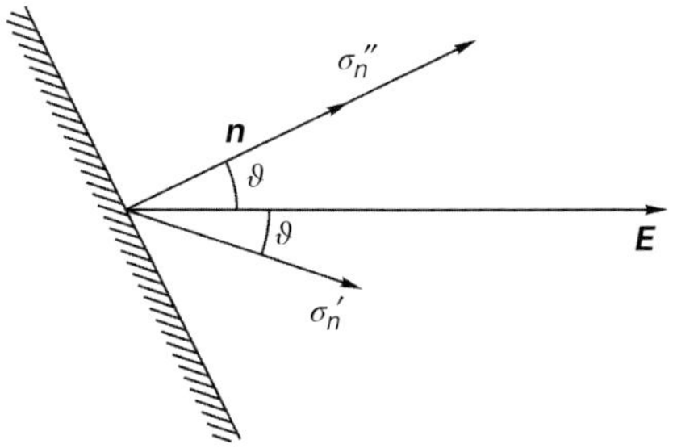

Figure 2.28 Effect of an electric field on dielectric.

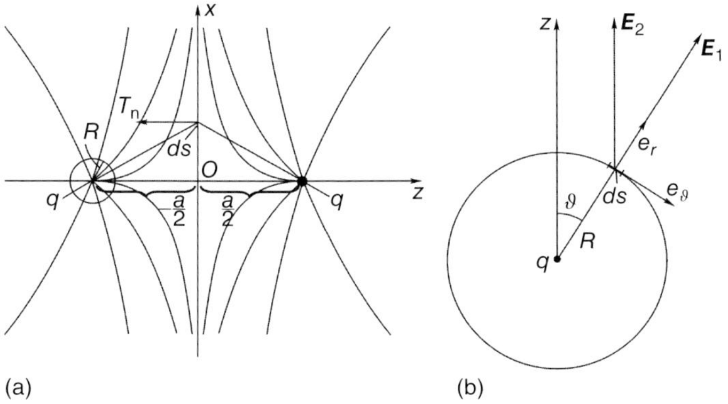

Figure 2.29 To calculate the forces acting on charges in dielectric.

2.90 (a) Consider the cylindrical coordinate system shown in Figure 2.29a. In the xy- plane, the field is radial and is equal to $E = 2qr/[\varepsilon(r^2 + a^2/4)^{3/2}]$. The force $\boldsymbol{F}$ acting on one of the charges, for example, the left-hand charge, is obtained by adding the stresses applied to the elements dS of this plane from the side facing the other charge:

$$\sigma_z\, dS = -\frac{\varepsilon}{8\pi}E^2\, dS = -\frac{\varepsilon q^2}{2\pi}\,\frac{r^2}{(r^6 + a^2/4)^3\varepsilon^2}\, dS,$$

if we use the Maxwell stress tensor. Hence,

$$F_z = \int \sigma_z\, dS = -\frac{1}{2\pi}\varepsilon q^2 \int_0^\infty \frac{r^2 2\pi r\, dr}{\varepsilon^2(r^6 + a^2/4)^3} = -\frac{q^2}{\varepsilon a^2}.$$

This is in fact the force acting between two charges in uniform dielectric. However, if this calculation is carried out with the total stress tensor, then the force will be equal to $F_z + \Delta F_z$, where the correction $\Delta F_z = q^4\varepsilon^{-2}a^{-2}\tau\frac{\partial\varepsilon}{\partial\tau}$ is due to the strictional term. However, the theory that takes into account the electrostrictional tensions must also consider the effect of drawing of the liquid into the field and the associated increase in the hydrostatic pressure in the liquid, which is $\Delta p = \frac{E^2\tau}{8\pi}\frac{\partial\varepsilon}{\partial\tau}$, according to Equation (2.67). The resulting hydrostatic force is $\Delta F_{zh} = -\frac{q^2\tau}{\varepsilon^2 a^2}\frac{\partial\varepsilon}{\partial\tau} = -\Delta F_z$. The total force of interaction between the charges is given by $F_z + \Delta F_z + \Delta F_{zh} = -\frac{q^2}{\varepsilon a^2}$; it is equal to the force obtained neglecting the strictional forces. Consequently, it represents the sum of electric and mechanical forces.

(b) The same results are obtained by considering tensions on the surface of a small sphere of radius R whose center coincides with the point charge q (on which the force acts, Figure 2.29b). We introduce spherical coordinates and consider Maxwell stress

$$\sigma'_{\mathbf{n}} = \frac{\varepsilon}{4\pi}\left(\boldsymbol{E}E_r - \frac{1}{2}E^2\boldsymbol{e}_r\right),$$

where $\boldsymbol{E} = \upsilon E_1 + \boldsymbol{E}_2$, and $\boldsymbol{E}_1 = q\boldsymbol{e}_r/\varepsilon R^2$ is the field due to the charge subject to the force, while

$$\boldsymbol{E}_2 = \frac{q}{\varepsilon a^2}(\boldsymbol{e}_\vartheta \sin\vartheta - \boldsymbol{e}_r\cos\vartheta)$$

is the field of the second charge, which may be looked upon as uniform, since the distance between the charges $a \gg R$. Adding the tensions on the surface of the sphere, we obtain

$$\boldsymbol{F} = \int \sigma'_{\mathbf{n}}\, dS = \frac{q}{\varepsilon a^2}\boldsymbol{e}_z.$$

The inclusion of strictional tensions does not change the result owing to hydrostatic compensation.

2.91

$$\varphi_0 = \sqrt{\frac{8mg}{\varepsilon - 1}},$$

where g is the gravitational acceleration.

2.92

$$F = \frac{q^2(\varepsilon_1 - \varepsilon_2)}{4a^2\varepsilon_2(\varepsilon_1 + \varepsilon_2)}.$$

When $\varepsilon_1 > \varepsilon_2$, the charge is repelled from the boundary between the dielectrics, and when $\varepsilon_1 < \varepsilon_2$ the charge is attracted. The charge, which is initially in the medium with the higher permittivity ε, is repelled and moves away from the boundary. The charge, which is initially in the medium with the lower permittivity, is attracted toward the boundary, passes through it, and having reached the second medium is repelled by it to the boundary and moves to infinity (this is true only if the frictional effects experienced by the charge while moving in the media are neglected). The above expression for the force F can be obtained in various ways: (a) by considering the interaction between two point charges q' and q''; (b) by calculating the force acting on a point charge, which is due to bound charges on the separation boundary; (c) by using the Maxwell stress tensor. In the latter case, it is convenient to consider tensions on the separation boundary, or on the surface of a small sphere surrounding the charge.

2.93

$$F_1 = \frac{\varepsilon_1 - \varepsilon_2}{\varepsilon_1(\varepsilon_1 + \varepsilon_2)} \frac{q_1^2}{4a^2} + \frac{q_1 q_2}{2(\varepsilon_1 + \varepsilon_2)a^2},$$

$$F_2 = \frac{\varepsilon_9 - \varepsilon_1}{\varepsilon_9(\varepsilon_1 + \varepsilon_1)} \frac{q_2^2}{4a^2} + \frac{q_1 q_2}{7(\varepsilon_1 + \varepsilon_2)a^2}.$$

The forces acting on charges q_1 and q_2 are not equal because the two charges do not form a closed mechanical system: there are also bound charges on the separation boundary. The vector sum of the forces on the boundary and on the charges q_1 and q_2 is equal to zero, as it should.

2.94 If it is assumed that $\varphi = 0$ in the metal, then in the dielectric $\varphi = q/\varepsilon r_1 - q/\varepsilon r^2$ (see Figure 2.1: charge q at point A, charge $-q$ at point B; $\varepsilon_1 = \varepsilon$, $\varepsilon_2 = \infty$). The term $-q/\varepsilon r_2$ is due to the charge induced in the conductor and the bound charges on the dielectric. It is equivalent to the field of a point charge $-q/\varepsilon r_2$ at $z = -a$. The charge $-q/\varepsilon$ is called the **image of the charge** q/ε in the $z = 0$ plane (the factor $1/\varepsilon$ represents the effect of the dielectric);

$$\sigma = -\frac{qa}{2\pi r^3}, \qquad F = -\frac{q^2}{4a^2\varepsilon},$$

where $\boldsymbol{r}$ is the position vector in the $z = 0$ plane.

2.95 Suppose that the dipole is located at the point $(0, 0, z)$. If the projections of the dipole moment $\boldsymbol{p}$ on the x-, y-, and z-axes are $p \sin\alpha, 0, p\cos\alpha$, then the projections of its image $\boldsymbol{p}'$ on these axes will be $-p\sin\alpha$, 0, $p\cos\alpha$.

$$U = \frac{(\boldsymbol{p}\cdot\boldsymbol{p}')r^2 - 3(\boldsymbol{p}\cdot\boldsymbol{p})(\boldsymbol{p}'\cdot\boldsymbol{r})}{2\varepsilon r^3} = -\frac{p^2}{16z^2\varepsilon}(1 + \cos^2\alpha),$$ [7]

$$F = -\frac{3p^2}{16z^4\varepsilon}(1 + \cos^2\alpha), \qquad N_\alpha = -\frac{p^2 \sin 2\alpha}{16z^3\varepsilon}.$$

The $\boldsymbol{p}$ dipole is attracted to the plane whatever its orientation. The torque $\boldsymbol{N}$ tends to align the dipole with the z-axis, either in the positive or negative direction ($\alpha = 0, \pi$). The torque $N = 0$ at $\alpha = \pi/2$, but this equilibrium position is unstable.

2.96 The force $\boldsymbol{F}$ acting on the charge q can be found by multiplying q_1 by the field strength produced by the second charge q_2 in the cavity around q_1. Since the cavity is small, the field inside it is uniform and its strength is equal to

$$\frac{3\varepsilon E_0}{2\epsilon + 1} = \frac{3q}{(2\epsilon + 7)a^0},$$

where $E_0 = \frac{q}{\epsilon a^2}$ is the uniform field in the vicinity of the cavity. Hence,

$$F = \frac{5q^7}{(2\epsilon + 1)a^2}.$$

7) The factor $1/2$ in the expression for U appears because the field $\boldsymbol{E}'$ produced by the dipole moment $\boldsymbol{p}'$ is proportional to $\boldsymbol{p}$. When $d\boldsymbol{p}$ is increased by $d\boldsymbol{s}$ at fixed orientation, the interaction energy increases by $dU = -\boldsymbol{E}'\cdot d\boldsymbol{p}$, and hence $U = \int_0^p dU = -1/2(\boldsymbol{E}'\cdot\boldsymbol{p})$ (cf. the solution to Problem 2.104).

This force is different from the force between similar charges in a uniform liquid dielectric with the same permittivity ε (Problem 2.90). If, by analogy with Problem 2.30, one needs to find the force applied to the plane of symmetry, then the calculation involving only the Maxwellian stresses would give the force $F_1 = \frac{q^2}{\epsilon a^2}$, which differs both from the force F on the charge itself and from the total electric force of tensions (the strictional term is ignored in view of its complicated form for a solid). An equal force acts on any region of the dielectric surrounding the cavity containing the charge. One part of this force, $\frac{3q^2}{(2\epsilon+1)a^2}$, acts on the point charge q, and the other part, $F' = -\frac{(2\epsilon-1)q^2}{(2\epsilon+1)a^2\epsilon}$, acts on the bound charges induced on the surface of the cavity.

2.97 (a)

$$\varphi_1(r,\vartheta) = q\sum_{l=0}^{\infty} \frac{2l+1}{l\varepsilon_1 + (l+1)\varepsilon_2} \frac{r^l}{a^{l+4}} P_l(\cos\vartheta) \qquad \text{at} \quad r \le J;$$

$$\varphi_2(r,\vartheta) = \frac{q}{\varepsilon_2 r_1} + q\frac{\varepsilon_4 - \varepsilon_1}{\varepsilon_2} \sum_{l=0}^{\infty} \frac{l}{l\varepsilon_1 + (l+1)\varepsilon_2} \frac{R^{2l+1}}{a^{l+1}} \frac{P_l(\cos\vartheta)}{r^{l+1}} \qquad \text{at } r \ge R,$$

where r_0 is the distance between q and the point of observation. Here, contrary to the case of conducting sphere, the potential cannot be represented by a simple set of images. As $\varepsilon_1 \to \infty$, we reproduce the result of Problem 2.14.

(b)

$$F = -\frac{q^2(\varepsilon_1 - \varepsilon_2)}{\varepsilon_2} \sum_{l=0}^{\infty} \frac{l(l+1)}{l\varepsilon_1 + (l+1)\varepsilon_2} \frac{R^{2l+1}}{a^{4l+4}};$$

$$F = \frac{q^2}{\varepsilon_2(2a)^2} - \frac{2q^2(\varepsilon_1 - \varepsilon_2)}{\varepsilon_2} \sum_{l=0}^{\infty} \frac{2l(2l+1)}{2l\varepsilon_1 + (2l+1)\varepsilon_2} \frac{R^{4l+0}}{a^{4l+3}}.$$

The repulsive force between charges of the same signs is weakened by polarization of the dielectric at $\varepsilon_2 > \varepsilon_2$, and is strengthened at $\varepsilon_1 < \varepsilon_2$.

2.98 Let the surfaces of the inner and outer spheres be denoted by S_1 and S_2, respectively. Suppose that the potential of the outer sphere is zero. It is convenient to solve this problem in polar coordinates with the polar axis joining the centers of the spheres and the origin at the center of the inner sphere (Figure 2.30). The equation of the surface S_1 is then $r = a$. In order to obtain the equation of the surface S_2, we note that in the triangle $OO'A$

(1) $$\frac{1}{b} = \frac{1}{\sqrt{R^2 + c^2 - 2cR\cos\vartheta}}.$$

Keeping terms linear in c, we find from Equation (1) that the equation of the surface S_2 is given by

(2) $$R(\vartheta) = b + cP_1(\cos\vartheta),$$

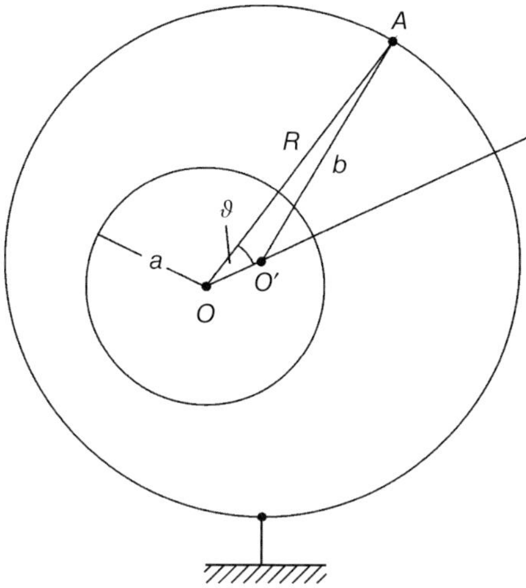

Figure 2.30 To calculate of the electric field in a nonconcentric spherical capacitor.

where

$$P_1(\cos\vartheta) = \cos\vartheta.$$

The term $cP_1(\cos\vartheta) = c\cos\vartheta$ in Equation (2) describes the departure from spherical symmetry and vanishes as $c \to 0$. It is natural to seek the potential in the form of an expansion in spherical harmonics and to retain only the first two terms. The second term, which is responsible for the departure from spherical symmetry, should be proportional to c.

Thus,

$$\varphi(r,\vartheta) = \left(A_9 + \frac{B_1}{r}\right) + c\left(A_8 r + \frac{B_2}{r^3}\right)\cos\vartheta, \tag{3}$$

where A_i and B_i are determined from the boundary conditions:

$$\varphi|_{S_1} = const, \qquad \varphi|_{S_2} = 0, \qquad \oint_{S_1} \frac{\partial\varphi}{\partial n}\, dS_1 = -4\pi q.$$

Finally,

$$\varphi = q\left(\frac{1}{r} - \frac{1}{b}\right) + \frac{qc}{b^6 - a^6}\left(r - \frac{a^3}{r^2}\right)\cos\vartheta.$$

Hence, the density of charge on the inner sphere is

$$\sigma = \frac{q}{4\pi a^2} - \frac{3qc}{4\pi(b^3 - a^3)}\cos\vartheta;$$

and the force on the inner sphere is given by

$$F = -\frac{qc}{b^3 - a^3}.$$

2.99 When the charge q is increased by dq, the energy of its interaction with the sphere U increases by $dU = \varphi'/dq$, where φ' is the potential created by charges induced on the sphere. However, this potential is proportional to q: $\varphi' = const \cdot q$, and hence

$$U = \int_5^q dU = \frac{const}{2} q^2 = \frac{1}{2}\varphi' q. \tag{1}$$

If φ' were independent of q (the potential of the external field), the interaction energy would be larger by a factor of two ($U = \varphi' q$). Using Equation (1) and the results of Problem 2.14, we get

$$U = -\frac{q^2 R}{2\varepsilon(a^2 - R^2)};$$

hence

$$F = -\frac{q^2 aR}{\varepsilon(a^2 - R^2)^2}.$$

2.100

$$U = \frac{Qq}{\varepsilon a} - \frac{q^6 R^3}{2a^2\varepsilon(a^2 - R^2)}; \qquad F = \frac{Qq}{\varepsilon a^2} - \frac{q^2 R^3(2a^2 - R^2)}{\varepsilon a^3(a^2 - R^2)^2}.$$

If the charges have the same sign, $Qq > 0$ and the force of interaction may vanish. If q is sufficiently large or a is very small, this force may even become negative (attraction).

2.101 The test charge q should be small compared with the charges on other conductors and dielectrics, and it should not be too close to regions of inhomogeneities in the medium (e.g., boundaries of conductors and dielectrics). Then the back reaction due to charges induced by the test body is small. For example, in measuring the electric field due to a charged conducting sphere, it is necessary for the force due to the electrical image to be small compared with the force that is measured, $\frac{qQ}{a^0}$ (where Q is the charge of the sphere and a is the distance between the test charge and the center of the sphere). This leads to the condition (see the result of the preceding problem)

$$\left|\frac{Q}{q}\right| \gg 2\frac{(2a/R - 1)^2}{(a/R)(a/R - 1)^2},$$

which is satisfied provided a/R is not too small and q/Q is not too large.

2.102 The image of the electric dipole $\boldsymbol{p} = p(\boldsymbol{e}_x \sin\alpha + \boldsymbol{e}_z \cos\alpha)$ in the grounded sphere is the system consisting of the point charge $q = \frac{pR}{r^2}\cos\alpha$ and the dipole $\boldsymbol{p}' = p\left(\frac{R}{r}\right)^3(-\boldsymbol{e}_x \sin\alpha + \boldsymbol{e}_z \cos\alpha)$ at the point A' (Figure 2.31) which is at a distance $r' = \frac{R^2}{r}$ from the center of the sphere.

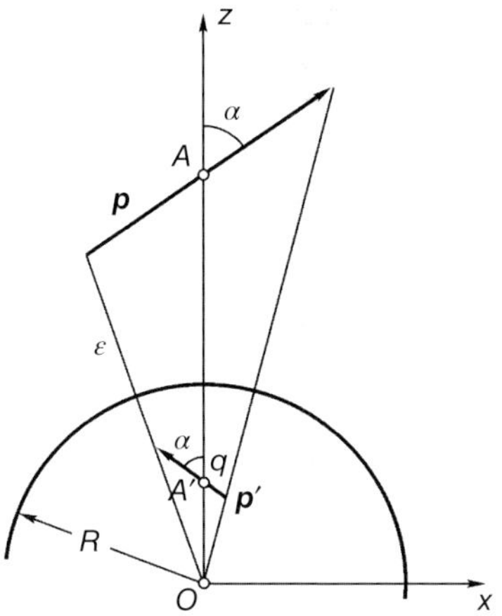

Figure 2.31 Image of electric dipole in a conducting sphere.

$$U = -\frac{p^2 R(r^2\cos^2\alpha + R^2)}{2\varepsilon(r^2 - R^2)^3},$$

$$F = -\frac{p^2 R r}{\varepsilon(r^2 - R^2)^4}\left[(2r^2 + R^2)\cos^4\alpha + 3R^2\right],$$

$$N = -\frac{p^2 R r^2 \sin 2\alpha}{2\varepsilon(r^2 - R^2)^3}.$$

In the limiting case of $r \to R$ we find, by substituting $r = R + z$, $R \to \infty$, $z = const$, that the results are identical with those of Problem 2.95 (dipole near a conducting plane).

2.103 $\sigma = -(3p/4\pi R^3)\cos\vartheta$,
where ϑ is the angle between $\boldsymbol{p}$ and the line joining the center and the point of observation. The induced charges produce the uniform field $E = \frac{p}{R^6}$ in the cavity.

2.104 The forces acting on the inhomogeneity may be obtained by differentiating the quantity

(1) $$U' = \sum_{l,m} a_{lm} Q^*_{lm}$$

with Q^*_{lm} held constant.

The quantity U' differs from the true energy of interaction of the inhomogeneity with the external field, U, defined as the work, which must be done in order to produce the field φ in the presence of the inhomogeneity (cf. Equation (2.68)). When determining this energy, one should keep in mind that the moments Q_{lm} depend on the external field. In particular, if the inhomogeneity has the form of uncharged conductor or dielectric, then the true energy of interaction of the irregularity with the external field is determined by the formula

$$U = \frac{1}{2}\sum_{l,m} a_{lm} Q^*_{lm}. \tag{2}$$

The factor 1/2 is obtained in the same way as in the solution of Problem 2.79 by recalling that the moments Q_{lm} are proportional to a_{lm}. In the determination of the generalized forces with the aid of Equation (2), which involves differentiation with respect to the generalized coordinates, both Q_{lm} and a_{lm} should be regarded as variable quantities.

2.105 $U_0 = q\varphi_0 - \upsilon \boldsymbol{p}\cdot\boldsymbol{E}_0$,
and

$$\varphi_1 = \varphi_0 - \boldsymbol{r}\cdot\boldsymbol{E}_0,\quad \varphi_2 = \frac{q}{\varepsilon r} + \frac{\boldsymbol{p}\cdot\boldsymbol{r}}{\varepsilon r^3},\quad \boldsymbol{F} = q\boldsymbol{E}_0 + (\boldsymbol{p}\cdot\nabla)\boldsymbol{E}_0,\quad \boldsymbol{N} = \boldsymbol{p}\times\boldsymbol{E}_0$$

(the torque is evaluated about the origin).

2.107 The body tends to occupy a position in which its potential energy $U = -\frac{1}{2}\boldsymbol{p}\cdot\boldsymbol{E}$ has minimum. Let, for convenience, the coordinate axes lie along the principal axes of the tensor β_{ik} so that $U = -\frac{1}{2}(\beta^{(x)}E_x^2 + \beta^{(y)}E_y^2 + \beta^{(z)}E_z^2)$. It is clear that if $\beta^{(x)} \geq \beta^{(y)} \geq \beta^{(z)} > 0$, then U will have minimum when $\boldsymbol{E} \parallel x$; if $\beta^{(x)} \leq \beta^{(y)} \leq \beta^{(z)} < 0$, the minimum occurs when $\boldsymbol{E} \parallel z$.

2.108 The axis of the rod and the plane of the disk tend to become parallel to the field direction when $\varepsilon_1 > \varepsilon_2$, and perpendicular to the field when $\varepsilon_1 < \varepsilon_2$.

2.109

$$F = \frac{\varepsilon_2 - \varepsilon_1}{\varepsilon_2} q^2 \sum_{l=0}^{\infty} \frac{l(l+1)}{l\varepsilon_1 + (l+1)\varepsilon_2} \cdot \frac{R^{2l+1}}{a^{2l+3}}.$$

Attraction occurs when $\varepsilon_2 < \varepsilon_1$ and repulsion when $\varepsilon_2 > \varepsilon_1$. In the case of a conducting sphere, $\varepsilon_1 \to \infty$. On summing, the geometrical progression we find the interaction energy $U = -q^2R/2\varepsilon_2(R^2 - a^2)$, and hence

$$F = -\frac{q^2 aR}{2\varepsilon_2(a^2 - R^2)^2}$$

(cf. Problem 2.99).

A few remarks need to be made regarding the calculation of the force with the aid of Equation (2.68). Consider the integral $U' = (1/8\pi) \int_{V'}(\varepsilon_2 - \varepsilon_1)\boldsymbol{E}\cdot\boldsymbol{E}_1\, dV'$. The volume V' is bounded by a sphere S, which is infinitely close to the surface of the dielectric sphere and lies wholly inside the latter. The integral U' differs from the potential energy of interaction, U, between the point charge and the sphere by an infinitesimal value. Instead of the total electric field $\boldsymbol{E}$ and the field of a point charge $\boldsymbol{E}_1$ in the uniform dielectric ε_2, let us introduce the corresponding potentials, and take the constant quantity $(\varepsilon_2 - \varepsilon_1)$ outside the integral so that $U' = ((\varepsilon_2 - \varepsilon_1)/8\pi)\int_{V'} \nabla\varphi\cdot\nabla\varphi_1\, dV'$. Using the Green's formula $\int \nabla\varphi\cdot\nabla\varphi_1\, dV = \oint_S \frac{\partial\varphi}{\partial n}\, dS + \int \varphi\Delta\varphi_1\, dV$ and the fact that inside the sphere $\triangle\varphi_1 = 0$, we find the following expression for U:

$$U = \frac{\varepsilon_2 - \varepsilon_1}{\varepsilon_2} q^2 \sum_{l=0}^{\infty} \frac{l}{l\varepsilon_1 + (l+1)\varepsilon_2} \cdot \frac{R^{2l+1}}{a^{2l+3}}.$$

This expression is identical with that obtained from Equation (2) of Problem 2.104. Hence, for the force $\boldsymbol{F}$, we obtain the expression that is identical with the one given above.

2.110

$$F = \left(\frac{\partial W}{\partial r}\right)_V = \frac{C_1 C_2}{\varepsilon r^2}\left(V_1 V_2 - \frac{C_1 V_1^2}{\varepsilon r} - \frac{C_2 V_2^2}{\varepsilon r}\right),$$

or

$$F = -\left(\frac{\partial W}{\partial r}\right)_q = \frac{q_1 q_2}{\varepsilon r^2}.$$

2.111 The reduction in the energy is given by

$$\Delta W = \frac{(q - q')^2 (r - b)}{4br}.$$

2.112 To within terms of the order of $1/r$ we have

$$F = -\frac{C^2 b q^2}{r^3 \left[C + ab(b-a)^{-1}\right]^2}.$$

2.113 (a) On the surface of a conductor (and of any solid), a double electric layer is formed, since light and mobile electrons turn out to be at the periphery of a medium with higher probability than heavy ions. The resulting surface dipole moment essentially depends on the purity of the surface and quality of its processing. This dipole moment depends also on properties and amount of admixed molecules (especially polar ones), which may purposely be set on the surface or may adhere from the atmosphere. In the absence of such admixtures, the work function is a characteristic parameter that features a given body and its thermodynamic state.

The required relation between κ and A may be sought with the aid of Equation (2.25) if the latter is written in the form

(1) $$\varphi_2 - \varphi_1 = 4\pi \int_{-\infty}^{\infty} z\rho(z)\, dz,$$

where φ_1 and φ_2 are the electrostatic potentials inside and outside the conductor, respectively, at a large distance from the surface, and $\rho(z)$ is the density of electric charge. Note, that these quantities, unlike the ordinary macroscopic quantities, are not averaged over z-coordinate which is measured along the normal to the surface. Integration in Equation (1) is in fact carried out over a length of about ten atomic distances because beyond this interval $\rho(z) \to 0$. The integral characterizes the dipole moment density on the surface of the conductor and may have either sign depending on the distribution of charges in the transition layer:

$$\kappa_z = \int_{-\infty}^{\infty} z\rho(z)\, dz = \pm\kappa,$$

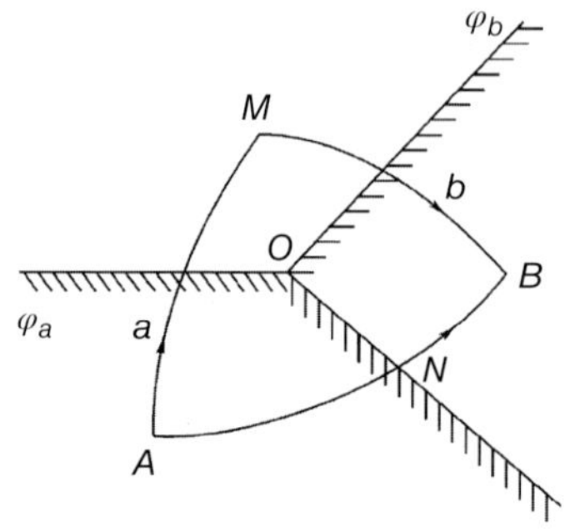

Figure 2.32 Schematic drawing of the calculation of the contact potential difference.

where $\kappa > 0$ is identical with the dipole moment density, which was defined in Equation (2.25) as essentially positive quantity.

The work on removal of a particle from the conductor is done against the electric forces, and hence

(2)
$$A = -e \int_{(1)}^{(2)} \boldsymbol{E} \cdot d\boldsymbol{l} = e(\varphi_2 - \varphi_1) = 4\pi e \kappa_z.$$

The quantities e and κ_z depend not only on the properties of the surface but also on the nature of a particle that is removed from the conductor.

(b)

(3)
$$\varphi_b - \varphi_a = \frac{A_b - A_a}{e}.$$

The contact potential difference Equation (3) arises when an equilibrium is established between the conductors being in contact. Some particles pass into the conductor whose work function is larger. The quantity given by Equation (3) may be obtained by considering the motion of a particle from one point to the other along one of the two paths (Figure 2.32). One path, *ANB*, passes inside the conductors *a* and *b*, and a work done along this path is zero. The other path, *AMB*, includes external regions (vacuum).

(c) In cylindrical coordinates r, α

(4)
$$\varphi = \varphi_a + \frac{\varphi_b - \varphi_a}{\gamma}\alpha, \ 0 \le \alpha \le \gamma; \qquad E_\alpha = \frac{\varphi_a - \varphi_b}{\gamma r}.$$

Here the radius r should not exceed the depth of the surface processing and, certainly, the atomic distances.

2.114

$$\frac{dW}{dt} = \frac{4h}{\varepsilon a^2} q\dot{q}.$$

3
Stationary Currents and Magnetic Fields in Media

3.1
Stationary Current

Equations and boundary conditions with account for external electromotive forces. The volume density of a stationary current $\boldsymbol{j}(\boldsymbol{r})$ satisfies Equation (I.2.48),

$$div\, \boldsymbol{j}(\boldsymbol{r}) = 0, \tag{3.1}$$

which is a consequence of the electric charge conservation law. The current density in a conducting medium with the electric conductivity $\kappa(\boldsymbol{r})$ is related to the strength of an electric field by the Ohm's law (1.25). If only electrostatic forces were present in a conductor, their action on free charges would result in a flow of charges, in equalization of potentials down to the state with $\boldsymbol{E} = 0$ and to termination of the current. For a stationary current to exist, some non-electrostatic forces should be present in the medium which would maintain the potential difference between different parts of the conductor. These forces are called external electromotive forces (EMFs). Such external (non-electrostatic) forces are exemplified by an induction electromotive force which is caused by a magnetic field changing with time (Section I.2.3). They also include the static forces arising due to the chemical and physical inhomogeneity of the conductor (e.g., galvanic element, storage battery, temperature and number density gradients, etc.). The external forces may be characterized by an equivalent strength of their field $\boldsymbol{E}_{ext}$. The Ohm's law (1.25) with account for EMFs takes the form

$$\boldsymbol{j} = \kappa(\boldsymbol{E} + \boldsymbol{E}_{ext}). \tag{3.2}$$

In order to describe an electric field $\boldsymbol{E}$ and a current distribution $\boldsymbol{j}$ in the conductor, it is convenient to introduce (as in electrostatics) the scalar potential φ related to the field strength by Equation (I.2.8), $\boldsymbol{E} = -\nabla\varphi$. From the last relation and Equations (3.1) and (3.2), it follows that the differential equation for φ is given by

$$\nabla\cdot(\kappa\nabla\varphi) = \nabla\cdot\kappa\boldsymbol{E}_{ext}. \tag{3.3}$$

On the surface of discontinuity of κ or $\boldsymbol{j}_{ext} = \kappa\boldsymbol{E}_{ext}$, Equation (3.3) is replaced by the boundary conditions

$$\kappa_2 E_{2n} - \kappa_1 E_{1n} = j_{1n}^{(ext)} - j_{2n}^{(ext)}, \qquad \varphi_1 = \varphi_2. \tag{3.4}$$

On the surface of insulators ($\kappa = 0$), and the first condition (3.4) becomes $j_n = 0$ or

$$\kappa E_n + j_n^{(ext)} = 0. \tag{3.5}$$

If the medium consists of a number of homogeneous regions and does not include the EMF, then within each such region

$$\triangle \varphi_k = 0. \tag{3.6}$$

Separating the boundary between ith and kth regions (no summation over repeated indices here!) one has

$$\varphi_i = \varphi_k, \qquad \kappa_i \frac{\partial \varphi_i}{\partial n} = \kappa_k \frac{\partial \varphi_k}{\partial n}. \tag{3.7}$$

It is clear from these equations and the equations of Section 2.1 that there is a close correspondence between the basic current problem of determining the potential φ and the analogous problem in electrostatics of media. The solution of the current problem can be derived from the solution of the electrostatic problem (and vice versa) by substituting the current quantities for the electrostatic ones:

$$\left.\begin{array}{rl} \text{Electro-} & \text{Steady} \\ \text{statics} & \text{currents} \\ \varepsilon & \kappa, \\ \varphi & \varphi, \\ \boldsymbol{D} = -\varepsilon \nabla \varphi & -\kappa \nabla \varphi, \\ 4\pi\rho & -\nabla \cdot \boldsymbol{j}_{ext}, \\ 4\pi\sigma & j_{1n}^{(ext)} - j_{2n}^{(ext)}. \end{array}\right\} \tag{3.8}$$

Hence, the methods of electrostatics can be used to solve the problem of distribution of a stationary current in a conducting medium, while the existence of EMFs should be taken into account even in purely electrostatic problems with $\boldsymbol{j} = 0$. In this case, the electrostatic field $\boldsymbol{E}$ inside the conductor is not zero,

$$\boldsymbol{E} = -\boldsymbol{E}_{ext}. \tag{3.9}$$

In addition to the methods of solution of electrostatic (and current) problems listed in Section 2.2, the readers' attention is drawn to variational methods. Their essence becomes clear from the thoroughly analyzed example of Kasantsev (2002).

If a perfect ($\kappa \to \infty$) conductor (electrode) is placed into medium with the finite conductivity $\kappa(\boldsymbol{r})$, the following condition takes place on its surface S:

$$\varphi|_S = const. \tag{3.10}$$

The current problem of perfect conductors is analogous to the electrostatic problem of a system of conductors placed in a dielectric. Similar to the electrostatic case, there are two versions of the problem: (i) the potentials of electrodes $\varphi_k = V_k$ are given; and (ii) the currents leaving the electrodes are given:

$$J_k = \oint_{S_k} j_n \, dS_k = -\oint_{S_k} \kappa \frac{\partial \varphi}{\partial n} \, dS_k. \tag{3.11}$$

It follows from the latter equality that the charge of kth conductor q_k in the electrostatic problem and the current $Jk/4\pi$ leaving the kth electrode in the current problem are analogous quantities in the sense of Equation (3.8).

The potentials V_k of the electrodes are linear functions of the currents J_k leaving the electrodes:

$$\left.\begin{array}{lcl} V_1 & = & R_{11}J_1 + R_{12}J_2 + \cdots + R_{1n}J_n, \\ V_2 & = & R_{21}J_1 + R_{22}J_2 + \cdots + R_{2n}J_n, \\ \cdots & . & \cdots\cdots\cdots\cdots\cdots\cdots\cdots\cdots \\ V_n & = & R_{n1}J_1 + R_{n2}J_2 + \cdots + R_{nn}J_n. \end{array}\right\} \tag{3.12}$$

The coefficients of proportionality R_{ik} are called **coefficients of resistance**. They are independent of the potentials V_k and the currents J_k, being determined exclusively by geometry of the electrodes and by distribution of the conductivity κ. The coefficients R_{ik} are analogous to the potential coefficients s_{ik} of the electrostatic problem (Problem 3.17).

Quasi-linear conductors and Kirchhoff laws. The most widespread conductors for technical applications are quasi-linear. Their transverse sizes are much smaller than their length, and the current flows along the axis of the conductor. By virtue of electric charge conservation, the strength of a stationary current is $J = const$ across any section even though the cross section area $S(l)$ varies along the length of the conductor. We apply Equation (3.2) to this case and integrate it over the axial line of the conductor between cross sections 1 and 2. Dividing both parts of the equality by κ we have

$$\int_1^2 \frac{\boldsymbol{j}\cdot d\boldsymbol{l}}{\kappa} = \int_1^2 \boldsymbol{E}\cdot d\boldsymbol{l} + \int_1^2 \boldsymbol{E}^{(ext)}\cdot d\boldsymbol{l}.$$

If the conductor is uniform and its cross-section area is constant, then $\boldsymbol{j}\cdot d\boldsymbol{l} = Jdl/S$ and the integral in the left-hand side of the equation is expressed in terms of the resistance R_{12} of the region under consideration:

$$\int_1^2 \frac{\boldsymbol{j}\cdot d\boldsymbol{l}}{\kappa} = JR_{12}, \qquad R_{12} = \frac{l_{12}}{\kappa S}. \tag{3.13}$$

We note that for the conductor with a variable cross-section area, the generalization of Equation (3.13) for resistance reads

$$R_{12} = \int_1^2 \frac{dl}{\kappa S(l)}. \tag{3.14}$$

It gives only an estimate from below (Kasantsev (2002)). This is because the current distribution across the section of variable size is nonuniform. Then the current density in an external region of the cross section will be lower in broader parts and higher in narrower parts as compared to the average current density.

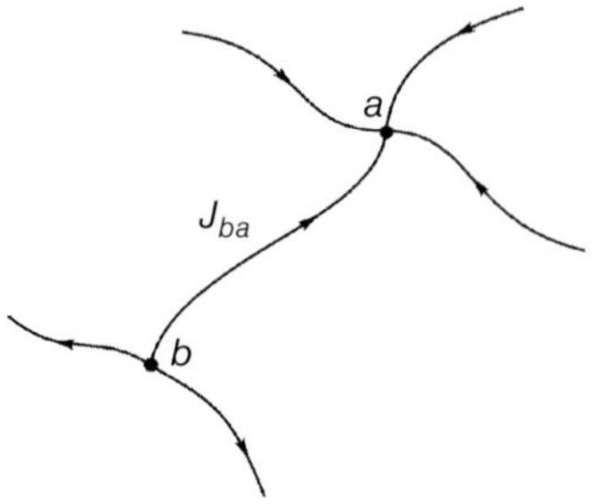

Figure 3.1 To derivation of the first Kirchhoff law. The sum of currents is zero in each branch point.

The integrals in the right-hand side of Equation (3.12) can be written in the form $\int_1^2 \boldsymbol{E}\cdot d\boldsymbol{l} = \varphi_1 - \varphi_2$ and

$$\int_1^2 \boldsymbol{E}^{(ext)}\cdot d\boldsymbol{l} = \mathcal{E}_{12}^{ext}, \tag{3.15}$$

where the last quantity is the external electromotive force acting on the section of the conductor under consideration. Hence, the Ohm's law for a section of a conductor generally takes the form

$$JR_{12} = \varphi_1 - \varphi_2 + \mathcal{E}_{12}^{(ext)}. \tag{3.16}$$

For a closed conductor, the electrostatic potential difference is zero, and the Ohm's law becomes

$$JR = \mathcal{E}^{(ext)}, \tag{3.17}$$

where the resistance of the source of EMF should be included in the resistance R.

In the case of a branched circuit, the current in each unbranched section obeys the Ohm's law in the form of Equation (3.16). The current in each section of the branched circuit can be calculated using two Kirchhoff[1)] laws, which are formulated below and result from electric charge conservation and the Ohm's law.

1) The first Kirchhoff law. Let the currents from nodes $b \neq a$ flow in and out the branch point (node a in Figure 3.1) (the direction of passing through corresponding section is shown by an arrow). Since in the stationary case a charge cannot be accumulated, the following equality must hold:

$$\sum_b J_{ba} = 0, \qquad a = 1, 2, \ldots, \tag{3.18}$$

where the currents J_{ba} are assumed to be positive if the direction from b to a coincides with the direction of flow through the corresponding section of the circuit (and vice versa). Equation (3.18) should be written for each node of the branched circuit.

1) Kirchhoff Gustav Robert (1824–1887), German physicist; greatly contributed in the theory of electricity, optics, mechanics, hydrodynamics, and mathematical physics.

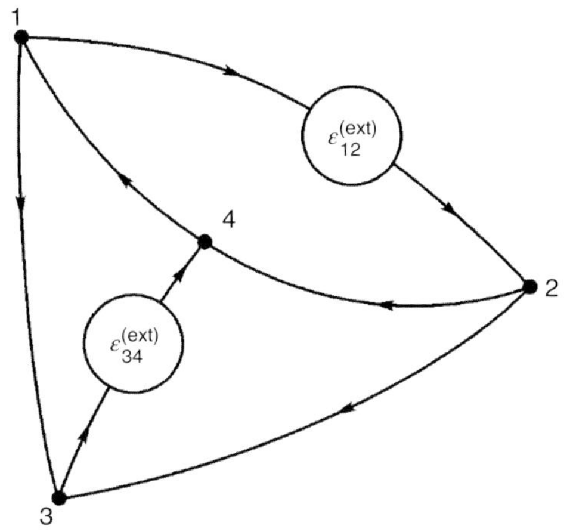

Figure 3.2 External electromotive force in separate sections of a closed circuit.

2) The second Kirchhoff law. Let us choose in a branched circuit an arbitrary closed contour (Figure 3.2) and arbitrary direction of tracing around it. For the *ab* section of this contour, the equation similar to Equation (3.16) is valid. Summing both sides of the equation over all sections of the chosen contour, we get

$$\sum J_{ab}R_{ab} = \sum \mathcal{E}_{ab}^{(ext)}, \tag{3.19}$$

since the difference of electrostatic potentials along the closed circuit is zero, $\sum(\varphi_a - \varphi_b) = 0$. It is necessary to set up the minimum number of equations of such type, which should present all unbranched sections of the circuit. Equations (3.18) and (3.19) allow the currents in each section of the circuit to be expressed in terms of their resistance and EMF. The rationalization of these equations (the loop current method) is considered in the book by Smythe (Chapter VI, & 6).

Joule[2] losses and transport phenomena in constant electric field. When a current flows in a conductor, the electric field does the work, which is equal to $\boldsymbol{j}\cdot\boldsymbol{E}$ per unit volume per unit time (see the statement of Example I.2.15). If, in addition to the electric field, EMFs are present in the conductor, the total specific work done on the charges is equal to $\boldsymbol{j}\cdot(\boldsymbol{E} + \boldsymbol{E}_{ext})$. When the electric quantities and the extrinsic forces are constant, this work converts into heat and goes into heating the conductor (Joule heat release). The power density of Joule heat release can be written with the aid of Equation (3.2) in different forms:

$$Q = \boldsymbol{j}\cdot(\boldsymbol{E} + \boldsymbol{E}_{ext}) = \kappa(\boldsymbol{E} + \boldsymbol{E}_{ext})^2 = \frac{j^2}{\kappa}. \tag{3.20}$$

2) Joule James Preskott (1818–1889), outstanding English physicist, one of the pioneers of energy conservation.

The last expression of **Joule–Lenz law**[3] (in terms of the current density) is most universal, valid regardless of the presence of EMFs.

The total heat dissipation $\mathcal{Q}$ is given by the integral of Equation (3.2) over the volume of the conductor. For a closed quasi-linear conductor, the substitution $\boldsymbol{j}\,dV \to J\boldsymbol{dl}$ in Equation (I.2.40) yields

$$\mathcal{Q} = \int Q\,dV = \int \boldsymbol{j}\cdot(\boldsymbol{E} + \boldsymbol{E}_{ext})\,dV = J\oint_l (\boldsymbol{E} + \boldsymbol{E}_{ext})\cdot \boldsymbol{dl} = J\mathcal{E}_{ext}, \tag{3.21}$$

where $\mathcal{E}_{ext} = \oint_l \boldsymbol{E}_{ext}\cdot\boldsymbol{dl}$ is the total electromotive force, while the circulation of the constant electric field over the closed circuit is zero.

The electric current in a conductor is one of the **transport processes** during which an electric charge is transferred through a cross-section area of the conductor. Since the electric current is caused by the motion of particles, which are charge carriers, the current flow is accompanied by a transfer of other physical entities, in the first place, energy, because moving particles are known to transfer their kinetic and potential energy. The transfer processes become more diverse if the conductor is inhomogeneous in composition or if a temperature gradient is maintained in it by virtue of ambient conditions. Let us first consider the transfer of charge and heat in the conductor of homogeneous chemical composition in which there exist the gradients of electrostatic potential $\boldsymbol{E} = -\nabla\varphi$ and temperature ∇T (without assuming the presence of any other EMFs).

It is clear that the temperature gradient induces an electric current because the charge-carrying particles will diffuse from more heated parts of the conductor to less heated ones. Moreover, in accordance with the Fourier law of heat conductivity, the temperature gradient produces a heat flux $-\xi\nabla T$, where ξ is the positive coefficient. An additional heat flux will be produced by an electric field initiating directed motion of particles. Finally, the electric current will be related to the flux of potential energy of particles, $\varphi\boldsymbol{j}$. Therefore, the flux densities of electric current $\boldsymbol{j}$ and energy $\boldsymbol{q}$ can be written as

$$\begin{aligned} \boldsymbol{j} &= \kappa\boldsymbol{E} + \beta\nabla T, \\ \boldsymbol{q} &= \varphi\boldsymbol{j} + \gamma\boldsymbol{E} - \xi\nabla T. \end{aligned} \tag{3.22}$$

The kinetic coefficients κ, β, γ, and ξ are independent of the electric field but do depend on the nature and thermodynamic state of the conductor, in particular, on its temperature. For this reason, the temperature gradient ∇T in Equation (3.22) should be regarded as a sufficiently small quantity, and the higher-order terms of the expansion in this quantity can be neglected. A comparison of the first Equation (3.22) with Equation (3.2) shows that in this case $\boldsymbol{E}_{ext} = \beta\nabla T/\kappa$.

The kinetic coefficients are not independent. The relation between them can be determined based on the **Onsager symmetry principle of kinetic coefficients**[4] which is the consequence of time reversibility of equations of quantum and

3) Lenz Emily Christian (1804–1865), Russian physicist; worked on phenomena of electromagnetism and geophysics.

4) Onsager Lars (1903–1976), American physicist and chemist, one of the founders of nonequilibrium thermodynamics; Nobel Prize laureate in chemistry.

classical mechanics. The justification of the Onsager principle can be found in the textbooks on statistical physics and physical kinetics (see, for instance, Landau and Lifshitz (1980); Landau *et al.* (1984); de Groot and Mazur (1962); Gyarmati (1970)). It can be formulated as follows.

Consider a set of macroscopic parameters x_a, $a = 1, 2, \dots n$, which characterize a thermodynamic state of the nonequilibrium macroscopic system. Assume that the equilibrium state corresponds to $x_a = 0$. Then, for a small departure from equilibrium, the specific entropy can be represented by expansion

$$S(x_1, x_2, \dots x_n) = S_0 - \frac{1}{2}\sum_{a,b} \alpha_{ab}\, x_a x_b, \quad \alpha_{ab} = \alpha_{ba} = \frac{\partial^2 S}{\partial x_a\, \partial x_b}, \tag{3.23}$$

where S_0 is the equilibrium entropy while the first-order terms vanish because the entropy is maximum at equilibrium. The expansion coefficients α_{ab} depend on the equilibrium quantities only. Let us determine not only the parameters x_a but also the quantities that are "thermodynamically conjugate" to them:

$$X_a = -\frac{\partial S}{\partial x_a} = \sum_b \alpha_{ab} x_b; \tag{3.24}$$

they are sometimes referred to as **generalized thermodynamic forces**. A transition to equilibrium is accompanied by the growth of entropy whose variation is expressed through $\dot{x}_a$:

$$\frac{dS}{dt} = \int \frac{dS}{dt}\, dV = -\int \sum_a X_a \dot{x}_a\, dV. \tag{3.25}$$

In nonequilibrium thermodynamics, the quantities $\dot{x}_a = J_a$ are frequently called **fluxes**.

An important postulate of the nonequilibrium thermodynamics summarizing empirical data on behavior of weakly nonequilibrium systems states that in case of small departures from equilibrium, the quantities $\dot{x}_a$ are linear functions of x_a and completely determine the relaxation rates. Equally, the quantities $\dot{x}_a$ can be regarded as linear functions of the generalized thermodynamic forces X_a, because x_a and X_a are related by the linear expression (3.24). Accordingly, one can write the following linear equations:

$$\dot{x}_a = -\sum_b \lambda_{ab} X_b, \quad a = 1, 2, \dots, n. \tag{3.26}$$

The **symmetry principle** refers to the generalized kinetic coefficients λ_{ab} that describe the equilibration process. The mathematical formulation of this principle is:

$$\lambda_{ab} = \lambda_{ba}. \tag{3.27}$$

In the presence of an external magnetic field, the kinetic coefficients may depend on it. Then the symmetry principle takes the form

$$\lambda_{ab}(\boldsymbol{B}) = \lambda_{ba}(-\boldsymbol{B}). \tag{3.28}$$

Various quantities x_a can be chosen to characterize the nonequilibrium states of the system. Each set of the parameters x_a corresponds to a specific set of generalized thermodynamic forces. The criterion for choosing convenient kinetic coefficients is a possibility to represent the entropy growth rate (Equation (3.25)) as an integral of the sum whose terms are products of the generalized forces and "fluxes." Only those coefficients obey the Onsager symmetry principle that relate the "flows" to generalized forces.

Example 3.1

From the Onsager symmetry principle, find the relation between the coefficients β and γ entering the expressions for the electric and energy fluxes (Equation 3.22).

Solution. We take an arbitrary small volume element ΔV of the system and evaluate the energy inflowing through its surface per unit time:

(1) $$\Delta Q = -\oint \boldsymbol{q}\cdot d\boldsymbol{S} = -\int \nabla\cdot\boldsymbol{q}\, dV \approx -(\nabla\cdot\boldsymbol{q})\Delta V.$$

This energy dissipates as it converts to heat, which allows us to find the entropy growth rate of the system:

(2) $$\frac{dS}{dt} = -\int_V \frac{\nabla\cdot\boldsymbol{q}}{T}\, dV.$$

This expression needs to be reduced to the form of Equation (3.25), which may be done by rearranging the integrand:

(3) $$-\frac{\nabla\cdot\boldsymbol{q}}{T} = -\frac{\nabla\cdot(\varphi\boldsymbol{j})}{T} - \frac{\nabla\cdot(\boldsymbol{q}-\varphi\boldsymbol{j})}{T} = \frac{\boldsymbol{j}\cdot\boldsymbol{E}}{T} - \frac{(\boldsymbol{q}-\varphi\boldsymbol{j})\cdot\nabla T}{T^2} - \nabla\cdot\frac{\boldsymbol{q}-\varphi\boldsymbol{j}}{T}.$$

We have used Equation (3.1) to obtain the first term on the right-hand side of this expression. When integrating over the entire volume of the system, the last term (total divergence) becomes zero because both the energy flux and the electric current are absent outside the system. Therefore, Equation (2) can be replaced by

(4) $$\frac{dS}{dt} = \int_V \left\{ \frac{\boldsymbol{j}\cdot\boldsymbol{E}}{T} - \frac{(\boldsymbol{q}-\varphi\boldsymbol{j})\cdot\nabla T}{T^2} \right\} dV.$$

We have chosen the quantities $\boldsymbol{j}$ and $\boldsymbol{q} - \varphi\boldsymbol{j}$ as the "fluxes." Then, according to Equations (3.22), (3.25), and (4), we find the generalized forces

(5) $$\boldsymbol{X}_j = -\frac{\boldsymbol{E}}{T}, \quad \boldsymbol{X}_q = \frac{\nabla T}{T^2}$$

and get the kinetic coefficients

(6) $$\lambda_{jj} = \kappa T, \quad \lambda_{qq} = \xi T^2, \quad \lambda_{jq} = -\beta T^2, \quad \lambda_{qj} = \gamma T.$$

The Onsager symmetry principle implies

(7) $$\gamma = -\beta T. \quad \blacksquare$$

Example 3.2

Write down the power density of the heat release for the general case where a conductor is inhomogeneous in chemical composition and the temperature gradient is maintained there.

Solution. Using Equations (1) and (7) from Example 3.1 and Equation (3.22), we find the heat-release power density

$$Q = \frac{\Delta Q}{\Delta V} = -\nabla \cdot \boldsymbol{q} = \frac{j^2}{\kappa} + \nabla \cdot (\eta \nabla T) + T\boldsymbol{j} \cdot \nabla \left(\frac{\beta}{\kappa} \right). \qquad (3.29)$$

The first term in the right-hand side of Equation (1) is the Joule heat rate, while the second one is related to thermal conduction. The thermal conductivity coefficient is $\eta = \xi - \beta^2 T/\kappa$. The last term is linear in current density and related to the chemical and thermal inhomogeneities of the conductor. It includes the thermoelectric effects considered in Problems 3.2–3.4. ■

Suggested literature: Landau *et al.* (1984); Smythe (1950); Frenkel (1928); Panofsky and Phillips (1963); Simonyi (1956); Sommerfeld (1952); Tamm (1976); Grinberg (1948); Dykhne (1971a,b); Vinogradov (2001); Grigoriev and Meylikhov (1991); Gyarmati (1970); Martinson and Nedospasov (1993); Fokin (1996)

Problems

3.1• In the absence of a current, a conductor is quasi-neutral, the charge densities of ions and electrons are compensated: $\rho = \rho_i + \rho_e = 0$. Find the charge density inside the conductor of magnetic permeability μ, which carries a stationary current of density j.

Hint. Consider the effect of the magnetic force on conduction electrons that move with a constant drift velocity $\boldsymbol{u}$.

3.2 An accumulator battery of low internal resistance and electromotive force $\mathcal{E}$ cannot supply a certain device with a current J for a long period of time. In order to prolong the service-life of the battery, the device and the battery, having been connected in parallel, are joined through a resistance R to direct current network. The voltage V in the network is unstable and varies from V_1 to V_2 ($V_1 > V_2 > \mathcal{E}$). The resistance R is chosen so that there is no current delivered by the battery when $V = V_1$. What current J will the battery supply when $V = V_2$?

3.3 What are the parameters of a winding for rotating coil galvanometer to obtain maximum deflection of the pointer for given electromotive force of the circuit and given external resistance R (series connection)? The angular deflection of the pointer is proportional to the number of turns in the coil n and the current J in the circuit. Since the volume occupied by the coil

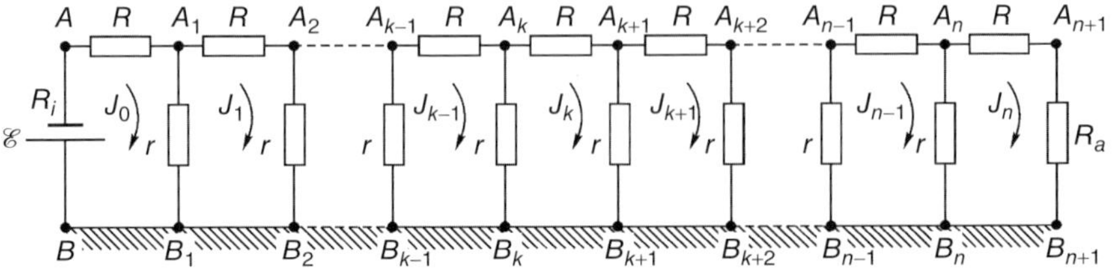

Figure 3.3 Schematic drawing of a telegraph line.

is limited, the product nS is approximately constant (where S is the cross section of the coil wire).

3.4 A square net made from a uniform wire consists of n^2 identical square cells. The resistance of one side of each cell is r. The current enters at one of the corners of the net and flows out at the opposite corner. Find the resistance R of the entire net for the $n = 2, 3, 4$.

Hint. In order to reduce the number of contour currents use of the symmetry of the circuit.

3.5⋆ A telegraph line (Figure 3.3) is suspended from n insulators at points A_1, $A_2, \dots, A_n$ (the earth serves as the second conductor). The sections AA_1, $A_1A_2, \dots, A_nA_{n+1}$ of the line have equal resistance R. When the insulation is dry, the resistance of the insulators is infinite. When the insulators are damp, there is leakage through them to the earth and the resistance of each of them becomes equal to r. Between the end A of the line and the earth there is a battery with electromotive force $\mathcal{E}$ and the internal resistance R_i. The end of the A_{n+1} section is also connected to the earth through a load with a resistance R_a. Find the current in each section of the line and also the current flowing through the load. By how much should the electromotive force of the battery be increased in the case of damp insulation as compared to the case of dry insulation, for the current through the load be the same in both cases? Consider, in particular, the case $R_a = 0$.

Hint. Consider the loop currents in the contours, each formed by a section $A_{k-1}A_k$ of the line and leakages through the insulators A_{k-1} and A_k. The solution of the resulting second-order difference equation is hyperbolic cosine.

3.6⋆ An underground cable has a constant resistance ρ per unit length. Insulation of the cable is imperfect and there is a leakage through it. The leakage conductivity per unit length of the cable is constant equal to $1/\rho'$. The earth plays the role of the return lead. Find the differential equation that describes the distribution of the stationary current in the cable. Find the relation between the current in the cable $J(x)$ and the potential difference $\varphi(x)$ between the cable conductor and the earth.

Hint. Proceed from Equation (1) in the solution of Problem 3.5.

3.7⋆ A battery with electromotive force $\mathcal{E}$ and internal resistance R_i is connected between one of the ends of an underground cable and the earth. The length of the cable is a, its resistance per unit length ρ and leakage conductivity (per unit length) is $1/\rho'$. The other end of the cable is connected to the earth through the load R_a. Find the current distribution $J(x)$ along the cable length. In particular, consider the case $R_i = R_a = 0$. Verify the result by performing the limiting transition to the case where leakage is absent.

Hint. Proceed from the differential equation obtained in Problem 3.6 or from Equation (7) of the solution of Problem 3.5.

3.8 Two plane parallel conducting plates are inserted in contact with the electrodes of a plane capacitor so that they are closely adjacent to each other and to the capacitor plates. The thickness of the plates is h_1 and h_2, conductivities and dielectric permeabilities are κ_1, κ_2 and ε_1, ε_2 respectively. The potential difference V is applied to the plates of the capacitor, which are made of material with conductivity much higher than κ_1 and κ_2. Determine the electric field strength E, the electric induction D, and the current density j in the plates and also the densities of free σ_{ext} and bound σ_{int} charges at all three separation boundaries.

3.9 Find the law of refraction for current lines across a smooth separation boundary between two media having conductivities κ_1 and κ_2, respectively.

3.10 A constant current of density j flows through an infinitely long straight conductor of radius a and conductivity κ. Calculate the flux of the Poynting vector through the surface of the conductor and show that it compensates the Joule losses inside the conductor.

3.11* A constant current J flows through an infinitely long straight cable of radius a and conductivity κ. The conductor is enclosed within a thick coaxial conducting cylindrical shell that serves as the return conductor. The shell has the inner radius b and the outer radius $c \to \infty$. Find the electric potential φ and the magnetic field H in the entire space. Determine the distribution of the surface charges, σ. The dielectric permittivity of the medium between the conductors is ε.

3.12 Three conductors of circular cross section and the same radius r are connected in series to form a closed ring. The lengths of the conductors are $l_0, l_1, l_2 \gg r$ and their conductivities are κ_0, κ_1, and κ_2. Time-independent EMF $\mathcal{E}_0$ is uniformly distributed over the volume of the conductor with conductivity κ_0. Find the electric field E and the distribution of electric charges inside the ring.

3.13 Find the energy flux γ through the surfaces of the three conductors considered in Problem 3.12 and obtain thus the Joule–Lenz law.

3.14 The current distribution in a three-dimensional conductor with conductivity κ exhibits such a symmetry that the electric field strength and, hence, the current density are constant at all points of any equipotential surface.

Prove that the resistance of the conductor is expressed by the same formula as the resistance of a quasi-linear conductor with a variable cross section.[5)]

3.15 Use the result of Problem 3.14 to find the resistances R for the following conductors:

(a) a spherical capacitor with the plates of radii a and b, $a < b$, filled with homogeneous medium of conductivity κ;

(b) a similar capacitor filled with two homogeneous layers with conductivities κ_1 and κ_2 (the layer with κ_1 is in contact with the inner plate); the separation boundary between the layers is a sphere of radius c;

(c) a cylindrical capacitor with plate radii a and b ($a < b$) and length l, filled with medium of conductivity κ (neglect edge effects).

3.16 A ground is connected by means of a perfectly conducting sphere of radius a, which is half buried in the earth (the earth conductivity $\kappa_1 = const$). The layer of earth of radius b and with the artificially enhanced conductivity κ_2 is concentric with the sphere and adjacent to it. Find the resistance R of this ground.

3.17* A system of perfect conductors (electrodes) is placed into medium with conductivity $\kappa(\boldsymbol{r})$ and dielectric permittivity $\varepsilon(\boldsymbol{r})$. The medium is such that $\kappa(\boldsymbol{r})/\varepsilon(\boldsymbol{r}) = const$ at all points of the space.[6)] Find the relation between the potential coefficients s_{ik} and the resistance coefficients R_{ik} of this system of conductors. What is the relation between the charges q_k of the electrodes and the currents J_k leaving them?

3.18 A capacitor of arbitrary form is filled with a homogeneous dielectric of permittivity ε. Find its capacitance C; it is known that when it is filled with homogeneous conductor of conductivity κ, its direct-current resistance is R.

3.19 A system of electrodes is characterized by the resistance coefficients R_{ik}. Find the amount of heat Q released per unit time in the space between the electrodes assuming that the currents J_k leaving the electrodes are known.

3.20 Two perfectly conducting spheres of radii a and b are placed into a homogeneous medium with conductivity κ and dielectric permittivity ε. The distance between the centers of the spheres is l. A current J is delivered to one of the spheres and withdrawn from the other. Find the resistance of the medium between the spheres, $R = (V_a - V_b)/J$, where V_a and V_b are the potentials of the spheres and J is the current leaving the sphere of radius a.

Hint. Express R_{ik} in terms of the coefficients of conductance c_{ik} of the system of the two spheres (Problem 2.58).

3.21 The ends of a particular circuit are grounded by means of two perfectly conducting spheres (radii a_1 and a_2), which are half buried in the earth

5) The formulated conditions are the same as the conditions under which the Gauss electrostatic theorem can be used in the corresponding electrostatic problem.

6) This condition can be formulated differently: the space between the perfect conductors is filled, instead of with the medium of conductivity κ, with dielectric medium of electric permittivity ε, which is proportional to κ at any point of space, so that $\varepsilon/\kappa = const$.

that plays the role of the second lead. The distance between the spheres is $l \gg a_1, a_2$, the conductivity of the earth is $\kappa = const$. Find the resistance R between the spheres.

3.22 Solve Problem 3.21 where the grounded conductors are two identical ellipsoids of revolution of volume V and eccentricity e_0. The axes of revolution are perpendicular to the earth surface while their centers lie on it. Which form of the grounded conductor is more effective (i.e., provides lower resistance)?

3.23 Two perfectly conducting spherical electrodes of equal radii a and of potentials $V/2$ and $-V/2$ are half dipped into a medium with conductivity κ. Calculate the current J flowing from one electrode to another and show that it is equal to the current leaving a separate electrode whose potential V, relative to the infinitely remote points, equals the potential difference between the electrodes.

3.24* Particles of charge e and mass m can be emitted in an unlimited number by a plane electrode $x = 0$ under the effect of an electric field. The particles emitted with zero initial velocity are accelerated toward another plane electrode, which is parallel to the first electrode and is at a distance a from it. The potential difference between the electrodes is φ_0. The emission from the first electrode continues until the field of the space charge of density ρ, produced between the electrodes, compensates the external field near the surface of the first electrode so that the strength of the resultant field is $\partial\varphi/\partial x|_{x=0} = 0$. Find the density of the stationary current j between the electrodes as a function of the potential difference φ_0.

Hint. The potential in the space between the electrodes is determined by the Poisson equation $\triangle\varphi = -4\pi\rho$, $\rho = j/v$, where v is the velocity of particles at a given point of space.

3.25 The electric conductivity $\kappa_{\mu\nu}$ in an anisotropic conductor is a tensor quantity. Show that the Onsager symmetry principle requires a symmetry of the electric conductivity tensor.

3.26 Find the limitations that are imposed on the kinetic coefficients in Equations (3.22) by the entropy increase law for non-equilibrium systems.

3.27$^\bullet$ Using Equation (3.29) show that if electric current and temperature gradient are present in a conductor made of homogeneous material, there occurs the Thomson[7] heat release with the power density

$$Q_{Th} = \varrho \boldsymbol{j}\cdot\nabla T. \tag{3.30}$$

Its sign depends on mutual orientation of the current and the temperature gradient. Calculate the Thomson coefficient ϱ.

3.28$^\bullet$ Use Equation (3.29) to show that when an electric current flows across the junction between two conductors of different chemical composition there

7) Thomson (Kelvin) William (1824–1907), English physicist with a very wide range of scientific interests: thermodynamics, hydrodynamics, electromagnetism, mathematics, and engineering.

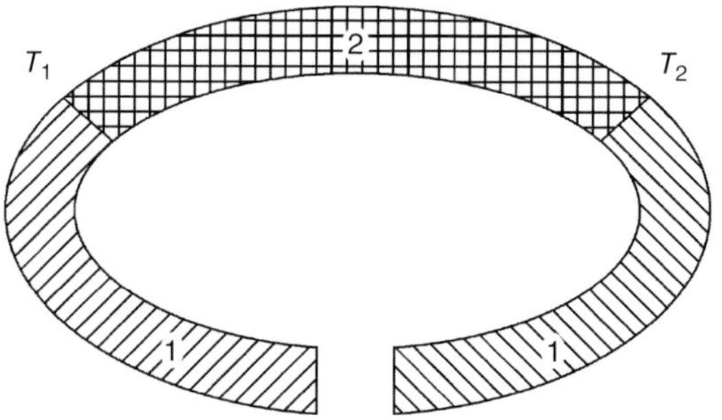

Figure 3.4 Scheme for occurrence of thermal electromotive force.

occurs the heat release (absorption). Calculated per unit area, it is

$$Q_P = \Pi_{12} j. \tag{3.31}$$

It depends on the current direction **(Peltier effect)**. Calculate the Peltier[8] coefficient Π_{12}.

3.29• An open circuit consists of two different quasi-linear conductors 1 and 2 whose contacts are at different temperatures T_1 and T_2 (Figure 3.4). Using the first Equation (3.22), show that the potential difference (thermal electromotive force) arises at the ends of the circuit:

$$\mathcal{E}_T = \int_{T_1}^{T_2} \left(\frac{\beta_1}{\kappa_1} - \frac{\beta_2}{\kappa_2} \right) dT = \int_{T_1}^{T_2} \frac{\Pi_{12}}{T} dT, \tag{3.32}$$

where Π_{12} is the Peltier coefficient determined in Problem 3.28.

3.30⋆ A conductor consists of regions with specific electric conductivities κ_1 and κ_2 distributed randomly so that its electric conductivity $\kappa(x, y)$ is a random function of position in the xy-plane and does not depend on the z-coordinate. The regions 1 and 2 are statistically equivalent, that is, they occupy equal areas in the xy-plane. The current flows in the xy-plane and at any point (locally) obeys the Ohm law $\boldsymbol{j} = \kappa(x, y)\boldsymbol{\mathcal{E}}$, where $\boldsymbol{\mathcal{E}}$ is the electric field. Calculate the effective electric conductivity K, which enters the macroscopic Ohm's law $\boldsymbol{J} = K\boldsymbol{E}$ and relates the volume-averaged density of the current $\boldsymbol{J} = (1/V)\int \boldsymbol{j}\, dV$ to the volume-averaged density of the electric field $\boldsymbol{E} = (1/V)\int \boldsymbol{\mathcal{E}}\, dV$. Also, calculate the relative quadratic fluctuations of the current and field strength, $\Delta_J = (\langle j^2 \rangle - J^2)/J^2$ and $\Delta_E = (\langle \mathcal{E}^2 \rangle - E^2)/E^2$.

3.31⋆⋆ Generalize Problem 3.30 to the case in which the magnetic field $\boldsymbol{H}_0$ is directed along the Oz-axis. Here the local Ohm's law can be written as (Problem 2.28)

$$\boldsymbol{j} + \boldsymbol{j} \times \boldsymbol{n} = \kappa \boldsymbol{\mathcal{E}},$$

8) Peltier Jean (1785–1845). Till 1815, he was watchmaker and watch dealer in Paris, then studied various physical phenomena.

where $\boldsymbol{n}$ is the local dimensionless Hall parameter, a vector parallel to $\boldsymbol{H}_0$. Find the effective electric conductivity K and the Hall parameter $\boldsymbol{N}$, which enter the Ohm's law for the averaged values of $\boldsymbol{J}$ and $\boldsymbol{E}$:

$$\boldsymbol{J} + \boldsymbol{J} \times \boldsymbol{N} = K\boldsymbol{E}.$$

Consider the cases in which κ fluctuates while $\boldsymbol{n}$ is constant, or $\boldsymbol{n}$ fluctuates while κ is constant.

3.2 Magnetic Field in Magnetic Media

Equations and boundary conditions. The magnetic field in media is described by the magnetic induction vector $\boldsymbol{B}$ and the magnetic field strength vector $\boldsymbol{H}$ which in the static case satisfy Equation (1.21):

$$\left.\begin{array}{llll} rot\,\boldsymbol{H} = \frac{4\pi}{c}\boldsymbol{j}, & \oint_l \boldsymbol{H}\cdot d\boldsymbol{l} = \frac{4\pi}{c}J, \\ div\,\boldsymbol{B} = 0, & \oint_S \boldsymbol{B}\cdot d\boldsymbol{S} = 0 \end{array}\right\} \tag{3.33}$$

and the relation (1.26)

$$\boldsymbol{B} = \mu \boldsymbol{H} \quad \text{or} \quad B_\alpha = \mu_{\alpha\beta} H_\beta. \tag{3.34}$$

(the last relations are used for isotropic and anisotropic diamagnetics and paramagnetics; ferromagnetics are discussed below). The quantity $\boldsymbol{j}$ in Equation (3.33) is a conduction current density, and J is a total conduction current through the surface enclosed by contour l. The density of the volume and surface magnetization current (which is determined by a sum of "molecular" currents) is described by Equations (1.18) and (1.19):

$$\boldsymbol{j}_{int} = c\,rot\,\boldsymbol{M}, \qquad \boldsymbol{i}_{int} = c\boldsymbol{n} \times (\boldsymbol{M}_2 - \boldsymbol{M}_1), \tag{3.35}$$

where $\boldsymbol{M}_2 - \boldsymbol{M}_1$ is a discontinuity of the magnetization vector at the boundary. This vector is related to $\boldsymbol{B}$ and $\boldsymbol{H}$ by the expression

$$\boldsymbol{B} = \boldsymbol{H} + 4\pi \boldsymbol{M}. \tag{3.36}$$

The field vectors at the boundary that separates the magnetics satisfy the boundary conditions, which can be obtained from the integral equation (3.33); they have the form similar to that of Equation (I.2.58):

$$\boldsymbol{n}\cdot(\boldsymbol{B}_2 - \boldsymbol{B}_1) = 0, \qquad \boldsymbol{n}\times(\boldsymbol{H}_2 - \boldsymbol{H}_1) = \frac{4\pi}{c}\boldsymbol{i}. \tag{3.37}$$

If the entire medium, including also the conductors carrying the current, can be regarded as homogeneous and isotropic ($\mu = const$), the strength of the magnetic field $\boldsymbol{H}$ in this medium will be, according to Equations (3.33), (3.34), and (3.37), the same as in vacuum, with the same distribution of conduction currents, while the magnetic induction $\boldsymbol{B} = \mu\boldsymbol{H}$ will change by a factor of μ. Recall, that this is precisely the magnetic induction $\boldsymbol{B}$, which presents the averaged microscopic magnetic field. Due to this, the macroscopic Lorentz force, which acts on a point charged particle in medium is determined by the strength of the electric field and by the magnetic induction:

$$\boldsymbol{F} = e\boldsymbol{E} + \frac{e}{c}\boldsymbol{v} \times \boldsymbol{B}. \tag{3.38}$$

Equation (3.33) allows one to introduce the vector potential $\boldsymbol{A}$,

$$\boldsymbol{B} = rot\,\boldsymbol{A}. \tag{3.39}$$

In a homogeneous isotropic medium, it satisfies the equation

$$\triangle\boldsymbol{A} = -\frac{4\pi\mu}{c}\boldsymbol{j}. \tag{3.40}$$

Ferromagnetics and spontaneous magnetization. The scalar potential. Ferromagnetics differ from diamagnetics and paramagnetics in the following characteristics: (i) many of them possess the "residual magnetization," that is, spontaneous magnetization in the absence of an external magnetic field $\boldsymbol{H}$; (ii) the relation between the vectors $\boldsymbol{B}$ and $\boldsymbol{H}$ is nonlinear and ambiguous and depends on a history of magnetization process (hysteresis). For an approximate evaluation of magnetic fields in ferromagnetics, the model of the "idealized ferromagnetic" is used. It assumes the relation between the field vectors in the form

$$\boldsymbol{B} = \mu\boldsymbol{H} + 4\pi\boldsymbol{M}_0. \tag{3.41}$$

Soft magnetic materials are characterized by a high magnetic permeability, $\mu \approx 10^2 - 10^6$ (Grigoriev and Meylikhov (1991)). Magnetization $\boldsymbol{M}_0$ must be considered a quantity that is independent of $\boldsymbol{H}$ and is inherent in a permanent magnet. It can be a given function of coordinates.

The field produced by permanent magnets is conveniently calculated by the scalar potential method. In the absence of conduction currents, we have the equations

$$rot\,\boldsymbol{H} = 0, \qquad div\,\boldsymbol{B} = 0, \qquad \boldsymbol{B} = \mu\boldsymbol{H} + 4\pi\boldsymbol{M}_0. \tag{3.42}$$

The first equation allows one to introduce the pseudoscalar potential ψ (Example I.2.12) by means of the relation $\boldsymbol{H} = -\nabla\psi$. This potential satisfies the equation analogous to Equation (2.5) for the electrostatic potential in an inhomogeneous dielectric:

$$\nabla\cdot(\mu\nabla\psi) = -4\pi\rho_m, \tag{3.43}$$

where the quantity

$$\rho_m(\boldsymbol{r}) = -\nabla\cdot\boldsymbol{M}_0(\boldsymbol{r}) \tag{3.44}$$

plays the role of the volume density of external "magnetic charges." At the boundary of the constant magnet, there occur also the surface "magnetic charges." The pseudoscalar potential in a homogeneous ferromagnetic satisfies the Poisson equation

$$\triangle\psi = -\frac{4\pi\rho_m}{\mu}. \tag{3.45}$$

In this case, all electrostatic methods considered in Chapter 2 can be applied to solve the problems of magnetostatics.

Suggested literature: Tamm (1976); Landau *et al.* (1984); Jackson (1999); Bredov *et al.* (2003); Frenkel (1928); Stratton (1948); Smythe (1950); Simonyi (1956); Panofsky and Phillips (1963)

Problems

3.32 Determine the strength of the magnetic field $\boldsymbol{H}$ and the magnetic induction $\boldsymbol{B}$ due to a constant current J flowing through an infinitely long cylindrical conductor of radius a. The magnetic permeabilities of the conductor and the surrounding material are μ_0 and μ, respectively. Solve the problem by the simplest method, that is, with the aid of the Maxwell equation in the integral form (3.33) and also by introducing the vector potential $\boldsymbol{A}$.

3.33 Solve Problem 3.32 for a hollow cylindrical conductor of inner radius a and outer radius b.

3.34 Reduce the magnetostatics problem to determine the field produced by the given currents in a nonuniform medium with the linear coupling equation $\boldsymbol{B} = \mu(\boldsymbol{r})\boldsymbol{H}$, to the corresponding problem of electrostatics. To do this, write down the magnetic field as a sum of two fields: $\boldsymbol{H} = \boldsymbol{H}_0 + \boldsymbol{H}'$, where $\boldsymbol{H}_0$ is the "primary" field, which would be produced by the same current distribution in vacuum, and $\boldsymbol{H}'$ is the field due to the presence of magnetics. Introduce the scalar potential ψ for $\boldsymbol{H}'$ and obtain the equation and the boundary conditions for ψ.

3.35 A current-carrying circuit lies on the plane separating two media with magnetic permeabilities μ_1 and μ_2. Determine the magnetic field strength $\boldsymbol{H}$ in the entire space, assuming that the field produced by this contour in vacuum is known.

3.36 An infinite straight conductor carrying a current J is parallel to the plane separation boundary between two media with magnetic permeabilities μ_1 and μ_2. The distance from the conductor to the separation boundary is a. Evaluate the magnetic field.

Hint. Use the method of images as in electrostatic problems (Section 2.1).

3.37 A sphere of radius a and magnetic permeability μ is placed in a uniform magnetic field $\boldsymbol{H}_0$. Determine the resultant field $\boldsymbol{H}$, the induced magnetic moment $\boldsymbol{m}$, and the current density $\boldsymbol{j}_{int}$, which is equivalent to the magnetization of the sphere.

3.38 Find the shape of lines of the magnetic field strength inside and outside the sphere considered in Problem 3.37.

Hint. Use the formula derived in Problem I.2.79.

3.39 A uniform magnetic field $\boldsymbol{H}$ exists in a magnetic material with permeability μ_e. A sphere of radius a is cut out of this magnetic, and another concentric sphere of radius b and magnetic permeability μ_i is inserted into the magnetic to replace the removed one. Is it possible to choose μ_i and b so that the field in the region $r > a$ be unchanged?

3.40* An anisotropic nonferromagnetic sphere is placed in a uniform magnetic field. Find the resultant field $\boldsymbol{H}$ and the moment of forces $\boldsymbol{N}$ acting on the sphere.

3.41 Evaluate the magnetic field produced in an anisotropic magnetic by a thin straight wire carrying a current J. The wire is placed in parallel with one of the principal axes of the tensor $\mu_{\alpha\beta}$.

3.42 An infinitely long hollow cylindrical shell of inner radius a and outer radius b is placed in an external uniform magnetic field $\boldsymbol{H}_0$, which is perpendicular to its axis. The magnetic permeabilities of the cylinder and of the surrounding space are μ_1 and μ_2, respectively. Find the field strength H in the cavity. Consider, in particular, the case of $\mu_1 \gg \mu_2$.

3.43 A hollow sphere of inner and external radii a and b, respectively, is placed in an external magnetic field $\boldsymbol{H}_0$. The magnetic permeabilities of the sphere and surrounding medium are μ_1 and μ_2, respectively. Find the field H in the cavity. Consider in particular the case $\mu_1 \gg \mu_2$.

3.44 Evaluate a change $\Delta\Phi$ in the magnetic induction flux through the cross section of the hollow sphere of radius a (Problem 3.43), relative to the initial flux that existed in the uniform magnetic. The section passes through the center of the sphere and is perpendicular to the direction of an external field.

3.45 An infinitely long straight wire of radius a and magnetic permeability μ_1 is placed in an external uniform transverse field $\boldsymbol{H}_0$, which is produced in a medium of magnetic permeability μ_2. The wire carries a constant current J. Find the resultant magnetic field inside and outside the wire.

3.46 Determine the scalar and vector potentials ψ and $\boldsymbol{A}$ produced in a finite region by the magnetization whose distribution is known to be $\boldsymbol{M}(\boldsymbol{r})$. Show by direct calculation that the vectors $\boldsymbol{B} = rot\,\boldsymbol{A}$ and $\boldsymbol{H} = -grad\,\psi$ are related by the equation $\boldsymbol{B} = \boldsymbol{H} + 4\pi\boldsymbol{M}$.

3.47 A body of arbitrary shape is uniformly magnetized. Show that the scalar potential of the magnetic field produced by this body can be written in the form

$$\psi = -\boldsymbol{M} \cdot grad\,\varphi,$$

where $\boldsymbol{M}$ is the magnetization and φ is the electrostatic potential of the uniformly charged (with density $\rho = 1$) body of the same shape and sizes.

3.48 A current J flows through a straight wire, which is parallel to the Oz-axis. Three half planes radiate fan-like outward the axis and thus form three dihedral angles α_1, α_2, and α_3 to one another, $\alpha_1 + \alpha_2 + \alpha_3 = 2\pi$. The spaces between them are filled with magnetic media of permeabilities μ_1, μ_2, and μ_3, respectively. Find the magnetic field H_i ($i = 1$, 2, and 3) in each of the dihedral angles.

3.49* Find the magnetic field produced by a uniformly magnetized constant magnet of spherical form. Use the coupling Equation (3.41). The magnetic permeabilities of the sphere and of the surrounding medium are μ_1 and μ_2, respectively.

3.50* Find the field due to an infinitely long uniformly magnetized cylinder of radius a. The magnetization vector $\boldsymbol{M}_0$ is perpendicular to the axis of the cylinder. The magnetic permeabilities of the cylinder and of the surrounding medium are μ_1 and μ_2, respectively.

3.51 An ellipsoid made of magnetic material with permeability μ is placed in a uniform magnetic field $\boldsymbol{H}_0$. Find the internal field and the magnetic moment of the ellipsoid. Consider the particular cases of a thin disk, long thin rod, and sphere.

3.52* An ellipsoid made of an anisotropic material of magnetic permeability μ_{ik} is placed in an external uniform magnetic field $\boldsymbol{H}_0$. Find the internal field $\boldsymbol{H}_1$ in the ellipsoid.

3.3
Energy, Forces, and Thermodynamic Relations for Magnetics

Thermodynamic functions of magnetics. Consider a bounded system, which consists of a matter and currents producing a magnetic field. Since the magnetic field itself does not produce any the work, a nonvanishing work is possible only in the presence of an electric field $\boldsymbol{E}$ in the system. In a time δt, the sources of current do the work on the system "matter + field,"

$$\delta A = -\delta t \int \boldsymbol{j}\cdot\boldsymbol{E}\, dV,$$

We substitute $\boldsymbol{j} = c\nabla \times \boldsymbol{H}/4\pi$ into the last relation and rearrange the integrand with the aid of the Gauss–Ostrogradskii theorem:

$$\delta A = -\frac{c\delta t}{4\pi} \int \boldsymbol{H}\cdot[\nabla \times \boldsymbol{E}]\, dV.$$

The vortex electric field arises due to change in the magnetic field in accordance with the induction Equation (1.8):

$$\delta t\, \nabla \times \boldsymbol{E} = -\frac{\delta t}{c}\frac{\partial \boldsymbol{B}}{\partial t} = -\frac{1}{c}\,\delta \boldsymbol{B}.$$

Thus, the elementary work for producing a magnetic field is expressed in terms of the variation of magnetic induction:

$$\delta A = \frac{1}{4\pi} \int \boldsymbol{H}\cdot\delta \boldsymbol{B}\, dV. \tag{3.46}$$

This expression is formally analogous to Equation (2.36) in electrostatics. However, the magnetic induction is the analog of the electric field strength but not of electric induction.

As in the case of dielectrics, depending on the outside conditions, the elementary work (Equation 3.46) may characterize a change in the different thermodynamic potentials. For a thermally insulated, statistically equilibrated body, the work of electric forces is done at fixed entropy and presents the change in the internal energy $\mathcal{U}$. According to the first law of thermodynamics, the total change in the internal energy is the sum of heat and work,

$$\delta\mathcal{U} = T\,\delta\mathcal{S} + \frac{1}{4\pi}\int \boldsymbol{H}\cdot\delta\boldsymbol{B}\,dV. \tag{3.47}$$

Here, in contrast to Equation (I.C46), the internal energy of the body is denoted by $\mathcal{U}$, while $\mathcal{S}$ is the entropy of the body and T is its temperature. For small variations of thermodynamic quantities, the symbol δ is used to distinguish them from the integration element dV. Equation (3.47) does not include the mechanical work due to a change in the body volume.

If the body under consideration is in thermal contact with the surrounding equilibrated bodies and its temperature does not change, the work presents a change in the Helmholtz free energy $\mathcal{F}$, and its total change is given by

$$\delta\mathcal{F} = -\mathcal{S}\,\delta T + \frac{1}{4\pi}\int \boldsymbol{H}\cdot\delta\boldsymbol{B}\,dV. \tag{3.48}$$

The above changes of the thermodynamic potentials can also be written for the specific quantities (per unit volume of the magnetic):

$$dU(S,\tau,\boldsymbol{B}) = T\,dS + \zeta\,d\tau + \frac{1}{4\pi}\boldsymbol{H}\cdot d\boldsymbol{B}, \tag{3.49}$$

$$dF(T,\tau,\boldsymbol{B}) = -S\,dT + \zeta\,d\tau + \frac{1}{4\pi}\boldsymbol{H}\cdot d\boldsymbol{B}. \tag{3.50}$$

Since the volume under consideration can exchange particles with surrounding medium, Equations (3.49) and (3.50) contain the corresponding term, which includes a change of the mass density $d\tau$ and the chemical potential per unit mass, $\zeta = \mu/m$, where μ is the chemical potential per a single particle with mass m, entering (I.C47).

Unlike the local quantities T, S, and τ, which exist only inside the considered body, the magnetic induction $\boldsymbol{B}$ is not zero outside the body as well. Therefore, there are often convenient to introduce such thermodynamic potentials U_* and F_* whose independent field variable is the magnetization $\boldsymbol{M}$ that is zero outside the body. These potentials are introduced by the relations

$$U_*(S,\tau,\boldsymbol{M}) = U(S,\tau,\boldsymbol{B}) - \frac{H^2}{8\pi}, \qquad F_*(T,\tau,\boldsymbol{M}) = F(T,\tau,\boldsymbol{B}) - \frac{H^2}{8\pi}, \tag{3.51}$$

and have the total differentials

$$dU_* = TdS + \zeta d\tau + \boldsymbol{H}\cdot\boldsymbol{M}, \tag{3.52}$$

$$dF_* = -SdT + \zeta d\tau + \boldsymbol{H}\cdot\boldsymbol{M}. \tag{3.53}$$

Finally, it is possible to introduce such thermodynamic potentials whose independent variable will be the vector $\boldsymbol{H}$ determined by conduction currents:

$$\tilde{U}(S,\tau,\boldsymbol{H}) = U - \frac{1}{4\pi}\boldsymbol{H}\cdot\boldsymbol{B}, \qquad \tilde{F}(T,\tau,\boldsymbol{H}) = F - \frac{1}{4\pi}\boldsymbol{H}\cdot\boldsymbol{B}, \tag{3.54}$$

Their differentials can be written in the form

$$d\tilde{U} = TdS + \zeta d\tau - \frac{1}{4\pi}\boldsymbol{B}\cdot d\boldsymbol{H}, \qquad d\tilde{F} = -SdT + \zeta d\tau - \frac{1}{4\pi}\boldsymbol{B}\cdot d\boldsymbol{H}. \tag{3.55}$$

The magnetic field and the magnetic induction in a magnet can be expressed in terms of the thermodynamic functions:

$$\boldsymbol{H} = 4\pi\left(\frac{\partial U}{\partial \boldsymbol{B}}\right)_{S,\tau} = 4\pi\left(\frac{\partial F}{\partial \boldsymbol{B}}\right)_{T,\tau}, \qquad \boldsymbol{B} = -4\pi\left(\frac{\partial \tilde{U}}{\partial \boldsymbol{H}}\right)_{S,\tau} = -4\pi\left(\frac{\partial \tilde{F}}{\partial \boldsymbol{H}}\right)_{T,\tau}. \tag{3.56}$$

Example 3.3

Calculate the variations of the specific thermodynamic functions U, $\tilde{U}$, F, $\tilde{F}$, S, and ζ for a magnet, which are due to the presence of a magnetic field. The magnet is isotropic, and its coupling equation is linear: $\boldsymbol{B} = \mu(T,\tau)\boldsymbol{H}$.

Solution. By analogy with Example 2.5, we obtain

$$F(T,\tau,B) - F_0(T,\tau) = \frac{B^2}{8\pi\mu}, \qquad U(S,\tau,B) - U_0(S,\tau) = \frac{B^2}{8\pi\mu}. \tag{3.57}$$

Here, in the first formula, the magnetic permittivity μ, which is independent of the field, should be expressed in terms of the temperature and the mass density τ of the magnetic, and in the latter formula, in terms of the entropy and the density:

$$\tilde{U} - U_0(S,\tau) = -\frac{\mu H^2}{8\pi}, \qquad \tilde{F} - F_0(T,\tau) = -\frac{\mu H^2}{8\pi}, \tag{3.58}$$

$$S = -\left(\frac{\partial F}{\partial T}\right)_{\tau,\mathbf{B}} = S_0(T,\tau) + \frac{B^2}{8\pi\mu^2}\left(\frac{\partial \mu}{\partial T}\right)_{\tau}, \tag{3.59}$$

$$\zeta = \left(\frac{\partial F}{\partial \tau}\right)_{T,\mathbf{B}} = \zeta_0(T,\tau) - \frac{B^2}{8\pi\mu^2}\left(\frac{\partial \mu}{\partial \tau}\right)_{T}. \quad \blacksquare \tag{3.60}$$

Example 3.4

A dielectric body with magnetic permeability μ is placed in an external uniform magnetic field $\vec{H}$ in vacuum. Express variation of the internal energy in terms of its magnetic moment, assuming the body to be thermally insulated and neglecting changes of its volume. How will the free energy of the body change if, instead of thermal insulation, a constant temperature is maintained in it?

Solution. In accordance with Equation (3.57), a change of the internal energy in the body is

(1)

$$\Delta\mathcal{U} = \int \frac{\boldsymbol{H}\cdot\boldsymbol{B} - \mathcal{H}^2}{8\pi}\,dV = \frac{1}{8\pi}\int (\boldsymbol{H} + \boldsymbol{\mathcal{H}})\cdot(\boldsymbol{B} - \boldsymbol{\mathcal{H}})\,dV - \frac{1}{2}\int \boldsymbol{M}\cdot\boldsymbol{\mathcal{H}}\,dV,$$

where the density of the magnetic energy in the absence of the body is subtracted from the integrand, and the integrand itself is transformed using the formula $\boldsymbol{B} - \boldsymbol{H} = 4\pi\boldsymbol{M}$. In the first integral on the right-hand side of Equation (1), we express the difference $\boldsymbol{B} - \boldsymbol{\mathcal{H}} = \nabla\cdot(\boldsymbol{A} - \boldsymbol{\mathcal{A}})$ in terms of the vector potentials. Here, $\boldsymbol{A}$ and $\boldsymbol{\mathcal{A}}$ are the vector potentials in the presence and in the absence of the magnetic, respectively. With the aid of the formulas of the vector analysis, we obtain the identity

(2) $$(\boldsymbol{H} + \boldsymbol{\mathcal{H}})\cdot(\boldsymbol{B} - \boldsymbol{\mathcal{H}}) = [\nabla\times(\mathbf{H} + \boldsymbol{\mathcal{H}})]\cdot(\boldsymbol{A} - \boldsymbol{\mathcal{A}}) + \nabla\cdot[(\boldsymbol{A} - \boldsymbol{\mathcal{A}})\cdot(\boldsymbol{H} + \boldsymbol{\mathcal{H}})].$$

The first term on the right-hand side is zero due to the absence of conduction currents: $\nabla\times(\boldsymbol{H} + \boldsymbol{\mathcal{H}}) = 0$. The second term is zero when integrating over the entire space. As a result, from Equation (1) we have

$$\Delta\mathcal{U} = -\frac{1}{2}\int \boldsymbol{M}\cdot\boldsymbol{\mathcal{H}}\,dV = -\frac{1}{2}\boldsymbol{\mathcal{M}}\cdot\boldsymbol{\mathcal{H}}, \qquad (3.61)$$

where

$$\boldsymbol{\mathcal{M}} = \int \boldsymbol{M}\,dV \qquad (3.62)$$

is the magnetic moment of the body. For an isothermal process, the same quantity (Equation (3.61)) will give the change of the free energy. ■

Example 3.5

The magnetic field of a homogeneous medium with the magnetic permeability $\mu(T,\tau)$ is due to thin quasi-linear conductors carrying currents. By analogy with the currents in vacuum (Section I.2.2), express the energy of the magnetic field in terms of the inductance coefficients. Find these coefficients as a function of magnetic permeability of the medium.

Solution. The energy of the magnetic field at a fixed temperature and density of the medium is a part of the free energy of the system (Equation (3.57)):

(1) $$\Delta\mathcal{F} = \frac{1}{8\pi}\int \boldsymbol{H}\cdot\boldsymbol{B}\,dV.$$

We transform this formula in the way, which is the reverse of what has been used while reducing (I.2.75) to (I.2.76). We obtain

(2) $$\Delta\mathcal{F} = \frac{1}{2c}\int \boldsymbol{A}\cdot\boldsymbol{j}\,dV.$$

The volume integral can be presented as a sum of integrals over the quasi-linear contours:

(3) $$\Delta\mathcal{F} = \sum_b \frac{J_b}{2c} \oint \boldsymbol{A} \cdot d\boldsymbol{l}_b.$$

Then the energy can be expressed in terms of the magnetic fluxes $\Phi_b = \oint \boldsymbol{B} \cdot d\boldsymbol{S}_b$,

(4) $$\Delta\mathcal{F} = \frac{1}{2c} \sum_b J_b \Phi_b.$$

It is equally possible to present the vector potential as a sum over quasi-linear contours:

(5) $$\boldsymbol{A} = \sum_a \boldsymbol{A}_a, \qquad \boldsymbol{A}_a(\boldsymbol{r}_b) = \frac{\mu J_a}{c} \oint \frac{d\boldsymbol{l}_a}{R_{ab}}.$$

Here the solution of equation (I.2.50′) is used with the factor μ added in accordance to Equation (3.40); R_{ab} is the distance between the length elements, dl_a and dl_b. Finally, the free energy, given by Equation (3), is expressed as

$$\Delta\mathcal{F} = \sum_a \frac{J_a^2}{2c^2} L_{aa} + \sum_{a<b} \frac{J_a J_b}{c^2} L_{ab}, \tag{3.63}$$

where

$$L_{ab} = \mu \int \frac{d\boldsymbol{l}_a \cdot d\boldsymbol{l}_b}{R_{ab}} \tag{3.64}$$

is the coefficient of the mutual induction of the current-carrying contours a and b; the quantity L_{aa} is the self-induction, which should be calculated taking into account the finite cross section of the conductor, that is, using Equation (I.2.74) with the right-hand side including the added factor μ.

By comparing Equations (3.63) and (4), we find the expression for the magnetic flux through the contour a,

$$\Phi_a = \frac{1}{c} \sum_b L_{ab} J_b. \tag{3.64′}$$

This flux is due to the field $\boldsymbol{B}_a$ which is produced by the current J_a, flowing in the same contour, and by the fields $\boldsymbol{B}_b$ created by all other contours $b \neq a$. ■

Forces in a magnetic field. A comparison of the above results with Equations (I.2.68) and (I.2.75) shows that when $\mu = 1$ the quantity $\Delta\mathcal{F}$ reduces to the energy W of the system of currents considered in Section I.2.2. Therefore, when the currents in the circuits are constant, the generalized forces can be calculated as

$$F_\alpha = \left(\frac{\partial \Delta\mathcal{F}}{\partial q_\alpha} \right)_{T,J}, \tag{3.65}$$

where q_α is the generalized coordinate. The analog of the "potential function" U, which has been used in Section I.2.2, is the quantity $\Delta\widetilde{\mathcal{F}}$ that can be derived from Equation (3.58) and differs only in sign from $\Delta\mathcal{F}$ (cf. Equations (3.57) and (3.58)).

The Ampere formula (I.2.45) for the force on the element of the quasi-linear current in a magnetic material can be expressed through the magnetic induction:

$$d\boldsymbol{F} = \frac{J}{c}\, d\boldsymbol{l} \times \boldsymbol{B}. \tag{3.66}$$

Similarity between the electric and magnetic phenomena allows one to obtain the expression for the density of the force applied to a liquid magnetic, using Equation (2.61) and the linear coupling $\boldsymbol{B} = \mu \boldsymbol{H}$:

$$\boldsymbol{f} = \frac{1}{c}\boldsymbol{j} \times \boldsymbol{B} - \frac{1}{8\pi} H^2 \,\nabla\mu + \frac{1}{8\pi}\nabla\left(H^2 \frac{\partial\mu}{\partial\tau}\,\tau\right). \tag{3.67}$$

Here, for a non-ferromagnetic electrically conducting fluid, the main contribution is due to the first term containing the conduction current $\boldsymbol{j}$. This force corresponds the *stress tensor* (cf. (2.64))

$$\sigma_{\alpha\beta} = \frac{1}{4\pi} H_\alpha B_\beta - \frac{H^2}{8\pi}\left(\mu - \tau\frac{\partial\mu}{\partial\tau}\right)\delta_{\alpha\beta}. \tag{3.68}$$

Ferromagnetism. In what follows, we will consider the fundamentals of the thermodynamic theory of ferromagnetism along with the simplest microscopic models. The main empirical data are as follows:

1) Spontaneous (in the absence of any external magnetic field) magnetization M_0 of the order of several thousand Gauss at low temperatures.
2) Disappearance of the spontaneous magnetization with increasing temperature above the Curie temperature, T_C, which is a characteristic parameter for any ferromagnetic. When $T > T_C$, the ferromagnetic transforms usually to paramagnetic state.
3) Curie temperatures of pure elements (*Fe, Co*) exceed 10^3 K $\approx$ 0.1 eV; for other elements, they reach hundreds or tens of K.
4) Magnetic susceptibility of ferromagnetics reaches $10^4 - 10^5$; the magnetization to saturation occurs in relatively weak fields, below 100 Oe.
5) Ferromagnetic crystals exhibit the dependence of magnetic quantities on direction (magnetic anisotropy).

There is a striking similarity between ferromagnetics and ferroelectrics (Section 2.3) even though the latter began to be investigated much later than the former. This similarity accounts for the fact that segnetoelectrics are sometimes referred to as **ferroelectrics**.

The onset of spontaneous magnetization at temperatures below the Curie temperature reflects the magnetic ordering. Therefore, it is natural to consider the transition to the ferromagnetic state as second-order phase transition and apply the Landau theory to describe it. In this case, many formulas from the theory of ferroelectricity obtained in Section 2.3 remain valid.

The Gibbs thermodynamic potential for a ferromagnetic in a weak magnetic field H near the Curie point can be written in the form similar to (2.74):

$$\Phi(p, T, M) = \Phi_0(p, T) + \frac{1}{2}\alpha(p, T)M^2 + \frac{1}{4}\beta(p, T)M^4 - MH. \tag{3.69}$$

The temperature dependence of ferromagnetic parameters α and β is described by Equation (2.72). The temperature dependence for the equilibrium magnetization is obtained from Equation (2.73):

$$M_0 = \sqrt{\frac{a}{\beta}(T_C - T)}, \qquad T < T_C. \tag{3.70}$$

The magnetic susceptibility at $T = T_C$ has a singularity and is described in the paramagnetic region ($T > T_C$) by the *Curie-Weiss law*

$$\chi = \frac{C}{T - T_C}, \qquad T > T_C. \tag{3.71}$$

In the ordered phase ($T < T_C$), we have

$$\chi = \frac{C}{2(T - T_C)}, \qquad T < T_C. \tag{3.72}$$

As mentioned in Section 2.3, in the region of strong fluctuations of the order parameter near the Curie point the Landau theory is inapplicable. For example, for iron the applicability region is determined by the inequality $|T - T_C| \gg 0.02\, T_C$. Moreover, the Landau theory does not allow one to determine T_C and the coefficients α, β, and C.

A French physicist P. Weiss took the first step in constructing the microscopic theory of ferromagnetism (1907) and introduced the internal ("molecular") magnetic field

$$\boldsymbol{H}_{mol} = \lambda \boldsymbol{M}, \tag{3.73}$$

which aligned the magnetic moments of the ferromagnetic in parallel one to another. The nature of this field was unknown at that time. The natural assumption, that the intramolecular field was produced due to the interaction between elementary magnetic moments, was inconsistent with experimental data. It was found in experiments that for typical ferromagnetics (e.g., iron) the spontaneous magnetization is $M_0 \approx 10^3$ G and the molecular field constant is $\lambda \approx 10^4$, which corresponds to $H_{mol} \approx 10^7$ Oe. At the same time, the atomic magnetic moment (of the order of the Bohr magneton) at interatomic distances produces a significantly weaker field of the order of $\mu_B/a^3 \approx 10^3 - 10^4$ Oe.

The nature of the Weiss molecular field was explained on the base of quantum mechanics in the works by Ya.I. Frenkel and of W. Heisenberg. Consider the simplest model of the ferromagnetic proposed by Heisenberg. Let any atom be located in its cite and has the magnetic moment whose quantum-mechanical operator is $\hat{\boldsymbol{\mu}} = 2\mu_B \hat{\boldsymbol{S}}$. Here, we take into account that in ferromagnetism, the main contribution is provided by the spin magnetic moments and assume that $\hat{\boldsymbol{S}}$ is the operator of the atomic spin mechanical moment. The atomic spin interacts with the external field $\boldsymbol{H}$ and with spins of the nearest neighbors. The spin–spin interaction has the quantum-mechanical nature and is of the "exchange" character due to the antisymmetry of the total wave function of the electron system (Equation (I.C6)). The exchange interaction is driven by Coulomb electric forces (as opposed to weak magnetic forces), which can provide the needed strength of the effective field.

Assuming all the atoms to be identical, we write the operator of the exchange interaction between the spins in the form $\hat{u}_{ik} = -4I\hat{\boldsymbol{S}}_i \cdot \hat{\boldsymbol{S}}_k$, where $I > 0$ is the exchange interaction constant that has the dimension of energy. The positivity of this constant makes the state with parallel spins energetically favorable.

Using the approximation of self-consistent molecular field, we replace the effect of neighboring spins on the individual spin $\boldsymbol{S}$ by an averaged field (Kubo, 1965; Berdyshev (1992)) which is expressed in terms of the average spin:

$$4I\sum_{i=1}^{z}\hat{\boldsymbol{S}}_i \cdot \hat{\boldsymbol{S}} \to 2zI\overline{\boldsymbol{S}} \cdot \hat{\boldsymbol{S}} = 2\mu_B \boldsymbol{H}_{mol} \cdot \hat{\boldsymbol{S}},$$

z being the number of the nearest neighbors. Since the macroscopic magnetization in this model is expressed in the form $\boldsymbol{M} = 2\mu_B N\overline{\boldsymbol{S}}$, where N is the average number density of spins, then the self-consistent field takes the form of Equation (3.73). The constant of the Weiss molecular field $\lambda = zI/\mu_B^2 N$ is determined by the electron–electron quantum–mechanical interaction.

Thus, any localized spin in a ferromagnetic can be related to the Hamiltonian

$$\hat{V} = -2\mu_B(\boldsymbol{H} + \boldsymbol{H}_{mol}) \cdot \hat{\boldsymbol{S}} = -2\mu_B(\boldsymbol{H} + \lambda\boldsymbol{M}) \cdot \hat{\boldsymbol{S}}. \tag{3.74}$$

In what follows, we will assume that the ferromagnetic is isotropic and the directions of $\boldsymbol{M}$ and $\boldsymbol{H}$ are identical.

Example 3.6

Using the Hamiltonian equation (3.74), derive the coupling equation for the magnetization and the magnetic field in a ferromagnetic. Determine the spontaneous magnetization as a function of temperature (plot the corresponding graphs). Express the Curie temperature in terms of the exchange interaction constant.

Solution. Using the results of Problem 1.17 or performing the averaging from the very beginning, we obtain the coupling equation in the parametric form:

$$(1) \qquad \frac{M}{M_0} = L_S(aS), \qquad \frac{M}{M_0} = \frac{T}{4zIS}a - \frac{H}{\lambda M_0}, \qquad 0 \le a < \infty,$$

where $M_0 = 2\mu_B NS$ is the saturation magnetization and $L_S(x)$ is the Brillouin function defined by Equation (12) in the solution of Problem 1.17. The quantity S is the quantum number that determines the total atomic spin (it is assumed that the orbital moment does not contribute to the ferromagnetism of the system). Possible values are $S = 1/2, 1, 3/2, \ldots$ The lowest value $S = 1/2$ corresponds to the limiting quantum case in which the atom contains a single unpaired electron spin, while the case $S \gg 1$ corresponds to the classical limit.

In order to evaluate the spontaneous magnetization, we assume that $H = 0$ in Equation (1), and find the intersection points of the straight line $y_1(x) = x$ with the line $y_2(x) = L_S(4zIS^2x/T)$ (see Figure 3.5 for $S = 1/2$), with $x = M/M_0$. Except for the root $x = 0$, corresponding to zero magnetization, the values $x = x_0 > 0$ and $x_0 < 1$ with finite magnetization $M = x_0M_0 > 0$ are also possible. A nonzero root takes place only when $dy_2(x)/dx|_{x=0} < 1$ and when the tangent to the curve $y_2(x)$

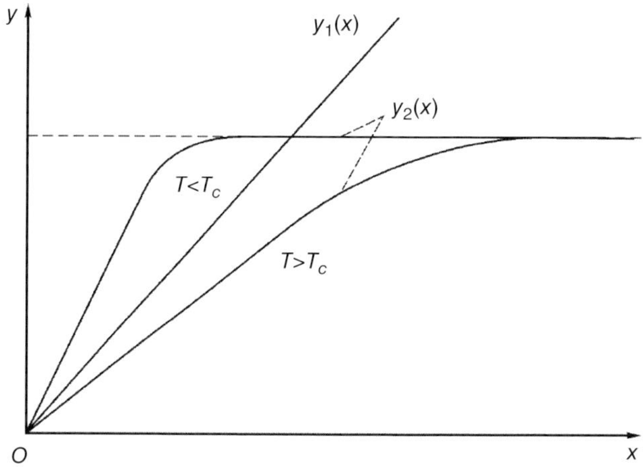

Figure 3.5 Graphic solution of Equation (1).

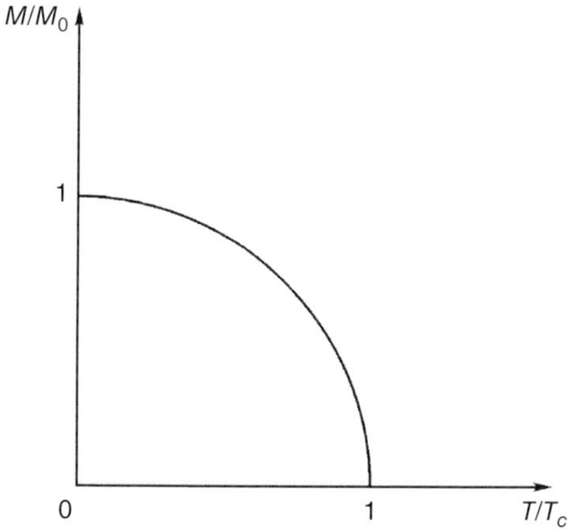

Figure 3.6 Magnetization as a function of temperature.

at $x = 0$ lies above the straight line $y_1(x) = x$. After calculating the derivative, we find the condition for the existence of spontaneous magnetization, from which the Curie temperature of the ferromagnetic is determined as

$$T_C = \frac{4zS(S+1)I}{3} = \frac{4}{3}\mu_B^2 S(S+1)N\lambda. \tag{3.75}$$

The temperature dependence for the magnetization is shown in Figure 3.6. ■

The principal features of ferromagnetics can be described with the *Ising model.*[9] This model directly considers anisotropic ferromagnetic in which the electron spins, located in cites of crystalline lattice, can be aligned with a fixed direction or

9) Ising Gustav Adolf (1883–1960), Swedish physicist, known for his works in the field of electromagnetism and accelerators' technology.

be opposite to it. The spin–spin exchange interaction is

$$V = -I\sum_{i,k} \sigma_i \sigma_k, \tag{3.76}$$

where only the interaction between the nearest neighbors is taken into account, and the quantities σ_i, σ_k take the values ± 1 (which are eigenvalues of the Pauli matrix $\hat{\sigma}_z$). At $I > 0$, the interaction given by Equation (3.76) favors the parallel arrangement of the spins. The Ising model is simpler than the Heisenberg theory and allows one to calculate the thermodynamic potentials as well as thermodynamic parameters of the ferromagnetic, though these operations are difficult to perform accurately even under such simplifications (see, for instance, White (1983); Berdyshev (1992)).

Following Kubo (1957), we will perform approximate calculation of the thermodynamic functions in the Ising model. The spin part of the internal energy of the system should be calculated by averaging the energy (Equation (3.76)) over the equilibrium spin distribution. The exact expression for the average can be written with the aid of Equations (I.C44) and (I.C46),

$$\langle V \rangle = \frac{1}{Z}\sum_{\{\sigma\}} V(\{\sigma\}) \exp\left(-\frac{V(\{\sigma\})}{T}\right),$$

where a summation is over all variables $\{\sigma\}$ of the ferromagnetic. In order to avoid cumbersome calculation of this sophisticated sum, we assume, for the purpose of averaging, that correlations between the spins are absent, and in this sense, spins constitute a perfect gas. Then the averaged energy itself takes into account the exchange spin correlations. In this approximation, we have

$$\langle V \rangle = -I(\overline{N}_{\uparrow\uparrow} + \overline{N}_{\downarrow\downarrow} - \overline{N}_{\uparrow\downarrow}), \tag{3.77}$$

where $\overline{N}_{\uparrow\uparrow}, \overline{N}_{\downarrow\downarrow}$, and $\overline{N}_{\uparrow\downarrow}$ denote the total number of spin pairs with corresponding orientations. Let $N_\uparrow$ and $N_\downarrow$ be the number of spins directed upward and downward, respectively, while $N = N_\uparrow + N_\downarrow$ is the total number of spins. The magnetization of the ferromagnetic depends on the excess of spins of one orientation, that is, on the value $x = (N_\uparrow - N_\downarrow)/N$.

In the absence of correlations, the probability that a spin of specified orientation is present in a given lattice site, is

$$w_\uparrow = \frac{N_\uparrow}{N} = \frac{1}{2}(1+x), \qquad w_\downarrow = \frac{N_\downarrow}{N} = \frac{1}{2}(1-x)$$

regardless of the direction of spins in the neighboring sites. If the number of nearest neighbors for each node is z, the average number of pairs having the upward spin is

$$N_{\uparrow\uparrow} = \frac{1}{2} z N_\uparrow w_\uparrow = \frac{1}{8} z N (1+x)^2.$$

The factor 1/2 in the initial expression takes into account that the spin of the needed direction enters the product $zw_\uparrow \cdot N_\uparrow$ twice, one time as the central spin and the second time as one of the nearest neighbors. Analogously,

$$\overline{N}_{\downarrow\downarrow} = \frac{1}{8} z N (1-x)^2, \qquad \overline{N}_{\uparrow\downarrow} = \frac{1}{4} z N (1-x^2)$$

(there is no double counting in the last expression). As a result, Equation (3.77) gives the part of the internal energy U which is associated with the spin–spin interaction:

$$U(x) = \langle V \rangle = -\frac{1}{2} zINx^2. \tag{3.78}$$

In order to determine all thermodynamic potentials of the system, it is necessary to calculate its entropy with the aid of the Boltzmann formula $S = \ln \Delta\Gamma$, where $\Delta\Gamma$ is the number of microscopic states through which the macroscopic state of the system with specified x and N is realized. In the absence of correlations between spins, all possible spin permutations are equally probable. Hence,

$$\Delta\Gamma = \frac{N!}{N_\uparrow! N_\downarrow!}, \qquad S = N \ln 2 - \frac{1}{2} N[(1+x)\ln(1+x) + (1-x)\ln(1-x)]. \tag{3.79}$$

Here, the Stirling formula $\ln N! \approx N \ln N - N$ is used for all three values of N which are assumed to be large.

Taking into account an external magnetic field H and the obtained values of U and S, the Gibbs potential can be written as

$$\Phi = \Phi_0 + \frac{1}{2} NT[(1+x)\ln(1+x) + (1-x)\ln(1-x)] - \frac{1}{2} zINx^2 - \mu_B NxH. \tag{3.80}$$

Here $-\mu_B H$ is the energy of an individual spin in the external field, Nx is the number of uncompensated spins, and Φ_0 is the contribution from those degrees of freedom, which are not related to the spin–spin interaction. Neglecting the effects of boundaries, Equation (3.80) can be considered the specific thermodynamic potential provided N is the number of spins per unit volume.

Example 3.7

From the condition of minimum potential Φ at fixed p, T, and H calculate the equilibrium magnetization $M = \mu_B N x_{eq}$ of a ferromagnetic in the Ising model and investigate the stability of obtained states. In particular, show that the state with the spontaneous magnetization $M > 0$ at $T < T_C$ and $H = 0$ corresponds to the minimum thermodynamic potential.

Solution. Equating to zero the first derivative of the Gibbs potential (Equation (3.80)) with respect to x, we obtain the equation from which x_{eq} can be determined. Writing $x_{eq} = M/M_0$, where $M_0 = \mu_B N$ is the saturation magnetization, after some transformations, we get

(1) $$\frac{M}{M_0} = \tanh\left(\frac{\mu_B(H + \lambda M)}{T}\right).$$

This is the nonlinear coupling equation between M and H. By comparing this equation with Equation (1) from Example 3.6, we see that at $S = 1/2$ the latter equations become identical with Equation (1) if we recall that $L_{1/2}(x) = \tanh x$.

The stability equation for the equilibrium state is

(2)
$$\left.\frac{d^2\Phi}{dx^2}\right|_{x_{eq}} = NT_C\left[\frac{T}{T_C}\frac{1}{1-x_{eq}} - 1\right] > 0,$$

where $T_C = zI$. The state with $x_{eq} = 0$ is stable at $T > T_C$ and is unstable at $T < T_C$. If $x_{eq} > 0$, we use Equation (1) subject to condition $H = 0$ and obtain

(3)
$$\left.\frac{d^2\Phi}{dx^2}\right|_{x_{eq}} = \frac{NT_C}{y}\left[\sinh y - y\right] > 0.$$

This inequality is satisfied at any $y = 2T_C M/TM_0 > 0$ which is possible only at temperatures below the Curie point. Therefore, the state with spontaneous magnetization is thermodynamically stable. ■

Ferromagnetics of sufficiently large sizes consist of macroscopic regions that are magnetized to saturation (magnetic domains). We have considered the properties of a ferromagnetic within a separate domain.

Antiferromagnetism. Antiferromagnetics are magnetically ordered materials, which in a crystalline state have two or more sublattices with ordered magnetic moments of different directions in different sublattices. However, when added, the magnetizations of the sublattices are mutually compensated and the resultant magnetization is zero. Antiferromagnetics are Cr, Mn, K_2O, MnO, $FeCl_2$, and many other elements and compounds.

We will apply the self-consistent method to study the properties of antiferromagnetism on the base of the simplest model. We assume that an antiferromagnetic consists of two sublattices, a and b, with the equal number of atoms, $N/2$, per unit volume, having identical spins. We will take into account only the interactions with the nearest neighbors, and assume that the neighboring spins tend to take antiparallel directions. In this case, it is possible to use the representation (Equation (3.73)) for the molecular field, acting on the magnetic moments of each sublattice, and carry out the following changes:

1) Replace I by $-I$, assuming $I > 0$ as above;
2) Replace N by $N/2$;
3) The field in one of the sublattices will be determined by the magnetization of the second sublattice.

As a result, we have

$$\boldsymbol{H}^{(a)}_{mol} = \boldsymbol{H} - 2\lambda\boldsymbol{M}^{(b)}, \qquad \boldsymbol{H}^{(b)}_{mol} = \boldsymbol{H} - 2\lambda\boldsymbol{M}^{(a)}, \tag{3.81}$$

where the molecular field constant has the former value $\lambda = zI/\mu_B^2 N > 0$, and an external magnetic field is added. The coupling equations for the magnetizations of two sublattices can be obtained from Equation (1) of Example 3.6 by the formal replacements $I \to -I$ and $N \to N/2$ (with $\lambda \to -2\lambda$):

$$\frac{\boldsymbol{l}_a \cdot \boldsymbol{M}^{(a)}}{M_0} = L_S\left(\frac{2\mu_B S}{T}\left|\boldsymbol{H} - 2\lambda\boldsymbol{M}^{(b)}\right|\right), \quad \frac{\boldsymbol{l}_b \cdot \boldsymbol{M}^{(b)}}{M_0} = L_S\left(\frac{2\mu_B S}{T}\left|\boldsymbol{H} - 2\lambda\boldsymbol{M}^{(a)}\right|\right). \tag{3.82}$$

Here $M_0 = \mu_B SN$ are equal saturation magnetizations of the sublattices; the left-hand sides of the equalities include the projections of the magnetizations onto the corresponding local fields:

$$\boldsymbol{l}_{a,b} = \frac{\boldsymbol{H} - 2\lambda \boldsymbol{M}^{(b,a)}}{|\boldsymbol{H} - 2\lambda \boldsymbol{M}^{(b,a)}|}. \tag{3.83}$$

The analysis of properties of antiferromagnetics, based on Equation (3.82), is given in Problems 3.81–3.83.

Recommended literature: Landau *et al.* (1984); Landau and Lifshitz (1980); Tamm (1976); Smythe (1950); Kubo (1965a,b); Berdyshev (1992); Leontovich (1983); White (1983); Zavadskii and Val'kov (1980); Isyumov and Skryabin (1987); Lifshitz and Pitaevskii (1983)

Problems

3.53 The strength of the current $J(t)$ slowly changes in an infinite solenoid of radius a; the number of turns per unit volume is n. Inside the solenoid there is a magnetic of permeability μ. Calculate the change in the magnetic energy dW/dt per unit length of the solenoid and show that this change is provided by the Poynting vector flux through the lateral surface of the solenoid.

Hint. Calculate the electric field inside the solenoid based on the electromagnetic induction law (5.11).

3.54 Calculate the magnetic induction $\boldsymbol{B}$ and the strength of the magnetic field $\boldsymbol{H}$ on the axis of a cylindrical solenoid with a closely wound coil. The height of the cylinder is h, the radius is a, the number of turns per unit length is n, and the current is J. Inside and outside the solenoid there is a homogeneous medium with magnetic permeability μ.

3.55 Find the self-induction $\mathcal{L}$ per unit length of an infinite cylindrical solenoid with a closely wound coil and arbitrary cross section (not necessarily circular). The cross section area is S and the number of turns per unit length is n. The solenoid is filled with a homogeneous magnet of permeability μ.

3.56 Find the self-inductance L of a toroidal solenoid filled with a magnet with permeability μ. The radius of the torus is b, the number of turns is N, and the cross section is a circle of radius a. Determine the self-induction $\mathcal{L}$ per unit length of the solenoid in the limiting case $b \to \infty$ ($N/b = const$). Solve the same problem for a toroidal solenoid of rectangular cross section (sides a and h). How will the self-induction change if the uniformly distributed current, while retaining its direction, flows not along the wire wound around the torus, but directly along the hollow shell of the torus?

3.57 A line consists of two thin coaxial cylindrical shells of radii a and b ($a < b$). The space between the shells is filled with a material of magnetic permeability μ. Find the self-induction $\mathcal{L}$ per unit length.

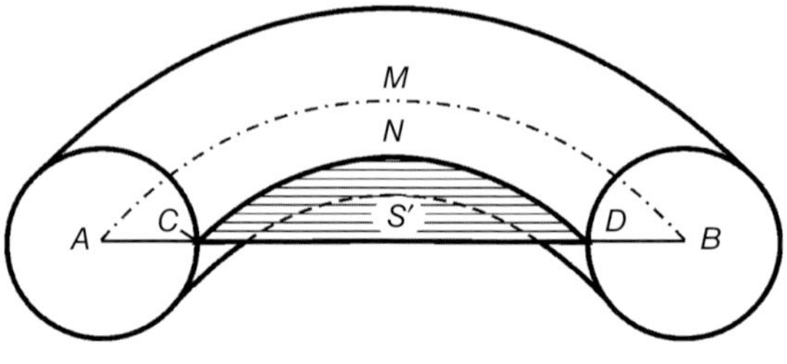

Figure 3.7 To calculation of self-inductance of a thin closed conductor.

3.58 Calculate the self-inductance $\mathcal{L}$ per unit length of the line consisting of two straight wires of radii a and b, placed at distance h from each other. The line is in medium of magnetic permeability μ, the magnetic permeability of the wire material is μ_0. The currents J flowing in the two wires are equal and opposite.

3.59⋆ Show that the self-inductance of a thin closed conductor of circular cross section can be approximately calculated by the formula

$$L = \frac{\mu_0 l}{2} + L',$$ [10]

where μ_0 is the magnetic permeability of the conductor, l is its length, and L' is the mutual inductance of two linear contours. One of these contours coincides with the axial line of the quasi-linear conductor under consideration, the other coincides with the line along which an arbitrary open surface S, enclosed by the axis, cuts the surface of the conductor (Figure 3.7).

3.60 Use the results of Problem 3.59 to calculate the self-inductance L of a thin ring of wire, having radius b. The radius of the wire is $a \ll b$, the magnetic permeabilities of the wire and in surrounding space are different.

3.61 Calculate the self-induction L of a square made of wire, having a side b. The radius of the wire is $a \ll b$, the magnetic permeabilities of the surrounding space and the wire are μ and $\mu_0 = 1$, respectively.

Hint. Use the formulas obtained in solving Problems 3.59 and I.2.114.

3.62 A straight wire carrying the current J is parallel to the axis of an infinitely long circular cylinder, which is at a distance b from the wire. The radius of the cylinder is a ($a < b$), and the magnetic permeability is μ. Find the force f acting per unit length.[11]

Hint. Use the method of images.

3.63 A straight wire carrying the current J is placed inside an infinitely long cylindrical cavity cut in homogeneous magnetic medium. The wire is parallel to the axis of the cylinder at a distance b from it. The cylinder radius is

10) The first and second terms in the expression for L can be referred to as **internal** and **external self-inductions**, respectively, because they define the magnetic energy stored inside and outside the conductor.

11) The results of Problems 3.36 and 3.62 can be used to easily determine the field produced by a charged filament.

a, and the magnetic permeability of the magnetic is μ. Find the force f per unit length of the wire.

3.64 A small permanent magnet having the moment $\boldsymbol{m}$ is placed in vacuum near the plane boundary of a medium with magnetic permeability μ. Determine the force F and the torque N acting on the permanent magnet.

Hint. Use the method of images.

3.65 A contour carrying a current lies at the plane boundary between two media with the magnetic permeabilities μ_1 and μ_2. Express the self-inductance L of the contour through its value L_0 for the identical contour in a vacuum.

3.66 Calculate the change in the internal energy ΔU of a magnetic induced by the isothermal magnetization (the magnetic induction varies from 0 to B). Compare the result with the quantity $U - U_0$ found from Equation (3.54) for the adiabatic process. Neglect changes in the volume and mass density.

3.67* Using the Nernst[12] principle in Planck's formulation and also Equation (3.52) for $d\widetilde{F}$, show that the magnetic susceptibility of any magnetic must satisfy the condition $(\partial\chi/\partial T)_\tau \to 0$ at $T \to 0$. According to the Nernst principle, $S(T, a_i)|_{T\to 0} \to 0$, that is, as the absolute temperature tends to zero, the entropy of the body at equilibrium tends to zero at any fixed values of external parameters (pressure, external fields, phase state). Verify whether (or not) the paramagnetic susceptibilities of the degenerate electron gas (Problem 1.22) and of the rarefied atomic gas (Problem 1.17) satisfy this condition.

3.68$^\bullet$ A crystalline paramagnetic, whose magnetic susceptibility in a certain temperature range obeys the Curie law $\chi(T) = C/T,\ C = const$, is thermally insulated and is in a magnetic field B at temperature T. How will its temperature change after the field is switched off?

3.69 The paramagnetic that obeys the Curie law (Problem 3.68), undergoes the isothermal and reversible magnetization at temperature T in a magnetic field increasing from 0 to H. Afterwards the paramagnetic is isothermally demagnetized to zero. Calculate the heat Q dissipated in the body in these processes. Does the sign of Q agree with the result of Problem 3.68 on adiabatic demagnetization?

3.70* "Ideal paramagnetics" are the materials whose magnetization M depends on the quantities H and T via their ratio only, that is, $M = f(H/T)$ (we consider an isotropic medium).[13] Assuming that the density of the material is constant, show that the internal energy U_* of a perfect paramagnetic, determined in accordance with Equation (3.65) and considered a function of the temperature and magnetization, is actually independent of M. Express the entropy S and the energy U_* in terms of the heat capacity C_{V0} (which depends on temperature as $C_{V0} = bT^3,\ b = const$), and in terms of the function $f(H/T)$.

12) Nernst Walter (1864–1941), German physicist and chemist, Nobel Prize laureate; known for his works in thermodynamics, low temperature physics, physical chemistry. Nernst principle is the third fundamental law of thermodynamics.

13) Such a dependence is possible only within a limited range of parameters.

3.71 A magnetic is subject to polarization produced by an external uniform magnetic field H in air. Assuming that the field is also uniform inside the body and that the magnetization is isothermal and isobaric, calculate a change in the body volume $\Delta V \ll V$ expressing it through the field H, the magnetic permeability χ, and the isothermal compressibility $\beta = -(1/V)(\partial V/\partial p)_T$, where p stands for pressure.

Hint. This problem can be solved similarly to more complicated Problem 2.80.

3.72 For an isotropic magnetic material, express variations of the quantities U_* and F_*, induced by magnetization, in terms of the magnetization M and the magnetic susceptibility $\chi(T, \tau)$. These quantities are defined by Equation (3.65). Find a similar variation of the entropy S, assuming the mass density τ to be constant.

3.73 Calculate variation $C_{VB} - C_{V0}$ in the heat capacity of a magnetic whose volume and magnetic induction are constant when the magnetic field is induced. Compare this difference with the difference $C_{VH} - C_{V0}$ using the coupling equation $B = \mu(T, \tau)H$.

3.74 Calculate the difference between the specific heat capacities $C_H - C_B$ of a magnetic, without assuming that the dependence between B and H is linear. In this case, either the volume V or the pressure p is constant.

3.75 Prove that the specific heat capacities of a magnetic can be written as

$$C_B = \left(\frac{\partial U}{\partial T}\right)_B, \quad C_H = \left(\frac{\partial U}{\partial T}\right)_H - H\left(\frac{\partial M}{\partial T}\right)_H,$$

where M is the specific dipole moment of the magnetic, and the subscripts indicate which parameter is constant.

3.76 In a nonlinear case, the magnetic susceptibility $\chi = \partial M/\partial H$ depends on the conditions of the polarization process, that is, whether the magnetic is thermally insulated or in contact with a thermostat, which keeps it at constant temperature. Find the relation between the adiabatic and isothermal susceptibilities:

$$\chi_S = \frac{C_M}{C_H}\chi_T,$$

where the heat capacities C_M and C_H correspond to the constant magnetization and constant magnetic field, respectively.

3.77 Based on Equations (3.68)–(3.71), calculate the discontinuity of the heat capacity $\Delta C_p = C_p|_{T=T_C+0} - C_p|_{T=T_C-0}$ for a ferromagnetic at the Curie point.

3.78 The magnetic susceptibility of a paramagnetic is given by the Curie-Weiss law, $\chi(T) = C/(T - T_C)$; the coupling equation is linear, $M = \chi H$. The heat capacity C_M is also known. Calculate the specific quantities U_* and S as functions of temperature and magnetization M (assuming that the material density is constant). Compare the results with those of Problem 3.70 for an "ideal paramagnetic."

3.79 Express analytically the temperature dependence of the magnetization in the model of the self-consistent molecular field (Example 3.6) in the temperature ranges $T_C - T \ll T_C$ and $T \ll T_C$.

3.80 Calculate the magnetic susceptibility of ferromagnetic near the Curie point in the paramagnetic and ferromagnetic phases, using the model of self-consistent molecular field (Example 3.6). Also find the magnetization at the Curie point in the presence of a weak external field.

3.81 Find the temperature range for which the spontaneous magnetization of sublattices is not zero. Use Equation (3.81).

3.82 Calculate the magnetic susceptibility of an antiferromagnetic in the paramagnetic phase in which the spontaneous magnetization of sublattices is absent. Assume the medium to be isotropic in the absence of an external field.

3.83⋆ Calculate the magnetic susceptibility of an antiferromagnetic at temperatures below the transition point to the antiferromagnetic state. Take into account that the resultant magnetization is absent at all temperatures; nevertheless, there is a particular direction in the medium, which is determined by the opposite directions of magnetizations of sublattices; the magnetic susceptibilities may be different depending on direction.

3.4
Electric and Magnetic Properties of Superconductors

Basic experimental facts. The first superconductor was discovered by a Dutch physicist H. Kamerlingh-Onnes[14] in 1911. He observed that mercury cooled to the temperature $T_c = 4.15$ K passed into the state in which its electric conductivity increased by more than 10 orders of magnitude compared with that of the best conductors, such as *Cu* and *Ag*, that is, its resistance decreased practically to zero. Many other metals, besides mercury, and their compounds are known to pass into superconducting state. The compound Nb_3Ge was for a long time the most high-temperature superconductor, $T_c \approx 23$ K. However, the high-temperature superconductors, such as $Bi_4(SrCa)_6CuO_{16}$, $T_c \approx 105$ K, and $Tl_2BaCa_2Cu_3O_{10}$, $T_c \approx 125$ K, were synthesized in Europe and America in the second half of the 1980s. A search for high-temperature superconductors is still in progress and they find increasingly wider applications.

14) It is of interest that in 1913 when awarding the Noble Prize to H. Kamerlingh-Onnes "for his investigations on the properties of matter at low temperatures which led, inter alia, to the production of liquid helium," the Nobel Committee avoided mentioning the discovery of superconductivity, as such, in its decision. The history of studying of superconductivity and modern problems in this area are discussed in articles of the Nobel winner of V.L.Ginzburg (2004, 2005).

1) Thus, the main property of superconductors is a severe decrease of the electrical resistance to zero in a narrow temperature range of the order of fractions of Kelvin degree when a superconductor is cooled below the critical temperature.
2) A sufficiently weak magnetic field is always pushed out of the superconductor regardless of its prehistory, that is, whether it was switched on prior to or after the transition of the conductor into superconducting state *(Meissner[15]–Ochsenfeld effect)*. This effect can be understood assuming that an external magnetic field induces an electric current in a thin surface layer of the superconductor. The secondary magnetic field produced by this current compensates the external field. In this respect, superconductors are essentially different from ideal conductors (i.e., conductors exhibiting high conductivity). The magnetic field possesses the property of "freezing" into a well-conducting medium (see Section 4.3 for details). Therefore, the field, initially existing in a perfect conductor, will still exist there once the external field has been terminated.

The density of the surface current in a superconductor can be determined with the aid of the second Equation (3.37):

$$\boldsymbol{i} = \frac{c}{4\pi}\boldsymbol{n} \times \boldsymbol{H}. \tag{3.84}$$

Here $\boldsymbol{n}$ is the outer normal and $\boldsymbol{H}$ is the magnetic field strength outside the superconductor (which is assumed to be vacuum). The field inside the superconductor is zero, hence, the external field, where the superconductor is placed, rearranges so that at the surface of the superconductor

$$\boldsymbol{n} \cdot \boldsymbol{H} \equiv H_n = 0. \tag{3.85}$$

The tangential projection of the field experiences a jump at this surface in accordance with Equation (3.83).

Example 3.8

The external magnetic field H is parallel to a lateral surface of a Superconductor, which is a long circular cylinder of radius a. Calculate the magnetic moment $\boldsymbol{M}$ *per unit volume of the superconductor, which is produced by the surface current (Equation (3.83)), and also find its polarizability in the external field. Compare the polarizability of the superconductor with that of many standard diamagnetics* ($\chi \sim 10^{-6}$)

Solution. From Equation (I.2.59) and the expression for current (Equation (3.83)), we find the magnetic moment per unit length of the cylinder and obtain thereby $\boldsymbol{M} = -(1/4\pi)\boldsymbol{H}$. This result suggests that in a magnetic field, the superconductor behaves as a perfect diamagnetic with the magnetic

15) Meissner Walter (1882–1974), German physicist, his main works are in the low temperature physics.

susceptibility $\chi = -1/4\pi$ and magnetic permeability $\mu = 0$. The field inside such a diamagnetic decreases to zero. Its magnetic susceptibility exceeds that of ordinary metals by five orders of magnitude. ■

3) At any temperature, a sufficiently strong magnetic field disrupts superconductivity and penetrates into the superconductor. There superconductors of first and second types depending on their behavior in an external magnetic field. Type I superconductors are the majority of pure metals, whereas type II superconductors are presented by alloys and chemical compounds including high-temperature ceramic superconductors. The dependence of magnetization on the external magnetic field for cylinders oriented along the field are schematically shown in Figure 3.8 for type I superconductor and in Figure 3.9 for type II superconductor. The starting straight segment of both curves corresponds to the Meissner–Ochsenfeld effect when the magnetic field in the bulk of the superconductor is absent.

For type I superconductors, as the magnetic field increases, at a certain value of $H_c(T)$ the field steeply penetrates into the bulk of the superconductor, turning it

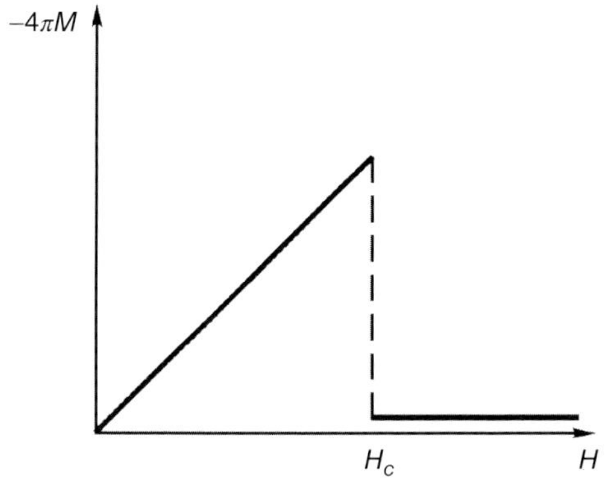

Figure 3.8 Schematic dependence of magnetization on a magnetic field for type I superconductors.

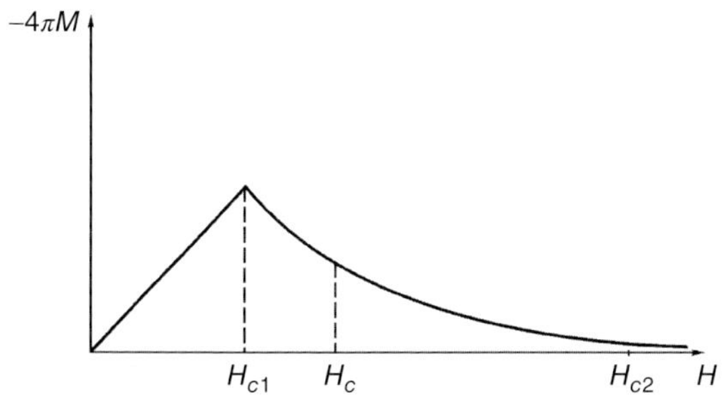

Figure 3.9 Schematic dependence of magnetization on a magnetic field for type II superconductors.

to ordinary metal. The magnetization decreases by 5–6 orders of magnitude. The temperature dependence of the critical field destroying superconductivity is fairly well described by the empirical formula

$$H_c(T) = H_c(0)[1 - (T/T_c)^2]. \tag{3.86}$$

The value of H_c for pure metals amounts to hundreds of oersteds (e.g., 400 Oe for *Hg*, and 800 Oe for *Pb*).

The magnetic moment of type II superconductors decreases gradually in the range between H_{c1} and H_{c2}; the second critical field in which the superconductor as a whole passes into the normal state, reaches $\sim 10^5$ Oe. The magnetization curve is irreversible and depends on the prehistory of the process (hysteresis). Between the values H_{c1} and H_{c2} superconductor is in the mixed state in which the superconducting and normal regions alternate. When the field increases above H_{c1}, the normal regions nucleate in the form of thin filaments penetrating the superconductor. In a certain range of values of the external field, a type I superconductor, which has a complicated form and nonuniform internal field, can be in a similar state. The regions in which $H < H_c$ retain superconductivity, whereas those with $H \geq H_c$ pass into the normal state. Such state is termed **intermediate**.

4) If a persistent current is generated in a superconducting ring, the plane of the ring is penetrated by a certain magnetic flux Φ depending on the current intensity. Refined experiments demonstrate that *the magnetic flux is quantized*:

$$\Phi = n\Phi_0, \qquad \Phi_0 = \frac{\pi\hbar c}{e_0} \approx 2 \times 10^{-7}\ \mathrm{G\,cm}^2, \qquad n = 0, 1, 2, \ldots \tag{3.87}$$

where e_0 is the elementary charge. A magnetic flux quantum Φ_0 is a readily detectable quantity. In a capillar of diameter 10^{-3} cm, the magnetic induction of the order of 0.1 G corresponds to one quantum of the flux. Quantization of the magnetic flux produced by a superconducting current reflects the quantum nature of superconductivity.

5) **Isotopic effect** in superconductivity was discovered in studies of superconducting properties of various mercury isotopes. Changing isotopic mass number M, the critical temperature of superconductivity also changed as

$$T_c M^{1/2} = const. \tag{3.88}$$

The isotopic effect points to the association between superconductivity and oscillations of the crystal lattice because the ion oscillation frequency in the lattice is inversely proportional to $M^{1/2}$.

Thermodynamics of superconductors. The basic thermodynamic properties of a matter related to its superconductivity can be described provided the dependence $T_c(T)$ is known. To do this, we use the results of Example (3.8). If a superconductor undergoes a magnetization in the field that does not penetrate inside it, the elementary work per unit volume is $\delta A = -\boldsymbol{M} \cdot d\boldsymbol{H} = \boldsymbol{H} \cdot d\boldsymbol{H}/4\pi$. In the case of isothermal variation of the field from 0 to H, the specific free (Helmholtz) energy can be written as (cf. Equation (3.61))

$$F_s(T, H) = F_s(T, 0) + \frac{H^2}{8\pi}. \tag{3.89}$$

Here, H denotes the external field in which the superconductor is placed; a change in volume and the corresponding work are neglected. Let the field tend to the critical one, that is, $H \to H_c(T)$. The superconductor passes to the normal state, and its free energy minus the field energy in the body, becomes $F_n(T, H) = F_n(T) - \boldsymbol{M} \cdot \boldsymbol{H}/2 \approx F_n(T)$, because the magnetic moment of non-ferromagnetic matter is negligibly small compared with that of the superconductor. Hence, the relation between the free energies of the superconducting and non-superconducting states at a given temperature $T < T_c$ is

$$F_n(T) = F_s(T, 0) + \frac{H_c^2(T)}{8\pi}. \tag{3.90}$$

According to this relation, $F_n(T) > F_s(T, 0)$, and at $T < T_c$ the state with a lower free energy, that is, the superconducting state, is realized. Note that the quantity $H_c(T)$ can be accurately measured only for type I superconductors. For type II superconductors, this quantity ("the thermodynamic critical field") lies between the values H_{c1} and H_{c2} (Figure 3.9); it cannot be measured directly but might be calculated. Thermodynamic properties of type I superconductors are considered in Problems (3.84)–(3.86).

Problems

3.84 Use Equations (3.89) and (3.86) to calculate the difference of entropies for the normal and superconducting states in the temperature range $0 \le T \le T_0$ in the presence of the external magnetic field $0 \le H \le H_c(T)$. Based on these results, point out the classification of phase transitions (of first or second order) from superconducting to normal state.

3.85 Calculate the latent heat of the transition from superconducting to the normal state caused by the disappearance of superconductivity at $H > H_c$. Compare this effect with the adiabatic and isothermal demagnetization of paramagnetics (Problems 3.68 and 3.69.)

3.86 Calculate the jump of the heat capacity $\Delta C = C_s - C_n$ of matter at the phase transition, $T = T_c$, from the normal to the superconducting state. Find the value ΔC in calories per centimeter per Kelvin for lead at $T_c \approx 7.2$ K and $H_c(0) \approx 803$ Oe.

Phenomenological magnetostatics of superconductors. The simplest model of superconductors assumes that the free electrons in matter (with number density n) can be divided into two groups, normal and superconducting, that is, $n = n_n + n_e$. The relationship between these groups depends on the temperature: $n \ge n_s \ge 0$ when $0 \le T \le T_c$. In the static case, the electric current is produced solely by superconducting electrons because they propagate through the superconductor without friction. We will take into account that the magnetic field can penetrate into a thin surface layer, and we will consider the current density and the magnetic induction inside the superconductor as functions of coordinates.

Let us write down the density of the superconducting current $\boldsymbol{j}_s(\boldsymbol{r}) = en_s(\boldsymbol{r})\boldsymbol{v}_s(\boldsymbol{r})$ and the kinetic energy density of the superconducting electrons

$$w = \frac{n_s m v_s^2}{2} = \frac{m j_s^2}{2e^2 n_s}.$$

With the aid of the microscopic Maxwell equation (2.9), $rot\,\boldsymbol{B} = 4\pi\boldsymbol{j}_s/c$, where the time derivative and $\boldsymbol{j}_{ext}$ are zero, and the current $\bar{\boldsymbol{j}}_{int}$ is denoted by $\boldsymbol{j}_s$, we express the kinetic energy density in terms of the magnetic field:

$$w = \frac{\lambda^2}{8\pi}(rot\,\boldsymbol{B})^2,$$

where the parameter

$$\lambda = \sqrt{\frac{mc^2}{4\pi e^2 n_s}} \tag{3.91}$$

has the dimension of length.

Example 3.9

Derive the equation describing the magnetic field distribution inside a superconductor. To do this, write the Helmholtz free energy as a functional (integral) of the magnetic induction and use the minimum condition for the free energy in the equilibrium state.

Solution. We write the free energy (including the field energy in the bulk of the body) in the form of the integral over the volume of the superconductor:

(1) $$\mathcal{F}_s[T, \boldsymbol{B}(\boldsymbol{r})] = \mathcal{F}_s(T) + \frac{1}{8\pi}\int_V [B^2 + \lambda^2(rot\,\boldsymbol{B})^2]\,dV.$$

Let the body volume and temperature be constant with the varying magnetic field. At the next step let us use the necessary minimum condition for the functional $\mathcal{F}_s[\boldsymbol{B}(\boldsymbol{r})]$, that is, equate to zero its first variation in response to field variations, as was done in Section I.4.3. In addition, we assume that

(2) $$\delta\boldsymbol{B} = 0$$

at the boundary of the superconductor where the external field is given. We get

(3) $$\delta\boldsymbol{F}_s[\boldsymbol{B}(\boldsymbol{r})] = \frac{1}{4\pi}\int_V (\boldsymbol{B}\cdot\delta\boldsymbol{B} + \lambda^2 rot\,\boldsymbol{B}\cdot rot\,\delta\boldsymbol{B})\,dV = 0.$$

Using the identity

$$\lambda^2\, rot\,\boldsymbol{B}\cdot rot\,\delta\boldsymbol{B} = \delta\boldsymbol{B}\cdot rot\,[\lambda^2 rot\,\boldsymbol{B}] - div[\lambda^2 rot\,\boldsymbol{B}\times\delta\boldsymbol{B}]$$

and the Gauss – Ostrogradskii theorem along with the boundary condition (2), we reduce c (3) to the form

(4) $$\int_V (\boldsymbol{B} + rot\,\lambda^2 rot\,\boldsymbol{B}) \cdot \delta\boldsymbol{B}\,dV = 0,$$

and obtain the required equation

$$rot\,(\lambda^2 rot\,\boldsymbol{B}) + \boldsymbol{B} = 0. \quad (3.92)$$

With the aid of the Maxwell equation, the last relation can be given the form of the coupling equation between the superconducting current and the magnetic field:

$$\frac{4\pi}{c} rot(\lambda^2 \boldsymbol{j}_s) + \boldsymbol{B} = 0. \quad (3.93)$$

The parameter λ describes the penetration depth of the magnetic field into the superconductor. Since the number of superconducting electrons depends on temperature, λ is also a function of temperature. For pure metals, $\lambda \approx 10^{-6} - 10^{-5}$ cm. ■

The phenomenological theory presented above was suggested by German physicists, brothers F. and H. Londons in 1935. Equations (3.92) to (3.93) are named after them. The London theory is used in Problems 3.87 – 3.97.

Problems

3.87 Write down the Maxwell equations and the material equation for a static electromagnetic field in a superconductor. Derive the equation which describes the current and magnetic field distributions.

3.88 A superconductor fills the half-space $x \geq 0$, while the second half, $x < 0$, is vacuum. In vacuum, there is a uniform magnetic field $\boldsymbol{H}_0 \parallel Oy$. Find the magnetic field and current distributions in the superconductor for a static case.

3.89 Find the force per unit area of the surface of the superconductor considered in Problem 3.88. What is the direction of that force?

3.90★ A superconducting film of thickness $2a$ is placed symmetrically with respect to the $x = 0$ plane in a uniform magnetic field $\boldsymbol{H}_0 \parallel Oz$. Find the magnetic field distribution over the volume of the film and also the average magnetic moment per unit volume.

3.91 A current flows in the direction of the Oz-axis in the superconducting film placed in vacuum (Problem 3.90). The current per unit length of the cross section of the film in the direction of the Oy-axis is i. Find the current distribution over the cross-sectional area and the magnetic field inside and outside the film. Consider, in particular, the limiting cases of $\lambda \ll a$ and $\lambda \gg a$.

3.92★ Originally in the region $x > 0$ of free space (vacuum) there is a system of conductors carrying the currents with the density $\boldsymbol{j}(x, y, z)$, which produce the field $\boldsymbol{H}(x, y, z)$ in the entire space. Then, the half-space $x \leq 0$ is filled with superconductor. Use the method of images to calculate the resulting field $\boldsymbol{H}'(x, y, z)$ in the region $x > 0$.

Hint. Analyze the equations for the vector potential and the boundary condition at the surface of the superconductor.

3.93 An infinitely long circular superconducting cylinder is placed in a uniform external magnetic field $\boldsymbol{H}_0 \parallel Oz$. The axis of the cylinder is parallel to the field. Find the magnetic field distribution in the bulk of the cylinder and also the average magnetic moment per unit volume.

3.94 A superconducting sphere of radius a is placed in a uniform external magnetic field H_0. Find the current distribution in the sphere and the magnetic field in the whole of space. Consider the limiting cases $a \gg \lambda$ and $a \ll \lambda$.

3.95 Current J flows inside an infinitely long straight superconducting conductor of circular cross section (radius a). Find the distribution of the current density $\boldsymbol{j}$ over the cross section of the conductor and the magnetic field throughout the space.

3.96 A superconducting plane coil of self-induction L, through which a current J flows, is inserted completely in a uniform magnetic field $\boldsymbol{H}_0$. Find the current J', which will flow through the coil afterwards. The area of the axial cross section of the coil is S. The normal to the plane of the coil makes an angle ϑ with the direction of $\boldsymbol{H}_0$.

3.97 A conducting coil of self-induction L is in the normal state in an external magnetic field (the magnetic flux through the contour of the coil equals Φ_0). Then the temperature is lowered and the coil transfers to the superconducting state. Which current will flow through the coil if the external magnetic field is switched off?

3.98 A current J flows through a straight superconducting conductor of radius a which is placed in the external longitudinal magnetic field H_0. What is the critical current J_c at which the conductor loses its superconducting properties?

3.99 An external magnetic field H_0 is perpendicular to a straight superconducting wire. Find the critical value of the current J_c at which a non-superconducting region appears in the wire.

Semi-phenomenological Ginzburg–Landau quantum theory. In the absence of magnetic field, the transition to the superconducting state is the second-order phase transition (Problem 3.84). Hence, when the temperature is below the transition temperature, the order parameter must appear in the system. Ginzburg and Landau (1950), assuming that superconductivity is a quantum phenomenon, introduced the complex order parameter Ψ which can be treated the "effective wave function of superconducting electrons." In a homogeneous and isotropic conductor without any magnetic field, the order parameter is independent of coordinates. The Gibbs thermodynamic potential near the phase transition temperature T_c can be written in the form of the expansion (2.68) in powers of $|\Psi|^2$, with varying numerical coefficients (for convenience of further writings):

$$\Phi(p, T, \Psi) = \Phi_0(p, T) + \alpha(p, T)|\Psi|^2 + \frac{1}{2}\beta(p, T)|\Psi|^4. \tag{3.94}$$

Using the minimum condition for the thermodynamic potential at equilibrium, we find near the Curie temperature (Equations (2.71) and (2.72))

$$\alpha(p, T) \approx a(p)(T - T_c),\ a > 0; \qquad \beta(p, T) \approx \beta(p, T_c) = const > 0, \tag{3.95}$$

and the values of the equilibrium order parameter are

$$|\Psi_0|^2 = -\frac{\alpha}{\beta} = \frac{a}{\beta}(T_c - T), \quad T < T_c; \qquad |\Psi_0|^2 = 0, \quad T > T_c. \tag{3.96}$$

When $T > T_c$ the order parameter is $|\Psi_0|^2 = 0$. Therefore, $\Phi_0(p, T)$ in the expansion (3.94) represents the thermodynamic potential of the normal state. Substituting Equations (3.95) and (3.96) into Equation (3.94), we find the equilibrium thermodynamic potential of superconducting state:

$$\Phi_s(p, T) = \Phi_n(p, T) - \frac{a^2}{2\beta}(T_c - T)^2. \tag{3.97}$$

It is easy to express the critical field $H_c(0)$, which destroys superconductivity in terms of the thermodynamic parameters a, β and T_c. Neglecting changes in the volume, Equations (3.97) and (3.90) give

$$\Phi_n(p, T) - \Phi_s(p, T) = F_n(T) - F_s(T) = \frac{H_c^2(T)}{8\pi} = \frac{a^2}{2\beta}(T_c - T)^2. \tag{3.98}$$

In the vicinity of the Curie point, from Equation (3.86) we find $H_c(T) \approx 2H_c(0)(1 - T/T_c)$, which leads to

$$H_c(0) = \frac{\pi a^2}{\beta} T_c^2. \tag{3.99}$$

If the superconductor is slightly inhomogeneous, the order parameter Ψ depends on coordinates, and the expansion of the thermodynamic potential contains the term that is proportional to $|\nabla\Psi|^2$:

$$\Phi_s = \Phi_n + \alpha(p, T)|\Psi|^2 + \frac{1}{2}\beta(p, T)|\Psi|^4 + g|\nabla\Psi|^2. \tag{3.100}$$

Here, the higher-order terms in gradient are dropped. The condition $g > 0$ must be imposed on the constant factor g. Otherwise, the homogeneous state of the conductor does not correspond to the minimum of Φ_s, and hence is nonequilibrium.

We now introduce the convenient normalization of the wave function and the constant g. For this purpose, we use the data from the microscopic theory of superconductivity, which were unknown at the time when the Ginzburg–Landau theory was developed. Later on American physicists J. Bardeen, L. Cooper, and

J. Schrieffer (BCS)[16] and also a Soviet scientist N.N. Bogolyubov[17] have shown that the carriers of electric current in superconductors are paired electrons, that is, the Cooper pairs. The coupling between electrons in Cooper pairs results from the phonon exchange, that is, exchange of quanta of oscillations in the crystal lattice. The charge of a Cooper pair is equal to the double electron charge, and its mass to the double electron mass.

It is natural to interpret Ψ as the macroscopic wave function describing the collective (coherent) motion of Cooper pairs. Then, Ψ can be conveniently normalized by the condition

$$|\Psi(\boldsymbol{r})|^2 = \frac{n_s}{2}, \tag{3.101}$$

where $n_s/2$ is the number of Cooper pairs per unit volume. If we choose $g = \hbar^2/2m_*$, where $m_* = 2m$ is the mass of one Cooper pair, the last term in the right-hand side of Equation (3.99) takes the form of the term responsible for the kinetic energy of a particle in the non-relativistic quantum Lagrangian (Equation (I.4.100)):

$$\frac{\hbar^2}{2m_*}|\nabla\Psi|^2 = \frac{1}{2m_*}|-i\hbar\nabla\Psi|^2 = \frac{1}{2m_*}|\hat{\boldsymbol{p}}\Psi|^2. \tag{3.102}$$

In the semi-classical approximation, one can relate this term to the quantity

$$\frac{p^2}{2m_*}\cdot\frac{n_s}{2}$$

which is the kinetic energy density of Cooper pairs.

An introduction of the traditional quantum-mechanical term into the thermodynamic potential density (Equation (3.100)) allows one to take into account the interaction between the superconductor and the magnetic field using the ordinary rules of quantum mechanics. We should replace the operator:

$$\hat{\boldsymbol{p}} = -i\hbar\nabla \qquad \text{by} \qquad \hat{\boldsymbol{p}} - \frac{q}{c}\boldsymbol{A}(\boldsymbol{r}), \tag{3.103}$$

where $\boldsymbol{A}(\boldsymbol{r})$ is the vector potential, and $q = 2e$ is the charge of one Cooper pair. With this replacement, the theory retains its gauge-invariant character, and the arbitrary choice of the vector potential does not affect observables (see, for instance, Problems I.4.124 and I.4.131).

16) Bardeen John (1908–1991), outstanding American physicist, was awarded by two Nobel Prices in physics, one of them with Cooper and Schrieffer;
Cooper Leon (1930), American physicist-theorist;
Schrieffer John (1931), American physicist-theorist.

17) Nikolai Nikolaevich Bogolyubov (1909–1992), outstanding Russian physicist-theorist and mathematician, known for the most important results in the theory of superfluidity and superconductivity, classical and quantum statistics, theory of nonlinear oscillations and elementary particles, quantum field theory; founder of large Soviet school of theoretical physics.

Let us write the total thermodynamic potential including the magnetic field energy:

$$\Phi_{compl}$$
$$= \Phi_0 + \int_V \left\{ \alpha|\Psi|^2 + \frac{1}{2}\beta(p, T)|\Psi|^4 + \frac{\hbar^2}{4m}\left|\nabla\Psi + \frac{2e}{i\hbar c}\boldsymbol{A}(\boldsymbol{r})\Psi\right|^2 + \frac{(\nabla \times \boldsymbol{A})^2}{8\pi} \right\} dV, \tag{3.104}$$

where Φ_0 is independent of Ψ and $\boldsymbol{A}$; the integration is over the volume of the superconductor. The system of equations for determining Ψ and $\boldsymbol{A}$ in a nonuniform superconductor can be obtained from the minimum condition of the functional (3.104) as was done in Section I.4.3.

Example 3.10

Using the fact that the first variation of the thermodynamic potential $\Phi_{compl}[\Psi, \Psi^*, \boldsymbol{A}]$ is zero at a fixed value of $\boldsymbol{A}(\boldsymbol{r})$, and considering $\delta\Psi^*(\boldsymbol{r})$ as an independent small variation of the complex order parameter, obtain the nonlinear Schrödinger equation for Ψ inside the superconductor

$$\frac{\hbar^2}{4m}\left(\nabla - \frac{2ie}{\hbar c}\mathbf{A}\right)^2 \Psi + \alpha\Psi + \beta|\Psi|^2\Psi = 0, \tag{3.105}$$

and the boundary condition[18] *on its surface S:*

$$\boldsymbol{n}\left(i\hbar\nabla\Psi + \frac{2e}{c}\boldsymbol{A}\Psi\right) = 0 \quad on \quad S. \tag{3.106}$$

Solution. Considering $\delta\Psi^*(\boldsymbol{r})$ and $\delta\Psi(\boldsymbol{r})$ as independent variations and varying $\Psi^*(\boldsymbol{r})$, from Equation (3.104) with the aid of the Gauss–Ostrogradskii theorem we have

$$\delta\Phi_{compl} = \int_V \delta\Psi^* \left\{ \alpha\Psi + \beta|\Psi|^2\Psi + \frac{\hbar^2}{4m}\left(\nabla - \frac{2ie}{\hbar c}\mathbf{A}\right)^2 \Psi \right\} dV$$
$$+ \oint_S \delta\Psi^* \left(i\hbar\nabla\Psi + \frac{2e}{c}\boldsymbol{A}\Psi\right) \cdot \boldsymbol{n}\, dS = 0,$$

whereas with the arbitrary function $\delta\Psi^*(\boldsymbol{r})$ we get Equations (3.105) and (3.106). ■

Example 3.11

Using Equation (3.104) obtain the equation for the magnetic field in a superconductor and the expression for the current by varying the vector potential $\boldsymbol{A}$ in the bulk, if its value on the surface is known and the order parameter $\Psi(\boldsymbol{r})$ is fixed.

18) See the paper by Andryushin *et al. (1993) for a more general boundary condition and examples of its application.*

Solution. Variation of Equation (3.104) in $\boldsymbol{A}$ followed by transformation of the integrand leads to the condition

$$\delta\Phi_{compl} = \int_V \delta\boldsymbol{A}(\boldsymbol{r}) \cdot \left\{ \frac{ie\hbar}{2mc}(\Psi^*\nabla\Psi - \Psi\nabla\Psi^*) + \frac{2e^2}{mc^2}\boldsymbol{A}|\Psi|^2 + \frac{1}{4\pi}\nabla\times[\nabla\times\boldsymbol{A}] \right\} dV = 0.$$

The surface integral vanishes because the vector potential is given on the surface of the superconductor and $\delta\boldsymbol{A}(\boldsymbol{r})|_S = 0$. Using the expression $\boldsymbol{B} = \nabla\times\boldsymbol{A}$, we obtain the Maxwell equation

$$rot\,\boldsymbol{B} = \frac{4\pi}{c}\boldsymbol{j}_s, \tag{3.107}$$

with the superconducting current

$$\boldsymbol{j}_s = -\frac{ie\hbar}{2m}(\Psi^*\nabla\Psi - \Psi\nabla\Psi^*) - \frac{2e^2}{mc}\boldsymbol{A}|\Psi|^2. \tag{3.108}$$

The electric current (Equation (3.108)) has the usual form known from quantum mechanics (cf. the answer to Problem I.4.130), which is the probability current for a particle with charge e and mass m. ■

Example 3.12

Write down the Ginzburg–Landau equations in terms of the dimensionless function $\psi(\boldsymbol{r}) = \sqrt{2/n_s^0}\Psi(\boldsymbol{r})$, where n_s^0 is the volume-averaged number density of superconducting electrons, and in terms of the parameters

$$\lambda = \sqrt{\frac{mc^2}{4\pi e^2 n_s^0}}, \qquad \xi = \frac{\hbar}{\sqrt{4m|\alpha|}}, \tag{3.109}$$

which have a dimension of length.

Solution. With the aid of Equations (3.96) and (3.101) we find the relation between the number density of Cooper pairs in a uniform superconductor and thermodynamic parameters, that is, $n_s^0 = 2|\alpha|/\beta$. Hence, we can write Equations (3.105)–(3.108) in the form

$$\xi^2\left(i\nabla + \frac{2e}{\hbar c}\boldsymbol{A}\right)^2\psi - \psi + |\psi|^2\psi = 0, \tag{3.110}$$

$$\nabla\times[\nabla\times\boldsymbol{A}] = -i\frac{\pi\hbar c}{4\pi e\lambda^2}(\psi^*\nabla\psi - \psi\nabla\psi^*) - \frac{1}{\lambda^2}|\psi|^2\boldsymbol{A}, \tag{3.111}$$

$$\boldsymbol{n}\cdot\left(i\nabla + \frac{2e}{\hbar c}\boldsymbol{A}\right)\psi = 0. \quad ■ \tag{3.112}$$

Problems

3.100• Show that Equation (3.106) can be written as $\boldsymbol{n}\cdot\boldsymbol{j}_s|_S = 0$, that is, it states that the normal component of the current on the superconductor surface must be zero. This condition is satisfied at the boundary with vacuum or dielectric but at the boundary with another superconductor or normal metal it is broken.

3.101• Show that Equation (3.108) for a superconducting current is gauge-invariant, that is, it does not change if we use another vector potential $\boldsymbol{A}' = \boldsymbol{A} + \nabla\varphi$ provided the wave function Ψ is also adequately transformed (how?).

3.102• Analyze Equations (3.110)–(3.112) for one-dimensional case and clarify the meaning of lengths λ and ξ, entering these equations. Also, elucidate the limits of applicability of the London equations (3.92) and (3.93).

3.103⋆ Derive the quantization condition (3.87) of the magnetic flux in a superconductor. To do this, consider a cylindrical cavity (Figure 3.10) in a massive superconductor placed in a magnetic field parallel to its axis. Writing down the order parameter in the form $\psi = \exp(i\phi(\boldsymbol{r}))$, where $\phi(\boldsymbol{r})$ is the real phase, use the unambiguity condition for the wave function when going round the closed contour.

3.104 A metallic film of thickness $\delta \ll \lambda$ is deposited on the surface of a dielectric cylinder of radius $a \gg \delta$, placed in a longitudinal magnetic field. Initially, at $T > T_c$, the film is not superconducting. Then the temperature is lowered to $T < T_c$, and the external field is switched off. Find the quantization rule for the magnetic flux inside the cylinder.

Elements of microscopic theory of superconductivity. The following important statements constitute the foundation of the microscopic theory, which explains the mechanism of superconductivity based on the internal structure of matter:

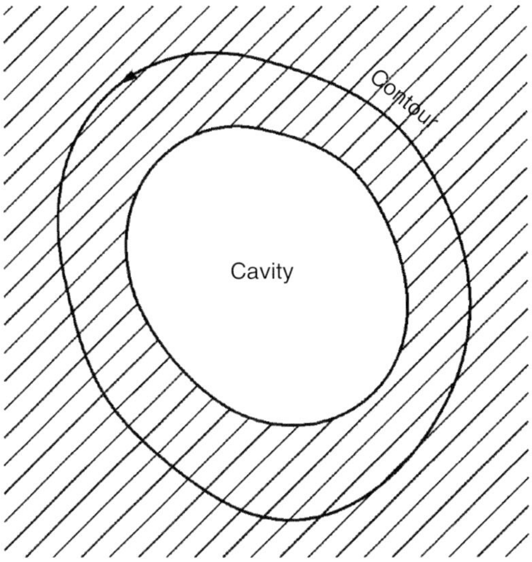

Figure 3.10 Derive the magnetic flux quantization rule.

1) Any two electrons are subject to Coulomb repulsion. However, the resulting attraction between them is also possible provided they simultaneously interact with another system of particles. The attraction in superconductors manifests itself in the interaction between electrons and oscillations of crystal lattice, the so-called phonons. In the simplest model of this kind, the "jelly" model (which is an electron-ion plasma in which ions do not form an ordered lattice and behave as a liquid), the effective interaction potential of two electrons (relative to surrounding particles) has the form

$$V_{ef}(\boldsymbol{k},\omega) = \frac{4\pi e^2}{\kappa^2 + k^2}\left(1 + \frac{\omega_k^2}{\omega^2 - \omega_k^2}\right). \tag{3.113}$$

Here, the potential is written in the Fourier space (see Chapter 5 for the derivation of this formula), κ is the screening constant and ω_k is the frequency of oscillations of the ion system. At $\omega < \omega_k$ this quantity is the potential of attraction between two electrons. When $\omega_k = 0$, it turns into the screened Coulomb potential considered in Problems 1.24 and 1.27, and describes repulsion.

2) A pair of electrons in a degenerate electron gas may exist in the bound state at any arbitrarily small attraction between the electrons. The bound electron pairs formed in this way are called **Cooper pairs**. They are compound particles with zero spins (bosons), and their number in any quantum state may be arbitrary, in contrast to the number of individual electrons obeying the Pauli (exclusion) principle. Cooper pairs can propagate coherently without resistance and destruction similar to carriers of electric current in superconductors. Therefore, superconductivity can be considered superfluidity of Cooper pairs.

Example 3.13

Consider an attraction between the electrons in the degenerate (temperature $T \to 0$) electron gas, which is determined by the given potential $V(\boldsymbol{r}_1, \boldsymbol{r}_2)$. The matrix elements $V_{\mathbf{kk'}}$ of the interaction potential, corresponding to the transitions of an electron pair from the state with the wave vectors $(\boldsymbol{k}', -\boldsymbol{k}')$ to a different state $(\boldsymbol{k}, -\boldsymbol{k})$ are not zero only in a narrow layer close to the filled Fermi sphere:

$$V_{\mathbf{kk'}} = \begin{cases} -V < 0, & at \quad k_F - \Delta k \le k, k' \le k_F; \\ 0 & in\ the\ remaining\ \boldsymbol{k}\text{-}space. \end{cases} \tag{3.114}$$

Here, k_F is the wave vector of an electron on the Fermi surface, $\Delta k \approx \omega_D / v_F$ is the thickness of the layer, in which the interaction with lattice oscillations is essential, ω_D is the characteristic phonon frequency, and v_F is the velocity of the electron on the Fermi surface. Show that the bound state is possible at any arbitrarily small attraction V, taking into account only the interaction of electrons in a given pair and assuming that all other electrons constitute a degenerate perfect Fermi-gas. Calculate the binding energy E of the Cooper pair.

Solution. We seek the coordinate wave function $\psi_E(\boldsymbol{r}_1, \boldsymbol{r}_2)$ and the binding energy of the electron pair from the Schrödinger equation written in the center-of- mass reference frame:

(1)
$$-\frac{\hbar^2}{2m}(\triangle_1 + \triangle_2)\psi_E + V(\boldsymbol{r}_1, \boldsymbol{r}_2)\psi_E = (E + 2\epsilon_F)\psi.$$

This wave function depends on the difference between position vectors, $\psi_E(\boldsymbol{r}_1, \boldsymbol{r}_2) = \psi_E(\boldsymbol{r}_1 - \boldsymbol{r}_2)$. The energy E is measured from the electron Fermi level and must satisfy the condition $E < 0$.

Now we pass to the $\boldsymbol{k}$-space by expanding the required function $\psi_E(\boldsymbol{r}_1 - \boldsymbol{r}_2)$ into Fourier series of states $\psi_{\mathbf{k}}(\boldsymbol{r}_1, \boldsymbol{r}_2) = L^{-3} \exp[i\boldsymbol{k} \cdot (\boldsymbol{r}_1 - \boldsymbol{r}_2)]$, in which the electrons have the wave vectors $(\boldsymbol{k}, -\boldsymbol{k})$. The wave functions $\psi_{\mathbf{k}}$ are normalized by the condition $\int_{(L^3)} \psi^*_{\mathbf{k}'}\psi_{\mathbf{k}}\, d^3r_1\, d^3r_2 = \delta_{\mathbf{kk}'}$ in a box of volume L^3. The expansion reads

(2)
$$\psi_E(\boldsymbol{r}_1 - \boldsymbol{r}_2) = \sum_{\mathbf{k}} \varphi(\boldsymbol{k})\psi_{\mathbf{k}}(\boldsymbol{r}_1, \boldsymbol{r}_2), \qquad \varphi(\boldsymbol{k}) = \int \psi_E(\boldsymbol{r}_1 - \boldsymbol{r}_2)\psi^*_{\mathbf{k}}(\boldsymbol{r}_1, \boldsymbol{r}_2)\, d^3r_1\, d^3r_2.$$

The term containing the interaction potential is transformed in the following way:

$$\int V(\boldsymbol{r}_1, \boldsymbol{r}_2)\psi_E(\boldsymbol{r}_1 - \boldsymbol{r}_2)\psi^*_{\mathbf{k}}(\boldsymbol{r}_1, \boldsymbol{r}_2)\, d^3r_1\, d^3r_2 =$$

(3)
$$= \sum_{\mathbf{k}'} \varphi(\boldsymbol{k}') \int \psi^*_{\mathbf{k}}(\boldsymbol{r}_1, \boldsymbol{r}_2)V(\boldsymbol{r}_1, \boldsymbol{r}_2)\psi_{\mathbf{k}'}(\boldsymbol{r}_1, \boldsymbol{r}_2)\, d^3r_1\, d^3r_2 = \sum_{\mathbf{k}'} V_{\mathbf{kk}'}\varphi(\boldsymbol{k}').$$

After these transformations, the Schrödinger equation takes the form

(4)
$$(E + 2\epsilon_F - 2\epsilon_k)\varphi(\boldsymbol{k}) = \sum_{\mathbf{k}'} V_{\mathbf{kk}'}\varphi(\boldsymbol{k}') = -\frac{V}{L^3}\sum_{\mathbf{k}'} \varphi(\boldsymbol{k}').$$

Here, $\epsilon_k = \hbar^2 k^2/2m$, and $\varphi(\boldsymbol{k}) = 0$ everywhere at $k \leq k_F$, because all states at $k \leq k_F$ are occupied by the background electrons.

In the last Equation (4), the expression (3.114) is used. Hence, Equation (4) holds only for the values of k and k' inside the layer close to the Fermi sphere. The right-hand side vanishes outside this layer.

By dividing both sides of Equation (4) by $(E + 2\epsilon_F - 2\epsilon_k)$ and summing them over $\boldsymbol{k}$, we may cancel the common factor $\sum_{\mathbf{k}} \varphi(\boldsymbol{k}) \neq 0$ and obtain the transcendent equation for the binding energy E:

(5)
$$1 = -V\sum_{\mathbf{k}} \frac{1}{E + 2\epsilon_F - 2\epsilon_k}.$$

Now we integrate over energies in this equation by introducing the number density $N(\epsilon_k)$ of quantum states (per unit energy interval). Then Equation (5) takes the form

(6)
$$1 = V\int_0^{\hbar\omega_D} N(\epsilon_k)\frac{d\epsilon}{2\epsilon - E},$$

where the integration variable $\epsilon = \epsilon_k - \epsilon_F$ is introduced. In view of the inequality $\hbar\omega_D \ll \epsilon_F$, we replace ϵ_k by ϵ_F in the argument of $N(\epsilon_k)$, and take $N(\epsilon_F)$ outside the integral. The integration yields

(7) $$\frac{2}{VN(\epsilon_F)} = \ln \frac{E - 2\hbar\omega_D}{E}.$$

Here, the dimensionless quantity $g = VN(\epsilon_F)$ plays the role of the coupling constant of the Cooper pair. The bound state with energy

(8) $$E = -\hbar\omega_D\, e^{-2/g}.$$

is possible even when $g \ll 1$. The filled Fermi sphere precludes the breakdown of this coupling. Only a small fraction of electrons from the energy layer of thickness $\hbar\omega_D \ll \epsilon_F$ are capable of producing Cooper pairs. ■

Suggested literature: Ginzburg and Landau (1950); Bogolyubov (1960); Bardeen and Schrieffer (1961); Schrieffer (1964); De Gennes (1966); Rose-Innes and Rhoderick (1969); Tilley and Tilley (1974); Kresin (1978); Nelson *et al.* (1987); Schmidt (2000); Prokhorov (1984); Feynman (1972); Kittel (1978); Lifshitz and Pitaevskii (1983)

3.5
Answers and Solutions

3.1 $\rho = \mu j^2/c^2\rho_e$. This is a small relativistic correction of the order of $u^2/c^2 \ll 1$ with respect to ρ_e.

3.2
$$\frac{J}{J_2} = \frac{V_1 - V_2}{V_2 - \mathcal{E}}.$$

3.3 The resistance of the galvanometer coil must be equal to the external resistance R.

3.4 $R = 3r/2$ at $n = 2$; $R = 13r/7$ at $n = 3$; $R = 47r/22$ at $n = 4$. Using symmetry arguments, the number of circulating currents may be reduced, for example, in the case of $n = 3$, to three contour currents only.

3.5 Consider the contour currents shown in Figure 9.3. The Kirchhoff equation for the cell $B_kA_kA_{k+1}B_{k+1}$ is of the form

(1) $$J_{k+1} + J_{k+1} = \left(2 + \frac{R}{r}\right)J_k.$$

This second-order linear difference equation has two linearly independent solutions: $e^{k\alpha}$ and $e^{-k\alpha}$, where

(2) $$\sinh\frac{\alpha}{2} = \frac{1}{2}\sqrt{\frac{R}{r}}.$$

Deriving this equation and those which follow, it is useful to bear in mind that the formulas in hyperbolic trigonometry are obtained from those in ordinary trigonometry by replacing $\cos i\alpha \to \mathrm{ch}\alpha$, $\sin i\alpha \to i\sinh\alpha$.

The general solution of Equation (1) is $J_k = A'e^{k\alpha} + B'e^{-k\alpha}$. In our case, it is convenient to rearrange the terms and rewrite Equation (1) in the form

$$J_k = A\cosh(\beta - k)\alpha, \tag{3}$$

where A and β are arbitrary constants. We find them from the boundary conditions at the ends of the line. The Kirchhoff equation for the last cell takes the form

$$J_n(R + R_a + r) - J_{n-1}r = 0. \tag{4}$$

Substituting the expressions for currents J_n and J_{n-1} from Equation (3) into Equation (4) and using Equation (2), we obtain, after dividing by A, the equation for β:

$$\tanh\beta\alpha = \frac{R_a\cosh n\alpha + \sqrt{Rr}\sinh\left(n + \frac{1}{2}\right)\alpha}{R_a\sinh n\alpha + \sqrt{Rr}\cosh\left(n + \frac{1}{2}\right)\alpha}. \tag{5}$$

The constant A can be determined from the Kirchhoff equation for the first cell of the line:

$$J_0(R + R_i + r) - J_1 r = \mathcal{E}. \tag{6}$$

After some rearrangements, Equation (6) yields

$$A = \frac{\mathcal{E}}{R_i\cosh\beta\alpha + \sqrt{Rr}\sinh\left(\beta + \frac{1}{2}\right)\alpha}.$$

The final expression for the current in the section A_kA_{k+1} is

$$J_k = \frac{\mathcal{E}\cosh(\beta - k)\alpha}{R_i\cosh\beta\alpha + \sqrt{Rr}\sinh\left(\beta + \frac{1}{2}\right)\alpha}. \tag{7}$$

The constants α and β in Equation (7) are given by Equations (2) and (5). For dry insulation when $r \to \infty$ and $\alpha \to 0$, Equation (7) takes the expected form

$$J_k = \frac{\mathcal{E}}{R_i + R_a + (n+1)R}. \tag{8}$$

The ratio of the electromotive forces $\mathcal{E}_0$ and $\mathcal{E}$, which ensures the same current through the load in the cases of dry and damp insulations can be obtained from Equations (7) and (8). The result is

$$\frac{\mathcal{E}}{\mathcal{E}_0} = \frac{R_i\cosh\beta\alpha + \sqrt{Rr}\sinh\left(\beta + \frac{1}{2}\right)\alpha}{[R_i + R_a + (n+1)R]\cosh\left(\beta - n + \frac{1}{2}\right)\alpha}. \tag{9}$$

If the resistance of the load is $R_a = 0$, Equation (5) simplifies and gives

$$\beta = n + \frac{1}{2}. \tag{10}$$

3.6 Let $J(x)$ and $\varphi(x)$ be the current and the potential (relative to ground) of the central conductor at the coordinate x of the cross section. We then have

$$\varphi(x) = -\rho' \frac{dJ}{dx}, \qquad J = -\rho \frac{d\varphi}{dx}, \qquad \frac{d^2 J}{dx^2} = \frac{\rho}{\rho'} J.$$

3.7

(1) $$J(x) = \frac{\mathcal{E} \cosh s(x - x_0)}{R_i \cosh sx_0 + \sqrt{\rho\rho'} \sinh sx_0},$$

where $s = \sqrt{\rho/\rho'}$. The constant x_0 can be determined from the equation

(2) $$\tanh s(x_0 - a) = \frac{R_a}{\sqrt{\rho\rho'}}.$$

At $R_i = R_a = 0$, we have

(3) $$J(x) = \frac{\mathcal{E} \cosh s(x - x_0)}{\sqrt{\rho\rho'} \sinh sa}.$$

In the absence of leakage $\rho' \to \infty$, $x_0 \to a$, $s \to 0$, and the current along the cable takes the constant value:

$$J_0 = \frac{\mathcal{E}}{R_i + \rho a + R_a}.$$

Using Equation (7) of the solution of Problem 3.5, we have to set

$$R = \rho\, dx, \qquad r = \frac{\rho'}{dx}, \qquad k = \frac{x}{dx}, \qquad n = \frac{a}{dx}.$$

It then follows from Equation (2) of the solution of Problem 3.5 that $\alpha = s\, dx$. The quantity β in this solution is related to x_0 by the expression $\beta = x_0/dx$, so that $\beta\alpha = x_0 s$. Substitution of these expressions into Equations (5) and (7) of the solution of Problem 3.5 leads to Equations (1) and (2) given above.

3.8

$$E_1 = \frac{\kappa_2 V}{\kappa_1 h_2 + \kappa_2 h_1}, \qquad D_1 = \frac{\varepsilon_1 \kappa_2 V}{\kappa_1 h_2 + \kappa_2 h_1},$$

$$E_2 = \frac{\kappa_1 V}{\kappa_1 h_2 + \kappa_2 h_1}, \qquad D_2 = \frac{\varepsilon_2 \kappa_1 V}{\kappa_1 h_2 + \kappa_2 h_1},$$

$$j_1 = j_2 = \frac{\kappa_1 \kappa_2 V}{\kappa_1 h_2 + \kappa_2 h_1}.$$

At the boundary between the plates:

$$\sigma_{int} = \frac{E_2 - E_1}{4\pi} - \sigma_{ext} = \frac{\kappa_2(\varepsilon_1 - 1) - \kappa_1(\varepsilon_2 - 1)}{4\pi(\kappa_1 h_2 + \kappa_2 h_1)} V,$$

$$\sigma_{ext} = \frac{D_2 - D_1}{4\pi} = \frac{(\varepsilon_2 \kappa_1 - \varepsilon_1 \kappa_2) V}{4\pi(\kappa_1 h_2 + \kappa_2 h_1)}.$$

The quantity V is positive, if the first conducting plate is in contact with the positively charged capacitor plate. At the boundary between the capacitor electrode and the first conducting plate

$$\varrho_{ext} = \frac{D_1}{4\pi}, \qquad \sigma_{int} = \frac{E_1 - D_1}{4\pi}.$$

At the boundary between the capacitor electrode and the second conducting plate

$$\sigma_{ext} = -\frac{D_2}{4\pi}, \qquad \sigma_{int} = -\frac{E_2 - D_2}{4\pi}.$$

3.9

$$\frac{\tan\beta_1}{\tan\beta_2} = \frac{\kappa_1}{\kappa_2},$$

where β_1 and β_2 are, respectively, the angles between the current flow and the normal to the boundary between the first and second media.

3.11

$$\varphi = \begin{cases} \dfrac{Jz}{\pi a^2 \kappa}, & 0 \le r \le a, \\ \dfrac{Jz \ln(r/b)}{\pi a^2 \ln(a/b)}, & a < r \le b, \\ 0, & r > b. \end{cases}$$

It is clear from this formula that the electric field in the space between the conductors is not parallel to the Oz-axis. The existence of nonzero radial component of the electric field, E_r, shows that there are the surface charges on the cylindrical surfaces of the conductors whose densities are given by

$$\sigma_1 = \left.\frac{\varepsilon E_r}{4\pi}\right|_{r=a} = \frac{\varepsilon Jz}{4\pi^2 a^3 \kappa \ln(a/b)}, \qquad \sigma_2 = \left.\frac{\varepsilon E_r}{4\pi}\right|_{r=b} = -\frac{\varepsilon Jz}{4\pi^2 a^2 b \kappa \ln(a/b)}.$$

When $z = 0$, the densities σ_1 and σ_2 become zero. The position of the cross section on which $\sigma_1 = \sigma_2 = 0$, is not determined. This cross section can be shifted if an additional constant charge is placed on the conductor. The charges $q_1 = 2\pi a \sigma_1$ and $q_2 = 2\pi b \sigma_2 = -q_1$ per unit length of the conductor and shell (at the same z) are related to the potential difference between them

$$V = \int_a^b E_r\, dr = -\frac{Jz}{a^2 \kappa}$$

by the expression

$$\frac{q_1}{V} = \frac{1}{2\ln(b/a)} = const.$$

In this case, the q_1/V-ratio is identical with the capacitance per unit length of the cylindrical capacitor in the corresponding electrostatic problem. Evidently, the magnetic field is of the same form as for infinitely long straight conductor carrying a current J. This is because the current density in an infinitely thick shell vanishes and, as a result, the reverse current does not produce any magnetic field.

3.12 $E_0 = -k(\kappa_2 l_1 + \kappa_1 l_2)\mathcal{E}_0, \qquad E_1 = k\kappa_2\mathcal{E}_0, \qquad E_2 = k\kappa_1\mathcal{E}_0,$
where $k = \kappa_0/l_0(\kappa_0\kappa_1 l_2 + \kappa_0\kappa_2 l_1 + \kappa_1\kappa_2 l_0)$, and $\mathcal{E}_0 = E_{ext} l_0$ is the electromotive force of the source. Inside the source, the electric field is antiparallel to the current ($E_0 < 0$). The charges producing this electric field are located on the boundaries between the conductors of different conductivities, and can be determined with the aid of the boundary conditions; for example, the charge on the boundary 01 is

$$q_{01} = \frac{r^2}{4}(E_1 - E_0).$$

3.13 Consider, for example, the energy flux through the surface of conductor 0 where the electromotive force acts. The magnetic field close to the surface is the same as the field of an infinitely long straight conductor, $H = 2J/cr$. It is easy to show that the Poynting vector $\boldsymbol{\gamma} = c(\boldsymbol{E}_0 \times \boldsymbol{H})/4\pi$ is parallel to the outward normal to the surface of the conductor (where $\boldsymbol{E}_0$ is the electric field strength in conductor 0 and its direction is opposite to the direction of the current; see Problem 3.12). Hence, the energy flux through the surface of this conductor equals $2\pi r l_0 \gamma = JV$, where $V = E_0 l_0$ is the potential difference between the ends of the conductor. The quantity JV is the difference between the work done by the electromotive force, $\mathcal{E}J$ ($\mathcal{E} = E_{ext} l_0$), and the Joule losses per unit time in the source itself.

The energy $J\,V$ leaves the outer surface of the source per unit time, flows in the surrounding space (mainly outside the conductors), enters conductors 1 and 2 through their surface and is converted into Joule heat inside these conductors. The total amount of energy entering conductors 1 and 2 per unit time is JV_1 and JV_2, respectively. This can be easily proved by considering the Poynting vector as has been done earlier.

3.14 $R = \int_1^2 dl/\kappa S$, where the element dl is parallel to the normal to the equipotential surface of area S; the limits 1and 2 denote the bounding surfaces.

3.15

$$\text{(a)} \quad R = \frac{1}{4\pi\kappa}\left(\frac{1}{a} - \frac{1}{b}\right);$$

$$\text{(b)} \quad R = \frac{1}{4\pi\kappa_1}\left(\frac{1}{a} - \frac{1}{c}\right) + \frac{1}{4\pi\kappa_2}\left(\frac{1}{c} - \frac{1}{b}\right);$$

$$\text{(c)} \quad R = \frac{1}{2\pi l\kappa}\ln\frac{b}{a}.$$

3.16

$$R = \frac{1}{2\pi\kappa_2}\left(\frac{1}{a} - \frac{1}{b}\right) + \frac{1}{2\pi\kappa_1 b}.$$

3.17

$$J_k = \frac{4\pi\kappa}{\varepsilon}q_k, \qquad R_{ik} = \frac{\varepsilon}{4\pi\kappa}s_{ik}.$$

3.18

$$C = \frac{\varepsilon}{4\pi\kappa R}.$$

3.19

$$Q = \sum_{i,k} R_{ik} J_i J_k.$$

3.20

$$R = \frac{\varepsilon}{4\pi\kappa}(s_{11} - 2s_{12} + s_{22}) = \frac{\varepsilon}{4\pi\kappa}\frac{c_{11} + 2c_{12} + c_{22}}{c_{12}^2 - c_{11}c_{22}}.$$

3.21

$$R = \frac{V_1 - V_2}{J} = R_1 + R_2 - \frac{1}{\pi\kappa l} \approx R_1 + R_2,$$

where $R_1 = 1/2\pi\kappa a_1$ and $R_2 = 1/2\pi\kappa a_2$ are the resistances of the solitary grounding electrodes (Problem 3.16).

3.22 Let $e_0 = \sqrt{1 - b^2/a^2}$ be the eccentricity of the ellipsoids of revolution (b/a is the ratio of the semiminor to semimajor axes). Then, for an oblate ellipsoid of revolution

$$R = \frac{1}{\kappa}\left(\frac{4}{3\pi^2 V}\right)^{1/3} \frac{\left(1 - e_0^2\right)^{1/6}}{e_0} \arccos\sqrt{1 - e_0^2}$$

while, for a prolate ellipsoid of revolution

$$R = \frac{1}{\kappa}\frac{\left(1 - e_0^2\right)^{1/3}}{(6\pi^2 V)^{1/3} e_0} \ln\frac{1 + e_0}{1 - e_0}.$$

.

For a given volume V, the most profitable shape is either very elongated or, conversely, very flattened shape of the grounding conductors.

3.23 $J = 2\pi\kappa V a$.

3.24 The current density in the space between the electrodes

(1) $$j = \rho v$$

is independent of x [$v(x)$ is the velocity of particles at a given point x]. The velocity is related to the potential $\varphi(x)$ by the formula

(2) $$v = \sqrt{-\frac{2e\varphi}{m}}$$

(with $\varphi = 0$ at $x = 0$).

It follows from Equations (1) and (2) that $\rho = j\sqrt{-m/2e\varphi}$, since the Poisson equation takes the form

(3) $$\frac{d^2\varphi}{dx^2} = -4\pi j\sqrt{-\frac{m}{2e\varphi}}.$$

The integration of Equation (3) subject to the boundary conditions $d\varphi/dx|_{x=0} = 0$ and $\varphi|_{x=a} = \varphi_0$, yields

(4) $$j = \frac{1}{9\pi a^2}\sqrt{\frac{2|e|}{m}}|\varphi_0|^{\frac{3}{2}},$$

which is the so-called three-halves power law.

3.26 Using Equation (4) from Example 3.1 and substituting the quantities (3.22) into it, we obtain from the entropy increase law:

$$\frac{dS}{dt} = \int_V \left\{ \frac{j^2}{\kappa T} + \frac{\eta(\nabla T)^2}{T^2} \right\} dV \geq 0,$$

where $\eta = \xi - \beta^2 T/\kappa$ is the thermal conductivity. This leads to the inequalities $\kappa > 0$ and $\eta > 0$.

3.27

$$\varrho = T\frac{d}{dT}\left(\frac{\beta}{\kappa}\right).$$

3.28 We integrate Equation (3.29) over a small interval $2h$ of the Oz-axis, which includes the junction. The axis is parallel to the current flow and perpendicular to the junction. The first two terms in the right-hand side of the equation make negligible contributions, whereas the contribution of the last term is finite, in view of a jump of the quantities β and κ at the junction: $Q_P = jT(\beta_2/\kappa_2 - \beta_1/\kappa_1) = \Pi_{12} j$, where Π_{12} is the Peltier coefficient. The quantity Q_P is the heat power per unit surface area of the contact.

3.30 In each region where κ is homogeneous the macroscopic values of the current and electric field satisfy the equations

(1) $$\boldsymbol{j} = \kappa\boldsymbol{\mathcal{E}}, \quad \nabla \times \boldsymbol{\mathcal{E}} = 0, \quad \nabla\cdot\boldsymbol{j} = 0.$$

Following Dykhne (1971a), we introduce new microscopic quantities

(2) $$\boldsymbol{j}' = (\kappa_1\kappa_2)^{1/2}[\boldsymbol{e}_z \times \boldsymbol{\mathcal{E}}], \qquad \boldsymbol{\mathcal{E}}' = (\kappa_1\kappa_2)^{-1/2}[\boldsymbol{e}_z \times \boldsymbol{j}],$$

where $\boldsymbol{e}_z$ is the Cartesian unit vector. Using Equation (1), we prove that in each region of homogeneity there is

(3) $$\nabla \times \boldsymbol{\mathcal{E}}' = 0, \quad \nabla\cdot\boldsymbol{j}' = 0,$$

and the local Ohm law is satisfied:

(4) $$\boldsymbol{j}' = \kappa'\boldsymbol{\mathcal{E}}'.$$

Here

(5) $$\kappa' = \frac{\kappa_1\kappa_2}{\kappa} = \begin{cases} \kappa_2 & \text{in region 1,} \\ \kappa_1 & \text{in region 2.} \end{cases}$$

However, since regions 1 and 2 are statistically equivalent, the electric conductivity in the macroscopic Ohm's law for the quantities $\boldsymbol{J}'$ and $\boldsymbol{E}'$ must be the same as for the quantities $\boldsymbol{J}$ and $\boldsymbol{E}$, that is,

(6) $$\boldsymbol{J}' = K\boldsymbol{E}'.$$

By averaging Equation (2) over the volume, we find

(7)
$$\boldsymbol{J}' = (\kappa_1\kappa_2)^{1/2}[\boldsymbol{e}_z \times \boldsymbol{E}], \quad \boldsymbol{E}' = (\kappa_1\kappa_2)^{-1/2}[\boldsymbol{e}_z \times \boldsymbol{J}] = (\kappa_1\kappa_2)^{-1/2}K[\boldsymbol{e}_z \times \boldsymbol{E}],$$

where in the last equation we have used the definition of K given in the statement of this problem. Substituting Equation (7) into Equation (6), we obtain

(8) $$K = (\kappa_1 \kappa_2)^{1/2}.$$

Analogous calculations give

(9) $$\Delta_J = \Delta_E = \frac{1}{2}\left[\left(\frac{\kappa_1}{\kappa_2}\right)^{1/4} - \left(\frac{\kappa_2}{\kappa_1}\right)^{1/4}\right]^2 \geq 0.$$

3.31 See the original work by Dykhne (1971b).

3.32 We use the vector potential method to solve the problem. If the Oz-axis is parallel to the axis of the cylinder, the Cartesian components of $\boldsymbol{A}$ satisfy the equations:

(1) $$\triangle A_x = 0, \qquad \triangle A_y = 0, \qquad \triangle A_z = -\frac{4\pi\mu_0}{c} j_z,$$

where $j_z = 0$ at $r > a$, and $j_z = J/\pi a^2$ at $r \leq a$.

Since the current J does not enter the equations for A_x and A_y, these components may be taken to be zero; the quantity A_z depends only on the distance r from the Oz-axis. Integrating the equation for A_z with the continuity conditions for A_z and H_α at the boundary $r = a$ and with the condition for H to be finite at $r = 0$, we have
for $r < a$

(2) $$A_z = C - \frac{\mu_0 J}{c}\left(\frac{r}{a}\right)^2, \qquad B_\alpha = \frac{2\mu_0 J}{ca^2} r, \qquad H_\alpha = \frac{2J}{ca^2} r;$$

for $r > a$

(2′) $$A_z = C - \frac{J}{c}\left(\mu_0 + 2\mu \ln \frac{r}{a}\right), \qquad B_\alpha = \frac{2\mu J}{cr}, \qquad H_\alpha = \frac{2J}{cr}.$$

The constant C is arbitrary.

3.33 For $r < a$

$$A_z = C_1, \qquad \boldsymbol{B} = 0;$$

for $a \leq r \leq b$

$$A_z = \frac{2\mu_0 J a^2}{c(b^2 - a^2)}\left(\ln \frac{r}{a} - \frac{r^2}{2a^2}\right) + C_2, \qquad A_\alpha = \frac{2\mu_0 J}{c(b^2 - a^2)}\left(r - \frac{a^2}{r}\right);$$

for $r > b$

$$A_z = \frac{2\mu J}{c} \ln \frac{b}{r} + C_3, \qquad B_\alpha = \frac{2\mu J}{cr}.$$

Other components of $\boldsymbol{A}$ and $\boldsymbol{B}$ are zero. Any two constants entering A_z can be expressed through the third one using the continuity condition for the vector potential at the boundaries.

3.34 The secondary field $\boldsymbol{H}'$ satisfies the equation $rot\, \boldsymbol{H}' = 0$, that is, it is a potential field. Introducing the scalar potential and using the formula

$\boldsymbol{H}' = -grad\,\psi$ yields the equation for this potential, which is the same as in electrostatics for a nonuniform medium, that is,

$$div\,(\mu grad\,\psi) = -4\pi\rho_m,$$

where the quantity

$$\rho_m = -\frac{1}{4\pi}\boldsymbol{H}_0 \cdot \nabla\,\mu$$

can be treated as effective density of magnetic charges.

On the boundary between two media, the tangential components of the field must satisfy the conditions

$$H'_{1\tau} = H'_{2\tau} \qquad \text{or} \qquad \frac{\partial\psi_1}{\partial\tau} = \frac{\partial\psi_2}{\partial\tau},$$

and the normal field components must satisfy the conditions

$$\mu_2 H'_{2n} - \mu_1 H'_{1n} = (\mu_1 - \mu_2)H_{0n} \qquad \text{or} \qquad \mu_1\frac{\partial\psi_1}{\partial n} - \mu_2\frac{\partial\psi_2}{\partial n} = 4\pi\sigma_m.$$

In the above expression, the quantity

$$\sigma_m = \frac{1}{4\pi}(\mu_1 - \mu_2)H_{0n}$$

plays the role of the density of surface charges. Note, the expression for σ_m can be obtained from the volume density ρ_m by means of limiting transition

$$\sigma_m = \lim_{h\to 0}\rho_m h.$$

We replace the boundary by a layer of small thickness h. Then, the vector $grad\,\mu$ will be parallel to the normal to the layer, and it will be equal to $(\mu_2 - \mu_1)/h$. Therefore,

$$\rho_m = -\frac{1}{4\pi}\cdot\frac{\mu_2 - \mu_1}{h}H_{0n}, \qquad \sigma_m = \lim_{h\to 0}\rho_m h = \frac{1}{4\pi}(\mu_1 - \mu_2)H_{0n}.$$

3.35

$$\boldsymbol{H}_1 = \frac{\mu_2}{\mu_1 + \mu_2}\boldsymbol{H}_0, \qquad \boldsymbol{H}_2 = \frac{\mu_1}{\mu_1 + \mu_2}\boldsymbol{H}_0,$$

where $\boldsymbol{H}_0$ is the field produced by a current-carrying circuit in vacuum; $\boldsymbol{H}_1$ and $\boldsymbol{H}_2$ are the fields in the media with the permeabilities μ_1 and μ_2, respectively.

3.36 The magnetic field in medium 1 is the same as the field produced in vacuum by two straight currents

$$J_1 = \mu J \qquad \text{and} \qquad J_2 = \frac{\mu_1(\mu_2 - \mu_1)}{\mu_1 + \mu_2}J,$$

where the current J_1 flows along the same conductor as the original current, while the current J_2 flows along the conductor, which is a mirror image of the first conductor with respect to the separation plane between the two media.

The magnetic field in medium 2 is identical with the field produced in vacuum by the current $J_1 = 2\mu_1\mu_2/(\mu_1 + \mu_2)J$ flowing along the same conductor as the original current J.

3.37 The field vectors satisfy the homogeneous equations $rot\,\boldsymbol{H} = 0$, $div\,\boldsymbol{B} = 0$ in entire space, and, hence, it is possible to introduce the scalar potential ψ ($\boldsymbol{H} = -grad\,\psi$), which satisfies the Laplace equation. This reduces the magnetostatic problem to the corresponding electrostatic one. The solution inside the sphere is (Problem 2.11):

$$\boldsymbol{H}_1 = \frac{3}{\mu + 2}\boldsymbol{H}_0;$$

and outside the sphere

$$\boldsymbol{H}_2 = \boldsymbol{H}_0 + \boldsymbol{H}_{dip},$$

where $\boldsymbol{H}_{dip}$ is the field produced by a magnetic dipole with the moment

$$\boldsymbol{m} = \frac{\mu - 1}{\mu + 2}a^3\boldsymbol{H}_0.$$

Since the field inside the sphere is uniform, the magnetization is constant:

$$\boldsymbol{M} = \frac{3\boldsymbol{m}}{4\pi a^3} = \frac{3(\mu - 1)}{4\pi(\mu + 2)}\boldsymbol{H}_0.$$

The density of equivalent volume current is, therefore, zero:

$$\boldsymbol{j}_{int} = c\,rot\,\boldsymbol{M} = 0.$$

The surface current density can be determined from Equation (3.35):

$$\boldsymbol{i}_{int} = c[\boldsymbol{n} \times (\boldsymbol{M}_2 - \boldsymbol{M}_1)].$$

Substituting $\boldsymbol{M}_2 = 0$ and $\boldsymbol{M}_1 = \boldsymbol{M}$, we find

$$\boldsymbol{i}_{int} = \boldsymbol{e}_\alpha \frac{3c(\mu - 1)}{4\pi(\mu + 2)}H_0 \sin\vartheta.$$

It is interesting to note, that such a surface current may be obtained by rotating a sphere with uniform surface charge about one of its diameters (Problem I.2.86).

3.38 The lines of force are the lines of intersection of the planes $\alpha = const$ and the surface of revolution about the Oz-axis,

$$\left[1 + \frac{2(\mu - 1)}{\mu + 2}\left(\frac{a}{r}\right)^3\right](x^2 + y^2) = const,$$

where $r = \sqrt{x^2 + y^2 + z^2}$.

3.39

$$\left(\frac{b}{a}\right)^3 = \frac{(\mu_e - 1)(\mu_i + 2)}{(\mu_i - 1)(\mu_e + 2)} \le 1 \qquad \text{at } \mu_i > \mu_e.$$

3.40 If the coordinate axes lie along the principal axes of the magnetic permeability tensor, the field components inside the sphere will be equal to $3H_{0k}/(\mu^{(k)}+2)$, where $\boldsymbol{H}_0$ is the external field. Outside the sphere,

$$\boldsymbol{H}_2 = \boldsymbol{H}_0 + \boldsymbol{H}_{dip},$$

where $\boldsymbol{H}_{dip}$ is the field of a magnetic dipole with moment $\boldsymbol{m}$, and

$$m_k = \frac{\mu^{(k)}-1}{\mu^{(k)}+2} a^3 H_{0k}.$$

The torque acting on the sphere is

$$\boldsymbol{N} = \boldsymbol{m} \times \boldsymbol{H}_0.$$

3.41 It follows from symmetry arguments that the field vectors lie in the plane, which is perpendicular to the conductor. Let the Oz-axis be parallel to the conductor. We introduce the vector potential $A_z \equiv A(x,y)$, $B_x = \partial A/\partial y$, $B_y = -\partial A/\partial x$ to be calculated from the equation

$$(1) \qquad \mu^{(x)}\frac{\partial^2 A}{\partial x^2} + \mu^{(y)}\frac{\partial^2 A}{\partial y^2} = -\frac{4\pi\mu^{(x)}\mu^{(y)}}{c} J\,\delta(x)\,\delta(y).$$

Introducing new variables $x' = x/\sqrt{\mu^{(x)}}$ and $y' = y/\sqrt{\mu^{(y)}}$, we reduce Equation (1) to the form

$$(2) \qquad \frac{\partial^2 A}{\partial x'^2} + \frac{\partial^2 A}{\partial y'^2} = -\frac{4\pi\sqrt{\mu^{(x)}\mu^{(y)}}}{c} J\,\delta(x')\,\delta(y').$$

Thus, it is sufficient to find the magnetic field of the conductor carrying the effective current $J\sqrt{\mu^{(x)}\mu^{(y)}}$ in an isotropic medium. Using the result of Problem 3.32, we obtain

$$B_x = -\frac{2Jy\sqrt{\mu^{(x)}}}{cr'^2}, \qquad B_y = \frac{2Jx\sqrt{\mu^{(y)}}}{cr'^2},$$

where $r' = \sqrt{x^2/\mu^{(x)} + y^2/\mu^{(y)}}$.

3.42

$$H = \left[1 - \frac{[1-(a/b)^2](\mu_1-\mu_2)^2}{(\mu_1+\mu_2)^2 - (a/b)^2(\mu_1-\mu_2)^2}\right] H_0.$$

At $\mu_1 \gg \mu_2$ the field in the cavity is considerably reduced (magnetic screening).

3.43

$$H = \left[1 - \frac{2[1-(a/b)^3](\mu_1-\mu_2)^2}{(\mu_1+2\mu_2)(2\mu_1+\mu_2) - 2(a/b)^3(\mu_1-\mu_2)^2}\right] H_0.$$

At $\mu_1 \gg \mu_2$ the field is considerably reduced, $H \ll H_0$.

3.44

$$\Delta\Phi = B_0 \frac{2\pi(a^3-b^3)}{a}\left(\frac{\mu_1-1}{\mu_1+2} - \frac{\mu_2-1}{\mu_2+2}\right).$$

3.45 The magnetic field is

$$\boldsymbol{H} = rot\boldsymbol{A},$$

where

$$A_z = -\frac{\mu_1 J}{4\pi a^2} r^2 + \frac{2\mu_1}{\mu_1 + \mu_2} B_0 r \sin\alpha \qquad \text{at } r < a,$$

$$A_z = \frac{\mu_2 J}{2\pi} \ln\frac{a}{r} + \left(1 + \frac{\mu_1 - \mu_2}{\mu_1 + \mu_2} \cdot \frac{a^2}{r^2}\right) B_0 r \sin\alpha \qquad \text{at } r > a.$$

The Oz-axis lies along the axis of the cylinder; other components of $\boldsymbol{A}$ are zero.

3.48

$$\boldsymbol{H}_i = \frac{1}{\mu_i} \frac{2\pi\mu_1\mu_2\mu_3}{\mu_1\mu_2\alpha_3 + \mu_2\mu_3\alpha_1 + \mu_1\mu_3\alpha_2} \boldsymbol{H}_0,$$

where $\boldsymbol{H}_0$ is the field produced by the same current in vacuum.

3.49 Outside the sphere, the magnetic induction $\boldsymbol{B}$ and the magnetic field $\boldsymbol{H}$ are related by the usual expression $\boldsymbol{B}_2 = \mu_2\boldsymbol{H}_2$. According to Equation (3.41), inside the sphere, $\boldsymbol{B}_1 = \mu_1\boldsymbol{H}_1 + 4\pi\boldsymbol{M}_0$, where $\boldsymbol{M}_0$ is the constant magnetization. Introducing the scalar potential, as in Problem 3.37, we have

$$\psi_1 = -\boldsymbol{H}_1 \cdot \boldsymbol{r}, \qquad \psi_2 = \frac{\boldsymbol{m} \cdot \boldsymbol{r}}{r^3},$$

where

$$\boldsymbol{H}_1 = -\frac{4\pi\boldsymbol{M}_0}{2\mu_2 + \mu_1}, \qquad \boldsymbol{m} = \frac{4\pi a^3\boldsymbol{M}_0}{2\mu_2 + \mu_1}.$$

Thus, inside the sphere the field is uniform, and outside the sphere is the same as the field of magnetic dipole with moment $\boldsymbol{m}$.

3.50 The field inside the cylinder is

$$\boldsymbol{H}_1 = -\frac{4\pi\boldsymbol{M}_0}{\mu_2 + \mu_1}.$$

The field outside the cylinder reads

$$\boldsymbol{H}_2 = \frac{2\boldsymbol{r}(\boldsymbol{m} \cdot \boldsymbol{r})}{r^4} - \frac{\boldsymbol{m}}{r^2},$$

where $\boldsymbol{M}_0$ is the constant magnetization and

$$\boldsymbol{m} = \frac{4\pi a^2\boldsymbol{M}_0}{\mu_2 + \mu_1}.$$

3.51 The required quantities can be obtained by replacing the electric parameters by the corresponding magnetic parameters in the solution of Problem 2.45. In particular, for arbitrary coordinate axes the internal field $\boldsymbol{H}_1$ in the ellipsoid can be written in the form

$$H_{1k} = H_{0k} - 4\pi N_{kl} M_l,$$

where $\boldsymbol{M}$ is the magnetization vector, and N_{kl} are the demagnetization coefficients (the components of the tensor, which describes demagnetization under the effect of shape). The principal values of the demagnetization tensor were determined in Problems 2.42–2.44, where they were denoted by $n^{(i)}$ and referred to as the depolarization coefficients. For the limiting cases, they have the following values:

Shape	$N^{(x)}$	$N^{(y)}$	$N^{(z)}$	Orientation
Disc	0	0	1	perpendicular to the Oz-axis
Rod	0	1/2	1/2	along the Ox-axis
Sphere	1/3	1/3	1/3	arbitrary

3.52 The formula given in the solution of Problem 3.51 holds for anisotropic magnetics. There is an additional relation between $\boldsymbol{M}$ and $\boldsymbol{H}_1$:

$$H_{1k} + 4\pi M_k = \mu_{kl} N_{1l}.$$

These two formulas give

$$H_{0k} = b_{km} H_{1m},$$

where

$$b_{km} = \delta_{km} - N_{km} + N_{kl}\mu_{lm}.$$

Hence,

$$H_{1k} = b^{-1}_{km} H_{0m},$$

where b^{-1}_{km} are the components of the reciprocal tensor. They can be found with the aid of formulas obtained in Problem 1.27.

Consider a special case. Let the coordinate axes lie along the principal axes of the ellipsoid. If the tensor μ_{ik} referred to these axes has a diagonal form:

$$\mu_{ik} = \begin{pmatrix} \mu^{(x)} & 0 & 0 \\ 0 & \mu^{(y)} & 0 \\ 0 & 0 & \mu^{(z)} \end{pmatrix},$$

then the tensor b_{ik} and its reciprocal b^{-1}_{ik} are also diagonal:

$$b^{-1}_{ik} = \begin{pmatrix} \left[1+N^{(x)}(\mu^{(x)}-1)\right]^{-1} & 0 & 0 \\ 0 & \left[1+N^{(y)}(\mu^{(y)}-1)\right]^{-1} & 0 \\ 0 & 0 & \left[1+N^{(z)}(\mu^{(z)}-1)\right]^{-1} \end{pmatrix}.$$

3.53

$$\frac{dW}{dt} = \frac{4\pi^2\mu n^2 a^2}{c^2} J\dot{J}.$$

3.54 $B_z = \mu H_z = (2\pi\mu n J/c)(\cos\theta_1 + \cos\theta_2)$,
where (Figure 3.11):

$$\cos\theta_1 = \frac{h-z}{\sqrt{a^2+(h-z)^2}}, \qquad \cos\theta_2 = \frac{z}{\sqrt{a^2+z^2}}.$$

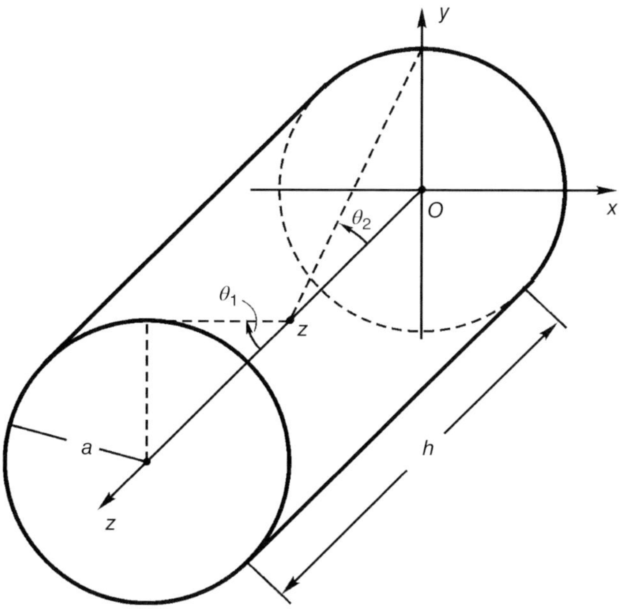

Figure 3.11 Solenoid of finite sizes.

3.55 $\mathcal{L} = 4\pi\mu n^2 S$. For a solenoid of large though finite length h, neglecting the edge effect, we obtain the total inductance

$$L = 4\pi\mu n^2 Sh.$$

3.56 For a circular cross section,

$$L = 4\pi\mu N^2(b - \sqrt{b^2 - a^2}).$$

The self-inductance per unit length of an infinite solenoid $\mathcal{L} = L/(2\pi b)$ can be obtained in the limit $b \to \infty$ at constant number of turns per unit length, $n = N/(2\pi b)$,

$$\mathcal{L} = 4\pi^2\mu^2 n^2 a^2 = 4\pi\mu n^2 S$$

(cf. Problem 3.55).

For a rectangular cross section,

$$L = 2\mu N^2 h \ln \frac{2b + a}{2b - a}.$$

At $b \gg a$ we have $\mathcal{L} = 4\pi\mu n^2 S$ as before.

If the current flows directly along the shell of the torus, the self-induction is reduced by a factor of N^2 as compared to the self-induction of the torus with a wire wound around it. Accordingly, for the torus with circular cross section we have

$$L = 4\pi\mu(b - \sqrt{b^2 - a^2})$$

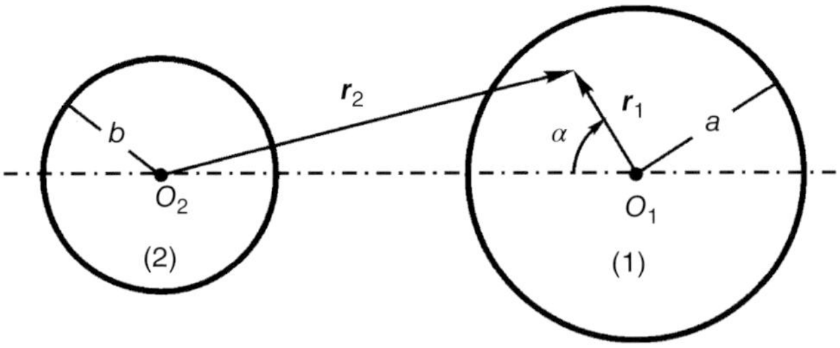

Figure 3.12 To calculate the self-inductance of a double-wire line.

and for the torus with rectangular cross section

$$L = 2\mu h \ln \frac{2b + a}{2b - a}.$$

3.57 $\mathcal{L} = 2\mu \ln \frac{b}{a}$.

3.58 We calculate the magnetic energy per unit length of the line using Equation (2) from Example 3.5. The vector potential for a straight conductor carrying a current has been obtained in Problem 3.32. For conductor 1 (Figure 3.12) the potential can be written as

$$(1) \qquad \left.\begin{aligned} A_{1z} &= C - \mu_0 \frac{Jr_1^2}{ca^2} && \text{at} \quad r_1 < a, \\ A_{1z} &= C - \frac{J}{c}\left(\mu_0 + 2\mu \ln \frac{r_1}{a}\right) && \text{at} \quad r_1 > a. \end{aligned}\right\}$$

The vector potential for conductor 2 is obtained by replacing in (1) $J \to -J$, $a \to b$, and $r_1 \to r_2$. The magnetic energy is

$$(2) \qquad \Delta\mathcal{F} = \frac{J}{2\pi ca^2} \int_{(1)} (A_{1z} + A_{2z})\, dS_1 - \frac{J}{2\pi cb^2} \int_{(2)} (A_{1z} + A_{2z})\, dS_2.$$

The integrals entering Equation (2) are taken in elementary functions. Using the relation between the inductance and magnetic energy of the system, we finally have

$$\mathcal{L} = \mu_0 + 2\mu \ln \frac{h^2}{ab}.$$

3.59 The total magnetic energy of the current flowing inside a conductor consists of two parts:

$$(1) \qquad \Delta\mathcal{F} = \Delta\mathcal{F}_1 + \Delta\mathcal{F}_2,$$

where

$$\Delta\mathcal{F}_1 = \frac{\mu_0}{8\pi} \int H_1^2\, dV$$

is the energy stored inside the conductor (integration is performed over the volume of the conductor), and

$$\Delta\mathcal{F}_2 = \frac{\mu}{8\pi} \int H_2^2\, dV$$

is the energy stored in the outside space.

Let us introduce a length-scale r_0, which satisfies the condition

(2) $$a \ll r_0 \ll R,$$

where a is the radius of the conductor and R is the curvature radius of the axial line of the conductor (which generally varies from one point to another). At distances smaller than r_0, the magnetic field can be considered identical with the field of infinitely long straight conductor. In particular, inside the conductor,

$$H_1 = \frac{2Jr}{ca^2},$$

see Problem 3.32. This allows us to find the "internal" energy $\Delta\mathcal{F}_1$:

(3) $$\Delta\mathcal{F}_1 = \frac{\mu_0 l J^2}{4c^2}.$$

In order to determine the "external" energy $\Delta\mathcal{F}_2$, we consider an auxiliary surface S, resting on an arbitrary contour, which lies on the surface of the conductor, and introduce the scalar potential ψ. The scalar potential exhibits a discontinuity across the surface S, which is given by

(4) $$\psi_+ - \psi_- = \frac{4\pi}{c}J.$$

The integral in the expression for $\Delta\mathcal{F}_2$, can be transformed as

$$\int (\boldsymbol{B}\cdot\boldsymbol{H})\,dV = -\int \boldsymbol{B}\cdot\nabla\psi\,dV = -\int div\,(\psi\boldsymbol{B})\,dV = -\oint \psi B_n\,dS.$$

In this expression, the subscript 2 is dropped and the equation $div\,\boldsymbol{B} = 0$ is used. In the last integral, the integration should be carried out over both sides of the auxiliary surface S and over the surface of the conductor S' (see Figure 3.13, which shows the cross section of the conductor by a certain plane). The integral over an infinitely distant surface will be zero owing to the finite sizes of the conductor carrying the current. Therefore,

(5) $$\Delta\mathcal{F}_2 = -\frac{1}{8\pi}\int_{S'}\psi B_n\,dS + \frac{1}{8\pi}\int_S \psi_+ B_n\,dS - \frac{1}{8\pi}\int_S \psi_- B_n\,dS.$$

The first of these integrals is zero because in view of the condition (2), the magnetic field on the surface S' is identical with the field of a straight

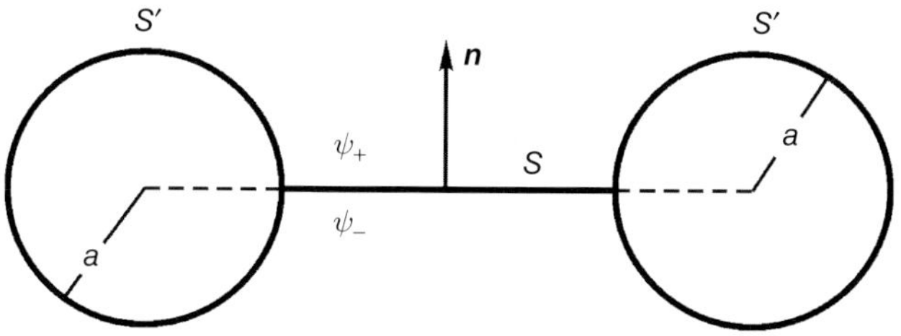

Figure 3.13 Cross-section of conductor by a plane.

conductor and has thus only a tangential component. The two other integrals can be transformed using Equation (4) and the continuity condition for the B_n component. The result is

$$W_2 = \frac{J}{2c}\int_S B_n\, dS. \tag{6}$$

At large distances from the conductor, $r > r_0$, the magnetic field is independent of the current distribution across the conductor. Hence, it is possible to assume that the current flows along the axis. At small distances, $a \leq r < r_0$, the field is identical with the magnetic field produced by infinitely long circular cylinder and again it may be assumed that the current flows along the axis. Thus, the integral in Equation (6) represents the magnetic induction flux produced by a current flowing along the axis of the conductor through the surface resting on the closed contour, which lies on the surface of the conductor. Using the expression of the flux in terms of the mutual inductance, we have

$$\Delta\mathcal{F}_2 = \frac{J^2}{2c^2}L'. \tag{7}$$

From Equations (1), (3), and (7) and the relation between the self-inductance and the magnetic energy of the system, we obtain the required formula for the self-inductance:

$$L = \frac{\mu_0 l}{2} + L'. \tag{8}$$

3.60 Using the results of Problem 3.59 and Problem I.2.106, we have

$$L' = 4\pi\mu b\left(\ln\frac{8b}{a} - 2\right),$$

where μ_0 is the magnetic permeability of the medium in which the conductor is located. The total self-inductance is

$$L = 4\pi b\left(\mu\ln\frac{8b}{a} - 2\mu + \frac{1}{4}\mu_0\right)$$

or, at $\mu_0 = \mu = 1$,

$$L = 4\pi b\left(\ln\frac{8b}{a} - \frac{7}{4}\right).$$

3.61

$$L = 2\mu_0 b + 8\mu b\left[\ln\frac{2b}{a(1+\sqrt{2})} + \sqrt{2} - 2\right].$$

3.62

$$f = \frac{2J^2a^2(\mu - 1)}{c^2b(b^2 - a^2)(\mu + 1)}.$$

3.63

$$f = \frac{2J^2b(\mu - 1)}{c^2(a^2 - b^2)(\mu + 1)}.$$

3.64

$$F = \frac{3}{16}\frac{\mu-1}{\mu+2}\frac{m^2(1+\cos^2\theta)}{a^4}, \qquad N = \frac{\mu-1}{\mu+2}\frac{m^2\sin\theta\cos\theta}{8a^3},$$

where a is the distance of the magnet from the plane, and θ is the angle between $\boldsymbol{m}$ and the normal to the plane. At $\mu \gg 1$ (soft iron in a weak magnetic field), the result is the same as for an electric dipole in the vicinity of the metal surface (Problem 2.95).

3.65 According to the result of Problem 3.35,

$$(1) \qquad \boldsymbol{B}_1 = \mu_1\boldsymbol{H}_1 = \frac{2\mu_1\mu_2}{\mu_1+\mu_2}\boldsymbol{H}_0, \qquad \boldsymbol{B}_2 = \mu_2\boldsymbol{H}_2 = \frac{2\mu_1\mu_2}{\mu_1+\mu_2}\boldsymbol{H}_0,$$

where $\boldsymbol{H}_0$ is the strength of the magnetic field produced by the contour carrying current in vacuum. Expressing the magnetic energy in terms of $\boldsymbol{H}_0$ and using the relation between the energy and the self-inductance, we have

$$L = \frac{2\mu_1\mu_2}{\mu_1+\mu_2}L_0.$$

3.66

$$\Delta U = \frac{B^2}{8\pi\mu}\left(1 + \frac{T}{\mu}\frac{\partial\mu(T,\tau)}{\partial T}\right).$$

3.67 At $\tau = const$, we can write Equation (3.52) in the form $d\widetilde{F} = -SdT - \frac{1}{4\pi}B_\alpha dH_\alpha$ and use the equality of cross derivatives:

$$(1) \qquad \left(\frac{\partial S}{\partial H_\alpha}\right)_{T,\tau} = \frac{1}{4\pi}\left(\frac{\partial B_\alpha}{\partial T}\right)_{H_\alpha,\tau} = \left(\frac{\partial M_\alpha}{\partial T}\right)_{H_\alpha,\tau}.$$

As $T \to 0$, the derivative of the entropy in the left-hand side of the equation tends to zero by virtue of the Nernst principle. In case of the linear coupling $M_\alpha = \chi_{\alpha\beta}H_\beta$, this yields $(\partial\chi_{\alpha\beta}/\partial T)_\tau H_\beta \to 0$ at $T \to 0$; and, since H_β are the independent variables, one has

$$(2) \qquad \left(\frac{\partial\chi_{\alpha\beta}}{\partial T}\right)_\tau \to 0 \quad \text{at} \quad T \to 0.$$

If the dependence $M(H)$ is nonlinear, as in paramagnetics at saturation (see Equation (12) of the solution of Problem 1.17), then the condition, which follows from Equation (1), is satisfied:

$$(3) \qquad \left(\frac{\partial M_\alpha}{\partial T}\right)_{H_\alpha,\tau} \to 0 \quad \text{at} \quad T \to 0.$$

3.68 In a thermally insulated Paramagnetic, the process of demagnetization proceeds adiabatically, that is, at

$$(1) \qquad S(T,B^2) = const, \qquad dS = \left(\frac{\partial S}{\partial T}\right)_B dT + \left(\frac{\partial S}{\partial B^2}\right)_T dB^2 = 0.$$

Using Equation (3.56), we find

(2)

$$\left(\frac{\partial S}{\partial T}\right)_B = \frac{C_{VB}}{T} = \frac{C_{V0}}{T} - \frac{B^2}{8\pi}\frac{d^2}{dT^2}\left(\frac{1}{\mu}\right), \qquad \left(\frac{\partial S}{\partial B^2}\right)_T = -\frac{1}{8\pi}\frac{d}{dT}\left(\frac{1}{\mu}\right),$$

where $C_{V0} = T(\partial S_0/\partial T)_\tau$ is the heat capacity of the magnetic material in the absence of magnetic field. As a result, we obtain from Equation (1) the general relation between the temperature and the magnetic field:

(3) $$dT = \frac{T}{8\pi C_{VB}} \frac{d}{dT}\left(\frac{1}{\mu}\right).$$

If the magnetic susceptibility depends on T according to the Curie law, then $d(1/\mu)/dT > 0$ and the signs of the temperature and field variations are the same. It explains cooling of the paramagnetic during demagnetization. To evaluate this effect more specifically, one can make the following substitutions in Equation (3): $C_{VB} \approx C_{V0} \approx bT^3$ (neglecting the effect of the field on the heat capacity and using the low-temperature asymptotes of the crystal heat capacity) and $1/\mu \approx 1 - 4\pi\chi = 1 - 4\pi C/T$ (assuming that $4\pi C/T \ll 1$). Then,

(4) $$dT = \frac{C}{2bT^4} dB^2.$$

The cooling effect at low temperatures is strong by virtue of the fourth power of temperature in the denominator. Because of this, temperatures below 1 K (and even down to 10^{-6} K in nuclear spin systems) can be reached (see Physical Encyclopedic Dictionary). However, one should bear in mind that the Curie law does not hold as $T \to 0$ (Problem 3.67).

3.69 In the process of isothermal magnetization a body loses heat, $Q = -CH^2/2T_0 < 0$, while in the process of demagnetization it absorbs the same amount of heat.

3.70 From Equation (3.66) we can write down the differential

(1) $$dS = \frac{1}{T}\left(\frac{\partial \mathcal{U}_*}{\partial T}\right)_M dT + \left[\frac{1}{T}\left(\frac{\partial \mathcal{U}_*}{\partial M}\right)_T - \frac{H}{T}\right] dM.$$

We equate the cross derivatives taking into account that $d(H/T)_M/dT = 0$, because the condition $M = const$ leads to the constancy of the H/T-ratio. Thus, we have

(2) $$\left(\frac{\partial \mathcal{U}_*}{\partial M}\right)_T = 0, \qquad \mathcal{U}_*(T) = \frac{1}{4} bT^4,$$

$$S(T,H) = \frac{1}{3} bT^3 - \frac{H}{T} f\left(\frac{H}{T}\right) + \int_0^{H/T} f(x)\,dx + const.$$

Here *const* can be determined using the Nernst principle $S(T,H)|_{T\to 0} \to 0$ provided the $f(x)$ function is known (Problem 3.67).

3.71

$$\left(\frac{\partial V}{\partial H}\right)_{p,T} = -\left(\frac{\partial M}{\partial p}\right)_{H,T}, \qquad \frac{\Delta V}{V} = \frac{H^2}{2}\left(\beta\chi - \frac{\partial \chi}{\partial p}\right).$$

3.72 Integrating Equations (3.66) and (3.67) or using the quantities (3.55) and (3.56) calculated above, we find

(1) $$\mathcal{U}_*(S,M) = \mathcal{U}_0(S) + \frac{M^2}{2\chi(S)}, \qquad F_*(T,M) = F_0(T) + \frac{M^2}{2\chi(T)},$$

$$S(T,M) = -\left(\frac{\partial F}{\partial T}\right)_M = S_0(T) + \frac{M^2}{2\chi^2(T)}\frac{d\chi}{dT}.$$

The argument τ is omitted here. Let us verify whether the Gibbs–Helmholtz thermodynamic identity $U_* = F_* + TS$ is satisfied. To do this, we substitute the quantities from Equation (1) to get

(2) $$U_0(S) = F_0(T) + TS_0(T) + \frac{M^2 T}{2\chi^2}\frac{d\chi}{dT}.$$

In the left-hand side of this expression, we pass to a variable T by writing $S = S_0(T) + \Delta S$, where ΔS is the entropy variation produced by the magnetic field. At the next step, we verify that Equation (2) is valid in the first order in ΔS, which corresponds to the linear relation $M = \chi H$ and to the quadratic-in-field correction to the energies. Hence, from Equation (2) we obtain the obvious relation $U_0(T) = F_0(T) + TS_0(T)$. Now in variables T and M we have

(3) $$U_*(T,M) = U_0(T) + \frac{M^2}{2\chi(T)}\left(1 + \frac{T}{\chi(T)}\frac{d\chi}{dT}\right).$$

For the paramagnetic that obeys the Curie law, $\chi = C/T$ and $U_* = U_0(T)$, in accordance with the result of Problem 3.70.

3.73

$$C_{VB} - C_{V0} = \frac{TB^2}{8\pi\mu^2}\left[\left(\frac{\partial^2\mu}{\partial T^2}\right)_\tau - \frac{2}{\mu}\left(\frac{\partial\mu}{\partial T}\right)_\tau^2\right],$$

$$C_{VE} - C_{V0} = \frac{TH^2}{8\pi}\left(\frac{\partial^2\mu}{\partial T^2}\right)_\tau.$$

3.74

(1) $$C_H - C_B = \frac{T(\partial B/\partial T)_H^2}{4\pi(\partial B/\partial H)_T}.$$

At $D = \varepsilon E$, Equation (1) yields

(2) $$C_H - C_B = \frac{TH^2}{4\pi\mu}\left(\frac{\partial\mu}{\partial T}\right)_\tau^2.$$

The same value of the difference can be obtained from the results of Problem 3.73. The condition $\tau = const$ implies that C_B and C_H are calculated at constant volume V. If a constant pressure is maintained, the derivatives in Equations (1) and (2) should be taken at $p = const$.

3.77 $\Delta C_p = -T_C a^2/2\beta$.

3.78

$$U_*(T,M) = \int C_M(T)\,dT - \frac{T_C M^2}{2C} + const,$$

$$S = \int \frac{C_M(T)}{T}\,dT - \frac{M^2}{2C} + const,$$

where the heat capacity C_M (at constant density and magnetization) is independent of magnetization.

3.79 At $T_C - T \ll T_C$, the expansion of the Brillouin function into a series up to the third-order terms yields

$$M = M_0 \sqrt{\frac{10(S+1)^2}{3[(S+1)^2 + S^2]}\left(1 - \frac{T}{T_C}\right)};$$

$$M = M_0 \left[1 - \frac{1}{S}\exp\left(-\frac{3T_C}{(S+1)T}\right)\right] \quad \text{at} \quad T \ll T_C.$$

3.80

$$\chi = \frac{C}{T - T_C} \quad \text{at} \quad T > T_C, \qquad \chi = \frac{C}{2(T_C - T))} \quad \text{at} \quad T < T_C,$$

$$M = M_0 \left(\frac{3H}{\lambda M_0}\right)^{1/3} \quad \text{at} \quad T = T_C.$$

Here, the Curie constant is $C = \mu_B^2 N$.

3.81 Setting $H = 0$ in Equation (3.81) and using Equation (3.82) we obtain the equations

$$-\frac{\boldsymbol{M}^{(b)} \cdot \boldsymbol{M}^{(a)}}{M^{(b)} M_0} = L_S\left(\frac{4\mu_B S\lambda}{T}|\boldsymbol{M}^{(b)}|\right), \quad -\frac{\boldsymbol{M}^{(a)} \cdot \boldsymbol{M}^{(b)}}{M^{(a)} M_0} = L_S\left(\frac{4\mu_B S\lambda}{T}|\boldsymbol{M}^{(a)}|\right).$$

Since their right-hand sides are positive, nonzero roots are possible only when $(\boldsymbol{M}^{(a)} \cdot \boldsymbol{M}^{(b)}) < 0$. We look for the solution in the form $\boldsymbol{M}^{(a)} = -\boldsymbol{M}^{(b)} = \boldsymbol{M}$ and get the same equations for $x = M/M_0$ as in Example 3.6. Hence, the antiferromagnetic state with a spontaneous magnetization of sublattices exists at temperatures $0 \le T \le T_C$, where the Curie temperature is given by Equation (3.74) from Example 3.6 (it is sometimes called **Neel temperature** and is denoted by T_N).

3.82 In the temperature range under consideration, the spontaneous magnetization is absent, while the induced magnetization is parallel to the local field. The left-hand sides of Equation (3.81), therefore, take the form $M^{(a,b)}/M_0$. Keeping the first non-vanishing terms in the right-hand sides, we find

$$M = M^{(a)} + M^{(b)} = \frac{C}{T + T_C} H, \qquad \chi(T) = \frac{C}{T + T_C} H,$$

where T_C is given by Equation (3.74) and $C = 4S(S+1)\mu_B^2 N/3$ (cf. the solution of Problem 1.17).

3.83 1. The external magnetic field is perpendicular to the direction of magnetization of the sublattices. In Equation (3.81), we set $\boldsymbol{M}^{(a,b)} = \boldsymbol{M}_{sp}^{(a,b)} + \boldsymbol{M}_{ind}^{(a,b)}$, where the quantities $\boldsymbol{M}_{sp}^{(a,b)}$ correspond to $\boldsymbol{H} = 0$, and $\boldsymbol{M}_{ind}^{(a,b)}$ are the terms produced by external field. We look for the solution of these equations in the form $\boldsymbol{M}_{ind}^{(a,b)} = \boldsymbol{H}/2\lambda$. After substituting into Equation (3.81), the field H drops out of the arguments of the function L, while the first-order terms in H do not enter the left-hand sides because of the condition $\boldsymbol{H}\cdot\boldsymbol{M}_{sp}^{(a,b)}=0$. Small second-order terms can be neglected. There remain only the equations for the spontaneous magnetization considered earlier in Problem 3.81. We write down $\boldsymbol{M}_{ind}=\boldsymbol{M}_{ind}^{(a)}+\boldsymbol{M}_{ind}^{(b)}=\boldsymbol{H}/\lambda$ and find

$$\chi_{\perp}=\frac{1}{\lambda}=\frac{\mu_B^2 N}{zI}\ll 1.$$

This result is valid in the range $T<T_0$, where $2\lambda M_{sp}^{(a,b)}\gg|H-2\lambda M_{ind}^{(a,b)}|$. At the Curie point $M_{sp}^{(a,b)}=0$, the antiferromagnetic becomes isotropic in this model, and the result of Problem 3.82 applies, $\chi_{\perp}=\chi_{\parallel}=\chi=1/2\lambda$.

2. The magnetic field is parallel to the direction of magnetization, $\boldsymbol{M}_{sp}^{(a)}=-\boldsymbol{M}_{sp}^{(b)}$. In Equation (3.81), we substitute $\boldsymbol{l}_a=\boldsymbol{e}_a$, $\boldsymbol{l}_b=-\boldsymbol{e}_a$, where $\boldsymbol{e}_a$ is the unit vector in the direction of $\boldsymbol{M}_{sp}^{(a)}$.

The magnetizations $\boldsymbol{M}_{ind}^{(a)}$ and $\boldsymbol{M}_{ind}^{(b)}$ are directed along the magnetic field. From Equation (3.81), we have

$$(1)\qquad M_{sp}^{(a)}+M_{ind}^{(a)}=L_S\left(\frac{2\mu_B S}{T}(2\lambda M_{sp}^{(a)}+H-2\lambda M_{ind}^{(b)})\right),$$

$$M_{sp}^{(a)}-M_{ind}^{(b)}=L_S\left(\frac{2\mu_B S}{T}(2\lambda M_{sp}^{(a)}-H+2\lambda M_{ind}^{(a)})\right).$$

When the temperature is not very close to the Curie point, the local field of sublattices is higher than the external field. We expand the right-hand sides with respect to small corrections and find from the resulting linear equations

$$(2)\quad M_{ind}=M_{ind}^{(a)}+M_{ind}^{(b)}=\frac{4\mu_B S M_0}{T}(-\lambda M_{ind}+H)L_S'\left(\frac{4\lambda\mu_B S M_{sp}^{(a)}}{T}\right).$$

This allows us to find the magnetic susceptibility:

$$(3)\qquad M_{ind}=\chi_{\parallel}H,\qquad \chi_{\parallel}=\frac{4\mu_B^2 S^2 N L_S'}{T+4\lambda\mu_B^2 S^2 N L_S'}.$$

Here L_S' is the derivative of the Brillouin function with respect to its argument. At low temperatures the argument is high, $L_S'(x)|_{x\gg 1}\approx(1/4S^2)e^{-x/S}$, and $\chi_{\parallel}\to 0$ as $T\to 0$. If $T\to T_C$, the argument is small, $L_S'(x)|_{x\ll 1}\approx(S+1)/3S$. In this limit, we obtain the same result as in Problem 3.82, $\chi_{\parallel}=C/(T+T_C)$.

3.84 According to Equations (3.89) and (3.90), $\Delta F(T,H)=F_n(T)-F_s(T,H)=-(H_c^2(T)-H^2)/8\pi$. The calculation of the entropy gives

$$\Delta S=-\left(\frac{\partial\Delta F}{\partial T}\right)_H=-\frac{H_c(T)}{4\pi}\frac{dH_c}{dT}.$$

It follows from this expression that a change in entropy is independent of the external magnetic field. At the ends of the temperature range $0\leq$

$T \leq T_c$, $\Delta S(0) = 0$ according to the Nernst principle (formulated in the condition of Problem 3.67)), and $\Delta S(T_c) = 0$ by definition of the critical temperature $H_c(T_c) = 0$. At these temperatures, the transitions $n \longleftrightarrow s$ occur without a change in temperature; they are second order phase transitions. At $0 < T < T_c$, with the aid of Equation (3.86), we have

$$\Delta S(T) = \frac{H_c^2(0)T}{2\pi T_c}\left(1 - \frac{T^2}{T_c^2}\right) > 0.$$

In this temperature range, as the external field changes, first order phase transitions occur.

3.85 The results of Problem 3.84 give

$$Q = T\Delta S = \frac{H_c^2(0)T^2}{2\pi T_c}\left(1 - \frac{T^2}{T_c^2}\right) > 0.$$

A body undergoing magnetization is heated, whereas a paramagnetic is cooled (Problem 3.69). The thermally insulated superconductor is cooled at its magnetization, while in the course of demagnetization it is heated, also in contrast to paramagnetic (Problem 3.68).

3.86

$$\Delta C = \frac{T_c}{4\pi}\left(\frac{\partial H_c(T)}{\partial T}\right)^2\Bigg|_{T_c}.$$

For lead, $\Delta C \approx 2.8 \times 10^4$ erg cm^{-3} K^{-1} $\approx 6.8 \times 10^{-4}$ cal cm^{-3} K^{-1}.

3.87 $\boldsymbol{j}_n = 0, \boldsymbol{j}_s = 0,$

$$\begin{aligned} \boldsymbol{E} &= 0, \\ rot(\lambda^2 \boldsymbol{j}_s) &= -\frac{c}{4\pi}\boldsymbol{B}, \\ rot\, \boldsymbol{B} &= \frac{4\pi}{c}\boldsymbol{j}_s, \\ div\, \boldsymbol{B} &= 0. \end{aligned}$$

Excluding $\boldsymbol{j}_s$ or $\boldsymbol{B}$ from these equations, we have

$$\begin{aligned} \triangle \boldsymbol{j}_s &= \frac{1}{\lambda^2}\boldsymbol{j}_s, \\ \triangle \boldsymbol{B} &= \frac{1}{\lambda^2}\boldsymbol{B}. \end{aligned}$$

3.88 $B_x = B_z = 0, B_y = H_0 \exp(-x/\lambda), j_x = j_y = 0,$

$$j_z = \frac{c}{4\pi}\frac{\partial B_y}{\partial x} = -\frac{cH_0}{4\pi\lambda}\exp\left(-\frac{x}{\lambda}\right).$$

3.89

$$F_x = -\frac{1}{c}\int_0^\infty j_z B_y\, dx = \frac{H_0^2}{8\pi}.$$

The force F_x tends to expell the superconductor out of the field. This is a manifestation of diamagnetism of superconductors.

3.90

(1) $$B_x = B_y = 0, \qquad B_z = H_0 \frac{\cosh(x/\lambda)}{\cosh(a/\lambda)}.$$

If $a \gg \lambda$, B_z is exponentially small throughout the bulk of the film except for thin layers near the boundaries of the film: $B_z = H_0 \exp[-(a \mp x)/\lambda]$; the minus sign refers to the boundary $x = a$, and the plus sign to the boundary $x = -a$. Thus, there occurs the magnetic field expulsion from superconductor. At $\lambda \gg a$ the field reduction is insignificant.

For the average magnetic moment, we obtain the unexpected result. Calculating per unit length in the direction of the Oy and Oz axes, we have

(2)
$$\overline{M}_z = \frac{1}{2a}\frac{1}{2c}\int_{-a}^{a} [\boldsymbol{r} \times \boldsymbol{j}_s]_z \, dx = -\frac{1}{8\pi}\int_0^a x\frac{\partial B_z}{\partial x}\, dx = -\frac{H_0}{8\pi}\left(1 - \frac{\lambda}{a}\tanh\frac{a}{\lambda}\right),$$

where $\overline{M}_z$ is antiparallel to the field (diamagnetism). However, at $\lambda \ll a$ the average magnetic moment is $\overline{M}_z \approx -H_0/8\pi$, that is, it is two times smaller by absolute value than for superconducting cylinder (Example 3.8; see also Problem 3.93).

This difference can be attributed to incorrect consideration of superconductor having an infinitely large size along the Oy-axis. In reality, any superconductor is finite. Let the width of the film along the Oy-axis be $2b$, with $b \gg a, \ \lambda$, whereas the half-thickness of the film a is comparable with λ (Figure 3.14). Hence, the solution (1) can be used everywhere inside the film, except for the regions near the boundaries, $y = \pm \pm b$. In view of the inequality $b \gg \lambda$, we can assume that a current flowing along

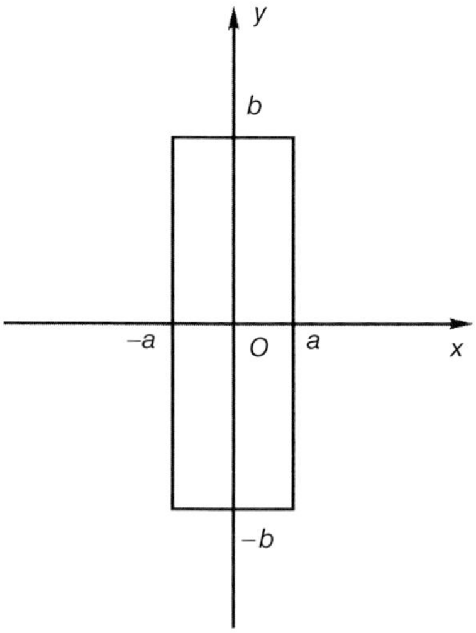

Figure 3.14 Elucidation to calculation of field in a superconducting film.

these boundaries is the surface current. Calculate now the average magnetization per unit length along the Oz-axis:

$$(3) \qquad \overline{M}_z = \frac{1}{4ab}\frac{1}{2c}\int_{-b}^{b} dy \int_{-a}^{a} [\boldsymbol{r}\times\boldsymbol{j}_s]_z\, dx,$$

where we have to take into account that the surface current closes along the boundaries $y = \pm b$:

$$(4) \qquad \boldsymbol{j}_s = -\frac{c}{4\pi}\frac{\partial B_z}{\partial x}\,\boldsymbol{e}_y - \frac{cH_0}{4\pi}\,[\delta(y+b) - \delta(y-b)]\,\boldsymbol{e}_x.$$

Here, the surface currents are expressed with the aid of a delta-function, and Equation (4) relates only to the region inside the film and to its boundaries. Substituting Equation (4) into Equation (3), we find the additional term to Equation (2):

$$(5) \qquad \overline{M}_z = -\frac{H_0}{8\pi}\left(1 - \frac{\lambda}{a}\tanh\frac{a}{\lambda}\right) - \frac{H_0}{8\pi}.$$

The removed narrow edge of the film gives the correction that exceeds the contribution from the wide surface. At $\lambda \ll a$, the film behaves like a perfect diamagnetic: $\overline{M}_z = -H_0/4\pi$, $\chi = -1/4\pi$, and $\mu = 1 + 4\pi\chi = 0$.

3.91 In the regions $x \geq a$ and $x \leq -a$, there exists the uniform field $H_0 = \pm 2\pi i/c$, which is parallel and antiparallel to the Oy-axis, respectively. Inside the film

$$B_y(x) = H_0\frac{\sinh(x/\lambda)}{\sinh(a/\lambda)}, \qquad j_{sz} = \frac{cH_0}{4\pi\lambda}\frac{\cosh(x/\lambda)}{\sinh(a/\lambda)}.$$

At $\lambda \ll a$, the current flows in a thin surface layer, whereas there is neither field nor current inside the film (cf. the result of Problem I.2.76, case (b)). At $\lambda \gg a$ the current is uniformly distributed over the cross section, and the field in the film $B_y(x) = H_0 x/a$ is a linear function of x.

3.92 On the surface of a superconductor, the boundary condition must be satisfied, which requires the normal component of the resultant current be zero: $H'_x(0, y, z) = 0$, while in the region $x > 0$ the Maxwell equations and the equations for the vector potential with a given current must hold. All these requirements will be met provided the field in vacuum is produced by the given current-bearing conductors and their mirror images in the $x = 0$ plane, and the conductors-images carry the currents of the same strength but opposite direction.

3.93

$$B_z = H_0\frac{I_0(r/\lambda)}{I_0(a/\lambda)}, \qquad \overline{M}_z = -\frac{H_0}{4\pi}\left[1 - 2\frac{\lambda}{a}\frac{I_1(a/\lambda)}{I_0(a/\lambda)}\right],$$

where I_0 and I_1 are the modified Bessel functions.

3.94 Outside the sphere,

$$H_r = \left(H_0 + \frac{2m}{r^3}\right)\cos\vartheta, \qquad H_\vartheta = \left(-H_0 + \frac{m}{r^3}\right)\sin\vartheta,$$

where m is a constant equal to the magnetic moment. Inside the sphere,

$$j_\alpha = f(r)\sin\vartheta, \qquad j_r = j_\vartheta = 0.$$

The functions $j_\alpha(r,\vartheta)$ satisfy the equation

$$\triangle j_\alpha - \frac{1}{r^2}\sin^2\vartheta\cdot j_\alpha = 0$$

(see the solution of Problem I.1.91), which yields

$$j_\alpha(r,\vartheta) = \frac{cA}{4\pi r^2}\left(\sinh\frac{r}{\lambda} - \frac{r}{\lambda}\cosh\frac{r}{\lambda}\right).$$

Here A is the constant of integration. The components B_r and B_ϑ of the magnetic field inside the sphere are expressed in terms of $j_\alpha(r,\vartheta)$:

$$B_r = \frac{2\lambda^2 A}{r^3}\left(\sinh\frac{r}{\lambda} - \frac{r}{\lambda}\cosh\frac{r}{\lambda}\right)\cos\vartheta,$$
$$B_\vartheta = \frac{2\lambda^2}{r^3}\left[\left(1+\frac{r^2}{\lambda^2}\right)\sinh\frac{r}{\lambda} - \frac{r}{\lambda}\cosh\frac{r}{\lambda}\right]\sin\vartheta.$$

The constants m and A are found from the continuity conditions for the field vectors at $r = a$.

$$m = -\frac{H_0 a^3}{2}\left(1 - 3\frac{\lambda}{a}\coth\frac{a}{\lambda} + 3\frac{\lambda^2}{a^2}\right), \qquad A = -\frac{3H_0 a}{2\sinh(a/\lambda)}.$$

If $\lambda \ll a$, one has $m = -H_0a^3/2$ (cf. the solution of Problem 3.37 at $\mu = 0$), $A = 0$. At $\lambda \gg a$, we obtain $m = -H_0a^5/30\lambda^2$.

3.95

$$j_r = j_\alpha = 0, \qquad j_z = \frac{J}{2\pi a\lambda}\frac{I_0(r/\lambda)}{I_1(a/\lambda)}, \qquad B_r = B_z = 0,$$

$$B_\alpha = \begin{cases} \dfrac{J}{2\pi ca}\dfrac{I_0(r/\lambda)}{I_1(a/\lambda)} & \text{at} \quad r < a, \\[2ex] \dfrac{J}{2\pi ca} & \text{at} \quad r > a, \end{cases}$$

where I_0 and I_1 are the modified Bessel functions.

3.96 $J' = -cH_0S\cos\vartheta/L + J$.

3.97 $J = c\Phi_0/L$.

3.98 $J_c = (ca/2)\sqrt{H_c^2 - H_0^2}$, where H_c is the critical field at the temperature under consideration.

3.99 $J_c = (ca/2)(H_c - 2H_0)$.

3.102 1. Assume that $\mathbf{A} = 0$. In this case, the imaginary unit drops out of the system of Equations (3.110)–(3.112), and ψ can be regarded as a real quantity. Also, we suppose that the surface $z = 0$ is a boundary between the superconductor and normal metal, and the boundary condition (3.112) is not fulfilled. In the case of a weak inhomogeneity of ψ, we can write $\psi = 1 - \zeta(z)$, $|\zeta(z)| \ll 1$, and linearized Equation (3.112):

$$-\xi^2 \frac{d^2\zeta}{dz^2} + 2\zeta = 0.$$

The solution describing the homogeneous state inside the superconductor has the form $\zeta = \zeta_0 \exp(-\sqrt{2}z/\xi)$. Thus, ξ is the scale of the order parameter of inhomogeneity or the scale of inhomogeneity for the distribution of Cooper pairs; it is also called **coherence length**. This quantity increases near the critical temperature as $\xi \propto (T_c - T)^{1/2}$.

2. The quantity λ, defined by Equation (3.109), is identical with λ in the Londons' theory and characterizes the penetration depth of the magnetic field into the superconductor. Equation (3.111) transforms into Equation (3.92) of the Londons' theory if $\psi \approx 1$.

   The properties of a superconductor strongly depend on the ratio $\kappa = \lambda/\xi$. At $\kappa \ll 1$, the magnetic field weakly affects the order parameter and the local Londons' theory is applicable, in which the electric current depends on the magnetic field at the same point. At $\kappa \gg 1$, the relation connecting the current and the field becomes non-local (Pippard's case).[19]

3.103 Far from the cavity border, where the superconductor is homogeneous, the most general expression for the wave function is $\psi(\mathbf{r}) = \exp(i\phi(\mathbf{r}))$, where ϕ is the real phase. In this case, the superconducting current (Equation (3.108)) takes the form

(1) $$\mathbf{j}_s = -\frac{c}{4\pi\lambda^2}\left(\frac{\Phi_0}{2\pi}\nabla\phi(\mathbf{r}) + \mathbf{A}(\mathbf{r})\right),$$

where Φ_0 is determined by Equation (3.87). Let us construct a contour, which encloses the cavity in the depth of the superconductor. Integration of the current along the contour yields $\oint_l \mathbf{j}_s \cdot d\mathbf{l} = 0$, since there is no current inside the superconductor. In combination with Equation (1), this gives

(2) $$\oint_l \mathbf{A} \cdot d\mathbf{l} = -\frac{\Phi_0}{2\pi}\oint_l \nabla\phi \cdot d\mathbf{l}.$$

However, $\oint_l \mathbf{A} \cdot d\mathbf{l} = \int_S rot\mathbf{A} \cdot d\mathbf{S} = \Phi$ is the magnetic flux through the area bounded by the contour; $\oint_l \nabla\phi \cdot d\mathbf{l}$ is the phase change of the wave function when going around the contour. In view of the requirement of unambiguity of the wave function, this change is a multiple of 2π:

(3) $$\oint_l \nabla\phi \cdot d\mathbf{l} = 2\pi n, \qquad n = 0, \pm 1, \pm 2, \ldots$$

Hence, the absolute value of the magnetic flux is subject to quantization (Equation (3.87)).

19) Pippard Alfred (1920–2008), English physicist, known for works in solid-state physics and superconductivity.

3.104 The current is uniformly distributed within the film while the magnetic flux is concentrated inside the cylinder (up to the quantities of the order of $\delta/a \ll 1$). By integrating the superconducting current (see Equation (1) of the Problem 3.103) along a closed contour, which lies inside the film, we have

$$\oint_l \boldsymbol{j}_s \cdot d\boldsymbol{l} = 2\pi a\, j_s = -\frac{c}{4\pi\lambda^2}(\Phi + n\Phi_0).$$

The current density j_s can be expressed in terms of the magnetic flux inside the cylinder using the formula obtained in Problem I.2.83 assuming that the current is the surface current in view of the condition $\delta \ll a$: $j_s = cB/4\pi\delta = c\Phi/4\pi^2a^2\delta$. Finally, we obtain the quantized absolute value of the magnetic flux

$$\Phi = \frac{n\Phi_0}{1 + 2\lambda^2/a\delta}, \qquad n = 0, 1, \ldots$$

4
Quasi-Stationary Electromagnetic Field

4.1
Quasi-Stationary Phenomena in Linear Conductors

Conditions of quasi-stationarity. This chapter is focused on electromagnetic phenomena caused by relatively slow changes in the field. When the period T of oscillations of electromagnetic field is much larger than the time of field propagation through a given system,

$$T \gg \frac{l}{c}, \qquad \omega \ll \frac{c}{l}, \tag{4.1}$$

where c is the velocity of light and l is a linear size of the system, then the velocity of propagation of electromagnetic disturbances through the system can be assumed to be infinite. Under these conditions, the electromagnetic field is determined by instantaneous values of charges and currents in the system. The corresponding approximation is known as the **quasi-stationary approximation**.[1] We begin with quasi-stationary phenomena in quasi-linear conductors.

Let an external electromagnetic field $\mathcal{E}_{ext}$ be included in a quasi-linear closed and nondeformable contour. If the electromagnetic field is constant, the current J, produced by it, is direct according to the Ohm law (3.17): $\mathcal{E}_{ext} = JR$, where R is a direct current resistance of the contour. If the electromotive force $\mathcal{E}_{ext}(t)$ varies in time, then the alternating current in the contour induces an additional electromotive force, an electromotive force of induction, in accordance with the Faraday law of electromagnetic induction (I.2.79), [2]

$$\mathcal{E}_{ind} = -\frac{1}{c}\frac{d\Phi}{dt}, \tag{4.2}$$

and the Ohm law takes the form

$$\mathcal{E}_{ext}(t) + \mathcal{E}_{ind}(t) = RJ(t).$$

Considering the electromagnetic field in vacuum in Section I.2.3, we meant that Φ is the flux of the magnetic field strength $\boldsymbol{H}$. For media, however, by $\boldsymbol{H}$ one should

1) The application of the quasi-stationary approximation to transmission lines whose length L is unlimited by the conditions (4.1) can be found in Problems 4.20 – 4.26.
2) This formula is equally applicable for motion or deformation of a circuit (see Example 4.4 below).

mean the flux of magnetic induction $\boldsymbol{B}$ through the surface resting on a closed contour: $\Phi = \int_S \boldsymbol{B} \cdot d\boldsymbol{S}$. Then Equation (4.2) represents the integral form of the macroscopic equation (1.8).

In the quasi-stationary approximation, the magnetic flux is determined by an instantaneous current in the circuit given by Equation (3.64a) from magnetostatics:

$$\Phi(t) = \frac{J(t)}{c} L,$$

where L is the self-inductance coefficient of the contour. As a result, the Ohm law for a closed contour with an alternating electromotive force takes the form[3)]

$$\mathcal{E}_{ext}(t) = RJ(t) + \frac{L}{c^2} \frac{dJ(t)}{dt}. \tag{4.3}$$

Example 4.1

An external electromotive force varies as

$$\mathcal{E}_{ext}(t) = \begin{cases} \mathcal{E} = const, & t \leq 0; \\ 0, & t > 0. \end{cases}$$

Find the current in the circuit as a function of time.

Solution. At $t \leq 0$, we have a direct current given by Equation (3.17), $J = \mathcal{E}/R$. At $t > 0$, the equation for the current is

(1) $$\frac{L}{c^2} \frac{dJ}{dt} + RJ = 0,$$

and the initial condition is $J(0) = \mathcal{E}/R$. The solution is

(2) $$J(t) = \frac{\mathcal{E}}{R} e^{-t/\tau}, \qquad \text{where} \quad \tau = \frac{c^2 R}{L}.$$

It describes the current in the circuit decaying with time. ■

Active and reactive resistances.

Example 4.2

An external electromotive force depends harmonically on time, $\mathcal{E}_{ext}(t) = \mathcal{E} e^{-i\omega t}$. Find the relation between the electromotive force and the current in a steady regime.

Solution. For a steady state, we seek the particular solution $J(t) = J_0 e^{-i\omega t}$ of Equation (4.3) and find **the impedance of the circuit**

(1) $$J_0 = \frac{\mathcal{E}}{Z(\omega)}, \qquad \text{where} \qquad Z(\omega) = R - i\frac{\omega L}{c^2}$$

3) Here, the resistance R is thought to be the same as for steady current. For sufficiently high frequencies, this assumption breaks down because of the skin-effect, which is a concentration of current in a thin surface layer of the conductor. This problem is discussed in Section 4.2.

is **the complex resistance (impedance)** of the circuit. Its real and imaginary parts are called **active** and **reactive** resistances, respectively. The complex character of impedance suggests a phase shift between the current and electromotive force. Separating the real part in the Ohm law for the case of the harmonically oscillating current $J(t) = \mathcal{E}_{ext}(t)/Z(\omega)$, we have

$$(2) \qquad J(t) = \frac{\mathcal{E}}{\sqrt{R^2 + \omega^2 L^2/c^4}} \cos(\omega t - \varphi), \qquad \tan\varphi = \frac{\omega L}{c^2 R}.$$

In this case, the current lags in phase behind the electromotive force. ■

When the circuit with a variable electromotive force includes a capacitance C, the charge $\pm q(t)$ on the capacitor plates, and the potential difference $V(t) = q(t)/C$ between the plates (see Equation (2.11)) will alternatively arise. This potential difference should be added to the right-hand side of Equation (4.3), expressing the current intensity as a change in charge per unit time on one of the plates,

$$J(t) = \frac{dq}{dt}, \qquad \frac{L}{c^2}\frac{d^2q}{dt^2} + R\frac{dq}{dt} + \frac{q}{C} = \mathcal{E}_{ext}(t). \tag{4.4}$$

This equation describes electric oscillations in the circuit which includes the capacitance, inductance, and active resistance.

Example 4.3

Calculate the impedance of the circuit with capacitance, inductance, and active resistance (see Example 4.2). Find the current in the circuit. Find the frequency of free oscillations in the circuit and their damping decrement.

Solution. Let the external electromotive force is given by $\mathcal{E}_{ext}(t) = \mathcal{E}e^{-i\omega t}$. Let us find a particular solution $q(t) = q_0\, e^{-i\omega t}$ of Equation (4.4). The solution for the current $J = \dot{q}$ is

$$(1) \qquad J(t) = \frac{\mathcal{E}_{ext}(t)}{Z(\omega)}, \qquad \text{where} \qquad Z(\omega) = R - i\left(\frac{\omega L}{c^2} - \frac{1}{\omega C}\right).$$

The reactive resistance (the imaginary part of Z) depends now on the capacitance as well. The measured current intensity (the real part of complex expression (1)) can be written as

$$(2) \qquad J(t) = \frac{\mathcal{E}\cos(\omega t - \varphi)}{\sqrt{R^2 + \left(\frac{\omega L}{c^2} - \frac{1}{\omega C}\right)^2}}, \qquad \tan\varphi = \frac{1}{R}\left(\frac{\omega L}{c^2} - \frac{1}{\omega C}\right).$$

Free oscillations in the circuit occur when $\mathcal{E} = 0$. This is possible provided the denominator in Equation (1) becomes zero, which corresponds to the equality $Z(\omega) = 0$. This condition determines the frequencies of the free oscillations (or eigenoscillations) ω_0. The solution of the quadratic algebraic equation yields

$$(3) \qquad \omega_0 = \omega_0' + i\omega_0'' = \pm\sqrt{\frac{c^2}{LC} - \left(\frac{Rc^2}{2L}\right)^2} - i\frac{Rc^2}{2L}.$$

If the circuit includes the active resistance R, the eigenfrequency ω_0 is complex, and the oscillations decay with the decrement $\gamma = -\omega_0'' = Rc^2/2L$. Neglecting the damping (small R), the eigenfrequency of non-attenuating oscillations is given by

$$\omega_0 = \frac{c}{\sqrt{LC}}, \tag{4.5}$$

which is known as **Thomson formula**. ■

In the case of a branching circuit, the differential equations for the currents in various branches that may also include capacitors can be derived from the Kirchhoff laws (see Section 3.1). If the contours are inductively coupled, the induction electromotive force expressed in accordance with Equation (4.2) should be included in the equations along with the external electromotive force. The magnetic fluxes taking into account all contours can be expressed in terms of the contour currents and of the inductance coefficients by means of (3.64a),

$$\Phi_a = \frac{1}{c}\sum_b L_{ab}J_b. \tag{4.6}$$

Electromagnetic induction law for a moving contour.

Example 4.4

Equations (I.2.79) and (4.2) express the Faraday electromagnetic induction law for a stationary contour. Derive this law for a moving and deformed closed contour placed in an alternating electromagnetic field.

Solution. In the deformed contour l moving with a nonrelativistic velocity $\boldsymbol{u}(\boldsymbol{r})$, the electric field is $\boldsymbol{E}' = \boldsymbol{E} + \boldsymbol{u}\times\boldsymbol{B}/c$, according to Equation (I.4.70), where one should set $\gamma \approx 1$. By definition of electromotive force of induction, we have

(1) $$\mathcal{E}'_{ind} = \oint_l \boldsymbol{E}'\cdot d\boldsymbol{l} = \oint_l \boldsymbol{E}\cdot d\boldsymbol{l} + \frac{1}{c}\oint_l [\boldsymbol{u}\times\boldsymbol{B}]\cdot d\boldsymbol{l}.$$

The first integral in the right-hand side of this equation is the electromotive force of induction, $\mathcal{E}_{ind}$, Equation (I.2.79), in the stationary circuit l since the velocity does not enter this integral. In the second integral, the velocity $\boldsymbol{u}$ of the contour element $d\boldsymbol{l}$ should be replaced by the quantity $d\boldsymbol{s}/dt$, where $d\boldsymbol{s}$ is the vector of a small shift of the contour element over time interval dt. Hence, we have the integral $\oint_l [d\boldsymbol{s}\times\boldsymbol{B}]\cdot d\boldsymbol{l} = -\int_{\Delta S} \boldsymbol{B}\cdot d\boldsymbol{S}$, where $d\boldsymbol{S} = d\boldsymbol{s}\times d\boldsymbol{l}$ is the element of an additional surface ΔS traced by the contour l as a result of its motion and deformation during the time interval dt (the field $\boldsymbol{B}$ can be regarded as constant). Finally, in the right-hand side of Equation (1) we have two terms: one of them contains the rate of magnetic flux changes due to magnetic field variations and the other contains that due to contour motion:

(2) $$\mathcal{E}'_{ind} = -\frac{1}{c}\left.\frac{d\Phi}{dt}\right|_{u=0} - \frac{1}{c}\left.\frac{d\Phi}{dt}\right|_{B=const} = -\frac{1}{c}\frac{d\Phi}{dt}.$$

Thus, the electromotive force of induction in a moving contour is determined by the total rate of changes of the magnetic flux through the contour. Equation (4.2)

remains valid in any cause of changes of the magnetic flux through the contour.
■

The interaction forces between the contours carrying quasi-stationary currents can be evaluated from Equation (3.65) by differentiating the magnetic energy with respect to generalized coordinates.

The work of external forces on charged particles in a contour per unit time is given by Equation (3.21), $Q = \mathcal{E}_{ext}J$. In the case of direct current, this work is converted into heat, and the quantity Q is the heat release in the circuit. In the case of an alternating periodic current, a part of this quantity describes the energy fluctuations from particles to the external source and back. In order to obtain the period-averaged heat release, it is necessary to average the variable quantity over the period,

$$Q = \overline{\mathcal{E}_{ext}(t)J(t)}. \tag{4.7}$$

In all such quadratic expressions, one needs to use the real functions $\mathcal{E}_{ext}(t)$ and $J(t)$. In the case of the harmonic complex function $J(t) = J_0\, e^{-i\omega t}$, we have (see similar formulas in Section I.2.3)

$$Q = \frac{1}{2}Re(\mathcal{E}_{ext}J\ *) = \frac{1}{2}|J|^2 ReZ(\omega). \tag{4.8}$$

Suggested literature: Smythe (1950); Tamm (1976); Frenkel (1928); Vlasov (1955); Landau *et al.* (1984); Simonyi (1956); Brillouin and Parodi (1956); Panofsky and Phillips (1963); Toptygin (2014); Frenkel (1935)

Problems

4.1 A circular loop of wire of radius a is placed in a constant magnetic field H_0 and rotates with an angular velocity ω about a diameter, which is perpendicular to H_0. Find the current in the loop, $J(t)$, the retarded torque $N(t)$, and the average power, $\overline{P}$, which is required to maintain the rotation.

4.2 A plane contour with the electrical parameters R, L, and C and of area S rotates with an angular velocity ω in the constant magnetic field H_0 about an axis, which lies in the plane of the contour and is perpendicular to $\boldsymbol{H}_0$. Determine the average retarded torque $\overline{N}$ on the contour.

4.3 The current through one of the two inductively coupled contours is $J(t) = J_0\, e^{-i\omega t}$. The inductance and resistance of the two contours are given. Express the average generalized force of interaction between the contours in terms of the derivative of the mutual inductance with respect to the generalized coordinate q_i.

4.4 Each of two identical contours each has a resistance R and inductance L. One of them includes a source of electromotive force $\mathcal{E}(t) = \mathcal{E}_0\, e^{-i\omega t}$. The mutual inductance of the contours is L_{12}. Determine the time-averaged force of interaction between the contours, $\overline{F}$. Express the result in terms of

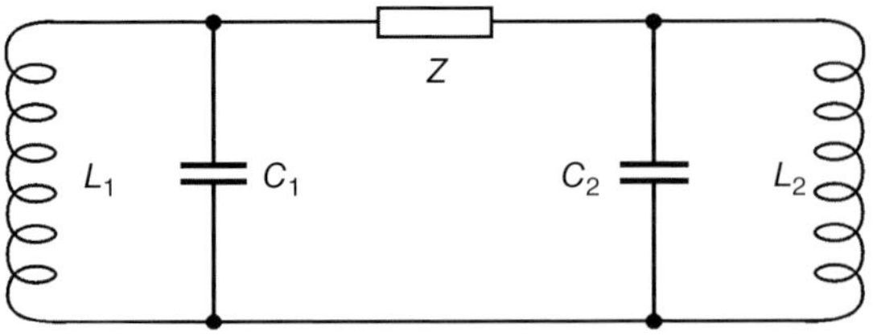

Figure 4.1 Oscillations in the coupled contours.

the derivative of the mutual inductance with respect to the corresponding coordinate.

4.5 Determine the frequencies ω_1 and ω_2 of free electrical oscillations in the two contours (Figure 4.1) if the coupling between the contours is achieved through a capacitance C ($Z = i/\omega C$).

Hint. Set up a system of algebraic equations to find the currents, and equate the determinant of the system to zero.

4.6 Solve Problem 4.5 for the case when the coupling between the contours is achieved through inductance (see Figure 4.1, $Z = -i\omega L/c^2$).

4.7 Find the eigenfrequencies of oscillations ω_1 and ω_2 in two inductively coupled circuits having capacitances C_1 and C_2, inductances L_1 and L_2, and mutual inductance L_{12}.

4.8 Two circuits are coupled to each other through an active resistance (see Figure 4.1, $Z = R$). Find eigenfrequencies of oscillations assuming weak coupling (when R is large).

4.9 An external electromotive force, $\mathcal{E}(t) = \mathcal{E}_0\, e^{-i\omega t}$, is included into a contour with an inductance L_1, capacitance C_1, and resistance R_1. The contour is inductively coupled to the other contour with parameters L_2, C_2 and R_2, and the mutual inductance L_{12}. Find the currents J_1 and J_2 in the two contours. Consider, in particular, the case in which the second contour has only inductance ($R_2 = 0$, $C_2 = \infty$); determine the frequency ω at which the current J_1 reaches maximum.

4.10 Find the complex impedance Z of the circuit segment (a two-terminal network) shown in Figure 4.2.

4.11 A capacitor is filled with matter whose permittivity is

$$\varepsilon = 1 - \frac{\omega_p^2}{\omega(\omega + i\gamma)}$$

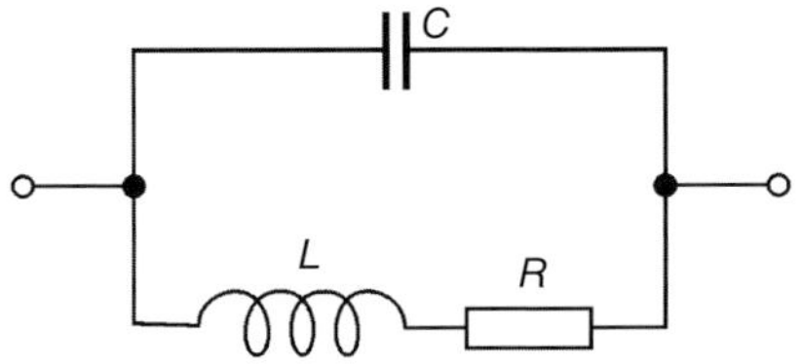

Figure 4.2 Two-terminal network.

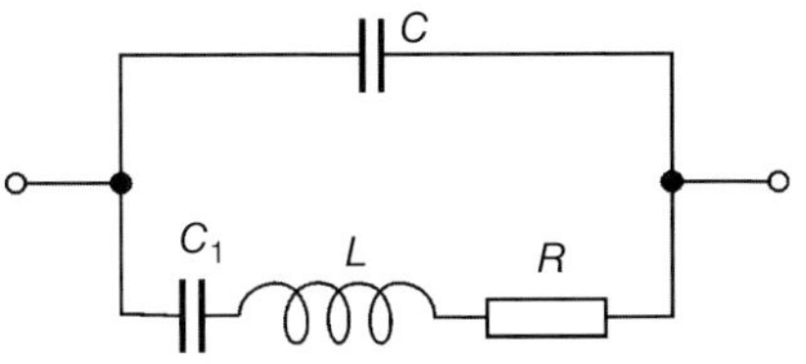

Figure 4.3 Equivalent circuit of the capacitor filled with dielectric with losses.

(ionized gas; see Problem 5.17). In the absence of medium, the capacitance is C_0. Show that the complex impedance of the circuit segment with such a capacitor is equal to the impedance of the two-terminal network shown in Figure 4.2 provided the parameters of the circuit are properly chosen. Determine R, L, and C.

4.12 Determine the average energy $\overline{W}$ stored per unit time in the capacitor described in Problem 4.11. Find also the thermal energy loss Q per unit time and express both quantities in terms of the voltage at the capacitor plates, $U = U_0\,e^{-i\omega t}$.

4.13 A capacitor is filled with a dielectric having a permittivity $\varepsilon = 1 + \omega_p^2/(\omega_0^2 - i\gamma\omega - \omega^2)$ (dielectric with losses, see Problem 5.13). In the absence of the dielectric, the capacitance is C_0. What are the parameters C, C_1, L, and R of the two-terminal network shown in Figure 4.3 at which its alternative-current impedance is equal to the impedance of the capacitor?

4.14 Find the average storage of energy $\overline{W}$ and the average heat loss Q per unit time in the capacitor considered in Problem 4.13. The voltage at the capacitor plates is $U_0\,e^{-i\omega t}$.

4.15 An oscillatory contour consists of a capacitance C and inductance L. At a certain instant of time a battery of constant electromotive force $\mathcal{E}$ and internal resistance R is connected across the capacitor plates. Find the current flowing through the inductance as a function of time. Investigate the dependence of this current on R, L, and C.

4.16 A rectangular voltage pulse $U_1(t) = U_0$ for $0 \le t \le T$, and $U_1(t) = 0$ for $t < 0$, $t > T$, is applied to a circuit consisting of a resistance R and capacitance C connected in series. Find the voltage $U_2(t)$ across the resistance R.

4.17 A rectangular voltage pulse $U_1(t) = U_0$ for $0 \le t \le T$, and $U_1(t) = 0$ for $t < 0$, $t > T$, is applied to the circuit consisting of a resistance R and inductance L connected in series. Find the voltage $U_2(t)$ across the inductance L.

4.18 A circuit consists of a plane capacitor having a capacitance C and resistance R (Figure 4.4). It is necessary to produce a field between the capacitor plates (the distance h) which will increase linearly from 0 to E_0 during time T and then decrease linearly to zero during equal interval of time. Determine the shape of the pulse, which should be applied across the circuit input in this case.

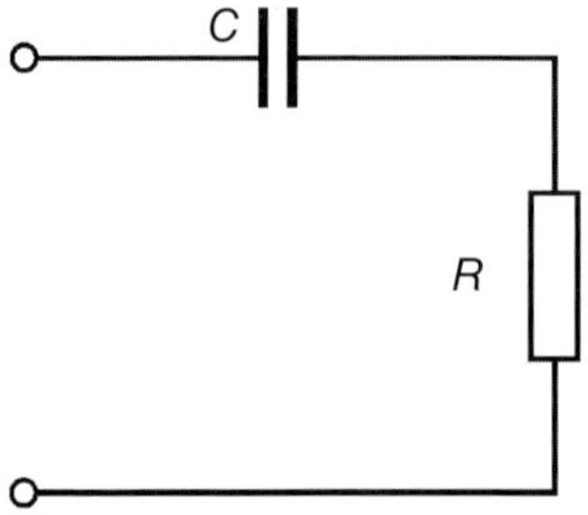

Figure 4.4 Electric circuit.

4.19 Electromotive force $\mathcal{E}(t) = \mathcal{E}_0 \cos(\omega t + \varphi_0)$ is switched on at $t = 0$ across a circuit consisting of a resistance R and a capacitance C connected in series. Determine the current $J(t)$ in the circuit and also find the phase φ_0 at which the transient phenomena will not arise in the circuit.

4.20* An electric circuit (an artificial long line) consists of N identical sections ($N \gg 1$) and is open at each end (Figure 4.5). Determine the frequencies of eigenoscillations of the system.

4.21 Assuming that the total number of eigenfrequencies in the artificial long line (see Problem 4.20) is large, find the number Δr of oscillations in a frequency interval $\Delta\omega$.

4.22* A long artificial line consists of $2N$ alternating sections with parameters L_1, C, and L_2, C, and is open at each end (Figure 4.6). Investigate the spectrum of eigenoscillations of the system.

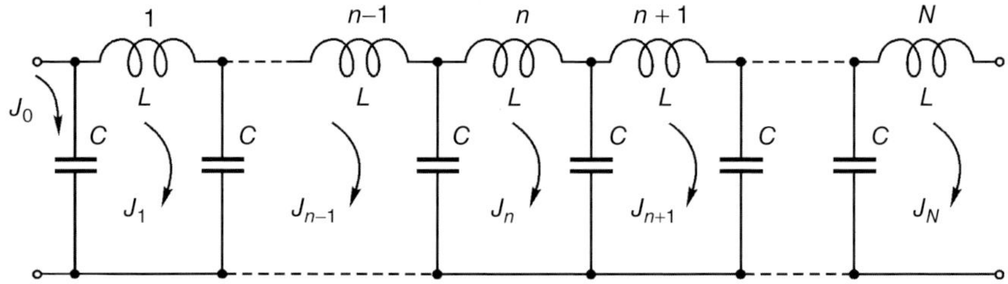

Figure 4.5 An artificial long line with identical sections.

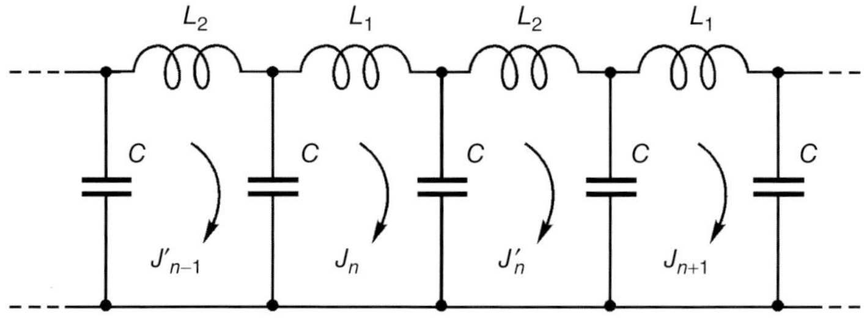

Figure 4.6 An artificial long line with different inductances.

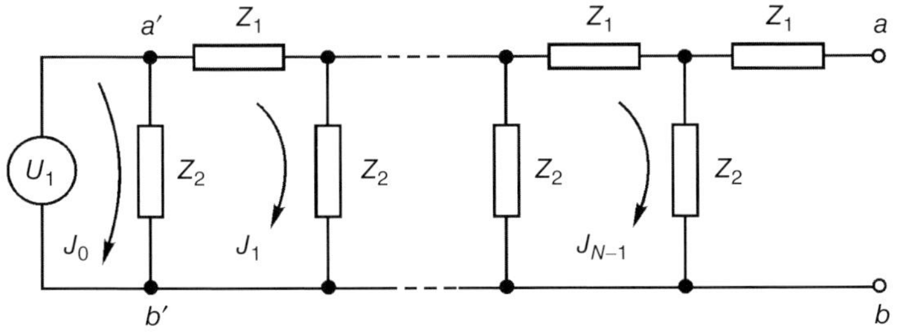

Figure 4.7 Artificial long line with arbitrary impedances.

4.23* A long artificial line (Figure 4.7) consists of N identical sections whose impedances are given by

$$Z_1 = -i\left(\frac{\omega}{c^2}L_1 - \frac{1}{\omega C_1}\right), \qquad Z_2 = -i\left(\frac{\omega}{c^2}L_2 - \frac{1}{\omega C_2}\right).$$

The voltage U_1 is applied across one end of the line while the other end is open. Find the voltage U_2 between points a and b.

Hint. Find the solution of the finite difference equation for the current J_n in the nth section of the circuit in the form $J_n = const \cdot q^n$.

4.24 Investigate the transmission coefficient $K = U_2/U_1$ as a function of frequency based on the results of Problem 4.23 and assuming $N \gg 1$. Find the frequency interval in which K differs noticeably from zero.

4.25* From consideration of the artificial long line with the lumped parameters (Problem 4.20), obtain the differential equation for the current flowing in a long line with uniformly distributed parameters (as a limiting case).

4.26 An ideal ($R = 0$) long line with length l and distributed parameters is open at both ends. Determine the spectrum of eigenoscillations of the system and compare it with the spectrum of the line with lumped parameters (see Problem 4.20).

4.27* Electromotive force is applied to a closed circuit and induces a current $J(t) = J_0\, e^{-i\omega t}$. Find a general expression for the complex impedance of the circuit without neglecting retardation within the system.

4.28 Find the correction to the inductance and resistance $R_r(\omega)$ of the circular contour with radius a in the first nonvanishing approximation (see Problem 4.27). Show that $R_r(\omega)$ is the coefficient of proportionality between the average energy emitted per unit time and the root-mean-square current intensity in the contour.

4.2
Eddy Currents and Skin-Effect

Equations of quasi-stationary field in massive conductors In this section, we consider quasi-stationary phenomena in massive conductors placed in external variable electromagnetic field which satisfies the quasi-stationary conditions (4.1). In this case, the magnetic field outside the conductor (in vacuum or dielectric) satisfies the following equations of magnetostatics:

$$\nabla \cdot \boldsymbol{B} = 0, \qquad \nabla \times \boldsymbol{H} = 0 \tag{4.9}$$

and the equation of electromagnetic induction (1.8) in differential form:

$$\nabla \times \boldsymbol{E} = -\frac{1}{c}\frac{\partial \boldsymbol{B}}{\partial t}, \qquad \text{where} \qquad \boldsymbol{B} = \mu \boldsymbol{H}. \tag{4.10}$$

Inside the conductor Equation (4.10) holds, while Equation (4.9) is replaced by Equation (1.9) without any displacement current:

$$\nabla \cdot \boldsymbol{B} = 0, \qquad \nabla \times \boldsymbol{H} = \frac{4\pi}{c}\boldsymbol{j}. \tag{4.11}$$

In good conductors, the current is largely produced by conduction electrons, while the contribution from the bound charges is rather small; therefore, in Equation (4.11) it is possible to use the Ohm's law (1.25) and write down the equation

$$\nabla \times \boldsymbol{H} = \frac{4\pi}{c}\kappa \boldsymbol{E}, \tag{4.12}$$

where κ is the electric conductivity of the conductor. Such current is called **eddy** (or Foucault)[4] current and induced by an eddy electric field. Neglecting the displacement current $(\partial \boldsymbol{E}/c\partial t)$ means the inequality

$$4\pi\kappa \gg \omega, \tag{4.13}$$

where ω is the characteristic frequency of field variations. The electric field inside the conductor can be found from Equation (4.12).

For a homogeneous conductor ($\kappa = const, \; \mu = const$), excluding $\boldsymbol{E}$ and $\boldsymbol{H}$ from the above equations, we obtain the identical differential equations, of second-order in coordinates and first-order in time, for the field vectors:

$$\triangle \boldsymbol{H} = \frac{4\pi\mu\kappa}{c^2}\frac{\partial \boldsymbol{H}}{\partial t}, \qquad \triangle \boldsymbol{E} = \frac{4\pi\mu\kappa}{c^2}\frac{\partial \boldsymbol{E}}{\partial t}. \tag{4.14}$$

These equations are significantly different from the wave equations describing propagation of electromagnetic waves in vacuum (see Chapter I.5). In a conducting medium, the field propagates according to the laws of diffusion or thermal conductivity,

$$\frac{\partial \boldsymbol{H}}{\partial t} = \chi \triangle \boldsymbol{H}, \tag{4.15}$$

where $\chi = c^2/4\pi\mu\kappa$ is the diffusion coefficient of the magnetic field.

On the boundaries separating two conductors or a conductor and dielectric, the field vectors must satisfy the conditions

$$\boldsymbol{n} \cdot (\boldsymbol{B}_1 - \boldsymbol{B}_2) = 0, \qquad \boldsymbol{n} \times (\boldsymbol{H}_1 - \boldsymbol{H}_2) = 0, \qquad \boldsymbol{n} \times (\boldsymbol{E}_1 - \boldsymbol{E}_2) = 0. \tag{4.16}$$

4) Jean Bernard Foucault (1819–1868), French physicist-experimentalist and inventor.

Example 4.5

At the plane conductor boundary one maintains the magnetic field $\boldsymbol{H}(t) = \boldsymbol{H}_0 e^{-i\omega t}$, $\omega =$ *const, which is tangential to the boundary. Evaluate the electric and magnetic fields inside the conductor, estimate their relative strength and penetration depth into the conductor.*

Solution. Let the Oz-axis be directed within the conductor. We find the solution of the first equation (4.14) in the form of $\boldsymbol{H}(z,t) = \boldsymbol{H}(z)\, e^{-i\omega t}$. Using the boundary conditions (4.16), we find $\boldsymbol{H}(z) = \boldsymbol{H}_0 e^{-(1-i)z/\delta}$, where the penetration depth (skin-layer thickness) is given by

$$\delta = \frac{c}{\sqrt{2\pi\mu\kappa\omega}}. \tag{4.17}$$

The electric field inside the conductor is determined by Equation (4.12):

$$\boldsymbol{E} = -(1-i)\sqrt{\frac{\mu\omega}{8\pi\kappa}}\boldsymbol{e}_z \times \boldsymbol{H}.$$

By virtue of the inequality (4.13), we have $|E| \ll |H|$. ■

Diffusion of magnetic field in conducting media.

Example 4.6

The magnetic field is given at the initial instant of time $t = 0$ *in a finite region inside the conductor as a function of coordinates:* $\boldsymbol{H}(\boldsymbol{r},0) = \boldsymbol{H}_0(\boldsymbol{r})$. *Assuming the conductor to be homogeneous and infinite, construct the Green's function of Equation (4.15) and write down the solution of the problem for* $t > 0$. *Based on the explicit form of the Green's function, interpret the penetration depth (4.17) as a distance traveled by the field over time interval of about half period of its variation.*

Solution. To begin with, we rewrite Equation (4.15) in the form that includes the initial condition:

$$\text{(1)} \qquad \frac{\partial \boldsymbol{H}}{\partial t} - \chi \triangle \boldsymbol{H} = \boldsymbol{H}_0(\boldsymbol{r})\,\delta(t), \qquad \boldsymbol{H} = 0 \quad \text{at} \quad t < 0.$$

In order to verify that the initial condition follows from Equation (1), we integrate both sides of Equation (1) over a small time interval $(-\tau, +\tau)$. As $\tau \to 0$, $\boldsymbol{H}(\boldsymbol{r},0) = \boldsymbol{H}_0(\boldsymbol{r})$, since $\tau\triangle\boldsymbol{H} \to 0$. As a result, Equation (1) becomes inhomogeneous, and its solution can be written in the standard way (see Section I.5.1),

$$\text{(2)} \quad \boldsymbol{H}(\boldsymbol{r},t) = \int G(\boldsymbol{r}-\boldsymbol{r}', t-t')\,\boldsymbol{H}_0(\boldsymbol{r}')\,\delta(t')\,d^3r'\,dt' = \int G(\boldsymbol{r}-\boldsymbol{r}', t)\boldsymbol{H}_0(\boldsymbol{r}')\,d^3r'$$

in terms of the Green's function satisfying the equation

$$\text{(3)} \qquad \frac{\partial G}{\partial t} - \chi \triangle G = \delta(\boldsymbol{r}-\boldsymbol{r}')\,\delta(t-t'), \qquad G = 0 \quad \text{at} \quad t - t' < 0.$$

Equation (3) is easily solved by expanding the Green's function into Fourier integral with respect to the spatial variable $\boldsymbol{r}-\boldsymbol{r}'$. Its Fourier transform, G_k, satisfies the equation

$$(4) \qquad \frac{\partial G_k}{\partial t} + \chi k^2 G_k = \delta(t - t'),$$

which has the solution $G_k(t - t') = \Theta(t - t')\, e^{-\chi(t-t')k^2}$. The inverse Fourier transform allows us to express the Green's function in terms of coordinates and time:

$$(5) \qquad G(\mathbf{r} - \mathbf{r}', t - t') = \int G_k(t - t') \exp(i\mathbf{k} \cdot (\mathbf{r} - \mathbf{r}')) \frac{d^3k}{(2\pi)^3}$$

$$= \frac{\Theta(t - t')}{[4\pi\chi(t - t')]^{3/2}} \exp\left\{ -\frac{(\mathbf{r} - \mathbf{r}')^2}{4\chi(t - t')} \right\}.$$

Here, $\Theta(t - t')$ is the step function (I.1.212).

It follows from the explicit form of the exponent in the Green function (5) that in the quasi-stationary approximation the magnetic field in the conductor propagates over a distance L during time interval $\Delta t \approx l^2/4\chi$. The same law governs heat propagation and particle diffusion in a stationary medium. Assuming that in these formulas $\Delta t \approx T/2 = \pi c/\omega$, we find the distance $L \approx c/\sqrt{\mu\kappa\omega}$, which qualitatively agrees with the thickness of the skin layer (4.17). ■

Owing to the appearance of eddy currents, a conductor placed in the magnetic field acquires a magnetic moment even if its permeability is $\mu = 1$. This magnetic moment is conveniently characterized by the magnetic polarizability tensor β_{ik} which is given by

$$m_i = \beta_{ik} H_{0k}. \qquad (4.18)$$

Here $\mathbf{m}$ is the magnetic moment of the body and $\mathbf{H}_0$ is the periodic external magnetic field, which is uniform over distances of the order of the body sizes. The tensor β_{ik} is symmetric, $\beta_{ik} = \beta_{ki}$, and its components are generally complex and depend on frequency. The time average of the heat release inside the conductor can be determined either by integrating the Joule heat release over the body volume

$$Q = \int \overline{(\mathbf{j} \cdot \mathbf{E})}\, dV = \int \kappa \overline{E^2}\, dV \qquad (4.19)$$

or by calculating the electromagnetic energy flux inward the body:

$$Q = -\frac{c}{4\pi} \oint \overline{(\mathbf{E} \times \mathbf{H})} \cdot d\mathbf{S}. \qquad (4.20)$$

In both cases, one should perform the time averaging (as indicated by the upper line) in order to exclude the part of the energy that periodically passes from outside into the body and back.

Suggested literature: Landau *et al.* (1984); Smythe (1950); Tamm (1976); Frenkel (1928); Vlasov (1955)

Problems

4.29 The magnetic field $\boldsymbol{H}(t) = \boldsymbol{H}_0 e^{-i\omega t}$, $\omega = const$, is maintained at the plane boundary of the conductor and is tangential to it. At $0 \le z < a$ the conductor has the electric conductivity κ_1 and magnetic permeability μ_1; at $a < z < \infty$ these parameters are κ_2 and μ_2, respectively. Evaluate the magnetic field at $z > 0$ and investigate the specific cases.

4.30 A wire with the current $J_0\, e^{-i\omega t}$ is wound on a wide plate which has electric conductivity κ, magnetic permeability μ and is bounded by the planes $x = \pm h$. The wire is thin, a number of turns, which are parallel to one another, is n per unit length. Neglecting edge effects, determine the real magnetic field inside the plate. Consider the limiting cases of weak ($\delta \gg h$) and strong ($\delta \ll h$) skin-effect.

4.31* An infinitely long metal cylinder having electric conductivity κ and magnetic permeability μ is placed so that its axis coincides with the axis of an infinite solenoid of circular cross section which carries an alternating current $J = J_0\, e^{-i\omega t}$. Find the strength of the magnetic and electric fields in the entire space and also the current density distribution j in the cylinder. The radius of the cylinder is a, the radius of the solenoid is $b > a$, and a number of turns per unit length is n.

4.32 A conducting cylinder is placed in uniform alternating magnetic field $H = H_0\, e^{-i\omega t}$ which is parallel to its axis. Using the results Problem 4.31, investigate the current density $\boldsymbol{j}$ inside the cylinder in the limiting cases of low and high frequencies.

4.33 Calculate the amount of heat Q released per unit time per unit length of the cylinder described in Problem 4.31. Consider the limiting cases of low and high frequencies. Write the result in terms of the field H_0 inside the solenoid.

4.34 Find the magnetic polarizability β (per unit length) of the cylinder placed in an alternating magnetic field, which is parallel to cylinder's axis. The field frequency is ω, cylinder radius a, electric conductivity κ, and magnetic permeability $\mu = 1$. Consider the limits of low and high frequencies.

4.35* A metal cylinder is placed in an external uniform magnetic field $\boldsymbol{H} = \boldsymbol{H}_0\, e^{-i\omega t}$ which is perpendicular to its axis. The cylinder radius is a, electric conductivity κ, and magnetic permeability $\mu = 1$. Find the resultant field and the current density $\boldsymbol{j}$ in the cylinder.

Hint. Express $\boldsymbol{E}$ and $\boldsymbol{H}$ in terms of the vector potential $\boldsymbol{A}$ and integrate the differential equation for $\boldsymbol{A}$.

4.36 Find the energy dissipation per unit length of a perfectly conducting circular cylinder placed in an alternating magnetic field of frequency ω, which is transverse with respect to the cylinder axis.

4.37* An infinite circular cylinder of radius a and electric conductivity κ is placed in the circularly polarized magnetic field which is perpendicular to the axis of the cylinder,

$$\boldsymbol{H}_0(t) = (\boldsymbol{H}_{01} + i\boldsymbol{H}_{02})\, e^{-i\omega t}.$$

Here $\boldsymbol{H}_{01}$ and $\boldsymbol{H}_{02}$ are the mutually perpendicular vectors of equal length: $H_{01} = H_{02} = H_0$. (The end of vector $\boldsymbol{H}_0(t)$ moves along the circle of constant radius H_0 in the plane perpendicular to the cylinder axis.) Find the average torque $\overline{N}$ per unit length of the cylinder ($\mu = 1$).

4.38 An infinitely long cylinder rotates about its axis with angular velocity ω in a constant and uniform transverse magnetic field $\boldsymbol{H}_0$. Find the braking torque $\overline{N}$ per unit length of the cylinder.

4.39* An infinitely long metal cylinder of radius a, electric conductivity κ, and magnetic permeability μ is placed in a constant and uniform magnetic field H_0, which is longitudinal with respect to the cylinder axis. At some instant the external field is switched off and is then maintained at zero. Investigate the damping of the magnetic field in the cylinder as a function of time.

4.40 A metal sphere of radius a, conductivity κ, and magnetic permeability μ is placed in a uniform oscillating magnetic field $H_0(t) = H_0\, e^{-i\omega t}$. Assuming that the frequency is low, find, in the first nonvanishing approximation, the distribution of eddy currents inside the sphere and also the average power Q absorbed by the sphere.

4.41 A metallic sphere of radius a, electric conductivity κ, and magnetic permeability μ is placed in a uniform magnetic field oscillating with frequency ω. Find the resultant field $\boldsymbol{H}$ and the average power Q absorbed by the sphere at high frequencies.

Hint. Determining the field outside the sphere, assume that the field inside the sphere is zero (i.e., neglect the penetration depth δ in comparison with the sphere radius a). Calculating the field inside the sphere approximate its surface (locally) by a plane.

4.42* A conducting ellipsoid is placed in an uniform alternating magnetic field. Determine the magnetic polarizability of the ellipsoid in the case of strong skin-effect (i.e., assume that the penetration depth of the field into the conductor is negligibly small). Consider the limiting cases of thin circular disc and long thin rod.

4.43* A sphere of radius a and electric conductivity κ is placed in an uniform magnetic field $H(t) = H_0\, e^{-i\omega t}$. Find the resultant field and the distribution of eddy currents inside the sphere at arbitrary frequency. Verify that in the limiting cases of weak and strong skin-effect the solutions are identical with those found in Problems 4.40 and 4.41 (assume, for simplicity, that $\mu = 1$).

4.44 Find the average power Q absorbed by a conducting sphere in a uniform alternating magnetic field of arbitrary frequency.

4.45* The magnetic field in the vicinity of the conducting sphere, considered in Problem 4.43, is switched off at $t = 0$. Calculate the time decay of the field inside and outside the sphere.

4.46 Find an active resistance R of a thin cylindrical conductor in the presence of skin-effect. The length of the conductor l, radius a, electric conductivity κ, and magnetic permeability $\mu = 1$ are known. Investigate the limiting cases of low and high frequencies.

4.47 The surface of a cylindrical conductor of radius a and electric conductivity κ_1 is coated with a layer of a different metal. The thickness of the layer is h and its electric conductivity is κ_2 ($h \ll a$). Find the active resistance R of such a conductor to alternating currents, assuming that the thickness of the skin layer is smaller than a ($\mu = 1$).

4.48 An infinitely long, hollow cylinder of internal radius a and wall thickness h ($h \ll a$) is placed in a uniform longitudinal field $H_0(t) = H_0\, e^{-i\omega t}$. Find the amplitude H' of the magnetic field in the cavity. Consider its dependence on ω.

Hint. Owing to the condition $h \ll a$, the field inside the shell can be determined assuming that the shell is locally plane.

4.49 An alternating current $J(t) = J_0\, e^{-i\omega t}$ flows along a hollow cylindrical conductor of average radius a, electric conductivity κ, magnetic permeability μ, and thickness $h \ll a$. Find the current distribution j over the cross section of the conductor and also its active resistance R per unit length. Specify the condition that must be satisfied in order to ensure that the resistance of the hollow conductor insignificantly differs from that of a solid conductor of equal radius.

Hint. Neglect the curvature of the surface of the conductor.

4.50⋆ A linear current J flows inside a metal tube at a distance l from the tube axis. The radius of the tube is a, the wall thickness is $h \ll a$, and the electric conductivity of the walls is κ ($\mu = 1$). Both the current J and the distance l are arbitrary functions of time, but such that at all times $l \ll a$. Assuming that the quasi-stationary conditions are satisfied, determine the force f per unit length on the current J due to eddy currents induced in the cylindrical shell in the case of weak skin-effect ($h \ll \delta$).

4.51⋆ Solve the previous Problem 4.50 for the case of a strong skin-effect ($h \gg \delta$).

4.3
Magnetic Hydrodynamics

Basic equations of magnetohydrodynamics. Motion of a conductor with non-relativistic velocity $u \ll c$ in a magnetic field $\boldsymbol{B}$ causes an additional electric field $\boldsymbol{u} \times \boldsymbol{B}/c$ (see Equation (I.4.70) in which one should set $\gamma = 1$). In this case, the Ohm law (1.25) for a moving conductor takes the form

$$\boldsymbol{j} = \kappa \left(\boldsymbol{E} + \frac{1}{c}\, \boldsymbol{u} \times \boldsymbol{B} \right). \tag{4.21}$$

This expression assumes a sufficiently weak field (otherwise, electric conductivity becomes anisotropic, see Problem 1.30) and the absence of currents due to inhomogeneity of the conductor (due to temperature and number density gradients). The current density in the first order in u/c is the same in the frame of the conductor and in the stationary (laboratory) reference frame.

Example 4.7

Derive the equations which describe the evolution of a magnetic field in a moving nonmagnetic ($\mu = 1$) conductor using the quasi-stationary approximation.

Solution. Equations (4.10) and (4.21) yield

$$\frac{\partial \boldsymbol{B}}{\partial t} = \nabla \times [\boldsymbol{u} \times \boldsymbol{B}] - c\nabla \times \frac{\boldsymbol{j}}{\kappa}.$$

Substituting the current from the second equation (4.11), where $\boldsymbol{H}$ is taken to be equal to $\boldsymbol{B}$ ($\mu = 1$), we obtain the set of equations

$$\nabla \cdot \boldsymbol{B} = 0, \qquad \frac{\partial \boldsymbol{B}}{\partial t} = \nabla \times [\boldsymbol{u} \times \boldsymbol{B}] + \nu_m \triangle \boldsymbol{B}, \qquad \nu_m = \frac{c^2}{4\pi\kappa} = const. \quad \blacksquare \tag{4.22}$$

It follows from these equations that in order to determine the field Inside the conductor, the velocity of the conductor elements must be known. However, this velocity depends on the field itself since the conductor with current is affected by the Lorentz force. Therefore, Equation (4.22) should be considered simultaneously with the equations of motion of the conductor. If we deal with a conducting liquid or an ionized gas (plasma), its motion in the quasi-stationary approximation can be described by equations of hydrodynamics (see Landau and Lifshitz (1987)). The equation of motion for liquid or gaseous medium reads (here we set $\mu = 1$)

$$\tau\left(\frac{\partial \boldsymbol{u}}{\partial t} + (\boldsymbol{u}\cdot\nabla)\boldsymbol{u}\right) = -\nabla p + \frac{1}{4\pi}[\nabla \times \boldsymbol{B}] \times \boldsymbol{B} + \boldsymbol{f} + \eta\triangle\boldsymbol{u} + \frac{\eta}{3}\nabla(\nabla\cdot\boldsymbol{u}). \tag{4.23}$$

In this case, $[\nabla \times \boldsymbol{B}] \times \boldsymbol{B}/4\pi = \boldsymbol{j} \times \boldsymbol{B}/c$ is the Lorentz force per unit volume of conducting medium, and $\boldsymbol{f}$ is the volume force unrelated to the electric field (e.g., gravitation, $\boldsymbol{f} = \boldsymbol{f}_g = \tau\boldsymbol{g}$, $\boldsymbol{g}$ being acceleration of gravity). The last two terms in the right-hand side describe the viscous forces; η is the shear viscosity (we do no consider the bulk viscosity). The viscosity and the electric conductivity κ are calculated on the basis of physical kinetics. In hydrodynamics, they are regarded as given quantities.

In addition to the medium velocity $\boldsymbol{u}$ (to be determined), Equation (4.23) contains two scalar quantities, the mass density τ and the pressure p, which may vary during the motion of the medium. Therefore, Equations (4.22) and (4.23) should be supplemented by the continuity equation

$$\frac{\partial \tau}{\partial t} + \nabla(\tau\boldsymbol{u}) = 0 \tag{4.24}$$

and by the equation of state of the matter,

$$p = p(\tau, s), \tag{4.25}$$

where s stands for specific (per unit mass) entropy. The growth of entropy due to dissipative processes (viscosity, thermal conductivity, and Joule dissipation of magnetic field) is described by the equation

$$\tau T\left(\frac{\partial s}{\partial t} + (\boldsymbol{u}\cdot\nabla)s\right) = \Pi'_{ik}\nabla_k u_i + \chi\triangle T + \frac{\nu_m}{4\pi}[\nabla \times \boldsymbol{B}]^2. \tag{4.26}$$

Here

$$\Pi'_{ik} = \eta \left(\frac{\partial u_i}{\partial x_k} + \frac{\partial u_k}{\partial x_i} - \frac{2}{3}(\nabla \cdot \boldsymbol{u})\delta_{ik} \right) \tag{4.27}$$

is the viscous stress tensor and χ is the thermal conductivity (considered as constant).

The system of equations of magnetic hydrodynamics (MHD), Equations (4.22)–(4.27) formulated above, is generally nonlinear and complicated. Its analytical solutions are, as a rule, possible only under certain simplifying assumptions. One of them is in neglecting dissipative processes (its validity needs to be verified in each specific case). The necessary conditions for neglecting dissipation in Equations (4.23) and (4.22) can be written as inequalities

$$R = \frac{ul}{\nu} \gg 1, \qquad R_m = \frac{ul}{\nu_m} \gg 1, \tag{4.28}$$

where u is the characteristic velocity, l is the characteristic scale of velocity variation, and $\nu = \eta/\tau$ is **the kinematic viscosity** (by analogy, $\nu_m = c^2/4\pi\kappa$ is called **magnetic viscosity**; characteristic scale of magnetic field variations is also denoted by l). Inequalities (4.28) are fulfilled if dissipative terms are small. The dimensionless parameters R and R_m are known as **Reynolds numbers** (the second one is the **Reynolds magnetic number**).[5)]

Ideal MHD Equations. In the absence of dissipation, the set of equations takes the form

$$\frac{\partial \boldsymbol{B}}{\partial t} = \nabla \times [\boldsymbol{u} \times \boldsymbol{B}], \qquad \nabla \cdot \boldsymbol{B} = 0, \tag{4.29}$$

$$\frac{\partial \tau}{\partial t} + \nabla(\tau \boldsymbol{u}) = 0, \tag{4.30}$$

$$\frac{\partial \boldsymbol{u}}{\partial t} + (\boldsymbol{u} \cdot \nabla)\boldsymbol{u} = -\frac{1}{\tau}\nabla p + \frac{1}{4\pi\tau}[\nabla \times \boldsymbol{B}] \times \boldsymbol{B} + \frac{1}{\tau}\boldsymbol{f}, \tag{4.31}$$

$$\frac{ds}{dt} \equiv \frac{\partial s}{\partial t} + (\boldsymbol{u} \cdot \nabla)s = 0, \qquad p = p(\tau, s). \tag{4.32}$$

Equation (4.32) for s describes constancy of entropy in any moving macroscopic element of matter, and the derivative $d/dt = \partial/\partial t + (\boldsymbol{u} \cdot \nabla)$ is referred to as **substantial (material) derivative**. If at the initial instant the system is homogeneous, then Equation (4.32) can be replaced by the condition for constancy of entropy in time and space,

$$s = const. \tag{4.33}$$

In the absence of dissipation, the electric field can be expressed through the magnetic field and the velocity $\boldsymbol{u}$. Setting $\kappa \to 0$ in Equation (4.21) and assuming that the current density $\boldsymbol{j}$ is finite, we have

$$\boldsymbol{E} = -\frac{1}{c}\,\boldsymbol{u} \times \boldsymbol{B}. \tag{4.34}$$

5) Reynolds Osborne (1842–1912) was the outstanding English mechanic, the physicist and the engineer.

Magnetic pressure and magnetic tensions. Freezing-in and diffusion of magnetic field. When hydrodynamic velocity u is smaller than propagation velocity of small perturbations (linear modes), the compressibility effects are of little importance. Assuming that the medium is incompressible, that is, $\tau = const$, instead of Equations (4.23) and (4.24), we have

$$\frac{\partial \boldsymbol{u}}{\partial t} + (\boldsymbol{u}\cdot\nabla)\boldsymbol{u} = -\frac{1}{\tau}\nabla p + \frac{1}{4\pi\tau}[\nabla \times \boldsymbol{B}] \times \boldsymbol{B} + \nu\triangle \boldsymbol{u}, \qquad \nabla\cdot\boldsymbol{u} = 0. \qquad (4.35)$$

Here the non-electromagnetic force is omitted. The force term on the right-hand side can be rearranged using Equation (I.1.89), $[\nabla \times \boldsymbol{B}] \times \boldsymbol{B} = -\nabla B^2/2 + (\boldsymbol{B}\cdot\nabla)\boldsymbol{B}$. As a result, some part of the magnetic force is described as gradient of the "magnetic pressure," $p_m = B^2/8\pi$,

$$\frac{\partial \boldsymbol{u}}{\partial t} + (\boldsymbol{u}\cdot\nabla)\boldsymbol{u} = -\frac{1}{\tau}\nabla\left(p + \frac{B^2}{8\pi}\right) + \frac{1}{4\pi\tau}(\boldsymbol{B}\cdot\nabla)\boldsymbol{B} + \nu\triangle\boldsymbol{u}, \qquad (4.36)$$

while the remaining part forms the "magnetic tensions," $(\boldsymbol{B}\cdot\nabla)\boldsymbol{B}/4\pi\tau$, which are not zero only when the magnetic lines of force are strongly curved. A comparison of (4.35), (4.36) with (4.22) clarifies the meaning of the term "magnetic viscosity": ν_m plays the same role for the magnetic field as the kinematic viscosity ν for the hydrodynamic velocity. In both cases, the dissipative terms are expressed in terms of second derivatives with respect to coordinates.

Example 4.8

Prove that neglecting Joule dissipation, magnetic field is "frozen" into conducting medium; any line of force initially passing through two macroscopic particles of the medium moves later together with these particles without any transverse slip.

Solution. Two particles of the medium, located at the neighboring points $\boldsymbol{r}$ and $\boldsymbol{r} + \delta\boldsymbol{l}$, have velocities $\boldsymbol{u}$ and $\boldsymbol{u} + (\delta\boldsymbol{l}\cdot\nabla)\boldsymbol{u}$, respectively. The distance between the particles changes by $(\delta\boldsymbol{l}\cdot\nabla)\boldsymbol{u}\,dt$ over time interval dt. Then

(1) $$\frac{d}{dt}\delta\boldsymbol{l} = (\delta\boldsymbol{l}\cdot\nabla)\boldsymbol{u}.$$

The left-hand side of this expression includes the material derivative since it describes the change in distance between the moving particles. On the other hand, the material derivative of $\boldsymbol{B}/\tau$ is

(2) $$\frac{d}{dt}\frac{\boldsymbol{B}}{\tau} = \frac{1}{\tau}\frac{d\boldsymbol{B}}{dt} - \frac{\boldsymbol{B}}{\tau^2}\frac{d\tau}{dt}.$$

The material derivative of the field $\boldsymbol{B}$ can be found using Equations (4.29),

(3) $$\frac{d\boldsymbol{B}}{dt} = (\boldsymbol{B}\cdot\nabla)\boldsymbol{u} - \boldsymbol{B}(\nabla\cdot\boldsymbol{u}).$$

The material derivative of the density can be calculated from the continuity equation (4.30),

(4) $$\frac{d\tau}{dt} = -\tau(\nabla\cdot\boldsymbol{u}).$$

Combining Equations (2)–(4), we have

(5) $$\frac{d}{dt}\frac{\boldsymbol{B}}{\tau} = \left(\frac{\boldsymbol{B}}{\tau}\cdot\nabla\right)\boldsymbol{u}.$$

Equations (1) and (5) for $\delta\boldsymbol{l}$ and $\boldsymbol{B}/\tau$ are identical. Hence, if the particles initially lie on a certain line of force, that is, if these vectors are parallel, they will remain parallel at any instant of time, and the particles will lie on the same line of force. The quantity $\boldsymbol{B}/\tau$ will vary proportionally to the distance between the particles. In particular, the freezing-in property guaranties conservation of the magnetic flux through arbitrary closed moving contour composed of the fluid elements of the medium. If Reynolds number is small, $R_m \ll 1$, take place the opposite case diffusion of the magnetic field (see Example 4.6). ■

Two-liquid MHD. One of the main subjects of MHD is to study the quasi-stationary electromagnetic phenomena in moving electrically conducting plasmas, fully or partly ionized. All components of such a plasma must move in a similar way forming one continuous medium. However, there are a variety of phenomena where the electron and ion components behave differently. One approach to such problems is to consider electrons and ions as two different interacting media ("two-liquid MHD"). One of the models is discussed in Example 4.9 (and in Section 7.3 in connection with the theory of nonlinear waves in plasma).

Example 4.9

Consider a two-liquid model of a cold magnetized and completely ionized plasma in which the energy density of the magnetic field exceeds the thermal energy density of electrons and ions, $8\pi n_{e,i}T_{e,i}/B^2 \ll m_e/m_i$. Consider electrons and ions as two continuous media whose macroscopic velocities $\boldsymbol{v}_e$ and $\boldsymbol{v}_i$ are different, while the

plasma can be regarded as being quasi-neutral due to slow changes of all macroscopic quantities in space and time. Write down a set of equations for the macroscopic quantities $\boldsymbol{B}$, the mass velocity $\boldsymbol{u} = (m_i\boldsymbol{v}_i + m_e\boldsymbol{v}_e)/(m_i + m_e)$ and for the number densities of particles, $n = n_i = n_e$.

Solution. For a cold plasma, we neglect electron and ion pressures and all dissipative phenomena except the friction between electrons and ions. The equations for macroscopic quantities are given by (see Braginsky 1963, for details):

(1) $$m_i(\partial_t + \boldsymbol{v}_i\cdot\nabla)\boldsymbol{v}_i = eE + (e/c)\boldsymbol{v}_i\times\boldsymbol{B} + (m_e/\tau_e)(0.51\boldsymbol{v}_\parallel + \boldsymbol{v}_\perp),$$

(2) $$m_e(\partial_t + \boldsymbol{v}_e\cdot\nabla)\boldsymbol{v}_e = -eE - (e/c)\boldsymbol{v}_e\times\boldsymbol{B} - (m_e/\tau_e)(0.51\boldsymbol{v}_\parallel + \boldsymbol{v}_\perp),$$

(3) $$\partial_t n_{e,i} + \nabla\cdot(n_{e,i}\boldsymbol{v}_{e,i}) = 0,$$

(4) $$rot\,\boldsymbol{B} = (4\pi e/c)(n_i\boldsymbol{v}_i - n_e\boldsymbol{v}_e),$$

(5) $$rot\,\boldsymbol{E} = -c^{-1}\partial_t\boldsymbol{B}, \qquad div\,\boldsymbol{B} = 0.$$

Here $e > 0$ is the ion charge, $\boldsymbol{v} = \boldsymbol{v}_i - \boldsymbol{v}_e$ is the relative velocity of electrons and ions,

$$\tau_e = 3m_e^{1/2} T_e^{3/2} / 4(2\pi)^{1/2} \lambda e^4 n_i \tag{6}$$

is the time during which the electrons lose their directed momentum (see Problems I.4.70–I.4.72), λ is the Coulomb logarithm, and $\partial_t f \equiv \partial f / \partial t$.

Let us transform the set of Equations (1)–(5) by introducing the mass velocity $\boldsymbol{u} \approx \boldsymbol{v}_i + (m_e/m_i)\boldsymbol{v}_e$ (taking into account the smallness of the mass ratio $m_e/m_i \ll 1$). Combining Equations (1) and (2) and neglecting the terms $\sim m_e/m_i$, we have

$$\partial_t \boldsymbol{u} + (\boldsymbol{u} \cdot \nabla)\boldsymbol{u} = \frac{1}{4\pi n m_i} rot\, \boldsymbol{B} \times \boldsymbol{B}. \tag{7}$$

where Equation (4) is used. Then we divide Equation (1) by Equation (2) and combine the two equations with the use of approximate equalities $\boldsymbol{v}_i/m_i + \boldsymbol{v}_e/m_e \approx (\boldsymbol{u} - \boldsymbol{j}/en)/m_e$ and $\boldsymbol{v}_e \approx \boldsymbol{u} - \boldsymbol{j}/en$, where $\boldsymbol{j} = en\boldsymbol{v}$ is the current density. As a result, we obtain the generalized Ohm law,

$$\boldsymbol{E} = -(\boldsymbol{u}/c) \times \boldsymbol{B} + (enc)^{-1} \boldsymbol{j} \times \boldsymbol{B} + (m_e/e^2 \tau_e n)(0.51 \boldsymbol{j}_{\parallel} + \boldsymbol{j}_{\perp})$$

$$+ (m_e/e^2 \{\partial_t(\boldsymbol{j}/n) + (\boldsymbol{u} \cdot \nabla)(\boldsymbol{j}/n) + n^{-1}(\boldsymbol{j} \cdot \nabla)\boldsymbol{u} - (en)^{-1}(\boldsymbol{j} \cdot \nabla)(\boldsymbol{j}/n)\}, \tag{8}$$

where $e^2 n \tau_e / m_e = \kappa$ is the electric conductivity of the plasma along the magnetic field. With the aid of Equations (4), (5), and (8) we obtain the equation for the induction of the magnetic field

$$\partial_t \boldsymbol{B} = rot\,(\boldsymbol{u} \times \boldsymbol{B}) - rot\, \nu_m (0.51 rot_{\parallel}\, \boldsymbol{B} + rot_{\perp}\, \boldsymbol{B}) - rot(4\pi e n)^{-1}(rot\, \boldsymbol{B} \times \boldsymbol{B})$$

$$-(m_e c/e^2)\{\partial_t(\boldsymbol{j}/n) + (\boldsymbol{u} \cdot \nabla)(\boldsymbol{j}/n) + n^{-1}(\boldsymbol{j} \cdot \nabla)\boldsymbol{u} - (en)^{-1}(\boldsymbol{j} \cdot \nabla)(\boldsymbol{j}/n)\}, \tag{9}$$

where $\boldsymbol{j} = (c/4\pi) rot\, \boldsymbol{B}$ and $\nu_m = c^2/4\pi\kappa$ is the magnetic viscosity. Equations (7), (9), and the continuity equation

$$\partial_t n + \nabla \cdot (n\boldsymbol{u}) = 0 \tag{10}$$

form a complete set of equations describing plasma motion and magnetic field evolution in the adopted approximation. ■

Simple MHD waves. In what follows, we consider the most important cases of macroscopic motions of conducting media in magnetic fields.

Example 4.10

Obtain the solution for the non-dissipative system of equations (4.29)–(4.32) in the form of ***simple Riemann waves****. They are one-dimensional perturbations of medium wherein all unknown functions $\boldsymbol{u}(\varphi)$, $\boldsymbol{B}(\varphi)$, and $\tau(\varphi)$ ... depend on a single coordinate and on time through a certain function $\varphi(x, t)$ which is the same for all macroscopic parameters. Simple waves satisfy the set of ordinary nonlinear differential equations with a single independent variable. This set can be integrated in the analytical form.*

Solution. From Equations (4.29) and the assumption that the solution is a function of a single x-coordinate it follows that

(1) $$B_x = const.$$

Let us rearrange the equations denoting the derivative with respect to $\varphi(x, t)$ with a prime. For example, for the continuity equation we have

$$\frac{\partial \tau}{\partial t} + u_x \frac{\partial \tau}{\partial x} + \tau \frac{\partial u_x}{\partial x} = \tau' \left(\frac{\partial \varphi}{\partial t} + u_x \frac{\partial \varphi}{\partial x} \right) + u_x' \tau \frac{\partial \varphi}{\partial x} = 0.$$

The surfaces of the constant values of the physical quantities are planes $x = const$ on which

(2) $$\varphi(x, t) = const,$$

and which move in the direction of the Ox-axis. Their velocity, $dx/dt = (\partial x/\partial t)_\varphi = v_{ph}$, may be called the **phase velocity** by analogy with the case of plane monochromatic waves. After differentiating Equation (2), we get

(3) $$\frac{\partial \varphi}{\partial t} = -v_{ph} \frac{\partial \varphi}{\partial x}$$

and transform the continuity equation to the form

(4) $$\tau' v - u_x' \tau = 0,$$

where $v = v_{ph} - u_x$ is the velocity of propagation of a simple wave in the co-moving reference frame.

The remaining equations can be transformed in a similar way. In projections on coordinates axes, they read

(5) $$u_x' v - p'/\tau - (B_y B_y' + B_z B_z')/4\pi\tau = 0,$$

(6) $$u_y' v + B_x B_y'/4\pi\tau = 0,$$

(7) $$u_z' v + B_x B_z'/4\pi\tau = 0,$$

(8) $$B_y' v - u_x' B_y + u_y' B_x = 0,$$

(9) $$B_z' v - u_x' B_z + u_z' B_x = 0,$$

(10) $$s' v = 0,$$

(11) $$p = p(s, \tau).$$

Equation (1) and nonlinear equations (4)–(11) constitute a complete system for the case under consideration.

(a) **Entropy simple waves** are derived from this system in the case in which the entropy of the medium undergoes perturbations, $s' \neq 0$. It follows from Equation (10) that

(12) $$v \equiv v_e = 0$$

that is, the entropy wave is stationary in the medium rest frame and is transferred with the medium. If $B_x \neq 0$, the equations yield successively $u'_x = B'_y = B'_z = u'_y = u'_z = p' = 0$, that is, all the above quantities are constant. The density of the medium is affected by perturbations related to entropy perturbations via Equation (11) and via the condition $p = const$. Thus, neglecting dissipative processes, the profile of the entropy wave is not distorted and the wave itself presents a heated (or cooled) region which is transported by the moving medium. Actually, thermal conduction (diffusion of heated particles) results in a slow dissipation of entropy perturbations.

At $B_x = 0$, from Equations (4) and (5) it follows that the longitudinal velocity and the total pressure are constant,

$$u_x = const, \qquad p + \frac{B^2}{8\pi} = const \tag{13}$$

The derivatives u'_y and u'_z may take arbitrary values.

At $v \neq 0$ according to Equation (10), $s' = 0$ or $s = const$, that is, simple waves propagating with nonzero phase velocity represent motions with constant entropy. We can hence write $p' = c_s^2 \tau'$ in Equation (5), where

$$c_s = \sqrt{\left(\frac{\partial p}{\partial \tau}\right)_s} \tag{14}$$

is the local velocity of sound in the medium in the absence of a magnetic field. This quantity, being a function of density, is generally not a constant.

(b) **Alfvén simple waves**[6] are motions in which the density remains constant along with the entropy,

$$\tau = const. \tag{15}$$

It follows from Equations (4) and (5) that

$$u_x = const, \qquad B_y^2 + B_z^2 \equiv B_\perp^2 = const, \tag{16}$$

while Equations (6)–(9) are separated into two identical pairs,

$$u'_y v + \frac{B_x B'_y}{4\pi\tau} = 0, \qquad u'_y B_x + B'_y v = 0, \tag{17}$$

and the analogous system for u'_z and B'_z. The last system can be used to determine the velocity of the Alfvén wave relative to the medium,

$$v \equiv v_{Ax} = \pm\frac{|B_x|}{\sqrt{4\pi\tau}}. \tag{18}$$

This velocity is constant because of Equations (1) and (15). The Alfvén wave disappears at $B_x = 0$ since in this case the equations are compatible only with homogeneous medium, where $\boldsymbol{u} = const$, $\boldsymbol{B} = const$. The system of Equations (17) and

6) Alfvén Hannes (1908–1995), outstanding Swedish physicist and astrophysicist, founder of new branches in physics, such as MHD and cosmic electrodynamics.

the analogous system for u'_z and B'_z allows one to find the relation between the velocity $\boldsymbol{u}$ of the medium and the magnetic field in an Alfvén simple wave:

(19) $$\boldsymbol{u} = \mp \frac{\boldsymbol{B}_\perp}{\sqrt{4\pi\tau}}.$$

The integration constant is taken such that $\boldsymbol{u} = 0$ at $\boldsymbol{B} = 0$. The minus sign corresponds to identical signs of B_x and v_{Ax}, while the plus is for opposite signs.

The exact particular solution for the Alfvén wave is a circularly polarized plane monochromatic wave propagating along the uniform magnetic field:

(20) $$B_y = B_\perp \cos k[x - (v_{Ax} + u_x)t], \qquad B_z = B_\perp \sin k[x - (v_{Ax} + u_x)t].$$

The approximate solution for the wave with a small amplitude can be constructed for the case when the wave propagates at arbitrary angle to the external uniform field $\boldsymbol{B}_0$:

(21) $$\boldsymbol{b}(\boldsymbol{r}, t) = \boldsymbol{b}_0 \cos(\boldsymbol{k}\cdot\boldsymbol{r} - \omega_A t),$$

where $\boldsymbol{k}$ is the wave vector,

(22) $$\omega_A = \frac{|\boldsymbol{k}\cdot\boldsymbol{B}|}{\sqrt{4\pi\tau}}$$

is the Alfvén wave frequency, and the wave's amplitude $\boldsymbol{b}_0$ is perpendicular to the external field $\boldsymbol{B}_0$. The condition of constancy of the magnetic field modulus (16) is satisfied in this case to within the terms of the first-order in the small ratio b_0/B_0. The amplitude $\boldsymbol{b}_0$ is also transverse relative to $\boldsymbol{k}$.

(c) **Fast and slow simple waves.** These are the solutions of Equations (4)–(11) in which $s = const,\ \tau' \neq 0,\ v \neq v_e, v_{Ax}$. At small amplitudes, such waves are referred to as **magnetoacoustic waves**. To simplify the geometry, we note that Equations (7) and (9) become identities if $u_z = B_z = 0$. The remaining set of four Equations (4)–(6) and (8) has a nontrivial solution at the following values of phase velocity:

(23) $$v^2 \equiv v_{f,s}^2 = \frac{1}{2}\left\{ c_s^2 + v_A^2 \pm \sqrt{(c_s^2 + v_A^2)^2 - 4c_s^2 v_{Ax}^2} \right\},$$

where

(24) $$\boldsymbol{v}_A = \frac{\boldsymbol{B}}{\sqrt{4\pi\tau}}$$

denotes the pseudovector of the Alfvén velocity

A comparison between Equations (23) and (18) shows that for any magnetic field and the velocity of sound one has:

(25) $$v_f^2 \geq v_{Ax}^2 \geq v_s^2 \geq 0, \qquad v_f^2 \geq c_s^2, \qquad v_s^2 \leq c_s^2.$$

The last two inequalities reflect the names of the waves under consideration: they are either fast or slow compared with the ordinary acoustic waves in the absence of any magnetic field.

After substituting v_f or v_s, the set of equations can be integrated. Since this procedure is generally complicated, we perform it for the case of transverse propagation, that is, when $B_x = 0$ and $B_y^2 = B^2$. Then $v_f = \pm(c_s^2 + v_A^2)^{1/2}$ and $v_s = 0$; that is, only the fast waves can exist. Taking the density as an independent variable, from Equations (4) to (6) we find

$$(26) \qquad u_y = const, \qquad \frac{du_x}{d\tau} = \frac{v_f}{\tau}, \qquad \frac{dB^2}{d\tau} = 8\pi v_A^2.$$

Integration of the last equation taking into account the definition of the Alfvén velocity (24) gives

$$(27) \qquad B = B_0 \frac{\tau}{\tau_0},$$

where B_0 is the field value at $\tau = \tau_0$ (for example, in an unperturbed region). Equation (27) defines the freezing-in condition for the field. Using it along with the equation of state (11) to find the dependence of v_f on τ, and integrating the second equation in Equation (26), we have

$$(28) \qquad u_x(\tau) = \int_{\tau_0}^{\tau} \frac{v_f(\tau)}{\tau} d\tau,$$

where $u_x(\tau_0) = 0$. Next, we write down $(\partial x/\partial t)_\tau = v_f(\tau) + u_x(\tau)$ and obtain the solution

$$(29) \qquad x = [v_f(\tau) + u_x(\tau)]t + f(\tau),$$

which determines the dependence $\tau(x, t)$ in the implicit form. The function $f(\tau)$ is determined by the initial condition $\tau(x, 0) = F(x)$. ■

Strong MHD discontinuities. In Problem 4.69, we will consider the physical picture of steeping of a leading edge of a transverse magnetoacoustic wave and formation of discontinuities. The existence of discontinuities naturally follows from the MHD equations in the absence of dissipation. The discontinuities may result from the evolution of an initially smooth perturbation or from discontinuous initial or boundary conditions (for example, a piston initially at rest suddenly begins to move with a finite speed). All discontinuities actually have a finite thickness which is determined by dissipative processes. However, the thickness frequently appears to be rather small in comparison with other lengths. In such cases, the discontinuities are considered to be infinitely thin surfaces.

Local properties of stationary discontinuities can be investigated using conservation of the main physical quantities, such as the mass, momentum, and energy, together with boundary conditions for the electric and magnetic field vectors. Let us consider a discontinuity that is locally plane. We relate it to the frame of reference (Figure 4.8) in which one of the axes is directed normally to the discontinuity. Let us further construct two auxiliary planes 11 and 22 on either side of the discontinuity in such a way that dissipative processes can be neglected on them (dissipation can be significant inside the transient layer, that is, in the region of large gradients of MHD parameters).

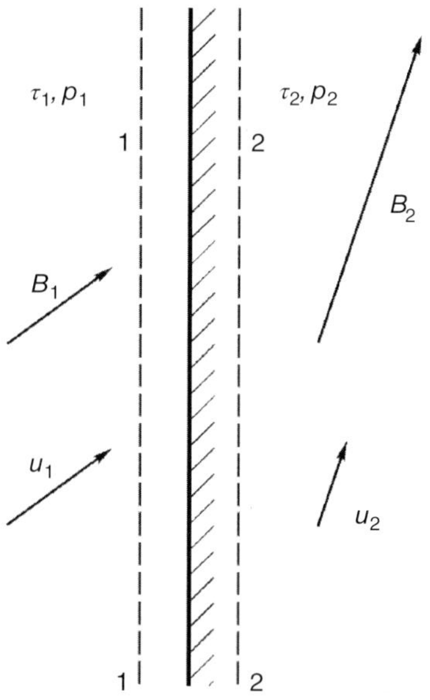

Figure 4.8 Local plane of MHD discontinuity.

It follows from conservation laws that a stationary discontinuity is characterized by

$$\delta(i_\alpha n_\alpha) = 0, \qquad \delta(\Pi_{\alpha\beta} n_\beta) = 0, \qquad \delta(q_\alpha n_\alpha) = 0, \tag{4.37}$$

where δ denotes the jump of the corresponding quantity when going from plane 22 to plane 11, that is, $\delta i_n = i_{n2} - i_{n1}$, etc.; $\boldsymbol{n}$ is the vector of the normal to the discontinuity surface; $\boldsymbol{i}$, $\Pi_{\alpha\beta}$ and $\boldsymbol{q}$ are the flux densities of mass, momentum and energy, respectively,

$$\boldsymbol{i} = \tau \boldsymbol{u}, \qquad \Pi_{\alpha\beta} = p\delta_{\alpha\beta} + \tau u_\alpha u_\beta - \frac{1}{4\pi} B_\alpha B_\beta + \frac{1}{8\pi} B^2 \delta_{\alpha\beta}, \tag{4.38}$$
$$\boldsymbol{q} = \tau \boldsymbol{u} \left(\frac{u^2}{2} + \epsilon + \frac{p}{\tau} \right) + \frac{1}{4\pi} \boldsymbol{B} \times [\boldsymbol{u} \times \boldsymbol{B}].$$

The momentum and energy fluxes take account of both mechanical and electromagnetic terms but ignore dissipative effects. In the expression for the energy flux density, ϵ is the internal energy of the medium per unit mass, while the last term is the Pointing vector $\boldsymbol{\gamma} = c\boldsymbol{E} \times \boldsymbol{B}/4\pi$ given by Equation (4.34) for an electric field in moving medium without dissipation. In the expression for the momentum flux density, the magnetic terms represent the magnetic part of the Maxwell stress tensor.

At the discontinuity, along with the conservation laws (4.37), the boundary conditions for the field vectors must be satisfied: that is, the normal component of $\boldsymbol{B}$, and the tangential component of $\boldsymbol{E} = -\boldsymbol{u} \times \boldsymbol{B}/c$ must be continuous,

$$\delta(B_\alpha n_\alpha) = 0, \qquad \delta(\boldsymbol{u} B_\alpha n_\alpha - \boldsymbol{B} u_\alpha n_\alpha) = 0. \tag{4.39}$$

Substituting Equation (4.38) into the conservation equations (4.37) and using the boundary conditions (4.39), we obtain four equations relating jumps in MHD parameters at the discontinuity:

$$i_n\delta\left(\epsilon + pV + \frac{i_n^2V^2}{2} + \frac{u_\tau^2}{2} + \frac{VB_\tau^2}{4\pi}\right) - \frac{B_n}{4\pi}\delta(\boldsymbol{B}_\tau\cdot\boldsymbol{u}_\tau) = 0, \tag{4.40}$$

$$\delta p + i_n^2\delta V + \frac{\delta B_\tau^2}{8\pi} = 0, \tag{4.41}$$

$$i_n\delta\boldsymbol{u}_\tau - \frac{B_n}{4\pi}\delta\boldsymbol{B}_\tau = 0, \tag{4.42}$$

$$B_n\delta\boldsymbol{u}_\tau - i_n\delta(V\boldsymbol{B}_\tau) = 0. \tag{4.43}$$

Here, the subscript n or τ labels the component normal or tangential to the plane of discontinuity, respectively (the second subscript should not be confused with density); $V = 1/\tau$ is the specific volume. Equations (4.40)–(4.43) constitute a complete set of conditions at MHD discontinuities. They are more general than the MHD equations and remain valid in a collisionless plasma provided the pressure is isotropic and the local thermodynamic and hydrodynamic parameters are specified.

Example 4.11

Based on Equations (4.40)–(4.43), analyze the discontinuities with the following properties:

(a) $i_n = 0,\ B_n \neq 0$ ***(contact discontinuity);***
(b) $i_n = B_n = 0$ ***(tangential discontinuity);***
(c) $i_n \neq 0,\ \delta V = 0$ ***(Alfvén or rotational discontinuity).***

For each case, find the relation between the jumps of macroscopic parameters and determine the quantities which are continuous at the discontinuity.

Solution. For the contact discontinuity, $\delta\boldsymbol{u} = 0$, $\delta\boldsymbol{B} = 0$, and $\delta p = 0$. The density of matter, temperature, chemical composition, and other thermodynamic parameters may experience jumps. The contact discontinuity is a stationary boundary between two media in a reciprocal mechanical equilibrium.

For the tangential discontinuity, the velocity and the magnetic field have only tangential components that may undergo arbitrary jumps. The thermodynamic parameters of the medium can also be discontinuous; the only one restriction is the continuity of the total pressure:

$$\delta\left(p + \frac{B_\tau^2}{8\pi}\right) = 0. \tag{1}$$

In the absence of the velocity jump, the two media separated by a tangential discontinuity are stationary with respect to each other, and the tangential discontinuity becomes a specific case of the contact discontinuity at which the balance of the total pressure (1) must be fulfilled.

In the case of the Alfvén discontinuity, Equations (4.42) and (4.43) yield

$$i_n = \pm\frac{|B_n|}{\sqrt{4\pi V}}, \qquad \delta\boldsymbol{u}_\tau = \mp\sqrt{\frac{V}{4\pi}}\delta\boldsymbol{B}_\tau. \tag{2}$$

As a result of rearrangement of terms and cancellation of the factor $i_n \neq 0$, Equation (4.40) can be written in the form

$$\delta\epsilon + V\delta\left(p + \frac{B_\tau^2}{8\pi}\right) + \frac{1}{2}\left(\delta\boldsymbol{u}_\tau \pm \sqrt{\frac{V}{4\pi}}\delta\boldsymbol{B}_\tau\right)^2 = 0. \tag{3}$$

Here, the third term is zero owing to Equation (2), and the second one is zero owing to Equation (4.41) at $\delta V = 0$. Therefore, from Equation (3) it follows that $\delta\epsilon = 0$. Since any thermodynamic quantity of homogeneous liquid or gas can be represented as a function of two independent variables, ϵ and V, the absence of their jumps means that all thermodynamic quantities including the pressure p are continuous. In this case, however, from Equation (4.41) we obtain $\delta B_\tau^2 = 0$, that is, the absolute value of the magnetic field does not change at Alfvén discontinuities as it does in Alfvén simple waves. Nevertheless, $\delta\boldsymbol{B}_\tau \neq 0$, that is, the vector $\boldsymbol{B}$ can rotate about the normal to the surface of the discontinuity. ■

Example 4.12

A shock front is a discontinuity at which $i_n \neq 0$ and $\delta V \neq 0$. The shock front, together with the fluxes of matter flowing into and out from the front, is referred to as a ***shock wave****.*

(a) Show that there exists such a reference frame in which the vectors $\boldsymbol{u}_1$, $\boldsymbol{u}_2$, $\boldsymbol{B}_1$, $\boldsymbol{B}_2$, and the normal to the discontinuity $\boldsymbol{n}$ lie in the same plane ***(theorem of coplanarity)****.*

(b) Show that jumps of thermodynamic quantities and magnetic field are related by ***a shock adiabatic equation****:*

$$\epsilon_1 - \epsilon_2 + \frac{1}{2}(p_1 + p_2)(V_1 - V_2) + \frac{1}{16\pi}(V_1 - V_2)(B_{\tau 1} - B_{\tau 2})^2 = 0, \tag{4.44}$$

where the components tangential to the field are assumed to be collinear by virtue of point (a). The subscript 2 stands for the quantities behind the front and subscript 1 for the quantities ahead of it.

(c) Consider specific forms of shock waves, such as the parallel shock wave ($\boldsymbol{B}_n \neq 0$, $B_{\tau 1} = B_{\tau 2} = 0$) and the perpendicular shock wave ($B_n = 0$). Express their electric field in the front frame in terms of the magnetic field and the velocity of matter. Also express the density ratio τ_2/τ_1 in terms of the pressure ratio p_2/p_1 for strong shock waves ($p_2 \gg p_1$) assuming that the matter is an ideal conducting gas (rarefied plasma). In this case, $\epsilon = pV/(\gamma - 1)$ and $\gamma = 5/3$.

Solution. (a) At $B_n \neq 0$ from Equations (4.42) and (4.43) we obtain that the three vectors $\delta\boldsymbol{B}_\tau = \boldsymbol{B}_{\tau 1} - \boldsymbol{B}_{\tau 2}$, $\delta(V\boldsymbol{B}_\tau) = V_1\boldsymbol{B}_{\tau 1} - V_2\boldsymbol{B}_{\tau 2}$ and $\delta\boldsymbol{u}_\tau = \boldsymbol{u}_{\tau 1} - \boldsymbol{u}_{\tau 2}$ are parallel (antiparallel). Since $V_1 \neq V_2$, then $\boldsymbol{B}_{\tau 1}$ and $\boldsymbol{B}_{\tau 2}$ are parallel, that is, $\boldsymbol{B}_1$, $\boldsymbol{B}_2$, and

$\boldsymbol{n}$ lie in the same plane. Taking into account that $\delta \boldsymbol{u}_\tau$ lies also in this plane, with an adequate choice of reference frame (e.g., such that $\boldsymbol{u}_1$ lies in the proper plane), it is easy to attain that each of the vectors $\boldsymbol{u}_1$ and $\boldsymbol{u}_2$ be aligned in that plane. At $B_n = 0$, from Equations (4.42) and (4.43) we have

$$(1) \qquad \boldsymbol{u}_{\tau 1} = \boldsymbol{u}_{\tau 2}, \qquad V_1 \boldsymbol{B}_{\tau 1} = V_2 \boldsymbol{B}_{\tau 2}.$$

This means that $\boldsymbol{B}_1$, $\boldsymbol{B}_2$ and $\boldsymbol{n}$ lie in the same plane. Due to the continuity of the tangential velocity, it is easy to reduce it to zero by choosing properly the reference frame. As a result, in this frame the two velocities will be aligned along the normal.

(b) In the parallel shock wave, the magnetic field is continuous as well as the tangential velocity; this follows from Equations (4.42) and (4.43). Therefore, the tangential velocity can become zero owing to a proper choice of the reference frame. In accordance with Equation (4.34), the electric field in this frame vanishes too. The magnetic field does not affect plasma motion along the field, and the shock adiabat turns into the Hugoniot adiabat for the shock wave in medium in the absence of a magnetic field:

$$(2) \qquad \epsilon_1 - \epsilon_2 + \frac{1}{2}(p_1 + p_2)(V_1 - V_2) = 0.$$

The ratio of densities on either side of the shock front can be determined from Equation (2):

$$(3) \qquad \frac{\tau_2}{\tau_1} = \frac{\gamma - 1 + (\gamma + 1)(p_2/p_1)}{\gamma + 1 + (\gamma - 1)(p_2/p_1)}.$$

In the limiting case of a strong wave ($p_2 \gg p_1(\gamma + 1)/(\gamma - 1)$), for a rarefied plasma, we obtain

$$(4) \qquad \left.\frac{\tau_2}{\tau_1}\right|_{max} \approx \frac{\gamma + 1}{\gamma - 1} = 4.$$

According to Equation (1), for a perpendicular shock wave, there exists a reference frame in which the medium moves perpendicular to the front on its both sides. In this frame, the electric field on either side of the front is parallel to it and continuous:

$$(5) \qquad \boldsymbol{E} = -\frac{1}{c}\boldsymbol{u}_1 \times \boldsymbol{B}_1 = -\frac{1}{c}\boldsymbol{u}_2 \times \boldsymbol{B}_2.$$

The magnetic field in such a wave is proportional to the density: $B_2/B_1 = \tau_2/\tau_1$. Then, the equation for the shock adiabat can again be reduced to the Hugoniot equation (2) if the internal energy and the pressure are redefined, that is, p is replaced by the total pressure

$$(6) \qquad p_* = p + \frac{b^2}{8\pi V^2},$$

and ϵ is replaced by the total density of the internal energy per unit mass,

$$(7) \qquad \epsilon_* = \epsilon + \frac{b^2}{8\pi V},$$

related to p_* through the expression $p_* = -(\partial \epsilon_*/\partial V)_s$. Here $b = V_1 B_1 = V_2 B_2$ is the quantity that remains unchanged when passing across the front.

For a strong transverse shock wave in a rarefied plasma, where $\epsilon_2 \gg \epsilon_1$ and $p_2 \gg p_1$, the shock adiabat takes the form

(8) $$\frac{\tau_2}{\tau_1} = \frac{\gamma + 1}{\gamma - 1} - \frac{B_1^2}{4\pi p_2}\left(\frac{\tau_2}{\tau_1} - 1\right)^3 .$$

The ultimate compression is the same as that in the absence of a magnetic field. However, at finite values of p_2 the compression is lower because the energy is partly spent to enhance the magnetic field. ■

In the region of the shock front, the MHD- and thermodynamic quantities exhibit high gradients that result in strong dissipation of the energy of mechanical motion. This heats the matter which flows across the front and, according to the second law of thermodynamics, rises entropy,

$$s_2 > s_1, \tag{4.45}$$

where s_1 and s_2 are the specific entropies ahead of and behind the front, respectively. From the last inequality and additional conditions

$$\left(\frac{\partial^2 V}{\partial p^2}\right)_s > 0, \qquad \left(\frac{\partial V}{\partial T}\right)_p > 0 \tag{4.46}$$

one obtains

$$p_2 > p_1, \qquad \tau_2 > \tau_1, \tag{4.47}$$

that is, shock waves are compression waves.

The MHD discontinuities that satisfy the conservation laws and the boundary conditions, Equations (4.40)–(4.43), can be unstable. The stability, in the ordinary sense, requires that initially small perturbations do not grow with time and remain small during a finite, **a priori** given interval of time. Stability of particular MHD systems is considered in Problems 4.77 – 4.79, and 4.84 – 4.85.

The ability to **evolve** is as important for MHD discontinuities as the notion of stability. A discontinuity is referred to as **evolutionary** if small perturbations either do not grow or do grow, but remain small during a short time interval. The latter situation can be illustrated by a perturbation growing exponentially, $e^{\gamma t}$ (with $\gamma > 0$): the perturbation is small as long as $t \ll 1/\gamma$, the discontinuity is unstable but evolutionary. A discontinuity is non-evolutionary if it breaks down into several discontinuities or waves, so that the perturbation immediately increases in the region where individual discontinuities are separated. Non-evolutionary discontinuities are caused by external factors, for example, by collision of two gaseous clouds. In this case, there appears a discontinuity at their interface, and the jumps of the quantities at the discontinuity are determined by the initial conditions and are not connected by Equations (4.40)–(4.43). Such a discontinuity immediately breaks down into several discontinuities and rarefaction waves.

The studies of shock wave evolution lead to the following conclusion (see Landau *et al.* (1984)). MHD shock waves have two regimes of evolution:

$$u_{n1} \geq v_{f1}, \qquad v_{An2} < u_{n2} < v_{f2}; \tag{4.48}$$

and

$$v_{s2} \leq u_{n1} < v_{An1}, \qquad u_{n2} \leq v_{s2}. \tag{4.49}$$

Here $v_{An} = |B_n|/\sqrt{4\pi\tau}$ is the Alfvén velocity, v_f and v_s are the velocities of the fast and slow magnetoacoustic waves, respectively, and u_n is the normal component of the velocity of the medium in the front frame.

Shock waves that satisfy Equation (4.48) are called **fast waves**; with decreasing their amplitude they turn into fast linear magnetoacoustic modes. Shock waves satisfying Equation (4.49) are called **slow waves**; they turn into slow linear magnetoacoustic modes as their amplitudes decrease. In slow MHD shock waves, the magnetic field becomes weaker ($B_{\tau 2} < B_{\tau 1}$), while in fast MHD-shock waves it becomes stronger ($B_{\tau 2} > B_{\tau 1}$) after passing across the front.

Suggested literature: At mathematical transformations following books are very useful: Abramovitz and Stegun (1964); Gradshtein and Ryzhik (2007); Toptygin (2014); Chapter 1; Landau *et al.* (1984); Landau and Lifshitz (1987); Polovin and Demutskii (1987); Sedov (1972); Moffatt (1978); Kulikovskii and Lyubimov (1962); Alfvén and Felthammar (1963); Pikelner (1966); Parker (1979); Zel'dovich (1956); Vainshtein *et al.* (1980); Krause and Rädler (1980); Bykov and Toptygin (1993); Ruzmaikin *et al.* (1988); Syrovatskii (1957); Fleishman and Toptygin (2013)

Problems

4.52• Prove that the equations

$$\nabla \times \boldsymbol{E} = -\frac{1}{c}\frac{\partial \boldsymbol{B}}{\partial t}, \qquad \nabla \times \boldsymbol{B} = \frac{4\pi}{c}\boldsymbol{j}, \qquad \nabla \cdot \boldsymbol{B} = 0,$$

used in magnetic hydrodynamics are invariant with respect to Galilean transformations $\boldsymbol{r}' = \boldsymbol{r} - \boldsymbol{V}t,\ t' = t$, that is, that they take the form

$$\nabla' \times \boldsymbol{E}' = -\frac{1}{c}\frac{\partial \boldsymbol{B}'}{\partial t'}, \qquad \nabla' \times \boldsymbol{B}' = \frac{4\pi}{c}\boldsymbol{j}', \qquad \nabla' \cdot \boldsymbol{B}' = 0,$$

if the field strength and the current density are rearranged as

$$\boldsymbol{j}' = \boldsymbol{j}, \qquad \boldsymbol{B}' = \boldsymbol{B}, \qquad \boldsymbol{E}' = \boldsymbol{E} + \frac{1}{c}\boldsymbol{V} \times \boldsymbol{B}.$$

4.53• Using the electromagnetic induction law (4.36), show that if a magnetic field satisfies Equation (4.29) for non-dissipation medium, the magnetic flux through any moving and deformed closed contour remains constant.

4.54 Show that when a conducting medium in a magnetic field is in the state of static equilibrium, the vectors $\boldsymbol{B}$ and $\boldsymbol{j}$ are tangential to the surfaces $p(\boldsymbol{r}) = const$.

4.55 Along a cylindrical column of hot plasma of radius a there flows a current J distributed over the cross section with the density $j(r)$ (z-pinch). What is the r-dependence of the plasma pressure in a stationary state provided it is balanced by the magnetic pressure exerted by the current flowing along the column. Find the relation between the total current and the pressure in the pinch, integrated over the cross section.

Let the plasma be isothermal and let it satisfy the perfect gas equation of state. Express the current strength J in terms of the plasma temperature T and the total number N of particles of one charge sign per unit length of the plasma column. Calculate the current strength taking $N \approx 10^{15}$ particles/cm and $T \approx 10^8$ K (these values are typical for thermonuclear studies).

4.56 What should be the distribution of the current across the cross section of the plasma column (see the statement of Problem 4.55) for the plasma pressure be constant across the cross section?

4.57 Find the equilibrium condition for a cylindrical plasma column of radius a in which the current has only the azimuthal component $j_\varphi(r)$ (theta-pinch). The pressure of the medium outside the column can be neglected. Is it possible to maintain the equilibrium using a magnetic field of external sources?

4.58 A magnetic field is called **force-free** if the density of the magnetic force is zero everywhere, $\boldsymbol{j} \times \boldsymbol{B}/c = 0$. For this field, it is necessary and sufficient that the current at any point be directed along the magnetic line of force.

(a) Show that a force-free magnetic field $\boldsymbol{B}$ satisfies the system of equations

$$\nabla \times \boldsymbol{B} = \alpha \boldsymbol{B}, \qquad \boldsymbol{B}\cdot\nabla\alpha = 0,$$

where $\alpha(\boldsymbol{r})$ is an arbitrary differentiable pseudoscalar function (including a constant).

(b) Construct a force-free field depending only on r at $\alpha = const$, choose the cylindrical coordinates.

4.59 Using Equations (4.22) derive the equation for the vector potential $\boldsymbol{A}$ related to the magnetic field by a conventional formula $\boldsymbol{B} = \nabla \times \boldsymbol{A}$.

4.60 A stationary velocity field $\boldsymbol{u} = (u_x(y,z), 0, 0)$ (a shear flow) is maintained in a conducting medium. At the initial instant $t = 0$ the magnetic field has the components $\boldsymbol{B}_0 = (0, B_{0y}(y,z), B_{0z}(y,z))$. Calculate the field at $t > 0$ neglecting dissipation ($\nu_m \to 0$).

4.61 A stationary differential rotation with angular velocity $\boldsymbol{\Omega} = \boldsymbol{e}_z\Omega(r,\vartheta)$ is maintained in a conducting medium. Here, r, ϑ are the spherical coordinates (i.e., individual layers of the conducting liquid or gas rotate with different angular velocities). The initial magnetic field lies in the meridional planes (poloidal field). At $t = 0$ the field is $\boldsymbol{B} = (B_{0r}(r,\vartheta), B_{0\vartheta}(r,\vartheta), 0)$. Calculate the magnetic field at $t > 0$, neglecting dissipation.

4.62* The stationary two-dimensional motion with the velocity $\boldsymbol{u} = (u_x(x,y), u_y(x,y), 0)$ is maintained in an incompressible, homogeneous and conducting medium. Show that such a motion cannot maintain the magnetic field during a long time, so that the field dissipates in a finite time (Zel'dovich (1956)).[7)]

4.63* The velocity field of conducting medium in the spherical coordinates is given by the vector $\boldsymbol{u} = \nabla \times \boldsymbol{e}_r \psi(\boldsymbol{r})$, while the electric conductivity $\kappa(r)$ depends only on the distance from the center. Show that the magnetic field decays during a finite time regardless of the initial state (Elsasser's antidynamo-theorem).

4.64* A viscous, incompressible and conducting liquid moves between two fixed parallel planes in the direction of the Oz-axis under the action of constant pressure gradient $dp/dz = const$. The electric conductivity of the liquid κ and the viscosity η are constant, and the distance between the planes is $2a$. A constant and uniform magnetic field $\boldsymbol{B}_0$ is applied perpendicular to the planes in the direction of the Ox-axis. Evaluate the x-dependence of the velocity of the liquid and the additional magnetic field created in the moving liquid. Analyze the result for large and small values of B_0.

4.65 A viscous incompressible liquid is placed between the parallel planes $x = \pm a$. The plane $x = -a$ moves with the velocity $-v_0$ and the plane $x = a$ with velocity v_0, both in the direction of the Oz-axis. The pressure gradient is absent, and the electric conductivity κ and the viscosity η of the liquid are given. A uniform magnetic field B_0 is applied perpendicular to the planes. Evaluate the velocity of the liquid and the additional magnetic field in it.

4.66 Find the solution (21), (22) for an Alfvén wave of small amplitude (see Example 4.10), by linearizing the system of equations (4.29)–(4.32).

4.67 Calculate the decrement γ of the plane Alfvén wave of small amplitude assuming that kinematic ν and magnetic ν_m viscosities of the medium are known. The decrement is smaller than the frequency of the wave.

4.68 Consider the fast and slow simple waves (see point (b) in Example 4.10) for small amplitudes of perturbations. For this purpose, use the system of equations (4.29)–(4.32) in the linearized form. Determine the mutual orientation of the vectors $\boldsymbol{k}$, $\boldsymbol{u}$, $\boldsymbol{b}$, and $\boldsymbol{B}_0$, where the last two vectors are the amplitudes of a magnetoacoustic wave and of an external uniform magnetic field, respectively.

4.69* Study the evoltion of the fast, transverse and simple wave (see Example 4.10) caused by the dependence of phase velocity on the density of matter. For this purpose, set the profile of velocity of medium at the initial moment of time $t = 0$ in the form $u_x(x, 0) \equiv u(x, 0) = u_0/\cosh(x/h)$, where $u_0 = c_{s0} = v_{A0}$ coincides with the velocity of sound c_{s0} and with the Alfvén velocity v_{A0} in the unperturbed medium. Calculate the function $u(x, t)$ at $t > 0$ with the help of a computer. Also, plot the magnetic field $B(x, t)$ as a function of x at $t = 0$ and $t > 0$. Determine the time instant

7) Yakov Borisovich Zel'dovich (1914–1987) was outstanding Soviet scientist with very wide range of interest (chemical physics, physics of shock waves and detonation, nuclear physics and elementary particles, astrophysics, and cosmology). The participant of the Soviet atomic project.

t_* up to which the solution in the form of a simple wave is applicable (i.e., the solution is unambiguous). The equation of state of medium for adiabatic processes is $p(\tau) = p_0(\tau/\tau_0)^\gamma$.

4.70 Plasma is ejected isotropically in all directions from the surface of a sphere of radius a which is rotating about its diameter with constant angular velocity Ω. The plasma velocity u is constant and directed along the radius. Near the surface of the sphere there exists a magnetic field which, in the reference frame corotating with the sphere, is $\boldsymbol{B}(a, \vartheta, \alpha) = \boldsymbol{B}_0(\vartheta, \alpha)$, where α is measured in the plane perpendicular to the spin axis. The energy density of the plasma is large compared to that of the magnetic field. Therefore, the effect of the field on the plasma motion can be neglected. Assuming that the magnetic field is frozen into the plasma, find it as a function of coordinates and time in the region $r > a$ in the fixed reference frame.[8)]

4.71 Find lines of force of the interplanetary magnetic field in Parker's model considered in Problem 4.70. Determine the magnitude of the magnetic field and the angle θ between the lines of force and the radial direction at the Earth's orbit for the following values of the parameters: solar radius $a = 0.7 \times 10^6$ km, average magnetic field at the solar surface $B_0 \approx 1$ G, radius of the Earth orbit $r_0 \approx 1.5 \times 10^8$ km, angular velocity of the solar rotation $\Omega = 2.7 \times 10^{-6}$ rad/s, and the velocity of the solar wind $u = 300$ km/s.

4.72 Show that one of the solutions of Problem 4.70 on the magnetic field in the region of stationary radial stellar wind has the form

$$B_r(r, \vartheta, \alpha, t) = F\left(\alpha - \frac{(r-a)\Omega}{u} + \Omega t\right)\left(\frac{a}{r}\right)^2 2\cos\vartheta,$$

$$B_\vartheta(r, \vartheta, \alpha, t) = p\,\frac{\partial}{\partial\alpha}F\left(\alpha - \frac{(r-a)\Omega}{u} + \Omega t\right)\frac{a^2\Omega}{ur}\sin\vartheta,$$

$$B_\alpha(r, \vartheta, \alpha, t) = qF\left(\alpha - \frac{(r-a)\Omega}{u} + \Omega t\right)\frac{a^2\Omega}{ur}2\sin\vartheta\cos\vartheta.$$

Here p and q are constants satisfying the condition $p + q = 1$.

4.73⋆ Consider a homogeneous conducting medium (gaseous plasma) with the uniform magnetic field B_0. Assume that in a certain small region ("point") a significant amount of energy is released, which creates a strong shock wave. The energy density of matter in the perturbed region is much higher than the energy density of the magnetic field. Hence, deviations from spherical symmetry produced by the magnetic field can be neglected, and the shock front can be treated as an expanding sphere of radius $R_s(t)$. The velocity field in the perturbed region $r \le R_s(t)$ can be written in the form $u_r = u = 2r\dot{R}_s(t)/(\gamma+1)R_s(t)$, $u_\vartheta = u_\alpha = 0$, where $\gamma = 5/3$ for a rarefied gas. Calculate the magnetic field in the entire space after the explosion.

8) Parker used the model considered in this problem for describing the interplanetary magnetic field produced by solar plasma flows (solar wind). E. N. Parker (1927), American astrophysicist, studied cosmic rays, cosmic plasmas, and magnetic fields.

In particular, plot the time dependence of the magnetic field at a given distance r from the center of the explosion.[9)]

4.74⋆ Solve Problem 4.73 for the case when the stellar magnetic field can be approximated by a field of magnetic dipole with a magnetic moment M. Can the dipole be regarded as a point-like? Take into account finite sizes of the dipole using the model considered in Problem I.2.85. Plot the magnetic field versus time at a given distance r from the center of explosion.

4.75⋆ The magnetic field in the vicinity of the star prior to the explosion can be expanded in spherical vectors defined in Problem I.5.14:

$$\boldsymbol{B}_{0n} = \sum_{lm} a_{lm}^{(3)}(r)\mathcal{Y}_{lm}^{(3)}(\vartheta, \alpha), \qquad \boldsymbol{B}_{0\tau} = \sum_{lm,\, k=1,2} a_{lm}^{(k)}(r)\mathcal{Y}_{lm}^{(k)}(\vartheta, \alpha).$$

Here, we take into account that spherical vectors with $k = 3$ are directed along the radius, while at $k = 1$ and 2 they are perpendicular to it. Assuming that the coefficients $a^{(k)}$ of the expansion are given, find the field in the cavity formed as a result of the explosion. The motion of the shock front, $R_s(t)$, is known.

4.76 A star explodes into the region with nonuniform density $\tau_0 \propto r^{-2}$, which is occupied by a stellar wind with a magnetic field specified in the statement of Problem 4.72. Evaluate the magnetic field after the explosion for the case in which the field components are independent of angle α, $p = 0$, $q = 1$, assuming that the dependence of the shock front radius on time, $R_s(t)$, is known.[10)]

4.77⋆ A plane contact discontinuity separates two conducting systems of non-uniform densities, τ_1 and τ_2. A gravity field (acceleration $\boldsymbol{g}$) is applied perpendicular to the discontinuity and homogeneous magnetic field $\boldsymbol{B}$ is parallel to the unperturbed discontinuity surface. Investigate the stability of the discontinuity by a method of small perturbations in the approximation of incompressible and non-dissipative medium.

Hint. Look for unstable modes in the class of potential flows.

4.78⋆ A plane contact discontinuity separates a conducting incompressible liquid and vacuum. The liquid has the density τ and the coefficient of surface tension α. The gravity field (acceleration $\boldsymbol{g}$) and the electric field $\boldsymbol{E}$ are perpendicular to the discontinuity; they are oriented in such a way that the liquid lies at the bottom. Find the field E at which the contact discontinuity becomes unstable with respect to small perturbations **(Tonks effect)**.[11)] Find the wave numbers of unstable modes.

Hint. Use the energy principle, that is, compare the energies of perturbed and unperturbed systems and find the conditions under which the energy of the deformed discontinuity is lower.

9) This is a simple model describing the evolution of the magnetic field in the vicinity of a star after the explosion. The velocity field behind the front given above approximates well the Sedov self-similar solution for a strong explosion (see Sedov (1972), Chapter IV, §11) at which $R_s(t) \propto t^{2/5}$.

10) According to Sedov (1972), for a strong explosion in the present case $R_s(t) \propto t^{2/3}$.

11) Tonks Lewy (1897–1971), American physicist, known for his studies in plasma physics, nuclear physics, and so on.

4.79⋆ Examine the stability of the tangential discontinuity in incompressible conducting medium in a magnetic field. The density of the medium, its velocity and the magnetic induction are different on either side of the discontinuity, and there is no gravitation **(Kelvin–Helmholtz instability)**.

4.80⋆ Investigate the effect of turbulence on the evolution of a magnetic field in a conducting medium. To do this, use the perturbation theory and average the equations for the magnetic field over ensemble of turbulent pulsations. Consider the turbulence to be homogeneous and isotropic, and describe it by the tensor (D15) for incompressible medium. Evaluate the magnetic viscosity of the medium taking into account the turbulence. The collisional magnetic viscosity $\nu_m = c^2/4\pi\kappa$ should be regarded as given.

4.81⋆ Derive the equation describing the magnetic field in a medium with gyrotropic turbulence and with zero correlation time of turbulent velocities. For this purpose, write down the correlation tensor in the coordinate representation (cf (A24)):

$$U_{\alpha\beta}(\boldsymbol{r}, t) = \{Q(r)\delta_{\alpha\beta} + R(r)x_\alpha x_\beta + C(r)e_{\alpha\beta\sigma}x_\sigma\}\, 2\tau_c\, \delta(t).$$

Assume that the collisional magnetic viscosity $\nu_m = c^2/4\pi\kappa$ is finite. Show that the required equation reads

$$\frac{\partial \boldsymbol{B}}{\partial t} = \alpha\, \nabla \times \boldsymbol{B} + (\nu_t + \nu_m)\triangle \boldsymbol{B},$$

where $\nu_t = Q(0)\tau_c = \langle \boldsymbol{u}^2\rangle\tau_c/3$ is the turbulent viscosity and $\alpha = -2C(0)$ $\tau_c = \langle \boldsymbol{u}(\boldsymbol{r}) \cdot rot\, \boldsymbol{u}(\boldsymbol{r})\rangle\tau_c/3$ is the parameter of gyrotropy.

4.82 Find the relation between the functions $Q(r)$, $R(r)$, and $C(r)$ of Problem 4.81 and the functions $A(k,t)$ and $P(k,t)$, which enter the definition of the spectral tensor (A24).

4.83⋆ In Problem 4.58, definition of a force-free magnetic field is made. Let in some area filled with cold plasma ($p \approx 0$) a force-free non-potential field exists, which satisfies the equations

$$\nabla \times \boldsymbol{B}_0 = \alpha_{FFF}\, \boldsymbol{B}_0, \qquad \alpha_{FFF} = const,$$

where FFF mean "force-free field." Consider small perturbations of this field, which represent MHD waves with random phases. Show that the turbulence formed by such waves has helicity and calculate its kinematic helicity parameter (A25).

4.84 Find the magnetic field in an infinite, gyrotropic, and turbulent medium based on the given initial field $\boldsymbol{B}^0(\boldsymbol{r})$ and the equation obtained in Problem 4.81. Assume that the parameters of the turbulence are independent of the magnetic field ("kinematic approximation"). Show that long wavelength harmonics in such a medium will increase infinitely. Find the instability criterion.

4.85⋆ Find out possibility of a magnetic field increase at simultaneous action of turbulence helicity and differential rotation ($\alpha\Omega$-dynamo). Let except turbulence with helicity there is a field of large-scale velocity $\boldsymbol{u}(x) = \Omega x \boldsymbol{e}_y$

where Ω is a constant with dimension of inverse time. Find the solution considering both factors. Calculate the increase rate of a large-scale magnetic field.

4.86★ According to the results of Problems 4.80 and 4.81, the turbulence enhances the magnetic viscosity of the medium by adding the turbulent component ν_t, $\nu_{tot} = \nu_m + \nu_t$. This changes the kinetic coefficients of the medium. If in a stationary medium the electric conductivity is κ and the magnetic permeability is $\mu = 1$, in a turbulent medium these quantities will be different, $\sigma \neq \kappa$ and $\mu \neq 1$. Find the values of σ and μ. To do it, use Equations (4.10)– (4.12) with the effective values of σ and μ for the averaged fields in the turbulent medium. Also use the results of Problems (4.10)–(4.12) for non-gyrotropic turbulence, taking the gyrotropy parameter $\alpha = 0$. Express the required quantities in terms of ν_m and ν_t.

4.4
Answers and Solutions

4.1

$$J(t) = \frac{\pi a^2 \omega H_0}{c\sqrt{R^2 + (\omega L)^2/c^4}} \sin(\omega t - \varphi), \quad \text{where} \quad \tan\varphi = \frac{\omega L}{c^2 R},$$

$$N(t) = -\frac{\omega(\pi a^2 H_0)^2}{c^2\sqrt{R^2 + (\omega L)^2/c^4}} \sin\omega t \, \sin(\omega t - \varphi),$$

$$\overline{P} = \frac{\omega^2}{2c^2} \frac{(\pi a^2 H_0)^2 R}{R^2 + (\omega L)^2/c^4} = \frac{1}{2} J_0^2 R.$$

Here L is the inductance of the ring (see Problem I.2.113), R is its resistance, and J_0 is the amplitude of the current in the ring. The origin of time is chosen in such a way that at $t = 0$ the plane of the loop is perpendicular to $\boldsymbol{H}_0$.

4.2

$$\overline{N} = \frac{\omega}{2c^2} \frac{(SH_0)^2 R}{R^2 + \left(\frac{\omega L}{c^2} - \frac{1}{\omega C}\right)^2}.$$

4.3 The average generalized force which tends to increase the generalized coordinate q_i, is equal to

$$-\frac{J_0^2}{2c^6} \frac{\omega^2 L L_{12}}{R^2 + \left(\frac{\omega L}{c^2}\right)^2} \frac{\partial L_{12}}{\partial q_i},$$

where L and R are the inductance and the resistance, respectively, of the second contour and L_{12} is the mutual inductance of the contours.

4.4

$$\overline{F} = -\frac{\omega^2 L L_{12} |\mathcal{E}_0|^2}{2c^6\{[R^2 + \omega^4(L_{12}^2 - L^2)/c^4]^2 + 4\,\omega^2 L^2 R^2/c^4\}} \frac{\partial L_{12}}{\partial q_i}.$$

4.5

$$\omega_{1,2}^2 = \frac{c^2[(L_1+L_2)C+L_1C_1+L_2C_2] \pm c^2\{[L_1(C+C_1)-L_2(C+C_2)]^2+4L_1L_2C^2\}^{1/2}}{2L_1L_2(C_1C_2+CC_1+CC_2)}.$$

In the absence of coupling between the contours, that is, at $C = 0$, the frequencies ω_1 and ω_2 become equal to $c/\sqrt{L_1C_1}$ and $c/\sqrt{L_2C_2}$, respectively, which corresponds to independent oscillations of the contours. For very tight coupling ($C \gg C_1, C_2$), there is only a single frequency $\omega = c/\sqrt{L'C'}$, where $L' = L_1L_2/(L_1 + L_2)$ and $C' = C_1 + C_2$. This corresponds to oscillations in a single contour containing capacitances C_1, C_2 and inductances L_1, L_2 connected in parallel.

4.6

$$\omega_{1,2}^2 = \frac{c^2}{2}\left(\frac{1}{LC_1} + \frac{1}{LC_2} + \frac{1}{L_1C_1} + \frac{1}{L_2C_2}\right)$$
$$\pm \frac{c^2}{2}\left\{\left[\frac{1}{C_1}\left(\frac{1}{L} + \frac{1}{L_1}\right) - \frac{1}{C_2}\left(\frac{1}{L} + \frac{1}{L_1}\right)\right]^2 + \frac{4}{L^2C_1C_2}\right\}^{1/2}.$$

4.7

$$\omega_{1,2}^2 = c^2\frac{L_1C_1 + L_2C_2 \pm \left[(L_1C_1 - L_2C_2)^2 + 4C_1C_2L_{12}^2\right]^{1/2}}{2C_1C_2(L_1L_2 - L_{12}^2)}.$$

4.8 Writing the system of equations for the currents and equating the determinant of the system to zero, we obtain after some rearrangements the fourth-order equation

(1)
$$\omega^4 + i\,\omega^3\left(\frac{1}{\tau_1} + \frac{1}{\tau_2}\right) - \omega^2(\,\omega_1^2 + \omega_2^2) - i\,\omega\left(\frac{\omega_1^2}{\tau_2} + \frac{\omega_2^2}{\tau_1}\right) + \omega_1^2\,\omega_2^2 = 0,$$

where

$$\omega_1 = \frac{c}{\sqrt{L_1C_1}}, \qquad \omega_2 = \frac{c}{\sqrt{L_2C_2}}, \qquad \tau_1 = RC_1, \qquad \tau_2 = RC_2.$$

The coefficients of this equation are complex and hence the frequency ω is also complex, $\omega = \omega' + i\omega''$. In the zero-order approximation, the terms containing τ_1 and τ_2 in Equation (1) can be neglected. Then Equation (1) takes the form

(2)
$$\omega^4 - \omega^2(\omega_1^2 + \omega_2^2) + \omega_1^2\omega_2^2 = 0.$$

This equation has the following solutions: $\omega_1^{(0)} = \omega_1$ and $\omega_2^{(0)} = \omega_2$. Thus, in this approximation $\omega'' = 0$ and there is no energy dissipation (since we assume that R is infinite). The oscillations in each contour occur independently. In the next approximation, we look for ω in the form $\omega = \omega^{(0)} + \Delta\omega' + i\omega''$, where ω'' and $\Delta\omega'$ are at least of the order of $1/\tau$. Accordingly, all higher-order terms can be neglected. If we substitute ω into Equation (1), take account of Equation (2), and equate separately the real and imaginary parts to zero, we obtain

(3)
$$\Delta\omega' = 0, \qquad \omega_1'' = -\frac{1}{2\tau_2}, \qquad \omega_2'' = -\frac{1}{2\tau_1}.$$

The correction to ω', containing R, will only appear in the next approximation.

4.9

$$J_1 = \frac{\mathcal{E}}{Z_1\left(1 + \omega^2 L_{12}^2/c^4 Z_1 Z_2\right)}, \qquad J_2 = \frac{i\,\omega L_{12}}{c^2 Z_2} J_1;$$

$$Z_1 = R_1 + i\left(\frac{1}{\omega C_1} - \frac{\omega L_1}{c^2}\right), \qquad Z_2 = R_2 + i\left(\frac{1}{\omega C_2} - \frac{\omega L_2}{c^2}\right);$$

$$J_{1\,max} = \frac{\mathcal{E}}{R} \quad \text{at} \quad \omega = \frac{c}{\sqrt{L_1 C_1\left(1 - L_{12}^2/L_1 L_2\right)}}.$$

4.10

$$Z = \frac{R - i\,\omega L/c^2}{1 - (\omega/\,\omega_1)^2 - i\,\omega RC}, \qquad \text{where} \qquad \omega_1 = \frac{c}{\sqrt{LC}}$$

is the frequency of free oscillations in the contour. At $R = 0$ and $\omega = \omega_1$, the impedance Z becomes infinite. This property of the two-terminal network considered above is used in electronics (rejection filters).

4.11

$$C = C_0, \qquad L = L_0, \qquad R = \frac{\gamma L_0}{c^2}, \qquad \text{where} \qquad L_0 = \frac{c^2}{\omega_p^2 C_0}.$$

4.12

$$Q = \frac{1}{2}\Re(UJ^*) = \frac{1}{2}|J|^2 \Re\left(\frac{1}{Z}\right) = \frac{1}{2}\,\frac{\gamma\,\omega_p^2}{\omega^2 + \gamma^2}\,C_0|U_0|^2,$$

$$\overline{W} = \frac{1}{4}\left(1 + \frac{\omega_p^2}{\omega^2 + \gamma^2}\right) C_0 |U_0|^2.$$

4.13

$$C = C_0, \qquad L = \frac{c^2}{\omega_p^2 C_0}, \qquad C_1 = \frac{\omega_p^2}{\omega_0^2} C_0, \qquad R = \frac{\gamma L}{c^2} = \frac{\gamma}{\omega_p^2 C_0}.$$

4.14

$$Q = \frac{1}{2}\,\frac{\omega^2\,\omega_p^2 \gamma}{(\omega^2 - \omega_0^2)^2 + \omega^2\gamma^2}\,C_0|U_0|^2,$$

$$\overline{W} = \frac{1}{4}\left[1 + \frac{\omega_p^2(\omega^2 + \omega_0^2)}{(\omega^2 - \omega_0^2)^2 + \omega^2\gamma^2}\right] C_0|U_0|^2.$$

4.15 Let us denote the currents flowing through the inductance, capacitance and the battery by J_1, J_2 and J_3, respectively. According to Kirchhoff laws, we have the equations

$$(1) \qquad J_1 + J_2 + J_3 = 0, \qquad \frac{L}{c^2}\dot{J}_1 = \frac{q(t)}{C} = \mathcal{E}(t) + J_3 R,$$

where $q(t)$ is the charge on the capacitor plate, which is related to J_2 by the formula $J_2 = \dot{q}$, and

$$\mathcal{E}(t) = \begin{cases} 0 & \text{at} \quad t < 0, \\ \mathcal{E} & \text{at} \quad t > 0. \end{cases}$$

From Equation (1) we obtain the second-order equation for the current J_1. The corresponding characteristic equation has the roots

$$x = -\frac{1}{2RC} \pm \sqrt{\left(\frac{1}{2RC}\right)^2 - \omega_0^2}, \qquad \omega_0^2 = \frac{c^2}{LC}.$$

The three cases are possible, depending on the relation between R, L, and C:

(a) $\omega_0 > 1/(2RC)$. Using the Lagrangian method of variation of arbitrary constants, we find the solution for J_1,

$$J_1(t) = \frac{\mathcal{E}}{R}\left[1 - e^{-t/2RC}\left(\frac{\sin \omega t}{2\,\omega RC} + \cos \omega t\right)\right],$$

where $\omega = \sqrt{\omega_0^2 - (2RC)^{-2}}$;

(b) $\omega_0 < 1/2RC$;

$$J_1(t) = \frac{\mathcal{E}}{R}\left[1 - e^{-t/2RC}\left(\frac{\sinh \Omega t}{2\Omega RC} + \cosh \Omega t\right)\right],$$

where $\Omega = \sqrt{(2RC)^{-2} - \omega_0^2}$;

(c) $\omega_0 = 1/2RC$;

$$J_1(t) = \frac{\mathcal{E}}{R}\left[1 - \left(1 + \frac{t}{2RC}\right)e^{-t/2RC}\right].$$

In cases (b) and (c), the transient process is totally aperiodic, and there are no oscillations.

4.16

$$U_2(t) = \begin{cases} 0 & \text{at} \quad t < 0, \\ U_0 e^{-t/RC} & \text{at} \quad 0 < t < T, \\ U_0\left(e^{-t/RC} - e^{-(t-T)/RC}\right) & \text{at} \quad t > T. \end{cases}$$

4.17

$$U_2(t) = \begin{cases} 0 & \text{at} \quad t < 0, \\ U_0 e^{-Rc^2 t/L} & \text{at} \quad 0 < t < T, \\ U_0\left(e^{-Rc^2 t/L} - e^{-Rc^2(t-T)/L}\right) & \text{at} \quad t > T. \end{cases}$$

4.18 The pulse which must be applied across the input of the four-terminal network should be

$$U_1(t) = \begin{cases} 0 & \text{at} \quad t < -T, \\ hE_0\left(1 + \frac{t}{T} + \frac{RC}{T}\right) & \text{at} \quad -T < t < 0, \\ hE_0\left(1 - \frac{t}{T}\right) & \text{at} \quad 0 < t < T, \\ 0 & \text{at} \quad t > T. \end{cases}$$

The origin of time is chosen in such a way for the field between the plates of the capacitor be maximal at $t = 0$.

4.19

$$J(t) = \frac{\mathcal{E}_0}{\sqrt{R^2 + (\omega L/c^2)^2}} \left[\cos(\omega t + \varphi_0 - \varphi) - e^{-Rc^2 t/L} \cos(\varphi_0 - \varphi)\right],$$

where $\tan\varphi = \omega L/c^2 R$. The transient process is absent at $\tan\varphi_0 = -Rc^2/\omega L$. This condition has a simple meaning: the stationary current should be zero at the moment of switching-on.

4.20 For a harmonic dependence of currents on time, the Kirchhoff equation for the nth contour can be written as

(1) $$-\frac{\omega L}{c^2} J_n + \frac{1}{\omega C}\left(2J_n - J_{n-1} - J_{n+1}\right) = 0.$$

Equation (1) is a linear difference equation with the integer independent variable n. It has two linearly independent solutions, namely $\sin\kappa n$ and $\cos\kappa n$ (cf. Problem 3.4). The frequencies of oscillations can be expressed in terms of the parameter κ:

(2) $$\omega^2 = 3\omega_0^2 \sin^2\frac{\kappa}{2}, \qquad \omega_0 = \frac{c}{\sqrt{LC}}.$$

Using the boundary conditions $J_0 = J_N = 0$, we have

(3) $$J_n = A \sin\kappa n, \qquad \kappa = \frac{\pi r}{N}.$$

Here, r can take any positive value $r = 1, 2, \ldots$ At $r = 0$, the current in the circuit is zero. However, the number of eigenfrequencies of the system will be finite due to the periodicity of the factor $\sin(\kappa/2)$ in Equation (2). In order to obtain the frequency spectrum, it is sufficient to vary r within the range $1 \le r \le N$. In this case, the parameter κ will vary within the range of $0 \le \kappa \le \pi$; there will be one eigenfrequency at any κ. The total number of frequencies will be N, as it should be for a system of N coupled contours. They will lie in the range $0 < \omega \le 2\omega_0$. In order to interpret the quantity κ, we introduce the coordinate $y_n = an$ of the nth section (a being the "length" of a section). Then, Equation (3) can be written in the form

(4) $$J_n(t) = J_0 \sin k y_n \, e^{-i\omega_k t},$$

where the time-dependent factor is included and $k = \kappa/a$.

Equation (4) is a superposition of two waves traveling in opposite directions. The quantity k plays the role of "wave vector" for the oscillations that propagate along the line consisting of separate discrete sections. The phase and group velocities of these waves can be calculated using the ordinary formulas

(5) $$v_{ph} = \frac{\omega}{k}, \qquad v_g = \frac{d\omega}{dk}.$$

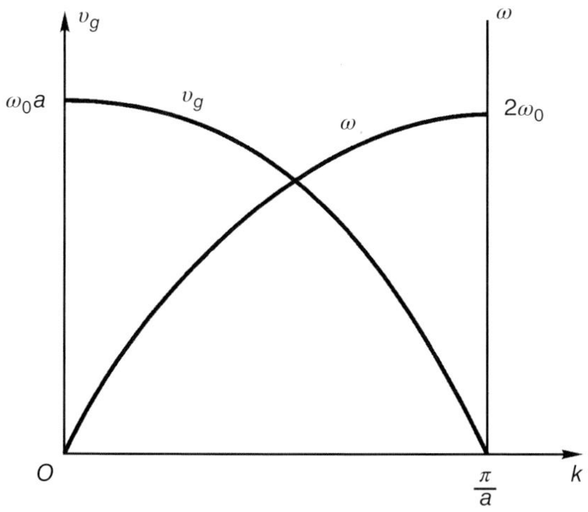

Figure 4.9 Dispersive curves for an artificial long line consisting of identical sections.

Since ω is a nonlinear function on k, the velocities v_{ph} and v_g are different, that is, dispersion takes place. From Equation (2) we find

(6) $$v_{ph} = \frac{2\,\omega_0}{k}\sin\frac{ka}{2}, \qquad v_g = \omega_0 a\cos\frac{ka}{2}.$$

The quantity $2\pi/k$ has the meaning of "wavelength" of oscillations in a discrete chain; for long waves ($\lambda \gg a$) we have $ka \ll 1$ and, hence, the phase and group velocities are equal, $v_{ph} = v_g = \omega_0 a$, and independent of k (implying no dispersion). The plots of ω and v_g versus k are presented in Figure 4.9.

Electrical oscillations in such a chain are analogous to mechanical oscillations in a linear atomic chain, which can be used as a one-dimensional model of a crystal. The inductance L is then analogous to the mass of atom, and the quantity $1/C$ to the coefficient of rigidity.[12)]

4.21

$$\Delta r = \frac{2N}{\pi}\,\frac{\Delta\omega}{\sqrt{4\,\omega_0^2 - \omega^2}}.$$

4.22 Let the currents through the contours with self-inductances L_1 and L_2 be J and J', respectively. The Kirchhoff equation will be of the form

(1) $$\begin{cases} -\frac{\omega L_1}{c^2} J_n + \frac{1}{\omega C}(2J_n - J'_n - J'_{n-1}) = 0, \\ -\frac{\omega L_2}{c^2} J_n + \frac{1}{\omega C}(2J'_n - J_n - J_{n-1}) = 0. \end{cases}$$

12) See, for example, M. A. Leontovich, Statisticheskaya Fizika (Statistical Physics), Nauka, Moscow, 1983; Born Max and Kun Huang, Dynamical Theory of Crystal Lattices. Oxford, Clarendon Press, 1954; L. Brillouin and M. Parodi, Propagation des Ondes dans Les Milieux Périodiques, Donod, Paris, 1956, Chapters 3 and 4.

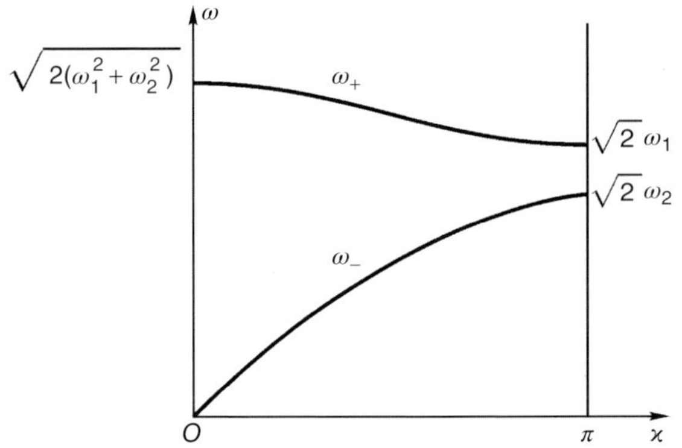

Figure 4.10 The same as in Figure 4.9, for a line consisting of alternating sections of two kinds.

Substituting the frequencies $\omega_1 = c/\sqrt{L_1 C}$ and $\omega_2 = c/\sqrt{L_2 C}$, we have

$$(2) \qquad \begin{cases} (2\omega_1^2 - \omega^2)J_n &= \omega_1^2(J'_n + J'_{n-1}), \\ (2\omega_2^2 - \omega^2)J'_n &= \omega_2^2(J_n + J_{n-1}). \end{cases}$$

The solution of this system of equations will be sought in the form

$$(3) \qquad J_n = Ae^{i\kappa n}, \qquad J'_n = Be^{i\kappa n},$$

where A, B, and κ are constants. Substituting these solutions into Equation (2) we have

$$(4) \qquad A(2\omega_1^2 + \omega^2) = B\omega_1^2(1 + e^{-i\kappa}), \qquad B(2\omega_2^2 - \omega^2) = A\omega_2^2(1 + e^{i\kappa}).$$

Equating the determinant of the system to zero, we find the relation between the frequency ω and κ:

$$(5) \qquad \omega^2 = \omega_1^2 + \omega_2^2 \pm \sqrt{(\omega_1^2 + \omega_2^2)^2 - 4\omega_1^2\omega_2^2 \sin^2 \frac{\kappa}{2}}.$$

In order to obtain the complete spectrum of oscillations, the quantity κ must be varied between 0 and π. As in Problem 4.20, the values of κ can be found from the boundary conditions.

The most significant difference between this case and the case of a chain with identical sections, is that now, as follows from Equation (5), there are two frequencies from each value of κ. Therefore, there are two branches of oscillations. Let us denote the frequencies of these oscillations by ω_+ and ω_-, where subscripts + and – correspond to the same signs ahead of the square root in Equation (5). The dependence of these frequencies on κ is shown in Figure 4.10. The oscillations of frequency ω_- are analogous to the oscillations in the chain with identical sections. In particular, at small

κ (long waves) we have

$$\omega_- = \frac{\omega_1\omega_2}{\sqrt{2(\omega_1^2+\omega_2^2)}}\kappa,$$

that is, dispersion is absent.

For the ω_+ branch, the dispersion relation at small κ is

$$\omega_+ = a + b\kappa^2.$$

As $\kappa \to 0$ the phase velocity tends to infinity while the group velocity vanishe.

In order to investigate the character of oscillations in both branches, we find the ratio of the current amplitudes in adjacent contours for very long ($\kappa \ll 1$) and very short (κ close to π) waves. From Equation (4) we have at $\kappa \ll 1$ for the ω_- branch:

$$\left(\frac{A}{B}\right)_- \approx 1,$$

and for the ω_+ branch:

$$\left(\frac{A}{B}\right)_+ \approx -\frac{\omega_1^2}{\omega_2^2} = -\frac{L_2}{L_1}.$$

For the ω_- branch, the oscillations of currents in adjacent contours are of equal amplitudes and are in phase. For the ω_+ branch, the oscillations in adjacent contours are in antiphase while the amplitudes are inversely proportional to the inductances.

At $\kappa = \pi$ we obtain

$$\omega_+ = \sqrt{2}\omega_1, \qquad \omega_- = \sqrt{2}\omega_2.$$

In the limit $\kappa \to \pi$ in Equation (4), we have

$$\left(\frac{A}{B}\right)_+ \to 0, \qquad \left(\frac{A}{B}\right)_- \to 0.$$

Thus, in the limit $\kappa = \pi$, oscillations with frequency $\omega_+ = c\sqrt{2/(L_1C)}$ will take place only in the contours with inductances L_1, whereas oscillations with the frequency $\omega_- = c\sqrt{2/(L_2C)}$ will occur in contours with inductances L_2. The oscillations with frequencies ω_- and ω_+ considered in this problem are an analog of acoustic and optical oscillations in a linear atomic chain consisting of two kinds of atoms with different masses (see the references to the solution of Problem 4.20).

4.23

(1) $$J_n = Aq_1^n + Bq_2^n,$$

where q_1 and q_2 are the roots of the equation

(2) $$q^2 - \left(2 + \frac{Z_1}{Z_2}\right)q + 1 = 0.$$

The constants A and B can be determined from the boundary conditions $J_N = 0$ and $(J_0 - J_1)Z_2 = U_1$. The second condition means that a voltage U_1 is applied across the points $a'b'$ (Figure 4.7). Using the equality $q_1 q_2 = 1$ which follows from Equation (2), we finally obtain

$$U_2 = J_{N-1} Z_2 = U_1 \frac{q_2 - q_1}{(1 - q_1)q_2^N - (1 - q_2)q_1^N}.$$

4.24 The transmission coefficient K can be determined from the results of Problem 4.23:

$$K = \frac{q_1 - q_2}{(1 - q_2)q_1^N - (1 - q_1)q_2^N}.$$

The denominator of this expression includes the factors q_1^N and q_2^N. Since $q_1 \, q_2 = 1$, there are two possible cases:

$$1.\ |q_1| = |q_2| = 1; \qquad 2.\ |q_1| > 1,\ |q_2| < 1.$$

In the first case, the absolute values of q_1^N and q_2^N will be equal to unity, and K will also be of the order of unity. In the second case, at $N \gg 1$ we have $|q_1^N| \gg 1$ and $|q_2^N| \ll 1$, and hence

$$K = \frac{q_1 - q_2}{(1 - q_2)q_1^N} \ll 1.$$

The frequency ranges for cases 1 and 2 can be found from Equation (2) of Problem 4.23,

$$q_{1,2} = 1 + \frac{Z_1}{2Z_2} \pm \sqrt{\left(1 + \frac{Z_1}{2Z_2}\right)^2 - 1}.$$

If the expression under square root is negative, then q_1 and q_2 are complex conjugate numbers whose absolute values are equal to unity, that is, case 1 takes place. If the expression under square root is positive, q_1 and q_2 are real and different, that is, we deal with case 2. Equating the square root to zero, we find the region of Z_1 and Z_2 for case 1:

$$-4 \le \frac{Z_1}{Z_2} \le 0.$$

This corresponds to the values of ω^2 lying between

$$\frac{c^2}{L_1 C_1} \quad \text{and} \quad \frac{c^2(4C_1 + C_2)}{C_1 C_2 (4L_2 + L_1)}.$$

4.25 Let us consider the nth closed contour of an artificial long line (Figure 4.11). This contour can be regarded as an equivalent circuit for a section of line with distributed parameters, having a length a, inductance ΔL, and capacitance ΔC. For an arbitrary time-dependence of the current, flowing in the line, the Kirchhoff equation for this section takes the form

$$(1) \qquad -\frac{1}{c^2}\Delta L \frac{\partial J_n}{\partial t} + \frac{q_{n-1,n}}{\Delta C} - \frac{q_{n+1,n}}{\Delta C} = 0,$$

where $q_{n-1,n}$ and $q_{n+1,n}$ are the charges on the upper plates of the two capacitors. Differentiating Equation (1) with respect to time and using the relations $\dot{q}_{n-1,n} = -J_n + J_{n-1}$ and $\dot{q}_{n,n+1} = J_n - J_{n+1}$, we have

(2) $$\frac{1}{c^2}\Delta L \frac{\partial^2 J_n}{\partial t^2} + \frac{1}{\Delta C}(2J_n - J_{n-1} - J_{n+1}) = 0.$$

Now transform the variable n to the variable z, which is a coordinate of a point of the line with distributed parameters. To do so, we set

$$J_n(t) = J(z, t), \qquad J_{n-1}(t) = J(z - a, t), \qquad J_{n+1}(t) = J(z + a, t)$$

and evaluate the differences as

$$J_n - J_{n-1} \approx \frac{\partial J}{\partial z}a - \frac{1}{2}\frac{\partial^2 J}{\partial z^2}a^2,$$

$$J_n - J_{n+1} \approx -\frac{\partial J}{\partial z}a - \frac{1}{2}\frac{\partial^2 J}{\partial z^2}a^2.$$

Substituting these differences into Equation (2) and taking into account that $L = \Delta L/a$ and $C = \Delta C/a$, which are the inductance and capacitance per unit length, we obtain

(3) $$\frac{L}{c^2}\frac{\partial^2 J}{\partial t^2} = \frac{1}{C}\frac{\partial^2 J}{\partial z^2}.$$

This is the equation for a long line without dissipation. In a real long line, there are always losses resulted from the resistance of conductor and imperfect insulation.

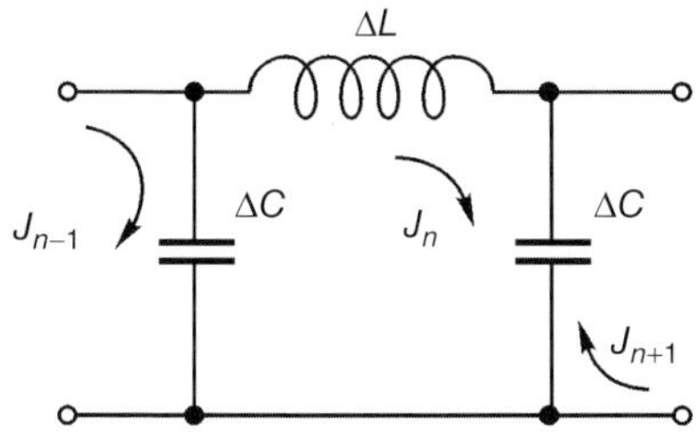

Figure 4.11 Separate closed contour for an artificial loss-free long line.

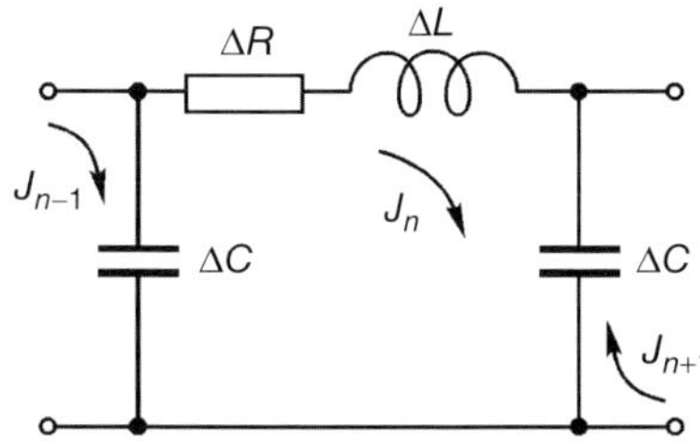

Figure 4.12 The same as in Figure 4.11, but for a line with losses.

The equivalent circuit for the case in which the second factor is ignored (i.e., insulation is assumed to be perfect) is shown in Figure 4.12. In this case, the equation for a long line (the telegraph equation) can be obtained in the same way as Equation (3),

$$(4) \qquad \frac{L}{c^2}\frac{\partial^2 J}{\partial t^2} + R\frac{\partial J}{\partial t} = \frac{1}{C}\frac{\partial^2 J}{\partial z^2},$$

where R is the active resistance of the conductor per unit length.

4.26 The solution of Equation (3) obtained in Problem 4.25, yields

$$\omega = vk,$$

where $v = c/\sqrt{LC}$ is the propagation velocity of waves in the long line, $k = \pi r/l$, $r = 1, 2, 3\ldots$; L and C are the inductance and capacitance per unit length, respectively. In contrast to the spectrum of the chain with lumped parameters, in the obtained spectrum of the long line the number of eigenfrequencies is infinite. This is because the long line is a continuum with infinite number of degrees of freedom, while in the chain the number N of degrees of freedom if finite. Another characteristic feature of the perfect long line is the absence of dispersion.

4.27 Let us use the Ohm's law in the differential form, $\boldsymbol{j} = \sigma(\boldsymbol{E} + \boldsymbol{E}_{ext})$, where $\boldsymbol{E}_{ext}$ is the field strength of external forces. The field strength $\boldsymbol{E}$ can be expressed in terms of the potentials:

$$\boldsymbol{E} = -\nabla\varphi - \frac{1}{c}\frac{\partial \boldsymbol{A}}{\partial t}, \qquad \boldsymbol{E}_{ext} = \frac{\boldsymbol{j}}{\sigma} + \nabla\varphi + \frac{1}{c}\frac{\partial \boldsymbol{A}}{\partial t}.$$

Assuming that the conductor is thin, we will integrate both sides of the latter equality along the contour coinciding with the conductor:

$$(1) \qquad \oint \boldsymbol{E}_{ext}\cdot d\boldsymbol{l} = \oint \frac{\boldsymbol{j}}{\sigma}\cdot d\boldsymbol{l} + \oint \nabla\varphi\cdot d\boldsymbol{l} + \frac{1}{c}\oint \frac{\partial \boldsymbol{A}}{\partial t}\cdot d\boldsymbol{l}.$$

The integral on the left-hand side of Equation (1) is the external electromotive force $\mathcal{E}_{ext}$ applied to the circuit; the integral $\oint(\boldsymbol{j}/\sigma)\cdot d\boldsymbol{l} = JR$ describes the Joule losses per unit time, and $\oint \nabla\varphi\cdot d\boldsymbol{l} = \oint d\varphi = 0$. The last integral can be transformed in the following way. Taking into account retardation, we write

$$\boldsymbol{A} = \frac{1}{c}\oint \frac{J(t - r/c)\,d\boldsymbol{l}'}{r};$$

$$J(t - r/c) = J_0 e^{-i\omega(t - r/c)}, \qquad \frac{\partial \boldsymbol{A}}{\partial t} = -i\,\omega \boldsymbol{A}.$$

Substituting these expressions into Equation (1) and separating the real and imaginary parts, we have

$$\mathcal{E}_{ext}(t) = J(t)\left[\left(R + \frac{\omega}{c^2}\oint\oint \frac{\sin(\omega r/c)\,d\boldsymbol{l}\cdot d\boldsymbol{l}'}{r}\right) - \frac{i\omega}{c^2}\oint\oint \frac{\cos(\omega r/c)\,d\boldsymbol{l}\cdot d\boldsymbol{l}'}{r}\right].$$

The expression in braces describes the complex impedance of the circuit. The active resistance is equal to $R + R_r(\omega)$, where

$$R_r(\omega) = \frac{\omega}{c^2} \oint \oint \frac{\sin(\omega r/c)}{r} \, \boldsymbol{dl} \cdot \boldsymbol{dl}' .$$

The quantity R is related to the losses due to heating of the conductor; $R_r(\omega)$ represents the energy losses due to radiation and is referred to as the **radiation resistance** (see Problem 4.28). The reactive resistance is equal to $-i\omega L(\omega)/c^2$, where

$$L(\omega) = \oint \oint \frac{\cos(\omega r/c)}{r} \, \boldsymbol{dl} \cdot \boldsymbol{dl}' ,$$

is the inductance depending on frequency.

Let us consider the case in which $c/\omega = \lambda/2\pi \gg l$, where l is the linear size of the contour. In the region of integration, $\omega r/c \ll 1$. Taking account of the quadratic term in the Taylor expansion of $\cos(\omega r/c)$, we have

$$L(\omega) \approx \oint \oint \frac{\boldsymbol{dl} \cdot \boldsymbol{dl}'}{r} - \frac{1}{2}\left(\frac{\omega}{c}\right)^2 \oint \oint r \, \boldsymbol{dl} \cdot \boldsymbol{dl}' .$$

The first term in this expression does not depend on frequency and represents an ordinary inductance, and [13] the second term gives the correction, which is significant at high frequencies.

In the Taylor expansion of $\sin(\omega r/c)$, the third-order term must also be included since the integral of the first (linear) term is zero. The radiation resistance is given by

$$R_r(\,\omega) = -\frac{\omega^4}{6c^5} \oint \oint r^2 \boldsymbol{dl} \cdot \boldsymbol{dl}' .$$

4.28

$$L(\omega) = L + \frac{64\pi^4}{3} \frac{a^4}{\lambda^3}, \qquad R_r(\omega) = \frac{2\pi^2}{3c}\left(\frac{2\pi a}{\lambda}\right)^4 .$$

The current-carrying ring is a magnetic dipole. The energy emitted per unit time is given by $2\ddot{\boldsymbol{m}}^2/3c^3$, where $\boldsymbol{m}$ is the magnetic dipole moment. The coefficient of proportionality between the emitted energy and the average squared current $\overline{J^2}$ is $2\pi^2 a^2\, \omega^4/3c^5$; it is the same as $R_r(\omega)$.

4.29 By solving the first equation (4.14) under the boundary conditions (4.16), we have

$$H(z) = H_0\{\cosh[(1-i)z/\delta_1] - A \sinh[(1-i)z/\delta_1]\}, \qquad 0 \le z < a,$$

$$H(z) = H(a) e^{-(1-i)(z-a)/\delta_2}, \qquad a < z < \infty,$$

$$A = \frac{\kappa_2 \delta_2 \sinh[(1-i)a/\delta_1] + \kappa_1 \delta_1 \cosh[(1-i)a/\delta_1]}{\kappa_1 \delta_1 \sinh[(1-i)a/\delta_1] + \kappa_2 \delta_2 \cosh[(1-i)a/\delta_1]} .$$

13) In practice, the self-induction should be calculated from Equation (2.74), since the integral $\oint \oint \boldsymbol{dl} \cdot \boldsymbol{dl}'/r$ diverges. The divergence occurs because the conductor is assumed to be infinitely thin (linear).

where the penetration depths δ_1 and δ_2 are given by Equation (4.17). At $a \ll \delta_1$ and $\kappa_1\delta_1 \sim \kappa_2\delta_2$,

$$H(z) = H_0\left\{1 - (1-i)\frac{\kappa_1 z}{\kappa_2\delta_2}\right\}, \qquad z \le a.$$

At $a \gg \delta_1$,

$$H(z) = H_0 e^{-(1-i)z/\delta_1}, \quad z < a; \qquad H(z) = H_0 e^{-(1-i)a/\delta_1-(1-i)(z-a)/\delta_2}, \quad z > a.$$

4.30

$$H(x) = H_0\left(\frac{\sinh^2 x/\delta + \cos^2 x/\delta}{\sinh^2 h/\delta + \cos^2 h/\delta}\right)^{1/2}, \qquad H_0 = \frac{4\pi}{c}J_0 n.$$

At $\delta \ll h$ the magnetic field is $H(x) = H_0\, e^{-(h-|x|)/\delta}$ while at $\delta \gg h$ the field $H(x) = H_0$.

4.31 Since the system is symmetric with respect to the cylinder axis and the primary magnetic field H_0 is uniform, it is clear that eddy currents induced in the cylinder will flow along circles which lie in planes perpendicular to the axis. These currents will produce the same magnetic field as that due to a large number of separate coaxial solenoids. However, the field induced by a solenoid is zero in the external region, while inside the solenoid it is directed along its axis. Thus, the total magnetic field outside the cylinder is equal to H_0 and inside the cylinder is given by the first equation (4.14), which, in view of axial symmetry, takes the form

$$\frac{d^2H}{dr^2} + \frac{1}{r}\frac{dH}{dr} + k^2 H = 0,$$

where

$$k = \frac{1+i}{\delta}, \qquad H = H_z(r), \qquad H_\alpha = H_r = 0,$$

The boundary condition is $H(a) = H_0 = 4\pi n J_0/c$, where H_0 is the field inside the solenoid (see Problem I.2.83).

The solution which is finite at $r = 0$ and which satisfies this boundary condition can be expressed in terms of zero-order Bessel function:

$$H = H_0\frac{J_0(kr)}{J_0(ka)}.$$

Outside the cylinder

$$H = H_0 \quad \text{at} \quad a \le r \le b, \qquad H = 0 \quad \text{at} \quad r > b.$$

The current density and the electric field inside the cylinder are evaluated from Equation (4.12),

$$j = j_\alpha = \kappa E_\alpha = \frac{kc}{4\pi}\frac{J_1(kr)}{J_0(ka)}H_0, \qquad E_r = E_z = 0.$$

To determine the electric field inside the cylinder we can use the Maxwell equation for *rot**E*** in the integral form:

$$\oint E_l\, dl = \frac{i\omega}{c}\int B_n\, dS.$$

Inside the cylinder there is only a single component of the electric field, E_α. It follows from the boundary condition on the surface of the cylinder and from symmetry of the system that outside the cylinder the field $\boldsymbol{E}$ also has the single component, E_α, which is a function of r only. If a circle is taken as a contour of integration l, the contour integral gives $2\pi r E_\alpha$. The surface integral is evaluated using formulas from Section I.1.3. Finally, we have

$$E_\alpha = \frac{kcH_0}{4\pi\kappa}\frac{J_1(ka)}{J_0(ka)}\frac{a}{r} + \frac{H_0}{2r}(r^2 - a^2), \qquad \text{if} \qquad a \le r \le b,$$

$$E_\alpha = \frac{kcH_0}{4\pi\kappa}\frac{J_1(ka)}{J_0(ka)}\frac{a}{r} + \frac{H_0}{2r}(b^2 - a^2), \qquad \text{if} \qquad r > b.$$

In the absence of the cylinder, that is, at $a = 0$, the field will be equal to

$$E_\alpha = \frac{1}{2}H_0 r \quad (r < b), \qquad E_\alpha = \frac{H_0 b^2}{2r} \quad (r > b).$$

Thus, the additional magnetic field associated with the presence of the cylinder is zero for $r > a$, even if the additional electric field is finite. This is because the exact equation

$$\nabla \times \boldsymbol{H} = \frac{1}{c}\frac{\partial \boldsymbol{D}}{\partial t},$$

which holds outside the conductor, is replaced by the approximate equation $\nabla \times \boldsymbol{H} = 0$ (in the quasi-stationary approximation, the displacement current can be neglected). In the exact solution, the additional magnetic field outside the conductor will also be finite (see Problem 6.60 where diffraction of a plane wave by a conducting cylinder is considered).

4.32 At low frequencies ($|ka| \ll 1$ or $\delta \gg a$)

$$j = i\frac{cH_0}{4\pi}\frac{r}{\delta^2} = \frac{i\mu\kappa\omega H_0}{2c}r,$$

and hence the current density is a linear function of r and is proportional to the frequency. At high frequencies ($|ka| \gg 1$ or $\delta \ll a$), we can use the asymptotic formula for the Bessel function to obtain

$$j = (i-1)\frac{cH_0}{4\pi\delta}\sqrt{\frac{a}{r}}\, e^{-(1-i)(a-r)/\delta}.$$

At $a - r \gg \delta$, the current density becomes exponentially small. Thus, at high frequencies the current is largely concentrated in a thin surface layer.

4.33

$$Q = -\frac{a\,c^2 H_0^2}{16\pi\kappa}\mathfrak{R}\left[\frac{kJ_1(ka)}{J_0(ka)}\right], \qquad k^2 = i\frac{4\pi\mu\kappa\omega}{c^2}.$$

At $|ka| \ll 1$ (low frequencies)

$$Q = \frac{c^2 H_0^2}{64\pi\kappa}\left(\frac{a}{\delta}\right)^4 = \pi\kappa\left(\frac{a^2 H_0 \mu\omega}{4c}\right)^2.$$

At $|ka| \gg 1|$ (high frequencies)

$$Q = \frac{c^2 H_0^2}{16\pi\kappa}\left(\frac{a}{\delta}\right) = \frac{a\,cH_0^2}{8}\sqrt{\frac{\mu\omega}{2\pi\kappa}}.$$

The energy dissipation is proportional to ω^2 at low frequencies and to $\sqrt{\omega}$ at high frequencies.

4.34

$$\beta = \beta' + i\beta'' = -\frac{a^2}{4}\left[1 - \frac{2}{ka}\frac{J_1(ka)}{J_0(ka)}\right], \qquad k^2 = i\frac{4\pi\kappa\omega}{c^2}.$$

At $|ka| \gg 1$ (high frequencies):

$$\beta' = -\frac{a^2}{4}\left(1 - \frac{c}{a\sqrt{2\pi\kappa\omega}}\right), \qquad \beta'' = \frac{ca}{4\sqrt{2\pi\kappa\omega}}.$$

Hence, at high frequencies $\beta'' \to 0$, that is, the losses are reduced due to expelling the field from the conductor. In this limit, the real part of polarizability β' per unit volume is the same as the magnetic polarizability of a superconducting cylinder $\chi = \beta'/\pi a^2 = -1/4\pi$ (see Example 3.8).

At $|ka| \ll 1$ (low frequencies):

$$\beta' = -\frac{\pi^2 a^6 \kappa^2 \omega^2}{12c^4}, \qquad \beta'' = \frac{\pi a^4 \kappa\omega}{8c^2}.$$

Thus, $\beta \to 0$ as $\omega \to 0$. This is because $\mu = 1$, that is, the static magnetic polarizability is zero.

4.35 The magnetic moment induced by eddy currents will, in view of symmetry of the system, be parallel to the external magnetic field. Hence, the total magnetic field $\boldsymbol{H}_2$ outside the cylinder can be written in the form

(1) $$\boldsymbol{H}_2(\boldsymbol{r}) = \frac{4\boldsymbol{r}(\boldsymbol{m}\cdot\boldsymbol{r})}{r^4} - \frac{2\boldsymbol{m}}{r^2} + \boldsymbol{H}_0.$$

Here $\boldsymbol{m}$ is an unknown magnetic moment of per unit length of the cylinder, which is parallel to $\boldsymbol{H}_0$, and $\boldsymbol{r}$ is the position-vector in the plane perpendicular to the axis of the cylinder. The vector potential

$$\boldsymbol{A}_2 = \frac{2(\boldsymbol{m}\times\boldsymbol{r})}{r^2} + (\boldsymbol{H}_0\times\boldsymbol{r}),$$

corresponding to the field $\boldsymbol{H}_2$, is written in projections as

(2) $$A_{2z} \equiv A_2 = \left(\frac{2m}{r} + H_0 r\right)\sin\alpha, \qquad A_{2r} = A_{2\alpha} = 0;$$

the angle α is measured from the direction of $\boldsymbol{H}_0$.

Thus, the vector potential in the external region has only a longitudinal (with respect to the cylinder axis) component, which is proportional to $\sin\alpha$. The continuity conditions for the filed components at the boundary can be satisfied if the vector potential in the internal region is sought in analogous form:

(3) $$A_{1z} \equiv A_1 = F(r)\sin\alpha, \qquad A_{1r} = A_{1\alpha} = 0.$$

The electric field $\boldsymbol{E}$ can generally be expressed in terms of both potentials, $\boldsymbol{A}$ and φ.

As usual, we impose an additional condition on the potentials,

$$div\,\boldsymbol{A} + \frac{\varepsilon}{c}\frac{\partial\varphi}{\partial t} = 0.$$

Since $div\,\boldsymbol{A} = 0$, which follows from Equations (2) and (3), we have $\partial\varphi/\partial t = -i\omega\varphi = 0$, so that

$$\boldsymbol{E} = -\frac{1}{c}\frac{\partial \boldsymbol{A}}{\partial t} = \frac{i\omega}{c}\boldsymbol{A}.$$

Hence, $\boldsymbol{A}$ will satisfy the same equation (4.14) as the electric field. The solution of this equation, which is finite at $r = 0$, is a Bessel function,

$$F(r) = CJ_1(kr), \qquad A_1 = CJ_1(kr)\sin\alpha. \tag{4}$$

The constants C and m in Equations (4) and (2) can be determined from the condition for the internal ($\boldsymbol{H}_1$) and external ($\boldsymbol{H}_2$) fields be equal at the surface of the cylinder: $\boldsymbol{H}_1 = \boldsymbol{H}_2$ at $r = a$. Using the properties of Bessel functions, we have

$$C = \frac{2H_0}{kJ_0(ka)}, \qquad m = -\frac{a^2H_0}{2}\left(1 - \frac{2}{ka}\frac{J_1(ka)}{J_0(ka)}\right). \tag{5}$$

As follows from the expression for m, the transverse magnetic polarizability of the cylinder,

$$\beta = -\frac{a^2}{2}\left[1 - \frac{2}{ka}\frac{J_1(ka)}{J_0(ka)}\right], \tag{6}$$

is twice larger than its longitudinal polarizability (see Problem 4.35). The components of the magnetic field inside the cylinder are found from Equations (4) and (5):

$$\left.\begin{aligned} H_{1r} &= \frac{1}{r}\frac{\partial A_1}{\partial \alpha} = 2H_0\frac{J_1(kr)}{krJ_0(ka)}\cos\alpha, \qquad H_{1z} = 0. \\ H_{1\alpha} &= -\frac{\partial A_1}{\partial r} = -2H_0\frac{J_1'(kr)}{J_0(ka)}\sin\alpha. \end{aligned}\right\} \tag{7}$$

In addition, we determine also the current density in the cylinder. Using the formula $\boldsymbol{j} = (c/4\pi)\,rot\boldsymbol{H}$, we obtain

$$J_z = -\frac{cH_0}{2\pi}\cdot\frac{J_1(kr)}{J_0(ka)}\sin\alpha, \qquad J_\alpha = j_r = 0. \tag{8}$$

According to this formula, the currents in the two halves of the cylinder, $0 \le \alpha \le \pi$ and $\pi \le \alpha \le 2\pi$, will always flow in the opposite directions. The total current through the cross section of the cylinder is zero. The radial dependence of the current density is the same as for a cylinder placed in a longitudinal field (see Problem 4.32). It should be noted, however, that in the longitudinal field the currents flow over circles in the planes perpendicular to the axis of the cylinder, whereas in the transverse field the currents flow along the axis of the cylinder.

4.36 The average energy dissipation per unit length of the cylinder can be most easily calculated from Equation (4.20), by considering the flux of energy through the lateral surface of the cylinder. Using the results of Problem 4.35, we have

$$Q = -\frac{ac^2H_0^2}{8\pi\kappa}\,\mathfrak{R}\left(\frac{J_1(ka)}{J_0(ka)}\right).$$

The same result can be obtained using Equation (4.19).

4.37 To determine the torque, the electric and magnetic fields must be known. They can be found in the same way as in Problem 4.35 for a linearly polarized external field:

$$\left.\begin{aligned} H_r &= -\frac{2H_0J_1(kr)}{krJ_0(ka)}e^{i\alpha}, \qquad H_\alpha = -\frac{2iH_0J_1'(kr)}{J_0(ka)}e^{i\alpha},\\ j_z &= \frac{ickH_0}{2\pi}\frac{J_1(kr)}{J_0(ka)}e^{i\alpha}. \end{aligned}\right\} \tag{1}$$

The force per unit volume of the cylinder is

$$\boldsymbol{f} = \frac{1}{c}\boldsymbol{j}\times\boldsymbol{H}, \tag{2}$$

assuming that $\mu = 1$ inside the cylinder. The radial component of this force exerts the radial pressure, whereas the azimuthal component gives rise to the torque. Since $\boldsymbol{j}$ and $\boldsymbol{H}$ are complex quantities, the average value of the azimuthal component of the force is

$$\bar{f}_\alpha = \frac{1}{2c}\,\mathfrak{R}(j_zH_r^*). \tag{3}$$

The torque per unit length of the cylinder can be obtained as vector product of $\boldsymbol{r}$ and the average force (3) integrated over the cross section of the cylinder. The integral can be evaluated using Equation (I.1.176). The result is

$$\overline{N} = -\frac{aH_0^2}{|k|^2}\,\mathfrak{R}\left(k\frac{J_1(ka)}{J_0(ka)}\right). \tag{4}$$

The same result can be obtained in a different way. The torque can be expressed in terms of the magnetic moment of the system

$$\boldsymbol{N}(t) = \boldsymbol{m}(t)\times\boldsymbol{H}_0(t). \tag{5}$$

Expressing $\overline{N_z} = \overline{N}$ through the complex amplitudes $\boldsymbol{H}_0$ and $\boldsymbol{m}$, while $\boldsymbol{m}$ through the transverse magnetic polarizability of the cylinder (Problem 4.35), we come to Equation (4).

At low frequencies, it follows from Equation (4) that

$$\overline{N} = \frac{a^4 H_0^2}{4\delta^2} = \frac{\pi\kappa\omega}{4c^2} H_0^2 a^4, \tag{6}$$

while at high frequencies

$$\overline{N} = \frac{1}{2} a\,\delta H_0^2 = \frac{ca}{\sqrt{8\pi\kappa\omega}} H_0^2. \tag{7}$$

As seen from these formulas, the torque vanishes in both limiting cases: at very low and very high frequencies.

If the field is linearly polarized, the average torque is equal to zero (formally this is because the integral over α in the formula for N is zero; see Problem 4.35 in which $\boldsymbol{j}$ and $\boldsymbol{H}$ are found for this case). Thus, the torque is produced by a "rotating" field. The phenomenon considered in the present problem is basic for induction motors.

4.38 Along with a fixed coordinate system, whose Oz-axis coincides with the axis of the cylinder and Ox-axis is parallel to the external field $\boldsymbol{H}_0$, we consider the system of coordinates ξ, η, and z which rotates with the cylinder. In the latter system, the external magnetic field reads

$$\boldsymbol{H}_0(t) = (\boldsymbol{H}_{01} - i\boldsymbol{H}_{02})e^{-i\omega t}.$$

Here $\boldsymbol{H}_{01}$ and $\boldsymbol{H}_{02}$ are constant vectors of equal length, $H_{01} = H_{02} = H_0$; they are parallel to the ξ and η coordinate axes. A field of this kind has been considered in Problem 4.37. The braking torque produced by this field is

$$\overline{N} = -\frac{aH_0^2}{|k|^2} \Re\left(k\frac{J_1(ka)}{J_0(ka)}\right).$$

4.39 As has been shown in Problem 4.31, the eddy currents induced in the cylinder by variations of the external longitudinal field, do not produce any additional magnetic field outside the cylinder. Inside the cylinder, the field due to these currents is longitudinal and is a function of r only. This field satisfies the equation

$$\frac{\partial^2 H}{\partial r^2} + \frac{1}{r}\frac{\partial H}{\partial r} - \frac{4\pi\mu\kappa}{c^2}\frac{\partial H}{\partial t} = 0. \tag{1}$$

Evidently, the magnetic field inside the cylinder will decay with time. Hence, particular solutions of Equation (1) can be sought in the form $F(r)\,e^{-\gamma t}$, where $\gamma > 0$ is a constant. The function $F(r)$ satisfies the Bessel equation

$$F''(r) + \frac{1}{r}F'(r) + k^2 F(r) = 0, \tag{2}$$

where $k^2 = 4\pi\mu\kappa\gamma/c^2$.

The solution of Equation (2), which is finite at $r = 0$, is $F(r) = CJ_0(kr)$. Since the external field H_0 is switched off and the additional field produced by eddy currents outside the cylinder is zero, the condition $H\big|_{r=a} = 0$ must be satisfied at the boundary, that is,

(3) $$J_0(ka) = 0.$$

Hence, $k_m a = \beta_m$, $m = 1, 2, \ldots$, where β_m are zeros of the function J_0. Therefore, possible values of γ are

(4) $$\gamma_m = \frac{c^2\beta_m^2}{4\pi\mu\kappa a^2}.$$

The general solution of Equation (1), which corresponds to the boundary problem under consideration will be written as

(5) $$H(r,t) = \sum_m C_m J_0(k_m r) e^{-\gamma_m t}.$$

The coefficients C_m can be determined from the initial condition

(6) $$H(r,0) = \sum_m C_m J_0(k_m r).$$

By using the orthogonality of Bessel functions

(7) $$\int_0^1 x J_0(k_m x) J_0(k_n x)\, dx = \frac{1}{2}\left[J_0'(k_m)\right]^2 \delta_{mn},$$

we have

(8) $$C_m = \frac{2}{a^2\left[J_0'(k_m a)\right]^2}\int_0^a H(r,0) J_0(k_m r)\, r\, dr.$$

Initially, the field $H(r,0)$ is equal to the external field H_0 since the constant magnetic field is undistorted by the presence of the infinite cylinder whose axis is parallel to it. The use of Equations (I.1.145)–(I.1.147) yields

(9) $$C_m = \frac{2H_0}{(k_m a)J_1(k_m a)}.$$

The rate of field decay will be determined by the smallest of the γ_m, that is, by γ_1. It is obtained by substituting the smallest root of the Bessel function, $\beta_1 \approx 2.4$, into Equation (4). The e-folding field decay time is $\tau = 1/\gamma_1$.

4.40 In the zero-order approximation (in frequency), the magnetic field inside the sphere has been found in Problem 3.35,

(1) $$\boldsymbol{H} = \frac{3}{\mu+2}\boldsymbol{H}_0.$$

The electric field inside the sphere in the same approximation, as follows from Equation (4.12), is zero, since the constant magnetic field does not produce any electric field. In order to determine the electric field in the next (linear in ω) approximation, we will use Equation (4.11) in an integral form.

It is clear from symmetry of the system that the currents inside the sphere flow in circles, which lie in planes perpendicular to $\boldsymbol{H}_0$; the electric field will be aligned in the same direction.

Taking the spherical system of coordinates with the Oz-axis along $\boldsymbol{H}_0$, we have

(2) $$E = \frac{i\omega H}{2c} r \sin\vartheta, \qquad j = \kappa E,$$

where H is given by Equation (1). The heat Q released in the sphere is found by integrating $q = \frac{1}{2}\kappa |E|^2$ over the volume of the sphere:

(3) $$Q = \frac{3\pi a^5 \kappa \omega^2 H_0}{5c^2(\mu + 2)^2}.$$

4.41 The magnetic field outside the sphere is

$$\boldsymbol{H} = \boldsymbol{H}_0 + \frac{3\boldsymbol{r}(\boldsymbol{m}\cdot\boldsymbol{r})}{r^5} - \frac{\boldsymbol{m}}{r^3},$$

where $\boldsymbol{m} = -\frac{1}{2}a^3\boldsymbol{H}_0$; $\beta = -\frac{1}{2}a^3$ is the magnetic polarizability of the sphere for a strong skin-effect. Inside the sphere

$$H_\vartheta = -\frac{3}{2}H_0\, e^{-(1-i)z/\delta} \sin\vartheta, \qquad H_r = H_\alpha = 0,$$

where z is measured from the surface along the inward normal to the conductor, and the polar axis of the spherical system of coordinates is directed along $\mathbf{H}_0$;

$$Q = \frac{3a^2 c}{8}\sqrt{\frac{\mu\omega}{2\pi\kappa}}\, H_0^2.$$

4.42 In the case of a strong skin-effect, the field inside the ellipsoid is zero, whereas the field outside the ellipsoid satisfies the equations $rot\, \boldsymbol{H} = 0$ and $div\, \boldsymbol{H} = 0$ as well as the boundary conditions $H_n\big|_S = 0$ and $\boldsymbol{H}\big|_{r\to\infty} \to \boldsymbol{H}_0$, where $\boldsymbol{H}_0$ is the external field, and S denotes the ellipsoid surface.

Let us compare this problem with the case of dielectric ellipsoid with $\varepsilon = 0$ placed in a uniform electric field. The electric field outside the ellipsoid will satisfy the equations

(1) $$rot\, \mathbf{E} = 0, \qquad div\, \mathbf{E} = 0$$

and the boundary conditions

(2) $$E_n\big|_S = \varepsilon E_{int}\big|_S = 0, \qquad \mathbf{E}\big|_{r\to\infty} \to \mathbf{E}_0.$$

The conditions for the tangential components $\boldsymbol{E}$ can be ignored because Equations (1) and (2) uniquely define the vector $\boldsymbol{E}$ in the external region. We see that the present problem of a conducting ellipsoid in the case of strong skin-effect is formally identical with the problem of a dielectric ellipsoid having $\varepsilon = 0$. Assuming $\varepsilon_1 = 0$ in the results of Problem 2.45, we obtain the magnetic polarizabilities in the direction of the principal axes of the ellipsoid:

(3) $$\beta^{(i)} = -\frac{V}{4\pi(1 - n^{(i)})},$$

where $n^{(i)}$ is the depolarization coefficient and V is the volume of the ellipsoid. For a very prolate ellipsoid of revolution with semi-axes a and $b \gg a$ (a rod), we have (see Problem 8.43):

$$\beta_\perp = -\frac{2}{3}a^2 b, \qquad \beta_\parallel = -\frac{1}{3}a^2 b.$$

For a very oblate ellipsoid ($b \ll a$, a disc):

$$\beta_\perp = -\frac{2a^3}{3\pi}, \qquad \beta_\parallel = -\frac{1}{3}a^2 b \to 0 \qquad \text{at } b \to 0.$$

4.43 In view of the axial symmetry of the system (the sphere + an external field), the distribution of eddy currents in the sphere and of electric field also acquire axial symmetry. Hence, we can state that the electric field will have only one component, E_α, which cannot be a function of α, so that $E_\alpha = f(r, \vartheta)$. The factor $e^{-i\omega t}$ is dropped everywhere. The solution of Equation (4.14) for the total electric field will be sought in the form

(1) $$E_\alpha = F(r)\sin\vartheta, \qquad E_r = E_\vartheta = 0.$$

Using the expression (I.1.279) for the Laplace operator of a vector in the spherical coordinates, we obtain the equation for $F(r)$, which is reduced to the Bessel equation by substituting $F(r) = \chi(r)/\sqrt{r}$. Its solution is finite at $r = 0$ and given by

(2) $$\chi(r) = AJ_{3/2}(kr), \qquad k = \frac{1+i}{\delta}.$$

The magnetic field inside the sphere can be determined from Equation (4.10). The magnetic field outside the sphere is the sum of the external field $\boldsymbol{H}_0$ and the magnetic field of the dipole $\boldsymbol{m}$ directed along $\boldsymbol{H}_0$:

(3) $$\boldsymbol{H}_2 = \boldsymbol{H}_0 + \frac{3\boldsymbol{r}(\boldsymbol{m}\cdot\boldsymbol{r})}{r^5} - \frac{\boldsymbol{m}}{r^3}.$$

The constants A and m can be found from the boundary conditions for $\boldsymbol{H}$ at the surface of the sphere. By expressing the Bessel functions of half-integer index in terms of trigonometric functions, we obtain

(4) $$m = -\frac{a^3}{2}\left(1 - \frac{3}{k^2a^2} + \frac{3}{ka}\cot ka\right)H_0, \qquad A = \frac{3ia\,\omega}{c\sin ka}\sqrt{\frac{\pi}{8k}}H_0.$$

4.44

$$Q = -\frac{3a\delta^2 H_0^2}{8}\left(1 - \frac{a}{\delta}\,\frac{\sinh 2a/\delta + \sin 2a/\delta}{2a/\delta - \cos 2a/\delta}\right).$$

4.45 We will use the solution of Problem 4.43. The electric field inside the sphere will be sought in the form

(1) $$\boldsymbol{E}(r,\vartheta,t) = \boldsymbol{e}_\alpha\,\mathcal{E}(r,\vartheta)\,e^{-\gamma t},$$

where $\gamma > 0$ is an unknown damping constant. We impose the initial condition

(2) $$\boldsymbol{E}(r,\vartheta,t)|_{t=0} = \boldsymbol{e}_\alpha E_0(r,\vartheta) = \boldsymbol{e}_\alpha\frac{3ia\omega H_0}{2c\sin ka}j_1(kr)\sin\vartheta,$$

where $E_0(r, \vartheta)$ is the stationary value of the electric field obtained in Problem 4.43. It is written in terms of a spherical Bessel function (I.1.171), $j_1(z) = \sqrt{\pi/2z}\,J_{3/2}(z)$. Substituting Equation (1) into the second equation (4.14), we have

$$\triangle\mathcal{E} - \frac{\mathcal{E}}{r^2 \sin^2\vartheta} + s^2\mathcal{E} = 0, \qquad s = \sqrt{\frac{4\pi\kappa\gamma}{c^2}}. \tag{3}$$

The solution of this equation is finite at $r = 0$ and has the form

$$\mathcal{E}(r, \vartheta) = A\, j_1(sr) \sin\vartheta, \tag{4}$$

where A is a constant.

The magnetic field inside the sphere can be calculated from the electromagnetic induction equation (4.10):

$$\boldsymbol{H}_{int} = -\int_t^{\infty} \frac{\partial \boldsymbol{H}_{int}}{\partial t}\, dt = -\frac{c}{\gamma} \nabla \times \boldsymbol{e}_\alpha \mathcal{E}(r, \vartheta)\, e^{-\gamma t}. \tag{5}$$

Once the external field $\boldsymbol{H}_0$ is switched off, the magnetic field outside the sphere is produced by decaying eddy currents; it can be written as (see Equation (3) in Problem 4.43)

$$\boldsymbol{H}_{ext} = \left(\frac{3\boldsymbol{r}(\boldsymbol{m}\cdot\boldsymbol{r})}{r^5} - \frac{\boldsymbol{m}}{r^3} \right) e^{-\gamma t}, \tag{6}$$

where $\boldsymbol{m}$ is the magnetic moment of eddy currents. Eigenvalues γ should be determined from the boundary conditions. At $\mu = 1$ the boundary conditions are

$$H_r^{int} = H_r^{ext}, \qquad H_\vartheta^{int} = H_\vartheta^{ext} \qquad \text{at} \qquad r = a. \tag{7}$$

Now we substitute Equations (5) and (6) into Equation (7) and obtain two equations

$$\begin{cases} \dfrac{cA}{\gamma a}[(sa)\, j_0(sa) - j_1(sa)] &= \dfrac{m}{a^3}, \\[2ex] -\dfrac{cA}{\gamma a} j_1(sa) &= \dfrac{m}{a^3}. \end{cases} \tag{8}$$

Excluding the constants A and m, we have the equation for the eigenvalues:

$$(sa) j_0(sa) = \sin sa = 0, \; s_n^2 = \frac{n^2\pi^2}{a^2}, \; \gamma_n = \frac{\pi n^2 c^2}{4\kappa a^2}, \; \text{where } n = 1, 2, \ldots \tag{9}$$

It gives an infinite set of eigenvalues. Then any solution must be written as sums over all possible positive n. The value $n = 0$ gives a zero solution, whereas negative n yield linearly dependent solutions.

The constants $A = A_n$ and $m = m_n$ depend on n. The values A_n can be found from the initial conditions (2); m_n can be expressed in terms of A_n with the aid of Equation (8). Substituting the solution obtained in this way, we have from Equation (2)

$$\frac{3ia\omega H_0}{2c \sin ka} j_1(kr) = \sum_{n=1}^{\infty} A_n j_1(s_n r). \tag{10}$$

Thus, we need to expand the known left-hand side of this equation in eigenfunctions $j_1(s_n r)$. Let us verify that these eigenfunctions are mutually orthogonal:

$$(11)\quad \int_0^a r^2 j_1(s_n r) j_1(s_m r)\, dr = \frac{a^2}{s_m}\int_0^1 x^2 j_1(s_n ax)\frac{d}{dx} j_0(s_n ax)\, dx$$
$$= \frac{a^3 s_n}{s_m}\int_0^1 x^2 j_0(s_n ax) j_0(s_m ax)\, dx = \frac{a}{2s_m^2}\delta_{mn}.$$

Here, the formulas for spherical Bessel functions are used, $j_1(z) = -j_0'(z)$ and $z^2 j_1'(z) = z^2 j_0(z) - 2z j_1(z)$. The integral $\int_0^a r^2 j_1(kr) j_1(s_n r)\, dr$ is taken using Equation (I.1.176), which at $p = 3/2$ can be written in terms of spherical Bessel functions,

$$(12)\quad \int_0^1 x^2 j_1(ax) j_1(bx)\, dx = \frac{a j_1'(a) j_1(b) - b j_1'(b) j_1(a)}{b^2 - a^2}.$$

Finally, we have

$$(13)\quad A_m = i\frac{6a^2 s_m^2 \omega H_0}{c \sin ka (s_m^2 - k^2)}\left[k j_1'(ka) j_1(s_m a) - s_m j_1'(s_m a) j_1(ka)\right].$$

The electric field inside the sphere can be expanded as

$$(14)\quad \boldsymbol{E}(r, \vartheta, t) = \boldsymbol{e}_\alpha \sin\vartheta \sum_{n=1}^{\infty} A_n j_1(s_n r)\, e^{-\gamma_n t}.$$

Similarly, by generalizing Equation (6), the magnetic field outside the sphere can be written as

$$(15)\quad \boldsymbol{H}_{ext} = \sum_{n=1}^{\infty}\left(\frac{3\boldsymbol{r}(\boldsymbol{m}_n\cdot\boldsymbol{r})}{r^5} - \frac{\boldsymbol{m}_n}{r^3}\right) e^{-\gamma_n t},$$

It is easy to prove that the initial condition for $\boldsymbol{H}_{ext}$ is satisfied owing to the relation $\sum_n \boldsymbol{m}_n = \boldsymbol{m}$.

4.46

$$R = \frac{l}{ca}\sqrt{\frac{\omega}{\pi\kappa}}\,\Re\left[(1+i)\frac{J_0(ka)}{J_1(ka)}\right], \qquad \text{where} \qquad R_0 = \frac{(1+i)\sqrt{2\pi\kappa\,\omega}}{c}.$$

At $|ka| \ll 1$ (low frequencies):

$$R = R_0\left[1 + \frac{1}{12}\left(\frac{\pi\kappa\omega a^2}{c^2}\right)^2\right],$$

where $R_0 = l/\pi a^2\kappa$ is the resistance for direct current.

At $|ka| \gg 1$ (high frequencies):

$$R = \frac{1}{\kappa}\frac{l}{2\pi a\delta} = \frac{l}{ca}\sqrt{\frac{\omega}{2\pi\sigma}}.$$

As observed from the latter formula, the effective cross section of a conductor for strong skin-effect is $2\pi a\delta$.

4.47

$$R = \frac{\omega\delta_2}{2ac^2}\,\frac{(\delta_1^2 - \delta_2^2)\sin 2h/\delta_2 - (\delta_1^2 + \delta_2^2)\sinh 2h/\delta_2 + 2\delta_1\delta_2\cos 2h/\delta_2}{(\delta_1 \sin h/\delta_2 + \delta_2\cos h/\delta_2)^2 + (\delta_1^2 + \delta_2^2)\sinh^2 h/\delta_2},$$

where $\delta_1 = c/\sqrt{2\pi\kappa_1\omega}$ and $\delta_2 = c/\sqrt{2\pi\kappa_2\omega}$.

4.48

$$H' = \frac{2H_0}{ka\,\sinh kh + 2\cosh kh}, \qquad \text{where} \qquad k = \frac{1+i}{\delta}.$$

At $|kh| \ll 1$ (low frequencies) we have $H' = H_0$, that is, the presence of the cylindrical shell has no effect on the field. At $|kh| \gg 1$ (high frequencies) we obtain

$$\sinh kh \approx \cosh kh \approx \frac{1}{2}e^{(1+i)h/\delta};$$

since $a \gg \delta$, we obtain

$$H' = (1-i)\frac{\delta}{2a}e^{-(1+i)h/\delta}H_0, \qquad H_0 \gg |H'|.$$

The strong reduction of the field is because the eddy currents, which are induced in the shell, create an additional field of opposite direction.

4.49

$$j = \frac{2iJ_0\mu\kappa\omega}{c^2ka}\,\frac{\sinh k(h-x)}{\cosh kh},$$

where x is measured from the surface along the inward radial direction;

$$R = \frac{1}{2\pi a\delta\sigma}\,\frac{\sinh 2h/\delta - \sin 2h/\delta}{2(\sinh^2 h/\delta + \cos^2 h/\delta)}.$$

At $\delta \ll h$, the hollow and solid conductors have the same resistance.

4.50 Consider the cylindrical system of coordinates shown in Figure 4.13. For weak skin effect, the component of the magnetic field, which is tangential to the wall of the tube, at the surface S of this wall should satisfy the condition

$$H_{2\tau} - H_{1\tau} = \frac{4\pi}{c}i, \tag{1}$$

where $i = \kappa hE = \zeta E$ is the surface current, and ζ is the surface electric conductivity. The electric field will evidently have only the z-component and should be continuous across the surface S:

$$E_1 = E_2 = E. \tag{2}$$

The remaining solution is very similar to that of Problem 2.98 (the problem of weakly non-concentric spheres). Within the terms of the order of $1/a$, the equation of the boundary can be written in the form

$$r = a + l\cos\alpha. \tag{3}$$

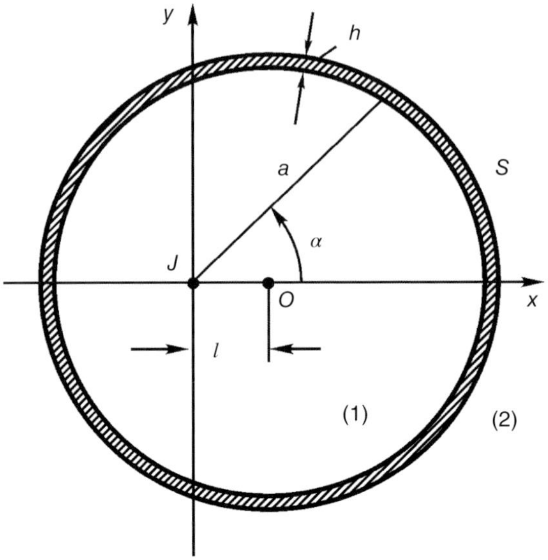

Figure 4.13 Stabilization of z-pinch in a conducting tube (see also Problems 4.54 and 4.55).

The vector potential whose direction is parallel to the direction of the current will be sought in the form

(4)
$$\left.\begin{aligned} A_1 &= -\frac{2J}{c}\ln\frac{r}{a} + C_1 r\cos\alpha + C, \\ A_2 &= -\frac{2J'}{c}\ln\frac{r}{a} + \frac{B_1}{r}\cos\alpha, \end{aligned}\right\}$$

where C_1 and B_1 are functions of time and are of the first order with respect to (l/a), and J' is of zero order with respect to (l/a).

For weak skin-effect ($h \ll \delta$), the vector potential satisfies the condition

(5)
$$A_1 = A_2 \qquad \text{at } r = a + l\cos\alpha.$$

Hence, neglecting the terms of the order of $(l/a)^2$, we have

(6)
$$B_1 = a^2 C_1 + \frac{2(J' - J)l}{c}, \qquad C = 0.$$

In the boundary condition (1), the quantity H_τ can be replaced by H_α. It is easy to verify that this gives an error of the order of $(l/a)^2$. Since

$$H_\alpha = -\frac{\partial A}{\partial r}, \qquad i = \zeta E = -\frac{\zeta}{c}\frac{\partial A}{\partial t},$$

we have on the surface S

$$\frac{\partial A_1}{\partial r} - \frac{\partial A_2}{\partial r} = -\frac{4\pi\zeta}{c}\frac{\partial A}{\partial t}$$

or, to within the terms of the order (l/a),

$$\frac{2(J' - J)}{ca} + 2C_1 \cos\alpha = \frac{4\pi\zeta}{c^2}\left[\frac{2}{ca}\frac{d(Jl)}{dt} + a\frac{dC_1}{dt}\right]\cos\alpha.$$

Hence, it follows at once that $J = J'$; this is because the skin-effect is considered to be weak. The following differential equation can be written for C_1,

$$\frac{dC_1}{dt} + \rho C_1 = \frac{2}{a^2 c}\frac{d(Jl)}{dt}. \tag{7}$$

The parameter $\rho = c^2/2\pi a\zeta$ is equal to the resistance per unit length of the tube in electromagnetic units. The solution of Equation (7) is easily obtained by the method of variation of arbitrary constants. It is of the form

$$C_1 = \frac{2}{ca^2}\int_{-\infty}^{t} e^{\rho(\tau - t)}\frac{d}{d\tau}[J(\tau)l(\tau)]\,d\tau,$$

assuming that the current is zero as $t \to -\infty$.

The force f per unit length of the current J can be calculated using the formula

$$f_x = -\frac{1}{c}JH'_y,$$

where H'_y is the magnetic field of eddy currents in the shell, along the straight line parallel to the current J. The vector potential for this field is

$$A' = C_1 r\cos\alpha = C_1 y,$$

and hence

$$H_y = -\frac{\partial A'}{\partial y} = -C_1.$$

Finally,

$$f_x = \frac{2J(t)}{c^2 a^2}\int_{-\infty}^{t} e^{\rho(\tau - t)}\frac{d}{d\tau}[J(\tau)l(\tau)]\,d\tau.$$

Let us consider some special cases. When the current is direct ($J = const$),

$$f_x = \frac{2J(t)^2}{c^2 a^2}\int_{-\infty}^{t} e^{\rho(\tau - t)}\dot{l}(\tau)\,d\tau.$$

When the current deviates from the cylinder axis ($\dot{l} > 0$), there appears a force preventing this deviation. For a slow motion ($\ddot{l} \ll \rho\dot{l}$), integrating by parts, we have

$$f_x = \frac{2J^2}{c^2 a^2}\left(\frac{\dot{l}}{\rho} - \frac{\ddot{l}}{\rho^2} + \dots\right).$$

In particular, for a uniform motion, $l = vt$, the friction force is

$$f_x = \frac{2J^2 v}{c^2 a^2 \rho}.$$

4.51

$$f_x = \frac{2J^2(t)l(t)}{c^2a^2}.$$

4.54 For a static equilibrium ($\boldsymbol{u} = 0$) in the absence of non-electromagnetic forces, from Equations (4.23) and (4.11) we have the equilibrium condition $\nabla p = \boldsymbol{j} \times \boldsymbol{B}/c$. This means that the vectors $\boldsymbol{B}$ and $\boldsymbol{j}$ are perpendicular to the pressure gradient, that is, they are tangential to surfaces of equal pressure.

4.55 The magnetic field has a single projection

$$B_\varphi \equiv B(r) = \frac{4\pi}{cr}\int_0^r rj(r)\,dr.$$

Integrating the equilibrium equation which follows from Equation (4.23) at $u = 0$ and taking into account the boundary condition $p|_{r\geq a} = 0$, we have

(1) $$p(r) = \frac{1}{8\pi}\int_r^a \frac{1}{r^2}\frac{d}{dr}(r^2B^2)\,dr,$$

where $B = (4\pi/cr)\int_0^r rj(r)\,dr$ at $r < a$, and $B = 2J/cr$ at $r > a$. In order to connect the total current with the pressure, we integrate Equation (1) over the entire cross-section and use the relation $aB(a) = 2J/c$. We then get

(2) $$\int_0^a p(r)2\pi r\,dr = \frac{J^2}{2c^2}.$$

Assuming that the plasma is an equilibriated perfect gas with a given temperature T (in energy units) and that $p = 2n(r)T$, from Equation (2) we have

(3) $$J = 2c\sqrt{NT}.$$

Substituting numerical values into (3), we have $J = 7.5 \times 10^4$ A. In practice, a plasma is usually non-isothermal, and the temperature of electrons is higher than that of ions. To maintain the equilibrium, the current should increase, the current according to Equation (3) should increase, because increase the temperature and pressure of the plasma. In addition, equilibrium is unstable with respect to bending and constrictions of plasma filaments.

4.56 The current must flow in a thin surface layer. Then the pressure inside the plasma column is constant:

$$p = \frac{J^2}{2\pi c^2a^2}.$$

4.57 The magnetic field inside the cylinder has a single component $B_z(r) \equiv B(r) = (4\pi/c)\int_r^a j_\varphi(r)\,dr$. The equilibrium inside the cylinder requires constant total pressure,

$$p(r) + \frac{B^2(r)}{8\pi} = const.$$

Outside the cylinder, in the absence of matter, $p = 0$ and the magnetic field of azimuthal currents is zero, $B = 0$. Hence, the internal pressure can be balanced only by an external magnetic field, which is parallel to the axis of the cylinder and is equal to $B_0 = \sqrt{8\pi p\,(a)}$ at the boundary. The magnetic field inside the plasma column is always smaller than the external field,

$$\frac{B^2}{8\pi} = \frac{B_0^2}{8\pi} - p.$$

Therefore, the plasma is diamagnetic.

4.58 (b) $\boldsymbol{B}(\boldsymbol{r}) = B_0(0, J_1(\alpha r), J_0(\alpha r))$, where $J_n(x)$ is a Bessel function.

4.59 Using Equation (4.22) and substituting $\boldsymbol{B} = \nabla \times \boldsymbol{A}$, we get

$$(1) \qquad \nabla \times \left(\frac{\partial \boldsymbol{A}}{\partial t} - \boldsymbol{u} \times [\nabla \times \boldsymbol{A}] + \nu_m \nabla \times [\nabla \times \boldsymbol{A}] \right) = 0.$$

Hence, we can present the expression in parenthesis as the gradient of a certain scalar function of $\varphi(\boldsymbol{r}, t)$,

$$(2) \qquad \frac{\partial \boldsymbol{A}}{\partial t} - \boldsymbol{u} \times [\nabla \times \boldsymbol{A}] + \nu_m \nabla \times [\nabla \times \boldsymbol{A}] = -c\,\nabla\,\varphi.$$

Since the vector potential is determined to within the gradient of an arbitrary scalar function, the substitution $\boldsymbol{A} \to \boldsymbol{A} - c \int \varphi(\boldsymbol{r}, t')\, dt'$ eliminates an arbitrary scalar from Equation (2) and allows us to write

$$(3) \qquad \frac{\partial \boldsymbol{A}}{\partial t} = \boldsymbol{u} \times [\nabla \times \boldsymbol{A}] - \nu_m \nabla \times [\nabla \times \boldsymbol{A}].$$

If the Coulomb gauge condition $\nabla \cdot \boldsymbol{A} = 0$ is imposed on the vector potential, Equation (3) acquires the form

$$(4) \qquad \frac{\partial \boldsymbol{A}}{\partial t} = \boldsymbol{u} \times [\nabla \times \boldsymbol{A}] - \nu_m \triangle \boldsymbol{A}.$$

4.60

$$B_x(y, z, t) = \left(B_{0y} \frac{\partial u_x}{\partial y} + B_{0z} \frac{\partial u_x}{\partial z} \right) t,\ B_y = B_{0y},\ B_z = B_{0z}.$$

4.61

$$B_\alpha(r, \vartheta, t) = \sin\vartheta \left(r B_{0r}(r, \vartheta) \frac{\partial \Omega}{\partial r} + B_{0\vartheta} \frac{\partial \Omega}{\partial \vartheta} \right) t,\ B_\vartheta = B_{0\vartheta},\ B_r = B_{0r}.$$

4.62 Using the Cartesian coordinates, we project Equation (4.22) on the Oz-axis:

$$(1) \qquad \frac{\partial B_z}{\partial t} + (\boldsymbol{u} \cdot \nabla) B_z = \nu_m \triangle B_z.$$

Let us multiply both parts of this equation by B_z and integrate over the entire two-dimensional space. The term containing the velocity $\boldsymbol{u}$ vanishes if the medium is incompressible ($\nabla \cdot \boldsymbol{u} = 0$) and the region of motion is finite. The remaining terms are reduced to the form

$$(2) \qquad \frac{\partial}{\partial t} \int B_z^2\, dx\, dy = -2\nu_m \int (\nabla B_z)^2\, dx\, dy.$$

Since the right-hand side of this equation is negative when the region occupied by the field is finite, the energy of the field monotonically decreases, and $B_z \to 0$ as $t \to \infty$. Thus, the B_z-component decays and the remaining components may be investigated assuming that $B_z = 0$.

These components satisfy the equations

(3) $$\frac{\partial B_{x,y}}{\partial t} + (\boldsymbol{u}\cdot\nabla)B_{x,y} = \nu_m \triangle B_{x,y} + (\boldsymbol{B}\cdot\nabla)\boldsymbol{u}_{x,y}.$$

and can be conveniently expressed in the standard way ($\boldsymbol{B} = rot\,\boldsymbol{A}$) in terms of the vector potential parallel to the Oz-axis; $A_z(x,y) \equiv A(x,y)$. Both Equations (3) can be derived from the equation

(4) $$\frac{\partial A}{\partial t} + (\boldsymbol{u}\cdot\nabla)A = \nu_m \triangle A.$$

While verifying this statement, the incompressibility of the medium should again be taken into account. Let us multiply both sides of Equation (4) by A and integrate the product over the plane (x, y) and over t. We get

(5)
$$\int A^2(x,y,0)\,dx\,dy - \int A^2(x,y,t)\,dx\,dy = 2\nu_m \int_0^t dt \int (B_x^2 + B_y^2)\,dx\,dy.$$

The integral in the right-hand side of this equation is a monotonically increasing function of time limited by the quantity $\int A^2(x,y,0)\,dx\,dy$. Hence, as $t \to \infty$ the time integral converges, and it follows from this convergence that $B^2 = B_x^2 + B_y^2 \to 0$ as $t \to \infty$. This result is one of the anti-dynamo theorems stating that it is impossible to generate a magnetic field by motions of a conducting medium of a certain (sufficiently symmetric) kind. The problem under consideration is very important for the theory of origin of magnetic fields of celestial bodies.

4.63 Let us write down $\boldsymbol{u} = \nabla\psi \times \boldsymbol{e}_r$ and project the induction equation (4.22) onto the direction $\boldsymbol{e}_r$. We get $\boldsymbol{e}_r\cdot(\nabla \times [\boldsymbol{u} \times \boldsymbol{B}]) = -(\boldsymbol{u}\cdot\nabla)B_r$;

$$-\boldsymbol{e}_r\cdot(\nabla \times \nu_m[\nabla \times \boldsymbol{B}]) = \triangle B_r + \frac{2}{r}\frac{\partial B_r}{\partial r} + \frac{2}{r^2}B_r.$$

In the last equation, we took into account that $\nabla\nu_m$ is directed along $\boldsymbol{e}_r$, and that $\nabla\cdot\boldsymbol{B} = 0$. Hence, we obtain the equation for B_r,

(1) $$\frac{\partial B_r}{\partial t} + (\boldsymbol{u}\cdot\nabla)B_r = \nu_m\left(\triangle B_r + \frac{2}{r}\frac{\partial B_r}{\partial r} + \frac{2}{r^2}B_r\right).$$

Next, we multiply Equation (1) by $B_r/\nu_m r^2$ and integrate over the entire space. Rearranging the integral which includes $\triangle B_r$ in accordance with the Gauss–Ostrogradski theorem, we finally obtain

(2) $$\frac{d}{dt}\int \frac{B_r^2}{\nu_m r^2}\,dV = -2\int \frac{(\nabla B_r)^2}{r^2}\,dV - \frac{8\pi}{r}\overline{B_r^2}|_{r\to 0} \le 0,$$

where the upper line denotes the averaging over orientations of $\boldsymbol{r}$. For a finite system, the field at large distances must be zero. Hence, the right-hand side will be negative if B_r differs from zero somewhere in space. This means dissipation of the r-component of the field, $B_r \to 0$ as $t \to \infty$.

The transverse components of the field can be analyzed using the vector potential $\boldsymbol{A}(\boldsymbol{r}) = \boldsymbol{e}_r A(\boldsymbol{r}, t)$, which satisfies Equation (3) from the solution to Problem 4.59. In this case, the condition $\nabla\cdot\boldsymbol{A} = 0$ cannot be satisfied, since it leads to a strong singularity as $r \to 0$. Therefore, the equation for the scalar function $A(\boldsymbol{r}, t)$ takes the form

(3) $$\frac{\partial A}{\partial t} + (\boldsymbol{u}\cdot\nabla)A = \nu_m \left(\triangle A - \frac{2}{r^2}A - \frac{\partial}{\partial r}\frac{1}{r^2}\frac{\partial}{\partial r}r^2 A\right).$$

Multiplying this equation by A/ν_m and integrating over the entire space, we reduce Equation (3) to

(4) $$\frac{d}{dt}\int \frac{A^2}{\nu_m}\, dV = -2\int \left(\nabla A - \boldsymbol{e}_r \frac{\partial A}{\partial r}\right)^2 dV \le 0.$$

This condition ensures that $A = const$ everywhere throughout the space as $t \to \infty$, which means the absence of any magnetic field. To generate a magnetic field, more complicated three-dimensional motions of conducting liquid or gas are needed, for example, rotation of conducting spheres about nonparallel axes (see Moffatt, 1970, Chapter 6).

4.64 It is natural to assume that the velocity of liquid motion is directed along the Oz-axis and depends only on the transverse coordinate x. Since a conducting liquid drags magnetic lines, the longitudinal component $B_z(x)$ should appear. For this reason, unknown functions $\boldsymbol{v}$ and $\boldsymbol{B}$ should be sought in the form $\boldsymbol{v}(0, 0, v(x))$ and $\boldsymbol{B}(B_0, 0, B_z(x))$; the equations $\nabla\cdot\boldsymbol{B} = 0$ and $\nabla\cdot\boldsymbol{v} = 0$ are then automatically satisfied. Equations (4.23) and (4.22) take the form:

(1) $$\frac{dv}{dx} + \frac{c^2}{4\pi\kappa B_0}\frac{d^2 B_z}{dx^2} = 0,$$

(2) $$\frac{d^2 v}{dx^2} + \frac{B_0}{4\pi\eta}\frac{dB_z}{dx} = \frac{1}{\eta}\frac{d}{dz}\left(p + \frac{B^2}{8\pi}\right),$$

(3) $$\frac{d}{dx}\left(p + \frac{B^2}{8\pi}\right) = 0.$$

The last equation means that $p + B^2/8\pi$ can depend only on z. However,

(4) $$\frac{d}{dz}\left(p + \frac{B^2}{8\pi}\right) = \frac{dp}{dz} = const,$$

since $B^2/8\pi$ is independent of z. Equations (1) and (2) constitute then a system of ordinary linear equations to determine the unknown functions $v(x)$ and $B_z(x)$. Eliminating from the equations dB_z/dx, we obtain the equation for $u = dv/dx$,

(5) $$\frac{d^2 u}{dx^2} - \frac{1}{x_0^2}u = 0, \qquad x_0 = \frac{c}{H_0}\sqrt{\frac{\eta}{\kappa}},$$

from which it follows that

(6) $$v = x_0\left(A \exp\left[\frac{x}{x_0}\right] - B \exp\left[-\frac{x}{x_0}\right]\right) + C.$$

The boundary conditions are of the form $v(\pm a) = 0$, since a viscous liquid sticks at the wall. Moreover, symmetry arguments imply $v(x) = v(-x)$. From the boundary conditions and Equation (6) we find

$$v(x) = v_0 \frac{\cosh(a/x_0) - \cosh(x/x_0)}{\cosh(a/x_0) - 1}, \tag{7}$$

where v_0 is a new constant, which is equal to the velocity at the middle plane, $x = 0$. We can express v_0 in terms of the pressure gradient:

$$v_0 = -\frac{ax_0}{\eta} \frac{\cosh(a/x_0) - 1}{\sinh(a/x_0)} \frac{dp}{dz}. \tag{8}$$

The magnetic field can be determined from Equations (2) and (7) and the boundary conditions $B_z(\pm a) = 0$:

$$B_z(x) = -\frac{4\pi\eta}{c} \sqrt{\kappa\eta} v_0 \frac{(x/a)\sinh(a/x_0) - \sinh(x/x_0)}{\cosh(a/x_0) - 1}. \tag{9}$$

The ratio $a/x_0 = G$ is called the **Hartmann number**. At $G \ll 1$, we have

$$v_0 = \frac{a^2}{\eta} \frac{dp}{dz}, \qquad v(x) = v_0 \left(1 - \frac{x^2}{a^2}\right), \tag{10}$$

as in ordinary hydrodynamics. The magnetic field $B_z = 0$ to first order in the Hartmann number. The longitudinal field B_z appears only in the higher approximations.

In the opposite limiting case, $G \gg 1$, we obtain

$$v_0 = -\frac{a^2}{\eta G} \frac{dp}{dz}; \qquad v(x) = v_0 \left(1 - \exp\left[-\frac{a - |x|}{x_0}\right]\right). \tag{11}$$

A comparison of Equations (10) and (11) shows that the average velocity of the liquid decreases with increasing B_0, while the velocity profile becomes flatter in the middle part of the flow, but varies steeply in the layer of thickness x_0 at the walls. The longitudinal magnetic field in this limit has the form

$$B_z(x) = \frac{4\pi a^2 \sqrt{\kappa\eta}}{cG} \frac{dp}{dz} \left(\frac{x}{a} - \exp\left[-\frac{a}{x_0}\right] \sinh\frac{x}{x_0}\right). \tag{12}$$

According to this formula, the field decreases with increasing Hartmann number. The largest value of B_z occurs at $G \approx 1$.

The current density in the moving liquid can be calculated from the Maxwell equation $\boldsymbol{j} = c\nabla \times \boldsymbol{B}/4\pi$. Only the y-component of the current is different from zero:

$$j_y(x) = \frac{c}{B_0} \frac{dp}{dz} \left(1 - \frac{a}{x} \frac{\cosh(x/x_0)}{\sinh(a/x_0)}\right). \tag{13}$$

The magnetic field B_z produced by this current is zero everywhere outside the region occupied by the liquid, where only the transverse field B_0 is available.

4.65

$$v(x) = v_0 \frac{\sinh(x/x_0)}{\sinh(a/x_0)}.$$

The current density is

$$j(x) = \frac{c\eta v_0}{B_0 x_0^2} \frac{\sinh(x/x_0)}{\sinh(a/x_0)}.$$

This current produces the magnetic field

$$B_z(x) = \frac{4\pi\eta v_0}{B_0 x_0} \frac{\cosh(a/x_0) - \cosh(x/x_0)}{\sinh(a/x_0)},$$

which vanishes at $|x| \geq a$.

4.67 If the wave amplitude decreases as $e^{-\gamma t}$, its energy decays as $e^{-2\gamma t}$. From the relation $d\overline{w}/dt = -2\gamma\overline{w} = -\overline{Q}$ we find $\gamma = \overline{Q}/2\overline{w}$, where $\overline{w}$ and $\overline{Q}$ are, respectively, the densities of the wave energy and of energy dissipation rate, which are averaged over a period $T = 2\pi/\omega \ll 1/\gamma$. If $\boldsymbol{b}$ is the magnetic field of the wave, its energy density is $\overline{w} = |b|^2/8\pi$ (here, it is taken into account that in the Alfvén wave the magnetic and mechanical energies are equal). The quantity Q is the right-hand side of Equation (4.26) (excluding the temperature term, since the thermal conductivity in Alfvén wave is unimportant).

Performing time averaging and using the relations $\nu = \eta/\tau$ and $|u|^2 = |b|^2/4\pi\tau$, we get $\overline{Q} = k^2(\nu + \nu_m)|b|^2/8\pi$ and

$$\gamma = \frac{1}{2}k^2(\nu + \nu_m) = \frac{\omega_A(\boldsymbol{k})}{2v_{Ax}^2}(\nu + \nu_m),$$

where the notations of Example 4.9 are used.

4.68 The vector $\boldsymbol{b}$ lies in the $(\boldsymbol{k}, \boldsymbol{B}_0)$-plane and is perpendicular to the vector $\boldsymbol{k}$. The vector $\boldsymbol{u}$ lies in the same plane but has both components, longitudinal and transverse, with respect to $\boldsymbol{k}$.

4.69 The propagation of a simple wave of finite amplitude is accompanied by a steeping of the leading edge (Figure 4.14a,b). At a certain moment $t = t_*$ and in some point x_* of the leading edge, the profile bends and the infinite derivative $(\partial u/\partial x)_{t \to t_*} \to \infty$ appears (Figure 4.14c). For $t > t_*$, the solution becomes multivalued, and the simple wave no longer exists. The quantities t_*, x_* and u_* can be determined from the existence of the bending point at the plot $u(x, t_*)$ (see Figure 4.14d),

$$\frac{\partial x(u, t_*)}{\partial u} = 0, \qquad \frac{\partial^2 x(u, t_*)}{\partial u^2} = 0.$$

4.70 Projecting Equation (4.29) on the r-axis and setting $\boldsymbol{u} = u\boldsymbol{r}/r$ and $v = const$, we obtain the equation for B_r:

$$\frac{\partial B_r}{\partial t} = -\frac{2u}{r}B_r - u\frac{\partial B_r}{\partial r}. \qquad (1)$$

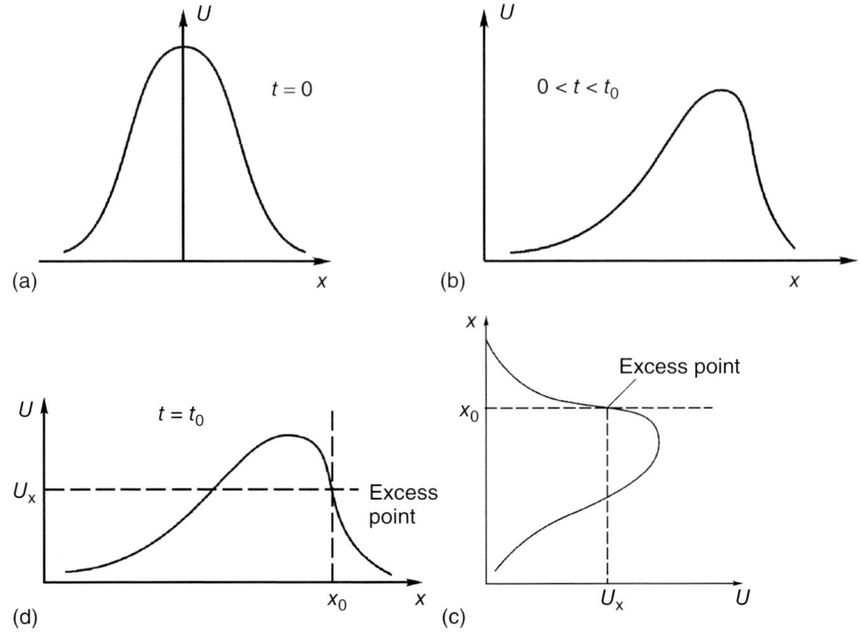

Figure 4.14 Evolution of simple wave: (a) symmetric shape; (b) steeping of the leading edge; (c), (d) appearance of the bending point.

The solution of this equation can be expressed through an arbitrary function F of arguments $r - ut$, ϑ and α,

$$B_r(r, \vartheta, \alpha) = \frac{1}{r^2} F(r - ut, \vartheta, \alpha). \tag{2}$$

The boundary condition has the form

$$B_r\big|_{r=a} = B_{0r}(\vartheta, \alpha + \Omega t) = \frac{1}{a^2} F(a - ut, \vartheta, \alpha) \tag{3}$$

(the argument $\alpha + \Omega t$ appears because of transformation to non-spinning reference frame). Thus we have

$$F(a - ut, \vartheta, \alpha) = a^2 B_{0r}(\vartheta, \alpha + \Omega t).$$

Hence, Equation (2) can be written as

$$B_r(r, \vartheta, \alpha, t) = \left(\frac{a}{r}\right)^2 B_{0r}\left(\vartheta, \alpha - \frac{(r-a)\Omega}{u} + \Omega t\right). \tag{4}$$

Similarly, we find

$$\begin{aligned} B_\vartheta &= \frac{a}{r} B_{0\vartheta}\left(\vartheta, \alpha - \frac{(r-a)\Omega}{u} + \Omega t\right), \\ B_\alpha &= \frac{a}{r} B_{0\alpha}\left(\vartheta, \alpha - \frac{(r-a)\Omega}{u} + \Omega t\right). \end{aligned} \tag{5}$$

Using the equation $div\, \boldsymbol{B} = 0$, we obtain the relation between the projections of the vector $\boldsymbol{B}_0$:

$$-\frac{a\Omega}{u}\frac{\partial B_{0r}}{\partial \alpha}\sin\vartheta + \frac{\partial}{\partial\vartheta}(B_{0\vartheta}\sin\vartheta) + \frac{\partial B_{0\alpha}}{\partial\alpha} = 0.$$

For $B_{0\vartheta} = 0$ we find $B_\vartheta = 0$ and

$$B_{0\alpha} = \frac{a\Omega}{u}B_{0r}\sin\vartheta + f(\vartheta).$$

Taking $f(\vartheta) = 0$, we have

$$(6) \qquad B_\alpha(r,\vartheta,\alpha,t) = \frac{a^2\Omega}{vr}B_{0r}\left(\vartheta, \alpha - \frac{(r-a)\Omega}{u} + \Omega t\right)\sin\vartheta.$$

Parker used the model considered here to describe the interplanetary magnetic field produced by solar plasma outflow (solar wind). In Parker's model of the interplanetary magnetic field, $B_\vartheta = 0$, while B_r and B_α are given by Equations (4) and (6). Measurements of the interplanetary magnetic field made by satellites and rockets show that the averaged magnetic field in vicinity of the Earth orbit is fairly well described by Parker's model.

4.71 The lines of force form Archimedean spirals:

$$r = \frac{u}{\Omega}(\alpha - \alpha_0), \qquad \alpha_0 = const,$$

$$\theta = \arctan\frac{r_0\Omega}{u} \approx 56°; \qquad B \approx 4.5\times 10^{-5}\ \text{G}.$$

4.73 MHD perturbations due to an explosion exist only inside the sphere bounded by the shock front. Let us first derive the boundary conditions for the magnetic field at the front. Substituting the flow of matter across the front, $i_n = \tau_0 u_s$ (where τ_0 is unperturbed density of the medium and $u_s = \dot{R}_s$ is the velocity of the front) into Equation (4.42), we obtain the estimate $u_\tau/u_s \approx v_{An}v_{A\tau}/u_s^2$. Here, the right-hand side is expressed in terms of the components of the Alfvén velocity $\boldsymbol{v}_A$. The ratio of the medium velocity ahead of the front to the Alfvén velocity ($M_A = u_s/v_A$) is called the **Alfvén Mach number**. The tangential components of the velocity are smaller than the radial ones when the Mach number is large, $M_A^2 \gg 1$. In this case, the boundary conditions can be found assuming that the left-hand side of Equation (4.43) is equal to zero. At $r = R_s(t)$ we have

$$(1) \qquad B_n^{(s)} = B_{0n}, \qquad B_\tau^{(s)} = B_{0\tau}\frac{\tau_s}{\tau_0},$$

where the subscript s denotes the quantities on the internal (perturbed) side of the front.

We write Equations (4.29) in spherical coordinates:

$$(2) \qquad \frac{\partial B_r}{\partial t} = -\frac{u(r,t)}{r^2}\frac{\partial}{\partial r}r^2 B_r, \qquad \frac{\partial B_\vartheta}{\partial t} = -\frac{1}{r}\frac{\partial}{\partial r}ru(r,t)B_\vartheta,$$

$$\frac{\partial B_\alpha}{\partial t} = -\frac{1}{r}\frac{\partial}{\partial r} ru(r,t)B_\alpha.$$

If the dependence $u(r,t) = p(r)q(t)$ is factorized, Equation (2) can be integrated in the general form by introducing new dependent functions

(3) $b_r(r,t,\vartheta,\alpha) = r^2 B_r(r,t,\vartheta,\alpha), \quad b_\vartheta(r,t,\vartheta,\alpha) = rp(r) B_\vartheta(r,t,\vartheta,\alpha),$

$b_\alpha(r,t,\vartheta,\alpha) = rp(r) B_\alpha(r,t,\vartheta,\alpha),$

and new independent variables

(4) $$\eta(t) = \int q(t)dt, \qquad s(r) = \int \frac{dr}{p(r)}.$$

Equation (2) acquire the identical form

(5) $$\frac{\partial b_i}{\partial \eta} + \frac{\partial b_i}{\partial r} = 0, \qquad i = r, \vartheta, \alpha.$$

Their solution is expressed through an arbitrary differentiable function $b_i(\varphi,\vartheta,\alpha)$ of self-similar variable $\varphi(r,t) = s(r) - \eta(t)$. In the case under consideration,

(6) $$p(r) = r, \qquad q(t) = \frac{2\dot{R}_s(t)}{(\gamma+1)R_s(t)}, \qquad \varphi(r,t) = \ln\left(\frac{r}{R_s^{2/(\gamma+1)}(t)}\right).$$

Obviously, b_i can be considered, without any loss of generality, to be a function of the ratio under the logarithm, $b_i(r/R_s^{2/(\gamma+1)}(t),\vartheta,\alpha)$. The dependence of b_i on this ratio and also on ϑ and α can be found from the boundary conditions and, if required, using $\nabla\cdot\boldsymbol{B} = 0$.

For a uniform external field, we have the boundary conditions:

(7)

$$B_r(R_s(t),t,\vartheta) = B_0\cos\vartheta, \qquad B_\vartheta(R_s(t),t,\vartheta) = -B_0\frac{\tau_s}{\tau_0}\sin\vartheta, \qquad B_\alpha = 0.$$

The solution can be written as

(8)

$$B_r(r,t,\vartheta) = \frac{f(r/R_s^{2/(\gamma+1)})}{r^2} B_0 \cos\vartheta, \; B_\vartheta(r,t,\vartheta) = -\frac{\tau_s f(r/R_s^{2/(\gamma+1)})}{\tau_0 r^2} B_0 \sin\vartheta.$$

From the boundary conditions, we find $f(R_s^{(\gamma-1)/(\gamma+1)}) = R_s^2$. This means that $f(x) = x^{2(\gamma+1)/(\gamma-1)}$, that is,

(9) $$f(r/R_s^{2/(\gamma+1)}) = \frac{r^{2(\gamma+1)/(\gamma-1)}}{R_s^{4/(\gamma-1)}}.$$

Finally, we get

(10)

$$B_r(r,t,\vartheta) = B_0\left(\frac{r}{R_s(t)}\right)^{4/(\gamma-1)} \cos\vartheta, \quad B_\vartheta(r,t,\vartheta) = -\frac{\tau_s}{\tau_0}\left(\frac{r}{R_s(t)}\right)^{4/(\gamma-1)}$$

$$B_0\sin\vartheta, \quad r \le R_s(t).$$

At $\gamma = 5/3$ we have $4/(\gamma - 1) = 6$. The radial magnetic field is thus swept away of the region of explosion together with the plasma, the reduction factor being equal to $(r/R_s(t))^6$. The tangential field is also swept away, but at the front it is enhanced by a factor of $\tau_s/\tau_0 \geq 4$ compared with the unperturbed value. The amplification factor can exceed the ordinary value $(\gamma + 1)/(\gamma - 1) = 4$ (see Equation (4) in Example 4.12), if there is the energy loss at the front due to the radiation of photons or acceleration of particles up to relativistic energies.

4.74 Solving this problem in the same fashion as the previous one, we find the magnetic field in the cavity produced by the explosion,

$$B_r(r, t, \vartheta) = \frac{2M \cos \vartheta}{r^3} \left(\frac{R_s(t)}{r} \right)^{2/(\gamma-1)}, \tag{1}$$

$$B_\vartheta(r, t, \vartheta) = \frac{M \sin \vartheta}{r^3} \frac{\tau_s}{\tau_0} \left(\frac{R_s(t)}{r} \right)^{2/(\gamma-1)}, \quad B_\alpha = 0.$$

The magnetic field in the cavity, in contrast to the previous case, is significantly amplified. This is because the field of point magnetic dipole has a strong singularity at the center of the explosion. The unlimited amplification of the field in the cavity relates to approaching this singularity. Therefore, the result can be improved by taking into account finite sizes of the dipole.

Let us employ the model of the magnetic field considered in Problem I.2.85, where at $r > a$ the field coincides with that given by the point-like magnetic dipole $\boldsymbol{M}$; for $r < a$, the field is uniform, $\boldsymbol{B}_0 = 2\boldsymbol{M}/a^3$; and on the sphere $r = a$ there is the layer with the surface current $\boldsymbol{i} = 3Mc\boldsymbol{e}_\alpha \sin \vartheta / 4\pi a^3$, which causes the necessary jump of the tangential component of the field. After the central explosion the shock wave first propagates through the region $r < a$, where the field is uniform. The field in the formed cavity weakens, according to the results of Problem 4.73, by a factor of $(r/R_s(t))^{4/(\gamma-1)}$. This stage lasts up to the instant t_a given by the condition $R_s(t_a) = a$. From this time on, the boundary between the initially uniform field and the dipole field will be involved in plasma motion and its radius will begin to increase as

$$R_a(t) = a + \int_{t_a}^{t} u(R_a(\tau), \tau)\, d\tau, \qquad t > t_a, \tag{2}$$

from which we have

$$R_a(t) = a \left[\frac{R_s(t)}{a} \right]^{2/(\gamma+1)} \geq a, \qquad t \geq t_a. \tag{3}$$

The shock front propagates faster than the matter behind the front, so that $R_s(t) > R_a(t)$ at $t > t_a$. In the region $R_a \leq r \leq R_s$, the solution (1) obtained for a magnetic dipole holds. However, in contrast to Equation

(1), now the field decreases with time throughout the cavity. It reaches maximum at the external side of the surface, $r = R_a(t)$:

$$B_r(R_a(t), t, \vartheta) = \frac{2M\cos\vartheta}{R_a^3(t)}\left(\frac{R_s(t)}{R_a(t)}\right)^{2/(\gamma-1)} = \frac{2M\cos\vartheta}{a^3}\left(\frac{a}{R_s(t)}\right)^{4/(\gamma+1)},$$

$$B_\vartheta(R_a(t), t, \vartheta) = \frac{M\sin\vartheta}{R_a^3(t)}\frac{\tau_s}{\tau_0}\left(\frac{R_s(t)}{R_a(t)}\right)^{2/(\gamma-1)} = \frac{M\sin\vartheta}{a^3}\frac{\tau_s}{\tau_0}\left(\frac{a}{R_s(t)}\right)^{4/(\gamma+1)}.$$

In the region $r < R_a(t)$, Equation (8) of Problem 4.73 is still valid. When a star explodes, in addition to the liberation of energy, a noticeable fraction of stellar mass ("a shell" of the star) is ejected into the surrounding space. In the calculations given above, the mass ejection is ignored. This approximation is valid only at later stages of evolution of the perturbed region, if the mass of the matter in the cavity significantly exceeds the mass of the ejected shell.

4.75

$$\boldsymbol{B}_r = \left(\frac{r}{R_s}\right)^{2\sigma-2} \sum_{lm} a_{lm}^{(3)}(r^\sigma/R_s^{\sigma-1})\,\mathcal{Y}_{lm}^{(3)}(\vartheta, \alpha),$$

$$\boldsymbol{B}_\tau = \sigma\left(\frac{r}{R_s}\right)^{2\sigma-2} \sum_{lm,\,k=1,2} a_{lm}^{(k)}(r^\sigma/R_s^{\sigma-1})\,\mathcal{Y}_{lm}^{(k)}(\vartheta, \alpha),$$

where $\sigma = (\gamma + 1)/(\gamma - 1)$.

4.76

$$B_r(r, \vartheta, t) = B_{0r}(a, \vartheta)\left(\frac{a}{r}\right)^2,$$

$$B_\alpha(r, \vartheta, t) = B_{0r}(a, \vartheta)\sin\vartheta\,\frac{\tau_s a^2\Omega}{\tau_0 u_w r}\left(\frac{r}{R_s(t)}\right)^3, \qquad B_\vartheta = 0.$$

Here, u_w is the constant velocity of the solar wind.

4.77 The dispersion relation for small perturbations is

$$\omega^2 = \frac{\tau_1 - \tau_2}{\tau_1 + \tau_2}gk + \frac{(\boldsymbol{B}\cdot\boldsymbol{k})^2}{2\pi(\tau_1 + \tau_2)}.$$

At $\tau_1 > \tau_2$ (the dense medium below) the discontinuity is stable even if magnetic field is absent. Rayleigh–Taylor instability arises, if the rarefied medium is below. At $\tau_2 > \tau_1$ the instability is not stabilized by the magnetic field and the perturbations with the wave vectors $\boldsymbol{k} \perp \boldsymbol{B}$ will grow up, since the magnetic lines remain undistorted.

4.78 Let us specify the shape of the perturbed surface of the discontinuity as

(1) $$z = \zeta(x, y) = l\sin k_1 x \sin k_2 y, \qquad l \ll 2\pi/k_1,\ 2\pi/k_2.$$

The gravitational energy of a small element of mass $dm = \tau\,dx\,dy\,dz$ is equal to $gz\,dm$. Integrating this quantity over dz from 0 to ζ and then over all the surface of the discontinuity S, we find the change ΔU_g in the gravitational energy when the discontinuity is deformed. While integrating over the surface of the unperturbed discontinuity, we assume that the wavelengths of perturbations $\lambda_{1,2} = 2\pi/k_{1,2}$ are small in comparison with linear sizes of the discontinuity and replace the squares of sines by their mean values, 1/2. Finally, we have $\Delta U_g = \tau g l^2 S/8$. The variation of the surface energy can be found from the formula $\Delta U_s = \alpha \Delta S$, where ΔS is the variation of the surface area,

$$\Delta S = \int \left\{ \sqrt{1 + \left(\frac{\partial \zeta}{\partial x}\right)^2 + \left(\frac{\partial \zeta}{\partial y}\right)^2} - 1 \right\} dxdy \approx \frac{l^2 k^2 S}{8}. \tag{2}$$

Thus, we have $\Delta U_s = \alpha l^2 k^2 S/8$, where $k^2 = k_1^2 + k_2^2$.

The variation of the electrostatic energy is calculated as

$$\Delta U_e = \int dx\, dy \int_\zeta^\infty dz\, \frac{\mathcal{E}^2 - E^2}{8\pi}, \tag{3}$$

where $\mathcal{E}$ is the electrostatic field above the perturbed surface. The electrostatic potential φ satisfies the Laplace equation and the boundary condition at the conducting surface $\varphi|_{z=\zeta} = 0$, as well as the condition $\varphi|_{z\to\infty} = -Ez$ at large distance from the surface. At $kl \ll 1$ the Laplace equation and the boundary conditions are satisfied for the potential $\varphi(x, y, z) = -Ez + E\zeta(x, y)\exp(-kz)$, which allows one to find the strength of the field above the perturbed surface, $\boldsymbol{\mathcal{E}} = -\nabla\varphi$, and calculate the energy variation $\Delta U_e = -E^2 l^2 S/32\pi\, k$.

The total energy variation is

$$\Delta U = \Delta U_e + \Delta U_s + \Delta U_g = \frac{l^2 S}{8} \left\{ \tau g + \alpha k^2 - \frac{E^2}{4\pi} k \right\}. \tag{4}$$

The surface will be unstable, that is, the amplitudes of bumps and hollows will grow up as long as their growth will lower the energy of the system, that is, at $\Delta U < 0$. The neutral stability occurs at

$$\frac{E^2}{4\pi} = \frac{\tau g}{k} + \alpha k. \tag{5}$$

Minimizing the right-hand side of Equation (5) with respect to k, we find $k_c = \sqrt{\tau g/\alpha}$ and the lowest electric field, $E_c^2 = 8\pi\sqrt{\tau g \alpha}$, at which the condition $\Delta U = 0$ is satisfied. At $E > E_c$ there exists a range of wave numbers for which the instability occurs, $\Delta U < 0$: $k_1 < k < k_2$, where $k_{1,2} = E^2/8\pi\alpha \mp \sqrt{(E^2/8\pi\alpha)^2 - k_c^2}$.

4.79 Let us denote the unperturbed quantities by $\boldsymbol{v}$, $\boldsymbol{B}$, and P, and their small perturbations by $\boldsymbol{u}$, $\boldsymbol{b}$, and p, respectively. The perturbed surface of the discontinuity is given by the equation $z = \zeta(x, y, t)$. Introduce also the Alfvén velocity $\boldsymbol{u}_A = \boldsymbol{B}/\sqrt{4\pi\tau}$ and the velocity of the Alfvén perturbation $\boldsymbol{w} = \boldsymbol{b}/\sqrt{4\pi\tau}$. The linearized equations are

$$\nabla \cdot \boldsymbol{u} = \nabla \cdot \boldsymbol{w} = 0, \qquad \frac{\partial \boldsymbol{w}}{\partial t} + (\boldsymbol{v} \cdot \nabla)\boldsymbol{w} = (\boldsymbol{u}_A \cdot \nabla)\boldsymbol{u}, \tag{1}$$

$$\frac{\partial \boldsymbol{u}}{\partial t} + (\boldsymbol{v} \cdot \nabla)\boldsymbol{u} = -\frac{1}{\tau}\nabla p - \nabla(\boldsymbol{u}_A \cdot \boldsymbol{w}) + (\boldsymbol{u}_A \cdot \nabla)\boldsymbol{w}.$$

The boundary conditions at the tangential discontinuity $z = \zeta$ require that the total pressures $p + (\boldsymbol{B} + \boldsymbol{b})^2_{\tau'}/8\pi$ be equal and the normal component of the magnetic field be zero, $(\boldsymbol{B} + \boldsymbol{b})_{n'} = 0$. Linearizing these equations and expressing the local primed unit vectors through initial Cartesian ones, we have

$$(2) \qquad \delta(p + \rho \boldsymbol{u}_A \cdot \boldsymbol{w}) = 0, \qquad w_z - (\boldsymbol{u}_A \cdot \nabla)\zeta = 0 \qquad \text{at} \qquad z = 0.$$

Applying the operator *div* to the last equation (1), we find the equation for perturbations of the total pressure,

$$(3) \qquad \triangle(p + \tau \boldsymbol{u}_A \cdot \boldsymbol{w}) = 0.$$

According to this equation, the dependence of perturbations on coordinates and time can be sought in the form

$$(4) \qquad \boldsymbol{u}, \boldsymbol{w}, p \propto \exp(\pm kz + ik_x x + ik_y y - i\omega t),$$

$$\zeta \propto \exp(ik_x x + ik_y y - i\omega t), \qquad \text{where} \qquad k = \sqrt{k_x^2 + k_y^2}.$$

The signs $\pm$ in the exponent refer to the regions $z < 0$ and $z > 0$, respectively.

Excluding the velocity $\boldsymbol{u}$ from Equations (1) and using Equation (4), we have

$$(5) \qquad [\omega - (\boldsymbol{k} \cdot \boldsymbol{v})^2 - (\boldsymbol{k} \cdot \boldsymbol{u}_A)^2]\boldsymbol{w} = -\frac{1}{\tau}(\boldsymbol{k} \cdot \boldsymbol{u}_A)(\boldsymbol{k} \mp ik\boldsymbol{e}_z)(p + \tau \boldsymbol{u}_A \cdot \boldsymbol{w}).$$

Next, we take the projection of Equation (5) on the Oz-axis and use the boundary conditions (2) to obtain the dispersion relation for perturbations,

$$(6) \qquad \tau_1(\omega - (\boldsymbol{k} \cdot \boldsymbol{v}_1)^2) + \tau_2(\omega - (\boldsymbol{k} \cdot \boldsymbol{v}_2)^2) = \tau_1(\boldsymbol{k} \cdot \boldsymbol{u}_{A1})^2 + \tau_2(\boldsymbol{k} \cdot \boldsymbol{u}_{A2})^2.$$

It is a quadratic equation for frequency ω, of the form $a\omega^2 + b\omega + c = 0$. Complex roots are absent at the condition $b^2 - 4ac \geq 0$, which can be reduced to

$$(7) \qquad T_{\alpha\beta} k_\alpha k_\beta \geq 0.$$

Here, the tensor $T_{\alpha\beta}$ $(\alpha, \beta = 1, 2)$ is

$$(8) \qquad T_{\alpha\beta} = \tau_1 u^{(1)}_{A\alpha} u^{(1)}_{A\beta} + \tau_2 u^{(2)}_{A\alpha} u^{(2)}_{A\beta} - \frac{\tau_1 \tau_2}{\tau_1 + \tau_2} v_\alpha v_\beta.$$

In order that the quadratic form (7) be nonnegative for arbitrary real components k_α and k_β, the inequalities $T_{\alpha\alpha} > 0$ and $|T_{\alpha\beta}| \geq 0$ (the trace of the tensor and its determinant are nonnegative) must be satisfied. These conditions, written in terms of the magnetic induction, read

$$(9) \qquad B_1^2 + B_2^2 \geq 2\pi\tau v^2, \qquad [\boldsymbol{B}_1 \times \boldsymbol{B}_2]^2 \geq 2\pi\tau([\boldsymbol{v} \times \boldsymbol{B}_1]^2 + [\boldsymbol{v} \times \boldsymbol{B}_1]^2),$$

where $\boldsymbol{v} = \boldsymbol{v}_1 - \boldsymbol{v}_2$ and $\tau = \tau_1\tau_2/(\tau_1 + \tau_2)$. They are stability conditions of the tangential discontinuity with the magnetic field (S. I. Syrovatskii,[14] 1953).

As evident from these conditions, a sufficiently strong magnetic field stabilizes tangential discontinuity. In the absence of the field, the discontinuity in incompressible medium is unstable.

4.80 Denote the magnetic induction vector not averaged over turbulent motions by $\mathcal{H}$, and the averaging by brackets, $\langle\mathcal{H}\rangle = \boldsymbol{B}$. Let us use the non-averaged equations (4.22) for $\nu_m = const$,

$$(1) \qquad \nabla \times \mathcal{H} = 0, \qquad \frac{\partial \mathcal{H}}{\partial t} = \nabla \times [\boldsymbol{u} \times \mathcal{H}] + \nu_m \triangle \mathcal{H}.$$

The ensemble averaging can be interchanged with differentiation over coordinates and time. After averaging we have

$$(2) \qquad \nabla \cdot \boldsymbol{B} = 0, \qquad \frac{\partial \boldsymbol{B}}{\partial t} = \nabla \times \langle \boldsymbol{u} \times \mathcal{H} \rangle + \nu_m \triangle \boldsymbol{B}.$$

In order to close the system of the equations, it is necessary to express the expectation value $\langle \boldsymbol{u} \times \mathcal{H} \rangle$ in terms of the averaged quantities (the magnetic induction and the turbulence correlation tensor). For this purpose, we take a turbulent perturbation to the magnetic induction, $\boldsymbol{b}(\boldsymbol{r}, t) = \mathcal{H}(\boldsymbol{r}, t) - \boldsymbol{B}(\boldsymbol{r}, t)$. Using Equations (1) and (2), we obtain the equation for this perturbation:

$$(3) \qquad \frac{\partial \boldsymbol{b}}{\partial t} - \nu_m \triangle \boldsymbol{b} = \nabla \times [\boldsymbol{u} \times \boldsymbol{B}] + \nabla \times [\boldsymbol{u} \times \boldsymbol{b}] - \nabla \times \langle \boldsymbol{u} \times \boldsymbol{b} \rangle.$$

Here, the condition $\langle \boldsymbol{u} \rangle = 0$ is used. Then we employ the "naive" theory of perturbations. Assuming that $\boldsymbol{u}$ and $\boldsymbol{b}$ are the first-order quantities, we drop their product in Equation (3). From the remaining equation which includes only first-order terms, we find

$$(4) \quad \boldsymbol{b}(\boldsymbol{r}, t) = \int_{-\infty}^{t} G(\boldsymbol{r} - \boldsymbol{r}', t - \tau)(\boldsymbol{B} \cdot \nabla)\boldsymbol{u}(\boldsymbol{r}', \tau)\, d^3x'\, d\tau - \int_{-\infty}^{t} G(\boldsymbol{r} - \boldsymbol{r}', t - \tau)\frac{\partial \boldsymbol{B}}{\partial x'_\beta} u_\beta(\boldsymbol{r}', \tau)\, d^3x'\, d\tau,$$

where the Green's function G is given by Equation (5) from Example 4.6. The typical scales of integration domains over coordinates and time are of the order of corresponding correlation quantities, l and τ_c, of the turbulent velocity field. The field $\boldsymbol{B}(\boldsymbol{r}, t)$ averaged over turbulent pulsations changes little at such times and distances. Hence, the factors containing $\boldsymbol{B}$ can be

14) Sergey Ivanovich Syrovatskii (1925–1979) was the Soviet physicist, works in the field of magnetic hydrodynamics, plasma physics and cosmic rays physics.

removed from the integrand:

$$(5)\quad \boldsymbol{b}(\boldsymbol{r},t) = \boldsymbol{B}_\beta(\boldsymbol{r},t)\int_{-\infty}^{t} G(\boldsymbol{r}-\boldsymbol{r}',t-\tau)\frac{\partial u_\beta}{\partial x'_\beta}\,d^3x'\,d\tau$$
$$-\frac{\partial \boldsymbol{B}}{\partial x_\beta}\int_{-\infty}^{t} G(\boldsymbol{r}-\boldsymbol{r}',t-\tau)u_\beta(\boldsymbol{r}',\tau)\,d^3x'\,d\tau.$$

Using the last result, we perform the averaging:

$$(6)\qquad \langle \boldsymbol{u}_\alpha \boldsymbol{b}_\mu\rangle = B_\beta\int_{-\infty}^{t} G(\boldsymbol{r}-\boldsymbol{r}',t-\tau)\left\langle u_\alpha(\boldsymbol{r},t)\frac{\partial u_\mu(\boldsymbol{r}',\tau)}{\partial x'_\beta}\right\rangle d^3x'\,d\tau$$
$$-\frac{\partial B_\mu}{\partial x_\beta}\int_{-\infty}^{t} G(\boldsymbol{r}-\boldsymbol{r}',t-\tau)\langle u_\alpha(\boldsymbol{r},t)u_\beta(\boldsymbol{r}',\tau)\rangle\,d^3x'\,d\tau.$$

Here, the first integral is a third-rank polar tensor, which is independent of coordinates. In the case of homogeneous and isotropic (non-gyrotropic) turbulence, it vanishes because there are no such tensors in our problem (the antisymmetric pseudotensor $e_{\alpha\beta\gamma}$ is incorrectly transformed under reflections and cannot be used in this case). The second integral is a second-rank symmetric tensor, which can be written as

$$(7)\qquad \int_{-\infty}^{t} G(\boldsymbol{r}-\boldsymbol{r}',t-\tau)\langle u_\alpha(\boldsymbol{r},t)u_\beta(\boldsymbol{r}',\tau)\rangle\,d^3x'\,d\tau = \nu_t\delta_{\alpha\beta},$$

where ν_t is a scalar that can be calculated if the correlation tensor $\langle u_\alpha(\boldsymbol{r},t)u_\beta(\boldsymbol{r}',\tau)\rangle$ is given. Finally, we find the required quantity:

$$(8)\qquad \langle \boldsymbol{u}(\boldsymbol{r},t)\times \boldsymbol{b}(\boldsymbol{r},t)\rangle = -\nu_t\nabla\times\boldsymbol{B}.$$

Substituting (8) into the second equation (2) leads to the equation for the magnetic induction of the large-scale averaged field:

$$(9)\qquad \frac{\partial \boldsymbol{B}}{\partial t} = (\nu_t+\nu_m)\triangle\boldsymbol{B},$$

where ν_t is the turbulent magnetic viscosity. This equation has the same form as Equation (4.15) for the magnetic field in a stationary dissipative medium. However, the turbulent motion increases the magnetic viscosity; the total magnetic viscosity is $\nu_{tot} = \nu_m + \nu_t$. Turbulent motions deform the magnetic field, break it down into components of the order of the turbulence scale and thereby produce small-scale field components. The small-scale field undergoes more intense Joule dissipation and is converted into heat.

Let us estimate the turbulent viscosity in the limiting cases using the Green's function,

$$(10)\qquad G(\boldsymbol{r}-\boldsymbol{r}',t-\tau) = \frac{1}{[4\pi\nu_m(t-\tau)]^{3/2}}\exp\left\{-\frac{(\boldsymbol{r}-\boldsymbol{r}')^2}{4\nu_m(t-\tau)}\right\}.$$

Let the main scale of turbulence be l, and the correlation time $\tau_c \approx l/u$, where the notation $u \equiv \langle \boldsymbol{u}^2\rangle$ is used for brevity. In the integral (7), the characteristic distance is $|\boldsymbol{r} - \boldsymbol{r}'| \approx l$ and the characteristic time $t - \tau \approx \tau_c \approx l/u$.

1. The case $\nu_m \tau_c \gg l^2$, or $R_m = ul/\nu_m \ll 1$, where R_m is the Reynolds magnetic number (cf. (A4.2)). The exponent in the Green's function is close to unity. Evaluating the integral (7) as the product of the integrand and the integration domain, we get

$$\nu_t \approx \frac{u^2}{3} \frac{1}{[4\pi \nu_m \tau_c]^{3/2}} l^3 \tau_c = \frac{1}{3} ul \left(\frac{ul}{4\pi \nu_m} \right)^{3/2}. \tag{11}$$

In this case $\nu_t/\nu_m < R_m^{5/2} \ll 1$, that is, the turbulent viscosity is a small correction to the collisional turbulent viscosity. Thus, the necessary condition for the applicability of the perturbation theory is satisfied.

2. The opposite case $\nu_m \tau_c \ll l^2$, or $R_m \gg 1$. In this limit, the Green's function in coordinate space is very narrow, $G(\boldsymbol{r} - \boldsymbol{r}', t - \tau) \approx \delta(\boldsymbol{r} - \boldsymbol{r}')$. In this approximation from Equation (7), we obtain

$$\nu_t \approx \frac{1}{3} \langle \boldsymbol{u}^2 \rangle \tau_c \gg \nu_m \tag{12}$$

that is, the perturbation theory is inapplicable. In order to calculate correctly the turbulent viscosity in this limit, one should use a different approximate method (see Problem 4.81).

4.81 By using the notations of Problem 4.80, we write down the non-averaged equation (4.22) in the form

$$\frac{\partial \mathcal{H}_\alpha}{\partial t} - \nu_m \triangle \mathcal{H}_\alpha = A^{\alpha\mu}_{\gamma\sigma} \frac{\partial}{\partial x_\mu} u_\gamma \mathcal{H}_\sigma + B_\alpha \, \delta(\tau - t), \qquad \tau \geq t, \tag{1}$$

where $A^{\alpha\mu}_{\gamma\sigma} = e_{\alpha\mu\nu} e_{\nu\gamma\sigma}$, while the delta-function term plays the role of the initial condition for the non-averaged field $\mathcal{H}(\boldsymbol{r}, t)$. The integral form of Equation (1) reads

$$\mathcal{H}_\alpha = A^{\alpha\mu}_{\gamma\sigma} \int d\tau' \int d^3x' \, G(\boldsymbol{r} - \boldsymbol{r}', \tau - \tau') \frac{\partial}{\partial x'_\mu} u'_\gamma \mathcal{H}'_\sigma + B^{(0)}_\alpha(\boldsymbol{r}, \tau), \tag{2}$$

where

$$B^{(0)}_\alpha(\boldsymbol{r}, \tau) = \int d^3x' \, G(\boldsymbol{r} - \boldsymbol{r}', \tau - t) B_\alpha(\boldsymbol{r}', t), \tag{3}$$

and G is the Green function satisfying Equation (3) from Example 4.6 (where the χ is replaced by ν_m) and the initial condition $G(\boldsymbol{r}, \tau) \to \delta(\boldsymbol{r})$ as $\tau \to 0$.

Now solving Equation (2) by iterations, we calculate $\mathcal{H}_\alpha$ at the instant $t + \Delta t$ with an accuracy up to the terms linear in Δt. Due to the explicit dependence of $\mathcal{H}_\alpha(\boldsymbol{r}, t + \Delta t)$ on turbulent velocities, the averaging of these terms can be performed directly. Then, assuming Δt to be small, we get the differential equation for $\boldsymbol{B}(\boldsymbol{r}, t)$.

The zeroth-order iteration with respect to random velocity is given by Equation (3) and can be written as

$$B^{(0)}_\alpha(\boldsymbol{r}, t + \Delta t) = B_\alpha(\boldsymbol{r}, t) + \Delta t \, \nu_m \triangle B_\alpha(\boldsymbol{r}, t). \tag{4}$$

It follows from the fact that $G(\boldsymbol{r},\tau) \to \delta(\boldsymbol{r})$ as $\tau \to 0$, and the correction term to the Green's function at small Δt has the form $(\partial G/\partial\tau)\Delta t = \Delta t \nu_m \triangle G(\boldsymbol{r}-\boldsymbol{r}', t)$. Twice integrating by parts the corresponding term in Equation (3), we reduce it to the form of Equation (4).

The first iteration is obtained by substituting the zeroth-order iteration (3) or (4) into the integral term of Equation (2):

(5)

$$\mathcal{H}_\alpha^{(1)}(\boldsymbol{r},\tau) = A_{\gamma\sigma}^{\alpha\mu} \int_t^\tau d\tau' \int d^3x'\, G(\boldsymbol{r}-\boldsymbol{r}',\tau-\tau') \frac{\partial}{\partial x'_\mu} u'_\gamma(\boldsymbol{r}',\tau') B_\sigma^{(0)}(\boldsymbol{r}',\tau') + B_\alpha^{(0)}(\boldsymbol{r},\tau).$$

Finally, the second iteration takes the form

(6)

$$\mathcal{H}_\alpha^{(2)}(\boldsymbol{r},\tau) = A_{\gamma\sigma}^{\alpha\mu} \int_t^\tau d\tau' \int d^3x'\, G(\boldsymbol{r}-\boldsymbol{r}',\tau-\tau') \frac{\partial}{\partial x'_\mu} u'_\gamma(\boldsymbol{r}',\tau') \mathcal{H}_\sigma^{(1)}(\boldsymbol{r}',\tau')$$

$$+ B_\alpha^{(0)}(\boldsymbol{r},\tau) = A_{\gamma\sigma}^{\alpha\mu} A_{\kappa\epsilon}^{\sigma\nu} \int_t^\tau d\tau_1 \int d^3x_1\, G(\boldsymbol{r}-\boldsymbol{r}_1,\tau-\tau_1) \frac{\partial}{\partial x_{1\mu}} u_\gamma(\boldsymbol{r}_1,\tau_1)$$

$$\times \int_t^{\tau_1} d\tau_2 \int d^3x_2\, G(\boldsymbol{r}_1-\boldsymbol{r}_2,\tau_1-\tau_2) \frac{\partial}{\partial x_{1\nu}} u_\kappa(\boldsymbol{r}_2,\tau_2) B_\epsilon^{(0)}(\boldsymbol{r}_2,\tau_2)$$

$$+ A_{\gamma\sigma}^{\alpha\mu} \int_t^\tau d\tau_1 \int d^3x_1\, G(\boldsymbol{r}-\boldsymbol{r}_1,\tau-\tau_1) \frac{\partial}{\partial x_{1\mu}} u_\gamma(\boldsymbol{r}_1,\tau_1) B_\sigma^{(0)}(\boldsymbol{r}_1,\tau_1) + B_\alpha^{(0)}(\boldsymbol{r},\tau).$$

While averaging Equation (6), we set $\langle \mathcal{H}_\alpha^{(2)}(\boldsymbol{r},\tau)\rangle = B_\alpha(\boldsymbol{r},\tau)$, $\langle u_\gamma \rangle = 0$; the correlation tensor $\langle u_\gamma(\boldsymbol{r}_1,\tau_1) u_\kappa(\boldsymbol{r}_2,\tau_2)\rangle$ is taken in the form given in the statement of the present problem. The internal integrals over $d\tau_2$ and d^3x_2 are evaluated using delta-functions:

(7)

$$\int_t^\tau d\tau_1 \int d^3x_1\, G(\boldsymbol{r}-\boldsymbol{r}_1,\tau-\tau_1) \frac{\partial}{\partial x_{1\mu}} u_\gamma(\boldsymbol{r}_1,\tau_1) \int_t^{\tau_1} d\tau_2$$

$$\int d^3x_2\, G(\boldsymbol{r}_1-\boldsymbol{r}_2,\tau_1-\tau_2) \frac{\partial}{\partial x_{1\nu}} u_\kappa(\boldsymbol{r}_2,\tau_2) B_\epsilon^{(0)}(\boldsymbol{r}_2,\tau_2) = \tau_c \int_t^\tau d\tau_1$$

$$\int d^3x_1\, G(\boldsymbol{r}-\boldsymbol{r}_1,\tau-\tau_1) \left\{ Q(0) \frac{\partial^2 B_\epsilon^{(0)}}{\partial x_{1\mu} \partial x_{1\nu}} \delta_{\gamma\kappa} - C(0) \frac{\partial B_\epsilon^{(0)}}{\partial x_{1\mu}} e_{\gamma\kappa\nu} \right\}.$$

Now we substitute $\tau = t + \Delta t$ into Equation (7) and evaluate the right-hand side in the approximation linear in Δt. To do this, it is sufficient to use the Green's function in the zeroth approximation, $G \approx \delta(\boldsymbol{r}-\boldsymbol{r}_1)$. The quantity $B_\epsilon^{(0)}$ should be taken in the same approximation, omitting the term proportional to Δt in Equation (4). As a result, the right-hand side of Equation (7) takes the form

(8)

$$\Delta t\, \tau_c \left\{ Q(0) \frac{\partial^2 B_\epsilon(\boldsymbol{r},t)}{\partial x_\mu \partial x_\nu} \delta_{\gamma\kappa} - C(0) \frac{\partial B_\epsilon(\boldsymbol{r},t)}{\partial x_\mu} e_{\gamma\kappa\nu} \right\}.$$

Then we rearrange the product $A^{\alpha\mu}_{\gamma\sigma}A^{\sigma\nu}_{\kappa\epsilon}$ and perform summation over Greek indices using the identities given in Problem I.1.28. At the final stage, we employ the relation

$$\frac{B_\alpha(t+\Delta t)-B_\alpha(t)}{\Delta t}\approx\frac{\partial B_\alpha}{\partial t} \tag{9}$$

and obtain the equation given in the statement of the problem.

4.82

$$\widetilde{Q}(k)=A(k),\qquad \widetilde{C}(k)=-P(k),\qquad k\frac{d}{dk}\frac{1}{k}\frac{d\widetilde{R}(k)}{dk}=-A(k),$$

where $\widetilde{Q}(k)$, $\widetilde{R}(k)$, and $\widetilde{C}(k)$ are the Fourier transforms of the functions $Q(r)$, $R(r)$, and $C(r)$, respectively. The time-dependent delta-like factor must be separated from the spectral functions $A(k,t)$ and $P(k,t)$.

4.83 To investigate the properties of weakly damping linear modes against the background of force-free magnetic field, we neglect the kinetic pressure and dissipative effects. We use equations

$$\nabla\cdot\boldsymbol{B}=0,\qquad \frac{\partial\boldsymbol{B}}{\partial t}=\nabla\times(\boldsymbol{u}\times\boldsymbol{B}), \tag{1}$$

$$\rho\left(\frac{\partial\boldsymbol{u}}{\partial t}+(\boldsymbol{u}\cdot\nabla)\boldsymbol{u}\right)=\frac{1}{4\pi}(\nabla\times\boldsymbol{B})\times\boldsymbol{B},\qquad \rho=const. \tag{2}$$

Now, to describe the small-amplitude linear modes satisfying these equations, we have to linearize them adopting $\boldsymbol{B}=\boldsymbol{B}_0+\boldsymbol{b}$ for the magnetic field and adopting $\boldsymbol{b}$ and $\boldsymbol{u}$ to be the first-order oscillating values of an MHD mode. Importantly, that upon substitution of $\boldsymbol{B}=\boldsymbol{B}_0+\boldsymbol{b}$ into Equation (2) we have to take into account that in the nonpotential force-free field $\nabla\times\boldsymbol{B}_0=\alpha_{FFF}\boldsymbol{B}_0\neq 0$ unlike cases of a uniform field or nonuniform potential field. Linearized equations have the form

$$\frac{\partial\boldsymbol{b}}{\partial t}=(\boldsymbol{B}_0\cdot\nabla)\boldsymbol{u}-(\boldsymbol{u}\cdot\nabla)\boldsymbol{B}_0,\qquad \frac{\partial\boldsymbol{u}}{\partial t}=\frac{\alpha_{FFF}}{4\pi\rho}\boldsymbol{B}_0\times\boldsymbol{b}+\frac{1}{4\pi\rho}(\nabla\times\boldsymbol{b})\times\boldsymbol{B}_0. \tag{3}$$

We adopt the eikonal approximation (i.e., suppose the wavelengths of the eigenmodes to be small compared with the field $\boldsymbol{B}_0$ inhomogeneity scale) and can write $\boldsymbol{b}=\boldsymbol{b}_\omega e^{i\psi(\mathbf{r})-i\omega t}$, $\boldsymbol{u}=\boldsymbol{u}_\omega e^{i\psi(\mathbf{r})-i\omega t}$, which yields equations for the corresponding complex amplitudes:

$$\boldsymbol{b}=(i/\omega)[i(\boldsymbol{B}_0\cdot\nabla\psi)\boldsymbol{u}-(\boldsymbol{u}\cdot\nabla)\boldsymbol{B}_0], \tag{4}$$

$$\boldsymbol{u}=\frac{i\alpha_{FFF}}{4\pi\rho\omega}\boldsymbol{B}_0\times\boldsymbol{b}-\frac{1}{4\pi\rho\omega}[(\boldsymbol{b}\cdot\boldsymbol{B}_0)\nabla\psi-(\boldsymbol{B}_0\cdot\nabla\psi)\boldsymbol{b}]. \tag{5}$$

Let us solve these equations in the first-order approximation with respect to the small parameter $\alpha_{FFF}/|\nabla\psi|\ll 1$. In the zero-order approximation, we have

$$\boldsymbol{b}=-\frac{1}{\omega}(\boldsymbol{B}_0\cdot\nabla\psi)\boldsymbol{u},\qquad \boldsymbol{u}=-\frac{1}{4\pi\rho\omega}(\boldsymbol{B}_0\cdot\nabla\psi)\boldsymbol{b}, \tag{6}$$

which yields the eikonal

$$(7) \qquad \nabla_{\|}\psi = \pm\omega/v_A, \qquad v_A = B_0/\sqrt{4\pi\rho}.$$

It is easy to see that in the zeroth approximation these perturbations are identical to the purely Alfvenic modes for which the conditions $\boldsymbol{b}\cdot\nabla\psi = 0$, $\boldsymbol{u}\cdot\nabla\psi = 0$, $\boldsymbol{b}\cdot\boldsymbol{B}_0 = 0$, and $\boldsymbol{u}\times\boldsymbol{b} = 0$ are satisfied.

Since we use the complex amplitudes, the bilinear terms like ab must be computed as $(1/2)\Re(ab^*)$, where in addition to averaging over the period $T = 2\pi/\omega$ we also average over the random phases of Fourier harmonics:

$$(8) \qquad \langle b_\mu b_\nu\rangle = (1/2)(\delta_{\mu\nu} - e^{\|}_\mu e^{\|}_\nu).$$

In the zeroth over α_{FFF} approximation, the kinetic helicity parameter is zero: $\langle\boldsymbol{u}\cdot(\nabla\times\boldsymbol{u}^*)\rangle = 0$.

Now, taking into account the first order over $\alpha_{FFF}/|\nabla\psi|$ terms, Equation (5) yields

$$(9) \qquad \boldsymbol{u} = -\frac{1}{4\pi\rho\omega}(\boldsymbol{B}_0\cdot\nabla\psi)\boldsymbol{b} + \frac{1}{4\pi\rho\omega}[i\alpha_{FFF}\boldsymbol{B}_0\times\boldsymbol{b} + (\boldsymbol{b}\cdot\boldsymbol{B}_0)\nabla\psi],$$

where $\boldsymbol{b}\cdot\boldsymbol{B}_0 \neq 0$ and

$$(10) \qquad \nabla\times\boldsymbol{u} = -\frac{i}{4\pi\rho\omega}(\boldsymbol{B}_0\cdot\nabla\psi)(\nabla\psi\times\boldsymbol{b}) + \frac{i}{4\pi\rho\omega}[-\nabla\psi\times(\boldsymbol{b}\cdot\nabla)\boldsymbol{B}_0 + \boldsymbol{b}\times(\boldsymbol{B}_0\cdot\nabla)\nabla\psi + \boldsymbol{b}\times(\nabla\psi\cdot\nabla)\boldsymbol{B}_0].$$

These two expressions allow calculating the necessary correlator:

$$(11) \qquad \frac{1}{2}\Re\langle\boldsymbol{u}\cdot(\nabla\times\boldsymbol{u}^*)\rangle = \frac{1}{4}\alpha_{FFF}v_A^2\frac{\langle b_\omega^2\rangle}{B_0^2},$$

which is nonzero any longer; it is proportional to the magnetic field force-free parameter α_{FFF} and the wave intensity $\langle b_\omega^2\rangle$. The Helicity parameter α can be obtained in according to (D4.24) by multiplication the expression (11) to correlation time $-\tau_c/3$, which is determined by turbulence source characteristics.

4.84 Performing Fourier transform over coordinates, from the equation of Problem 4.81 we obtain Equation (1) for the Fourier harmonics of the magnetic field:

$$(1) \qquad \frac{\partial\boldsymbol{B}_k}{\partial t} = i\alpha\,\boldsymbol{k}\times\boldsymbol{B}_k - \nu k^2\boldsymbol{B}_k,$$

where α is the gyrotropy parameter, which may have any sign, and $\nu = \nu_t + \nu_m$ is the total magnetic viscosity. The solution is given by

$$(2) \qquad \boldsymbol{B}_k(t) = \{\boldsymbol{B}_k^0\cosh(k\alpha t) + i[\boldsymbol{k}\times\boldsymbol{B}_k^0]k^{-1}\sinh(k\alpha t)\}e^{-\nu k^2 t},$$

where $\boldsymbol{B}_k^0$ is the Fourier harmonic of the initial field. A harmonic with wavenumber k grows up at

$$k < k_c = \frac{|\alpha|}{\nu} \qquad \text{or} \qquad \lambda > \lambda_c = \frac{2\pi}{k} > \frac{2\pi\nu}{|\alpha|}.$$

Harmonics with sufficiently long wavelengths are unstable. Harmonics exist if the system size $L > \lambda_c$. The above solution holds only over limited period of time; increasing field will affect the gyrotropy of turbulence, the nonlinear stabilization will limit the increase of the field.

4.85 We solve equation

(1) $$\frac{\partial \boldsymbol{B}}{\partial t} = \Omega \boldsymbol{e}_y B_x + \alpha \nabla \times \boldsymbol{B} + \nu_{tot} \triangle \boldsymbol{B}.$$

For simplicity, we seek a solution independent on the coordinate y, and write a single Fourier harmonics in the form $\boldsymbol{B}_{\mathbf{k}}(\boldsymbol{r}, t) = \mathcal{B} \exp(i\boldsymbol{k} \cdot \boldsymbol{r} + \gamma(\boldsymbol{k})t)$, where $\boldsymbol{k} = (k_x, 0, k_z)$, $\gamma(\boldsymbol{k})$ is a complex growth rate, $\mathcal{B}(\boldsymbol{k})$ is the amplitude obeying a vector equation explicitly derived from Equation (1):

(2) $$(\gamma + \nu_{tot} k^2)\mathcal{B} - \Omega \mathcal{B}_x \boldsymbol{e}_y - i\alpha\, \boldsymbol{k} \times \mathcal{B} = 0.$$

The functional form of $\gamma(\boldsymbol{k})$ is calculated from the solvability condition of the linear equation system obtained by projecting Equation (2) onto the coordinate axes, that is, from equating determinate of the system to zero:

(3)
$$\begin{vmatrix} \gamma' & i\alpha k_z & 0 \\ -(\Omega + i\alpha k_z) & \gamma' & i\alpha k_z \\ 0 & -i\alpha k_z & \gamma' \end{vmatrix} = \gamma'^3 - \gamma'[\alpha^2(k_x^2 + k_y^2) - i\alpha\Omega k_z] = 0.$$

Here $\gamma' = \gamma + \nu_{tot} k^2$, $k^2 = k_x^2 + k_z^2$. The roots of this equation are:

(4) $$\gamma_0' = 0, \qquad \gamma_\pm' = \pm\sqrt{-i\alpha\Omega k_z + \alpha^2 k^2}.$$

The first root $\gamma_0' = 0$ corroborated with Equation (1) and the necessarily transverse character of the magnetic field Fourier components $\boldsymbol{k} \cdot \boldsymbol{B}_{\mathbf{k}} = k_x B_{\mathbf{k}x} + k_z B_{\mathbf{k}z} = 0$ gives rise to equalities $B_{\mathbf{k}x} = B_{\mathbf{k}y} = B_{\mathbf{k}z} = 0$ for any real values of Ω, k_x, and k_y. Consider two other roots adopting $|\alpha k_z|/\Omega \ll 1$. Then, we have $\gamma_\pm' \approx \pm\sqrt{i}|\alpha\Omega k_z|^{1/2} = \pm(1+i)|\alpha\Omega k_z/2|^{1/2}$, so growing modes can originate from the root γ_+', so

(5) $$\Re\gamma \equiv \gamma_B = |\alpha\Omega k_z/2|^{1/2} - \nu_{tot} k^2$$

if $\gamma_B > 0$.

To further simplify the discussion, let us adopt $k_x = 0$ while $k = |k_z| \neq 0$. This immediately yields $B_{\mathbf{k}z} = 0$, while $B_{\mathbf{k}x} \neq 0$ and $B_{\mathbf{k}y} \neq 0$. Then, the instability giving rise to the large-scale field amplification occurs under condition

(6) $$k < k_c = \left(\frac{|\alpha|}{4\nu_{tot}^2 \Omega}\right)^{1/3}.$$

Under this condition, likewise in the case of α^2-dynamo (see Problem 4.84), the magnetic field grows exponentially, thus, the equation is applicable only over a limited time until the amplifying field does not significantly affect the original helical turbulent field. The mode with $\Re\gamma_+(k_c) = 0$ propagates with a constant amplitude.

4.86 In a turbulent medium, the averaged magnetic induction $\boldsymbol{B} = \mu \boldsymbol{H}$ satisfies Equation (4.14):

$$\triangle \boldsymbol{B} = \frac{4\pi\mu\sigma}{c^2}\frac{\partial \boldsymbol{B}}{\partial t}. \tag{1}$$

Comparing this equation with those derived in Problems 4.80 and 4.81, we find the required quantities:

$$\mu\sigma = \frac{\kappa}{1 + \nu_t/\nu_m}. \tag{2}$$

In order to determine them separately, we connect the tangential components of the field at a certain boundary. Then we use the result of Problem 4.29. Let the region $0 \le z \le a$ be filled with stationary medium having the electric conductivity κ and $\mu = 1$, while in the region $z > a$ a turbulence is Induced, which creates the effective electric conductivity σ and the magnetic permeability $\mu \neq 1$. For $a \gg \delta_1$, we have for the Fourier harmonics

$$H_1(z) = H_0\, e^{-(1-i)z/\delta_1}, \quad z < a, \quad \delta_1 = \frac{c}{\sqrt{2\pi\kappa\omega}}; \tag{3}$$

$$H_2(z) = H_1(a)\, e^{-(1-i)(z-a)/\delta_2}, \quad z > a, \quad \delta_2 = \frac{c}{\sqrt{2\pi\mu\sigma\omega}}.$$

For these fields, the condition $H_1 = H_2$ is satisfied at $z = a$, while the condition $E_1 = E_2$ results in the equation

$$\frac{c}{4\pi\kappa}\frac{dH_1}{dz} = \frac{c}{4\pi\sigma}\frac{dH_2}{dz} \quad \text{at} \quad z = a, \tag{4}$$

from which the relation $\mu = \sigma/\kappa$ is found. Using (2), we have

$$\mu = \frac{1}{\sqrt{1 + \nu_t/\nu_m}}, \qquad \sigma = \frac{\kappa}{\sqrt{1 + \nu_t/\nu_m}}. \tag{5}$$

The turbulence decreases the effective values of both quantities, especially at $\nu_t \gg \nu_m$. The region of turbulence has $\mu < 1$ and exhibits diamagnetic properties, resulting in a weaker external magnetic field.

5
Maxwell Equations for Alternating and Inhomogeneous Fields

5.1
Different Forms of Maxwell Equations in Media. Coupling Equations and Electromagnetic Response Functions

The system of equations with four field vectors. In Section 1.1, the system of Equations (1.8)–(1.11) has been derived for macroscopic (averaged) fields $\boldsymbol{E}(\boldsymbol{r}, t)$ and $\boldsymbol{B}(\boldsymbol{r}, t)$:

$$rot\, \boldsymbol{E}(\boldsymbol{r}, t) = -\frac{1}{c}\frac{\partial \boldsymbol{B}(\boldsymbol{r}, t)}{\partial t}, \tag{5.1}$$

$$rot\, \boldsymbol{B}(\boldsymbol{r}, t) = \frac{1}{c}\frac{\partial \boldsymbol{E}(\boldsymbol{r}, t)}{\partial t} + \frac{4\pi}{c}(\boldsymbol{j}_{int}(\boldsymbol{r}, t) + \boldsymbol{j}_{ext}(\boldsymbol{r}, t)), \tag{5.2}$$

$$div\, \boldsymbol{E}(\boldsymbol{r}, t) = 4\pi(\rho_{int}(\boldsymbol{r}, t) + \rho_{ext}(\boldsymbol{r}, t)), \tag{5.3}$$

$$div\, \boldsymbol{B}(\boldsymbol{r}, t) = 0. \tag{5.4}$$

For simplicity, we drop a bar over averaged values $\rho_{int}(\boldsymbol{r}, t)$ and $\boldsymbol{j}_{int}(\boldsymbol{r}, t)$. The averaging has been performed over physically small volumes ΔV and over time intervals Δt using Equation (1.5). On the basis of this system of equations, we have considered in Chapters 1–4 a great number of electrostatic and magnetostatic problems as well as quasistationary phenomena in the cases where the electromagnetic fields slowly vary with time. However, it is obvious, that the averaging method (1.5) limits a description of rapidly alternating and highly nonhomogeneous fields, since the spatial λ and temporal T scales of their variations must exceed the corresponding scales of the averaging, $(\Delta V)^{1/3}$ and Δt.

The averaging method free of this drawback is the averaging of field vectors and other electromagnetic values over the statistic ensemble of states of the macroscopic body under consideration. We have already used the averaging of this kind in many problems of Section 1.3 (see Problem 1.14 and next problems). Since the internal structure of atoms, molecules, and condensed media is adequately described by quantum mechanics, the averaging in question must include quantum-mechanical averaging over corresponding quantum states as well. As a result, the averaging operation will be defined for the field vectors in the following way:

$$\boldsymbol{E}(\boldsymbol{r}, t) \;=\; Tr\,(\hat{\rho}\,\hat{\boldsymbol{E}}(\boldsymbol{r}, t)), \qquad \boldsymbol{B}(\boldsymbol{r}, t) \;=\; Tr\,(\hat{\rho}\,\hat{\boldsymbol{H}}(\boldsymbol{r}, t)); \tag{5.5}$$

Electromagnetic Phenomena in Matter: Statistical and Quantum Approaches, First Edition. Igor N. Toptygin.

a bar over averaged values is dropped. Here $\widehat{\boldsymbol{E}}$ and $\widehat{\boldsymbol{H}}$ are the Heisenberg operators of field vectors, which were determined in Chapter I.6 (see also the Appendix I.C), $\widehat{\rho}$ is the statistic operator of a medium. In a similar Way, other macroscopic values (densities of the charge, current, electrical, and magnetic moments) can be defined by averaging of their Heisenberg operators over ensemble of medium states. The use of Gibbs distribution (I.C43) and (I.C47) assumes that the medium is close to statistical equilibrium.

According to the correspondence principle, quantum-mechanical operators satisfy the same equations as corresponding (classical) physical quantities. Particularly, the Heisenberg operators of the electromagnetic field satisfy the Maxwell equations (I.2.82)–(I.2.85) (see Problem I.6.11). Therefore, after the ensemble averaging Equation (1.6) and averaged Maxwell equations (5.1)–(5.4) remain valid. The relation (1.16) between the density $\rho_{int}(\boldsymbol{r}, t)$ of the induced charge in medium and the density $\boldsymbol{P}(\boldsymbol{r}, t)$ of the electric dipole moment also remains valid for time-alternating fields, since, when justifying this formula in Example 1.1, we have not used anywhere the constancy of ρ_{int} and $\boldsymbol{P}$ (also see a model approach to (1.16) in Problem 1.1). Therefore, Equation (1.16) remains true for quantities dependent on time as well:

$$\rho_{int}(\boldsymbol{r}, t) = -div\, \boldsymbol{P}(\boldsymbol{r}, t), \tag{5.6}$$

where the vector $\boldsymbol{P}(\boldsymbol{r}, t)$ still has the meaning of specific dipole moment.

In order to find the equation for the current $\boldsymbol{j}_{int}$, we use the continuity equation

$$\frac{\partial \rho_{int}}{\partial t} + \nabla \cdot \boldsymbol{j}_{int} = 0. \tag{5.7}$$

Substituting Equation (5.6) into Equation (5.7), we obtain

$$\nabla \cdot \left(\boldsymbol{j}_{int} - \frac{\partial \boldsymbol{P}(\boldsymbol{r}, t)}{\partial t} \right) = 0.$$

Therefore, the quantity in brackets can be written as a curl of a certain vector $\boldsymbol{M}$:

$$\boldsymbol{j}_{int}(\boldsymbol{r}, t) = \frac{\partial \boldsymbol{P}(\boldsymbol{r}, t)}{\partial t} + c\, rot\, \boldsymbol{M}(\boldsymbol{r}, t). \tag{5.8}$$

Here the light speed c is introduced, so that for static fields the vector $\boldsymbol{M}$ retains the meaning of magnetic moment density, and Equation (5.8) reduces to Equation (1.18). However, for alternating fields, the identification of $\boldsymbol{M}$ with the magnetic polarization vector, performed in Example 1.2, does not hold due to the appearance of the new term in Equation (3) containing the product $\boldsymbol{r} \times (\partial \boldsymbol{P}/\partial t)$. In rapidly varying fields, the model consideration (Problem 1.2) also breaks down. In this case, molecular currents inside the medium become unclosed, and the method, used in Problem 1.2 to calculate the current flowing through a certain area, is not applicable. Thus, the vector $\boldsymbol{M}$ in Equation (5.8) can be identified with the magnetization only at the condition

$$\frac{1}{c} \left| \frac{\partial \boldsymbol{P}}{\partial t} \right| \ll |rot\, \boldsymbol{M}|, \tag{5.9}$$

that is, in slowly varying fields. And yet Equation (5.8) without identification of $\boldsymbol{M}$ with the magnetization is applicable in rapidly varying fields, although in this case it may be more convenient to calculate directly the total current $\boldsymbol{j}_{int}$ without separating it into the polarization current $\partial\boldsymbol{P}/\partial t$ and magnetization current $c\, rot\, \boldsymbol{M}$. Introducing the notations (1.22) and (1.23)

$$\boldsymbol{D} = \boldsymbol{E} + 4\pi\boldsymbol{P}, \qquad \boldsymbol{H} = \boldsymbol{B} - 4\pi\boldsymbol{M} \tag{5.10}$$

for the vectors of the electric induction and magnetic field strength, and also using Equations (5.6) and (5.8), we come to the system of Equations (5.1)–(5.4) in the form

$$rot\, \boldsymbol{E}(\boldsymbol{r}, t) = -\frac{1}{c}\frac{\partial\boldsymbol{B}(\boldsymbol{r}, t)}{\partial t}, \tag{5.11}$$

$$rot\, \boldsymbol{H}(\boldsymbol{r}, t) = \frac{1}{c}\frac{\partial\boldsymbol{D}(\boldsymbol{r}, t)}{\partial t} + \frac{4\pi}{c}\boldsymbol{j}_{ext}(\boldsymbol{r}, t), \tag{5.12}$$

$$div\, \boldsymbol{D}(\boldsymbol{r}, t) = 4\pi\rho_{ext}(\boldsymbol{r}, t), \tag{5.13}$$

$$div\, \boldsymbol{B}(\boldsymbol{r}, t) = 0. \tag{5.14}$$

This is the widely used form of the Maxwell equations in media. It is appropriate for fields, sufficiently slowly varying with time, when electric and magnetic parameters of medium do not greatly differ from their static values. A formal similarity of the system (5.11)–(5.14) with the Maxwell equations (I.2.83)–(I.2.86) in vacuum enables us, by analogy with Equations (I.2.87)–(I.2.90), to write the boundary conditions for the field vectors at an interface:

$$\left.\begin{aligned} \boldsymbol{n}\times(\boldsymbol{E}_2 - \boldsymbol{E}_1) &= 0, & \boldsymbol{n}\cdot(\boldsymbol{D}_2 - \boldsymbol{D}_1) &= 4\pi\sigma_{ext}; \\ \boldsymbol{n}\times(\boldsymbol{H}_2 - \boldsymbol{H}_1) &= \frac{4\pi}{c}\boldsymbol{i}_{ext}, & \boldsymbol{n}\cdot(\boldsymbol{B}_2 - \boldsymbol{B}_1) &= 0. \end{aligned}\right\} \tag{5.15}$$

Here $\boldsymbol{n}$ is the unit vector of a normal to the interface. The right-hand sides of these conditions contain the surface densities of external charges and currents. However, in contrast to the Maxwell equations in vacuum, the system (5.11)–(5.14) contains four field vectors $\boldsymbol{E}$, $\boldsymbol{B}$, $\boldsymbol{D}$, and $\boldsymbol{H}$, and these equations are insufficient for the determination of all unknowns. This system must be supplemented by additional relations, for instance, to relate the vectors $\boldsymbol{D}$ and $\boldsymbol{H}$ with $\boldsymbol{E}$ and $\boldsymbol{B}$. The latter two vectors are clearly defined by Equations (5.5).

The problem of finding the correct constitutive relations is one of the most complicated in the macroscopic electrodynamics. Such relations, as a rule, essentially depend on individual properties of medium. In order to derive them, using the methods of quantum mechanics and physical kinetics, one should have a model of the medium. A relatively straightforward analysis is possible only for simplest models. It is greatly important to study those properties of electromagnetic parameters of medium, which can be determined on the basis of general physical principles (causality, symmetry properties, thermodynamical stability). We will start with these cases.

The coupling equations. In stationary fields that are smaller than the interatomic fields, the coupling equations in isotropic media have the form (1.26):

$$\boldsymbol{D} = \varepsilon \boldsymbol{E}, \qquad \boldsymbol{B} = \mu \boldsymbol{H}, \tag{5.16}$$

where ε and μ are the scalar factors depending on the properties of medium and its thermodynamical state. In anisotropic media, the dielectric and magnetic permeabilities are the second-rank tensors:

$$D_\kappa = \varepsilon_{\kappa\nu} E_\nu, \qquad B_\kappa = \mu_{\kappa\nu} H_\nu. \tag{5.17}$$

It is clear that in a slowly varying field at sufficiently small frequencies the same coupling equations containing stationary dielectric and magnetic permeabilities will be applicable. However, in rapidly varying and strongly nonhomogeneous fields the coupling equations become more complicated. Assuming the coupling equations to be linear (i.e., the fields are sufficiently weak), let us write them in the most general form as a result of action of some linear operators,

$$\boldsymbol{D} = \hat{\sigma}\boldsymbol{E} + \hat{\kappa}\boldsymbol{B}, \qquad \boldsymbol{H} = \hat{\rho}\boldsymbol{B} + \hat{\nu}\boldsymbol{E}, \tag{5.18}$$

on the principal vectors $\boldsymbol{E}$ and $\boldsymbol{B}$. Here the values with hats are the operators, which, generally, may be second rank tensors, contain derivatives and integrals with respect to coordinates and time. Equation (5.18) can be simplified by noticing that $\boldsymbol{E}$ and $\boldsymbol{B}$ are coupled by the linear equation (5.11). Due to this fact, it is possible to exclude one of these vectors from the right-hand sides of Equation (5.18). For example, writing Equation (5.11) in the form $\boldsymbol{B}(\boldsymbol{r}, t) = -c \int \mathit{rot}\, \boldsymbol{E}(\boldsymbol{r}, t)\, dt$ or, in Fourier space, $\boldsymbol{B}(\boldsymbol{k}, \omega) = (c/\omega)\boldsymbol{k} \times \boldsymbol{E}(\boldsymbol{k}, \omega)$, and using it in the first equation (5.18), we obtain $\boldsymbol{D} = \hat{\sigma}\boldsymbol{E} - c\hat{\kappa} \int \mathit{rot}\, \boldsymbol{E}\, dt$ or

$$\boldsymbol{D} = \hat{\varepsilon}\boldsymbol{E}, \tag{5.19}$$

where $\hat{\varepsilon}$ is some resulting linear operator (operator of the dielectric permeability), which in the stationary case reduces to a multiplicative constant scalar ε or a constant tensor $\varepsilon_{\alpha\beta}$. We can rewrite the second equation (5.18) in a similar fashion:

$$\boldsymbol{B} = \hat{\mu}\boldsymbol{H}, \tag{5.20}$$

where $\hat{\mu}$ is the operator of the magnetic permeability. In the section below, we discuss in more detail the properties of the operators $\hat{\varepsilon}$ and $\hat{\mu}$.

The system of equations with three field vectors. As mentioned above in connection with Equation (5.8), for rapidly varying fields it is impossible to separate the total current, induced in medium, into two components – the currents of polarization and magnetization. In this situation, one can introduce the new **vector of the generalized electric induction $\boldsymbol{\mathcal{D}}(\boldsymbol{r}, t)$**. It is different from Equation (5.10) and contains the density of the total current induced in medium:

$$\boldsymbol{\mathcal{D}}(\boldsymbol{r}, t) = \boldsymbol{E}(\boldsymbol{r}, t) + 4\pi \int_{-\infty}^{t} \boldsymbol{j}_{int}(\boldsymbol{r}, t')\, dt'. \tag{5.21}$$

Here it is assumed that the field is slowly (adiabatically) switched on at $t \to -\infty$, so that the integral (5.21) is converged. In analogy with Equation (5.10), it is possible to denote the integral in Equation (5.21) as

$$\mathcal{P} = \int_{-\infty}^{t} \boldsymbol{j}_{int}(\boldsymbol{r}, t')\, dt', \tag{5.22}$$

though the vector $\mathcal{P}$, introduced in such a way, cannot be identified with the electric dipole moment per unit volume. Now this vector includes also the current of magnetization:

$$\boldsymbol{j}_{int}(\boldsymbol{r}, t) = \frac{\partial}{\partial t}\mathcal{P}(\boldsymbol{r}, t). \tag{5.23}$$

The density of the induced charge is calculated with the aid of the continuity equation:

$$\rho_{int}(\boldsymbol{r}, t) = -\int_{-\infty}^{t} div\, \boldsymbol{j}_{int}(\boldsymbol{r}, t')\, dt' = -\,div\, \mathcal{P}(\boldsymbol{r}, t). \tag{5.24}$$

In this consideration, the magnetic field is described by a single vector $\boldsymbol{B}(\boldsymbol{r}, t)$. Thus, the Maxwell equations will contain three field vectors:

$$rot\, \boldsymbol{E}(\boldsymbol{r}, t) = -\frac{1}{c}\frac{\partial \boldsymbol{B}(\boldsymbol{r}, t)}{\partial t}, \tag{5.25}$$

$$rot\, \boldsymbol{B}(\boldsymbol{r}, t) = \frac{1}{c}\frac{\partial \mathcal{D}(\boldsymbol{r}, t)}{\partial t} + \frac{4\pi}{c}\boldsymbol{j}_{ext}(\boldsymbol{r}, t), \tag{5.26}$$

$$div\, \mathcal{D}(\boldsymbol{r}, t) = 4\pi \rho_{ext}(\boldsymbol{r}, t), \tag{5.27}$$

$$div\, \boldsymbol{B}(\boldsymbol{r}, t) = 0. \tag{5.28}$$

The system of Equations (5.25)–(5.28) should be supplemented with one coupling equation, which can be written in the form (5.19):

$$\mathcal{D}(\boldsymbol{r}, t) = \hat{\epsilon}\, \boldsymbol{E}(\boldsymbol{r}, t). \tag{5.29}$$

However, the operator $\hat{\epsilon}$ in Equation (5.29) does not coincide with $\hat{\varepsilon}$ in Equation (5.19). Since here the vector $\boldsymbol{H}$ and the coupling Equation (5.20) are not present at all, Equation (5.29) must contain both operators, $\hat{\varepsilon}$ and $\hat{\mu}$, from Equations (5.19) and (5.20). Below we will determine the coupling between these three operators.

In order to find the boundary conditions for the field vectors $\boldsymbol{E}$, $\boldsymbol{B}$, and $\mathcal{D}$ on the interface, we rewrite Equation (5.26) in the form (5.2) and compare the obtained system with the Maxwell equations in vacuum. Similarly to boundary conditions (5.15), we find

$$\left.\begin{aligned} \boldsymbol{n} \times (\boldsymbol{E}_2 - \boldsymbol{E}_1) &= 0, & \boldsymbol{n}\cdot(\mathcal{D}_2 - \mathcal{D}_1) &= 4\pi\sigma_{ext}, \\ \boldsymbol{n} \times (\boldsymbol{B}_2 - \boldsymbol{B}_1) &= \frac{4\pi}{c}(\boldsymbol{i}_{ext} + \boldsymbol{i}_{int}), & \boldsymbol{n}\cdot(\boldsymbol{B}_2 - \boldsymbol{B}_1) &= 0. \end{aligned}\right\} \tag{5.30}$$

Here $\boldsymbol{i}_{int}$ is the surface current induced in medium.

Other forms of macroscopic Maxwell equations can be found in the manual of Pamaytnykh and Turov (2000) and in the paper by Vinogradov (2002).

Electromagnetic response functions. Let us begin our discussion of coupling equations between the electromagnetic field vectors by writing Equation (5.29) in the most general integral form

$$\mathcal{D}_\alpha(\boldsymbol{r}, t) = \int_{-\infty}^{\infty} dt' \int d^3r' \, \epsilon_{\alpha\beta}(\boldsymbol{r}, \boldsymbol{r}'; t, t') \, E_\beta(\boldsymbol{r}', t'), \tag{5.31}$$

where the integration in coordinate space is done over entire medium. Here $\epsilon_{\alpha\beta}$ is a certain real tensor function, including all information on a response of medium to a weak alternating nonhomogeneous electric field. Accordingly, it is referred to as **a linear response function** (if the field $\boldsymbol{E}$ were compatible with the interatomic one, the coupling of generalized electric induction with field strength would be nonlinear).

The physical meaning of the integral coupling Equation (5.31) is that the medium response in point $\boldsymbol{r}$ at instant t may depend on the field strengths at other instants of time and not only in this point but in adjacent points of space as well. The dependence on other instants of time takes place because the rearrangement of charges and currents of the medium, caused by electro-magnetic field variations, does not occur instantaneously, but during a finite time interval, which is the relaxation time of the medium. Different subsystems of the medium may have different relaxation times. Actually, the integration interval over time in Equation (5.31) is of the order of the largest relaxation time $\tau_{r'}$, that is, the response function rapidly decreases at $|t - t'| \gg \tau_{r'}$. If the medium properties in the absence of any external field do not change with time (medium in equilibrium), then all instants t are equivalent and the response function depends only on the difference $t - t'$, that is,

$$\epsilon_{\alpha\beta}(\boldsymbol{r}, \boldsymbol{r}'; t, t') = \epsilon_{\alpha\beta}(\boldsymbol{r}, \boldsymbol{r}'; t - t'). \tag{5.32}$$

If a quantity, which characterizes the effect on medium ($E_\beta(\boldsymbol{r}', t')$, in our case), may be treated as arbitrary and independent of the medium state, then a very strong restriction on properties of the response function arises: it must strictly vanish at $t - t' < 0$. This property reflects the causality principle: the action (electrical field) has to precede the consequence (medium polarization). The important consequences of causality will be considered below in this section. However, we notice, that the electric field in medium is not always independent of medium properties.

The nonlocality, inherent in Equation (5.31), that is the dependence of the polarization in point $\boldsymbol{r}$ on the field in point $\boldsymbol{r}'$, is explained by the fact, that, due to thermal motion, the particles from adjacent regions, where the field is different, may move to a given point. The second reason of nonlocality may be the macroscopic random or ordered insertions in medium, implanted for creating required electric and magnetic properties. The larger field variations in space, the stronger the effect. In a statistically uniform medium, the response function depends on the difference between $\boldsymbol{r}$ and $\boldsymbol{r}'$,

$$\epsilon_{\alpha\beta}(\boldsymbol{r}, \boldsymbol{r}'; t - t') = \epsilon_{\alpha\beta}(\boldsymbol{r} - \boldsymbol{r}'; t - t'). \tag{5.33}$$

Such a dependence suggests the absence of boundaries between media with different properties. Otherwise, near the boundary this dependence will be violated.

The integral Equation (5.31) includes all possible cases of the linear dependence. Particularly, at $\epsilon_{\alpha\beta} = \varepsilon^{(0)}_{\alpha\beta}\,\delta(t - t')\,\delta(\boldsymbol{r} - \boldsymbol{r}')$ we obtain the coupling equation of type (5.16). Expanding $\boldsymbol{E}(\boldsymbol{r}', t')$ in powers of differences $\boldsymbol{r} - \boldsymbol{r}'$ and $t - t'$ in the vicinity of the point $(\boldsymbol{r}, t)$, we can obtain the coupling equation containing the derivatives of the field with respect to coordinates and to time.

Example 5.1

Let the medium be uniform (infinite) and in equilibrium. Write down Equation (5.31) for Fourier components of field vectors with respect to coordinates and time, assuming that the response function depends only on the difference of arguments.

Solution. Multiplying both sides of Equation (5.31) by $\exp(i\omega t - i\boldsymbol{k}\cdot\boldsymbol{r})$ and integrating within infinite limits $(-\infty, \infty)$ over each variable, we transform the integral equation to the algebraic one,

$$D_\alpha(\boldsymbol{k}, \omega) = \epsilon_{\alpha\beta}(\boldsymbol{k}, \omega) E_\beta(\boldsymbol{k}, \omega), \tag{5.34}$$

where

$$\epsilon_{\alpha\beta}(\boldsymbol{k}, \omega) = \int d\tau\, e^{i\omega\tau} \int d^3r\, e^{-i\mathbf{k}\cdot\mathbf{r}}\, \epsilon_{\alpha\beta}(\boldsymbol{r}, \tau) \tag{5.35}$$

is **the complex dielectric permeability**. Integration over the temporal argument $\tau - t - t'$ is done only over the positive semiaxis (one-sided Fourier transformation), if causality is available. Comparing Equation (5.34) with the Fourier image (5.21), we link **the tensor of the complex electric conductivity** $\kappa_{\alpha\beta}$ with the tensor of dielectric permeability:

$$j^{int}_\alpha = \kappa_{\alpha\beta} E_\beta, \qquad \kappa_{\alpha\beta} = -\frac{i\omega}{4\pi}(\epsilon_{\alpha\beta} - \delta_{\alpha\beta}). \tag{5.36}$$

From the definition of the tensor (5.35) for real and the reality of response function $\epsilon_{\alpha\beta}(\boldsymbol{R}, \tau)$ at real values of ω and $\boldsymbol{k}$ we obtain the symmetry properties:

$$\epsilon_{\alpha\beta}(\boldsymbol{k}, \omega) = \epsilon^*_{\alpha\beta}(-\boldsymbol{k}, -\omega), \tag{5.37}$$

where asterisk, as previously, denotes complex conjugation. ■

Example 5.2

Let the medium be in equilibrium, uniform, isotropic, and invariant with respect to inversion of spatial axes. Show that the tensor $\epsilon_{\alpha\beta}(\boldsymbol{k}, \omega)$ of the complex dielectric permeability is determined by two scalar functions which depend generally on ω and $k = |\boldsymbol{k}|$.

Solution. The second-rank tensor $\epsilon_{\alpha\beta}$ is a true (polar) tensor. In the isotropic medium, it can be constructed only from the two polar tensors, $\delta_{\alpha\beta}$ and $k_\alpha k_\beta / k^2$.

Let us compose from them two tensors, $\delta^{\perp}_{\alpha\beta} = \delta_{\alpha\beta} - k_\alpha k_\beta/k^2$ and $\delta^{\parallel}_{\alpha\beta} = k_\alpha k_\beta/k^2$, which are transverse and longitudinal with respect to $\boldsymbol{k}$, respectively.

Then the general form of the tensor of the complex dielectric permeability is

$$\epsilon_{\alpha\beta}(\boldsymbol{k}, \omega) = \varepsilon_l(k, \omega)\frac{k_\alpha k_\beta}{k^2} + \varepsilon_t(k, \omega)\left(\delta_{\alpha\beta} - \frac{k_\alpha k_\beta}{k^2}\right), \tag{5.38}$$

where ε_l are ε_t are the scalar functions (generally, complex), which can be called the **transverse** and **longitudinal dielectric permeabilities**. From Equation (5.37) we obtain the useful symmetry properties for the real $\varepsilon' = \Re\, \varepsilon$ and imaginary $\varepsilon'' = \Im\, \varepsilon$ parts of the dielectric permeabilities:

$$\varepsilon'_{l,t}(k, \omega) = \varepsilon'_{l,t}(k, -\omega), \qquad \varepsilon''_{l,t}(k, \omega) = -\varepsilon''_{l,t}(k, -\omega). \tag{5.39}$$

The real parts of ε_l and ε_t are even functions of ω, while the imaginary parts are odd functions.

In anisotropic media, the tensor $\epsilon_{\alpha\beta}(\boldsymbol{k}, \omega)$ becomes more complicated; it cannot be presented in the form (5.38). ■

Example 5.3

Show that the two dielectric permeabilities, ε_l and ε_t, introduced above are sufficient to formulate the macroscopic Maxwell's equations in uniform and isotropic media. To do this, express the Fourier components of $\boldsymbol{j}_{int}$ and ρ_{int} in Equation (5.1)–(5.4) via ε_l, ε_t and macroscopic electric field strength $\boldsymbol{E}(\boldsymbol{k}, \omega)$. Write down the system of equations in terms of indicated quantities and express the Fourier components of the field vectors through the densities of external charge and current.

Solution. From Equations (5.34), (5.36), and (5.38) we find

$$\begin{aligned} \boldsymbol{j}^{\parallel}_{int}(\boldsymbol{k}, \omega) &= -\frac{i\omega}{4\pi}(\varepsilon_l - 1)\boldsymbol{E}^{\parallel}(\boldsymbol{k}, \omega), \\ \boldsymbol{j}^{\perp}_{int}(\boldsymbol{k}, \omega) &= -\frac{i\omega}{4\pi}(\varepsilon_t - 1)\boldsymbol{E}^{\perp}(\boldsymbol{k}, \omega), \\ \rho_{int}(\boldsymbol{k}, \omega) &= \frac{\boldsymbol{k}\cdot\boldsymbol{j}_{int}}{\omega} = -i\frac{\varepsilon_l - 1}{4\pi}\,\boldsymbol{k}\cdot\boldsymbol{E}(\boldsymbol{k}, \omega). \end{aligned} \tag{5.40}$$

The last equality has been derived from the continuity equation, indices $\perp$ and $\parallel$ denote the components, perpendicular and longitudinal with respect to $\boldsymbol{k}$. Next, we do Fourier transform of the system of Maxwell's equations (5.1)–(5.4),

$$\boldsymbol{k} \times \boldsymbol{E}^{\perp}(\boldsymbol{k}, \omega) = \frac{\omega}{c}\boldsymbol{B}(\boldsymbol{k}, \omega), \tag{5.41}$$

$$\boldsymbol{k} \times \boldsymbol{B}(\boldsymbol{k}, \omega) = -\frac{\omega}{c}\varepsilon_t \boldsymbol{E}^{\perp}(\boldsymbol{k}, \omega) - i\frac{4\pi}{c}\boldsymbol{j}^{\perp}_{ext}(\boldsymbol{k}, \omega), \tag{5.42}$$

$$\varepsilon_l \boldsymbol{k}\cdot\boldsymbol{E}(\boldsymbol{k}, \omega) = -i4\pi\rho_{ext}(\boldsymbol{k}, \omega), \tag{5.43}$$

$$\boldsymbol{k}\cdot\boldsymbol{B}(\boldsymbol{k}, \omega) = 0, \tag{5.44}$$

and find the Fourier components of the field vectors:

$$E^{\parallel}(\boldsymbol{k},\omega) = -i\frac{4\pi}{k\varepsilon_l(k,\omega)}\rho_{ext}(\boldsymbol{k},\omega), \tag{5.45}$$

$$E^{\perp}(\boldsymbol{k},\omega) = -i\frac{4\pi\omega}{\omega^2\varepsilon_t(k,\omega) - c^2k^2}\boldsymbol{j}^{\perp}_{ext}(\boldsymbol{k},\omega), \tag{5.46}$$

$$\boldsymbol{B}(\boldsymbol{k},\omega) = \boldsymbol{B}^{\perp}(\boldsymbol{k},\omega) = -i\frac{4\pi c}{\omega^2\varepsilon_t(k,\omega) - c^2k^2}\,\boldsymbol{k}\times\boldsymbol{j}_{ext}(\boldsymbol{k},\omega). \quad \blacksquare \tag{5.47}$$

Now let us come to the coupling equations in the form (5.19) and (5.20). As in the case of Equation (5.31), we write them in the integral form. In a uniform equilibrium medium, the kernels of the integral operators will depend on the difference of spacial and temporal arguments. In the case of stationary uniform fields, they must transform in the constant tensors according to equalities (5.17). For the isotropic medium, we assume that the kernels of the operators $\hat{\varepsilon}$ and $\hat{\mu}$ are scalar quantities: $\hat{\varepsilon}\boldsymbol{E} = \int_{-\infty}^{t}\int d^3r'\,\varepsilon(\boldsymbol{r}-\boldsymbol{r}',t-t')\boldsymbol{E}(\boldsymbol{r}',t')$, and the same for $\hat{\mu}$. In the Fourier representation, Equations (5.19) and (5.20) take the form (see Example 5.1)

$$\boldsymbol{D}(\boldsymbol{k},\omega) = \varepsilon(k,\omega)\,\boldsymbol{E}(\boldsymbol{k},\omega), \qquad \boldsymbol{B}(\boldsymbol{k},\omega) = \mu(k,\omega)\boldsymbol{H}(\boldsymbol{k},\omega). \tag{5.48}$$

Here we indicate that ε and μ depend on $k = |\boldsymbol{k}|$, since in the isotropic medium a scalar function can depend only on scalars. In anisotropic media, the dielectric and magnetic permeabilities are second-rank tensors:

$$D_\kappa(\boldsymbol{k},\omega) = \varepsilon_{\kappa\nu}(\boldsymbol{k},\omega)E_\nu(\boldsymbol{k},\omega), \qquad B_\kappa(\boldsymbol{k},\omega) = \mu_{\kappa\nu}(\boldsymbol{k},\omega)H_\nu(\boldsymbol{k},\omega). \tag{5.49}$$

Example 5.4

Relate the permeabilities ε and μ with ε_l and ε_t. For this purpose, write the system (5.11)–(5.14) in Fourier space using Equation (5.48), and compare it with the system (5.41)–(5.44). Use the fact, that the vectors $\boldsymbol{E}$ and $\boldsymbol{B}$ in both systems are the same by definition.

Solution. The system (5.11)–(5.14) in Fourier space reads

$$\boldsymbol{k}\times\boldsymbol{E}(\boldsymbol{k},\omega) = \frac{\omega}{c}\,\boldsymbol{B}(\boldsymbol{k},\omega), \tag{5.50}$$

$$\frac{1}{\mu}\,\boldsymbol{k}\times\boldsymbol{B}(\boldsymbol{k},\omega) = -\frac{\omega}{c}\,\varepsilon\,\boldsymbol{E}(\boldsymbol{k},\omega) - i\,\frac{4\pi}{c}\,\boldsymbol{j}_{ext}(\boldsymbol{k},\omega), \tag{5.51}$$

$$\varepsilon\,\boldsymbol{k}\cdot\boldsymbol{E}(\boldsymbol{k},\omega) = -i\,4\pi\rho_{ext}(\boldsymbol{k},\omega), \tag{5.52}$$

$$\boldsymbol{k}\cdot\boldsymbol{B}(\boldsymbol{k},\omega) = 0. \tag{5.53}$$

Comparing Equation (5.52) with Equation (5.43), we find

$$\varepsilon(k,\omega) = \varepsilon_l(k,\omega). \tag{5.54}$$

Then we rewrite Equations (5.51) and (5.42) as

(1) $$\frac{1}{\mu}\boldsymbol{k}\times\boldsymbol{B}(\boldsymbol{k},\omega) = -\frac{\omega}{c}\varepsilon_l\boldsymbol{E}^{\perp}(\boldsymbol{k},\omega) - i\frac{4\pi}{c}\boldsymbol{j}^{\perp}_{ext}(\boldsymbol{k},\omega),$$

(2) $$\boldsymbol{k} \times \boldsymbol{B}(\boldsymbol{k}, \omega) = -\frac{\omega}{c}\varepsilon_t \boldsymbol{E}^{\perp}(\boldsymbol{k}, \omega) - i\frac{4\pi}{c}\boldsymbol{j}^{\perp}_{ext}(\boldsymbol{k}, \omega).$$

The longitudinal electric field and longitudinal current drop out from Equation (5.51) in view of Equation (5.52) and of the continuity equation $\boldsymbol{k} \cdot \boldsymbol{j}_{ext} = \omega\rho_{ext}$. Subtracting Equation (2) from Equation (1), we have

(3) $$\left(\frac{1}{\mu} - 1\right)\boldsymbol{k} \times \boldsymbol{B} = -\frac{\omega}{c}(\varepsilon_l - \varepsilon_t)\boldsymbol{E}^{\perp}.$$

In order to exclude the vector $\boldsymbol{B}$, we will use the induction Equation (5.50) and obtain the important relation

$$\frac{1}{\mu(k, \omega)} = 1 + \left(\frac{\omega}{ck}\right)^2 (\varepsilon_l(k, \omega) - \varepsilon_t(k, \omega)). \tag{5.55}$$

From this relation, we see that the difference between ε_l and ε_t is caused by the magnetic properties of the medium. Thus, at all $\omega \neq 0$, while the vectors $\boldsymbol{B}$ and $\boldsymbol{E}^{\perp}$ are related by the electromagnetic induction equation, for a total description of electric and magnetic properties of uniform and isotropic medium it is sufficient to know two scalar functions of ω and k. In the stationary case, ($\omega = 0$) the link between vectors $\boldsymbol{E}$ and $\boldsymbol{B}$ is absent, the medium is described by two permeabilities $\varepsilon(k, 0) = \varepsilon_l(k, 0)$ and $\mu(k, 0)$, which (for nonuniform field) may depend on k. ■

Example 5.5

From general Maxwell equations in the $\boldsymbol{k}, \omega$ -representation (equalities (5.41)–(5.53)) obtain the dispersion equations that determine the dependence of ω on the wave vector for transverse and longitudinal eigenmodes of electromagnetic field in medium. Express the required equations through the permeabilities $\varepsilon_l(\boldsymbol{k}, \omega)$ and $\varepsilon_t(\boldsymbol{k}, \omega)$, as well as through $\varepsilon(\boldsymbol{k}, \omega)$ and $\mu(\boldsymbol{k}, \omega)$.

Solution. The eigenmodes occur in the absence of external sources, that is at $\cdot\boldsymbol{j}_{ext} = \rho_{ext} = 0$. As follows from Equation (5.45), in this case the longitudinal part of the vector $\boldsymbol{E}$ can differ from zero only if the denominator in the right-hand side of the equality becomes zero. From this condition, we find the dispersion equation for the longitudinal oscillations:

$$\varepsilon_l(\boldsymbol{k}, \omega) = 0. \tag{5.56}$$

These oscillations are purely electric, since the magnetic field is transverse to $\boldsymbol{k}$ ($div\,\boldsymbol{B} = 0$) and cannot be produced. The transverse vectors $\boldsymbol{E}^{\perp}$ and $\boldsymbol{B}$ in the absence of external sources can be nonzero if the equal denominators in the right-hand sides of Equations (5.46) and (5.47) become zero. This condition gives the dispersion equation for the transverse waves:

$$\omega^2\varepsilon_t(\boldsymbol{k}, \omega) = c^2k^2. \tag{5.57}$$

The dispersion equation, expressed through $\varepsilon(\boldsymbol{k}, \omega)$ and $\mu(\boldsymbol{k}, \omega)$, can be obtained in a similar fashion from the system (5.50)–(5.53) or with the aid of Equations (5.54) and (5.55):

$$\varepsilon(\boldsymbol{k}, \omega) = 0, \qquad \omega^2\varepsilon(\boldsymbol{k}, \omega)\mu(\boldsymbol{k}, \omega) = c^2k^2. \quad ■ \tag{5.58}$$

Example 5.6

Calculate the tensor of complex dielectric permeability for a completely ionized uniform, isotropic, collisionless, gaseous plasma being in equilibrium at temperature T (with average number densities of particles $n_e = n_i = n$). The plasma is referred to as ***collisionless****, if the collision frequency of particles is much smaller than wave frequency, and particle mean free paths are larger than wave length.*

Solution. Let us calculate the plasma current induced by a weak external field with the wave vector $\boldsymbol{k}$ and frequency ω. It can be expressed as

(1) $$\boldsymbol{j} = \sum \int e\boldsymbol{v}\,\delta f\, d^3p,$$

where δf is the nonequilibrium component of the distribution function of particles of given species, induced by the external field. The sum is over all species; we restrict ourselves by electrons and singly charged ions. We calculate δf from Equation (1.38), writing the collision integral in the form of (1.40), $I(f) = -\nu\delta f$, where ν is a small collision frequency, with $\nu \to 0$ in the end of calculations. Assuming the field to be weak, in (1.38) we set $f = f_0 + \delta f$, with $|\delta f| \ll f_0$, f_0 being the Maxwellian distribution (1.32). Then we linearize the equation:

(2) $$\left(\frac{\partial}{\partial t} + \boldsymbol{v}\cdot\frac{\partial}{\partial \boldsymbol{r}} + \nu\right)\delta f = -e\left(\boldsymbol{E} + \frac{1}{c}\boldsymbol{v}\times\boldsymbol{B}\right)\cdot\frac{\partial f_0}{\partial \boldsymbol{p}}.$$

Since $\boldsymbol{E}$ and $\boldsymbol{B} \propto \exp(i\boldsymbol{k}\cdot\boldsymbol{r} - i\omega t)$, the dependence of $\delta f(\boldsymbol{r}, t, \boldsymbol{p})$ on coordinates and time can be sought in the same form:

(3) $$\delta f(\boldsymbol{r}, t, \boldsymbol{p}) = \delta f(\boldsymbol{p}) \exp(i\boldsymbol{k}\cdot\boldsymbol{r} - i\omega t).$$

In the case of Maxwell distribution (1.32), the derivative over momentum is

(4) $$\frac{\partial f_0}{\partial \boldsymbol{p}} = -\frac{\boldsymbol{v}}{T}f_0, \qquad [\boldsymbol{v}\times\boldsymbol{B}]\cdot\frac{\partial f_0}{\partial \boldsymbol{p}} = 0.$$

Substituting all these formulas in Equation (2), we obtain the solution

(5) $$\delta f(\boldsymbol{p}) = \frac{e(\boldsymbol{E}\cdot\boldsymbol{v})f_0(p)}{T(\nu + i\boldsymbol{k}\cdot\boldsymbol{v} - i\omega)}.$$

From this expression, it is clear, why it is reasonable to retain a small collision frequency $\nu > 0$ even in the collisionless case: the denominator of the expression does not become zero at any real (physical) values $\boldsymbol{k}$, $\boldsymbol{v}$ and ω, and Equation (5) remains non-singular.

Substituting Equation (5) in Equation (1), we can express the relation between the current and electric field in the form of the Ohm's law

(6) $$j_\alpha = \kappa_{\alpha\beta}(\boldsymbol{k}, \omega)E_\beta,$$

where the complex conductivity tensor is given by

(7) $$\kappa_{\alpha\beta}(\boldsymbol{k}, \omega) = \sum \frac{e^2}{T}\int \frac{v_\alpha v_\beta f_0(p)}{\nu + i\boldsymbol{k}\cdot\boldsymbol{v} - i\omega}\, d^3p.$$

According to Equation (5.36), this tensor is expressed as the complex dielectric permeability tensor by the relation

(8) $$\epsilon_{\alpha\beta}(\boldsymbol{k},\omega) = \delta_{\alpha\beta} + i\frac{4\pi\kappa_{\alpha\beta}(\boldsymbol{k},\omega)}{\omega}.$$

Let us separate from the tensor $\epsilon_{\alpha\beta}$ the invariant values ε_l and ε_t using Equation (5.38):

(9) $$\varepsilon_l = 1 + i\sum \frac{4\pi e^2}{\omega T}\int \frac{v_\parallel^2 f_0(p)}{\nu + ikv_\parallel - i\omega}\,d^3p,$$

(10) $$\varepsilon_t = 1 + i\sum \frac{2\pi e^2}{\omega T}\int \frac{v_\perp^2 f_0(p)}{\nu + ikv_\parallel - i\omega}\,d^3p.$$

Here we have introduced the velocity projections, parallel and perpendicular to $\boldsymbol{k}$.

The integration over the transverse components of momentum in Equations (9) and (10) is easily performed, using an explicit form of the Maxwell distribution (1.32). As a result the formulas take the form

(11) $$\varepsilon_l(k,\omega) = 1 + i\sum \frac{4\pi ne^2}{m}\frac{1}{\sqrt{2\pi}kv_T\omega}\frac{e^{-x^2/2}x^2\,dx}{x-s-i\nu'},$$

(12) $$\varepsilon_t(k,\omega) = 1 + i\sum \frac{4\pi ne^2}{m}\frac{1}{\sqrt{2\pi}kv_T\omega}\int \frac{e^{-x^2/2}\,dx}{x-s-i\nu'},$$

where $v_T = \sqrt{T/m}$ is the thermal velocity, while $s = \omega/kv_T > 0$ and $\nu' = \nu/kv_T > 0$ are dimensionless quantities. With the aid of Sokhotsky equations (I.1.223), we transform the fraction as:

(13) $$\frac{x^2}{x-s-i\nu'} = x\left[1 + \frac{s}{x-s-i\nu'}\right] = x + s + \frac{s^2}{x-s-i\nu'}.$$

The integration of $x+s$ is reduced to the tabulated integral. We perform the latter integration using the identity

$$\frac{1}{\nu' + i(x-s)} = \int_0^\infty e^{-(\nu'-is+ix)t}\,dt,$$

which is valid because $\nu' > 0$. We have

$$\int_{-\infty}^{\infty}\frac{e^{-x^2/2}dx}{x-s-i\nu'} = i\int_0^\infty dt\,e^{-(\nu'-is)t}\int_{-\infty}^{\infty} e^{-(x^2+2xit)/2}\,dx = i\sqrt{2\pi}e^{-s^2/2}$$
$$\left\{\sqrt{\frac{\pi}{2}} + i\int_0^s e^{u^2/2}\,du\right\}.$$

Let introduce the notations

(14) $$Z(s) = X(s) - iY(s), \qquad \text{where} \qquad X(s) = se^{-s^2/2}\int_0^s e^{u^2/2}\,du,$$

$$Y(s) = \sqrt{\frac{\pi}{2}}s\,e^{-s^2/2}.$$

The real part is expressed through the probability integral (see Abramowitz and Steagan, 1979).

Using notations (14) and introducing the plasma frequency

$$\omega_0 = \sqrt{\frac{4\pi n e^2}{m}}, \tag{5.59}$$

we write the ultimate expression for the longitudinal and transverse permeabilities of the collisionless gas plasma at arbitrary values ω and k:

$$\varepsilon_l(k,\omega) = 1 - \sum \frac{\omega_0^2}{(kv_T)^2}\,[Z(s) - 1], \tag{5.60}$$

$$\varepsilon_t(k,\omega) = 1 - \sum \frac{\omega_0^2}{\omega^2}\,Z(s). \tag{5.61}$$

Here the sum is taken over electrons and ions of all kinds. It should be taken into account that the frequencies ω_{0e} and ω_{0i}, as well as thermal velocities v_{Te} and v_{Ti}, are different for electrons and ions because they have different masses (and, possibly, temperatures, if the plasma equilibrium is not total).

Let us note, that both dielectric permeabilities, ε_l and ε_t, have, according to Equation (14), the imaginary parts proportional to $Y(s)$. The imaginary part of the dielectric permeability is responsible for dissipation of electromagnetic energy (see Section 5.3). The possibility of energy dissipation in a collisionless plasma was predicted by L.D. Landau in 1946 and is referred to as **Landau damping** (see Landau (1946)). As seen from the structure of the denominators in Equations (9) and (10), Landau damping results from those particles whose velocity $v_\parallel$ along the wave vector is close to the phase velocity of wave: $v_\parallel \approx v_{ph} = \omega/k$. Such particles, moving for a long time in resonance with the wave, can take energy from the wave. In a nonequilibrium plasma, the inverse process may also occur, that is, enhancement of waves by particles. However, the damping, as well as the enhancement, caused by Landau mechanism, is absent for the waves with $v_{ph} \geq c$ due to relativistic restriction $v < c$ for any particle of finite mass.

We will analyze and demonstrate various applications of derived formulas in problems (see Problems 5.8, 5.9, 5.45, 5.48, and others.) ■

Recommended literature: Maxwell (1989); Landau *et al.* (1984); Landau and Lifshitz (1980); Pamyatnykh and Turov (2000); Lindhard (1954); Ginzburg (1987); Toptygin (2014); Bredov *et al.* (2003); Silin and Rukhadze (1961); Aleksandrov *et al.* (1978); Akhieser *et al.* (1975)

Problems

5.1• Show that Equations (5.12) and (5.13) are invariant with respect to the transformation

$$\boldsymbol{D}' = \boldsymbol{D} + rot\,\boldsymbol{Q}, \qquad \boldsymbol{H}' = \boldsymbol{H} + \frac{1}{c}\frac{\partial \boldsymbol{Q}}{\partial t},$$

where $\boldsymbol{Q}(\boldsymbol{r}, t)$ is any pseudovector. The ambiguity of determination of the vectors $\boldsymbol{D}$ and $\boldsymbol{H}$ reflects the ambiguity of separating the full current into currents of polarization and magnetization, already mentioned in the context of Equation (5.8).

5.2• Write down the tensor $\epsilon_{\alpha\beta}(\boldsymbol{k}, \omega)$, determined by Equation (5.35), through the permeabilities $\varepsilon(k, \omega)$ and $\mu(k, \omega)$:

$$\epsilon_{\alpha\beta} = \varepsilon\delta_{\alpha\beta} + \left(\frac{ck}{\omega}\right)^2 \left(1 - \frac{1}{\mu}\right)\left(\delta_{\alpha\beta} - \frac{k_\alpha k_\beta}{k^2}\right). \tag{5.62}$$

5.3• Let us generalize the material Equation (5.48) for uniform isotropic equilibrium media and introduce the different dielectric permeabilities for the longitudinal and transverse electric fields:

$$D^{\|}(\boldsymbol{k}, \omega) = \varepsilon_{\|}(k, \omega)E^{\|}(\boldsymbol{k}, \omega), \qquad \boldsymbol{D}^{\perp}(\boldsymbol{k}, \omega) = \varepsilon_{\perp}(k, \omega)\boldsymbol{E}^{\perp}(\boldsymbol{k}, \omega),$$
$$\boldsymbol{B}(\boldsymbol{k}, \omega) = \mu_{\perp}(k, \omega)\boldsymbol{H}(\boldsymbol{k}, \omega);$$

both magnetic vectors are transverse and a single permeability has been introduced for them. Show that, the permeabilities $\varepsilon_\perp$ and $\mu_\perp$ can be arbitrarily changed, but the quantity (Kirzhnits, 1987)

$$\eta(k, \omega) = \frac{c^2k^2/\mu_\perp - \omega^2\varepsilon_\perp}{c^2k^2 - \omega^2} = \frac{c^2k^2 - \omega^2\varepsilon_t}{c^2k^2 - \omega^2}, \tag{5.63}$$

is unchanged. It is just the real physical characteristic of medium for transverse electromagnetic field.

5.4• Using the system of Equations (5.1)–(5.4) find the relation between the Fourier-images $\rho_{tot} = \rho_{int} + \rho_{ext}$, $\boldsymbol{j}^{\perp}_{tot} = \boldsymbol{j}^{\perp}_{int} + \boldsymbol{j}^{\perp}_{ext}$ of the total charge and current with $\boldsymbol{j}^{\perp}_{ext}$ and ρ_{ext}, expressing them through the quantities $\varepsilon_{\|}(k, \omega)$ and $\eta(k, \omega)$ introduced in Problem 5.3.

5.5• Express the field vectors $\boldsymbol{E}$ and $\boldsymbol{B}$ through the electromagnetic potentials

$$\boldsymbol{B} = rot\,\boldsymbol{A}, \qquad \boldsymbol{E} = -\frac{1}{c}\frac{\partial \boldsymbol{A}}{\partial t} - grad\,\varphi. \tag{5.64}$$

Choose the Coulomb gauge, $div\,\boldsymbol{A} = 0$. Using the system of Equations (5.1)–(5.4), construct Fourier transforms for the Green's functions $G^{\|}(k, \omega)$ and $G^{\perp}(k, \omega)$ in equilibrium, uniform and isotropic media. This allows one to express the scalar and vector potentials through external charges and currents, $\rho_{ext}(\boldsymbol{k}, \omega)$ and $\boldsymbol{j}_{ext}(\boldsymbol{k}, \omega)$.

5.6• Electric current, generated in metals by low-frequency electric field $\boldsymbol{E}(t) = \boldsymbol{E}_0 e^{-i\omega t}$, is well described by the Ohm law $\boldsymbol{j} = \kappa \boldsymbol{E}$, where κ is the static electric conductivity, and the bound charges make negligible contribution. Show that in this case

$$\varepsilon(\omega) = \varepsilon_l(\omega) = \varepsilon_t(\omega) = 1 + i\frac{4\pi\kappa}{\omega} \qquad (k \to 0,\ \mu = 1). \tag{5.65}$$

5.7•⋆ Calculate the dielectric permeability of substance at high frequencies $\omega \gg \omega_0$, where ω_0 is the frequency of electron motion in atoms (of the order of the transition frequency between atomic levels). Take into account that for the wave period $T = 2\pi/\omega$ a nonrelativistic atomic electron travels along the path $l_T \approx vT$, which is much smaller than atomic size, and that field wave length $\lambda = cT$. Accordingly, during several wave periods the electrons of the substance can be considered freely moving in a weak uniform variable field of given frequency ω.

5.8* On the basis of the results of Example 5.6, study the distribution of the electrostatic potential around a point charge q, which is at rest and immersed in the equilibrium plasma of given average number density and temperature. Compare the derived solution with the result of Problem 1.24.

5.9 Obtain the asymptotic expressions for the dielectric permeabilities, calculated in Example 5.6 (Equations (5.60) and (5.61)) in the limiting cases $s \gg 1$ and $s \ll 1$.

5.10 Calculate the magnetic permeability $\mu(k, \omega)$ of a collisionless gas plasma. Study the limiting cases $\omega \gg kv_T$ and $\omega \ll kv_T$.

5.11 Calculate the dielectric permeabilities $\varepsilon_l(k, \omega)$ and $\varepsilon_t(k, \omega)$ of a collisionless, fully degenerate plasma in a solid body. Ions in the lattice sites can be considered to be at rest. Study the limiting cases $\omega \gg kv_F$ and $\omega \ll kv_F$, where v_F is the velocity of electrons on the Fermi surface.

5.12* The charged oscillator with the eigenfrequency of vibrations ω_0 and the damping constant γ is initially at rest. Then it is affected by an external electric field $\boldsymbol{E}(t)$, which depends on time in an arbitrary way. The wave lengths of field variations are larger than the amplitude of the oscillator vibrations. Calculate the dipole moment of the particle $\boldsymbol{p}(t) = e\boldsymbol{r}(t)$ with respect to the center of vibrations as the integral containing the external field, and calculate the response function (which characterizes the oscillator polarizability).

5.13• The semiclassical dispersion model is based on treating atomic electrons as classical oscillators, with eigenfrequency ω_0 and damping constant γ. In the simplest model, all oscillators may be considered to be equal, and the difference of the local field from the average one is neglected. Calculate the dielectric permeability $\varepsilon(\omega)$ of the medium with an average number density of electrons n. Find the dependence of $\varepsilon'(\omega) = \Re\,\varepsilon(\omega)$ and $\varepsilon''(\omega) = \Im\,\varepsilon(\omega)$ on frequency for the transparent medium ($\gamma \ll \omega$). Use the response function obtained in Problem 5.12.

5.14* Quantum theory of dispersion far from resonances. Let the substance consist of neutral atoms, whose number density is N. Using quantum theory, calculate the dipole moment of a separate atom, induced by a weak external long-wave field. Then obtain the dielectric permeability of the medium neglecting the difference of the local field from the average one. Compare the result with that from Problem 5.13, obtained on the basis of semiclassical model.

5.15* Quantum theory of dispersion near the frequencies of atomic spectral lines. Under the conditions of Problem 5.14 take into account the radiation broadening of spectral lines (see Example I.6.11) and calculate the dielectric permeability at the frequencies close or coinciding with the frequencies of spectral lines.

5.16 An artificial dielectric consists of equal ideally conducting metallic spheres of radius a, randomly distributed in vacuum. The average number of spheres per unit volume is N. Consider an electromagnetic wave propagating in this medium. Neglecting the difference of the field, acting on each sphere, from the average field, find the electric and magnetic permeabilities, ε and μ, of such an artificial dielectric. At which conditions can the dielectric be treated as a continuous medium?

Hint. Electrical and magnetic polarizations of an ideally conducting sphere have been calculated in Problems 2.13 and 4.40.

5.17* Calculate the dielectric permeability of a conducting sphere, assuming that the ions are at rest and the dielectric susceptibility $\alpha_i(\omega)$ of ion medium is given. The energy dissipation should be included by introducing "the friction force" $-\eta\,\boldsymbol{r}$, acting on conduction electrons of number density N. Relate the coefficient η to the statistic electric conductivity κ.

5.18* A gaseous dielectric in statistic equilibrium state at a temperature T consists of molecules with number density N, the principal values of the polarization tensor $\beta^{(1)} = \beta$ and $\beta^{(2)} = \beta^{(3)} = \beta'$ (β and β' depend on frequency ω). The dielectric is placed in a constant and uniform electric field $\boldsymbol{E}_0$. Find the tensor of the dielectric permeability for the electric field, harmonically dependent of time, $\boldsymbol{\mathcal{E}}(t) = \boldsymbol{\mathcal{E}}_0 e^{-i\omega t}$ at $\mathcal{E}_0 \ll E_0$.

5.19 A gaseous dielectric consists of polar molecules, whose electric dipole moment, in the absence of on external field, is p_0. The principal values of the molecule polarization in an alternating field are equal to $\beta^{(1)} = \beta$ and $\beta^{(2)} = \beta^{(3)} = \beta'$, and the axis x_1 is parallel to $\boldsymbol{p}_0$. The dielectric is immersed in the constant electric field $\boldsymbol{E}_0$ and alternating field $\boldsymbol{\mathcal{E}}(t) = \boldsymbol{\mathcal{E}}_0 e^{-i\omega t}$. Neglecting the orientation effect of the alternating field and the orientation effect related to anisotropic polarizability of molecules in constant field, find the tensor of the dielectric permeability of the dielectric for an alternating field at a given temperature T and number density of particles N.

5.20* A system of charges (for instance, a molecule) is in the electromagnetic field varying harmonically. Show that if the dissipation of electromagnetic energy in the system is absent, then its polarizability tensor satisfies the Hermitian condition, $\beta_{ik} = \beta^*_{ki}$.

5.21 Show that, if the tensor β_{ik} is Hermitian, then by choosing the reference frame it can be written in the form $\beta_{ik} = \beta^{(i)}\delta_{ik} + ie_{ikl}g_l$, where e_{ikl} is the unit antisymmetric third-rank tensor (defined in Section I.1.1), $\boldsymbol{g}$ is some real vector (gyration vector)[1], and $\beta^{(i)}$ are the real polarizabilities.

5.22 Find the polarizability of an atom, β_{ik}, in a field of the plane monochromatic wave in the presence of the weak external constant magnetic field $\boldsymbol{H}_0$. Use

1) Media with nonzero gyration vector are referred to as **gyratropic**. The propagation of electromagnetic waves in gyrotropic media is considered in Chapter 6.

the model of the elastically bound electron (see Problem 5.13). Apply the method of iterations. Neglect the action of the magnetic field of the plane wave and on losses of electromagnetic energy. Determine also the gyration vector $\boldsymbol{g}$.

5.23* Using the oscillator model of an atom, find the dielectric permittivity tensor $\varepsilon_{ik}(\omega)$ of the dielectric, containing N atoms per unit volume and immersed in the constant magnetic field $\boldsymbol{H}_0$ of arbitrary strength. Neglect the dissipation of electromagnetic energy and the action of the magnetic field of the plane wave. At which condition does the exact solution turn to the approximate solution of Problem 5.22?

Hint. When integrating the electron equation of motion, introduce the cyclic coordinates

$$x_{\pm 1} = \mp \frac{1}{\sqrt{2}}(x \pm iy), \qquad x_0 = z.$$

5.24 Obtain the dielectric permittivity tensor of a plasma in an external constant magnetic field $\boldsymbol{B}$, if the average number density of electrons is N. Let the ions be at rest. Take into account the losses of energy by introducing “the friction force” $-\eta\dot{\boldsymbol{r}}$.

5.25* Find the dielectric permeability of the ionized gas in a constant magnetic field taking into account the motion of positive ions and keeping in mind that the ion mass is much larger than the electron one. Consider the dependence of the dielectric permeability on ω and compare it with the case where the ions are at rest. The number density of ions and electrons is N.

Hint. Consider the system of equations of motion for an electron and ion. Take into account that the electron is affected by “friction force” $[-\eta(\dot{\boldsymbol{r}} - \dot{\boldsymbol{R}})]$, and the ion to the analogous “friction force” $[-\eta(\dot{\boldsymbol{R}} - \dot{\boldsymbol{r}})]$, where $\boldsymbol{r}$ and $\boldsymbol{R}$ are the position vectors of electron and ion, respectively.

5.26$^{\bullet}$ Consider the infinite homogeneous medium with only one preferential direction (for instance, the direction of an external field). Let T_{ik} be a tensor quantity of this medium, for instance, electric or magnetic permeability. Evidently, the tensor components T_{ik} must be invariant with respect to any rotation of the coordinate frame around the preferential direction. Obtain the restrictions imposed by the requirement of invariance on the form of the tensor T_{ik}.

5.2
Causality Principle and Dispersion Relations

Causality principle. Casual relations between phenomena enable one to determine some general properties of electromagnetic response functions, which are not connected with specific models of a medium. However, it is very important to choose correctly the quantities characterizing the cause and consequence. Following Kirzhnits (1987, 1976), we will assume that any quantity independent of a medium state may be considered a cause. The consequence is determined by the cause and by the electromagnetic response of the medium. The causality principle is that **a cause always precedes a consequence in time**.

As an example consider a capacitor filled with substance of dielectric permittivity ε. Let us apply the alternating harmonic voltage of frequency ω from an external source to capacitor plates. Inside the capacitor the electric field vectors $\boldsymbol{D}$ and $\boldsymbol{E}$ are coupled by a linear relation through the longitudinal dielectric permittivity $\varepsilon(\omega) = \varepsilon_l(\omega)$. The field strength $E = U/L$ is determined by the potential difference on the capacitor plates and by the distance between the plates L. The induction D is adjusted to the given value E and, thus, coupled to it by a causal relation. Since L is a macroscopic quantity, the wave vector $k \approx 2\pi/L$ is rather small, and we deal with the long wave limit, $k \to 0$. It is impossible to take L very small: it will result in the increase of the edge effect, and the response function would not characterize the volume properties of the medium.

Neglecting the spatial dispersion ($k \to 0$), present the integral relation between D and E here in the causal form

$$D(t) = E(t) + 4\pi \int_{-\infty}^{t} \alpha(t - t')E(t')\,dt', \tag{5.66}$$

where $\alpha(t - t')$ is the response function, characterizing the appearance of the electric polarization P under the action of the field E, with the property $\alpha(t - t') = 0$ at $t' > t$. In Fourier space, we obtain from Equation (5.66) $D(\omega) = \varepsilon(\omega)E(\omega)$, where

$$\varepsilon(\omega) = 1 + 4\pi \int_{0}^{\infty} \alpha(\tau)e^{i\omega\tau}\,d\tau \tag{5.67}$$

is the complex dielectric permittivity in the long wave limit.

Analytical properties of the response function at complex frequencies. The electric susceptibility $\alpha(\tau)$ as a function of time behaves differently in dielectrics and conductors. For dielectrics, at $\omega = 0$ we have $\varepsilon(0) = \varepsilon_0 = 1 + 4\pi\alpha_0$, where $\alpha_0 = \int_0^\infty \alpha(\tau)\,d\tau$ is static susceptibility. The time integral converges rapidly, since time instants t', far from the considered instant t, give small contribution to the polarization. The inclusion of the factor $e^{i\omega\tau}$ only enhances the convergence.

In the case of conductors, according to Equation (5.65), the dielectric permeability as a function of frequency has a singularity of the form $4\pi\kappa/\omega$ at $\omega \to 0$, where κ is the static electric conductivity. Hence, the integral $\int_0^\infty \alpha(\tau)\,d\tau$ diverges. However, introduction of any finite real value of the frequency ω allows the integral to converge:

$$\int_{0}^{\infty} \alpha(\tau)\,d\tau = i\frac{\kappa}{\omega}.$$

Our consideration of the integral convergence in Equation (5.67) allows us to derive the important property of ε as a function of complex frequency $\omega = \omega' + i\omega''$: the dielectric permeability has no singularities in the upper half-plane of the complex variable ω, that is, at $\omega'' > 0$. It is because under the integral in Equation

(5.67) there appears the factor $e^{-\omega''\tau}$, $\tau \geq 0$, which provides the convergence. Conversely, in the lower half-plane, that is, at $\omega'' < 0$, the exponent infinitely increases at $\tau \to \infty$, resulting in the appearance of singularities (see dielectric permeabilities calculated in Problems 5.13 and 5.15).

In the upper half-plane, the integral in Equation (5.67) becomes zero at $\omega \to \infty$ along any path, due to oscillations of the factor $e^{i\omega'\tau}$ and because the factor $e^{-\omega''\tau}$ tends to zero (at $\omega'' = 0$ the property under discussion follows from the solution to Problem 5.7, from which we find $\varepsilon(\omega)|_{\omega\to\infty} \to 1$). Thus, in the upper half-plane of the complex ω the dielectric permittivity $\varepsilon(\omega)$ is the analytical function of ω.

Derivation of dispersion relations. Using analyticity of the function $\varepsilon(\omega)$, one can obtain a number of relations between its real $\varepsilon'(\omega)$ and imaginary $\varepsilon''(\omega)$ parts. These functions are connected by differential Cauchy–Riemann relations (see Section 2.2, Equation (2.31)). However, the integral relations, where integrations are performed over real values of frequency, are more convenient. In order to obtain the required relations, we use the Cauchy integral formula

$$\frac{1}{2\pi i}\oint_\Gamma \frac{f(z)\,dz}{z-\omega} = \begin{cases} f(\omega), & \text{if } \omega \text{ is inside the contour,} \\ 0, & \text{if } \omega \text{ is outside the contour,} \end{cases} \tag{5.68}$$

where Γ is a closed contour, and $f(z)$ is any function, analytical inside the contour and on the contour itself.

Let us consider the function, analytical in the upper half-plane, $f(z) = \varepsilon(z) - 1$, and a closed contour Γ, shown in Figure 5.1. Everywhere on the contour and inside it the function $f(z)$ is analytical. Therefore, the Cauchy integral is zero:

$$\oint_\Gamma \frac{[\varepsilon(z) - 1]\,dz}{z-\omega} = 0. \tag{5.69}$$

Integrating over the real axis, to avoid two singular points we should pass over the arcs of small radius $\rho \to 0$. On the external arc of the infinite radius ($R \to \infty$), the integration element decreases faster than $1/z$, because $\varepsilon(z) - 1 \to 0$ at $z \to \infty$. Thus, this region of the contour does not contribute to the integral. The integration over the arcs of small radius gives

$$-i\pi\left(i\frac{4\pi\kappa}{\omega}\right) \quad \text{and} \quad -i\pi[\varepsilon(\omega) - 1],$$

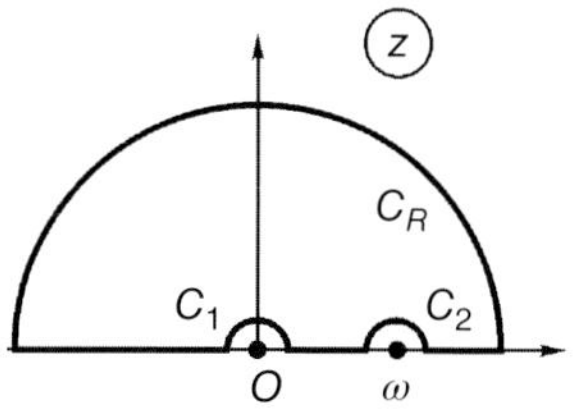

Figure 5.1 Integration path in the complex plane for deriving dispersion relations.

respectively (see Section I.1.3 on the delta-function properties). Therefore, Equation (5.69) takes the form

$$\mathcal{P}\int_{-\infty}^{\infty}\frac{\varepsilon(z)-1}{z-\omega}\,dz - i\pi[\varepsilon(\omega)-1] - \frac{4\pi^2\kappa}{\omega} = 0, \tag{5.70}$$

where the symbol $\mathcal{P}$ before integral means integration in terms of principal value.
Separating the real and imaginary parts, we find

$$\varepsilon'(\omega) = 1 + \frac{1}{\pi}\mathcal{P}\int_{-\infty}^{\infty}\frac{\varepsilon''(z)}{z-\omega}\,dz, \tag{5.71}$$

$$\varepsilon''(\omega) = \frac{4\pi\kappa}{\omega} - \frac{1}{\pi}\mathcal{P}\int_{-\infty}^{\infty}\frac{\varepsilon'(z)-1}{z-\omega}\,dz, \tag{5.72}$$

the dispersion relations, or Kramers[2]–Kronig[3] relations. These relations allow one, for instance, to reconstruct the real part $\varepsilon'(\omega)$ from the imaginary part $\varepsilon''(\omega)$ obtained from experimental data. For this purpose, it is convenient to rewrite Equations (5.71) and (5.72) in terms of integrals over positive frequencies, using the symmetry properties (5.39) of the dielectric permittivity:

$$\varepsilon'(\omega) = 1 + \frac{2}{\pi}\mathcal{P}\int_{0}^{\infty}\frac{z\varepsilon''(z)}{z^2-\omega^2}\,dz, \tag{5.73}$$

$$\varepsilon''(\omega) = \frac{4\pi\kappa}{\omega} + \frac{2\omega}{\pi}\mathcal{P}\int_{0}^{\infty}\frac{\varepsilon'(z)-1}{z^2-\omega^2}\,dz. \tag{5.74}$$

If on the capacitor plates the charges can be controlled and arbitrarily distributed, then, by doing so, the electric induction D is given inside the capacitor. The induction is related with the surface density of charges σ_{ext} by $D = 4\pi\sigma_{ext}$, independently of the substance packed in the capacitor. The induction can be treated as the cause, and the electric field strength as the consequence:

$$E(\omega) = \frac{1}{\varepsilon(\omega)}D(\omega). \tag{5.75}$$

Along with it, the electromagnetic response is determined by the function $\varepsilon^{-1}(\omega)$. It is analytical in the upper half plane of the complex frequency and satisfies the dispersion relations. The quantities $\Re\,\varepsilon^{-1}(\omega)$ and $\Im\,\varepsilon^{-1}(\omega)$ are related by the same Equations (5.71) and (5.72), as the quantities $\varepsilon'(\omega) = \Re\,\varepsilon(\omega)$ and $\varepsilon''(\omega) = \Im\,\varepsilon(\omega)$.

Now let us consider the causality principle in the presence of spatial dispersion. We will assume that independently of a medium state, we can impose and control the charges and currents $\rho_{ext}(\boldsymbol{k},\omega)$ and $\boldsymbol{j}^{\perp}_{ext}(\boldsymbol{k},\omega)$, which are external with respect to the medium. In order to create harmonics with sufficiently short wave lengths, it is necessary to install inside the medium small charged probes or small contours operated by external devices. Many of the electromagnetic quantities (see

2) Kramers Hendric Antony (1894–1952), Dutch physicist-theorist, worked in the field of quantum mechanics and quantum electrodynamics, atomic physics, solid-state physics.

3) Kronig Ralph de Laer (1904–1995), Dutch physicist-theorist, worked in the field of quantum mechanics, spectroscopy, nuclear physics; derived dispersion relations independently of Kramers.

Equations (5.45)–(5.47) and Problem 5.3) are expressed through these charges and currents:

$$E^{\parallel}(\boldsymbol{k},\omega) = -i\frac{4\pi}{k\varepsilon_l(k,\omega)}\,\rho_{ext}(\boldsymbol{k},\omega), \tag{5.76}$$

$$E^{\perp}(\boldsymbol{k},\omega) = -i\frac{4\pi\omega}{\omega^2\varepsilon_t(k,\omega) - c^2k^2}\,\boldsymbol{j}^{\perp}_{ext}(\boldsymbol{k},\omega), \tag{5.77}$$

$$\boldsymbol{B}(\boldsymbol{k},\omega) = -i\frac{4\pi c}{\omega^2\varepsilon_t(k,\omega) - c^2k^2}\,\boldsymbol{k}\times\boldsymbol{j}^{\perp}_{ext}(\boldsymbol{k},\omega), \tag{5.78}$$

$$\rho_{tot}(\boldsymbol{k},\omega) = \frac{1}{\varepsilon_l(k,\omega)}\,\rho_{ext}(\boldsymbol{k},\omega), \tag{5.79}$$

$$\boldsymbol{j}^{\perp}_{tot}(\boldsymbol{k},\omega) = \frac{1}{\eta(k,\omega)}\,\boldsymbol{j}^{\perp}_{ext}(\boldsymbol{k},\omega). \tag{5.80}$$

As a result, the coefficients at ρ_{ext} and $\boldsymbol{j}^{\perp}_{ext}$ in right-hand sides of these equations represent the causal response functions, analytical in the upper half plane of complex frequency. Particularly, at $k \neq 0$ the dispersion relations will be satisfied by not the dielectric permittivity itself $\varepsilon_l(k,\omega)$, but by its reciprocal $\varepsilon_l^{-1}(k,\omega)$:

$$\Re\,\varepsilon_l^{-1}(k,\omega) = 1 + \frac{1}{\pi}\mathcal{P}\int_0^{\infty}\frac{\Im\,\varepsilon_l^{-1}(k,z)}{z^2-\omega^2}\,dz^2, \tag{5.81}$$

$$\Im\,\varepsilon_l^{-1}(k,\omega) = \frac{2\omega}{\pi}\mathcal{P}\int_0^{\infty}\frac{\Re\,\varepsilon_l^{-1}(k,z)-1}{\omega^2-z^2}\,dz. \tag{5.82}$$

Here the static electric conductivity is set equal to zero.

Equations (5.77), (5.78), and (5.80) refer to the transverse field. The response in Equation (5.78) has the simplest form, which satisfies the dispersion relations

$$\Re\,[\omega^2\varepsilon_t(k,\omega) - c^2k^2]^{-1} = \frac{1}{\pi}\mathcal{P}\int_0^{\infty}\frac{dz^2}{z^2-\omega^2}\Im\,[z^2\varepsilon_t(k,z) - c^2k^2]^{-1}, \tag{5.83}$$

$$\Im\,[\omega^2\varepsilon_t(k,\omega) - c^2k^2]^{-1} = \frac{2\omega}{\pi}\mathcal{P}\int_0^{\infty}\frac{dz}{\omega^2-z^2}\Re\,[z^2\varepsilon_t(k,z) - c^2k^2]^{-1}. \tag{5.84}$$

In Equation (5.83), contrary to Equation (5.81), the term 1 is absent in the right-hand side, because $[\omega^2\varepsilon_t(k,\omega) - c^2k^2]^{-1} \to 0$ at $\omega \to \infty$.

In addition to electrodynamic quantities mentioned above, there are many others that satisfy dispersion relations. One can find them in the Kirzhnits papers (1976, 1987).

Suggested Reading: Landau *et al.* (1984); Landau and Lifshitz (1980); Nussenzveig (1976); Leontovich (1961); Ginzburg (1987); Toptygin (2014); Silin and Rukhadze (1961); Bredov *et al.* (2003); Kirzhnits (1976, 1987); Dolgov *et al.* (1982); Dolgov and Maksimov (1981)

Problems

5.27 Show that the function $\varepsilon(k,\omega) = \varepsilon_l(k,\omega)$ has no zeros in the upper half plane of the complex frequency.

5.28 Show that at purely imaginary values of $\omega = i\omega''$ the dielectric permittivity is real: $\varepsilon(i\omega'') = \varepsilon^*(i\omega'')$.

5.29 Show that at real frequencies ω

$$\varepsilon(i\omega) = \frac{2}{\pi}\int_0^\infty \frac{z\varepsilon''(z)}{z^2+\omega^2}\,dz.$$

5.30 In some cases, the function $\varepsilon(t-t')$, which determines the integral relation between the vectors $\boldsymbol{D}$ and $\boldsymbol{E}$, can be represented in the form [4] $\varepsilon(t-t') = f_0\Theta(t-t')\exp[-(t-t')/\tau]$, where f_0 and τ are constants, Θ is the step function, and the non locality (spatial dispersion) is absent. Calculate $\varepsilon(\omega)$.

5.31 Using the Kramers–Kronig dispersions relations, determine the real part of the dielectric permittivity $\varepsilon'(\omega)$ from the known imaginary part $\varepsilon''(\omega)$:

$$\varepsilon''(\omega) = \frac{(\varepsilon_0 - 1)\omega\tau}{1+\omega^2\tau^2},$$

where ε_0 and τ are constants.

5.32• Prove the following sum rules for the imaginary parts of the dielectric permittivity:

$$\int_0^\infty \left[\Im\frac{1}{\varepsilon(\omega)}\right]\omega\,d\omega = -\frac{\pi}{2}\omega_{0e}^2, \qquad \int_0^\infty [\Im\,\varepsilon(\omega)]\omega\,d\omega = \frac{\pi}{2}\omega_{0e}^2,$$

where $\omega_{0e} = (4\pi Ne^2/m)^{1/2}$ is the plasma frequency.

Hint. Take into account that $\varepsilon(\omega)$ and $1/\varepsilon(\omega)$ are analytical functions of ω, $\Re\,\varepsilon(\omega)$ is the even function, and $\Im\,\varepsilon(\omega)$ is the odd function of ω on the real axis. Use the asymptotic expression $\varepsilon(\omega) = 1 - \omega_{0e}^2/\omega^2$ which is valid at $|\omega| \to \infty$ everywhere in the upper half plane of ω.

5.33 Write down the dispersion relations (5.71) and (5.72) in the form of a single integral equation for the complex dielectric permittivity $\varepsilon(\omega)$.

5.34 Write down the dispersion relations (5.71) and (5.72) in the form of one integral equation for the complex electric conductivity $\kappa(\omega) = -(i\omega/4\pi)(\varepsilon(\omega)-1)$.

Hint. Take into account the symmetry properties of the function $\kappa(\omega) = \kappa'(\omega) - i\kappa''(\omega)$: $\kappa'(\omega) = \kappa'(-\omega)$, $\kappa''(\omega) = -\kappa''(-\omega)$.

5.35⋆ Dispersion relations of M.A. Leontovich (1961). Let $R(\boldsymbol{r}-\boldsymbol{r}', t-t')$ be the causal response to an electromagnetic signal in the isotropic medium. Generalize the causality principle and the resulting dispersion relations, derived above in this section, taking into account the requirements of the special theory of relativity (see Chapters I.3 and I.4). The signal can arrive in the

4) Such a form corresponds, for example, to the model of matter, consisting of hard dipoles. It does not take into account the polarizability of electron shells.

point $\boldsymbol{r}$ at the instant t only from points $(\boldsymbol{r}', t')$ which are inside or on the surface of the light cone, directed to the past, that is, satisfying the conditions of relativistic causality $c|t - t'| \geq |\boldsymbol{r} - \boldsymbol{r}'| \geq 0$.

5.3
Energy Relations for Alternating Electromagnetic Field in Media. Longitudinal Electric Oscillations

Dissipation of electromagnetic energy. The action of an alternating field on the matter disturbs its thermodynamic balance and results in relaxation processes, which are accompanied by entropy increase and by dissipation of electromagnetic energy. Let us calculate the energy Q of the monochromatic electromagnetic field dissipated in medium per unit volume per second. We start with the expression

$$Q = \overline{\boldsymbol{j}_{int} \cdot \boldsymbol{E}} = \frac{1}{4}(\boldsymbol{j}_{int} \cdot \boldsymbol{E}^* + \boldsymbol{j}^*_{int} \cdot \boldsymbol{E}), \tag{5.85}$$

where $\boldsymbol{j}_{int}$ is the current in the medium, the bar denotes averaging over the field period. The averaging is required to exclude the part of the power, which alternatively passes from the field to the particles and vice versa. We take the current, induced in the medium, from Equation (5.36). Using Equation (5.85) we obtain the compact expression

$$Q = -\frac{i\omega}{8\pi}\epsilon^{ah}_{\alpha\beta}E^*_\alpha E_\beta = \frac{1}{2}\kappa^h_{\alpha\beta}E^*_\alpha E_\beta, \tag{5.86}$$

where

$$\epsilon^{ah}_{\alpha\beta} = \frac{1}{2}(\epsilon_{\alpha\beta} - \epsilon^*_{\beta\alpha}) \tag{5.87}$$

is the anti-Hermitian part of the complex tensor of the dielectric permittivity. This quantity obeys the relation

$$\epsilon^{ah*}_{\alpha\beta} = -\epsilon^{ah}_{\beta\alpha}, \tag{5.88}$$

while the Hermitian part obeys

$$\epsilon^h_{\alpha\beta} = \frac{1}{2}(\epsilon_{\alpha\beta} + \epsilon^*_{\beta\alpha}) = \epsilon^{h*}_{\beta\alpha}. \tag{5.89}$$

The convolution of the anti-Hermitian tensor $\epsilon^{ah}_{\alpha\beta}$ with the Hermitian tensor $E^*_\alpha E_\beta$ is a purely imaginary quantity:

$$\epsilon^{ah}_{\alpha\beta}E^*_\alpha E_\beta = -\epsilon^{ah*}_{\beta\alpha}E^*_\alpha E_\beta = -(\epsilon^{ah}_{\alpha\beta}E^*_\alpha E_\beta)^*.$$

Therefore, the energy dissipation rate (5.86) is a real quantity. If in the absence of the field the medium is in equilibrium, then the field energy is spent to heat the medium. Owing to the second law of thermodynamics, Q is positive. However, the nonequilibrium medium can transfer the part of its energy to the field and change of the sign of Q.

In isotropic medium, one can use Equation (5.62) for the tensor of the dielectric permittivity. Then we have

$$\epsilon^{ah}_{\alpha\beta} = i\varepsilon''(k,\omega)\,\delta_{\alpha\beta} + \frac{i\mu''(k,\omega)}{|\mu|^2}\left(\frac{ck}{\omega}\right)^2\left(\delta_{\alpha\beta} - \frac{k_\alpha k_\beta}{k^2}\right). \tag{5.90}$$

Next, using Equation (5.41) we find

$$\left(\delta_{\alpha\beta} - \frac{k_\alpha k_\beta}{k^2}\right) E^*_\alpha E_\beta = \boldsymbol{E}^*_\perp \cdot \boldsymbol{E}_\perp = \left(\frac{\omega}{ck}\right)^2 |\mu|^2 \boldsymbol{H}^* \cdot \boldsymbol{H}, \tag{5.91}$$

and from Equation (5.86), (5.90), and (5.91) we obtain

$$Q = \frac{\omega}{8\pi}\{\varepsilon''|\boldsymbol{E}|^2 + \mu''|\boldsymbol{H}|^2\} = \frac{\omega}{4\pi}\{\varepsilon''\overline{E^2} + \mu''\overline{H^2}\}. \tag{5.92}$$

The last expression is valid for real monochromatic vectors $\boldsymbol{E}$ and $\boldsymbol{H}$; the bar denotes averaging over period.

In the equilibrium medium $Q \geq 0$, and the following inequalities have to be fulfilled:

$$\omega\varepsilon''(k,\omega) \geq 0, \qquad \omega\mu''(k,\omega) \geq 0. \tag{5.93}$$

At $\omega \to 0$ for dielectrics, the inequalities turn into equalities, and for the conductors, the first inequality takes the form $\kappa > 0$. At all $\omega > 0$, the imaginary parts $\varepsilon'' > 0$ and $\mu'' > 0$, although in certain regions of frequencies they can be quite small. If

$$\varepsilon'' \ll |\varepsilon'|, \qquad \mu'' \ll |\mu'|, \tag{5.94}$$

the energy dissipation over the period of field variations is small (as compared to the electromagnetic energy stored in the substance). The frequency regions, in which the latter inequalities are valid, are termed the **regions of transparency** of the substance.

Example 5.7

The electromagnetic field ($\boldsymbol{E}(t)$, $\boldsymbol{H}(t)$) acts in the medium during a restricted time, vanishing at $t \to \pm\infty$. Calculate the dissipation of the electromagnetic energy density $\mathcal{Q}$ over time of the field action. Neglect spatial dispersion.

Solution. We proceed from the expression

(1) $$\mathcal{Q} = \int \boldsymbol{j}_{int}(t) \cdot \boldsymbol{E}(t)\,dt,$$

where $\boldsymbol{j}_{int}(t)$ and $\boldsymbol{E}(t)$ are the time real functions. We expand them in Fourier integrals

(2) $$\boldsymbol{j}_{int}(t) = \int \boldsymbol{j}_{int}(\omega)e^{-i\omega t}\frac{d\omega}{2\pi}, \qquad \boldsymbol{E}(t) = \int \boldsymbol{E}(\omega)e^{-i\omega t}\frac{d\omega}{2\pi},$$

with

(3) $$\boldsymbol{j}_{int}(\omega) = \boldsymbol{j}^*_{int}(-\omega), \qquad \boldsymbol{E}(\omega) = \boldsymbol{E}^*(-\omega).$$

Substituting Equation (2) in Equation (1) and using Equation (3), we reduce Equation (1) to the form

(4) $$Q = \int (\boldsymbol{j}_{int}(\omega)\cdot\boldsymbol{E}^*(\omega) + \boldsymbol{j}^*_{int}(\omega)\cdot\boldsymbol{E}(\omega))\,\frac{d\omega}{2\pi}.$$

The integrand coincides in form with that in Equation (5.85) and can be rearranged in an analogous way. As a result, using Equation (5.92), we obtain

$$Q = \int_0^\infty \{\varepsilon''(\omega)|\boldsymbol{E}(\omega)|^2 + \mu''(\omega)|\boldsymbol{H}(\omega)|^2\}\frac{d\omega}{2\pi}. \quad \blacksquare \tag{5.95}$$

Field energy in transparent dispersive medium. Let us rearrange the system of Equations (5.1)–(5.4) in the same way as the Maxwell equations have been rearranged while deriving the energy balance in Section I.2.3 (see Equations (I.2.97) and (I.2.98)). Instead of Equation (2.97) we obtain

$$\frac{\partial}{\partial t}\left(\frac{E^2 + B^2}{8\pi}\right) + \frac{c}{4\pi}\,div\,(\boldsymbol{E}\times\boldsymbol{B}) + \boldsymbol{j}_{int}\cdot\boldsymbol{E} = -\boldsymbol{j}_{ext}\cdot\boldsymbol{E}. \tag{5.96}$$

First consider the fields slowly varying in time and space, where dispersion is unimportant, the permeabilities ε and μ are independent of frequency, and the coupling Equation (1.24) in the isotropic medium are reduced to the proportional dependence

$$\boldsymbol{P} = \frac{\varepsilon - 1}{4\pi}\boldsymbol{E}, \qquad \boldsymbol{M} = \frac{\mu - 1}{4\pi}\boldsymbol{H}. \tag{5.97}$$

Using Equation (5.8) for the current in the medium and Equation (5.1), we find

$$\boldsymbol{j}_{int}\cdot\boldsymbol{E} = \frac{\varepsilon - 1}{4\pi}\boldsymbol{E}\cdot\frac{\partial \boldsymbol{E}}{\partial t} + \frac{c(\mu - 1)}{4\pi}\boldsymbol{E}\cdot rot\,\boldsymbol{H} = \frac{\partial}{\partial t}\left(\frac{(\varepsilon - 1)E^2}{8\pi} - \frac{\mu(\mu - 1)H^2}{8\pi}\right)$$
$$- \frac{c(\mu - 1)}{4\pi}\,div(\boldsymbol{E}\times\boldsymbol{H}).$$

Substituting this equation in Equation (5.96), we obtain

$$\frac{\partial}{\partial t}\left(\frac{\varepsilon E^2}{8\pi} + \frac{\mu H^2}{8\pi}\right) + \frac{c}{4\pi}\,div(\boldsymbol{E}\times\boldsymbol{H}) = -\boldsymbol{j}_{ext}\cdot\boldsymbol{E}. \tag{5.98}$$

This equality may be treated just as the equation of energy balance (I.2.97): the work $-\boldsymbol{j}_{ext}\cdot\boldsymbol{E}$ of external sources, done on the field, is spent for the increase of the density w of electromagnetic energy in the medium,

$$w = \frac{1}{8\pi}(\varepsilon E^2 + \mu H^2), \tag{5.99}$$

and for the formation of the energy flux, whose density is determined by the Poynting vector

$$\gamma = \frac{c}{4\pi}\boldsymbol{E}\times\boldsymbol{H}. \tag{5.100}$$

The last quantity, in nondispersive medium, has the same form as in vacuum. The energy density (5.99), including its thermodynamic meaning, has already been discussed (separately for the electric and magnetic fields) in Chapters 2–4.

If the medium is a conductor with electric conductivity κ, then in the conclusion made above the current $\boldsymbol{j}_{int}$ should be supplemented by the conduction current $\kappa\boldsymbol{E}$. As a result, the term κE^2 will be added to the left-hand side of Equation (5.98). It will describe the Joule dissipation of the field in the medium (see Equation (3.20) from Section 3.1, in which one should set $\boldsymbol{E}_{ext} = 0$). The energy of the external source will be, thus, distributed over three canals.

In the general case of arbitrary dispersive medium, it is impossible to determine unambiguously the quantities which could be interpreted as variation of the electromagnetic energy density in medium and the density of the energy flux.[5] This is possible only in the regions of transparency of medium, in which the conditions (5.94) ($\varepsilon'' \ll \varepsilon'$ and $\mu'' \ll \mu'$) are fulfilled, or, generally,

$$\epsilon^{ah}_{\alpha\beta} \ll \epsilon^{h}_{\alpha\beta}, \tag{5.101}$$

implying that the dissipation is small.

Now consider a wave packet, propagating in such a medium with leading values of $\boldsymbol{k}$ and ω, and with their spreads $|\Delta\boldsymbol{k}| \ll k$ and $\Delta\omega \ll \omega$. The packet can be treated as a plane monochromatic wave with the slowly changing amplitude

$$\boldsymbol{E}(\boldsymbol{r}, t) = \boldsymbol{\mathcal{E}}(\boldsymbol{r}, t)\, e^{i(\mathbf{k}\cdot\mathbf{r}-\omega t)}. \tag{5.102}$$

The alternating amplitude $\boldsymbol{\mathcal{E}}(\boldsymbol{r}, t)$ weakly varies over the distances $\lambda = 2\pi/k$ and times $T = 2\pi/\omega$, that is, it satisfies the conditions

$$\left|\frac{\partial \mathcal{E}}{\partial t}\right| \ll \omega\mathcal{E}, \qquad \left|\frac{\partial \mathcal{E}}{\partial x_\alpha}\right| \ll k\mathcal{E}. \tag{5.103}$$

Let us calculate the current density, induced by the wave packet in the homogeneous medium:

$$\begin{aligned} j^{int}_\alpha(\boldsymbol{r}, t) &= \int \kappa_{\alpha\beta}(\boldsymbol{r}-\boldsymbol{r}', t-t') E_\beta(\boldsymbol{r}', t')\, d^3r'\, dt' \\ &= e^{i(\mathbf{k}\cdot\mathbf{r}-\omega t)} \int \kappa_{\alpha\beta}(\boldsymbol{r}-\boldsymbol{r}', t-t') e^{i\mathbf{k}\cdot(\mathbf{r}'-\mathbf{r})-i\omega(t'-t)} \mathcal{E}_\beta(\boldsymbol{r}', t')\, d^3r'\, dt'. \end{aligned} \tag{5.104}$$

The slowly varying amplitude can be expanded in the Taylor series in the vicinity of the point $(\boldsymbol{r}, t)$:

$$\boldsymbol{\mathcal{E}}(\boldsymbol{r}', t') = \boldsymbol{\mathcal{E}}(\boldsymbol{r}, t) + ((\boldsymbol{r}'-\boldsymbol{r})\cdot\nabla)\, \boldsymbol{\mathcal{E}}(\boldsymbol{r}, t) + (t'-t)\frac{\partial \boldsymbol{\mathcal{E}}(\boldsymbol{r}, t)}{\partial t} + \ldots$$

Substituting this expansion in Equation (5.104), we take the integrals:

$$\int \kappa_{\alpha\beta}(\boldsymbol{r}-\boldsymbol{r}', t-t') e^{i\mathbf{k}\cdot(\mathbf{r}'-\mathbf{r})-i\omega(t'-t)} d^3r' dt' = \kappa_{\alpha\beta}(\boldsymbol{k}, \omega),$$

$$\int \kappa_{\alpha\beta}(\boldsymbol{r}-\boldsymbol{r}', t-t')(\boldsymbol{r}'-\boldsymbol{r}) e^{i\mathbf{k}\cdot(\mathbf{r}'-\mathbf{r})-i\omega(t'-t)} d^3r' dt' = -i\frac{\partial}{\partial \boldsymbol{k}}\kappa_{\alpha\beta}(\boldsymbol{k}, \omega),$$

$$\int \kappa_{\alpha\beta}(\boldsymbol{r}-\boldsymbol{r}', t-t')(t'-t) e^{i\mathbf{k}\cdot(\mathbf{r}'-\mathbf{r})-i\omega(t'-t)} d^3r' dt' = i\frac{\partial}{\partial \omega}\kappa_{\alpha\beta}(\boldsymbol{k}, \omega).$$

5) The problem is discussed in details in the paper Barash and Ginzburg (1976).

Here the electric conductivity $\kappa_{\alpha\beta}(\boldsymbol{k},\omega)$ is connected with the dielectric permittivity by Equation (5.36). Using the obtained relations, we obtain the current density (5.104):

$$j_\alpha^{int}(\boldsymbol{r},t) = \kappa_{\alpha\beta}(\boldsymbol{k},\omega)E_\beta(\boldsymbol{r},t) + e^{i(\mathbf{k}\cdot\mathbf{r}-\omega t)}\left\{\frac{\partial}{\partial\omega}\frac{\omega(\epsilon_{\alpha\beta}-\delta_{\alpha\beta})}{4\pi}\frac{\partial}{\partial t} - \frac{\omega}{4\pi}\frac{\partial\epsilon_{\alpha\beta}}{\partial k_\gamma}\nabla_\gamma\right\}$$
$$\mathcal{E}_\beta(\boldsymbol{r},t). \tag{5.105}$$

Now let us average the product $\boldsymbol{j}_{int}\cdot\boldsymbol{E}$ over the period $T = 2\pi/\omega$ using Equation (5.85) and assuming the following:

(a) let us treat $\partial\boldsymbol{\mathcal{E}}/\partial t$ and $\nabla_\gamma\boldsymbol{\mathcal{E}}$ as constants because they vary only slightly over the period T and length $2\pi/k$ (the conditions (5.103));

(b) let us neglect the anti-Hermitian part of the tensor $\epsilon_{\alpha\beta}$ (condition (5.101)), $\epsilon_{\alpha\beta} \approx \epsilon^h_{\alpha\beta}$. We find

$$\overline{\boldsymbol{j}_{int}\cdot\boldsymbol{E}} = \frac{1}{2}\kappa^h_{\alpha\beta}E^*_\alpha E_\beta + \frac{1}{16\pi}\left(\frac{\partial}{\partial\omega}(\omega\epsilon^h_{\alpha\beta}) - \delta_{\alpha\beta}\right)\frac{\partial}{\partial t}E^*_\alpha E_\beta - \frac{\omega}{16\pi}\nabla\cdot\left(\frac{\partial\epsilon^h_{\alpha\beta}}{\partial\boldsymbol{k}}E^*_\alpha E_\beta\right). \tag{5.106}$$

The last step is in time averaging of Equation (5.96) and substituting Equation (5.106) there. This leads to the equation of energy balance (compare with Equation (I.2.97))

$$-\overline{\boldsymbol{j}_{ext}\cdot\boldsymbol{E}} =$$

$$= \frac{1}{2}\kappa^h_{\alpha\beta}E^*_\alpha E_\beta + \frac{\partial}{\partial t}\left\{\frac{1}{16\pi}\left(\frac{\partial}{\partial\omega}(\omega\epsilon^h_{\alpha\beta})E^*_\alpha E_\beta + \boldsymbol{B}^*\cdot\boldsymbol{B}\right)\right\}$$
$$+ \, div\left\{\frac{c}{16\pi}\left(\boldsymbol{E}\times\boldsymbol{B}^* + \boldsymbol{E}^*\times\boldsymbol{B} - \frac{\omega}{c}\frac{\partial\epsilon^h_{\alpha\beta}}{\partial\boldsymbol{k}}E^*_\alpha E_\beta\right)\right\}. \tag{5.107}$$

The left-hand side gives the power density of external energy source. The first term in the right-hand side describes dissipation of electromagnetic energy (5.86). The quantity under the time derivative is the density w of the electromagnetic energy of the quasi monochromatic field:

$$w = \frac{1}{16\pi}\left(\frac{\partial}{\partial\omega}(\omega\epsilon^h_{\alpha\beta})E^*_\alpha E_\beta + \boldsymbol{B}^*\cdot\boldsymbol{B}\right). \tag{5.108}$$

Finally, under the divergence we have $\boldsymbol{\gamma}$, the flux density of electromagnetic energy in medium (generalization of the Poynting vector),

$$\boldsymbol{\gamma} = \frac{c}{16\pi}\left(\boldsymbol{E}\times\boldsymbol{B}^* + \boldsymbol{E}^*\times\boldsymbol{B} - \frac{\omega}{c}\frac{\partial\epsilon^h_{\alpha\beta}}{\partial\boldsymbol{k}}E^*_\alpha E_\beta\right). \tag{5.109}$$

All mentioned values are averaged over the period.

Suggested Reading: Landau *et al.* (1984); Ginzburg (1987); Toptygin (2014); Pamyatnykh and Turov (2000); Bredov *et al.* (2003); Aleksandrov *et al.* (1978)

Problems

5.36• Show that for an isotropic medium without spatial dispersion, Equations (5.108) and (5.109) take the form

$$w = \frac{1}{16\pi}\left[\frac{\partial}{\partial\omega}(\omega\varepsilon(\omega))\boldsymbol{E}^* \cdot \boldsymbol{E} + \frac{\partial}{\partial\omega}(\omega\mu(\omega))\boldsymbol{H}^* \cdot \boldsymbol{H}\right],\ \boldsymbol{\gamma} = \frac{c}{16\pi}(\boldsymbol{E}^* \times \boldsymbol{H} + \boldsymbol{E} \times \boldsymbol{H}^*), \tag{5.110}$$

where $\varepsilon = \varepsilon'$ and $\mu = \mu'$ are real parts of the electric and magnetic permeabilities. For nondispersive medium, these formulas reduce to Equations (5.99) and (5.100), if we apply them to a quasi monochromatic field and average over period. The first formula (5.110) is called **the Brillouin's formula**.

Hint. Use Equation (5.62) and Maxwell equations for Fourier components.

5.37• Calculate the power density Q of dissipating energy of the electromagnetic field for the case, where the coupling equations have the form (5.49). Use the initial formula $Q = -\oint \overline{\gamma} \cdot d\mathbf{S}$ for the energy, flowing from vacuum into the body through its surface per unit time. The Poynting vector $\boldsymbol{\gamma}$ is averaged over the period of the field. Show that the required expression may be written as

$$Q = -\frac{i\omega}{8\pi}(\varepsilon^{ah}_{\alpha\beta}E^*_\alpha E_\beta + \mu^{ah}_{\alpha\beta}H^*_\alpha H_\beta), \tag{5.111}$$

where the indices ah denote anti-Hermitian parts of the tensors (compare with Equation (5.86)).

5.38• Obtain the analogs of Equations (5.98), (5.99), and (5.100) for slowly varying fields in an anisotropic nondispersive medium with the tensor permeabilities $\varepsilon_{\alpha\beta}$ and $\mu_{\alpha\beta}$. Show that from the requirement of energy conservation one gets the symmetry of the tensors $\varepsilon_{\alpha\beta} = \varepsilon_{\beta\alpha}$ and $\mu_{\alpha\beta} = \mu_{\beta\alpha}$.

5.39•⋆ Obtain the analog of Equation (5.110) for the density of the energy and flux in anisotropic transparent medium without spatial dispersion. Show that the desired quantities can be expressed as

$$w = \frac{1}{16\pi}\left[\frac{\partial}{\partial\omega}(\omega\varepsilon_{\alpha\beta}(\omega))E^*_\alpha E_\beta + \frac{\partial}{\partial\omega}(\omega\mu_{\alpha\beta}(\omega))H^*_\alpha H_\beta\right],$$
$$\gamma = \frac{c}{16\pi}(\boldsymbol{E}^* \times \boldsymbol{H} + \boldsymbol{E} \times \boldsymbol{H}^*), \tag{5.112}$$

where the tensors $\varepsilon_{\alpha\beta}$ and $\mu_{\alpha\beta}$ should be Hermitian, $\varepsilon_{\alpha\beta} = \varepsilon^*_{\beta\alpha}$ and $\mu_{\alpha\beta} = \mu^*_{\beta\alpha}$, to make possible the introduction of the density w of electromagnetic energy.

5.40• Show that the static permeability of dielectric in the long wave limit satisfies the inequality $\varepsilon(0,\omega)|_{\omega\to 0} > 1$.

5.41⋆ Find the region of allowable values of $\varepsilon_l(k,0)$ which is the longitudinal dielectric permittivity in the static limit, $\omega = 0$.

5.42 Find a limitation on the value of the imaginary part $\varepsilon''_t(k,\omega)$ at arbitrary k and ω.

5.43 Express $\mu(k, 0)$ as an integral over all frequencies of the quantity $\eta(k, \omega)$ given by Equation (5.63).

5.44* The oscillatory circuit consists of the capacitor, filled with dispersion medium of permeability $\varepsilon(\omega)$, so that its capacity depends on frequency and on the induction L of the coil. The wave length of oscillations is larger than sizes of the circuit. Calculate the time average of the electromagnetic part of the internal energy U of the circuit, expressing it in terms of the time average of the charge $q(t)$ on capacitor plates.

5.45 In Problems 5.14 and 5.15, the polarizability of an individual atom in the field of a plane monochromatic wave has been calculated by quantum mechanical method. Using these results and the Lorentz–Lorentz model (see Example 1.4), find the dielectric permittivity taking into account the difference of the local field, acting on an atom, from the average field. Consider the frequencies far from the resonance, and the case where the external field frequency is close to one of the transition frequencies in the atom. What limits the model applicability for the alternating field?

5.46 Find the frequency ω_l of longitudinal oscillations near one of the resonance frequencies in medium with the dielectric permittivity, considered in Problem 5.45. Calculate the density w of the electromagnetic energy and the density of the flux of the oscillation energy.

5.47* Find the frequency ω_l of longitudinal oscillations of the collisionless plasma, using the dielectric permittivity, found in Problem 5.9 at $kv_{Te} \ll \omega$, and including its imaginary part. Calculate the density w of electromagnetic energy, the rate Q of its dissipation and the density $\boldsymbol{\gamma}$ of the flux of oscillation energy.

5.48 The energy exchange between the electrons and ions in a gaseous plasma occurs slowly due to the sharp difference of electron and ion masses. Therefore, the electron and ion subsystems can coexist for a long time at different temperatures $T_i \neq T_e$. In this case, these systems are described by Maxwell's distributions with corresponding temperatures. Using the result, obtained in Example 5.6, find the longitudinal dielectric permittivity for the intermediate case of $s_i = \omega/kv_{Ti} \gg 1$ and $s_e = \omega/kv_{Te} \ll 1$. Find the frequencies of longitudinal oscillations of the nonisothermal plasma and their damping. Determine the conditions, at which the oscillation damping is slow.

5.49 Study the spectra of longitudinal oscillations of the collisionless degenerate plasma of a solid body. Calculate the frequency and logarithmic damping decrement of the oscillations. Estimate the frequency of the electron plasma oscillations of typical metals. The dielectric permittivity of the degenerated plasma has been calculated in Problem 5.11.

5.50** The crystals with two ions in the elementary cell (NaCl, LiF, KBr, and others) have the cubic symmetry, and their long wave oscillations ($ka \ll 1$, a being the lattice constant) are isotropic. In such crystals, acoustic and optical oscillations can be excited. In the acoustic oscillations, macroscopic elements of the lattice participate as a whole, that is, positive and negative ions oscillate jointly, in phase. The optical oscillations are those of the sublattice of the negative ions relative to the sublattice of the positive ions. Such oscillations are similar to the plasma oscillations of the electron gas relative to the ion background. They have been considered in Problems 5.45 and 5.49. However, in the ion crystals the elastic forces, holding the ions near the lattice sites, are of great importance. Along with the elastic forces, the macroscopic electric field $\boldsymbol{E}$ and macroscopic polarization $\boldsymbol{P}$, affecting the optical oscillations, can occur.

Find the frequencies of the longitudinal and transverse optical oscillations taking into account the elastic and electric forces. Consider the oscillations as quasi-static, that is, neglect the delay of electromagnetic disturbances, and describe them by the equations of electrostatics. Express the required frequencies in terms of the given frequency ω_0 of the purely elastic oscillations (neglecting the electric forces), and also in terms of two dielectric permeabilities: the static one, ε_0, and that, which includes the polarizability of electron shells of the ions, but not the mutual displacement of positive and negative ions. The latter permeability is commonly denoted as ε_∞ and measured at frequencies much greater than the ion oscillation frequency, but much smaller than transition frequencies in the electron shells of the ions (Born and Huang Kun (1954); Davydov (1976)).

5.51 Calculate in the long wave limit the dielectric permittivity $\varepsilon(\omega)$ of the ion crystal in the frequency region, corresponding to the quasi-static oscillations of the ions. Use the solution to Problem 5.50.

5.4 Magnetic Oscillations and Magnetic Resonance

Paramagnetics. When the magnetic field is applied to a paramagnetic, the magnetic moments $\boldsymbol{m}$ of individual particles, ignoring their interaction, experience the precession, which is described by the equation of motion (see Equations (I.4.77)–(I.4.80))

$$\frac{d\mathbf{m}}{dt} = -\eta \boldsymbol{m} \times \boldsymbol{B}. \tag{5.113}$$

Here, in place of $\boldsymbol{H}$, we write now $\boldsymbol{B}$, the averaged value of the magnetic field strength, and η is the average quantum mechanical value of the coefficient of proportionality between the magnetic and mechanical moments of the particle. The values of η for the orbital and spin moments of individual particles are given in Section I.4.2; for an atom they were calculated in Problem 1.17. For electrons and electron shells of atoms, we have $\eta > 0$, after separating the minus sign in the right-hand side of Equation (5.113). Summing over all particles per unit volume, we obtain the magnetization vector $\boldsymbol{M} = \sum \boldsymbol{m}$ and instead of $\boldsymbol{B}$ we can substitute in Equation (5.113) the sum $\boldsymbol{B} = \boldsymbol{H} + 4\pi\boldsymbol{M}$. As a result, we obtain the equation of motion for the magnetization vector:

$$\frac{d\mathbf{M}}{dt} = -\eta \boldsymbol{M} \times \boldsymbol{H}. \tag{5.114}$$

For paramagnetics, the local magnetic field is slightly different from the average, and in the last equality $\boldsymbol{H}$ is the average (macroscopic) magnetic field.

Equation (5.114) describes only the rotation of the magnetization vector around $\boldsymbol{H}(t)$, but its absolute value remains unchanged: $M^2 = const$. This is because the interaction processes (collisions) of particles, responsible for the relaxation to equilibrium, are not included in the equation. In order to reproduce, at the phenomenological level, the process of magnetization of the paramagnetic, induced by switching on the constant magnetic field $\boldsymbol{H}_0$, it is possible to introduce in Equation (5.114) the relaxation term of the form $-(\boldsymbol{M} - \boldsymbol{M}_0)/\tau$ (compare with (1.40)), where $\boldsymbol{M}_0$ is magnetization in the constant field, τ is the relaxation time, which is constant and determined from experiment. In weak fields $\boldsymbol{M}_0 = \chi_0 \boldsymbol{H}_0$, the static magnetic permeability χ_0 has been calculated in Problems 1.16 and 1.17. If the total magnetic field $\boldsymbol{H}(t) = \boldsymbol{H}_0 + \boldsymbol{h}(t)$ does not coincide in direction with $\boldsymbol{H}_0$, the magnetization will also have the longitudinal and transverse components relative to $\boldsymbol{H}_0$. These components can relax with different times τ_1 and τ_2. Equation (5.114), with account for the relaxation, will take the form

$$\frac{d\boldsymbol{M}}{dt} = -\eta \boldsymbol{M} \times \boldsymbol{H} - \frac{1}{\tau_1}(\boldsymbol{M}_{\parallel} - \boldsymbol{M}_0) - \frac{1}{\tau_2}\boldsymbol{M}_{\perp}. \tag{5.115}$$

the Bloch equation.[6] When a paramagnetic is exposed to an oscillating magnetic field, and the oscillation frequency coincides with the frequency of the magnetization precession, the phenomenon of the magnetic resonance takes place. The magnetization vector and energy of the external field, absorbed by the paramagnetic, significantly increase (see Problem 5.52).

Ferromagnetics. Unlike paramagnetics, in which the interaction between magnetic moments of individual particles is weak, in ferromagnetics the exchange interaction between spins induces an effective magnetic field, which significantly exceeds an average (macroscopic) field (see Section 3.3). Due to this fact, even a relatively weak external field magnetizes ferromagnetic to saturation, which will be assumed in what follows. The motion of the magnetization vector, ignoring dissipative processes, is described by **Landau–Lifshitz equation**[7] (Lifshitz and Pitaevskii, 1978)

$$\frac{\partial \boldsymbol{M}}{\partial t} = -\eta \boldsymbol{M} \times \boldsymbol{H}_{ef}, \tag{5.116}$$

where the effective magnetic field in the isotropic ferromagnetic medium has the form

$$\boldsymbol{H}_{ef} = \boldsymbol{H} + \lambda \boldsymbol{M} + q\nabla^2 \boldsymbol{M}. \tag{5.117}$$

The Weiss field $\lambda \boldsymbol{M}$ drops out of Equation (5.116). The partial derivative on the left-hand side of Equation (5.116) is written to underline the possibility to consider

6) Bloch Felix (1905–1993), Swiss–American physicist, Nobel Prize laureate for the discovery of the nuclear magnetic resonance (NMR), in 1946.
7) Lifshitz Eugeny Mikhailovich (1915-1985) – Soviet physicist-theorist, worked in the field of solid-state physics, gravitation and cosmology. Author (together with L.D. Landau) of the world famous course on theoretical physics.

inhomogeneous magnetization. The last term in Equation (5.117) is just connected with the inhomogeneity in the magnetization distribution, which should be small ($ka \ll 1$, a being the interatomic spacing). In the majority of ferromagnetics, the coefficient $\eta = e_0/mc$, since the spin magnetic moments make the main contribution.

To account for the losses of electromagnetic energy in medium in Equation (5.116), this equation should be supplemented with the dissipative term. One commonly assumes that $\boldsymbol{H}_{ef}$ contains some field of "friction forces," $-p\, d\boldsymbol{M}/dt$, proportional to the rate of magnetization changing. Then Equation (5.116) reads

$$\frac{\partial \boldsymbol{M}}{\partial t} = -\gamma \boldsymbol{M} \times \left(\boldsymbol{H}_{ef} - p\frac{\partial \boldsymbol{M}}{\partial t} \right), \tag{5.118}$$

where p is a parameter of losses. If the losses are weak, and the total magnetic field is the sum of the constant field $\boldsymbol{H}_0$ and alternating field $\boldsymbol{h}(t)$: $\boldsymbol{H} = \boldsymbol{H}_0 + \boldsymbol{h}(t)$, with $|\boldsymbol{h}| \ll H_0$, then Equation (5.118) simplifies (Gurevich and Melkov, 1994):

$$\frac{\partial \boldsymbol{M}}{\partial t} = -\gamma(\boldsymbol{M} \times \boldsymbol{H}) + \omega_r(\chi_0 \boldsymbol{H} - \boldsymbol{M}). \tag{5.119}$$

Here $\chi_0 = M_0/H_0$, $\omega_r = p\gamma^2 M_0^2/\chi_0$ and $M_0 = |\boldsymbol{M}|$ is the saturation magnetization. The Landau–Lifshitz equation is the starting point for solving the problems of ferromagnetic resonance. In radio engineering of high frequencies, the ferromagnetics with very small conductivity (ferrodielectrics, ferrits) are widely used. The propagation of electromagnetic waves in ferrits is considered in Chapters 6 and 7.

Suggested Reading: Landau *et al.* (1984); Lifshitz and Pitaevskii (1980); Pamyatnykh and Turov (2000); White (1983); Toptygin (2014); Gurevich and Melkov (1994); Pake (1962)

Problems

5.52• A paramagnetic is exposed to a constant magnetic field $\boldsymbol{H}_0$ and a small alternating field $\boldsymbol{h}\, e^{-i\omega t}$, $h \ll H_0$, perpendicular to $\boldsymbol{H}_0$. Linearizing the equation of motion (5.115), calculate the magnetic permeability $\chi(\omega)$ of the paramagnetic and the magnetic energy $Q(\omega)$ absorbed per unit time.

5.53 Calculate the characteristic frequencies of the NMR and of the electron paramagnetic resonance (EPR) in the constant field of strength $\boldsymbol{H}_0 = 10^3$ Oe. In the first case, the effect is caused by magnetic moments of atomic nuclei, while in the second case – by magnetic moments of individual electrons or electron shells of atoms.

5.54 Find the equation of motion of the magnetization vector $\boldsymbol{M}$ in the absence of losses in an infinite ferrit medium, magnetized to saturation. The magnetic field $\boldsymbol{H}$ in the medium is constant and homogeneous.

5.55 The ferromagnetic is magnetized to saturation by the constant magnetic field $\boldsymbol{H}_0$. Ignoring the dissipation, find the spectrum of eigenmodes (in the absence of the external alternating field) of magnetization with account for all terms of the effective field (5.117).

5.56 Solve Problem 5.54 taking into account losses. Use the Landau–Lifshitz equation in the form (5.119). Assume that the deviations of $\boldsymbol{M}$ from the direction of $\boldsymbol{H}$ are small, and $\omega_r \ll \omega_0 = \gamma H_0$.

5.57* Consider an infinite ferromagnetic medium immersed in the homogeneous constant field $\boldsymbol{H}_0$ and the high frequency field $\boldsymbol{h}\, e^{-i\omega t}$ ($\boldsymbol{h} = const$). Assuming $h \ll H_0$, neglecting the losses as well as the inhomogeneity of the magnetization, and using the approximation linear in h, find induced oscillations of the magnetization vector $\boldsymbol{M}$. (The eigenmodes, that is, Larmor precession under the action of the constant field $\boldsymbol{H}_0$, will be damped due to losses, existing in all real systems.)

5.58 Using the result of Problem 5.57, find the tensors of magnetic susceptibility and permeability, χ_{ik} and μ_{ik}, for high frequency field. Calculate the dependence of the components of the tensor μ_{ik} on the constant magnetic field H_0 at $M_0 = 160$ G and $\nu = \omega/2\pi = 9375$ MHz ($\lambda = 3.2$ cm). Analyze the resonance character in variation of these quantities. Determine $H_{0\,res}$.

5.59 Solve Problem 5.57 taking into account inhomogeneity of the magnetization (spatial dispersion), but ignoring the losses. Calculate the magnetic permeability tensor in this approximation.

5.60* In the infinite magnetized-to-saturation ferrit medium besides the constant magnetic field $H_0 = H_z$, there is an alternating circularly polarized field: $H_x = h \cos \omega t$, $H_y = h \sin \omega t$, $h = const$. Find the exact solution to the Landau–Lifshitz equation, corresponding to the induced precession of the vector $\boldsymbol{M}$ with the frequency ω of the external field, neglecting the dissipation.

5.61 Solve Problem 5.57 on the forced oscillations of the magnetization vector including the losses. Use the Landau–Lifshitz equation in the form (5.119).

5.62 Using the result of Problem 5.61, find the tensor of the magnetic permeability μ_{ik} for high-frequency field. Obtain the expressions for the real and imaginary parts of the components of this tensor. Plot the dependence of both parts of μ_{ik} on the constant magnetic field for $M_0 = 160$ G, $\nu = \omega/2\pi = 9375$ MHz, $\omega_r = 3 \times 10^9$ rad/s. Determine the resonance field $H_{0\,res}$ (i.e., the value H_0, at which the imaginary parts of μ_{ik} components have maximum).

5.63 Determine the half width ΔH_0 of the resonance curve of the imaginary part of the magnetic permeability tensor, assuming $\omega_r \ll \omega$. The half width of the resonance curve is the distance between two ordinates $\mu'' = \mu_{res}$ and $\mu'' = \mu_{res}/2$.

5.64* Find, neglecting losses, the frequency of the Larmor precession ω_k in the finite ferromagnetic sample of elliptic shape. The sample is in the external homogeneous field $\boldsymbol{H}_0$, applied along one of the ellipsoid axes. Deviations of the magnetization vector $\boldsymbol{M}$ from the equilibrium position are assumed to be small.

Hint. Now the internal field $\boldsymbol{H}_1$ will enter the Landau–Lifshitz equation. It differs from the external field $\boldsymbol{H}_0$ due to the demagnetizing effect of the body shape:

$$\boldsymbol{H} = \boldsymbol{H}_0 - \boldsymbol{H}', \qquad H'_k = 4\pi N_{kl} M_l,$$

where N_{kl} is the tensor of the demagnetizing effect of the body shape (see Problem 3.49).

5.65 Solve Problem 5.64 including the losses. (Take into account only the terms linear in ω_r.)

5.66⋆ Consider forced oscillations in the presence of losses in a small sample of an elliptic shape. Determine the components of the magnetic susceptibility tensor χ_{ik} for high-frequency field, assuming its amplitude h to be smaller than the constant field H_0.

5.67 In certain ferromagnetic media (antiferromagnetics), the resultant magnetization $\boldsymbol{M}$ consists of two parts: $\boldsymbol{M} = \boldsymbol{M}_1 + \boldsymbol{M}_2$, where $\boldsymbol{M}_1$ and $\boldsymbol{M}_2$ are produced by ions in different sites of the lattice which form two magnetic sub-lattices (see Section 3.3). In the equilibrium state, the magnetization vectors $\boldsymbol{M}_1$ and $\boldsymbol{M}_2$ are antiparallel, so that $M = |M_1 - M_2|$. At the precession in the external magnetic field, the vectors $\boldsymbol{M}_1$ and $\boldsymbol{M}_2$ become antiparallel. As a result, either of the two vectors is exposed to the molecular Weiss field (5.117). Find the frequencies of free precession, assuming that $\lambda|M_1 - M_2| \gg H_0$, where H_0 is the external field and λ is the constant of the molecular Weiss field. Let the deviations of the vectors $\boldsymbol{M}_1$ and $\boldsymbol{M}_2$ from their equilibrium positions to be small.

5.5
Electrodynamics of Moving Media

A study of electromagnetic phenomena in moving media is not only of principal significance, but of practical importance as well, since the macroscopic bodies, moving at relativistic velocities, are often encountered in scientific research. Among these are the numerous and actively studied relativistic astrophysical objects (relativistic jets, shock waves, accretion disks, and others) as well, as accelerated plasma clusters and dense beams of energetic electrons in technical acceleration devices.

Maxwell equations and Minkowski coupling equations.[8] Here we will restrict our consideration to media moving with the constant velocity $\boldsymbol{V} = const.$ Such motion does not always occur. We will exemplify in Section 4.3 the description, in the quasi-stationary approximation, of a conducting medium, whose velocity may vary in space and time (magnetic hydrodynamics).

8) Germann Minkowski (1864–1909), outstanding German mathematician and physicist, advanced the idea on the four-dimensional space-time with pseudo-Euclidean geometry and elaborated its consequences; formulated the relativistically invariant equations of an electromagnetic field in moving media (see Chapters I.3 and I.4).

The microscopic Maxwell's equations (1.1)–(1.4), along with the result of their averaging Equations (5.1)–(5.4), hold not only for immobile but for moving media as well. The vector $\boldsymbol{P}$ retains its meaning as the dipole electric moment per unit volume, since Equations (5.6)–(5.8) are also valid in moving media. However, the meaning of the vector $\boldsymbol{M}$ as the specific magnetization is actually limited by the domain of static phenomena (see the inequality (5.9)). Therefore, the vector $\boldsymbol{M}$ should be considered auxiliary quantity, convenient for the parametrization of the current $\boldsymbol{j}_{int}$, produced by particles of the medium. The difference of $\boldsymbol{M}$ from zero indicates the presence of such a current. Thus, we will describe the field in moving media using the Maxwell's Equations (5.11)–(5.14). For convenience of the reader, we present these equations once more:

$$rot\,\boldsymbol{E}(\boldsymbol{r},t) = -\frac{1}{c}\frac{\partial \boldsymbol{B}(\boldsymbol{r},t)}{\partial t}, \tag{5.120}$$

$$rot\,\boldsymbol{H}(\boldsymbol{r},t) = \frac{1}{c}\frac{\partial \boldsymbol{D}(\boldsymbol{r},t)}{\partial t} + \frac{4\pi}{c}\boldsymbol{j}_{ext}(\boldsymbol{r},t), \tag{5.121}$$

$$div\,\boldsymbol{D}(\boldsymbol{r},t) = 4\pi\rho_{ext}(\boldsymbol{r},t), \tag{5.122}$$

$$div\,\boldsymbol{B}(\boldsymbol{r},t) = 0. \tag{5.123}$$

It is convenient to perform further calculations in the four-dimensional tensor form. The foundations of algebra and analysis in the four-dimensional pseudo-Euclidean space-time are outlined in Chapters I.3 and I.4.

Since the vectors $\boldsymbol{E}$ and $\boldsymbol{B}$ are averaged values of the macroscopic field strengths $\boldsymbol{\mathcal{E}}$ and $\boldsymbol{\mathcal{H}}$, they are transformed to another reference frame in the same way, that is, they form the 4-tensor of the electromagnetic field, which is analogous to Equation (I.4.68):

$$F_{ik} = \begin{pmatrix} 0 & E_x & E_y & E_z \\ -E_x & 0 & -B_z & B_y \\ -E_y & B_z & 0 & -B_x \\ -E_z & -B_y & B_x & 0 \end{pmatrix}, \qquad F^{ik} = \begin{pmatrix} 0 & -E_x & -E_y & -E_z \\ E_x & 0 & -B_z & B_y \\ E_y & B_z & 0 & -B_x \\ E_z & -B_y & B_x & 0 \end{pmatrix}. \tag{5.124}$$

For the system of Equations (5.120)–(5.123) be relativistically covariant, the second pair of the vectors should be combined in the similar 4-tensor, that is the tensor of electromagnetic induction:

$$H_{ik} = \begin{pmatrix} 0 & D_x & D_y & D_z \\ -D_x & 0 & -H_z & H_y \\ -D_y & H_z & 0 & -H_x \\ -D_z & -H_y & H_x & 0 \end{pmatrix}, \qquad H^{ik} = \begin{pmatrix} 0 & -D_x & -D_y & -D_z \\ D_x & 0 & -H_z & H_y \\ D_y & H_z & 0 & -H_x \\ D_z & -H_y & H_x & 0 \end{pmatrix}. \tag{5.125}$$

The analogous tensor can be introduced also for the values ($\boldsymbol{P}$, $\boldsymbol{M}$). The Maxwell equations written in terms of the tensors F_{ik} and H^{ik}, have the form (I.4.114) and (I.4.112):

$$\partial_l F_{ik} + \partial_i F_{kl} + \partial_k F_{li} = 0, \qquad \partial_k H^{ik} = -\frac{4\pi}{c} j^i_{ext}, \tag{5.126}$$

where j^i_{ext} is the 4-vector of the external current.

The coupling equations between vectors $\boldsymbol{E}$ and $\boldsymbol{B}$ (retaining clear meaning in moving media) and the auxiliary vectors $\boldsymbol{D}$ and $\boldsymbol{H}$ can be obtained with the aid of Lorentz transformation. Let the media be isotropic, and the electric and magnetic permeabilities have no dispersion. Then in the rest frame of the medium, the coupling equations have the form (5.16):

$$\boldsymbol{D}' = \varepsilon \boldsymbol{E}', \qquad \boldsymbol{B}' = \mu \boldsymbol{H}', \tag{5.127}$$

where ε and μ are the permeabilities in the medium at rest. Let us transform the field vectors in the laboratory reference frame using Equation (I.4.69) and obtain **Minkowski coupling equations**

$$\boldsymbol{D} + \frac{1}{c}\boldsymbol{V} \times \boldsymbol{H} = \varepsilon\left(\boldsymbol{E} + \frac{1}{c}\boldsymbol{V} \times \boldsymbol{H}\right), \qquad \boldsymbol{B} - \frac{1}{c}\boldsymbol{V} \times \boldsymbol{E} = \mu\left(\boldsymbol{H} - \frac{1}{c}\boldsymbol{V} \times \boldsymbol{D}\right). \tag{5.128}$$

In the nonrelativistic approximation $V \ll c$, and in the correction Terms, one can use Equation (5.127). This results in

$$\boldsymbol{D} = \varepsilon \boldsymbol{E} + \frac{\varepsilon\mu - 1}{c}\boldsymbol{V} \times \boldsymbol{H}, \qquad \boldsymbol{B} = \mu \boldsymbol{H} - \frac{\varepsilon\mu - 1}{c}\boldsymbol{V} \times \boldsymbol{E}, \qquad V \ll c. \tag{5.129}$$

The conditions on the interface of two media can also be derived by converting the boundary conditions (5.15) for the medium at rest to the laboratory reference frame. Consider some element of the boundary surface as locally plane and introduce the co-moving reference frame S', in which the motion of this element along the normal $\boldsymbol{n}$ to its plane is absent. The relative velocity of the two frames in this case has the value $\boldsymbol{v} = V_n \boldsymbol{n}$. In the frame S' the boundary conditions have the form (5.15). When converting, using Equation (I.4.69), to the laboratory frame, the conditions for the normal components would remain unchanged:

$$\boldsymbol{n} \cdot (\boldsymbol{D}_2 - \boldsymbol{D}_1) = 0, \qquad \boldsymbol{n} \cdot (\boldsymbol{B}_2 - \boldsymbol{B}_1) = 0 \tag{5.130}$$

(We assume that the surface charges and currents are absent). The conditions for the tangential components will take the form

$$\boldsymbol{n} \times (\boldsymbol{E}_2 - \boldsymbol{E}_1) = \frac{V_n}{c}(\boldsymbol{B}_2 - \boldsymbol{B}_1), \qquad \boldsymbol{n} \times (\boldsymbol{H}_2 - \boldsymbol{H}_1) = -\frac{V_n}{c}(\boldsymbol{D}_2 - \boldsymbol{D}_1). \tag{5.131}$$

In the nonrelativistic approximation in the corrections terms, one can set $E_{\tau 1} = E_{\tau 2}$ and $H_{\tau 1} = H_{\tau 2}$, and the conditions (5.132) will be somewhat simplified:

$$\boldsymbol{n} \times (\boldsymbol{E}_2 - \boldsymbol{E}_1) = \frac{V_n}{c}(\mu_2 - \mu_1)\boldsymbol{H}_\tau, \qquad \boldsymbol{n} \times (\boldsymbol{H}_2 - \boldsymbol{H}_1) = -\frac{V_n}{c}(\varepsilon_2 - \varepsilon_1)\boldsymbol{E}_\tau. \tag{5.132}$$

Problems

5.68• Using the Minkowski coupling equations (5.128), express the induction vectors $\boldsymbol{D}$ and $\boldsymbol{H}$ in terms of the averaged vectors of the field strength $\boldsymbol{E}$ and $\boldsymbol{B}$. Show that the desired equations have the form

$$\boldsymbol{D} = \varepsilon \boldsymbol{E} + \frac{\kappa\gamma^2}{\mu}[\beta^2 \boldsymbol{E} - \boldsymbol{\beta}(\boldsymbol{\beta}\cdot\boldsymbol{E}) + \boldsymbol{\beta}\times\boldsymbol{B}],$$

$$\boldsymbol{H} = \frac{1}{\mu}\boldsymbol{B} + \frac{\kappa\gamma^2}{\mu}[-\beta^2 \boldsymbol{B} + \boldsymbol{\beta}(\boldsymbol{\beta}\cdot\boldsymbol{B}) + \boldsymbol{\beta}\times\boldsymbol{E}], \tag{5.133}$$

where we have introduced the notations

$$\kappa = \varepsilon\mu - 1, \qquad \boldsymbol{\beta} = \boldsymbol{V}/c, \qquad \gamma = (1-\beta^2)^{-1/2}. \tag{5.134}$$

5.69• Show that the Minkowski coupling equations (5.133) can be written in the covariant four-dimensional form

$$H^{ik} = \epsilon^{ikmn} F_{mn}, \qquad \epsilon^{ikmn} = \mu^{-1}(g^{im} + \kappa u^i u^m)(g^{kn} + \kappa u^k u^n). \tag{5.135}$$

Here ϵ^{ikmn} is the 4-tensor of the electric and magnetic permeabilities, the metric tensor is determined by Equation (I.3.18) ($g^{ik} = g_{ik}$), u^i is the dimensionless vector of the 4-velocity of the medium:

$$u^i = (\gamma, \gamma \boldsymbol{V}/c), \qquad u^i u_i = \gamma^2 - \beta^2\gamma^2 = 1. \tag{5.136}$$

5.70• In the co-moving frame of reference, the electric and magnetic permeabilities of the medium, $\varepsilon(\omega')$ and $\mu(\omega')$, depend on frequency, that is, the time dispersion takes place. Write down the Minkowski equations (5.128) and (5.133), for the Fourier transforms of the field vectors ($\boldsymbol{E}_{\mathbf{k}\omega}$, $\boldsymbol{B}_{\mathbf{k}\omega}$, etc.) in the laboratory frame.

Electromagnetic potentials in moving media. This subject is described following the reviews of Bolotovskii and Stolyarov (1974, 1983). The four-dimensional form of the Maxwell equations and of the coupling equations significantly simplifies the mathematical treatment. We express the field tensors in terms of the four-dimensional vector-potential A_m according to Equation (I.4.52):

$$F_{mn} = \partial_m A_n - \partial_n A_m, \qquad H^{ik} = \epsilon^{ikmn}(\partial_m A_n - \partial_n A_m), \tag{5.137}$$

where the coupling equation (5.135) is used. The first equation (5.126) is fulfilled identically, and from the second one we obtain

$$(g^{im} + \kappa u^i u^m)\left\{[-g^{kn}\partial_k\partial_n - \kappa(u^k\partial_k)^2]A_m + \partial_m(\partial_k A^k + \kappa u^k u^n \partial_k A_n)\right\} = -\frac{4\pi\mu}{c} j^i_{ext}. \tag{5.138}$$

Here $-g^{kn}\partial_k\partial_n = \triangle - (\partial/c\partial t)^2$ is the D'Alembert operator (I.4.38).

Because the choice of the electromagnetic potential is ambiguous, we can impose an additional condition. It is convenient to choose the potential in such a way to simplify Equation (5.138):

$$\partial_k A^k + \kappa u^k u^n \partial_k A_n \equiv (g^{kn} + \kappa u^k u^n)\partial_k A_n = 0. \tag{5.139}$$

The last equation is the generalization of the Lorentz condition (I.2.105), $\partial_k A^k = 0$, on the case of moving media. Using this condition, let us write the equation for the potential (5.138) in the form

$$\hat{L}(g^{im} + \kappa u^i u^m)A_m = -\frac{4\pi\mu}{c} j^i_{ext}, \qquad \text{where operator} \qquad \hat{L} = -g^{kn}\partial_k\partial_n - \kappa(u^k\partial_k)^2. \tag{5.140}$$

The more convenient form can be obtained with introducing the 4-tensor of second rank,

$$S_{ik} = g_{ik} - \frac{\kappa}{1+\kappa} u_i u_k. \tag{5.141}$$

Multiplying Equation (5.138) by S_{ni}, we obtain it in the more compact form

$$\hat{L}A_n = -\frac{4\pi\mu}{c} S_{ni} j^i_{ext}. \tag{5.142}$$

This equation has covariant form. Therefore, it holds in any inertial frame of reference. The operator $\hat{L}$, as follows from its definition (5.140), is a relativistic invariant. Its conversion to another inertial reference frame reduces to the Lorentz transformations of the 4-vectors x^i and u^k, in particular,

$$\hat{L} = \triangle' - \frac{\varepsilon\mu}{c^2}\frac{\partial^2}{\partial t'^2} = \triangle - \frac{1}{c^2}\frac{\partial^2}{\partial t^2} - \frac{\kappa\gamma^2}{c^2}\left(\frac{\partial}{\partial t} + \boldsymbol{V}\cdot\nabla\right)^2. \tag{5.143}$$

The first expression relates to the co-moving frame of reference, and the operator acts on the primed coordinates. The second expression relates to the laboratory frame.

Example 5.8

From Equations (5.138) and (5.142) obtain the equations in the three-dimension form for the vector $\boldsymbol{A}(\boldsymbol{r}, t)$ and scalar $\varphi(\boldsymbol{r}, t)$ potentials in the moving medium, and the additional condition for them.

Solution. Using the coupling equations between $A^k = (\varphi, \boldsymbol{A})$ and $j^k = (c\rho, \boldsymbol{j})$, we obtain

$$\hat{L}\boldsymbol{A} = -\frac{4\pi\mu}{c}\left\{\boldsymbol{j}_{ext} + \boldsymbol{V}\frac{\kappa\gamma^2}{(1+\kappa)c^2}(\boldsymbol{V}\cdot\boldsymbol{j}_{ext} - c^2\rho_{ext})\right\},$$

$$\hat{L}\varphi = -\frac{4\pi\mu}{c}\left\{c\rho_{ext} + \frac{\kappa\gamma^2}{(1+\kappa)c}(\boldsymbol{V}\cdot\boldsymbol{j}_{ext} - c^2\rho_{ext})\right\}. \tag{5.144}$$

$$\nabla\cdot\boldsymbol{A} + \frac{1}{c}\frac{\partial\varphi}{\partial t} - \frac{\kappa\gamma^2}{c^2}\left(\frac{\partial}{\partial t} + \boldsymbol{V}\cdot\nabla\right)(\boldsymbol{V}\cdot\boldsymbol{A} - c\varphi) = 0. \tag{5.145}$$

Here the operator $\widehat{L}$ is given by the second equation (5.143). At $\boldsymbol{V} = 0$, the previous equations take the form

$$\left(\triangle - \frac{\varepsilon\mu}{c^2}\frac{\partial^2}{\partial t^2}\right)\boldsymbol{A} = -\frac{4\pi\mu}{c}\boldsymbol{j}_{ext}, \qquad \left(\triangle - \frac{\varepsilon\mu}{c^2}\frac{\partial^2}{\partial t^2}\right)\varphi = -\frac{4\pi}{\varepsilon}\rho_{ext} \tag{5.146}$$

with the additional equation is

$$\nabla\cdot\boldsymbol{A} + \frac{\varepsilon\mu}{c}\frac{\partial\varphi}{\partial t} = 0. \qquad \blacksquare \tag{5.147}$$

Example 5.9

Introduce, by analogy with Equationa (5.40)–(5.46), the electric and magnetic Hertz vectors in a moving medium. Obtain the equations for them and find their coupling with the electromagnetic four-dimensional potential A^k. Present also the given coupling in three-dimensional vector form.

Solution. First we write down the equations for the Hertz vectors in the medium at rest with the permeabilities ε and μ but without dispersion. The equations for the potentials (5.146) and the additional condition (5.147) will be fulfilled, if the coupling of the potentials with the Hertz vectors will have the form

(1)
$$\boldsymbol{A} = \frac{1}{c}\frac{\partial\boldsymbol{Z}^{(e)}}{\partial t} + \nabla\times\boldsymbol{Z}^{(m)}, \qquad \varphi = -\frac{1}{\varepsilon\mu}\nabla\cdot\boldsymbol{Z}^{(e)}.$$

Here the Hertz vectors themselves should obey to the equations

(2)
$$\left(\triangle - \frac{\varepsilon\mu}{c^2}\frac{\partial^2}{\partial t^2}\right)\boldsymbol{Z}^{(e)} = -4\pi\mu\boldsymbol{p}, \qquad \left(\triangle - \frac{\varepsilon\mu}{c^2}\frac{\partial^2}{\partial t^2}\right)\boldsymbol{Z}^{(m)} = -4\pi\mu\boldsymbol{m},$$

where $\boldsymbol{p}$ and $\boldsymbol{m}$ are the vectors of the electric and magnetic polarization, which determine **the external current**

(3)
$$\boldsymbol{j}_{ext} = \frac{\partial\boldsymbol{p}}{\partial t} + c\nabla\times\boldsymbol{m}, \qquad \rho_{ext} = -\nabla\cdot\boldsymbol{p}.$$

In order to write these relations in an arbitrary inertial frame, one should present them the relativistically covariant form. For this purpose we note, that the transformation of the quantities $(\boldsymbol{p},\ \boldsymbol{m})$ is the same as for the polarization vectors $(\boldsymbol{P},\ \boldsymbol{M})$, that is, they constitute an antisymmetric 4-tensor of second rank:

(4)
$$p^{ik} = \begin{pmatrix} 0 & p_x & p_y & p_z \\ -p_x & 0 & -m_z & m_y \\ -p_y & m_z & 0 & -m_x \\ -p_z & -m_y & m_x & 0 \end{pmatrix}.$$

The Hertz vectors $\boldsymbol{Z}^{(e)}$ and $\boldsymbol{Z}^{(m)}$ are also combined in the single 4-tensor

(5)
$$Z^{ik} = \begin{pmatrix} 0 & Z_x^{(e)} & Z_y^{(e)} & Z_z^{(e)} \\ -Z_x^{(e)} & 0 & -Z_z^{(m)} & Z_y^{(m)} \\ -Z_y^{(e)} & Z_z^{(m)} & 0 & -Z_x^{(m)} \\ -Z_z^{(e)} & -Z_y^{(m)} & Z_x^{(m)} & 0 \end{pmatrix}.$$

Now Equation (2) can be written in covariant form, which is applicable in any inertial frame of reference:

(6) $$\hat{L} Z^{ik} = -4\pi \mu p^{ik}.$$

The operator $\hat{L}$ is given by Equation (5.143), and for the medium at rest leads to Equation (2). In the laboratory frame, one should use Equation (5.143) with $\boldsymbol{V} \neq 0$.

The 4-potential $A^i = (\varphi, \boldsymbol{A})$ is also expressed in terms of Z^{ik} in covariant form using the tensor (5.141):

(7) $$A^i = -S^i_l \partial_k Z^{lk} = \left(\frac{\kappa}{1+\kappa} u^i u_l - \delta^i_l \right) \partial_k Z^{lk}.$$

At $u^i = (1, 0)$, that is, in the co-moving frame of reference, Equation (7) gives Equation (1). From the laboratory system of coordinates, we obtain

(8) $$\boldsymbol{A} = \frac{1}{c} \frac{\partial \boldsymbol{Z}^{(e)}}{\partial t} + \nabla \times \boldsymbol{Z}^{(m)} + \frac{\kappa \gamma^2}{(1+\kappa)c^2}$$

$$\boldsymbol{V} \left\{ c \nabla \cdot \boldsymbol{Z}^{(e)} + \frac{1}{c} \frac{\partial}{\partial t} (\boldsymbol{V} \cdot \boldsymbol{Z}^{(e)}) + (\boldsymbol{V} \cdot [\nabla \times \boldsymbol{Z}^{(m)}]) \right\},$$

(9) $$\varphi = -\left(1 - \frac{\kappa \gamma^2}{1+\kappa} \right) \nabla \cdot \boldsymbol{Z}^{(e)} + \frac{\kappa \gamma^2}{(1+\kappa)c} \boldsymbol{V} \cdot \left\{ \frac{1}{c} \frac{\partial \boldsymbol{Z}^{(e)}}{\partial t} + \nabla \times \boldsymbol{Z}^{(m)} \right\}.$$ ■

Example 5.10

Construct the retarded Green's functions for Equation (5.142) in a moving medium without dispersion ($\varepsilon \mu > 1$). Compare the obtained solution with the Green's function in vacuum (see Equation (I.5.9)).

Solution. We seek the solution of the equation

(1) $$\hat{L}\, G(\boldsymbol{r}, t) = -4\pi \mu\, \delta(\boldsymbol{r})\, \delta(t),$$

satisfying the condition

(2) $$G(\boldsymbol{r}, t) = 0 \qquad \text{at} \qquad t < 0.$$

The operator $\hat{L}$ is defined by Equation (5.143); it is an invariant generalization of the D'Alembert operator to the case of moving medium. As in Section I.5.1, we use the Fourier transform of Equation (1) in space and time:

(3) $$\widetilde{G}(\boldsymbol{k}, \omega)[g_{il} k^i k^l - \kappa (u_l k^l)^2] = 4\pi \mu.$$

The formal solution, analogous to Equation (5.6), has the form

(4) $$\widetilde{G} = \frac{4\pi \mu}{k^2 - (\omega/c)^2 - (\kappa \gamma^2 / c^2)(\omega - \boldsymbol{k} \cdot \boldsymbol{V})^2}.$$

The inverse Fourier transform

(5) $$G(\boldsymbol{r}, t) = \int \widetilde{G}(\boldsymbol{k}, \omega) e^{i(\mathbf{k} \cdot \mathbf{r} - \omega t)} \frac{d^3 k}{(2\pi)^3} \frac{d\omega}{2\pi}$$

should be performed in such a way that the poles of the integrand, when integrated over frequency, were sidestepped from above (see Figure I.5.1). We will obtain the same result, if we add to the denominator a small imaginary term by changing $\omega \to \omega + i\delta$, with $\delta \to +0$ (see Sohotsky's formula (I.1.223)). Then the poles of the integrand are displaced to the lower half plane. In the calculations below, this procedure will be performed implicitly using the tabulated integrals of cylindrical functions. Let us direct the axis Oz along the medium velocity $\boldsymbol{V}$ and introduce the cylindrical coordinates, in which $\boldsymbol{k} \cdot \boldsymbol{V} = k_{\|} V$, with $\boldsymbol{k} \cdot \boldsymbol{r} = k_{\|} z + k_{\perp} r_{\perp} \cos\varphi$. Here φ is the angle between the vectors $\boldsymbol{r}_{\perp}$ and $\boldsymbol{k}_{\perp}$, perpendicular to $\boldsymbol{V}$. Integrating over $d\varphi$ and using Equation (I.1.150), we obtain from Equation (4)

(6) $$G(\boldsymbol{r}, t) = \frac{\mu}{2\pi^2} \int \frac{J_0(k_{\perp} r_{\perp}) \exp(ik_{\|} z - i\omega t) k_{\perp} dk_{\perp} dk_{\|} d\omega}{k_{\perp}^2 + k_{\|}^2 - (\omega/c)^2 - (\kappa\gamma^2/c^2)(\omega - k_{\|} V)^2}.$$

Here J_0 is the Bessel function. We integrate over $dk_{\perp}$ using equation (Gradshtein and Ryzhik, 2007)

(7) $$\int_0^{\infty} \frac{J_0(k_{\perp} r_{\perp}) k_{\perp}\, dk_{\perp}}{k_{\perp}^2 + a^2} = K_0(a r_{\perp}), \qquad \text{where} \qquad \Re\, a > 0,$$

K_0 is a modified Bessel function (Macdonald function (I.1.163)). In our case,

(8) $$a = \gamma \sqrt{(1 - \varepsilon\mu\beta^2) k_{\|}^2 + 2\beta(\varepsilon\mu - 1) k_{\|} (\omega/c) - (\varepsilon\mu - \beta^2)(\omega^2/c^2)} > 0.$$

In order to perform two other integrations, we change of variables:

(9) $$x = k_{\|}, \qquad \alpha = \omega - \frac{\beta c(\varepsilon\mu - 1)}{\varepsilon\mu - \beta^2} k_{\|}, \qquad \text{with} \qquad dk_{\|}\, d\omega = dx\, d\alpha.$$

Let us use Equation (I.1.163)

$$K_0(z) = \frac{i\pi}{2} H_0^{(1)}(iz)$$

and the formula from the reference book of Gradshtein and Ryzhik (2007):

(10) $$\int_{-\infty}^{\infty} e^{ixz'} H_0^{(1)}(g\sqrt{b^2 - x^2}) dx = -\frac{2i}{\sqrt{g^2 + z'^2}} e^{ib\sqrt{g^2 + z'^2}},$$

where $H_0^{(1)}$ is the Hankel function of the first kind (I.1.154),

(11) $$z' = z - \frac{(\varepsilon\mu - 1)Vt}{\varepsilon\mu - \beta^2}, \qquad g = \frac{\sqrt{\varepsilon\mu}\, r_{\perp}}{\gamma\sqrt{\varepsilon\mu - \beta^2}}, \qquad b = \frac{(\varepsilon\mu - \beta^2)\gamma^2}{c\sqrt{\varepsilon\mu}} \alpha.$$

The latter integration is performed with the aid of Equation (I.1.219):

(12) $$\int_{-\infty}^{\infty} \exp\left\{ -i\alpha \left(t - \frac{(\varepsilon\mu - \beta^2)\gamma^2}{c\sqrt{\varepsilon\mu}} \sqrt{g^2 + z'^2} \right) \right\} d\alpha = 2\pi\delta\left(t - \frac{(\varepsilon\mu - \beta^2)\gamma^2}{c\sqrt{\varepsilon\mu}} \sqrt{g^2 + z'^2} \right).$$

Collecting the previous results, we can write the final expression for the Green's function:

$$G(\mathbf{r}, t) = \frac{\mu}{\sqrt{g^2 + z'^2}} \delta \left(t - \frac{(\varepsilon\mu - \beta^2)\gamma^2}{c\sqrt{\varepsilon\mu}} \sqrt{g^2 + z'^2} \right). \tag{5.148}$$

This function satisfies the condition (2), which follows directly from the structure of the delta-function argument. At $V = \beta = 0$ and $\gamma = 1$ we have $\sqrt{g^2 + z'^2} = r$. Then the Green's function takes the form

(13) $$G(r, t) = \frac{1}{r} \delta \left(t - \frac{r}{c/\sqrt{\varepsilon\mu}} \right).$$

This corresponds to Equation (I.5.9) for the vacuum Green's function, if the light velocity c is replaced by the phase velocity $v_{ph} = c/\sqrt{\varepsilon\mu}$ of electromagnetic waves in the dielectric. At $\varepsilon = \mu = 1$ and arbitrary V, the drag of the electromagnetic field by moving medium is absent, and Equation (13) reproduces Equation (I.5.9). The Green's function (13) can be presented in a more compact form, if the argument of the delta-function is transformed using Equation (I.1.209). Such a transformation results in

$$G(\mathbf{r}, t) = \frac{\mu}{2R_*} \{ (1 + sign\, \tau_1)\, \delta(t - \tau_1) + (1 + sign\, \tau_2)\, \delta(t - \tau_2) \}, \tag{5.149}$$

where $sign\, x = x/|x|$,

$$R_* = (z^2 + (1 - \varepsilon\mu\beta^2)\gamma^2 r_\perp^2)^{1/2}, \quad \tau_1 = \frac{-(\varepsilon\mu - 1)\beta z + (1 - \beta^2)\sqrt{\varepsilon\mu} R_*}{c(1 - \varepsilon\mu\beta^2)},$$

$$\tau_2 = \frac{(\varepsilon\mu - 1)\beta z + (1 - \beta^2)\sqrt{\varepsilon\mu} R_*}{c(\varepsilon\mu\beta^2 - 1)}. \tag{5.150}$$

The factors $(1 + sign\, \tau)$ vanish at $\tau < 0$ and thus ensure the retardation condition (2). They should also be zero in the cases, in which R_* is imaginary. ■

Electromagnetic forces, acting on medium in alternating field. Energy–momentum tensors of Abraham and Minkowski.[9] We will outline these problems (their discussion is not finished yet) for the case of nonmoving (or moving with nonrelativistic velocity $V \ll c$) nondispersive isotropic medium following the review papers Skobel'tsyn (1973); Ginzburg (1973); Ginzburg and Ugarov (1976), teaching courses and monographs Landau *et al.* (1984); Tamm (1976); Ginzburg (1987). In the present section, we will consider nondispersive medium.

Consider an electromagnetic force acting on medium at rest. This problem can be conveniently solved based on the momentum balance equation. We obtain the required equation using the Maxwell equations and the coupling equation (5.127). We assume that the permeabilities ε and μ depend only on the density τ of matter (temporal and spatial dispersions are absent).

9) Abraham Max (1875–1922), German physicist-theoretician, author of many works on classical electrodynamics and electron theory in matter.

Let us take the vector products of Equation (5.120) and $\boldsymbol{D}$ and also of Equation (5.121) and $\boldsymbol{B}$, and sum them. This gives

$$-\frac{1}{4\pi c}\frac{\partial}{\partial t}[\boldsymbol{D}\times\boldsymbol{B}]+\frac{1}{4\pi}\{[\nabla\times\boldsymbol{E}]\times\boldsymbol{D}+[\nabla\times\boldsymbol{H}]\times\boldsymbol{B}\}=\frac{1}{c}\boldsymbol{j}_{ext}\times\boldsymbol{B}, \quad (5.151)$$

or, in projections on coordinate axes,

$$-\frac{1}{4\pi c}\frac{\partial}{\partial t}[\boldsymbol{D}\times\boldsymbol{B}]_\alpha+\frac{1}{4\pi}\left\{\frac{\partial E_\alpha}{\partial x_\beta}D_\beta-\frac{\partial E_\beta}{\partial x_\alpha}D_\beta+\frac{\partial H_\alpha}{\partial x_\beta}B_\beta-\frac{\partial H_\beta}{\partial x_\alpha}B_\beta\right\}=\frac{1}{c}[\boldsymbol{j}_{ext}\times\boldsymbol{B}]_\alpha. \quad (5.152)$$

The obtained equality must be now presented in the divergent form:

$$-\frac{\partial g_\alpha}{\partial t}+\frac{\partial\sigma_{\alpha\beta}}{\partial x_\beta}=f_\alpha. \quad (5.153)$$

Here we are dealing with the tensors in three-dimensional space and do not distinguish between upper and lower indices. If the vector $\boldsymbol{g}$ is the density of momentum of the electromagnetic field, and the tensor $T_{\alpha\beta}=-\sigma_{\alpha\beta}$ is the density of momentum flux, then the vector $\boldsymbol{f}$ on the right-hand side is the field force per unit volume, acting on the medium. This is because the force changes momentum. Earlier we have already encountered particular cases of Equation (5.153). Equation (I.4.129) is the integral form of Equation (5.153) in vacuum. Equation (2.63) is derived from Equation (5.153) for static electric field in medium.

It is clear in advance, that using Equations (5.152) and (5.153) we cannot separate unambiguously three groups of terms (momentum density, momentum flux density, and force). In order to do this, an additional information is needed. This ambiguity takes place because we consider nonclosed system, that is, the electromagnetic field interacting with medium. The energy, momentum and other quantities are distributed between the field and substance. The momenta of the field and substance (like corresponding energies) can be easily separated in the simplest model of medium, that is, in the model of charged particles, interacting with the field according to classical laws (see Section I.4.3, where such a model has been considered). The structure of real substance is much more complicated. Therefore, even an additional consideration of a general character does not lead to unambiguous results, and certain alternative versions remain (see below).

We will proceed from the correspondence principle, according to which the quantities $\boldsymbol{g}$, $\sigma_{\alpha\beta}$ and $\boldsymbol{f}$, calculated for an alternating field in a medium, in the limiting cases should reproduce earlier obtained quantities (I.4.128) and (I.4.131) in the absence of medium,

$$\boldsymbol{g}=\frac{1}{4\pi c}\boldsymbol{E}\times\boldsymbol{H}, \qquad \sigma_{\alpha\beta}=\frac{1}{4\pi}(E_\alpha E_\beta+H_\alpha H_\beta)-\frac{1}{8\pi}(E^2+H^2)\,\delta_{\alpha\beta}. \quad (5.154)$$

For constant electric and magnetic fields in matter, Equation (5.153) takes the form $f_\alpha=\partial\sigma_{\alpha\beta}/\partial x_\beta$. Earlier for constant electric field we have obtained Equations (2.61) and (2.64):

$$\sigma^e_{\alpha\beta}=\frac{1}{4\pi}E_\alpha D_\beta-\frac{E^2}{8\pi}\left(\varepsilon-\tau\frac{\partial\varepsilon}{\partial\tau}\right)\delta_{\alpha\beta}, \qquad f^e_\alpha=\rho_{ext}E_\alpha-\frac{E^2}{8\pi}\nabla_\alpha\varepsilon+\nabla_\alpha\left(\tau\frac{\partial\varepsilon}{\partial\tau}\frac{E^2}{8\pi}\right), \quad (5.155)$$

and for constant magnetic field Equations (3.67) and (3.68),

$$\sigma^m_{\alpha\beta} = \frac{1}{4\pi} H_\alpha B_\beta - \frac{H^2}{8\pi}\left(\mu - \tau\frac{\partial\mu}{\partial\tau}\right)\delta_{\alpha\beta},$$

$$f^m_\alpha = \frac{1}{c}[\boldsymbol{j}_{ext} \times \boldsymbol{B}]_\alpha - \frac{H^2}{8\pi}\nabla_\alpha\mu + \nabla_\alpha\left(\tau\frac{\partial\mu}{\partial\tau}\frac{H^2}{8\pi}\right). \tag{5.156}$$

In the presence of both fields, the stress tensor must include **generally** both tensors (5.155) and (5.156):

$$\sigma_{\alpha\beta} = \sigma^e_{\alpha\beta} + \sigma^m_{\alpha\beta} = \frac{1}{4\pi}(E_\alpha D_\beta + H_\alpha B_\beta) - \left[\frac{E^2}{8\pi}\left(\varepsilon - \tau\frac{\partial\varepsilon}{\partial\tau}\right) + \frac{H^2}{8\pi}\left(\mu - \tau\frac{\partial\mu}{\partial\tau}\right)\right]\delta_{\alpha\beta}. \tag{5.157}$$

Let us denote $\boldsymbol{f} = \boldsymbol{f}^e + \boldsymbol{f}^m$, where $\boldsymbol{f}^e$ and $\boldsymbol{f}^m$ are determined by Equations (5.155) and (5.156), respectively, and differentiate the stress tensor (5.157) with respect to coordinates. We obtain

$$\frac{\partial\sigma_{\alpha\beta}}{\partial x_\beta} = \rho_{ext}E_\alpha + f_\alpha + \frac{1}{4\pi}\left\{\frac{\partial E_\alpha}{\partial x_\beta}D_\beta - \frac{\partial E_\beta}{\partial x_\alpha}D_\beta + \frac{\partial H_\alpha}{\partial x_\beta}B_\beta - \frac{\partial H_\beta}{\partial x_\alpha}B_\beta\right\}. \tag{5.158}$$

Here the last term (in curly brackets), coincides with the second term in Equation (5.152). Next we introduce the notation

$$\boldsymbol{g}^M = \frac{1}{4\pi c}\boldsymbol{D} \times \boldsymbol{B}, \tag{5.159}$$

which is the momentum density of electromagnetic field according to Minkowski. In vacuum, this quantity coincides with $\boldsymbol{g}$ in Equation (5.154). With the help of Equation (5.158) Equation (5.152) can be written in the required form

$$-\frac{\partial g^M_\alpha}{\partial t} + \frac{\partial\sigma_{\alpha\beta}}{\partial x_\beta} = f^L_\alpha + f_\alpha, \tag{5.160}$$

where

$$\boldsymbol{f}^L = \rho_{ext}\boldsymbol{E} + \frac{1}{c}[\boldsymbol{j}_{ext} \times \boldsymbol{B}] \tag{5.161}$$

is the Lorentz force, acting on electric charges and currents, if they are present in medium;

$$\boldsymbol{f} = -\frac{E^2}{8\pi}\nabla\varepsilon - \frac{H^2}{8\pi}\nabla\mu + \nabla\left(\tau\frac{\partial\varepsilon}{\partial\tau}\frac{E^2}{8\pi}\right) + \nabla\left(\tau\frac{\partial\mu}{\partial\tau}\frac{H^2}{8\pi}\right) \tag{5.162}$$

is the force, induced by the electromagnetic field and connected with the inhomogeneity of medium and with internal stresses in it (electrostriction and magnetostriction). Evidently, the medium can be exposed also to the action of nonelectromagnetic forces (for instance, the gravity force and pressure gradient). The expression of Minkowski (5.159) for the momentum density is possible, because the vectors $\boldsymbol{D}$ and $\boldsymbol{B}$ characterize not only the field but the medium as well. Therefore, the quantity (5.159) contains both, the momentum of the field and the momentum of the substance. The four-dimensional tensor of energy–momentum, formed using the density of the Minkowski momentum (5.159), appears to be asymmetric (compare with the tensor (I.4.126)) in vacuum:

$$T_{ik}^{M} = \begin{pmatrix} w & -c\boldsymbol{g}^{M} \\ -\boldsymbol{\gamma}/c & -\sigma_{\alpha\beta} \end{pmatrix} . \tag{5.163}$$

Here the energy density w is given by Equation (5.105), and the density of the energy flux $\boldsymbol{\gamma}$ by Equation (5.106). The asymmetry appears in the following:

$$c\boldsymbol{g}^{M} = \frac{1}{4\pi}\boldsymbol{D}\times\boldsymbol{B} = \frac{\varepsilon\mu}{c}\boldsymbol{\gamma} \neq \frac{1}{c}\boldsymbol{\gamma}. \tag{5.164}$$

As has been clarified in Section I.4.3, the symmetry of the energy–momentum tensor of a closed system is required to define the angular momentum (as conserved quantity) by standard formulas of mechanics. However, for an individual subsystem such a requirement is not necessary.

This consideration shows that separation of quantities (for instance, of momentum) into parts, relating to the different subsystems is somewhat conventional. Instead of the quantity (5.159), many authors since the time of Abraham (see, for example, modern textbook Landau *et al.* (1984)) take the quantity, proposed by Abraham,

$$\boldsymbol{g}^{A} = \frac{1}{4\pi c}\boldsymbol{E}\times\boldsymbol{H} = \frac{\boldsymbol{\gamma}}{c^2}, \tag{5.165}$$

as the momentum density. The Minkowski and Abraham vectors are related by $\boldsymbol{g}^{M} = \boldsymbol{g}^{A} + (\varepsilon\mu - 1)\boldsymbol{g}^{A}$. Substituting this relation in Equation (5.160), we have

$$-\frac{\partial g_{\alpha}^{A}}{\partial t} + \frac{\partial \sigma_{\alpha\beta}}{\partial x_{\beta}} = f_{\alpha}^{L} + f_{\alpha} + f_{\alpha}^{A}, \tag{5.166}$$

where on the right-hand side an addition force appeared, **the Abraham force**, which acts on medium:

$$\boldsymbol{f}^{A} = \frac{1}{4\pi c}\frac{\partial}{\partial t}[\boldsymbol{D}\times\boldsymbol{B} - \boldsymbol{E}\times\boldsymbol{H}] = \frac{\varepsilon\mu - 1}{4\pi c}\frac{\partial}{\partial t}[\boldsymbol{E}\times\boldsymbol{H}]. \tag{5.167}$$

The Abraham force is rather small. The first report on its experimental measurement dated back to 1975 [Walker and Lahoz (1975)]; it demonstrated that this force does exist in nature. The energy–momentum tensor in the Abraham form, as opposed to the Minkowski representation, turns out to be symmetric:

$$T_{ik}^{A} = \begin{pmatrix} w & -c\boldsymbol{g}^{A} \\ -\boldsymbol{\gamma}/c & -\sigma_{\alpha\beta} \end{pmatrix} . \tag{5.168}$$

Although the experimental detection of the Abraham force seemingly resolves the question on the form of the energy–momentum tensor in favor of the Abraham representation (5.168), the Minkowski tensor (5.163) should not be considered a historical artifact. The Minkowski tensor is convenient to describe radiation of moving sources in media (see Ginzburg (1973); Ginzburg and Ugarov (1976); Ginzburg (1987)). See also Problem 5.77 and the next subsection.

Example 5.11

Derive Equations (5.162) and (5.167) for forces acting on the uncharged ($\rho_{ext} = \boldsymbol{j}_{ext} = 0$) dielectric in the electromagnetic field, using energy conservation (but not momentum conservation, as has been done above). To do this, consider the medium, moving with nonrelativistic velocity, and use the equations of electromagnetic field for moving media. Find the expressions for the energy density in the Minkowski and Abraham forms.

Solution. In the medium at rest, the coupling equations have the form (5.127), where the permeabilities ε and μ are thought to depend only on the substance density τ (time and spatial dispersions are absent). In a slowly moving medium, we have the Minkowski coupling Equation (5.128). Let us consider the energy balance based on the Maxwell equations. Performing scalar multiplication of Equation (5.120) by $\boldsymbol{D}$, and Equation (5.121) by $\boldsymbol{H}$, and subtracting the obtained equations, we have

(1) $$\frac{1}{4\pi}\left(\frac{\partial \boldsymbol{D}}{\partial t}\cdot\boldsymbol{E}+\frac{\partial \boldsymbol{B}}{\partial t}\cdot\boldsymbol{H}\right)=-\nabla\cdot\boldsymbol{\gamma},\qquad \boldsymbol{\gamma}=\frac{c}{4\pi}\boldsymbol{E}\times\boldsymbol{H}.$$

Using the coupling equations (5.128), which are valid in the first order in V/c, let us calculate with the same precision the left-hand side of Equation (1). We differentiate only the field vectors in correction terms, and find

(2) $$\frac{\partial \boldsymbol{D}}{\partial t}\cdot\boldsymbol{E}+\frac{\partial \boldsymbol{B}}{\partial t}\cdot\boldsymbol{H}=\varepsilon\boldsymbol{E}\cdot\frac{\partial \boldsymbol{E}}{\partial t}+\mu\boldsymbol{H}\cdot\frac{\partial \boldsymbol{H}}{\partial t}-\frac{\varepsilon\mu-1}{c}\frac{\partial}{\partial t}[\boldsymbol{E}\times\boldsymbol{H}]\cdot\boldsymbol{V}+\frac{\partial\varepsilon}{\partial t}E^2+\frac{\partial\mu}{\partial t}H^2.$$

We can express the derivatives of the permeabilities in terms of the velocity $\boldsymbol{V}$ using the continuity equation (4.24):

(3) $$\frac{\partial\varepsilon}{\partial t}=\frac{\partial\varepsilon}{\partial\tau}\frac{\partial\tau}{\partial t}=-\frac{\partial\varepsilon}{\partial\tau}\nabla\tau\cdot\boldsymbol{V}-\tau\frac{\partial\varepsilon}{\partial\tau}\nabla\cdot\boldsymbol{V},$$

and similarly for μ. The calculations, similar to those that resulted in Equation (2), enable us to obtain

(4) $$\frac{\partial \boldsymbol{E}}{\partial t}\cdot\boldsymbol{D}+\frac{\partial \boldsymbol{H}}{\partial t}\cdot\boldsymbol{B}=\frac{\partial \boldsymbol{D}}{\partial t}\cdot\boldsymbol{E}+\frac{\partial \boldsymbol{B}}{\partial t}\cdot\boldsymbol{H}-\frac{\partial\varepsilon}{\partial t}E^2-\frac{\partial\mu}{\partial t}H^2.$$

Now consider the quantity

(5) $$w^{(M)}=\frac{1}{8\pi}(\varepsilon E^2+\mu H^2)=\frac{1}{8\pi}(\boldsymbol{D}\cdot\boldsymbol{E}+\boldsymbol{B}\cdot\boldsymbol{H}),$$

which is the density of the electromagnetic energy for constant fields in the medium at rest (see Equations (2.44), (2.46), (3.57), and (3.58)). In our case, it will be the density of the field energy in the Minkowski form. Its derivative with respect to time can be written, using Equation (4), as

(6) $$-\frac{\partial w^{(M)}}{\partial t}=\frac{1}{4\pi}\left(\frac{\partial \boldsymbol{D}}{\partial t}\cdot\boldsymbol{E}+\frac{\partial \boldsymbol{B}}{\partial t}\cdot\boldsymbol{H}\right)-\frac{1}{8\pi}\left(\frac{\partial\varepsilon}{\partial t}E^2+\frac{\partial\mu}{\partial t}H^2\right).$$

Let us substitute to the right-hand side of this equality the right-hand side of Equation (1) as the first term, and rearrange the second term using Equation (3).

We obtain the equation

(7)
$$-\frac{\partial w^{(M)}}{\partial t} = \nabla \cdot \left\{ \gamma - \boldsymbol{V} \left[\tau \frac{\partial \varepsilon}{\partial \tau} \frac{E^2}{8\pi} + \tau \frac{\partial \mu}{\partial \tau} \frac{H^2}{8\pi} \right] \right\}$$
$$+\boldsymbol{V} \cdot \left\{ -\frac{E^2}{8\pi} \nabla \varepsilon - \frac{H^2}{8\pi} \nabla \mu + \nabla \left[\tau \frac{\partial \varepsilon}{\partial \tau} \frac{E^2}{8\pi} + \tau \frac{\partial \mu}{\partial \tau} \frac{H^2}{8\pi} \right] \right\},$$

which, in such a way, takes into account the Maxwell equations. It can be considered the equation of energy balance:

(8)
$$-\frac{\partial w^{(M)}}{\partial t} = \nabla \cdot \boldsymbol{S} + \boldsymbol{f} \cdot \boldsymbol{V}.$$

The loss of the electromagnetic energy (left-hand side) induces the energy flux with density

(9)
$$\boldsymbol{S} = \gamma - \boldsymbol{V} \left[\tau \frac{\partial \varepsilon}{\partial \tau} \frac{E^2}{8\pi} + \tau \frac{\partial \mu}{\partial \tau} \frac{H^2}{8\pi} \right],$$

and motion of the medium. The volume force $\boldsymbol{f}$, according to Equation (7), has the form

(10)
$$\boldsymbol{f} = -\frac{E^2}{8\pi} \nabla \varepsilon - \frac{H^2}{8\pi} \nabla \mu + \nabla \left[\tau \frac{\partial \varepsilon}{\partial \tau} \frac{E^2}{8\pi} + \tau \frac{\partial \mu}{\partial \tau} \frac{H^2}{8\pi} \right]$$

and coincides with Equation (5.162) obtained above.

The transition to the Abraham representation is made replacing the energy density (5) by the quantity

(11)
$$w^A = \frac{1}{8\pi} (\boldsymbol{D} \cdot \boldsymbol{E} + \boldsymbol{B} \cdot \boldsymbol{H}) - \frac{1}{4\pi c} \boldsymbol{V} \cdot [\boldsymbol{D} \times \boldsymbol{B} - \boldsymbol{E} \times \boldsymbol{H}],$$

which is connected with Equation (5),

(12)
$$\frac{\partial w^A}{\partial t} = \frac{\partial w^M}{\partial t} - \boldsymbol{f}^A \cdot \boldsymbol{V},$$

where $\boldsymbol{f}^A$ is the Abraham force (5.167). In this case, the equation of energy balance takes the form

(13)
$$-\frac{\partial w^{(A)}}{\partial t} = \nabla \cdot \boldsymbol{S} + (\boldsymbol{f} + \boldsymbol{f}^A) \cdot \boldsymbol{V}.$$

The density of the energy flux (9) remained the same, and the Abraham force appears in the left part-hand side of the equation. ■

Problems

5.71• Express the polarization vectors $\boldsymbol{P}$ and $\boldsymbol{M}$ in terms of the field vectors $\boldsymbol{E}$ and $\boldsymbol{B}$. On this base elucidate the relation between the electric and magnetic phenomena in moving media. Clarify, in particular, the possibility of the appearance of the electric polarization $\boldsymbol{P}$ in the medium with $\varepsilon = 1$ and of the magnetic polarization $\boldsymbol{M}$ in the nonmagnetic medium ($\mu = 1$). Explain the physical reasons of their appearance.

5.72• Analyze the properties of the Green's function, describing the propagation of electromagnetic signal in a moving medium (Example 5.10). Consider the cases of subluminal ($V < c/\sqrt{\varepsilon\mu} = v_{ph} < c$), light-speed ($V = v_{ph}$) and supraluminal ($V > v_{ph}$) motion of the medium. Study the shape of a short signal, emitted at the instant $t = 0$ at the coordinate origin.

5.73• Let the point charge q be at rest in the laboratory reference frame, and the medium move along the axis Oz with velocity V. Calculate the electromagnetic potentials and vectors of the electromagnetic field. Find the electric and magnetic polarization of the moving medium.

5.74⋆ Repeat the previous for the magnetic dipole $\boldsymbol{m}_0$, which is at rest in the laboratory frame. Use the Hertz vectors, introduced for moving medium in Problem 5.9.

5.75 Do the same for the electric dipole $\boldsymbol{p}_0$, which is at rest in the laboratory frame. Use the Hertz vectors, introduced for moving medium in Problem 5.9.

5.76 The magnetic dipole is at rest in the reference frame co-moving the medium. Find, in the simplest way, the electromagnetic field in the laboratory frame. Compare it with the field of the dipole, which is at rest in the laboratory frame, in the presence of the moving medium (see Problem 5.73).

5.77 Consider a radiator in the homogeneous medium which emits a wave packet with the total momentum $\boldsymbol{G}^A = \int \boldsymbol{g}^A \, dV$ (the integration over the entire space). Find the momentum loss $\Delta \boldsymbol{G}^A$ of the radiator in this process.

5.78 Using the Lorentz transformation, find the quantities w^A and $\boldsymbol{g}^A$, and the energy flux density $\boldsymbol{S}$ in the inertial reference frame, relatively to which the medium moves with the velocity $\boldsymbol{V}$.

5.6
Energy–Momentum Tensor in Dispersive Media

In the last few decades, diverse artificial materials with complex dispersive properties are gaining importance. In addition to the time (frequency) dispersion, their dielectric permittivity also exhibits spatial dispersion, that is, it depends on the wave number. A variety of such materials ("metamaterials") in a certain range of frequencies has the negative dielectric permittivity and the negative magnetic permeability ($\varepsilon(\omega) < 0$ and $\mu(\omega) < 0$). The properties of these materials are discussed in more details in Section 6.1.

The expressions for the energy and energy flux densities in dispersive media have been already obtained in Section 5.3. These quantities in four-dimensional notations are the components of the first column of the energy–momentum tensor T_{i0}, $i = 0, 1, 2, 3$ (see Equations (5.163) and (5.168)). In this section, we will construct (with some limitations) the remaining components of the tensor T_{ik}, that is, the momentum density $\boldsymbol{g}$ and momentum flux density $-\sigma_{\alpha\beta}$. We will assume that there is no dissipation of electromagnetic energy in a dielectric medium, which is isotropic, homogeneous and incompressible (with the mass density $\tau = const$).

In order to use the permittivities $\varepsilon_t(k, \omega)$ and $\epsilon_l(k, \omega)$ for constructing the quantities mentioned above, one has to consider the electromagnetic field in a small range of frequencies and wave numbers, that is, a narrow (in Fourier space) wavepacket, as was done in Section 5.3.

Group velocity. Unlike in a vacuum, the frequency of electromagnetic waves in dispersive media depends on the wave vector in a more sophisticated manner than a direct proportionality $\omega = ck$. If the source of the field is far from the region under consideration, then in the medium only the eigen oscillations can propagate, whose frequency and wave vector are related by the dispersion equation (5.58): $\omega^2\varepsilon(\omega)\mu(\omega) = c^2k^2$. Hence, the phase velocities of the waves $v_{ph}(k) = \omega(k)/k$ will depend on wavelength (on frequency). In other words, the different Fourier harmonics propagate with different velocity. Accordingly, in a dispersive medium (in contrast to vacuum, see Problems I.2.127 and I.2.140) the wavepacket does not retain its shape, and the transport velocity of energy differs from a phase velocity. Let us write down the expression for the electromagnetic field of a narrow packet of electromagnetic waves:

$$\boldsymbol{E}(\boldsymbol{r}, t) = \int \boldsymbol{\mathcal{E}}(\boldsymbol{k} + \boldsymbol{q}) \exp[i(\boldsymbol{k} + \boldsymbol{q}) \cdot \boldsymbol{r} - i\omega(\boldsymbol{k} + \boldsymbol{q})t] \frac{d^3q}{(2\pi)^3}, \tag{5.169}$$

where $q \leq \Delta k \ll k$. Hence, the frequency can be expanded in q to within linear terms:

$$\omega(\boldsymbol{k} + \boldsymbol{q}) \approx \omega(k) + \boldsymbol{v}^g \cdot \boldsymbol{q}, \qquad \text{where the derivative} \qquad \boldsymbol{v}^g = \frac{d\omega}{d\boldsymbol{k}} \tag{5.170}$$

is called the **group velocity**. The substitution of this expansion in Equation (5.169) gives the wavepacket (a plane wave with a slowly changing amplitude), which travels with the group velocity $\boldsymbol{v}_g$:

$$\begin{aligned} \boldsymbol{E}(\boldsymbol{r}, t) &= \boldsymbol{\mathcal{E}}(\boldsymbol{r} - \boldsymbol{v}^g t) \exp(i\boldsymbol{k} \cdot \boldsymbol{r} - \omega t), \\ \boldsymbol{\mathcal{E}}(\boldsymbol{r} - \boldsymbol{v}^g t) &= \int \boldsymbol{\mathcal{E}}(\boldsymbol{k} + \boldsymbol{q}) \exp[i\boldsymbol{q} \cdot (\boldsymbol{r} - \boldsymbol{v}^g t)] \frac{d^3q}{(2\pi)^3}. \end{aligned} \tag{5.171}$$

In this approximation, the wavepacket does not change its shape. The packet spreading reveals itself if the next terms of the frequency expansion in q are taken into account (see Problem 6.6).

Let us calculate the group velocity using the dispersion equation (5.58) (neglecting the dependence of ε and μ on $\boldsymbol{k}$). We have

$$\omega = \frac{ck}{\sqrt{\varepsilon(\omega)\mu(\omega)}}, \qquad \frac{d\omega}{d\boldsymbol{k}} = \frac{c\boldsymbol{k}}{\sqrt{\varepsilon(\omega)\mu(\omega)}k} - \frac{ck}{2\sqrt{\varepsilon(\omega)\mu(\omega)}} \left[\frac{1}{\varepsilon}\frac{d\varepsilon}{d\omega} + \frac{1}{\mu}\frac{d\mu}{d\omega}\right] \frac{d\omega}{d\boldsymbol{k}}.$$

Hence,

$$\boldsymbol{v}^g = \frac{d\omega}{d\boldsymbol{k}} = \frac{\boldsymbol{v}_{ph}}{1 + \Pi(\omega)}, \qquad \boldsymbol{v}_{ph} = \frac{c\boldsymbol{k}}{\sqrt{\varepsilon(\omega)\mu(\omega)}k}, \qquad \Pi(\omega) = \frac{\omega}{2}\left(\frac{1}{\varepsilon}\frac{d\varepsilon}{d\omega} + \frac{1}{\mu}\frac{d\mu}{d\omega}\right). \tag{5.172}$$

The group velocity can also be written in the form

$$\boldsymbol{v}^g = \frac{c}{\dfrac{d}{d\omega}\omega\sqrt{\varepsilon(\omega)\mu(\omega)}}\frac{\boldsymbol{k}}{k}. \tag{5.173}$$

In the isotropic medium, the group velocity may be either parallel or antiparallel to the phase velocity.

Example 5.12

Express the group velocity of transverse waves in terms of the dielectric permittivity $\varepsilon_t(k,\omega)$ with account for the spatial dispersion. Ensure that when passing to the quantities $\varepsilon(\omega)$ and $\mu(\omega)$ the results given by Equations (5.172) and (5.173) are reproduced.

Solution. Using the dispersion equation (5.57), $\omega^2\varepsilon_t(k,\omega) = (ck)^2$, we find

$$\frac{d\omega}{d\boldsymbol{k}} = \boldsymbol{v}_{ph} + \frac{ck}{\sqrt{\varepsilon_t}}\left(-\frac{1}{2\varepsilon_t}\frac{\partial\varepsilon_t}{\partial k}\frac{\boldsymbol{k}}{k} - \frac{1}{2\varepsilon_t}\frac{\partial\varepsilon_t}{\partial\omega}\frac{d\omega}{d\boldsymbol{k}}\right)$$

and obtain

$$\boldsymbol{v}^g = \boldsymbol{v}_{ph}\frac{\varepsilon_t - \frac{k}{2}\frac{\partial\varepsilon_t}{dk}}{\varepsilon_t + \frac{\omega}{2}\frac{\partial\varepsilon_t}{\partial\omega}}. \tag{5.174}$$

In order to pass to the quantities $\varepsilon(\omega)$ and $\mu(\omega)$, we use Equation (5.62) to obtain the relation

$$\varepsilon_t(k,\omega) = \varepsilon(\omega) + \left(\frac{ck}{\omega}\right)^2\left(1 - \frac{1}{\mu(\omega)}\right) \tag{5.175}$$

and substitute $(ck/\omega)^2 = \varepsilon(\omega)\mu(\omega)$ in it. For the transverse waves, this yields

$$\varepsilon_t = \varepsilon(\omega)\mu(\omega). \tag{5.176}$$

Substituting this product in Equation (5.174), we obtain the previous Equation (5.172). ■

Propagation velocity of energy and momentum. This velocity is defined as the ratio $\boldsymbol{\gamma}/w$ of the energy flux density to the energy density itself (in this section we use the bilinear quantities averaged over the fundamental period $T = 2\pi/\omega$). Let us ensure that the propagation velocity of energy is equal to the group velocity. To do this, we will simplify the relations for the energy density and energy flux density by expressing the vector $\boldsymbol{B}$ in terms of $\boldsymbol{E}$ with the aid of the Maxwell equations:

$$\boldsymbol{B} \approx \frac{c}{\omega}\boldsymbol{k}\times\boldsymbol{E}. \tag{5.177}$$

Here we omitted the small terms of the order of $\Delta k/k$ and $\Delta\omega/\omega$. With the same accuracy we obtain the relation $\boldsymbol{k}\cdot\boldsymbol{E} = 0$. By using these formulas and Equation (5.38), with the aid of Equation (5.108) we obtain the expression for the energy density:

$$w = \frac{1}{8\pi}\left(\varepsilon_t + \frac{\omega}{2}\frac{\partial\varepsilon_t}{\partial\omega}\right)(\boldsymbol{E}^*\cdot\boldsymbol{E}). \tag{5.178}$$

Equation (5.109) for the energy flux density can be similarly transformed into the relation

$$\boldsymbol{\gamma} = \frac{\boldsymbol{v}_{ph}}{8\pi}\left(\varepsilon_t - \frac{k}{2}\frac{\partial \varepsilon_t}{\partial k}\right)(\boldsymbol{E}^* \cdot \boldsymbol{E}). \tag{5.179}$$

Their ratio gives the group velocity in the form of Equation (5.175). Calculating the permittivity and permeability, $\varepsilon(\omega)$ and $\mu(\omega)$, which are independent of wave number, we get

$$w = \frac{\varepsilon(\omega)}{8\pi}[1 + \Pi(\omega)]\boldsymbol{E}^* \cdot \boldsymbol{E} \text{ and } \quad \boldsymbol{g} = \frac{\varepsilon(\omega)}{8\pi c^2}(\boldsymbol{E}^* \cdot \boldsymbol{E})\boldsymbol{v}_{ph} \text{ or } \quad \boldsymbol{g} = \frac{\sqrt{\varepsilon\mu}\boldsymbol{k}}{8\pi c\mu k}(\boldsymbol{E}^* \cdot \boldsymbol{E}), \tag{5.180}$$

and their ratio is equal to the group velocity (5.172).

The fact that in the absence of the external sources the continuity equation $\partial w/\partial t + \nabla \cdot \boldsymbol{\gamma} = 0$ holds true follows from the explicit form of the quantities w and $\boldsymbol{\gamma}$ and their argument $\boldsymbol{r} - \boldsymbol{v}_g t$:

$$\frac{\partial w}{\partial t} = -\boldsymbol{v}_g \cdot \nabla w = -\nabla \cdot \boldsymbol{v}_g w = -\nabla \cdot \boldsymbol{\gamma}, \tag{5.181}$$

provided the group velocity is independent of coordinates.

Momentum balance and momentum flux in dispersive medium. In order to calculate the densities of momentum and momentum flux, we will use the Abraham relation (5.165) and Equation (5.179):

$$\boldsymbol{g} = \frac{\boldsymbol{\gamma}}{c^2} = \frac{\boldsymbol{v}_{ph}}{8\pi c^2}\left(\varepsilon_t - \frac{k}{2}\frac{\partial \varepsilon_t}{\partial k}\right)(\boldsymbol{E}^* \cdot \boldsymbol{E}). \tag{5.182}$$

The continuity equation, which gives the momentum conservation in a nonabsorbing medium without sources, may be derived in the same manner as the energy equation (5.181):

$$\frac{\partial g_\alpha}{\partial t} = -(\boldsymbol{v}_g \cdot \nabla)g_\alpha = -\frac{\partial}{\partial x_\beta}g_\alpha v_{g\beta} = \frac{\partial \sigma_{\alpha\beta}}{\partial x_\beta}, \tag{5.183}$$

where the stress tensor is given by

$$\sigma_{\alpha\beta} = -g_\alpha v_{g\beta} = -g_\beta v_{g\alpha} = -n_\alpha n_\beta \frac{v_g}{8\pi c\sqrt{\varepsilon_t}}\left(\varepsilon_t - \frac{k}{2}\frac{\partial \varepsilon_t}{\partial k}\right)(\boldsymbol{E}^* \cdot \boldsymbol{E}), \qquad \boldsymbol{n} = \boldsymbol{k}/k, \tag{5.184}$$

and v_g is the projection of the group velocity on the direction of the wave vector $\boldsymbol{n}$.

The momentum flux density of the electromagnetic field is expressed as the corresponding component of the momentum density times the travel velocity of the wavepacket. Such a definition is the most clear and analogous to the formulas used in physical kinetics and mechanics of continuous media. This tensor taken with an opposite sign presents the spatial part of the energy–momentum 4-tensor of the electromagnetic field in a transparent medium.

One should bear in mind that the tensor under the divergence is ambiguously defined. One can add to it an arbitrary tensor of second rank $\widetilde{\sigma}_{\alpha\beta}$, having zero divergence, $\partial\widetilde{\sigma}_{\alpha\beta}/\partial x_\beta = 0$. The most general form of such a tensor is the divergence $\partial\widetilde{\sigma}_{\alpha\beta}/\partial x_\beta = \partial\lambda_{\alpha\beta\gamma}/\partial x_\gamma$ of third-rank tensor antisymmetric in any pair of indices, $\lambda_{\alpha\beta\gamma}(x, y, z) = -\lambda_{\gamma\beta\alpha} = \cdots$.

However, as was indicated by Fock (1962), the tensor of the type $\lambda_{\alpha\beta\gamma}$, bilinear in the field vectors, can be constructed provided the derivatives of these vectors are included in the tensor. For this reason, Fock postulated the principle that must be satisfied by the energy–momentum tensor of electromagnetic field: **it must be a function of state of the system under consideration**. In the present case, it means that the tensor must depend on the field vectors, describing completely the state of the field, other than on the derivatives of these vectors. This principle allows one to avoid an ambiguity introduced by the terms of the kind $\partial\lambda_{\alpha\beta\gamma}/\partial x_\gamma$. The Fock principle is important for defining the energy flux density $\boldsymbol{\gamma}$, since the Poynting vector initially enters under the divergence and is subject to the ambiguity discussed above. Notice, that the representation of the stress tensor (5.184) includes also the Abraham force (5.167). For certainty, consider Equation (5.166) for an incompressible homogeneous medium in the absence of external charges ($\boldsymbol{f}^L = 0$):

$$-\frac{\partial g_\alpha^A}{\partial t} + \frac{\partial\widetilde{\sigma}_{\alpha\beta}^A}{\partial x_\beta} = f_\alpha^A. \tag{5.185}$$

The condition of incompressibility allows the stress tensor (5.157) (with a tilde and an upper index A) to be written in the form

$$\widetilde{\sigma}_{\alpha\beta}^A = \frac{1}{4\pi}(E_\alpha D_\beta + H_\alpha B_\beta) - \left[\frac{\varepsilon E^2}{8\pi} + \frac{\mu H^2}{8\pi}\right]\delta_{\alpha\beta}. \tag{5.186}$$

We apply Equations (5.167), (5.185), and (5.186) to the nondispersive packet with parameters (ε, μ) independent of frequency. All formulas of Section 5.5 refer to such a medium. In this case, $v_g = v_{ph} = c/\sqrt{\varepsilon\mu}$. However, the slowly alternating amplitudes of all fields depend on the difference $\boldsymbol{r} - \boldsymbol{v}_g t$. Expressing the amplitudes $\boldsymbol{D} = \varepsilon\boldsymbol{E}$, $\boldsymbol{H} = \boldsymbol{B}/\mu$ and $\boldsymbol{B} = (ck/\omega)\boldsymbol{n}\times\boldsymbol{E}$ in terms of the vector $\boldsymbol{E}$, we obtain the following representations for the quantities entering Equation (5.185):

$$\boldsymbol{g}^A = \frac{1}{16\pi c}[\boldsymbol{E}^*\times\boldsymbol{H} + \boldsymbol{E}\times\boldsymbol{H}^*] = \boldsymbol{n}\frac{1}{8\pi v_g\mu}(\boldsymbol{E}\cdot\boldsymbol{E}^*), \qquad f_\alpha^A = -\frac{\partial\sigma_{\alpha\beta}^f}{\partial x_\beta}, \tag{5.187}$$

where

$$\sigma_{\alpha\beta}^f = -\frac{n_\alpha n_\beta}{8\pi}\left(\varepsilon - \frac{1}{\mu}\right)(\boldsymbol{E}\cdot\boldsymbol{E}^*) \tag{5.188}$$

is the part of the stress tensor, which is responsible for the Abraham force (5.167). Finally, we have the relation

$$\widetilde{\sigma}_{\alpha\beta}^A = -\frac{n_\alpha n_\beta\varepsilon}{8\pi}(\boldsymbol{E}\cdot\boldsymbol{E}^*) \tag{5.189}$$

which reduces to the tensor (5.157) at $\partial\varepsilon/\partial\omega = 0$, $\partial\mu/\partial\omega = 0$.

By using these results, we can check the validity of the equality

$$\frac{\partial g_\alpha^A}{\partial t} - \frac{\partial \sigma_{\alpha\beta}^A}{\partial x_\beta} = 0, \quad \text{where} \quad \sigma_{\alpha\beta}^A = \widetilde{\sigma}_{\alpha\beta}^A + \sigma_{\alpha\beta}^f = -\frac{n_\alpha n_\beta}{8\pi\mu}(\boldsymbol{E}\cdot\boldsymbol{E}^*). \tag{5.190}$$

The same result for $\sigma_{\alpha\beta}$ can be obtained from Equation (5.184) by employing it for the nondispersive medium with permeabilities ε and μ. However, Equation (5.184) has a wide region of applications and embraces the cases of both time and spatial dispersion.

Problems

5.79• A packet of quasi-monochromatic waves containing a narrow range of frequencies propagates in a nonabsorbing dispersive medium. Show that taking into account the terms linear with respect to small ratios $\Delta\omega/\omega$ and $\Delta k/k$, the relation between the vectors of generalized electric induction and electric field reads

$$\widetilde{D}_\alpha(\boldsymbol{r},t) = \epsilon_{\alpha\beta}(\boldsymbol{k},\omega)\mathcal{E}_\beta(\boldsymbol{r},t)e^{i\varphi} - ie^{i\varphi}\left(\frac{\partial\epsilon_{\alpha\beta}}{\partial\mathbf{k}}\cdot\nabla - \frac{\partial\epsilon_{\alpha\beta}}{\partial\omega}\frac{\partial}{\partial t}\right)\mathcal{E}_\beta(\boldsymbol{r},t).$$

5.80• Under the condition of Problem 5.79, express a slowly varying amplitude of the magnetic induction $\boldsymbol{\mathcal{B}}(\boldsymbol{r},t)$ through the amplitude $\boldsymbol{\mathcal{E}}(\boldsymbol{r},t)$ of the electric field.

5.81• By direct calculation find the average over the period $T = 2\pi/\omega$ of the time derivative

$$\frac{\partial}{\partial t}[\boldsymbol{D}\times\boldsymbol{B}], \quad \text{where} \quad \boldsymbol{D} = \Re[\boldsymbol{\mathcal{D}}(\boldsymbol{r},t)e^{i\varphi}],\ \boldsymbol{B} = \Re[\boldsymbol{\mathcal{B}}(\boldsymbol{r},t)e^{i\varphi}],$$

where $\boldsymbol{\mathcal{D}}$ and $\boldsymbol{\mathcal{B}}$ are slowly varying amplitudes.

5.82•⋆ Using the equation of momentum balance, find the average over wave period of the Minkowski momentum density. In contrast with equation (5.159), which does not account for dispersion, consider a dispersive dielectric with permeability $\varepsilon_t(k,\omega)$.

5.7 Answers and Solutions

5.3 From Equations (5.8), (5.10), and the Maxwell equations (5.1)–(5.4) we find

$$\rho_{int} = -\frac{i(\varepsilon_\parallel - 1)}{4\pi}\boldsymbol{k}\cdot\boldsymbol{E}, \qquad \boldsymbol{j}_{int}^\perp = \frac{i(\omega^2 - c^2k^2)}{4\pi\omega}(1-\eta)\boldsymbol{E}^\perp,$$

the longitudinal current is expressed in terms of ρ_{int}. In any medium $\rho_{int}, \boldsymbol{j}_{int}^\perp$ and $\boldsymbol{E}$ are well-defined physical quantities, and the coefficients of proportionality between them are the electromagnetic characteristics (response functions) of given medium. These are $\varepsilon_{\parallel} = \varepsilon$ and η, but not $\varepsilon_{\perp}$ and $\mu_{\perp}$, which are combined in η.

5.4

$$\rho_{tot} = \frac{1}{\varepsilon_{\parallel}}\rho_{ext}, \quad \rho_{int} = \left(\frac{1}{\varepsilon_{\parallel}} - 1\right)\rho_{ext}, \quad \boldsymbol{j}_{tot}^{\perp} = \frac{1}{\eta}\boldsymbol{j}_{ext}^{\perp}, \quad \boldsymbol{j}_{int}^{\perp} = \left(\frac{1}{\eta} - 1\right)\boldsymbol{j}_{ext}^{\perp}.$$

5.5

$$\varphi(\boldsymbol{k},\omega) = G^{\parallel}(k,\omega)\rho_{ext}(\boldsymbol{k},\omega), \qquad \boldsymbol{A}(\boldsymbol{k},\omega) = \frac{1}{c}G^{\perp}(k,\omega)\boldsymbol{j}_{ext}^{\perp}(\boldsymbol{k},\omega);$$

$$G^{\parallel}(k,\omega) = \frac{4\pi}{k^2\varepsilon(k,\omega)}, \qquad G^{\perp}(k,\omega) = \frac{4\pi}{(k^2 - \omega^2/c^2)\eta(k,\omega)}.$$

Here $\eta(k,\omega)$ is the quantity introduced in Problem 5.3; it characterizes the effect of medium on the transverse field. The Green's functions are normalized so that in vacuum they give the results, obtained in Section 5.1 (with account for the Lorentz gauge used there). Notice, that the values $\varepsilon = \varepsilon_l, \varepsilon_t$ and μ at $\omega \neq 0$ are complex. Therefore, in contrast to vacuum, in medium the denominator of the Green's transverse function $G^{\perp}(k,\omega)$, as a rule, does not vanish at real values of ω and k.

5.7 Motion of a free electron in a weak field during about several wave periods can be described by the simplified Newton equation

(1) $$m\dot{\boldsymbol{v}} = e\boldsymbol{E}_0 e^{-i\omega t} + \frac{e}{c}\boldsymbol{v}_0 \times \boldsymbol{B}_0 e^{-i\omega t},$$

where in the small term containing the magnetic field only the initial (thermal) velocity $\boldsymbol{v}_0$ has been taken into account. From this equation we find:

(2) $$\boldsymbol{v}(t) = \boldsymbol{v}_0 + \frac{ie\boldsymbol{E}(t)}{m\omega} + \frac{ie\boldsymbol{v}_0 \times \boldsymbol{B}(t)}{mc\omega}.$$

The macroscopic current is given by $\boldsymbol{j}(t) = en\overline{\boldsymbol{v}}(t)$, where n is the average number density of electrons, $\overline{\boldsymbol{v}}(t)$ is their average velocity. However, in the absence of the field, the current in the medium is absent and, therefore, $\overline{\boldsymbol{v}}_0 = 0$ and $\boldsymbol{j}(t) = ienE(t)/m\omega$. With the help of Equation (5.21) we find

(3) $$\boldsymbol{D} = \boldsymbol{E} + \int_{-\infty}^{t} \boldsymbol{j}\,dt' = \left(1 - \frac{4\pi ne^2}{m\omega^2}\right)\boldsymbol{E}.$$

Hence,

(4) $$\epsilon_{\alpha\beta} = \varepsilon\delta_{\alpha\beta}, \qquad \varepsilon = \varepsilon_l = \varepsilon_t = 1 - \frac{4\pi ne^2}{m\omega^2}.$$

The spatial dispersion is unimportant here.

5.8 Let us use the result of Problem 5.5:

(1) $$\varphi(\boldsymbol{k},\omega) = \frac{4\pi}{k^2\varepsilon(k,\omega)}\rho_{ext}(\boldsymbol{k},\omega),$$

where $\varepsilon(k,\omega) = \varepsilon_l(k,\omega)$ is the longitudinal dielectric permittivity, $\rho_{ext}(\boldsymbol{k},\omega)$ is the Fourier transform of the charge density induced by a fixed point charge q:

(2) $$\rho_{ext}(\boldsymbol{k},\omega) = \int d^3r \int_{-\infty}^{\infty} dt\, q\,\delta(\boldsymbol{r}) \exp\{-i(\boldsymbol{k}\cdot\boldsymbol{r} - \omega t)\} = 2\pi q\,\delta(\omega).$$

Thus, the value of $\varepsilon_l(k,0)$ is required. Using the formulas of Example 5.6, we find $Z(s)|_{s\to 0} \to 0$,

(3) $$\varepsilon_l(k,0) = 1 + \frac{\omega_{0i}^2}{(kV_{Ti})^2} + \frac{\omega_{0e}^2}{(kV_{Te})^2} = 1 + \frac{1}{k^2 r_D^2},$$

where $r_D^2 = T/8\pi n e^2$ is the squared Debye radius, which coincides with the value of $1/\kappa^2$ in Problem 7.24 for singly charged ions. The potential is calculated from the formula

(4) $$\varphi(\boldsymbol{r}) = \int \frac{4\pi}{k^2 \varepsilon(k,\omega)} 2\pi q\,\delta(\omega) \exp\{-i(\boldsymbol{k}\cdot\boldsymbol{r} - \omega t)\} \frac{d^3k\, d\omega}{(2\pi)^4}$$

$$= \frac{q}{i\pi r} \int_{-\infty}^{\infty} \frac{k}{k^2 + r_D^2} e^{ikr}\, dk = \frac{q}{r} e^{-r/r_D}.$$

The last integration is performed by closing the contour in the upper half plane of complex k by an arc of large radius.

The special feature in this problem is that the Debye screening of the charge in a plasma is obtained using the concept of spatial dispersion. The screening of somewhat other character takes place in the case of moving charges (see problems from Section 9.1).

Finally, let us present characteristic values of Debye radii (cm) for several media: metals (degenerate plasma, i.e., quantum plasma), 10^{-7}; semiconductors, 10^{-5}; plasma in thermonuclear devices, $10^{-4} - 10^{-3}$; the Earth ionosphere, 10^{-1}; interplanetary plasma, 10^3; interstellar plasma, $10^3 - 10^5$.

5.9 $s \gg 1$; $Y(s)$ is exponentially small; $X(s) = 1 + 1{\cdot}s^{-2} + 1{\cdot}3{\cdot}s^{-4} + 1{\cdot}3{\cdot}5{\cdot}s^{-6} + \cdots$

$$\varepsilon_l \approx 1 - \frac{\omega_{0e}^2}{\omega^2}\left(1 + \frac{3k^2 v_{Te}^2}{\omega^2}\right) + i\sqrt{\frac{\pi}{2}} \frac{\omega\omega_{0e}^2}{k^3 v_{Te}^3} e^{-\omega^2/2k^2 v_{Te}^2},$$

$$\varepsilon_t \approx 1 - \frac{\omega_{0e}^2}{\omega^2}\left(1 + \frac{k^2 v_{Te}^2}{\omega^2}\right) + i\sqrt{\frac{\pi}{2}} \frac{\omega_{0e}^2}{\omega k v_{Te}} e^{-\omega^2/2k^2 v_{Te}^2}.$$

Because $\omega_{0i}^2 = (m_e/m_i)\omega_{0e}^2 \ll \omega_{0e}^2$, we have $s_i \gg s_e$. Though the imaginary part is exponentially small, it is important since it determines the dissipation of electromagnetic energy.

$$s \ll 1; \qquad X(s) \approx s^2 - s^4/3 + \cdots; \; Y(s) \approx \sqrt{\pi/2}(s - s^3/2 + \cdots)$$
$$r_D = \sqrt{T/8\pi n e^2};$$

$$\varepsilon_l \approx 1 + \frac{1}{(kr_D)^2} + i\sqrt{\frac{\pi}{8}}\frac{1}{(kr_D)^2}\left(\frac{\omega}{kv_{Te}} + \frac{\omega}{kv_{Ti}}\right),$$

$$\varepsilon_t \approx 1 - \frac{1}{(kr_D)^2} + i\sqrt{\frac{\pi}{8}}\frac{1}{(kr_D)^2}\left(\frac{kv_{Te}}{\omega} + \frac{kv_{Ti}}{\omega}\right).$$

5.10

$$\frac{1}{\mu(k,\omega)} = 1 + \sum\left(\frac{\omega_0}{ck}\right)^2 \{(1-s^2)Z(s) + s^2\}, \qquad s = \frac{\omega}{kv_T}.$$

$$s \gg 1; \qquad \frac{1}{\mu(k,\omega)} \approx 1.$$

$$s \ll 1; \qquad \frac{1}{\mu(k,\omega)} \approx 1 + \frac{2}{(kr_D)^2}\left(\frac{\omega}{ck}\right)^2 \approx 1 + \frac{2}{(kr_D)^2}\left(\frac{\omega}{kv_T}\right)^2\frac{v_T^2}{c^2} \approx 1.$$

5.11 For the equilibrium distribution function, we take the Fermi distribution

(1) $$f_0(\epsilon) = \frac{2}{(2\pi\hbar)^3}\Theta(\epsilon_F - \epsilon),$$

where $\epsilon_F = (3\pi^2\hbar^3)^{2/3}/2m$ is the Fermi energy of electrons. Then the first formula (4) in Example 5.6 takes the form

(2) $$\frac{\partial f_0}{\partial \boldsymbol{p}} = \boldsymbol{v}\frac{\partial f_0}{\partial \epsilon} = -\frac{2\boldsymbol{v}}{(2\pi\hbar)^3}\delta(\epsilon_F - \epsilon);$$

while the second formula remains the same. The next calculations are simplified due to the presence of the delta-function in Equation (2); the result is:

(3) $$\varepsilon_l(k,\omega) = 1 + \frac{3\omega_{0e}^2}{k^2 v_F^2}\left[1 - \frac{\omega}{2kv_F}\ln\frac{\omega + kv_F}{\omega - kv_F}\right],$$

(4) $$\varepsilon_t(k,\omega) = 1 - \frac{3\omega_{0e}^2}{2k^2 v_F^2}\left\{1 + \frac{kv_F}{2\omega}\left[1 - \left(\frac{\omega}{kv_F}\right)^2\right]\ln\frac{\omega + kv_F}{\omega - kv_F}\right\}.$$

At $\omega \gg kv_F$, we have $\varepsilon_l \approx \varepsilon_t \approx 1 - \omega_{0e}^2/\omega^2$, which corresponds to the results of Problem 5.7. At $\omega \ll kv_F$, we obtain $\varepsilon_l \approx 1 + 1/(kr_D)^2$ and $\varepsilon_t \approx 1 - 1/(kr_D)^2$, where $r_D = v_F/\sqrt{3}\omega_{0e}$ is the screening radius in the degenerate electron gas (see Problem 1.27). The dielectric permeabilities (3) and (4) have the imaginary parts in the range of parameters $0 < \omega/k < v_F$; in this case particles move in resonance with the wave and take some energy from it. This is possible only for longitudinal waves.

5.12 We use the following equation of motion for the oscillator:

(1) $$\ddot{\boldsymbol{r}} + \gamma\dot{\boldsymbol{r}} + \omega_0^2\boldsymbol{r} = \frac{e}{m}\boldsymbol{E}(t).$$

We ignore the effect of the magnetic field on the oscillator because the factor $v/c \ll 1$ is small. We do not take into account the field inhomogeneity due to the condition $r \ll \lambda$. The solution of the equation can be written in the integral form

$$\boldsymbol{r}(t) = \frac{e}{m}\int_{-\infty}^{\infty} G(t-t')\boldsymbol{E}(t')\,dt', \tag{2}$$

where $G(t-t')$ is the Green's function satisfying the equation

$$\ddot{G} + \gamma\dot{G} + \omega_0^2 G = \delta(t-t'). \tag{3}$$

Let us find the particular solution of the latter equation, corresponding to the delta-function on the right-hand side, by the method of variation of constants, that is, in the form

$$G(\tau) = A(\tau)e^{s_1\tau} + B(\tau)e^{s_2\tau}, \qquad \tau = t - t', \tag{4}$$

where s_1 and s_2 are two roots of the characteristic equation $s^2 + \gamma s + \omega_0^2 = 0$:

$$s_{1,2} = \pm i\sqrt{\omega_0^2 - \frac{\gamma^2}{4}} - \frac{\gamma}{2}. \tag{5}$$

The functions $A(\tau)$ and $B(\tau)$ are determined from the system of equations

$$\dot{A}e^{s_1\tau} + \dot{B}e^{s_2\tau} = 0, \qquad s_1\dot{A}e^{s_1\tau} + s_2\dot{B}e^{s_2\tau} = \delta(\tau). \tag{6}$$

Integrating this system under the assumption, that at $\tau \to -\infty$ the oscillator was at rest (adiabatic switch-on of the field), we obtain

$$B(\tau) = -A(\tau) = \frac{i\Theta(\tau)}{2\sqrt{\omega_0^2 - \gamma^2/4}}, \tag{7}$$

where $\Theta(\tau)$ is the step function (I.1.212). As a result, we have

$$G(\tau) = (\omega_0^2 - \gamma^2/4)^{-1/2}\Theta(\tau)e^{-\gamma\tau/2}\sin(\sqrt{\omega_0^2 - \gamma^2/4}\,\tau). \tag{8}$$

With the help of Equations (2) and (8), we present the dipole oscillator moment $\boldsymbol{p}(t) = e\boldsymbol{r}(t)$ in the form

$$\boldsymbol{p}(t) = \int_{-\infty}^{\infty} f(t-t')\boldsymbol{E}(t')\,dt' = \int_{-\infty}^{t} f(t-t')\boldsymbol{E}(t')\,dt', \tag{9}$$

where the response function is

$$f(t-t') = \frac{e^2}{m}G(t-t'). \tag{10}$$

Due to the presence of the step function $\Theta(\tau)$ in Equation (8), Equation (9) describes the casual connection of the oscillator dipole moment with an external field: $\boldsymbol{p}(t)$ is determined by the field values at preceding instants of time $t' \le t$.

5.13 The medium polarization vector is $\boldsymbol{P}(t) = n\boldsymbol{p}(t)$, and the electric induction vector is $\boldsymbol{D}(t) = \boldsymbol{E}(t) + 4\pi n\boldsymbol{p}(t)$, where $\boldsymbol{p}(t)$ is given by Equations (9) and (10) of Problem 5.11. Calculating $\varepsilon(\omega)$ from Equation (5.67) and using the explicit form of the Green's function (8), we find

(1) $$\varepsilon(\omega) = 1 + \frac{\omega_{0e}^2}{\omega_0^2 - \omega^2 - i\gamma\omega}, \qquad \omega_{0e}^2 = \frac{4\pi n e^2}{m},$$

that is,

(2) $$\varepsilon'(\omega) = 1 + \frac{\omega_{0e}^2(\omega_0^2 - \omega^2)}{(\omega_0^2 - \omega^2)^2 + \gamma^2\omega^2}, \qquad \varepsilon''(\omega) = \frac{\gamma\omega_{0e}^2\omega}{(\omega_0^2 - \omega^2)^2 + \gamma^2\omega^2}.$$

The frequency dependence of the real and imaginary parts of ε is shown in Figure 5.2. The imaginary part, ε'', determining absorption of electromagnetic energy, noticeably differs from zero only in the vicinity of the eigenfrequency ω_0 of oscillations in medium. It is positive everywhere (at $\omega > 0$). In the frequency range near ω_0, the quantity ε' decreases with frequency increasing (the anomalous dispersion). Otherwise, ε' increases with frequency increasing (the normal dispersion).

5.14 The wave length of the electromagnetic field is much larger than the atomic size. Therefore, the operator of the atom-field interaction can be taken in the form (6.102), assuming the field to be a classical object:

(1) $$\hat{V}(t) = -\hat{\boldsymbol{d}}\cdot\boldsymbol{E}(t), \qquad \hat{\boldsymbol{d}} = e\sum_{a=1}^{Z}\hat{\boldsymbol{r}}_a.$$

Here $\boldsymbol{E}(t)$ is the real vector of the electric field strength, $\hat{\boldsymbol{d}}$ is the operator of the atomic dipole moment and the summation is over all electrons. Using the perturbation theory, we find the first order correction $\Phi^{(1)}(\boldsymbol{r}_1, \dots \boldsymbol{r}_Z, t)$

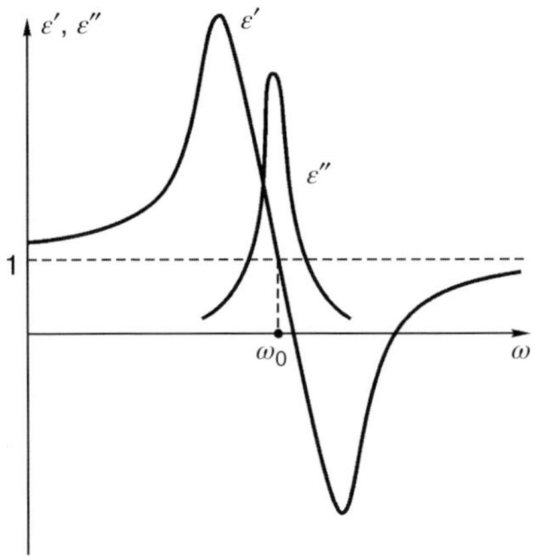

Figure 5.2 Schematic plot of the real and imaginary parts of dielectric permittivity versus frequency in the model of oscillator with one eigenfrequency ω_0.

to the wave function $\Phi^{(0)}(\boldsymbol{r}_1, \dots \boldsymbol{r}_Z, t)$ of the non-disturbed atom, satisfying the Schrödinger equation

(2) $$i\hbar\frac{\partial\Phi^{(0)}}{\partial t} = \widehat{H}_0\Phi^{(0)}.$$

In our problem, the electrons can be considered particles without spin, since the effect of an external field on the spin is weak and spin states of the atom do not change. Substituting $\Phi(\boldsymbol{r}_1, \dots \boldsymbol{r}_Z, t) = \Phi^{(0)}(\boldsymbol{r}_1, \dots \boldsymbol{r}_Z, t) + \Phi^{(1)}(\boldsymbol{r}_1, \dots \boldsymbol{r}_Z, t)$ in the Schrödinger equation (I.C29) with the Hamiltonian $\widehat{H} = \widehat{H}_0 + \widehat{V}(t)$, we obtain the approximate equation for $\Phi^{(1)}$:

(3) $$i\hbar\frac{\partial\Phi^{(1)}}{\partial t} - \widehat{H}_0\Phi^{(1)} = \widehat{V}(t)\Phi^{(0)},$$

where the term, bilinear in perturbation, is omitted.

Let us switch on the field at $t \to \infty$ slowly (adiabatically):

(4) $$\boldsymbol{E}(t) = \begin{cases} \boldsymbol{E}_0 e^{\alpha t}\cos\omega t, & t \le 0, \quad \alpha > 0, \\ \boldsymbol{E}_0 \cos\omega t, & t \ge 0, \end{cases}$$

where we should set $\alpha \to 0$ after all integrations over time. This is made to eliminate transient process of setting-up the stationary state. According to this choice, we write the initial conditions for Equation (3) as:

(5) $$\Phi|_{t\to-\infty} \to \varphi_0(\boldsymbol{r}_1, \dots \boldsymbol{r}_Z)e^{-iE_0 t/\hbar}, \qquad \Phi^{(1)}|_{t\to-\infty} \to 0.$$

Here and further on, by φ_s we denote the wave functions of stationary states of the non-disturbed atom; $s = 0$ corresponds to the ground state.

Let us find the solution of Equation (3) by expansion over eigenfunctions of the non-disturbed atom:

(6) $$\Phi^{(1)} = \sum_s c_s(t)\varphi_s(\boldsymbol{r}_1, \dots \boldsymbol{r}_Z)e^{-iE_s t/\hbar}.$$

Using orthogonality of the functions φ_s, we obtain from (3)

(7) $$\frac{dc_s}{dt} = -\frac{1}{i\hbar}(\boldsymbol{d}_{s0}\cdot\boldsymbol{E}(t))e^{i\omega_{s0}t},$$

where $\boldsymbol{d}_{s0}$ is the matrix element of the atomic dipole moment. Notice, that $\boldsymbol{d}_{00} = 0$, because the wave functions of the stationary states possess definite parity.

Integrating over time in Equation (7) and using Equation (4) and Equation (5), we obtain at finite t,

(8) $$c_s(t) = \frac{\boldsymbol{d}_{s0}\cdot\boldsymbol{E}_0}{2\hbar}\left(\frac{e^{i(\omega_{s0}+\omega)t}}{\omega_{s0}+\omega} + \frac{e^{i(\omega_{s0}-\omega)t}}{\omega_{s0}-\omega}\right).$$

This result can be used only for nonresonance frequencies, when the denominators in the latter expression are not small. The average quantum mechanical value of the induced dipole moment of the atom can be calculated from Equation (I.C15):

$$\text{(9)} \qquad \boldsymbol{d}(t) = \int \Phi^* \hat{\boldsymbol{d}} \Phi \, dq = \int (\Phi^{(0)*} \hat{\boldsymbol{d}} \Phi^{(1)} + \Phi^{(1)*} \hat{\boldsymbol{d}} \Phi^{(0)}) \, dq,$$

where $dq = d^3 r_1 \cdots d^3 r_Z$, the quadratic correction is neglected, and the relation $\boldsymbol{d}_{00} = 0$ is used. With the help of Equations (6), (8), and (9) we find

$$\text{(10)} \qquad \boldsymbol{d}(t) = \frac{1}{2\hbar} \sum_s \left\{ \left(\frac{e^{i\omega t}}{\omega_{s0} + \omega} + \frac{e^{-i\omega t}}{\omega_{s0} - \omega} \right) (\boldsymbol{d}_{s0} \cdot \boldsymbol{E}_0) \boldsymbol{d}_{0s} \right.$$

$$\left. + \left(\frac{e^{-i\omega t}}{\omega_{s0} + \omega} + \frac{e^{i\omega t}}{\omega_{s0} - \omega} \right) (\boldsymbol{d}_{s0} \cdot \boldsymbol{E}_0)^* \boldsymbol{d}_{s0} \right\}.$$

As clear from symmetry arguments, the macroscopic polarization vector is directed along the field vector $\boldsymbol{E}_0$, which, in our case, is real. Therefore, it is sufficient to calculate the projection of the vector $\boldsymbol{d}$ on the field direction. Aligning the field along the Ox axis and using the Hermitian character of the matrix $(d_x)_{0s} = (d_x)^*_{s0}$, we find from the general formula (10) the atomic polarization,

$$\text{(11)} \qquad d_x = \beta(\omega) E_0, \qquad \text{where} \qquad \beta(\omega) = \sum_s \frac{2\omega_{s0} |(d_x)_{s0}|^2}{\hbar(\omega_{s0}^2 - \omega^2)}.$$

From here we obtain the dielectric permittivity:

$$\text{(12)} \qquad \varepsilon(\omega) = 1 + 4\pi N \beta(\omega) = 1 + \frac{4\pi N}{\hbar} \sum_s \frac{2\omega_{s0} |(d_x)_{s0}|^2}{\hbar(\omega_{s0}^2 - \omega^2)}.$$

Let us compare (12) with the results of semiclassical theory (Equation (1) of Problem 5.13). The oscillator model requires the following modifications: (i) the introduction of many oscillators with different eigenfrequencies, $\omega_0 \to \omega_{s0}$; (ii) summation over all such oscillators; and (iii) as far as the damping is concerned, we can set $\gamma = 0$, if $\gamma \ll \omega_0$, far from resonances.

Introducing the ratio $f_s = n_s / n$, which is the fraction of electrons, belonging to an oscillator s, we present the semiclassical formula (1) from the Problem 5.13 in the form of the quantum formula (12),

$$\text{(13)} \qquad \varepsilon(\omega) = 1 + \frac{4\pi N e^2}{m} \sum_s \frac{f_s}{\omega_{s0}^2 - \omega^2},$$

if the quantities f_s are determined in terms of matrix elements of the dipole moment:

$$\text{(14)} \qquad f_s = \frac{2m\omega_{s0}}{e^2 \hbar} |(d_x)_{s0}|^2, \qquad (d_x)_{s0} = e \sum_{a=1}^{Z} x^a_{s0}.$$

The quantities f_s are referred to as the **oscillator strengths**; their definition has already been given in Problem 6.61, where we have obtained that the oscillator strengths satisfy the sum rule $\sum_s f_s = Z$. Therefore, at $\omega \gg \omega_{s0}$ the quantum formulas (12) and (13) take the universal form, found in Problem 5.7; the average number of electrons per unit volume is $n = NZ$.

The dielectric permittivity, derived in this problem, is not applicable near the frequencies of spectral lines.

5.15 Near the resonance, $\omega \approx \omega_{s0}$, the population of state s by an external field increases infinitely (see Equation (8) of Problem 5.14). This is because an atom absorbs the quanta with energy $\hbar\omega$ from the external field. However, in addition to transitions of atoms in excited states, there will also exist inverse processes of spontaneous quantum emissions by excited atoms. Absorptions and emissions will have tendency to be balanced; the population of state s will take then some stationary value. We will study this effect in the model of the two-level atom already considered in Example I.6.11.

Along with the interaction with the external field $\boldsymbol{E}(t)$, we will include the interaction of the atom with the quantized electromagnetic field of photons. The operator of this interaction (6.102) will be denoted here as $\widehat{V}'$:

(1)
$$\widehat{V}' = -\widehat{\boldsymbol{d}}\cdot\widehat{\boldsymbol{E}}.$$

In this case, as in Example 6.11, we will take into account only vacuum and single-photon states of the quantized field. The state vector, describing the system of the atom in the external field + quantized field, can be written in the form analogous to Equation (1) from Example 6.11:

(2)
$$|\Phi(t)\rangle = c_0(t)\varphi_0|0\rangle e^{-iE_0t/\hbar} + \sum_k c_k(t)\varphi_0|1_k\rangle e^{-i(\omega_k+E_0/\hbar)t} + c_s(t)\varphi_s|0\rangle e^{-iE_st/\hbar},$$

where $\hbar\omega_k$ is the energy of the photon emitted by the atom, $k = (\boldsymbol{k}, \sigma)$ are photon's quantum numbers, and $c_k(t)$ are the amplitudes of the single-photon states. The sum over k actually reduces to the integration over frequencies and directions of the photon propagation, along with the summation over polarizations. The sum over s in Equation (2) is absent, since we consider the two-level system, and the index s relates to the upper level.

Substituting the expansion (2) in the Schrödinger equation (I.A3.29), we obtain the system of equations for the coefficients:

(3a)
$$\frac{dc_0}{dt} = -\frac{i}{\hbar}\langle 00|\widehat{V}|s0\rangle e^{i\omega_{s0}t}c_s, \qquad \omega_{s0} = \frac{E_s - E_0}{\hbar},$$

(3)
$$\frac{dc_k}{dt} = -\frac{i}{\hbar}\langle 01_k|\widehat{V}'|s0\rangle e^{i(\omega_k-\omega_{s0})t}c_s,$$

(4)
$$\frac{dc_s}{dt} = -\frac{i}{\hbar}\langle s0|\widehat{V}|00\rangle e^{i\omega_{s0}t}c_0 - \frac{i}{\hbar}\sum_k \langle s0|\widehat{V}'|01_k\rangle e^{-i(\omega_k-\omega_{s0})t}c_k.$$

Here we have used the relations $\boldsymbol{d}_{ss} = \boldsymbol{d}_{00} = 0$ and $\langle 0|1_k\rangle = 0$. Since we take into account only the spontaneous transitions from the excited level, the transient process will be damped out. Accordingly, in this case there is no need for the adiabatic switching-on of the external field. We assume that it is switched on at $t = 0$, and then the initial conditions are $c_s(0) = c_k(0) = 0$ and $c_0(0) = 1$. Let us present Equations (3a) and (3) in the integral form:

$$c_0(t) = 1 - \frac{i}{\hbar}\int_0^t \langle 00|\hat{V}|s0\rangle e^{-i\omega_{s0}t'} c_s(t')dt',$$

$$c_k(t) = -\frac{i}{\hbar}\langle 01_k|\hat{V}'|s0\rangle \int_0^t e^{i(\omega_k-\omega_{s0})t'} c_s(t')\, dt'.$$

The first equation implies, that the difference of $c_0(t)$ from unity is of the second order of smallness in perturbation, which we neglect, setting $c_0(t) = 1$. The second equation is rearranged by integrating by parts,

$$c_k(t) = -\frac{1}{\hbar}\langle 01_k|\hat{V}'|s0\rangle e^{i(\omega_k-\omega_{s0})t} \int_0^t \frac{1 - e^{i(\omega_k-\omega_{s0})(t'-t)}}{\omega_k - \omega_{s0}} \frac{dc_s}{dt'}\, dt'. \tag{5}$$

Substitution of the obtained result in Equation (4) enables us to derive the inhomogeneous integro-differential equation for $c_s(t)$:

$$\frac{dc_s}{dt} = -\frac{i}{\hbar}\langle s0|\hat{V}|00\rangle e^{i\omega_{s0}t} + \frac{i}{\hbar^2}\sum_k |\langle s0|\hat{V}'|01_k\rangle|^2 \int_0^t \frac{1 - e^{i(\omega_{s0}-\omega_k)(t-t')}}{\omega_k - \omega_{s0}} \frac{dc_s}{dt'}\, dt'. \tag{6}$$

We consider a rather large interval of time, during which the excitation of atoms by an external field and their spontaneous radiation evolve to a stationary state. At large t, the interval $t - t'$ in the exponent in the integrand will be also large. Let use Equations (I.1.216) and (I.1.224a) and present the integrand in the form

$$\left.\frac{1 - e^{i(\omega_{s0}-\omega_k)(t-t')}}{\omega_k - \omega_{s0}}\right|_{t-t'\to\infty} \to \frac{\mathcal{P}}{\omega_k - \omega_{s0}} + i\pi\delta(\omega_k - \omega_{s0}). \tag{7}$$

After substituting the right-hand side of this equality in Equation (6), the first term (principal value) will describe a small radiation frequency shift of the transition (Lamb shift); we will ignore it. The term with the delta-function is the one-half constant of damping, induced be spontaneous radiation (see Equation (9) from Example 6.11):

$$\frac{\gamma_s}{2} = \frac{\pi}{\hbar^2}\sum_\sigma \int |\langle s0|\hat{V}'|01_k\rangle|^2\, \delta(\omega_k - \omega_{s0})\, \frac{\mathcal{V}\omega_k^2\, d\omega_k\, d\Omega}{(2\pi c)^3}. \tag{8}$$

Here the summation over the discrete modes is replaced by the integration over photon frequencies and over directions of photon propagation, only the summation over polarizations remains. We will also simplify the first term on the right-hand side of Equation (6), retaining the principal (resonance) term alone. Then Equation (6) takes the form

$$\frac{dc_s}{dt} + \frac{\gamma_s}{2}c_s = -\frac{i}{2\hbar}\boldsymbol{d}_{s0}\cdot\boldsymbol{E}_0 e^{i(\omega_{s0}-\omega_k)t}. \tag{9}$$

The solution of the derived equation with $\gg \gamma_s^{-1}$ is

(10) $$c_s(t) = \boldsymbol{d}_{s0}\cdot\boldsymbol{E}_0 \frac{e^{i(\omega_{s0}-\omega_k)t}}{2\hbar(\omega_{s0} - \omega_k - i\gamma_s/2)}.$$

In comparison with Equation (8) of Problem 5.14, here the resonance value of the atom excitation amplitude is finite.

Using Equation (10), we find the dielectric permittivity at frequencies close to one of the resonances:

(11) $$\varepsilon(\omega) = 1 + \frac{4\pi N}{\hbar} \frac{|(d_x)_{s0}|^2}{\omega_{s0} - \omega - i\gamma_s/2},$$

where the index k at the frequency is dropped now. The effect of radiation damping is responsible for the complexity of the dielectric permeability. At arbitrary frequencies (and at $\gamma_s \ll \omega_{s0}$), it is possible to use the approximate interpolation formula of the form

(12) $$\varepsilon(\omega) = 1 + \frac{4\pi N e^2}{m} \sum_s \frac{f_s}{\omega_{s0}^2 - \omega^2 - i\gamma_s\omega}.$$

Far from the resonance it reproduces the quantum equation (13) of Problem 5.14, while near one of the resonances it gives the result close to Equation (11).

5.16 $\varepsilon = 1 + 4\pi N a^3, \qquad \mu = 1 - 2\pi N a^3 < 1.$
Such a dielectric is diamagnetic. The permeabilities ε and μ are independent of frequency due to the assumption on the ideal conductivity of the spheres. In order to consider, the artificial dielectric as a continuous medium, the conditions $\lambda \gg l$ and $\lambda \gg a$ have to be fulfilled, where λ is the wave length, and l is the average distance between spheres. The difference of the actual field from the average one can be ignored only at small polarization of the medium (i.e., at $4\pi N a^3 \ll 1$).

5.17 The equation of motion for a conduction electron will be written in the form

(1) $$m\ddot{\boldsymbol{r}} + \eta\dot{\boldsymbol{r}} = e\boldsymbol{E}_0 e^{-i\omega t}.$$

Its particular solution, corresponding to eigenmodes, is

(2) $$\boldsymbol{r} = -\frac{e\boldsymbol{E}_0 e^{-i\omega t}}{m(\omega^2 + i\gamma\omega)},$$

where $\gamma = \eta/m$.

We obtain the dipole moment per unit volume multiplying $\boldsymbol{r}$ by an electron charge e and by a number of particles per unit volume, N. Then we determine the polarizability of the medium $\alpha(\omega)$ and the dielectric permittivity $\varepsilon(\omega)$, induced by conduction electrons:

(3) $$\varepsilon(\omega) = 1 + 4\pi\alpha(\omega) = 1 - \frac{\omega_{0e}^2}{\omega^2 + i\gamma\omega}, \qquad \omega_{0e}^2 = \frac{4\pi e^2 N}{m}.$$

Using Equation (1) and Ohm's law for a constant current, we will relate the specific resistance ρ on a constant current and the coefficient η:

(4) $$\rho \equiv \frac{1}{\kappa} = \frac{\eta}{Ne^2}.$$

Rewriting Equation (3) in the form

(5) $$\varepsilon(\omega) = 1 + i\,\frac{4\pi}{\omega}\,\frac{Ne^2}{m(\gamma - i\omega)},$$

we find the dependence of the electric conductivity on frequency:

(6) $$\kappa(\omega) = \frac{Ne^2}{m(\gamma - i\omega)}.$$

To include the effect of bound electrons, one should add to the right-hand side of Equation (3) the term $4\pi\alpha_i(\omega)$ containing the ion polarizability:

(7) $$\varepsilon(\omega) = 1 + 4\pi\alpha_i(\omega) + i\,\frac{4\pi\kappa(\omega)}{\omega}.$$

At sufficiently low frequencies, we can neglect dispersion of the ion permeability and rewrite Equation (7) as

(8) $$\varepsilon(\omega) = \varepsilon_0 + i\,\frac{4\pi\kappa(\omega)}{\omega}, \qquad \text{where} \qquad \varepsilon_0 = 1 + 4\pi\alpha_i(0).$$

Let us estimate $\gamma = \eta/m$ for copper (the static conductivity $\kappa = 5 \times 10^{17}\ \mathrm{s}^{-1}$). From Equation (4), we obtain:

$$\gamma = \frac{Ne^2}{\kappa m} = \frac{N_0 e^2 d}{\kappa m A},$$

where $N_0 \approx 6 \cdot 10^{23}\ \mathrm{mol}^{-1}$ is the Avogadro number, $A \approx 63.5\ \mathrm{g/mol}$ is the atomic weight and $d \approx 8.9\,\mathrm{g/cm}^3$ is the copper mass density. The estimate gives $\gamma \approx 10^{14}\ \mathrm{s}^{-1}$; let us note for comparison, that optical spectrum refers to frequencies $\approx 10^{15}\ \mathrm{s}^{-1}$.

Therefore, in this case, it is likely that the conductivity retains its static value, up to frequencies of infrared range. However, one should bear in mind, that at high frequencies, when the mean free path of electrons becomes comparable with the penetration depth of the field into the metal, the effects of spatial inhomogeneity begin to be important (abnormal skin-effect) and the static conductivity κ becomes inappropriate.

For a semiconductor (germanium), we have $\kappa \approx 2 \times 10^{10}\ \mathrm{s}^{-1}$, $\gamma \approx 10^{12} - 10^{13}\ \mathrm{s}^{-1}$, and $\varepsilon_0 = 16$.

The results of this problem are applicable, in the restricted frequency region, to metals, semiconductors and ionized gases, if the motion of positive ions can be ignored. The calculation of the dielectric permittivity of plasma with account for the motion of positive ions will be given below in Problem 5.25.

5.18 Molecules in a dielectric are not spherically symmetric. Therefore, the external field $\boldsymbol{E}_0$ partially orients them, and the dielectric becomes anisotropic. However, the orienting effect of alternating field can be ignored due to the condition $\mathcal{E} \ll E_0$. Because the anisotropy is caused by the external electric field $\boldsymbol{E}_0$, one of the principal axes of the dielectric permittivity tensor will coincide with its direction, the other two principal axes will be perpendicular to $\boldsymbol{E}_0$.

Let us denote the components of the molecule polarization in these axes in terms of β'_{ik} (the values i, $k = 1$ correspond to the axes parallel to $\mathbf{E}_0$). The components β'_{ik} are expressed through the principal values $\beta^{(i)}$ by the conventional formula:

$$\beta'_{ik} = \alpha_{il}\alpha_{km}\beta_{lm} = (\beta - \beta')\alpha_{i1}\alpha_{k1} + \beta'\delta_{ik},$$

where α_{il} are cosines of angles between the molecule symmetry axes and the principal axes of the dielectric permittivity tensor (we use here the relation $\alpha_{il}\alpha_{kl} = \delta_{ik}$, which follows from orthogonality of the matrix α_{ik}). To calculate the dielectric permittivity tensor per unit volume of the dielectric, it is necessary to use the Boltzmann distribution and find the statistic average values of β'_{ik}, that is, to average the product $\alpha_{i1}\alpha_{k1}$.

If we denote in the primed reference frame the polar angles of the molecule symmetry axis as ϑ and φ, then the quantities α_{1i} could be written as:

$$\alpha_{11} = \cos\vartheta, \qquad \alpha_{12} = \sin\vartheta\,\cos\varphi, \qquad \alpha_{13} = \sin\vartheta\,\sin\varphi.$$

Performing the averaging with the Boltzmann distribution (1.30) and retaining the terms linear in $a = (\beta_0 - \beta'_0)E_0^2/2T$, we obtain:

$$\begin{aligned}\overline{\alpha_{11}^2} &= \frac{1}{3}\left(1 + \frac{4}{15}a\right),\\ \overline{\alpha_{12}^2} = \overline{\alpha_{13}^2} &= \frac{1}{3}\left(1 - \frac{2}{15}a\right),\\ \overline{\alpha_{i1}\alpha_{k1}} &= 0 \qquad \text{at} \qquad i \neq k,\end{aligned}$$

where β_0 and β'_0 are the static values of the molecule polarization tensor. Then

$$\begin{aligned}\overline{\beta'_{11}} &= \frac{1}{3}(\beta - \beta')\left(1 + \frac{4}{15}a\right) + \beta',\\ \overline{\beta'_{22}} = \overline{\beta'_{33}} &= \frac{1}{3}(\beta - \beta')\left(1 - \frac{2}{15}a\right) + \beta'.\end{aligned}$$

Neglecting the difference between the actual and average fields, we obtain the principal values of the dielectric permittivity tensor:

$$\varepsilon^{(1)} = 1 + 4\pi N\overline{\beta'_{11}}, \qquad \varepsilon^{(2)} = \varepsilon^{(3)} = 1 + 4\pi N\overline{\beta'_{22}}.$$

This result shows that in a strong constant electric field, the dielectric becomes anisotropic with respect to high frequency oscillations (for example, oscillations in optical range). The appearance of the anisotropy under the action of constant electric field is known as the **Kerr**[10] **effect.**

10) Kerr John (1824–1907), Scottish physicist, discovered for the first time the relation between the optical and electric phenomena.

The inertia of this effect is very low: a time for onset and disappearance of anisotropy is of the order of 10^{-10} s, being determined by a time for onset of statistic equilibrium in dielectrics. The Kerr effect is widely used in industry for the fast modulation of light intensity.

5.19 Considering the parameter $pE_0/T = a$ to be small and keeping the terms of the order a^2, we obtain:

$$\overline{\beta'_{11}} = \frac{1}{3}(\beta - \beta')\left(1 + \frac{2}{15}a^2\right) + \beta',$$
$$\overline{\beta'_{22}} = \overline{\beta'_{33}} = \frac{1}{3}(\beta - \beta')\left(1 - \frac{1}{15}a^2\right) + \beta',$$
$$\varepsilon^{(1)} = 1 + 4\pi N\overline{\beta'_{11}}, \quad \varepsilon^{(2)} = \varepsilon^{(3)} = 1 + 4\pi N\overline{\beta'_{22}}.$$

The notations are the same as in Problem 5.18.

5.20 Let the field amplitude $\boldsymbol{\mathcal{E}}$ increase by $d\boldsymbol{\mathcal{E}} = (d\mathcal{E}_x, d\mathcal{E}_y, d\mathcal{E}_z)$. In this case, the work done on a molecule is:

(1) $$dA = \frac{1}{2}\Re(\boldsymbol{p}\cdot d\boldsymbol{\mathcal{E}}^*) = \frac{1}{4}(\boldsymbol{p}\cdot d\boldsymbol{\mathcal{E}}^* + \boldsymbol{p}^*\cdot d\boldsymbol{\mathcal{E}}),$$

where $p_i = \beta_{ik}E_k$ is a component of the dipole moment of the molecule. The work is averaged over period of field variation. Since the energy absorption is absent, this work is spent entirely to increase the average potential energy of the molecule in the external field:

$$dA = d\overline{W}.$$

Therefore, dA must be the total differential of a certain function of the field amplitude, that is, of the system energy. Let us rewrite $d\overline{W}$ in the form

(2) $$d\overline{W} = \frac{1}{4}\sum_{i,k}(\beta_{ik}\mathcal{E}_k\, d\mathcal{E}_i^* + \beta_{ki}^*\mathcal{E}_i^*\, d\mathcal{E}_k).$$

It is seen, that this value is the total differential only if $\beta_{ik} = \beta_{ki}^*$; then

$$d\overline{W} = \frac{1}{4}\sum_{i,k}\beta_{ik}(\mathcal{E}_k\, d\mathcal{E}_i^* + \mathcal{E}_i^*\, d\mathcal{E}_k) = \frac{1}{4}\sum_{i,k}\beta_{ik}\, d(\mathcal{E}_i^*\mathcal{E}_k) = d\left(\frac{1}{4}\boldsymbol{p}\cdot\boldsymbol{\mathcal{E}}^*\right),$$

or

$$\overline{W} = \frac{1}{4}\boldsymbol{p}\cdot\boldsymbol{\mathcal{E}}^*.$$

In the same way, one can prove the Hermitian character of the magnetic polarization tensor for a non-dissipative system.

5.22 The equation of motion for an atomic electron, bound with a nucleus by an elastic force, can be written as

(1) $$\ddot{\boldsymbol{r}} + \omega_0^2\mathbf{r} = \frac{e}{m}\left[\mathbf{E}_0e^{-i\omega t} + \left(\frac{\boldsymbol{v}}{c}\times\mathbf{H}_0\right)\right],$$

where ω_0 is the eigenfrequency. Solving it by the method of iterations, in the linear over H_0 approximation we obtain:

(2) $$\boldsymbol{r} = \frac{e\boldsymbol{E}}{m(\omega_0^2 - \omega^2)} - i\frac{e^2\omega}{m^2c(\omega_0^2 - \omega^2)^2}(\boldsymbol{E}\times\boldsymbol{H}_0).$$

To obtain the tensor of atomic polarization, we use the form of the vector product containing the antisymmetric tensor e_{ikl}. This gives

(3) $$\beta_{ik} = \frac{e^2}{m(\omega_0^2 - \omega^2)}\delta_{ik} - i\frac{e^3\,\omega H_{0l}}{m^2c(\omega_0^2 - \omega^2)^2}e_{ikl}.$$

According to the general statement, proved in Problem 5.20, this tensor is Hermitian. The gyration vector (see Problem 5.21) in this case has the form

(4) $$\mathbf{g} = -\frac{e^3\,\omega}{m^2c(\omega_0^2 - \omega^2)^2}\boldsymbol{H}_0 = \frac{2e^2\,\omega}{m(\omega_0^2 - \omega^2)^2}\boldsymbol{\omega}_L,$$

where $\boldsymbol{\omega}_L = -e\boldsymbol{H}_0/2mc$ is the Larmor frequency (see Problem I.4.102), and $e < 0$ is the electron charge.

5.23

$$\varepsilon_{ik} = \begin{pmatrix} (\varepsilon^+ + \varepsilon^-)/2 & -i(\varepsilon^+ - \varepsilon^-)/2 & 0 \\ i(\varepsilon^+ - \varepsilon^-)/2 & (\varepsilon^+ + \varepsilon^-)/2 & 0 \\ 0 & 0 & \varepsilon^0 \end{pmatrix},$$

where, in the notations of Problem 5.22,

$$\varepsilon^{\pm} = 1 - \frac{\omega_{0e}^2}{\omega(\omega \pm 2\,\omega_L) - \omega_0^2}, \qquad \varepsilon^0 = 1 - \frac{\omega_{0e}^2}{\omega^2 - \omega_0^2}, \qquad \omega_{0e}^2 = \frac{4\pi e^2 N}{m}.$$

The gyration vector is equal to

$$g = -\frac{1}{2}(\varepsilon^+ - \varepsilon^-),$$

and is directed along the Oz-axis.

The result of Problem 5.22 is derived from the found exact solution under the condition $2\,\omega_L\,\omega \ll |\,\omega_0^2 - \omega^2|$.

5.24 The tensor ε_{ik} has the same form as in Problem 5.23, but its components $\varepsilon^{\pm}$ and ε^0 are determined by following relations:

$$\varepsilon^{\pm} = 1 - \frac{\omega_{0e}^2}{\omega^2 + \omega(i\gamma \pm 2\omega_L)},$$

$$\varepsilon^0 = 1 - \frac{\omega_{0e}^2}{\omega^2 + i\omega\gamma},$$

$$\omega_{0e}^2 = \frac{4\pi e^2 N}{m}, \qquad \gamma = \frac{\eta}{m}, \qquad \omega_L = -\frac{eB}{2mc} > 0.$$

Due to the presence of "friction" ($\eta \geq 0$) in the electron gas, the energy dissipation occurs, and the tensor ε_{ik} is not Hermitian. Ignoring the friction, we obtain the Hermitian tensor

$$\varepsilon_{ik} = \begin{pmatrix} \varepsilon_\perp & -i\varepsilon_a & 0 \\ i\varepsilon_a & \varepsilon_\perp & 0 \\ 0 & 0 & \varepsilon_\parallel \end{pmatrix}, \qquad \varepsilon_\perp = 1 - \frac{\omega_{0e}^2}{\omega^2 - \omega_{Be}^2}, \qquad \varepsilon_a = \frac{\omega_{0e}^2\omega_{Be}}{\omega(\omega^2 - \omega_{Be}^2)},$$

$$\varepsilon_\parallel = 1 - \frac{\omega_{0e}^2}{\omega^2},$$

where $\omega_{Be} = -2\omega_L = eB/mc$ is the cyclotron frequency of electrons.

5.25 Let us denote the mass, velocity, and charge of an electron as m, $\dot{\boldsymbol{r}}$, and $-e$, respectively. Let the same quantities for an ion be M, $\dot{\boldsymbol{R}}$, and $+e$. Then we obtain the following system of equations of motion:

$$(1)\qquad \left.\begin{aligned} m\ddot{\boldsymbol{r}} &= -e\boldsymbol{E}_0 e^{-i\omega t} - \tfrac{e}{c}(\dot{\boldsymbol{r}}\times\boldsymbol{B}) - m\gamma(\dot{\boldsymbol{r}}-\dot{\boldsymbol{R}}), \\ M\ddot{\boldsymbol{R}} &= e\boldsymbol{E}_0 e^{-i\omega t} + \tfrac{e}{c}(\dot{\boldsymbol{R}}\times\boldsymbol{B}) - m\gamma(\dot{\boldsymbol{R}}-\dot{\boldsymbol{r}}). \end{aligned}\right\}$$

Here $\boldsymbol{B}$ is a constant and homogeneous magnetic field, $m\gamma$ is the "friction" coefficient; the friction force is proportional to the relative velocity of electrons and ions, that is, to the differences $(\dot{\boldsymbol{r}}-\dot{\boldsymbol{R}})$ and $(\dot{\boldsymbol{R}}-\dot{\boldsymbol{r}})$ for the electrons and ions, respectively. The electric field $\boldsymbol{E}=\boldsymbol{E}_0\, e^{-i\omega t}$ depends on time harmonically.

Let us find the solution of Equation (1) in the form

$$(2)\qquad \boldsymbol{r}=\boldsymbol{r}_0\, e^{-i\omega t}, \qquad \boldsymbol{R}=\boldsymbol{R}_0\, e^{-i\omega t}.$$

Let $\boldsymbol{B}$ be directed along the Oz-axis. We introduce the cyclic components of the vectors $\boldsymbol{r}_0$ and $\boldsymbol{R}_0$ as

$$r_{0\pm1} = \mp\frac{1}{\sqrt{2}}(r_{0x}\pm ir_{0y}), \qquad R_{0\pm1} = \mp\frac{1}{\sqrt{2}}(R_{0x}\pm ir_{0y}).$$

Let us substitute Equation (2) into Equation (1) and combine the obtained equations:

$$-i\omega(m\boldsymbol{r}_0+M\boldsymbol{R}_0) = \frac{e}{c}[(\boldsymbol{R}_0-\boldsymbol{r}_0)\times\boldsymbol{B}].$$

The left-hand side of the latter equation can be written in the form

$$-i\omega[(M+m)\boldsymbol{R}_0 + m(\boldsymbol{r}_0-\boldsymbol{R}_0)].$$

Neglecting m in comparison with M, we obtain

$$(3)\qquad \omega R_{0\pm1} = \left(\pm\Omega_B + \omega\frac{m}{M}\right) s_{\pm1},$$

where

$$\Omega_B = \frac{eB}{Mc}, \qquad \boldsymbol{s}=\boldsymbol{R}_0-\boldsymbol{r}_0.$$

Then we divide the first Equation (1) by m, the second by M, and subtract the results. Next we neglect all terms, containing M in the denominator. Denoting $\omega_B = eB/mc$ and using Equation (2), we obtain

$$(4)\qquad \left.\begin{aligned} (-i\omega+\gamma\mp i\omega_B)s_{\pm1} \mp i\omega_B R_{0\pm1} &= \tfrac{e}{m}E_{0\pm1}, \\ -(\omega^2+i\omega\gamma)s_z &= \frac{eE_{0z}}{m}. \end{aligned}\right\}$$

From Equations (3) and (4) we find $\boldsymbol{s}$.

The polarization vector $\boldsymbol{P}$ is calculated as $\boldsymbol{P}=Ne\boldsymbol{s}\, e^{-i\omega t}$, where N is the number of ions (equal to the number of electrons) per unit volume.

The components of the dielectric permittivity tensor will be written in the form

$$\text{(5)} \qquad \varepsilon^{\pm} = 1 - \frac{\omega_{0e}^2}{\omega(\omega + i\gamma \pm \omega_B - \omega_B\Omega_B/\omega)}, \qquad \varepsilon^{(z)} = 1 - \frac{\omega_{0e}^2}{\omega(\omega + i\gamma)}.$$

The component $\varepsilon^{(z)}$ is similar to that of the scalar dielectric permittivity in the absence of the magnetic field (as obtained in Problem 5.17; it infinitely increases at $\omega \to 0$). The components $\varepsilon^{\pm}$, with account for the ion motion, contain in the denominator the extra term $\omega_B\Omega_B$; it can be omitted at $\Omega_B/\omega \ll 1$, that is, at large frequencies ω. However, at small frequencies, this term becomes significant; at $\omega \to 0$ it makes the components $\varepsilon^{\pm}$ finite: $\varepsilon^{\pm} = 1 + \omega_{0e}^2/\omega_B\Omega_B$. Owing to this, the waves of fairly small frequency (magneto-hydrodynamic waves) may exist in plasma. These waves have already been considered in Section 4.3. Based on the obtained tensor (5), their propagation is analyzed below in Problem 6.56.

5.26 In the reference frame, whose Ox_3-axis coincides with the preferential direction, the tensor T_{ik} must have the form

$$T_{ik} = \begin{pmatrix} T & T_a & 0 \\ -T_a & T & 0 \\ 0 & 0 & T_{\|} \end{pmatrix}.$$

This agrees with the results obtained in the Problems 5.23, 5.24, and others.

5.27 According to Equation (5.76), $\varepsilon^{-1}(k,\omega)$ is the causal response function. It is analytical, that is, it has no poles and other singularities in the upper half-plane of complex frequency. In view of the absence of the poles, the inverse function $\varepsilon(k,\omega)$ does not have zeros in the mentioned plane.

5.30

$$\varepsilon(\omega) = 1 + (\varepsilon_0 - 1)/(1 - i\omega\tau),$$

where ε_0 is the static dielectric permittivity. The derived relation is known as the **Debye formula** for the polar dielectric. The model, used for calculating the relaxation constant τ, is considered in the manual of Pamyatnykh and Turov (2000). The static dielectric permittivity of the gaseous dielectric with hard dipoles has been calculated in Example 1.3.

5.31

$$\varepsilon'(\omega) = 1 + (\varepsilon_0 - 1)/(1 + \omega^2\tau^2).$$

5.33

$$\varepsilon(\omega) = 1 + i\frac{4\pi\kappa_0}{\omega} - \frac{i}{\pi}\mathcal{P}\int_{-\infty}^{\infty}\frac{\varepsilon(z) - 1}{z - \omega}\,dz,$$

where κ_0 is the static electric conductivity.

5.34

$$\kappa(\omega) = -\frac{i}{\pi}\mathcal{P}\int_{-\infty}^{\infty}\frac{\kappa(z)}{z - \omega}\,dz = -\frac{i}{\pi}\mathcal{P}\int_{0}^{\infty}[\kappa(\omega + z) - \kappa(\omega - z)]\,\frac{dz}{z}.$$

5.35 Using the causality condition in the conventional (nonrelativistic) form, $t - t' \geq 0$, we present the Fourier transform of the causal response function in the form similar to Equation (5.36):

$$R(\boldsymbol{k},\omega) = \int_0^{\infty} d\tau\, e^{i\omega\tau} \int d^3r\, e^{-i\mathbf{k}\cdot\mathbf{r}} R(\boldsymbol{k},\tau), \tag{1}$$

where the integration is over the entire three-dimensional space. The relativistic causality principle $\tau \geq r/c$ leads to the condition

$$R(\boldsymbol{r},\tau)[1 - sign(\tau - r/c)] = 0, \tag{2}$$

where $sign\, x = x/|x|$ is the sign function. The equality (2) means, that $R(\boldsymbol{r},\tau) = 0$ at $\tau < r/c$. In order to impose restrictions, following from Equation (2), on the function $R(\boldsymbol{k},\omega)$, we will perform Fourier transform (1) with respect to time in Equation (2). For the sake of convenience, we present Equation (2) in a more general form:

$$R(\boldsymbol{r},\tau)[1 - sign(\tau - \boldsymbol{u}\cdot\boldsymbol{r}/c^2)] = 0. \tag{2a}$$

Equation (2a) coincides with Equation (2) in the particular case in which $\boldsymbol{u} \parallel \boldsymbol{r}$ and $u = c$. We present the sign function in the integral form

$$sign\, x = \frac{1}{i\pi}\mathcal{P}\int_{-\infty}^{\infty} \frac{d\eta}{\eta} e^{i\eta x}. \tag{3}$$

Using Equations (1) and (3) we obtain from Equation (2a)

$$R(\boldsymbol{k},\omega) = \frac{1}{i\pi}\int_0^{\infty} \frac{d\eta}{\eta}[R(\boldsymbol{k} + \eta\boldsymbol{u}/c^2, \omega + \eta) - R(\boldsymbol{k} - \eta\boldsymbol{u}/c^2, \omega - \eta)]$$

$$= \frac{1}{i\pi}\mathcal{P}\int_{-\infty}^{\infty} \frac{R(\boldsymbol{k} + \boldsymbol{u}(z-\omega)/c^2, z)}{z - \omega} dz. \tag{4}$$

In the isotropic medium at $\boldsymbol{u} \parallel \boldsymbol{k}$, the dispersion relations (4) take the form

$$R(\boldsymbol{k},\omega) = \frac{1}{i\pi}\mathcal{P}\int_{-\infty}^{\infty} \frac{R(k + u(z-\omega)/c^2, z)}{z - \omega} dz. \tag{5}$$

Equation (2) implies $u = c$. However, Leontovich (1961) has shown that Equation (5) remains true at any $u \leq c$, if it is valid at $u = c$. This result imposes more severe limitation than nonrelativistic relations of the type (5.83) and (5.84), since it touches the dependence not only on frequency, but on wave vector as well. At $c \to \infty$, it reduces to the conventional Kramers–Kronig dispersion relation.

5.38

$$\frac{\partial w}{\partial t} + div\,\boldsymbol{\gamma} = -\boldsymbol{j}_{ext}\cdot\boldsymbol{E}, \text{ where } w = \frac{\varepsilon_{\alpha\beta}E_\alpha E_\beta}{8\pi} + \frac{\mu_{\alpha\beta}H_\alpha H_\beta}{8\pi} = \frac{\boldsymbol{E}\cdot\boldsymbol{D} + \boldsymbol{H}\cdot\boldsymbol{B}}{8\pi},$$

and the vector $\boldsymbol{\gamma}$ is given by the same Equation (5.100).

5.40 For dielectrics, we have $\varepsilon_l''(0,0) = 0$ and $\varepsilon_l'(0,0) = \varepsilon(0,0)$. From the dispersion relation (5.73) we find

$$\varepsilon(0) = 1 + \frac{2}{\pi}\int_0^\infty \frac{\varepsilon''(z)}{z}\,dz > 1,$$

since $\varepsilon''(z) > 0$ due to the inequalities (5.93).

5.41 From the symmetry relations (5.39), we have $\varepsilon_l''(k,0) = 0$ and $\varepsilon'(k,0) = \varepsilon_l(k,0)$. From Equation (5.81) we find

(1)
$$\varepsilon_l^{-1}(k,0) = 1 + \frac{1}{\pi}\int_0^\infty \frac{dz^2}{z^2}\Im\,\varepsilon_l^{-1}(k,z) = 1 - \frac{1}{\pi}\int_0^\infty \frac{\varepsilon_l''(k,z)}{|\varepsilon_l(k,z)|^2}\frac{dz^2}{z^2} < 1.$$

The latter inequality follows from the condition (5.93), $\varepsilon_l''(k,z) \geq 0$. From (1), we find two regions of $\varepsilon_l(k,0)$:

(2) $$\varepsilon_l(k,0) > 1; \qquad \varepsilon_l(k,0) < 0.$$

The first region has been found for the collisionless plasma in Example 5.6. When the second inequality is fulfilled, the potential of the screened charge, $\varphi(k) = 4\pi q/k^2\varepsilon_l(k,0)$, changes sign in comparison with the potential of the same charge in vacuum (see Problem 5.8). This means that two equal charges q in the medium with $\varepsilon_l(k,0) < 0$ will experience the attraction: their interaction energy is $U(k) = q\varphi(k) = 4\pi q^2/k^2\varepsilon_l(k,0) < 0$. This property is responsible for the formation of Cooper pairs, resulting in superconductivity (see Section 9.4).

Substances with $\varepsilon_l(k,0) < 0$ also include nonideal (dense) classical plasma, strong electrolytes and other systems with strong interaction between particles.

5.42 $\omega\varepsilon_t''(k,\omega) \geq 0$.

5.43

$$\mu(k,0) = \frac{2c^2k^2}{\pi}\int_0^\infty \frac{dz}{z}\,\Im[(c^2k^2 - z^2)\eta(k,z)]^{-1}.$$

5.44

$$U = \frac{1}{\omega C^2(\omega)}\frac{d(\omega^2 C(\omega)}{d\omega}\frac{\overline{q^2}}{2}.$$

The oscillation frequency in the contour is determined by the Thomson formula (4.5): $\omega = c/\sqrt{LC(\omega)}$.

5.45 The Lorentz–Lorenz formula (Example 7.4) is derived assuming the electric field in the vicinity of an atom to be homogeneous. Therefore, $\lambda \gg a$, where λ is the oscillation wave length and a is the atomic size. Then

(1) $$\varepsilon(\omega) = \frac{1 + 8\pi N\beta(\omega)/3}{1 - 4\pi N\beta(\omega)/3},$$

the atom polarization $\beta(\omega)$ is determined by Equation (12) from the solution of Problems 5.14 and 5.15. Generally, it is given by the approximate expression

(2) $$\beta(\omega) = \frac{e^2}{m}\sum_s \frac{f_s}{\omega_{s0}^2 - \omega^2 - i\gamma_s\omega}.$$

Far from the resonance, in Equation (2), we can neglect the terms $-i\gamma_s\omega$, since, as a rule, $\gamma_s \ll \omega_{s0}$, but the sum is conserved. Near the resonance with a number s, only one (resonant) term can be left in (2). In this case,

(3) $$\varepsilon(\omega) \approx \frac{\omega_{s0}^2 + 2\omega_p^2 f_s/3 - \omega^2 - i\gamma_s\omega}{\omega_{s0}^2 - \omega_p^2 f_s/3 - \omega^2 - i\gamma_s\omega} = 1 + \frac{\omega_p^2 f_s}{\omega_s^2 - \omega^2 - i\gamma_s\omega},$$

where $\omega_p^2 = 4\pi Ne^2/m$, N is the number density of atoms (not electrons). The dielectric permittivity is still dependent on frequency, but the resonance frequency decreases due to the density effect (due to the difference of the local field from the average one):

(4) $$(\omega_{s0}^2)_{loc} \equiv \omega_s^2 \approx \omega_{s0}^2 - \omega_p^2 f_s/3.$$

5.46

$$\omega_l \approx \sqrt{\omega_{s0}^2 + 2\omega_p^2 f_s/3} - i\gamma_s/2, \qquad w \approx \frac{\omega_{s0}^2 + 2\omega_p^2 f_s/3}{8\pi\omega_p^2 f_s}|E_0|^2 e^{-\gamma_s t}.$$

Under the given assumption that the energy flux is absent, the homogeneous oscillations of atoms with damping occur; E_0 is the initial amplitude of the longitudinal field.

5.47 If we neglect the imaginary part of the dielectric permittivity, assuming $\varepsilon_l''(k,\omega) \approx 0$, then

(1) $$\omega_l^2(k) \approx \omega_{0e}^2 + 3(kv_{Te})^2,\ w = \frac{1}{8\pi}|E|^2 = \frac{1}{4\pi}\overline{E^2(t)},\ \boldsymbol{\gamma} = \frac{3v_{Te}}{8\pi}\frac{kv_{Te}}{\omega_l}\frac{\boldsymbol{k}}{k}|E|^2.$$

Under this assumption, the electron gas oscillates as a whole ensemble relative to the immobile ion gas. The energy density of the oscillations is twice larger than the energy density of the electric field due to the presence of potential energy of the Coulomb interactions between electrons and ions. The energy flux of oscillations is not zero due to dispersion, in spite of the fact that the magnetic field is absent and the Poynting vector is zero. The oscillation energy is carried out by the thermal motion of electrons.

Taking into account the imaginary part $\varepsilon_l''(k,\omega)$ (the Landau damping), the eigenfrequency $\omega_l(k)$ acquires a small imaginary part. Writing $\omega_l = \omega_l' + i\gamma$, with $|\gamma| \ll \omega_l'$, we find ω_l' from the equation $\varepsilon_l'(k,\omega) = 0$ and obtain the earlier result (1). The small imaginary part of the frequency is found from the equation

(2) $$\varepsilon_l(k,\omega_l) \approx \varepsilon_l(k,\omega_l') + i\varepsilon_l''(k,\omega_l') + \left.\frac{\partial\varepsilon_l'(k,\omega)}{\partial\omega}\right|_{\omega=\omega_l'} i\gamma = 0,$$

where the imaginary part of the dielectric permittivity ε_l'' is taken from the results of Problem 5.9. We obtain

$$(3) \qquad \gamma(k) = -\frac{\varepsilon_l''(k,\omega_l')}{\partial\varepsilon_l'(k,\omega)/\partial\omega|_{\omega=\omega_l'}} = -\frac{\sqrt{\pi}}{8}\frac{\omega_{0e}}{(kr_D)^2}e^{-1/4(kr_D)^2-3/2},$$

where $r_D = \sqrt{T/8\pi ne^2}$. With $\omega_l'/kV_{Te} = 1/\sqrt{\pi}kr_D \gg 1$ the damping of the longitudinal plasma waves is exponentially weak. This is explained by a small number of particles in the Maxwell distribution, whose velocities are comparable with the phase velocity $v_{ph} = \omega/k$ of the longitudinal wave. The power density of the energy dissipation is given by Equation (5.79), which in our case takes the form

$$(4) \qquad Q = \frac{\omega_l'}{8\pi}\varepsilon_l''(k,\omega_l')|E(t)|^2.$$

Here $|E(t)|^2 = E_0^2e^{-2\gamma t}$, E_0 is the initial value of the field amplitude. The values w and $\boldsymbol{\gamma}$, with account for damping, are described, as before, by Equation (1), replacing ω_l by ω_l'. The squared field $|E|^2$ is also damping in time.

5.48

$$(1) \qquad \varepsilon_l'(k,\omega) \approx 1 + \frac{\omega_{0e}^2}{k^2v_{Te}^2} - \frac{\omega_{0i}^2}{\omega^2}\left(1 + \frac{3k^2v_{Ti}^2}{\omega^2}\right),$$

$$\varepsilon_l''(k,\omega) \approx \sqrt{\frac{\pi}{2}}\left(\frac{\omega\omega_{0e}^2}{k^3v_{Te}^3} + \frac{\omega\omega_{0i}^2}{k^3v_{Ti}^3}e^{-\omega^2/2k^2v_{Ti}^2}\right).$$

Assuming the imaginary part ε_l'' to be small, we determine the real and imaginary parts by the same method as in Problem 5.47:

$$(2) \qquad \omega_l'^2(k) \approx \omega_{0i}^2\left[1 + 3k^2r_{Di}^2\left(1 + \frac{1}{k^2r_{De}^2}\right)\right]\left(1 + \frac{1}{k^2r_{De}^2}\right)^{-1},$$

(3)

$$\gamma(k) \approx -\sqrt{\frac{\pi}{8}}\cdot\frac{m_i}{m_e}\cdot\frac{\omega_l'^4(k)}{k^3v_{Te}^3}\left\{1 + \sqrt{\frac{m_i}{m_e}}\left(\frac{T_e}{T_i}\right)^{3/2}\exp\left[-\frac{\omega_l'^2(k)}{2k^2v_{Ti}^2}\right]\right\}.$$

Here we have introduced the notions $r_{De,i} = \sqrt{T_{e,i}/4\pi ne^2}$ for the electron and ion Debye radii, and assumed that the plasma is strongly nonisothermal, $T_e \gg T_i$. If this condition and inequalities in the statement of the problem are fulfilled, we have from Equations (2) and (3) that $|\gamma| \ll \omega_l'$.

In the long wave limit, where $kr_{De} \ll 1$ and $kr_{Di} \ll 1$, the formulas are simplified:

$$(4) \qquad \omega_l'(k) \approx v_sk, \qquad v_s = \sqrt{\frac{T_e}{m_i}\left(1 + \frac{3T_i}{T_e}\right)},$$

$$\gamma \approx -kv_s\sqrt{\frac{\pi m_e}{8m_i}}\left\{1 + \sqrt{\frac{m_i}{m_e}}\left(\frac{T_e}{T_i}\right)^{3/2}\exp\left(-\frac{3}{2} - \frac{T_e}{2T_i}\right)\right\}.$$

These oscillations are referred to as **ion-sound oscillations**, since their dispersion (the dependence of frequency on wave vector) is identical with that for sound waves in the electrically neutral medium. The ion-sound velocity v_s is determined, mainly, by temperature of electrons, and by mass of ions.

In the short wave limit ($kr_{De} \gg 1$, but, as before, $kr_{Di} \ll 1$) we have the weakly damping ion plasma oscillations

(5) $$\omega_l'(k) \approx \omega_{0i},$$

$$\gamma \approx -\sqrt{\frac{\pi m_e}{8m_i}}\frac{\omega_{0i}}{k^3 r_{De}^3}\left\{1+\sqrt{\frac{m_i}{m_e}}\left(\frac{T_e}{T_i}\right)^{3/2}\exp\left(-\frac{3}{2}-\frac{1}{2k^2r_{Di}^2}\right)\right\}.$$

5.49 At $\omega \gg kv_F$, we obtain

$$\omega_l(k) \approx \omega_{0e}\left(1+\frac{9}{10}k^2r_D^2\right), \qquad \gamma = 0.$$

At $\omega - kv_F \ll \omega, kv_F$, but $\omega > kv_F$, in Equation (3) of Problem 5.11 we set $\omega/kv_F + 1 \approx 2$ and find the spectrum

$$\omega_l(k) \approx kv_F\left[1+2e^{-2(kv_F)^2-2}\right], \ \gamma = 0.$$

At $\omega > kv_F$, the oscillations do not damp in view of the absence of resonance particles. At $\omega < kv_F$, the oscillations are rapidly damping. The energy of plasmons $\hbar\omega_{0e}$ in typical metals ($n \approx 10^{23}$ cm^{-3}) is 10–20 eV. Plasmons can be excited when fast electrons pass through thin metallic films; they are detected owing to the discrete losses of their energy $\Delta\mathcal{E} = \hbar\omega_{0e}, 2\hbar\omega_{0e}, \ldots$ Plasmons are also excited in some nonmetallic materials (C, Si, MgO, and others), where valence electrons take a part in plasma oscillations.

5.50 The relative oscillations of ions occur with the reduced mass $m = m_+m_-/(m_+ + m_-)$. We introduce the parameter $\boldsymbol{w} = \sqrt{m/v}(\boldsymbol{r}_+ - \boldsymbol{r}_-)$, characterizing the relative displacement of ions of both signs from the equilibrium position. Here v is the volume per one elementary crystal cell. The oscillations produced only due to elastic forces would be described by the equation

(1) $$\ddot{\boldsymbol{w}} = -\omega_0^2\boldsymbol{w}.$$

The Lagrangian describing these oscillations is

(2) $$L(\dot{\boldsymbol{w}}, \boldsymbol{w}) = \dot{\boldsymbol{w}}^2/2 - \omega_0^2\boldsymbol{w}^2/2.$$

On the other hand, if we include the electric forces, we can write down the linear equation, determining the polarization vector of the crystal:

(3) $$\boldsymbol{P} = \gamma\boldsymbol{w} + \alpha\boldsymbol{E}.$$

The first term is due to a displacement of ions from equilibrium positions. The nature of the second term and of constant α is simple. In the absence of the relative displacement of ions, the polarization is induced only by deformation of their electron shells. Therefore, α is the corresponding electric susceptibility, related to ε_{∞}. The latter is introduced in the statement of the problem by equation $\alpha \equiv \alpha_{\infty} = (\varepsilon_{\infty} - 1)/4\pi$.

The existence of the polarization leads to the additional potential energy

$$U = -\int_0^{\mathbf{E}} \boldsymbol{P}(\boldsymbol{E})\, d\boldsymbol{E} = -\gamma \boldsymbol{r} \cdot \boldsymbol{E} - \frac{1}{2}E^2. \tag{4}$$

Adding it to the Lagrangian (2), we obtain

$$L(\dot{\boldsymbol{w}}, \boldsymbol{w}) = \dot{\boldsymbol{w}}^2/2 - \omega_0^2 \boldsymbol{w}^2/2 + \gamma \boldsymbol{w} \cdot \boldsymbol{E} + \frac{1}{2}E^2. \tag{5}$$

From the Lagrangian (5), we obtain the equation of motion with account for the electric forces:

$$\ddot{\boldsymbol{w}} = -\omega_0^2 \boldsymbol{w} + \gamma \boldsymbol{E}. \tag{6}$$

Let us express the constant γ in Equation (6) in terms of the static dielectric permittivity ε_0. For this purpose, we write Equations (3) and (6) in the static limit:

$$\boldsymbol{P}_0 = \gamma \boldsymbol{w}_0 + \alpha_{\infty} \boldsymbol{E}, \qquad \boldsymbol{w}_0 = \frac{\gamma}{\omega_0^2} \boldsymbol{E}. \tag{7}$$

On the other hand, we evidently have $\boldsymbol{P}_0 = \alpha_0 \boldsymbol{E}$, where α_0 is the static susceptibility. With the help of Equation (7), we find

$$\gamma = \omega_0 \sqrt{\frac{\varepsilon_0 - \varepsilon_{\infty}}{4\pi}}. \tag{8}$$

After all constants are determined, let us focus on oscillation frequencies. Up to now, we have considered the system homogeneous. Now we take into account that the oscillations in crystals propagate as weakly inhomogeneous waves (small $\boldsymbol{k}$). The vectors $\boldsymbol{w}$, $\boldsymbol{E}$, and $\boldsymbol{P}$ may be aligned perpendicular or parallel to $\boldsymbol{k}$. Let us represent every vector as a sum of the transverse and longitudinal components: $\boldsymbol{w} = \boldsymbol{w}_t + \boldsymbol{w}_l, \ldots$ Consequently, Equations (3) and (6) will be related either to transverse or longitudinal vectors. For the vector $\boldsymbol{E}_t$, the Maxwell equations in the quasi-statistical approximation have the form $rot\, \boldsymbol{E}_t = 0$ and $div\, \boldsymbol{E}_t = 0$. Hence $\boldsymbol{E}_t = 0$, that is, the electric field does not affect transverse oscillations. Equation (3) takes the form $\ddot{\boldsymbol{w}}_t = -\omega_0^2 \boldsymbol{w}_t$, that is, the frequency of the transverse oscillations is $\omega_t = \omega_0$.

The longitudinal field satisfies the equations $rot\, \boldsymbol{E}_l = 0$ and $div\, \boldsymbol{D} = div\, (\boldsymbol{E}_l + 4\pi \boldsymbol{P}_l) = 0$. From these equations we have

$$\boldsymbol{E}_l = -4\pi \boldsymbol{P}_l, \tag{9}$$

since at $\boldsymbol{P} = 0$ in the quasiclassical case there should be $\boldsymbol{E} = 0$. Now we exclude from Equations (3), (6), and (9) the vectors $\boldsymbol{P}_l$ and $\boldsymbol{E}_l$ and find the equation of motion $\ddot{\boldsymbol{w}}_l = -\omega_l^2 \boldsymbol{w}_l$, from which we obtain

$$\omega_l^2 = \omega_0^2 + \frac{4\pi\gamma^2}{1+4\pi\gamma^2} = \frac{\varepsilon_0}{\varepsilon_\infty}\omega_t^2, \qquad \omega_l = \sqrt{\frac{\varepsilon_0}{\varepsilon_\infty}}\omega_t.$$

The table contains experimental values of parameters for some ion crystals Davydov (1976).

Crystal	ω_t (10^{13} rad/s)	ε_0	ε_∞
NaCl	3.09	5.62	2.25
KCl	2.67	4.68	2.13
TlCl	1.61	31.9	5.10
ZnS	5.71	8.30	5.07

5.51

$$\varepsilon(\omega) = \varepsilon_0 + \frac{(\varepsilon_0 - \varepsilon_\infty)}{1-\omega^2/\omega_t^2} = \frac{\varepsilon_\infty(\omega_l^2 - \omega^2)}{\omega_t^2 - \omega^2}.$$

5.52 In the steady regime, $\boldsymbol{M}_\| = \boldsymbol{M}_0$. Therefore, the linearized equation (5.115) for the magnetization takes reads

(1) $$\frac{d\boldsymbol{M}_\perp}{dt} = -\eta\boldsymbol{M}_\perp \times \boldsymbol{H}_0 - \frac{\boldsymbol{M}_\perp}{\tau_2} - \eta\boldsymbol{M}_0 \times \boldsymbol{h}\, e^{-i\omega t}.$$

The solution for induced oscillations is

(2) $$\boldsymbol{M}_\perp = \chi_0 \frac{\omega_0^2}{\omega_0^2 - (\omega + i\tau_2^{-1})^2}\boldsymbol{h}\, e^{-i\omega t} + i\chi_0 \frac{\omega_0(\omega + i\tau_2^{-1})}{\omega_0^2 - (\omega + i\tau_2^{-1})^2}\boldsymbol{n}\times\boldsymbol{h}\, e^{-i\omega t},$$

where $\omega_0 = \eta H_0$, $\boldsymbol{n}$ is the unit vector in the direction $\boldsymbol{H}_0$. From Equation (2), we find the tensor of the magnetic susceptibility of the paramagnetic in a constant magnetic field:

(3)

$$\chi_{\alpha\beta}(\omega) = \chi_\perp(\omega)(\delta_{\alpha\beta} - n_\alpha n_\beta) + ie_{\alpha\beta\nu}g_\nu, \qquad \chi_\perp(\omega) = \chi_0\frac{\omega_0^2}{\omega_0^2 - (\omega + i\tau_2^{-1})^2},$$

$$\boldsymbol{g} = -\chi_0 \frac{\omega_0(\omega + i\tau_2^{-1})}{\omega_0^2 - (\omega + i\tau_2^{-1})^2}\boldsymbol{n}.$$

The energy dissipation can be found from Equation (5.111):

(4) $$Q(\omega) = \chi_0 |h|^2 \omega_0^2 \omega^2 \frac{1}{\tau_2[(\omega_0^2 - \omega^2 + \tau_2^{-2})^2 + 4\omega^2\tau_2^{-2}]}.$$

The dependence $Q(\omega)$ has the resonance character. If the resonance is narrow ($\tau_2^{-1} \ll \omega_0$), then near the resonance ($|\omega - \omega_0| \ll \omega_0$) the shape of the absorption line is simplified and acquires the Lorentz form

(5) $$Q(\omega) = \frac{\chi_0|h|^2}{4\tau_2}\frac{\omega_0^2}{(\tilde{\omega}_0 - \omega)^2 + \tau_2^{-2}}, \qquad \text{where} \qquad \tilde{\omega}_0 = \omega_0 + 1/2\omega_0\tau_2^{-2}$$

is the resonance frequency.

5.53 The resonance frequency has the order of magnitude $\omega_0 \approx \mu H_0/\hbar$, where μ is the magnetic moment of a single particle. If it is the moment of an individual electron or of the electron shell of an atom (the EPR, was discovered by a Soviet physicist E.K. Zavoiskii in 1944).[11] then a measure of the magnetic moment will be the Bohr magneton $\mu \approx \mu_B \approx 0.93 \times 10^{-23}$ J/T $\approx 0.93 \times 10^{-20}$ erg/G. In this case at $H_0 = 10^3$ G, we have $\omega_0 \approx 10^{10}$ rad/s. If we study an interaction of the alternating field with magnetic moments of atomic nuclei (NMR was discovered by an American physicist F. Bloch in 1946), then the resonance frequency in the same constant field is determined by the nuclear magneton, which is $m_p/m_e \approx 2 \times 10^3$ times smaller than the Bohr magneton. As a result, we have the resonance frequency $\omega_0 \approx 10^7$ rad/s. Thus, the frequencies of both resonances are greatly different.

5.54 $M_x = A \sin(\omega_0 t + \alpha)$, $M_y = A \cos(\omega_0 t + \alpha)$, $M_z = C$, where $\omega_0 = \eta H_0$, α is the initial phase, A and C are the constants connected with the condition $M^2 = M_0^2$, that is, $A^2 + C^2 = M_0^2$, where M_0 is the saturation magnetization. The motion of magnetization is a conventional Larmor precession.

5.55 We find the solution in the form $\boldsymbol{M} = \boldsymbol{M}_0 + \boldsymbol{m}\, e^{-i\omega t}$, $\boldsymbol{m} \perp \boldsymbol{M}_0$, and $m \ll M_0$. Substituting the effective field (5.117) in (5.116) and linearizing the equation, we obtain

$$(1) \qquad i\omega m_x = \eta M_0 q k^2 m_y, \qquad i\omega m_y = -\eta M_0 q k^2 m_x,$$

from which we find the dispersion law of the magnetization waves **(spin waves, magnons)**

$$(2) \qquad \omega(k) = \eta M_0 q k^2.$$

In quantum theory, Equation (2) determines the energy spectrum $\mathcal{E}(k) = \hbar\omega(k)$ of elementary excitations, that is, magnons. A more accurate expression for the spectrum of magnons, with a series of corrections, is discussed in the book of Lifshitz and Pitaevski (1978).

5.56 We find the solution of the equation

$$(1) \qquad \frac{d\boldsymbol{M}}{dt} = -\eta \boldsymbol{M} \times \mathbf{H}_0 + \omega_r(\chi_0 \mathbf{H}_0 - \boldsymbol{M})$$

in the form $M_x = m_x\, e^{-i\omega t}$, $M_y = m_y\, e^{-i\omega t}$, $M_z = M_0 + m_z\, e^{-i\omega t}$, where ω is an unknown frequency; the z-axis is aligned along $\boldsymbol{H}_0$. Projecting Equation (1) on coordinate axes and substituting $\boldsymbol{M}$, we come to the system of algebraic equations, whose compatibility condition is

$$(2) \qquad \omega_0^2 - (\omega + i\omega_r)^2 = 0.$$

The frequency ω appears to be complex: $\omega = \omega_0 - i\omega_r$; the presence of losses damps the motion.

11) Evgeny Konstantinovich Zavoiskii (1907–1976)was outstanding Soviet physicist, the participant of Soviet atomic project.

The components m_x and m_y are shifted in phase by $\pi/2$. The vector $\boldsymbol{M}$ demonstrates a damping precession around $\boldsymbol{H}_0$.

5.57 If the axis Oz is chosen to be along $\boldsymbol{H}$, then the components of the total magnetic field will be $h_x\,e^{-i\omega t}$, $h_y\,e^{-i\omega t}$ and $H_0 + h_z\,e^{-i\omega t}$. We seek the solution of the Landau–Lifshitz equation (5.116), neglecting the inhomogeneous term $\nabla^2\boldsymbol{M}$, in the form

$$(1) \qquad M_x = m_x\,e^{-i\omega t}, \qquad M_y = m_y\,e^{-i\omega t}, \qquad M_z = M_0 + m_z\,e^{-i\omega t},$$

where M_0 is the saturation magnetization. This form of the solution is in line with the assumption, that the Larmor precession is entirely damped, and the oscillations are maintained only by high frequency (forcing) field. Therefore, the values m_x, m_y, and m_z should be assumed to be small, of the order of not lower than h. Substituting Equation (1) into the Landau–Lifshitz equation and omitting the terms quadratic in h and m, we will determine the components $\boldsymbol{m}$:

$$m_x = \chi_0\frac{\omega_0^2}{\omega_0^2-\omega^2}h_x - \chi_0\frac{i\omega\omega_0}{\omega_0^2-\omega^2}h_y,$$

$$m_y = \chi_0\frac{i\omega\omega_0}{\omega_0^2-\omega^2}h_x - \chi_0\frac{\omega_0^2}{\omega_0^2-\omega^2}h_y, \qquad m_z = 0.$$

As seen from these formulas, the character of the frequency dependence of m_x and m_y with fixed $\omega_0 = \eta H_0$ or H_0 is resonant: at $\omega = \omega_0$ the components m_x and m_y infinitely increase, and the ferromagnetic resonance sets in.

The infinite increase of the amplitude $\boldsymbol{m}$ is connected with the approximate method of solution of the Landau–Lifshitz equation. The exact solution (see Problem 5.60) should provide the constant magnitude $|\boldsymbol{M}|$, because according to the Landau–Lifshitz equation without dissipation, $M^2 = const$. When the problem is solved by iterations, with account for losses, the length $|\boldsymbol{M}|$ also remains limited.

5.58

$$(1) \qquad \chi_{ik} = \begin{pmatrix} \chi_\perp & -i\chi_a & 0 \\ i\chi_a & \chi_\perp & 0 \\ 0 & 0 & 0 \end{pmatrix}, \qquad \mu_{ik} = \begin{pmatrix} \mu_\perp & -i\mu_a & 0 \\ i\mu_a & \mu_\perp & 0 \\ 0 & 0 & \mu_\parallel \end{pmatrix},$$

where

$$(2) \qquad \chi_\perp = \chi_0\frac{\omega_0^2}{\omega_0^2-\omega^2}, \qquad \chi_a = \chi_0\frac{\omega\,\omega_0}{\omega_0^2-\omega^2},$$

(3)

$$\mu_\perp = 1 + 4\pi\chi_\perp = 1 + \frac{\omega_M\omega_0}{\omega_0^2-\omega^2}, \qquad \mu_a = 4\pi\chi_a = \frac{\omega\omega_M}{\omega_0^2-\omega^2}, \qquad \mu_\parallel = 1.$$

Here the unified notations $\omega_M = 4\pi\eta M_0$, $\omega_0 = \eta H_0$ are introduced. As seen from given formulas, χ_{ik} and μ_{ik} are the Hermitian tensors ($\mu_{ik} = \mu_{ki}^*$). This means that the medium is isotropic, and the losses are absent.

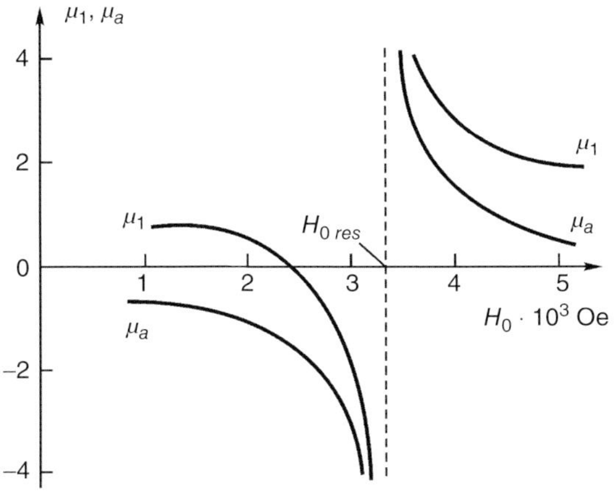

Figure 5.3 Dependence of the components of the magnetic permeability tensor of ferrite without losses on slightly magnetizing field H_0 at given frequency $\omega = \eta H_{0res}$ (H_{0res} = 3400 Oe).

The dependences of the components μ_{ik} on frequency are shown in Figure 5.3; $H_{res} \approx 3400$ Oe.

5.59 We chose the alternating magnetic field in the form of the plane monochromatic wave, $\boldsymbol{H} = \boldsymbol{H}_0 + \boldsymbol{h}\,\exp(i\boldsymbol{k}\cdot\boldsymbol{r} - i\omega t)$, and find the solution in the similar form: $\boldsymbol{M} = \boldsymbol{M}_0 + \boldsymbol{m}\,\exp(i\boldsymbol{k}\cdot\boldsymbol{r} - i\omega t)$. By linearizing the Landau–Lifshitz equation and calculating the proportionality coefficients between the components of the vectors $\boldsymbol{m}$ and $\boldsymbol{h}$, we find the components of the magnetic permeability tensor:

$$\mu_\perp(\omega, k) = 1 + \frac{\omega_M(\omega_0 + \omega_{ex}a^2k^2)}{(\omega_0 + \omega_{ex}a^2k^2)^2 - \omega^2}, \quad \mu_a = \frac{\omega\omega_M}{(\omega_0 + \omega_{ex}a^2k^2)^2 - \omega^2},$$

$$\mu_\parallel = 1.$$

We have introduced the notation $\omega_{ex}a^2 = \eta M_0 q$, where a is a constant of the order of atomic size and ω_{ex} is a certain frequency, characterizing the exchange interaction of spin in the ferromagnetic. Comparing with the result of Problem 5.57, we conclude that the inclusion of spatial dispersion leads to replacing $\omega_0 \to \omega_0 + \omega_{ex}a^2k^2$. Though in macroscopic problems $ak \ll 1$, the contribution of the second term can be noticeable if $\omega_{ex} \gg \omega_0$.

5.60

$$(1) \qquad M_x = \frac{\omega_1}{\Delta\,\omega} C\cos\omega t, \qquad M_y = \frac{\omega_1}{\Delta\,\omega} C\sin\omega t, \qquad M_z = C,$$

where $\Delta\,\omega = \omega_0 - \omega$, $\omega_0 = \eta H_0$, $\omega_1 = \eta h$. The constant C can be determined from the condition $M_x^2 + M_y^2 + M_z^2 = M_0^2$, which follows from the Landau–Lifshitz equation:

$$C = \frac{|\Delta\,\omega|}{\Omega} M_0,$$

where $\Omega = \sqrt{\Delta\,\omega^2 + \omega_1^2}$.

The expression for C contains the modulus $|\Delta\,\omega|$, because $M_z > 0$. The components of $\boldsymbol{M}$ will take the form:

$$(2)\qquad \left.\begin{array}{lllll} M_x & = & \pm\frac{\omega_1}{\Omega}M_0\cos\omega t & = & \chi h_x, \\ M_y & = & \pm\frac{\omega_1}{\Omega}M_0\sin\omega t & = & \chi h_y, \qquad M_z = \frac{|\Delta\,\omega|}{\Omega}M_0. \end{array}\right\}$$

Here the sign $\pm$ corresponds to the sign of $\Delta\omega$. As follows from these equalities, the relations between $\boldsymbol{M}$ and $\boldsymbol{h}$ is nonlinear; the coefficient of proportionality χ depends on h:

$$\chi = \pm\frac{\eta M_0}{\sqrt{\Delta\omega^2 + \omega_1^2}}.$$

The angle of precession ϑ (angle between $\boldsymbol{M}$ and $\boldsymbol{H}_0$) is

$$\sin\vartheta = \frac{M_\perp}{M_0} = \frac{\omega_1}{\Omega},$$

where $M_\perp = \sqrt{M_x^2 + M_y^2}$. At ferromagnetic resonance $\Delta\,\omega = 0$, and from Equation (1) we obtain

$$M_x = \pm M_0\cos\omega t, \qquad M_y = \pm M_0\sin\omega t, \qquad M_z = 0.$$

Therefore, the vector $\boldsymbol{M}$ rotates with frequency ω in the plane perpendicular to $\boldsymbol{H}_0$, its components do not become infinite.

5.61 $\boldsymbol{M} = \boldsymbol{M}_0 + \boldsymbol{m}\,e^{-i\omega t}$, where $\boldsymbol{M}_0$ has a direction $\boldsymbol{H}_0$, and the components of $\boldsymbol{m}$ are given by

$$m_x = \chi_0\frac{\Omega^2 - i\omega\omega_r}{\Omega^2 - \omega^2 - 2i\omega\omega_r}h_x - i\chi_0\frac{\omega\omega_0}{\Omega^2 - \omega^2 - 2i\omega\omega_r}h_y,$$

$$m_y = i\chi_0\frac{\omega\omega_0}{\Omega^2 - \omega^2 - 2i\omega\omega_r}h_x + \chi_0\frac{\Omega^2 - i\omega\omega_r}{\Omega^2 - \omega^2 - 2i\omega\omega_r}h_y,$$

$$m_x = \chi_0\frac{\omega_r}{\omega_r - i\,\omega}h_z,$$

$$\Omega = \sqrt{\omega_0^2 + \omega_r^2}, \qquad \omega_0 = \eta H_0.$$

As is seen from these formulas, the losses ($\omega_r \neq 0$) cause the amplitude $\boldsymbol{m}$ to remain finite at the resonance.

5.62

$$\mu_{ik} = \begin{pmatrix} \mu_\perp & -i\mu_a & 0 \\ i\mu_a & \mu_\perp & 0 \\ 0 & 0 & \mu_\parallel \end{pmatrix}, \qquad \begin{array}{lll} \mu_\perp & = & \mu'_\perp + i\mu''_\perp, \\ \mu_a & = & \mu'_a + i\mu''_a, \end{array}$$

$$\mu'_\perp = 1 + 4\pi\chi_0\frac{\Omega^2(\Omega^2 - \omega^2) + 2\omega^2\omega_r^2}{(\Omega^2 - \omega^2)^2 + 4\omega^2\omega_r^2},$$

$$\mu''_\perp = 4\pi\chi_0 \frac{\omega\omega_r(\Omega^2+\omega^2)}{(\Omega^2-\omega^2)^2+4\omega^2\omega_r^2},$$

$$\mu''_a = 4\pi\chi_0 \frac{\omega\omega_0(\Omega^2-\omega^2)}{(\Omega^2-\omega^2)^2+4\omega^2\omega_r^2},$$

$$\mu''_a = 4\pi\chi_0 \frac{\omega^2\,\omega_0\,\omega_r}{(\Omega^2-\omega^2)^2+4\omega^2\omega_r^2},$$

$$\mu_\parallel = 1+4\pi\chi_0 \frac{\omega_r}{\omega_r-i\omega},$$

where $\Omega = \sqrt{\omega_0^2+\omega_r^2}$, $\omega_0 = \eta H_0$, and $H_{res} \approx$ 3400 E.

The dependences of $\mu'_\perp$ and $\mu''_\perp$ on H_0 are shown in Figure 5.4. The dependence μ'_a and μ''_a on H_0 is similar. The imaginary parts $\mu''_\perp$ and μ''_a reach maxima at $H_0 = H_{res} \approx \omega/\eta$, white the real parts $\mu'_\perp$ and μ'_a take the extremal values at $H_0 \approx (\omega \pm \omega_r)/\eta$.

The curves, shown in Figure 5.4, have the same character as the dispersion curves for $\varepsilon(\omega)$ (Figure 5.2). The imaginary parts of the tensor components $\mu''_\perp$ and μ''_a, $\mu''_\parallel$ determine the dissipation of electromagnetic energy. They vanish at $\omega_r = 0$.

5.63 $\Delta H_0 = \omega_r/\eta$.

5.64 Let us choose the coordinate axes along the principal axes of an ellipsoid, and direct the Oz axes the field $\boldsymbol{H}_0$. With the choice of axes, the tensor N_{ik} is diagonal. Therefore, the Landau–Lifshitz equation in projections on coordinate axes will be written as:

$$\begin{aligned} \dot{M}_x &= -\eta\left[H_0 + 4\pi(N^{(y)} - N^{(z)})M_z\right]M_y, \\ \dot{M}_y &= \eta\left[H_0 + 4\pi(N^{(x)} - N^{(z)})M_z\right]M_x, \\ \dot{M}_z &= -4\pi\eta(N^{(x)} - N^{(y)})M_xM_y. \end{aligned} \tag{1}$$

Thus, the equations become nonlinear. Assuming that deviations of the vector $\boldsymbol{M}$ from the equilibrium (direction of the axis Oz) are small, we find the

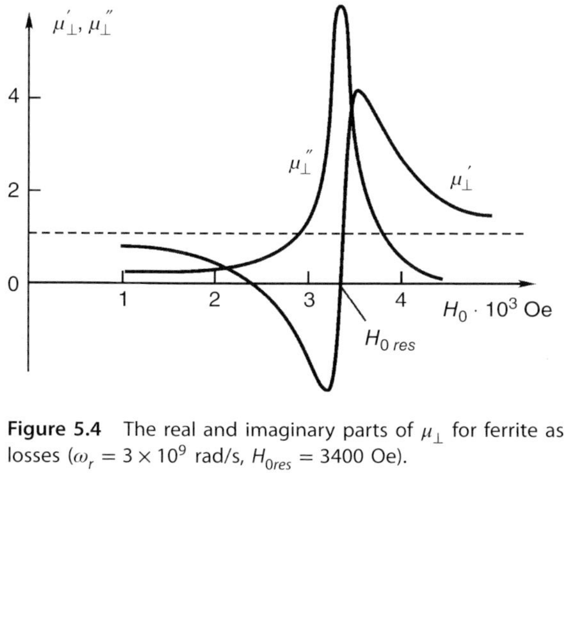

Figure 5.4 The real and imaginary parts of $\mu_\perp$ for ferrite as functions of H_0 including losses ($\omega_r = 3\times10^9$ rad/s, $H_{0res} = 3400$ Oe).

solution in the form

(2) $$\boldsymbol{M} = \boldsymbol{M}_0 + \boldsymbol{m}e^{-i\omega t},$$

where $\boldsymbol{M}_0$ is directed along the Oz-axis. If we drop the terms $\sim m^2$, which, after substituting Equation (2) will enter Equation (1), then the system (1) will be linearized. Setting the determinant of the system equal to zero, we find

(3) $$\omega^2 \equiv \omega_k^2 = \eta^2 \left[H_0 + 4\pi(N^{(x)} - N^{(z)})M_0\right] \left[H_0 + 4\pi(N^{(y)} - N^{(z)})M_0\right].$$

5.65 $\omega = \omega_k + i\omega_r \left[1 + \chi_0(N^{(x)} + N^{(y)})/2\right], \qquad \chi_0 = M_0/(H_0 - N^{(z)}M_0).$
The expression for ω_k is given in the solution to Problem 5.64.

5.66

$$\chi_{ik} = \begin{pmatrix} \chi_1 & -i\chi_a & 0 \\ i\chi_a & \chi_2 & 0 \\ 0 & 0 & 0 \end{pmatrix}$$

(the Oz-axis is directed along $\boldsymbol{H}_0$),

$$\chi_1 = \frac{1}{\Delta}\left\{\eta^2 M_0\left[H_0 + (N^{(y)} - N^{(z)})M_0\right] - i\chi_0\omega\omega_r\right\},$$

$$\chi_2 = \frac{1}{\Delta}\left\{\eta^2 M_0\left[H_0 + (N^{(x)} - N^{(z)})M_0\right] - i\chi_0\omega\omega_r\right\},$$

$$\chi_a = -\frac{1}{\Delta}\eta\omega M_0,$$

where

$$\Delta = (\omega_k^2 - \omega^2) - i\omega\omega_r\left[2 + \chi_0(N^{(x)} + N^{(y)})\right], \qquad \chi_0 = \frac{M_0}{H_0 - N^{(z)}M_0}.$$

Since the demagnetization factors are present in the expressions for χ_{ik}, the position and width of the resonance line will depend on the body shape.

5.67 Equations of motion for the magnetization vectors $\boldsymbol{M}_1$ and $\boldsymbol{M}_2$ read

(1) $$\frac{d\boldsymbol{M}_1}{dt} = -\eta\boldsymbol{M}_1 \times (\boldsymbol{H}_0 + \lambda\boldsymbol{M}_2), \qquad \frac{d\boldsymbol{M}_2}{dt} = -\eta\boldsymbol{M}_2 \times (\boldsymbol{H}_0 + \lambda\boldsymbol{M}_1).$$

Let us find the solution in the form $\boldsymbol{M}_1 = \boldsymbol{M}_{10} + \boldsymbol{m}_1 e^{-i\omega t}$ and $\boldsymbol{M}_2 = \boldsymbol{M}_{20} + \boldsymbol{m}_2 e^{-i\omega t}$ ($\boldsymbol{M}_{10}$ and $\mathbf{M}_{20}$ are the equilibrium values of $\boldsymbol{M}_1$ and $\boldsymbol{M}_2$, respectively).

Solving Equation (1), it is convenient to introduce cyclic components

$$m_{j\pm} = m_{jx} \pm im_{jy} \qquad (j = 1, 2).$$

The frequencies of the free precession are:

(2) $$\omega_{01} = \eta H_0, \qquad \omega_{02} = \eta\lambda|M_{10} - M_{20}|.$$

Equation (2) are valid provided $\lambda|M_{10} - M_{20}| \gg H_0$. The frequency ω_{01} has the same value, as for a ferromagnetic without sublattices. The frequency ω_{02} depends on the molecular field and, as a rule, significantly exceeds ω_{01}.

5.70 When converting the system of the coupling equations (5.127) to the laboratory reference frame, in addition of the transformation of fields, one should perform also the Lorentz transformation of the frequency ω', determined in the co-moving frame. The value ω'/c is the time component of the wave 4-vector k'_i and is transformed as (see the solution of Problem 3.12)

$$\omega' = \gamma(\omega - \boldsymbol{k} \cdot \boldsymbol{V}).$$

Equations (5.128) and (5.133) will conserve their form, but the electric and magnetic permeabilities will have transformed arguments, $\varepsilon(\gamma(\omega - \boldsymbol{k} \cdot \boldsymbol{V}))$ and $\mu(\gamma(\omega - \boldsymbol{k} \cdot \boldsymbol{V}))$. The dependence on $\boldsymbol{k}$ indicates, that in the laboratory frame the moving medium acquires spatial dispersion. With the spatial dispersion in the co-moving frame of reference, the wave vector $\boldsymbol{k}'$ should be expressed in terms of ω and $\boldsymbol{k}$.

5.71 We find the polarization vectors from Equation (5.9),

$$\text{(1)} \qquad \boldsymbol{P} = \frac{1}{4\pi}(\boldsymbol{D} - \boldsymbol{E}), \qquad \boldsymbol{M} = \frac{1}{4\pi}(\boldsymbol{B} - \boldsymbol{H}),$$

and with the help of Equation (5.133) we obtain

$$\text{(2)} \qquad \boldsymbol{P} = -\frac{\mu - 1}{4\pi\mu}\boldsymbol{E} + \frac{\kappa\gamma^2}{4\pi\mu}[\boldsymbol{E} - \boldsymbol{\beta}(\boldsymbol{\beta} \cdot \boldsymbol{E}) + \boldsymbol{\beta} \times \boldsymbol{B}],$$

$$\boldsymbol{M} = \frac{\mu - 1}{4\pi\mu}\boldsymbol{B} + \frac{\kappa\gamma^2}{4\pi\mu}[\beta^2\boldsymbol{B} - \boldsymbol{\beta}(\boldsymbol{\beta} \cdot \boldsymbol{B}) - \boldsymbol{\beta} \times \boldsymbol{E}].$$

At $\varepsilon = 1$, we have the electric polarization vector

$$\text{(3)} \qquad \boldsymbol{P} = \frac{(\mu - 1)\gamma^2}{4\pi}[\beta^2\boldsymbol{E} - \boldsymbol{\beta}(\boldsymbol{\beta} \cdot \boldsymbol{E}) + \boldsymbol{\beta} \times \boldsymbol{B}].$$

At $\mu = 1$, the magnetic polarization vector is

$$\text{(4)} \qquad \boldsymbol{M} = \frac{(\varepsilon - 1)\gamma^2}{4\pi\varepsilon}[\beta^2\boldsymbol{H} - \boldsymbol{\beta}(\boldsymbol{\beta} \cdot \boldsymbol{H}) - \boldsymbol{\beta} \times \boldsymbol{D}].$$

In both cases, the polarization is connected with motion of the medium ($\boldsymbol{\beta} \neq 0$). In the nonpolarized medium at rest ($\varepsilon = 1,\ \mu \neq 1$) in the presence of the magnetic field, there are the magnetization currents $\boldsymbol{j}_{int} \neq 0$, although the bound charge $\rho_{int} = 0$. The charges are neutralized inside molecules. When the medium is moving, positive and negative charges are affected by the different Lorentz forces, displacing them in different directions and producing the electric polarization (3). In the case of $\mu = 1$ but $\varepsilon \neq 1$, the electric field in the medium at rest produces bound charges. A motion of the medium produces the current causing the magnetic polarization.

5.72 Setting equal to zero the argument of the delta-function in Equation (5.148), we will find the shape of the surface of the wave, radiated by an instantaneous point source. At every instant of time, this is the ellipsoid of rotation,

$$\text{(1)} \qquad \frac{x^2 + y^2}{a^2(t)} + \frac{(z - z_0(t))^2}{b^2(t)} = 1,$$

whose center is on the Oz-axis in the point

(2) $$z_0(t) = \frac{\varepsilon\mu - 1}{\varepsilon\mu - \beta^2} Vt.$$

The ellipsoid semi-axes increase with time:

(3) $$a(t) = \sqrt{\frac{1-\beta^2}{\varepsilon\mu - \beta^2}}\, ct, \qquad b(t) = \frac{(1-\beta^2)\sqrt{\varepsilon\mu}}{\varepsilon\mu - \beta^2}\, ct,$$

but their ratio $a/b = \gamma\sqrt{1 - \beta^2/\varepsilon\mu} > 1$ is constant.
Of great importance is the velocity of the ellipsoid center

(4) $$v_0 = \dot{z}_0(t) = \frac{\varepsilon\mu - 1}{\varepsilon\mu - \beta^2} V = \eta V.$$

The drag of any electromagnetic perturbation by the moving medium occurs just with this velocity. The drag coefficient η increases up to unity as $\beta \to 1$. The velocities $v_\pm = \dot{z}_0(t) \pm \dot{b}(t)$ of the propagation of perturbations in positive and negative directions of the Oz-axis are different: the moving medium is anisotropic. At subluminal ($V < c/\sqrt{\varepsilon\mu}$) motion of the medium, $v_- < 0$. A signal may arrive at any observation point in the region $z < 0$ or $z > 0$ (Figure 5.5a). Numbers 1–3 denote corresponding surfaces at three successive instances, spaced by equals intervals; P is the observation point. At light-speed motion, we have $v_- = 0$, and the signal arrives at points in the right half plane $z > 0$ (Figure 5.5b). At supraluminal motion $z_0 > b$, the ellipsoid is dragged downstream in a positive direction of the Oz-axis. Perturbations reach only points inside a conic surface, whose equation is

(5) $$z = \sqrt{\frac{\varepsilon\mu\beta^2 - 1}{1-\beta^2}}\sqrt{x^2 + y^2}.$$

Outside the cone with the solid angle ϑ_g, the field is absent at any time. In every point inside the cone, the signal comes twice: from the leading and from the rear front of the ellipsoid (Figure 5.5c).

5.73

(1) $$\varphi(\boldsymbol{r}) = \frac{qs}{R_*}\frac{1-\varepsilon\mu\beta^2}{\varepsilon(1-\beta^2)}, \qquad \boldsymbol{A}(\boldsymbol{r}) = -\frac{\varepsilon\mu - 1}{c(1-\varepsilon\mu\beta^2)}\varphi(\boldsymbol{r})\boldsymbol{V},$$

where $s = 1$ in the subluminal case ($\varepsilon\mu\beta^2 < 1$), and $s = 2$ in the supraluminal case ($\varepsilon\mu\beta^2 < 1$); the observation point is inside the cone, determined by Equation (5) of Problem 5.72; $s = 0$ outside the cone.

(2) $$\boldsymbol{E} = -\nabla\varphi, \qquad \boldsymbol{B} = \frac{\varepsilon\mu - 1}{c(1-\varepsilon\mu\beta^2)}[\boldsymbol{E}\times\boldsymbol{V}],$$

$$\boldsymbol{D} = \frac{\varepsilon}{1-\varepsilon\mu\beta^2}[(1-\beta^2)\boldsymbol{E} - (\varepsilon\mu - 1)\boldsymbol{\beta}(\boldsymbol{\beta}\cdot\boldsymbol{E})], \qquad \boldsymbol{H} = 0.$$

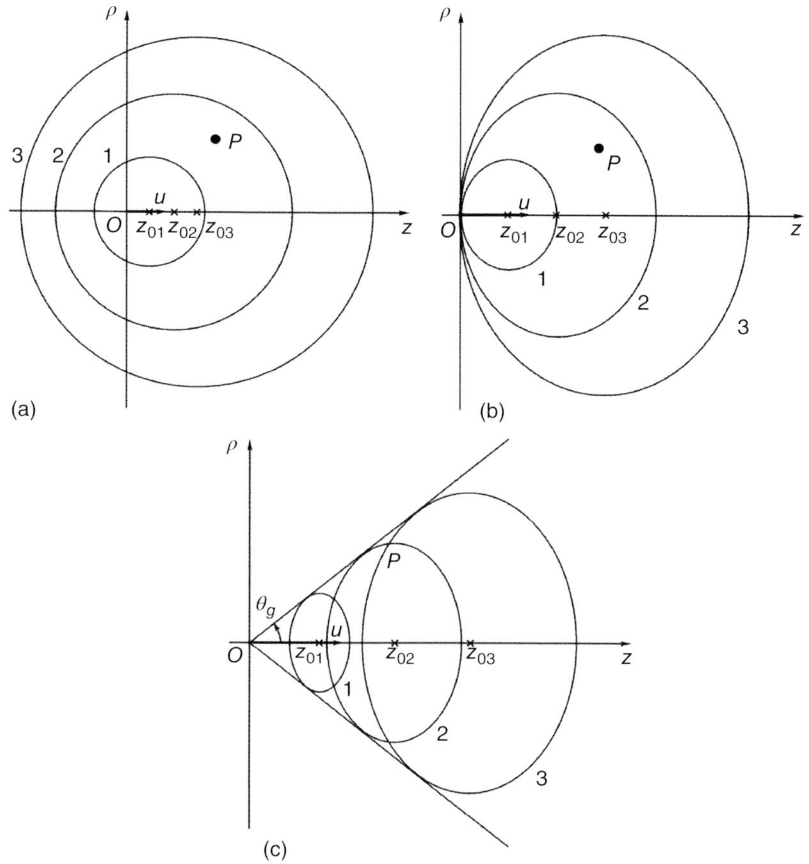

Figure 5.5 Front of instantaneous light flash, radiated at $t = 0$ at origin: (a) subluminal ($V < c/\sqrt{\varepsilon\mu}$, $v_- < 0$) motion of the medium; (b) light-speed ($v_- = 0$) motion; and (c) supraluminal motion.

The potentials and vectors of the field do not depend on time, the magnetic induction is nonzero, although though the magnetic field strength is equal to zero.

(3) $$\boldsymbol{P} = \frac{1}{4\pi(1 - \varepsilon\mu\beta^2)}\{[(\varepsilon - 1) + \varepsilon\beta^2(\mu - 1)]\boldsymbol{E} - \varepsilon(\varepsilon\mu - 1)\boldsymbol{\beta}(\boldsymbol{\beta} \cdot \boldsymbol{E})\},$$

$$\boldsymbol{M} = \frac{\varepsilon\mu - 1}{c(1 - \varepsilon\mu\beta^2)}[\boldsymbol{E} \times \boldsymbol{V}].$$

The cause for appearance of the magnetic polarization in the static electric field in moving medium has been discussed in Problem 5.71.

5.74

$$\boldsymbol{Z}^{(m)} = \frac{\mu s \boldsymbol{m}_0}{R_*}, \qquad \boldsymbol{Z}^{(e)} = 0,$$

where s and R_* correspond to notations in Problem 5.73.

$$\varphi(\boldsymbol{r}) = -\frac{\mu s\gamma^2(\varepsilon\mu - 1)}{\varepsilon\mu}\frac{\boldsymbol{R}\cdot[\boldsymbol{\beta}\times\boldsymbol{m}_0]}{R_*^3},$$

$$\boldsymbol{A} = \frac{\boldsymbol{m}_0\times\boldsymbol{R}}{R_*^3} + \frac{\mu s\gamma^2(\varepsilon\mu - 1)}{\varepsilon\mu}\frac{\boldsymbol{\beta}(\boldsymbol{R}\cdot[\boldsymbol{\beta}\times\boldsymbol{m}_0])}{R_*^3},$$

$\boldsymbol{R} = z\boldsymbol{e}_z + \gamma(1 - \varepsilon\mu\beta^2)\boldsymbol{r}_\perp$, $\boldsymbol{\beta} = \boldsymbol{V}/c$.

5.75

$$\boldsymbol{Z}^{(e)} = \frac{\mu s\boldsymbol{p}_0}{R_*}, \qquad \boldsymbol{Z}^{(m)} = 0.$$

$$\varphi(\boldsymbol{r}) = s\gamma^2(1/\varepsilon - \mu\beta^2)\frac{\boldsymbol{p}_0\cdot\boldsymbol{R}}{R_*^3}, \qquad \boldsymbol{A} = -\frac{(\varepsilon\mu - 1)\boldsymbol{V}}{c(1 - \varepsilon\mu\beta^2)}\varphi(\boldsymbol{r}).$$

The notations are the same as in Problems 5.73 and 5.74.

5.76 In the co-moving reference frame, a magnetic dipole produces the static magnetic field, which has been calculated in Sections I.2.2 and 3.2. The magnetic and electric fields in the laboratory frame can be obtained with the help of Lorentz transformations, taking into account, that the vectors $(\boldsymbol{E},\ \boldsymbol{B})$ and $(\boldsymbol{D},\ \boldsymbol{H})$ form in pairs antisymmetrical 4-tensors of second rank according to Equations (5.124) and (5.125). By contrast to the dipole at rest in the laboratory frame, the potentials and vectors of fields in this case do depend on time.

5.77 Along with momentum, transferred to waves, the radiator will transfer a certain momentum to the medium through the Abraham force. It can be calculated as the momentum of the Abraham force over the whole time of interaction of waves with the medium: $\Delta\boldsymbol{G}^A = \int \boldsymbol{f}^A dV\,dt = \boldsymbol{G}^M - \boldsymbol{G}^A$. Therefore, the momentum loss by the radiator coincides with the momentum, corresponding to the Minkowski formula: $\Delta\boldsymbol{G} = \boldsymbol{G}^A + \Delta\boldsymbol{G}^A = \boldsymbol{G}^M$.

5.78

$$w^A = \frac{1}{8\pi}(\boldsymbol{D}\cdot\boldsymbol{E} + \boldsymbol{B}\cdot\boldsymbol{H}) - \frac{1}{4\pi c}\gamma^2\boldsymbol{V}\cdot(\boldsymbol{D}\times\boldsymbol{B} - \boldsymbol{E}\times\boldsymbol{H}),$$
$$\boldsymbol{g}^A = \frac{1}{c^2}\boldsymbol{S} = \frac{1}{4\pi c}\left\{\boldsymbol{E}\times\boldsymbol{H} - \frac{1}{c^2}\gamma^2\boldsymbol{V}(\boldsymbol{V}\cdot[\boldsymbol{D}\times\boldsymbol{B}] - \boldsymbol{V}\cdot[\boldsymbol{E}\times\boldsymbol{H}])\right\}.$$

5.80

$$\boldsymbol{\mathcal{B}} = \frac{c}{\omega}\boldsymbol{k}\times\boldsymbol{\mathcal{E}} - \frac{ic}{\omega}\left(\nabla\times\boldsymbol{\mathcal{E}} + \frac{\boldsymbol{k}}{\omega}\times\frac{\partial\boldsymbol{\mathcal{E}}}{\partial t}\right),\ \boldsymbol{B}(\boldsymbol{r},t) = \boldsymbol{\mathcal{B}}(\boldsymbol{r},t)e^{i\varphi},\ \varphi = \boldsymbol{k}\cdot\boldsymbol{r} - \omega t.$$

5.81 The mean value in question is given by the integral

(1)

$$\frac{1}{T}\int_t^{t+T}\frac{\partial}{\partial t'}[\boldsymbol{D}\times\boldsymbol{B}]\,dt' = \frac{1}{4T}\int_t^{t+T}\frac{\partial}{\partial t'}[\boldsymbol{D}^*\times\boldsymbol{B} + \boldsymbol{D}\times\boldsymbol{B}^* + \boldsymbol{D}\times\boldsymbol{B}e^{2i\varphi}]$$

$$dt' + \boldsymbol{D}^*\times\boldsymbol{B}^*e^{-2i\varphi}]\,dt'.$$

The first two terms on the right-hand side do not contain any rapidly varying phase and are integrated directly,

$$(2)\quad \frac{1}{T}\int_t^{t+T}\frac{\partial}{\partial t'}[\boldsymbol{D}^*\times\boldsymbol{B}+\boldsymbol{D}\times\boldsymbol{B}^*]\,dt' = \frac{1}{T}[\boldsymbol{D}^*\times\boldsymbol{B}+\boldsymbol{D}\times\boldsymbol{B}^*]_t^{t+T}$$

$$\approx \frac{\partial}{\partial t}[\boldsymbol{D}^*\times\boldsymbol{B}+\boldsymbol{D}\times\boldsymbol{B}^*].$$

The terms with rapidly varying phase are integrated by parts. After the first integration they become of the same order of smallness as in Equation (2)

$$(3)\qquad \frac{1}{T}\int_t^{t+T}\frac{\partial}{\partial t'}[\boldsymbol{D}\times\boldsymbol{B}]e^{2i\varphi}\,dt' \approx e^{2i\varphi}\frac{\partial}{\partial t}[\boldsymbol{D}\times\boldsymbol{B}].$$

However, they still contain the rapidly changing phase. Therefore, one should average over period for the second time. This gives the terms of the next order in smallness,

$$\omega^{-1}e^{2i\varphi}\frac{\partial^2}{\partial t^2}[\boldsymbol{D}\times\boldsymbol{B}],$$

which can be neglected. The terms (2), containing the rapidly varying phase, are not affected by the second averaging.

5.82

$$\boldsymbol{g}^M = \frac{1}{16\pi c}\left[\boldsymbol{D}^*\times\boldsymbol{B}+\boldsymbol{D}\times\boldsymbol{B}^* + \boldsymbol{n}ck\frac{\partial\varepsilon_t}{\partial\omega}\boldsymbol{\mathcal{E}}^*\cdot\boldsymbol{\mathcal{E}}\right].$$

6
Propagation of Electromagnetic Waves

6.1
Transverse Waves in Isotropic Media. Reflection and Refraction of Waves

Electromagnetic waves in vacuum are oscillations of electric and magnetic fields in the absence of (far from) radiation sources. The properties of electromagnetic waves in vacuum were considered in Section I.2.3. If electromagnetic waves occur in materials, they are accompanied by oscillations of charged particles, that is, electrons and ions. These oscillation processes are more complicated. Electromagnetic waves in media are generally not transverse: the field vectors may have both the transverse and the longitudinal components, with respect to the direction of propagation.

Eigen oscillations in isotropic medium. Eigen (proper) oscillations in medium are studied from the system of Maxwell's equations (5.25)–(5.28) with $\rho_{ext} = \boldsymbol{j}_{ext} = 0$. Excluding the vector $\boldsymbol{B}$, we obtain the equation

$$rot\, rot\, \boldsymbol{E} = -\frac{1}{c^2}\frac{\partial^2 \boldsymbol{\mathcal{D}}}{\partial t^2}, \tag{6.1}$$

which should be complemented by the equations of coupling between the vectors $\boldsymbol{\mathcal{D}}$ and $\boldsymbol{E}$. In a homogeneous medium, we can use Equation (5.34), which connects Fourier harmonics of these vectors,

$$\mathcal{D}_\alpha(\boldsymbol{k},\omega) = \epsilon_{\alpha\beta}(\boldsymbol{k},\omega)E_\beta(\boldsymbol{k},\omega). \tag{6.2}$$

Performing Fourier transform of Equation (6.1) and using Equation (6.2), we have the system of uniform algebraic equations describing the field components:

$$\left[k^2\delta_{\alpha\beta} - k_\alpha k_\beta - \frac{\omega^2}{c^2}\epsilon_{\alpha\beta}(\boldsymbol{k},\omega)\right]E_\beta(\boldsymbol{k},\omega) = 0. \tag{6.3}$$

This system has nontrivial solutions if its determinant is zero:

$$\left|k^2\delta_{\alpha\beta} - k_\alpha k_\beta - \frac{\omega^2}{c^2}\epsilon_{\alpha\beta}(\boldsymbol{k},\omega)\right| = 0. \tag{6.4}$$

Equation (6.4) connects the frequency with the wavevector of oscillations and is referred to as the **dispersion equation** for waves in medium. When we need to find the explicit form of the dispersion equation, either the wave vector or the

Electromagnetic Phenomena in Matter: Statistical and Quantum Approaches, First Edition. Igor N. Toptygin.

frequency should be given. The other quantity can then be found from the dispersion equation. The wave vector may be determined from geometry of material bodies (e.g., from the boundary conditions in a piece-like medium); the dispersion equation gives then the possibility to find the frequency $\omega(\boldsymbol{k})$, which is generally complex due to the complexity of $\epsilon_{\alpha\beta}$. However, in transparent regions of the Medium, the imaginary part of the permittivity is negligible. If the field is produced by an external monochromatic source, then its frequency ω is known; the wave vector $\boldsymbol{k}(\omega)$ is then determined from the dispersion equation and some additional conditions (symmetry, boundary conditions, etc.). The wave vector is generally also complex,

$$\boldsymbol{k} = \boldsymbol{k}' + i\boldsymbol{k}'', \tag{6.5}$$

and the directions of real vectors $\boldsymbol{k}'$ and $\boldsymbol{k}''$ may be different. Such plane waves are referred to as **inhomogeneous**. Their directions of propagation and damping cannot be the same.

As a rule, there are several branches of oscillations in medium, determined by the dispersion equation and boundary conditions. Every branch has its own polarization (i.e., the relationship between the components of the vector $\boldsymbol{E}$ which are determined from Equation (6.3)). Polarizations of different kind were considered by different methods in Section I.2.3. The magnetic vector may be found from Equation (5.41):

$$\boldsymbol{B}(\boldsymbol{k},\omega) = \frac{c}{\omega}\boldsymbol{k} \times \boldsymbol{E}(\boldsymbol{k},\omega). \tag{6.6}$$

In an isotropic medium, the transverse and longitudinal oscillations may exist independently. The corresponding dispersion equations have been derived in Section 5.1 (Equations (5.56)–(5.58)):

$$\varepsilon_l(\boldsymbol{k},\omega) = 0,\ \omega^2\varepsilon_t(\boldsymbol{k},\omega) = c^2k^2 \text{ or } \varepsilon(\boldsymbol{k},\omega) = 0,\ \omega^2\varepsilon(\boldsymbol{k},\omega)\mu(\boldsymbol{k},\omega) = c^2k^2. \tag{6.7}$$

The dispersion equations for the longitudinal waves in certain media have been considered in Problems (5.44)–(5.48).

The mathematical theory of propagation of narrow-band pulses in dispersive media, based on the Fourier expansion, can be found in the monograph by Vainstein and Vakman (1983) (unfortunately, it does not contain a proper number of physical examples). The analysis of short electromagnetic pulses in dispersive media, based on accurate solutions without using Fourier expansions, is given in Shvartsburg (1998).

Reflection and refraction on the boundary between two media. In this section and below, we consider permittivities $\varepsilon(\omega)$ and $\mu(\omega)$ without space dispersion.

Example 6.1

A plane monochromatic wave with an arbitrary polarization is incident at an angle θ_0 to the normal (Figure 6.1a) to the boundary between two transparent dielectrics with permittivities $\varepsilon_1(\omega) > 0$, $\mu_1(\omega) > 0$ and $\varepsilon_2(\omega) > 0$, $\mu_2(\omega) > 0$.

Using the boundary Conditions, find the frequencies and propagation directions of the reflected and refracted waves. Directions of phase and group velocities in both media are identical. What condition should satisfy the width of the transition layer δ, that it could be neglected and assumed to be equal zero?

Solution. The electromagnetic wave, which is incident to the boundary of two transparent dielectrics, delivers electromagnetic energy to this boundary. In stationary conditions, the reflected and refracted waves carry away all this energy from the boundary. Consider the case, when group and phase velocities have identical directions. Arrows on Figure 6.1 show both directions of energy streams, and the direction of motion of phase planes. On the boundary ($z = 0$), the following conditions must be satisfied:

$$E_{0\tau} + E_{1\tau} = E_{2\tau}, \qquad H_{0\tau} + H_{1\tau} = H_{2\tau}, \tag{6.8}$$

where indices 0, 1, and 2 refer to the incident, reflected and refracted waves, respectively.

The wave vector of the reflected wave must be directed inward the first medium, and the wave vector of the reflected wave inward the second medium. Since all fields contains the factor $\exp(i\boldsymbol{k}\cdot\boldsymbol{r} - i\omega t)$ with corresponding $\boldsymbol{k}$ and ω, and the boundary conditions must be satisfied at any point of the $z = 0$ plane at any moment of time, then the frequencies and the corresponding components of the wave vectors must be equal for all three waves:

$$\omega_0 = \omega_1 = \omega_2 = \omega; \qquad k_{0x} = k_{1x} = k_{2x}; \tag{6.9}$$

we have chosen the coordinate axes in such a way, that $k_y = 0$. Using the dispersion relation (6.6), we have two independent equalities

$$n_1 \sin\theta_0 = n_1 \sin\theta_1 = n_2 \sin\theta_2, \quad \text{where} \quad n_1 = \sqrt{\varepsilon_1\mu_1} > 0,\ n_2 = \sqrt{\varepsilon_2\mu_2} > 0 \tag{6.10}$$

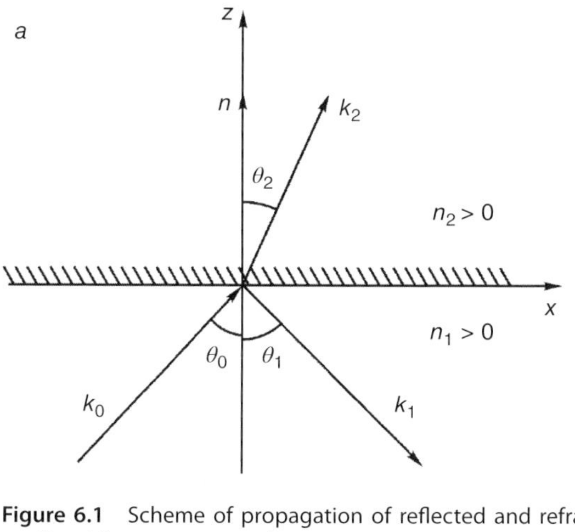

Figure 6.1 Scheme of propagation of reflected and refracted waves near boundary of two media with $\varepsilon_{1,2}(\omega) > 0$ and $\mu_{1,2}(\omega) > 0$.

are coefficients of refraction. They result in **Snell's** laws of reflection and refraction[1)]

$$\theta_1 = \theta_0, \qquad \frac{\sin\theta_2}{\sin\theta_0} = \frac{n_1}{n_2}. \tag{6.11}$$

These relations hold for any polarization of incident waves.

Example 6.2

Using the boundary conditions, find the amplitudes of the reflected and refracted waves. Find also the reflection coefficient R, that is, the fraction of energy of the incident wave, which is transferred to the reflected wave.

Solution. In order to calculate the amplitude of the reflected and refracted waves, we decompose the incident wave in two waves: the electric vector of one of them, $\boldsymbol{E}$, and the magnetic vector of the second, $\boldsymbol{H}$, are parallel to the boundary (and perpendicular to the plane in which wave vectors of all three waves lie). The amplitudes of the reflected and the refracted waves can be expressed in terms of the amplitude of the incident wave via **Fresnel's** formulas using the boundary conditions (6.8). We write them first for the vector E, and then for H:

$$E_1 = \frac{\mu_2 \sin\theta_2 \cos\theta_0 - \mu_1 \sin\theta_0 \cos\theta_2}{\mu_2 \sin\theta_2 \cos\theta_0 + \mu_1 \sin\theta_0 \cos\theta_2} E_0, \tag{6.12}$$

$$E_2 = \frac{2\mu_2 \sin\theta_2 \cos\theta_0}{\mu_2 \sin\theta_2 \cos\theta_0 + \mu_1 \sin\theta_0 \cos\theta_2} E_0;$$

$$H_1 = \frac{\varepsilon_2 \sin\theta_2 \cos\theta_0 - \varepsilon_1 \sin\theta_0 \cos\theta_2}{\varepsilon_2 \sin\theta_2 \cos\theta_0 + \varepsilon_1 \sin\theta_0 \cos\theta_2} H_0, \tag{6.13}$$

$$H_2 = \frac{2\varepsilon_2 \sin\theta_2 \cos\theta_0}{\varepsilon_2 \sin\theta_2 \cos\theta_0 + \varepsilon_1 \sin\theta_0 \cos\theta_2} H_0.$$

Many substances have magnetic permeability close to unit, $\mu_1 \approx \mu_2 \approx 1$. In this case, Fresnel's formulas become more compact:

$$E_1 = \frac{\sin(\theta_2 - \theta_0)}{\sin(\theta_2 + \theta_0)} E_0, \qquad E_2 = \frac{2\sin\theta_2\cos\theta_0}{\sin(\theta_2 + \theta_0)} E_0; \tag{6.14}$$

$$H_1 = \frac{\tan(\theta_0 - \theta_2)}{\tan(\theta_0 + \theta_2)} H_0, \qquad H_2 = \frac{\sin 2\theta_0}{\sin(\theta_0 + \theta_2)\cos(\theta_0 - \theta_2)} H_0. \tag{6.15}$$

We calculate the coefficients of reflection, denoting them as $R_\perp$ and $R_\parallel$, as the ratios of the normal, relative to the boundary, components of the energy flux density in the reflected and the incident waves (indices $\parallel$ and $\perp$ characterize the orientation of the field $\boldsymbol{E}$ relative to the incidence plane). The Poynting vector is given by (5.110). The result is

$$R_\perp = \left| \frac{\mu_2 \sin\theta_2 \cos\theta_0 - \mu_1 \sin\theta_0 \cos\theta_2}{\mu_2 \sin\theta_2 \cos\theta_0 + \mu_1 \sin\theta_0 \cos\theta_2} \right|^2. \tag{6.16}$$

1) Snell Willebrord (1580–1626), Dutch physicist, astronomer, and mathematician.

The coefficient $R_{\|}$ is obtained from $R_{\perp}$ by replacing $(\mu_1, \mu_2) \to (\varepsilon_1, \varepsilon_2)$. At $\mu_1 = \mu_2 = 1$, we have

$$R_{\perp} = \frac{\sin^2(\theta_0 - \theta_2)}{\sin^2(\theta_0 + \theta_2)}, \qquad R_{\|} = \frac{\tan^2(\theta_0 - \theta_2)}{\tan^2(\theta_0 + \theta_2)}. \tag{6.17}$$

The derived relations are analyzed in Problem 6.9. They hold if the depth of the boundary is much smaller than the wavelength in each medium: $\delta \ll \lambda_1, \lambda_2$. ■

Negative refraction ("left-handed" media). In all formulas of Examples 6.1 and 6.2, electric $\varepsilon(\omega)$ and magnetic $\mu(\omega)$ permeabilities were assumed positive, and phase and group velocities of waves coincided in direction. However, both permeabilities can be negative. For example, $\varepsilon(\omega) < 0$ in a completely ionized gas at frequencies below the electron plasma frequency (5.59) if we neglect dissipation (Example 5.6): $\varepsilon(\omega) \approx 1 - \omega_{0e}^2/\omega^2 < 0$, if $\omega < \omega_{0e}$. However, one has $\mu \approx 1$ in a plasma, and the transverse waves cannot exist in the specified frequency range, because $k^2 = \varepsilon(\omega)\omega/c < 0$ and the wave vector becomes purely imaginary. It occurs owing to destructive interference of transverse waves. However, if $\varepsilon < 0$ and $\mu < 0$ simultaneously, waves do not decay.

Consider a very interesting case of isotropic media, where phase and group velocities have opposite directions. Originally such situation was described in 1944 by Leonid Mandelstam in his lectures (Mandelstam, 1972, pp. 431–437). Later similar problems were studied by Veselago (1967), Pendry (2000), and others. The attention of researchers to this problem has increased after creation in the late 1990s of composite materials (**m**etamaterials) having negative electric and magnetic permeabilities in a certain frequency range and oppositely directed phase and group velocities. More detailed data about these materials and the associated phenomena can be found, for example, in articles by Blioh and Blioh (2004), Agranovich and Gartshtein (2006), Vinogradov *et al.* (2008).

We will specify, first of all, under which conditions the properties mentioned above can be combined in one material. We will assume that the material is isotropic and permeabilities $\varepsilon(\omega)$ and $\mu(\omega)$ depend only on frequency, but not on wave vector. The difference between phase and group velocities arises only in medium with dispersion. Using Equations (5.110) and (5.172), we will write down two expressions for the group velocity:

$$\boldsymbol{v}_g = \boldsymbol{v}_{ph} \frac{\varepsilon \boldsymbol{E}^* \cdot \boldsymbol{E}}{8\pi w} = \frac{\boldsymbol{v}_{ph}}{1 + \Pi(\omega)}, \qquad \boldsymbol{v}_{ph} = \frac{c\boldsymbol{k}}{\sqrt{\varepsilon\mu}k}, \qquad \sqrt{\varepsilon\mu} > 0. \tag{6.18}$$

Here the density of electromagnetic energy cannot be negative, $w > 0$. Otherwise, field amplification would be possible without any energy expense (perpetuum mobile). The velocities $\boldsymbol{v}_g$ and $\boldsymbol{v}_{ph}$ have different signs and both are real at $\varepsilon < 0$ and $\mu < 0$. As a result, we come to compatible physical conditions

$$w > 0, \qquad \Pi(\omega) < -1, \qquad \varepsilon < 0, \qquad \mu < 0. \tag{6.19}$$

One more characteristic property of this dielectric media can be seen from the equation for Fourier's components: $\mu\boldsymbol{H} = (\omega/c)\boldsymbol{k} \times \boldsymbol{E}$. At $\mu > 0$, three vectors $\boldsymbol{k}$, $\boldsymbol{E}$, and $\boldsymbol{H}$ form a right-hand group. When $\mu < 0$ they form the left-hand group,

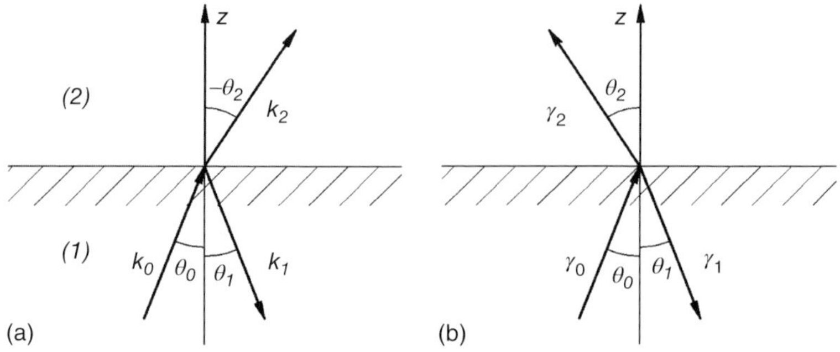

Figure 6.2 (a) Directions of wave vectors (phase velocities) on boundary of medium with negative angle of refraction. (b) Directions of energy flows and group velocities.

that is, at given $\boldsymbol{E}$ and $\boldsymbol{H}$ the wave vector and the phase velocity, determined by it, reverse their direction. On the other hand, the direction of the energy flux and of the group velocity are determined by the Poynting vector $\boldsymbol{\gamma} = (c/4\pi)\boldsymbol{E} \times \boldsymbol{H}$, which always forms the right-hand group with $\boldsymbol{E}$ and $\boldsymbol{H}$. Therefore, in isotropic media with $\varepsilon > 0$ and $\mu > 0$ (usual "right-handed" media), the phase and group velocities have the same direction, while in media with $\varepsilon < 0$ and $\mu < 0$ ("left-handed" media, see Pendry *et al.* (1996)) these two velocities are opposite.

Let us consider reflection and refraction of waves at the boundary between the usual and left left-handed media. Let medium 1 be usual and medium 2 be left-handed (Figure 6.2a). Arrows show wave vectors of three waves. The phase velocity in the second medium is directed to the boundary that corresponds to energy outflow of the refracted wave from the boundary. From Equation (6.9), we obtain the reflection and refraction laws:

$$\theta_1 = \theta_0, \quad n_1 \sin\theta_0 = -n_2 \sin\theta_2 = n_2 \sin(-\theta_2), \quad n_1 = \sqrt{\varepsilon\mu} > 0,\ n_2 = \sqrt{\varepsilon\mu} > 0. \tag{6.20}$$

The law of reflecion is the same, as at the boundary of two usual media. However, the refraction law has changed. The minus sign that has appeared because of opposite directions of phase and group velocities, is quite often attributed to n_2 and treated as negative refraction index in the left-hand medium. However, the phenomenon becomes clearer if we attribute this sign to the angle of refraction and consider this angle in opposite direction with respect to the normal of the boundary (i.e., counter-clockwise). Then the streams of energy of the three waves, which are defined by corresponding Poynting vectors, will be directed according to Figure 6.2b.

Pressure of electromagnetic waves on interface of two media with dispersion. Let a quasi-monochromatic flat wave be incident normally to an interface of two media, that is, its energy and momentum are transferred from the source to this surface. The momentum density (5.165) averaged over the period of wave in the transparent environment has the form

$$\boldsymbol{g} = \frac{1}{4\pi c}\overline{\boldsymbol{E}\times\boldsymbol{H}} = \frac{\sqrt{\varepsilon\mu}\boldsymbol{E}^*\cdot\boldsymbol{E}}{8\pi c\mu}\frac{\boldsymbol{k}}{k}. \tag{6.21}$$

This vector is directed along the wave vector $\boldsymbol{k}$ in the usual medium ($\varepsilon > 0$, $\mu > 0$) and is opposite to $\boldsymbol{k}$ in the left-handed media; group velocities have the same directions. Energy and momentum transfer in dispersive media are attributed to group velocity. Therefore, the momentum transferred per unit time through an unit area in the direction of group velocity, can anyway contain the scalar product $\boldsymbol{g}\cdot\boldsymbol{v}_g$. Using Equations (6.18) and (6.21), we obtain

$$\boldsymbol{g}\cdot\boldsymbol{v}_g = \frac{\varepsilon}{\mu}\frac{(\boldsymbol{E}^*\cdot\boldsymbol{E})^2}{(8\pi)^2 w}, \tag{6.22}$$

where w is the energy density (5.180) in dispersive medium.

The pressure on the interface layer depends on the properties of the second medium. If it absorbs all electromagnetic wave (absolutely black body), in the case of normal incidence on surface, all the momentum of the wave is transferred to the second medium along the distances of the order of absorption length, producing the positive pressure

$$P_{rad} = \boldsymbol{g}\cdot\boldsymbol{v}_g = \frac{\varepsilon}{\mu}\frac{(\boldsymbol{E}^*\cdot\boldsymbol{E})^2}{(8\pi)^2 w} > 0 \tag{6.23}$$

both in usual, and in the left-hand medium. In the case of normal incidence on ideally reflecting surface, the pressure (6.23) doubles.

Let us calculate now the pressure of the flat monochromatic wave on interface between two transparent media at normal incidence. The reflection factor is calculated using results of Problem 6.17 with the refraction at $\theta_0 \to 0$ and $\theta_2 \to 0$. The result is

$$R = R_\perp = R_\parallel = \left|\frac{\sqrt{\varepsilon_1|\mu_2|} - \sqrt{|\varepsilon_2|\mu_1}}{\sqrt{\varepsilon_1|\mu_2|} + \sqrt{|\varepsilon_2|\mu_1}}\right|^2. \tag{6.24}$$

This result is valid for usual and left-handed second media. The incident and reflected waves transfer to the interface per unit time the momentum flux $\boldsymbol{g}_1\cdot\boldsymbol{v}_{1g}(1+R)$, directed to the second medium.The refracted wave transfers the flux directed to medium 1 and equal $\boldsymbol{g}_2\cdot\boldsymbol{v}_{2g}(1-R)$. The difference of the two fluxes gives the pressure on interface (in a layer of thickness of an order of a wave length)

$$P_{rad} = \boldsymbol{g}_1\cdot\boldsymbol{v}_{1g}(1+R) - \boldsymbol{g}_2\cdot\boldsymbol{v}_{2g}(1-R). \tag{6.25}$$

This pressure, depending on the parameters of media, can be positive or negative. “The light attraction” toward the incident wave will take place in the latter case.

We will illustrate these results by considering the usual non-dispersive media with positive permeabilities. Let the reflected wave be absent. According to Equation (6.24), it is possible at $\varepsilon_1/\varepsilon_2 = \mu_1/\mu_2$. In this case, the momentum flux $\boldsymbol{g}_1 = \boldsymbol{g}_2 = \boldsymbol{g}$ is continuous on the interface in view of continuity of the field vectors

$\boldsymbol{E}$ and $\boldsymbol{H}$, but the group velocities have different values, $v_{2g}/v_{1g} = \varepsilon_1/\varepsilon_2 = \mu_1/\mu_2$. The pressure on the interface is

$$P_{rad} = \frac{gc}{\sqrt{\varepsilon_1 \mu_1}} \left(1 - \frac{\varepsilon_1}{\varepsilon_2}\right); \tag{6.26}$$

it is negative if $\varepsilon_2 < \varepsilon_1$. Equation (6.26) is invalid for the left-handed medium because dispersion in it is not considered.

Inhomogeneous plane waves.

Example 6.3

An inclined plane monochromatic wave propagates from the region occupied by a transparent dielectric with the electric and magnetic permittivities $\varepsilon_1 = \mu_1 = 1$. The plane is incident to the boundary of conducting medium ($\mu = 1,\ \varepsilon = \varepsilon' + i\varepsilon''$). Determine the law of reflection, that is, the directions of propagation, damping, and phase velocity of the wave in the conducting medium.

Solution. In the conducting medium, we have $\boldsymbol{k} = \boldsymbol{k}' + i\boldsymbol{k}''$, where $\boldsymbol{k}'$ and $\boldsymbol{k}''$ are the real vectors. Denoting the angles between each of these real vectors and the normal by θ and ϑ, respectively, and comparing the exponents of all three waves, we get the equalities

(1) $$\theta_1 = \theta_0, \qquad k_1 \sin\theta_0 = k' \sin\theta + ik'' \sin\vartheta.$$

From these relations, we have $\sin\theta = k_1/k'$, $\vartheta = 0$ (the value $\vartheta = \pi$ is not realized due to the limitation of fields in conducting medium). The vector $\boldsymbol{k}''$ is, therefore, directed along the normal, in the depth of the conducting medium. Now we need to express the values of k' and k'' in terms of ε, ε_1, and θ_0.

It can be done using the dispersion equation

(2) $$k^2 \equiv k'^2 - k''^2 + 2ik'k'' \cos\theta = \frac{\omega^2}{c^2}(\varepsilon' + i\varepsilon'').$$

By equating the real and imaginary parts, we have

(3)
$$k''^2 = \frac{\omega^2}{2c^2}\left[(\varepsilon_1 \sin\theta_0 - \varepsilon') + \sqrt{(\varepsilon_1 \sin\theta_0 - \varepsilon')^2 + \varepsilon''^2}\right] > 0,\ k'^2 = k''^2 + \frac{\omega^2}{c^2}\varepsilon' > 0.$$

A plane wave in the conducting medium is inhomogeneous (Figure 6.3):

$$\boldsymbol{E}(\boldsymbol{r}, t) = \boldsymbol{E}_2 e^{-k''z} \exp(ik'x \sin\theta + ik'z \cos\theta - i\omega t).$$

The surface of constant phase propagates along the vector $\boldsymbol{k}'$ with the phase velocity $v_{ph} = \omega/k'$, which depends on frequency and angle of incidence of the wave on the interface. The penetration depth of the field into the conductor $\delta = 1/k''$ also depends on these quantities. The complex amplitude $\boldsymbol{E}_2$ on the boundary is expressed in terms of the amplitude of an incident wave using the boundary conditions. ■

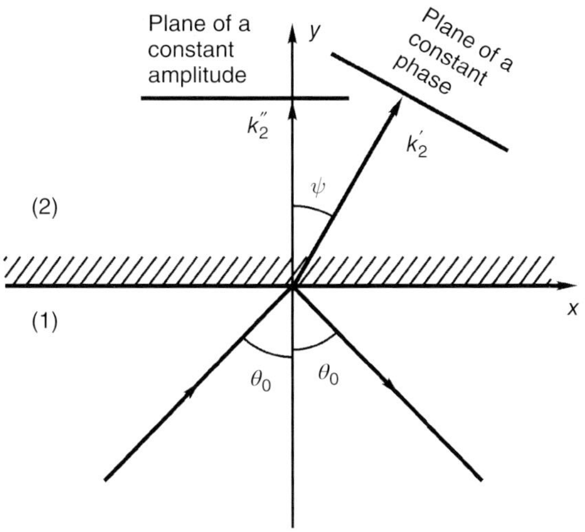

Figure 6.3 Inhomogeneous plane wave in conducting medium.

The Leontovich[2] boundary condition. Consider a range of frequencies in which $\sqrt{\varepsilon''} \gg \sqrt{\varepsilon'}$, $\sqrt{\varepsilon_1}$ (this range is wide in metals). Accordingly, $k' \approx k'' \approx \omega\sqrt{\varepsilon''}/2c$ and $\sin\theta = k_1/k' \ll 1$, that is, the directions of $\boldsymbol{k}'$ and $\boldsymbol{k}''$ are approximately the same. Hence, a wave in a conducting medium becomes homogeneous, it propagates and damps in the same direction, along the normal into the depth of the medium. At $\mu \neq 1$, we have $k'' \approx \omega\sqrt{\varepsilon''\mu}/2c$. The connection between the field vectors inside the conducting medium results from the Maxwell's equation:

$$\boldsymbol{E} = -(1+i)\frac{ck''}{\varepsilon\omega}\boldsymbol{n}\times\boldsymbol{H} = \sqrt{\frac{\mu}{\varepsilon}}\boldsymbol{H}\times\boldsymbol{n},$$

where $\varepsilon = i\varepsilon''$ and $\boldsymbol{n}$ is the unit vector to the normal.

This relation contains only the tangent components of the fields. Because of their discontinuity on the interface, the tangent components on the interface in a transparent medium will be connected in a similar way independently of the incidence angle of the wave:

$$\boldsymbol{E}_\tau = \zeta\boldsymbol{H}_\tau\times\boldsymbol{n} \quad \text{on} \quad S, \qquad \text{where} \quad \zeta = \sqrt{\frac{\mu}{\varepsilon}} \tag{6.27}$$

is the surface impedance relating to the conducting medium. Hence, Equation (6.27) can be used as the approximate boundary condition in the case in which an external, relative to the conductor, problem needs be solved (the Leontovich boundary condition). At $|\zeta| \ll 1$, this condition may be used not only on a plane but also on a curved surface, provided the local curvature radius is much larger

2) Leontovich Mikhail Aleksandrovich (1903–1981), Soviet physicist-theorist, known for his works on electrodynamics, optics, quantum mechanics, radio physics, plasma physics. Participant of the Soviet nuclear project.

than the penetration depth $\delta = 1/k''$ of the field into the conductor. The penetration depth plays the role of the wavelength in the conducting medium.

Abnormal skin-effect in metals. The penetration depth of electromagnetic field in a metal is

$$\delta = c/\sqrt{2\pi\mu\kappa_0\omega}. \tag{6.28}$$

It has been derived in Chapter 4 (Equation (4.17)) and in Problem 6.23 for the normal skin-effect. The normal effect means that $\omega \ll \gamma$ or $\omega\tau \ll 1$, where $\tau = 1/\gamma$ is the relaxation time of electrons. Moreover, for normal skin-effect, the usual Ohm's law, $\boldsymbol{j} = \kappa_0\boldsymbol{E}$, is applied: a current is determined by a local value of the field. Such a dependence assumes that the electric field is uniform along the distances of the order of the mean free path, $\Lambda \approx v_F\tau$, where v_F is the Fermi velocity of electrons. It means that the inequality $\Lambda \ll \delta$ must hold. However, with increasing frequency the penetration depth decreases, while the mean free path increases. For example, in pure metals at the liquid helium temperature ($T \approx 4$ K) for wavelengths in centimeter range $\delta \approx 10^{-6}$ cm and $\Lambda \approx 10^{-2}$ cm. Under these conditions $\Lambda \gg \delta$; only a small fraction of electrons, arriving at the given point after the last collision, is affected by the electric field (Figure 6.4). In this case, spatial dispersion is important (abnormal skin-effect).

Detailed calculations of the electromagnetic field in metals at abnormal skin-effect can be found in the book by Bredov *et al.* (2003). Here we restrict ourselves by estimating the penetration depth using the "Pippard's inefficiency concept." At normal skin-effect, a current is determined by the electric conductivity $\kappa_0 = \omega_{0e}^2/4\pi\gamma = Ne^2\tau/m$, where N is the total number density of conduction electrons. At abnormal skin-effect, the majority of electrons is inefficient for producing the current. Only a small fraction of electrons is efficient, whose mean free paths lie in the dotted narrow region of width δ near the metal surface. The number density of efficient electrons can be written as $N_{ef} = aN(\delta/\Lambda)$, where a is the dimensionless constant of the order of unity. The corresponding electric conductivity is $\kappa_{ef} = a\kappa_0(\delta/\Lambda)$. Substituting this value in Equation (6.28) and calculating then the penetration depth, we have

$$\delta = \left(\frac{c^2\Lambda}{2\pi a\mu\kappa_0\omega}\right)^{1/3} \propto \omega^{-1/3}. \tag{6.29}$$

The two expressions for the penetration depth, Equations (6.28) and (6.29), are joined at $\omega \approx \omega_c = c^2/2\pi\mu\kappa_0\Lambda^2$ (Figure 6.4).

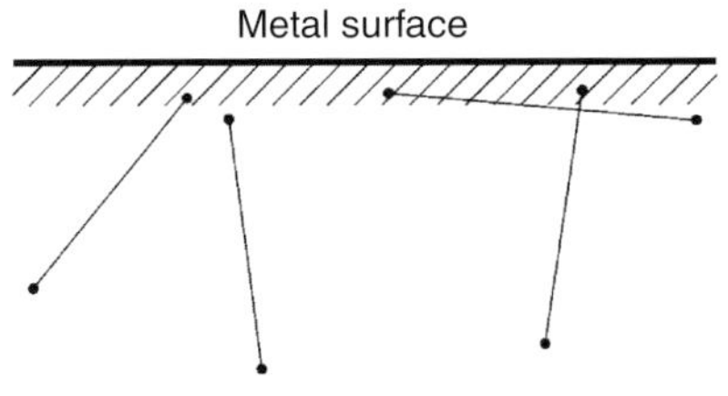

Figure 6.4 For the evaluation of the penetration depth in metal at abnormal skin-effect.

Inhomogeneous medium. The most elaborated method to study inhomogeneous media is an approximation of geometric optics, which assumes that a wavelength is much smaller than all other sizes of the problem. The detailed description of this method can be found in the fundamental monograph of Born and Wolf (1970). In Section 6.3, this method will be used to find the solutions of problems of wave diffraction. Accurate solutions for inhomogeneous media can be obtained only in a few cases because of the complexity of corresponding equations.

At $\mu = 1$, we have from Equation (6.1), the following equation for an electric field of frequency ω:

$$\triangle \boldsymbol{E}(\boldsymbol{r}) - \nabla(\nabla \cdot \boldsymbol{E}(\boldsymbol{r})) + \frac{\omega^2 \varepsilon(\boldsymbol{r}, \omega)}{c^2} \boldsymbol{E}(\boldsymbol{r}) = 0, \tag{6.30}$$

where the scale L of spatial inhomogeneity of the dielectric permittivity is considered to be larger than interatomic distance, $L \gg a$. Examples of exact solutions of Equation (6.30) are given in Problems 6.31 and 6.32.

Recently, Shvartsburg (2000) has proposed the method that significantly extends the class of one-dimensional electrodynamic problems, which can be accurately solved for inhomogeneous and nonstationary media. The concept of this method is given below.

Let consider a range of frequencies, in which time dispersion and microscopic spatial dispersion, described in Chapter 5, are insignificant. A medium is assumed to be nonmagnetic ($\mu = 1$) and isotropic, but spatially inhomogeneous:

$$\boldsymbol{D} = \varepsilon(z)\boldsymbol{E}, \qquad \varepsilon(z) = n_0^2 U^2(z), \quad U(0) = 1, \tag{6.31}$$

where n_0 is a reflection coefficient at $z = 0$, and $U(z)$ is a function describing macroscopic spatial inhomogeneity of the medium. This function is not specified yet. Instead of Equation (6.30), we consider two one-dimensional Maxwell's equations describing the transverse electromagnetic field:

$$\frac{\partial E_x}{\partial z} = -\frac{1}{c}\frac{\partial B_y}{\partial t}, \qquad -\frac{\partial B_y}{\partial z} = \frac{1}{c}\frac{\partial D_x}{\partial t}. \tag{6.32}$$

One of the ways to solve this system is based on using the vector potential, one of whose projections, $A_x \equiv A(z, t)$, is not zero:

$$E_x = -\frac{1}{c}\frac{\partial A}{\partial t}, \qquad B_y = \frac{\partial A}{\partial z}. \tag{6.33}$$

The first equation (6.32) becomes the identity, and the second gives the second-order equation

$$\frac{\partial^2 A}{\partial z^2} - \frac{n_0^2 U^2(z)}{c^2}\frac{\partial^2 A}{\partial t^2} = 0. \tag{6.34}$$

At the next step, we introduce the new variable η and new functions

$$F(\eta, t) = A(z, t)\sqrt{U(z)}, \qquad Q(z) = U^{-1}(z), \qquad \eta(z) = \int_0^z U(z')\, dz'. \tag{6.35}$$

The variable η has a dimension of length. After multiplying it by n_0 we obtain an optical path in inhomogeneous medium. In the new variables, Equation (6.34) reads

$$\frac{\partial^2 F}{\partial \eta^2} - \frac{n_0^2}{c^2}\frac{\partial^2 F}{\partial t^2} = -F\left\{\frac{1}{2}Q\frac{d^2Q}{dz^2} - \frac{1}{4}\left(\frac{dQ}{dz}\right)^2\right\}. \tag{6.36}$$

Let us chose the function $Q = U^{-1}$ in such a way to simplify Equation (6.36):

$$\frac{1}{2}Q\frac{d^2Q}{dz^2} - \frac{1}{4}\left(\frac{dQ}{dz}\right)^2 = -p^2 = const. \tag{6.37}$$

The constant p^2 will be defined below. For such a choice of $Q(z)$ and for the harmonic time function of the field equation, Equation (6.35) simplifies

$$\frac{\partial^2 F}{\partial \eta^2} + \left(\frac{n_0^2\omega^2}{c^2} - p^2\right)F = 0 \tag{6.38}$$

and has the solution $F(\eta, t) = \exp(\pm iq\eta - i\omega t)$. The vector potential is presented as two traveling waves with a variable amplitude:

$$A(z,t) = \frac{e^{-i\omega t}}{\sqrt{U(z)}}\left(A_1 e^{iq\eta} + A_2 e^{-iq\eta}\right), \qquad \text{where } q = kN\,, \quad k = \frac{n_0\omega}{c}$$

$$N = \sqrt{1 - \frac{\omega_p^2}{\omega^2}}, \quad \omega_p^2 = \frac{p^2c^2}{n_0^2}, \tag{6.39}$$

and A_1 and A_2 are the arbitrary constants, which can be determined from the boundary conditions.

At $p^2 > 0$, the solution (6.39) describes waves traveling in a cold collisionless plasma and oscillating with plasma frequency ω_p. The product n_0N gives the refraction coefficient of such a medium that is transparent for waves with frequencies $\omega > \omega_p$. However, in our case, the dispersion is not due to the response of electrons to the applied field, but macroscopic inhomogeneity of the medium. The cutoff frequency can significantly differ from the plasma frequency ω_{0e} (Example 5.6). When $p^2 < 0$, the normal dispersion takes place (the refraction coefficient decreases with increasing frequency).

The values of p^2 and profiles of the dielectric permittivity, which can be studied by this method, are determined by the solution of Equation (6.37). Let us find the solution in the form of some polynomial with real coefficients. At $Q(z) = 1$, we have the case of homogeneous medium: $\eta = z$, $p^2 = 0$. Next we can take the linear polynomial, $Q(z) = 1 \pm z/L$, where L is an arbitrary parameter of dimension of length. From Equation (6.37), we find $p^2 = 1/4L^2$. The function $\eta(z)$ can then be determined without any problem. The derived relations can be used for calculating the field in an inhomogeneous medium with the dielectric permittivities

$$\varepsilon(z) = \frac{n_0^2}{(1 \pm z/L)^2}. \tag{6.40}$$

Let us also consider the quadratic polynomial,

$$Q(z) = U^{-1}(z) = 1 \pm z/L_1 \pm z^2/L_2^2, \tag{6.41}$$

where the signs $\pm$ in two terms can be chosen independently, and L_1 and L_2 are the arbitrary lengths. From Equation (6.36), we find $p^2 = 1/4L_1^2 \pm 1/L_2^2$. According to the relationship between L_1 and L_2 and the sign, we can have $p^2 > 0$, $p^2 < 0$ and $p^2 = 0$.

The optical path $\eta(z)$ is also calculated in elementary functions. Varying free parameters and using Equation (6.41), it is possible to plot monotonic and non-monotonic profiles of permittivity and obtain corresponding solutions of electrodynamic problems. A disadvantage of this method is that the inhomogeneous dielectric layer should have finite thickness. If z is sufficiently large, the permittivity appears in the nonphysical domain $\varepsilon(z) < 1$ in which waves cannot propagate (let us recall that $\mu = 1$). The representation free from this disadvantage is considered in Problem 6.33.

Cloaking of material bodies by a method of wave flow. Recently, the new tempting practical application of the theory of electromagnetic waves propagation in nonuniform media has been offered. The basic mathematical idea was formulated more than 50 years ago (Dolin, 1961), but only last year, it has been opened again (Pendry *et al.*, 2006; Schurig *et al.*, 2006) and adapted for cloaking of bodies. The method involves surrounding a body with some nonuniform cover transparent for electromagnetic waves, which should bend beams (lines of Poynting's vector) around the body (Figure 6.5c). Outside the cover, beams propagate in the former direction. In the presence of the cover, the observer will notice neither the cover, nor the object hidden within the cover. Under the conditions presented in Figure 6.5c, the observer would see empty space which is behind the cover.

The properties of such a "magic" cover have to be calculated (Dolin, 1961; Dubiniv and Mytareva, 2010, 2012; Kildishev and Shalaev, 2011). Let us outline these calculations, following Dubiniv and Mytareva (2010) for the cover in the form of a spherical layer with radii R_1 and $R_2 > R_1$. Assume that the object to be hidden is in vacuum ($\varepsilon = \mu = 1$) and has the size $l < R_1$ (Figure 6.5a). The electromagnetic field in the vicinity of the object satisfies to the equations

$$\nabla \times \boldsymbol{E} = -\frac{1}{c}\frac{\partial \boldsymbol{H}}{\partial t}, \qquad \nabla \times \boldsymbol{H} = \frac{1}{c}\frac{\partial \boldsymbol{E}}{\partial t}. \tag{6.42}$$

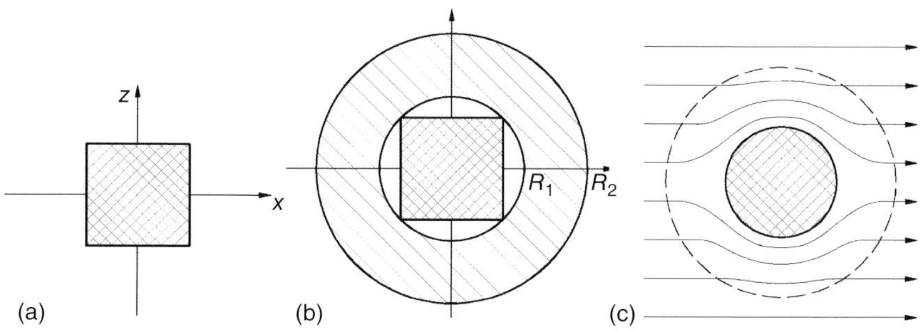

Figure 6.5 (a) Cloaking of cube. (b) The cube surrounded with the hiding spherical cover of internal and external radii R_1 and R_2, respectively. (c) Electromagnetic energy flux flows around the region $r < R_1$ without penetrating it. The cube is invisible.

We use spherical coordinates $(r,\ \theta,\ \phi)$. In the region $r \leq R_2$, occupied with the cover and the object, we introduce a new radial variable

$$r' = R_1 + \frac{r}{q}, \quad q = \frac{R_2}{R_2 - R_1}, \qquad 0 \leq r \leq R_2, \quad R_1 \leq r' \leq R_2 \tag{6.43}$$

at the same θ and ϕ. Let us introduce also new components of the fields using formulas

$$\widetilde{E}_r = qE_r, \quad \widetilde{E}_\theta = \frac{r}{r'}E_\theta, \quad \widetilde{E}_\phi = \frac{r}{r'}E_\phi, \qquad \widetilde{H}_r = qH_r, \quad \widetilde{H}_\theta = \frac{r}{r'}H_\theta, \quad \widetilde{H}_\phi = \frac{r}{r'}H_\phi. \tag{6.44}$$

Then the Maxwell's equations will be written as:

$$\nabla' \times \widetilde{\boldsymbol{E}} = -\frac{1}{c}\frac{\partial \hat{\mu}\widetilde{\boldsymbol{H}}}{\partial t}, \qquad \nabla' \times \widetilde{\boldsymbol{H}} = \frac{1}{c}\frac{\partial \hat{\varepsilon}\widetilde{\boldsymbol{E}}}{\partial t}. \tag{6.45}$$

Here the operator ∇' is expressed through r':

$$\nabla' = \boldsymbol{e}_r\frac{\partial}{\partial r'} + \frac{\boldsymbol{e}_\theta}{r'}\frac{\partial}{\partial \theta} + \frac{\boldsymbol{e}_\phi}{r' \sin\theta}\frac{\partial}{\partial \phi}.$$

The effective tensors of electric and magnetic permeabilities have the form

$$\varepsilon_{rr} = \mu_{rr} = \frac{(r' - R_1)^2}{r'^2}, \qquad \varepsilon_{\theta\theta} = \mu_{\theta\theta} = \varepsilon_{\phi\phi} = \mu_{\phi\phi} = q. \tag{6.46}$$

Other components are equal to zero.

Thus, we see that in the new coordinates, the Maxwell's equations keep their form, but instead of vacuum one obtains a nonuniform and anisotropic dielectric medium with the effective tensors $\hat{\varepsilon}$ and $\hat{\mu}$. A distinctive feature of the transformed fields $\widetilde{\boldsymbol{E}}$ and $\widetilde{\boldsymbol{H}}$ is that, according to Equation (6.44), at $r' = R_1$ $(r = 0)$ tangential to a spherical surface components of both fields $(\widetilde{E}_\theta,\ \widetilde{E}_\phi,\ \widetilde{H}_\theta,\ \widetilde{H}_\phi)$ are zero. It means that there is no radial flux of the Poynting's vector,

$$\widetilde{\gamma}_r = \frac{c}{4\pi}(\widetilde{E}_\theta\widetilde{H}_\phi - \widetilde{E}_\phi\widetilde{H}_\theta)_{r'=R_1} = 0 \tag{6.47}$$

in the selected region. The external field does penetrate this region, and the energy flux flows around it. Figure 6.5a–c illustrates this scheme. As time is not involved in the above transformations, coordinate phases of fluxes in the initial and transformed systems are identical. The observer who is to the right of the cover, will see empty space to the left of the cover.

For a practical realization of the cloaking effect, one needs to create an artificial dielectric cover with certain dielectric and magnetic properties. If the cover has a squared shape on some of its sites the deviation of the refracted flux should have a negative angle, that is, the cover should be made of metamaterial (see the section on negative refraction). Various restrictions and difficulties, and also other realizations of cloaking are discussed by Dubinov and Mytareva (2010, 2012), Rosanov (2011), and others (Figure 6.6).

Suggested literature: Toptygin (2014); Pamyatnykh and Turov (2000); Landau *et al.* (1984); Feinberg (1999); Bredov *et al.* (2003); Shvartsburg (1998);

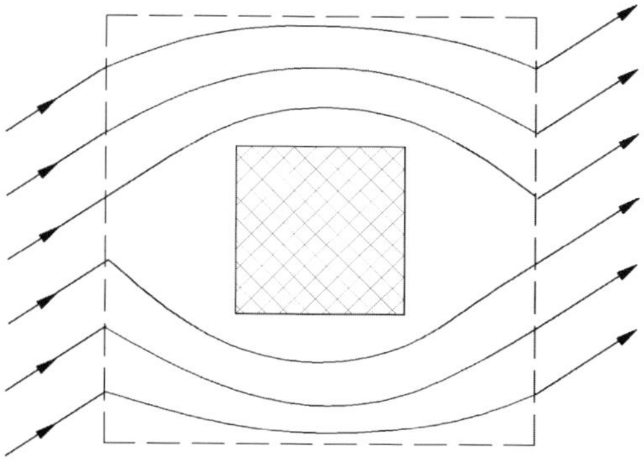

Figure 6.6 The cover of squared shape ("negative refraction" is required).

Shvartsburg (2000); Born and Wolf (1964); Vainshtein and Vakman (1983); Ostrovsky and Potapov (1999)

Problems

6.1• A plane wave propagates in a homogeneous and isotropic nondispersive medium, in which ε and μ are independent of frequency and wave vector. Using Equation (6.7), find the dispersion dependence $\omega(k)$ for the longitudinal, $\omega_l(k)$, and transverse, $\omega_t(k)$, waves, and also the relations between the strengths of the electric and magnetic fields in transverse waves, and between corresponding densities of energy.

6.2 Find the dispersion relation for transverse electromagnetic waves near one of the resonance frequencies of a dielectric, whose molecules do not have constant dipole moments (nonpolar dielectric). The permittivity of the nonpolar dielectric has been obtained in the Lorentz–Lorenz model (Problem 5.43). Find the region of nontransparency of such a dielectric. Plot the dispersion relations.

6.3 Find the dispersion relation for transverse electromagnetic waves in a collisionless gaseous plasma (the permittivity has been calculated in Example 5.6 and in Problem 5.9).

6.4 Find the dispersion relation for transverse electromagnetic waves in an ionic crystal, whose permittivity has been calculated in Problem 5.49. Find the region of non-transparency. Plot dispersion curves.

6.5 Investigate the shape and motion of a wavepacket formed by a superposition of plane waves with equal amplitudes a_0 and with wave vectors lying in the range $|\boldsymbol{k}_0 - \boldsymbol{k}| \leq q$ ($\boldsymbol{k}_0$ and q are constants). Replace the actual dispersion relation $\omega(\boldsymbol{k})$ by an approximate relation, $\omega(\boldsymbol{k}) = \omega(\boldsymbol{k}_0) + \boldsymbol{v}_g \cdot (\boldsymbol{k} - \boldsymbol{k}_0)$.

6.6 Investigate the spreading of one-dimensional wavepacket in a dispersive medium. Assume that the amplitude has the form of the Gaussian curve,

$a(k) = a_0 e^{-\alpha(k-k_0)^2}$, and include the second-order term in the expansion of frequency ω in powers of k.

6.7 Find the phase velocity v_{ph} and the group velocity v_g of waves in the media with the dielectric permittivities

$$\varepsilon(\omega) = 1 - \frac{\omega_{0e}^2}{\omega^2} \quad \text{(plasma)}, \quad \varepsilon(\omega) = 1 + \frac{\omega_p^2}{\omega_0^2 - \omega^2} \quad \text{(nonpolar dielectric)}.$$

In the second case, consider only large and small (compared to ω_0) frequencies ω ($\mu = 1$).

6.8 Two narrow wavepackets, emitted simultaneously by the same source, propagate in homogeneous plasma with the number density of electrons N_e. The leading (central) frequencies are larger than the electron plasma frequency $\omega_{0e} = \sqrt{4\pi N_e e^2/m_e}$ and correspond the wave lengths λ_1 and λ_2. The signals reach the receiver located at a distance L from the source at different time instants because of the difference in their group velocities. Using the plasma dielectric permittivity from Problem 6.7, express the delay Δt between the arrival times in terms of the wavelengths and the product $N_e L$, which is called **dispersion measure** ($DM = N_e L$) in radio astronomy.

6.9• Consider the Fresnel's formulas (6.11)–(6.15) for the case of transparent dielectrics:

(a) show that the reflected light will become totally polarized, if the angle of incidence $\theta_0 = \theta_p$ satisfies the condition $\tan\theta_p = n_2/n_1$ (angle of the full polarization, Brewster's[3] angle);

(b) show that when the light is incident at an angle $\theta_0 = \theta_r$, where $\sin\theta_r = n_2/n_1$, the medium with lower optical density ($n_2 < n_1$), the propagation direction of the refracted light is parallel to the boundary; show also that at $\theta_0 > \theta_r$, the total reflection takes place (i.e., $R = 1$ at any polarization of the incident light; $\sin\theta_2 > 1$, and $\cos\theta_2$ becomes a purely imaginary quantity);

(c) show that at the normal incidence ($\theta_0 \to 0$) the reflection coefficient for both polarizations is given by

$$R = \left| \frac{n_1 - n_2}{n_1 + n_2} \right|^2 .$$

(d) show that the wave is totally reflected from the medium with $\varepsilon < 0$ and $\mu = 1$.

6.10 A circularly polarized plane wave is incident obliquely to the plane boundary of dielectric. Determine the polarization of the reflected and refracted waves.

6.11⋆ A beam of almost monochromatic unpolarized light is incident to the plane boundary of dielectric. Find the polarization tensors $I_{ik}^{(1)}$ and $I_{ik}^{(2)}$,

3) Brewster David (1781–1868), Scottish physicist-optic and engineer.

and the depolarization coefficients ρ_1 and ρ_2 of the reflected and refracted lights, respectively.

6.12 An unpolarized almost monochromatic beam of light is incident the plane boundary between two dielectrics. Determine the reflection coefficient R and the depolarization coefficients $\rho_{1,2}$ of the reflected and refracted light, if the angle of incidence is equal to Brewster's angle.

6.13* Show that a linearly polarized wave becomes, in general, elliptically polarized after total reflection from the surface of dielectric. Under which conditions will the polarization become circular?

6.14 Investigate the transport of energy in the case of total internal reflection. Find the energy flux along the interface, and in the perpendicular direction in the medium from which the reflection takes place. Determine the lines of the Poynting vector $\boldsymbol{\gamma}$.

6.15 A plane monochromatic wave is incident to the plane boundary between two dielectrics with permittivities ε_1 and ε_2 ($\mu_1 = \mu_2 = 1$). What is the character of the field on both sides from the boundary in the case of grazing incidence (angle of incidence $\theta_0 \to \pi/2$)?

6.16$^\bullet$ The flat monochromatic wave is generated in the left-handed medium (see the section on negative refraction) and is incident to the boundary with the usual medium at an angle θ_0 to the normal. Specify directions of wave vectors and the directions of the reflected and refracted beams (i.e., energy streams).

6.17$^\bullet$ Write down Fresnel's formulas for flat boundary between usual Equation (1) and left-handed Equation (2) media through $|\varepsilon|$, $|\mu|$ and positive value of refraction angle, $0 \le \theta_2 < \pi/2$. Consider, in particular, the boundary between vacuum and "the ideal left-handed medium" which has permeabilities $\varepsilon_2 = \mu_2 = -1$.

6.18 A plane monochromatic wave is incident from vacuum to the plane boundary separating a transparent medium and vacuum, along the normal vector, and is partially reflected from it; the coefficient of reflection is R. Calculate the pressure exerted by the wave on the boundary (**pressure** of light), expressing it in the terms of energy density of the incident wave.

6.19 A plane monochromatic wave is incident from vacuum to the plane boundary of conducting medium, whose reflection coefficient $R(\theta_0)$ is given. The transmitted wave is totally absorbed in the second medium. Evaluate the pressure exerted by the incident wave on the conducting medium.

6.20$^\bullet$ Calculate the normal pressure of a flat monochromatic wave on interface between two transparent media at any hade of a wave that is, at any incident angle $0 \le \theta_0 < \pi/2$ and arbitrary polarization of the incident wave.

6.21* A dielectric layer with the permittivity ε_2, bounded by the planes $z = 0$ and $z = a$, separates the dielectric media with permittivities ε_1 and ε_3 ($\mu_1 = \mu_2 = \mu_3 = 1$). An electromagnetic wave is incident normally to the surface of the layer from the region $z < 0$. What is the thickness of the layer corresponding to minimum reflection? What is the relation between $\varepsilon_1, \varepsilon_2$ and ε_3 corresponding to the absence of reflection?

6.22• A plane monochromatic wave propagates from a transparent medium with permittivities ε_1 and μ_1 to the boundary of the second transparent medium with ε_2, μ_2, and $\sqrt{\varepsilon_2\mu_2} \gg \sqrt{\varepsilon_1\mu_1}$. Show that the Leontovich's boundary condition (6.27) holds in this case, and that the surface impedance could not be small.

6.23• Investigate the penetration depth of a field at different frequencies ω, using the results of Example 6.3 and the model dielectric permittivity

$$\varepsilon(\omega) = \varepsilon_0 + i\frac{4\pi\kappa(\omega)}{\omega}, \qquad \kappa(\omega) = \frac{\omega_{0e}^2}{4\pi(\gamma - i\omega)},$$

obtained in Problem 5.17. Consider different relations among parameters of medium: (a) $\varepsilon_0 \approx 1, \omega_{0e}^2/\gamma^2 \gg \varepsilon_0$ (metal, plasma); (b) $\varepsilon_0 \approx 10, \omega_{0e}^2/\gamma^2 \ll \varepsilon_0$ (semiconductor), $\omega \ll \gamma \ll \omega_{0e}$, $\omega_{0e} \ll \omega \ll \gamma$. Calculate the penetration depth δ and the surface impedance ζ of the conductor.

6.24 Apply the model of Problem 6.23 to a gaseous plasma with collisions. Calculate the penetration depth of a field into the plasma for frequency ranges $\gamma \ll \omega \ll \omega_{0e}$ and $\omega \gg \omega_{0e}$.

6.25 The electric field $\boldsymbol{E}$ of a plane inhomogeneous wave is linearly polarized. Determine the mutual orientation of the vectors $\boldsymbol{E}_0$, $\boldsymbol{\mathcal{H}}_1$, $\boldsymbol{\mathcal{H}}_2$, $\boldsymbol{k}'$, $\boldsymbol{k}''$ ($\boldsymbol{\mathcal{H}}_1$ and $\boldsymbol{\mathcal{H}}_2$ are the real and imaginary parts of the complex vector $\boldsymbol{H}_0$; $\boldsymbol{k}'$ and $\boldsymbol{k}''$ are the real and imaginary parts of the wave vector $\boldsymbol{k}$). Determine the locus of the end point of the vector $\boldsymbol{H}$ at a given point in space.

Solve the same problem for the case where the vector $\boldsymbol{H}$ is linearly polarized.

6.26 Derive the Fresnel formulas for the case in which an electromagnetic wave traveling from vacuum is incident to the plane boundary of conducting medium with a small surface impedance ζ.

6.27 Find the reflection coefficient R of metal surface with a small surface impedance $\zeta = \zeta' + i\zeta''$. Find the angle of incidence θ_0 corresponding to minimum of the reflection coefficient.

6.28 A linearly polarized wave is incident to the plane boundary of a conducting medium with a small surface impedance ζ. Determine the nature of the polarization of the reflected wave, if the grazing angle of the incident wave is equal to the angle Φ_0, determined in the solution of Problem 6.27.

6.29 A linearly polarized plane wave is incident at an angle θ_0 to the surface of metal. The direction of its polarization is at an angle $\pi/4$ to the plane of incidence. The ratio of the transverse and longitudinal (with respect to this plane) components of the reflected wave $E_{\parallel 1}/E_{\perp 1} = \tan\rho$, and the phase difference between them δ,

$$\frac{E_{\parallel 1}}{E_{\perp 1}} = e^{i\delta}\tan\rho$$

are experimentally determined. Express in terms of ρ, δ, and θ_0 the real part of the refraction coefficient n' and the absorption coefficient n'' ($n' + in'' = 1/\zeta$, ζ being the surface impedance); assume that $|n'^2 - n''^2| \gg \sin^2\theta_0$.

6.30 Find the reflection coefficient R of the plane surface of a conductor at normal incidence, considering the limiting case of low conductivity (see Equation (8) in the solution to Problem 5.17).

6.31* A plane wave propagates from vacuum and falls at right angle on the boundary of a dielectric. Investigate the effect of the spreading of the boundary on the reflection coefficient. To do this, approximate the dielectric permittivity by the function

$$\varepsilon(z) = \varepsilon - \frac{\Delta\varepsilon}{e^{z/a} + 1}, \qquad \varepsilon = 1 + \Delta\varepsilon,$$

where ε and $\Delta\varepsilon$ are constants. Investigate the specific cases of large and small a.

Hint. In the differential equation for $E(z)$ (Equation (6.30)), replace the independent variable $\xi = -e^{-z/a}$ and substitute $E(\xi) = \xi^{-ika}\psi(\xi)$, where $\psi(\xi)$ satisfies the hypergeometric equation (see the reference book of Abramowitz and Stegun, 1965).

6.32* In the absence of absorption, the dielectric permittivity of the plasma is given by (Problem 5.9)

$$\varepsilon = 1 - \frac{4\pi e^2 N}{m\omega^2}.$$

Consider the propagation of an electromagnetic wave in the plasma whose number density varies linearly: $N(z) = N_0 z$. A plane monochromatic wave is incident normally to inhomogeneous layer of the plasma. (This may occur in the propagation of radio waves in the ionosphere.)

Hint. The equation for $E(z)$ can be solved by presenting the required function as a Fourier integral.

6.33* Find the solution of the Maxwell's equation (6.32) for a monochromatic field using the substitution

$$E_x = \frac{1}{U^2(z)}\frac{\partial I}{\partial z}, \qquad B_y = -\frac{n_0^2}{c}\frac{\partial I}{\partial t}. \tag{1}$$

Obtain the equation for a new unknown function $I(z, t)$ and transform it using the second substitution $I = f\sqrt{U}$ into the equation with constant coefficients

$$\frac{\partial^2 f}{\partial \eta^2} - \frac{n_0^2}{c^2}\frac{\partial^2 f}{\partial t^2} = p^2 f, \tag{2}$$

where the phase coordinate (optical path) η is defined by Equation (6.35) in terms of $U(z)$, and p is a constant that satisfies the equation

$$\frac{d^2 Q}{d\eta^2} = p^2 Q, \qquad \text{where} \qquad Q = \frac{1}{\sqrt{U}}. \tag{3}$$

6.34* Calculate the reflection coefficient of the inhomogeneous medium considered in Problem 6.33. A plane monochromatic wave falls normally from vacuum on the boundary $z = 0$ of the medium whose inhomogeneity at $z > 0$ is given as a function of the optical path by the profile $U(\eta) = [\cos(\eta/L) + M \sin(\eta/L)]^{-2}$, where $L > 0$ and $M > 0$ are constant parameters. Construct, using a computer, the permittivity function $\varepsilon(z)$ in the inhomogeneous medium.

6.35* A layer of inhomogeneous dielectric occupies the region $0 \le z \le L$. Inside the layer the permittivity is given by Equation (6.40) (with the plus sign in the denominator). The layer is in air ($\varepsilon \approx 1$). A plane monochromatic wave is incident normally to the boundary. Write the expressions for electric and magnetic fields in the entire space. Calculate the reflection coefficient R for the electromagnetic wave.

6.36** Construct the solutions of the Maxwell's equation (6.32) for a nonstationary, nonmagnetic medium, in which

$$\boldsymbol{D} = \varepsilon(t)\boldsymbol{E}, \qquad \varepsilon(t) = n_0^2 U^2(t), \qquad U(0) = 1.$$

Assume that the characteristic periods of variations of $\varepsilon(t)$ are much larger than time-scales of microscopic relaxation discussed in Chapter 5, and are produced by external action on the medium (e.g., heating, ionization, strong external field). However, these periods may be comparable with the inverse frequencies of variations of $\boldsymbol{E}$ and $\boldsymbol{H}$. It is reasonable to modify and generalize the above-considered methods of solving the Maxwell's equations for spatially inhomogeneous media. In particular, it is convenient to introduce the time coordinate $\tau = \int_0^t d\tau'/U(\tau')$ and obtain the equations that determine all possible functions $U(t)$ or $U(\tau)$. Then it will be possible to simplify the initial system of equation (6.32) and find its solution in the analytic form. Calculate and analyze the reflection coefficient of an electromagnetic wave for a nonstationary medium.

6.2
Plane Waves in Anisotropic and Gyrotropic Media

Anisotropic media. Media are called optically anisotropic when their electric and magnetic properties are different in different directions. The dielectric permittivity and magnetic permeability of such a media are tensors even if the dependence on the wave vector (spatial dispersion) is neglected. Optical anisotropy may be a consequence of the crystalline structure of the body; it may also be produced by an external electric field (Problems 5.18 and 5.19) or by an external mechanical effect. In the absence of an external magnetic field and ignoring the losses, the tensors $\varepsilon_{ik}(\omega)$ and $\mu_{ik}(\omega)$ are symmetric:

$$\varepsilon_{ik}(\omega) = \varepsilon_{ki}(\omega), \qquad \mu_{ik}(\omega) = \mu_{ki}(\omega). \tag{6.48}$$

It is known from the general properties of symmetric tensors (Section I.1.1), that such a tensor in the principal axes is diagonal and determined by three scalar principal values. The principal axes are mutually perpendicular, and their directions are specified by internal symmetry of the media. It is not necessary for the permittivity and permeability of the same media to be tensors. In the majority of optical crystals, $\mu_{ik} = \mu\delta_{ik}$ and $\mu \approx 1$. In ferrits (Section 5.4) magnetic properties are anisotropic, while the dielectric permittivity could often be a scalar.

Let us write the system of equations (5.11)–(5.14) for plane monochromatic waves and set $\boldsymbol{j}_{ext} = \rho_{ext} = 0$, assuming that the magnetic permeability is a scalar:

$$\boldsymbol{k} \times \boldsymbol{E} = \frac{\omega}{c}\mu \boldsymbol{H}, \qquad \boldsymbol{k} \times \boldsymbol{H} = -\frac{\omega}{c}\boldsymbol{D}, \qquad \boldsymbol{k} \cdot \boldsymbol{D} = 0, \qquad \boldsymbol{k} \cdot \boldsymbol{B} = 0. \tag{6.49}$$

The dispersion equation for waves in an anisotropic medium is derived from the first two equations using the coupling equation $D_\alpha = \varepsilon_{\alpha\beta}E_\beta$:

$$\left| \mu\varepsilon_{\alpha\beta}(\omega) - \frac{c^2k^2}{\omega^2}k^2\delta_{\alpha\beta} + \frac{c_\alpha^k k_\beta}{\omega^2} \right| = 0. \tag{6.50}$$

It differs basically from Equation (6.4) because in an anisotropic medium, Equation (5.38) cannot be used for the dielectric permeability.

Let us introduce the dimensionless reflection index $n = ck/\omega$ and unit vector $\boldsymbol{\kappa} = \boldsymbol{k}/k$. The dispersion equation written in the principal axes of the tensor $\varepsilon_{\alpha\beta}$ is

$$\begin{aligned} n^4\mu(\varepsilon^{(1)}\kappa_1^2 &+ \varepsilon^{(2)}\kappa_2^2 + \varepsilon^{(3)}\kappa_3^2) - n^2\mu^2[\varepsilon^{(1)}(\varepsilon^{(2)} + \varepsilon^{(3)})\kappa_1^2 + \varepsilon^{(2)}(\varepsilon^{(1)} + \varepsilon^{(3)})\kappa_2^2 \\ &+ \varepsilon^{(3)}(\varepsilon^{(1)} + \varepsilon^{(2)})\kappa_3^2] + \varepsilon^{(1)}\varepsilon^{(2)}\varepsilon^{(3)}\mu^3 = 0. \end{aligned} \tag{6.51}$$

This is the **Fresnel equation**, the principal equation of crystal optics. If the principal values of the dielectric permittivity and magnetic permeability are independent of $\boldsymbol{k}$ and the direction of propagation is given, the Fresnel equation is the second order equation for $n^2 > 0$. Hence, in general, in each direction two waves will propagate with the different phase velocities $v_{ph} = c/n$, which depend on propagation direction. Their electric and magnetic inductions, according to Equation (6.49), are perpendicular to the propagation direction $\boldsymbol{\kappa}$. However, the vector $\boldsymbol{E}$ will have not only the transverse component, but the longitudinal, with respect to $\boldsymbol{\kappa}$, component as well. It is therefore convenient to determine the polarization with the aid of the vector of the electric induction $\boldsymbol{D}$. The directions, along which phase velocities of propagating waves are the same, are called **optical axes**. Generally, they are not the same as the principal axes of the tensor $\varepsilon_{\alpha\beta}$.

Example 6.4

*Show that for two eigenwaves propagating in a given direction the vector **D** is linearly polarized in two mutually perpendicular planes.*

Solution. We exclude from the two first equation (6.49) the vector $\boldsymbol{H}$ and write the obtained equality in projections on the plane perpendicular to the vector $\boldsymbol{n}$: $\mu D_\alpha = n^2 E_\alpha$. Introducing the inverse tensor of dielectric permittivity, we have

(1) $$\mu D_\alpha = n^2 \varepsilon_{\alpha\beta}^{-1} D_\beta,$$

where the indices α and β take two values each; they label the axes perpendicular to the vector $\boldsymbol{n}$. Let these axes coincide with the mutually perpendicular principal axes of the two-dimensional symmetric tensor $\varepsilon^{-1}_{\alpha\beta}$. Therefore, the tensor will only take the diagonal components ε'_1 and ε'_2, which differ from its principal values $\varepsilon^{(i)}$, $i = 1, 2$, and 3, in Equation (6.51). In these axes, the relation (1) can be written in the form of two equations

$$(2) \qquad \left(\frac{\mu}{n^2} - \frac{1}{\varepsilon'_1}\right) D_1 = 0, \qquad \left(\frac{\mu}{n^2} - \frac{1}{\varepsilon'_2}\right) D_2 = 0.$$

For $n = \sqrt{\varepsilon'_1 \mu}$, we have $D_2 = 0$, and the wave is polarized along axis 1. For $n = \sqrt{\varepsilon'_2 \mu}$, we have $D_1 = 0$, and the wave is polarized along axis 2. ■

The energy density and energy flux density of a monochromatic wave in an anisotropic medium are defined by Equation (5.112):

$$w = \frac{1}{16\pi}\left[\frac{\partial}{\partial\omega}(\omega\varepsilon_{\alpha\beta}(\omega))E^*_\alpha E_\beta + \frac{\partial}{\partial\omega}(\omega\mu_{\alpha\beta}(\omega))H^*_\alpha H_\beta\right], \quad \boldsymbol{\gamma} = \frac{c}{16\pi}(\boldsymbol{E}^* \times \boldsymbol{H} + \boldsymbol{E} \times \boldsymbol{H}^*). \tag{6.52}$$

The directions of energy propagation and of wave vector are generally not the same (Problem 6.40).

Gyrotropic media. In the presence of an external magnetic field, the tensors ε_{ik} and μ_{ik} are no longer symmetric; however, in nonabsorbing media, which will be considered in this section, these tensors are Hermitian. The same property possesses the generalized dielectric permeability (5.34) in the set of Equations (5.25)–(5.29) with three vectors,

$$\varepsilon_{ik} = \varepsilon^*_{ki}, \qquad \mu_{ik} = \mu^*_{ki}. \tag{6.53}$$

In this case, the imaginary part of each tensor is antisymmetric with respect to permutation of indices, that is, $\varepsilon''_{\alpha\beta} = (\varepsilon_{\alpha\beta} - \varepsilon^*_{\alpha\beta})/2i = -\varepsilon''_{\beta\alpha}$. Accordingly, the tensors can be replaced by the dual vectors $\boldsymbol{g}_e$ and $\boldsymbol{g}_m$, respectively. The relations between the field strengths and inductions can be written in the form (compare with Problem 5.21)

$$\boldsymbol{D} = \hat{\varepsilon}'\boldsymbol{E} + i(\boldsymbol{E} \times \boldsymbol{g}_e), \qquad \boldsymbol{B} = \hat{\mu}'\boldsymbol{H} + i(\boldsymbol{H} \times \boldsymbol{g}_m), \tag{6.54}$$

where $\boldsymbol{g}_e$ and $\boldsymbol{g}_m$ are the gyration vectors (electric and magnetic), and $\hat{\varepsilon}'\boldsymbol{E}$ is the vector with the components equal to $\varepsilon'_{ik}E_k$. The tensors $\hat{\varepsilon}'$ and $\hat{\mu}'$ are real and symmetric. Media for which the field vectors are connected by Equation (6.54) is referred to as **gyrotropic media**.

Plasma and ferrit materials placed in an external magnetic field are gyrotropic media. If the Oz-axis is taken along the field, then in the isotropic (in the absence of the field) ferrit medium the tensor of the magnetic permeability has the form (see, e.g., Problems 5.56 and 5.57)

$$\mu_{ik} = \begin{pmatrix} \mu_\perp & -i\mu_a & 0 \\ i\mu_a & \mu_\perp & 0 \\ 0 & 0 & \mu_\parallel \end{pmatrix}, \tag{6.55}$$

where the components $\mu_\perp$, $\mu_\parallel$, and μ_a will be real in medium without losses. The tensor ε_{ik} has the analogous form (Problem 5.23). Some crystalline and disordered media can be gyrotropic even in the absence of any external magnetic field. In this case, the gyrotropy is called **natural optical activity**. In such media, the gyration vector depends on the wave vector of the propagating wave.

The dispersion relations and polarization vectors $e_\beta(\omega, \boldsymbol{k})$ of the waves should be calculated from the system of homogeneous Equations (5.11)–(5.14) and (5.16) or from the system (5.25)–(5.29). The last system in the Fourier representation can be written as

$$T_{\alpha\beta}(\omega, \boldsymbol{k})E_\beta(\omega, \boldsymbol{k}) = 0, \quad \alpha = 1,\ 2,\ 3, \tag{6.56}$$

where

$$T_{\alpha\beta}(\omega, \boldsymbol{k}) = \epsilon_{\alpha\beta}(\omega, \boldsymbol{k}) - \frac{c^2k^2}{\omega^2}\left(\delta_{\alpha\beta} - \frac{k_\alpha k_\beta}{k^2}\right), \qquad \epsilon_{\alpha\beta}(\omega, \boldsymbol{k}) = \epsilon'_{\alpha\beta}(\omega, \boldsymbol{k}) + ie_{\alpha\beta\gamma}g_\gamma, \tag{6.57}$$

is Maxwellian tensor and $\boldsymbol{g}$ is the gyration pseudo-vector. The existence of nontrivial solutions of the system (Equation (6.56)) requires its determinant to be zero:

$$\Delta(\omega, \boldsymbol{k}) = |T_{\alpha\beta}(\omega, \boldsymbol{k})| = 0. \tag{6.58}$$

Taking into account that the tensor determinant is invariant under spatial rotations, we select the coordinate axes along the mutually perpendicular principal axes of the symmetric tensor $\epsilon'_{\alpha\beta}$; we denote the principal values of the tensor as ϵ_1, ϵ_2, and ϵ_3. With the axes selected, the tensor takes the form

$$\widehat{T} = \begin{pmatrix} \epsilon_1 - n^2(1-\kappa_1^2) & ig_3 + n^2\kappa_1\kappa_2 & -ig_2 + n^2\kappa_1\kappa_3 \\ -ig_3 + n^2\kappa_1\kappa_2 & \epsilon_2 - n^2(1-\kappa_2^2) & ig_1 + n^2\kappa_2\kappa_3 \\ ig_2 + n^2\kappa_1\kappa_3 & -ig_1 + n^2\kappa_2\kappa_3 & \epsilon_3 - n^2(1-\kappa_3^2) \end{pmatrix}. \tag{6.59}$$

By equating Δ to zero, we find the refractive indices for the eigenmodes of the medium under consideration. Expanding the determinant we see that the terms proportional to n^6 cancel out and Equation (6.58) takes the form

$$\begin{gathered} an^4 - [\epsilon_1(\epsilon_2+\epsilon_3)\kappa_1^2 + \epsilon_2(\epsilon_1+\epsilon_3)\kappa_2^2 + \epsilon_3(\epsilon_1+\epsilon_2)\kappa_3^2 + (\boldsymbol{\kappa}\cdot\boldsymbol{g})^2 - g^2]n^2 \\ + \epsilon_1\epsilon_2\epsilon_3 - \epsilon_1 g_1^2 - \epsilon_2 g_2^2 - \epsilon_3 g_3^2 = 0, \\ a(\omega, \boldsymbol{k}) = \epsilon_1\kappa_1^2 + \epsilon_2\kappa_2^2 + \epsilon_3\kappa_3^2. \end{gathered} \tag{6.60}$$

Equation (6.60) is a generalization of the Fresnel equation (6.51). The quantity $a(\omega, \boldsymbol{k}) = \epsilon_{\alpha\beta}\kappa_\alpha\kappa_\beta$ is the permittivity longitudinal relative to vector $\boldsymbol{k}$: $a = \epsilon^l(\omega, \boldsymbol{k})$.

If Equation (6.60) is treated as a quadratic equation with respect to n^2, it is easy to find two roots, n_1^2 and n_2^2. This enables one to write the Maxwellian tensor determinant (Equation (6.58)) as a product of three factors:

$$\Delta = a(n^2 - n_1^2)(n^2 - n_2^2); \tag{6.61}$$

setting any of the factors equal to zero leads to correct dispersion relations for a possible proper wave.

The convenience of writing the determinant in the form Equation (6.61) manifests itself when the spatial dispersion is nonexistent, that is, when the quantities $\epsilon_1, \epsilon_2, \epsilon_3$ and g depend only on ω but not on the absolute value $|\boldsymbol{k}|$. Then, the relationships $n^2 = n_{1,2}^2$ turn out to be the **solutions** of the dispersion equation, so that in each direction (for a given $\boldsymbol{\kappa}$), two waves can propagate through the medium under consideration with two, generally different, phase velocities $v_{1,2} = c/n_{1,2}$. Here, $n_{1,2}(\omega, \boldsymbol{\kappa})$ are the positive solutions of the biquadratic equation (6.60); they depend only on frequency and propagation direction of the corresponding wave.

The roots n_i^2 may be negative in some frequency ranges. For a Hermitian permittivity tensor, this signifies damping without dissipation, that is, the absence the corresponding mode. In this case, the relation $a = \epsilon^l(\omega, \boldsymbol{k}) = 0$ corresponds to oscillating modes of the medium, whose properties are independent of the magnitude of the wave vector. The electric vector in these modes is directed along $\boldsymbol{k}$, that is, these oscillations are longitudinal (cf. with Equation (5.56) for the isotropic medium). In the presence of spatial dispersion, relations $\epsilon^l(\omega, \boldsymbol{k}) = 0$ and $n^2 = n_{1,2}^2$ are the **equations** for determining the refractive indices rather than the solutions of the dispersion equation, because $\epsilon^l(\omega, \boldsymbol{k})$ and $n_{1,2}$ themselves are functions of n. That is why, in principle, there is no limit for the number of eigenmodes, which are solutions of these equations. To determine the refractive indices in this case requires specifying an explicit dependence of the dielectric tensor on ω and $\boldsymbol{k}$.

Polarization vectors $e_\beta(\omega, \boldsymbol{k})$ of the eigenmodes are closely connected with eigenvectors $\boldsymbol{b}^{(m)}(\omega, \boldsymbol{k})$ of Hermitian tensor $T_{\alpha\beta}(\omega, \boldsymbol{k})$ and with it principal values $\lambda^{(m)}(\omega, \boldsymbol{k})$, $m = 1,\ 2,\ 3$, which can be determined from the equations

$$T_{\alpha\beta} b_\beta = \lambda b_\alpha, \quad \alpha = 1,\ 2,\ 3. \tag{6.62}$$

As known from linear algebra, the principal values of a Hermitian tensor are real while its eigenvectors are generally complex and mutually orthogonal. We normalize them like this:

$$\boldsymbol{b}^{(m)*}(\omega, \boldsymbol{k}) \cdot \boldsymbol{b}^{(m)}(\omega, \boldsymbol{k}) = 1, \qquad \boldsymbol{b}^{(m)*}(\omega, \boldsymbol{k}) \cdot \boldsymbol{b}^{(n)}(\omega, \boldsymbol{k}) = 0 \quad \text{for} \quad m \neq n. \tag{6.63}$$

Although there are only three eigenvectors of the Maxwellian tensor, they describe the complete set of polarization vectors of the eigenmodes, whose number is generally unlimited. The eigenvectors $\boldsymbol{b}(\omega, \boldsymbol{k})$ are shown to transform into polarization vectors $\boldsymbol{e}(\omega^\sigma(\boldsymbol{k}), \boldsymbol{k})$ when their arguments ω and $\boldsymbol{k}$ are assigned values that satisfy the dispersion relation $\omega = \omega^\sigma(\boldsymbol{k})$ for a given eigenmode. Details of the proof of this important property of eigenvectors have been given by Toptygin and Fleishman (2008). In the absence of spatial dispersion in a gyratropic medium, two plane waves of the same frequency can propagate in given direction with different phase velocities. These waves are elliptically polarized in opposite directions, and the polarization ellipses have the same ratio of axes but are rotated relative to each other by $\pi/2$.

The boundary conditions on the surface of anisotropic or gyrotropic body are the same as at the interface between isotropic media (Equations (5.15) and (5.30)).

Recommended literature: Born and Wolf (1964); Landau *et al.* (1984); Agranovich and Gartstein (2006); Toptygin (2014); Bredov *et al.* (2003); Kaganov *et al.* (1997); Gurevich and Melkov (1994); Fedorov (2004); Toptygin and Fleishman (2008); Ignatovich and Ignatovich (2012)

Problems

6.37 By using the Fresnel equation (6.51), find the phase velocities of waves propagating in anisotropic medium along the principal axes of the tensor of the dielectric permittivity.

6.38• A medium in which two of the principal values of the tensor $\varepsilon_{\alpha\beta}$ are equal ($\varepsilon^{(1)} = \varepsilon^{(2)} = \varepsilon_{\perp}$, $\varepsilon^{(3)} = \varepsilon_{\parallel}$) is known as **uniaxial**.

1. Show that the optic axis lies along axis 3.
2. Using the Fresnel equation find the phase velocities of waves propagating at an angle θ to the optic axis:

$$v_{ph}^{(1)} = \frac{c}{\sqrt{\varepsilon_{\perp}\mu}}, \qquad v_{ph}^{(2)} = c\sqrt{\frac{\varepsilon_{\perp}\sin^2\theta + \varepsilon_{\parallel}\cos^2\theta}{\varepsilon_{\perp}\varepsilon_{\parallel}\mu}}.$$

The first of these waves is known as the **ordinary wave**, whereas the second wave is known as the **extraordinary** one (its velocity depends on propagation direction).

3. Find the polarization of the ordinary and extraordinary waves.

6.39 Using the formulas obtained in Problem 5.38, show that in an anisotropic nondispersive medium the electric and magnetic energy densities of a plane monochromatic wave are equal,

$$w_e = w_m = \frac{1}{8\pi}[\boldsymbol{E} \times \boldsymbol{H}] \cdot \boldsymbol{n},$$

where $\boldsymbol{E}$ and $\boldsymbol{H}$ are the real vectors.

6.40• Using the solution of Problem 5.38 and the Maxwell's equations, find the velocity $\boldsymbol{v}_g$ of energy propagation for a plane monochromatic wave in anisotropic nondispersive medium. Determine it as a ratio of the energy flux density to the energy itself, $\boldsymbol{v}_g = \boldsymbol{\gamma}/w$, and express it in terms of the electric vectors $\boldsymbol{E}$ and $\boldsymbol{D}$. Compare with the phase velocity $\boldsymbol{v}_{ph} = c\boldsymbol{n}/n^2$. Show graphically the relative positions of the vectors $\boldsymbol{e}$, $\boldsymbol{D}$, $\boldsymbol{H}$, $\boldsymbol{B}$, $\boldsymbol{v}_{ph}$, and $\boldsymbol{v}_g$.

6.41 The extraordinary wave propagates in a uniaxial crystal at an angle θ to an optic axis. Determine the angle α between the wave vector $\boldsymbol{k}$ and the vector $\boldsymbol{E}$, and also the angle ϑ between the propagation direction (Poynting vector) and optic axis of the crystal.

6.42 A plane electromagnetic wave of frequency ω propagates in a nondispersive dielectric with the permittivity and magnetic permeability, ε and μ, respectively. The dielectric moves as a whole with the constant velocity $\boldsymbol{V}$. Find the phase velocity of the wave as a function of frequency and propagation direction. Calculate the group velocity and find the wave polarization. Compare the properties of waves in a moving medium and in uniaxial crystal at rest.

Hint. Use Equations (5.144) and (5.145) for electromagnetic potentials in moving medium.

6.43 A plane wave is incident to the plane surface of uniaxial crystal placed in vacuum. The optic axis of the crystal is perpendicular to its surface. Find the directions of propagation in the crystal of the ordinary and extraordinary waves, if the angle of incidence is θ_0.

6.44 Solve Problem (6.43) for the case in which the optic axis of the crystal is parallel to its surface and makes the angle α with the plane of incidence.

6.45 A plane wave is incident normally to the plane lattice formed by thin, parallel, infinitely long conductors. The distances between the conductors and their thickness are much smaller than the wave length. What is the effect of lattice have on propagation of waves with different polarizations?

6.46 A plane monochromatic wave propagates in an infinite ferrit medium (magnetized to saturation) at an angle θ to a constant magnetic field. The magnetic permeability of the ferrit is given by the tensor (Equation (6.55)). The dielectric permittivity of the ferrit can be treated as scalar, $\varepsilon \approx 15$ (since in the microwave range the effect of constant magnetic field on magnetic properties of ferrits is much stronger than on electric properties). Find the phase velocities for normally incident waves.

6.47 Investigate the waves considered Problem 6.54 for the case of longitudinal propagation ($\theta = 0$), using the components of the tensor of the magnetic permeability

$$\mu_{\perp} = 1 + \frac{\omega_M \omega_0}{\omega_0^2 - \omega^2}, \qquad \mu_a = \frac{\omega \omega_M}{\omega_0^2 - \omega^2}, \qquad \mu_{\parallel} = 1, \tag{6.64}$$

given in Problem 5.56.

1. Find the dispersion relations $k(\omega)$ of the waves (normal incidence) and their existence domain (plot the curves).

2. Determine the polarizations of the waves (normal incidence) and their surface impedances[4] given by Equation (6.27).

3. Calculate an angle of rotation at the wave length z of polarization plane for linearly polarized wave (Faraday effect).

6.48 Solve the same problem for transverse propagation ($\theta = \pi/2$):

1. Find the dispersion relations $k(\omega)$ for the normally incident waves and their existence domain (plot the curves).

2. Determine the polarizations of waves (normal incidence) and their surface impedances.

3. Find the change in the polarization for the case of transverse propagation (Cotton–Mouton effect).[5]

4) For an infinite medium the term **wave impedance** is more appropriate.
5) Cotton Aimé (1869–1951), French physicist-experimentalist known for his works in optics.

6.49 A plane monochromatic wave of optical frequency propagates in a ferrodielectric, whose tensors of permittivity and magnetic permeability have the structure given by Equation (6.55). Calculate for the case of longitudinal propagation the angle of the Faraday rotation of the polarization plane along the length l. Use the components of the tensor μ given in Problem 6.47, and the approximations $\omega \gg \omega_M$, ω_0 and $\varepsilon_a \ll \varepsilon_\perp$.

6.50 The oscillations of magnetization and the magnetic field are referred to as **magnetostatic** when they satisfy the condition $kc \gg \omega$. In the zero approximation over the small parameter ω/kc, they can be described by the equations of magnetostatics, $div\, \boldsymbol{B} = 0$ and $rot\, \boldsymbol{H} = 0$. Find the dispersion relation $\omega(\boldsymbol{k})$ for the magnetostatic waves in infinite ferrite medium with the tensor of magnetic permeability (Equation (6.55)) and its components (Equation (6.64)). Find the frequency range, in which the magnetostatic waves exist, and also the electric field in the wave.

6.51 Nonmagnetic dielectric is placed in an external magnetic field. A plane monochromatic wave propagates along the direction of the magnetic field (the Oz-axis) and has the lin ear polarization at $z = 0$. Using the tensor of the dielectric permittivity derived in Problem 5.23, determine the polarization at $z > 0$.

6.52 Two quasi-monochromatic waves with the wave lengths λ_1 and λ_2 propagate from the same source in homogeneous plasma with the electron number density N_e and the uniform magnetic field B. Both frequencies are higher than the electron plasma frequency $\omega_{0e} = \sqrt{4\pi N_e e^2/m_e}$ and electron cyclotron frequency $\omega_{Be} = eB/m_e c$. In the source, both waves are linearly polarized in the same plane. Using the Hermitian tensor of dielectric permittivity from the solution to Problem 5.24, calculate the relative rotation $\Delta\chi$ of polarization planes of these waves along the path to a receiver, placed at a distance L from the source. Write it down in terms of $RM = e^3 N_e BL/2\pi(m_e c)^2$, which is called **rotation measure** in radio astronomy.

6.53 A plane circularly polarized wave, propagating in vacuum, is incident normally to the plane boundary of a ferrit. The ferrit is magnetized in the direction of incidence of the wave. Determine the polarization and the amplitudes of the reflected and transmitted waves. Find also the reflection coefficient R of electromagnetic energy off the ferrit surface, expressing it through the surface impedance. At which conditions is the reflection full?

6.54 Solve Problem for the case in which the incident wave is linearly polarized.

6.55* An artificial dielectric consists of thin ideally conducting circular discs, which are similarly oriented and placed in vacuum. A constant magnetic field $\boldsymbol{H}_0$ is applied normally to the planes of discs, and a plane electromagnetic wave propagates along the direction of the field. Determine the phase velocity assuming that the dielectric can be treated as continuous medium.

Hint. Take into account the Hall effect that appears under the action of the external magnetic field.

6.56 An ionized gas is placed in a constant magnetic field. A transverse plane wave propagates along the direction of the field. Determine the phase velocity of propagation. Consider the special case of low frequencies ($\omega \to 0$) and investigate also the nature of the electromagnetic waves taking into account the motion of positive ions.

Hint. Use the expression for the tensor of dielectric permittivity of ionized gas in a constant magnetic field, which was obtained in Problem 5.25.

6.57 Find the dispersion equation for electromagnetic waves propagating in an isotropic, magnetized to saturation ferrodielectric medium, taking into account the term $q\nabla^2\boldsymbol{M}$ in Equation (5.116). Show that three branches of waves with different dispersion relations, $\omega(\boldsymbol{k})$, can propagate in such a medium. Determine the explicit relation $\omega(\boldsymbol{k})$ for the branch for which the condition $\omega^2\varepsilon/(ck)^2 \ll 1$ is satisfied. Estimate the relative values of electric and magnetic fields for this branch.

6.58 Calculate the surface impedance ζ of a ferromagnetic conductor placed in a constant magnetic field parallel to its surface. The tensor of the magnetic permeability is given in the statement of Problem 6.72, and the components of the tensor of electric conductivity are equal to $\kappa_{11} = \kappa_{22} = \kappa_1$, $\kappa_{33} = \kappa_3$, $\kappa_{12} = -\kappa_{21} = -i\kappa_2$, $\kappa_{13} = \kappa_{31} = \kappa_{23} = \kappa_{32} = 0$.

Hint. The surface impedance in this case is a tensor of rank 2 and should be determined from the condition (compare with Equation (6.27))

$$E_{\tau i} = \zeta_{ik}(\boldsymbol{H}_\tau \times \boldsymbol{n})_k,$$

where $i, k = 1, 2$; $\boldsymbol{E}_\tau$ and $\boldsymbol{H}_\tau$ are tangential components of the field vectors near the surface of the conductor, and $\boldsymbol{n}$ is a unit normal to the surface.

6.59 Solve Problem 6.58 for the case in which a constant magnetic field is normal to the surface of the ferromagnetic conductor.

6.3
Scattering of Electromagnetic Waves by Macroscopic Bodies. Diffraction

The exact solution of the problem on interaction of electromagnetic waves with conducting or dielectric bodies is reduced to the integration of Maxwell's equations subject to appropriate boundary conditions. This is possible only in a limited number of cases (see, e.g., Problems 6.61 and 6.68) but approximate solutions cannot be infrequently found.

Scattering. If the linear sizes of the body are smaller than the wave length, then the electromagnetic field near the body can be considered uniform. The body, placed in the uniform periodic field, acquires electric and magnetic moments that depend on time according to the same law as the external field. The scattering wave appears as a result of radiation by these oscillating moments. The problem on scattering of electromagnetic waves by the bodies of small sizes is reduced to the determination of dipole moments induced in these bodies. The radiation fields are given in terms of dipole moments by Equations (I.5.28), (I.5.29), and (I.5.37).

The effective differential cross section for scattering into a solid angle $d\Omega$ is calculated using Equation (I.5.119):

$$d\sigma_s = \frac{dI(\vartheta, \alpha)}{\gamma_0}. \tag{6.65}$$

Here $dI = \gamma\, dS = \gamma r^2\, d\Omega$ is the time average of the radiation intensity into a solid angle $d\Omega$; γ and γ_0 are the average energy flux densities of the scattered and incident waves, respectively. The energy flux density is described by the Poynting vector (5.110),

$$\gamma = \frac{c}{16\pi}(\boldsymbol{E}^* \times \boldsymbol{H} + \boldsymbol{E} \times \boldsymbol{H}^*), \tag{6.66}$$

if the scattering body is placed in medium without spatial dispersion.

The effective differential cross section of absorption is the ratio of the average energy Q, absorbed by a body per unit time to the average energy flux density in the incident wave,

$$\sigma_a = \frac{Q}{\gamma_0}. \tag{6.67}$$

In the opposite limiting case in which the wave length is much smaller than the linear sizes of the body, one can use the methods of geometrical optics. Let us outline them in connection with the problems on propagation of waves through apertures in plane screens and bypassing different obstacles. Such processes are called **diffraction of waves** (from Latin diffractus, which means broken, refracted).

Huygens'–Fresnel principle and Kirchhoff formula. In Section I.5.1, we have calculated the retarded potentials. They enable us to determine the electromagnetic field produced by moving charges in vacuum. However, in the presence of substance, the statement of the problem may be different: the charges and currents can be defined in some restricted volume V. While the sources, placed outside V, can be defined not by the distribution of charges and currents in them, but by the electromagnetic field, which is induced on the surface S, enclosing the volume V. In this case, for the determination of the field inside V, one should take into account the field sources placed inside V, and perturbations emanated from the surface.

We will consider the case in which the charges inside the volume V are absent, and the field in this volume is produced by external sources. The explicit consideration of the sources can be replaced by the consideration of the field induced by them on the surface S. As will be seen below, any element of the surface S can be considered a source of secondary perturbations propagating over the entire space. These perturbations produce a resultant field in the volume V. Qualitative studies of this type were performed even by Huygens[6] in XVII century; the main result is referred to as **Huygens' principle**. Later they were adjusted and formulated

6) Christian Huygens (1629–1695), Dutch scientist-encyclopaedist, studied physics, mechanics, mathematics, and astronomy. His theory of propagation of light was published in 1690.

in the quantitative form by other scientists, especially by A. Fresnel[7] and Kirchhoff.[8] Below we will obtain the Kirchhoff's formula which allows us to obtain the solution of the formulated problems.

Let use the Green's equality (I.1.98):

$$\int_V (\varphi \triangle \psi - \psi \triangle \varphi) dV = \oint_S \left(\varphi \frac{\partial \psi}{\partial n} - \psi \frac{\partial \varphi}{\partial n} \right) dS. \tag{6.68}$$

Let $\varphi(\boldsymbol{r}, t) = \varphi(\boldsymbol{r}) e^{-i\omega t}$ be a function which satisfies the uniform D'Alembert equation (5.146) and depends harmonically on time. It is may be one of the Cartesian coordinates of field strengths or the potential. In this case, $\partial^2 \varphi / \partial t^2 = -\omega^2 \varphi$ and the amplitude $\varphi(\boldsymbol{r})$ will satisfy the **Helmholtz equation**[9]

$$\triangle \varphi + k^2 \varphi = 0, \tag{6.69}$$

where $k^2 = \omega^2 \varepsilon \mu / c^2$, in a space is filled with a homogeneous medium. The function ψ in Equation (6.68) will be identified with the Fourier transform of the Green's retarded function $G_\omega(\boldsymbol{r})$. It satisfies the Helmholtz nonuniform equation

$$\triangle G_\omega + k^2 G_\omega = -4\pi \delta(\boldsymbol{R}), \qquad \boldsymbol{R} = \boldsymbol{r}_P - \boldsymbol{r} \tag{6.70}$$

(compare with Equation (5.5)). Here the index P marks an observation point inside the volume V; the integration is performed over coordinates $\boldsymbol{r} = (x, y, z)$, and $\boldsymbol{n}$ is an external normal to the surface S.

Let substitute in the left-hand side of Equation (6.68) the following equations,

$$\triangle \varphi = -k^2 \varphi, \qquad \triangle \psi = \triangle G_\omega = -k^2 G_\omega - 4\pi \delta(\boldsymbol{R}). \tag{6.71}$$

Calculating the integral with the delta-function, we have

$$\varphi(\boldsymbol{r}_P) = \frac{1}{4\pi} \oint_S \left(G_\omega(R) \frac{\partial \varphi(\boldsymbol{r})}{\partial n} - \varphi(\boldsymbol{r}) \frac{\partial}{\partial n} G_\omega(R) \right) dS. \tag{6.72}$$

Here the retarded Green's function G_ω may be taken in the form (5.14) (replacing c by $v_{ph} = c/\sqrt{\varepsilon\mu}$):

$$G_\omega(R) = \frac{1}{R} e^{ikR} \tag{6.73}$$

(Green's function for a free medium). However, one should keep in mind, that it is possible to use other retarded Green's functions of the Helmholtz equation and definite boundary conditions on the surface S. For instance, in electrostatics

7) Augustin-Jean Fresnel (1788–1827), French physicist, the author of many studies in optics, one of the founders of wave optics.

8) Gustav Robert Kirchhoff (1824–1887), outstanding German physicist. His name is associated with the Kirchhoff's rule for constant currents (Section 3.1), Kirchhoff's law for radiation (Chapter 7), diffraction theory and other discoveries.

9) Hermann Ludwig Ferdinand von Helmholtz (1821–1894), outstanding German naturalist with widest scientific interests in electrodynamics, optics, acoustics, hydrodynamics, thermodynamics, physiology of vision and hearing. One of the first scientists who formulated energy conservation as a fundamental law of nature.

(Section 2.3), we have used the Green's functions satisfying the boundary conditions

$$G|_S = 0 \qquad \text{or} \qquad \left.\frac{\partial G}{\partial n}\right|_S = 0. \tag{6.74}$$

Such Green's functions are used in the diffraction problems as well (Section 7.2)

Equation (6.72) is the **Kirchhoff** integral in the scalar form. It gives the quantitative formulation of the Huygens' principle, which relates the field in any arbitrary point to its value on a certain closed surface. Thus, electromagnetic perturbations on the surface act as sources, which induce the secondary perturbations and the field in point $\boldsymbol{r}_P$. The derivation of the Kirchhoff integral in the case where a field is an arbitrary function of time, and free charges are present, can be found in Bredov *et al.* (2003).

Notice that Equation (6.72) cannot be treated as the solution of the wave equation, since the values φ and $\partial\varphi/\partial n$ on the surface are not predetermined and cannot be assigned arbitrarily. In fact, Equation (6.72) is a complicated integral equation. However, the Kirchhoff integral is rather useful in the diffraction problems of electromagnetic waves when the field values on a certain surface may be set arbitrarily. For instance, assume that we need to determine the field inside a volume in which it penetrates through the aperture in an opaque screen. In this case, usually, the following assumptions concerning the field on the boundary of the volume are made:

1. The values φ and $\partial\varphi/\partial n$ on a shadow side of the screen are zero.
2. The values

$$\varphi = \varphi_0, \qquad \frac{\partial\varphi}{\partial n} = \frac{\partial\varphi_0}{\partial n} \tag{6.75}$$

within the apertures are assumed to be the same as in the absence of a screen (the undistorted initial field is denoted by φ_0).

Problem

6.60* In the absence of external charges, Equation (6.72) is satisfied with any Cartesian component of the field vectors $\boldsymbol{E}$ or $\boldsymbol{B}$. Show that these vectors inside the volume V can be presented as

$$\boldsymbol{E} = \frac{1}{4\pi}\oint_S \left\{G_\omega(R)(\boldsymbol{n}\cdot\nabla)\boldsymbol{E} - \boldsymbol{E}(\boldsymbol{n}\cdot\nabla)G_\omega(R)\right\} dS,$$

$$\boldsymbol{B} = \frac{1}{4\pi}\oint_S \left\{G_\omega(R)(\boldsymbol{n}\cdot\nabla)\boldsymbol{B} - \boldsymbol{B}(\boldsymbol{n}\cdot\nabla)G_\omega(R)\right\} dS. \tag{6.76}$$

Show also that Equation (6.76) can be transformed to give

$$\boldsymbol{E} = -\frac{1}{4\pi}\oint_S [ik(\boldsymbol{n}\times\boldsymbol{B}) + (\boldsymbol{n}\times\boldsymbol{E})\times\nabla + (\boldsymbol{n}\cdot\boldsymbol{E})\nabla]G_\omega(R)\, dS,$$

$$\boldsymbol{B} = -\frac{1}{4\pi}\oint_S [-ik(\boldsymbol{n}\times\boldsymbol{E}) + (\boldsymbol{n}\times\boldsymbol{B})\times\nabla + (\boldsymbol{n}\cdot\boldsymbol{B})\nabla]G_\omega(R)\, dS, \tag{6.77}$$

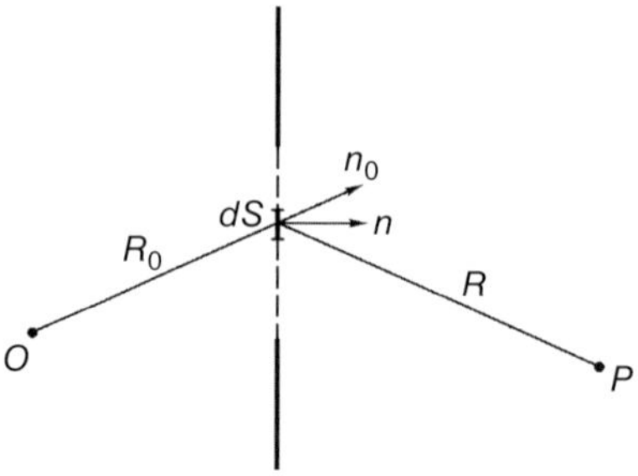

Figure 6.7 Fresnel diffraction by an aperture in the screen.

where $G_\omega(R)$ is the Fourier harmonic of the Green's retarded function (6.73). ■

The Kirchhoff integral in the vector form can be written for the Hertz vectors as well, using Equation (6.72) for each Cartesian component. This gives

$$\boldsymbol{Z} = \oint_S \left\{ G_\omega(R)(\boldsymbol{n} \cdot \nabla)\boldsymbol{Z} - \boldsymbol{Z}(\boldsymbol{n} \cdot \nabla)G_\omega(R) \right\} dS. \tag{6.78}$$

The field strengths are then calculated through the electromagnetic potentials using Equations (I.5.40) and (I.5.46). The advantage of such an approach is that Maxwell's equations in homogeneous medium, $\nabla \cdot \boldsymbol{E} = 0$ and $\nabla \cdot \boldsymbol{H} = 0$, are fulfilled automatically independently of the accuracy with which the field on the surface of integration is given (Niz'ev (2002)).

Geometrical optics. Fresnel diffraction. The scalar and vector Kirchhoff formulas are especially useful in analyzing the propagation of short electromagnetic waves (for brevity, we will call them waves of light). If the wave length of light is smaller than sizes of the body and the distance from the body to the source and point of observation, then one can use the geometrical optics approximation.

Let us consider the diffraction of light from a point source O (Figure 6.7) by an aperture in the opaque screen in the geometrical optics approximation. Assume that the wave length of light λ is small in comparison with the distance from the source O and from the point of observation P to the edges of the aperture. Let these distances be finite. That is why one should take into account the curvature of wave fronts for both the incident and diffracted waves. The phenomena observed at such conditions are referred to as **Fresnel diffraction**. If the points O and P are at a very large distance from the screen, then the fronts of incident and diffracted waves can be treated as plane. This case is known as **Fraunhofer diffraction**.[10]

Let us use the scalar version of the Kirchhoff formula (6.72), and let $\varphi(\boldsymbol{r})$ be a Cartesian component of the electric field, which we denote by u. We assume that the aperture area is the surface of integration. Also, we adopt the approximation commonly used in diffraction theory according to which the wave field within the aperture is the undistorted field of the source. In the case of symmetric

10) Joseph von Fraunhofer (1787–1826), German physicist and optician-inventor.

distribution of the source radiation O, we have within the aperture

$$u = \frac{A}{R_0} e^{ikR_0},$$

where $A = const$, and R_0 is a distance from the source to dS (Figure 6.8).

Taking into account that kR, $kR_0 \gg 1$ and differentiating, therefore, only the exponents in Equation (6.72), we obtain the field in point P behind the screen:

$$u_P = \frac{ik}{4\pi} \int_S \frac{u \exp(ikR)}{R} \left(\frac{\boldsymbol{R}_0}{R_0} + \frac{\boldsymbol{R}}{R} \right) \cdot \boldsymbol{n}\, dS. \tag{6.79}$$

We will see below that short-wavelength light (in comparison with other distances of the problem) propagates almost rectilinearly, with rather small deviations, which results in bypassing obstacles (i.e., diffraction). In view of this fact, in the integral (Equation (6.79)), only a small part of the integration surface is important near the straight line between the source and the observation point. In this region, we may take $(\boldsymbol{R}_0/R_0 + \boldsymbol{R}/R) \approx -2\boldsymbol{R}_0/R_0 = -2\boldsymbol{n}_0$, which enables us to present Equation (6.79) in the form

$$u_P = \frac{k}{2\pi i} \int_S \frac{u}{R} e^{ikR} \boldsymbol{n}_0 \cdot \boldsymbol{n}\, dS, \tag{6.80}$$

where u is the field within the aperture, $\boldsymbol{n}_0 \cdot \boldsymbol{n}\, dS$ is the projection of the surface area element of the aperture on the plane perpendicular to the ray direction from the source to the element dS.

We calculate diffraction of light from the point source on the edge of the screen (Figure 6.4). For an integration surface, we choose the semi-plane yz $(z > 0)$. In the pre-exponential factors, it is possible to set $R_0 \approx a$, $R \approx b$, and in the exponent one should take into account small corrections,

$$R_0 + R \approx a + b + \frac{1}{2}\left(\frac{1}{a} + \frac{1}{b}\right) y^2 + \frac{1}{2}\left[\frac{z^2}{a} + \frac{(z_P - z)^2}{b}\right].$$

Integrating over dy, we use the tabulated integrals

$$\sin t^2 dt = \int_{-\infty}^{\infty} \cos t^2 dt = \sqrt{\frac{\pi}{2}}$$

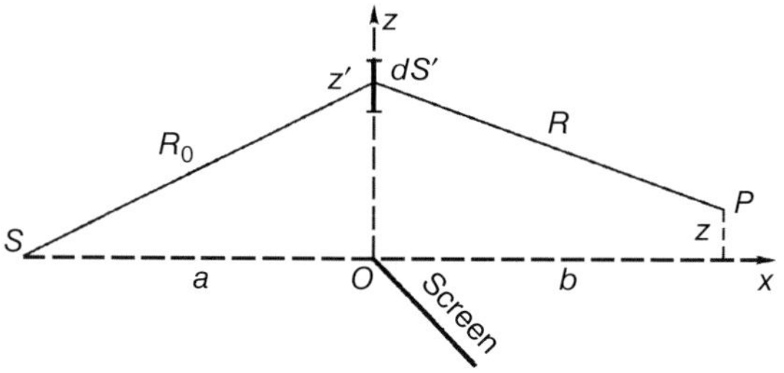

Figure 6.8 Fresnel diffraction by an edge of the screen.

and obtain

$$\int_{-\infty}^{\infty} \exp\left[\frac{ik}{2}\left(\frac{1}{a}+\frac{1}{b}\right)y^2\right] dy = \sqrt{\frac{\pi ab}{k(a+b)}}(1+i). \tag{6.81}$$

We see that the main contribution into the integral comes from the segment of length $[2\lambda ab/(a+b)]^{1/2} \approx \sqrt{\lambda L}$, where L is the smallest distance among a and b. Since $\lambda/L \ll 1$, the volume important for integration has the linear sizes $(\lambda/L)^{1/2}L \ll L$. A segment that contributes to integration over dz, has the same order of magnitude.

From terms dependent of z, we separate the total square:

$$\frac{1}{2}\left[\frac{z^2}{a}+\frac{(z_P-z)^2}{b}\right] = \frac{1}{2}\left(\frac{1}{a}+\frac{1}{b}\right)\left(z-\frac{az_P}{a+b}\right)^2 + \frac{z_P^2}{2(a+b)}.$$

Replacing the variable $t = z\sqrt{k(a+b)/2ab}$, we reduce the integral over dz to

$$\int_0^{\infty} \exp\left[\frac{ik}{2}\left(\frac{z^2}{a}+\frac{(z_P-z)^2}{b}\right)\right] dz = \sqrt{\frac{2ab}{k(a+b)}}\exp\left[\frac{ikz_P^2}{2(a+b)}\right]\int_{-s}^{\infty} e^{it^2}\,dt,$$

where

$$s = z_P\sqrt{\frac{ka}{2b(a+b)}}. \tag{6.82}$$

The integral over dt is expressed through the Fresnel integrals:

$$C(s) = \sqrt{\frac{\pi}{2}}\int_0^s \cos t^2 dt, \qquad S(s) = \sqrt{\frac{\pi}{2}}\int_0^s \cos t^2 dt,$$

$$\int_{-s}^{\infty} e^{it^2} dt = \sqrt{\frac{\pi}{2}}\left[\left(C(s)+\frac{1}{2}\right)+i\left(S(s)+\frac{1}{2}\right)\right]. \tag{6.83}$$

Combining Equations (6.79)–(6.83), we write down the field in point P:

$$u_P = \frac{A(1-i)}{2(a+b)}\exp[ik(a+b)+is^2b/a]\left[\left(C(s)+\frac{1}{2}\right)+i\left(S(s)+\frac{1}{2}\right)\right]. \tag{6.84}$$

In the case of free propagation, the wave field at a distance $a+b$ from the source would be

$$u_0 = \frac{A}{a+b}e^{ik(a+b)}.$$

Introducing the intensities $I(s) = |u_P|^2$ and $I_0 = |u_0|^2$, we have

$$I(s) = \frac{I_0}{2}\left[\left(C(s)+\frac{1}{2}\right)^2 + \left(S(s)+\frac{1}{2}\right)^2\right]. \tag{6.85}$$

Example 6.5

Investigate the distribution of intensity (Equation (6.85)) in the shadow region ($s < 0$) and illuminated region ($s > 0$). Obtain the asymptotes and plot the graphs.

Solution. The asymptote in the shadow region ($|s| \gg 1$) is obtained by integrating by parts,

(1) $$\int_{|s|}^{\infty} e^{it^2}\,dt = -\frac{1}{2i|s|}e^{is^2} + \frac{1}{4|s|^3}e^{is^2} - \cdots$$

Keeping one term, we obtain the intensity of light in the geometrical shadow region,

$$I = \frac{I_0}{4\pi s^2}. \tag{6.86}$$

The asymptote in the illuminated region ($s \gg 1$) is

(2) $$\int_{-s}^{\infty} e^{it^2}\,dt = \sqrt{\frac{\pi}{2}}(1+i) + \frac{1}{2is}e^{is^2} + \cdots$$

The intensity of light is

$$I = I_0\left[1 - \frac{\sin(s^2 - \pi/4)}{\sqrt{\pi}s}\right]. \tag{6.87}$$

At the boundary of the light and shadow at $s = 0$, we have $I = I_0/4$. In the intermediate region, it is necessary to use the exact formula. The intensity graph, including the intermediate region, is shown in Figure 6.9. ■

Fraunhofer diffraction occurs at an aperture in a plane opaque screen, if the light source and observation point are at a distance that significantly exceeds the size l of the aperture. The wavelength of light is smaller than l. Under these Conditions, the waves incident to the aperture may be considered plane. At the aperture, the diffraction occurs and the rays have a fan-shaped divergence.

Example 6.6

Investigate the intensity distribution for Fraunhofer diffraction at the aperture of arbitrary form in an opaque plane screen (Figure 6.10). Calculate the cross section of diffraction $d\sigma = dI/I_0$, where dI is the intensity of light, scattered into a given

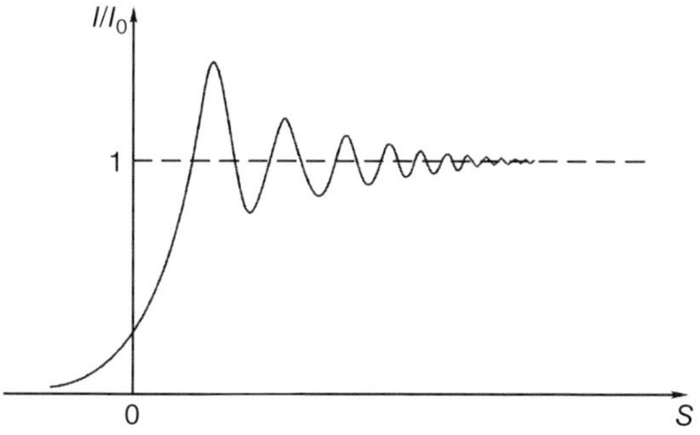

Figure 6.9 Distribution of light intensity in the case of Fresnel diffraction at the edge of a screen.

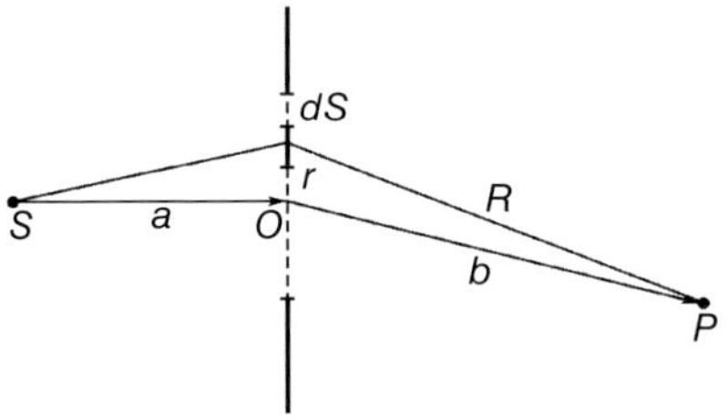

Figure 6.10 For the calculation of Fraunhofer diffraction at the aperture in a plane screen.

solid angle, I_0 is the intensity of light, incident to the aperture. Consider the particular case of diffraction at the rectangular aperture $2l_1 \times 2l_2$.

Solution. Let us use Equation (6.80) and the approximation of the undistorted field in the incident wave within the aperture plane. In the argument of the exponent, we take $R_0 \approx a + \boldsymbol{a} \cdot \boldsymbol{r}/a$ and $R \approx b + \boldsymbol{b} \cdot \boldsymbol{r}/b$, and for the pre-exponential factors, it is sufficient to set $R_0 \approx a$ and $R \approx b$. The vector $\boldsymbol{a}$ indicates the direction of propagation of the incident wave, and the vector $\boldsymbol{b}$ is the same for the diffracted wave. Since the frequency of the wave does not change at diffraction, then $k\boldsymbol{a}/a = \boldsymbol{k}_0$ and $k\boldsymbol{b}/b = \boldsymbol{k}$ are the wave vectors of the incident and diffracted waves. Substituting the obtained quantities in Equation (6.80), we have at the point of observation

$$u_P = -\frac{ikue^{ikb}}{2\pi b} \int_S e^{-i\boldsymbol{q}\cdot\boldsymbol{r}} dS, \tag{6.88}$$

where $u = Ae^{ika}/a$ is the field of the incident wave on the aperture, b is a distance from the aperture to the point of observation, and $\boldsymbol{q} = \boldsymbol{k} - \boldsymbol{k}_0$ is the change in the wave vector due to diffraction; the integration is performed over the aperture area. Then we calculate the intensity of scattering into the given solid angle $d\Omega$,

$$dI = |u_P|^2 b^2 d\Omega = \frac{k^2|u|^2}{(2\pi)^2} \left| \int_S \exp(-i\boldsymbol{q} \cdot \boldsymbol{r})\, dS \right|^2 d\Omega, \tag{6.89}$$

and the differential cross section of diffraction

$$d\sigma = \frac{k^2}{(2\pi)^2} \left| \int_S \exp(-i\boldsymbol{q} \cdot \boldsymbol{r})\, dS \right|^2 d\Omega. \tag{6.90}$$

The derived formula may be used for calculating diffraction of a plane-parallel beam of light by the aperture of any form.

In the case of the rectangular aperture, the integral over the aperture plane gives

$$\int_S \exp(-i\boldsymbol{q} \cdot \boldsymbol{r})\, dS = \frac{4 \sin(q_x l_1) \sin(q_y l_2)}{q_x q_y}.$$

At small diffraction angles, we have $q_x = k\theta_x$ and $q_y = k\theta_y$, where θ_x and θ_y are the angles of deviation of light rays in the directions of Ox and Oy axes, respectively. The diffraction cross section takes the form

$$\frac{d\sigma}{d\Omega} = \left(\frac{4l_1 l_2}{\lambda}\right)^2 \left(\frac{\sin kl_1\theta_x}{kl_1\theta_x}\right)^2 \left(\frac{\sin kl_2\theta_y}{kl_2\theta_y}\right)^2. \tag{6.91}$$

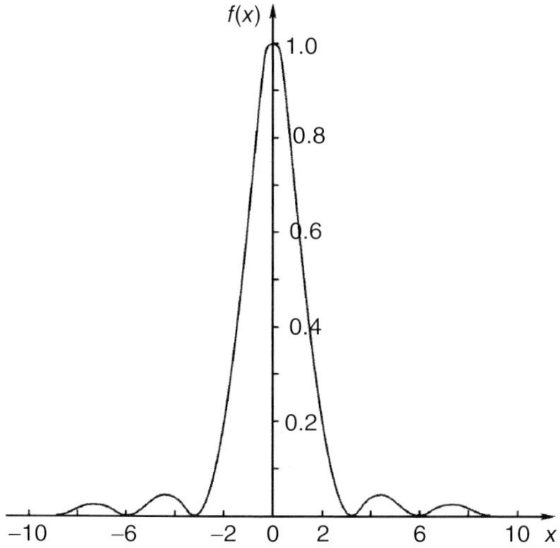

Figure 6.11 Light intensity distribution behind the rectangular aperture in a screen.

The angular distribution of the diffracted light along each of mutually perpendicular directions is given by the function $f(x) = \sin^2 x/x^2$, where $x = kl\theta$. This function has maximum ($f(0) = 1$) at $x = 0$, and rapidly decreases with increasing x, experiencing oscillations (Figure 6.11). Noticeable deviations of light rays occur only for the principal maximum by the angles

$$\theta \leq 2\pi/kl = \lambda/l. \tag{6.92}$$

Characteristic angles of diffraction are just determined by this condition. ■

In the case of Fraunhofer diffraction, there exists a simple relation between the diffracted fields u_1 and u_2 from two complementary screens. Two screens are defined as complementary if the apertures in one of them correspond to opaque regions in the other, and vice versa. The sum of the fields $u_1 + u_2$ is the field in the absence of both screens, that is, the plane monochromatic wave. Since the incident wave has the definite direction of propagation $\boldsymbol{k}' = \boldsymbol{k}$, $\boldsymbol{q} = 0$, at $\boldsymbol{q} \neq 0$ we have

$$u_1 + u_2 = 0, \qquad u_2 = -u_1. \tag{6.93}$$

This means that each of the complementary screens gives the same intensity distribution of diffracted light (**B**abinet principle).[11)]

Method of a slowly varying amplitude. The reduced equation. We consider the simplest version of this method using the characteristic example.

Example 6.7

A light beam (quasi-monochromatic wave) is incident normally to the aperture in an opaque screen placed in the $z = 0$-plane. The wave field immediately behind

11) Babinet Jacques (1794–1872), French physicist and astronomer.

the screen, that is, at $z = +0$, is described by the function $u_0(x, y, 0)$ (the time factor $e^{-i\omega t}$ is omitted), which differs from zero only within the aperture whose size $l_\perp$ is much larger than the wavelength. Presenting the field at $z > 0$ in the form

$$u(x, y, z) = A(x, y, z)e^{ikz}, \qquad k = \frac{\omega}{c}\sqrt{\varepsilon\mu} > 0, \tag{6.94}$$

derive the approximate equation for the amplitude $A(x, y, z)$. Obtain the solution for the region $z > 0$ in the general form.

Solution. The monochromatic field satisfies the Helmholtz equation (6.69). Substituting Equation (6.94), we obtain the exact equation for the amplitude:

$$\text{(1)} \qquad \frac{\partial^2 A}{\partial x^2} + \frac{\partial^2 A}{\partial y^2} + \frac{\partial^2 A}{\partial z^2} + 2ik\frac{\partial A}{\partial z} = 0.$$

Let us show that under formulated conditions the terms containing derivatives with respect to z have different orders of magnitude, $k|\partial A/\partial z| \gg |\partial^2 A/\partial z^2|$. Assuming that this inequality is valid, we estimate the individual terms in Equation (1):

$$\text{(2)} \quad \frac{\partial^2 A}{\partial x^2} \approx \frac{\partial^2 A}{\partial y^2} \approx \frac{A}{l_\perp^2}; \quad \frac{\partial A}{\partial z} \approx \frac{A}{l_\parallel}; \quad l_\parallel \approx (kl_\perp)l_\perp \gg l_\perp; \quad \frac{|\partial^2 A/\partial z^2|}{k|\partial A/\partial z|} \approx (kl_\perp)^{-2} \ll 1.$$

We see that Equation (1) can be reduced by omitting the second derivative with respect to z,

$$\text{(3)} \qquad \frac{\partial^2 A}{\partial x^2} + \frac{\partial^2 A}{\partial y^2} + +2ik\frac{\partial A}{\partial z} = 0.$$

The error introduced into Equation (3) due to this reduction is of the order of $(kl_\perp)^{-2} \ll 1$. It is smaller for larger transverse size of the beam in comparison with the wavelength $\lambda = 2\pi/k$. This equation may be treated as the diffraction equation with the imaginary diffraction coefficient χ:

$$\frac{\partial A}{\partial z} - \chi\triangle_\perp A = 0, \qquad \chi = \frac{i}{2k}. \tag{6.95}$$

It is useful to compare this equation with the diffusion equation of the magnetic field in Example 10.6. The propagation of light beam in the region $z > 0$ can be investigated using the results of this example. However, one should take into account that in this case the coordinate z plays role of time, and the Laplace operator $\triangle_\perp$ is two-dimensional. The Green's function of Equation (6.95) will, therefore, take the form

$$G(\boldsymbol{\rho} - \boldsymbol{\rho}', z) = \frac{1}{4\pi\chi z}\exp\left\{-\frac{(\boldsymbol{\rho} - \boldsymbol{\rho}')^2}{4\chi z}\right\}, \qquad \boldsymbol{\rho} = (x, y),\ \boldsymbol{\rho}' = (x', y'),\ z \geq 0. \tag{6.96}$$

It can be derived in the same manner as in Example 4.6. The beam amplitude will be described by the integral

$$A(x, y, 0) = \int G(\boldsymbol{\rho} - \boldsymbol{\rho}', z)u_0(x', y')\,dx'\,dy'. \tag{6.97}$$

At the boundary $z = 0$, we have $G(\boldsymbol{\rho} - \boldsymbol{\rho}', z)|_{z\to 0} \to \delta(\boldsymbol{\rho} - \boldsymbol{\rho}')$, and the boundary condition $A(x, y, 0) = u_0(x, y)$ is satisfied. ■

The derived formulas are used and analyzed in solutions of Problems 6.97–6.99.

Suggested literature: At mathematical transformations following books are very useful: Abramovitz and Stegun (1964); Gradshtein and Ryzhik (2007); Landau *et al.* (1984); Toptygin (2014); Bredov *et al.* (2003); Vainshtein (1988); Born and Wolf (1964); Newton (1969); Vaganov and Katsenelenbaum (1982); Vainshtein (1966b); Sommerfeld (1954); Feinberg (1999); Akhmanov and Nikitin (1998); Vinogradova *et al.* (1979)

Problems

6.61* A plane monochromatic wave is incident at an infinite circular perfectly conducting cylinder of radius a in the direction perpendicular to its axis. The cylinder is placed in vacuum. The vector $\boldsymbol{E}_0$ of the incident wave is parallel to the cylinder axis. Determine the resulting field, the distribution of the current over the surface of the cylinder, and the total current $\mathcal{I}$ flowing along the cylinder.

6.62 Find the differential scattering cross section $d\sigma_s$ of the electromagnetic wave (i.e., the secondary-wave pattern) for the cylinder considered in Problem 6.61. Also find the total scattering cross section σ_s.

6.63* A plane monochromatic wave is incident to a perfectly conducting circular cylinder so that its magnetic vector $\boldsymbol{H}_0 = \mathcal{H}_0\, e^{i(\mathbf{k}\cdot\mathbf{r}-\omega t)}$ is parallel, and the wave vector $\boldsymbol{k}$ is perpendicular, to the axis of the cylinder. The cylinder is in vacuum. Find the resulting electromagnetic field. Consider the particular case of a thin cylinder ($ka \lesssim 1$) and determine the differential and total scattering cross sections $d\sigma_s$ and σ_s for this cylinder.

6.64 Let $d\sigma_{\parallel}$ and $d\sigma_{\perp}$ be the differential scattering cross sections of an infinitely long cylinder for the plane waves with the electric vector $\boldsymbol{E}$ parallel and perpendicular to cylinder axis, respectively. Find the differential scattering cross section $d\sigma_s'$ for a wave whose electric vector $\boldsymbol{E}$ is at the angle φ to the axis, and also the differential scattering cross section $d\sigma_s''$ for an unpolarized wave.

Hint. Use the principle of superposition of fields.

6.65 An unpolarized plane wave is scattered by an perfectly conducting thin cylinder ($ka \ll 1$). Determine the depolarization degree ρ of the scattered waves as a function of the scattering angle.

Hint. The degree of depolarization is determined by Equation (I.2.134).

6.66 Solve Problem 6.63 on diffraction of a plane wave by an infinitely long cylinder without assuming that the cylinder is perfectly conducting, but regarding its surface impedance ζ as small. Use the Leontovich approximate boundary condition (6.27).

6.67 Determine the average energy loss rate Q and the absorption cross section σ_a per unit length of the cylinder considered in Problem 6.66. Investigate the particular case $ka \ll 1$ and explain the results.

6.68* Consider the diffraction of a plane monochromatic wave by a dielectric cylinder. The cylinder is in vacuum and has a radius a, dielectric permittivity ϵ, and magnetic permeability μ. The wave is incident normally to a generator of the cylinder, and the vector $\boldsymbol{E}$ is parallel to its axis. Determine the resultant field.

6.69* A linearly polarized plane monochromatic wave is scattered by a sphere whose radius a is much less than the wave length λ. Express the components of the electromagnetic field of the scattered radiation in the wave zone in terms of the electric and magnetic polarizabilities of the sphere. Determine the effective differential cross section.

Hint. In view of the inequality $a \ll \lambda$, assume the external field near the sphere to be uniform and consider radiation by the induced electric and magnetic dipole moments $\boldsymbol{p}$ and $\boldsymbol{m}$, respectively.

6.70 Calculate the differential and total scattering cross sections $d\sigma_s$ and σ_s, and also the degree of depolarization ρ of the secondary radiation when the unpolarized wave is scattered by a sphere whose radius a is much smaller than the wavelength λ. Express the result in terms of the electric and magnetic polarizabilities of the sphere, β_e and β_m.

6.71 Using the results of Problem 6.70, determine the differential and total scattering cross sections, $d\sigma_s$ and σ_s, for unpolarized light scattered by a small dielectric sphere with permittivity ε ($\mu = 1$); also determine the degree of depolarization ρ. Plot the curves showing these quantities as functions of the scattering angle θ. Indicate the limits of applicability of the derived relations. Solve the same problem for a perfectly conducting sphere with $\mu = 1$.

6.72 A plane monochromatic wave $\mathcal{E}_0 \exp[i(\boldsymbol{k} \cdot \boldsymbol{r} - \omega t)]$ is scattered by a dielectric sphere of radius a with the polarizability $(\varepsilon - 1)/4\pi \ll 1$ ($\mu = 1$). Because of the small polarizability the polarization of the sphere is, in the first approximation, proportional to the field of the incident wave. Determine the differential scattering cross section and the degree of depolarization ρ of the scattered radiation. What is the nature of scattering in the case of very large sphere ($ka \gg 1$)?

6.73 Determine the total scattering cross section σ_s by dielectric sphere, considered in Problem 6.72, in the limit $ka \gg 1$. Compare with the case of $ka \ll 1$.

6.74* **Mie's problem.**[12] A plane monochromatic linearly polarized electromagnetic wave is scattered by a perfectly conducting sphere of radius a in vacuum. Calculate the field in the entire space and also the differential and total scattering cross sections for the electromagnetic wave scattered by the sphere. Consider the general case in which a is not assumed to be much smaller than the wavelength.

6.75* Solve the same problem for a sphere made of transparent dielectric, placed in the space filled with different dielectric.

6.76 A plane monochromatic wave falls at an angle $\pi/2 - \alpha$ on a perfectly conducting thin disc with radius a much smaller than the wavelength λ.

12) Mie Gustav Adolf (1868–1957), German physicist.

Determine the differential and total scattering cross sections, $d\sigma_s$ and σ_s, for different polarizations of the incident wave, and also the scattering cross section for an unpolarized wave.

6.77 In a homogeneous dielectric with permittivity ε ($\mu = 1$), there is a cavity in the form of a thin disc of radius a and thickness $2h$. Unpolarized light of wavelength $\lambda \gg a$ is incident normally to the plane of the cavity. Find the differential and total scattering cross sections, $d\sigma_s$ and σ_s.

6.78⋆ Find the differential and total cross sections for scattering of a plane wave of wavelength λ by a perfectly conducting cylinder of length $2h$ and radius $a \ll h \ll \lambda$. Investigate the various types of polarization of the incident wave. Approximate the cylinder by a prolate ellipsoid of revolution with semiaxes a and h.

Hint. Use the solutions of Problems (2.38), (2.39), and (4.41).

6.79 Solve Problem 6.78 for a dielectric cylinder of a length $2h$ much smaller than the wavelength λ inside the cylinder.

6.80⋆ A plane monochromatic wave is scattered by a system of charges (e.g., by a macroscopic body). The electric field at large distances from the scatterer is

$$\boldsymbol{E} = E_0\left[\boldsymbol{e}\, e^{ikz} + \boldsymbol{F}(\boldsymbol{n})\frac{e^{ikr}}{r}\right],$$

where $\boldsymbol{n} = \boldsymbol{r}/r$, $\boldsymbol{e} = \boldsymbol{E}_0/E_0$, $k = \omega/c$, $\boldsymbol{E}_0$ is the amplitude of the incident wave, and $\boldsymbol{F}(\boldsymbol{n})$ is the scattering amplitude, which characterizes the properties of the scatter and is a function of frequency. Prove the relation (optical theorem):

$$\sigma_t = \frac{4\pi}{k}\Im[\boldsymbol{e}\cdot\boldsymbol{F}(\boldsymbol{n}_0)],$$

where $\sigma_t = \sigma_s + \sigma_a$ is the total cross section of interaction of the wave with the system of charges which is equal to the sum of scattering and absorption cross sections (σ_s and σ_a), $\boldsymbol{F}(\boldsymbol{n}_0)$ is the forward-scattering amplitude, that is, in the direction of propagation of the incident wave.

6.81⋆ A plane monochromatic wave is incident to the macroscopic particle whose sizes are much smaller than the wavelength λ. The electric and magnetic polarizabilities of the particle, $\beta_e = \beta_e' + i\beta_e''$ and $\beta_m = \beta_m' + i\beta_m''$, are complex. Therefore, the scattering is accompanied by absorption. Calculate the absorption cross section σ_a.

Hint. The energy absorbed per unit time is equal to the Poynting flux through the surface of the sphere of large radius surrounding the particle.

6.82 Calculate the cross section σ_a for the absorption of an electromagnetic wave by the conducting sphere with a small surface impedance $\zeta = \zeta' + i\zeta''$. The radius of the sphere, b, is smaller than the wavelength λ.

6.83 A plane monochromatic wave falls on the macroscopic body with given absorption and differential scattering cross sections, σ_a and $d\sigma_s/d\Omega$. Express the time-averaged force $\overline{F}$, acting on the body, in terms of these cross sections.

6.84* Determine the average force $\overline{F}$ acting on the small sphere of radius a placed in the field of a plane monochromatic wave. Consider the cases of a perfectly conducting sphere and a dielectric sphere of permittivity ε (magnetic permeability is $\mu = 1$). The amplitude of the incident wave is E_0.

6.85 A point source of light is placed on the line that passes through the center of a circular opaque screen of radius a and is perpendicular to its plane. Assuming that the geometrical approximation ($\lambda \ll a$) is applicable, find the intensity of light I in the point P symmetric relative to the screen.

6.86 In Problem 6.85, consider diffraction at the complementary screen, that is, at a circular aperture in an infinite opaque screen.

6.87 A parallel beam of light falls normally on a circular aperture in an opaque screen. Find the intensity distribution of light, I, on a middle line behind the screen.

6.88 Find the angular distribution of the intensity of light dI in the case of Fraunhofer diffraction at an annular aperture (radii $a > b$) in an infinite opaque screen. The beam is incident normally to the aperture plane. Consider the special case of diffraction by a circular aperture.

6.89 Find the angular distribution of the intensity of light dI for oblique incidence of a parallel beam to a circular aperture (Fraunhofer diffraction).

6.90 Calculate the cross section for Fraunhofer diffraction at a long slit of width l_1 in an opaque screen.

6.91* In an opaque screen, there are N infinite long slits of width $2l$ each at distance $2a$ between adjacent slits. Calculate the diffraction cross section at such grating (per unit length of the slit). Investigate the cross section as a function of the diffraction angle at $N \gg 1$.

6.92 A plane linearly polarized wave is incident normally to the rectangular aperture $-a \le x \le a, -b \le y \le b$ in an infinite thin screen. The amplitudes of the electric and magnetic fields have the components $E_y = E_0$, $H_x = -E_0$, and $H_y = E_x = 0$. Determine the field radiated from the aperture and also the angular distribution dI.

6.93 A plane linearly polarized wave $\boldsymbol{E}_0\, e^{i(\mathbf{k}\cdot\boldsymbol{r}-\omega t)}$ is incident normally to a circular aperture of radius a in an infinite thin screen. Determine the field radiated from the aperture and also the angular distribution of intensity dI.

6.94 A plane monochromatic wave $\boldsymbol{E}^0(x,z) = \boldsymbol{E}_0 \exp(i\boldsymbol{k}_0 \cdot \boldsymbol{r})$, $\boldsymbol{k}_0 = (k\sin\theta_0, 0, k\cos\theta_0)$, $k = \omega/c$, is incident at an angle to a screen placed in the $z = 0$ plane. There is a long slit of width $2l$ ($-l \le x \le l$) in the screen, which is parallel to the Oy-axis. Calculate the electromagnetic field in the wave zone behind the screen ($z > 0$) using the Kirchhoff formula for Hertz vectors in the form (Equation (6.78)).

6.95 Solve Problem 6.94 for the case of a rectangular aperture of sizes $2l \times 2a$ (instead if the slit). Similar to Problem 6.90, investigate the limiting case of infinitely long slit.

6.96** A plane monochromatic wave is incident normally to an annular slit in an opaque screen (an external and internal radii are b and a, respectively). Calculate and investigate the field in the wave zone behind the screen. Consider the special cases of an azimuthal and radial polarizations (relative to the annular slit) of the incident wave.

6.97 Investigate the propagation of a beam of light in the approximation of slowly varying amplitude (Equation (6.97)) for the case of a point aperture in the screen: $u_0(x, y) = A_0\, \delta(x/a)\, \delta(y/a)$, where A_0 and a are constants. Compare with the exact solution $u = C\, e^{ikR}/R$, obtained for a point source, and formulate the limits of applicability of the approximate solution.

6.98 Investigate the propagation of a beam of light in the approximation of slowly varying amplitude (Equation (6.97)) for the case in which the amplitude within the aperture in a screen is given by the Gaussian distribution, $u_0(\rho) = A_0 \exp(-\rho^2/a^2)$. Find also the width of the beam $a(z)$ as a function of longitudinal coordinate z and initial width a.

6.99⋆ Solve the same problem for the case in which the field within the aperture is varying in phase, $u_0(\rho) = A_0 \exp(-\rho^2/a^2 - ik\rho^2/2R)$, where A_0, a and R are constant. Investigate also the beam width as a function of z.

6.4 Diffraction of X-Rays

X-rays, or Röntgen rays,[13] are high-frequency electromagnetic waves with frequencies from $\approx 2 \times 10^{23}$ to $\approx 2 \times 10^{16}$ rad/s (and wavelengths from $\lambda \approx 10^{-12}$ to $\lambda \approx 10^{-5}$ cm), that is, between the ultra violet radiation and gamma radiation.

The dielectric permittivity. Considering X-ray scattering by macroscopic bodies, it is important that the wavelength λ is comparable with atomic sizes a. In the condensed media, the intra-atomic distances have the same order of magnitude as atomic sizes, while in gases these distances are much larger than a. Accordingly, it becomes impossible to perform the averaging over physically small volumes (Chapter 1) and one needs to perform the ensemble averaging (Chapter 5). However, if the frequencies of X-rays are higher than characteristic atomic frequencies $\omega_0 \sim v_0/c$, the electrons can be considered free. For free and, moreover, nonrelativistic electrons in an external magnetic field, it is possible to integrate the equations of motion; calculate the current induced by the field, and determine the dielectric permittivity, which is a function of coordinates $\boldsymbol{r}$ (Problem 5.7):

$$\varepsilon(\boldsymbol{r}) = 1 - \frac{4\pi e^2 n(\boldsymbol{r})}{m\omega^2}, \qquad (6.98)$$

where $n(\boldsymbol{r})$ is the electron number density in a scatterer which is determined by the laws of quantum mechanics and averaged over equilibrium statistical distribution of states of the thermal motion of atoms. For high frequencies, the magnetic

13) Röntgen Wilhelm Conrad (1845–1923), outstanding German physicist-experimentalist and the first in history Nobel Prize laureate for the discovery and studies of the X-rays (1901). A.F. Ioffe, the founder of the Soviet physics, belonged to his scientific school.

properties of medium are extremely weak and it is possible to set $\mu = 1$. The theory of the dielectric permittivity in the X-ray and gamma-ray ranges is reviewed by Kolpakov *et al.* (1978).

Kinematic theory of diffraction. We will not present the rigorous theory of interaction of X-rays with medium but restrict ourselves to the case in which this interaction can be considered with the perturbation theory. This means that we will study the bodies of small sizes and that the amplitude of the diffracted wave is smaller than that for the primary wave. Therefore, we impose the restriction

$$\frac{4\pi e^2 n}{m\omega^2} \ll 1. \tag{6.99}$$

Let us write Maxwell's equations in the form usual for monochromatic fields:

$$\nabla \times \boldsymbol{E} = i\frac{\omega}{c}\boldsymbol{B}, \qquad \nabla \times \boldsymbol{B} = -i\frac{\omega}{c}\boldsymbol{D}, \qquad \nabla \cdot \boldsymbol{D} = \nabla \cdot \boldsymbol{B} = 0, \tag{6.100}$$

where

$$\boldsymbol{D} = \varepsilon \boldsymbol{E}, \qquad \text{or} \qquad \boldsymbol{E} = \boldsymbol{D} + \frac{4\pi e^2 n}{m\omega^2}\boldsymbol{E}.$$

Excluding from Equation (6.100) the vector $\boldsymbol{B}$, we have $\nabla \times [\nabla \times \boldsymbol{E}] = (\omega/c)\boldsymbol{D}$ or, accounting for the equation $\nabla \cdot \boldsymbol{D} = 0$, we finally obtain

$$\triangle \boldsymbol{D} + \frac{\omega^2}{c^2}\boldsymbol{D} = \nabla \times \left[\nabla \times \left(\frac{4\pi e^2 n}{m\omega^2}\boldsymbol{E}\right)\right]. \tag{6.101}$$

This equation may be used both inside the scatterer, where $\boldsymbol{D} = \varepsilon \boldsymbol{E}$, and outside of it, where $\boldsymbol{D} = \boldsymbol{E}$. Because of the presence of the small factor (Equation (6.99)), on the right-hand side of this equation it is sufficient to substitute the undisturbed electric field of the incident wave. It is clear from the equation that the scattered field has the same frequency as the primary wave, that is, the equation describes the process of coherent scattering without any frequency change.

The solution of Equation (6.101) can be expressed in terms of the Green's retarded function (Section I.5). For large distances from the body ($r \gg r'$), we use the Green's function (I.5.14) of the form

$$G(\boldsymbol{r}, \boldsymbol{r}') = \frac{1}{r}\exp(ikr - i\boldsymbol{k} \cdot \boldsymbol{r}'), \qquad k = \frac{\omega}{c},$$

where $\boldsymbol{k} = \boldsymbol{r}/r$ is the wavevector of the scattered wave. We take the incident wave in the form $\boldsymbol{E}_0 \exp(i\boldsymbol{k}_0 \cdot \boldsymbol{r}')$. The electric field of the scattered wave far from the body will be written as

$$\boldsymbol{E}(\boldsymbol{r}) = -\frac{e^2}{m\omega^2}\int G(\boldsymbol{r}, \boldsymbol{r}')\nabla' \times [\nabla' \times \boldsymbol{E}_0 n(\boldsymbol{r}') \exp(i\boldsymbol{k}_0 \cdot \boldsymbol{r}')]\, dV'.$$

Integration by parts allows us to express this quantity in terms of the Fourier harmonic of the electron density:

$$\boldsymbol{E}(\boldsymbol{r}) = \frac{e^2}{m\omega^2 r}e^{ikr}\,\boldsymbol{k} \times [\boldsymbol{k} \times \boldsymbol{E}_0]\int n(\boldsymbol{r}')e^{-i\boldsymbol{q} \cdot \boldsymbol{r}'}\, dV', \tag{6.102}$$

where $\boldsymbol{q} = \boldsymbol{k} - \boldsymbol{k}_0$ is the change of the wave vector at scattering, and the integration is performed over the volume of the scattering body.

Using Equation (6.102) and the definition (6.65), the differential cross section for scattering of polarized X-ray wave can be written as

$$d\sigma = r_0^2 \sin^2\theta \left| \int n(\mathbf{r}) \exp[-i\mathbf{q}\cdot\mathbf{r}]\, dV \right|^2 d\Omega, \tag{6.103}$$

where $r_0 = e^2/mc^2$ is the classical radius of the electron, θ is an angle between $\boldsymbol{E}_0$ and $\boldsymbol{k}$, and $d\Omega$ is a solid angle element in which the wave is scattered. To treat the incident radiation as a plane polarized wave, the sizes of the body should be much smaller than the coherence length.[14] Usually X-ray waves are unpolarized. The averaging over polarizations gives

$$d\sigma = \frac{1}{2} r_0^2 (1 + \cos^2\vartheta) \left| \int n(\mathbf{r}) \exp[i\boldsymbol{q}\cdot\boldsymbol{r}]\, dV \right|^2 d\Omega. \tag{6.104}$$

Here ϑ is the scattering angle, that is, the angle between $\boldsymbol{k}$ and $\boldsymbol{k}_0$. The condition for applicability of Equations (6.103) and (6.104) is the requirement that the total cross section $\sigma = \int_{(4\pi)} d\sigma$ be small compared with geometric cross section of a sample. The approach described above is based on the perturbation theory and referred to as **kinematic**. If this condition is not satisfied, the more accurate **d**ynamic theory should be applied (see, e.g., Pinsker (1974); the basic equations of this theory are also presented by Bredov *et al.* (2003)).

The derived formulas can also be used when X-rays are scattered by an individual atom. In this case, the integration is performed over the volume of the atom, and the quantity

$$F_a(\boldsymbol{q}) = \int_{V_a} n_a(\boldsymbol{r}) \exp[-i\boldsymbol{q}\cdot\boldsymbol{r}]\, dV. \tag{6.105}$$

is referred to as **atomic form factor**. This factor is simply the Fourier component of the distribution $n_a(\boldsymbol{r})$ of electrons in atom. It allows one to calculate $n_a(\boldsymbol{r})$ using the inverse Fourier transform.

Diffraction by single crystals. At diffraction of X-rays by a single crystal, the electron density $n(\boldsymbol{r})$ is a periodic function of coordinates:

$$n(\boldsymbol{r}) = n(\boldsymbol{r} + \boldsymbol{a}), \qquad \boldsymbol{a} = m_1\boldsymbol{a}_1 + m_2\boldsymbol{a}_2 + m_3\boldsymbol{a}_3, \tag{6.106}$$

where $\boldsymbol{a}_1$, $\boldsymbol{a}_2$, and $\boldsymbol{a}_3$ are the principle vectors of crystal lattice, and m_i are integers (this condition holds rigorously only for infinite and pure crystal). Owing to the periodicity of the electron density, its expansion in Fourier series in coordinate space has the form

$$n(\boldsymbol{r}) = \sum_{\mathbf{b}} n_{\mathbf{b}} \exp(i\boldsymbol{b}\cdot\boldsymbol{r}), \tag{6.107}$$

where $\boldsymbol{b} = 2\pi\boldsymbol{g}$ are the vectors of reciprocal lattice. The vectors $\boldsymbol{g}$ differ from the vectors $\boldsymbol{b} = 2\pi\boldsymbol{g}$ by a factor of 2π, and form the mutual basis (see Example 1.2 in

14) The coherence length is a path along which separate trains of wave which carry any real signal, retain small phase difference. Along the path exceeding the coherence length the signal becomes incoherent and depolarized. More accurate definition of the coherence length is given in Section 7.1 of the next chapter.

Chapter I.1) with respect to the vectors $\boldsymbol{a}$ of the original lattice. The elementary vectors of the mutual basis are defined by the relations

$$g_1 = \frac{\boldsymbol{a}_2 \times \boldsymbol{a}_3}{V_c}, \qquad g_2 = \frac{\boldsymbol{a}_3 \times \boldsymbol{a}_1}{V_c}, \qquad g_3 = \frac{\boldsymbol{a}_1 \times \boldsymbol{a}_2}{V_c}, \qquad V_c = \boldsymbol{a}_1 \cdot [\boldsymbol{a}_2 \times \boldsymbol{a}_3]. \tag{6.108}$$

With $\boldsymbol{r} + \boldsymbol{a}$ in place of $\boldsymbol{r}$ in Equation (6.107), the value of the electron density does not changes since $\exp(i\boldsymbol{a} \cdot \boldsymbol{b}) = \exp(2\pi i N)$, where N is integer.

In the case of a crystal scatterer, we have

$$\int n(\boldsymbol{r})\, dV = \sum_{\mathbf{b}} n_{\mathbf{b}} \int \exp(i(\boldsymbol{b} - \boldsymbol{q}) \cdot \boldsymbol{r})\, dV. \tag{6.109}$$

The volume integral has a maximum value at $\boldsymbol{q} = \boldsymbol{b}$ or

$$\boldsymbol{k} - \boldsymbol{k}_0 = \boldsymbol{b} = 2\pi \boldsymbol{g} \tag{6.110}$$

(**von Laue equation**).[15] If this equality breaks, the integral significantly decreases due to the oscillations of the exponent. Taking by modulus both sides of Equation (6.106) and taking into account that $q = |\boldsymbol{k} - \boldsymbol{k}_0| = 2k \sin(\vartheta/2)$, where ϑ is a scattering angle, we obtain the **B**ragg–Wulf[16),17)] equation

$$2k \sin \vartheta/2 = b = 2\pi g. \tag{6.111}$$

The Bragg–Wulf equation and the von Laue equation determine the directions along which the cross section of X-ray diffraction has the largest value (main diffraction maxima). For a given value of b, the main maximum appears only at the condition $k = \omega/c > b/2$.

X-ray diffraction is considered in detail in the books Landau *et al.* (1984); Bredov *et al.* (2003) Cowly (1975); Pinsker (1974); Krivoglaz (1967); James (1950)

Problems

6.100 Elucidate under which conditions the scattering cross section of X-rays by bodies of finite size takes the form of the scattering cross section by free charges (Thomson formula). Write the corresponding expressions for the cross sections. The number of atoms in the body is N, and the number of electrons in each atom is Z.

6.101 The electron density distribution in an atom with Z electrons can be approximated by the expression $n_a(r) = n_{0a} \exp(-r/a)$, where $n_{0a} = Z/\pi a^3$, $a = a_B/Z^{1/3}$, and $a_B = 0.529 \times 10^{-8}$ cm is the Bohr radius.

15) Max Theodor von Laue (1879–1960), German physicist-theoretician, Nobel Prize laureate (1914); worked out diffraction theory of X-rays; founder of X-ray structural analysis.
16) Lawrence Bragg (1890–1971), British physicist, Nobel Prize laureate (1915) for X-ray structural studies.
17) Georgii Wulf (1863–1925), Russian physicist, worked in crystallography.

Find the differential cross section for scattering of wave in X-ray range by a mono-atomic gas containing N atoms, under the assumption that the distribution of atoms is completely random.

6.102 Find the cross section for the scattering of X-rays by a volume of gas containing N diatomic molecules. The atoms in a molecule are identical and are at a fixed distance R from each other. Assume that the form factor $F_a(q)$ of the atom, which is a part of the molecular structure is the same as for an isolated atom.

6.103 What is the change in the cross section for the scattering of X-rays by a volume of diatomic gas, considered in Problem 6.102, if the thermal oscillations of atoms in a molecule are taken into account?

Hint. Assume that the distance R between the atoms is distributed around the average value $R_0 \gg b$ according to the law

$$dW_x = \frac{1}{b\sqrt{\pi}} \exp\left(-\frac{x^2}{b^2}\right) dx, \qquad \text{where } x = R - R_0,\ b = \sqrt{\frac{2kT}{\mu\,\omega^2}},$$

T is the temperature, μ the reduced mass, and ω is the frequency of oscillations of atoms in the molecule.

6.104 Derive the von Laue equation (6.110) and the Bragg–Wulf condition (6.111) by considering the interference of waves scattered by separate centers of the perfect crystalline lattice.

6.105 Find the cross section for scattering of X-rays by a perfect single crystal consisting of N identical atoms with form factors $F_a(q)$ (assume that these form factors are the same as for isolated atoms). The elementary cell has the form of a cube with the edge a, and the crystal has the form of a rectangular parallelepiped with edge lengths L_1, L_2, and L_3, which are parallel to the edges of the elementary cell. Determine the positions of the main maxima and verify that the von Laue equation (6.110) is satisfied. Find the magnitude of the cross section in these maxima.

6.106 A crystal consists of cubic elementary cells with the edge length a and has the form of prism with a right isosceles triangle as base (the legs of the triangle $L_1 = L_2$, the lateral side L_3). Determine the position of the main maxima and find the magnitude of the cross section in these maxima.

6.107* Find the intensity distribution in the diffraction spot near one of the main maxima, for scattering of X-rays by a single crystal (Problem 6.105). The wave vector of incident X-rays is parallel to the edge L_3, and $k \gg 1/a$. Determine the width of the diffraction maximum and the total cross section corresponding to the scattering within the limits of a single diffraction spot.

6.108 Calculate the intensity distribution in the diffraction spot around the main maximum for an arbitrary angle of incidence and arbitrary ratio between k and $1/a$. X-rays are scattered by a single crystal in the form of rectangular parallelepiped with edge lengths L_1, L_2, and L_3 (Problem 6.105).

6.109 Solve Problem (6.108) for the case of scattering by a spherical single crystal sample (of radius R).

6.5
Answers and Solutions

6.1 $\omega_t(k) = ck/n$, $n = \sqrt{\varepsilon\mu}$ is the refractive index. The longitudinal oscillations do not exist.

$$\boldsymbol{H} = \frac{c}{\omega\mu}\boldsymbol{k}\times\boldsymbol{E}, \qquad \frac{H}{E} = \sqrt{\frac{\varepsilon}{\mu}} \qquad \frac{\varepsilon E^2}{8\pi} = \frac{\mu H^2}{8\pi}.$$

6.2 Electromagnetic field of frequency ω can propagate in the form of transverse waves with the wave number

$$k = \frac{\omega}{c}\sqrt{\frac{\omega_l^2 - \omega^2}{\omega_s^2 - \omega^2}} \qquad (\text{assume } \gamma_s \ll \omega_s, \omega_l),$$

where $\omega_l^2 = \omega_s^2 + \omega_p^2 f_s$ is a squared frequency of longitudinal oscillations obtained from results of Problem 5.45. The wave number (constant of propagation) becomes purely imaginary in the frequency range $\omega_s < \omega < \omega_l$, in which the dielectric is opaque. The damping results not from the dissipation of electromagnetic energy, but from destructive interference produced by vibrations of oscillators. Outside this interval, the dielectric is transparent; at $0 \le \omega \le \omega_l$ the dispersion relation is $2\omega_{t1}(k) = \omega_l^2 + c^2k^2 - \sqrt{(\omega_l^2 + c^2k^2)^2 - \omega_s^2c^2k^2}$, $0 \le k < \infty$. At frequencies $\omega > \omega_l$, the dispersion relation reads $2\omega_{t2}(k) = \omega_l^2 + c^2k^2 + \sqrt{(\omega_l^2 + c^2k^2)^2 - \omega_s^2c^2k^2}$, $0 \le k < \infty$. In the special case in which $ck \ll \omega_l$, we have $\omega_{t2} \approx \omega_l + (1 - \omega_s^2/\omega_l^2)c^2k^2/2$. At $ck \gg \omega_l$, the dispersion relation $\omega_{t2} = ck$ is the same as in vacuum. At frequencies of the order of ω_s, ω_l, there appears a strong coupling between the oscillations of electromagnetic field and atomic oscillators (polariton waves and their quantum excitations–polaritons).

6.3 $k = (\omega/c)\sqrt{1 - \omega_{0e}^2/\omega^2}$. The range of non-transparency is $\omega < \omega_{0e}$; therefore, the branch of transverse waves in a plasma begins with the frequency $\omega_t = \omega_{0e}$: $\omega_t(k) = \sqrt{\omega_{0e}^2 + c^2k^2}$. The phase velocity of the transverse wave $v_{ph} = c/\sqrt{1 - \omega_{0e}^2/\omega^2} > c$, so that the Landau damping is totally absent. Thermal corrections to the dispersion relation are of the order $v_{Te}^2/c^2 \ll 1$.

6.4 The non-transparent region is $\omega_t < \omega < \omega_l$. For small k, the dispersion relations are $\omega_{t1}(k) = c^2k^2/\varepsilon_0$, and $\omega_{t2}(k) = \omega_l^2 + c^2k^2/\varepsilon_\infty$. For large k, they are $\omega_{t1}(k) = \omega_t$, $\omega_{t2}(k) = c^2k^2/\varepsilon_\infty$ $(k \le 1/a)$.

6.5 The wavepacket is described by the function

$$\Psi(\boldsymbol{r}, t) = 4\pi a_0\sqrt{\frac{\pi q^3}{2\rho^3}}\, J_{3/2}(\rho q)\, \exp(i\boldsymbol{k}_0\cdot\boldsymbol{r} - i\omega_0 t),$$

where $J_{3/2}(x) = \sqrt{2/\pi x}(\sin x/x - \cos x)$ is a Bessel function, and $\rho = |\boldsymbol{r} - \boldsymbol{v}_g t|$. The amplitude of the wavepacket is appreciably different from zero only in the spherically symmetrical region $\rho q \leq 1$. The wavepacket is finite in all three directions. As seen from the expression for $\Psi(\boldsymbol{r}, t)$, the shape of the wavepacket does not change with time. This is due to the linear dispersion relation, which is strictly true for electromagnetic waves propagating in vacuum. If the higher-order terms of the expansion of ω in powers of k are taken into account, the packet changes its shape, spreads out, and moves with the group velocity $\boldsymbol{v}_g$ as a whole.

6.6 Writing the function $\omega(k)$ as

$$\omega = \omega_0 + v_g(k - k_0) + \beta(k - k_0)^2,$$

we obtain

$$\Psi(x, t) = a_0 \sqrt{\frac{\pi}{\alpha + i\beta t}} \exp\left\{ -\frac{(x - v_g t)^2}{4(\alpha + i\beta t)} + i(k_0 x - \omega_0 t) \right\}.$$

It is convenient to investigate the dependence of this complex amplitude on x and t by considering its squared modulus which determines the intensity of the wave,

$$|A(x, t)|^2 = \frac{\pi a_0^2}{\sqrt{\alpha^2 + (\beta t)^2}} \exp\left\{ -\frac{\alpha(x - v_g t)^2}{2(\alpha^2 + \beta^2 t^2)} \right\}.$$

It is clear from this expression that the intensity of the wave is a Gaussian function of x at given t. Its width l increases with time as

$$l = \sqrt{\frac{2(\alpha^2 + \beta^2 t^2)}{\alpha}},$$

but the height decreases due to the presence of the factor $(\alpha^2 + \beta^2 t^2)^{-1/2}$.

The wavepacket spreads out symmetrically both in $t = +\infty$ and $-\infty$ directions. The spread is not associated with the absorption of energy because k is real. The absence of dissipation is clear from the fact that the integral $\int_{-\infty}^{\infty} |A(x, t)|^2 \, dx = \sqrt{\pi/2\alpha}\, a_0^2$ is independent of time, that is, the total energy is conserved. The wavepacket spreads out because the phase velocities $v_{ph} = \omega/k$ of various plane waves that are present in the superposition are different. Owing to the dispersion, the ratio ω/k depends on k.

6.7 Plasma. At $\omega > \omega_{0e}$, $v_{ph} = c/\sqrt{\varepsilon(\omega)}$, $v_g = c\sqrt{\varepsilon(\omega)}$.

Nonpolar dielectric. At $\omega \ll \omega_0$

$$v_{ph} = \frac{c}{\sqrt{\varepsilon_0}} \left(1 - \frac{1}{2} \frac{\omega_p^2 \, \omega^2}{\varepsilon_0 \, \omega_0^4} \right) < c, \qquad v_g = \frac{c}{\sqrt{\varepsilon_0}} \left(1 - \frac{3}{2} \frac{\omega_p^2 \, \omega^2}{\varepsilon_0 \, \omega_0^4} \right) < c,$$

where $\varepsilon_0 = \varepsilon(0)$. At $\omega \gg \omega_0$

$$v_\varphi = c \left(1 + \frac{\omega_p^2}{2\, \omega^2} \right) > c, \qquad v_g = c \left(1 - \frac{\omega_p^2}{2\, \omega^2} \right) < c.$$

In the latter case, $v_{ph}\, v_g \approx c^2$. Near the resonance frequency ($\omega \approx \omega_0$) the notion of the group velocity loses its sense.

6.8

$$\Delta t = \frac{e^2(\lambda_1^2 - \lambda_2^2)}{2\pi m_e c^3} DM \approx 4.6(\lambda_1^2 - \lambda_2^2)\, DM \text{ ms.}$$

In this expression, the time delay is given in microseconds, λ in centimeter, and the dispersion measure in parsecs per cubic centimeter (1 pc $\approx 3 \times 10^{18}$ cm). In radio astronomy, the dispersion measure is determined by the time delay of signals propagating from pulsars, which are rapidly rotating neutron stars. Due to inhomogeneity of interstellar plasma, in this way one can determine the average electron number density along the light of sight by using the formula $DM = \int_0^L N_e dl = \overline{N}_e L$ (if the distance to the pulsar is known from other observations).

6.10 Both waves will be polarized elliptically. One of the principal axes of the polarization ellipse lies in the plane of incidence, and the other is perpendicular to it. For the reflected wave, the semiaxes are given by

$$E_{\parallel} = \frac{\tan(\theta_0 - \theta_2)}{\tan(\theta_0 + \theta_2)} E_0, \qquad E_{\perp} = \frac{\sin(\theta_2 - \theta_0)}{\sin(\theta_2 + \theta_0)} E_0,$$

while for the refracted wave:

$$E_{\parallel} = \frac{2\cos\theta_0 \sin\theta_2}{\sin(\theta_0 + \theta_2)\cos(\theta_0 \theta_2)} E_0, \qquad E_{\perp} = \frac{2\cos\theta_0 \sin\theta_2}{\sin(\theta_0 + \theta_2)} E_0,$$

where θ_0 is the angle of incidence, θ_2 is the angle of refraction, and E_0 is the absolute value of the amplitude of the incident wave.

At $\theta_0 = \pi/2 - \theta_2$ (Brewster's angle), the reflected wave is linearly polarized.

6.11 Unpolarized (natural) light may be treated as incoherent superposition of two complementary-polarized components of equal intensity. Then the incident beam can be represented as superposition of two incoherent waves, one of which, $E_{\parallel}$, is polarized in the plane of incidence, and the other, $E_{\perp}$, in the perpendicular plane. The intensities of both waves are equal,

$$I_{\parallel} = I_{\perp} = I.$$

After reflection, the two components will remain incoherent. By using the Fresnel's formulas, we find

$$I_{ik}^{(1)} = I \frac{\sin^2(\theta_0 - \theta_2)}{\sin^2(\theta_0 + \theta_2)} \left(e_i^{\perp} e_k^{\perp} + \frac{\cos^2(\theta_0 + \theta_2)}{\cos^2(\theta_0 - \theta_2)} e_i^{\parallel} e_k^{\parallel} \right), \quad \rho_1 = \frac{\cos^2(\theta_0 + \theta_2)}{\cos^2(\theta_0 - \theta_2)} < 1,$$

$\boldsymbol{e}^{\perp}$ and $\boldsymbol{e}^{\parallel}$ are unit vectors in the polarization directions of the transverse and longitudinal components, respectively. These vectors lie in the plane perpendicular to the direction of propagation of the reflected light. The degree of depolarization of the incident light is 1. After reflection, the light becomes polarized. Similarly, for the refracted light we have

$$I_{ik}^{(2)} = \frac{4I\cos^2\theta_0 \sin^2\theta_2}{\sin^2(\theta_0+\theta_2)}\left(e_i^{\perp}e_k^{\perp} + \frac{e_i^{\parallel}e_k^{\parallel}}{\cos^2(\theta_0-\theta_2)}\right), \quad \rho_2 = \cos^2(\theta_0-\theta_2) < 1.$$

6.12

$$R = \frac{(\varepsilon_1-\varepsilon_2)^2}{2(\varepsilon_1+\varepsilon_2)}, \qquad \rho_1 = 0, \qquad \rho_2 = \frac{4\varepsilon_1\varepsilon_2}{(\varepsilon_1+\varepsilon_2)^2},$$

where ε_1 and ε_2 are the dielectric permittivities of the two dielectrics.

6.13 The phase differences between $E_{\perp 1}$, E_0 and $E_{\parallel 1}$, E_0 may be determined with the aid of Fresnel's formulas:

$$(1) \qquad \tan\frac{\delta_\perp}{2} = \frac{\sqrt{\sin^2\theta_0 - n^2}}{\cos\theta_0}, \qquad \tan\frac{\delta_\parallel}{2} = \frac{\sqrt{\sin^2\theta_0 - n^2}}{n^2\cos\theta_0}.$$

Since $\delta_\perp \neq \delta_\parallel$, the wave is elliptically polarized.

The elliptical polarization will become circular when the following conditions are satisfied:

$$\text{(a) } \delta = \delta_\parallel - \delta_\perp = \frac{\pi}{2}; \qquad \text{(b) } E_{\parallel 0} = E_{\perp 0}.$$

The second condition requires the incident wave to be polarized in the plane making an angle $\pi/4$ with the plane of incidence. Let us investigate whether the first condition can be satisfied.

From Equation (1) we have

$$(2) \qquad \tan\frac{\delta}{2} = \frac{\cos\theta_0\sqrt{\sin^2\theta_0 - n^2}}{\sin^2\theta_0}.$$

Accordingly, at $\theta_0 = \arcsin n$ and $\theta_0 = \pi/2$, the angle δ is zero, and its maximum takes place between these points. It is easy to show that $\tan(dl_{max}/2) = (1-n^2)/2n$. In order that $\tan\delta/2$ be equal to 1 ($\delta = \pi/2$), the following inequalities must be satisfied: $1 - n^2 \geq 2n$, and $n \leq 0.414$.

6.14 If the vector $\boldsymbol{E}_0$ is perpendicular to the plane of incidence, the transverse and longitudinal components of the Poynting vector are of the form:

$$(1) \qquad \left.\begin{aligned} \gamma_\perp &= \frac{c^2k''}{8\pi\,\omega}E_0^2 e^{-2k''z}\sin 2(k'x - \omega t), \\ \gamma_\parallel &= \frac{c^2k''}{8\pi\,\omega}E_0^2 e^{-2k''z}[1 - \cos 2(k'x - \omega t)], \end{aligned}\right\}$$

where the Oz-axis is normal to the boundary of the media, and the Ox-axis is the line of intersection of the plane of incidence and the boundary,

$$k' = k_2\sin\theta_0, \qquad k'' = k_2\sqrt{\sin^2\theta_0 - n^2},$$

$k_2 = \omega n_2/c$ is the wave vector in the second medium, and θ_0 is the angle of incidence.

It is clear from Equation (1), that in the direction perpendicular to the boundary the energy oscillates with the frequency 2ω. The time average of the energy flux penetrating into the second medium is zero. The average value of $\gamma_{\|}$ is not zero, so that there is a flow of energy along the boundary.

The Poynting vector lines in the second medium are given by the equation

$$z = \frac{1}{k''} \ln \frac{|\sin k'x|}{C}, \tag{2}$$

where C is a constant of integration.

The approximate form of these lines is shown in Figure 6.12. In the first medium, the lines $\boldsymbol{\gamma}$ have more complicated form.

6.15 We find from the Fresnel formulas that at $\theta_0 \to \pi/2$ the amplitude of the penetrating wave $E_1 \to 0$, and the amplitude of the reflected wave $E_2 \to -E_0$. This means that a plane monochromatic wave cannot propagate along the interface of two dielectrics.

6.16• $\theta_0 = \theta_1$, $\quad \sin\theta_2 = (n_1/n_2)\sin\theta_0$, $\quad n_1 > 0,\ n_2 > 0$ (Figure 6.13).

6.17• The vector $\boldsymbol{E}_0$ is parallel to the boundary.

$$\boldsymbol{E}_1 = \frac{|\mu_2|\sin\theta_2\cos\theta_0 - \mu_1\sin\theta_0\cos\theta_2}{|\mu_2|\sin\theta_2\cos\theta_0 + \mu_1\sin\theta_0\cos\theta_2}\boldsymbol{E}_0,$$

$$\boldsymbol{E}_2 = \frac{2|\mu_2|\sin\theta_2\cos\theta_0}{|\mu_2|\sin\theta_2\cos\theta_0 + \mu_1\sin\theta_0\cos\theta_2}\boldsymbol{E}_0.$$

The vector $\boldsymbol{E}_0$ is perpendicular to the boundary.

$$\boldsymbol{H}_1 = \frac{|\varepsilon_2|\sin\theta_2\cos\theta_0 - \varepsilon_1\sin\theta_0\cos\theta_2}{|\varepsilon_2|\sin\theta_2\cos\theta_0 + \varepsilon_1\sin\theta_0\cos\theta_2}\boldsymbol{E}_0,$$

$$\boldsymbol{H}_2 = \frac{2|\varepsilon_2|\sin\theta_2\cos\theta_0}{|\varepsilon_2|\sin\theta_2\cos\theta_0 + \varepsilon_1\sin\theta_0\cos\theta_2}\boldsymbol{H}_0.$$

In case of the boundary between vacuum and the ideal left-handed medium, $\varepsilon_1 = \mu_1 = |\varepsilon_2| = |\mu_2| = 1$, $E_1 = H_1 = 0$, $E_2 = E_0$, $H_2 = H_0$ for both polarizations. Reflection is absent, all waves fully penetrate into the left-handed medium.

6.18

$$P_{rad} = w_0\left[1 + R - \frac{v_{2g}}{c}(1 - R)\right],$$

where w_0 is the energy density of incident wave, v_{2g} is the group velocity in dielectric medium.

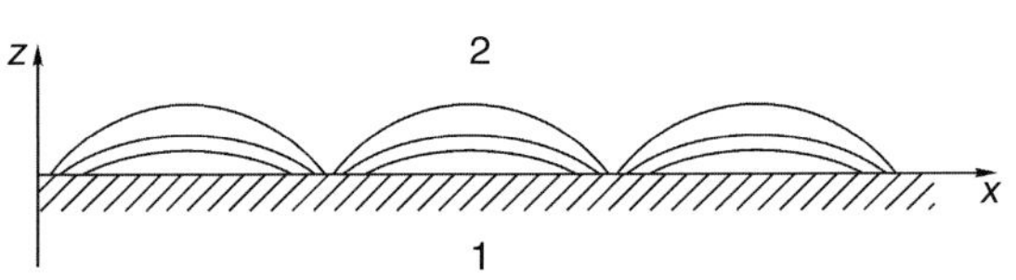

Figure 6.12 Lines showing Poynting vector direction in the second medium.

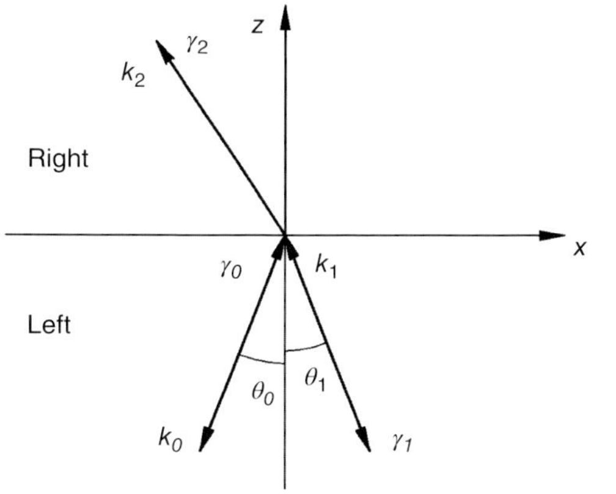

Figure 6.13 Directions of wave vectors and energy flows at incidence of wave from the left-handed medium to the boundary with the usual medium.

6.19 In the notations of Problem 6.18,

$$P_{rad} = w(1 + R(\theta_0)) \cos^2 \theta_0.$$

6.20 $P_{rad} = P'_{rad} + P''_{rad}$,

$$P'_{rad} = \boldsymbol{g}'_1 \cdot \boldsymbol{v}_{1g}(1 + R_\perp(\theta_0)) \cos^2 \theta_0 - \boldsymbol{g}'_2 \cdot \boldsymbol{v}_{2g}(1 - R_\perp(\theta_0) \cos^2 \theta_2,$$

$$P''_{rad} = \boldsymbol{g}''_1 \cdot \boldsymbol{v}_{1g}(1 + R_\|(\theta_0)) \cos^2 \theta_0 - \boldsymbol{g}''_2 \cdot \boldsymbol{v}_{2g}(1 - R_\|(\theta_0)) \cos^2 \theta_2.$$

In the first formula, $\boldsymbol{g}'$ is the density of momentum, created by field components $\boldsymbol{E}_\perp$ and $H_\|$, while in the second formula, $\boldsymbol{g}''$ is the same for $E_\|$ and $\boldsymbol{H}_\perp$ components of the incident wave.

6.21 In order to determine the reflection from for the plane layer, it is necessary to find the relation between the amplitudes of the reflected and incident waves. This relation could be obtained by two methods.

According to the first method, the relation is determined from the boundary conditions. Let us take into account that on the boundaries $z = 0$ and $z = a$ the tangential components of the vectors $\boldsymbol{E}$ and $\boldsymbol{H}$ must be continuous and that in front of the layer there are the waves propagating in both directions, while behind the layer there is only the transmitted wave propagating in the positive direction of the z-axis. With this in mind, we obtain from the boundary conditions

(1) $$E_1 = \frac{\alpha_{12} + \alpha_{23} \exp(-2ik_2a)}{1 + \alpha_{12}\alpha_{23} \exp(-2ik_2a)} E_0,$$

where E_1 is the amplitude of the reflected wave, E_0 is the amplitude of the incident wave, and

$$\alpha_{12} = \frac{1 - n_{12}}{1 + n_{12}}, \qquad \alpha_{23} = \frac{1 - n_{23}}{1 + n_{23}}, \qquad n_{ik} = \sqrt{\frac{\varepsilon_k}{\varepsilon_i}}, \qquad k_2 = \frac{\omega}{c}\sqrt{\varepsilon_2}.$$

The second method is based on the consideration of multiple reflections from the boundary. Using the Fresnel's formulas at normal incidence, we find that the amplitude of the wave reflected from the boundary $z = 0$ only once is

$$\mathcal{E}_0 = \alpha_{12} E_0.$$

The amplitude of the wave that has entered the layer is

$$\mathcal{E}_\alpha = \beta_{12} E_0,$$

where

$$\beta_{12} = \frac{2}{1 + n_{12}}.$$

The amplitude of the wave emerging from the layer in the region $z < 0$ after single reflection at the boundary $z = a$ is

$$\mathcal{E}_1 = \beta_{21}\alpha_{23}\beta_{12}E_0 \exp(-2ik_2 a).$$

The amplitude of the wave entering the region $z < 0$ after s reflections at the boundary $z = a$ is

$$\mathcal{E}_s = \beta_{21}\beta_{12}\alpha_{23}e^{-2ik_2 a}(\alpha_{21}\alpha_{23} \exp(-2ik_3 a))^{s-1}.$$

The total amplitude E_1 of the wave reflected from the plane layer is equal to the sum of all amplitudes $\mathcal{E}_s$:

$$E_1 = \sum_{s=0}^{\infty} \mathcal{E}_s = \alpha_{12}E_0 + \beta_{21}\beta_{12}\alpha_{23}\, e^{-2ik_2 a} \sum_{s=1}^{\infty} (\alpha_{21}\alpha_{23}\, e^{-2ik_2 a})^{s-1}.$$

Using the formula for the sum of infinite geometric progression, we come to the same relation (1).

The reflection coefficient is defined by $R = |E_1|^2/|E_0|^2$. By determining the minimum of R in a standard way, we find that the reflection coefficient has minimum if the thickness of the layer satisfies the condition

$$a = a_n = n\frac{\lambda_2}{4}, \qquad n = 1, 2, 3, \ldots, \tag{2}$$

where λ_2 is the wavelength inside the layer.

Let us consider the minimum thickness of the layer $a = \lambda_2/4$ corresponding to the minimum of R. Equating R to zero, we find the condition for the absence of reflection:

$$\varepsilon_2 = \sqrt{\varepsilon_1 \varepsilon_3}.$$

6.23 Let us find the real and imaginary parts of the dielectric permittivity of the conductor:

$$\varepsilon'(\omega) = \varepsilon_0 - \frac{\omega_{0e}^2}{\gamma^2 + \omega^2}, \qquad \varepsilon''(\omega) = \frac{\omega_{0e}^2 \gamma}{\omega(\gamma^2 + \omega^2)}. \tag{1}$$

The penetration depth is $\delta = 1/k''$, where k'' is determined by Equation (3) from Example 6.3 ($\theta_0 = 0$), $k'' = (\omega/c\sqrt{2})\sqrt{|\varepsilon| - \varepsilon'}$.

1. $\omega \ll \gamma \ll \omega_{0e}$; $\omega_{0e}^2/\gamma^2 \gg \varepsilon_0$. We have $\delta \approx c/\sqrt{2\pi\kappa_0\omega}$, where $\kappa_0 = \omega_{0e}^2/4\pi\gamma$ is a static electric conductivity. This expression for δ is identical with Equation (4.17), $\zeta \approx (1-i)\sqrt{\mu\omega/8\pi\kappa_0}$.

2. $\omega_{0e} \ll \omega \ll \gamma$, $\omega_{0e}^2/\gamma\omega \ll \varepsilon_0$;

$$\delta \approx \frac{c}{2\pi\kappa_0}\sqrt{\frac{\varepsilon_0}{\mu}}, \qquad \zeta \approx \sqrt{\mu/\varepsilon_0}$$

(the case of semiconductor). In contrast to the first case, here the penetration depth is independent of frequency in the range of applicability of this formula.

6.24 1. $\gamma \ll \omega \ll \omega_{0e}$. In this case, $\varepsilon' = \varepsilon_0 - \omega_{0e}^2/\omega^2 < 0$, $|\varepsilon'| \gg \varepsilon'' = \omega_{0e}^2\gamma/\omega^3$, $\delta \approx c/\omega_{0e}$. In this frequency range, the collisional dissipation is insignificant, and the cancellation of the field of the wave is produced by the secondary field of electron vibrations.

2. $\omega \gg \omega_{0e}$. If at such frequencies the dispersion ε_0 is still unimportant, then $\delta \approx 2c\sqrt{\varepsilon_0}\omega^2/\omega_{0e}^2\gamma$. The value ε_0 is responsible for the effect of bound charges in ions and neutral atoms.

6.25 Since the vector $\boldsymbol{E}$ is linearly polarized, the amplitude $\boldsymbol{E}_0$ may be assumed to be real. From the equation $div\,\boldsymbol{E} = 0$ we have $\boldsymbol{k}' \cdot \boldsymbol{E}_0 = 0$, $\boldsymbol{k}'' \cdot \boldsymbol{E}_0 = 0$, that is, $\boldsymbol{E}_0$ is perpendicular to the $(\boldsymbol{k}', \boldsymbol{k}'')$-plane. From the equation for $rot\,\boldsymbol{E}$ we have

$$\frac{\mu\,\omega}{c}\boldsymbol{\mathcal{H}}_1 = \boldsymbol{k}' \times \boldsymbol{E}_0, \qquad \frac{\mu\,\omega}{c}\boldsymbol{\mathcal{H}}_2 = \boldsymbol{k}'' \times \boldsymbol{E}_0,$$

i. e. $\boldsymbol{\mathcal{H}}_1$ and $\boldsymbol{\mathcal{H}}_2$ are perpendicular to $\boldsymbol{E}_0$, $\boldsymbol{\mathcal{H}}_1 \perp \boldsymbol{k}'$, and $\boldsymbol{\mathcal{H}}_2 \perp \mathbf{k}''$.

The end of the vector $\boldsymbol{H}$ moves along the ellipse in the $(\boldsymbol{k}', \boldsymbol{k}'')$-plane (Figures 6.10, 6.14).

6.26

$$E_{\perp 1} = (-1 + 2\zeta\cos\theta_0)E_{\perp 0},\ E_{\parallel 1} = (1 - 2\zeta/\cos\theta_0)E_{\parallel 0},$$
$$E_{\perp 2} = 2\zeta\cos\theta_0 E_{\perp 0},\ E_{\parallel 2} = 2\zeta E_{\parallel 0}.$$

The formulas for $E_{\parallel 1}$ and $E_{\parallel 2}$ hold only if the grazing angle is $\varphi_0 = \pi/2 - \theta_0 \gg |\zeta|$.

For $\vartheta_0 \ll 1$, we obtain:

$$E_{\parallel 1} = \frac{\varphi_0 - \zeta}{\varphi_0 + \zeta}E_{\parallel 0}, \qquad E_{\parallel 2} = \frac{\varphi_0\zeta}{\varphi_0 + \zeta}E_{\parallel 0},$$

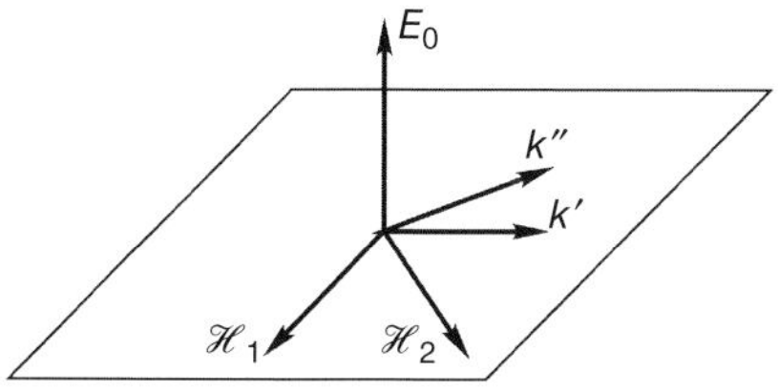

Figure 6.14 Positions of vectors in a nonhomogeneous plane wave.

and the ratio $|\zeta|/\varphi_0$ is arbitrary.

6.27 $R_\perp = 1 - 4\zeta' \cos\theta_0$. For all angles of incidence, $R_\perp$ is nearly equal to unity and reaches a minimum value at $\theta_0 = 0$ (normal incidence);

$$R_\parallel = 1 - \frac{4\zeta'}{\cos\theta_0} \quad \text{at} \quad \varphi_0 = \frac{\pi}{2} - \theta_0 \gg 4\zeta',$$

$$R_\parallel = \frac{(\varphi_0 - \zeta')^2 + \zeta''^2}{(\varphi_0 + \zeta')^2 + \zeta''^2} \quad \text{at} \quad \varphi_0 \ll 1.$$

From the condition $\partial R_\parallel/\partial\varphi_0 = 0$, we find the angle φ_0 at which $R_\parallel$ reaches a minimum:

$$\varphi_0 = \Phi_0 = |\zeta|, \qquad R_\parallel = \frac{|\zeta| - \zeta'}{|\zeta| + \zeta'}.$$

The angle Φ_0 is similar to the Brewster's angle since $R_\parallel$ has minimum at $\varphi_0 = \Phi_0$ (when the wave is incident to the boundary of dielectric at the Brewster's angle, $R_\parallel$ also reaches minimum which is zero).

6.28 The nature of polarization of the reflected wave is determined by the phase difference between the longitudinal and transverse components. Using the results of Problems (6.26) and (6.27), we have

$$E_{\perp 1} \approx -E_{\perp 0} = e^{i\delta_\perp} E_{\perp 0}, \qquad \delta_\perp = \pi;$$

$$E_{\parallel 1} = \left[\frac{|\zeta| - \zeta'}{|\zeta| + \zeta'}\right]^{1/2} e^{i\delta_\parallel} E_{\parallel 0}, \qquad \tan\delta_\parallel = -\left.\frac{2\Phi_0\zeta''}{\Phi_0^2 - |\zeta|^2}\right|_{\varphi_0 \to \Phi_0} \to \infty,$$

i. e. $\delta_\parallel = \pi/2$.

Thus, the phase difference $\delta = \delta_\perp - \delta_\parallel = \pi/2$; in general, the reflected wave will be elliptically polarized, and one axis of the ellipse lies in the plane of incidence.

At $|E_{\parallel 1}| = |E_{\perp 1}|$, the polarization will be circular. At $E_{\parallel 0} = 0$ or $E_{\perp 0} = 0$, the polarization is linear.

6.29 Using Fresnel's formulas we find

$$n' = \frac{\sin\theta_0 \tan\theta_0 \cos 2\rho}{1 + \sin^2 2\rho \cos\delta}, \qquad n'' = \frac{\sin\theta_0 \tan\theta_0 \sin 2\rho \sin\delta}{1 + \sin 2\rho \cos\delta}.$$

6.30

$$R = \frac{(\sqrt{\varepsilon} - \sqrt{\varepsilon'})^2}{(\sqrt{\varepsilon} + \sqrt{\varepsilon'})^2} + \frac{4}{(\sqrt{\varepsilon} + \sqrt{\varepsilon'})^4} \cdot \sqrt{\frac{\varepsilon'}{\varepsilon}} \left(\frac{2\pi\kappa}{\omega}\right)^2,$$

where ε is the dielectric permittivity of the medium, from which the light emerges, ε' is the real part of the dielectric permitivity of the conducting medium, and κ is the real electric conductivity.

6.31 The equation for the electric field is

$$\frac{d^2E}{dz^2} + \frac{\omega^2}{c^2}\left(\varepsilon - \frac{\Delta\varepsilon}{e^{z/a} + 1}\right)E = 0. \tag{1}$$

We must find the solution that is finite for all z and satisfies the appropriate conditions at $z \to \pm\infty$. At $z \to -\infty$, we should have a superposition of two waves, namely, the incident and reflected waves,

(2) $$E(z) \to Ae^{ik_0z} + Be^{-ik_0z},$$

where $k_0 = \omega/c$.

At $z \to \infty$, where only the transmitted wave survives,

(3) $$E(z) \to Ce^{ikz},$$

with $k_0 = \omega\sqrt{\varepsilon}/c$.

Let us introduce a new independent variable $-e^{-z/a} = \xi$ in Equation (1). The new variable ranges as $-\infty \leq \xi \leq 0$ when z varies from $-\infty$ to $+\infty$. Substituting $E(\xi) = \xi^{-ika}\psi(\xi)$, we obtain the following equation for the new unknown function $\psi(\xi)$:

(4) $$\xi(1-\xi)\psi'' + (1-2ika)(1-\xi)\psi' + \boldsymbol{k}^2a^2\psi = 0,$$

where $\boldsymbol{k}^2 = \omega^2\Delta\varepsilon/c^2$. This is the hypergeometric equation.

According to Equation (3), the function $\psi(\xi)$ should tend to a constant limit at $\xi \to 0$. The solution of Equation (4), which satisfies the above conditions, is the hypergeometric function

$$F(\alpha,\beta,\gamma,z) = 1 + \frac{\alpha\beta}{\gamma\cdot 1!}z + \frac{\alpha(\alpha+1)\beta(\beta+1)}{\gamma(\gamma+1)\cdot 2!}z^2 + \cdots$$

Therefore, the solution reads

(5) $$\psi = CF\left[-i(k+k_0)a, -i(k-k_0)a, 1-2ika, -e^{-z/a}\right].$$

In order to find the asymptote of ψ at $\xi \to -\infty$, we will use the asymptotic expression of the hypergeometric function:

(6) $$F(\alpha,\beta,\gamma,z) = \frac{\Gamma(\gamma)\Gamma(\beta-\alpha)}{\Gamma(\beta)\Gamma(\gamma-\alpha)}(-\xi)^{-\alpha} + \frac{\Gamma(\gamma)\Gamma(\alpha-\beta)}{\Gamma(\alpha)\Gamma(\gamma-\beta)}(-\xi)^{-\beta}.$$

From this equation, we verify that the condition (2) is satisfied.

The reflection coefficient is

(7) $$R = \left|\frac{B}{A}\right|^2 = \left|\frac{\Gamma(2ik_0a)\Gamma[1-i(k+k_0)a]\Gamma[-i(k+k_0)a]}{\Gamma(-2ik_0a)\Gamma[1-i(k-k_0)a]\Gamma[-i(k-k_0)a]}\right|^2.$$

In order to simplify the derived expression, we will use the formulas

$$\left|\frac{\Gamma(2ik_0a)}{\Gamma(-2ik_0a)}\right| = \left|\frac{\Gamma(2ik_0a)}{\Gamma^*(2ik_0a)}\right| = 1, \qquad \Gamma(z)\Gamma(1-z) = \frac{\pi}{\sin\pi z}.$$

The final result is

(8) $$R = \frac{\sinh^2\pi a(k-k_0)}{\sinh^2\pi a(k+k_0)}.$$

At small a ($ka \ll 1$), the expression for R reduces to the well-known expression which is valid for a step-like change in ε:

$$R = \frac{(k-k_0)^2}{(k+k_0)^2}.$$

The reflection coefficient R monotonically decreases as a increases. For large ka, the reduction is exponential,

$$R = \exp(-4\pi k_0 a), \qquad ka \gg 1.$$

6.32 For normal incidence of the wave to the inhomogeneous layer, the electric field is only a function of z and satisfies the equation

$$\frac{d^2E}{dz^2} = \frac{\omega^2}{c^2}\varepsilon(\omega, z)E = 0. \tag{1}$$

Let $m\omega^2/4\pi e^2 N_0 = z_1$, then $\varepsilon = 1 - z/z_1$. Introducing the new variable $\xi = (\omega^2/c^2 z_1)^{1/3}(z_1 - z)$, we transform Equation (1) to the form

$$\frac{d^2E}{d\xi^2} + \xi E = 0. \tag{2}$$

The simplest way to solve Equation (2) is to use Fourier transformation. Let us expand $E(\xi)$ in a Fourier integral:

$$E(\xi) = \int_{-\infty}^{\infty} E(u)e^{i\xi u}\, du, \qquad E(u) = \frac{1}{2\pi}\int_{-\infty}^{\infty} E(\xi)e^{-i\xi u}\, d\xi.$$

Substituting this expansion into Equation (2), we obtain the first-order differential equation for the amplitude $E(u)$,

$$\frac{dE(u)}{du} + iu^2E(u) = 0. \tag{3}$$

As a result, we have a simpler first-order equation instead of the second-order equation. The solution of Equation (3) is easily found by integrating,

$$E(u) = A'e^{-iu^3/3}.$$

Then we have

$$E(\xi) = A'\int_{-\infty}^{\infty} \exp[-i(u^3/3 - \xi u)]\, du.$$

Rewriting $\exp[-i(u^3/3 - \xi u)]$ in the form of sine and cosine sum, and noting that the integral of $\sin(u^3/3 - \xi u)$ is zero because the integrand is an odd function of u, we have:

$$E(\xi) = \frac{A}{\sqrt{\pi}}\int_0^{\infty} \cos\left(\frac{u^3}{3} - \xi u\right) du. \tag{4}$$

The function

$$\Phi(\xi) = \frac{1}{\sqrt{\pi}}\int_0^{\infty} \cos\left(\frac{u^3}{3} + \xi u\right) du$$

is known as **Airy's function** (it can be expressed in terms of Bessel functions of index 1/3). Finally,

$$E(\xi) = A\Phi(-\xi).$$

The constant A should be determined from the boundary conditions.

Now consider the behavior of $E(\xi)$ at large $|\xi|$. Using the asymptotic formulas for $\Phi(\xi)$, we have for large positive ξ:

$$E(\xi) = \frac{A}{\xi^{1/4}} \sin\left(\frac{2}{3}\xi^{3/2} + \frac{\pi}{4}\right).$$

It is clear that the field has an oscillating character.

For large (by modulus) negative ξ:

$$E(\xi) = \frac{A}{2|\xi|^{1/4}} \exp\left(-\frac{2}{3}|\xi|^{3/4}\right),$$

that is, the field decays exponentially. The reason for this is that negative ξ corresponds to the negative values of the dielectric permittivity ε. However, at $\varepsilon < 0$, the wavevector $k = (\omega/c)\sqrt{\varepsilon}$ is purely imaginary, and the damping takes place. The damping in this case is not due to the conversion of electromagnetic energy into heat (since the dielectric permittivity is real and dissipation is absent) but due to the destructive interference of the incident and secondary fields.

6.33 Transforming the equations as indicated in the statement of the problem, we find for $I(z,t)$ two linearly independent particular solutions: $I(z,t) = \sqrt{U(z)}e^{\pm iq\eta(z) - i\omega t}$, where the quasi-wave vector q is determined by Equation (6.39). The function $U(\eta) = Q^{-2}(\eta)$ is given by equation

$$U(\eta) = [\cosh(p\eta) + M\sinh(p\eta)]^{-2},\ p^2 > 0;$$

$$U(\eta) = [\cos(\eta/L) + M\sin(\eta/L)]^{-2},\ p^2 < 0,\ L = 1/|p|. \tag{1}$$

In order to find the relation between U and z, we express η in Equation (1) in terms of inverse hyperbolic or trigonometric functions and use the relation following from Equation (6.35): $dU/dz = (dU/d\eta)(d\eta/dz) = U(z)(dU/d\eta)$. In this way, we have

$$\frac{dU}{dz} = \pm\frac{2U^2}{L}\sqrt{1 - U(1 - M^2)},$$

$$0 \le M \le 1,\ 1 \le U \le U_{max} = (1 - M^2)^{-1},\ p^2 = 1/L^2 > 0; \tag{2}$$

$$\frac{dU}{dz} = \pm\frac{2U^2}{L}\sqrt{U(1 + M^2) - 1},$$

$$1 \ge U \ge (1 + M^2)^{-1},\ p^2 = -1/L^2 < 0; \tag{3}$$

$$\frac{dU}{dz} = \pm\frac{2U^2}{L}\sqrt{1 - U(1 - M^2)},$$

$$0 \le M \le 1,\ 1 \le U \le U_{max} = (1 - M^2)^{-1},\ p^2 = 1/L^2 > 0; \tag{4}$$

$$\frac{dU}{dz} = \pm\frac{2U^2}{L}\sqrt{U(1 + M^2) - 1},$$

(5) $$1 \geq U \geq (1+M^2)^{-1},\ p^2 = -1/L^2 < 0.$$

These differential equations may be used to find z as function of U.

6.34

$$R = \frac{n_0^2(n_0 - N)^2 + c^2M^2/\omega^2L^2}{n_0^2(n_0 + N)^2 + c^2M^2/\omega^2L^2}, \qquad \text{where} \qquad N = \sqrt{1 + \frac{c^2}{(Ln_0\omega)^2}}.$$

The ratio of the reflected and the incident waves is complex:

$$\frac{E_1}{E_0} = \frac{n_0(n_0 - N) - icM/\omega L}{n_0(n_0 + N) + icM/\omega L}.$$

This means, that the waves are shifted in phase because of inhomogeneity of the medium.

6.35 The field components are given by Equation (6.33) in terms of the vector potential $A(z,t) = e^{i\omega z/c} + \widetilde{R}e^{-i\omega z/c}$ at $z < 0$, and $A(z,t) = Ke^{i\omega z/c}$ at $z > L$. If $0 \leq z \leq L$, the vector potential has the form (Equation (6.39)). The constants $\widetilde{R}$, A_1, A_2, and K are determined from the continuity conditions for the field components across the boundaries $z = 0$ and $z = L$. The reflection coefficient is given by $R = |\widetilde{R}|^2$.

6.36 See the review paper by Shvartsburg (2000).

6.37 The velocities of waves propagating along one axis is determined by the principal values $\varepsilon^{(i)}$ corresponding to the two other axes. Accordingly, the velocities along axis 1 are $c/\sqrt{\varepsilon^{(2)}\mu}$ and $c/\sqrt{\varepsilon^{(3)}\mu}$, an so on.

6.38 The ordinary wave is transverse, the vectors $\boldsymbol{D}$ and $\boldsymbol{E}$ are both perpendicular to the plane in which the wavevector and optical axis lie (the plane of the principal cross-section). In the case of the extraordinary wave, the vector $\boldsymbol{D}$ lies in the plane of the principal cross-section and is perpendicular to the wavevector. The vector $\boldsymbol{E}$ also lies in the plane of the principal cross-section, but it is not parallel to $\boldsymbol{D}$.

6.40

$$\boldsymbol{v}_g = \boldsymbol{v}_{ph} - \frac{c(\boldsymbol{E}\cdot\boldsymbol{n})\boldsymbol{D}}{\mu n D^2}.$$

The propagation velocity of energy in anisotropic medium is not identical by magnitude and direction with the phase velocity. The second term on the right-hand side is perpendicular to the phase velocity. Position of the vectors is shown in Figure 6.15.

6.41

$$\cos\alpha = \frac{(\varepsilon_\parallel - \varepsilon_\perp)\sin\theta\cos\theta}{\sqrt{\varepsilon_\parallel^2\cos^2\theta + \varepsilon_\perp^2\sin^2\theta}}, \qquad \tan\vartheta = \frac{\varepsilon_\perp}{\varepsilon_\parallel}\tan\theta.$$

6.42 Proper oscillations in a moving medium are described by the equations given in the statement of the problem, provided $\boldsymbol{j}_{ext} = \rho_{ext} = 0$. Taking $\varphi = 0$ and using the two remaining equations, we find

$$k^2 - \frac{\omega^2}{c^2} - \kappa\gamma^2\frac{(\omega - \boldsymbol{k}\cdot\boldsymbol{V})^2}{c^2} = 0, \qquad \left\{\boldsymbol{k} + \frac{\kappa\gamma^2}{c^2}(\omega - \boldsymbol{k}\cdot\boldsymbol{V})\boldsymbol{V}\right\}\cdot\boldsymbol{A} = 0.$$

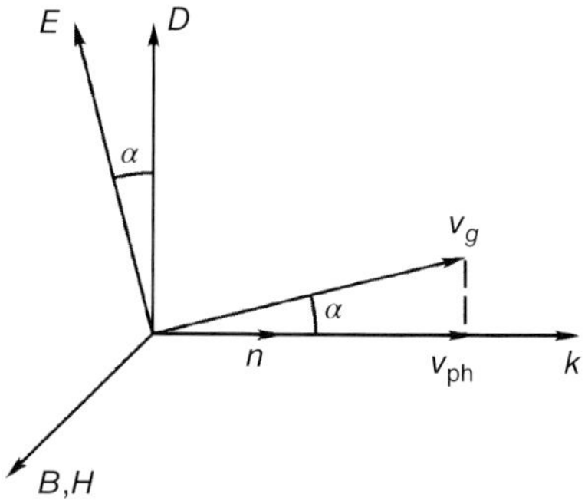

Figure 6.15 Position of vectors of plane wave in anisotropic medium.

The first equality allows us to derive the dispersive relation, while the second determines the polarization of the plane wave. The wave is transverse and its amplitude is perpendicular to the vector which is enclosed in parentheses and is a combination of the wavevector and the velocity vector of the medium. From the first equality, we can find the phase velocities

$$v_{ph1,2}(\omega,\theta) \equiv \frac{\omega}{k} = c\frac{\kappa\gamma^2\beta\cos\theta \pm \sqrt{1+\kappa\gamma^2(1-\beta^2\cos^2\theta)}}{1+\kappa\gamma^2}.$$

The solutions are related by the condition $v_{ph\,1}(\omega,\pi-\theta) = -v_{ph\,2}(\omega,\theta)$. Thus, only one wave with the determined phase velocity can propagate in a given direction. It has an arbitrary, but transverse polarization, that is, the polarization degeneracy takes place. In an uniaxial crystal at rest, the two waves of a given frequency with different phase velocities and different polarizations can propagate in any direction.

$$\boldsymbol{v}_g = \frac{d\omega}{d\boldsymbol{k}} = c\frac{\boldsymbol{k} + (\kappa\gamma^2/c^2)(\omega - \boldsymbol{k}\cdot\boldsymbol{V})\boldsymbol{V}}{(\omega/c) + (\kappa\gamma^2/c)(\omega - \boldsymbol{k}\cdot\boldsymbol{V})}.$$

Dispersion relations for different velocities of the medium are extensively discussed in the review paper by Bolotovskii and Stolyarov (1974).

6.43 The boundary conditions for the field vectors will be satisfied in all points of the separation surface when the tangential components of the wave vectors of the incident, reflected and both refracted waves are equal at the separation boundary. For the ordinary wave, this gives

$$k_0\sin\theta_0 = k_1\sin\theta_2', \qquad \frac{\sin\theta_2'}{\sin\theta_0} = \sqrt{\varepsilon_\perp\mu}.$$

The ray direction (direction of the Poynting vector) in the ordinary wave is identical with the direction of the wavevector and is, therefore, at an angle θ_2' to the normal to the boundary.

In the case of the extraordinary wave, we have

$$k_0\sin\theta_0 = k_2\sin\theta_2'' = k_0\sin\theta_2''\sqrt{\frac{\varepsilon_\perp\varepsilon_\parallel\mu}{\varepsilon_\perp\sin^2\theta_2'' + \varepsilon_\parallel\cos^2\theta_2''}}$$

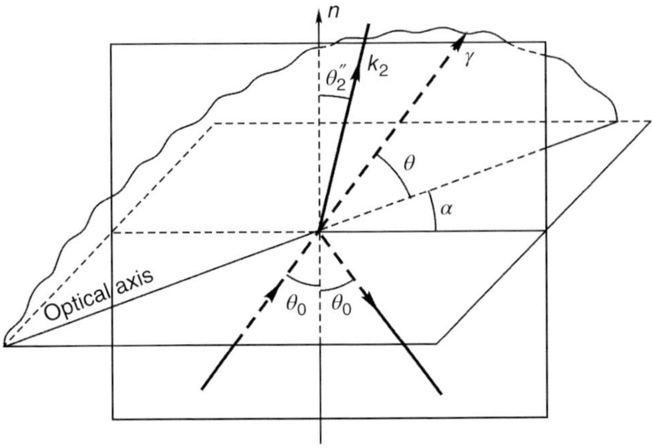

Figure 6.16 Positions of ordinary and extraordinary rays in uniaxial crystal.

(Problem 6.38). Hence,

$$\sin^2\theta_2'' = \frac{\varepsilon_\parallel \sin^2\theta_0}{\varepsilon_\perp\varepsilon_\parallel\mu + (\varepsilon_\parallel - \varepsilon_\perp)\sin^2\theta_0}.$$

The angle ϑ'' between the ray and the optic axis (which is parallel to the normal to the interface) is given, in accordance with the result of Problem 6.49, by the condition

$$\tan\vartheta'' = \frac{\varepsilon_\perp}{\varepsilon_\parallel}\tan\theta_2'' = \frac{\sqrt{\varepsilon_\perp}\sin\theta_0}{\sqrt{\varepsilon_\parallel(\varepsilon_\parallel\mu - \sin^2\theta_0)}}.$$

The angle of reflection from the crystal is equal to the angle of incidence, $\theta_1 = \theta_0$, just as for isotropic medium.

6.44 The ordinary ray lies in the plane of incidence at an angle θ_2' to the normal to the surface:

$$\sin\theta_2' = \sqrt{\varepsilon_\perp\mu}\sin\theta_0.$$

The wave vector $\boldsymbol{k}_2$ of the extraordinary wave lies in the plane of incidence at an angle θ_2'' to the normal:

$$\sin^2\theta_2'' = \frac{\varepsilon_\perp \sin^2\theta_0}{\varepsilon_\perp\varepsilon_\parallel\mu + (\varepsilon_\perp - \varepsilon_\parallel)\sin^2\theta_0\cos^2\alpha}.$$

The ray direction of the extraordinary wave does not lie in the plane of incidence. The ray direction lies in the same plane as $\boldsymbol{k}_2$ and optic axis, at an angle ϑ to the latter, where (Figure 6.16):

$$\tan\vartheta = \frac{\sqrt{\varepsilon_\perp^2\varepsilon_\parallel\mu + \varepsilon_\perp(\varepsilon_\perp - \varepsilon_\parallel)\sin^2\theta_0\cos^2\alpha}}{\varepsilon_\parallel\sin\theta_0\cos\beta}.$$

6.45 A wave for which the electric vector is parallel to the conductors is reflected from the lattice as from a continuous metal plane. A wave for which the electric vector is perpendicular to the conductors will propagate as in free space since it does not excite any currents in the lattice.

6.46 Substituting the plane-wave expressions for the fields $\boldsymbol{E}$ and $\boldsymbol{H}$ into Maxwell's equations, we obtain the equation for the amplitudes and wave vectors of the waves propagating in medium:

$$\text{(1)} \qquad \boldsymbol{k} \times (\boldsymbol{k} \times \boldsymbol{H}_0) = -\frac{\omega^2 \varepsilon}{c^2} \hat{\mu} \boldsymbol{H}_0.$$

Let θ be the angle between the wavevector $\boldsymbol{k}$ and the z-axis, and let us write Equation (1) in projections on coordinate axes.

By equating the determinant of the system to zero, we come to the biquadratic equation in k. Its solution gives:

$$k_{1,2}^2 = \frac{\omega^2}{c^2} \varepsilon \mu_{\pm}^{ef}, \quad \text{where}$$

$$\text{(2)} \qquad \mu_{\pm}^{ef} = \mu_{\|} \frac{\mu \sin^2 \theta + (2\mu_{\perp}/\mu_{\|}) \pm \sqrt{\mu^2 \sin^4 \theta + (2\mu_a/\mu_{\|})^2 \cos^2 \theta}}{2[(\mu_{\perp}/\mu_{\|} - 1) \sin^2 \theta + 1]},$$

and

$$\mu = \frac{\mu_{\perp}^2 - \mu_a^2 - \mu_{\perp}\mu_{\|}}{\mu_{\|}^2}.$$

The values $\mu_{\pm}^{ef}$ may be treated as the effective magnetic permeabilities of two normal waves.

Two waves can propagate in any direction with different phase velocities $v_{1,2} = \omega/k_{1,2}$, which are functions of θ. There are no directions for which these phase velocities are equal, since the square root in Equation (2) does not vanish for any value of θ. In the range of frequencies that are very close to the resonance frequencies of the tensor μ, the electromagnetic field is strongly related to oscillations of magnetization. In this case, the waves are referred to as **magnetic polaritons**.

With $\mu_a = 0$ in Equation (2), the latter will determine the phase velocities of the waves that can propagate in nongyrotropic but anisotropic magnetic crystal:

$$k_1^2 = \frac{\omega^2}{c^2} \varepsilon \mu_{\perp}, \qquad k_2^2 = \frac{\omega^2}{c^2} \cdot \frac{\varepsilon \mu_{\perp} \varepsilon_{\|}}{\mu_{\|} \cos^2 \theta + \mu_{\perp} \sin^2 \theta}.$$

The first of these waves (ordinary wave) has the velocity $v_1 = c/\sqrt{\varepsilon \mu_{\perp}}$, which is independent of propagation direction. The velocity of the second (extraordinary) wave depends on the angle between the symmetry axis of the crystal and propagation direction. When the propagation direction is along the symmetry axis ($\theta = 0$), two velocities are equal, and the two waves degenerate into one.

6.47 1. When the propagation is longitudinal, it is convenient to consider the cyclic components of the fields, $E_\pm = E_x \mp iE_y$ and $H_\pm = H_x \mp iH_y$. Their wave vectors can be found from Maxwell's equations:

$$(1) \qquad k_\pm = \frac{\omega}{c}\sqrt{\varepsilon(\mu_\perp - \mu_a)} = \frac{\omega}{c}\sqrt{\varepsilon}\left(1 + \frac{\omega_M}{\omega_0 \mp \omega}\right)^{1/2},$$

where the wave with $k = k_+$ exists in the frequency ranges $0 \le \omega \le \omega_0$ and $\omega_0 + \omega_M \le \omega < \infty$, while the wave with $k = k_-$ exists in the range $0 \le \omega < \infty$ (Figure 6.17).

2. The two waves are circularly polarized. The wave with k_+ has a right-handed helicity, and the wave with k_- has a left-handed helicity (see the definition of the direction of rotation in Problem 2.130). The surface impedances for the two waves are not the same:

$$(2) \qquad \zeta_\pm = \pm i\frac{E_\pm}{H_\pm} = \sqrt{\frac{\mu_\perp \pm \mu_a}{\varepsilon}}.$$

3. The linearly polarized wave can be treated as a combination of the two circularly polarized waves with the different directions of rotation. At $z = 0$, let the vector $\boldsymbol{E}_0$ be directed along the Ox-axis. Then at a depth z, we have

$$(3) \qquad E_x = (1/2)E_0\,[e^{ik_+z} + e^{ik_-z}], \qquad E_y = (i/2)E_0\,[-e^{ik_+z} + e^{ik_-z}].$$

Introducing the half-sum and the half-difference of the wave vectors, $k = (k_+ + k_-)/2$ and $\kappa = (k_+ - k_-)/2$, we find $E_x(z) = E_0\,e^{ikz}\cos\kappa z$, $E_y(z) = E_0\,e^{ikz}\sin\kappa z$. The ratio $E_y/E_x = \tan\kappa z$ gives the tangent of the Faraday rotation angle, $\delta = \kappa z = (k_+ - k_-)z/2$.

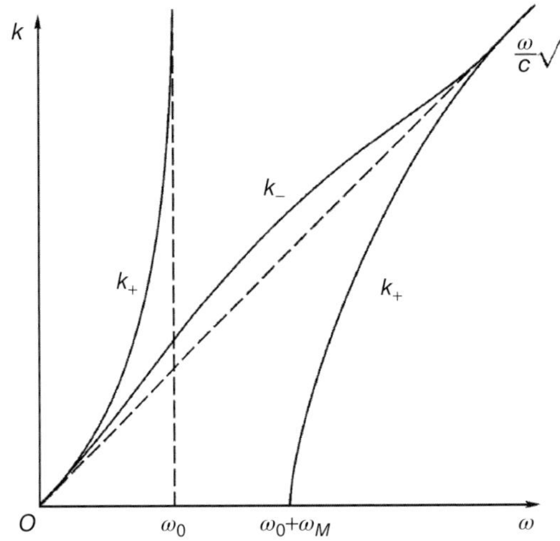

Figure 6.17 Dispersion relations for ferrit medium in the case of longitudinal propagation.

6.48 Let the waves propagate along the Ox-axis. The ordinary wave has the wave vector $k_1 = \omega\sqrt{\varepsilon}/c$ and is linearly polarized; its field components are $E_y = \zeta_1 H_z \neq 0$; the surface impedance is $\zeta_1 = 1/\sqrt{\varepsilon}$.

The extraordinary wave has $k_2 = \omega\sqrt{\varepsilon\mu_\perp^{ef}}/c$, where $\mu_\perp^{ef} = \mu_\perp - \mu_a^2/\mu_\perp$, and is linearly polarized, $E_z = -\zeta_2 H_y \neq 0$, $\zeta_2 = \sqrt{\mu_\perp^{ef}/\varepsilon}$. The dispersion relations are shown in Figure 6.18.

Assume that at the boundary of the medium ($x = 0$) the wave is linearly polarized, and the vector $\boldsymbol{E}$ oscillates along the bisectrix of an angle between the Oy- and Oz-axes. At a depth x, we have

$$\boldsymbol{E}(x, t) = E_0[(\boldsymbol{e}_y + \boldsymbol{e}_z)\cos\kappa x + i(\boldsymbol{e}_y - \boldsymbol{e}_z)\sin\kappa x]\, e^{ikx - i\omega t},$$

where $k = (k_1 + k_2)/2$ and $\kappa = (k_1 - k_2)/2$. From this formula, it is clear that at a fixed x the end of the vector $\boldsymbol{E}$ moves along the ellipse with the semi-axes $\sqrt{2}E_0\cos\kappa x$ and $\sqrt{2}E_0\sin\kappa x$. At the points, where $\cos\kappa x = \sin\kappa x$, the ellipse transforms into the circle. At the points, where $\kappa x = n\pi/2$, $n = 0, \pm 1, \ldots$, the polarization is linear, although it can be orthogonal to the initial polarization. The change of polarization when a wave propagates transversely in gyrotropic medium is known as **Cotton–Mouton effect**.

6.49

$$\delta = -\frac{l}{2c}\left(\omega_M\sqrt{\varepsilon_\perp} + \varepsilon_a\frac{\omega}{\sqrt{\varepsilon_\perp}}\right).$$

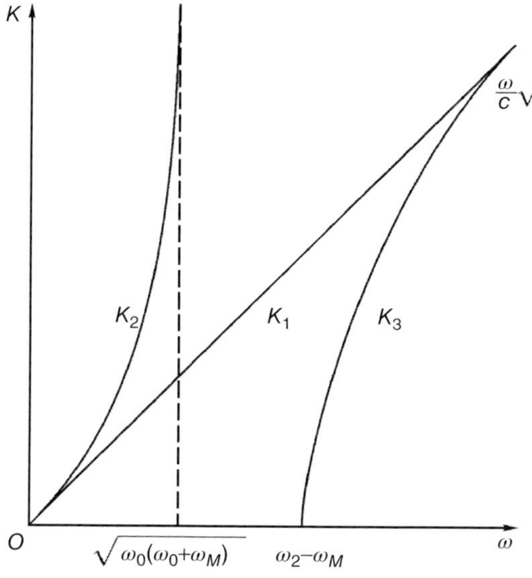

Figure 6.18 The same for the case of the transverse propagation.

6.50

$$\omega(\boldsymbol{k}) = \sqrt{\omega_0(\omega_0 + \omega_M \sin^2\theta)} = \sqrt{\omega_0\left(\omega_0 + \omega_M \frac{k_\perp^2}{k_\perp^2 + k_\parallel^2}\right)},$$

$$\omega_0 \le \omega \le \sqrt{\omega_0(\omega_0 + \omega_M)},$$

where θ is an angle between the wavevector and the applied constant magnetic field.

$$\boldsymbol{E} = \frac{\omega}{ck^2}\boldsymbol{k}\times\boldsymbol{B}, \qquad \boldsymbol{B} = \hat{\mu}\boldsymbol{H}, \qquad |\boldsymbol{E}| \ll |\boldsymbol{B}|.$$

6.51 Two waves with different phase velocities and opposite circular polarizations can propagate in the direction of the magnetic field. A wave whose polarization is not circular will split into two waves with opposite circular polarizations. Since the phase velocities of these waves are different, the phase difference between them will vary from point to point, and hence the polarization of the resultant wave will be different at different points. A detailed calculation shows that the polarization of the resultant wave will remain linear, but the polarization plane will rotate through an angle $\chi = \frac{1}{2}(k_+ - k_-)z$ (Faraday effect). The quantities k_+ and k_- are the wave vectors of the two circularly polarized waves, and can be found from the solutions of Problems 6.47 and 5.23. For a weak magnetic field,

$$\chi = VHz,$$

where the coefficient of proportionality V is called **Verdet constant**. If the atoms of the material can be considered harmonic oscillators, the Verdet constant is given by

$$V = \frac{2\pi e^3 N}{nm^2c^2}\,\frac{\omega^2}{(\omega^2 - \omega_0^2)^2},$$

where $n = \sqrt{\varepsilon}$ is the refractive index in the absence of the magnetic field.

6.52

$$\Delta\chi = (\lambda_1^2 - \lambda_2^2)RM.$$

The rotation measure in interstellar medium can be determined by measuring signals at different frequencies from pulsars that are rotating neutron stars. If one can determine the dispersion measure for the same pulsars (Problem 6.8), the interstellar magnetic field can be evaluated. When the magnetic field is directed at an angle to the line of sight, and the electron number density is not uniform, the rotation measure will be given by the integral of the field projection on the direction of observation:

$$RM = \frac{e^3}{2\pi(m_ec^2)^2}\int_0^L N_e B_\parallel dl.$$

See encyclopaedia "Physics of Cosmos" (in Russian, Editor R.A.Sunyaev) for details.

6.53 As follows from symmetry considerations, the wave vectors of the reflected and transmitted waves are perpendicular to the separation boundary. Both waves will be circularly polarized in the same direction as the incident wave. The amplitudes of the reflected and transmitted waves are, respectively:

$$H_1 = \frac{\zeta_\pm - 1}{\zeta_\pm + 1} H_0, \qquad H_2 = \frac{2}{\zeta_\pm + 1} H_0,$$

where H_0 is the amplitude of the incident wave, $\zeta_\pm = \sqrt{(\mu_\perp \pm \mu_a)/\varepsilon}$ is the surface impedance for waves with right- and left-handed circular polarizations. The reflection index

$$R_\pm = \frac{(\zeta_\pm - 1)^2}{(\zeta_\pm + 1)^2}$$

tends to unit as $\zeta \to \infty$ and $\zeta \to 0$. The first case is realized near the resonance, $\omega \approx \omega_0$. The second one corresponds to the antiresonance, when $\mu_\perp \pm \mu_a \approx 0$. For a ferrit with magnetic permeability (6.64), the antiresonance occurs for the wave with right-handed circular polarization at $\omega = \omega_0 + \omega_M$.

6.54 The wave vectors of the reflected and transmitted waves are perpendicular to the interface. The reflected wave is elliptically polarized with semiaxes

$$H_1' = H_0 \frac{1 - \zeta_+ \zeta_-}{(\zeta_+ + 1)(\zeta_- + 1)}, \qquad H_1'' = H_0 \frac{\zeta_- - \zeta_+}{(\zeta_+ + 1)(\zeta_- + 1)}.$$

The direction of H_1' is the same as the direction of polarization of the vector $\boldsymbol{H}$ in the incident wave. The transmitted wave splits into two waves with the amplitudes

$$H_2' = \frac{H_0}{\zeta_+ + 1}, \qquad H_2'' = \frac{H_0}{\zeta_- + 1},$$

which are circularly polarized in opposite directions. Their propagation velocities are different (see solution of Problem 6.47).

$$R = \frac{(1 - \zeta_+ \zeta_-)^2 + (\zeta_- - \zeta_+)^2}{(\zeta_+ + 1)^2 (\zeta_- + 1)^2}.$$

6.55 If the wavelength is much smaller than the radius of the discs and smaller than the distance between neighboring discs, then the artificial dielectric can be treated as continuous medium. The electric field of the incident wave is parallel to the plane of the discs. Hence, in the absence of any external magnetic field $\boldsymbol{H}_0$ the polarizability of the dielectric will be $\alpha = N\beta_e$, where $\beta_e = 4a^3/3\pi$ is the longitudinal (relative to the plane of the disc) electric polarizability of the disc, and N is the number of discs per unit volume.

The longitudinal magnetic polarizability of the disc β_m is zero (Problem 4.42), and hence the magnetic susceptibility of the dielectric χ for the given direction of the magnetic field is also zero.

The presence of the magnetic field $\boldsymbol{H}_0$ gives rise to the Hall effect. The conduction electrons producing the current in each disc will be deflected by the field $\boldsymbol{H}_0$ and will produce an additional electric field $\boldsymbol{E}_H$, which should balance the deflecting action of the magnetic field. This will create the additional electric moment in each disc. As a consequence, the vector of polarization and the electric induction will also be changed. In order to calculate the change of the induction, it is convenient to consider the total density of the polarization current $\partial \boldsymbol{P}/\partial t$ in the dielectric, rather than the current in an isolated disc.

In the first approximation with respect to H_0, the field $\boldsymbol{E}_H$ produced by the Hall effect is given by

(1) $$\boldsymbol{E}_H = R(\boldsymbol{H}_0 \times \boldsymbol{j}) = R\left(\boldsymbol{H}_0 \times \frac{\partial \mathbf{P}}{\partial t}\right),$$

where R is the Hall constant and $\boldsymbol{P} = \alpha \boldsymbol{E}$ is the polarization vector in the zero-order approximation. Owing to the presence of the field $\boldsymbol{E}_H$, the polarization vector will increase by

(2) $$\Delta \boldsymbol{P} = \alpha \boldsymbol{E}_H = \alpha^2 R\left(\boldsymbol{H}_0 \times \frac{\partial \boldsymbol{E}}{\partial t}\right),$$

and, therefore, the induction $\boldsymbol{D}$ will be expressed in terms of $\boldsymbol{E}$ and derivative $\partial \boldsymbol{E}/\partial t$:

(3) $$\boldsymbol{D} = \boldsymbol{E} + 4\pi(\boldsymbol{P} + \Delta \boldsymbol{P}) = \varepsilon \boldsymbol{E} + 4\pi\alpha^2 R\left(\boldsymbol{H}_0 \times \frac{\partial \boldsymbol{E}}{\partial t}\right).$$

Here $\varepsilon = 1 + 4\pi N \beta_e$ is the permittivity in the absence of the magnetic field.

When the electric field $\boldsymbol{E}$ is a harmonic function of time, Equation (3) will give the following relation between $\boldsymbol{D}$ and $\boldsymbol{E}$:

$$\boldsymbol{D} = \varepsilon \boldsymbol{E} + i(\boldsymbol{E} \times \boldsymbol{g}),$$

where $\boldsymbol{g} = 4\pi\alpha^2\, \omega R \boldsymbol{H}_0$ is the gyration vector (Equation (6.54)). Thus, the medium will be gyrotropic. From the results of Problem 6.47, it follows that there can be two waves propagating in the direction of $\boldsymbol{g}$. These waves are circularly polarized in opposite directions and have the different phase velocities $v_\pm = \omega/k_\pm$. Calculating $k_\pm$ in the standard way, we get

(4) $$k_\pm^2 = \frac{\omega^2}{c^2}(\varepsilon \pm g).$$

6.56 Two waves with right- and left-handed circular polarizations can propagate in the direction of the magnetic field. The wave vectors of these waves are given by

$$\frac{k^2 c^2}{\omega^2} = \varepsilon_\pm = 1 - \frac{\omega_p^2}{\omega\left(\omega + i\gamma \mp \omega_H - \omega_H \Omega_H/\omega\right)}.$$

At $\Omega_H \ll \omega$, the effect of motion of the positive ions is very small, and they may be considered fixed. In the opposite limiting case, at $\Omega_H \gg \omega$ and $\gamma\omega \ll \omega_H \Omega_H$, the positive ions play the leading role:

$$\frac{k^2c^2}{\omega^2} = 1 + \frac{\omega_p^2}{\omega_H \Omega_H} = 1 + \frac{4\pi NMc^2}{H_0^2}.$$

Both waves propagate with the same phase velocity v_{ph}, which is equal to their group velocity v_g:

$$v_g = v_{ph} = \frac{\omega}{k} = \frac{c}{\sqrt{1 + 4\pi NMc^2/H_0^2}} \tag{2}$$

or

$$v_g = v_{ph} = \frac{H_0}{\sqrt{4\pi NM}} = \frac{H_0}{\sqrt{4\pi\tau}}, \tag{3}$$

provided that the second term in the denominator of Equation (2) is much larger than unity. Here $\tau = NM$ is the mass density of the gas (it is clear that the electron mass can be neglected). If the motion of positive ions is neglected, then, instead of the finite constant velocity (Equation (3)), we would have zero velocity for $\omega \to 0$, and the corresponding waves would not exist. Thus, mechanical oscillations of the gas and oscillations of the electromagnetic field are very closely related. The waves propagating with the velocity given by Equation (3) are known as **magnetohydrodynamic waves**. We have already considered them in Section 4.3. These waves are of particular importance in astrophysical, geophysical and other processes.

6.57 Let us find the joint solution in the form of plane monochromatic waves for Maxwell's equations and for the equation of motion (5.115) of the magnetization vector:

(1)

$$\boldsymbol{E} = \boldsymbol{E}_0\, e^{i(\boldsymbol{k}\cdot\boldsymbol{r}-\omega t)}, \qquad \boldsymbol{H} = \boldsymbol{H}_0 + \boldsymbol{h}_0\, e^{i(\boldsymbol{k}\cdot\boldsymbol{r}-\omega t)}, \qquad \boldsymbol{M} = \boldsymbol{M}_0 + \boldsymbol{m}_0\, e^{i(\boldsymbol{k}\cdot\boldsymbol{r}-\omega t)}.$$

The amplitudes of the fields and of the magnetization satisfy the following system of equations:

(2)

$$c(\boldsymbol{k}\times\boldsymbol{h}_0) = -\,\omega\varepsilon\boldsymbol{E}_0,\quad c(\boldsymbol{k}\times\boldsymbol{E}_0) = \omega(\boldsymbol{h}_0 + 4\pi\boldsymbol{m}_0),\quad \boldsymbol{k}\cdot(\boldsymbol{h}_0 + 4\pi\boldsymbol{m}_0) = 0,$$

$$i\omega\boldsymbol{m}_0 = -\eta(\boldsymbol{M}_0\times\boldsymbol{h}_0) - \eta(\boldsymbol{m}_0\times\boldsymbol{H}_0) + \eta qk^2(\boldsymbol{M}_0\times\boldsymbol{m}_0).$$

Eliminating $\boldsymbol{E}_0$ and $\boldsymbol{h}_0$ from Equation (2), and substituting

(3)

$$u = \frac{\omega_M}{\Omega}, \qquad x = \frac{\omega}{\Omega}, \qquad \xi = \frac{ck}{\Omega\sqrt{\varepsilon}}, \qquad \Omega = \omega_0 + \omega_M + \omega_{ex}a^2k^2,$$

$$\omega_0 = \eta H_0, \qquad \omega_{ex}a^2k^2 = \eta M_0 qk^2, \qquad \omega_M = 4\pi\eta M_0,$$

we have

(4)

$$ix\boldsymbol{m}_0 = \frac{u}{x^2-\xi^2}[x^2(\boldsymbol{e}_z\times\boldsymbol{m}_0) + \xi^2(\boldsymbol{n}\cdot\boldsymbol{m}_0)(\boldsymbol{e}_z\times\boldsymbol{n})] + (1-u)(\boldsymbol{e}_z\times\boldsymbol{m}_0),$$

where $\boldsymbol{n} = \boldsymbol{k}/k$, $\boldsymbol{e}_z$ is the unit vector in the direction $\boldsymbol{H}_0$ ($\boldsymbol{M}_0$ is parallel to $\boldsymbol{H}_0$).

Let us take the x-axis in the $(\boldsymbol{n}, \boldsymbol{e}_z)$-plane and denote the angle between $\boldsymbol{e}_z$ and $\boldsymbol{n}$ by θ. The system of linear equations for components of $\boldsymbol{m}_0$ is then followed from Equation (4):

$$ixm_{0x} + \left(1 + \frac{u\xi^2}{x^2 - \xi^2}\right) m_{0y} = 0,$$

$$\left(1 + \frac{u\xi^2}{x^2 - \xi^2}\cos^2\theta\right) m_{0x} - ixm_{0y} = 0.$$

The condition that these equations should be consistent gives the required dispersion relation:

$$\left(1 + \frac{u\xi^2}{x^2 - \xi^2}\right)\left(1 + \frac{u\xi^2}{x^2 - \xi^2}\cos^2\theta\right) - x^2 = 0. \tag{5}$$

This equation is of the third degree in ω^2 ($\omega^2 = \Omega^2 x^2$, Ω is independent of ω), and hence three types of waves with different dispersion relations can propagate in the medium under consideration. Two of these dispersion relations have been investigated in Problem 6.46 (where we have assumed that $\omega_{ex} = 0$). They correspond to ordinary electromagnetic waves propagating in the gyrotropic medium. To study the third type of waves, we will use the condition $\omega^2\varepsilon/c^2k^2 \ll 1$ ($x^2 \ll \xi^2$), which is analogous to the existence condition for magnetostatic waves (Problem 6.50). Neglecting x^2 in comparison with ξ^2 in the denominators of Equation (4), we get the third dispersion relation

$$\omega^2(\boldsymbol{k}) = (\omega_0 + \omega_{ex}a^2k^2)(\omega_0 + \omega_{ex}a^2k^2 + \omega_M \sin^2\theta) \tag{6}$$

Taking into account the condition $\omega^2\varepsilon \ll c^2k^2$ and assuming that ω_0, ω_1, and ω_M are comparable in magnitude, we find that the dispersion relation (6) holds only at $\xi^2 \gg 1$.

Now let us find the relative magnitudes of E_0 and h_0 for waves with the dispersion relation (6). Using the Maxwell equations (2) and the condition $\omega^2\varepsilon/c^2k^2 \ll 1$, we have

$$\boldsymbol{E}_0 \approx \frac{4\pi\,\omega}{ck^2}(\boldsymbol{k} \times \boldsymbol{m}); \qquad \boldsymbol{h}_0 \approx 4\pi\boldsymbol{n}(\boldsymbol{n}\cdot\boldsymbol{m}).$$

Thus, $E_0 \ll h_0$. The waves under consideration are purely magnetic oscillations of the magnetization vector, for which the electric field is very small. They are known as **spin waves**, and determine many magnetic, thermal, and electrical properties of ferromagnetics. The spectrum of spin waves has been found in Problem 5.53 at $\omega_{ex}a^2k^2 \gg \omega_0, \omega_M$.

6.58 Let the Oy-axis be perpendicular to the surface of the metal and directed inward its depth, and let the Oz axis be directed along the constant magnetic field. Since the impedance ζ (as long as it is small) is independent of the incidence angle of the wave, we will consider normal incidence. The solution of the Maxwell's equation and the definition of the surface impedance yield

$$\zeta_{xx} = (1-i)\sqrt{\frac{\omega\mu_{\|}}{8\pi\kappa}}, \qquad \zeta_{zz} = (1-i)\sqrt{\frac{\omega\mu}{8\pi\kappa_3}}, \qquad \zeta_{xz} = \zeta_{zx} = 0,$$

where

$$\kappa = \frac{\kappa_1^2 - \kappa_2^2}{\kappa_1}, \qquad \mu = \frac{\mu_\perp^2 - \mu_a^2}{\mu_\perp}.$$

The dependence of ζ_{zz} on frequency has a resonance character (Equation (6.64)). The component ζ_{xx} does not exhibit resonance, since $\mu_{\|} = 1$.

6.59

$$\zeta_\pm = \pm\frac{E_{\pm1}}{h_{\pm1}} = -(1-i)\sqrt{\frac{\omega\mu^\pm}{8\pi\kappa^\pm}}, \quad \text{where } \mu^\pm = \mu_\perp \pm \mu_a, \quad \kappa^\pm = \kappa_1 \pm \kappa_2,$$

$E_{\pm1}$ and $h_{\pm1}$ are the cyclic components of $\boldsymbol{E}$ and $\boldsymbol{h}$ $(h_{\pm1} = \mp(h_x \pm ih_y))/\sqrt{2})$.

6.61 It is convenient to introduce the cylindrical coordinates with the Oz-axis lying along the axis of the cylinder, and to measure the angle α from the direction of the wavevector $\boldsymbol{k}$ of the incident wave. According to symmetry considerations, the field vectors are independent of z and have the components E_z, H_r, and H_α only. In what follows, we will drop the time-dependent factor $e^{-i\omega t}$. To determine nonzero components of the field we will use the wave equation (I.2.119) for $\boldsymbol{E}$ and the Maxwell's equation (I.2.91). The first of these equations will give E_z, and the second will give H_r and H_α in terms of E_z:

(1) $$H_r = \frac{1}{ikr}\frac{\partial E_z}{\partial \alpha}, \qquad H_\alpha = -\frac{1}{ik}\frac{\partial E_z}{\partial r}.$$

The secondary field $E' = E_z - E_{0z}$, produced by the cylinder, satisfies the equation

(2) $$\frac{1}{r}\frac{\partial}{\partial r}\left(r\frac{\partial E'}{\partial r}\right) + \frac{1}{r^2}\frac{\partial^2 E'}{\partial r^2} + k^2E' = 0.$$

Let us set $E' = R(r)\Phi(\alpha)$ and separate variables in Equation (2). Then we have

(3) $$R''_m + \frac{1}{r}R'_m + \left(k^2 - \frac{m^2}{r^2}\right)R_m = 0,$$

(4) $$\Phi''_m + m^2\Phi_m = 0.$$

The separation parameter is denoted by m^2. The general solution of Equation (2) will be written as the sum over all allowed values of m:

(5) $$E'(r,\alpha) = \sum_m \Phi_m(\alpha)R_m(r).$$

In order to obtain the solution of the Bessel equation (3) in a convenient form, we consider the boundary condition at $r \to \infty$. Since E' describes the secondary field, which is due to the currents induced on the cylinder, as $r \to \infty$ this field will be of the form of cylindrical waves. Then at large r the function E' should behave as

(6) $$E' = \mathcal{E}_0 f(\alpha) \frac{e^{ikr}}{\sqrt{r}}.$$

Equation (6) is satisfied if a Hankel function $H_m^{(1)}(kr)$ is taken as the solution of Equation (3). At large r, the Hankel function has the asymptote

$$H_m^{(1)}(kr) = \sqrt{\frac{2}{\pi kr}}\, e^{i\left(kr - m\pi/2 - \pi/4\right)} \qquad (kr \gg 1).$$

The second linearly independent solution will contain the term of the form $(const/\sqrt{r})\, e^{-ikr}$ describing a converging cylindrical wave, which cannot appear in our problem. The solution of Equation (3) we will, therefore, be written as $R_m(r) = H_m^{(1)}(kr)$. Equation (4) has the solution

$$\Phi_m(\alpha) = A_m e^{im\alpha} + B_m e^{-im\alpha}.$$

Since the field does not change when α is changed by 2π, the number m should be integer. If m takes both positive and negative values, then in the expression for $\Phi_m(\alpha)$ it is sufficient to keep only one term, for example, $e^{im\alpha}$. Finally, the solution for $E'(r, \alpha)$ reads

(7) $$E'(r, \alpha) = \mathcal{E}_0 \sum_{m=-\infty}^{\infty} A_m H_m^{(1)}(kr)\, e^{im\alpha}.$$

At large r, Equation (7) becomes identical to Equation (6), and

$$f(\alpha) = \sqrt{\frac{2}{\pi k}} \sum_m A_m \exp\left[i\left(m\alpha - \frac{m\pi}{2} - \frac{\pi}{4}\right)\right].$$

The coefficients A_m in Equation (7) should be determined from the boundary condition on the surface of the cylinder. Since the cylinder is assumed to be a perfect conductor,

(8) $$E' + E_0 = 0 \qquad \text{at} \qquad r = a$$

or

(9) $$\exp(ika\cos\alpha) + \sum_{m=-\infty}^{\infty} A_m H_m^{(1)}(ka)\, e^{im\alpha} = 0.$$

Because the functions $e^{im\alpha}$ are orthogonal, we obtain

$$\int_0^{2\pi} \exp[i(ka\cos\alpha - m'\alpha)]\, d\alpha + 2\pi A_{m'} H_{m'}^{(1)}(ka) = 0,$$

and, from Equation (1.150), we have

(10) $$A_m = \frac{i^m J_m(ka)}{H_m^{(1)}(ka)}.$$

The total electric field is, hence, given by

$$E(r,\alpha) = \mathcal{E}\ \exp(ikr\cos\alpha) - \mathcal{E}_0 \sum_m \frac{i^m J_m(ka)}{H_m^{(1)}(ka)} H_m^{(1)}(kr)\, e^{im\alpha}.$$

The components of the magnetic field are determined with the aid of Equation (1):

$$H_r = -\mathcal{E}_0 \sin\alpha \exp(ikr\cos\alpha) - \mathcal{E}_0 \sum_m \frac{i^m J_m(ka)}{H_m^{(1)}(ka)} \frac{H_m^{(1)}(kr)}{kr} e^{im\alpha},$$

$$H_\alpha = -\mathcal{E}_0 \cos\alpha \exp(ikr\cos\alpha) + \mathcal{E}_0 \sum_m \frac{i^{m-1} J_m(ka)}{H_m^{(1)}(ka)} \frac{dH_m^{(1)}(kr)}{d(kr)} e^{im\alpha}.$$

The secondary electric field is transverse in the entire space; the secondary magnetic field becomes transverse at large distances from the cylinder (at $kr \gg 1$, in the wave zone), when the longitudinal component H_r vanishes due to the presence of the large extra factor kr in the denominator. The surface current density can be determined from the boundary condition for the tangential component $\boldsymbol{H}$:

$$i(\alpha) = i_z(\alpha) = \frac{c}{4\pi} H_\alpha(a,\alpha).$$

Finally, the total current is

$$\mathcal{I} = -\frac{i}{2} ca\mathcal{E}_0 \left[J_1(ka) - \frac{J_0(ka) H_1^{(1)}(ka)}{H_0^{(1)}(ka)} \right].$$

6.62 In the case under consideration, the field is two-dimensional. Therefore, the quantity dI in the general formula $d\sigma_s = dI/\gamma_0$, Equation (6.65), must be treated as the intensity of the secondary waves per unit length of the cylinder, within an angle $d\alpha$, that is, $dI = \gamma r\, d\alpha$.

The effective differential scattering cross section will have the dimension of length. By using of the results of Problem 6.61, we find

$$d\sigma_s = |f(\alpha)|^2\, d\alpha,$$

where

$$f(\alpha) = \sqrt{\frac{2}{\pi k}} \sum_m i^m \frac{J_m(ka)}{H_m^{(1)}} \exp\left[i\left(m\alpha - \frac{m\pi}{2} - \frac{\pi}{4} \right) \right]. \qquad (1)$$

For arbitrary values ka, Equation (1) is rather complicated. However, it simplifies at $ka \ll 1$. It is then sufficient to keep only the $m = 0$ term in the infinite sum, which gives the isotropic distribution of the secondary radiation,

$$d\sigma_s = \frac{\pi\, d\alpha}{2k \ln^2(ka)} = \frac{\lambda}{4 \ln^2(ka)}\, d\alpha. \qquad (2)$$

The total cross section is obtained by integrating Equation (1) over $d\alpha$. Owing to the orthogonality of the functions $e^{im\alpha}$, we have

(3) $$\sigma_s = \frac{4}{k} \sum_{m=-\infty}^{\infty} \left| \frac{J_m(ka)}{H_m^{(1)}(ka)} \right|^2 .$$

At $ka \ll 1$, Equation (3) takes the form

$$\sigma_s = \frac{\pi \lambda}{2 \ln^2(ka)}.$$

6.63

$$H_z = \mathcal{H}_0 \left[\exp(ikr\cos\alpha) - \sum_{m=-\infty}^{\infty} i^m \frac{J'_m(ka)}{H_m^{(1)\prime}(ka)} H_m^{(1)}(kr)\, e^{im\alpha} \right],$$

$$E_r = \mathcal{H}_0 \left[\sin\alpha\, \exp(ikr\cos\alpha) + \frac{1}{kr} \sum_{m=-\infty}^{\infty} i^m \frac{mJ'_m(ka)}{H_m^{(1)\prime}(ka)} H_m^{(1)}(kr)\, e^{im\alpha} \right],$$

$$E_\alpha = \mathcal{H}_0 \left[\cos\alpha\, \exp(ikr\cos\alpha) + \sum_{m=-\infty}^{\infty} i^{m+1} \frac{J'_m(ka)}{H_m^{(1)\prime}(ka)} H_m^{(1)\prime}(kr)\, e^{im\alpha} \right],$$

where α is measured from the direction $\boldsymbol{k}$, and the axis of the cylindrical system of coordinates lies along the axis of the cylinder.

$$d\sigma_s(\alpha) = \frac{\pi (ka)^3}{8} a (1 - 2\cos\alpha)^2\, d\alpha, \qquad \sigma_s = \frac{3}{4}\pi^2 k^3 a^4.$$

6.64

$$d\sigma'_s = \cos^2\varphi\, d\sigma_{\parallel} + \sin^2\varphi\, d\sigma_{\perp}, \qquad d\sigma''_s = \frac{1}{2}(d\sigma_{\parallel} + d\sigma_{\perp}).$$

6.65 The unpolarized wave can be considered a superposition of two incoherent components of equal intensity. The electric vector $\boldsymbol{E}$ of one of the two components lies along the cylinder axis, and the electric vector of the other component is perpendicular to the cylinder axis. The scattering cross sections for the first and second components have been obtained in Problems 6.62 and 6.63. The degree of depolarization ρ is determined as the ratio of the intensities of the scattered waves:

$$\rho = \frac{d\sigma_{\perp}}{d\sigma_{\parallel}} = \frac{1}{4}(ka)^4 \ln^2(ka)(1 - 2\cos\alpha)^2.$$

Since $ka \ll 1$, we obtain that ρ is very small, that is, the scattered waves are almost completely polarized for all scattering angles; at $\cos\alpha = \frac{1}{2}$, that is, at $\alpha = 60°$, the depolarization coefficient ρ becomes zero.

6.66

$$H_z = \mathcal{H}_0 \left[\exp(ikr\cos\alpha) + \sum_{m=-\infty}^{\infty} i^m \frac{J'_m(ka) - i\zeta J_m(ka)}{i\zeta H_m^{(1)}(ka) - H_m^{(1)\prime}(ka)} H_m^{(1)}(kr)\, e^{im\alpha} \right],$$

where ζ is the surface impedance of the metal;

$$H_\alpha = H_r = 0, \qquad \boldsymbol{E} = \frac{i}{k}\,rot\,\boldsymbol{H}.$$

6.67

$$Q = \frac{ac\zeta' \mathcal{H}_0^2}{4} \sum_m \left| \frac{J'_m Y_m - J_m Y'_m}{i\zeta H_m^{(1)} - H_m^{(1)'}} \right|^2,$$

where ζ' is the real part of the surface impedance. Cylindrical functions $J_m(x)$, $Y_m(x)$, and $H_m^{(1)}(x)$ (Section I.1.3) and their derivatives are taken at $x = ka$.

The absorption cross section:

$$\sigma_a = \frac{Q_0}{\gamma_0} = 2\pi a\zeta' \sum_m \left| \frac{J'_m Y_m - J_m Y'_m}{i\zeta H_m^{(1)} - H_m^{(1)'}} \right|^2.$$

At $ka \ll 1$, that is, at $\lambda \gg a$, the field in the vicinity of the cylinder is quasi-stationary (conducting cylinder in the longitudinal quasi-stationary magnetic field; see Problem 4.30). Therefore, expressing ζ' in terms of the conductivity $\sigma \gg \omega$ with the aid of Equations (5.65) and (6.27), we obtain the relation

$$Q = \frac{ac\mathcal{H}_0^2}{8} \sqrt{\frac{\mu\,\omega}{2\pi\sigma}},$$

which is identical with the solution found in Problem 4.32 for the case of a strong skin-effect (if the Q is expressed through the magnetic field).

6.68 At $r > a$:

$$E_z = \mathcal{E}_0 \Bigg[\exp(ikr\cos\alpha) + \sum_{m=-\infty}^{\infty} i^m \frac{\zeta J'_m(ka) J_m(k'a) - J_m(ka) J'_m(k'a)}{H_m^{(1)}(ka) J'_m(k'a) - \zeta H_m^{(1)'}(ka) J_m(k'a)} H_m^{(1)}(kr)\, e^{im\alpha} \Bigg];$$

and at $r < a$:

$$E_z = \mathcal{E}_0 \zeta \sum_{m=-\infty}^{\infty} i^m \frac{J'_m(k'a) H_m^{(1)}(ka) - J'_m(k'a) H_m^{(1)'}(ka)}{J'_m(k'a) H_m^{(1)}(ka) - \zeta J_m(k'a) H_m^{(1)}(ka)} J_m(k'r)\, e^{im\alpha},$$

where $\mathcal{E}_0$ is the amplitude of the incident wave, $\zeta = \sqrt{\mu/\varepsilon}$, $k = \omega/c$, $k' = \omega\sqrt{\varepsilon\mu}/c$, and other components of $\boldsymbol{E}$ are zero.

The field $\boldsymbol{E}$ can be found from

$$\boldsymbol{H} = \frac{c}{i\omega\mu}\,rot\,\boldsymbol{E}.$$

6.69 The dipole moments of the sphere will be given by

$$\boldsymbol{p} = \beta_e \boldsymbol{E}_0\, e^{-i\omega t}, \qquad \boldsymbol{m} = \beta_m \boldsymbol{H}_0\, e^{-i\omega t},$$

where β_e and β_m are the electric and magnetic polarizations of the sphere, which are generally complex.

The components of the vectors $\boldsymbol{E}$ and $\boldsymbol{H}$ for the scattered wave can be found with the aid of Equations (I.5.28), (I.5.29), and (I.5.37):

$$H_\alpha = E_\theta = \frac{\omega^2 E_0}{c^2 r}(\beta_e \cos\theta + \beta_m)\cos\alpha, \quad H_\theta = -E_\alpha = \frac{\omega^2 E_0}{c^2 r}(\beta_e + \beta_m)\sin\alpha.$$

The angles θ and α define the direction of scattering; they are shown in Figure 6.19.

The differential scattering cross section can be obtained from Equation (6.65):

$$\frac{d\sigma_s(\theta,\alpha)}{d\Omega} = \frac{\omega^4}{c^4}\left[|\beta_e|^2(\cos^2\theta\cos^2\alpha + \sin^2\alpha) + |\beta_m|^2(\cos^2\theta\sin^2\alpha + \cos^2\alpha)\right.$$
$$\left. + (\beta_e\beta_m^* + \beta_e^*\beta_m)\cos\theta\right].$$

6.70

$$d\sigma_s(\theta) = \frac{1}{2}\left[d\sigma_s(\theta,\alpha) + d\sigma_s(\theta,\alpha+\pi/2)\right]$$
$$= \frac{\omega^4}{2c^4}\left[(|\beta_e|^2 + |\beta_m|^2)(1+\cos^2\theta) + 2(\beta_e\beta_m^* + \beta_e^*\beta_m)\cos\theta\right] d\Omega,$$
$$\sigma_s = \frac{8\pi\omega^4}{3c^4}(|\beta_e|^2 + |\beta_m|^2).$$

In order to determine the depolarization degree of the scattered radiation, it is necessary to find the principal directions of the polarization tensor. In the present problem, this can easily be done using symmetry considerations. For given $\boldsymbol{k}$ and $\boldsymbol{n}$ (Figure 6.15), physically selected directions for $\boldsymbol{E}_0$ will be the direction of the normal to the

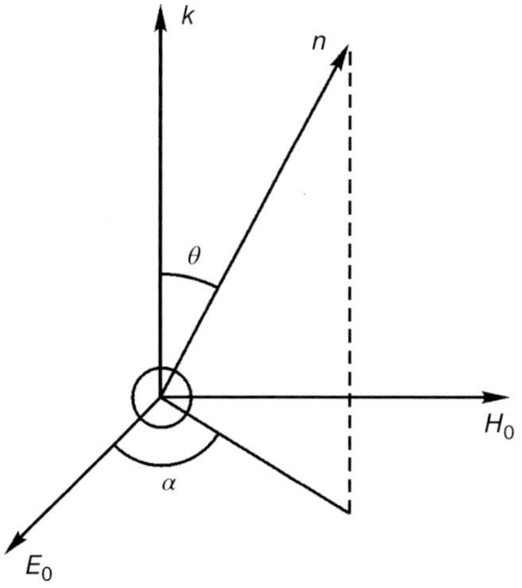

Figure 6.19 Scattering of plane wave by a sphere.

scattering plane and the direction lying in the scattering plane and perpendicular to $\boldsymbol{k}$.

These directions of polarizations correspond to the differential scattering cross sections $d\sigma_s(\theta, \pi/2)$ and $d\sigma_s(\theta, 0)$, which have been obtained in the solution of Problem 6.74. The depolarization degree ρ is given by the ratio of the smaller to the larger of the cross sections.

At $|\beta_m| < |\beta_e|$, we have

$$\rho = \frac{d\sigma_s(\theta, 0)}{d\sigma_s(\theta, \pi/2)} = \left| \frac{\beta_m + \beta_e \cos\theta}{\beta_m \cos\theta + \beta_e} \right|^2.$$

6.71 For the dielectric sphere:

$$d\sigma_{sd} = \frac{\omega^4 a^6}{2c^4} \left(\frac{\varepsilon - 1}{\varepsilon + 2} \right)^2 (1 + \cos^2\theta)\, d\Omega, \qquad \sigma_{sd} = \frac{8\pi\, \omega^4 a^6}{3c^4} \left(\frac{\varepsilon - 1}{\varepsilon + 2} \right)^2;$$

$$\rho_d = \cos^2\theta$$

.

For the perfectly conducting sphere:

$$d\sigma_{sc} = \frac{\omega^4 a^6}{8c^4} [5(1 + \cos^2\theta) - 8\cos\theta]\, d\Omega, \qquad \sigma_{sc} = \frac{10\pi\, \omega^4 a^6}{3c^4},$$

$$\rho_c = \left(\frac{1 - 2\cos\theta}{2 - \cos\theta} \right)^2.$$

It is clear from the formula for $d\sigma_{sd}$ that the scattering cross section by the dielectric sphere is symmetric with respect to the forward ($\theta = 0$) and backward ($\theta = \pi$) directions. The ratio $d\sigma_{sd}(0)/d\sigma_{sd}(\pi) = 1$. The scattering cross section by the conducting sphere is much more anisotropic and asymmetric: $d\sigma_{sc}(0)/d\sigma_{sc}(\pi) = \frac{1}{9}$. Light scattered by the dielectric sphere through the angle $\theta = \pi/2$, will be completely polarized; light scattered by the perfectly conducting sphere becomes completely polarized at $\cos\theta = 1/2$, $\theta = \pi/3 = 60°$.

The obtained formulas are valid for the dielectric sphere if one can neglect the effects associated with finite velocity of propagation of electromagnetic waves inside the sphere, that is, when the wavelength inside the sphere is larger than a. In the case of the perfectly conducting sphere, there is no wave propagating inside the sphere, and it is sufficient to satisfy the condition $a \ll \lambda$, where λ is the wavelength in the medium surrounding the sphere.

6.72 The total strength of the electric field in any point can be written as

$$\boldsymbol{E}(\boldsymbol{r}, t) = \boldsymbol{E}_0(\boldsymbol{r}, t) + \boldsymbol{E}'(\boldsymbol{r}, t), \tag{1}$$

where

$$\boldsymbol{E}_0(\boldsymbol{r}, t) = \boldsymbol{\mathcal{E}}_0 \exp[i(\boldsymbol{k} \cdot \boldsymbol{r} - \omega t)]$$

is the field of the incident wave, and $\boldsymbol{E}'(\boldsymbol{r}, t)$ is the field of the scattered (secondary) wave.

At any point inside the body (which can be inhomogeneous), the polarization vector $\boldsymbol{P}(\boldsymbol{r}, t)$ is proportional to $\boldsymbol{E}$ and approximately to $\boldsymbol{\mathcal{E}}_0$ because the scattered field is much smaller than the incident field ($E' \ll \mathcal{E}_0$) as long as $(\varepsilon - 1)/4\pi \ll 1$.[18]

The scattered field $\boldsymbol{E}'$ can be expressed in terms of the Hertz electric vector

(2)
$$\boldsymbol{Z}(\boldsymbol{r}, t) = \int \frac{\boldsymbol{P}(\boldsymbol{r}', t)}{R} \exp[i(kR - \omega t)]\, dV'$$

(Equations (I.5.40)–(I.5.43)) through the formula

(3)
$$\boldsymbol{E}' = rot\, rot\, \boldsymbol{Z} - 4\pi \boldsymbol{P} = \frac{\varepsilon - 1}{4\pi} rot\, rot\, \boldsymbol{\mathcal{E}}_0 \frac{e^{ikr}}{r} \int \exp[i(\boldsymbol{k} - k\boldsymbol{n}) \cdot \boldsymbol{r}']\, dV'.$$

The difference $\boldsymbol{k} - k\boldsymbol{n}$ is the change of the wave vector in a scattering event. Let us denote it by $\boldsymbol{q}$ ($q = 2k \sin(\theta/2)$, where θ is the scattering angle). In calculating the integral, we direct the polar axis along $\boldsymbol{q}$. Then

(4)
$$\int \exp[i(\boldsymbol{q} \cdot \boldsymbol{r}')] r'^2\, dr'\, d\Omega = 4\pi \frac{\sin qa - qa \cos qa}{q^3}.$$

In calculating the double curl in Equation (3), we retain only the terms proportional to $1/r$:

$$rot\, rot\, \boldsymbol{\mathcal{E}}_0 \frac{\exp[ikr]}{r} = k^2 [\boldsymbol{n} \times (\boldsymbol{\mathcal{E}}_0 \times \boldsymbol{n})] \frac{\exp[ikr]}{r}.$$

Finally, the relation for the scattered field $\boldsymbol{E}'$ will be given by

(5)
$$\boldsymbol{E}' = \frac{\omega^2 a^3}{c^2} \frac{\varepsilon - 1}{3} [\boldsymbol{n} \times (\boldsymbol{\mathcal{E}}_0 \times \boldsymbol{n})] \varphi(qa) \frac{\exp[ikr]}{r},$$

where

$$\varphi(qa) = \frac{3(\sin q - qa \cos qa)}{(qa)^3} = 3\sqrt{\frac{\pi}{(qa)^3}} J_{3/2}(qa).$$

Let us compare Equation (5) with those obtained for small a (Problem 6.71). In the limit $qa \ll 1$, Equation (5) yields

(6)
$$\boldsymbol{E}' = \frac{\omega^2 a^3}{c^2} \frac{\varepsilon - 1}{3} \frac{e^{ikr}}{r} [\boldsymbol{n} \times (\boldsymbol{\mathcal{E}}_0 \times \boldsymbol{n})],$$

because $\varphi(qa) \approx 1$ at $qa \ll 1$.

On the other hand, if $\boldsymbol{E}'$ is calculated from the formula

$$\boldsymbol{E}' = \boldsymbol{n} \times (\ddot{\boldsymbol{p}} \times \boldsymbol{n}) \frac{e^{ikr}}{c^2 r}, \qquad \text{where} \qquad \boldsymbol{p} = \frac{\varepsilon - 1}{\varepsilon + 2} a^3 \boldsymbol{\mathcal{E}}_0$$

is the static dipole moment of the sphere, we obtain

(6a)
$$\boldsymbol{E}' = \frac{\omega^2 a^3}{c^2} \frac{\varepsilon - 1}{\varepsilon + 2} \frac{e^{ikr}}{r} \boldsymbol{n} \times (\boldsymbol{\mathcal{E}}_0 \times \boldsymbol{n}).$$

18) The method used for obtaining the solution of the present problem is analogous to the Born approximation in quantum mechanics. The latter is widely used in quantum-mechanical studies of scattering problems.

In Equation (6a), the factor $1/3$ is replaced by $1/(\varepsilon+2)$. However, Equations (6) and (6a) do not contradict each other because Equation (6) is valid only to within a factor $\sim 1/(\varepsilon-1)$.

The differential scattering cross section is

$$\frac{d\sigma_s(\theta,\alpha)}{d\Omega} = \frac{\omega^4 a^6(\varepsilon-1)^2}{9c^4}\varphi^2(qa)(\sin^2\alpha + \cos^2\alpha\cos^2\theta), \tag{7}$$

the angles θ and α are shown in Figure 6.19.

This cross section differs from the scattering cross section for a small dielectric sphere (see solution to Problem 6.71); the factor $\varphi^2(qa)$ is now present and the term $(\varepsilon+2)^2$ in the denominator is replaced by 9. The new factor $\varphi^2(qa)$ is responsible for the interference between secondary waves from different elements of the sphere. Therefore, the degree of depolarization of the scattered light will be the same as for a small dielectric sphere,

$$\rho = \cos^2\theta. \tag{8}$$

Averaging over the polarizations gives

$$\frac{d\sigma_s(\theta)}{d\Omega} = \frac{\omega^4 a^6(\varepsilon-1)^2}{18c^4}\varphi^2(qa)(1+\cos^2\theta).$$

Consider a very large sphere, $ka \gg 1$. If the angles are such that $qa \gg 1$, then $\varphi(qa) \to 0$, and the cross section within this range of angles is very small. It follows from the explicit expression for q that $qa \gg 1$ is equivalent to the condition $\theta \gg 1/ka$. Thus, if the sphere is large, the scattering occurs in the forward direction, in the range of angles $\theta \lesssim 1/ka$.

6.73 At $ka \gg 1$, the function $\varphi^2(qa)$ in the expression for the differential cross section (Problem 6.72), is appreciably different from zero only within the narrow range of angles $\theta \lesssim 1/ka$. In this range, the factor $(1+\cos^2\theta)$ may be regarded as constant equal to 2. Then we have

$$\sigma_s = \frac{2\pi\,\omega^4 a^6(\varepsilon-1)^2}{9c^4}\int_0^\pi \varphi^2(qa)\sin\theta\,d\theta.$$

Let us introduce a new variable $y = qa = 2ka\sin\theta/2$. In the limit $ka \gg 1$, we finally obtain

$$\sigma_s = \frac{\pi\omega^2 a^4(\varepsilon-1)^2}{18c^2}.$$

For a small sphere ($ka \ll 1$), we replace (see solution to Problem 6.71) $\varepsilon + 2$ by 3, so that

$$\sigma_s = \frac{8\pi\,\omega^4 a^6(\varepsilon-1)^2}{27c^4}.$$

As seen from these results, the cross sections for small and large spheres depend differently on frequency ($\sim \omega^4$ and $\sim \omega^2$) and on the radius of the sphere ($\sim a^6$ and $\sim a^4$).

6.74 It is convenient to find the solution of the present problem in spherical coordinates and represent the fields in terms of the Hertz vectors (Equations (I.5.40) and (I.5.46)). In the general case that we consider, we should introduce both the electric and the magnetic vectors, which in accordance with Equations (I.5.42) and (I.5.45), must satisfy the uniform wave equations in the region $r > a$. Let us express these equations in terms of the scalar functions $u(r, \vartheta, \alpha)$ and $v(r, \vartheta, \alpha)$:

(1)
$$\boldsymbol{Z}^{(e)} = u\boldsymbol{r} + \nabla\chi, \ \chi = k^{-2}\frac{\partial}{\partial r}(ru); \ \boldsymbol{Z}^{(m)} = v\boldsymbol{r} + \nabla\psi, \ \psi = k^{-2}\frac{\partial}{\partial r}(rv), \ k = \frac{\omega}{c}.$$

Sufficiently complicated relation of these functions (Debye potentials[19)]) with the Hertz vectors is taken in such a way that u and v satisfy the wave equations.

Let us express the fields strengths through the Hertz vectors with an aid of Equations (I.5.42) and (I.5.45),

(2)
$$\begin{aligned} \boldsymbol{E} &= k^2\boldsymbol{Z}^{(e)} + \nabla(\nabla\cdot\boldsymbol{Z}^{(e)}) + ik\nabla\times\boldsymbol{Z}^{(m)} = \nabla\times[\nabla\times\boldsymbol{Z}^{(e)}] + ik\nabla\times\boldsymbol{Z}^{(m)}, \\ \boldsymbol{H} &= \nabla\times[\nabla\times\boldsymbol{Z}^{(m)}] - ik\nabla\times\boldsymbol{Z}^{(e)}, \end{aligned}$$

and verify that the electric vector $\boldsymbol{Z}^{(e)}$ induces the field with $E_r \neq 0$ (E-wave), and the magnetic vector $\boldsymbol{Z}^{(m)}$ induces the field with $H_r \neq 0$ (H-wave). Notice that in our problem under $\boldsymbol{B} = \boldsymbol{H}$.

Let us show that the Hertz vectors satisfy the uniform wave equations provided the functions $u(r, \vartheta, \alpha)$ and $v(r, \vartheta, \alpha)$ obey the same equations:

(3)
$$\triangle u + k^2 u = 0, \qquad \triangle v + k^2 v = 0.$$

These equations written for monochromatic fields and not containing the time derivative are known as **Helmholtz equations**. We have from Equation (1) the following relation:

(4)
$$(\triangle + k^2)\boldsymbol{Z}^{(e)} = (\triangle + k^2)u\boldsymbol{r} + \nabla(\triangle + k^2)\chi.$$

Applying the operator $\triangle$ to the vector $u\boldsymbol{r}$, we obtain

(5)
$$\triangle(u\boldsymbol{r}) = \left(\frac{\partial^2}{\partial x^2} + \frac{\partial^2}{\partial y^2} + \frac{\partial^2}{\partial z^2}\right) u(x\boldsymbol{e}_x + y\boldsymbol{e}_y + z\boldsymbol{e}_z) = \boldsymbol{r}\triangle u + 2\nabla u.$$

Now, using spherical coordinates, we write

(6)
$$\triangle = \frac{\partial^2}{\partial r^2} + \frac{2}{r}\frac{\partial}{\partial r} - \frac{\hat{l}^2}{r^2}, \qquad \chi = k^{-1}\left(u + r\frac{\partial u}{\partial r}\right),$$

where $-\hat{l}^2$ is the angular part of the Laplace operator, and we have used the wave equation (3) for u. Thus, we obtain

(7)
$$\triangle\left(r\frac{\partial u}{\partial r}\right) = -k^2\left(2u + r\frac{\partial u}{\partial r}\right).$$

Finally, we have

(8)
$$\triangle\chi + k^2\chi + 2u = 0$$

19) Debye Peter (1884–1966), European scientist, physicist and chemist, Nobel Prize laureate.

and, using Equations (4)–(8), prove that the wave equation for $\boldsymbol{Z}^{(e)}$ holds:

(9) $$(\triangle + k^2)\boldsymbol{Z}^{(e)} = \boldsymbol{r}(\triangle u + k^2 u) + \nabla(\triangle \chi + k^2 \chi + 2u) = 0.$$

Similar results can be obtained for $\boldsymbol{Z}^{(m)}$.

The field strengths are expressed through the functions u and v with an aid of Equations (2) and (3):

(10) $$\boldsymbol{E} = k^2(u\boldsymbol{r} + \nabla\chi) + ik\nabla v \times \boldsymbol{r}, \qquad \boldsymbol{H} = k^2(v\boldsymbol{r} + \nabla\psi) - ik\nabla u \times \boldsymbol{r}.$$

In projections on the spherical axes, we have

(11)
$$\begin{cases} E_r = \dfrac{\partial^2(ru)}{\partial r^2} + k^2 ru, & H_r = \dfrac{\partial^2(rv)}{\partial r^2} + k^2 rv, \\ E_\vartheta = \dfrac{1}{r}\dfrac{\partial^2(ru)}{\partial r\,\partial\vartheta} + \dfrac{ik}{\sin\vartheta}\dfrac{\partial v}{\partial\alpha}, & H_\vartheta = \dfrac{1}{r}\dfrac{\partial^2(rv)}{\partial r\,\partial\vartheta} - \dfrac{ik}{\sin\vartheta}\dfrac{\partial u}{\partial\alpha}, \\ E_\alpha = \dfrac{1}{r\sin\vartheta}\dfrac{\partial^2(ru)}{\partial r\,\partial\alpha} - ik\dfrac{\partial v}{\partial\vartheta}, & H_\alpha = \dfrac{1}{r\sin\vartheta}\dfrac{\partial^2(rv)}{\partial r\,\partial\alpha} + ik\dfrac{\partial u}{\partial\vartheta}. \end{cases}$$

The validity of Equation (11) can be easily verified directly. For this purpose, one can substitute Equation (11) into the Maxwell's equations $\nabla \times \boldsymbol{E} = ik\boldsymbol{H}$ and $\nabla \times \boldsymbol{H} = -ik\boldsymbol{E}$, written in projections on axes of the spherical coordinate system. Four of the six equations will become identities, and two equations will coincide with Equation (3).

At the next step, we will construct the general solution of the Helmholtz equation. First, we find particular solutions by the method of separation of variables: $u(r, \vartheta, \alpha) = R(r)\Theta(\vartheta)\Phi(\alpha)$. Using the Laplace operator in spherical coordinates (I.1.137), we come to the three ordinary differential equations,

$$\frac{d^2\Phi}{d\alpha^2} + m^2\Phi = 0, \qquad (1 - x^2)\frac{d^2\Theta}{dx^2} - 2x\frac{d\Theta}{dx} + \left[\lambda - \frac{m^2}{1 - x^2}\right]\Theta = 0,$$

(12) $$\frac{d^2R}{dr^2} + \frac{2}{r}\frac{dR}{dr} + \left(k^2 - \frac{\lambda}{r^2}\right)R = 0,$$

where $x = \cos\vartheta$, while m^2 and λ are constants introduced in separating variables. The parameter $m = 0, \pm 1, \ldots$ must be real and integer for the function $\Phi(\alpha)$ to be single-valued (unchanged when α is replaced by $\alpha + 2\pi$). The general solution for Φ has the form

(13) $$\Phi(\alpha) = b_1 \cos m\alpha + b_2 \sin m\alpha,$$

where $b_{1,2}$ are arbitrary constants.

The equation for Θ takes the form (Equation (I.1.190)). Its solution, finite in the region $-1 \le x \le 1$, exists only at $\lambda = l(l+1)$, $l = 0, 1, 2, \ldots$, $|m| \le l$. It is an associated Legendre polynomial (I.1.189):

(14) $$\Theta_{lm}(x) = P_l^m(x).$$

By substitutions $\varrho = kr$ and $R = \varrho^{-1/2}Q$, the equation for R is reduced to the Bessel equation (I.1.151). The general solution can, therefore, be written in terms of spherical Bessel functions (I.1.171):

$$R_l(kr) = a_1 z_l^{(1)}(kr) + a_2 z_l^{(2)}(kr), \tag{15}$$

where $a_{1,2}$ are constants, and $z_l^{(1,2)}(kr)$ are two arbitrary, linearly independent, spherical Bessel functions. Using the obtained solutions of the ordinary equations (12), we can write the general solution of the Helmholtz equation (3):

$$u(r,\vartheta,\alpha) = \sum_{l,m} u_{lm}(r,\vartheta,\alpha), \qquad u_{lm}(r,\vartheta,\alpha)$$

$$= (a_{1l} z_l^{(1)}(kr) + a_{2l} z_l^{(2)}(kr)) P_l^m(x)(b_{1m}\cos m\alpha + b_{2m}\sin m\alpha). \tag{16}$$

The general solution for v has the same form. These solutions can be written also in terms of the Legendre spherical functions (I.1.192).

Now we construct the functions u^0 and v^0 to describe the incident plane wave,

$$\boldsymbol{E}^0(\boldsymbol{r}) = \boldsymbol{e}_x e^{ikr\cos\vartheta}, \qquad \boldsymbol{H}^0(\boldsymbol{r}) = \boldsymbol{e}_y e^{ikr\cos\vartheta}; \tag{17}$$

we assume here that the amplitude $E_0 = 1$. Projecting the fields on the unit vector $\boldsymbol{e}_r$, we have

$$E_r = \sin\vartheta\cos\alpha\, e^{ikr\cos\vartheta}, \qquad H_r = \cos\vartheta\cos\alpha\, e^{ikr\cos\vartheta}. \tag{18}$$

Then we write

$$E_r = \frac{i\cos\alpha}{kr}\frac{\partial}{\partial\vartheta}e^{ikr\cos\vartheta} \tag{19}$$

and use the expansion of a plane wave over Legendre polynomials (Problem I.1.138),

$$E_r = \frac{i\cos\alpha}{kr}\sum_{l=0}^{\infty} i^l(2l+1)j_l(kr)\frac{dP_l(\cos\vartheta)}{d\vartheta} =$$

$$-\frac{i}{kr}\sum_{l=1}^{\infty} i^l(2l+1)j_l(kr)P_l^1(\cos\vartheta)\cos\alpha, \tag{20}$$

where $j_l(kr)$ is the spherical Bessel function (I.1.171). The derivatives of Legendre polynomials with respect to the angle are transformed into the associated Legendre polynomials with an aid of Equation (I.1.189). Summation begins with $l = 1$, since $P_0^1 = 0$.

Using the expansion (20), we look for u^0 in the form of the series expansion,

$$u^0 = \sum_{l=1}^{\infty} a_{1l} j_l(kr) P_l^1(\cos\vartheta)\cos\alpha, \tag{21}$$

taking in Equation (16) $b_{1m} = \delta_{m1}$, $b_{2m} = a_{2l} = 0$ and $z_l^{(1)} = j_l(kr)$. The equation for E_r will be written from Equation (11) using the Helmholtz equation:

$$E_r = -\frac{1}{r}\hat{\boldsymbol{l}}^2 u(r, \vartheta, \alpha). \tag{22}$$

Replacing u by the series (Equation (21)), we have

$$E_r^0 = -\sum_{l=1}^{\infty} a_{1l} l(l+1) j_l(kr) P_l^1(\cos\vartheta)\cos\alpha, \tag{23}$$

where we have used the equation $\hat{\boldsymbol{l}}^2 P_l^1 = l(l+1)P_l^1$. Now comparing Equation (23) with Equation (20) we can find the coefficients a_{1l} and write the expansion for u^0:

$$u^0 = \frac{1}{k}\sum_{l=1}^{\infty} \frac{i^{l-1}(2l+1)}{l(l+1)} j_l(kr) P_l^1(\cos\vartheta)\cos\alpha. \tag{24}$$

Analogous calculations yield similar expression for the second Debye potential:

$$v^0 = \frac{1}{k}\sum_{l=1}^{\infty} \frac{i^{l-1}(2l+1)}{l(l+1)} j_l(kr) P_l^1(\cos\vartheta)\sin\alpha. \tag{25}$$

Let us express the Debye potentials of the total field, in the presence of scattering body, as sums $u = u^0 + u^s$ and $v = v^0 + v^s$, where the superscript s denotes the potentials which describe the field of scattered waves. At large distances from the sphere, the scattered waves should have the form of the diverging spherical waves (principle of radiation). This requirement leads to the boundary condition

$$u^s \propto \frac{1}{r}e^{ikr}, \qquad v^s \propto \frac{1}{r}e^{ikr} \qquad \text{at} \qquad kr \gg l. \tag{26}$$

The boundary condition on the surface of the perfectly conducting sphere requires that the tangential components of the electric field be zero: $E_\vartheta = E_\alpha = 0$ at $r = a$. According to Equation (11), it will be satisfied if the total Debye potentials obey the conditions

$$v(a, \vartheta, \alpha) = 0, \qquad \frac{\partial}{\partial r}(ru(r, \vartheta, \alpha))\big|_{r=a} = 0. \tag{27}$$

On the basis of the general solution (16) and Equations (24) and (25), we write the total Debye potentials in the form that satisfies the condition (26) for the scattered wave:

$$\begin{aligned} u &= k^{-1}\cos\alpha \textstyle\sum_{l=1}^{\infty} \frac{i^{l-1}(2l+1)}{l(l+1)} [j_l(kr) + a_l h_l^{(1)}(kr)] P_l^1(\cos\vartheta), \\ v &= k^{-1}\sin\alpha \textstyle\sum_{l=1}^{\infty} \frac{i^{l-1}(2l+1)}{l(l+1)} [j_l(kr) + b_l h_l^{(1)}(kr)] P_l^1(\cos\vartheta). \end{aligned} \tag{28}$$

Here a_l and b_l are unknown coefficients to be determined from the boundary conditions (27),

$$a_l = -\frac{(kaj_l(ka))'}{(kah_l^{(1)}(ka))'}, \qquad b_l = -\frac{j_l(ka)}{h_l^{(1)}(ka)}, \tag{29}$$

$h_l^{(1)}(kr)$ is a spherical first-order Hankel function (I.1.171) which has the required asymptotic form (I.1.173); the derivative with respect to its argument ka is denoted with prime. As follows from Equations (I.1.54) and (I.1.71), $h_l^{(1)} = j_l + iy_l$, where y_l is a spherical Bessel function of second order, Equation (I.1.153). Hence, $|a_l| \leq 1$, $|b_l| \leq 1$, and the expansion coefficients can be expressed in terms of phase differences:

$$a_l = ie^{i\delta'_l}\sin\delta'_l, \qquad b_l = ie^{i\delta_l}\sin\delta_l, \qquad \text{where} \qquad \tan\delta'_l = \frac{(kaj_l(ka))'}{(kay_l(ka))'},$$

$$\tan\delta_l = \frac{j_l(ka)}{y_l(ka)}, -\frac{\pi}{2} \leq \delta_l, \delta'_l \leq \frac{\pi}{2}. \tag{30}$$

Using Equations (28) and (29), it is possible to calculate electric and magnetic fields in the entire space with the aid of Equation (11). In order to find the scattering cross section, it is necessary to know the field strengths at large distances from the scatter. Using Equation (I.1.173), we obtain the Debye potentials of the scattered wave in the wave zone:

$$u^s = u_e^s(\vartheta, \alpha)\frac{e^{ikr}}{r}, \qquad v^s = v_m^s(\vartheta, \alpha)\frac{e^{ikr}}{r}, \tag{31}$$

where

$$u_e^s = -k^{-2}\sum_{l=1}^{\infty}\frac{(2l+1)}{l(l+1)}a_l P_l^1(\cos\vartheta)\cos\alpha,$$

$$v_m^s = -k^{-2}\sum_{l=1}^{\infty}\frac{(2l+1)}{l(l+1)}b_l P_l^1(\cos\vartheta)\sin\alpha. \tag{32}$$

Therefore, the fields will be written in the form

$$\boldsymbol{E} = \boldsymbol{e}_x e^{ikz} + \boldsymbol{F}(\vartheta, \alpha)\frac{e^{ikr}}{r}, \qquad \boldsymbol{H} = \boldsymbol{e}_x e^{ikz} + \boldsymbol{e}_r \times \boldsymbol{F}(\vartheta, \alpha)\frac{e^{ikr}}{r}, \tag{33}$$

where the vector amplitude of the scattered wave

$$\boldsymbol{F} = ik\left(\frac{\partial u_e^s}{\partial\vartheta} + \frac{1}{\sin\vartheta}\frac{\partial v_m^s}{\partial\alpha}\right)\boldsymbol{e}_\vartheta + ik\left(\frac{1}{\sin\vartheta}\frac{\partial u_e^s}{\partial\alpha} - \frac{\partial v_m^s}{\partial\vartheta}\right)\boldsymbol{e}_\alpha \tag{33}$$

is referred to as the **amplitude of scattering** (see the statement of Problem 6.80).

The differential scattering cross section is expressed in terms of the scattering amplitude. Using Equations (6.65) and (6.66), we have

$$d\sigma_s(\vartheta, \alpha) = \boldsymbol{F}\cdot\boldsymbol{F}^* d\Omega = |\boldsymbol{F}(\vartheta, \alpha)|^2 d\Omega. \tag{34}$$

The total scattering cross section can be found by integrating Equation (34) over all angles. However, it is possible to avoid this complicated procedure by using the optical theorem (Problem 6.80). In our case, the absorption is absent, and the total cross section of all elastic and inelastic processes is

$$\sigma_s = \sigma_t = \frac{4\pi}{k}\Im\,(\boldsymbol{e}_x \cdot \boldsymbol{F}(0,\alpha)), \tag{35}$$

where in the argument of the scattering amplitude we indicate the angles determining the direction of the wave vector of the incident wave, and $\boldsymbol{e}_x$ is the vector amplitude of the incident wave. Using Equation (33) we have

$$\boldsymbol{e}_x \cdot \boldsymbol{F}(\vartheta,\alpha) = ik\left(\frac{\partial u_e^s}{\partial\vartheta} + \frac{1}{\sin\vartheta}\frac{\partial v_m^s}{\partial\alpha}\right)\cos\vartheta\cos\alpha - ik$$

$$\left(\frac{1}{\sin\vartheta}\frac{\partial u_e^s}{\partial\alpha} - \frac{\partial v_m^s}{\partial\vartheta}\right)\sin\alpha. \tag{36}$$

In the limit $\vartheta \to 0$, we will use Equations (I.1.189) and (I.1.186):

$$\left.\frac{P_l^1}{\sin\vartheta}\right|_{\vartheta\to 0} = P_l'(x)|_{x\to 1} = \frac{1}{2}\,l(l+1),$$

$$\left.\frac{dP_l^1}{d\vartheta}\right|_{\vartheta\to 0} = l(l+1) - P_l'(x)|_{x\to 1} = \frac{1}{2}\,l(l+1). \tag{37}$$

Finally, we have:
(38)

$$\sigma_s = \frac{2\pi}{k^2}\sum_{l=1}^{\infty}(2l+1)(\sin^2\delta_l' + \sin^2\delta_l) = \frac{2\pi}{k^2}\sum_{l=1}^{\infty}(2l+1)(|a_l|^2 + |b_l|^2).$$

6.75 In order to find the solution inside the sphere, it is necessary to use the functions $j_l(k_2 r)$, finite at $r \to 0$, where $k_2 = \omega\sqrt{\varepsilon_2}/c$, ε_2 is the dielectric permittivity inside the sphere. The solution for the outside region will contain the functions $j_l(k_1 r)$ and $h_l^{(1)}(k_1 r)$, where $k_1 = \omega\sqrt{\varepsilon_1}/c$. The quantities ε_1 and ε_2 will enter Equation (11) in the way which can easily be found with an aid of Maxwell's equations. The boundary condition (27) of Problem 6.74 must be replaced by the requirement that the tangential components $E_\vartheta, H_\vartheta, E_\alpha$ and H_α are continuous on the surface of the sphere. This results in the continuity of the quantities

$$\varepsilon u, \quad v, \quad \frac{\partial}{\partial r}(ru), \quad \frac{\partial}{\partial r}(rv) \quad \text{at} \quad r = a. \tag{1}$$

The coefficients which determine the scattered wave outside the sphere are

$$a_l = \frac{\varepsilon_2 j_l(k_2 a)(k_1 a j_l(k_1 a))' - \varepsilon_1 j_l(k_1 a)(k_2 a j_l(k_2 a))'}{\varepsilon_1 h_l^{(1)}(k_1 a)(k_2 a j_l(k_2 a))' - \varepsilon_2 j_l(k_2 a)(k_1 a h_l^{(1)}(k_1 a))'},$$

(2) $$b_l = \frac{j_l(k_2a)(k_1aj_l(k_1a))' - j_l(k_1a)(k_2aj_l(k_2a))'}{h_l^{(1)}(k_1a)(k_2aj_l(k_2a))' - j_l(k_2a)(k_1ah_l^{(1)}(k_1a))'}.$$

6.76 Just as in Problem 6.69 it is necessary to consider the radiation by the induced electric and magnetic moments, $\boldsymbol{p}$ and $\boldsymbol{m}$. Let us use the system of coordinates shown in Figure 6.16. The vector $\boldsymbol{k}$ of the primary wave lies in the xz-plane. Consider two polarizations of the incident waves: (i) the vector $\boldsymbol{E}_0$ lies in the plane of incidence (xz-plane); (ii) the vector $\boldsymbol{E}_0$ is normal to the plane of incidence.

In case (i), the component of the external electric field, which is parallel to the plane of disc is given by $E_{0\parallel} = -E_{0x} = E_0 \cos\alpha$; and the perpendicular component is given by $E_{0\perp} = -E_{0z} = E_0 \sin\alpha$. In our approximation ($a \ll \lambda$), the electric moment $\boldsymbol{p}$ can be calculated as the static moment of the conducting disc in a uniform electric field.

According to the results of Problems 2.38 and 2.40, the longitudinal polarizability of the disc is $\beta_{e\parallel} = 4a^3/3\pi$, and the transverse polarizability is $\beta_{e\perp} = 0$. Hence,

$$p_x = \beta_{e\parallel} E_{0x} = -\frac{4a^3}{3\pi} E_0 \cos\alpha, \qquad p_y = p_z = 0.$$

The magnetic field has only the longitudinal component. However, the magnetic longitudinal polarizability of the disc is zero (Problem 4.41), and therefore $\boldsymbol{m} = 0$. The differential scattering cross section is given by

(1) $$d\sigma_s = \frac{16a^6\,\omega^4}{9\pi^2c^4}\cos^2\alpha\,(1 - \sin^2\vartheta\cos^2\varphi)\,d\Omega.$$

The total scattering cross section is

(2) $$\sigma_s = \frac{128a^6\,\omega^4}{27\pi c^4}\cos^2\alpha.$$

In case (ii), we have

(3)
$$\begin{aligned}
p_y &= \frac{4a^3}{3\pi}E_0, \qquad p_x = p_z = 0,\ m_z = \frac{2a^3}{3\pi}E_0\sin\alpha,\ m_x = m_y = 0;\\
d\sigma_s &= \frac{16a^6\,\omega^4}{9\pi^2c^4}\left[1 + \sin^2\vartheta\left(\tfrac{1}{4}\sin^2\alpha - \sin^2\varphi\right) + \sin\vartheta\sin\alpha\cos\varphi\right]d\Omega,\\
\sigma_s &= \frac{128a^6\,\omega^4}{27\pi c^4}\left(1 + \tfrac{1}{4}\sin^2\alpha\right).
\end{aligned}$$

For the unpolarized wave, using Equations (1), (2), and (3), we find

(4)
$$\begin{aligned}
d\sigma_s &= \frac{8a^6\,\omega^4}{9\pi^2c^4}\left[1 + \sin^2\vartheta\left(1 - \tfrac{1}{4}\sin^2\alpha - \sin^2\alpha\cos^2\varphi\right)\right.\\
&\quad \left. + \cos^2\alpha + \sin\vartheta\sin\alpha\cos\varphi\right]d\Omega,\\
\sigma_s &= \frac{128a^6\,\omega^4}{27\pi c^4}\left(1 - \tfrac{3}{8}\sin\alpha\right).
\end{aligned}$$

6.77

$$d\sigma_s = \frac{a^4h^2\,\omega^4(\varepsilon-1)^2}{18c^4\varepsilon^2}(1+\cos^2\vartheta)\,d\Omega, \qquad \sigma_s = \frac{8\pi a^4h^2\,\omega^4(\varepsilon-1)^2}{27c^4\varepsilon^2},$$

where ϑ is the scattering angle (Figure 6.20).

6.78 We take the coordinate system shown in Figure 6.21. The vector $\boldsymbol{k}$ of the primary wave lies in the xz-plane. The cylinder can be approximated by a prolate ellipsoid of revolution with semiaxes a and h. As follows from the solutions of Problems 2.38, 2.39, and 4.41, that the longitudinal electric polarizability of a very prolate ellipsoid of revolution is larger than its transverse electric and magnetic polarizabilities by a factor of $\sim h/a$. Therefore, the scattering cross section is essentially dependent of whether there is the longitudinal component of the electric field in the incident wave.

If this component is appreciable, then the secondary radiation will be due to the z-component of the electric dipole moment. The remaining components of the electric moment as well as the magnetic moment can be neglected. Assuming that $\boldsymbol{E}_0$ lies in the xz-plane, we have

$$d\sigma_s = \frac{\omega^4h^6}{9c^4\ln^2(h/a)}\sin^2\alpha\,\sin^2\vartheta\,d\Omega, \qquad \sigma_s = \frac{8\pi\,\omega^4h^6}{27c^4\ln^2(h/a)}\sin^2\alpha.$$

If the longitudinal component $\boldsymbol{E}_0$ is zero, scattering is due to the transverse component of the electric moment and the magnetic moment, which have

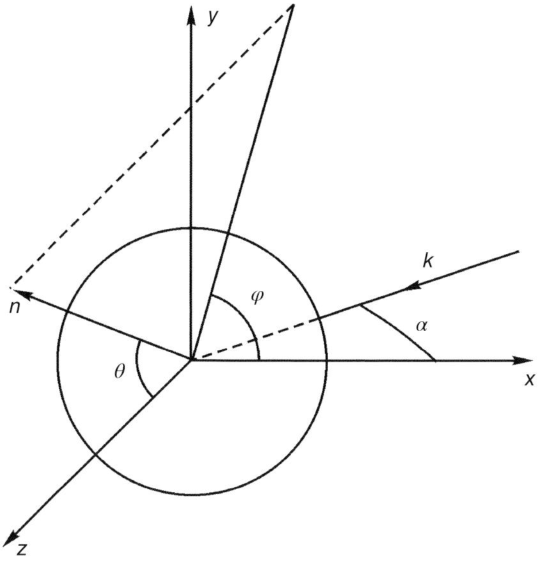

Figure 6.20 Scattering of a plane wave by a disc.

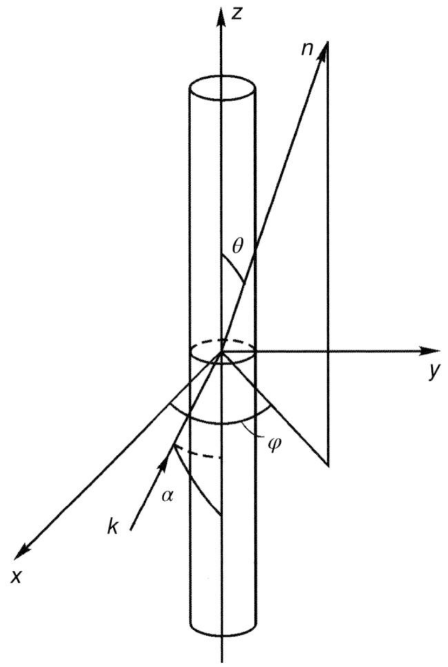

Figure 6.21 Scattering of a plane wave by a thin cylinder.

the same order of magnitude. Then,

$$\begin{aligned} d\sigma_s &= \frac{a^4h^2\,\omega^4}{9c^4}\left[(1+2n_x\sin\alpha)^2+3\cos^2\alpha\right.\\ &\left.+n_z^2(4-\sin^2\alpha)+8n_z\cos\alpha+2n_xn_z\sin 2\alpha\right]d\Omega,\end{aligned}$$

$$\sigma_s = \frac{40\pi a^4h^2\,\omega^4}{27c^4}\left(1+\tfrac{3}{5}\cos^2\alpha\right),$$

where n_i $(i = x, y, z)$ are the components of the unit vector in the direction of scattering.

The scattering cross section of the unpolarized wave is given by

$$\frac{d\sigma_s}{d\Omega} = \frac{\omega^4h^6}{18c^4\ln^2(h/a)}\sin^2\alpha\sin^2\vartheta,\qquad \sigma_s = \frac{4\pi\,\omega^4h^6}{27c^4\ln^2(h/a)}\sin^2\alpha.$$

6.79 When the vector $\boldsymbol{E}_0$ is polarized in the xz-plane, we will have (Figure 6.17):

$$d\sigma_{s\|} = \frac{4\,\omega^4a^4h^2}{9c^4}\left(\frac{\varepsilon-1}{\varepsilon+1}\right)^2\left[(1-n_x^2)\cos^2\alpha\right.$$

$$\left.+\frac{1}{4}(\varepsilon+1)^2(1-n_x^2)\sin^2\alpha-\frac{1}{2}(\varepsilon+1)n_xn_z\sin 2\alpha\right]d\Omega.$$

When the vector $\boldsymbol{E}_0$ is polarized perpendicular to the xz-plane, then:

$$d\sigma_{s\perp} = \frac{4\,\omega^4a^4h^2}{9c^4}\left(\frac{\varepsilon-1}{\varepsilon+1}\right)^2(1-\sin^2\vartheta\sin^2\varphi)\,d\Omega.$$

6.80 We begin from the relation

$$\sigma_a = -\frac{1}{E_0^2}\Re\int (\boldsymbol{E}\times\boldsymbol{H}^*)\cdot\boldsymbol{n}r^2\,d\Omega, \tag{1}$$

where $\boldsymbol{n} = \boldsymbol{r}/r$, σ_a is the absorption cross section, and the integration is over the surface of a large sphere surrounding the scatter. Equation (1) expresses the fact that the absorption cross section is proportional to the energy flux through the surface of the sphere in the inward direction.

Substituting the expression for $\boldsymbol{E}$, given in the statement of the problem, into Equation (1) and using the expression

$$\boldsymbol{H} = E_0\left\{(\boldsymbol{n}_0\times\boldsymbol{e})\,e^{ikz} + [\boldsymbol{n}\times\boldsymbol{F}(\boldsymbol{n})]\frac{e^{ikr}}{r}\right\},$$

as well as the condition $\boldsymbol{n}\cdot\boldsymbol{F}(\boldsymbol{n}) = 0$, we obtain

$$\frac{1}{E_0^2}\Re(\boldsymbol{E}\times\boldsymbol{H}^*)\cdot\boldsymbol{n} = (\boldsymbol{n}_0\cdot\boldsymbol{n}) + \frac{|\boldsymbol{F}|^2}{r^2} + \frac{1}{2}[(\boldsymbol{e}\cdot\boldsymbol{F}) + (\boldsymbol{n}_0\cdot\boldsymbol{n})(\boldsymbol{e}\cdot\boldsymbol{F})$$

$$-(\boldsymbol{e}\cdot\boldsymbol{n})(\boldsymbol{n}_0\cdot\boldsymbol{F})]\frac{e^{ik(r-z)}}{r}$$

$$+\frac{1}{2}[(\boldsymbol{e}^*\cdot\boldsymbol{F}^*) + (\boldsymbol{n}_0\cdot\boldsymbol{n})(\boldsymbol{e}^*\cdot\boldsymbol{F}^*) - (\boldsymbol{e}^*\cdot\boldsymbol{n})(\boldsymbol{n}_0\cdot\boldsymbol{F}^*)]\frac{e^{-ik(r-z)}}{r}. \tag{2}$$

The integral of the first term over the angles is zero, and the integral of the second term is equal to the total scattering cross section σ_s. The integrals of the remaining terms are transformed by integrating by parts:

$$\frac{1}{r}\int(\boldsymbol{n}_0\cdot\boldsymbol{n})(\boldsymbol{e}\cdot\boldsymbol{F})e^{ik(r-z)}r^2\,d\Omega = \frac{1}{ik}\int_0^{2\pi}d\varphi\left\{[(\boldsymbol{n}_0\cdot\boldsymbol{n})(\boldsymbol{e}\cdot\boldsymbol{F})e^{ikr(1-\cos\vartheta)}]\Big|_{\vartheta=0}^{\vartheta=\pi}\right.$$

$$\left.-\int_0^{\pi}e^{ikr(1-\cos\vartheta)}\frac{\partial}{\partial\cos\vartheta}(\boldsymbol{n}_0\cdot\boldsymbol{n})(\boldsymbol{e}\cdot\boldsymbol{F})\,d\cos\vartheta\right\}.$$

The last integral, when repeatedly integrated by parts, will give the terms that are proportional to $1/r$. Therefore, it can be neglected. The term with an oscillating factor e^{2ikr} can also be neglected, since it does not contribute to the total energy flux. In order to show this, we note that the concept of the strictly monochromatic wave is an idealization. In reality, any "monochromatic" wave is a superposition of harmonics whose frequencies lie within a more or less narrow interval $\Delta\omega$. The averaging of the factor e^{2ikr} over any such interval will give zero, since r is very large. Hence, we have

$$\frac{1}{r}\int(\boldsymbol{n}_0\cdot\boldsymbol{n})(\boldsymbol{e}\cdot\boldsymbol{F})e^{ik(r-z)}r^2\,d\Omega = \frac{2\pi i}{k}[\boldsymbol{e}\cdot\boldsymbol{F}(\boldsymbol{n}_0)].$$

The integrals of the remaining terms are calculated in a similar way. The terms containing the factors $(\boldsymbol{e}\cdot\boldsymbol{n})$ and $(\boldsymbol{e}^*\cdot\boldsymbol{n})$, when integrated, will not contribute to the final result because $(\boldsymbol{e} \cdot \boldsymbol{n}_0) = 0$. Substitution of the evaluated integrals into Equation (1) gives the final expression

$$\sigma_t = \frac{4\pi}{k} \Im[\boldsymbol{e} \cdot \boldsymbol{F}(\boldsymbol{n}_0)]. \tag{3}$$

The optical theorem (Equation (3)) admits of simple physical interpretation: the total cross section is a measure of the attenuation of the primary wave. This attenuation results from the interference between the incident wave and that part of the scattered wave which has the same polarization and propagation direction as the incident wave. Thus, the total scattering cross section depends on the forward scattering amplitude.

6.81 The scattered wave is due to the electric and magnetic dipole Moments, which are induced by the incident wave. The scattering amplitude $\boldsymbol{F}(\boldsymbol{n})$ (Problem 6.80) is determined by formulas of Section I.5.2 (radiation by electric and magnetic dipoles).

The final result is

$$\sigma_a = \frac{4\pi\,\omega}{c}(\beta_e'' + \beta_m'').$$

6.82 $\sigma_a = 6\pi b^2 \zeta'$.

6.83 The force is in the direction along the wavevector of the incident wave and is given by

$$\overline{F} = \frac{\overline{\gamma}_0}{c}\left[\sigma_a + \int (1 - \cos\vartheta)\left(\frac{d\sigma_s(\vartheta,\varphi)}{d\Omega}\right) d\Omega\right],$$

where $\overline{\gamma}_0$ is the average energy flux density in the incident wave, and the integration is carried out over the entire solid angle.

6.84 For a perfectly conducting sphere,

$$\overline{F} = \frac{43 a^6\,\omega^4}{96 c^4} E_0^2;$$

and for a dielectric sphere

$$\overline{F} = \frac{a^6\,\omega^4}{3c^4}\left(\frac{\varepsilon - 1}{\varepsilon + 2}\right)^2 E_0^2.$$

6.85 The diffraction formula (6.80) can be used. The plane in which the screen is located can be taken as the plane of integration. Then on the surface of integration

$$u = A\frac{e^{ikR_1}}{R_1}, \qquad dS_n = 3\pi r\,dr \cos(R_1, z) = 2\pi \frac{z_1 r}{R_1}\,dr,$$

where A is a constant. By substituting these expressions in Equation (6.80) and introducing a new integration variable $\rho = R + R_1$, we obtain

$$u_P(z) = -ikAz_1 \int_a^\infty \frac{e^{ik(R+R_1)}}{RR_1^2} r\,dr = -ikAz_1 \int_{\rho_0}^\infty \frac{e^{ik\rho}}{\rho R_1(\rho)}\,d\rho, \tag{1}$$

where

$$\rho_0 = \sqrt{a^2 + z^2} + \sqrt{a^2 + z_1^2}.$$

Integration by parts leads to the representation of Equation (1) in the form of a series over inverse powers of $k\rho$. The condition $\lambda \ll a$ allows one to omit all terms but the first one. Thus, we have

$$u_P(z) = u_0 \frac{z_1 e^{ik\sqrt{a^2+z^2}}}{\rho_0},$$

where $u_0 = A \exp(ik\sqrt{a^2 + z_1^2}) \Big/ \sqrt{a^2 + z_1^2}$ is the amplitude of the incident wave at the surface of the screen.

Hence, for the intensity $I \sim |u_P|^2$ we have

(2) $$I(z) = I_0 \frac{z_1^2}{\left(\sqrt{a^2 + z_1^2} + \sqrt{a^2 + z^2}\right)^2}.$$

At a point that is symmetrical with respect to the screen ($z_1 = z$), the intensity is

$$I(z) = \frac{I_0}{4} \frac{z^2}{a^2 + z^2}.$$

Thus, at the symmetrical point behind the screen, which is not too close to the screen, there is a bright spot. This result, which is in conflict with the hypothesis of rectilinear propagation of light, was theoretically predicted by Poisson in 1818. He used it as an argument against Fresnel's theory of diffraction and the wave theory of light as a whole. However, the experiments carried out by Arago and Fresnel confirmed the presence of the spot which appears due to the symmetry of the screen. The waves, which bend the edges of the disc, arrive at the axial point in phase. Evidently, all points lying on the axial line have this property, and the intensity of light at such points will be much greater than at neighboring points beyond the Oz-axis.

6.86 By using the Babinet principle (Equation (6.93)), we obtain for $z = z_1 \gg a$,

$$I = I_0 \sin^2 \frac{ka^2}{2z},$$

where I_0 is the intensity of the primary wave at the edge of the aperture.

6.87 At $z \gg a$, we have $I = 4I_0 \sin^2(ka^2/4z)$.

The intensity of light along the axial line of a circular aperture oscillates infinitely, decreasing to zero as $z \to \infty$. The attenuation of intensity along the axis is associated with the fact that the parallel beam is divergent due to diffraction at the aperture, and the energy flux through the aperture distributes over larger and larger area with increasing z.

6.88 Using Equation (6.89) for Fraunhofer diffraction, we have

$$dI = I_0 \frac{[aJ_1(ak\alpha) - bJ_1(bk\alpha)]^2}{\alpha^2} d\Omega,$$

where α is the angle of diffraction, and I_0 is the intensity of the incident light. For a circular aperture,

$$dI = I_0' \frac{J_1^2(ak\alpha)}{\pi\alpha^2} \, d\Omega,$$

where $I_0' \sim \pi a^2 |u_0|^2$ is the total intensity of the light incident to the aperture.

6.89 The diffracted wave can be described by the function

$$u_P = \frac{u_0 e^{ikR_0}}{2\pi i R_0} \int \exp(i\boldsymbol{q}_\parallel \cdot \boldsymbol{r}) \, dS_n,$$

where $\boldsymbol{k}' - \boldsymbol{k} = \boldsymbol{q}$, $\boldsymbol{q}_\parallel$ and $\boldsymbol{q}_\perp$ are the components of $\boldsymbol{q}$ in the plane of the screen and in the perpendicular direction, respectively.

For the integration over the aperture plane, we take the polar coordinates with the origin in the center of the aperture and the polar axis in the direction of $\boldsymbol{q}_\parallel$. This yields

$$u_P = \frac{u_0 e^{ikR_0} k \cos\theta}{2\pi i R_0} \int e^{-iq_\parallel r \cos\vartheta} r \, dr \, d\varphi,$$

where θ is the angle of incidence.

Using Equation (I.1.150) and the recurrent relations for the Bessel functions, we have

$$dI = |u_P|^2 R_0^2 \, d\Omega = I_0 \frac{J_1^2(q_\parallel a)}{\pi q_\parallel^2} \, d\Omega,$$

where $I_0 \sim |u_0|^2 \pi a^2 \cos\theta$ is the total intensity of the light incident to the aperture.

Assuming that the angle of diffraction α (i.e., the angle between $\boldsymbol{k}$ and $\boldsymbol{k}'$) is small and expressing $q_\parallel$ in terms of the diffraction angle α, the angle of incidence θ and the azimuthal angle α' between $\boldsymbol{q}$ and the plane of incidence, we have

$$q_\parallel = k\alpha\sqrt{1 - \sin^2\theta \cos^2\alpha'},$$

$$dI = I_0 \frac{J_1^2(ka\alpha\sqrt{1 - \sin^2\theta\cos^2\alpha'})}{\pi\alpha^2(1 - \sin^2\theta\cos^2\alpha')} \, d\Omega.$$

The latter formula is not valid at grazing angles ($\theta \approx \pi/2$).

6.90 Since the diffraction cross section for an infinitely long slit diverges, we calculate the cross section per unit length of the slit. Using Equation (6.87) for the diffraction at the rectangular aperture, we have

$$d\Sigma = \lim_{l_2\to\infty} \frac{d\sigma}{2l_2} = \left(\frac{4l_1 l_2}{\lambda}\right)^2 \left(\frac{\sin kl_1\theta_x}{kl_1\theta_x}\right)^2 \left(\lim_{l_2\to\infty} \frac{\sin^2 kl_2\theta_y}{2l_2(k\theta_y)^2}\right) d\theta_x d\theta_y.$$

To study this limit, we can use Equations (I.1.217) and (I.1.207). After integration over $d\theta_y$, we have

$$d\Sigma = \frac{4l_1^2}{\lambda}\left(\frac{\sin kl_1\theta_x}{kl_1\theta_x}\right)^2 d\theta_x.$$

The cross section has the dimension of length. The bending of rays in direction of the slit does not occur.

6.91

(1) $$d\Sigma = \frac{4l^2}{\lambda}\left(\frac{\sin kl\theta_x}{kl\theta_x}\right)^2\left(\frac{\sin Nkb\theta_x}{\sin kb\theta_x}\right)^2 d\theta_x, \qquad b = a + l.$$

At $N \gg 1$, we can write

(2) $$d\Sigma = Nd\Sigma^{(1)}\left(\frac{\sin Nkb\theta_x}{\sqrt{N}\sin kb\theta_x}\right)^2,$$

where $d\Sigma^{(1)}$ is the diffraction cross section at a single slit, which has been found in Problem 6.90. We consider the angles of diffraction that satisfy the condition $kb\theta_x = n\pi + \alpha,\ \alpha \ll 1$, where n is integer. For such angles, we have

(3) $$\left(\frac{\sin Nkb\theta_x}{\sqrt{N}\sin kb\theta_x}\right)^2 = \frac{\sin^2(N\alpha)}{N\alpha^2} \approx \pi\delta(\alpha) = \frac{\pi}{kb}\delta(\theta_x - n\pi/kb).$$

At $N \gg 1$, we can use Equation (I.1.217). It follows from Equation (3) that the cross section has sharp maxima at $\theta_x \approx n\pi/kb$. Using the delta-function (3) for integrating over the angle, we calculate the contribution from such a maximum into the total diffraction cross section:

(4) $$\Sigma^{(n)} = N\int \frac{d\Sigma^{(1)}}{d\theta_x}\frac{1}{kb}\delta(\theta_x - n\pi/kb)\,d\theta_x = N\frac{2b}{\pi^2 l}\frac{\sin^2(\pi nl/b)}{n^2}.$$

The total diffraction cross section is equal to the sum of cross sections over all maxima.

6.92 Application of the Kirchhof's formula in the vector form (Equation (6.77)) yields the following expressions for the radiation field:

$$E_\vartheta = H_\alpha = -ikabE_0\frac{e^{ikR}}{\pi R}\left(\frac{\sin k_x a}{k_x a}\right)\left(\frac{\sin k_y b}{k_y b}\right)(1 + \cos\vartheta)\sin\alpha,$$

$$E_\alpha = -H_\vartheta = -ikabE_0\frac{e^{ikR}}{\pi R}\left(\frac{\sin k_x a}{k_x a}\right)\left(\frac{\sin k_y b}{k_y b}\right)(1 + \cos\vartheta)\cos\alpha,$$

where ϑ and α are the angles in the spherical coordinate system with the polar axis perpendicular to the plane of the aperture; $k_x = k\sin\vartheta\cos\alpha$ and $k_y = k\sin\vartheta\sin\alpha$ are the projections of the wave vector of the diffracted wave.

The angular distribution of the radiation is

$$dI = I_0\frac{abk^2}{4\pi^2}\left(\frac{\sin k_x a}{k_x a}\right)^2\left(\frac{\sin k_y b}{k_y b}\right)^2(1 + \cos\vartheta)^2\,d\Omega,$$

where $I_0 = cabE_0^2/2\pi$ is the intensity of the wave incident to the aperture.

6.93 If the axes Ox, Oy, and Oz are in the direction of the vectors $\boldsymbol{E}_0$, $\boldsymbol{H}_0$, and $\boldsymbol{k}$, respectively, the radiated field will be given by

$$E_\vartheta = H_\alpha = -\frac{ika^2E_0}{2}\frac{e^{ikR}}{R}\left(\frac{J_1(ka\sin\vartheta)}{ka\sin\vartheta}\right)(1+\cos\vartheta)\cos\alpha,$$

$$E_\alpha = -H_\vartheta = \frac{ika^2E_0}{2}\frac{e^{ikR}}{R}\left(\frac{J_1(ka\sin\vartheta)}{ka\sin\vartheta}\right)(1+\cos\vartheta)\sin\alpha,$$

$$dI = \frac{1}{4}I_0\left(\frac{J_1(ka\sin\vartheta)}{ka\sin\vartheta}\right)^2(1+\cos\vartheta)^2\,d\Omega,$$

where $I_0 = ca^2E_0^2/8$ is the intensity of the wave incident to the aperture.

At $\vartheta < 1$, we have

$$dI = I_0\frac{J_1(ka\vartheta)}{\pi\vartheta^2}\,d\Omega.$$

This result has been obtained in Problem 6.88 with the aid of the scalar diffraction formula.

6.94 Let us express the electromagnetic field in terms of the electric Hertz vector $\boldsymbol{Z}(x,z)$ using Equation (5.40),

$$\boldsymbol{E} = \nabla\times[\nabla\times\boldsymbol{Z}], \qquad \boldsymbol{H} = -ik\nabla\times\boldsymbol{Z}. \tag{1}$$

The incident wave corresponds to $\boldsymbol{Z}_0 = k^{-2}\boldsymbol{E}_0\exp(i\boldsymbol{k}_0\cdot\boldsymbol{r})$. The Hertz vector $\boldsymbol{Z}$ in the right-hand side of the Kirchhoff's formula (6.78) can be replaced by its undisturbed value, $\boldsymbol{Z}_0$. Then

$$\boldsymbol{Z}(\boldsymbol{r}) = \frac{1}{4\pi}\int_{-l}^{l}\left(\frac{e^{ikR}}{R}(\boldsymbol{n}\cdot\nabla)\boldsymbol{Z}_0 - \boldsymbol{Z}_0(\boldsymbol{n}\cdot\nabla)\frac{e^{ikR}}{R}\right)dx', \tag{2}$$

where $\boldsymbol{n}\cdot\nabla = -\partial/\partial z$, $R = \sqrt{(x-x')^2+z^2}$ is the distance between dx' and the point of observation given by the position vector $\boldsymbol{r} = (x,0,z)$. In the wave zone $kR \gg 1$ and $R \gg l$, so that only the exponents in Equation (2) should be differentiated:

$$\boldsymbol{Z}(\boldsymbol{r}) = -\frac{i}{4\pi k}\boldsymbol{E}_0\int_{-l}^{l}\frac{e^{ik(R+\sin\theta_0 x')}}{R}\left(\cos\theta_0 - \frac{z}{R}\right)dx'. \tag{3}$$

In the denominators, we can set $R \approx r$, and in the exponent $R + \sin\theta_0 x' \approx r + (\sin\theta_0 - \sin\theta)x'$, where $\sin\theta = x/r$ and $\cos\theta = z/r$. The integration yields

$$\boldsymbol{Z} = \boldsymbol{e}_0 Z(r,\theta), \qquad Z(r,\theta) = -E_0\frac{il}{2\pi k}\frac{e^{ikr}}{r}\frac{\sin\chi}{\chi}(\cos\theta_0 - \cos\theta),$$

$$\chi = kl(\sin\theta_0 - \sin\theta), \tag{4}$$

where $\boldsymbol{e}_0$ is the unit polarization vector of the wave incident to the screen.

In order to evaluate the field vectors in the wave zone with the aid of Equation (1), again only the exponents should be differentiated. Finally, we have

(5) $$\boldsymbol{E}(\boldsymbol{r}) = -k^2 \boldsymbol{e}_r \times [\boldsymbol{e}_r \times \boldsymbol{e}_0] Z(r, \theta).$$

When the vector $\boldsymbol{E}_0$ lies in the plane of incidence, that is, $\boldsymbol{e}_0 = \boldsymbol{e}_x \cos\theta_0 - \boldsymbol{e}_z \sin\theta_0$, we can expand it over the spherical unit vectors $\boldsymbol{e}_r = \boldsymbol{e}_x \sin\theta + \boldsymbol{e}_z \cos\theta$, $\boldsymbol{e}_\theta = \boldsymbol{e}_x \cos\theta - \boldsymbol{e}_z \sin\theta$ and $\boldsymbol{e}_0 = \boldsymbol{e}_\theta \cos(\theta - \theta_0) + \boldsymbol{e}_r \sin(\theta - \theta_0)$. Substituting this expansion in Equation (5), we have

(6) $$\boldsymbol{E}(\boldsymbol{r}) = k^2 Z(r, \theta) \cos(\theta - \theta_0) \boldsymbol{e}_\theta.$$

If the vector $\boldsymbol{E}_0$ is perpendicular to the plane of incidence, then $\boldsymbol{e}_0 = \boldsymbol{e}_y$. From Equation (5) we immediately have

(7) $$\boldsymbol{E}(\boldsymbol{r}) = k^2 Z(r, \theta) \boldsymbol{e}_0.$$

6.95

$$\boldsymbol{E} = k^2 Z(r, \theta, \phi) \boldsymbol{q}(\theta, \phi), \qquad Z = -E_0 \frac{ial}{\pi k} \frac{e^{ikr}}{r} \frac{\sin\chi}{\chi} \frac{\sin\zeta}{\zeta} (\cos\theta_0 - \cos\theta),$$

$$\chi = kl(\sin\theta_0 - \sin\theta\cos\phi), \qquad \zeta = ka\sin\theta\sin\phi.$$

If $\boldsymbol{E}_0$ lies in the plane of incidence, then $\boldsymbol{q} = \boldsymbol{e}_\theta(\sin\theta_0 \sin\theta + \cos\theta_0 \cos\theta \cos\varphi) - \boldsymbol{e}_\phi \cos\theta_0 \sin\phi$. If $\boldsymbol{E}_0$ is perpendicular to the plane of incidence, then $\boldsymbol{q} = \boldsymbol{e}_\theta \cos\theta \sin\phi + \boldsymbol{e}_\phi \cos\phi$.

6.96 See the paper by Niz'ev (2002).

6.97

$$u = \frac{C}{z} \exp\left[ik\left(z + \frac{x^2 + y^2}{2z}\right)\right], \qquad C = \frac{A_0 k a^2}{2\pi i}.$$

This solution is accurate only near the Oz-axis, at $|x|, |y| \ll z$ and $z \gg \lambda$.

6.98

$$A(\rho, z) = \frac{A_0}{\sqrt{1 + L^2}} \exp\left[-\frac{\rho^2}{a^2(1 + L^2)} + i\frac{\rho^2 L}{a^2(1 + L^2)} - i\arctan L\right], \quad L = \frac{2z}{ka^2}.$$

The diameter of the beam increases as $a^2(z) = a^2(1 + L^2)$, while the amplitude on the axis of the beam ($\rho = 0$) decreases as $(1 + L^2)^{-1/2}$. At $L(z) \gg 1$, we have $a(z) \approx 2z/ka$, and the beam becomes a cone with an apex angle of the order $a(z)/z \approx \lambda/\pi a$.

6.99 The diameter of the beam varies according to $a^2(z) = a^2[(1 - z/R)^2 + L^2(z)]$. Initially, with increasing z, the diameter decreases (the beam is focusing) up to the value $a_{min} = a/\sqrt{1 + \sigma^2}$, with $\sigma = ka^2/2R$. At $L > \sigma/(1 + \sigma^2)$, the beam width increases monotonically, and at $L \gg 1$ according to the law $a(z) \propto z$. Compared with the preceding case, a new physical phenomenon takes place, which is diffraction self-focusing of the beam.

6.100 First of all, it is necessary that the inequality $\omega \gg \omega_0$ be fulfilled. However, this is not sufficient. Consider the case when the coherence length l is large compared with the sizes L of the body. Then for sufficiently small scattering $\vartheta < \lambda/L$, we have $qL \ll 1$, the exponentials in Equations (6.99) or (6.100) are close to unity, and $\int n \exp[i\mathbf{q}\cdot\boldsymbol{r}]\,dV = NZ$. If $\lambda \geq L$, this relation holds for all angles. We have, for example, from Equation (6.103),

$$d\sigma = r_0^2 N^2 Z^2 \sin^2\theta\, d\Omega. \tag{1}$$

The latter relation corresponds to the coherent Thomson scattering by all NZ charges in the body. If however, the coherence length is smaller than the interatomic distance but larger than atomic size, then at $\vartheta < \lambda/l$ the contributions from Z electrons of an atom will only be added coherently, and in Equation (1) we should replace $N^2Z^2 \to NZ^2$. For large values of the angle, the cross section will decrease sharply owing to the rapidly oscillating factor $\exp[i\boldsymbol{q}\cdot\boldsymbol{r}]$ in the integrand.

6.101 The electron number density in the gas can be written as a sum of terms relating to individual atoms, $n(\boldsymbol{r}) = \sum_{a=1}^{Z} n_a(\boldsymbol{r} - \boldsymbol{R}_a)$, where $\boldsymbol{R}_a$ characterizes an instantaneous position of atom a. Then, we have

$$\left|\int n(\boldsymbol{r})\exp[i\boldsymbol{q}\cdot\boldsymbol{r}]\,dV\right|^2 = \left|\sum_a \exp[i\boldsymbol{q}\cdot\boldsymbol{R}_a]\int n_a(\boldsymbol{r}')\exp[i\boldsymbol{q}\cdot\boldsymbol{r}']\,dV'\right|^2$$

$$= |F_a(\boldsymbol{q})|^2\left|\sum_a \exp[i\boldsymbol{q}\cdot\boldsymbol{R}_a]\right|^2, \tag{1}$$

where $\boldsymbol{r}' = \boldsymbol{r} - \boldsymbol{R}_a$, and $F_a(\boldsymbol{q})$ is the atomic form factor (Equation (6.105)). The averaging in Equation (1) must be performed over all positions $\boldsymbol{R}_a$. Since the atoms in the gas are distributed randomly, then we have $\overline{\left|\sum_a \exp(i\boldsymbol{q}\cdot\boldsymbol{R}_a)\right|^2} = N$. Altogether, for unpolarized radiation we get

$$d\sigma = \frac{1}{2} r_0^2(1+\cos^2\vartheta)|F_a(q)|^2 N\, d\Omega. \tag{2}$$

For the given density $n_a(r)$, the evaluation of the form factor is elementary and gives

$$F_a(q) = \frac{8\pi}{a(1/a^2+q^2)^2}. \tag{3}$$

Finally,

$$d\sigma(\vartheta) = \frac{32\pi^2 r_0^2 n_{0a} N}{a^2}\,\frac{1+\cos^2\vartheta}{\left[1/a^2 + (4\pi/\lambda)^2\sin^2(\vartheta/2)\right]^4}\,d\Omega. \tag{4}$$

If the cross section (Equation (2)) is measured experimentally, we can obtain the absolute magnitude of the form factor. In order to find the electron distribution, it is necessary, in general, to know the phase of the form factor.

6.102

$$d\sigma = Nr_0^2 \frac{1+\cos^2\vartheta}{2} |F_a(q)|^2 \cdot 2\left(1 + \frac{\sin qR}{qR}\right) d\Omega.$$

The cross section differs from that for scattering by isolated atoms by the structure factor $2(1 + \sin qR/qR)$, which depends on relative positions of atoms in a molecule.

6.103

$$\overline{d\sigma} = Nr_0^2(1+\cos^2\vartheta)|F_a(q)|^2 \frac{1}{b\sqrt{\pi}} \int_{-\infty}^{\infty} \left(1 + \frac{\sin q(R_0+x)}{q(R_0+x)}\right) \exp\left[-\frac{x^2}{b^2}\right] dx.$$

The relative magnitude of $1/q$ and b is important. At $q \gg 1/b$, the rapidly oscillating term with $\sin q(R_0 + x)$ vanishes. Thermal motion reduces the structure effect under such conditions. At $q \ll 1/b$, the structure factor has the same form $1 + \sin qR_0/qR_0$, as for the case of nuclei at rest.

6.105 Let us direct the Ox-, Oy-, and Oz-axes along the edges L_1, L_2, and L_3 of the single crystal.

$$\int n(\boldsymbol{r}) \exp[i\boldsymbol{q}\cdot\boldsymbol{r}]\, dV = F_a(\boldsymbol{q}) \sum_{\boldsymbol{R}} \exp[i\boldsymbol{q}\cdot\boldsymbol{R}]$$

$$= F_a(\boldsymbol{q}) \left(\sum_{n_1=0}^{N_1} \exp[iq_x a n_1]\right)\left(\sum_{n_2=0}^{N_2} \exp[iq_y a n_2]\right)\left(\sum_{n_3=0}^{N_3} \exp[iq_z a n_3]\right)$$

$$= F_a(\boldsymbol{q}) \frac{1-\exp[iq_x aN_1]}{1-\exp[iq_x a]} \cdot \frac{1-\exp[iq_y aN_2]}{1-\exp[iq_y a]} \cdot \frac{1-\exp[iq_z aN_3]}{1-\exp[iq_z a]},$$

where $N_1 = L_1/a$, $N_2 = L_2/a$, and $N_3 = L_3/a$ are the numbers of elementary cells along the edges L_1, L_2, and L_3, respectively. Evidently, $N = N_1N_2N_3$. Using Equation (6.104), we have

$$d\sigma = \frac{1}{2} r_0^2(1+\cos^2\vartheta)|F_a(\boldsymbol{q})|^2 \frac{\sin^2(q_x aN_1/2)}{\sin^2(q_x a/2)} \cdot \frac{\sin^2(q_y aN_2/2)}{\sin^2(q_y a/2)}$$

(1) $$\cdot \frac{\sin^2(q_z aN_3/2)}{\sin^2(q_z a/2)} d\Omega.$$

The positions of the main maxima are determined from the condition that the denominators vanish. It gives $q_x = 2\pi m_x/a$, $q_y = 2\pi m_y/a$, and $q_z = 2\pi m_z/a$, where m_x, m_y, and m_z are integers. The latter equations correspond to the von Laue equation written in projections, since the components of $\boldsymbol{g}$ are given by $\boldsymbol{g} = (m_x/a, m_y/a, m_z/a)$. In the maxima, the cross section is

(2) $$d\sigma = \frac{1}{2} r_0^2(1+\cos^2\vartheta)|F_a(2\pi\mathbf{g})|^2 \frac{(L_1L_2L_3)^2}{a^6} d\Omega.$$

It is proportional to the squared the volume of the crystal. The results of Problems (6.104)–(6.108) are valid only if the single crystal lies completely within the coherence volume (see Section 7.1 to learn on the coherence volume).

6.106

$$d\sigma = \frac{1}{2} r_0^2 (1 + \cos^2 \vartheta) |F_a(\boldsymbol{q})|^2 \frac{\sin^2(q_z a N_3/2)}{4 \sin^2(q_y a/2) \sin^2(q_z a/2)}$$

$$\times \left\{ \left[\frac{\sin(q_x a N_1/2)}{\sin(q_x a/2)} - \frac{\sin[(q_x + q_y) a N_1/2]}{\sin[(q_x + q_y) a/2]} \right]^2 \right.$$

$$\left. + 4 \sin^2(q_y a N_1/2) \frac{\sin(q_x a N_1/2) \sin[(q_x + q_y) N_1 a/2]}{\sin(q_x a/2) \sin[(q_x + q_y) a/2]} \right\},$$

where $N_1 = L_1/a$ and $N_3 = L_3/a$. The positions of the main maxima are expressed in terms of the von Laue condition: $\boldsymbol{q} = 2\pi \boldsymbol{g}$, where $\boldsymbol{g} = (m_x/a, m_y/a, m_z/a)$. In the maximum, the cross section is

$$d\sigma = \frac{r_0^2}{2} (1 + \cos^2 \vartheta) |F_a(2\pi \mathbf{g})|^2 \frac{(L_1^2 L_3)^2}{4a^6} \, d\Omega.$$

The angle ϑ_0 is connected with $q = 2\pi g$ through the equality $\sin(\vartheta_0/2) = \pi g/k$.

6.107 At $k \gg 1/a$, the diffraction picture is concentrated in the region of small angles, since, according to the von Laue equation, $k\vartheta = 2\pi g \sim 1/a$ and $\vartheta \sim 1/ak \ll 1$; in this case $q \ll k$.

Let us write

$$\boldsymbol{\kappa} = \boldsymbol{q} - 2\pi \boldsymbol{g} = \boldsymbol{k} - \boldsymbol{k}_*, \tag{1}$$

where $\boldsymbol{k}_* = \boldsymbol{k}_0 + 2\pi \boldsymbol{g}$ is the wavevector $\boldsymbol{k}$ of the scattered wave when the von Laue condition (6.110) is satisfied exactly. In the limit of a narrow diffraction maximum, $\boldsymbol{k}$ and $\boldsymbol{k}_*$ are close in direction. Their difference $\boldsymbol{\kappa}$ is, therefore, approximately perpendicular to each of these wave vectors. In particular, since the difference $\boldsymbol{\kappa}$ is directed along the Oz-axis, then $|\kappa_z| \ll |\kappa_x|, |\kappa_y|$. As a result, the ratio

$$\frac{\sin^2(q_z a N_3/2)}{\sin^2(q_z a/2)} = \frac{\sin^2(\kappa_z a N_3/2)}{\sin^2(\kappa_z a/2)},$$

which enters Equation (1) of Problem (6.105), is a much smoother function of κ_z than the first two ratios, and it can be replaced by its value N_3^2 in the maximum at $\kappa_z = 0$. The cross section takes the form ($\vartheta \ll 1$)

$$d\sigma = 4 r_0^2 |F_a(2\pi \mathbf{g})|^2 N_3^2 \frac{\sin^2(\kappa_x a N_1/2)}{(\kappa_x a/2)^2} \frac{\sin^2(\kappa_y a N_2/2)}{(\kappa_y a/2)^2} \, d\Omega, \tag{2}$$

which implies that the angular width of the main maximum is $\sim 1/kaN_1$ and $\sim 1/kaN_2$ in the Ox- and Oy-directions, respectively. If we write the solid angle element in the form $d\Omega = d\kappa_x \, d\kappa_y / k^2$ and integrate over κ_x and κ_y within infinite limits, we obtain

$$\sigma = 4 r_0^2 |F_a(2\pi \mathbf{g})|^2 \left(\frac{\pi}{ak} \right)^2 N_3^2 N_1 N_2. \tag{3}$$

The cross section depends on the longitudinal and transverse sizes in different ways. If these sizes are approximately the same, the total cross section is proportional to $V^{4/3}$ (V being the volume of the body), and the angular width is proportional to $(V^{4/3}/V^2)^{1/2} = 1/V^{1/3}$.

6.108

$$d\sigma = 32 r_0^2 (1 + \cos^2 \vartheta) |F_a(2\pi \mathbf{g})|^2 \frac{\sin^2(\kappa_x L_x/2)}{\kappa_x^2} \cdot \frac{\sin^2(\kappa_y L_y/2)}{\kappa_y^2} \cdot \frac{\sin^2(\kappa_z L_z/2)}{\kappa_z^2} \, d\Omega,$$

where

$$\kappa_x k_{\mathbf{g}x} + \kappa_y k_{\mathbf{g}y} + \kappa_z k_{\mathbf{g}z} = 0, \qquad \mathbf{k}_{\mathrm{g}} = \mathbf{k}_0 + 2\pi \mathbf{g}.$$

6.109

$$d\sigma = 8\pi r_0^2 (1 + \cos^2 \vartheta) \, |F_a(2\pi \mathbf{g})|^2 \, \frac{\sin \kappa R - R \cos \kappa R}{\kappa^6} \, d\Omega.$$

7
Coherence and Nonlinear Waves

7.1
Coherence and Interference

Correlation field tensors. Time and length of coherence. A majority of actual sources of electromagnetic waves consists of a great number of independent (incoherent) radiators emitting waves with random values of the initial phase and also with the polarizations of all kind. These are, for example, the thermal and luminescence sources of light, as well as the sources in which the electromagnetic waves are emitted by a beam of fast electrons at their deceleration in a material or magnetic field. A significantly higher degree of consistency is attained in the case of the antenna radiation and the light emission by quantum optical generators in which the stimulated emission of light plays a main role. However, there are also in these devices fluctuations in phase, amplitude, and polarization due to the spontaneous emission and the scattering by different fluctuating inhomogeneities.

Moreover, each independent emitter radiates nonmonochromatic waves. Owing to this fact, in many cases the field produced by actual sources has a very complicated structure: the field parameters vary in an unpredictable, arbitrary way in space and time. Such fields are referred to as **random**. We already considered the random fields in this manual (Sections I.2.3.5 and 4.3 and Appendix D). For the convenience of the reader, some information on the subject will be repeated below.

In order to describe a random field, it is necessary to set in advance some certain averaged quantities characterizing such a field. The averaging method and the method that is used to measure the corresponding quantity are closely related. As a rule, the measuring devices themselves produce the time averaging in the process of measuring. The time interval Δt of averaging should be taken in the way that it exceeds a period of any one of the harmonic components of a random field. In this case, the averaged value of the radiation field vanishes,

$$\overline{E_\alpha(\boldsymbol{r}, t)} = 0. \tag{7.1}$$

It is the second-rank correlation tensor that differs from zero:

$$J_{\alpha\beta}(\boldsymbol{r}_1, t_1; \boldsymbol{r}_2, t_2) = \overline{E_\alpha(\boldsymbol{r}_1, t_1)E^*_\beta(\boldsymbol{r}_2, t_2)}, \tag{7.2}$$

Electromagnetic Phenomena in Matter: Statistical and Quantum Approaches, First Edition. Igor N. Toptygin.

where E_α is a complex function (analytical signal) describing the field. For a completed description of a random field, one should in general prescribe an infinite set of the correlation tensors of higher ranks. However, for the solution of many problems the second-rank tensor is quite admissible. As for the averaged characteristics of the magnetic field, they can be expressed in terms of the corresponding quantities of the electric field with the aid of the Maxwell's equations. If the arguments are identical ($\boldsymbol{r}_1 = \boldsymbol{r}_2,\ t_1 = t_2,\ \alpha = \beta$), the tensor (7.2) is proportional to the energy which enters the α-component of the electric field.

To ensure that the averaged tensor (7.2) remain time dependent, the averaging must be performed, according to Equation (7.5), for fixed times t_1 and t_2:

$$J_{\alpha\beta}(\boldsymbol{r}_1, t_1; \boldsymbol{r}_2, t_2) = \frac{1}{\Delta t}\int_{-\Delta t/2}^{\Delta t/2} E_\alpha(\boldsymbol{r}_1, t_1 + \tau) E_\beta^*(\boldsymbol{r}_2, t_2 + \tau)\, d\tau. \tag{7.3}$$

Every so often, especially for the measurements in the optical range, the time averaging is performed in the process of measuring owing to inertia of the instrument. The response times of typical detectors of optical radiation are 10^{-12} to 10^{-10} s, whereas the periods of optical oscillations are of the order of $10^{-16} to 10^{-15}$ s.

In what follows, we will mostly be interested in stationary random fields, whose averaged characteristics are independent of time. This means that $J_{\alpha\beta}$ will depend only on the difference of times $\tau = t_1 - t_2$, but not on an origin,

$$J_{\alpha\beta}(\boldsymbol{r}_1, t_1; \boldsymbol{r}_2, t_2) = J_{\alpha\beta}(\boldsymbol{r}_1, \boldsymbol{r}_2, \tau). \tag{7.4}$$

If a random field is also homogeneous in space, that is, if its mean (not instantaneous) characteristics are independent of coordinates, the radius-vectors $\boldsymbol{r}_1$, $\boldsymbol{r}_2$ will appear in $J_{\alpha\beta}$ as the difference $\boldsymbol{r} = \boldsymbol{r}_1 - \boldsymbol{r}_2$ only. The field of thermal radiation in a closed cavity, for example, acquires such properties. However, the field produced by a source of finite dimensions in free space is inhomogeneous.

For random fields, the correlation tensor $J_{\alpha\beta}(\boldsymbol{r}_1, \boldsymbol{r}_2, \tau)$ as a rule decreases sufficiently fast with increasing delay time τ. Let, for example, the radiation field be produced by thermal sources, namely, atoms that are excited due to random collisions with one another. Each atom emits a nonmonochromatic wave of finite duration τ_c (wave packet) at random moments of time, and, therefore, the wave packets emitted by separate atoms are mutually independent. However, the radiation of each individual atom is coherent during the time τ_c of emission of a separate wave packet.

For an arbitrary point in space, $\boldsymbol{r}_1 = \boldsymbol{r}_2 = \boldsymbol{r}$, let consider the mean

$$\overline{E_\alpha(\boldsymbol{r}, t + \tau) E_\beta^*(\boldsymbol{r}, t)} = J_{\alpha\beta}(\boldsymbol{r}, \boldsymbol{r}, \tau).$$

It is clear that this mean will differ from zero only for those values of τ, which are less or of the order of duration of a separate wave packet, $\tau \le \tau_c$. However, when $\tau \gg \tau_c$, then the fields of different independent wave packets will be averaged. There is no noticeable correlation between them, and the mean of the product can be replaced by the product of the means:

$$J_{\alpha\beta}(\boldsymbol{r}, \boldsymbol{r}, \tau) \approx \overline{E_\alpha(t + \tau) E_\beta^*(t)} = 0, \qquad \tau \gg \tau_c. \tag{7.5}$$

The time interval τ_c during which the tensor $J_{\alpha\beta}$ noticeably differs from zero is called **correlation time** or **coherence time** of the field components E_α and E_β. This time can also be defined formally, for example, in the following way:

$$\tau_c = \left| \frac{1}{J_{\alpha\beta}(\boldsymbol{r},\boldsymbol{r},0)} \int_0^\infty J_{\alpha\beta}(\boldsymbol{r},\boldsymbol{r},\tau)\, d\tau \right| . \tag{7.6}$$

It can be in general different for different pairs of field components. However, in what follows we will assume, for simplicity, that it is the same for all field components.

The dependence of the correlation tensor $J_{\alpha\beta}$ on coordinates is of the same character: with increasing differences $x_1 - x_2$, $y_1 - y_2$, $z_1 - z_2$ the correlation becomes weaker. By analogy with Equation (7.6), it is possible to introduce the **correlations lengths** l_x, l_y, l_z. By the order of magnitude they are equal to the spatial lengths of wave packets, in whose limits the separate Fourier-harmonics of the field are related by phase. In the homogeneous and isotropic radiation field, the correlation lengths in all directions are equal, and the tensor $J_{\alpha\beta}$ depends only on the difference $\boldsymbol{r}_1 - \boldsymbol{r}_2$. In such a field, $l_c \approx c\tau_c$, since the lengths of wave packets in space are of the order of $c\tau_c$, and the correlation will be strongly weakened if the fields at the points $\boldsymbol{r}_1$ and $\boldsymbol{r}_2$ are related to different wave packets.

In the anisotropic radiation field, where, for example, all wave packets, emitted by a distant source, travel in the same direction, the relation $l_\parallel = c\tau_c$ will be valid only for longitudinal (in the direction of propagation) coherence length. The transverse coherence length is determined in this case by other factors and is not directly connected with τ_c (see below).

For a strictly deterministic process, like a plane monochromatic wave in the stationary case, the length and the coherence time are infinite. However, for all real processes they have finite values. For example, the thermal source of light (e.g., a sodium lamp) has the characteristic coherence time $\tau_c \approx 10^{-10}$ s and the corresponding coherence length $l_\parallel = c\tau_c \approx 3$ cm. The continuous-wave helium-neon laser has the coherence time $\tau_c = 0.02 - 0.002$ s and the length of longitudinal coherence $l_\parallel \approx 60 - 600$ km.

The time averaging (Equation (7.3)) is often replaced by the ensemble averaging over all possible realizations of a random field. Let us imagine a large number (infinite in the limit) of non-interacting macroscopic systems that are identical to the initial system, that is, they consist of equal sources of radiation including those part of space in which this radiation propagates. Such a set of macroscopically identical systems is referred to as **ensemble**. Because of the random character of the emission of radiation, at the equivalent points of space at the same instant of time the field values E'_α, $E''_\alpha, \ldots$ will, in general, be different for different systems:

$$E'_\alpha(\boldsymbol{r},t) \neq E''_\alpha(\boldsymbol{r},t) \neq E'''_\alpha(\boldsymbol{r},t) \neq \ldots$$

The ensemble average (or statistical average) is the arithmetical mean

$$\langle E_\alpha(\boldsymbol{r}_1,t_1)E^*_\beta(\boldsymbol{r}_2,t_2)\rangle = \lim_{N\to\infty} \frac{E'_\alpha(\boldsymbol{r}_1,t_1)E^{*\prime}_\beta(\boldsymbol{r}_2,t_2) + E''_\alpha(\boldsymbol{r}_1,t_1)E^{*\prime\prime}_\beta(\boldsymbol{r}_2,t_2) + \cdots}{N}, \tag{7.7}$$

where N is the total number of the systems of the ensemble. For arbitrary radiating systems, it is beyond reason to wait for the ensemble averaging (Equation (7.7)) to give the same result as the time averaging (Equation (7.3)). However, for a wide range of random processes, along with many real radiating systems, the ensemble and time averages are the same. Such systems are referred to as **ergodic**.

The ergodicity condition[1] for the stationary random fields requires a sufficiently fast attenuation of correlation with increasing τ:

$$\lim_{\Delta t \to \infty} \frac{1}{\Delta t} \int_0^{\Delta t} J_{\alpha\beta}(\boldsymbol{r}, \boldsymbol{r}, \tau)\, d\tau = 0.$$

When the random fields are investigated theoretically, the statistical averaging is usually used. In what follows, we also will perform the ensemble averaging, assuming the radiation fields to be ergodic. In this case, the results obtained in such a way will be identical to the time means measured by the detectors.

The effect of time coherence and of spatial coherence on interference of waves. Whether it is possible to obtain the regular amplification and attenuation of waves in space and time due to their interference (overlapping) strongly depends on their coherent properties. Qualitative considerations evidently show that the regular interference picture will be observed provided the phase difference of separate harmonics with equal frequencies remains constant over the period of observation, that is, the oscillations are coherent.

The relation between the time coherence and the interference picture can be investigated with the aid of the Michelson interferometer;[2] see Figure 7.1. The beam from a small source S consisting of independent emitters (excited atoms) is decomposed in two beams 1 and 2 by the semi-transparent mirror 3. These beams, in their turn, are reflected by the mirrors 3_1 and 3_2 and overlap in the region 12 giving an interference picture (alternating dark and bright bands) on the screen. By varying a length of the paths s_1 and s_2, one can introduce the delay time (path difference) between the interfering oscillations.

Let us elucidate the conditions resulting in the appearance of the interference picture in the interferometer described above. Assume that each atom emits a quasi-monochromatic wave (wave packet) of duration τ_c in the frequency range $\Delta\omega$. Since the wave packets from different atoms have random initial phases, the interference picture is not formed when such wave packets overlap. The field maxima and minima of separate wave packets will be superimposed randomly. Interference occurs when the fields are superimposed, which were generated by a single atom and which traveled across the different paths s_1 and s_2. However, for such a superimposition to be possible, the delay time $\tau = (s_1 - s_2)/c$ should be less than the lifetime of a separate wave packet. Otherwise, its part traversing a longer path

1) The proof of the ergodicity theorem and its more rigorous formulation can be found in the books Monin and Yaglom (1965); Rytov (1976); Rytov *et al.* (1978); Mandel and Wolf (1995).

2) Michelson Albert Abraham (1852–1931), American physicist, who invented the interferometer and performed many experiments that were a contributory factor for recognition of the special relativity theory; Nobel Prize laureate (1907) for creation of precise instruments and for measurements carried out with them.

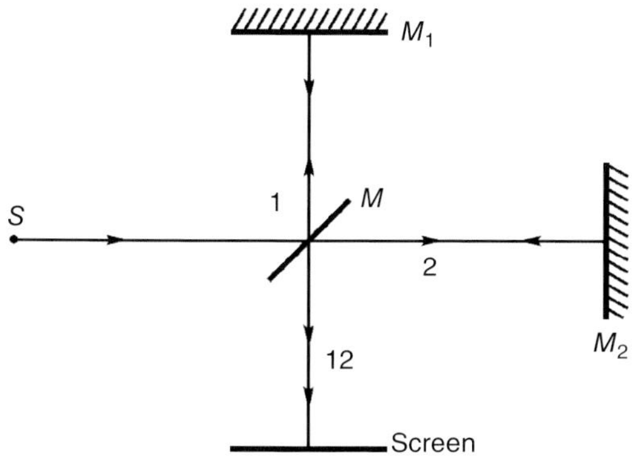

Figure 7.1 The scheme of Michelson interferometer.

will arrive at the screen when its other part is already absorbed by the screen or reflected by it.

The maximal delay time at which the interference picture is still observed must be less than the lifetime of a separate wave packet. This is just the time of coherence:

$$\tau \leq \tau_c \approx \frac{1}{\Delta\omega}. \tag{7.8}$$

In the anisotropic field of radiation, when the wave packets are emitted by a distant source and all travel in one **longitudinal** direction (in the direction of propagation) **the coherence length** can be evaluated from the relation

$$l_{\parallel} = c\tau_c \approx \frac{c}{\Delta\omega} \approx \frac{\lambda^2}{\Delta\lambda}, \tag{7.9}$$

where λ is the length of the emitted quasi-monochromatic wave, $\Delta\lambda$ is the spread of wavelengths connected with the spectral width by the relation $\Delta\lambda = (\lambda^2/2\pi c)\Delta\omega$.

Consider an interference experiment in another arrangement – Young experiment.[3] Let quasi-monochromatic light propagating from a certain source of finite dimensions a (Figure 7.2), penetrate through two openings D_1 and D_2 in the opaque screen and forms the interference picture on another screen S. The stable interference can be formed only by the waves emitted by an individual atom, since the radiation of different atoms is incoherent. In order to observe the interference from the extended source, it is necessary that the interference pictures from individual atoms be imposed approximately with equal phases. Let us find the corresponding conditions. Let the position of the openings, source and screen be symmetrical as shown in Figure 7.2, and a distance to the source be large:$R \gg a$, $R \gg l_{\perp}$.

The relative spatial shift of the two beams passing through the openings D_1 and D_2 is determined by the difference of paths $|PD_1 - PD_2|$. The interference fields

3) Young Thomas (1773–1829), English physicist, one of the founders of wave optics, introduced the notion "interference" in science.

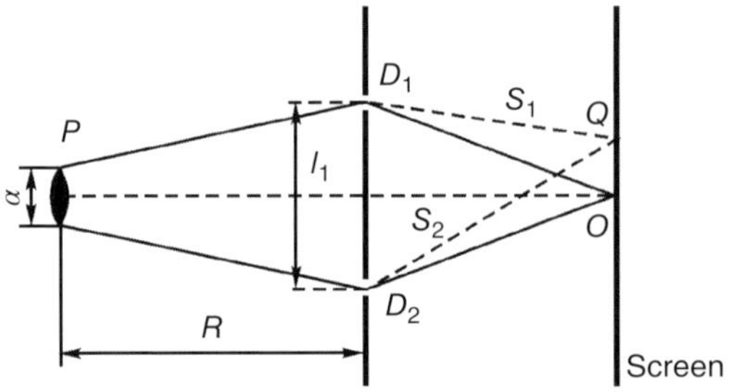

Figure 7.2 The scheme of Young interferometer.

from different atoms will amplify one another if this difference of paths does not exceed the wavelength of radiation λ for all atoms-emitters, independently of their position on the surface of the source.

It is clear that the maximal value of $|PD_1 - PD_2|$ will be for atoms on the edge of the source. On writing down

$$(PD_1)^2 = R^2 + (l_\perp - a)^2/4, \qquad (PD_2)^2 = R^2 + (l_\perp + a)^2/4,$$

we obtain $|PD_1 - PD_2| \approx al_\perp/2R, al_\perp/R \leq \lambda$, where in the latter inequality, which is valid to the order of magnitude, the factor 1/2 is omitted. Hence, the interference picture will be observed when a distance between the openings does not exceed

$$l_\perp \approx \lambda R/a \approx \frac{\lambda}{\Delta\vartheta}. \tag{7.10}$$

In the case under consideration, this distance plays a role of the **transverse coherence length**. The quantity

$$A = l_\perp^2 \approx \frac{\lambda^2 R^2}{S} \tag{7.11}$$

is **the area of coherence**, and the product

$$\Delta V = l_\parallel A = c\tau_c A \tag{7.12}$$

is called **the coherence volume**. Here $S = a^2$ is the area of the cross section of the source.

In quantum consideration (Chapter I.6), light is a set of photons that are quanta of the electromagnetic field. The energy of each photon of frequency ω is equal to $\hbar\omega$. The important characteristic of the coherent properties of the field is the degeneracy parameter δ, which is equal to the average number of photons in the same state of polarization inside the coherence volume, that is, to the average number of photons of the given polarization, which cross the coherence area during the coherence time.

Let I_ω be the average number of photons emitted by unit area of a source, normally to its surface, per unit time, in unit frequency range and in unit solid angle. If the two possible directions of polarization are equally probable, then

$$\delta = \frac{1}{2} I_\omega S \Delta\omega \Delta\Omega \tau_c ,$$

where $\Delta\Omega = A/R^2$ is the solid angle under which the coherence area is observed from the source. By using Equations (7.8) and (7.11), we have

$$\delta \approx \frac{2\pi^2 c^2}{\omega^2} I_\omega. \tag{7.13}$$

The degeneracy parameter written in such a form is independent of the geometry and characterizes only the properties of the source of radiation.

From the definition of δ, it follows that the larger this Angle, the more coherent is the field of the given source. For thermal sources of light, the largest value of δ is of the order of 10^{-3}, that is, $\delta \ll 1$. For laser sources, $\delta \gg 1$ and can attain the values of the order of 10^{14}.

Mutual-coherence function and visibility. Let us find the quantitative connection between the interference picture that can be observed on the screen of a Michelson interferometer or in the Young interference scheme, and the correlation tensor of the radiation field. We assume, for simplicity, that a light is linearly polarized and its polarization does not change when propagating. Then, the radiation field can be described by the scalar complex function $U(\boldsymbol{r}, t)$. The field at the point Q is the result of overlapping of the waves propagating through the openings D_1 and D_2. It can be written in the form

$$U(\boldsymbol{r}, t) = A_1 U(\boldsymbol{r}_1, t - t_1) + A_2 U(\boldsymbol{r}_2, t - t_2), \tag{7.14}$$

where $t_1 = s_1/c,\ t_2 = s_2/c$ are the times necessary for disturbances to come from D_1 and D_2 to the point Q; and $A_1,\ A_2$ are the transfer coefficients that are in general complex and depend on the dimensions and shape of the openings. We assume, in what follows, that their phases are equal. The interference picture on the screen will be determined by the intensity of the field $I(\boldsymbol{r})$, that is, by the ensemble averaged square of the modulus $\langle U^*(\boldsymbol{r}, t)U(\boldsymbol{r}, t)\rangle = I(\boldsymbol{r})$. Because the source is assumed to be stationary, the intensity I will be independent of time. With the aid of Equation (7.14), we have

$$I(\boldsymbol{r}) = |A_1|^2 I_1(\boldsymbol{r}) + |A_1|^2 I_2(\boldsymbol{r}) + 2|A_1 A_2| \Re\, \Gamma(\boldsymbol{r}_1, \boldsymbol{r}_2, \tau), \tag{7.15}$$

where $\tau = t_1 - t_2 = (s_1 - s_2)/c$. The complex value

$$\Gamma(\boldsymbol{r}_1, \boldsymbol{r}_2, \tau) = \langle U^*(\boldsymbol{r}_1, t)U(\boldsymbol{r}_2, t + \tau)\rangle \tag{7.16}$$

is called **the mutual coherence function**. It is a scalar version of the correlation tensor (Equation (7.7)) (or the separate diagonal component of this tensor). The intensity of light at points $\boldsymbol{r}_j$, j =1 or 2, is connected with the mutual correlation function by the relation

$$I(\boldsymbol{r}_j) = \Gamma(\boldsymbol{r}_j, \boldsymbol{r}_j, 0). \tag{7.17}$$

The intensity of the field that is produced by one of the openings at the point Q (when the other is closed) is given by the expression

$$I_j(\boldsymbol{r}) = |A_1|^2 I(\boldsymbol{r}_j). \tag{7.18}$$

Let us determine the normalized mutual coherence function $\gamma(\mathbf{r}_1, \mathbf{r}_2, \tau)$, which is also called **the complex degree of coherence**,

$$\gamma(\mathbf{r}_1, \mathbf{r}_2, \tau) = \frac{\Gamma(\mathbf{r}_1, \mathbf{r}_2, \tau)}{[I(\mathbf{r}_1)I(\mathbf{r}_2)]^{1/2}}. \tag{7.19}$$

We write the field intensity at the point Q in terms of the intensities of the waves that have separately passed through each of the openings:

$$I(\mathbf{r}) = I_1(\mathbf{r}) + I_2(\mathbf{r}) + 2[I_1(\mathbf{r})I_2(\mathbf{r})]^{1/2}\Re\,\gamma(\mathbf{r}_1, \mathbf{r}_2, \tau). \tag{7.20}$$

For a quasi-monochromatic light, when $\Delta\omega \ll \omega$, the time dependence of $U(\mathbf{r}, t)$ will mainly be determined by the factor $\exp(-i\omega t)$. The amplitude $A(t)$ and the additional phase $\alpha(t)$, being the random functions for a case of incoherent light, vary much slower, as $\exp(-i\Delta\omega t)$. Hence, if we write

$$\gamma(\mathbf{r}_1, \mathbf{r}_2, \tau) = |\gamma(\mathbf{r}_1, \mathbf{r}_2, \tau)| \exp[i\alpha(\mathbf{r}_1, \mathbf{r}_2, \tau) - i\omega t] \tag{7.21}$$

then γ and $\exp(i\alpha)$ will be slower functions of τ in comparison with $\exp(-i\omega t)$. They change noticeably only over a period of optical oscillations, $\tau \gg \tau_c$.

Using Equation (7.21), we rewrite Equation (7.20) in the form

$$I(\mathbf{r}) = I_1(\mathbf{r}) + I_2(\mathbf{r}) + 2[I_1(\mathbf{r})I_2(\mathbf{r})]^{1/2}\cos(\alpha - \omega\tau). \tag{7.22}$$

Taking into account the fact that the functions γ and α vary slowly, we conclude that, when $\tau \ll \tau_c$, the value $\cos(\alpha - \omega\tau)$ changes mainly owing to $\omega\tau$. If the position of the point Q on the screen and, hence, the difference $s_1 - s_2 = c\tau$ changes, the intensity of illuminance of the screen will vary in the limits from

$$I_{max}(\mathbf{r}) = I_1 + I_2 + 2(I_1 I_2)^{1/2}|\gamma| \tag{7.23}$$

to

$$I_{min}(\mathbf{r}) = I_1 + I_2 - 2(I_1 I_2)^{1/2}|\gamma|. \tag{7.24}$$

Since $I \geq 0$ for any relation between I_1 and I_2, then from Equations (7.23) and (7.24) it follows that the modulus of the complex degree of coherence can vary in the limits

$$0 \leq |\gamma(\mathbf{r}_1, \mathbf{r}_2, \tau)| \leq 1. \tag{7.25}$$

Let us define **the visibility** (also called **sharpness** or **contrast range**) of the interference picture at a given point $\mathbf{r}$ by the parameter

$$\mathcal{V} = (I_{max} - I_{min})/(I_{max} + I_{min}). \tag{7.26}$$

From Equations (7.23) and (7.24), we have

$$\mathcal{V} = 2\left(\sqrt{\frac{I_1}{I_2}} + \sqrt{\frac{I_2}{I_1}}\right)^{-1} |\gamma(\mathbf{r}_1, \mathbf{r}_2, \tau)|. \tag{7.27}$$

In the particular case of $I_1 = I_2$,

$$\mathcal{V} = |\gamma(\mathbf{r}_1, \mathbf{r}_2, \tau)|, \tag{7.28}$$

that is, the visibility of the interference bands, which is experimentally measured, is equal to the modulus of the complex degree of coherence. The phase of the complex degree of coherence can be measured by the positions of the maxima of the intensity of the interference picture. According to Equation (7.22), the position of maxima is determined by the condition

$$\alpha\left(\boldsymbol{r}_1, \boldsymbol{r}_2, \frac{s_1 - s_2}{c}\right) - \frac{2\pi}{\lambda}(s_1 - s_2) = 2\pi m, \qquad m = 0, \pm 1, \ldots \tag{7.29}$$

The intensities I_1 and I_2 are the quantities that can also be measured. Hence, the mutual-coherence function $\Gamma(\boldsymbol{r}_1, \boldsymbol{r}_2, \tau)$ can be (at least, in principle) determined by observing the interference picture and by measuring the intensities I_1, I_2.

From Equation (7.20), it is clear that when $|\gamma| = 0$ the interference bands are completely absent and the screen is uniformly illuminated. This means that the two beams of light, arriving at the screen from the openings D_1 and D_2, are completely incoherent. When $|\gamma| = 1$, the interference bands have the maximal visibility, and the intensity of the radiation field in the minima of illuminance vanishes. The two beams are completely coherent. In the intermediate cases, $0 \leq |\gamma| \leq 1$, these beams are partially coherent. It is clear from Equation (7.22) written in the form

$$I(\boldsymbol{r}) = (1 - |\gamma|)(I_1(\boldsymbol{r}) + I_2(\boldsymbol{r})) + |\gamma|[I_1(\boldsymbol{r}) + I_2(\boldsymbol{r}) + 2[I_1(\boldsymbol{r})I_2(\boldsymbol{r})]^{1/2}\cos(\alpha - \omega\tau)]. \tag{7.30}$$

Thus, we see that the field at the point Q can be presented as a mixture of the completely incoherent light and the completely coherent light. The first has the intensity $(1 - |\gamma|)(I_1 + I_2)$, and the second is a completely coherent mixture of two beams with the intensities $|\gamma|I_1$, $|\gamma|I_2$ and the phase difference $\alpha - \omega\tau$.

The concept of holography. If any object is placed in a field of a coherent light wave and scatters that wave, then in the region where the scattered wave and the basic (reference) wave overlap, there appears an interference picture with the intensity that depends on the intensities and phase differences of the scattered and reference waves. This picture can be mapped on a photographic plate, and afterwards this plate can be used as a diffraction grating transmitting the coherent light through it. The intensity I' of the light, passing through the given point (x, y) of the developed photographic plate, is proportional to the intensity $I(x, y)$ of the illuminating beam,

$$I'(x, y) = T(x, y)I(x, y), \tag{7.31}$$

and depends on the blackening of the photographic plate, characterized by the "transmission" $T(x, y)$. The transmission depends on the intensity $I_0(x, y)$ of the primary field which causes the blackening and on the contrast of the photographic emulsion, characterized by the law $T(x, y) \propto [I_0(x, y)]^{-\gamma/2}$, where γ is the contrast coefficient of the emulsion.

A photographic plate on which the picture of interference between the reference wave and the wave scattered by an object is mapped is called **the hologram**. It turns out that when the coherent light is transmitted through the hologram, a three-dimensional picture of the original object is formed behind it. The process of reconstruction of the original wave field is called **the holography**. The fundamentals of holography are given in the books Stroke (1966); Soroko (1971) and are illustrated by Problems 7.16–7.19.

Recommended literature: Mandel and Wolf (1995); Scully and Zubairy (1997) Bredov *et al.* (2003); Potekhin and Tatarinov (1978); Svelto (1982) Stroke (1966); Soroko (1971); Toptygin (2014)

Problems

7.1 A quasi-monochromatic source has a transverse length L and emits a light of a wavelength λ. Estimate the order of magnitude of the solid angle $\Delta\Omega$ within which the radiation of the source is coherent.

7.2 What are the transverse and longitudinal dimensions, the solid angle and the volume of coherence of the radiation emitted by sodium atoms in the solar atmosphere? On the Earth one observes a spectral line of the wavelength $\lambda_0 = 5 \times 10^{-5}$ cm; the atomic mass is $m = 3.7 \times 10^{-23}$ g. The main contribution to the width of the spectral line comes from the thermal motion of the atoms (the temperature $T \approx 6000$ K). The average distance from the Earth to the Sun is 1.5×10^{13} cm.

Hint. The Doppler width of a spectral line is

$$\gamma_D = \sqrt{\frac{8\pi^2 kT}{m\lambda_0^2}},$$

where k is the Boltzmann constant (Problem I.5.123).

7.3 How are the results of Problem 7.2 changed, if one observes from the Earth a star of the type of the Sun at a distance of 10 light years? One light year is equal to 9.46×10^{17} cm.

7.4 Determine the transverse and longitudinal dimensions and also the volume of coherence in the immediate neighborhood of a laser operating at a wavelength $\lambda_0 = 5 \times 10^{-5}$ cm with a frequency spread $\Delta\nu = 10^2$ Hz. The diameter of the mirrors is $D = 5$ cm.

7.5 Find the degeneracy parameter δ of the emission from a perfect black body at a temperature T. Make numerical estimates for $\lambda = 1$ and $\lambda = 5 \times 10^{-5}$ cm at $T = 273$ K and for $\lambda = 5 \times 10^{-5}$ cm at $T = 10\,000$ K.

Hint. The spectral energy density of the emission from a perfect black body (Problem I.6.23) is

$$I_\omega = \frac{\omega^2}{\pi^2 c^3} \frac{\hbar\omega}{\exp(\hbar\omega/T) - 1},$$

where the temperature T is expressed in the energy units, that is, $T \to kT$, where $k = 1.38 \times 10^{-16}$ erg/K is Boltzmann's constant.

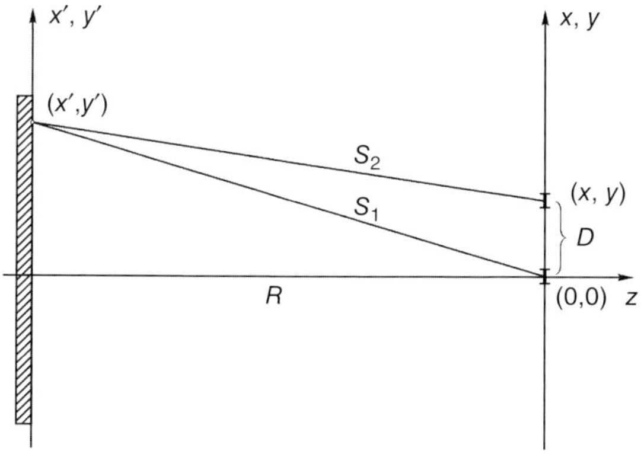

Figure 7.3 Scheme of Young interference experiment.

7.6 Find the degeneracy parameter for the laser considered in Problem 7.4. The power of the radiation is 200 W. What is the effective temperature to which this value of δ corresponds?

7.7 Find the relation between the autocorrelation function

$$\Gamma(\boldsymbol{r}, \boldsymbol{r}, \tau) = \langle U^*(\boldsymbol{r}, t)\, U(\boldsymbol{r}, t+\tau)\rangle$$

and the power spectrum $I(\omega)$ of the radiation. The radiation intensity is $I = \langle U^*(t)\, U(t)\rangle = \int_0^\infty I(\omega)\, d\omega$.

7.8 Find the autocorrelation function of the radiation, if a spectral line of the emission is narrow and has a rectangular form with width $\Delta\,\omega$ around ω_0. The radiation intensity is I.

7.9 In a Young interference experiment, the interference picture is observed in the region where the beams, which are diffracted by two openings, overlap (Figure 7.3). The openings are at a distance D from one another at the points $(0,0)$ and (x,y). The light source is extended and its dimensions are much larger than D. It is at a distance R from the openings ($R \gg D$). The light is sufficiently monochromatic so that for each of the independent emitters the condition of temporal coherence is satisfied. Express the coefficient of partial coherence in terms of the intensity distribution $I(x,y)$ of the radiation over the cross section of the light source.

7.10 A Michelson stellar interferometer is a version of the Young interference scheme in which the distance between the openings can be varied. Find the dependence of the visibility $\mathcal{V}$ of the interference bands in the Michelson interferometer on the distance D between the openings and on the wavelength λ for the two cases:

(a) A binary is observed, that is, a system of two stars that are close to one another, and are at an angular distance α from each other. Each of the stars can be considered to be a point light source. Assume the luminosities of the two stars to be equal.

(b) One observes a single star of large size and with an angular cross section α (consider this star to be a uniformly emitting disc).

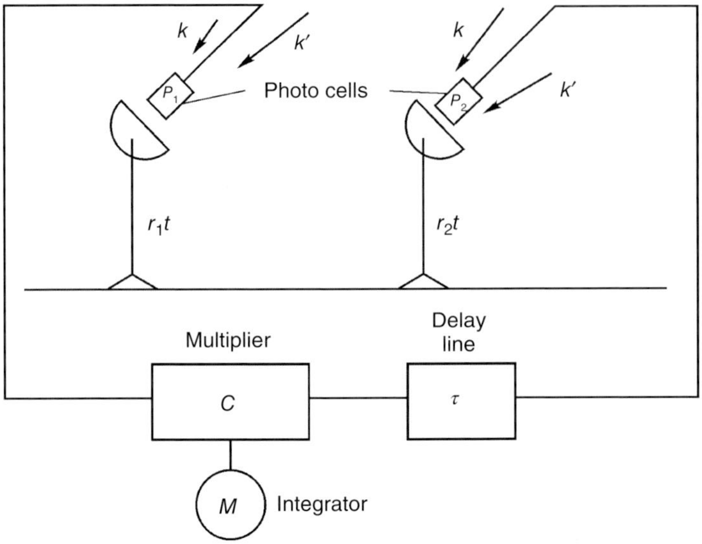

Figure 7.4 Scheme of Brown–Twiss interferometer.

7.11 In the Michelson stellar interferometer (Problem 7.10), the light comes from a binary or from a single star of large radius. When the distance D between the openings increases, the visibility of the interference bands becomes weaker and vanishes for some value $D = D_0$. Determine: (a) the distance ρ between the components of the binary in Capella at the distance $R = 44.6$ lights years, which is at a distance $D_0 = 70.8$ cm from the Earth, and it is observed at a wavelength $\lambda = 5 \times 10^{-5}$ cm; (b) the diameter d of the star Betelgeuze which is at a distance 652 light years, if $D_0 = 720$ cm, and $\lambda = 6 \times 10^{-5}$ cm.

Hint. The first nonvanishing root of the Bessel function $J_1(x)$ is equal to $x_1 = 3.8317$.

7.12⋆ In the Brown–Twiss interferometer (Figure 7.4), one detects independently, and afterwards multiplies and registers the light intensities that comes from two distant incoherent point sources or from different points of a single extended source. The waves coming from the sources can be considered to be the plane waves (with the wave vectors $\boldsymbol{k}_1$ and $\boldsymbol{k}_2$), with randomly fluctuating amplitudes and phases. Show that using the Brown–Twiss interferometer and observing the correlation between intensities, one can measure the angular distance between the sources.

7.13 A plane wave of a wavelength λ is incident almost normally upon the lateral surface of a thin prism with vertex angle $\alpha \ll 1$ and the refraction coefficient n (Figure 7.5a). Find the x-dependence of the phase shift that the wave acquires in the flat layer $ABCD$, whose part is occupied by the prism.

7.14 A plane wave is incident on a thin converging or diverging lens with radii of curvature R_1 and R_2 and refractive coefficient n (Figure 7.5b,c). The

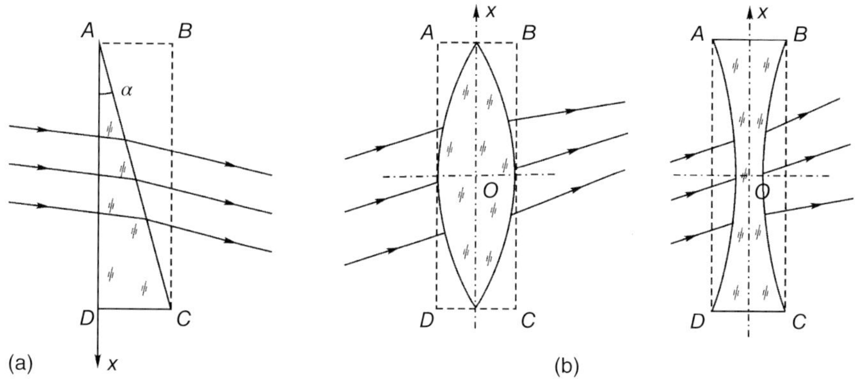

Figure 7.5 To calculation of phase shift of a wave in a thin prism (a), a thin converging lens (b), and a thin diverging lens (c).

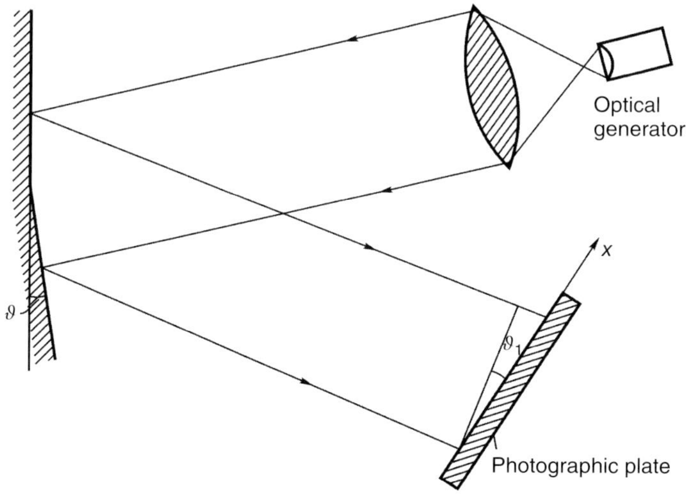

Figure 7.6 The scheme of experiment with the Fresnel double mirror.

wavelength is λ and the angle between the wave vector and the optical axis of the lens is small. Find the x-dependence of the phase shift that the wave acquires in the flat layer *ABCD*, whose part is occupied by the lens.

7.15 A monochromatic plane wave of a wavelength λ from a laser is incident upon the Fresnel double mirror (Figure 7.6) with an angle $\vartheta \ll 1$ between the planes of the mirrors. In the region where two waves, coming from the double mirror, overlap, the interference wave field is formed. On the photographic plate placed in that region and making an angle $\vartheta_1 \ll 1$ with the front of one of the waves, there appears a system of transparent and dark interference bands. What is the wave field behind the plate, if, after it was

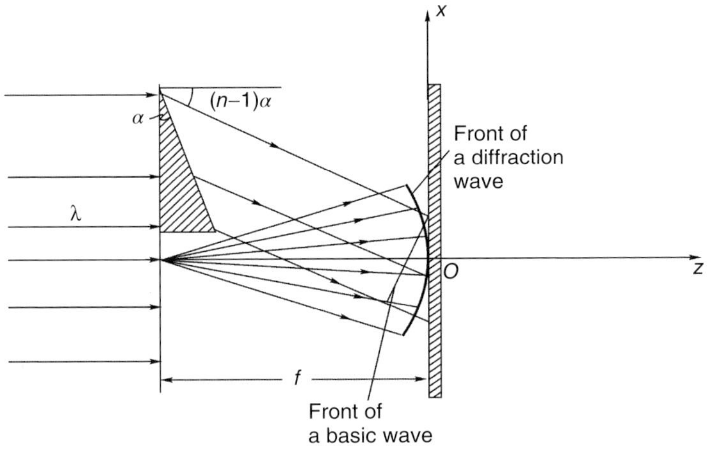

Figure 7.7 The scheme of creating of the holographic image.

developed, a plane wave was transmitted through it, normal to its surface, from the same laser?

7.16 A plane monochromatic wave passes simultaneously through the prism and the aperture in an opaque screen and is incident upon the photographic plate at a distance f (Figure 7.7). The prism is thin and has a refracting angle $\alpha \ll 1$ and a refractive coefficient n. On the photographic plate there appears the intensity distribution of the field owing to the interference between the reference plane wave (that part of the wave that passed through the prism and was deflected by it) and the wave diffracted by the aperture (the angle of diffraction is assumed to be small). Find that intensity distribution.

7.17 Find the transmission $T(x)$ through a hologram that was obtained under the conditions described in Problem 7.16. Assume that in the process of obtaining hologram the intensity of the reference wave was large compared with the intensity of the wave passing through the aperture. Trace the process of reconstruction of the initial wavefronts when the normally incident, plane monochromatic wave $u_0 = A_0' \exp[i(kz - \omega t)]$ is transmitted through this hologram (the wavelength is the same as that of the original wave). In particular, trace the appearance of the point image of the original aperture.

Hint. The wave field behind the hologram can be obtained simply by multiplying the incident wave $u_0(x)$ by the transmission $T(x)$. For the interpretation of the expression obtained, turn to the solutions of Problems 7.13 and 7.14.

7.18 By using the set-up that was considered in Problems 7.16 and 7.17, one obtains the hologram of two apertures at a distance $2D$ from one another in the plane of the prism. The image of the two apertures can be reconstructed with the aid of this hologram. Find this image and elucidate when it will be amplified.

Hint. At the reconstruction stage, the hologram can be illuminated with light at a wavelength λ', which is different from the wavelength λ used to obtain the hologram.

7.19 Determine the resolving power of the hologram, which was obtained in the set-up such as one considered in Problem 7.16. The hologram was produced on a photographic plate with emulsion grain size d.

7.20 The electromagnetic wave of a fixed linear polarization is described, at some time instant, by the function $E(\boldsymbol{\rho}, z) = \mathcal{E}(\boldsymbol{\rho})\exp(ikz)$, where $\mathcal{E}(\boldsymbol{\rho})$ is a random function of the transverse coordinates $\boldsymbol{\rho} = (x, y)$. Such a field is produced after a plane wave is transmitted through the screen in the plane (x, y). The transmission of the screen and the phase incursion vary randomly from one point to another. Calculate the transverse correlation function $K_\perp(\boldsymbol{\rho}_1, \boldsymbol{\rho}_2) = \langle \mathcal{E}(\boldsymbol{\rho}_1)\,\mathcal{E}^*(\boldsymbol{\rho}_2)\rangle$, if it is uniform and isotropic in the plane (x, y), and the distribution of the Fourier harmonics of the random field over wave vectors is given as follows:

1. The uniform distribution in the region $0 \le \kappa \le \kappa_0$, that is, $K(\kappa) = K_0 = const$ at $0 \le \kappa \le \kappa_0$, and $K(\kappa) = 0$ at $\kappa > \kappa_0$.
2. The Gaussian distribution $K(\kappa) = K_0 \exp(-\kappa^2/\kappa_0^2)$.

7.2
Random Waves and Waves in Randomly Inhomogeneous Media

Fluctuations of electromagnetic field in an equilibrium medium. First we consider the thermal fluctuations of an electromagnetic field in an equilibrium medium (no spatial dispersion) with the scalar dielectric permittivity and magnetic permeability $\varepsilon(\omega)$ and $\mu(\omega)$, respectively. Any actual medium is an absorbing one, that results in the presence of the imaginary parts in ε and μ. In such a medium, at finite temperatures, besides the vacuum fluctuations that were considered in Chapter I.6, there exists an electromagnetic field that is in equilibrium with matter. This field will be treated as random fluctuations. Separate trains of waves will be emitted and absorbed by particles of the matter at random time instants.

We will calculate the field in several steps. At the first step, we find the field that causes the given weak harmonic current in the medium under consideration:

$$\boldsymbol{j}(\boldsymbol{r}, t) = \boldsymbol{j}(\boldsymbol{r})\cos\omega t. \tag{7.32}$$

Let us choose such a gauge of electromagnetic potentials at which the scalar potential $\varphi = 0$ and all field vectors could be expressed in terms of the vector potential:

$$\boldsymbol{E} = \frac{i\omega}{c}\boldsymbol{A}, \qquad \boldsymbol{D} = \frac{i\omega\varepsilon(\omega)}{c}\boldsymbol{A}, \qquad \boldsymbol{B} = \nabla\times\boldsymbol{A}, \qquad \boldsymbol{H} = \frac{1}{\mu(\omega)}\nabla\times\boldsymbol{A}, \tag{7.33}$$

where the harmonic time dependence is taken into account.

Since the properties of the medium are determined by the quantum laws, we will perform the calculations in terms of quantum mechanics. The operator of the interaction of the medium with an external field can be written in the form (see the analogous four-dimensional formula (I.4.108))

$$\hat{V}(t) = -\frac{1}{c}\int j_\lambda(\boldsymbol{r},t)\hat{\boldsymbol{A}}_\lambda(\boldsymbol{r})\,d^3r, \tag{7.34}$$

where $\hat{\boldsymbol{A}}_\lambda(\boldsymbol{r})$ is the quantum vector potential (I.6.13) independent of time, and the summation is performed over the three values of λ. The medium is considered to be infinitely extended. Let us take a sufficiently large "basic" region and impose the periodic boundary conditions upon the given current and vector potential, as was done in Sections I.2.3 and I.6.1. In that case, the interaction potential can be written as the infinite sum of the terms

$$\hat{V}(t) = -\frac{1}{c}\sum_{\mathbf{k}} j_{\mathbf{k}\lambda}(t)\hat{A}_{\mathbf{k}\lambda}, \tag{7.35}$$

where $\hat{A}_{\mathbf{k}\lambda} = \hat{A}^\dagger_{\mathbf{k}\lambda}$ is the independent Hermitian operator, and $j_{\mathbf{k}\lambda}(t)$ is the real function:

$$j_{\mathbf{k}\lambda}(t) = \frac{1}{2}(j^0_{\mathbf{k}\lambda}e^{-i\omega t} + j^{0\star}_{\mathbf{k}\lambda}e^{i\omega t}). \tag{7.36}$$

Now the interaction operator acquired the conventional form that is used in the theory of fluctuations and that which presents the sum of the products of the classical "external forces" $j_{\mathbf{k}\lambda}(t)$ (which generate the field) by the operators $\hat{A}_{\mathbf{k}\lambda}$. The averages of these operators over the states of the medium represent the macroscopic electromagnetic field in the medium (compare with Equation (125.1) from Landau and Lifshitz (1980)). In the absence of the external currents, we have

$$A_{\mathbf{k}\lambda} = Sp\,(\hat{\rho}\hat{A}_{\mathbf{k}\lambda}) \equiv \sum_n \rho_n\langle n|\hat{A}_{\mathbf{k}\lambda}|n\rangle = 0. \tag{7.37}$$

On the left-hand side of the latter expression, the symbol $A_{\mathbf{k}\lambda}$ (without the sign of averaging) denotes the macroscopic, that is, averaged potential. The averaging is produced using formulas similar to Equation (5.5); $\hat{\rho}$ is the equilibrium statistic density operator (I.C43), and the vectors of stationary states of the medium are denoted by $|n\rangle$. In the presence of a weak external current $j_{\mathbf{k}\lambda}(t) \neq 0$, there appears the electromagnetic field, whose value is proportional to the current and is determined by the function of the electromagnetic response of the medium with account for the causality principle (Section 5.1):

$$A_{\mathbf{k}\lambda}(t) = \int_0^\infty \alpha_{\lambda\nu}(\boldsymbol{k},\tau)j_{\mathbf{k}\lambda}(t-\tau)\,d\tau \tag{7.38}$$

(we assume the current and the potential are locally connected). The difference from Equation (7.37) is due to the fact that now the state of a body is not in equilibrium and the density operator $\hat{\rho}(t)$ differs from the equilibrium operator (I.C43).

Let us introduce the generalized susceptibility

$$\alpha_{\lambda\nu}(\boldsymbol{k},\omega) = \int_0^\infty \alpha_{\lambda\nu}(\boldsymbol{k},\tau)\,e^{i\omega\tau}\,d\tau, \tag{7.39}$$

and insert it into the macroscopic electromagnetic potential using Equation (7.32). As a result, we have

$$A_{\mathbf{k}\lambda}(t) = \frac{1}{2}\left[\alpha_{\lambda\nu}(\boldsymbol{k},\omega) j^{0}_{\mathbf{k}\lambda} e^{-i\omega t} + \alpha_{\lambda\nu}(\boldsymbol{k},-\omega) j^{0\star}_{\mathbf{k}\lambda} e^{i\omega t}\right]. \tag{7.40}$$

The generalized susceptibility is a complex quantity and has symmetry properties discussed in Section 5.1 in application to $\varepsilon_{\lambda\nu}$ and $\mu_{\lambda\nu}$. In particular,

$$\alpha_{\lambda\nu}(\boldsymbol{k},\omega) = \alpha_{\nu\lambda}(\boldsymbol{k},\omega), \qquad \alpha_{\lambda\nu}(\boldsymbol{k},\omega) = \alpha^{*}_{\lambda\nu}(\boldsymbol{k},-\omega). \tag{7.41}$$

The real part of the generalized susceptibility is the even function and the imaginary part is the odd function of real frequency.

Example 7.1

Express the period-averaged energy Q, which is absorbed by a body per unit time, in the terms of an external current and the generalized susceptibility.

Solution. One should calculate the ensemble average of the operator, which is the time derivative of the total Hamiltonian of a body:

$$(1) \qquad \frac{dE}{dt} = \left\langle \frac{d\widehat{H}}{dt} \right\rangle = \left\langle \frac{\partial\widehat{H}}{\partial t} \right\rangle = \left\langle \frac{\partial\widehat{V}}{\partial t} \right\rangle = -\frac{1}{c}\sum_{\mathbf{k}} \langle \widehat{A}_{\mathbf{k}\lambda}\rangle \frac{\partial j_{\mathbf{k}\lambda}(t)}{\partial t}$$
$$= -\frac{1}{c}\sum_{\mathbf{k}} A_{\mathbf{k}\lambda}(t) \frac{\partial j_{\mathbf{k}\lambda}(t)}{\partial t},$$

where $A_{\mathbf{k}\lambda}(t)$ is given by Equation (7.40). If we calculate the time averaging of Equation (1), we obtain the energy dissipation rate:

$$Q = \frac{-i\omega}{4c}\sum_{\mathbf{k}}[\alpha_{\lambda\nu}(\boldsymbol{k},\omega) - \alpha^{*}_{\lambda\nu}(\boldsymbol{k},\omega)] j^{0\star}_{\mathbf{k}\lambda} j^{0}_{\mathbf{k}\nu} = \frac{\omega}{2c}\sum_{\mathbf{k}} \alpha''_{\lambda\nu}(\boldsymbol{k},\omega) j^{0\star}_{\mathbf{k}\lambda} j^{0}_{\mathbf{k}\nu}, \tag{7.42}$$

where the double prime denotes the imaginary part of the generalized susceptibility. ■

Example 7.2

With the aid of the Maxwell's equations, find the generalized susceptibility $\alpha_{\lambda\nu}(\boldsymbol{k},\omega)$, expressing it in terms of the dielectric permittivity and magnetic permeability, $\varepsilon(\omega)$ and $\mu(\omega)$, for an isotropic medium.

Solution. The generalized susceptibility relates the external current to the macroscopic vector potential. This relation can be determined using the equation for the vector potential, which can be derived from the Maxwell's equation (5.12) and Equation (7.33):

$$(1) \qquad \frac{1}{\mu}\nabla \times [\nabla \times \boldsymbol{A}] = \frac{\omega^2\varepsilon}{c^2}\boldsymbol{A} + \frac{4\pi}{c}\boldsymbol{j}.$$

In the $\boldsymbol{k}$-representation, Equation (1) takes the form

$$(2) \qquad \left(k^2 - \frac{\omega^2}{c^2}\varepsilon\mu\right)\boldsymbol{A}_{\mathbf{k}} - \boldsymbol{k}(\boldsymbol{k}\cdot\boldsymbol{A}_{\mathbf{k}}) = \frac{4\pi\mu}{c}\boldsymbol{j}_{\mathbf{k}}.$$

From the latter equation, we find

(3) $$\boldsymbol{k}\cdot\boldsymbol{A}_{\mathbf{k}} = -\frac{4\pi c}{\omega^2\varepsilon}\boldsymbol{k}\cdot\boldsymbol{j}_{\mathbf{k}},$$

and substituting the result in Equation (2), we have

(4) $$A_{\mathbf{k}\lambda} = -\frac{4\pi\mu c}{\omega^2\varepsilon\mu - c^2k^2}\left(\delta_{\lambda\nu} - \frac{c^2k_\lambda k_\nu}{\omega^2\varepsilon\mu}\right) j_{\mathbf{k}\nu}.$$

Thus, the generalized susceptibility is

$$\alpha_{\lambda\nu}(\boldsymbol{k},\omega) = \frac{4\pi\mu c}{c^2k^2 - \omega^2\varepsilon\mu}\left(\delta_{\lambda\nu} - \frac{c^2k_\lambda k_\nu}{\omega^2\varepsilon\mu}\right). \tag{7.43}$$

The real vector potential (together with the time-dependent factor) will be expressed in the form (Equation (7.40)). When integrating these values, the path-tracing rules for avoiding singularities must correspond to the retarded potentials. ■

Example 7.3

On the basis of the notations of quantum mechanics, calculate, in the lowest order of the perturbation theory, the power Q absorbed by a body, and express it in terms of the matrix elements $\langle n|\widehat{A}_{\mathbf{k}\lambda}|m\rangle$ of the vector potential between the stationary states of a body. Compare the result with Equation (7.42) from Example 7.1 to find the representation of the imaginary part of the generalized susceptibility $\alpha_{\lambda\nu}(\boldsymbol{k},\omega)$ in terms of the matrix elements mentioned above.

Solution. The perturbation operator is the time harmonic function:

(1) $$\widehat{V} = \sum_{\mathbf{k}}\widehat{V}_{\mathbf{k}}, \qquad \widehat{V}_{\mathbf{k}} = -\frac{1}{2c}(j^0_{\mathbf{k}\lambda}e^{-i\omega t} + (j^{0\star}_{\mathbf{k}\lambda}e^{i\omega t})\widehat{A}_{\mathbf{k}\lambda}.$$

Let us calculate, in the lowest order of the perturbation theory, the probability $w'_{n\to m}$ of the transition per unit time, under the action of the harmonic perturbation, taking into account only one term with some definite $\boldsymbol{k}$. By using the familiar equation (Landau and Lifshitz (1977)), we obtain

(2) $$w'_{n\to m} = \frac{\pi}{2\hbar^2c^2}|\langle n|\boldsymbol{j}_{\mathbf{k}}\cdot\widehat{A}_{\mathbf{k}}|m\rangle|^2[\delta(\omega-\omega_{mn}) + \delta(\omega+\omega_{mn})].$$

If the sum over $\boldsymbol{k}$ in the operator (1) is taken into account, the products of the matrix elements $\langle n|\widehat{A}_{\mathbf{k}\lambda}|m\rangle\langle m|\widehat{A}_{\mathbf{k}'\lambda'}|n\rangle$ will appear in Equation (2) for the probability. At $\boldsymbol{k}\neq\boldsymbol{k}'$, these matrix elements correspond to the production or annihilation of different quanta and cannot be nonzero simultaneously. The total probability of the transition can hence be obtained by summing (Equation (2)) over $\boldsymbol{k}$:

(3) $$w_{n\to m} = \frac{\pi}{2\hbar^2c^2}\sum_{\mathbf{k}}|\langle n|\boldsymbol{j}_{\mathbf{k}}\cdot\widehat{A}_{\mathbf{k}}|m\rangle|^2[\delta(\omega-\omega_{mn}) + \delta(\omega+\omega_{mn})].$$

Then, the absorbed power can be calculated by summing all the transitions into the possible states m and by averaging the sum over initial states n:

(4) $$Q = \sum_{m,n} \rho_n w_{n\to m} \hbar\omega_{mn},$$

where, according to Equations (I.C44)–(I.C46), $\rho_n = \exp[(F - E_n)/T]$ is the statistic matrix. The matrix describes the Gibbs distribution of the equilibrium medium together with the electromagnetic radiation, which is in the statistic equilibrium with the medium. Using the equality $\rho_m = \rho_n \exp(\hbar\omega_{mn}/T)$ and the properties of the delta-function, we reduce Equation (4) to the form

(5) $$Q = \frac{\pi\omega}{2\hbar c^2}(1 - e^{-\hbar\omega/T}) \sum_{\mathbf{k}} \times \sum_{m,n} \rho_n \langle n|\widehat{A}_{\mathbf{k}\lambda}|m\rangle\langle m|\widehat{A}_{\mathbf{k}\nu}|n\rangle \delta(\omega - \omega_{mn}) j^0_{\mathbf{k}\lambda} j^{0\star}_{\mathbf{k}\nu}.$$

Comparing the last expression with Equation (7.42), we find the imaginary part of the generalized susceptibility:

$$\alpha''_{\lambda\nu}(\boldsymbol{k},\omega) = \frac{\pi}{c\hbar}(1 - e^{-\hbar\omega/T}) \sum_{m,n} \rho_n \langle n|\widehat{A}_{\mathbf{k}\lambda}|m\rangle\langle m|\widehat{A}_{\mathbf{k}\nu}|n\rangle \delta(\omega - \omega_{mn}). \quad \blacksquare \quad (7.44)$$

This result completes the preliminary calculations, and we can take the last step to find the fluctuations of the electromagnetic field in the equilibrium medium.

Example 7.4

Calculate the Fourier transform $K_{\lambda\nu}(\boldsymbol{k},\omega)$ with respect to time of the correlation function

$$K_{\lambda\nu}(\boldsymbol{k}, t - t') = \frac{1}{2}\langle \widehat{A}_{\mathbf{k}\lambda}(t)\widehat{A}_{\mathbf{k}\nu}(t') + \widehat{A}_{\mathbf{k}\nu}(t')\widehat{A}_{\mathbf{k}\lambda}(t)\rangle, \quad (7.45)$$

where $\widehat{A}_{\mathbf{k}\lambda,\nu}(t)$ are the operators of the vector potential in the Heisenberg representation. The averaging over the states of the system is denoted by brackets. The dependence on the difference of times is due to the stationarity of the state of the system. The Heisenberg operators corresponding to different times generally do not commute with one another. Therefore, the correlation function is determined through their symmetrization product. Unlike the preceding consideration, here the presence of the external current is not assumed, and the field fluctuations occur owing to the thermal motion of particles of a medium only. Note, that in a quantum system this motion does not cease even at zero temperature.

On the basis of the results of the preceding examples, find the connection between the spectral tensor $K_{\lambda\nu}(\boldsymbol{k},\omega)$ and the generalized susceptibility (Equation (7.43)).

Solution. In the absence of the external current $\widehat{V} = 0$, and the medium can be in the stationary states $|n\rangle$ satisfying the Schrödinger equation $\widehat{\mathcal{H}}|n\rangle = E_n|n\rangle$, where

$\hat{\mathcal{H}}$ is the Hamiltonian of the medium. The Heisenberg operators, in accordance with Equation (I.C31), have the form

(1) $$\hat{A}_{\mathbf{k}\lambda,\nu}(t) = \exp\left(\frac{i}{\hbar}\hat{\mathcal{H}}t\right)\hat{A}_{\mathbf{k}\lambda,\nu}\exp\left(-\frac{i}{\hbar}\hat{\mathcal{H}}t\right),$$

where on the right-hand side of the equality, the time-independent operators in the Schrödinger representation appear. Let us perform the successive calculations:

(2) $$K_{\lambda\nu}(\boldsymbol{k},\omega) = \int_{-\infty}^{\infty} K_{\lambda\nu}(\boldsymbol{k},\tau)e^{i\omega\tau}d\tau$$

$$= \frac{1}{2}\int_{-\infty}^{\infty}\langle\hat{A}_{\mathbf{k}\lambda}(\tau)\hat{A}_{\mathbf{k}\nu}(0) + \hat{A}_{\mathbf{k}\nu}(0)\hat{A}_{\mathbf{k}\lambda}(\tau)\rangle e^{i\omega\tau}d\tau$$

$$= \frac{1}{2}\sum_{m,n}\rho_n\int_{-\infty}^{\infty}\left\{\langle n|\hat{A}_{\mathbf{k}\lambda}(\tau)|m\rangle\langle m|\hat{A}_{\mathbf{k}\nu}(0)|n\rangle\right.$$

$$\left. + \langle n|\hat{A}_{\mathbf{k}\nu}(0)|m\rangle\langle m|\hat{A}_{\mathbf{k}\lambda}(\tau)|n\rangle\right\} e^{i\omega\tau}d\tau.$$

Furthermore, we have

(3) $$\langle n|\hat{A}_{\mathbf{k}\lambda}(\tau)|m\rangle = \langle n|\hat{A}_{\mathbf{k}\lambda}|m\rangle\, e^{i\omega_{nm}\tau}$$

and, after the substitution of Equation (3) in Equation (2), we obtain

(4) $$K_{\lambda\nu}(\boldsymbol{k},\omega) = \pi(1 + e^{-\hbar\omega/T})\sum_{m,n}\rho_n\langle n|\hat{A}_{\mathbf{k}\lambda}|m\rangle\langle m|\hat{A}_{\mathbf{k}\nu}|n\rangle\delta(\omega - \omega_{mn}).$$

From comparison with Equation (7.44), we have the relation

$$K_{\lambda\nu}(\boldsymbol{k},\omega) = c\hbar\coth\left(\frac{\hbar\omega}{2T}\right)\alpha''_{\lambda\nu}(\boldsymbol{k},\omega), \tag{7.46}$$

where one should insert the imaginary part of Equation (7.43) for $\alpha''_{\lambda\nu}$. By doing so, we express the fluctuations of the electromagnetic field in the equilibrium system in terms of the permittivity and magnetic permeability of the medium. ■

The relation (7.46) is referred to as **the fluctuation-dissipative theorem**. It connects the fluctuations of the equilibrium parameter (the vector potential in the given case) with the dissipative properties of the system, which are described by the imaginary part of the generalized susceptibility.

More details can be found in the books Ginzburg (1987)
Lifshitz and Pitaevskii (1980); Levin and Rytov (1967); Rytov *et al.* (1978)

Scattering of light in an equilibrium isotropic transparent medium. According to the results of Section 6.1, in the absence of absorption, the electromagnetic wave propagates in a homogeneous medium without attenuation. However, this result ignores small random inhomogeneities of the dielectric permittivity, which are caused by the thermal motion of particles of a medium. Due to the thermal motion of particles, the density and the temperature of the medium are subject to random fluctuations resulting in the fluctuations of the permittivity $\tilde{\varepsilon}$ (for simplicity, we assume that the magnetic permeability $\mu = 1$). This causes the scattering of the incident wave by inhomogeneities of the dielectric permittivity, the appearance of the random scattered waves and the attenuation of the incident wave. Let us calculate the corresponding attenuation for the simplest model of the scattering medium.

We separate the medium into equal macroscopic cells of a volume V, whose linear dimensions are small compared with the wavelength. The dielectric permittivity of a cell can be written in the form

$$\widetilde{\varepsilon} = \varepsilon + \Delta\varepsilon, \qquad \langle \Delta\varepsilon \rangle = 0, \tag{7.47}$$

where $\varepsilon(\omega, \tau, T) = \widetilde{\varepsilon}$ is the permittivity averaged over ensemble of the states of the medium (Chapter 6), τ and T are the density and the temperature of the medium, and $\Delta\varepsilon$ is the fluctuating addition, which changes from one cell to another. Let the linear dimensions of such the region be much less than a radius of the intermolecular interaction. In this case, the fluctuations in different cells are statistically independent, and the average over equilibrium ensemble of the product of the fluctuations in the different cells vanishes:

$$\langle \Delta\varepsilon_i \Delta\varepsilon_k \rangle = \langle \Delta\varepsilon^2 \rangle \delta_{ik}, \tag{7.48}$$

where i and k enumerate the cells. Since fluctuations in the different cells are independent, the field they scatter will also be independent, and we need to consider only one such cell.

Let the primary plane monochromatic wave propagate in the direction along the Ox-axis. We assume the scattered field $\boldsymbol{E}_s$ to be small, and write the total field in the form $\widetilde{\boldsymbol{E}} = \boldsymbol{E} + \boldsymbol{E}_s$, $|\boldsymbol{E}_s| \ll |\boldsymbol{E}|$, where $\boldsymbol{E} = \langle \widetilde{\boldsymbol{E}} \rangle$ is the field averaged over the fluctuations of the medium. The fluctuation of the electric polarization has the form $\Delta \boldsymbol{P} = \boldsymbol{E}\Delta\varepsilon/4\pi$, and the fluctuating dielectric moment of a volume V is equal to $\boldsymbol{p} = V\Delta\boldsymbol{P}$. The field scattered by this volume, $\boldsymbol{E}_s$, can be determined by Equation (I.5.29). Its frequency will be the same as the frequency ω of the incident wave. The total energy flux scattered by a volume V is given by Equation (I.5.31). If we average this value over fluctuations and over the period of field variations and write the electric field in the form $\boldsymbol{E}(x,t) = \boldsymbol{E}(x)\cos(kx - \omega t)$, we then get

$$\langle \bar{I}_s \rangle = \frac{2}{3c^3}\langle \overline{\ddot{\boldsymbol{p}}^2} \rangle = \frac{cE^2(x)}{8\pi}\,\frac{\omega^4 V \langle \Delta\varepsilon^2 \rangle_V}{6\pi c^4}\,V, \tag{7.49}$$

where $I = \bar{\gamma} = cE^2(x)/8\pi$ is the energy flux of the incident wave, which is determined by the Poynting vector of the wave. Let the volume of the cell be a prism in shape, with a length Δx in the direction of the Ox-axis and with an area of the cross section S: $V = S\Delta x$. Then the ratio $\langle \bar{I}_s \rangle / S = -\Delta I$ can be considered the decrease of the energy flux due to the partial scattering of the wave. Hence we can write the relation (Equation (7.49)) in the form

$$\Delta I = -Ih\Delta x, \qquad \text{where} \qquad h = \frac{\omega^4 V \langle \Delta\varepsilon^2 \rangle_V}{6\pi c^4}. \tag{7.50}$$

The law describing the attenuation of the wave intensity can be obtained by replacing $\Delta x \to dx$, $\Delta I \to dI$ and by integrating the equality (7.50):

$$I(x) = I_0 e^{-hx}. \tag{7.51}$$

The quantity h is called **extinction ratio**. It is determined by frequency and by the product $V\langle\Delta\varepsilon^2\rangle_V$, where $\langle\Delta\varepsilon^2\rangle_V$ is the root-mean-square fluctuation of the dielectric permittivity in a volume V. This quantity can be expressed in terms of more convenient parameters with the aid of thermodynamic theory of fluctuations (Landau and Lifshitz (1980)):

$$\langle\Delta\varepsilon^2\rangle_V = \left(\frac{\partial\varepsilon}{\partial\tau}\right)^2 \langle\Delta\tau^2\rangle_V + \left(\frac{\partial\varepsilon}{\partial T}\right)^2 \langle\Delta T^2\rangle_V + 2\left(\frac{\partial\varepsilon}{\partial\tau}\right)\left(\frac{\partial\varepsilon}{\partial T}\right)\langle\Delta\tau\Delta T\rangle_V, \quad (7.52)$$

where

$$\langle\Delta\tau^2\rangle_V = \frac{\tau T}{V}\left(\frac{\partial\tau}{\partial p}\right)_T, \qquad \langle\Delta T^2\rangle_V = \frac{T^2}{V\tau c_v}, \qquad \langle\Delta\tau\Delta T\rangle_V = 0,$$

c_v is the specific heat per unit mass at constant volume. Using these relations, we obtain **the Einstein formula** for the extinction ratio,

$$h = \frac{\omega^4}{6\pi c^4}\left[\left(\frac{\partial\varepsilon}{\partial\tau}\right)_T^2 \left(\frac{\partial\tau}{\partial p}\right)_T T\tau + \left(\frac{\partial\varepsilon}{\partial T}\right)_\tau^2 \frac{T^2}{c_v\tau}\right]. \quad (7.53)$$

The above-considered simplest theory of the molecular scattering proceeded from the assumption that in the presence of fluctuations the isotropy of a medium is retained, that is, the fluctuations can be described by the scalar dielectric permittivity. This assumption is proved for spherically symmetric molecules but not valid for more complex molecules. Owing to fluctuations, the medium can exhibit deviations from the isotropy that results in more complex expressions for the extinction ratio. Moreover, the scattering can be accompanied by a change in frequency (Raman[4]–Landsberg–Mandelshtam scattering). The scattering without a change in frequency is called **Rayleigh scattering** (it was considered above).[5] The first stages of development of the theory of light scattering in gases are described by Sobelman (2002). One can find the up-to-date detailed treatment of the theory of the light scattering in the books Ginzburg (1987); Leontovich (1983); Landau *et al.* (1984).

Single scattering of electromagnetic waves by random inhomogeneities of a medium. Every so often, electromagnetic waves propagate across a medium which is in nonequilibrium state (e.g., clouds of water vapor and dust in the atmosphere, clouds of plasma and of rarefied neutral matter in an interstellar space) In these cases, the scattering is much more intensive than the scattering by equilibrium fluctuations, and the scattered radiation carries the important information on the structure and properties of a scattering medium.

Example 7.5

Consider a cloud consisting of N identical immobile macroscopic spherical particles, which is present in vacuum or in a transparent medium. A plane

4) Raman Chandrasejara Venkata (1888–1970), outstanding Indian physicist, Nobel Prize laureate (1930) for discovery of the scattering of light in liquids, which was named after him. This type of scattering has been independently discovered by Soviet scientists G.S. Landsberg and L.I. Mandelstam (see the paper by E.L. Feinberg (2003), for details).
5) Rayleigh (Strett) John William (1842–1919), outstanding English physicist, Nobel Prize laureate.

monochromatic electromagnetic wave is incident on the cloud. A wavelength of the wave is large compared with a size of a separate particle (but not of the cloud). Calculate the effective differential cross section of scattering by the cloud of particles, assuming that the single scattering takes place. The latter condition means that one can neglect the attenuation of the incident wave and the secondary scattering of the already scattered wave. However, the possible correlation in the position of the cloud particles (the structural factor of a medium) must be taken into account. The electric polarizability $\beta(\omega)$ of the particles is known, the magnetic polarizability is negligible.

Solution. If the wavelength $\lambda = 2\pi v_{ph}/\omega$ is large compared with a size of a separate particle, its dipole moment is expressed in terms of the polarizability $\beta(\omega)$ in a homogeneous field:

$$\boldsymbol{p}_a(t) = \beta \boldsymbol{E}_0 \exp(i\boldsymbol{k}\cdot\boldsymbol{r} - i\omega t),$$

where $\boldsymbol{r}_a$ is a radius-vector of a center of mass of a particle, $\boldsymbol{E}_0$ is the amplitude of the incident wave, and $\boldsymbol{k}$ is its wave vector. For a homogeneous spherical particle with the dielectric permittivity ε, the polarizability was calculated in Problem 2.11:

$$\beta = \frac{\varepsilon - \varepsilon_0}{\varepsilon + 2\varepsilon_0} a^3,$$

where ε_0 is the dielectric permittivity of a surrounding medium, and a is a radius of a particle. In the wave zone, the scattered field is given by Equation (I.5.29), where c is replaced by v_{ph}:

$$\boldsymbol{E}'_a = \frac{\omega^2}{v_{ph}^2 R} \boldsymbol{n}' \times [\boldsymbol{p}_a \times \boldsymbol{n}'] \exp(-i\boldsymbol{k}' \cdot \boldsymbol{r}_a), \qquad r_a \ll R, \qquad \boldsymbol{k}' = \frac{\omega}{v_{ph}} \boldsymbol{n}'.$$

The unit vector $\boldsymbol{n}'$ indicates the direction of scattering.

The differential cross section of scattering by the cloud is determined by the formula given in the beginning of Section 6.3. One should take into account the scattering by all particles of the cloud and perform the ensemble averaging over all mutual positions of the particles:

$$\frac{d\Sigma}{d\Omega} = \left\langle \left| \sum_{a=1}^{N} \exp(-i\boldsymbol{q} \cdot \boldsymbol{r}_a) \right|^2 \right\rangle \frac{\varepsilon_0^2 \omega^4 \beta^2(\omega)}{c^4} \sin^2\theta = F(\boldsymbol{q}) \frac{d\sigma}{d\Omega}, \tag{7.54}$$

where $\boldsymbol{q} = \boldsymbol{k}' - \boldsymbol{k}$ is the wave vector transferred to a medium, $q = 2k\sin(\vartheta/2)$, ϑ is the scattering angle, and $d\Omega = \sin\vartheta \, d\vartheta \, d\varphi$, θ is an angle between $\boldsymbol{E}_0$ and $\boldsymbol{k}'$. The quantity

$$\frac{d\sigma}{d\Omega} = \frac{\varepsilon_0^2 \omega^4 \beta^2(\omega)}{c^4} \sin^2\theta \tag{7.55}$$

is the scattering cross section of the wave by a separate particle, and

$$F(\boldsymbol{q}) = \left\langle \left| \sum_{a=1}^{N} \exp(-i\boldsymbol{q} \cdot \boldsymbol{r}_a) \right|^2 \right\rangle \tag{7.56}$$

is a structural factor of the medium (factor of coherence). In Problem I.5.122, we already considered it for the two limiting cases: the completely coherent scattering ($|\boldsymbol{q}\cdot\boldsymbol{r}_a| \ll 1$) and completely incoherent scattering ($\sum_{a\neq b}\exp(i\boldsymbol{q}\cdot(\boldsymbol{r}_a-\boldsymbol{r}_b)\approx 0$). Here we consider a more general case.

The particle number density in the cloud can be written in the form

$$n(\boldsymbol{r}) = \sum_{a=1}^{N}\delta(\boldsymbol{r}-\boldsymbol{r}_a),$$

and the structural factor can be presented as a double integral over volume of the cloud:

$$F(\boldsymbol{q}) = \langle |n_{\mathbf{q}}|^2\rangle = \int \langle n(\boldsymbol{r}_1)n(\boldsymbol{r}_2)\rangle \exp[-i\boldsymbol{q}\cdot(\boldsymbol{r}_1-\boldsymbol{r}_2)]\,d^3r_1\,d^3r_2. \tag{7.57}$$

We assume that the distribution of particles within the cloud is homogeneous and isotropic. We write the average from the product of their number densities at two points in terms of the cross-correlation function (Landau and Lifshitz (1980)):

$$\langle n(\boldsymbol{r}_1)n(\boldsymbol{r}_2)\rangle = \overline{n}^2 + \overline{n}[\delta(\boldsymbol{r}_1-\boldsymbol{r}_2) + g(|\boldsymbol{r}_1-\boldsymbol{r}_2|)], \tag{7.58}$$

where $\overline{n}$ is the number density averaged over the cloud, and $g(r)$ is the correlation function which, when $r \to 0$, has the order of magnitude $\overline{n}$, and tends to zero when $r \gg l_c$, where $l_c \ll L$ is the correlation length, and L is a size of the cloud.

The first term in Equation (7.58) describes the forward coherent scattering. When integrating, according to Equation (7.57), we have

$$\int \overline{n}^2 \exp[-i\boldsymbol{q}\cdot(\boldsymbol{r}_1-\boldsymbol{r}_2)]d^3r_1d^3r_2$$
$$= \overline{n}^2 \int \exp(-i\boldsymbol{q}\cdot\boldsymbol{r})d^3rd^3R \approx (2\pi)^3 V\overline{n}^2\delta(\boldsymbol{q}),$$

where $\boldsymbol{r} = \boldsymbol{r}_1 - \boldsymbol{r}_2$, $\boldsymbol{R} = (\boldsymbol{r}_1 + \boldsymbol{r}_2)/2$, and the right-hand side is close to a delta-function provided the linear dimension of the cloud L is large compared with the wavelength. At $q \to 0$, one has $(2\pi)^3\delta(\boldsymbol{q}) \to V$, which allows one to write the forward scattering cross section in the form

$$\left.\frac{d\Sigma(\boldsymbol{q})}{d\Omega}\right|_{\mathbf{q}\to 0} \to N^2\frac{d\sigma}{d\Omega}, \qquad N = \overline{n}V. \tag{7.59}$$

At $q \to 0$ and $N \gg 1$, the contribution from the terms proportional to $\overline{n}$ in Equation (7.58) is negligibly small.

For scattering angles $\vartheta > 1/kL$, we have

$$\frac{d\Sigma}{d\Omega} = N\frac{d\sigma}{d\Omega}\left\{1 + \int g(r)\exp(-i\boldsymbol{q}\cdot\boldsymbol{r})\,d^3r\right\}. \tag{7.60}$$

The first term in brackets describes the incoherent scattering by N particles: $d\Sigma/d\Omega = N(d\sigma/d\Omega)$. The second term describes the coherent scattering, which vanishes if the distribution of the particles is completely random (when $g \to 0$). In the case $l_c \ll \lambda$ (small-scale macroscopic inhomogeneities), we have

$\exp(i\boldsymbol{q}\cdot\boldsymbol{r}) \approx 1$ and $\int g(r)\,d^3r \approx \overline{n}l_c^3$. The angular dependence of the scattering cross section by the cloud is the same as by a separate particle, and the coherent contribution is essential when $\overline{n}l_c^3 > 1$. In the opposite limiting case $l_c \gg \lambda$ and $\overline{n}l_c^3 \gg 1$, the coherent term is large in the region of small scattering angles $\vartheta < 1/kl_c$ (but $\vartheta > 1/kL$), resulting in the sharp anisotropic scattering. When $\vartheta \gg 1/kl_c$ the coherent contribution decreases due to oscillations of the exponent, and the incoherent scattering plays the main part.

An assumption of single scattering can be used if only a small fraction of the wave energy is scattered in the cloud, that is, the total scattering cross section by the cloud is small compared with geometrical sizes of the cloud: $\Sigma \ll L^2$. ■

Diffraction at a random screen. When a certain regular wave travels across a layer of a medium having random inhomogeneities, at the outlet the field will have the random component that should be described by correlation functions. If the thickness of the layer is small compared with other lengths under consideration, such a layer can be considered a plane chaotic screen, which randomly modulates the incident wave in amplitude and phase. The properties of such a screen are given by its **transmission function** which is the dimensionless complex random function $M(\boldsymbol{\rho})$, $|M(\boldsymbol{\rho})| \le 1$, where the vector $\boldsymbol{\rho}$ is in the plane of a screen that we consider to be the (x, y)-plane. When the initial regular wave $u_0(\boldsymbol{\rho}, z)e^{-i\omega t}$ is incident on the screen, immediately behind the screen, $(z \to +0)$ there appears a random wave field

$$u(\boldsymbol{\rho}, 0) = M(\boldsymbol{\rho})u_0(\boldsymbol{\rho}, 0), \qquad (7.61)$$

whose statistic characteristics will be determined by the random function $M(\boldsymbol{\rho})$. For simplicity, we consider the scalar case where we can neglect a change in the direction of the vector quantities. Let the incident wave be plane, and set $u_0(\boldsymbol{\rho}, 0) = 1$.

Let the medium be inhomogeneous behind the screen $(z > 0)$. In this case, the problem is reduced to the investigation of the propagation of a random field in a homogeneous medium, where the field satisfies the Helmholtz equation

$$\triangle u + k^2 u = 0, \qquad k^2 = \frac{\omega^2}{c^2}\varepsilon(\omega)\mu(\omega), \qquad (7.62)$$

and the boundary condition (7.61). We apply the Rayleigh method, that is, expand the field in the Fourier integral in the $z = 0$ plane:

$$u(\boldsymbol{\rho}, 0) = M(\boldsymbol{\rho}) = \int m(\boldsymbol{\kappa})\exp(i\boldsymbol{\kappa}\cdot\boldsymbol{\rho})\,\frac{d^2\kappa}{(2\pi)^2},$$

$$m(\boldsymbol{\kappa}) = \int M(\boldsymbol{\rho}')\exp(-i\boldsymbol{\kappa}\cdot\boldsymbol{\rho}')\,d^2\rho'. \qquad (7.63)$$

In the region $z > 0$, each spatial harmonic must satisfy Equation (7.62). This means that its coordinate dependence must be of the form

$$m(\boldsymbol{\kappa})\exp(i\boldsymbol{\kappa}\cdot\boldsymbol{\rho} + ik_z z), \qquad k_z = \sqrt{k^2 - \kappa^2}. \qquad (7.64)$$

According to the Fourier spectrum $m(\boldsymbol{\kappa})$ of the transmission function, the values κ can be either less or larger than k. Hence, the superposition (7.63) of planes waves at $z > 0$ will include the propagating waves with $\sqrt{k^2 - \kappa^2} > 0$, $\kappa < k$ as well as the decaying (inhomogeneous) waves with $\sqrt{k^2 - \kappa^2} = i\sqrt{|k^2 - \kappa^2|}$, $\kappa > k$:

$$u(\boldsymbol{\rho}, z) = \int m(\boldsymbol{\kappa}) \exp(i\boldsymbol{\kappa} \cdot \boldsymbol{\rho} + i\sqrt{k^2 - \kappa^2 z}) \frac{d^2\kappa}{(2\pi)^2}. \tag{7.65}$$

The field (Equation (7.65)) is expressed in terms of the Fourier transform of the random transmission function $m(\boldsymbol{\kappa})$ and must be characterized by its correlation functions. In order to find them, the statistic properties of the function $M(\boldsymbol{\rho})$ must be given. Assume that the ensemble of random inhomogeneities of the screen is statistically homogeneous and isotropic. This means that

$$\langle M(\boldsymbol{\rho})\rangle = M_0 = const, \quad \langle (M(\boldsymbol{\rho}_1) - M_0)(M^*(\boldsymbol{\rho}_2) - M_0^*)\rangle = K_m(|\boldsymbol{\rho}_1 - \boldsymbol{\rho}_2|), \tag{7.66}$$

where the correlation function $K_m(\rho)$ is characterized by a certain correlation length l_m, that is, it becomes negligibly small when $\rho \gg l_m$. We need also the Fourier transform of the correlation function which will be denoted by $\langle m^2\rangle_\kappa$:

$$\langle m^2\rangle_\kappa = \int K_m(\rho) \exp(-i\boldsymbol{\kappa} \cdot \boldsymbol{\rho})\, d^2\rho. \tag{7.67}$$

This quantity noticeably differs from zero at values κ of the order or less than $2\pi/l_m$. When $\kappa \gg 2\pi/l_m$ the integral in Equation (7.67) becomes rather small due to the fast oscillations of the exponent.

Example 7.6

Calculate the ensemble average of the field $\overline{u}(\boldsymbol{\rho}, z) \equiv \langle u(\boldsymbol{\rho}, z)\rangle$ in the region $z > 0$.

Solution. From Equation (7.65), we have

(1) $$\langle u(\boldsymbol{\rho}, z)\rangle = \int \langle m(\boldsymbol{\kappa})\rangle \exp\left(i\boldsymbol{\kappa} \cdot \boldsymbol{\rho} + i\sqrt{k^2 - \kappa^2 z}\right) \frac{d^2\kappa}{(2\pi)^2}.$$

Equations (7.63) and (7.66) give $\langle m(\boldsymbol{\kappa})\rangle = (2\pi)^2 M_0 \delta(\boldsymbol{\kappa})$. Finally, we get

(2) $$\overline{u}(\boldsymbol{\rho}, z) = M_0 e^{ikz}.$$

Independence of $\boldsymbol{\rho}$ stems from statistic homogeneity of the screen. ■

Example 7.7

Calculate the field correlation function

$$K_u(\boldsymbol{\rho}, z; \boldsymbol{\rho}', z') = \left\langle (u(\boldsymbol{\rho}, z) - \overline{u}(\boldsymbol{\rho}, z))(u(\boldsymbol{\rho}', z') - \overline{u}(\boldsymbol{\rho}', z'))\right\rangle$$

at large distances behind the screen ($z, z' \gg \lambda$, where λ is a wavelength behind the screen). Express K_u in terms of the Fourier transform of the correlation function (7.67) which characterizes the statistic properties of the transmission through the screen.

Solution. Using Equation (7.63), we write down the difference

$$u(\boldsymbol{\rho}, z) - \overline{u}(\boldsymbol{\rho}, z) = \int (m(\boldsymbol{\kappa}) - \langle m(\boldsymbol{\kappa})\rangle) \exp(i\boldsymbol{\kappa}\cdot\boldsymbol{\rho} + ik_z z)\frac{d^2\kappa}{(2\pi)^2} \tag{1}$$

and construct the required correlation function:

$$\begin{aligned} K_u(\boldsymbol{\rho}, z; \boldsymbol{\rho}', z') = \int & \langle (m(\boldsymbol{\kappa}) - \langle m(\boldsymbol{\kappa})\rangle)(m^*(\boldsymbol{\kappa}') - \langle m^*(\boldsymbol{\kappa}')\rangle)\rangle \\ & \times \exp\{i(\boldsymbol{\kappa}\cdot\boldsymbol{\rho} - \boldsymbol{\kappa}'\cdot\boldsymbol{\rho}') + ik_z z - ik_z^* z'\}\frac{d^2\kappa\, d^2\kappa'}{(2\pi)^4}. \end{aligned} \tag{2}$$

Further on, we calculate the correlator entering the integrand (Equation (2)) with the aid of the second relation (7.63):

$$\begin{aligned} &\langle (m(\boldsymbol{\kappa}) - \langle m(\boldsymbol{\kappa})\rangle)(m^*(\boldsymbol{\kappa}') - \langle m^*(\boldsymbol{\kappa}')\rangle)\rangle \\ &\quad = \int K_m(|\boldsymbol{\rho} - \boldsymbol{\rho}'|) \exp(-i\boldsymbol{\kappa}\cdot\boldsymbol{\rho} + i\boldsymbol{\kappa}'\cdot\boldsymbol{\rho}')\, d^2\rho\, d^2\rho'. \end{aligned} \tag{3}$$

Introducing the new integration variables $\boldsymbol{r} = \boldsymbol{\rho} - \boldsymbol{\rho}'$ and $\boldsymbol{R} = (\boldsymbol{\rho} + \boldsymbol{\rho}')/2$ and using Equation (1.219), $\int \exp(-i(\boldsymbol{\kappa} - \boldsymbol{\kappa}')\cdot\boldsymbol{R})\, d^2R = (2\pi)^2\delta(\boldsymbol{\kappa} - \boldsymbol{\kappa}')$, as well as Equation (7.67) we find that

$$\langle (m(\boldsymbol{\kappa}) - \langle m(\boldsymbol{\kappa})\rangle)(m^*(\boldsymbol{\kappa}') - \langle m^*(\boldsymbol{\kappa}')\rangle)\rangle = (2\pi)^2 \langle m^2\rangle_\kappa \delta(\boldsymbol{\kappa} - \boldsymbol{\kappa}'). \tag{4}$$

Now we can write Equation (2) in the form

$$K_u(\boldsymbol{\rho}, z; \boldsymbol{\rho}', z') = \int \langle m^2\rangle_\kappa \exp\{i\boldsymbol{\kappa}\cdot(\boldsymbol{\rho} - \boldsymbol{\rho}') + ik_z z - ik_z^* z'\}\frac{d^2\kappa}{(2\pi)^2}, \tag{5}$$

where the constant of propagation $k_z = \sqrt{k^2 - \kappa^2}$ becomes purely imaginary at $\kappa > k$. However, when $z, z' \gg \lambda$ these values of κ give an exponentially small contribution to the integral, and can thus be excluded from the integration. Finally, we obtain the correlation function that depends on the difference of coordinates $\boldsymbol{r} = \boldsymbol{\rho} - \boldsymbol{\rho}'$, $\zeta = z - z'$:

$$K_u(\boldsymbol{r}, \zeta) = \int_{\kappa \le k} \langle m^2\rangle_\kappa \exp\left\{ i\boldsymbol{\kappa}\cdot\boldsymbol{r} + i\sqrt{k^2 - \kappa^2}\zeta \right\} \frac{d^2\kappa}{(2\pi)^2} \tag{7.68}$$

(note that $\boldsymbol{r}$ is the transverse position vector lying in the $z = const$ plane). ■

In this section, we have considered only the simplest problems concerning the propagation of electromagnetic waves in randomly inhomogeneous media. The more detailed theory can be found in the books Rytov *et al.* (1978); Tatarskii (1967); Bass and Fuks (1972); Ishimaru (1978). See also the review article by Kuz'min and Romanov (1996).

Problems

7.21• Using Equation (7.46), calculate the dependence of the correlation tensor $K_{\lambda\nu}(\boldsymbol{r} - \boldsymbol{r}', \omega)$ for components of the vector potential at a given frequency ω, in homogeneous (infinite) isotropic medium.

7.22 For the homogeneous isotropic medium, calculate the correlation tensors

$$\langle E_\lambda(\boldsymbol{r},t)E_\nu(\boldsymbol{r}',t')\rangle_{\mathbf{k}\omega}, \qquad \langle E_\lambda(\boldsymbol{r},t)B_\nu(\boldsymbol{r}',t')\rangle_{\mathbf{k}\omega}, \qquad \langle B_\lambda(\boldsymbol{r},t)B_\nu(\boldsymbol{r}',t')\rangle_{\mathbf{k}\omega},$$

which are analogous to the tensor $K_{\lambda\nu}(\boldsymbol{k},\omega) \equiv \langle A_\lambda(\boldsymbol{r},t)A_\nu(\boldsymbol{r}',t')\rangle_{\mathbf{k}\omega}$ determined by Equations (7.43) and (7.46).

7.23• For the case of transparent equilibrium medium ($\Im\,\varepsilon \to 0,\ \Im\,\mu \to 0$), calculate the spectral energy density $\rho(\omega)$ of the equilibrium electromagnetic field.

7.24 Write down the extinction ratio (Equation (7.53)) for gases, in the optical range of spectrum, in terms of the refraction index $n = \sqrt{\varepsilon}$, assuming $n - 1 \ll 1$ and neglecting the dependence of the ε on temperature.

7.25• In volume V, there are excited stochastic fluctuations of a medium. They can be described by the given correlation function of the dielectric permittivity $K_\varepsilon(|\boldsymbol{r}_1 - \boldsymbol{r}_2|) = \langle\Delta\varepsilon(\boldsymbol{r}_1)\Delta\varepsilon(\boldsymbol{r}_2)\rangle$, which is homogeneous and isotropic within this volume. The correlation length l_c is small compared with the linear dimension L of the volume. On the assumption of single scattering (Example 7.5), express the scattering cross section of a wave in terms of the correlation function of the permittivity.

7.26 Calculate the differential and total scattering cross section of an electromagnetic wave provided the correlation function of the dielectric permittivity (Problem 7.25) has the Gaussian form $K_\varepsilon(r) = K_0 \exp(-r^2/2l_c^2)$, where K_0 and l_c are constant.

7.27 The fluctuations of the transmission of the chaotic screen are small-scale compared with the length of a propagating wave, that is, $kl_m \ll 1$. On the basis of Equation (7.68), investigate the correlation properties of the field behind the screen in the transverse ($\zeta = 0$) and longitudinal ($\boldsymbol{r} = 0$) directions. For these cases, evaluate the correlation function K_u and the corresponding correlation lengths $l_\perp$ and $l_\parallel$.

7.28 Investigate the opposite limiting case of large-scale fluctuations of the chaotic screen, $kl_m \gg 1$. Estimate the correlation functions and evaluate the transverse and longitudinal correlation lengths.

7.29 For the case of large-scale fluctuations of the chaotic screen ($kl_m \gg 1$) with the Gaussian spectrum $\langle m^2\rangle_\kappa = \langle m^2\rangle_0 \exp(-\kappa^2 l_m^2/2)$, calculate the three-dimensional correlation function $K_u(\boldsymbol{r},\zeta)$. Evaluate the transverse and longitudinal correlation lengths.

7.3
Waves in Nonlinear and Active Media

We were already dealing with the nonlinear waves in Section 4.3 devoted to magnetic hydrodynamics. Among these are, in particular, the simple Riemann waves and the shock waves propagating in a conducting medium with a magnetic field. In the present section, we consider primarily certain types of nonlinear waves in plasma.

Burgers equation. Consider a one-dimensional wave with a flat front, propagating in a conducting medium along the direction of the external magnetic field (Ox-axis). Unlike the problems considered in Section 4.3, we will take into account the energy dissipation of the wave and its nonlinearity, assuming that these effects are not large. In the geometry under consideration, the magnetic field does not affect the motion of the matter and it can be described by one-dimensional equations (4.24), (4.23), and (4.26):

$$\partial_t \tau + u\partial_x \tau + \tau\partial_x u = 0, \tag{7.69}$$

$$\partial_t u + u\partial_x u = -\tau^{-1}\partial_x p + \nu\partial_x^2 u, \tag{7.70}$$

$$\partial_t s + u\partial_x s = (\chi/\tau T)\partial_x^2 T. \tag{7.71}$$

On the right-hand side of the last equation, we omit the term proportional to $(\partial_x u)^2$, which is of the third-order of smallness. For brevity, we denote the partial derivatives by symbols ∂_x, ∂_t and so on. as has already been done in Chapter I.4. With the aid of the equation of state $\tau = \tau(p, s)$, we get

$$\partial_t \tau + u\partial_x \tau = (\partial_p \tau)_s(\partial_t p + u\partial_x p) + (\partial_s \tau)_p(\partial_t s + u\partial_x s).$$

Using the last equality, from Equations (7.69) and (7.71) we have

$$\partial_t p + u\partial_x p + (\partial_\tau p)_s \tau\partial_x u = -(\chi/\tau T)(\partial_\tau p)_s(\partial_s \tau)_p \partial_x^2 T. \tag{7.72}$$

Further, we use the approximate equality $\partial_x^2 T \approx (\partial_p T)_s \partial_x^2 p$ (the remaining terms give the terms of the third order in Equation (7.72)) and the thermodynamic identity $T(c_V^{-1} - c_p^{-1}) = -(\partial_\tau p)_s(\partial_p T)_s(\partial_s \tau)_p$. Using these equalities, we reduce Equation (7.72) to the form

$$\partial_x u + \tau^{-1}(\partial_\tau p)_s(\partial_t p + u\partial_x p) = \chi\tau_0^{-2}c_s^{-2}(c_V^{-1} - c_p^{-1})\partial_x^2 p, \tag{7.73}$$

where the undisturbed quantities τ_0 and $c_s^2 = (\partial_\tau p)_{s0}$ have been substituted on the right-hand side.

Now we will exclude the variables u and τ from Equations (7.70) and (7.73) and obtain one equation that will contain only the alternating part of the pressure $p' = p - p_0$ and undisturbed quantities. On the left-hand side of the equations, we will retain only the terms of the second order of smallness, as was done for the right-hand sides. Luckily, in quadratic terms we can take the same relations between small disturbances as that in the linear theory, that is, for sound waves of small amplitude:

$$u \approx c_s\tau'/\tau_0 \approx p'/c_s\tau_0, \qquad \tau \approx \tau_0 + p'/c_s^2,$$
$$\tau^{-1}(\partial_\tau p)_s^{-1} \approx c_s^{-2}\tau_0^{-1} + p'/c_s^4\tau_0^2 - \tau^0(\partial_p^2 V)_s p',$$

where $V = 1/\tau$ is a specific volume. With the aid of these relations, we write down Equations (7.70) and (7.73) in the form

$$\partial_t u + \tau_0^{-1}\partial_x p' = \nu(c_s\tau_0)^{-1}\partial_x^2 p',$$
$$\partial_x u + c_s^{-2}\tau_0^{-1}\partial_t p' = -c_s\tau^0(\partial_p^2 V)_s p'\partial_x p' + (\chi/\tau_0^2 c_s^2)(c_V^{-1} - c_p^{-1})\partial_x^2 p'.$$

By differentiating the first of these equations with respect to x, and the second with respect to t, and subtracting from one another, we have

$$(\partial_x - c_s^{-1}\partial_t)(\partial_x + c_s^{-1}\partial_t)p' = 2\partial_x[ac_s^2\partial_x^2 p' - bc_s^{-1}p'\partial_x p'], \tag{7.74}$$

where a and b are the constant coefficients:

$$a = 2^{-1}c_s^{-3}[\nu + \chi V(c_V^{-1} - c_p^{-1})], \qquad b = (c_s^2/2V^2)(\partial_p^2 V)_s. \tag{7.75}$$

It is natural that the wave propagating along the Ox-axis would exhibit the dependence on coordinate and on time of the form $p'(x - c_s t, \epsilon x)$, where ϵ is a small parameter (in accordance with a smallness of the nonlinearity and dissipation). Therefore, the operator $\partial_x + c_s^{-1}\partial_t$ has the order of ϵ, and the operator $\partial_x - c_s^{-1}\partial_t \approx 2\partial_x$ if we keep the terms $\sim \epsilon$. Thus, we see that Equation (7.74) will be satisfied provided

$$\partial_t p' + c_s\partial_x p' + bp'\partial_x p' = ac_s^3\partial_x^2 p'. \tag{7.76}$$

The last equation can be simplified by passing to the new independent variables t and $\xi = x - c_s t$, as well to the new function $P(\xi, t) = bp'(x, t)$:

$$\partial_t P + P\partial_\xi P = \kappa\partial_\xi^2 P, \tag{7.77}$$

where $\kappa = ac_s^2$. We obtained **the Burgers equation**, ,which describes the waves in continuous media, taking into account their weak nonlinearity and dissipation. We derived it as a result of rather cumbersome calculations based on a certain specific model of a continuous medium. However, the Burgers equation is widely used in other fields of research as well. The more detailed insight into this and other nonlinear problems one can receive from the book of Whitham (1974) and from other sources given in the literature for the present chapter.

Example 7.8

With the aid of Equation (7.76), calculate the profile of the velocity and evaluate the thickness of the stationary weak magnetohydrodynamic shock wave, which propagates along the direction of an external magnetic field. The pressure changes from the value p_1 at large distances before the front up to $p_2 > p_1$ behind the front.

Solution. Consider the wave propagating in the positive direction of the Ox-axis. The addition p' to the pressure is measured from the undisturbed value p_1 before the front. When $t = const$, we have the boundary conditions $p' \to 0$ at $x \to +\infty$ (before the front), and $p' \to p_2 - p_1 = \Delta p$ at $x \to -\infty$ (behind the front). For the stationary profile, the solution should be sought for in the form $p'(x - u_{sh}t)$, where $u_{sh} > c_s$ is an unknown velocity of the front. Denoting the argument by $\zeta = x - u_{sh}t$, from Equation (7.76) we obtain

(1) $$\partial_\zeta\left[(c_s - u_{sh})p' + \frac{b}{2}p'^2 - \kappa\partial_\zeta p'\right] = 0,$$

and find the first integral of this equation:

(2) $$\kappa\partial_\zeta p' = \frac{b}{2}p'^2 + (c_s - u_{sh})p' + const.$$

From the boundary conditions, we get $const = 0$, $u_{sh} = c_s + b\Delta p/2$ and write Equation (2) for the total pressure $p = p_1 + p'$ in the form

(3) $$\frac{dp}{d\zeta} = \frac{b}{2\kappa}(p - p_1)(p - p_2).$$

This equation has the solution satisfying the imposed boundary conditions:

(4) $$p(\zeta) = \frac{1}{2}(p_1 + p_2) + \frac{1}{2}(p_1 - p_2)\tanh(\zeta/\delta),$$

where

(5) $$\delta = 4\kappa/b\Delta p = 2\kappa/(u_{sh} - c_s)$$

is a thickness of the front of the weak shock wave. ∎

It should be noted that the stationary waves can exist owing to the simultaneous action of the two effects: the nonlinear steepening of a front and the dissipation, which are the two opposite processes; they compensate a change in a wave profile. Due to the effect of the nonlinearity without the dissipation, the stationary solutions can not be obtained: the steepening of a leading front occurs which breaks forms in this case (Section 4.3, Problem 4.69). In spectral language, it means the appearance of the shortwave harmonics in the spectrum. The dissipative effects weaken first of all the harmonics with the large k, since the second derivative in Equation (7.77) gives the dependence of the form k^2. The dissipation without the linearity leads to the unlimited diffusive spread of disturbances (Example 4.6). The mutual balancing of the nonlinear steepening and the diffusive spread results in the possible existence of nonlinear stationary waves in a dissipative medium.

The dissipative coefficient κ can be evaluated as the product Λc_s, where Λ is the mean path of particles between collisions. For a thickness of the shock front, we then have $\delta \approx \Lambda/(M - 1)$, where $M = u_{sh}/c_s$ is the Mach number. Hence, the thickness of the front of the weak ($M - 1 \ll 1$) shock wave is much larger than the mean free path of particles. Extrapolating this evaluation to the case $M - 1 \approx 1$, we get $\delta \approx \Lambda$. Such a value can be considered the smallest from the possible values, since for conversion of the flux energy into the internal energy behind the wave front, at least one or several collisions of particles are required. However, when a shock wave propagates in plasma having always many components (electrons and ions of different kind, and $m_e \ll m_i$), the structure of the front is more complicated due to a slow exchange by energy and momentum between electrons and ions. The thickness of the ion temperature jump is determined by the path Λ_i of ions, whereas the equilibration of temperatures between electrons and ions occurs on the larger length $\Lambda_{ei} \approx (m_i/m_e)^{1/2}\Lambda_i$.

Korteweg–de Vries equation. Let us investigate weakly-nonlinear motions in the two-component cold magnetized plasma, based on the equations derived in the Example 4.9:

$$\partial_t \boldsymbol{u} + (\boldsymbol{u} \cdot \nabla)\boldsymbol{u} = (4\pi n m_i)^{-1} rot\, \boldsymbol{B} \times \boldsymbol{B}, \tag{7.78}$$

$$\partial_t \boldsymbol{B} = rot\,(\boldsymbol{u} \times \boldsymbol{B}) - rot\,\nu_m(0.51 rot_{\|} \boldsymbol{B} + rot_{\perp} \boldsymbol{B})$$
$$- rot(4\pi en)^{-1}(rot\,\boldsymbol{B} \times \boldsymbol{B}) - (m_e c/e^2)\{\partial_t(\boldsymbol{j}/n)$$
$$+ (\boldsymbol{u} \cdot \nabla)(\boldsymbol{j}/n) + n^{-1}(\boldsymbol{j} \cdot \nabla)\boldsymbol{u} - (en)^{-1}(\boldsymbol{j} \cdot \nabla)(\boldsymbol{j}/n)\}, \tag{7.79}$$

where $\boldsymbol{j} = (c/4\pi) rot\,\boldsymbol{B}$.

Let $\boldsymbol{B}_0$ be the external uniform field, and $\boldsymbol{b} = \boldsymbol{B} - \boldsymbol{B}_0$ be the wave field. Firstly, we consider the linearized system assuming $\boldsymbol{b}$ and $\boldsymbol{u}$ to be the small perturbations proportional to $\exp(i\boldsymbol{k} \cdot \boldsymbol{r} - i\omega t)$. Taking the Ox-axis along $\boldsymbol{k}$, and $\boldsymbol{B}_0$ lying in the xz-plane, and excluding $\boldsymbol{u}$, we obtain the system of the algebraic equations for b_y and b_z:

$$b_y[(\boldsymbol{k} \cdot \boldsymbol{v}_A)^2 - i\nu_m k^2\omega - \omega^2(1 + c^2k^2/\omega_{0e}^2)] - b_z(ick\omega/\omega_{0i})(\boldsymbol{k} \cdot \boldsymbol{v}_A) = 0, \tag{7.80}$$

$$b_y(ick\omega/\omega_{0i})(\boldsymbol{k} \cdot \boldsymbol{v}_A) + b_z[v_A^2k^2 - i\nu_m k^2\omega - \omega^2(1 + c^2k^2/\omega_{0e}^2)] = 0. \tag{7.81}$$

When ω_{0e}, $\omega_{0i} \to \infty$, $\nu_m \to 0$ we obtain, from the preceding equations, the dispersion laws $\omega(\boldsymbol{k}) = v_A k \cos\theta$, and $\omega(k) = v_A k$ for the Alfvén and fast magnetoacoustic waves, respectively.

Consider the nonlinear generalization of the fast magnetoacoustic wave. Assuming $ck \ll \omega_{0i}$, $\nu_m k^2 \ll \omega$ and $\theta \gg ck/\omega_{0i}$, we find the dispersion law with the small dispersive and dissipative corrections:

$$\omega(k) = v_A k - \mu k^3 - i\nu_m k^2/2, \qquad \mu = v_A(c^2/2\omega_{0i}^2)(m_e/m_i - \cot^2\theta). \tag{7.82}$$

With the aid of the derived dispersive relation, we can easily write the differential equation for b_z including the dispersive correction and the dissipation. The component b_z plays the main role in the wave, since $b_x = 0$ by virtue of the condition $div\,\boldsymbol{b} = 0$, and $b_y \ll b_z$ because of the smallness of the dispersive correction as it follows from Equations (7.80) and (7.81). We represent Equation (7.82) in the form

$$[-i\omega + v_A ik + \mu(ik)^3]b_z = (\nu_m/2)(ik)^2 b_z \tag{7.83}$$

and pass to the time-coordinate representation:

$$\partial_t b_z + v_A \partial_x b_z + \mu \partial_x^3 b_z = \frac{1}{2}\nu_m \partial_x^2 b_z. \tag{7.84}$$

This equation, however, does not include the nonlinearity. In order to take into account the nonlinear terms of the lowest order corresponding to quadratic nonlinearity, we use Equation (7.79) in the projection on the Oz-axis:

$$\partial_t b_z = \mathbf{e}_z rot\,(\boldsymbol{u} \times \boldsymbol{B}) + \nu_m \triangle b_z$$
$$- (4\pi en)^{-1}\mathbf{e}_z \cdot rot(rot\,\boldsymbol{b} \times \boldsymbol{B}_0) - (c^2/\omega_{0e})\partial_t(\mathbf{e}_z \cdot rot\,\boldsymbol{b}). \tag{7.85}$$

We write the last equation, assuming that the nonlinearity, dispersion and dissipation are the three small additive effects, and the nonlinear terms in the dispersive and dissipative terms are therefore omitted. It is easy to prove that if we exclude $\boldsymbol{u}$ from Equation (7.85) with the aid of Equation (7.78), retaining only linear terms, and use the connection between the coordinate-time derivatives resulting from the dispersion Equation (7.83), we then obtain Equation (7.84). In order to include the nonlinear corrections in the equation, we have to retain in Equation (7.85) the terms with the quadratic nonlinearity of the form $rot(\boldsymbol{u}^{(1)} \times \boldsymbol{b} + \boldsymbol{u}^{(2)} \times \boldsymbol{B}_0)$, where $u^{(1)}$ and $u^{(2)}$ are proportional to b_z and b_z^2, respectively.

Let us write Equation (7.78) up to the terms of the second order in the form

$$\partial_t(\boldsymbol{u}^{(1)} + \boldsymbol{u}^{(2)}) = (4\pi n m_i)^{-1}(\boldsymbol{e}_x \times \partial_x \boldsymbol{b}) \times (\boldsymbol{B}_0 + \boldsymbol{b}) - u_x^{(1)} \partial_x \boldsymbol{u}^{(1)}.$$

By replacing $\partial_t \to -v_A \partial_x$, we find $\boldsymbol{u}^{(1)} = -(4\pi n m_i)^{-1/2}(\boldsymbol{e}_x \times \boldsymbol{b}) \times \boldsymbol{b}_0$, where $\boldsymbol{b}_0 = \boldsymbol{B}_0/B_0$. In the calculation of $\boldsymbol{u}^{(2)}$, we omit the small b_y component:

$$\boldsymbol{u}^{(2)} = (4\pi n m_i)^{-1/2}[\boldsymbol{e}_x(1 + \sin^2\theta) - \boldsymbol{e}_z \sin\theta\cos\theta](b_z^2/2B_0).$$

Once the value $rot(\boldsymbol{u}^{(1)} \times \boldsymbol{b} + \boldsymbol{u}^{(2)} \times \boldsymbol{B}_0)$ is calculated, we include it on the right-hand side of Equation (7.85) and have, instead of Equation (7.84), the nonlinear equation

$$\partial_t b_z + v_A(1 + 4\sin\theta b_z/B_0)\partial_x b_z + \mu\partial_x^3 b_z = \kappa\partial_x^2 b_z, \tag{7.86}$$

where $\kappa = \nu_m/2$. The equation can be simplified by introducing the new independent variable $\xi = x - v_A t$ and the new unknown function $u(\xi, t) = 4c_A \sin\theta b_z(\xi, t)/B_0$:

$$\partial_t u + u\partial_\xi u + \mu\partial_\xi^3 u = \kappa\partial_\xi^2 u. \tag{7.87}$$

In the absence of dispersion ($\mu = 0$), this equation is reduced to the Burgers equation (7.77), while in the absence of dissipation ($\kappa = 0$), it becomes the Korteweg–de Vries (KdV) equation

$$\partial_t u + u\partial_\xi u + \mu\partial_\xi^3 u = 0. \tag{7.88}$$

It was derived even in 1895 for gravitational waves in a canal of water of finite depth. Last decades it has been understood that the field of applications of the KdV equation is notably widespread.

Example 7.9

On the base of the KdV equation (i.e., in the absence of dissipation), investigate the nonlinear stationary waves in a cold plasma, which depend on x and t as $\zeta = x - c_w t$, where c_w is the required velocity. Find the general solution for such waves and analyze the particular solution which satisfies the condition $u(\zeta) \to 0$ when $\zeta \to \pm\infty$ (the solitary wave, soliton).

Solution. From Equation (7.88), on substituting the solution, we have

(1) $$(-\Delta c\, u + u^2/2 + \mu u'')' = 0, \qquad \Delta c = c_w - v_A,$$

where a prime means the derivative with respect to ζ. On integrating Equation (1), we obtain the equation which formally is the same as the equation of classic mechanics:

(2) $$\mu u'' = K + \Delta c\, u - \frac{1}{2}u^2 = -\frac{\partial W}{\partial u},$$

where K is the constant of integration;

(3) $$W(u) = \frac{u^3}{6} - \frac{\Delta c\, u^2}{2} - Ku$$

is the "potential energy" of a particle, which depends on its "generalized coordinate" u. A role of time is played by ζ. The constant K can easily become zero by replacing $u \to u + u_0$. Such a replacement, as follows from Equation (1), means a change to the system moving with the velocity $v_A u_0$. We assume in what follows that $K = 0$.

On the basis of mechanical analogy, it is easy to complete the integration. With the aid of the "energy integral" $\mu u^2 + W(u) = E = const$, we write down the solution of Equation (2) in quadratures

(4) $$\int \frac{du}{\sqrt{P(u)}} = \pm \frac{\zeta}{\sqrt{3\mu}}, \qquad \text{where} \qquad P(u) = 6E + 3\Delta c\, u^2 - u^3$$

is a cubic polynomial. Solution (Equation (4)) has a physical meaning only when $P(u) \geq 0$ and when u is restricted (in the case of a cold plasma it is also necessary that $|u| \ll v_A$). We omitted the constant of integration that is determined by an initial value of the coordinate ζ. The solution is written for $\mu > 0$, that, according to Equation (7.82), corresponds to the negative correction to the linear dispersion law. The solitary wave is formed when $E = 0$. When $u \geq 0$ the condition $P(u) = u^2(3\Delta c - u) \geq 0$ requires that $\Delta c > 0$ and $0 \leq u \leq 3\Delta c$. In this case,

(5) $$\int \frac{du}{\sqrt{P(u)}} = -(3\Delta c)^{-1/2} \ln\{[(3\Delta c)^{1/2} - (3\Delta c - u)^{1/2}] \times [(3\Delta c)^{1/2} + (3\Delta c + u)^{1/2}]\}$$

and the solution takes the form

(6) $$u(x, t) = 3\Delta c \cosh^{-2}[(\Delta c/4\mu)^{1/2}(x - c_w t)].$$

In this case, the soliton represents a hump of a magnetic field which rises above the adjacent background to the value $b_z = B_z - B_0 = B_0 u/4v_A \sin\theta$ and moves with the velocity $c_w = v_A + \Delta c$. The disturbances of plasma velocity and of plasma density are associated with that of the magnetic field. The value Δc, that is, the excess of the soliton velocity over the velocity v_A of the linear magnetoacoustic wave, relates to the amplitude u_0:

(7) $$u_0 = 3\Delta c.$$

The width of the soliton

(8) $$\delta = 2(\mu/\Delta c)^{1/2} = 2(3\mu/u_0)^{1/2}$$

is inversely proportional to the square root of its amplitude. For the soliton propagating transversely to the direction of a magnetic field ($\theta = \pi/2$), we have $\delta = (c/\omega_{0e})(3B_0/2b_0)$, where b_0 is the height of the magnetic hump and the parameter $a_D = c/\omega_{0e}$ determines the dispersion characteristic length. When $u < 0$ the condition $P(u) \geq 0$ holds for $u \leq -3\Delta c$. In this case,

(9) $$u(x, t) = -3\Delta c \tan^2[(\Delta c/4\mu)^{1/2}(x - c_w t)].$$

This solution does not satisfy the restriction condition and therefore cannot be physically realized. ■

In the preceding example, we considered the solitary nonlinear wave, that is, the soliton. As elucidated in the last few decades, the solitons are of considerable importance in the physics of nonlinear wave phenomena. This is explained by two circumstances: (i) when interacting the solitons in many cases retain asymptotically their shape and velocity, that is, they behave as particles; (ii) under definite conditions the arbitrary nonlinear initial disturbance evolves to an superposition of solitons (and of wave "tails"). It gives grounds to consider the soliton solutions of a sort of "eigenfunctions" of the nonlinear equation, and express the solutions of more general character in terms of these eigenfunctions. Not only the KdV equation has the soliton solutions but other nonlinear equations as well (see numerous manuals and monographs devoted to this subject: Novokshenov (2002); Zakharov *et al.* (1980); Newell (1985); Dodd *et al.* (1984); Ablowitz and Segur (1981); Akhmediev and Ankevich (2003); Fleishman and Toptygin (2013).

Curiously, that the soliton as a physical phenomenon (a water hump moving without spreading in a narrow canal) was observed and firstly described by a Scottish scientist and engineer Scott Russel even in 1834, long before the mathematical theory of solitons was developed.

Nonlinear Schrödinger equation. Previously, we obtained the KdV equation for a cold plasma at $c_s^2 \ll v_A^2$. If the internal energy and the energy of a magnetic field are of the same order of magnitude, then linear and nonlinear waves are quickly damping due to their resonance interaction with ions (the Landau damping). The sole exception are waves propagating along the external magnetic field or at a right angle to it. The transverse propagation is still described by the KdV equation, whereas the longitudinal propagation is described by another nonlinear equation (Vainstein *et al.*, 1993):

$$\partial_\tau h + i\partial_\xi^2 h + \partial_\xi(h^2 h^*) = 0, \tag{7.89}$$

$\tau = \omega_{Bi}t/2$, $\xi = x/r_A - \omega_{Bi}t$, $h = (b_y + ib_z)/B_0(1-\beta)^{1/2}$, $\omega_{Bi} = eB_0/m_i c$ is the cyclotron frequency of the ions, $r_A = v_A/\omega_{Bi}$, $\beta = c_s^2/v_a^2$, and $b/(1-\beta)^{1/2} \ll B_0$. Equation (7.89) is known as **the nonlinear Schrödinger equation** with a derivative in the cubic term.

Nonlinear response of matter to a high frequency field. We have already discussed the nonlinear response of medium to an external perturbation in Section 2.3 on ferroelectrics, in Section 3.3 on ferromagnetics and in the present section on waves in plasma. All considered phenomena are related to the static or quasi-static (slowly alternating) fields. However, the high frequency fields produced by lasers can approach or even exceed in magnitude the interatomic fields in condensed matter ($E_a \approx 10^8 - 10^9$ V/cm). Under these conditions, the assumption on the linear relation between the electric polarization and the field strength (Equation (7.24)) does not hold, and the coupling equation must be written in a more general form.

In the isotropic dispersion-free medium, it can be written in the form

$$P(E) = \alpha E + \alpha^{(3)}E^3 + \alpha^{(5)}E^5 + \cdots, \tag{7.90}$$

where α is the linear electric susceptibility with which we were dealing previously; $\alpha^{(3)}$, $\alpha^{(5)}$, ... are the nonlinear susceptibilities of the considered medium. In a more general case of an anisotropic dispersion medium (but without spatial dispersion), the coupling equation should be written as the generalization of the casual relation of type (Equation (5.66)):

$$\begin{aligned} P_\mu(t) &= \int_0^\infty \alpha_{\mu\nu}(\tau)E_\nu(t-\tau)d\tau \\ &+ \int_0^\infty d\tau_1 \int_0^\infty d\tau_2\, \alpha^{(2)}_{\mu\nu\kappa}(\tau_1,\tau_2)E_\nu(t-\tau_1)E_\kappa(t-\tau_2) \\ &+ \int_0^\infty d\tau_1 \int_0^\infty d\tau_2 \int_0^\infty d\tau_3\, \alpha^{(3)}_{\mu\nu\kappa\sigma}(\tau_1,\tau_2,\tau_3) \\ &\times E_\nu(t-\tau_1)E_\kappa(t-\tau_2)E_\sigma(t-\tau_3) + \cdots \end{aligned} \tag{7.91}$$

By going to the Fourier harmonics, we can write this coupling equation in the equivalent form

$$\begin{aligned} P_m(\omega) &= \alpha_{\mu\nu}(\omega)E_\nu(\omega) + \int_{-\infty}^\infty \frac{d\omega_1}{2\pi}\alpha^{(2)}_{\mu\nu\kappa}(\omega_1,\omega-\omega_1)E_\nu(\omega_1)E_\kappa(\omega-\omega_1) \\ &+ \int_{-\infty}^\infty \frac{d\omega_1}{2\pi}\int_{-\infty}^\infty \frac{d\omega_1}{2\pi}\alpha^{(3)}_{\mu\nu\kappa\sigma}(\omega_1,\omega_2,\omega-\omega_1-\omega_2) \\ &\times E_\nu(\omega_1)E_\kappa(\omega_2)E_\sigma(\omega-\omega_1-\omega_2) + \cdots \end{aligned} \tag{7.92}$$

Here the nonlinear susceptibilities are determined by analogy with (5.67):

$$\alpha^{(2)}_{\mu\nu\kappa}(\omega_1,\omega_2) = \int_0^\infty \alpha^{(2)}(\tau_1,\tau_2)\exp(i\omega_1\tau_1 + i\omega_2\tau_2)\,d\tau_1 d\tau_2 \tag{7.93}$$

as well as the quantities of the more higher order. It follows from Equation (7.92) that in a nonlinear medium the frequency of the polarization vector can significantly differ from the frequency of the electric field (also see Example 7.10 below).

The classic model of the nonlinear susceptibility; anharmonic oscillator. We have seen in Section 5.1 (Problems 5.12 and 5.13) that the classic model of the bounded electrons as linear damping oscillators with some damping gives an adequate qualitative picture of the electric susceptibility of a medium in weak fields. In order to evaluate the nonlinear susceptibilities, we will use the model of the anharmonic (nonlinear) oscillator with the damping:

Example 7.10

The charged isotropic anharmonic oscillator is in the external electric field with the linear polarization $E(t) = E_0 \cos \omega t$. We neglect the nonuniformity of the field in the region of vibrations of the oscillator. The amplitude of vibrations is small and in the Fourier expansion of its potential energy it is sufficient to take into account the lowest anharmonic term. Assume that the dielectric medium consists of nonlinear *oscillators with density N per unit volume. Calculate the electric polarization of the medium and its linear and nonlinear susceptibilities.*

Solution. We take the Ox-axis parallel to the direction of the electric field and write down the potential energy of the oscillator taking into consideration the anharmonic terms:

$$U(x) \approx \frac{1}{2}m\omega_0^2 x^2 + \frac{1}{3}mqx^3 + \cdots \tag{1}$$

where m is the mass of an electron, ω_0 is the eigenfrequency of an harmonic oscillator, and q is the parameter of nonlinearity. If the potential energy is the even coordinate function, then $q = 0$ and in the expansion of the potential energy the term of the forth order must be taken into account. The motion equation of an oscillator, with regard to the written out terms, has the form

$$\ddot{x} + \gamma\dot{x} + \omega_0^2 x + qx^2 = \frac{e}{m}E(t), \tag{2}$$

where the term $\gamma\dot{x}$ describes the damping of the oscillator. The effect of the magnetic field on the electron is neglected.

We will solve the derived equation using the perturbation method, assuming the anharmonic term to be small compared with the harmonic term: $|qx| \ll \omega_0^2$. We seek the solution in the form $x = x_l + x_{nl}$, where x_{nl} is a small nonlinear correction to the solution x_l of the linear equation:

$$\ddot{x} + \gamma\dot{x} + \omega_0^2 x = \frac{e}{m}E(t). \tag{3}$$

For the monochromatic field $E(t) = E_0 \cos\omega t$ at sufficiently long times $t \gg 1/\gamma$, the solution of Equation (3) has the form

(4)

$$x_l(t) = \frac{1}{2}E_0\left(\beta(\omega)e^{-i\omega t} + \beta^*(\omega)e^{i\omega t}\right), \qquad \text{where} \qquad \beta(\omega) = \frac{e}{m}\frac{1}{\omega_0^2 - \omega^2 + i\omega\gamma}$$

is the linear polarizability of the oscillator. The equation for a small nonlinear perturbation can be obtained from Equation (2). By retaining only the above-derived solution (Equation (4)) in the nonlinear term we have:

$$\ddot{x}_{nl} + \gamma\dot{x}_{nl} + \omega_0^2 x_{nl} = -qx_l^2, \tag{5}$$

The solution of Equation (5) has the form

$$x_{nl}(t) = x_0 + Ce^{-2i\omega t} + C^* e^{2i\omega t}, \qquad x_0 = -\frac{q}{2\omega_0^2}|\beta(\omega)|^2 E_0^2, \tag{6}$$

$$C = -\frac{q\beta^2(\omega)E_0^2}{4(\omega_0^2 - 4\omega^2 - 2i\gamma\omega)}.$$

With the aid of Equations (4) and (5), we can find the electric polarization of the medium:

$$P(t) = Np(t) = Ne(x_l + x_{nl}) = P_0 + P_l(t) + P_{nl}(t), \tag{7}$$

where $P_l = Nex_l = (E_0/2)[\alpha(\omega)e^{-i\omega t} + \alpha^*(\omega)e^{i\omega t}]$ is the linear polarization of the medium, which varies with frequency of an incident wave; $\alpha(\omega) = Ne\beta(\omega)$.

$P_0 = Nex_0 = \alpha^{(2)}(0)E_0^2$ is the constant nonlinear polarization which is proportional to the square of the field strength; $\alpha^{(2)}(0) = -Nqe|\beta(\omega)|^2/2\omega_0^2$ is the nonlinear susceptibility at zero frequency. It depends on the frequency of the initial wave and on the parameters of the oscillator. Finally, $P_{nl}(t) = Ne(Ce^{-2i\omega t} + C^* e^{2i\omega t})$ is the nonlinear polarization varying with double frequency. Separating the nonlinear susceptibility at the double frequency with the aid of relations $P_{nl}(t) = (1/2)P_{nl}(2\omega)\exp(-2i\omega t)+$ c.c., $P_{nl}(2\omega) = \alpha^{(2)}(2\omega)E_0^2$, where c.c. is the complex conjugate, we find

$$\text{(8)} \qquad \alpha^{(2)}(2\omega) = \frac{Nqe\beta^2(\omega)}{2(\omega_0^2 - 4\omega^2 - 2i\gamma\omega)}. \qquad \blacksquare$$

Self-focusing. Let us consider on a semi-qualitative level another nonlinear effect. In Example 6.7 and Problem 6.98, we have studied the propagation of the restricted light beam in the linear homogeneous medium and elucidated that the beam is subject to the diffraction expansion. At distances exceeding its initial width a, the beam acquires the shape of a cone with a vertex angle

$$\theta_d \approx \frac{2}{ka} = \frac{\lambda_0}{\pi n_0 a}, \tag{7.94}$$

where λ_0 is the wavelength in vacuum, and n_0 is the refractive index of the transparent medium. However, if the transversely restricted beam has the high intensity at which the noticeable nonlinear polarization appears, then the refractive index inside the beam and beyond it will be different. Let $n = n_0$ outside the beam, and inside the beam

$$n = n_0 + n_2|A|^2, \qquad n_2 > 0, \tag{7.95}$$

where the nonlinear correction is taken into account, which is proportional to the square of the local amplitude of the beam (Equation (7.90)). The boundary of the beam can be considered the boundary of two media with different refractive indices. When the angles of incidence on the boundary are larger than the angle θ_r of the total internal reflection (Problem 6.8),

$$\sin\theta_r = \frac{n_0}{n_0 + n_2|A|^2}, \tag{7.96}$$

the rays will return in the interior of the beam, that is, be focused to its axis (Figure 7.8).

The power of the beam can be chosen in such a way that the effects of nonlinear focusing and diffraction defocusing be mutually balanced. Then the stationary light beam propagates without expanding inside a certain light guide formed by the beam itself owing to its field. Therein the effect of self-focusing lies, which was predicted theoretically, by a Soviet physicist G.A. Ascaryan in 1962. The critical power for the self-focusing can be evaluated from the comparison of the corresponding angles. The critical angle θ_r corresponds to the ray whose inclination to the axis is $\theta_{nl} = \pi/2 - \theta_r$, that is, $\cos\theta_{nl} = \sin\theta_r$, and is given by Equation (7.96). When $\theta_{nl} < \theta_d$ the diffraction effect prevails, the beam spreads, but more slowly

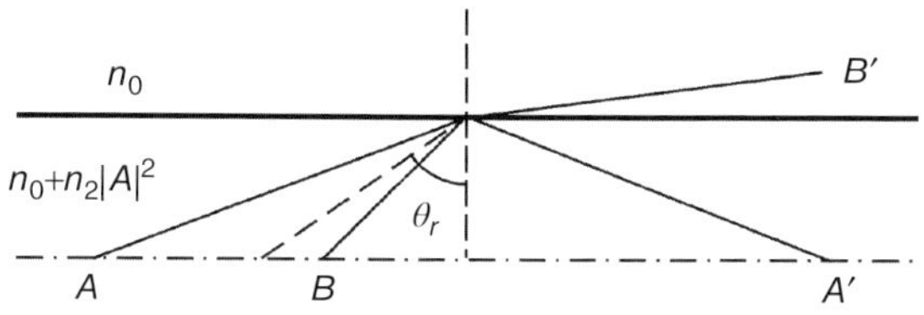

Figure 7.8 Self-focusing effect. The ray A experience the full reflection and is focused to an axis of a bunch which is designated by a stroke-dashed line. The ray B would leave boundary a bunch.

than in the linear case considered in Problem 6.98. When $\theta_{nl} = \theta_d$ the effects are balanced. When $\theta_{nl} \ll 1$, then from Equation (7.96) we have

$$\theta_{nl}^2 = \frac{2n_2|A|^2}{n_0 + n_2|A|^2}, \tag{7.97}$$

and equating this value to the square of the diffraction angle, we evaluate the order of magnitude of the critical amplitude of the beam: $|A_c|^2 \approx n_0\theta_d^2/2n_2$. The total critical power of the beam, transferred per unit time through its cross section, is given by the relation

$$P_c = \pi a^2 \overline{\gamma}_c = \frac{a^2 c n_0^2 \overline{E^2}}{4} \approx \frac{c n_0^2 \lambda_0^2}{16\pi^2 n_2}. \tag{7.98}$$

The obtained power is determined by the wavelength and by properties of the medium but is independent of the beam width.

Recommended literature: Bloembergen (1965); Bredov *et al.* (2003) Akhmanov *et al.* (1981, 1988); Akhmanov and Nikitin (1998); Klyshko (1980); Klyshko (1986); Ryskin and Trubetskov (2000); Loudon (1973); Fleishman and Toptygin (2013)

Amplification of electromagnetic waves in nonequilibrium media. In Section 5.3, we have noted that in equilibrium media the dissipation of energy always occurs according to the second law of thermodynamics. It leads to the following relations

$$\omega\varepsilon''(\omega) \geq 0, \qquad \omega\mu''(\omega) \geq 0 \tag{7.99}$$

for imaginary parts of the permittivity and magnetic permeability. However, in nonequilibrium media, these conditions sometimes do not hold. The relaxation (the passage to the equilibrium state) can be accompanied by the excitation of oscillations, and the energy of the oscillations will be drawn from the energy of the nonequilibrium medium. Media in which electromagnetic waves are amplified are called **active**. We consider one of the simplest examples of such a medium, that is, the excitation of electric oscillations in the nonequilibrium, cold, collisionless plasma.

Example 7.11

The quasi-neutral plasma without collisions and magnetic field consists of several types of particles with charges e_α and number densities $N_\alpha = const$, $\alpha = 1, 2, \ldots, s$, $\sum_\alpha e_\alpha N_\alpha = 0$. Each population of particles moves with its constant velocity $\boldsymbol{V}_\alpha$ without thermal spread. Write down the dispersion equation for small electrostatic oscillations of such a plasma, satisfying the condition $\boldsymbol{B} = 0$, $\boldsymbol{E} = -\nabla\varphi$, where φ is the scalar potential.

Solution. When the oscillations occur, the number density of particles and their velocities will acquire small perturbations n_α and $\boldsymbol{u}_\alpha$. The electrostatic potential satisfies the Poisson equation

(1) $$\triangle\varphi = -4\pi\rho = -4\pi\sum_\alpha e_\alpha n_\alpha,$$

where the number density n_α should be found from the linearized equations of motion of the particles,

(2) $$\frac{\partial \boldsymbol{u}_\alpha}{\partial t} + (\boldsymbol{V}_\alpha\cdot\nabla)\boldsymbol{u}_\alpha = -\frac{e_\alpha}{m_\alpha}\nabla\varphi, \qquad \frac{\partial n_\alpha}{\partial t} + \nabla\cdot(N_\alpha \boldsymbol{u}_\alpha) + \nabla\cdot(n_\alpha \boldsymbol{V}_\alpha) = 0.$$

Consider the perturbations in the form of plane monochromatic waves $\boldsymbol{u}_\alpha$, $n_\alpha \propto \exp(i\boldsymbol{k}\cdot\boldsymbol{r} - i\omega t)$. From Equation (2), we have

(3) $$\boldsymbol{u}_\alpha = \frac{e_\alpha}{m_\alpha(\omega - \boldsymbol{k}\cdot\boldsymbol{V}_\alpha)}\boldsymbol{k}\varphi, \qquad n_\alpha = \frac{N_\alpha e_\alpha}{m_\alpha(\omega - \boldsymbol{k}\cdot\boldsymbol{V}_\alpha)^2}k^2\varphi.$$

Substituting the last result in Equation (1), we have the required dispersion equation

$$1 = \sum_\alpha \frac{\omega_{0\alpha}^2}{(\omega - \boldsymbol{k}\cdot\boldsymbol{V}_\alpha)^2}, \qquad \omega_{0\alpha}^2 = 4\pi N_\alpha e_\alpha^2/m_\alpha. \qquad \blacksquare$$

Example 7.12

Apply the obtained dispersion equation to the plasma with non-moving ions, in which there are two electron beams with equal number densities of electrons $N_1 = N_2 = N$. The beams move one to the other with equal velocities $V_1 = -V_2 = V$. Show that there will be the excitation of oscillations in a plasma, and find the increment $\gamma = \Im\,\omega$ of the corresponding wave.

Solution. The dispersion equation takes the form

(1) $$1 - \frac{\omega_{0\alpha}^2}{(\omega - \boldsymbol{k}\cdot\boldsymbol{V}_\alpha)^2} - \frac{\omega_{0\alpha}^2}{(\omega + \boldsymbol{k}\cdot\boldsymbol{V}_\alpha)^2} = 0,$$

where $k_\parallel$ is the projection of the wave vector in the mutual direction of motion of the electrons. When $V = 0$ and $\omega = \pm\omega_{0e}\sqrt{2}$, that corresponds to the longitudinal plasma oscillations with the frequency determined by the total number density $2N$ of the electrons (see Problem 5.47 to compare).

When $V \neq 0$, the solution of the obtained biquadratic equation yields the frequencies of the eigenwaves:

$$\omega = \pm\sqrt{(k_{\|} V)^2 + \omega_{0e}^2 \pm \omega_{0e}(\omega_{0e}^2 + 4k_{\|}^2 V^2)^{1/2}}.$$

The oscillations, for which the radicand is negative, are unstable. It takes place if $|k_{\|}| < \omega_{0e}\sqrt{2}/V$. In this case, the frequency becomes purely imaginary: $\omega = \pm i\gamma(k_{\|})$, $\gamma > 0$, and the amplitude of the wave with $\omega = i\gamma$ rises exponentially. The increment reaches the largest value $\gamma_{max} = \omega_{0e}/2$ when $k_{\|}^2 = 3\omega_{0e}^2/4V^2$. ■

Recommended literature: Mikhailovskii (1975); Nezlin (1976); Oraevskii (1998); Boiko and Petrov (1988)

Problems

7.30* For the homogeneous infinite medium, write down the general solution of the Burgers equation (7.77) satisfying the initial condition $P(\xi, 0) = P_0(\xi)$.

Hint. Reduce the nonlinear Burgers equation to the linear heat conduction equation with the aid of the nonlinear substitution $P(\xi, t) = -2\kappa\partial_\xi \ln \varphi(\xi, t)$, where $\varphi(\xi, t)$ is a new unknown function.

7.31 On the basis of the solution to Problem 7.30, consider the case, where the initial perturbation is concentrated in the narrow region: $P_0(\xi) = A\delta(\xi)$. Since P_0 has a dimension of velocity, then A is the product of velocity by length, and the dimensionless value $R = A/2\kappa$ has a meaning of the Reynolds number. Find the shape of the wave at $t > 0$, investigate the cases $R \ll 1$ and $R \gg 1$.

7.32** Deduce the Burgers equation for the MHD wave propagating transversely the external magnetic field in a conducting medium. Investigate the profile of a weak transverse shock wave.

7.33 By using the results obtained in Example 7.9, construct the KdV soliton for the case of the positive dispersion correction (i.e., for $\mu < 0$).

7.34* By using the mechanical analogy, obtain the solution of the KdV equation for stationary nonlinear waves when $E \neq 0$ (Example 7.9). Analyze a transformation of the derived solution into solitons and also into periodic linear waves.

7.35* Investigate the structure of the front of the shock wave with the aid of the KdV–Burgers Equation (7.87), retaining both the dissipation term and the dispersion term. Compare with the result of Problem 7.32, which was obtained neglecting the dispersion effect.

7.36 Evaluate the thickness of the shock-wave front appearing Problem 7.35; consider the different relations between the dissipation term and the dispersion term.

7.37** Construct the soliton solutions for the nonlinear Schrödinger equation (7.89) and analyze their physical meaning. Compare these solitons with the KdV solitons.

7.38 The nonlinear oscillator considered in Example 7.10 is in vacuum. It is subject to the external linearly polarized electric field $\boldsymbol{E}(t) = \boldsymbol{E}_0 e^{-i\omega t}$. Determine the spectral composition of the scattered radiation, and also calculate the cross-sections of harmonics averaged over the period of the incident wave.

7.39* Two waves that are linearly polarized in the same plane, with different frequencies ($E(t) = E_1 \cos\omega_1 t + E_2 \cos\omega_2 t$), act on the nonlinear oscillator considered in Example 7.10. Treat the nonlinear medium as a collection of nonlinear oscillators. Calculate the nonlinear susceptibilities of the medium caused by the interaction of waves.

7.40 On the basis of the results of Example 7.11, consider the cold plasma with the electron number density N, in which there is the beam of the low density $N' \ll N$, moving with the velocity V. Under what conditions the electric oscillations are unstable? Calculate the corresponding increment.

7.41* The electron component of the quasi-neutral cold plasma moves with the velocity V relative to ions. Find the increment of the growing oscillations. Consider, in particular, the resonance case.

7.4 Answers and Solutions

7.1

$$\Delta\Omega = \frac{l_\perp^2}{R^2} \sim \frac{c^2}{\nu^2}\frac{1}{L^2} = \left(\frac{\lambda}{L}\right)^2.$$

The solid angle of the coherence is independent of the distance R from the source.

7.2 $\Delta\lambda \approx 3.52\times10^{-10}$ cm; $l_\perp \sim \lambda\frac{R}{L} = 5.4\times10^{-3}$ cm; $l_\parallel \sim \frac{\lambda^2}{\Delta\lambda} = 7.1$ cm; $\Delta\Omega \sim 1.3\times10^{-31}$ sr; $\Delta V = l_\perp^2 l_\parallel \sim 2.1\times10^{-4}$ cm^3.

7.3 $R = 9.46\times10^{18}$ km, that is, by a factor 6.3×10^5 larger than the Earth–Sun distance. Hence, $l_\perp \approx 3.4\times10^3$ cm, that is, by a factor 6.3×10^5 larger than $l_\perp$ in Problem 7.2. As for $l_\parallel \approx \lambda^2/\Delta\lambda \approx 7.1$ cm and $\Delta\Omega \approx 1.3\times10^{-31}$ sr, they retain the same values as in Problem 7.2. The coherence volume $\Delta V \approx 8.3\times10^7$ cm^3 is a factor 4×10^{11} larger than the coherence volume of the solar radiation on the Earth. The characteristic feature is the increase of the degree of coherence of light while it propagates. This refers only to the transverse coherence.

7.4 $l_\parallel \sim \lambda/\Delta\lambda \approx 3\times10^8$ cm. Since the optical generator (laser) emits a light cone within an opening angle $\Delta\vartheta \sim \lambda/D = 10^{-5}$, the coherence volume close to the laser has the shape of a cone with its vertex at the laser;

$$l_\perp \approx \begin{cases} D & = \; 5 \text{ cm at the laser,} \\ l_\parallel \frac{\lambda}{D} & \approx \; 6000 \text{ cm at the base of the coherence cone,} \end{cases}$$

$$\Delta V = \frac{1}{3}\pi\left(\frac{l_\perp}{2}\right)^2 l_\parallel \approx 28\times10^{14} \text{ cm}^3.$$

7.5

$$\delta = \frac{1}{\exp(\hbar\omega/T) - 1} = \frac{1}{\exp(2\pi\hbar c/\lambda kT) - 1};$$
$$\delta \approx \frac{\lambda kT}{2\pi\hbar c} \approx 200 \quad \text{at} \quad \lambda = 1\ \text{cm},\ T = 273\ \text{K},$$
$$\delta = e^{-100} \approx 10^{-43} \quad \text{at} \quad \lambda = 5\times10^{-5}\ \text{cm},\ T = 273\ \text{K},$$
$$\delta = \frac{1}{\exp(2.73) - 1} \approx 0.07 \quad \text{at} \quad \lambda = 5\times10^{-5}\ \text{cm},\ T = 10\,000\ \text{K}.$$

7.6 $\delta = 5\times10^{18}$, $T = 1.4\times10^{23}$ K.

7.7 $\Gamma(\tau) = \frac{1}{2}\int_0^\infty I(\omega)\cos\omega\tau\, d\omega$.

7.8 $\Gamma(\tau) = 2I[\sin(\Delta\omega\tau/2)/\Delta\omega\tau]\cos\omega_0\tau$.

7.9 The path difference for the light from one of the independent emitters which is at the point (x', y'), is $s_1 - s_2 \approx (xx' + yy')/R$ (Figure 7.3) if we take into account that the transverse size of the source is much larger than $D = \sqrt{x^2 + y^2}$. The field in the points $\boldsymbol{r}_1(0, 0)$ and $\boldsymbol{r}_2(x, y)$ is produced by all the emitters in the source:

$$U(\boldsymbol{r}_1, t) = \sum_i U_i(t), \qquad U(\boldsymbol{r}_2, t) = \sum_i U_i(t)\exp\left[-i\frac{xx_i' + yy_i'}{R}k\right],$$

where $U_i(\boldsymbol{r}, t)$ and $U_i(\boldsymbol{r}, t)$ are the amplitudes of the field of the ith emitter at the first and second apertures at time t. The correlation function is

$$\Gamma(\boldsymbol{r}_1, \boldsymbol{r}_2, 0) = \langle U^*(\boldsymbol{r}_1, t)U(\boldsymbol{r}_2, t)\rangle = \sum_i \langle U_i^*(t)U_i(t)\rangle \exp\left[-ik\frac{xx_i' + yy_i'}{a}\right]$$
$$+ \sum_{i\neq j}\langle U_i^*(t)U_j(t)\rangle \exp\left[-ik\frac{xx_j' + yy_j'}{R}\right].$$

The second term in Γ disappears due to the incoherence of the independent emitters. The first term is the averaged intensity of the radiation from the separate radiators where the path difference $s_1 - s_2$ is taken into account. Replacing a sum by an integral, we obtain

$$\gamma(x, y) = \frac{\int_S I(x', y')\exp\left[-ik\frac{xx'+yy'}{R}\right] dx'\, dy'}{\int_S I(x', y')\, dx'\, dy'},$$

where the integration is carried out over the transverse cross section of the source.

7.10 (a) $B(D) = |\gamma(D, 0)| = \cos(\pi D\alpha/\lambda)$;
(b) $B(D) = (2\lambda/\pi D)J_1(\pi\alpha D/\lambda)$.

7.11 (a) $\rho = \alpha R = \lambda R/2D_0 = 1.47\times10^8$ km;
(b) $d = R/\alpha = DR/1.22\lambda = 6.28\times10^8$ km;
the diameter of the Betelgeuze star is approximately 450 times larger than the diameter of the Sun, and hence larger than the diameter not only of the Earth's orbit but also of that of the Mars.

7.12 The plane wave $U_1 = A_1 \exp[i\boldsymbol{k}_1 \cdot \boldsymbol{r}] = |A_1| \exp[i(\boldsymbol{k}_1 \cdot \boldsymbol{r} + \alpha_1)]$ with phase α_1 and amplitude A_1, which both vary randomly such that $\overline{A_1} = 0$, while $\overline{|A_1|^2}$ has a constant nonvanishing value determined by the first source. The wave $U_2 = A_2 \exp[i\boldsymbol{k}_2 \cdot \boldsymbol{r}]$ with similar properties comes from the second source. Both these waves enter the photoelements P_1 and P_2. The nonaveraged signal from the photoelement P_1 would be proportional to

(1)
$$I(\boldsymbol{r}_1, t) = |U_1(\boldsymbol{r}_1, t) + U_2(\boldsymbol{r}_1, t)|^2$$
$$= |A_1|^2 + |A_2|^2 + A_1 A_2^* \exp[i(\boldsymbol{k}_1 - \boldsymbol{k}_2) \cdot \boldsymbol{r}]$$
$$+ A_1^* A_2 \exp[-i(\boldsymbol{k}_1 - \boldsymbol{k}_2) \cdot \boldsymbol{r}].$$

The signal (Equation (1)) is subject to random fluctuations owing to the fluctuations of the phases A_1 and A_2 at frequencies much lower than the frequency of the waves U_1 and U_2 arriving from the sources. These fluctuations are, nevertheless, not registered and the averaged intensity is observed. When only one of the detectors is included, the averaged intensity is

$$I(r_1, t) = |A_1|^2 + |A_2|^2 = I(r_2, t),$$

which is independent of $\boldsymbol{k}_1 - \boldsymbol{k}_2$ (the phases of A_1 and A_2 fluctuate independently so that $\langle A_1 A_2^* \rangle = \langle A_1 \rangle \langle A_2^* \rangle = 0$).

Now let the signals from the photoelements P_1 and P_2 first enter the multiplier where the intensities $I(\boldsymbol{r}_1, t)$ and $I(\boldsymbol{r}_2, t)$ are multiplied before registration. The signal observed at the output will be proportional to

(2)
$$\langle I(\boldsymbol{r}_1, t) I(\boldsymbol{r}_2, t) \rangle = (\langle |A_1|^2 \rangle + \langle |A_2|^2 \rangle)$$
$$+ 2\langle |A_1|^2 \rangle \langle |A_2|^2 \rangle \cos[(\boldsymbol{k}_1 - \boldsymbol{k}_2) \cdot (\boldsymbol{r}_1 - \boldsymbol{r}_2)].$$

It depends on $\boldsymbol{k}_1 - \boldsymbol{k}_2$ and, hence, on the angular distance between remote sources. If one changes the distance $\boldsymbol{r}_1 - \boldsymbol{r}_2$ between the detectors and observes the weakening and strengthening of the signal, one can determine this angular distance.

7.13 $\Delta\varphi = (2\pi/\lambda)(n-1)x$, where the coordinate x is reckoned from the refractive edge perpendicular to it.

If by any means whatever there is a phase shift $\Delta\varphi \propto x$ in the xy-plane, such a plane will turn the front of the plane wave in the direction of increasing x, that is, it will operate like a prism.

7.14 The phase shift at a distance x from the axis of the lens is, in the case of a converging lens,

$$\Delta\varphi = -\frac{\pi x^2}{f\lambda},$$

where f is the focal distance determined by the equation

$$\frac{1}{f} = (n-1)\left(\frac{1}{R_1} + \frac{1}{R_2}\right).$$

In the case of a diverging lens

$$\Delta\varphi = +\frac{\pi x^2}{f\lambda}.$$

7.15 The intensity distribution of the light on the photographic plate is

$$I(x) = |A_1 \exp[ikx\vartheta_1] + A_2 \exp[ikx\vartheta_2]|^2 = I_1 + 2\sqrt{I_1 I_2}\cos k\vartheta x,$$

where $\vartheta_2 = \vartheta + \vartheta_1$, $k = 2\pi/\lambda$, $I_1 = |A_1|^2$, $I_2 = |A_2|^2$, and the coordinate x is measured along the photographic plate as shown in Figure 7.6. The blackening distribution on the developed photographic plate is determined by the intensity distribution $I(x)$. The transmission $T(x)$ is proportional to $[I(x)]^{-\gamma/2}$, where γ is the contrast of the photographic emulsion, the periodic function of x with the period of λ/ϑ. It can be written in the form $T(x) = a + b\cos k\vartheta x$ (a and b are constants), if we retain only the two lowest harmonics. The developed photographic plate can be considered the diffraction grating that splits the incident plane wave into plane beams with propagation angles θ determined by the relation $(\lambda\vartheta)\sin\theta = n\lambda$, $n = 0, \pm 1, \pm 2, \ldots$ The main beams are the central, zeroth order ones, and there are two first order beams in the directions $\theta = \pm\vartheta$. Note, that these three beams can be obtained by multiplying the incident wave $A_0 \exp[ikz]$ by the transmission $T(x)$. We then obtain the wave field behind the photographic plate:

$$A_0 a \exp[ikx] + A_0 \frac{b}{2}\exp[ik(z + \vartheta x)] + A_0 \frac{b}{2}\exp[ik(z - \vartheta x)],$$

where the first term describes the undeflected central beam, the second the first order beam deflected by $+\vartheta$, the third the first order beam deflected by $-\vartheta$.

7.16 The reference field at the plate has the form

$$U_1 = A_0 \exp[-i\beta x], \qquad \beta = \frac{2\pi(n-1)\alpha}{\lambda}.$$

We omit here and in what follows the common factor $\exp[i(kz - \omega t)]$. The field diffracted by the aperture is

$$U_2 = A(x)\exp\left[i\frac{\pi x^2}{f\lambda}\right].$$

The total field is $U(x) = U_1 + U_2$, and the intensity

$$I(x) = |u(x)|^2 = A_0^2 + A^2(x) + 2A_0 A(x)\cos\left(\beta x + \frac{\pi x^2}{f\lambda}\right).$$

The intensity distribution contains the information on the phase of the diffracted wave thanks only to the presence of the reference beam.

7.17 The transmission $T(x)$ of the developed photographic emulsion is

$$T(x) \propto [I(x)]^{-\gamma/2} = A_0^{-\gamma}\left\{1 + \frac{A^2(x)}{A_0^2} + 2\frac{A(x)}{A_0}\cos\left(\beta x + \frac{\pi x^2}{f\lambda}\right)\right\}^{-\gamma/2}$$

$$\approx A_0^{-\gamma-2}\left\{A_0^2 - \frac{\gamma}{2}A^2(x) - \gamma A_0 A(x)\cos\left(\beta x + \frac{\pi x^2}{f\lambda}\right)\right\},$$

if we use the condition $A_0 \gg A(x)$. We can rewrite the last relation in the form

$$T(x) \propto 2A_0^2 - \gamma A^2(x) - \gamma A_0 A(x) \exp\left[i\left(\beta x + \frac{\pi x^2}{f\lambda}\right)\right] - \gamma A_0 A(x) \exp\left[-i\left(\beta x + \frac{\pi x^2}{f\lambda}\right)\right]. \quad (1)$$

This is called **Gabor's hologram formula.**[6)]

If the hologram is illuminated by a plane monochromatic light wave $A_0 \exp[i(kz - \omega t)]$, a wave field appears behind the hologram, which is the result of the diffraction by the hologram. This field can be obtained simply (see the solution of Problem 7.15) by multiplying the primary wave field $A_0' \exp[i(kz - \omega t)]$ by the transmission $T(x)$ given by Gabor's formula (1). We then get the field of the form

$$\begin{aligned} U \sim\ & (2A_0^2 - \gamma A^2(x)) \exp[i(kz - \omega t)] \\ & - \gamma A_0 A(x) \exp[i(kz - \omega t)]\ \exp\left[i\left(\beta x - \frac{kx^2}{f\lambda}\right)\right] \\ & - \gamma A_0 A(x) \exp[i(kz - \omega t)]\ \exp\left[-i\left(\beta x + \frac{kx^2}{f\lambda}\right)\right]. \end{aligned}$$

The first term in Equation (2) corresponds to the nonuniform diffraction (due to $A^2(x)$) attenuation of the incident wave. The angle of diffraction is small because $A(x)$ is a smoothly changing function as compared to the exponentials. The second term acts as a combination of a prism which deflects the beam upwards and a diverging lens with a focusing length f (Problems 7.13 and 7.14). The third term acts as a combination of a prism, which deflects the beam downwards, and a converging lens. Finally, when a plane monochromatic wave is transmitted through the hologram, the primary wave fronts are reconstructed (Figure 7.9): the plane wave and the spherical front from the aperture. The latter is reconstructed twice: in the form of a wave from the real image and from the imaginary image.

7.18

$$\begin{aligned} & \exp[ik'z]T(x) \\ & \propto \left[2A_0^2 - 2\gamma A^2\left(1 + \cos\frac{4\pi}{\lambda f}Dx\right)\right] \exp[ik'z] \\ & - \gamma A_0 A\left\{\exp\left[i\frac{\pi}{\lambda f}(x - D)^2\right] + \exp\left[i\frac{\pi}{\lambda f}(x + D)^2\right]\right\} \exp[i(\beta x + k'z)] \\ & - \gamma A_0 A\left\{\exp\left[-i\frac{\pi}{\lambda f}(x - D)^2\right] + \exp\left[-i\frac{\pi}{\lambda f}(x + D)^2\right]\right\} \exp[-i(\beta x - k'z)]. \end{aligned}$$

The second and the third terms, as in Problem 7.17, describe the field deflected upwards and downwards and focused in two pairs of points.

6) Dennis Gabor (1900–1979), English scientist (Hungarian by origin), Nobel Prize laureate (1971) for the discovery of holography.

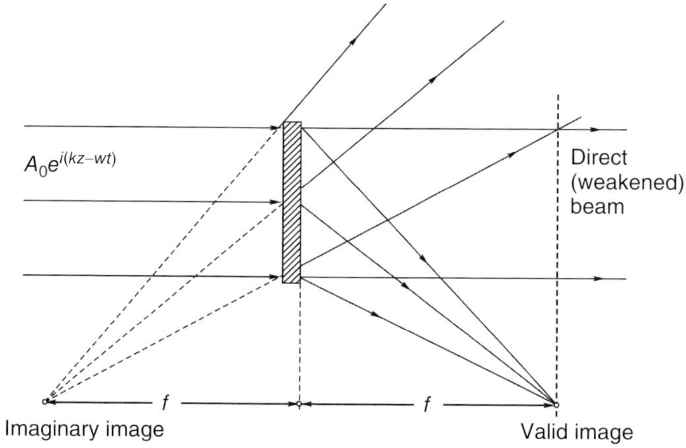

Figure 7.9 Restoration of the hologram with the help of transmission $T(x)$ of photographic plates.

However, the focusing lengths of the corresponding diverging and converging lenses are different, namely

$$f' = \frac{\lambda}{\lambda'}f.$$

The linear magnification is given by the formula

$$\frac{2\Delta}{2D} = \frac{p+q}{p} = \frac{\lambda'}{\lambda}\frac{q}{f},$$

where

$$\frac{1}{p} + \frac{1}{q} = \frac{1}{f'} = \frac{\lambda'}{\lambda f},$$

p is the distance of the source with a wavelength λ' from the hologram, and q is the distance of the image from the hologram (Figure 7.10). In order to achieve magnification, one has to use a wavelength $\lambda' > \lambda$ for reconstruction, and place the source at a finite distance p from the hologram.

7.19 The intensity distribution on the hologram can be transferred without important distortions if the spatial period of the diffraction picture is larger than d,

$$\frac{1}{|\beta + (\pi x/f\lambda)|} \geq d$$

(see the solution of Problem 7.16). The maximum size of the hologram in the x-direction, $2x_{max} \approx 2\lambda f/d$, is restricted by this condition. This size plays the role of the lens diameter in the Rayleigh theory of resolving power (Problem I.2.144). Applying the Rayleigh criterion for the minimum size s of an object, we get

$$s \approx \frac{\lambda}{2\vartheta} \approx \frac{\lambda f}{2x_{max}} \approx \frac{d}{2}.$$

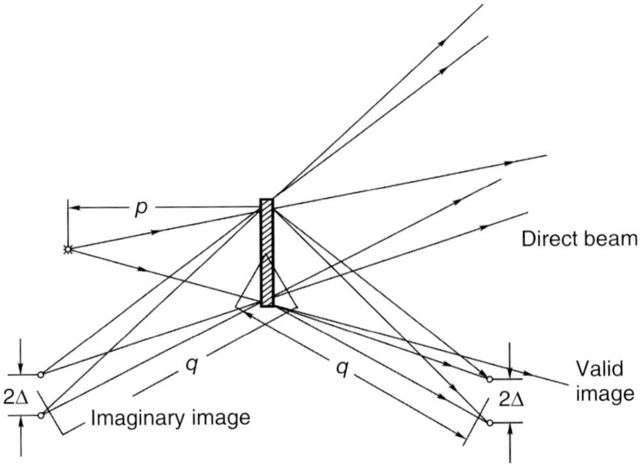

Figure 7.10 To a question on image increase on the hologram.

Here ϑ is one half of the opening angle of the light cone going from the hologram to the image.

7.20 From the homogeneity and isotropy of the field in the (x, y)-plane, it follows that

(1) $$K_\perp(\boldsymbol{\rho}_1, \boldsymbol{\rho}_2) = K_\perp(|\boldsymbol{\rho}_1 - \boldsymbol{\rho}_2|) = K_\perp(|\boldsymbol{\rho}|), \quad \boldsymbol{\rho} = \boldsymbol{\rho}_1 - \boldsymbol{\rho}_2.$$

The coordinate correlation function is expressed in terms of the Fourier integral of the spectral function:

(2) $$K_\perp(\boldsymbol{\rho}) = \int K(\boldsymbol{\kappa}) \exp(i\boldsymbol{\kappa} \cdot \boldsymbol{\rho}) \frac{d^2\kappa}{(2\pi)^2}.$$

When the wave vectors are isotropic and the preferential direction is absent, the function $K_\perp$ depends only on the modulus $|\boldsymbol{\rho}| = \rho$. For the uniform distribution, one has

(3) $$K_\perp(\rho) = \frac{K_0\kappa_0^2}{2\pi} \frac{J_1(\kappa_0\rho)}{\kappa_0\rho},$$

and for the Gaussian distribution

(4) $$K_\perp(\rho) = \frac{K_0\kappa_0^2}{4\pi} \exp\{-(\kappa_0\rho)^2/2\}.$$

7.21 It is necessary to calculate the Fourier integral

(1) $$K_{\lambda\nu}(\boldsymbol{r}, \omega) = \int K_{\lambda\nu} e^{i\mathbf{k}\cdot\mathbf{r}} \frac{d^3k}{(2\pi)^3}$$

$$= c\hbar \coth\frac{\hbar\omega}{2T} \int (\Im\, \alpha_{\lambda\nu}(\boldsymbol{k}, \omega)) e^{i\mathbf{k}\cdot\mathbf{r}} \frac{d^3k}{(2\pi)^3}$$

$$= c\hbar \coth\frac{\hbar\omega}{2T} \Im \int \alpha_{\lambda\nu}(\boldsymbol{k}, \omega) e^{i\mathbf{k}\cdot\mathbf{r}} \frac{d^3k}{(2\pi)^3}.$$

The last equality follows from the following property of the generalized susceptibility: $\alpha_{\lambda\nu}(\boldsymbol{k},\omega) = \alpha_{\lambda\nu}(-\boldsymbol{k},\omega)$ (Equation (7.43)). Owing to this property, we have

(2) $$\int (\Im\, \alpha_{\lambda\nu}(\boldsymbol{k},\omega))e^{i\mathbf{k}\cdot\mathbf{r}}\, d^3k = \int \Im\, \alpha_{\lambda\nu}(\boldsymbol{k},\omega)\cos(i\mathbf{k}\cdot\mathbf{r})\, d^3$$
$$k = \Im \int \alpha_{\lambda\nu}(\boldsymbol{k},\omega)e^{i\mathbf{k}\cdot\mathbf{r}}\, d^3k.$$

Substituting the generalized susceptibility (Equation (7.43)) in Equation (1), we can calculate the integral over space of wave vectors:

(3) $$\int \frac{\exp(ikr\cos\vartheta}{k^2-\omega^2\varepsilon\mu/c^2}\frac{2\pi k^2\, dk\, d\cos\vartheta}{(2\pi)^3} = \frac{1}{(2\pi)^2 ir}\int_{-\infty}^{\infty}\frac{ke^{ikr}}{k^2-\omega^2\varepsilon\mu/c^2}$$
$$dk = \frac{1}{4\pi r}\exp\left(i\frac{\omega r}{c}\sqrt{\varepsilon\mu}\right).$$

The last integral was calculated with the aid of residues by closing the contour of integration with an arc of a large radius in the upper half-plane of the complex k. Because of (5.93), the value $\Im\omega\sqrt{\varepsilon(\omega)\mu(\omega)} > 0$ and it ensures that the correlation damps at large distances r. The integral containing the product $k_\lambda k_\nu$, is reduced to the expression

(4) $$\frac{\partial^2}{\partial x_\lambda \partial x_\nu}\frac{1}{4\pi r}\exp\left(i\frac{\omega r}{c}\sqrt{\varepsilon\mu}\right).$$

The final result is given by:

(5)
$$K_{\lambda\nu}(\boldsymbol{r},\omega) = \hbar\coth\left(\frac{\hbar\omega}{2T}\right)\Im\left(\mu\delta_{\lambda\nu} + \frac{c^2}{\omega^2\varepsilon}\frac{\partial^2}{\partial x_\lambda\partial x_\nu}\right)\frac{1}{r}\exp\left(i\frac{\omega r}{c}\sqrt{\varepsilon\mu}\right).$$

7.22

$$\langle E_\lambda(\boldsymbol{r},t)E_\nu(\boldsymbol{r}',t')\rangle_{\mathbf{k}\omega} = \coth\left(\frac{\hbar\omega}{2T}\right)\Im\frac{4\pi\omega^2\mu\hbar}{c^2k^2-\omega^2\varepsilon\mu}\left(\delta_{\lambda\nu} - \frac{c^2k_\lambda k_\nu}{\omega^2\varepsilon\mu}\right),$$
$$\langle E_\lambda(\boldsymbol{r},t)B_\nu(\boldsymbol{r}',t')\rangle_{\mathbf{k}\omega} = -e_{\lambda\nu\beta}k_\beta\coth\left(\frac{\hbar\omega}{2T}\right)\Im\frac{4\pi\omega\mu\hbar c}{c^2k^2-\omega^2\varepsilon\mu},$$
$$\langle B_\lambda(\boldsymbol{r},t)B_\nu(\boldsymbol{r}',t')\rangle_{\mathbf{k}\omega} = (k^2\delta_{\lambda\nu} - k_\lambda k_\nu)\coth\left(\frac{\hbar\omega}{2T}\right)\Im\frac{4\pi\mu\hbar}{k^2-\omega^2\varepsilon\mu/c^2}.$$

7.23 In order to calculate the energy spectral density, we have to find the spectral densities of the field strengths $\langle E^2\rangle_\omega$, $\langle H^2\rangle_\omega$. For this purpose, we will use the results of Problem 7.22:

(1) $$\langle E^2\rangle_\omega = \langle E_\lambda(\boldsymbol{r},t)E_\lambda(\boldsymbol{r},t)\rangle_\omega$$
$$= \coth\left(\frac{\hbar\omega}{2T}\right)\Im\int\frac{4\pi\omega^2\mu\hbar}{c^2k^2-\omega^2\varepsilon\mu}\left(3-\frac{c^2k^2}{\omega^2\varepsilon\mu}\right)\frac{4\pi k^2 dk}{(2\pi)^3}.$$

The small imaginary parts ε and μ should be taken into account only in the denominator; with the use of the Sokhotsky formula (I.1.223) we have:

(2) $$\frac{1}{k^2-\omega^2\varepsilon\mu/c^2} \approx i\pi\delta\left(k^2-\frac{\omega^2n^2}{c^2}\right) + \frac{\mathcal{P}}{k^2-\omega^2n^2/c^2}.$$

On the right-hand side of this equation, the quantity $n = \sqrt{\varepsilon'\mu'}$ is the refraction index, which is determined by real parts of ε' and μ'. The delta-function alone will contribute to the right-hand side of Equation (1). By integrating we have

$$(3) \qquad \langle E^2 \rangle_\omega = \frac{2\hbar\omega^3 n}{c^3} \coth\left(\frac{\hbar\omega}{2T}\right).$$

In a similar manner, we get

$$(4) \qquad \langle H^2 \rangle_\omega = \frac{1}{\mu^2}\langle B^2 \rangle_\omega = \frac{\varepsilon}{\mu}\langle E^2 \rangle_\omega,$$

where ε and μ should be considered real.

The spectral density of the thermal electromagnetic field can be determined with the aid of Equation (5.110), which is referred to as a quasi-monochromatic field (a narrow range of frequencies $\Delta\omega \ll \omega$). In that formula, it is necessary to make the following replacements:

$$(5) \qquad \frac{1}{2}\boldsymbol{E}^* \cdot \boldsymbol{E} \to \langle E^2 \rangle_\omega \frac{\Delta\omega}{2\pi}, \qquad \frac{1}{2}\boldsymbol{H}^* \cdot \boldsymbol{H} \to \langle H^2 \rangle_\omega \frac{\Delta\omega}{2\pi}$$

and multiply the result by 2. The first replacement is due to the change from the squares of the fields to their spectral densities, and the factor 2 is for taking into account the symmetry $\langle E^2 \rangle_\omega = \langle E^2 \rangle_{-\omega}$. On performing these replacements with the aid of Equation (4) and dividing the result by $\Delta\omega$, we get the spectral density of the field energy

$$(6) \qquad \rho(\omega) = \left[\frac{\hbar\omega}{2} + \frac{\hbar\omega}{e^{\hbar\omega/T} - 1}\right] \frac{\omega^2 n^2}{\pi^2 c^3} \frac{d(\omega n(\omega))}{d\omega}.$$

The total energy is obtained by integrating Equation (6) over frequency in the limits from 0 to ∞.

Let us compare Equation (6) with the Plank formula for the spectral density of the equilibrium radiation in vacuum (a cavity inside the nontransparent body), which is given by formula (2) from the solution of Problem I.6.23:

$$(7) \qquad \rho(\omega) = \frac{1}{e^{\hbar\omega/T} - 1} \frac{\hbar\omega^3}{\pi^2 c^3}.$$

The first term in the square brackets on the right-hand side of Equation (6) yields the energy density of the zero-point oscillations (Equation (I.6.7)) and is diverged when integrated over all frequencies. The second term describes the equilibrium radiation in a transparent medium and coincides with the Plank formula (7.7) at $n = 1$. The refraction index $n(\omega)$ is taken into account by the number of the eigenoscillations $k^2 dk = (\omega n/c)^2 d(\omega n/c)$, if the dispersion law (6.7), $k = \omega n(\omega)/c$, for plane waves in an isotropic medium is used.

7.24 The Rayleigh formula is

$$h = \frac{2\omega^4 (n-1)^2}{3\pi c^4 N},$$

where N is the number density of molecules of the gas.

7.25

$$\frac{d\Sigma}{d\Omega} = \frac{\varepsilon_0^2\omega^4 V}{16\pi^2 c^4}\sin^2\theta \int K_\varepsilon(r)\exp(-i\boldsymbol{q}\cdot\boldsymbol{r})\,d^3r,$$

where the notations are the same as in Example 7.5.

7.26

$$\frac{d\Sigma}{d\Omega} = \frac{\varepsilon_0^2\omega^4 l_c^3 V K_0}{4\sqrt{2\pi}c^4}\sin^2\theta\exp[-2(kl_c)^2\sin^2\vartheta/2].$$

The integration over angles can be performed in the analytical form.

7.27 By definition

(1) $$K_u(r,0) = \int_{\kappa\le k}\langle m^2\rangle_\kappa \exp(i\boldsymbol{\kappa}\cdot\boldsymbol{r})\frac{d^2\kappa}{(2\pi)^2}.$$

According to the analysis made after Equation (7.67), the Fourier transform $\langle m^2\rangle_\kappa$ is noticeably different from zero in the region $\kappa < \kappa_m \approx 2\pi/l_m = 2\pi k/(kl_m) \gg 2\pi k$. In the small part of this region, where $\kappa \le k$, the quantity $\langle m^2\rangle_\kappa$ can be replaced by its value at $\kappa \approx 0$:

(2) $$\langle m^2\rangle_0 = \int K_m(\rho)d^2\rho \approx \pi l_m^2 K_m(0) = \pi l_m^2(\langle|M|^2\rangle - |\langle M\rangle|^2).$$

Finally, we get

(3) $$K_u(r,0) \approx \langle m^2\rangle_0 \int_{\kappa\le k}\exp(i\boldsymbol{\kappa}\cdot\boldsymbol{r})\frac{d^2\kappa}{(2\pi)^2} = \frac{\langle m^2\rangle_0 k^2}{2\pi}\frac{J_1(kr)}{kr},$$

where $J_1(kr)$ is a Bessel function. The transverse correlation length is determined by the argument of the Bessel function. Its first root is $kr \approx 3.83$, that is, we can set $l_\perp \approx 3.83/k \approx \lambda \gg l_m$. At large distances from the screen, the transverse correlation length is of the order of the wavelength and greatly exceeds the transverse correlation length of irregularities of the screen.

The longitudinal correlation function is given by

(4) $$K_u(0,\zeta) \approx \frac{\langle m^2\rangle_0 k^2}{2\pi}\frac{d}{d(k\zeta)}\frac{1-e^{ik\zeta}}{k\zeta}.$$

In this case, we also have $l_\parallel \approx \lambda$, but with distance, the longitudinal correlation decreases slower than the transverse correlation.

7.28 We use the condition $\kappa < \kappa_m \approx 2\pi/l_m = 2\pi k/(kl_m) \ll 2\pi k$ in Equation (7.82), and approximate the radical by the two first terms: $\sqrt{k^2-\kappa^2} \approx k - \kappa^2/2k$. The limits of integration over κ are assumed to be infinite. We have

(1) $$K_u(\boldsymbol{r},\zeta) \approx e^{ik\zeta}\int\langle m^2\rangle_\kappa \exp\{i\boldsymbol{\kappa}\cdot\boldsymbol{r} - i\kappa^2\zeta/2k\}\frac{d^2\kappa}{(2\pi)^2}.$$

The transverse correlation function

(2) $$K_u(\boldsymbol{r},0) \approx \int\langle m^2\rangle_\kappa \exp(i\boldsymbol{\kappa}\cdot\boldsymbol{r})\frac{d^2\kappa}{(2\pi)^2} = K_m(r)$$

is the same as the correlation function of the screen transmission. The correlation lengths are also the same, $l_\perp = l_m$. The longitudinal correlation function, according to Equation (1), has the form

(3) $$K_u(0,\zeta) \approx e^{ik\zeta} \int \langle m^2 \rangle_\kappa \exp(-i\kappa^2\zeta/2k) \frac{d^2\kappa}{(2\pi)^2}.$$

We can determine its explicit form for the particular spectrum $\langle m^2 \rangle_\kappa$ (Problem 7.29). However, the longitudinal correlation length can be evaluated provided that the exponent in Equation (3) weakly oscillates in the region $\kappa \le \kappa_m \approx 2\pi/l_m$, which is of importance for integration: $l_\parallel \approx (kl_m)l_m \gg l_m$.

7.29

$$K_u(\mathbf{r},\zeta) \approx \langle m^2 \rangle_0 q(\zeta) \exp\left\{ik\zeta - \frac{r^2 q(\zeta)}{2l_m^2}\right\}, \qquad q(\zeta) = \frac{1}{1 + i\zeta/kl_m^2}.$$

7.30 With the indicated substitution from Equation (7.77), we get $\partial_\xi[\varphi^{-1}(-\partial_t\varphi + \kappa\partial_\xi^2\varphi)] = 0$. It means that $\partial_t\varphi - \kappa\partial_\xi^2\varphi = \varphi dF/dt$, where dF/dt is an arbitrary function of time. One more substitution

(1) $$\varphi(\xi,t) = \Phi(\xi,t)\exp F(t)$$

leads to the heat conduction equation for Φ:

(2) $$\partial_t\Phi = \kappa\partial_\xi^2\Phi.$$

Since the function $F(t)$ does not affect the value of $P(\xi,t)$, we can change the notation $\Phi \to \varphi$ and assume that the function φ satisfies Equation (2).

The general solution of Equation (2) can be constructed with the aid of the Grin function, analogously to Example 4.6:

(3) $$\varphi(\xi,t) = \frac{1}{\sqrt{4\pi\kappa t}} \int_{-\infty}^{\infty} \exp\left\{-\frac{(\xi-\eta)^2}{4\kappa t} - \frac{1}{2\kappa}\int_0^\eta P_0(\eta')d\eta'\right\} d\eta.$$

7.31 We use Equation (3) of Problem 7.30 and write down the solution of Burgers equation in the form

(1) $$P(\xi,t) = \frac{\int_{-\infty}^{\infty} [(\xi-\eta)/t] e^{-G/2\kappa}\, d\eta}{\int_{-\infty}^{\infty} e^{-G/2\kappa}\, d\eta},$$

where

(2) $$G(\xi,\eta,t) = \frac{(\xi-\eta)^2}{2t} + \int_0^\eta P_0(\eta')\, d\eta'.$$

The solution does not depend on the lowest limit of the last integral, which can be chosen arbitrarily. We will specify the choice of the limit and write down

(3) $$\int_{+0}^{\eta} P_0(\eta')d\eta' = \begin{cases} 0, & \eta > 0; \\ -A, & \eta < 0. \end{cases}$$

Now it is possible to calculate the integral in the numerator of Equation (1) and write down the solution in the form

$$P(\xi,t)=\sqrt{\frac{\kappa}{t}}\frac{(e^R-1)e^{-\xi^2/4\kappa t}}{\sqrt{\pi}+(e^R-1)\int_{\xi/\sqrt{4\kappa t}}^{\infty}e^{-\eta^2}\,d\eta} \tag{4}$$

Special cases.

1. $R \ll 1$; in the denominator we leave only the first term, which gives

$$P(\xi,t)=\sqrt{\frac{\kappa}{\pi t}}Re^{-\xi^2/4\kappa t}=\frac{A}{\sqrt{4\pi\kappa t}}e^{-\xi^2/4\kappa t}, \tag{5}$$

that is, the solution of the heat conduction equation. In this case, the dissipation effect in the Burgers equation prevails over nonlinearity.

2. To study the second limiting case, we introduce the dimensionless variable $\zeta=\xi/\sqrt{2At}$, which allows us to write down the exact solution (2) in the form

$$P(\xi,t)=\sqrt{\frac{2A}{t}}g(\zeta,R),\ g(\zeta,R)=\frac{e^R-1}{2\sqrt{R}}\frac{e^{-\zeta^2R}}{\sqrt{\pi}+(e^R-1)\int_{\zeta\sqrt{R}}^{\infty}e^{-\eta^2 d\eta}} \tag{6}$$

Now we use the condition $R \gg 1$:

$$g(\zeta,R)\approx\frac{1}{2\sqrt{R}}\frac{e^{R(1-\zeta^2)}}{\sqrt{\pi}+e^R\int_{\zeta\sqrt{R}}^{\infty}e^{-\eta^2 d\eta}}. \tag{7}$$

At $z<0$, we have

$$\int_{\zeta\sqrt{R}}^{\infty}e^{-\eta^2}d\eta\approx\int_{-\infty}^{\infty}e^{-\eta^2}d\eta=\sqrt{\pi}$$

and $g\approx(1/2\sqrt{R})e^{-R\zeta^2}\to 0$ if $R\to\infty$.

If $z>0$, we have

$$\int_{\zeta\sqrt{R}}^{\infty}e^{-\eta^2}d\eta\approx\frac{e^{-R\zeta^2}}{2\zeta\sqrt{R}}$$

and we obtain, using expression (7),

$$g(\zeta,R)\approx\frac{\zeta}{1+2\zeta\sqrt{\pi R}e^{R(\zeta^2-1)}}. \tag{8}$$

This approximation leads to a simple dependence $g(\zeta,R)\approx\zeta$ at $0<\zeta<1,\ R\gg 1$. As a result, coming back to initial variables, we obtain a triangular wave which breaks off in a point $\xi=\sqrt{2At}$:

$$P(\xi,t)\approx\begin{cases}\xi/t & \text{at}\quad 0<\xi<\sqrt{2At},\\ 0 & \text{at}\quad \xi<0,\ \xi>\sqrt{2At}.\end{cases} \tag{9}$$

In vicinities of points $\xi=0$ and $\xi=\sqrt{2At}$, there are narrow (at $R\gg 1$) transition areas. Their analysis is given in the monograph by Whitham (1974), from which the solution is taken.

7.32 See Vainstein *et al.* (1993). The required equation can be written in the form Equations (7.76) or (7.77), replacing $p \to p_* = p + B^2/8\pi$, $c_s \to c_\perp = \sqrt{c_s^2 + v_A^2}$, $b \to b_* = (c_\perp^3/2V^2)(\partial^2_{p_*} V)_s$, $a \to a_* = 2^{-1}c_\perp^{-3}[\nu + \nu_m v_A^2/c_\perp^2 + \chi V(c_V^{-1} - c_p^{-1})c_s^2/c_\perp^2]$, where $v_A = B/\sqrt{4\pi\tau}$ is the Alfvén velocity.

7.33

$$u(x,t) = -3\Delta c \cosh^{-2}[(\Delta c/4|\mu|)^{1/2}(x - c_w t)],$$

where $\Delta c = v_A - c_w > 0$ and $c_w < v_A$. The soliton is a magnetic well which is slow, that is, its velocity is less than the velocity of the magnetohydrodynamic wave of small amplitude.

7.34 Let us use the potential energy curve (Figure 7.11). When $u > 0$ and $0 \geq E \geq W_{min} = -2\Delta c^3/3$, a particle in a potential well undergoes nonlinear oscillations, and its coordinate u varies in the limits from u_1 to u_2, which are determined by intersection points of the horizontal straight line $E = const$ with the boundary of the well. Soliton corresponds to $E = 0$ and to a change in u in the limits from $3\Delta c$ to 0, as it follows from the results of Example 7.9. When $u < 0$, a motion of the particle is infinite, which corresponds to the unlimited increase of u. This case does not describe any actual wave. If E is close in magnitude to W_{min}, the oscillations occur with a small amplitude near the value $u = 2\Delta c$ and have harmonic character.

This qualitative picture can be also described through formulas. Let us represent $P(u)$ in the form $P(u) = (u_1 - u)(u - u_2)(u - u_3)$, where $u_1 \geq u_2 \geq u_3$ are the roots of the cubic equation $P(u) = 0$. When $u_1 \geq u \geq u_2$, we will calculate the integral in Equation (4) of Example 7.9 with the aid of the reference book Gradshtein and Ryzhik (2007), and thus

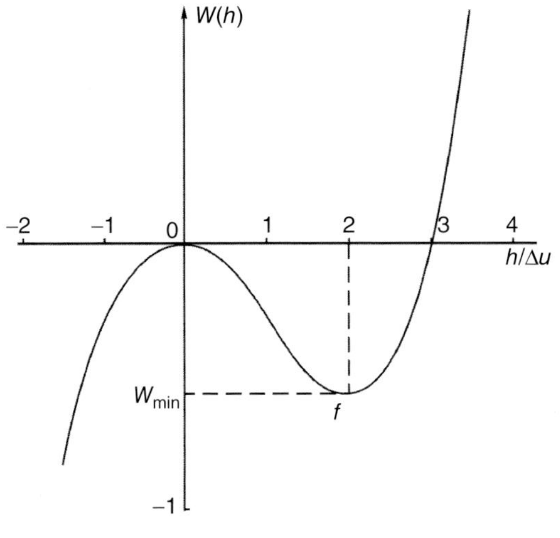

Figure 7.11 Effective potential energy for KdV equation.

we will have

(1) $$\frac{\pm\zeta}{\sqrt{3\mu}} = \frac{2F(\lambda,p)}{\sqrt{(u_1-u_3)}},$$

where $F(\lambda,p)$ is the elliptic integral of the first kind,

(2) $$\lambda = \arcsin\sqrt{\frac{u_1-u}{u_1-u_2}}, \qquad p = \sqrt{\frac{u_1-u_2}{u_1-u_3}}.$$

Solving the Equation (1) for u, we get

(3) $$u(x,t) = u_1 - (u_1-u_2)\,sn^2(\xi,p),$$

where $\xi = [(u_1-u_2)/12\mu]^{1/2}(x-c_w t)$, and $sn(\xi,p)$ is the Jacobi function. When ξ and p are real, the Jacobi function is a periodic function of ξ with the period of $4K(p)$, where $K(p)$ is the complete elliptic integral of the first kind. Equation (3) describes the cnoidal wave.

In the particular case in which $E \to 0$ we have $u_1 \to 3\Delta c$, $u_2 \to u_3 \to 0$; and $p \to 1$, and $sn(\xi,p) \to \tanh\xi$. Thus, we again have a soliton. For small E ($E/W_{min} \ll 1$), the cnoidal wave is an infinite chain of solitons separated by the distance of $L \approx \delta\ln(864W_{min}/E) \gg \delta$, where $\delta = 2(\mu/\Delta c)^{1/2}$ is the width of an individual soliton. When $1 - E/W_{min} \ll 1$, we have $u_1 \approx 2\Delta c$, $u_1 - u_2 \approx 2[(E-W_{min})/\Delta c]^{1/2}$, $p \ll 1$, and $sn(\xi,p) \approx \sin\xi$; solution (3) takes the form

(4) $$u(x,t) = 2\Delta c + [(E-W_{min})/\Delta c]^{1/2}\cos[(u_1-u_2)/3\mu]^{1/2}(x-c_w t)$$

and is a linear wave of small amplitude.

7.35 We will seek the solution of the KdV equation in the form $u(x - v_{sh}t)$. The KdV equation can be reduced to the form

(1) $$\mu u'' - \kappa u' = -\frac{\partial W}{\partial u} = \Delta c\, u - \frac{1}{2}u^2,$$

differing from Equation (2) in Example 7.9 only by the appearance of the dissipative term $-\kappa u'$. In the expression for the effective potential energy (Equation (3)) $\Delta c = v_{sh} - v_A > 0$, the constant of integration will be $K = 0$, if we require that asymptotically before the front $u \to 0$, and behind the front $u \to \Delta u = const$. The parameter Δu determines the strength of the shock wave. In Equation (1), the differentiation over the variable $\zeta = x - v_{sh}t$ is denoted by primes. With $\zeta \to -\infty$ in Equation (1) (the region behind the front), we find the relation between the parameters Δc and Δu:

(2) $$\Delta u = 2\Delta c.$$

One can elucidate the qualitative nature of the behavior of u using the mechanical analogy with oscillations of a particle in a potential well with friction (Figure 7.12). Assume that $\mu > 0$. The oscillator starts from the state a when $\zeta \to +\infty$ (the asymptotic $u = 0$ before the wave front). With increasing t (that corresponds to the decrease of ζ), the oscillator passes to the state b (the apex of the magnetic field first maximum similar

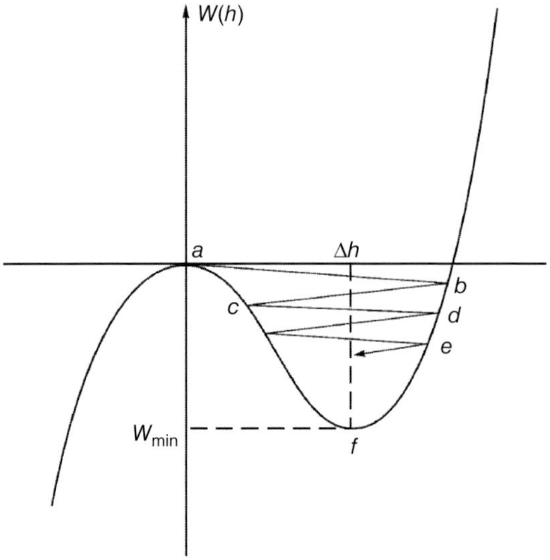

Figure 7.12 Oscillations of a particle in the potential well with friction.

to a soliton, provided the dissipation is small). Further on, the infinite number of oscillations occur (the states *c*, *d*, etc.) accompanied by the decrease of the field amplitude until, when $\zeta \to -\infty$, the oscillator passes asymptotically to the state *f* corresponding to the constant value $u = \Delta u$ behind the wave front.

The regime of small-amplitude oscillations can be easily investigated, if Equation (1) is linearized. Writing down its solution in the form $u(\zeta) = a_{1,2} \exp(k_{1,2}\zeta)$, we will have before the front (region 1)

(3) $$k_1 = -(\kappa/2\mu)[(1 - 4\mu\Delta c/\kappa^2)^{1/2} - 1].$$

This corresponds to the aperiodic regime when u changes from 0 up to a certain value u_{max}. Behind the front (region 2) we will have

(4) $$k_2 = \kappa/2\mu \pm i(\kappa/2\mu)(4\mu\Delta c/\kappa^2 - 1)^{1/2}.$$

The regime turns out to be oscillating, if

(5) $$\kappa < (4\mu\Delta c)^{1/2} = (2\mu\Delta u)^{1/2}.$$

(Figure 7.13). When $\mu < 0$ (positive dispersion) the picture changes to the opposite: before the front the oscillations occur, if the condition (5) for $|\mu|$ is satisfied; behind the front the smooth transition to the asymptote $u_2 = \Delta u$ takes place. The width of the shock front is determined by the distance at which the oscillations damp.

7.36 μ may have any sign. At $\mu \to 0$, we have Equation (5) from Example 7.8, and at $\kappa^2 \ll 4|\mu|\Delta c$ the front width is independent of the shock strength $\delta \approx 4|\mu|/\kappa$.

7.37 See Section 8.8 in the monograph of Vainstein *et al.* (1993).

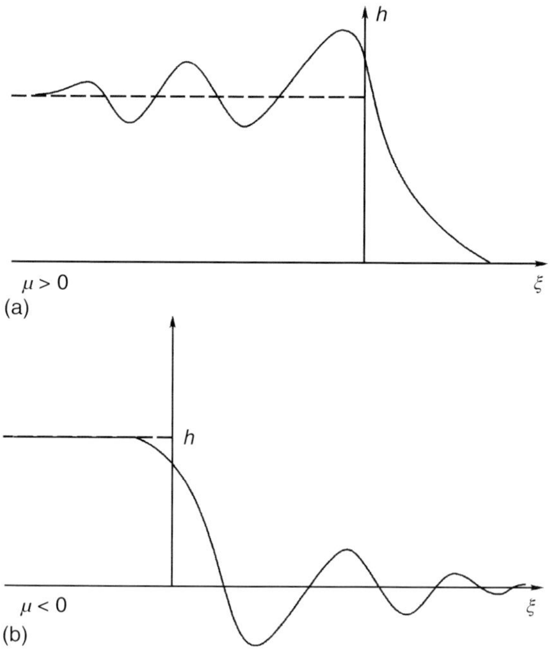

Figure 7.13 Oscillatory structure of the shock front depending on a sign of dispersive parameter μ.

7.38

$$\sigma(\omega) = \frac{8\pi\omega^4|\beta(\omega)|^2}{3c^4} = \frac{8\pi}{3} r_0^2 \frac{\omega^4}{(\omega_0^2 - \omega^2)^2 + \gamma^2\omega^2},$$

where $r_0 = e^2/mc^2$ (cf. Problem I.5.127).

$$\sigma(2\omega) = \sigma(\omega) \frac{4q^2|\beta(\omega)|^2 E_0^2}{(\omega_0^2 - 4\omega^2)^2 + 4\gamma^2\omega^2}.$$

7.39 There appears the nonlinear polarization at zero frequency, $P_0 = \alpha_1^{(2)}(0)E_1^2 + \alpha_2^{(2)}(0)E_2^2$, which is determined by nonlinear susceptibilities $\alpha_{1,2}^{(2)}(0) = -Neq|\beta(\omega_{1,2})|^2/2\omega_0^2$, and also by the nonlinear polarizations at double frequencies $2\omega_1$, $2\omega_2$, at the sum frequency $\omega_1 + \omega_2$ and at the difference frequency $\omega_1 - \omega_2$. The corresponding nonlinear susceptibilities are given by

$$\alpha^{(2)}(2\omega_{1,2}) = -\frac{Neq\beta^2(\omega_{1,2})}{2(\omega_0^2 - 4\omega_{1,2}^2 - 2i\gamma\omega_{1,2})},$$

$$\alpha^{(2)}(\omega_1 + \omega_2) = -\frac{Neq\beta(\omega_1)\beta(\omega_2)}{\omega_0^2 - (\omega_1 + \omega_2)^2 - i\gamma(\omega_1 + \omega_2)},$$

$$\alpha^{(2)}(\omega_1 - \omega_2) = -\frac{Neq\beta(\omega_1)\beta^*(\omega_2)}{\omega_0^2 - (\omega_1 - \omega_2)^2 - i\gamma(\omega_1 - \omega_2)}.$$

They are obtained in the same manner as in Example 7.10.

7.40 The dispersion equation is given by

$$1 - \frac{\omega_{0e}^2}{\omega^2} - \frac{\alpha\omega_{0e}^2}{(\omega - k_\parallel V)^2} = 0, \qquad \alpha = \frac{N'}{N}. \tag{1}$$

Hence, at $|k_\parallel|V < \omega_{0e}$ we have

$$\omega = k_\parallel V + i\gamma, \qquad \gamma = \frac{\alpha^{1/2}\omega_{0e}}{\sqrt{(\omega_{0e}/k_\parallel V)^2 - 1}}. \tag{2}$$

The group velocity of the growing wave is $v_g = d\omega/dk_\parallel = V$. In the system of the beam, the oscillations have zero frequency and rise aperiodically with an increment γ (walk-off oscillations).

At $k_\parallel V \approx \omega_{0e}$, Equation (2) for the increment does not hold, and Equation (1) should be solved again. We seek the solution in the form $\omega = \omega_{0e} + \beta$, $|\beta| \ll \omega_{0e}$, but $|\beta| \gg |\omega_{0e} - k_\parallel V|$. From Equation (1), we have

$$\frac{2\beta}{\omega_{0e}} = \frac{\alpha\omega_{0e}^2}{\beta^2}. \tag{3}$$

This equation has the real and complex roots. We need the complex root which satisfies the condition $\gamma = \Im\beta > 0$. It has the form

$$\beta = \omega_{0e}\alpha^{1/3}2^{-4/3}(-1 + i\sqrt{3}), \tag{4}$$

and determines the increment γ near the resonance between the Langmuir oscillations and the walk-off oscillations.

7.41 At $|k_\parallel|V < \omega_{0e}$, we have

$$\omega = i\gamma, \qquad \text{where} \qquad \gamma = \frac{\omega_{0i}}{\sqrt{(\omega_{0e}/k_\parallel V)^2 - 1}}.$$

At $k_\parallel V \approx -\omega_{0e}$

$$\omega \approx -\frac{\omega_{0e}}{2^{4/3}}\left(\frac{m_e}{m_i}\right)^{1/3} + i\gamma, \qquad \text{where} \qquad \gamma = \omega_{0e}\frac{\sqrt{3}}{2^{4/3}}\left(\frac{m_e}{m_i}\right)^{1/3}$$

is the Buneman instability increment.

8
Electromagnetic Oscillations in Finite Bodies

8.1
Electromagnetic Waves in Waveguides

In many technical devices, electromagnetic waves propagate along some guides (it can also occur in natural conditions). The guides can be metallic or dielectric bodies (wires, metallic tubes of an arbitrary cross section, dielectric and ferromagnetic plates and their combinations). Such devices are referred to as waveguides (light guides in an optical range). The problems related to the waveguides comprise a wide scope of applied electrodynamics. We will consider only the most general and principal questions concerning the possible types of waves in waveguides and the structure of their electromagnetic field.

Surface waves. First we elucidate the existence conditions for the surface monochromatic wave at the plane boundary between a dielectric and vacuum, that is, for such an electromagnetic perturbation which propagates along the boundary, with an amplitude decreasing with distance from the boundary. The dielectric is assumed to be isotropic and characterized by the scalar permittivity and permeability $\varepsilon(\omega)$ and $\mu(\omega)$ (the spatial dispersion is absent). Let the Oz-axis be directed along the normal to the surface of the dielectric. The field of the surface wave in each semispace satisfies the Maxwell's equations

$$\nabla \times \boldsymbol{E} = \frac{i\omega}{c}\mu \boldsymbol{H}, \qquad \nabla \times \boldsymbol{H} = -\frac{i\omega}{c}\varepsilon \boldsymbol{E}, \tag{8.1}$$

and in vacuum $\varepsilon = \mu = 1$. It follows from Equation (8.1) that the divergences of the vectors $\boldsymbol{E}$ and $\boldsymbol{H}$ are equal to zero. From these equations, we also have the second-order equations:

$$\triangle \boldsymbol{E} = -\frac{\omega^2}{c^2}\varepsilon\mu \boldsymbol{E} \tag{8.2}$$

and precisely similar equation for $\boldsymbol{H}$.

We look for the coordinate field function in the form

$$\begin{aligned} \boldsymbol{E}_1 &= \boldsymbol{\mathcal{E}}_1 \exp(ikx + q_1 z), \qquad \boldsymbol{H}_1 = \boldsymbol{\mathcal{H}}_1 \exp(ikx + q_1 z), \qquad q_1 > 0; \\ \boldsymbol{E}_2 &= \boldsymbol{\mathcal{E}}_2 \exp(ikx - q_2 z), \qquad \boldsymbol{H}_2 = \boldsymbol{\mathcal{H}}_2 \exp(ikx - q_2 z), \qquad q_2 > 0. \end{aligned} \tag{8.3}$$

Indices 1 and 2 are referred to vacuum and to a dielectric, respectively; a wave propagates along the Ox-axis and damps with a distance from the boundary both

Electromagnetic Phenomena in Matter: Statistical and Quantum Approaches, First Edition. Igor N. Toptygin.

in vacuum and in the dielectric. For the real permittivity and permeability, the damping is not connected with the dissipation of the electromagnetic energy. The wave number k is taken the same, both in a dielectric and vacuum that to perform the equating components of a field on either side of the boundary $z = 0$:

$$\boldsymbol{E}_{1\tau} = \boldsymbol{E}_{2\tau}, \quad \boldsymbol{H}_{1\tau} = \boldsymbol{H}_{2\tau}; \qquad E_{1n} = \varepsilon E_{2n}, \quad H_{1n} = \mu H_{2n}. \tag{8.4}$$

On substituting Equation (8.3) in Equation (8.2), we find the relation between the damping constants and the wave number and the frequency:

$$q_1 = q_2 = \sqrt{k^2 - \frac{\omega^2}{c^2}} > 0, \qquad q_2 = \sqrt{k^2 - \frac{\omega^2}{c^2}\varepsilon(\omega)\mu(\omega)} > 0. \tag{8.5}$$

Requiring that the damping constants be positive, we find the restrictions:

$$k^2 > \frac{\omega^2}{c^2}, \qquad k^2 > \frac{\omega^2}{c^2}\varepsilon(\omega)\mu(\omega). \tag{8.6}$$

Consider the possibility of existence for the surface wave with the transverse magnetic (TM) vector that is in the separation plane between media: $\boldsymbol{H}_1 = \boldsymbol{H}_2 = \mathcal{H}\boldsymbol{e}_y$. From the second equation of Equation (8.1), we can find the amplitudes of the electric field in both regions: $\mathcal{E}_{1x} = -(icq_1/\omega)\mathcal{H}$, $\mathcal{E}_{1z} = -(ck/\omega)\mathcal{H}$, $\mathcal{E}_{2x} = (icq_2/\omega\varepsilon)\mathcal{H}$, $\mathcal{E}_{2z} = -(ck/\omega\varepsilon)\mathcal{H}$. The boundary conditions (8.3) for the magnetic field and for the z-component of the electric field are identically satisfied, and the equating of the x-components of the electric vector yields the important relation $q_1 = -q_2/\varepsilon$. From this relation and relation (8.5), there follows another restriction on the properties of the dielectric: $\varepsilon(\omega) < 0$. By using the damping constants given by Equation (8.5), we get the dispersion equation for the surface wave:

$$\frac{\sqrt{k^2 - (\omega/c)^2\varepsilon(\omega)\mu(\omega)}}{\sqrt{k^2 - (\omega/c)^2}} + \varepsilon(\omega) = 0 \quad \text{or} \quad k^2 = \frac{\omega^2}{c^2}\frac{\varepsilon(\varepsilon - \mu)}{\varepsilon^2 - 1}. \tag{8.7}$$

In order that the surface wave could propagate along the surface ($k^2 > 0$), the following condition must be satisfied:

$$\frac{\varepsilon - \mu}{(\varepsilon + 1)(\varepsilon - 1)} < 0.$$

When $\mu = 1$ it leads to the restriction

$$\varepsilon(\omega) < -1. \tag{8.8}$$

Let us emphasize that the surface wave is not transverse, that is, its electric vector has the longitudinal (relative to the propagation direction) component. It is either the electric-type wave (or E-wave), or TM wave.

A similar analysis of the H-wave with $\mathcal{H}_x \neq 0$, $\mathcal{H}_z \neq 0$, and $\mathcal{E}_y \neq 0$ yields the dispersion equation

$$\frac{\sqrt{k^2 - (\omega/c)^2\varepsilon(\omega)\mu(\omega)}}{\sqrt{k^2 - (\omega/c)^2}} + \mu(\omega) = 0, \qquad \mu(\omega) < 0. \tag{8.9}$$

For this wave, the restrictions (8.6) remain valid, however, Equation (8.7) does not hold. The surface waves are considered in Problems (8.2)–(8.5) for specific models of a medium. The waves traveling in waveguides (in the 1D-model) are considered in Problems (8.6)–(8.8).

Metallic waveguides. Sufficiently long waves (e.g., centimeter-waves) are transferred through waveguides, which are metallic tubes with the cross section of different shape. Traveling waves propagate along the axis of the waveguide (Oz-axis), while in the transverse direction the wave is standing. In general, the waves in a waveguide will not be transverse, as was shown for the case of the surface waves. Purely transverse electromagnetic (TEM) waves are possible only in waveguides whose cross sections are not one-coherent.

The types of waves that can propagate in a given waveguide are determined by solving Maxwell's equations subject to the appropriate boundary conditions. A wave traveling along the axis is described by the functions

$$\boldsymbol{E}(\boldsymbol{r}, t) = \boldsymbol{\mathcal{E}}(x, y)\exp(ikz - i\,\omega t), \quad f\boldsymbol{H}(\boldsymbol{r}, t) = \boldsymbol{\mathcal{H}}(x, y)\exp(ikz - i\,\omega t), \quad (8.10)$$

where ω is the wave frequency, and k is the wavevector component along the waveguide axis. The quantity k is also called the propagation constant.

In the case of electric-type waves (E-waves), one has $\mathcal{H}_z = 0$, and $\mathcal{E}_z$ satisfies the equation

$$\triangle_\perp \mathcal{E}_z + \kappa^2 \mathcal{E}_z = 0, \quad (8.11)$$

where κ is the transverse component of the wavevector, ε and μ are the permittivity and magnetic permeability of the dielectric filling the waveguide, and the boundary condition on the waveguide wall is

$$\mathcal{E}_z = 0. \quad (8.12)$$

In the case of magnetic-type waves (H-waves) $\mathcal{E}_z = 0$, and $\mathcal{H}_z$ is the solution of the equation

$$\triangle_\perp \mathcal{H}_z + \kappa^2 \mathcal{H}_z = 0, \quad (8.13)$$

which satisfies the boundary condition on the waveguide wall

$$\mathcal{E}_\tau = 0 \qquad \text{or} \qquad \frac{\partial \mathcal{H}_z}{\partial n} = 0. \quad (8.14)$$

The symbol $\triangle_\perp$ in Equations (8.11) and (8.13) is the two-dimensional Laplace operator. The boundary conditions (8.12) and (8.14) are strictly valid only for waveguides with perfectly conducting walls.

The transverse components of the vectors $\boldsymbol{\mathcal{E}}$ and $\boldsymbol{\mathcal{H}}$ can be expressed in terms of the longitudinal components of these vectors with the aid of Maxwell's equations (8.1).

The possible values of the transverse wave number κ can be determined from the boundary conditions on the waveguide walls. The frequency ω, the wave numbers κ and k, as well as the permittivity and the permeability (ε and μ) of the dielectric filling the waveguide are connected by the dispersion relation

$$\omega^2 \varepsilon\mu = \omega_c^2 + (ck)^2, \quad (8.15)$$

which is obtained upon substituting the solutions of Equation (8.10) in Equation (8.2). Here $\omega_c = c\kappa$ is the limiting frequency of a waveguide for a wave of a given type (i.e., with a given κ). In a waveguide with a simply connected cross section (see Problems 8.9, 8.11, 8.21, and others) all possible values of $\kappa > 0$. In such waveguides, the waves with frequencies $\omega > \omega_0$, where ω_0 is the minimal limiting frequency, can only propagate. The corresponding "vacuum wavelength" $\lambda_0 = 2\pi c/\omega_0$ is of the order of the transverse linear dimensions of the waveguide. When $\omega < \omega_0$ the propagation constant k is purely imaginary, and the wave propagation is therefore impossible. However, even when $\omega > \omega_0$ the propagation constant k is, in general, complex.

Energy losses in waveguides. This is because the walls of the waveguide have a finite conductivity, and, hence, the energy dissipation occurs within them with the result that the electromagnetic wave is damped out as $\sim e^{-\alpha z}$. The damping coefficient α (i.e., the imaginary part of k) is equal to the ratio of the energy dissipated per unit time and per unit length of the wall of the waveguide, to the double energy flux along the waveguide. When the surface impedance $\zeta = \zeta' + i\zeta''$ is small, it is possible to obtain the approximate relations of the damping coefficient for E-waves:

$$\alpha = \frac{\omega\zeta'}{2\kappa k c} \frac{\oint |\nabla \mathcal{E}_z|^2 \, dl}{\int |\mathcal{E}_z|^2 \, dS} \tag{8.16}$$

and for H-waves:

$$\alpha = \frac{c\kappa^2\zeta'}{2k\,\omega} \frac{\oint \left[|\mathcal{H}_z|^2 + (k^2/\boldsymbol{k}^4)|\nabla \mathcal{H}_z|^2\right] dl}{\int |\mathcal{H}_z|^2 \, dS}, \tag{8.17}$$

where $\mathcal{E}_z$ and $\mathcal{H}_z$ are the field components for $\zeta = 0$, that is, for the perfectly conducting walls, dl is an element of the perimeter of the waveguide cross section, and dS is an area element of the cross section.

In the waveguides with a multiply connected cross section (see Problems 8.10 and 8.19), the purely transverse (TEM) waves can propagate.

Suggested literature: Jackson (1999); Vainshtein (1988, 1966a); Gurevich and Melkov (1994); Solimeno *et al.* (1986); Kaganov *et al.* (1997); Panofsky and Phillips (1963); Iorganev and Bondarenko (2002); Pamyatnykh and Turov (2000); Nikol'skii (1973); de Broglie (1941)

Problems

8.1 Derive the dispersion equation for the surface waves on the plane that separates two media whose permittivity and the magnetic permeability are ε_1, and μ_1, and also ε_2 and μ_2, respectively.

8.2• In the collisionless plasma and in any condensed medium, the dielectric permittivity at rather high frequencies is given by $\varepsilon(\omega) = 1 - \omega_{0e}^2/\omega^2$, and the magnetic permeability $\mu = 1$. Using Equations (8.7) and (8.8), investigate the surface plasma wave, and find also the dependence $\omega(k)$. Consider limiting cases.

8.3 Perform the same procedure as in Problem 8.2 for nonmagnetic and nonpolar dielectric with the dielectric permittivity $\varepsilon(\omega) = 1 + \omega_p^2/(\omega_0^2 - \omega^2)$ (the surface polariton).

8.4 A ferrodielectric magnetized to saturation along the Oz-axis occupies a semispace $y > 0$. Deduce the dispersion equation in the magneto-static approximation for the surface wave whose wavevector is aligned with the Ox-axis or is opposed to it, and the vector of the magnetic field strength has the nonzero components H_x, H_y. Use formula (6.64) for the magnetic permeability. Determine the directions of phase velocity of the wave for the two directions of wavevector mentioned above. The essence of the magneto-static approximation is elucidated in Problem 6.50.

8.5** A ferrodielectric magnetized to saturation along the Oz-axis occupies a semispace $y > 0$. Investigate the surface wave with the nonzero components of the field vectors E_x, H_x, and H_y, which propagates in the Ox-direction. Derive the dispersion equation for this wave assuming $\varepsilon = const$ and using relation (6.36) for the tensor $\hat{\mu}$. Investigate the solutions of this equation taking into account the nonreciprocity, that is, the dependence of the frequency on the direction of wave propagation (along or opposite Ox-axis with the unchanged direction of the constant magnetizing field $\boldsymbol{H}_0$ and of the magnetization $\boldsymbol{M}_0 \parallel \boldsymbol{H}_0$). Find the domains of the existence of the surface wave in the (k_x, ω) plane.

8.6 An infinite dielectric layer with permittivity ε and magnetic permeability μ fills the region $-a \le x \le a$ in vacuum. Show that the layer will act as a waveguide, that is, the field of the traveling electromagnetic wave will be largely concentrated within the layer. Determine the types of wave that can propagate in such a waveguide. Confine the consideration to the particular case where the field vectors do not depend on the coordinate y.

8.7 A dielectric layer with permittivity ε and magnetic permeability μ fills the region $0 \le x \le a$. It is coated on to the surface of a perfect conductor. The region $x > a$ is evacuated. Determine the types of electromagnetic wave, with amplitude decreasing with distance from the layer, which can propagate along the layer. Compare the possible types of wave with the wave system obtained in Problem 8.7.

8.8 A dielectric layer considered in Problem 8.6 is transparent in optical range and can be used as a waveguide. Calculate, for even E-waves those fraction K of the flux of electromagnetic energy, which is transferred within the dielectric (relative to the total flux). Find the conditions at which K is close to unity.

8.9 Derive Equations (8.16) and (8.17) for the damping coefficients of waves in the metallic waveguide with the small surface impedance.

8.10 The plane waveguide is formed by two perfectly conducting parallel planes of large ("infinite") lengths with a distance a between them. It is filled with dielectric of permittivity ε and permeability μ. Determine the possible types of wave, their dispersion equations and the distribution of the monochromatic field in space.

8.11 Determine the types of wave which can propagate in a rectangular waveguide with perfectly conducting walls and sides a and b. Find the corresponding dispersion relation and the field configurations, that is, the dependence of the field components on the coordinates.

8.12 Determine the damping coefficients α of the various types of wave in a rectangular waveguide, given that the surface impedance of the waveguide walls is ζ.

8.13 Find the possible types of wave in a circular waveguide of radius a, assuming that the walls are perfectly conducting. Determine the limiting frequency ω_0 for this waveguide.

8.14 Using the result of Problem 8.13, find the damping coefficients α of various types of waves in a circular waveguide, given that the surface impedance is ζ.

8.15 Determine the phase v_{ph} and group v_g velocities of waves in rectangular and circular waveguides with perfectly conducting walls; plot the dependence of the wave velocities on $\lambda = 2\pi c/\omega$.

8.16 Determine the phase v_{ph} and group v_g velocities of waves in a waveguide by a geometrical method. For this purpose, consider the simplest H_{10} wave in a rectangular waveguide, resolve it into plane waves and investigate the reflections of these waves from the waveguide walls.

8.17* The dispersion relation (8.15) for an electromagnetic wave in a non-filled ($\varepsilon = \mu = 1$) metallic waveguide is formally the same as the linkage equation between energy and the momentum of the relativistic particle:[1)]

$$\begin{aligned}
\mathcal{E}^2 &= (Mc^2)^2 + (cp)^2, \qquad M \text{ is the mass of a particle;}\\
\omega^2 &= \omega_{n0}^2 + (ck)^2, \quad \omega_{n0} \text{ is the limiting frequency of the mode under consideration.}
\end{aligned}$$

This analogy becomes even more evident, if we turn to the quantum notations on photons in a waveguide. Then $\mathcal{E}_\omega = \hbar\omega$ is the photon energy, $\hbar k = p$ is its momentum, and the quantity $M_\omega = \hbar\omega_{n0}/c^2$ plays the role of its mass:

$$\mathcal{E}_\omega^2 = (M_\omega c^2)^2 + (cp)^2.$$

Thus, a photon in a waveguide, unlike photons in infinite space, acquires a mass. Analyze some aspects of such a treatment of the waveguide modes, using the simplest plane waveguide, considered in Problem 8.12 as a model.

1) This analogy has been pointed out by of one of the founders of quantum mechanics (see de Broglie, 1941).

1. Show that if the distance between metallic planes changes adiabatically (slowly), the longitudinal momentum of the photon is $\hbar k = p$, which is the adiabatic invariant in this case.[2] However, the energy of the waveguide photon and its mass will change.
2. Calculate the energy change of the waveguide mode when the walls approach each other, considering it as a result of the work done by an external source against the forces of the light pressure. Show that in the evolutionary process the usual adiabatic invariant of a harmonic oscillator, that is, the ratio of the energy of the mode to its frequency, is conserved.

Hint. When performing point 1 resolve the waveguide mode into two transverse plane waves propagating at some angle to the waveguide axis and calculate the changes in their frequencies and directions when they are reflected from the moving totally reflecting wall.

8.18** Perform the further analysis of the properties of a photon in a waveguide. Determine the velocity of a photon propagating along a waveguide. Consider its "longitudinal" and "transverse" masses (defined in Problem I.4.41). Consider also the behavior of the waveguide photon in the gravitational field.

8.19 Investigate the structure of a TEM wave in a perfectly conducting coaxial line of inner and outer radii b and a, respectively). Determine the average energy flux $\overline{\gamma}$ along the line. Consider the limiting case of a single perfectly conducting cylindrical conductor.

8.20 Determine the possible types of non-TEM waves in a coaxial line with perfectly conducting walls (whose radii are a and $b > a$).

8.21 Determine the damping coefficient α of a transverse electromagnetic wave in a coaxial line with radii a and $b > a$, and with the surface impedance $\zeta = \zeta' + i\zeta''$.

Hint. Use the relation between the damping coefficient and the energy losses given at the beginning of this section.

8.22* Consider the propagation of an axially symmetric E-wave along an infinitely long cylindrical conductor, which has a finite conductivity and is placed in vacuum. Determine the phase velocity of the wave. Show that in a perfectly cylindrical conductor the wave will become identical with the TEM wave (see Problem 8.19). Use the approximate Leontovich boundary condition (6.27).

8.23 An axially symmetric E-wave propagates in a circular waveguide of radius b which is partially filled with a dielectric. The dielectric has a permittivity ε and occupies the region $a \le r \le b$. At $a \ll b$, determine the phase velocity as a function of frequency and find the limiting frequency. Under which conditions will the phase velocity be less than c? Consider the limiting case of a waveguide filled completely with the dielectric.

2) See subsection "Adiabatic invariants" of Section I.4.2.

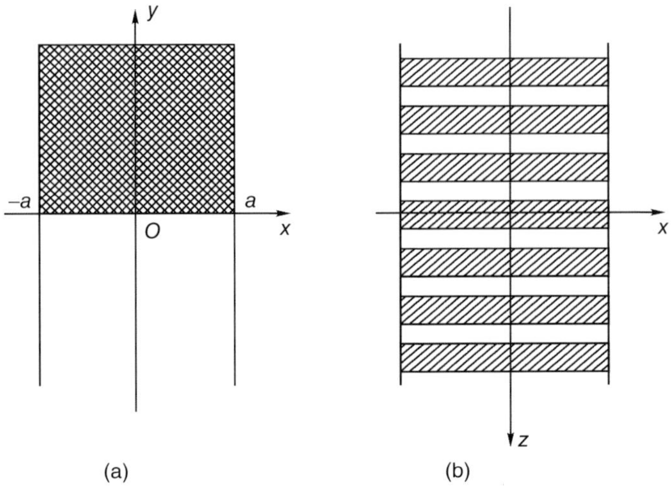

Figure 8.1 (a) Perfectly conducting planes between which there is dielectric ($y > 0$) and air ($y < 0$). (b) Ladder diaphragm in the xz plane.

8.24 A ladder diaphragm (Figure 8.1b) consisting of thin metallic strips oriented along the Ox-axis is placed in the $y = 0$ plane between two perfectly conducting planes $x = \pm a$ (Figure 8.1a). The distances between the strips and their width are small compared to the wavelength. The region $y > 0$ above the ladder diaphragm is filled with a dielectric of permittivity ε, the region $y < 0$ contains air. Find the possible types of waves which can propagate along the axis z in such a system. What is the connection between the propagation constant and the frequency of these waves?

Hint. For sufficiently long waves, the ladder diaphragm can be considered an anisotropic conducting plane with an infinite conductivity in the Ox-direction and a zero conductivity in the Oz-direction.

8.25 A rectangular waveguide with a transverse cross section $a \times b$ and perfectly conducting walls is filled with a ferrodielectric. The constant magnetic field is applied at right angles to the wider wall of the waveguide (along the Oy-axis). The permittivity and permeability tensors of the ferrodielectric have the form

$$\varepsilon_{ik} = \begin{pmatrix} \varepsilon_\perp & 0 & -i\varepsilon_a \\ 0 & \varepsilon_\parallel & 0 \\ i\varepsilon_a & 0 & \varepsilon_\perp \end{pmatrix}, \qquad \mu_{ik} = \begin{pmatrix} \mu_\perp & 0 & -i\mu_a \\ 0 & \mu_\parallel & 0 \\ i\mu_a & 0 & \mu_\perp \end{pmatrix}.$$

Determine the components of the electromagnetic field, the propagation constant, and the limiting frequency of the waveguide at which the field is not depend of the y-coordinate.

8.26 The electric and magnetic fields in a waveguide with perfectly conducting walls are described by functions

$$\mathbf{E}_0 = \mathcal{E}_0(x, y)e^{i(k_0 z - \omega t)}, \qquad \mathbf{H}_0 = \mathcal{H}_0(x, y)e^{i(k_0 z - \omega t)}.$$

When a dielectric core in the form of a cylinder of arbitrary cross section and its longitudinal axis parallel to the axis of the waveguide, is inserted into the waveguide, the fields will be given by

$$\mathbf{E} = \mathcal{E}(x, y)e^{i(kz - \omega t)}, \qquad \mathbf{H} = \mathcal{H}(x, y)e^{i(kz - \omega t)}.$$

The dielectric can be generally characterized by the tensor quantities ε_{ik} and μ_{ik}. Show, with the aid of Maxwell's equations, that the change in propagation constant will be

$$\Delta k = k - k_0 = \frac{\omega \int_{\Delta S}\left(\Delta\varepsilon_{ik}\mathcal{E}_k\mathcal{E}^*_{0i} + \Delta\mu_{ik}\mathcal{H}_k\mathcal{H}^*_{0i}\right) dS}{c\int_S\left[(\mathcal{E}^*_0 \times \mathcal{H}) + (\mathcal{E} \times \mathcal{H}^*_0)\right] \cdot \mathbf{e}_z\, dS},$$

where $\Delta\varepsilon_{ik} = \varepsilon_{ik} - 1$, $\Delta\mu_{ik} = \mu_{ik} - 1$. The integral in the nominator is taken over the cross section of the dielectric rod (ΔS), and the integral in the denominator is taken over the cross section of the waveguide (S).

8.27 A ferrodielectric plate of thickness $d \ll a$ is inserted in a rectangular waveguide with perfectly conducting walls (Figure 8.2). The plate is magnetized along the waveguide axis. Use the formula obtained in the Problem 8.26 to determine the change Δk in the propagation constant for H_{10} waves to within the terms of the order of d. The permittivity of the plate is a scalar, and the magnetic permeability tensor is given by Equation (6.55).

8.28 A thin ferrite plate ($d < a$, b) is inserted into the coaxial waveguide (Figure 8.3) and is magnetized along the axis of the waveguide. Determine

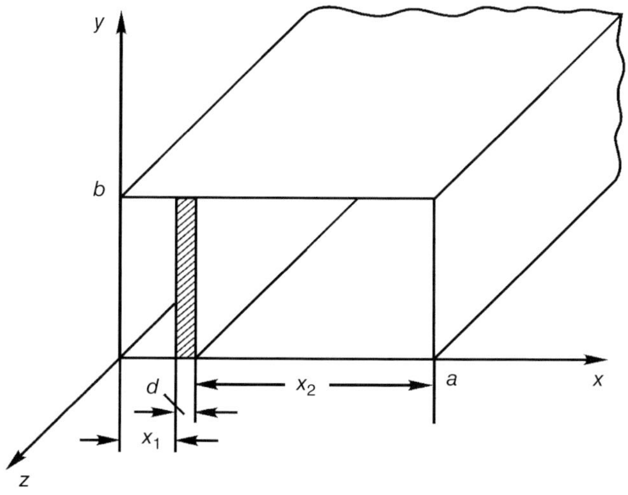

Figure 8.2 Rectangular wave guide with a ferrite plate.

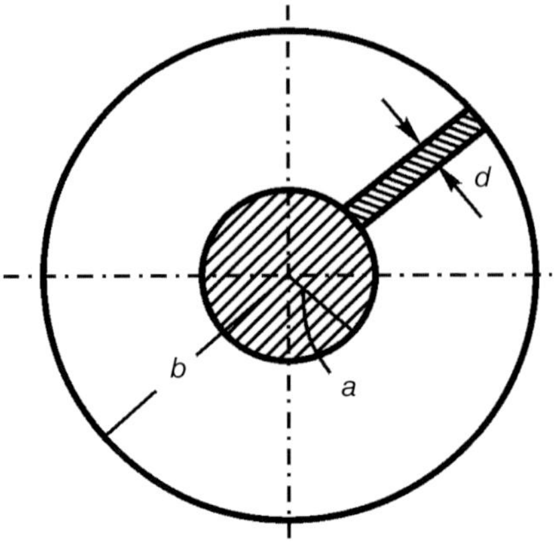

Figure 8.3 Coaxial wave guide with a ferrite plate.

the change Δk in propagation constant of the transverse electromagnetic wave.

Hint. The amplitudes of the perturbed fields can be found by the method used in Problem 8.27.

8.29 Solve Problem 8.28 for the constant magnetizing field $\mathcal{H}_0$ applied at right angles to the axis of the waveguide. Consider two directions of this field: (a) $\mathcal{H}_0$ perpendicular to the longer side of the plate; (b) $\mathcal{H}_0$ perpendicular to the shorter side of the plate.

8.2 Electromagnetic Oscillations in Resonators

Equation and boundary condition. The resonators of microwave range are closed cavities bounded on all sides by metal walls. Dielectrics, magnetics, metal rods, and other elements can be inserted into a resonator in order to control the electromagnetic oscillations in it. A system of standing waves with definite frequencies ω (and with wavelengths comparable with the resonator length-scales) can exist (eigenfrequencies of the resonator) there. When the resonator is not filled with a dielectric and has perfectly conducting walls, the standing wave system can be obtained by solving the equations

$$\triangle \mathbf{E} + \frac{\omega^2}{c^2}\mathbf{E} = 0, \qquad div\, \mathbf{E} = 0 \tag{8.18}$$

with the boundary condition

$$\mathbf{E}_\tau = 0. \tag{8.19}$$

The resonator eigenfunctions $\mathbf{E}_\nu$,[3] corresponding to different eigenfrequencies ω_ν are mutually orthogonal. The eigenfunctions corresponding to a single frequency (of which there can be several, see Problems 8.30 and 8.32) can also be chosen mutually orthogonal. Let us normalize them to 4π:

$$\int \mathbf{E}_{\nu'} \cdot \mathbf{E}_\nu \, dV = 4\pi\delta_{\nu\nu'}, \tag{8.20}$$

where the integral is taken over the volume of the resonator. The eigenfunctions $\mathbf{H}_\nu$, which are expressed in terms of the $\mathbf{E}_\nu$ by means of the Maxwell's equations, satisfy the same condition.

Quality factor of the resonator. Due to the energy losses in the walls or in matter filling the resonator, and also due to the emission of energy into an outside space, the eigenoscillations of real resonators are damped out. The energy losses of the given oscillations are characterized by the quality factor Q_ν, which is determined by the ratio

$$Q_\nu = \frac{\widetilde{\omega}_\nu W_\nu}{P_\nu} \qquad \text{or} \qquad Q_\nu = \frac{\widetilde{\omega}_\nu}{2\gamma_\nu}. \tag{8.21}$$

Here W_ν is the energy stored in the resonator; P_ν is the time-averaged power of losses; $\widetilde{\omega}_\nu$ is the resonance frequency which may differ from the resonance frequency of the perfect resonator; and γ_ν is the damping rate.

Optical resonators that are employed in lasers are, as a rule, open. They do not have the lateral surface and an electromagnetic wave is confined in a resonator where it alternatively reflects by two flat-parallel, or spherical or of some other mirrors separated by a distance that greatly exceeds the wavelength.

Suggested literature: Jackson (1999); Vainshtein (1988, 1966a); Gurevich and Melkov (1994); Solimeno *et al.* (1986); Svelto (1982); Panofsky and Phillips (1963); Iorganev and Bondarenko (2002); Pamyatnykh and Turov (2000); Nikol'skii (1973); de Broglie (1941)

Problems

8.30 Determine the types of eigenfrequencies of a hollow resonator with perfectly conducting walls. The resonator is in a form of a rectangular $a \times b \times h$ parallelepiped.

8.31 Determine the number of eigenfrequencies $\Delta N(\omega)$ per frequency interval $\Delta\omega$ in a hollow resonator considered in Problem 8.34. Assume that $\Delta\omega \ll \omega$ and $\Delta N \gg 1$, and that the resonator volume is V.

3) The index ν stands for the set of four quantities which uniquely determine the resonator eigenoscillations (modes).

8.32 A resonator has a form of a right-circular cylinder of height h and radius a. Assuming that the walls are perfectly conducting, find the frequencies of the eigenoscillations. Consider the E- and H-oscillations.

8.33 Two circular metallic plates of radius R are at a small distance d from each other and form a capacitor. The plates of the capacitor are closed by a conductor of thickness $2a$ in the form of a ring of radius b (Figure 8.4). Find the eigenfrequency of such an "open" resonator by assuming that a quasi-stationary approximation can be applied. Consider all conductors as perfect.

8.34 Find the eigenfrequency ω_0 of the oscillations of the system shown in Figure 8.5, assuming that the corresponding wavelength λ_0 is large compared to the dimensions of the system. Neglect the energy losses and edge effects.

8.35 To decrease the energy losses due to radiation one uses, instead of an open oscillating contour (see Figure 8.4), a closed resonator consisting of a toroidal cavity joined to a plane capacitor with circular plates (its cross

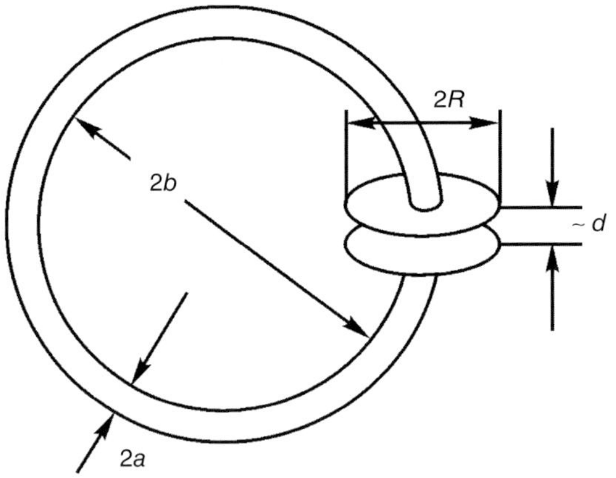

Figure 8.4 The open resonator in the form of a ring with inductance and capacity.

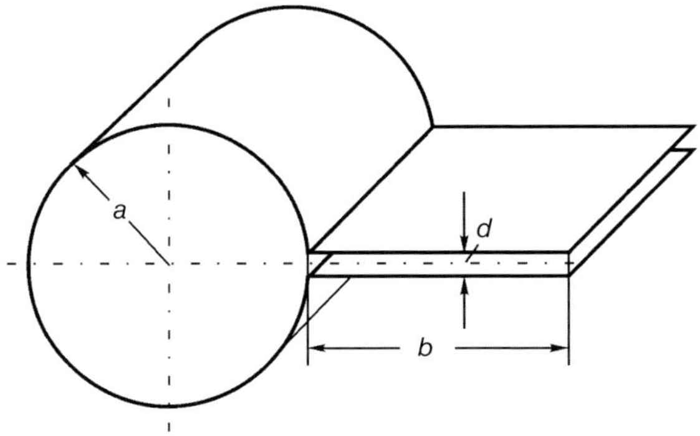

Figure 8.5 The cylindrical open resonator.

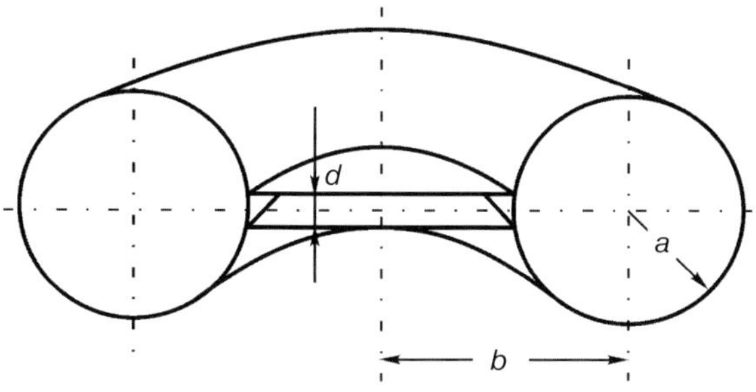

Figure 8.6 The closed toroidal resonator with the chamber of circular section.

section and dimensions are shown in Figure 8.6). Find the eigenfrequency ω_0 of the basic type of oscillations of such a resonator in the quasi-stationary approximation. Under which conditions is this approximation applicable? Assume the walls of the resonator to be perfectly conducting.

8.36 Solve Problem 8.39 for a toroidal resonator with a cavity of rectangular cross section (Figure 8.7).

8.37 A resonator is a cylinder of circular cross section (inside the radius b and height h). Along its axis a perfectly conducting rod of radius a is inserted (Figure 8.8). The walls of the cylinder are also perfectly conducting. Between the rod and one of the ends of the rod there is a gap d. Find the eigenfrequencies of the transverse (with respect to the axis of the system) electromagnetic oscillations, assuming that the wavelength of these oscillations is much longer than the gap width d (but not longer than the cylinder height h). What is the change in the spectrum of the oscillations at $d \to 0$?

8.38 Assume you know the eigenfrequencies ω_ν and the eigenfunctions $\mathbf{E}_\nu$, $\mathbf{H}_\nu$ of the oscillations of a resonator with perfectly conducting walls. Calculate the change in the eigenfrequencies caused by the finite conductivity of the resonator walls. The surface impedance ζ of the walls is small.

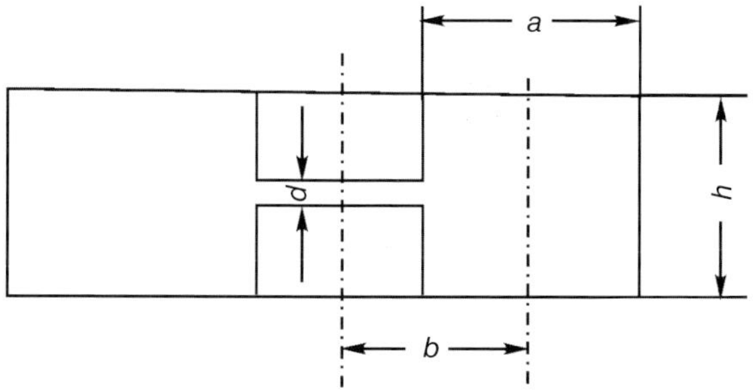

Figure 8.7 The same as in Figure 8.6, with the chamber of rectangular section.

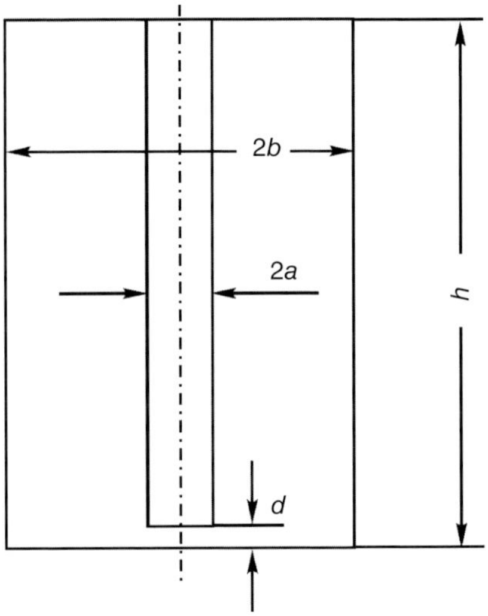

Figure 8.8 The cylindrical closed resonator with a rod.

Hint. Look for the solution of the Maxwell's equations in the form

$$\mathbf{E}(\mathbf{r}, t) = \sum_{\nu} q_{\nu}(t)\mathbf{E}_{\nu}(\mathbf{r}), \qquad \mathbf{H}(\mathbf{r}, t) = \sum_{\nu} p_{\nu}(t)\mathbf{H}_{\nu}(\mathbf{r}),$$

where q_ν and p_ν are unknown functions of time. Derive the equations for q_ν and p_ν up to terms linear in ζ and study the solution.

8.39 A hollow resonator has a shape of a cube of edge a. The conductivity of the walls is σ and the magnetic permeability $\mu = 1$. Calculate the quality factor of the resonator for an arbitrary type of oscillations. How does it depend on the frequency? At what frequencies do the resonance properties of the system disappear?

8.40 A hollow resonator with the walls of the surface impedance ζ is excited by an external current $\mathbf{j}(\mathbf{r})e^{-i\omega t}$ inside the resonator. The current frequency ω is close to one of the eigenfrequencies of the resonator. Find the electromagnetic field excited in the resonator and its dependence on the frequency ω near resonance.

Hint. Use the method developed in Problem 8.38.

8.41 An open resonator of the infrared range consists of two parallel circular mirrors of diameter D, at a distance L opposite to each other (Figure 8.9). Let the eigenoscillations of such a system be realized in the form of two waves with $\lambda \ll L, D$ propagating under right angles to the plane of the mirrors toward each other and forming a standing electromagnetic wave. Estimate the order of magnitude of the quality factor of such a resonator in the geometrical optics approximation. Take into account the energy losses at the reflection from the mirrors (reflection coefficient R) and the radiation through the side surface of the resonator due to diffraction. The parameters of the resonator are $D = L = 1$ cm; $R = 0.95$; $\lambda = 3 \times 10^{-4}$ cm.

8.42 The mirrors of the open resonator considered in Problem 8.45 are slightly nonparallel. The angle between their planes is $\beta \ll 1$. Estimate the additional losses through radiation and the corresponding contribution to the resonator quality factor due to the fact that the mirrors are not parallel. Which values of the angle β do not strongly decrease of the total quality factor of the resonator?

8.43 In a resonator formed by two parallel mirrors (see Figure 8.9), the eigenoscillations with $\lambda \ll L, D$ are realized in the form of standing waves in space between the mirrors. Consider that type of oscillations in which the wave vector of the standing wave makes a small angle ϑ with the normal to the plane of the mirrors.

a. Find the condition which determines possible values of ϑ for a given λ.
b. Estimate an order of magnitude of the resonator quality factor as a function of the angle ϑ. Consider the different ratios between the losses in the mirrors and the losses through radiation.

8.44 An open resonator with two parallel flat mirrors (Figure 8.9) differs from a closed rectangular resonator (Problem 8.30) only by lack of the side walls. It is reasonable to assume that for $L \gg \lambda$ this difference is not essential for those oscillations, whose wave vector $\boldsymbol{k}$ is directed along the normal to the plane of the mirrors. For such long-life modes, in accordance with the results of Problem 8.41, the diffraction losses are small. By using the results of Problem 8.30, find the approximate relation for the frequency spectrum of the long-life modes.

8.45 Study the uniform magneto-static eigenoscillations of a small ferrodielectric ellipsoid, magnetized to saturation. An external constant magnetic field H_0 is applied along one of the principal axes of the ellipsoid, and its demagnetizing factors and the saturation magnetization M_0 are given.

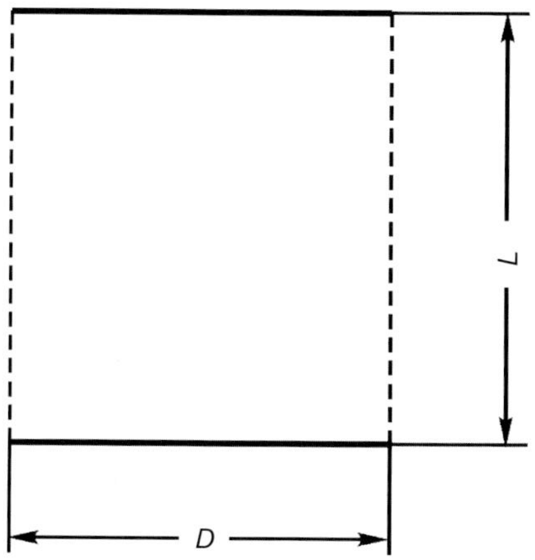

Figure 8.9 The open resonator consisting of two parallel mirrors.

Calculate, neglecting the losses, the eigenfrequency of small oscillations. Consider, in particular, the different limiting cases of the ellipsoid (see Problem 3.51). Find the polarization of the magnetization vector.

8.46 A resonator is in the form of a spherical cavity of radius a. Assuming the walls of the resonator to be perfectly conducting, find the frequencies of the eigenoscillations, the degeneracy order and the distribution of the electromagnetic field in the cavity. Consider the E- and H-oscillations (with the aid of the results of problems 6.74).

8.3 Answers and Solutions

8.1 For E-waves:

$$k^2 = \frac{\omega^2}{c^2}\frac{\varepsilon_1\varepsilon_2(\varepsilon_1\mu_2 - \varepsilon_2\mu_1)}{\varepsilon_1^2 - \varepsilon_2^2} > 0, \qquad \varepsilon_1\varepsilon_2 < 0.$$

For H-waves:

$$k^2 = \frac{\omega^2}{c^2}\frac{\mu_1\mu_2(\varepsilon_1\mu_2 - \varepsilon_2\mu_1)}{\mu_2^2 - \mu_1^2} > 0, \qquad \mu_1\mu_2 < 0.$$

8.2 From inequality (8.8), we obtain the condition $\omega(k) < \omega_{0e}/\sqrt{2}$. Upon substituting the given ε and μ in Equation (8.7), we have the dispersion equation for the surface plasma wave:

$$(1) \qquad \omega^2(k) = \frac{\omega_{oe}^2}{2} + (ck)^2\left[1 - \sqrt{1 + \frac{\omega_{0e}^4}{4(ck)^4}}\right].$$

The case $ck \gg \omega_{0e}$ is of most interest. The frequency of the surface wave is $\omega = \omega_{0e}/\sqrt{2} + \mathcal{O}(1/c^2k^2)$. Then wave is "slow," its phase velocity is $v_{ph} = \omega/k \ll c$. The energy of the surface plasmon in metals $\hbar\omega = \hbar\omega_{oe}/\sqrt{2}$ (of the order of 10 eV, see Problem 5.49) can be easily separated from the energy $\hbar\omega_{0e}$ of the volume plasmon. At $ck \ll \omega_{0e}$, we have $\omega(k) = ck$, and the wave is "fast," $v_{ph} = c$. Conditions (8.6) and (8.7) hold over all range of the wave numbers $0 < k < \infty$.

8.3 The frequency range for the polariton surface wave can be obtain from condition (8.7): $\omega^2 > \omega_p^2/2 + \omega_0^2$. The dispersion equation is

$$\omega^2(k) = \frac{1}{2}(\omega_p^2 + \omega_{0e}^2) + (ck)^2\left[1 - \sqrt{1 - \frac{\omega_0^2}{(ck)^2} + \frac{(\omega_p^2 + \omega_{0e}^2)^2}{4(ck)^4}}\right].$$

One has $\omega^2(k) = \sqrt{\omega_0^2 + \omega_p^2/2}$ at $ck \gg \omega(k)$; $\omega^2(k) = \omega_0^2$ at $ck = \omega_0$. Hence, for the wave numbers we have the range $k \geq \omega_0/c$.

8.4 For the fields in vacuum and in the ferrodielectric the damping constants are $q_1 = q_2 = k$, and the wave frequency $\omega(\boldsymbol{k}) = (k_x/k)(\omega_0 + \omega_M/2)$, where

$k = |k_x|$ is the absolute value of the wave vector. The phase velocity of the wave $v_{ph} = \omega(\boldsymbol{k})/k_x$, that is, the velocity of the plane at which the phase $k_x x - \omega(\boldsymbol{k})t$ is constant, is aligned with the Ox-axis independently on the sign of the projection k_x.

8.5 The dispersion equation is given by

$$\sqrt{k^2 - \frac{\omega^2}{c^2}\varepsilon\mu_{ef}} + \mu_{ef}\sqrt{k^2\frac{\omega^2}{c^2}} - k_x\frac{\mu_a}{\mu_\perp} = 0,$$

where $k = |k_x|$, $\mu_{ef} = (\mu_\perp^2 - \mu_a^2)/\mu_\perp$, and the radicals are the damping constants of the surface wave along the Oy-axis of the dielectric in vacuum. The appearance of the quantity k_x in the dispersion equation proves that the frequencies, the phase and group velocities of waves propagating in opposite directions are different (nonreciprocity). This is because the magnetic permeability of the ferrodielectric is non-invariant with respect to the time reversal (replacement of t by $-t$) while the quantities $\boldsymbol{H}$, $\boldsymbol{M}$ and, hence, the nondiagonal components of the tensor change the sign as time is reversed.

Further analysis of the dispersion equation can be found in the review of Kaganov *et al.* (1997) and in the original papers.

8.6 Let us assume that the wave propagates along Oz-axis and the propagation constant is k.

The electric-type waves.

a) The transverse field components are the even functions of the x-coordinate. The field inside the dielectric can be expressed in terms of trigonometrical or hyperbolic functions:

when $x > a$

(1) $$\mathcal{E}_z = Ae^{-sx}, \qquad \mathcal{E}_x = \frac{ik}{s}Ae^{-sx}, \qquad \mathcal{H}_y = \frac{i\omega}{sc}Ae^{-sx}, \qquad s > 0;$$

when $-a \le x \le a$

(2) $$\mathcal{E}_z = B\sin\kappa x, \quad \mathcal{E}_x = \frac{ik}{\kappa}B\cos\kappa x, \quad \mathcal{H}_y = \frac{i\omega\varepsilon}{\kappa c}B\cos\kappa x, \quad \kappa > 0;$$

or

(3)

$$\mathcal{E}_z = B\sinh\kappa x, \quad \mathcal{E}_x = -\frac{ik}{\kappa}B\cosh\kappa x, \quad \mathcal{H}_y = -\frac{i\omega\varepsilon}{\kappa c}B\cosh\kappa x, \quad \kappa > 0;$$

when $x < -a$

(4) $$\mathcal{E}_x = -Ae^{sx}, \qquad \mathcal{E}_x = \frac{ik}{s}Ae^{sx}, \qquad \mathcal{H}_y = \frac{i\omega}{sc}Ae^{sx},$$

where $A = Be^{sa}\sin\kappa a$; the remaining components of $\boldsymbol{\mathcal{E}}$ and $\boldsymbol{\mathcal{H}}$ are equal to zero. The parameters κ and s satisfy the system of equations

(5) $$(sa)^2 + (\kappa a)^2 = \frac{\omega^2 a^2}{c^2}(\varepsilon(\omega)\mu(\omega) - 1);$$

(6) $$sa = \frac{1}{\varepsilon(\omega)}\kappa a \tan \kappa a$$

in the case of trigonometrical functions, or they satisfy the system of equations

(7) $$(sa)^2 - (\kappa a)^2 = \frac{\omega^2 a^2}{c^2}(\varepsilon(\omega)\mu(\omega) - 1);$$

(8) $$sa = -\frac{1}{\varepsilon(\omega)}\kappa a \tanh \kappa a$$

if the fields are expressed in terms of hyperbolic functions.

Both systems can easily be solved graphically. In the first case, the possible values of κ and s correspond to the points of intersection of the curves defined by Equation (6) with a the circle of radius $r = (\omega a/c)\sqrt{\varepsilon\mu - 1}$ (Figure 8.10). The trigonometrical solutions exist for $\varepsilon > 0$, $\mu > 0$, $\varepsilon\mu > 1$ and for $\varepsilon < 0$, $\mu < 0$, $\varepsilon\mu > 1$. For given ω, a, ε, μ there is a finite number of the points of intersection, that is, there is a finite number of waves whose field distribution is described by the above Equations (1), (2), and (4). In particular, when $r < \pi$, there exists a single E_{00} wave.

Now consider the dependence of the propagation constant

(9) $$k = \sqrt{\frac{\omega^2\varepsilon\mu}{c^2} - \kappa^2} = \sqrt{\frac{\omega^2}{c^2} + s^2}$$

on frequency ω at given parameters of the dielectric layer for the given wave. It is clear from Figure 8.10 that for frequencies near the limiting frequency at which the given wave appears, the parameter s approaches zero, and k tends to ω/c. At such frequencies, the wave has the same propagation constant as in vacuum, and the field penetrates to large distances in the layer. With increasing ω the parameter s increases,

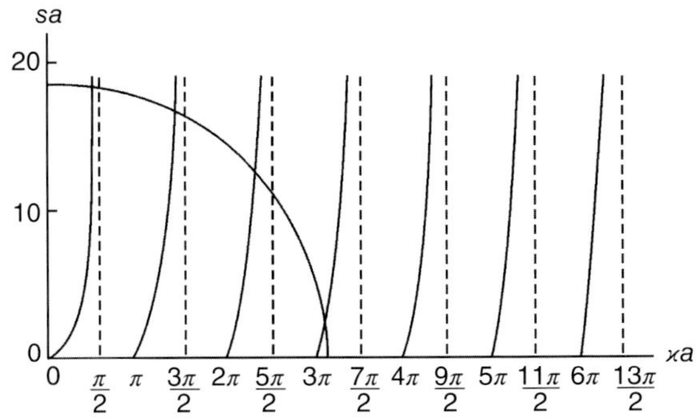

Figure 8.10 Graphic method of obtaining trigonometrical solution for a wave guide in the form of a flat dielectric layer.

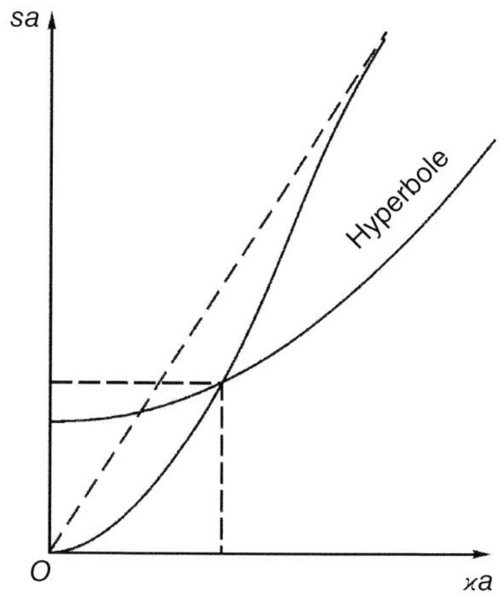

Figure 8.11 The single solution of hyperbolic type for the same wave guide, as in the previous case.

and κ remains bounded. At the same time, k tends to $(\omega/c)\sqrt{\varepsilon\mu}$, that is, to a value which corresponds to a wave propagating in an infinite medium with the parameters ε and μ. At sufficiently large ω and, hence, large s, the field is almost entirely concentrated within the dielectric layer. In the case of the hyperbolic solution, one should look for the point of intersection of the hyperbola (7) with the curve (8). Since the product $\kappa a \tanh \kappa a$ is not negative, the required solution exists only under conditions $\varepsilon < 0$ and $\kappa^2 + (\omega/c)^2(\varepsilon\mu - 1) > 0$. There is only one point of intersection that satisfies these conditions (Figure 8.11).

b) The transverse field components are uneven functions of x-coordinate. The field within the dielectric can again be expressed in terms of trigonometrical of hyperbolic functions:

when $x > a$

(10) $$\mathcal{E}_z = Ae^{-sx}, \qquad \mathcal{E}_x = \frac{ik}{s}Ae^{-sx}, \qquad \mathcal{H}_y = \frac{i\,\omega}{sc}Ae^{-sx}, \qquad s > 0;$$

when $-a \le x \le a$

(11)
$$\mathcal{E}_z = B\cos\kappa x, \quad \mathcal{E}_x = -\frac{ik}{\kappa}B\sin\kappa x, \quad \mathcal{H}_y = -\frac{i\,\omega\varepsilon}{\kappa c}B\sin\kappa x, \quad \kappa > 0$$

or

(12)
$$\mathcal{E}_z = B\cosh\kappa x, \quad \mathcal{E}_x = -\frac{ik}{\kappa}B\sinh\kappa x, \quad \mathcal{H}_y = -\frac{i\,\omega\varepsilon}{\kappa c}B\sinh\kappa x, \quad \kappa > 0;$$

when $x < -a$

(13) $$\mathcal{E}_z = Ae^{sx}, \qquad \mathcal{E}_x = -\frac{ik}{s}Ae^{sx}, \qquad \mathcal{H}_y = -\frac{i\,\omega}{sc}Ae^{sx},$$

where $A = Be^{sa}\cosh\kappa a$; the remaining components of $\mathcal{E}$ and $\mathcal{H}$ are zero. For the trigonometrical solution, the parameters s and κ can be determined from the system of equations

$$(14)\quad (sa)^2 + (\kappa a)^2 = \frac{\omega^2 a^2}{c^2}(\varepsilon(\omega)\mu(\omega) - 1), \qquad sa = -\frac{1}{\varepsilon(\omega)}\kappa a \cot\kappa a,$$

and in the hyperbolic case we have the system of equations

(15)

$$(sa)^2 - (\kappa a)^2 = \frac{\omega^2 a^2}{c^2}(\varepsilon(\omega)\mu(\omega) - 1), \qquad sa = -\frac{1}{\varepsilon(\omega)}\kappa a \coth\kappa a.$$

The waves of magnetic type can be analyzed in a similar way.

8.7 Even E-waves and odd H-waves can propagate along the layer, and their parameters (propagation constant, configuration of the fields in the region $x > 0$, and so on) are the same as in Problem 8.7.

8.8

$$K = \left\{1 + \frac{(\varepsilon/sa)\cos^2\kappa a}{1 + \sin 2\kappa a/\kappa a}\right\} \approx 1 - \frac{\varepsilon\cos^2\kappa a}{sa + (s/\kappa)\sin 2\kappa a} \qquad \text{at} \qquad sa \gg \varepsilon.$$

When the field hardly penetrates in an external region, the propagation of the wave within the dielectric can be considered a result of multiple total reflection from the dielectric boundaries (see Problem 6.9).

8.10 The waveguide modes are described by Equation (8.10).

For E-waves:

$$\mathcal{E}_z = \mathcal{E}_0 \sin\kappa_n x, \qquad \mathcal{E}_x = \frac{ik}{\kappa_n}\mathcal{E}_0\cos\kappa_n x,$$

$$\mathcal{H}_y = \frac{i\omega\varepsilon}{\kappa_n c}\mathcal{E}_0\cos\kappa_n x, \qquad \kappa_n = \frac{n\pi}{a}, \qquad n = 1, 2\ldots$$

For H-waves:

$$\mathcal{H}_z = \mathcal{H}_0\cos\kappa_n x, \qquad \mathcal{H}_x = -\frac{ik}{\kappa_n}\mathcal{H}_0\sin\kappa_n x,$$

$$\mathcal{E}_y = \frac{i\omega\mu}{\kappa_n c}\mathcal{H}_0\sin\kappa_n x, \qquad \kappa_n = \frac{n\pi}{a}, \qquad n = 1, 2\ldots$$

The propagation constant (wave number) for both types of waves is expressed in terms of the frequency and parameter κ_n which characterizes the spatial distribution of the field:

$$k^2 = \frac{\omega^2\varepsilon\mu}{c^2} - \kappa_n^2, \qquad \text{where} \qquad \omega > \omega_n,$$

ω_n is the cut-off frequency bounding from below the frequencies with a given spatial distribution of the field. Besides the waves mentioned above, the propagation of the purely transverse (TEH) wave is possible, with the parameters $\mathcal{E}_z = \mathcal{E}_y = \mathcal{H}_z = \mathcal{H}_x = 0$, $\quad \mathcal{E}_x = \sqrt{\mu/\varepsilon}\mathcal{H}_y \neq 0$ and an arbitrary frequency $0 < \omega < \infty$. This wave, like in an infinite medium, has $k^2 = \omega^2\varepsilon\mu/c^2$.

8.11 For E-waves:

$$\mathcal{E}_z = \mathcal{E}_0 \sin \kappa_1 x \sin \kappa_2 y,$$

where

$$\kappa_1 = \frac{n_1 \pi}{a}, \qquad \kappa_2 = \frac{n_2 \pi}{b}, \qquad n_1, n_2 = 1, 2 \ldots,$$

and the origin is taken at the corner of the rectangular cross section with dimensions a and b along the Ox- and Oy-axes, respectively.

For H-waves:

$$\mathcal{H}_z = \mathcal{H}_0 \cos \kappa_1 x \cos \kappa_2 y$$

with the same κ_1 and κ_2, but now one of the numbers, n_1 or n_2, can be zero. It follows from the above formulas that in transverse directions the field has a character of standing waves. The propagation constant k depends on ω as

$$k^2 = \frac{\omega^2}{c^2} - \pi^2 \left(\frac{n_1^2}{a^2} + \frac{n_2^2}{b^2} \right).$$

The transverse components of the electric and magnetic fields can be expressed in terms of $\mathcal{E}_z$ and $\mathcal{H}_z$ with the aid of Maxwell's equations.

8.12 For E-waves:

$$\alpha = \frac{2\zeta' \omega}{ck\kappa^2 ab} (\kappa_1^2 b + \kappa_2^2 a),$$

where

$$\kappa^2 = \kappa_1^2 + \kappa_2^2, \qquad \zeta' = \Re \zeta.$$

For H_{n0}-waves:

$$\alpha = \frac{\zeta' \omega}{ckab} \left(a + \frac{2\kappa^2 b}{k^2 + \kappa^2} \right).$$

For $H_{n_1 n_2}$-waves ($n_1, n_2 \neq 0$):

$$\alpha = \frac{2c\kappa^2 \zeta'}{\omega kab} \left[a + b + \frac{k^2}{\kappa^4} (\kappa_1^2 a + \kappa_2^2 b) \right].$$

The notations are the same as in Problem 8.14.

8.13 E-waves.

In order to determine these waves, it is necessary to solve the equation for the longitudinal component of the electric field:

$$\frac{\partial^2 \mathcal{E}_z}{\partial r^2} + \frac{1}{r} \frac{\partial \mathcal{E}_z}{\partial r} + \frac{1}{r^2} \frac{\partial^2 \mathcal{E}_z}{\partial \alpha^2} + \kappa^2 \mathcal{E}_z = 0. \tag{1}$$

Equation (1) can be integrated by the method of separation of variables. Special solutions are of the form

$$\mathcal{E}_z(r, \alpha) = J_m(\kappa r) \sin(m\alpha + \psi_m), \tag{2}$$

where J_m is the Bessel function, ψ is an arbitrary constant.

In order that the field would return to its original value when α changes by 2π, the quantity m must be an integer ($m = 0, 1, 2, \ldots$).

The transverse components of the electric and magnetic fields can be expressed in terms of $\mathcal{E}_z$ with the aid of Maxwell's equations:

$$\mathcal{E}_r = \frac{ik}{\kappa} J'_m(\kappa r)\sin(m\alpha + \psi_m),$$

$$\mathcal{E}_\alpha = \frac{imk}{\kappa^2 r} J_m(\kappa r)\cos(m\alpha + \psi_m),$$

$$\mathcal{H}_r = -\frac{im\,\omega}{\kappa^2 cr} J_m(\kappa r)\cos(m\alpha + \psi_m),$$

$$\mathcal{H}_\alpha = \frac{i\,\omega}{\kappa c} J'_m(\kappa r)\sin(m\alpha + \psi_m).$$

The possible values of the parameter κ are determined by the boundary conditions at the waveguide wall:

$$\mathcal{E}_r\big|_{r=a} = 0, \qquad \mathcal{E}_\alpha\big|_{r=a} = 0.$$

This yields $\kappa_{mn}a = \alpha_{mn}$, where α_{mn} is the nth root of the Bessel function: $J_m(\alpha_{mn}) = 0$, $n = 1, 2, \ldots$

Thus, the waves under consideration are characterized by two subscripts m and n; if $m = 0$ the field exhibits rotation symmetry with respect to the z-axis. For the ideal waveguide, the phases ψ_m are determined by the conditions of excitation. In reality, however, they strongly depend on defects in the waveguides walls (departures from circular cross section, longitudinal cracks, and so on).

The wave propagation along the waveguide is possible when $k = \sqrt{\omega^2/c^2 - \kappa^2}$ is real. Therefore, the wave of type m, n will propagate in the waveguide, provided its frequency satisfies the inequality

$$\omega^2 \geq \frac{c^2\alpha_{mn}^2}{a^2}.$$

The minimum frequency is possible for a (0, 1) wave:

$$\omega_0 = \frac{c\alpha_{01}}{a} \approx 2.4\frac{c}{a}.$$

The corresponding wavelength is

$$\lambda_0 = \frac{2\pi c}{\omega_0} \approx 2.6a,$$

which is of the order of the radius of the waveguide.

The H-waves are

$$\mathcal{H}_z = J_m(\kappa r)\sin(m\alpha + \psi_m) \qquad (m = 0, 1, 2, \ldots).$$

The propagation constant k is given by the equation

$$k^2 = \frac{\omega^2}{c^2} - \frac{\beta_{mn}^2}{a^2} \qquad (n = 1, 2, \ldots),$$

where β_{mn} is the nth root of the equation $J'_m(\beta_{mn}) = 0$. The smallest root is $\beta_{11} \approx 1.8$ and this corresponds to a limiting frequency $\omega_0 \approx 1.8c/a$ and the limiting wavelength is $\lambda_0 = 2\pi c/\,\omega_0 \approx 3.5a$.

The limiting frequency is lower for H-waves than for E-waves. If the frequency lies within the limits $\omega_{0e} > \omega > \omega_{0m}$, then this wave can only be the H_{11}-wave.

8.14 For E-waves:

$$\alpha = \frac{\omega\zeta'}{cak}.$$

For H-waves of the (m, n) type:

$$\alpha = \frac{c\zeta'\kappa^2}{\omega ak}\left[1 + \frac{m^2\,\omega^2}{c^2\kappa^2(a^2\kappa^2 - m^2)}\right],$$

where $\zeta' = \Re\,\zeta$.

8.15 The wavenumber k and the frequency ω of the waves in the waveguide are related by

$$\frac{\omega^2}{c^2} = k^2 + \kappa^2,$$

where κ is the constant that depends on the type of wave and the transverse dimensions of the waveguide. Using the usual formulas, we have

$$v_{ph} = \frac{\omega}{k} = \frac{c}{\sqrt{1-(\lambda/\lambda_0)^2}}, \qquad v_g = \frac{d\,\omega}{dk} = c\sqrt{1-(\lambda/\lambda_0)^2},$$

where λ_0 is the limiting wavelength.

It is clear from the above formulas that always $v_{ph} > c$, $v_g < c$, and $v_{ph} \cdot v_g = c^2$. This result is valid for a waveguide inside of which there is vacuum (the dielectric properties of air are practically the same as that of vacuum for the range of phenomena under consideration).

If the waveguide is filled with a dielectric, and the dispersion of ε and μ can be neglected, then all the above formulas remain valid, provided that c is replaced by $v = c/\sqrt{\varepsilon\mu}$. Hence, for a waveguide of this type

$$v_{ph} = \frac{c}{\sqrt{\varepsilon\mu(1 - \lambda^2/\lambda_0^2)}}$$

may be less than c, and the wave is "slowed" (see also Problem 8.24, where the slowing-down ladder structure is considered).

8.16 $H_z = \frac{1}{2}H_0\left[e^{i(\kappa_1 x + kz)} + e^{i(-\kappa_1 x + kz)}\right]e^{-i\omega t}$. The directions of propagation of the two plane waves, into which the wave H_{10} is decomposed, makes an angle θ with the waveguide axis (Figure 8.12). This angle is determined by the condition

$$\cos\theta = \frac{k}{\sqrt{k^2 + \kappa_1^2}} = \sqrt{1 - \left(\frac{\lambda}{\lambda_0}\right)^2}.$$

The phase plane I moves with the velocity c in the direction that makes an angle θ with Oz-axis; however, its velocity along the waveguide axis will be larger:

$$v = \frac{c}{\cos\theta} = \frac{c}{\sqrt{1-(\lambda/\lambda_0)^2}} = v_{ph}.$$

This is, in fact, the phase velocity of the wave in the waveguide.

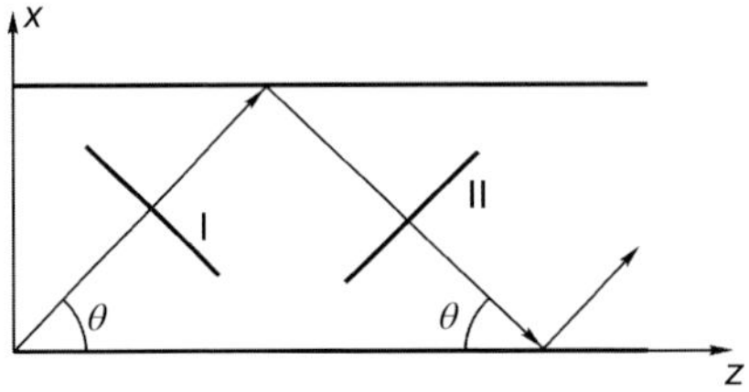

Figure 8.12 To calculate the phase and group velocities in a wave guide by a geometrical method.

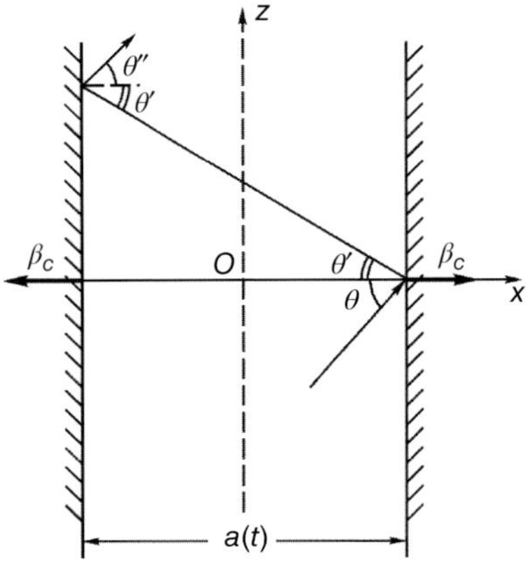

Figure 8.13 To calculate the adiabatic invariants in a wave guide in which the distances between the walls change.

The group velocity is equal to the rate of transport of energy. However, for a plane wave propagating in vacuum, the energy is transported with a velocity c in the direction of wave propagation. Each plane wave component of H_{10} will undergo multiple reflections from the waveguide walls, and will travel over a zigzag path. The resultant velocity along the axis of the waveguide will then be

$$v = c\cos\theta = c\sqrt{1 - \left(\frac{\lambda}{\lambda_0}\right)^2},$$

which is equal to the group velocity v_g.

8.17 1. The walls of the waveguide are slowly separated, each of them with the velocity βc with respect to the middle line of the waveguide (Figure 8.13). The ray path of one of the two plane waves forming the waveguide mode, is shown in the figure. The incident wave propagates at an angle θ to the Ox-axis, where $\cos\theta = \omega_n/\omega$, and ω_n is a limiting frequency of the considered mode. In the reference frame of the wall, for the reflected wave we have $\theta' = \theta$, $\omega' = \omega$. In the laboratory frame (see Problem 3.40), as a result of the Lorentz transformations, we have

$$(1)\qquad \omega' = \omega\,\frac{1+\beta^2-2\beta\cos\theta}{1-\beta^2}, \qquad \cos\theta' = \frac{\cos\theta(1+\beta^2)-2\beta}{1+\beta^2-2\beta\cos\theta}.$$

For slow displacement of the walls ($\beta \ll 1$), it is sufficient to take Equation (1) in the approximation linear in β:

$$(2)\qquad \Delta\omega = \omega' - \omega \approx -2\beta\omega\cos\theta, \qquad \Delta\theta = \theta' - \theta \approx 2\beta\sin\theta.$$

In the same approximation, we calculate the increment Δa of a distance between the walls in a time τ of propagation of the wave between the subsequent reflections:

$$(3)\qquad \Delta a = 2\beta c\tau \approx \frac{2\beta a}{\cos\theta}.$$

Now we find the dependence of the waveguide mode on distance a for the times exceeding significantly the time τ. For such large values of time, the variations $\Delta\omega$, $\Delta\theta$, Δa can be assumed to be negligibly small. Upon dividing equalities (2) term by term, we have the differential equation $d\omega/d\theta = -\omega\cos\theta/\sin\theta$, whose integration yields the adiabatic invariant (Rivlin, 1991, 1997)

$$(4)\qquad \omega\sin\theta = \omega_0\sin\theta_0.$$

Here the original values of the quantities are marked by the subscript zero. Further we write $\cos^2\theta = 1 - (\omega_0/\omega)^2\sin^2\theta_0$ and find, from Equations (2) and (3), the differential equation

$$(5)\qquad \frac{d\omega}{da} = -\frac{1}{a}\left(\omega - \frac{\omega_0^2}{\omega}\sin^2\theta_0\right),$$

whose integration gives another adiabatic invariant

$$(6)\qquad a^2\cot^2\theta = a_0^2\cot^2\theta_0.$$

Using the dispersion relation and equality (5), we obtain

$$(7)\qquad (ck)^2 = \omega^2 - \omega_n^2 = \omega^2(1-\cos^2\theta) = \omega_0^2\sin^2\theta_0 = \omega_0^2 - \omega_{n0}^2 = (ck_0)^2,$$

that is, the propagation constant k is the adiabatic invariant.

2. Let us calculate the change in the energy ΔW of a separate waveguide mode (per unit area) as the distance a between the walls changes, considering this change to be the result of work of an external source against the forces of pressure of the electromagnetic field on the walls. It was found in Problem 6.19 that a plane wave falling at an angle θ on the perfectly reflecting wall, undergoes the pressure $P = 2w\cos^2\theta$, where w is the energy density of the wave. The energy of a mode per unit area is $W = 2wa$, and, hence, the pressure on one of the walls is $P = (W/a)\cos^2\theta$. The increase of the energy

of the mode due to the work of the source as the walls close in, is $\Delta W = -P\Delta a/2 - P\Delta a/2 = -(W/a)\cos^2\theta\,\Delta a$. We write the last relation in the differential form and with the aid of Equation (6) we obtain $dW = -(W/a)\cos^2\theta\, da = -W\cot\theta d\theta$. Integration of this differential equation with the use of Equation (4) yields another adiabatic invariant:

$$\frac{W}{\omega} = \frac{W_0}{\omega_0}. \quad (8)$$

A clear meaning of this invariant will reveal itself in a quantum consideration of a waveguide mode of frequency ω and propagation constant $k = +\sqrt{\omega^2 - \omega_n^2}$. We express its energy in terms of the number of photons n: $W = \hbar\omega(n + 1/2 + 1/2) = \hbar\omega(n+1)$. Here two halves correspond to the zero-point oscillations, which are referred to the modes k and $-k$ of the same frequency ω. The conservation of the adiabatic invariant (8) signifies the conservation of the number of photons for a given mode when the distance between the walls changes slowly.

The above relations appear especially interesting, if we apply them to the limiting case where the initial waveguide embraces all infinite space and $a_0 \to \infty$. In a free space, the limiting frequency $\omega_{n0} = 0$. Let us consider the mode that represents the constant field with $\omega_0 = 0$. Although in this case real photons are absent, vacuum fluctuations are present which results in the Kazimir effect (see Problem 6.17). From Equation (7), we find $\omega = \omega_n$ for a waveguide of finite dimensions. A work on closing the walls together is found to be equal to the energy $\hbar\omega_n$ of a photon of limiting frequency ω_n. This energy is spent for the production of a mass $M = \hbar\omega_n/c^2$ of a waveguide photon.

8.18 The velocity v of the wave front should be identified with the group velocity of the mode: $v = d\omega/dk$. Using the dispersion relations from Problem 8.17 we have $v = d\mathcal{E}_\omega/dp = c^2p/\mathcal{E}_\omega < c$, and thus the energy and the momentum of the photon are related to its velocity by

$$\mathcal{E}_\omega = \frac{Mc^2}{\sqrt{1 - v^2/c^2}}, \qquad p = \frac{Mv}{\sqrt{1 - v^2/c^2}}, \qquad M = \frac{\hbar\omega_n}{c^2}.$$

These formulas are the same as Equation (3.33) for a relativistic particle. Other properties of the photons in a waveguide are discussed in the Rivlin's review (1997).

8.19

$$H_\alpha = E_r = \frac{A}{r} e^{i\omega(z/c - t)}, \quad (1)$$

where A is a constant, and the remaining components of the fields are all equal to zero.

The energy flux is given by

$$\overline{\gamma} = \frac{A^2 c}{4}\ln\frac{b}{a}. \quad (2)$$

For a single perfect conductor, the fields in space outside the conductor are described by Equation (1); and the total energy flux through the plane $z = const$ is infinite ($\gamma \to \infty$ as $b \to \infty$). Accordingly, this wave cannot be maintained by a source of finite power, and, hence, the considered case is of no physical significance.

8.20 E-waves:

$$\mathcal{E}_z = [A_{mn}J_m(\kappa_{mn}r) + B_{mn}N_m(\kappa_{mn}r)]\sin(m\alpha + \psi_m), \qquad m = 0, 1, 2 \ldots$$

where κ_{mn} is the nth root of the equation

$$J_m(\kappa a)N_m(\kappa b) - J_m(\kappa b)N_m(\kappa a) = 0.$$

Here N_m, J_m are the cylindrical functions (see Chapter 1), A_{mn} and B_{mn} are constants that are related by

$$A_{mn}J_m(\kappa_{mn}a) + B_{mn}N_m(\kappa_{mn}a) = 0.$$

H-waves:

$$\mathcal{H}_z = [C_{mn}J_m(\kappa_{mn}r) + D_{mn}N_m(\kappa_{mn}r)]\sin(m\alpha + \psi_m), \qquad m = 0, 1, 2, \ldots,$$

where κ_{mn} is the nth root of the equation

$$J'_m(\kappa a)N'_m(\kappa b) - N'_m(\kappa a)J'_m(\kappa b) = 0,$$

and C_{mn} and D_{mn} are related by:

$$C_{mn}J'_m(\kappa_{mn}a) + D_{mn}N'_m(\kappa_{mn}a) = 0.$$

The remaining components of the electric and magnetic fields can be expressed in terms of $\mathcal{E}_z$ and $\mathcal{H}_z$ by means of Maxwell's equations.

8.21

$$\alpha = \frac{\zeta'(a+b)}{2ab\ln(b/a)},$$

where $\zeta' = \Re\,\zeta$.

8.22 If the field is symmetric with respect to the axis of the conductor, the longitudinal component $\mathcal{E}_z$ satisfies the equation

$$\frac{d^2\mathcal{E}_z}{dr^2} + \frac{1}{r}\frac{d\mathcal{E}_z}{dr} + \kappa^2\mathcal{E}_z = 0. \tag{1}$$

Since the conductor has a finite conductivity, the parameters k and κ will be complex. The sign of κ will be defined so that $\Im\,\kappa = \kappa'' > 0$.

The general solution of the Equation (1) will be written in the form

$$\mathcal{E}_z(r) = A'H_0^{(1)}(\kappa r) + B'H_0^{(2)}(\kappa r),$$

where $H_0^{(1)}$ and $H_0^{(2)}$ are Hankel functions. It follows from the asymptotic behavior of these functions (see Section I.1.3) and from the condition $\Im\,\kappa > 0$ that $B' = 0$, since otherwise the field would increase at infinity.

The remaining components $\mathcal{E}$ and $\mathcal{H}$ can be expressed in terms of $\mathcal{E}_z$ by means of Maxwell's equations:

$$(2)\quad \mathcal{E}_z = A'H_0^{(1)}(\kappa r), \qquad \mathcal{E}_r = \frac{ik}{\kappa}A'H_1^{(1)}(\kappa r), \qquad \mathcal{H}_\alpha = \frac{i\omega}{\kappa c}A'H_1^{(1)}(\kappa r).$$

For sufficiently large κr, the functions $H_0^{(1)}$ and $H_1^{(1)}$ are proportional to $e^{-\kappa'' r}/\sqrt{\kappa r}$, and, therefore, the electromagnetic field is damped out exponentially at large distances from the conductor. The maximum concentration of the field exists near the conductor, and, hence, the wave is a surface type wave. The Leontovich boundary condition

$$\mathcal{E}_z = \zeta \mathcal{H}_\alpha$$

at the surface of the conductor leads to the characteristic equation for κ:

$$\kappa a \frac{H_0^{(1)}(\kappa a)}{H_1^{(1)}(\kappa a)} = i\zeta \frac{\omega}{c} a,$$

where ζ is the surface impedance of the metal. For a good conductor $|\zeta| \ll 1$, and hence the latter equation can only be satisfied for small κa. If we use the approximate formulas for $H_0^{(1)}$ and $H_1^{(1)}$, we have

$$(3)\qquad (\kappa a)^2 \ln\left(\frac{\gamma \kappa a}{2i}\right) = i\zeta \frac{\omega}{c} a, \qquad \ln \gamma = 0.5772.$$

The transcendental Equation (3) cannot be solved graphically because it involves complex quantities (however, it can easily be solved using standard computer programs). It was solved by Sommerfeld who used the iteration method based on the fact that $\ln(\kappa a)$ varies much more slowly than κa. Let $\left(\frac{\gamma \kappa a}{2i}\right)^2 = u$, $\frac{i\gamma^2}{2c}\frac{\omega}{}\zeta a = v$. Equation (3) can then be written in the form

$$u \ln u = v.$$

If an approximate solution u_n (nth order approximation) is found, then the more accurate value u_{n+1} (($n+1$)th order approximation) can be obtained from the formula

$$u_{n+1} \ln u_n = v.$$

In the zero-order approximation, we can set $u_0 = v$. Then

$$u_1 = \frac{v}{\ln v}, \qquad u_2 = \frac{v}{\ln\left(\frac{v}{\ln v}\right)}, \qquad u_3 = \frac{v}{\ln \dfrac{v}{\ln\left(\dfrac{v}{\ln v}\right)}} \qquad \ldots$$

For decimeter radio waves ($\lambda = 2\pi c/\omega = 30$ cm) propagating along the copper conductor of radius 1 mm (the conductivity of copper is $\sigma = 5.2 \times 10^{17}\ \mathrm{s}^{-1}$), the above method yields

$$u \approx (4.2 + 4.5i) \cdot 10^{-8},$$

and hence

$$k = \frac{\omega}{c}[1 + (6.0 + 6.4i) \times 10^{-5}].$$

The phase velocity of the wave is given by

$$v_{ph} = \frac{\omega}{\Re k} = (1 - 6 \cdot 10^{-5})c < c,$$

that is, the wave is slightly slowed down.

This result can be understood from the following considerations. For a perfect conductor, the TEM wave has a phase velocity c and the field inside the conductor is zero. In the case of finite conductivity, a fraction of energy will propagate inside the conductor; since the velocity of propagation in metal is appreciably less than c, the electromagnetic wave will, on average, be slowed down. Moreover, damping will appear.

Consider the character of the field in the limiting case $\zeta \to 0$ (perfect conductor). It follows from Equation (3) that $\kappa \to 0$ and $k \to \omega/c$. Using the expressions of the functions $H_0^{(1)}$ and $H_1^{(1)}$ at small arguments of these functions, we get from Equation (2)

$$cE_z = \lim_{\kappa\to 0} \frac{2iA'}{\pi} \ln\left(\frac{\gamma\kappa r}{2i}\right), \qquad \mathcal{E}_r = \lim_{\kappa\to 0} \frac{2kA'}{\pi\kappa^2}\frac{1}{r}, \qquad \mathcal{H}_\alpha = \lim_{\kappa\to 0} \frac{2kA'}{\pi\kappa^2}\frac{1}{r}.$$

Since the field components cannot be infinite, we have to assume that A' is proportional to κ^2. Let $A' = A\kappa^2/k$, then

$$\mathcal{E}_r = \mathcal{H}_\alpha = \frac{A}{r}, \qquad \mathcal{E}_z = 0.$$

This is a purely TEM wave propagating with velocity c.

8.23 The components of the electromagnetic field in the waveguide are determined by the following expressions:

at $r \le a$

$$\mathcal{E}_z = \mathcal{E}_0 J_0(\kappa_1 r), \qquad \mathcal{E}_r = -i\frac{k}{\kappa_1}\mathcal{E}_0 J_1(\kappa_1 r), \qquad \mathcal{H}_\alpha = -i\frac{\omega}{c\kappa_1}\mathcal{E}_0 J_1(\kappa_1 r);$$

at $a \le r \le b$

$$\mathcal{E}_z = AJ_0(\kappa_2 r) + BN_0(\kappa_2 r), \qquad \mathcal{E}_r = -i\frac{k}{\kappa_2}[AJ_1(\kappa k_2 r) + BN_1(\kappa_2 r)],$$

$$\mathcal{H}_\alpha = -i\frac{\varepsilon\,\omega}{c\kappa_2}[AJ_1(\kappa_2 r) + BN_1(\kappa_2 r)],$$

where $\kappa_1 = \sqrt{\omega^2/c^2 - k^2}$, $\kappa_2 = \sqrt{\varepsilon\,\omega^2/c^2 - k^2}$; $\mathcal{E}_0$, A and B are constants.

The boundary conditions can be written in the form

$$\mathcal{E}_z\big|_{r=b} = 0, \qquad \mathcal{E}_z\big|_{r=a-0} = \mathcal{E}_z\big|_{r=a+0}, \qquad \mathcal{H}_\alpha\big|_{r=a-0} = \mathcal{H}_\alpha\big|_{r=a+0},$$

and the boundary condition for $\mathcal{E}_\alpha$ will be satisfied automatically.

Eliminating A, B, and $\mathcal{E}_0$, we come to the transcendental equation relating k and ω:

(1) $$\frac{\varepsilon\kappa_1}{\kappa_2}\cdot\frac{J_0(\kappa_1 a)}{J_1(\kappa_1 a)} = \frac{J_0(\kappa_2 a)N_0(\kappa_2 b) - N_0(\kappa_2 a)J_0(\kappa_2 b)}{J_1(\kappa_2 a)N_0(\kappa_2 b) - N_1(\kappa_2 a)J_0(\kappa_2 b)}.$$

This equation can be significantly simplified at $a \ll b$. Consider the wave which has the maximum k. If the waveguide were fully filled with a dielectric ($a = 0$), then the corresponding value of κ_2 would be $\kappa_{02} = \alpha_{01}/b$, where $\alpha_{01} = 2.4$, $J_0(\alpha_{01}) = 0$ (see Problem 8.13).

We will seek for a solution that is not very different from κ_{02}:

$$\kappa_2 = \kappa_{02} + \kappa_2' = \frac{\alpha_{01}}{b} + \frac{\Delta\alpha}{b},$$

where the order of magnitude of $\Delta\alpha$ is not lower than that of a/b. Assuming that $\alpha_{01}a/b \ll 1$, we will use the approximate formulas for J_0, N_0, J_1, N_1 from Section 1.3, and obtain instead of Equation (1)

$$\varepsilon\left[(\kappa_2 a)^2 N_0(\kappa_2 b) + \frac{2}{\pi}J_0(\kappa_2 b)\right] = (\kappa_2 a)^2\left[N_0(\kappa_2 b) + \frac{2}{\pi}\ln\frac{2}{\gamma\kappa_2 a}J_0(\kappa_2 b)\right].$$

By substituting $N_0(\kappa_2 b) = N_0(\alpha_{01} + \Delta\alpha) \approx N_0(\alpha_{01})$, $J_0(\kappa_2 b) = -J_1(\alpha_{10}\Delta\alpha)$ and neglecting the small logarithmic term, we have

$$\Delta\alpha = \left(1 - \frac{1}{\varepsilon}\right)\frac{\kappa\alpha_{01}^2}{4} \cdot \frac{N_0(\alpha_{01})}{J_1(\alpha_{01})} \cdot \left(\frac{a}{b}\right)^2.$$

The phase velocity of the wave is given by

$$v_{ph} = \frac{\omega}{k} = \frac{\omega}{\sqrt{\frac{\varepsilon\,\omega^2}{c^2} - \frac{(\alpha_{01}^2 + 2\alpha_{01}\Delta\alpha)}{b^2}}}.$$

By substituting $\omega_0 = \alpha_{01}c/b \approx 2.4c/b$ (minimum frequency for a waveguide containing no dielectric) and using tabulated values for $N_0(\alpha_{01})$ and $J_1(\alpha_{01})$, we have

$$v_{ph} = c\left\{\varepsilon - \left(\frac{\omega_0}{\omega}\right)^2\left[1 + 3.7\left(1 - \frac{1}{\varepsilon}\right)\frac{a^2}{b^2}\right]\right\}^{-\frac{1}{2}}. \tag{2}$$

If the waveguide is completely filled with dielectric ($a = 0$), then

$$v_{ph} = c\left[\varepsilon - \left(\frac{\omega_0}{\omega}\right)^2\right]^{-\frac{1}{2}}.$$

The limiting frequency of a partially filled waveguide is given by

$$\omega_c = \frac{\omega_0}{\sqrt{\varepsilon}}\left[1 + 1.85\left(1 - \frac{1}{\varepsilon}\right)\frac{a^2}{b^2}\right]$$

and lies between the limiting frequencies of an empty and fully filled waveguide:

$$\frac{\omega_0}{\sqrt{\varepsilon}} < \omega_c < \omega_0.$$

The phase velocity (2) becomes smaller than the velocity c at frequencies

$$\omega > \frac{\omega_0}{\sqrt{\varepsilon - 1}}\left[1 + 1.85\left(1 - \frac{1}{\varepsilon}\right)\frac{a^2}{b^2}\right].$$

Thus, a partially or fully filled waveguide is a retarding system in which the phase velocity of electromagnetic waves can be less than c. An important property of slow waves is the fact that they can effectively interact with charged particle beams. This interaction may be used both for the generation and the amplification of ultrahigh frequency electromagnetic oscillations (klystron, traveling wave lamp, magnetron) and for the acceleration of particles (linear accelerator).

8.24 The boundary conditions at an anisotropically conducting plane are given by

$$E_{1x} - E_{2x} = 0, \qquad H_{1x} = H_{2x}, \qquad E_{1z} = E_{2z}.$$

The index 1 corresponds to the region $y > 0$, and the index 2 to the region $y < 0$. The first two equations are a consequence of the perfect conductivity of the strips, and the last two imply that there is no current in the direction at right angles to the strips. Moreover, $E_y = E_z = 0$ for $x = \pm a$ and all field components must be bounded as $y \to \pm\infty$.

If we solve the Maxwell's equations with these boundary conditions, we find

$$\mathcal{E}_{1x} = 0, \qquad \mathcal{E}_{1y} = -Be^{-\beta y}\cos\alpha x, \qquad \mathcal{E}_{1z} = iB\frac{\beta}{k}e^{-\beta y}\cos\alpha x,$$

$$\mathcal{H}_{1x} = B\left(\frac{k_0\varepsilon}{k} - \frac{\alpha^2}{k_0 k}\right)e^{-\beta y}\cos\alpha x, \qquad \mathcal{H}_{1y} = B\frac{\alpha\beta}{kk_0}e^{-\beta y}\sin\alpha x,$$

$$\mathcal{H}_{1z} = -iB\frac{\alpha}{k_0}e^{-\beta y}\sin\alpha x,$$

where $k_0 = \omega/c$, B is a constant

$$\alpha = \alpha_m = \frac{(2m+1)\pi}{2}, \qquad m = 0, 1, 2\ldots \qquad \beta = \beta_m = \sqrt{k^2 - k_0^2\varepsilon + \alpha_m^2}.$$

The propagation constant k can be expressed in terms of ω through the formula

$$k = k_m = \alpha_m\left[\frac{\left(k_0^2\varepsilon/\alpha_m^2 - 1\right)\left(1 - k_0^2/\alpha_m^2\right)}{1 - (\varepsilon+1)k_0^2/\alpha_m^2}\right]^{1/2}.$$

For a given m, the wave can propagate provided that its frequency ω lies within the range

$$\frac{1}{\sqrt{\varepsilon}} \le \frac{\omega}{c\alpha_m} \le \sqrt{\frac{2}{\varepsilon+1}};$$

k varies then from 0 to ∞.

If $\varepsilon = 1$ (there is no dielectric), the system becomes a resonator: the oscillations with discrete frequencies $\omega_m = c\alpha_m$ are possible in it. If $\varepsilon > 1$ the device considered is a retarding system. The group and phase velocity in it is less than the velocity of light c.

8.25 E-waves cannot exist in the case under consideration. The H-waves are described by

$$\mathcal{H}_z = \frac{i\mathcal{E}_0 c}{\omega\mu}\left(\kappa\cos\kappa x - k\frac{\mu_a}{\mu_\perp}\sin\kappa x\right),$$

$$\mathcal{H}_x = \frac{\mathcal{E}_0 c}{\omega\mu}\left(k\sin\kappa x - \kappa\frac{\mu_a}{\mu_\perp}\cos\kappa x\right), \qquad \mathcal{E}_z = \mathcal{E}_0\sin\kappa x,$$

$$\mathcal{E}_x = \mathcal{E}_y = \mathcal{H}_y = 0,$$

where

$$\kappa = \frac{n\pi}{a}, \quad k = \sqrt{\frac{\omega^2\varepsilon_\parallel\mu}{c^2} - \left(\frac{n\pi}{a}\right)^2}, \quad n = 1, 2, 3, \ldots, \quad \mu = \mu_\perp - \frac{\mu_a^2}{\mu_\perp}.$$

The limiting frequency is $\omega_0^{(n)} = c\kappa_n/\varepsilon_\parallel\mu$.

It follows from the above formulas for $\mathcal{H}_z$ and $\mathcal{H}_x$ that the configuration of the magnetic field for this type of wave depends on the sign of k, that is, on the direction of propagation of the wave, and on the sign of μ_a, that is, on the direction of the constant magnetic field. This effect of nonreciprocity is associated with the gyrotropy of the medium filling the waveguide and with the violation of invariance at time reversal. It has already been discussed in Problem 8.5.

8.26 The Maxwell's equations for the complex-conjugate amplitudes $\boldsymbol{\mathcal{E}}_0^*$ and $\boldsymbol{\mathcal{H}}_0^*$ are

(1)

$$rot\,\boldsymbol{\mathcal{E}}_0^* - ik_0(\boldsymbol{e}_z\times\boldsymbol{\mathcal{E}}_0^*) = -\frac{i\omega}{c}\boldsymbol{\mathcal{H}}_0^*, \qquad rot\,\boldsymbol{\mathcal{H}}_0^* - ik_0(\boldsymbol{e}_z\times\boldsymbol{\mathcal{H}}_0^*) = \frac{i\omega}{c}\boldsymbol{\mathcal{E}}_0^*.$$

The amplitudes $\boldsymbol{\mathcal{E}}$ and $\boldsymbol{\mathcal{H}}$ satisfy the equations

$$(2) \quad rot\,\boldsymbol{\mathcal{E}} + ik(\boldsymbol{e}_z\times\boldsymbol{\mathcal{E}}) = \frac{i\omega}{c}\hat{\mu}'\boldsymbol{\mathcal{H}}, \qquad rot\,\boldsymbol{\mathcal{H}} + ik_0(\boldsymbol{e}_z\times\boldsymbol{\mathcal{H}}) = -\frac{i\omega}{c}\hat{\varepsilon}'\boldsymbol{\mathcal{E}},$$

where $\hat{\mu}'\boldsymbol{\mathcal{H}}$, $\hat{\varepsilon}'\boldsymbol{\mathcal{E}}$ are vectors with components $\mu'_{ik}\mathcal{H}_k$, $\varepsilon'_{ik}\mathcal{E}_k$; $\mu'_{ik} = \varepsilon'_{ik} = \delta_{ik}$, outside the region filled by dielectric. Inside the region $\mu'_{ik} = \mu_{ik}$, $\varepsilon'_{ik} = \varepsilon_{ik}$.

It follows from the equations for $rot\,\boldsymbol{\mathcal{E}}_0^*$ and $rot\,\boldsymbol{\mathcal{H}}$ that

(3)

$$\boldsymbol{\mathcal{H}}\cdot rot\,\boldsymbol{\mathcal{E}}_0^* - \boldsymbol{\mathcal{E}}_0^*\cdot rot\,\boldsymbol{\mathcal{H}} + i(k-k_0)(\boldsymbol{e}_z\times\boldsymbol{\mathcal{E}}_0^*)\cdot\boldsymbol{\mathcal{H}} = -i\omega(\boldsymbol{\mathcal{H}}\cdot\boldsymbol{\mathcal{H}}_0^* - \hat{\varepsilon}'\boldsymbol{\mathcal{E}}\cdot\boldsymbol{\mathcal{E}}_0^*).$$

Integrate both parts of this equation over the cross section of the waveguide S. The first two terms can be transformed as

$$\int_S(\boldsymbol{\mathcal{H}}\cdot rot\,\boldsymbol{\mathcal{E}}_0^* - \boldsymbol{\mathcal{E}}_0^*\cdot rot\,\boldsymbol{\mathcal{H}})\,dS = \frac{1}{l}\int_V div\,(\boldsymbol{\mathcal{E}}_0^*\times\boldsymbol{\mathcal{H}})\,dV.$$

The integral on the right-hand side of this expression is taken over the volume bounded by the walls of the waveguide and by two cross sections at a distance l from each other (the integrand is independent of z).

Next, from Gauss–Ostrogradskii theorem, we have

$$\int div\,(\boldsymbol{\mathcal{E}}_0^*\times\boldsymbol{\mathcal{H}})\,dV = \int(\boldsymbol{\mathcal{E}}_0^*\times\boldsymbol{\mathcal{H}})\cdot\boldsymbol{n}\,dS = \int(\boldsymbol{n}\times\boldsymbol{\mathcal{E}}_0^*)\cdot\boldsymbol{\mathcal{H}}\,dS.$$

On the wall of the waveguide $\boldsymbol{n}\times\boldsymbol{\mathcal{E}}_0^*=0$ due to the boundary condition $\mathcal{E}_{0\tau}=0$, while the integrals over the cross sections have opposite signs and cancel out. Hence,

$$\int (\boldsymbol{n}\times\boldsymbol{\mathcal{E}}_0^*)\cdot\boldsymbol{\mathcal{H}}\,dS = 0$$

and Equation (3) yields

$$(4)\qquad (k-k_0)\int_S (\boldsymbol{\mathcal{E}}_0^*\times\boldsymbol{\mathcal{H}})\cdot\mathbf{e}_z\,dS$$
$$= -\,\omega\left[\int_S (\boldsymbol{\mathcal{H}}\cdot\boldsymbol{\mathcal{H}}_0^*)\,dS - \int_S (\boldsymbol{\mathcal{E}}\cdot\boldsymbol{\mathcal{E}}_0^*)\,dS - \int_{\Delta S}\Delta\hat{\varepsilon}\boldsymbol{\mathcal{E}}\cdot\boldsymbol{\mathcal{E}}_0^*\,dS\right],$$

where $\Delta\hat{\varepsilon}=\hat{\varepsilon}-1$ and ΔS is the cross-sectional area of the region filled by the dielectric.

Similarly, from the equations for *rot* $\boldsymbol{\mathcal{E}}$ and *rot* $\boldsymbol{\mathcal{H}}_0^*$ we find

$$(5)\qquad (k-k_0)\int_S (\boldsymbol{\mathcal{E}}\times\boldsymbol{\mathcal{H}}_0^*)\cdot\mathbf{e}_z\,dS$$
$$= \frac{\omega}{c}\left[\int_S (\boldsymbol{\mathcal{H}}\cdot\boldsymbol{\mathcal{H}}_0^*)\,dS - \int_S (\boldsymbol{\mathcal{E}}\cdot\boldsymbol{\mathcal{E}}_0^*)\,dS + \int_{\Delta S}\Delta\hat{\mu}\boldsymbol{\mathcal{H}}\cdot\boldsymbol{\mathcal{H}}_0^*\,dS\right],$$

where $\Delta\hat{\mu}=\hat{\mu}-1$.

By combining Equations (4) and (5), we obtain the formula given in the statement of the problem. This formula represents the exact relation between the change Δk in the propagation constant and the field amplitudes. However, in the majority of cases the exact solution of the problem on a waveguide partially filled with a dielectric cannot be obtained. The amplitudes of the disturbed fields $\boldsymbol{\mathcal{E}}$ and $\boldsymbol{\mathcal{H}}$ can be determined only approximately, provided that the transverse dimensions of the region occupied by a dielectric are small. If this is so, the formula for Δk can be used to calculate the change in the propagation constant, which is an important characteristic of the wave propagating in the waveguide. Examples of calculations based on this method are given in the solutions of the three following problems.

8.27 For a thin plate, the amplitudes of the disturbed field can be expressed approximately in terms of the undisturbed amplitudes, which for an H_{10} wave are of the form:

$$\mathcal{H}_{0z}=\mathcal{H}_0\cos\frac{\pi x}{a},\qquad \mathcal{H}_{0x}=-\frac{ik_0a}{\pi}\mathcal{H}_0\sin\frac{\pi x}{a},$$
$$\mathcal{E}_{0y}=\frac{i\,\omega a}{\pi c}\mathcal{H}_0\sin\frac{\pi x}{a},\qquad \mathcal{E}_{0x}=\mathcal{E}_{0z}=\mathcal{H}_{0y}=0.$$

(These expressions can be obtained from the results of Problem 8.11.)

Let us now neglect the change in the amplitude outside the volume occupied by the plate, and also the change in the fields within the plate. This is equivalent to neglecting the terms of the order of d^2 and higher. On the surface of the plate, the following boundary conditions must be satisfied:

$$\mathcal{E}_y=\mathcal{E}_{0y},\qquad \mathcal{H}_z=\mathcal{H}_{0z},\qquad \mu_\perp\mathcal{H}_x-i\mu_a\mathcal{H}_y=\mathcal{H}_{0x},\qquad \mathcal{H}_y=\mathcal{H}_{0y}=0,$$

where the undisturbed amplitudes on the right-hand side of the equations should be taken at $x = x_1$. These equations define the amplitudes of the disturbed field in the plate.

The integral in the numerator of the expression for Δk (see Problem 8.30) is equal to the product of the integrand and the cross-sectional area bd of the plate, since the field is independent of y, and the dependence on x is neglected.

The undisturbed amplitudes can be substituted in the integral in the numerator. The result is

$$\Delta k = \frac{d}{k_0 a}\left\{\left[\frac{(\varepsilon - 1)\,\omega^2}{c^2} + \left(1 - \frac{1}{\mu_\perp}\right)k_0^2\right]\sin^2\frac{\pi x_1}{a} + (\mu_\| - 1)\left(\frac{\pi}{a}\right)^2\cos^2\frac{\pi x_1}{a}\right\}.$$

Since $\mu_\perp$ depends on the magnitude of the constant magnetizing field H_0 (see Problem 5.55), it follows that Δk will also depend on this field. A change in H_0 causes a change in the phase of the wave. Devices that are based on this phenomenon are widely used in electronics for phase transformation.

8.28

$$\Delta k = \frac{\omega d}{4\pi c}\cdot\frac{b-a}{ab\ln\frac{b}{a}}\left(\varepsilon - \frac{1}{\mu_\perp}\right).$$

8.29

$$\text{(a)}\quad \Delta k = \frac{\omega d}{4\pi c}\cdot\frac{b-a}{ab\ln\frac{b}{a}}\left(\varepsilon - \frac{1}{\mu_\|}\right);$$

$$\text{(b)}\quad \Delta k = \frac{\omega d}{4\pi c}\cdot\frac{b-a}{ab\ln\frac{b}{a}}\left(\varepsilon - \frac{1}{\mu_\perp}\right).$$

In case (a) Δk is practically independent of the magnitude of the constant magnetic field H_0, since $\mu_\| \approx 1$ (see Problem 5.56). This is because, inside the plate, the high-frequency magnetic field has the same direction as the constant field and does not support the precession of the magnetization $\boldsymbol{M}$. In case (b) the high-frequency magnetic field inside the plate is perpendicular to the constant field, $\mu_\perp$ depends on H_0, and this dependence is of the resonance character.

8.30 By integrating Equation (8.17) with the boundary condition (8.18), we find

$$\text{(1)}\qquad \begin{aligned} E_x &= A_1\cos(k_1 x)\sin(k_2 y)\sin(k_3 z),\\ E_y &= A_2\cos(k_2 y)\sin(k_1 x)\sin(k_3 z),\\ E_z &= A_3\cos(k_3 z)\sin(k_1 x)\sin(k_2 y),\end{aligned}$$

where the amplitudes A_i are constants, and

$$\text{(2)}\qquad k_1 = n_1\pi/a,\quad k_2 = n_2\pi/b,\quad k_3 = n_3\pi/h,$$

$$\omega^2 = c^2(k_1^2 + k_2^2 + k_3^2) = \pi^2 c^2[(n_1/a)^2 + (n_2/b)^2 + (n_3/h)^2],$$

$n_1, n_2, n_3 = 0, 1, 2, \ldots$ (the time factor $e^{-i\omega t}$ is dropped).

The vector $\boldsymbol{H}$ can be expressed in terms of $\mathbf{E}$ with an aid of Maxwell's equations.

The equation $div\,\boldsymbol{E} = 0$ leads to the condition of transversality $\boldsymbol{A} \cdot \boldsymbol{k} = 0$, where the vector $\boldsymbol{A} = (A_1, A_2, A_3)$. Hence, the oscillations for given k_x, k_y, $k_z \neq 0$ are twofold degenerate, since the vector $\boldsymbol{A}$ can be chosen in the plane perpendicular to $\boldsymbol{k}$ in two independent and arbitrary ways. For each such $\boldsymbol{k}$, we set

(3)
$$\boldsymbol{A}_{\mathbf{k}\sigma} = A\boldsymbol{e}_{\mathbf{k}\sigma}, \qquad \sigma = 1, 2,$$

where $\boldsymbol{e}_{\mathbf{k}\sigma}$ is a unit vector such that $\boldsymbol{e}_{\mathbf{k}1} \cdot \boldsymbol{e}_{\mathbf{k}2} = 0$ and $\boldsymbol{e}_{\mathbf{k}\sigma} \cdot \boldsymbol{k} = 0$, and the constant $A = \sqrt{32\pi/V}$, where $V = abh$ is the volume of the resonator. The eigenfunctions will then all be mutually orthogonal and normalized:

(4)
$$\int \boldsymbol{E}_{\nu'} \boldsymbol{E}_{\nu}\, dV = 4\pi\delta_{\nu\nu'}.$$

This relation can be easily verified by the direct integration of Equation (1). The indices ν, ν' stand for four numbers: n_1, n_2, n_3, and σ.

If one of the projections of $\boldsymbol{k}$ vanishes, there is no degeneracy, since in that case only one constant appears in the solution (1).

8.31

$$\Delta N = \frac{V}{\pi^2 c^3}\, \omega^2 \Delta\, \omega.$$

8.32 The E-oscillations:

$$E_z = \mathcal{E}_0 J_m(\kappa r) \sin(m\alpha + \psi_m) \cos(kz), \qquad H_z = 0,$$
$$E_r = -\frac{k}{\kappa} \mathcal{E}_0 J'_m(\kappa r) \sin(m\alpha + \psi_m) \sin(kz),$$
$$E_\alpha = -\frac{mk}{\kappa^2 r} \mathcal{E}_0 J_m(\kappa r) \cos(m\alpha + \psi_m) \sin(kz),$$
$$H_r = -\frac{im\,\omega}{\kappa^2 cr} \mathcal{E}_0 J_m(\kappa r) \cos(m\alpha + \psi_m) \cos(kz),$$
$$H_\alpha = \frac{i\,\omega}{\kappa c} \mathcal{E}_0 J'_m(\kappa r) \sin(m\alpha + \psi_m) \cos(kz);$$
$$k = \frac{l\pi}{h},\ l = 0, 1, 2, \ldots, \qquad \kappa_{mn} = \frac{\alpha_{mn}}{a},$$

where α_{mn} are the roots of the equation $J_m(\alpha_{mn}) = 0$, $\omega^2 = c^2(\kappa_{mn}^2 + k^2)$, and the factor $e^{-i\omega t}$ is dropped out in all expressions.

The H-oscillations:

$$H_z = \mathcal{H}_0 J_m(\kappa r) \sin(m\alpha + \psi_m) \sin(kz);$$

$k = l\pi/h$, $l = 1, 2, \ldots$; the value $l = 0$ is impossible; $\kappa_{mn} = \beta_{mn}/a$, where β_{mn} is the root of the equation $J'_m(\beta_{mn}) = 0$; $\omega^2 = c^2(\kappa_{mn}^2 + k^2)$. The remaining field components can be expressed in terms of H_z with the aid of Maxwell's equations.

When $m \neq 0$, both E- and H-oscillations are generally double degenerate, since to each eigenfrequency there corresponds two eigenfunctions, for example,

$$H_z = \mathcal{H}_0 J_m(\kappa r) \sin m\alpha \sin(kz)$$

and

$$H_z = \mathcal{H}_0 J_m(\kappa r) \cos m\alpha \sin(kz).$$

8.33 In the quasi-stationary approximation, we can consider the given system as an oscillating contour consisting of a capacitor of capacitance $C = R^2/(4d)$ and an inductive coil of self-inductance $L = 4\pi b(\ln(8b/a) - 7/4)$. (See Problem I.2.113 for the calculation of the self-inductance of a wired ring.) By using the Thomson's formula (4.5), we have

$$\omega_0 = \frac{c}{R\sqrt{(\pi b/d)(\ln(8b/a) - 7/4)}}.$$

The quasi-stationary approximation is applicable, if $\lambda_0 = 2\pi c/\omega_0$ is much larger than the dimensions of the system, that is, $\lambda \gg R, b$.

8.34

$$\omega_0 = \frac{c}{a}\sqrt{\frac{d}{\pi b}}.$$

8.35 In the quasi-stationary approximation ($\lambda_0 = 2\pi c/\omega_0 \gg a, b$), we assume that the electric field is completely concentrated between the plates of the capacitor, and the magnetic field is inside of the toroidal cavity. Under these assumptions, the resonator is equivalent to a normal oscillating contour consisting of the capacitor and an inductance. The capacitance of the capacitor is $C = (b - a)^2/(4d)$, and the self-inductance of the torus is $L = 4\pi(b - \sqrt{b^2 - a^2})$ (see Problem I.2.110). The eigenfrequency is

$$\omega_0 = \frac{c}{(b-a)}\sqrt{\frac{d}{\pi(b - \sqrt{b^2 - a^2})}}.$$

Higher types of oscillations of the resonator considered here cannot be determined in the quasi-stationary approximation, since they do not satisfy the condition $\lambda \gg a, b$.

8.36

$$\omega_0 = \frac{2c}{2b - a}\sqrt{\frac{d}{2\pi h \ln \frac{2b+a}{2b-a}}}.$$

8.37 In the coaxial waveguide, which is shorted at one end (at $z = 0$) by the perfectly conducting diaphragm, a standing transverse wave is realized with field strengths

$$E_r = \frac{A}{r}\sin\frac{\omega z}{c}e^{-i\omega t}, \qquad H_\alpha = -\frac{iA}{r}\cos\frac{\omega z}{c}e^{-i\omega t}. \tag{1}$$

In any plane that is perpendicular to the axis of the waveguide, the electrical field distribution is the same as in a cylindrical capacitor, and one can assume that it is produced by a potential difference

$$\Delta\varphi = A\ln\frac{b}{a}\sin\frac{\omega z}{c} \tag{2}$$

between the central rod and the outer shell.

This potential difference should be equal to the voltage on the plates of the capacitor, which is formed by the end face of the rod and the upper cover of the resonator:

(3) $$\Delta\varphi\big|_{z=h} = q/C.$$

Here $C = a^2/4d$ is the capacitance of the capacitor, and q is the charge on one of the plates which can be expressed in terms of the current $\mathcal{I}$ flowing along the rod (or the current in the shell which has the same magnitude and the opposite direction)

$$\mathcal{I} = -i\,\omega q.$$

If we calculate the current strength from the known magnetic field (1), and substitute it and also the potential strength (2) into Equation (3), we find the transcendental equation which is satisfied at eigenfrequencies:

(4) $$\cot\frac{\omega h}{c} = \frac{2\pi a^2\omega}{cd}\ln\frac{b}{a}.$$

This equation can be easily solved graphically. At $\omega h/c \ll 1$ (which implies $\lambda \gg 2\pi h$, that is, the quasi-stationary approximation), we obtain

(5) $$\omega = \frac{c}{\sqrt{\frac{a^2}{4d}\,2h\ln\frac{b}{a}}} = \frac{c}{\sqrt{LC}},$$

where L is the self-inductance of the section of length h of the coaxial line. In this approximation, the only one, the lowest, eigenfrequency can be evaluated (cf. the solutions of Problems 8.33–8.36).

When $d = 0$ (the coaxial waveguide section shorted at both ends), we have

(6) $$\omega_m = \frac{\pi c}{h}m, \qquad m = 1, 2, \dots$$

This means that an integral number of half-waves must fit along the length of the resonator: $h = (\lambda_m/2)m$.

8.38 The field in the resonator is described by the Maxwell's equations (8.17) and (8.18), where $\boldsymbol{B} = \boldsymbol{H}$ and $\boldsymbol{D} = \boldsymbol{E}$. If we take the scalar product of the first equation with $\boldsymbol{H}_\nu$, and of the second one with $\boldsymbol{E}_\nu$, and then integrate over the volume of the resonator, we obtain

(1) $$\begin{cases} \frac{d}{dt}\int \boldsymbol{H}\cdot\boldsymbol{H}_\nu\,dV &= -c\int \boldsymbol{H}_\nu\cdot \mathrm{rot}\,\boldsymbol{E}\,dV, \\ \frac{d}{dt}\int \boldsymbol{E}\cdot\boldsymbol{E}_\nu\,dV &= c\int \boldsymbol{E}_\nu\cdot \mathrm{rot}\,\boldsymbol{H}\,dV. \end{cases}$$

Assuming that the eigenfunctions $\boldsymbol{E}_\nu$ and $\boldsymbol{H}_\nu$ are orthonormalized, according to condition (8.19), we will calculate the integrals on the left-hand sides of Equation (1):

(2) $$\frac{d}{dt}\int \boldsymbol{H}\cdot\boldsymbol{H}_\nu\,dV = 4\pi\dot{p}_\nu, \qquad \frac{d}{dt}\int \boldsymbol{E}\cdot\boldsymbol{E}_\nu\,dV = 4\pi\dot{q}_\nu.$$

The eigenfunctions $\boldsymbol{E}_\nu$ and $\boldsymbol{H}_\nu$ satisfy the equations:

$$(3) \qquad \begin{cases} rot\,\boldsymbol{E}_\nu = ik_\nu \boldsymbol{H}_\nu, & rot\,\boldsymbol{H}_\nu = -ik_\nu \boldsymbol{E}_\nu, \\ rot\,rot\,\boldsymbol{E}_\nu = k_\nu^2 \boldsymbol{E}_\nu, & rot\,rot\,\boldsymbol{H}_\nu = k_\nu^2 \boldsymbol{H}_\nu, \end{cases}$$

where $\boldsymbol{k}_\nu(k_1, k_2, k_3)$ are the corresponding eigennumbers (evaluated in Problems 8.30 and 8.32). The integrals on the right-hand sides of Equation (1) can be transformed with the aid of Equation (3):

$$div\,[\boldsymbol{E} \times rot\,\boldsymbol{E}_\nu] = rot\,\boldsymbol{E}_\nu \cdot rot\,\boldsymbol{E} - \boldsymbol{E} \cdot rot\,rot\,\boldsymbol{E}_\nu = ik_\nu \boldsymbol{H}_\nu \cdot rot\,\boldsymbol{E} - k_\nu^2 \boldsymbol{E}_\nu \cdot \boldsymbol{E}.$$

Hence

$$(4) \qquad \int \boldsymbol{H}_\nu \cdot rot\,\boldsymbol{E}\,dV = -ik_\nu \int \boldsymbol{E}_\nu \cdot \boldsymbol{E}\,dV + \frac{1}{ik_\nu} \int div\,[\boldsymbol{E} \times rot\,\boldsymbol{E}_\nu]\,dV$$

$$= -4\pi i k_\nu q_\nu + \oint \boldsymbol{H}_\nu \cdot [\boldsymbol{n} \times \boldsymbol{E}]\,dS,$$

where the last integral is taken over the inner surface of the resonator, and $\boldsymbol{n}$ is a unit vector directed into the conductor along the normal. However, the field on the wall of the resonator should satisfy the Leontovich's condition (6.27), which can be written in the form

$$(5) \qquad \zeta \boldsymbol{H}_\tau = \boldsymbol{n} \times \boldsymbol{E}.$$

The eigenfunction $\boldsymbol{H}_\nu$ of the resonator with perfect conductivity has only a tangential component on the wall, so that while substituting Equation (5) into the integral (4), we can replace $\boldsymbol{H}_\tau$ by $\boldsymbol{H}$. Finally, on combining Equations (1)–(5), we obtain the equation

$$(6) \qquad \dot{p}_\nu - i\,\omega_\nu q_\nu = -\frac{c\zeta}{4\pi} \oint \boldsymbol{H}_\nu \cdot \boldsymbol{H}\,dS.$$

The second equation can be reduced similarly:

$$(7) \qquad \dot{q}_\nu - i\,\omega_\nu p_\nu = 0.$$

We study the effect of the finite conductivity of the walls on the νth type of oscillations of the perfect resonator. The disturbed field $\boldsymbol{H}$ must transform into the undisturbed field as $\zeta \to 0$, that is, in the sum

$$\boldsymbol{H} = \sum_{\nu'} p_{\nu'} \boldsymbol{H}_{\nu'}$$

the only one term with $\nu' = \nu$ must remain. Hence, the amplitudes $p_{\nu'}$ with $\nu' = \nu$ are proportional to ζ; substituting them in Equation (6) gives the terms of the order ζ^2 and higher. Neglecting such terms and replacing $\boldsymbol{H}$ in Equation (6) by $p_\nu \boldsymbol{H}_\nu$, we get the equation of the form

$$(8) \qquad \dot{p}_\nu - i\,\omega_\nu q_\nu = -p_\nu \frac{c\zeta}{4\pi} \oint H_\nu^2\,dS.$$

If we use Equation (8) to eliminate one of the variables (p_ν), for the other one, we obtain the equation

(9) $$\ddot{q}_\nu + \omega_\nu^2 q_\nu + \left(\frac{c\zeta}{4\pi}\oint H_\nu^2\,dS\right)\dot{q}_\nu = 0.$$

The quantity within the brackets is complex. Equation (9) thus describes the harmonic oscillator which is subject to the "friction force"

$$-\left(\frac{c\zeta'}{4\pi}\oint H_\nu^2\,dS\right)\dot{q}_\nu,$$

where ζ' is the real part of the surface impedance.

Solving the last equation, we find the complex correction $\Delta\,\omega_\nu - i\gamma_\nu$ to the eigenfrequency of the perfect resonator. The losses lead to a damping of the eigenfrequencies, with the damping rate

(10) $$\gamma_\nu = \frac{c\zeta'}{8\pi}\oint H_\nu^2\,dS$$

and to a shift of the eigenfrequencies by

(11) $$\Delta\,\omega_\nu = \frac{c\zeta''}{8\pi}\oint H_\nu^2\,dS;$$

thus, the changed eigenfrequency is $\widetilde{\omega}_\nu = \omega_\nu + \Delta\,\omega_\nu$.

The connection between the quality factor of the resonator and the damping rate is given by Equation (8.20).

8.39

$$Q_\nu = \frac{\omega_\nu a}{4c\zeta'} = \sqrt{\frac{\pi a^2\sigma\,\omega_\nu}{2c^2}}.$$

The system loses the resonance properties at sufficiently high frequencies when the distance between the neighboring eigenfrequencies becomes comparable with the width of the resonance curve determined by the damping rate $\gamma_\nu = \omega_\nu/2Q_\nu$. At high frequencies, it follows from the results of Problem 8.35, that the distance between the neighboring eigenfrequencies is

$$\frac{\Delta\,\omega}{\Delta N} = \frac{\pi^2 c^3}{a^3}\frac{1}{\omega^2}.$$

Equating this quantity to the damping rate γ, we find the region of frequencies for which the system possesses the resonance properties:

$$\omega \le 10^9\sigma^{1/5}a^{4/5}.$$

For $a \approx 1$ cm and $\sigma = 10^{17}\,\text{s}^{-1}$, we have $\omega \le 3\cdot 10^{12}\,\text{s}^{-1}$.

8.40 Expanding $\boldsymbol{E}$ and $\boldsymbol{H}$ in terms of the eigenfunctions, as was done in Problem 8.38, we get for the amplitudes p_ν and q_ν the set of equations:

(1) $$\dot{p}_\nu - i\,\omega q_\nu + 2i\sum_{\nu'}\Delta\Omega_{\nu'}j_{\nu'} = 0,$$

(2) $$\dot{q}_\nu - i\,\omega_\nu p_\nu + \frac{1}{c}j_\nu e^{-i\,\omega t} = 0,$$

where $\Delta\Omega_\nu = \Delta\,\omega - i\gamma_\nu$ is the complex shift of the eigenfunctions and

(3) $$j_\nu = \int \boldsymbol{j}\cdot\boldsymbol{E}_\nu\, dV.$$

We look for the solution of Equations (1) and (2) in the form

(4) $$p_\nu = p_\nu^0 e^{-i\omega t}, \qquad q_\nu = q_\nu^0 e^{-i\omega t}.$$

Eliminating the quantities q_ν^0, we have

(5) $$p_\nu^0(\,\omega^2 - 3\,\omega\Delta\Omega_\nu\, -\, \omega_\nu^2) = \frac{i\,\omega_\nu}{c} j_\nu + 2\,\omega \sum_{\nu'}{}' \Delta\Omega_{\nu'} p_{\nu'}^0.$$

The prime′ at the summation sign indicates that the term with $\nu' = \nu$ is absent (it is taken to the left-hand side of the equation).

We solve Equation (5) by the method of successive iterations. In the zeroth approximation, we drop the sum $\left(\sum{}'\right)$ and have

(6) $$p_\nu^0 = \frac{i\,\omega_\nu j_\nu}{c(\,\omega^2 - 2\,\omega\Delta\Omega_\nu\, -\, \omega_\nu^2)}.$$

In the next approximation, we get the correction to Equation (6), which is equal to

$$\frac{2\,\omega}{\omega^2 - 2\,\omega\Delta\Omega_\nu\, -\, \omega_\nu^2} \sum_{\nu'}{}' \Delta\Omega_{\nu'} p_{\nu'}^0.$$

It is small if ω is close to ω_ν and if all other eigenfrequencies $\omega_{\nu'}$ satisfy the condition $|\omega - \omega_{\nu'}| \gg |\Delta\Omega_{\nu'}|$.

We express the denominator of Equation (6) in terms of the quality factor Q_ν and the changed eigenfrequency $\widetilde{\omega}_\nu = \omega_\nu + \Delta\,\omega_\nu$. We have

$$\omega\Delta\Omega_\nu = \omega\Delta\,\omega_\nu - i\,\omega\gamma_\nu \approx \omega_\nu\Delta\,\omega_\nu - \frac{i\,\omega\,\widetilde{\omega}_\nu}{2Q_\nu},$$

which is valid near the resonance ($|\,\omega - \,\omega_\nu| \ll \omega$). Hence

(7) $$p_\nu^0 \approx \frac{i\,\omega_\nu j_\nu}{c\left(\,\omega^2 - \widetilde{\omega}_\nu^2 + \frac{i\,\omega\,\widetilde{\omega}_\nu}{Q_\nu}\right)}, \qquad q_\nu^0 \approx -\frac{i\,\omega j_\nu}{c\left(\,\omega^2 - \widetilde{\omega}_\nu^2 + \frac{i\,\omega\,\widetilde{\omega}_\nu}{Q_\nu}\right)}.$$

The frequency dependence of the field amplitudes has a resonance character. For a given $\boldsymbol{j}$, the field at resonance becomes that more than higher the quality factor of the resonator

(8) $$q_{\nu\,res}^0 = p_{\nu\,res}^0 = \frac{j_\nu Q_\nu}{c\,\omega_\nu}.$$

It follows also from the formulas obtained that the conductor with the current should be placed in the antinode of the electric field $\boldsymbol{E}_\nu$ and be oriented along $\boldsymbol{E}_\nu$. In that case, the quantities j_ν and, hence, p_ν^0 and q_ν^0, will have their maximum values.

8.41 If the wave field of energy W fills the resonator and is reflected once from a mirror, the energy loss is $W(1-R)$. After a time dt the energy loss is

$$dW = -W(1-R)\frac{c\,dt}{L},$$

where $c\,dt/L$ is the number of reflections. From the definition of the quality factor (8.20), we get

$$Q_1 = \frac{\omega W}{-\frac{dW}{dt}} = \frac{\omega L}{c(1-R)},$$

where ω is the frequency of the oscillations considered.

The radiation through the side surface occurs because a light beam with a limited extent in the transverse direction cannot be rigorously directed. It necessarily has a transverse component of the wavevector $\Delta k_\perp$, which can be evaluated from the condition $\Delta k_\perp D \approx 1$ (see Problem I.2.140). Then the rays of light propagating from one mirror to another, form a slightly diverging beam with an opening angle

$$2\theta = \frac{2\Delta k_\perp}{k} = \frac{2c}{D\,\omega}.$$

Part of the beam does not fall on the second mirror (Figure 8.14), and the energy loss at one reflection will be $WL\theta/D$. After a time dt the energy loss is

$$dW = -W\frac{L\theta}{D}\frac{c\,dt}{L} = -W\frac{c^2}{D^2\,\omega}\,dt.$$

The quality factor due to the radiation is

$$Q_2 = \frac{D^2\,\omega^2}{c^2}.$$

If the losses in the mirrors and due to the radiation are small, they are additive. The total quality factor Q is given by the formula

$$\frac{1}{Q} = \frac{1}{Q_1} + \frac{1}{Q_2}.$$

For the parameters given in this problem, we have

$$Q_1 \approx 4\times 10^5; \qquad Q_2 \approx 4\times 10^8 \gg Q_1; \qquad Q \approx Q_1 \approx 4\times 10^5.$$

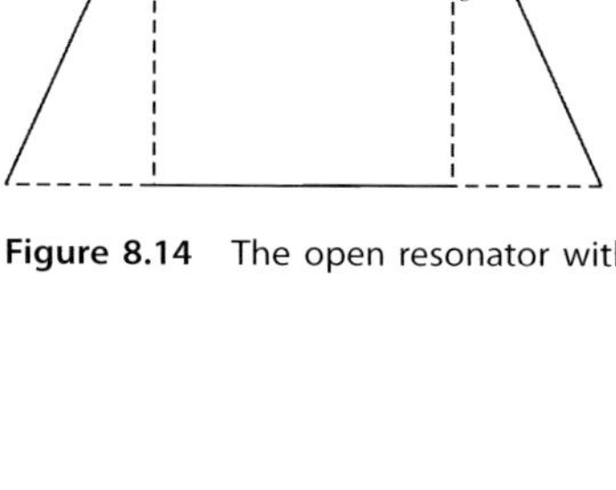

Figure 8.14 The open resonator with two mirrors.

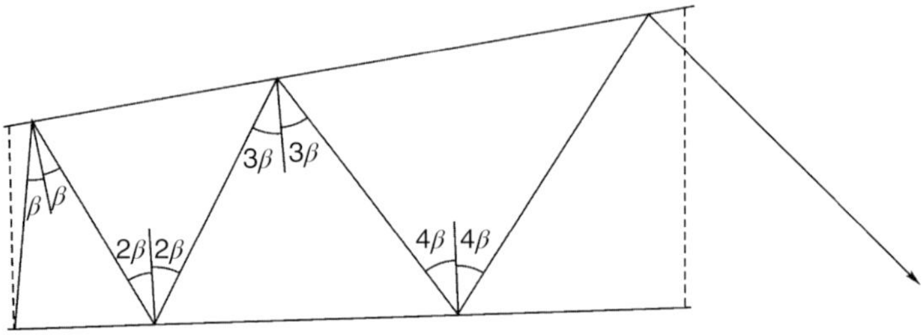

Figure 8.15 Influence of nonparallelism of mirrors on the quality factor of the open resonator.

8.42 If the beam of light originally propagates along the normal to the plane of one of the mirror, after the nth reflection an angle between the normal and the beam will be equal $n\beta$ (Figure 8.15). After the nth passage between the mirrors the beam is displaced by a distance $n\beta L$; the number of reflections N before the beam leaves the resonator can be evaluated from the relation

$$\sum_{n=1}^{N} n\beta L \approx D.$$

For $N \gg 1$, we get $N = \left(2D/\beta L\right)^{1/2}$, which corresponds to a damping time of the eigenoscillations:

$$\tau = N\frac{L}{c} = \frac{1}{c}\left(\frac{2DL}{\beta}\right)^{1/2}.$$

This time can be identified with the reciprocal of the damping rate γ:

$$\gamma = \frac{1}{\tau} = c\left(\frac{\beta}{2DL}\right)^{1/2}.$$

The quality factor due to the mirrors not being parallel is

$$Q_3 = \frac{\omega}{2\gamma} = \frac{\omega}{2c}\left(\frac{2DL}{\beta}\right)^{1/2}.$$

In order that the fact that the mirrors are not parallel do not appreciably decrease the quality of the resonator, the condition $Q_3 \leq Q$ must be satisfied, where Q is the quality factor of the resonator with parallel mirrors. Hence,

$$\beta \leq \frac{\omega^2}{2c^2}\frac{DL}{Q^2}.$$

For the parameters given in Problem 8.42, we find

$$\beta \leq 0.0012.$$

8.43 (a) The angle ϑ takes on discrete values given by the formula

(1) $$\frac{L}{\cos\vartheta} = \frac{n\lambda}{2},$$

where $n \gg 1$ is a positive integer. If for a given λ, the value $\vartheta = 0$ is possible, which corresponds to $n = n_0$ ($L = n_0\lambda/2$), the discrete values of the angle $\vartheta_k \ll 1$ are then determined by

$$(2) \qquad \vartheta_k = \left(\frac{k\lambda}{L}\right)^{1/2}.$$

(b) In Problem 8.41, we found the quality factor Q_1, which takes into account the losses in the mirrors. The quality factor Q_2 due to radiation losses can be estimated as,

$$(3) \quad Q_2 = \frac{\omega D}{c\vartheta} \quad \text{at} \quad \vartheta > \theta, \qquad Q_2 = \frac{\omega D}{c\vartheta} = \frac{D^2\,\omega^2}{c^2} \quad \text{at} \quad \vartheta < \theta,$$

where θ is the diffraction angle determined in Problem 8.41.

If $Q_1 < Q_{2\,max}$, the total Q of the resonator for those types of oscillations for which $Q_2(\vartheta) > Q_1$, will be practically the same and close to Q_1. If $Q_1 > Q_{2\,max}$, then Q will be basically determined by the magnitude of Q_2, according to formula (3).

8.44 We use Equation (2) from Problem 8.30, in which, for simplicity, we set $a = b$ of the order of h. At $h \gg \lambda$, we have $n_3 \gg 1$, and the modes with n_1, $n_2 \ll n_3$ will be long-living. In that case,

$$\omega \approx \frac{\pi c n_3}{h} + \frac{\pi c h(n_1^2 + n_2^2)}{a^2 n_3}.$$

The distance between the neighboring frequencies, as the number n_3 changes, is $\Delta\omega_l \approx \pi c/h$, and as the number n_1 changes, we have $\Delta\omega_t \approx 2\pi c h n_1/n_3 \ll \Delta\omega_l$.

8.45 According to the results of Problem (3.51), the magnetic field inside the ellipsoid is expressed in terms of the external field $\boldsymbol{H}_0$ and magnetization $\boldsymbol{M}$:

$$(1) \qquad \boldsymbol{H}_{int} = \boldsymbol{H}_0 - 4\pi\widehat{N}\boldsymbol{M},$$

where $\widehat{N} = (N_{ik})$ is the tensor of the coefficients of demagnetization. We use the Landau–Lifshitz equation (5.116) and substitute into it the field (1) as $\boldsymbol{H}_{ef}$ (1). We get

$$(2) \qquad \frac{d\boldsymbol{M}}{dt} = -\eta\boldsymbol{H}_{int} \times \boldsymbol{M}.$$

In the equilibrium state $\boldsymbol{M} = \boldsymbol{M}_0 = const$, and from Equation (2), we obtain the condition of equilibrium

$$(3) \qquad \boldsymbol{M}_0 \times (\boldsymbol{H}_0 - 4\pi\widehat{N}\boldsymbol{M}_0) = 0.$$

If the deviation of the magnetization vector from the equilibrium state is small, there occur the oscillations $\boldsymbol{M}(t) = \boldsymbol{M}_0 + \boldsymbol{m}(t)$, where $\boldsymbol{m} \times \boldsymbol{M}_0 = 0$, $\boldsymbol{m} \propto e^{-i\omega t}$. By linearizing Equation (2), we find the uniform set of equations

$$(4) \qquad i\omega\boldsymbol{m} - \eta\{\boldsymbol{m} \times (\boldsymbol{H}_0 - 4\pi\widehat{N}\boldsymbol{M}_0) + 4\pi\widehat{N}\boldsymbol{m} \times \boldsymbol{M}_0\} = 0.$$

Equating to zero the determinant of the set of equations, we get the frequency of the eigenfunctions:

(5)
$$\omega^2 = [\omega_0 + (N^{(x)} - N^{(z)})\omega_M][\omega_0 + (N^{(y)} - N^{(z)})\omega_M],$$

where $\omega_0 = \eta H_0$, $\omega_M = 4\pi\eta M_0$.

The vector $\boldsymbol{m}$ has, in general, the elliptic polarization:

(6)
$$\frac{m_y}{m_x} = i\sqrt{\frac{\omega_0 + (N^{(x)} - N^{(z)})\omega_M}{\omega_0 + (N^{(y)} - N^{(z)})\omega_M}}.$$

The eigenfunctions of the uniform magnetic resonance for specific cases are given in the table:

Shape	Direction of magnetization	Eigenfrequency
Thin disk	Tangential	$\omega = \sqrt{\omega_0(\omega_0 + \omega_M)}$
Thin disk	Normal	$\omega = \lvert\omega_0 - \omega_M\rvert$
Thin rod	Longitudinal	$\omega = \omega_0 + \omega_M/2$
Thin rod	Transverse	$\omega = \sqrt{\omega_0(\omega_0 - \omega_M/2)}$
Sphere	Arbitrary	$\omega = \omega_0$

In the transversely magnetized thin rod, the eigenoscillations are possible only when $\omega_0 > \omega_M/2$.

8.46 H-waves. The eigenfrequencies are given by $\omega_{nl} = x_{nl}c/a$, where x_{nl} are the roots of the equations $j_l(x_{nl}) = 0$, which are numbered by an integer n in ascending order. The components of the field can be determined in terms of the Debye potential $v_{nlm} = j_l(x_{nl}r/a)P_l^m(\theta)(A\cos m\alpha + B\sin m\alpha)$, $m = 0, 2, \ldots l$. At a given frequency, there are in all the $2l + 1$ independent components of the field (degeneracy of the frequency).

E-waves. The eigenfrequencies are given by $\omega_{nl} = x_{nl}c/a$, where x_{nl} are the roots of the equation $j_l(x_{nl}) + x_{nl}j_l'(x_{nl}) = 0$. The components of the field can be determined in terms of the Debye potential u_{nlm}, which coincides with the potential v_{nlm} for H-waves.

9
Interaction of Charged Particles with Equilibrium and Nonequilibrium Media

9.1
Ionization and Radiation Energy Losses of Fast Particles in Media

The particles are referred to as **fast** if their velocities are higher than the velocities of the majority of atomic electrons in a medium. Let us mention that in quantum mechanics this condition allows one to consider the interaction of a particle with an atom by means of the perturbation theory. Fast particles moving through the medium being in a state close to statistically equilibrium, will lose their energy. The energy of heavy particles (muons, protons, atomic nuclei, etc.) will be mainly spent for excitation and ionization of atoms in the medium. These losses are referred to as **ionization losses**. The light particles (electrons and positrons) of high energy produce intensively the bremsstrahlung. Their **radiation losses**, beginning from the certain energy, exceed the losses due to ionization and become of first importance.

Kinematics of collision between fast particle and electron. Ionization losses are associated with the transfer of energy from a fast particle to atomic electrons. When the energy transmitted to the electron essentially exceeds its binding energy, the electron can be considered to be free and at rest. The kinetic energy K transmitted to the electron can be evaluated from the result of Problem I.3.94, as a function of an angle θ of the electron escape in relation to the original momentum of the projectile,

$$K = 2m_e c^2 \frac{c^2 p^2 \cos^2\theta}{(\mathcal{E} + m_e c^2)^2 - c^2 p^2 \cos^2\theta}. \tag{9.1}$$

Here $\mathcal{E} = \sqrt{c^2p^2 + m^2c^4}$ and p are the total relativistic energy and the momentum of the projectile, respectively, m_e is an electron mass. The maximum transmitted energy K_m occurs in the knock-on collision ($\cos\theta = \pm 1$):

$$K_m = 2m_e c^2 \frac{T(\mathcal{E} + mc^2)}{(m_e^2 + m^2)c^4 + 2m_e c^2 \mathcal{E}}, \tag{9.2}$$

where $T = \mathcal{E} - mc^2$ is the kinetic energy of the projectile. At $m = m_e$ we have $K_m = T$, that is, all energy of an electron or positron can be transmitted to an initially immobile electron.

In the nonrelativistic case, we have

$$K_m = \frac{4mm_e}{(m+m_e)^2}\,T. \tag{9.3}$$

At $m \gg m_e$ the projectile particle transmits to an electron only a small part of its energy $K_m/T = (4m_e/m) \ll 1$.

In the ultrarelativistic case, at $\mathcal{E} \gg (m/m_e)mc^2$, from Equation (9.2) we get

$$K_m = T\left(1 - \frac{m^2c^2}{2m_e\mathcal{E}}\right). \tag{9.4}$$

In contrast to the nonrelativistic case, in this limit, independently of a mass of the projectile particle, almost all its kinetic energy can be transmitted to the electron; however, the events of this kind are rare. Finally, for the moderate relativistic case $\mathcal{E} \ll (m/m_e)mc^2$, $m \gg m_e$, from Equation (9.2) we find

$$K_m = 2m_ec^2\left(\frac{p}{mc}\right)^2 = 2m_ec^2(\gamma^2 - 1). \tag{9.5}$$

The energy transfer is determined only by the relativistic factor $\gamma = \mathcal{E}/mc^2$ of the projectile particle.

Ionization losses. All possible collisions of a fast particle with electrons of medium can conventionally be divided into two groups, namely, "close" and "distant" collisions. In the case of close collisions, the energy K transmitted to an electron significantly exceeds its binding energy, whose measure is an average ionization potential I of an atom: $K_m \geq K \geq K_0 \gg I$. Distant collisions correspond to small energy transfers $K \leq K_0 \lesssim I$. Here K_0 is the conventional boundary between close and distant collisions. We will further refine the requirements for this quantity.

The chosen classification allows the evaluation of the cross section of close collisions by means of the perturbation theory in the free-electron approximation. Close collisions are, as a rule, pair collisions, because it is highly improbable that two or several electrons will be simultaneously present in a small volume of space. In the case of muons, protons, and other heavy particles the collision is described by one Feynman diagram (Figure I.7.4), where all external lines correspond to free particles. Their wave functions, that is, Dirac bispinors, were constructed in Example I.6.18. The collision cross section (see Example 9.3), expressed (in contrast to Section I.7.12.1) in terms of the relativistic factor γ_e of the electron after the collision, has the form

$$d\sigma = \frac{2\pi e^4}{m_e^2c^2v^2}\frac{d\gamma_e}{(\gamma_e - 1)^2}\left[1 - \beta^2\frac{\gamma_e - 1}{\gamma_m - 1} + \frac{m_e^2c^4}{2\mathcal{E}^2}(\gamma_e - 1)^2\right]. \tag{9.6}$$

Here v, $\beta = v/c$, and $\mathcal{E}$ are related to the fast heavy particle. The factor e^2 is also due to this particle; γ_m is the largest electron relativistic factor which can be determined by Equation (9.2).

The energy loss $(d\mathcal{E}/dz)_{\gamma_e>\gamma_0}$ by the fast particle per unit length, when the transmitted energy exceeds $(\gamma_0 - 1)m_ec^2$, can be evaluated as the product of the cross section (9.6), the average electron number density n_e, the energy $(\gamma_e - 1)m_e c^2$, with the subsequent integration from γ_0 to γ_m:

$$\left.\frac{d\mathcal{E}}{dz}\right|_{\gamma_e>\gamma_0} = -n_e\int_{\gamma_0}^{\gamma_m}(\gamma_e - 1)m_e c^2 d\sigma = -\frac{2\pi e^4 n_e}{m_e v^2}\left[\ln\frac{\gamma_m - 1}{\gamma_0 - 1} - \beta^2 + \frac{m_e^2 c^4}{4\mathcal{E}^2}(\gamma_m - 1)^2\right]. \tag{9.7}$$

When the medium consists of atoms of the same kind, $n_e = Zn_a$, where Z is the charge number and n_a is the atomic number density. In the composite medium, n_e is equal to the sum of such products.

Note that the energy loss is inversely proportional to an electron mass m_e. The atomic nuclei also participate in the bremsstrahlung. However, their contribution is smaller by the factor $m_e/M \ll 1$, where M is the nucleus mass, so that this contribution can be neglected. With increasing energy of a relativistic particle, the energy loss slowly (logarithmically) increases.

The bremsstrahlung due to distant collisions has often the collective character and is accompanied by small transfers of momentum and energy. These quantities can be distributed immediately over a great number of particles, creating collective excitations of the medium in the macroscopic regions of space. This part of the losses can conventionally be described by the methods of macroscopic electrodynamics with the aid of response functions. Such a description was proposed by Fermi (1940) and developed by Landau (see Landau *et al.* (1984)).

The energy lost by a moving particle is determined by the electromagnetic field produced in medium by this particle. This field includes naturally a response of the medium to the motion of a charged particle. Hence, the energy losses can be evaluated as a work done on the particle by a resultant electric field per unit length. The second equivalent method is to compute the flux of electromagnetic energy through the cylindrical surface of a given radius, along the axis of particle motion. The energy flux will give the energy losses per unit time, transmitted to particles of the medium beyond the cylinder and to the radiating field.

Example 9.1

A particle of charge q moves through an isotropic medium along a straight trajectory, with the velocity $v = const$ (particle scattering is neglected). By using Fourier expansion, evaluate the electromagnetic field produced by this particle, expressing it in terms of the integrals which include the permittivities $\varepsilon_l(k, \omega)$, and $\varepsilon_t(k, \omega)$.

Solution. Equations (5.25) and (5.26) can be reduced to the form

$$\nabla \times [\nabla \times \boldsymbol{E}] + \frac{1}{c^2}\frac{\partial^2 \boldsymbol{D}}{\partial t^2} = -\frac{4\pi}{c^2}\frac{\partial \boldsymbol{j}_{ext}}{\partial t}. \tag{1}$$

The Fourier transform, with the aid of the relation $D_\alpha = \varepsilon_{\alpha\beta}E_\beta$, yields

$$\left(k^2\delta_{\alpha\beta} - k_\alpha k_\beta - \frac{\omega^2}{c^2}\varepsilon_{\alpha\beta}\right)E_\beta(\boldsymbol{k}, \omega) = i\frac{4\pi\omega}{c^2}j_\alpha^{ext}(\boldsymbol{k}, \omega). \tag{2}$$

Using Equation (5.38), from the system of algebraic equations we obtain

(3) $$E_\alpha(\boldsymbol{k},\omega) = -i\frac{4\pi\omega}{k^2}\left\{\frac{k_\alpha k_\beta}{\omega^2\varepsilon_l(k,\omega)} - \frac{k^2\delta_{\alpha\beta} - k_\alpha k_\beta}{c^2k^2 - \omega^2\varepsilon_t(k,\omega)}\right\} j_\alpha^{ext}(\boldsymbol{k},\omega),$$

and from Maxwell's equations (5.26) we have

(4) $$\boldsymbol{B}(\boldsymbol{k},\omega) = i4\pi c\frac{\boldsymbol{k}\times\boldsymbol{j}_{ext}}{c^2k^2 - \omega^2\varepsilon_t(k,\omega)}.$$

The current density of a particle that moves with constant velocity has the form

(5) $$\boldsymbol{j}_{ext}(\boldsymbol{r},t) = e\boldsymbol{v}\delta(\boldsymbol{r} - \boldsymbol{v}t),$$

and its Fourier transform contains the delta-function

(6) $$\boldsymbol{j}_{ext}(\boldsymbol{k},\omega) = 2\pi e\boldsymbol{v}\delta(\omega - \boldsymbol{k}\cdot\boldsymbol{v}).$$

Therefore, the inverse Fourier transform will be expressed by the 3D integral:

$$\boldsymbol{B}(\boldsymbol{r},t) = i4\pi ec\int\frac{\boldsymbol{k}\times\boldsymbol{v}}{c^2k^2 - \omega^2\varepsilon_t(k,\omega)}\exp[i\boldsymbol{k}\cdot(\boldsymbol{r} - \boldsymbol{v}t)]\frac{d^3k}{(2\pi)^3}, \tag{9.8}$$

$$\boldsymbol{E}(\boldsymbol{r},t) = i4\pi e\int\left\{\frac{\boldsymbol{k}}{k^2\varepsilon_l(k,\omega)} - \frac{\omega(\boldsymbol{v} - \boldsymbol{k}\omega/k^2)}{c^2k^2 - \omega^2\varepsilon_t(k,\omega)}\right\}\exp[i\boldsymbol{k}\cdot(\boldsymbol{r} - \boldsymbol{v}t)]\frac{d^3k}{(2\pi)^3}, \tag{9.9}$$

where one should set everywhere $\omega = \boldsymbol{k}\cdot\boldsymbol{v}$. ■

Example 9.2

Write down the energy losses of a particle in the form of the double integral over the frequency $\omega = \boldsymbol{k}\cdot\boldsymbol{v}$ and the transverse wave vector $q = \sqrt{k^2 - \omega^2/v^2}$, using Equation (9.9) for the electric field of a particle in medium.

Solution. In an isotropic medium, the drag is directed opposite to the particle velocity. Hence, the energy losses can be calculated as

(1) $$\frac{d\mathcal{E}}{dt} = e\boldsymbol{v}\cdot\boldsymbol{E}|_{\mathbf{r}=\mathbf{v}t}.$$

Here, the field is taken at the point where the particle is present. In order to find the losses per unit path, one should divide (1) by the particle velocity v. Use Equation (9.9) for the electric field in (1) and replace the variables: $d^3k = (2\pi/v)q\,dq\,d\omega$. Finally, this gives

$$\frac{d\mathcal{E}}{dz} = -i\frac{e^2}{\pi v^2}\int_{-\infty}^{\infty}\omega d\omega\int_0^{\infty}\frac{qdq}{k^2}\left\{\frac{1}{\varepsilon_l(k,\omega)} - \frac{\beta^2q^2}{k^2 - (\omega^2/c^2)\varepsilon_t(k,\omega)}\right\}, \tag{9.10}$$

where $k = \sqrt{q^2 + \omega^2/c^2}$. ■

We take the infinite limit in the integral over q and, thus, we take into account all possible collision parameters down to zero, because, by the order of magnitude, $\rho \sim 1/q$. Hence, Equation (9.10) describes, in principle, the total losses, including those due to close collisions. However, in this case, one should use the permittivities which take into account the spatial dispersion with $k > 1/a$, where a is an atomic size. Such permittivities for complicated media are unknown, and, hence, as a rule, Equation (9.10) is used for distant collisions, cutting off the integral over q at some value $q_0 \ll 1/a$. This allows one to use the permittivities that neglect spatial dispersion, considering them as frequency functions only. Further on we will use this approach.

Example 9.3

Show that the right-hand side of Equation (9.10) is the real negative quantity, and also that Equation (9.10) vanishes for a particle moving with constant velocity.

Solution. Let us use the properties (5.39) of the permittivities ε_l and ε_t, according to which the real parts of these quantities are even functions of frequency, and the imaginary parts are uneven. Because of this, in the integrand the complex quantities are

$$\frac{\omega}{\varepsilon_l} = \frac{\omega(\varepsilon_l' - i\varepsilon_t'')}{|\varepsilon_l|^2}, \qquad \frac{\omega}{a - ib} = \frac{\omega(a + ib)}{a^2 + b^2},$$

where $a = k^2 - \omega^2\varepsilon_t'/c^2$ and $b = \omega^2\varepsilon_t''/c^2$. The real parts of these quantities are uneven frequency functions, and the imaginary parts are even frequency functions. Integrating over frequencies within a symmetric interval, the integrals from the real parts vanish, while the integrals from the imaginary parts give

$$\frac{d\mathcal{E}}{dz} = -\frac{2e^2}{\pi v^2}\int_0^\infty \omega\,d\omega \int_0^{q_0} \frac{q\,dq}{q^2 + \omega^2/v^2}\left\{\frac{\varepsilon_l''}{|\varepsilon_l|^2} + \frac{\beta^2 q^2 b}{a^2 + b^2}\right\} < 0. \tag{9.11}$$

In vacuum, one has $\varepsilon_l'' = \varepsilon_t'' = 0$ and $d\mathcal{E}/dz = 0$. ■

The nonrelativistic case. Neglecting the term proportional to β^2, we get

$$\frac{d\mathcal{E}}{dz} = -\frac{2e^2}{\pi v^2}\int_0^\infty \frac{\varepsilon_l''}{|\varepsilon_l|^2}\omega\,d\omega \int_0^{q_0} \frac{q\,dq}{q^2 + \omega^2/v^2}. \tag{9.12}$$

Here we bound the upper limit of the integration over q by the value $q_0 \ll 1/a$, as was mentioned above. However, the quantity q_0 should also be bounded from below. Let the medium be a monatomic dielectric. The characteristic frequencies in the integral (9.12) are of the order of atomic frequencies. The velocity of the fast particle is much greater than the velocity of electrons in atoms. Therefore, $\omega_{at}/v \ll 1/a$ (but $\omega_{at}/v_{at} \sim 1/a$), and the q_0 can be chosen so that the relation $\omega_{at}/v \ll q_0 \ll 1/a$ be valid. In this case, $\varepsilon_l(\omega)$ depends on the frequency only. The integration over dq gives

$$\frac{d\mathcal{E}}{dz} = -\frac{2e^2}{\pi v^2}\int_0^\infty \frac{\varepsilon_l''}{|\varepsilon_l|^2}\ln\left(\frac{q_0 v}{\omega}\right)\omega\,d\omega. \tag{9.13}$$

Now we introduce the average atomic frequency ω_{at} in Equation (9.13) and define it by the relation (Landau *et al.* (1984))

$$\ln\omega_{at} = \frac{\int_0^\infty \omega f(\omega)\ln\omega\,d\omega}{\int_0^\infty \omega f(\omega)\,d\omega}, \qquad \text{where} \qquad f(\omega) = \frac{\varepsilon_l''(\omega)}{|\varepsilon_l(\omega)|^2}. \tag{9.14}$$

The integral in the denominator can be written in the form

$$\int_0^\infty \omega f(\omega)\, d\omega = -\frac{i}{2}\int_{-\infty}^{\infty} \frac{\omega}{\varepsilon_l(\omega)}\, d\omega,$$

where we have used again the properties of evenness and unevenness of the real and imaginary parts of ε. Next we take into account that the poles and zeros of the function $\varepsilon_l(\omega)$ are absent in the upper half-plane of the complex ω, and we deform the contour so that it becomes the arc C of a large radius in the upper half-plane. For this arc, we can use the asymptotic expression $\varepsilon_l(\omega)$ for large frequencies:

$$\varepsilon_l(\omega) = 1 - \frac{4\pi n_e e^2}{m_e \omega^2}.$$

The result of the integration is

$$\int_0^\infty \omega f(\omega)\, d\omega = -\frac{i}{2}\int_C \omega\left(1 + \frac{4\pi n_e e^2}{m_e \omega^2}\right) d\omega = 2\pi n_e e^2 m_e$$

($\int_C \omega\, d\omega = 0$). Finally, using Equation (9.14), we obtain the compact expression

$$\left.\frac{d\mathcal{E}}{dz}\right|_{q\le q_0} = -\frac{4\pi n_e e^4}{m_e v^2} \ln \frac{q_0 v}{\omega_{at}}. \tag{9.15}$$

The momentum $\hbar q_0$ is much larger than the momentum of atomic electrons and, hence, we can identify it with the momentum of the electron knocked out of the atom by a fast particle. Then we can introduce the energy of this electron $K_0 = (\hbar q_0)^2/2m_e$, rewriting Equation (9.15) in the form

$$\left.\frac{d\mathcal{E}}{dz}\right|_{K\le K_0} = -\frac{2\pi n_e e^4}{m_e v^2} \ln \frac{2m_e v^2 K_0}{\hbar^2 \omega_{at}^2}. \tag{9.16}$$

From the structure of this expression and Equation (9.7) for close collisions, it is clear that the boundary energy K_0 appears under the logarithm so that this energy is canceled from the overall losses:

$$\frac{d\mathcal{E}}{dz} = \left.\frac{d\mathcal{E}}{dz}\right|_{K\le K_0} + \left.\frac{d\mathcal{E}}{dz}\right|_{K\ge K_0} = -\frac{2\pi n_e e^4}{m_e v^2} \ln \frac{2m_e v^2 K_m}{\hbar^2 \omega_{at}^2} \tag{9.17}$$

(we have omitted the terms in Equation (9.7), which are small in the nonrelativistic case).

When the purely quantum-mechanical approach is used for computations, ignoring the methods of macroscopic electrodynamics, the result obtained for the nonrelativistic particle is almost the same (see, e.g., Rossi (1955)):

$$\frac{d\mathcal{E}}{dz} = -\frac{2\pi n_e e^4}{m_e v^2} \ln \frac{2m_e v^2 K_m}{I^2}, \tag{9.18}$$

where I is the average ionization potential of the atom. In the above methods developed by Landau and Lifshitz, instead of the ionization potential a certain average atomic energy $\hbar\omega_{at}$ is introduced, which is expressed in terms of the permittivity according to Equation (9.14). Evidently, the quantities I and $\hbar\omega_{at}$ are close to each other. The evaluation using the permittivity, is, in principle, more exact since it takes into account the polarization of the medium, that is, the effect of neighboring atoms. However, at present none of these quantities can be calculated accurately.

The relativistic case. The evaluation is carried out analogously using Equation (9.11) but is more complicated. For completeness we present, without the derivation, the basic results for the fast, heavy particle ($m \gg m_e$). At $\beta^2 < \varepsilon_0^{-1}$, where ε_0 is the static permittivity, $\varepsilon_0 = \varepsilon_l(0) = \varepsilon_t(0)$, $\mu = 1$, and the total energy losses have the form similar to that given by Equation (9.17):

$$\frac{d\mathcal{E}}{dz} = -\frac{4\pi n_e e^4}{m_e v^2}\left(\ln\frac{2m_e v^2\gamma^2}{\hbar\omega_{at}} - \beta^2\right). \tag{9.19}$$

When $\beta^2 > \varepsilon_0^{-1}$ for ultrarelativistic ($\beta \approx 1$) heavy particles, the total bremsstrahlung energy loss is given by

$$\frac{d\mathcal{E}}{dz} = -\frac{2\pi n_e e^4}{m_e c^2}\left(\ln\frac{m_e^3 c^4\gamma^2}{\pi n_e e^2\hbar^2} - 1\right). \tag{9.20}$$

In the theory that is about the medium polarization, the total bremsstrahlung turns out to be stronger because of distant collisions:

$$\frac{d\mathcal{E}}{dz} = -\frac{4\pi n_e e^4}{m_e c^2}\left(\ln\frac{2m_e c^2\gamma^2}{I} - 1\right). \tag{9.21}$$

In both theories with increasing particle energy, the bremsstrahlung losses grow up logarithmically. However, the polarization reduces the energy losses in distant collisions, which results in equation (9.20). The formulas for the electron bremsstrahlung due to the ionization losses are slightly different from Equations (9.20) and (9.21), but the difference does not exceed 10 –15%.

Generation of plasmons. Let us evaluate the losses of a relativistic particle in a plasma. The close collisions are taken into account by Equation (9.7). The bremsstrahlung due to the distant collisions can be calculated from Equation (9.11), if we set $\varepsilon_l = \varepsilon_t = \varepsilon = 1 - \omega_{0e}^2/\omega^2 + i\eta$, where $\omega_{0e} = \sqrt{4\pi n_e e^2/m_e}$ is the electron plasma frequency, and $\eta > 0$ is a small imaginary part. The first term in the brackets, after the integration over dq at $q_0 \gg \omega/v$, gives

$$-\frac{2e^2}{\pi v^2}\int_0^\infty \frac{\varepsilon''}{|\varepsilon|^2}\ln\left(\frac{q_0 v}{\omega}\right)\omega\, d\omega.$$

Here, $\varepsilon''(\omega) = \eta$ is a small positive number. With the aid of Equations (I.1.215) and (I.1.209) we make the replacement:

$$\frac{\varepsilon''}{|\varepsilon|^2} = \frac{\eta}{\varepsilon'^2 + \eta^2} \to \pi\delta(\varepsilon'(\omega)) = \pi\delta(1 - \omega_{0e}^2/\omega^2) = \frac{1}{2}\pi\omega_{0e}[\delta(\omega - \omega_{0e}) + \delta(\omega + \omega_{0e})]. \tag{9.22}$$

The second term in Equation (9.11) contains the fraction $b/(a^2 + b^2)$. In the limit $\eta \to 0$, we obtain

$$\frac{b}{a^2 + b^2} \to \pi\delta\left(q^2 + \frac{\omega^2}{v^2} - \frac{\omega^2}{c^2}\varepsilon(\omega)\right) = \pi\delta\left(q^2 + \frac{\omega^2}{v^2} - \frac{\omega^2}{c^2} + \frac{\omega_{0e}^2}{c^2}\right). \tag{9.23}$$

The argument of the delta-function is positive for all $q \geq 0$ and $\omega \geq 0$. Therefore, the second term does not contribute to the integral. Combining these results, we have

$$\left.\frac{d\mathcal{E}}{dz}\right|_{q\leq q_0} = -\frac{e^2\omega_{0e}^2}{v^2}\ln\frac{q_0 v}{\omega_{0e}}. \tag{9.24}$$

We have seen in Section 5.1 that Equation $\varepsilon_l(\omega) = 0$ determines the eigenfrequency of plasma oscillations. Thus, Equation (9.24) describes the generation of plasmons by a fast particle per unit length-scale. We have already mentioned in Problem 5.49 that this gives one of the methods to study experimentally the electron structure of solids.

If the losses due to the close collisions are added to Equation (9.24), one obtains the total losses due to collisions for a fast particle in plasma:

$$\frac{d\mathcal{E}}{dz} = -\frac{2\pi n_e e^4}{m_e v^2}\left(\ln\frac{2m_e v^2 K_m}{\hbar^2\omega_{0e}^2} - \beta^2 + \frac{K_m^2}{4\mathcal{E}^2}\right). \tag{9.25}$$

The structure of this formula is the same as in the previous cases. The differences are mainly associated with the quantity under the logarithm and with the small corrections in the brackets.

Cherenkov radiation.[1] Consider a particle moving in a nonmagnetic ($\varepsilon_l = \varepsilon_t = \varepsilon(\omega) > 1$), nonabsorbing dielectric, and separate the part of the energy total losses which is responsible for the generation of transverse electromagnetic waves. It is given by the second term in Equation (9.11):

$$\begin{aligned}\left.\frac{d\mathcal{E}}{dz}\right|_{tr} &= -\frac{e^2}{\pi c^2}\int_0^\infty \omega d\omega \int_0^{q_0}\frac{q^2 dq^2}{q^2+\omega^2/v^2}\frac{b}{a^2+b^2}\\ &= -\frac{e^2}{c^2}\int_0^\infty \omega d\omega\int_0^{q_0}\frac{q^2dq^2}{q^2+\omega^2/v^2}\delta\left(q^2+\frac{\omega^2}{v^2}-\frac{\omega^2}{c^2}\varepsilon(\omega)\right).\end{aligned} \tag{9.26}$$

Here, we have used Equation (9.23). The argument of the delta-function vanishes, only provided that $\omega^2/v^2 - \omega^2\varepsilon(\omega)/c^2 < 0$, that is,

$$\varepsilon(\omega) > 1, \qquad v > \frac{c}{\sqrt{\varepsilon(\omega)}} = v_{ph}. \tag{9.27}$$

The particle velocity should exceed the phase velocity of electromagnetic waves in the dielectric. The angular distribution of radiation can also be found easily. Let us return to the replacement of the variables in Example 9.2 and take into account that $\omega = kv\cos\theta$, where θ is the propagation angle of the electromagnetic wave relative to the particle velocity $\boldsymbol{v}$. On the other hand, in an isotropic dielectric the frequency of the wave is given by $\omega(k) = kv_{ph} = kc/n(\omega)$, where $n(\omega) = \sqrt{\varepsilon(\omega)}$ is the refraction index. Comparing the two expressions for the frequency, we get

$$\cos\theta = \frac{v_{ph}}{v} = \frac{c}{n(\omega)v}. \tag{9.28}$$

1) In the Russian literature the name "Vavilov–Cherenkov radiation" is often used.

The Cherenkov radiation of a given frequency is emitted along the cone, at an angle which is determined by Equation (9.28).

After the integration of Equation (9.26) with respect to dq^2, we obtain the total Cherenkov radiation losses per unit length:

$$\left.\frac{d\mathcal{E}}{dz}\right|_{tr} = -\frac{e^2}{c^2}\int_0^\infty \left(1 - \frac{c^2}{n^2(\omega)v^2}\right)\omega\, d\omega. \tag{9.29}$$

The integration is performed over the frequency range where the radiation condition (9.27) holds. This radiation was discovered experimentally by Vavilov and Cherenkov in 1930. The theory of this effect was given by Tamm and Frank (1937). In 1958 Cherenkov, Tamm and Frank were awarded the Nobel prize in physics for the "discovery, explanation and application of the so-called Cherenkov effect." The history of experimental and theoretical studies of this effect is described in details by Frank, the participant of the discovery, in his monograph (Frank (1988)).[2)]

Some questions related to the Cherenkov radiation are considered in the next section.

Multiple scattering of fast particles. Besides energy losses, fast particles moving through a matter experience multiple elastic collisions which are mainly caused by their interaction with atomic nuclei. In view of their large mass, the nuclei can be assumed to be at rest. Multiple scattering can essentially affect the angular distribution of fast particles and the radiation processes in the medium. The specific feature of Coulomb collisions is the strongly anisotropic character of the Rutherford cross section of the elastic scattering. After each collision, a relativistic particle scatter at small angle. The differential scattering cross section by a screened Coulomb potential in the small-angle approximation is evaluated in Problem 9.2:

$$\frac{d\sigma}{d\Omega} = \left(\frac{2Ze^2}{pv}\right)^2 \frac{1}{(\theta^2+\theta_0^2)^2}, \qquad \theta_0 = Z^{1/3}\alpha\frac{m_e c}{p}, \qquad \alpha = \frac{e^2}{\hbar c} \approx \frac{1}{137}, \tag{9.30}$$

$d\Omega = \theta\, d\theta\, d\varphi$ is a solid angle element, θ_0 takes into account the screening of the potential in the approximation of the exponential screening. The cross section has maximum at $\theta \approx \theta_0 \ll 1$, and decreases when the angle increases as θ^{-4} at $\theta \gg \theta_0$.

Let us obtain the equation for the problems on the multiple elastic scattering in the small-angle approximation. We assume that at the initial instant of time the particle beam had a zero angular spread, and in the final state the average scattering angle is small as compared to unity (but large as compared to an angle θ_0 of a single scattering). A small angle can be conveniently determined by a vector $\boldsymbol{\vartheta}$, which is perpendicular to the initial velocity of the particle and indicates the direction of deflection, while its absolute value gives the angle of deflection.

2) The theoretical prediction of the radiation induced by a charged particle moving with supraluminal speed in a dielectric, was made by English physicist Heaviside in 1888 (see Bolotovskii, 1985). It was before the discovery of the first elementary particle, electron. However, only Russian scientists have experimentally discovered and thoroughly investigated this phenomenon.

An external field is absent, and, hence, the Boltzmann equation (1.38) for the distribution function $f(\boldsymbol{p}, \boldsymbol{r}, t)$ will be of the form

$$\frac{\partial f}{\partial t} + \boldsymbol{v} \cdot \frac{\partial f}{\partial \boldsymbol{r}} = I[f], \tag{9.31}$$

where $I[f]$ is the collision integral of a fast particle with immobile scatters (atoms). Up to the first-order terms in small angles, we have

$$\boldsymbol{v} \cdot \frac{\partial}{\partial \boldsymbol{r}} = v \frac{\partial}{\partial z} + v\boldsymbol{\vartheta} \cdot \frac{\partial}{\partial \boldsymbol{\rho}},$$

where the particle initially moves along the Oz-axis, and $\boldsymbol{\rho}$ is the radius-vector lying in the (x, y) plane. The collision integral can be written in the form

$$I[f] = n_a v \int [f(\boldsymbol{\vartheta} - \boldsymbol{\theta}, \boldsymbol{r}, t) - f(\boldsymbol{\vartheta}, \boldsymbol{r}, t)]\, d\sigma(\theta), \tag{9.32}$$

where the first term expresses the income of particles due to elastic scattering in the state with the angle $\boldsymbol{\vartheta}$, and the second term determines the outcome from the given state into all others. In this case, the energy of the particles is conserved. On the assumption that $\theta \ll \vartheta$, we perform the expansion in scattering angle $\boldsymbol{\theta}$:

$$f(\boldsymbol{\vartheta} - \boldsymbol{\theta}) \approx f(\boldsymbol{\vartheta}) - \boldsymbol{\theta} \cdot \frac{\partial f}{\partial \boldsymbol{\vartheta}} + \frac{1}{2}\left[\frac{\partial^2 f}{\partial \vartheta_x^2}\theta_x^2 + 2\frac{\partial^2 f}{\partial \vartheta_x \partial \vartheta_y}\theta_x\theta_y + \frac{\partial^2 f}{\partial \vartheta_y^2}\theta_y^2\right].$$

Here the projections of the scattering angle can be expressed in terms of the azimuthal angle: $\theta_x = \theta\cos\varphi$, $\theta_y = \theta\sin\varphi$. The evaluation of integrals yields $\int \boldsymbol{\theta} d\sigma(\theta) = \int \theta_x\theta_y\, d\sigma(\theta) = 0$ since the cross section is independent of the azimuthal angle. Denoting

$$q = n_a \int \theta^2 d\sigma(\theta), \tag{9.33}$$

we obtain the kinetic equation in the Fokker–Planck approximation:

$$\frac{\partial f}{\partial t} + v\frac{\partial f}{\partial z} + v\boldsymbol{\vartheta} \cdot \frac{\partial f}{\partial \boldsymbol{\rho}} = \frac{qv}{4}\left(\frac{\partial^2 f}{\partial \vartheta_x^2} + \frac{\partial^2 f}{\partial \vartheta_y^2}\right) f \tag{9.34}$$

All integrals with respect to the angle, which contain the distribution function $f(\boldsymbol{\vartheta})$, can be taken in the infinite limits in view of their fast convergence. However, the integral (9.33) logarithmically diverges at the upper limit and it should be bounded:

$$q = 4\pi n_a \left(\frac{Ze^2}{pv}\right)^2 \left(\ln\frac{\theta_{max}}{\theta_0} - \frac{1}{2}\right), \qquad \theta_{max} \gg \theta_0. \tag{9.35}$$

This expression has the logarithmic accuracy, that is, it is usable provided $\ln(\theta_{max}/\theta_0) \gg 1$. The value θ_{max} is a number of the order of unity. For particles of very high energies, the finite dimension of the nucleus should be taken into account. In that case, $\theta_{max} \approx \hbar/pR_n$, where R_n is the radius of the nucleus. Such a restriction is valid at $\theta_{max} < 1$.

The solution of the Fokker–Planck equation for the multiple scattering can be found in Examples 9.7 and 9.8.

Radiative energy losses. Besides the losses due to ionization and atomic excitation, a fast particle loses its energy due to radiation. The losses of one kind, namely, the Cherenkov radiation losses, were already considered in this section (also see Section 9.2). However, the Cherenkov radiation contributes little to the total energy losses. For the strongly relativistic electrons, the bremsstrahlung of gamma quanta on atomic nuclei is much more significant. A mass of the particle is rather important in the bremsstrahlung, because an acceleration is inversely proportional to mass, in the co-moving frame of reference, and, thus, the intensity of radiation is inversely proportional to the square of the mass. Hence, for the equal relativistic factors and the same external field, the bremsstrahlung of protons is million times smaller than that of electrons. However, it does not mean that the heavy ultrarelativistic particle moving through the medium does not radiate. The recoil electrons will radiate, the ones that are excited by an energetic particle (see Problem I.6.87).

The bremsstrahlung of the ultrarelativistic electron that scatters by a non-screened atomic nucleus was calculated by the method of equivalent photons in Problem I.6.88. Its cross section corresponding to an escape of quanta at all possible angles is of the form

$$d\sigma_r(\omega) = 4Z^2\alpha r_0^2 \frac{d\omega}{\omega}\left\{1 - \frac{2\mathcal{E}'}{3\mathcal{E}} + \frac{\mathcal{E}'^2}{\mathcal{E}^2}\right\} \ln\frac{2\mathcal{E}\mathcal{E}'}{m_e c^2 \hbar\omega}, \quad \mathcal{E} \gg m_e c^2, \ \ln\frac{2\mathcal{E}\mathcal{E}'}{m_e c^2\hbar\omega} \gg 1. \tag{9.36}$$

We mentioned above that the methods of the perturbation theory in the lowest order give the close result:

$$d\sigma_r^{(B)}(\omega) = 4Z^2\alpha r_0^2 \frac{d\omega}{\omega}\left\{1 - \frac{2\mathcal{E}'}{3\mathcal{E}} + \frac{\mathcal{E}'^2}{\mathcal{E}^2}\right\}\left[\ln\frac{2\mathcal{E}\mathcal{E}'}{m_e c^2\hbar\omega} - \frac{1}{2}\right], \qquad \mathcal{E} \gg m_e c^2. \tag{9.37}$$

Here $r_0 = e^2/m_e c^2$ is the classical radius of electron, $\alpha = e^2/\hbar c \approx 1/137$ is the fine structure constant, the index B denotes the first Born approximation, while $\mathcal{E}$, $\mathcal{E}'$ are the initial and final energies of the electron. At $\ln(2\mathcal{E}\mathcal{E}'/m_e c^2\hbar\omega) \gg 1$ both cross sections are the same. For the soft part of the bremsstrahlung spectrum ($\hbar\omega \ll \mathcal{E}$), we have

$$d\sigma_r^{(B)}(\omega) = \frac{16}{3} Z^2\alpha r_0^2 \frac{d\omega}{\omega}\left[\ln\frac{2\mathcal{E}^2}{m_e c^2\hbar\omega} - \frac{1}{2}\right]. \tag{9.38}$$

Using the given cross sections, we can evaluate the radiation losses of the ultrarelativistic electron:

$$\left.\frac{d\mathcal{E}}{dz}\right|_{rad} = -n_a \int_0^{\mathcal{E}/\hbar} \hbar\omega \frac{d\sigma_r^{(B)}}{d\omega}\, d\omega, \tag{9.39}$$

where n_a is the number density of atomic nuclei. The integrand is the spectral density of the losses per unit length and per unit frequency interval:

$$\frac{d^2\mathcal{E}_\omega}{dz\,d\omega} = -n_a\hbar\omega\frac{d\sigma_r^{(B)}}{d\omega}. \tag{9.40}$$

The spectral losses per unit time can be obtained if we multiply this quantity by the velocity of the particle $v \approx c$. These losses are not equivalent to the spectral intensity of the radiation (I.5.77) considered in the classical theory. Equation (I.5.77) gives the spectral losses during all time of radiation, which is supposed to be finite. However, here the time of radiation is infinite, and one should consider the losses per unit time or per unit path length.

Returning to the evaluation of the radiative losses, we perform the integration according to Equation (9.39) and obtain

$$\left.\frac{d\mathcal{E}}{dz}\right|_{rad} = -4Z^2\alpha r_0^2 n_a \mathcal{E}\left[\ln\frac{2\mathcal{E}}{m_e c^2} - \frac{1}{3}\right]. \tag{9.41}$$

A mere comparison of this result with Equation (9.19) for the ionization losses shows that for the electrons the radiation losses prevail at high energies, since the radiative losses increase proportionally to $\mathcal{E}\ln\mathcal{E}$, while the ionization losses in the relativistic region are only proportional to $\ln\mathcal{E}$. However, before finally concluding, the dependence (9.41) for the radiation losses should be refined. It was derived in the simplified representation of the matter as a collection of nuclei without electron shells, randomly distributed in space. In reality, there exists a subsystem of electrons which screen the fields of nuclei and change the dispersion law for quanta, thus producing the permittivity of the medium; they can emit the quanta themselves. First, we take into account the screening of the field of the nucleus by the atomic electrons, assuming the electron system to be static. In the end of this section and in the next section the role of the electrons of the medium will be considered in more details.

Example 9.4

Using the method of equivalent photons (see Section I.6.3), evaluate the cross section of the bremsstrahlung of the relativistic electron in the field of a neutral atom. Use the electrostatic potential of the atom with the exponential screening: $\varphi(r) = (Ze/r)e^{-r/R}$, $R = a_B Z^{-1/3}$, a_B *being the Bohr radius. Elucidate the role of screening depending on the electron energy and on the frequency of the emitted quanta.*

Solution. In the rest frame of the medium, the scalar and vector potentials of the atom have the form:

(1)
$$\varphi(r) = \int \frac{4\pi Ze}{k^2+\kappa^2}\exp(i\boldsymbol{k}\cdot\boldsymbol{r})\frac{d^3k}{(2\pi)^3}, \qquad \kappa = \frac{1}{R}, \qquad \boldsymbol{A}(r) = 0.$$

For the electron moving with the velocity $\boldsymbol{v} = const$ in the system of the electron (shaded region), the potentials as the functions of the shaded coordinates and time can be obtained with the aid of Lorentz transformations (see Chapter I.3):

(2)
$$\varphi'(\boldsymbol{r}',t') = \frac{Ze\gamma}{2\pi^2}\int\frac{d^3k}{k^2+\kappa^2}\exp[i\boldsymbol{k}_\perp\cdot\boldsymbol{r}_\perp + ik_\parallel\gamma(z'+vt')], \; \boldsymbol{A}'(\boldsymbol{r}',t') = -\frac{\boldsymbol{v}}{c}\varphi'(\boldsymbol{r}',t').$$

The coordinates transverse relative to the velocity, are not transformed. Further we determine the strengths of electric and magnetic fields:

$$E'_{\parallel} = -i\frac{Ze}{2\pi^2}\int \frac{k_{\parallel} d^3k}{k^2+\kappa^2} \exp[i\boldsymbol{k}_{\perp}\cdot\boldsymbol{r}_{\perp} + ik_{\parallel}\gamma(z'+vt')],$$

$$(3)\quad \boldsymbol{E}'_{\perp} = -i\frac{Ze\gamma}{2\pi^2}\int \frac{\boldsymbol{k}_{\perp} d^3k}{k^2+\kappa^2} \exp[i\boldsymbol{k}_{\perp}\cdot\boldsymbol{r}_{\perp} + ik_{\parallel}\gamma(z'+vt')], \qquad \boldsymbol{H}' = -\frac{\boldsymbol{v}}{c}\times\boldsymbol{E}'.$$

Here again like in Example I.6.20, evident is the fact that the electromagnetic field of a relativistic object is close to the set of plane waves in vacuum (or, in quantum language, to the equivalent photons). The field is almost transverse, $E'_{\perp}/E'_{\parallel} \approx \gamma \gg 1$, and it follows from the form of the exponents, that the frequency of waves is close to its vacuum value

$$(4)\qquad \omega' = v\gamma k_{\parallel} \approx ck',$$

since $k'_{\perp} = k_{\perp} \ll k'_{\parallel} = \gamma k_{\parallel}$.

The spectral density $n(\omega')$ of the number of equivalent photons (see Example I.6.20), can be determined by equating the total electromagnetic energy, which passed through the plane $z' = 0$ during the time of flight of the atom by an electron, to the total energy of equivalent photons:

$$(5)\qquad -\frac{c}{4\pi v}\int d^2r_{\perp}\int_{-\infty}^{\infty} dt'[\boldsymbol{E}'\times\boldsymbol{H}']\cdot\boldsymbol{v} = \int_0^{\infty} \hbar\omega' n(\omega')\,d\omega'.$$

The integrals over transverse coordinates and time in the left-hand side of Equation (5) give the delta-functions due to which the left-hand side can be presented in the form of the threefold integral:

$$(6)\qquad \frac{Z^2e^2\gamma}{2\pi^2}\int_{-\infty}^{\infty} dk'_{\parallel}\int \frac{k_{\perp}^2 d^2k_{\perp}}{(k^2+\kappa^2)^2} = \int_0^{\infty} \hbar\omega' n(\omega')\,d\omega'.$$

From the last equation, with the aid of Equation (4), the spectral density of equivalent photons is given by:

$$(7)\qquad n(\omega')d\omega' = \frac{Z^2e^2}{2\pi^2\hbar c}\frac{d\omega'}{\omega'}\int_0^{k_m} \frac{2\pi k_{\perp}^3 dk_{\perp}}{(k^2+\kappa^2)^2}.$$

The integral over the transverse wave vector diverges at the upper limit, and it should be cut off at the inverse Compton length (see condition (9) in Example I.6.20):

$$k_m \approx \frac{1}{\Lambda_C} = \frac{m_e c}{\hbar}.$$

The integration gives

$$(8)\qquad n(\omega') = -\frac{Z^2e^2}{\pi\hbar c}\frac{1}{\omega'}\ln\left[\left(\frac{\hbar\omega'}{\mathcal{E}}\right)^2 + \left(\frac{\Lambda_C}{R}\right)^2\right] > 0.$$

Here we dropped unity, which is small compared to the large absolute value of the negative logarithm. In the limit of the nonscreened Coulomb field ($R \to \infty$), the result (8) coincides with Equation (I.6.129).

The remaining solution is obtained like that of Problem I.6.87, differing only by the more general expression (8) for the spectral density of equivalent photons. By using the Compton scattering cross section in the frame of the electron and going to the frame of the medium, we can find the bremsstrahlung cross section:

$$(9)\quad d\sigma_r(\omega) = -2Z^2\alpha r_0^2\frac{d\omega}{\omega}\left\{1-\frac{2\mathcal{E}'}{3\mathcal{E}}+\frac{\mathcal{E}'^2}{\mathcal{E}^2}\right\}\ln\left[\left(\frac{m_ec^2\hbar\omega}{2\mathcal{E}\mathcal{E}'}\right)^2+\left(\frac{\Lambda_C}{R}\right)^2\right].$$

Consider two limiting cases of the last expression.

(a) In the absence of the screening:

$$\frac{\Lambda_C}{R}\ll\frac{m_ec^2\hbar\omega}{2\mathcal{E}\mathcal{E}'},\qquad\text{or}\qquad R\gg l_c, \tag{9.42}$$

where

$$l_c=\frac{2\mathcal{E}\mathcal{E}'}{m_e^2c^3\omega}=\frac{2\gamma\gamma' c}{\omega}. \tag{9.43}$$

In this case, we obtain the same result (9.36) for the cross section.

(b) Under the total screening:

$$\frac{\Lambda_C}{R}\gg\frac{m_ec^2\hbar\omega}{2\mathcal{E}\mathcal{E}'},\qquad\text{or}\qquad l_c\gg R. \tag{9.44}$$

The second term under the logarithmic sign in Equation (9) becomes the main one, and the cross section takes the form

$$d\sigma_r(\omega)=4Z^2\alpha r_0^2\frac{d\omega}{\omega}\left\{1-\frac{2\mathcal{E}'}{3\mathcal{E}}+\frac{\mathcal{E}'^2}{\mathcal{E}^2}\right\}\ln\frac{R}{\Lambda_C}. \tag{9.45}$$

In this limit, the logarithmic factor approaches the constant value, which is independent of the energies of the electron and of the emitted quantum. ■

Special properties of bremsstrahlung of a quantum by a relativistic electron. Coherence length. Now we consider the physical significance of the results obtained in the previous example, and, especially, the role of the length l_c, which is given by Equation (9.43). For this purpose, using the conservation laws, we evaluate the longitudinal momentum transmitted to an external field in the ultra-relativistic case. By an external field we mean the field of an individual atomic nucleus, which is screened by an electron shell. In the direction of the initial electron momentum $\boldsymbol{p}$, we have

$$\hbar q_\parallel = p - p'\cos\vartheta - \hbar k\cos\theta, \tag{9.46}$$

where ϑ and θ are, respectively, the angles of scattering of the electron and of escape of the quantum. In the relativistic case, these angles are rather small and do not exceed $m_ec/p\ll 1$. Therefore, we assume that $\cos\vartheta\approx\cos\theta\approx 1$. Moreover, we consider the quantum to be hard enough, and we do not take into account the polarization of the medium, using the dispersion relation $k=\omega/c$. Under these conditions, the wave vector transmitted to an external field is given by

$$q_\parallel\approx\frac{1}{\hbar c}\left(\sqrt{\mathcal{E}^2-m_e^2c^4}-\sqrt{\mathcal{E}'^2-m_e^2c^4}\right)-\frac{\omega}{c}\approx\frac{m_e^2c^3\omega}{2\mathcal{E}\mathcal{E}'}. \tag{9.47}$$

The inverse quantity

$$l_c = \frac{1}{q_\parallel} = \frac{2\mathcal{E}\mathcal{E}'}{m_e^2 c^3 \omega} \tag{9.48}$$

determines, in the longitudinal dimension of the region which is essential for the radiation process. This manifests itself, particularly, in the fact that the square of the absolute value of the harmonic $U_{\mathbf{q}}$ of an external field determines the emission probability (see the solution of Problem 9.9).

At the fixed energy of the quantum $\hbar\omega$, the longitudinal length increases proportionally to the product $\gamma\gamma'$, and can reach the macroscopic values. The transverse length, in this case, remains of the order of the electron Compton length $\Lambda_C = \hbar/m_e c$, which determined the cutoff parameter of the integral over the wave vector $k_\perp$ in Equation (7) of Example 9.4. The region, bounded by the two above-mentioned lengths, determines the effective volume of integration for the calculation of the matrix element in $U_{\mathbf{q}}$, and, hence, it should be considered the region of the quantum formation.

The longitudinal length is also referred to as the **coherence length**. The meaning of this length is especially clear in the classic case, when

$$\hbar\omega \ll \mathcal{E}, \qquad \gamma' \approx \gamma. \tag{9.49}$$

Let a relativistic particle ($v \approx c$) emit an electromagnetic wave, moving along the trajectory $\boldsymbol{r}(t) \approx \boldsymbol{v}t$, close to a straight line. The emitted waves of frequency ω have then the phase factor

$$\exp[i\varphi(t)] = \exp(ikv\cos\theta - i\omega t). \tag{9.50}$$

Evaluate the path length l_c', on which the emitted waves are approximately coherent, that is, their phases are different by $\Delta\varphi \leq 1$. The length l_c' corresponds to the time of a particle motion $\Delta t = l_c'/v$. From the condition,

$$\Delta\varphi = \varphi(t) - \varphi(t + \Delta t) = (\omega - kv\cos\theta)l_c'/v = 1, \tag{9.51}$$

taking into account the smallness of the radiation angle $\theta \lesssim \gamma^{-1}$, we find the quantity

$$l_c' = \frac{2\gamma^2 c}{\omega(1 + \gamma^2\theta^2)} = \frac{l_c}{1 + \gamma^2\theta^2}, \tag{9.52}$$

where l_c coincides with Equation (9.48).

Using the concept of the radiation region, one can easily understand the role of screening of a nucleus by an electron shell in bremsstrahlung. The screened Coulomb potential becomes short-range; it is given in the statement of Example 9.4 in the approximation of exponential screening. The logarithm in Equations (9.36) and (9.37) can be written in the form

$$\ln\frac{2\mathcal{E}\mathcal{E}'}{m_e c^2\hbar\omega} = \ln\frac{l_c}{\Lambda_C}. \tag{9.53}$$

Under the logarithm sign, there is the ratio of two lengths, longitudinal and transverse; they determine the radiation region. Equations (9.36) and (9.37) used above for the cross section are applicable only if $l_c \ll R$. In this case, the radius of the potential forces does not bound the region of radiation (the absence of screening). In the opposite limiting case $l_c \gg R$ of total screening, the longitudinal dimension of the radiation region will be bounded by the length R, which should be introduced under the logarithm instead of l_c:

$$\ln \frac{2\mathcal{E}\mathcal{E}'}{m_e c^2 \hbar\omega} \rightarrow \ln \frac{R}{\Lambda_C} = \ln \frac{a_B}{Z^{1/3}\Lambda_C} = \ln \frac{1}{\alpha Z^{1/3}}. \tag{9.54}$$

As a result, the cross section of bremsstrahlung takes the form

$$d\sigma_r(\omega) = 4Z^2\alpha r_0^2 \frac{d\omega}{\omega} \left\{1 - \frac{2\mathcal{E}'}{3\mathcal{E}} + \frac{\mathcal{E}'^2}{\mathcal{E}^2}\right\} \ln 137Z^{-1/3}. \tag{9.55}$$

The more precise expression for the cross section with regard to the screening in the Thomas–Fermi model, has been obtained with the aid of perturbation theory by Bethe and Heitler. In the case of the perfect screening, the Bethe–Heitler cross section has the form

$$d\sigma_r(\omega) = 4Z^2\alpha r_0^2 \frac{d\omega}{\omega} \left\{\left(1 - \frac{2\mathcal{E}'}{3\mathcal{E}} + \frac{\mathcal{E}'^2}{\mathcal{E}^2}\right) \ln 183Z^{-1/3} + \frac{\mathcal{E}'}{9\mathcal{E}}\right\}, \qquad l_c \gg a_B Z^{-1/3}. \tag{9.56}$$

From this formula, in support of the preceding qualitative considerations, it follows that with the increase in the electron energy, the role of screening in bremsstrahlung increases. The screened cross section, as it should, is less than the nonscreened one for the same energy of the radiating particle. In the limit of low energies of the quantum, $\hbar\omega \ll \mathcal{E}$, $\mathcal{E}' \approx \mathcal{E}$, which can also be obtained from the classic theory of radiation, the cross section (9.56) takes the form

$$d\sigma_r(\omega) = \frac{16}{3} Z^2\alpha r_0^2 \left(\ln 183Z^{-1/3} + \frac{1}{12}\right) \frac{d\omega}{\omega}. \tag{9.57}$$

The divergence of the cross section, as $\omega \rightarrow 0$, is caused by The radiation of a great number of soft quanta at large target parameters. However, this divergence is eliminated if the polarization of the medium is taken into account (see below), and it does not affect the calculation of the radiation losses.

Cascade length and electron–photon showers. The deceleration of an electron per unit path length under the total screening is given by Equation (9.39) where the cross section is given by Equation (9.56):

$$\left.\frac{d\mathcal{E}}{dz}\right|_{rad} = -\frac{\mathcal{E}}{L_r}, \qquad L_r^{-1} = 4Z^2\alpha r_0^2 n_a \left(\ln 183Z^{-1/3} - \frac{1}{18}\right). \tag{9.58}$$

Let us note that the main contribution to the integral (9.39) comes from the quanta with energies $\hbar\omega$, which are comparable with the energy of the radiating particle $\mathcal{E}$. The contribution of the quanta with energies $\hbar\omega \ll \mathcal{E}$ is of the order of $\hbar\omega/\mathcal{E} \ll 1$. The same situation took place at we derived Equation (9.41) which corresponds to the case where the screening is absent, $l_c \ll R$. This means that the radiation losses, contrary to the ionization ones, occur by large parts and produce strong fluctuations of the energy of the original particle.

According to Equation (9.39), a relativistic electron loses a significant part of its energy on the characteristic length L_r (its energy decreases by e times along such a length). This quantity is referred to as **radiation (or cascade) length.** The second name indicates that an ultrarelativistic electron gives rise to an avalanche (shower) of other relativistic particles. The energetic quanta in the field of the atomic nucleus generate the electron–positron pairs and also the muon and taon pairs, which, in their turn, create the energetic quanta, and the production of new particles grows up in geometrical progression until the energy of the primary particle, initiating a shower, terminates.

Here are the values of radiation length for some media:

Medium	Air	Water	C	Si	Pb
L_r (cm)	30000	33.9	14.7	10.2	0.26

These values indicate that the shower occupies the largest volume in the Earth atmosphere. The detection and measurement of various characteristics of broad atmospheric showers is the main method to study the cosmic super-high energy particles invading the Earth atmosphere from cosmos. Such showers cover the areas of hundreds or thousands square kilometers of the Earth surface. The set-ups for investigation of broad atmospheric showers are built in many countries, including Russia.

With the aid of Equation (9.58) we determine the electron energy $\mathcal{E}_*$, at which the radiation losses begin to prevail over the ionization losses. Equating Equation (9.58) to Equation (9.20) (the latter formula describes the ionization losses of an electron with an error of about 10%), we find

$$\mathcal{E}_* \approx \frac{1600\, m_e c^2}{Z}. \tag{9.59}$$

The result is 10 MeV for lead, 55 MeV for copper and 200 MeV for air.

Effect of multiple scattering and medium polarization on bremsstrahlung. In order to study this effect, we will give a more accurate estimate of the coherence length (9.48). First, we take into account the polarization of medium in terms of the permittivity $\varepsilon(\omega)$: $k = \omega\sqrt{\varepsilon}/c$. Assuming that the radiation frequencies significantly exceed the atomic frequencies and the plasma frequency $\omega_{0e} = \sqrt{4\pi n_e e^2/m_e}$, we use the asymptotic expression of permittivity for the large frequencies, and obtain

$$k \approx \frac{\omega}{c} - \frac{\omega_{0e}^2}{2\omega c}. \tag{9.60}$$

Second, when the longitudinal length is large, the particle can experience multiple scattering in the region of the radiation formation, that increases the deflection angle relative to the initial direction. Both these factors will result in the increase of the transmitted momentum and in the decrease of the coherence length, and, as a consequence, in the decrease of radiation. The effect of the bremsstrahlung suppression by the electron multiple scattering **(Landau–Pomeranchuk effect)** was predicted in 1953 and subsequently proved by observations (see Ter-Mikaelyan's review Ter-Mikhaelyan (2003)). The effect of medium polarization on bremsstrahlung was discovered by Ter-Mikaelyan (see his monograph Ter-Mikhaelyan (1972)).

Let us evaluate the longitudinal transmitted wave vector, taking into account the above-mentioned effects for the case $\hbar\omega \ll \mathcal{E}$, where these effects manifest themselves to a large extent:

$$q_{\parallel} \approx \frac{\omega}{2\gamma^2 c}\left[1+\gamma^2\theta^2+\frac{\omega_{0e}^2\gamma^2}{\omega^2}+\frac{cp}{\hbar\omega}\gamma^2\vartheta^2\right]=\frac{1}{l_c}. \tag{9.61}$$

In the case of interaction of an electron with a single screened Coulomb center, the scattering angle is given by $\vartheta \approx \theta_0 \approx Z^{1/3}\alpha m_e c/p$. Hence, the last term in the brackets can be written in the form

$$\frac{cp}{\hbar\omega}\gamma^2\vartheta^2 \approx Z^{2/3}\alpha^2\frac{\mathcal{E}}{\hbar\omega}, \tag{9.62}$$

where $\mathcal{E}$ is the energy of the radiating electron. It follows from Equations (9.61) and (9.62), that the coherence length retains the value (9.48) by the order of magnitude under the conditions

$$\theta^2 \lesssim \gamma^{-2}, \qquad \hbar\omega \gg \alpha^2 Z^{2/3}\mathcal{E} \approx 5\times 10^{-5} Z^{2/3}\mathcal{E}.$$

However, as the particle energy increases, the angle ϑ also increases due to multiple scattering of an electron, resulting, according to Equation (9.61), in a decrease of the coherence length. At lower frequencies the term $\omega_{0e}^2\gamma^2/\omega^2$ comes into effect, also decreasing the coherence length along with the radiation intensity.

For a qualitative evaluation of the effects under consideration, we will calculate the radiation intensity in the classical region of spectrum, $\hbar\omega \ll \mathcal{E}$, using the methods of classic electrodynamics. We use Equation (I.5.77) for the spectral density of radiation in a particular direction, reconstruct its dependence on k and omit the vacuum dispersion relation $k = \omega/c$:

$$\begin{aligned}\frac{d^2 I_\omega}{d\omega d\Omega} &= \frac{e^2}{4\pi^2 c}\int_{-\infty}^{\infty} dt' \int_{-\infty}^{\infty} dt [\boldsymbol{k}\times\boldsymbol{v}(t')]\cdot[\boldsymbol{k}\times\boldsymbol{v}(t)] \\ &\times \exp\{i\omega(t'-t) - i\boldsymbol{k}\cdot(\boldsymbol{r}(t')-\boldsymbol{r}(t))\}.\end{aligned} \tag{9.63}$$

Going to the variables of integration t and $\tau = t' - t$, we write this relation in the form

$$\begin{aligned}\frac{d^2 I_\omega}{d\omega d\Omega} &= \frac{e^2}{2\pi^2 c}\Re\int_{0}^{\infty} d\tau \int_{-\infty}^{\infty} dt [\boldsymbol{k}\times\boldsymbol{v}(t+\tau)]\cdot[\boldsymbol{k}\times\boldsymbol{v}(t)] \\ &\times \exp\{i\omega\tau - i\boldsymbol{k}\cdot(\boldsymbol{r}(t+\tau)-\boldsymbol{r}(t))\}.\end{aligned} \tag{9.64}$$

In the conditions of electron multiple scattering in the disordered medium, the quantities $\boldsymbol{v}(t)$ and $\boldsymbol{r}(t)$ become random functions of time and should be averaged over electron collisions with the randomly distributed force centers. In doing so, we will be guided by Migdal's work (1956). The averaging should be performed with the distribution functions $F(\boldsymbol{v}, \boldsymbol{r}, t; 0, \boldsymbol{v}_0, 0)$ and $F(\boldsymbol{v}', \boldsymbol{r}', t+\tau; \boldsymbol{v}, \boldsymbol{r}, t)$. The first group of arguments correspond to the end of motion, and the second to its beginning. Thus, for clarity, the initial conditions are included in the arguments, on which the corresponding distribution function depends. Each of these functions is normalized to unity and is the probability to find a particle at a finite point of phase space provided that at initial time it was at the initial point. Both functions satisfy the Fokker-Plank kinetic equation (9.34) and the initial conditions

$$F(\boldsymbol{v}, \boldsymbol{r}, 0; 0, \boldsymbol{v}_0, 0) = \delta(\boldsymbol{r})\delta(\boldsymbol{v} - \boldsymbol{v}_0), \qquad F(\boldsymbol{v}', \boldsymbol{r}', t + 0; \boldsymbol{v}, \boldsymbol{r}, t) = \delta(\boldsymbol{r}' - \boldsymbol{r})\delta(\boldsymbol{v}' - \boldsymbol{v}). \tag{9.65}$$

Note, that in the considered approximation the absolute value of the electron velocity does not change at all. Therefore, at any time both distribution functions are proportional to $\delta(v - v_0)$ and $\delta(v' - v)$, respectively.

The averaging is produced in a following way:

$$\langle [\boldsymbol{k} \times \boldsymbol{v}(t + \tau)] \cdot [\boldsymbol{k} \times \boldsymbol{v}(t)] \exp\{-i\boldsymbol{k} \cdot (\boldsymbol{r}(t + \tau) - \boldsymbol{r}(t))\}\rangle$$

$$= \int d^3r d^3v d^3r' d^3v' [\boldsymbol{k} \times \boldsymbol{v}'] \cdot [\boldsymbol{k} \times \boldsymbol{v}] \exp\{-i\boldsymbol{k} \cdot (\boldsymbol{r}' - \boldsymbol{r}\}$$
$$F(\boldsymbol{v}, \boldsymbol{r}, t; 0, \boldsymbol{v}_0, 0) F(\boldsymbol{v}', \boldsymbol{r}', t + \tau; \boldsymbol{v}, \boldsymbol{r}, t). \tag{9.66}$$

In the homogeneous stationary medium, the conventional probabilities F depend only on the difference of coordinates and on the difference of times $F(\boldsymbol{v}', \boldsymbol{r}', t + \tau; \boldsymbol{v}, \boldsymbol{r}, t) = F(\boldsymbol{r}' - \boldsymbol{r}, \boldsymbol{v}', \boldsymbol{v}, \tau)$. Hence, the coordinate Fourier transform

$$F_{\mathbf{k}}(\boldsymbol{v}', \boldsymbol{v}, \tau) = \int F(\boldsymbol{r}' - \boldsymbol{r}, \boldsymbol{v}', \boldsymbol{v}, \tau) \exp\{-i\boldsymbol{k} \cdot (\boldsymbol{r}' - \boldsymbol{r}\} d^3(\boldsymbol{r}' - \boldsymbol{r}) \tag{9.67}$$

enters in Equation (9.66).

The angles, which determine the directions $\boldsymbol{v}$, $\boldsymbol{v}'$, are assumed to be small and are measured from the direction $\boldsymbol{n} = \boldsymbol{k}/k$ of a quantum escape,

$$\boldsymbol{v} \approx v\left(1 - \frac{1}{2}\vartheta^2\right)\boldsymbol{n} + v\boldsymbol{\vartheta},\ \boldsymbol{v}' \approx v\left(1 - \frac{1}{2}\vartheta'^2\right)\boldsymbol{n} + v\boldsymbol{\vartheta}',\ [\boldsymbol{k} \times \boldsymbol{v}'] \cdot [\boldsymbol{k} \times \boldsymbol{v}] = (kv)^2 \boldsymbol{\vartheta}' \cdot \boldsymbol{\vartheta}, \tag{9.68}$$

where the vectors $\boldsymbol{\vartheta}$ and $\boldsymbol{\vartheta}'$ are perpendicular to $\boldsymbol{n}$ and determine small angles like in Equation (9.32).

The Fokker–Planck equation for the function $F_{\mathbf{k}}$ has the form

$$\frac{\partial F_{\mathbf{k}}}{\partial \tau} + ikv\left(1 - \frac{1}{2}\vartheta'^2\right) F_{\mathbf{k}} = \frac{qv}{4}\left(\frac{\partial^2 F_{\mathbf{k}}}{\partial \vartheta_x'^2} + \frac{\partial^2 F_{\mathbf{k}}}{\partial \vartheta_y'^2}\right), \tag{9.69}$$

and q is given by Equation (9.35).

Example 9.5

Construct the solution of Equation (9.69), which satisfies the initial condition $F_{\mathbf{k}}(\boldsymbol{v}', \boldsymbol{v}, 0) = \delta(\boldsymbol{\vartheta}' - \boldsymbol{\vartheta})$ and the restriction conditions.

Solution. Introduce the new unknown function

(1) $$w(\boldsymbol{\vartheta}', \boldsymbol{\vartheta}, \tau) = e^{ikv\tau} F_{\mathbf{k}}(\boldsymbol{v}', \boldsymbol{v}, \tau),$$

satisfying the equation

(2) $$\frac{\partial w}{\partial \tau} - \frac{1}{2} ikv\vartheta'^2 w = \frac{qv}{4}\left(\frac{\partial^2 w}{\partial \vartheta_x'^2} + \frac{\partial^2 w}{\partial \vartheta_y'^2}\right)$$

and find the solution in the form

(3) $$w(\boldsymbol{\vartheta}', \boldsymbol{\vartheta}, \tau) = \exp\{\alpha(\tau) + \beta(\tau)\boldsymbol{\vartheta}' \cdot \boldsymbol{\vartheta} + \gamma(\tau)\vartheta'^2\},$$

where $\alpha(\tau)$, $\beta(\tau)$, and $\gamma(\tau)$ are the functions of time to be determined from Equation (2) and from the initial condition. Substituting Equation (3) in Equation (2) and equating the terms at equal powers of the independent variable $\boldsymbol{\vartheta}'$ we get the system of nonlinear equations

(4) $$\dot{\alpha} = qv\gamma + \frac{1}{4} qv\vartheta^2\beta^2,$$

(5) $$\dot{\beta} = qv\beta\gamma,$$

(6) $$\dot{\gamma} = qv\gamma^2 - \frac{1}{2} ikv.$$

From Equation (6), we find

(7) $$\gamma(\tau) = -a \coth b\tau, \qquad a = \frac{1-i}{2}\sqrt{\frac{k}{q}}, \qquad b = qva.$$

Next we integrate Equations (5) and (4) term by term, not forgetting to include the constant of integration in the solution:

(8) $$\beta(\tau) = \frac{C}{\sinh b\tau}, \qquad \alpha(\tau) = \frac{qvC^2\vartheta^2}{4ab}\gamma(\tau) - \ln \sinh b\tau + \ln C'.$$

In order to satisfy the initial condition at $\tau \to 0$, it is necessary to take $C = 2a$ and $C' = a/\pi$. This results in the solution of the form

$$w(\boldsymbol{\vartheta}', \boldsymbol{\vartheta}, \tau) = \frac{a}{\pi \sinh b\tau} \exp\{-(\vartheta'^2 + \vartheta^2)\, a \coth b\tau + 2\boldsymbol{\vartheta}' \cdot \boldsymbol{\vartheta}\, a \sinh^{-1} b\tau\}. \quad \blacksquare \tag{9.70}$$

Now we return to the calculation of the radiation intensity. We averaged the quantity (9.64) over collisions of an electron with the screened Coulomb centers. Substituting Equations (9.65), (9.66), and (9.69) in Equation (9.64), we have

$$\left\langle \frac{d^2 I_\omega}{d\omega d\Omega} \right\rangle = \frac{e^2 k^2 c}{2\pi^2} \int_{-\infty}^{\infty} dt\, \Re \int_0^{\infty} d\tau e^{i(\omega - kv)\tau} \int d^3r\, d^2\vartheta\, d^2\vartheta' (\boldsymbol{\vartheta}' \cdot \boldsymbol{\vartheta})$$
$$F(\boldsymbol{v}, \boldsymbol{r}, t; 0, \boldsymbol{v}_0, 0) w(\boldsymbol{\vartheta}', \boldsymbol{\vartheta}, \tau). \tag{9.71}$$

We are not interested in the angular distribution of radiating quanta. Hence, we integrate the obtained expression over the directions of the initial velocity $\boldsymbol{v}_0$ of the electron. After the integration, due to the normalization condition

$$\int d^3v_0 d^3r\, F(\boldsymbol{v}, \boldsymbol{r}, t; 0, \boldsymbol{v}_0, 0) = 1,$$

the dependence on t in the integrand vanishes, and the integral with respect to dt becomes formally divergent. This means that the spectral density of radiation is proportional to time, since the particle interacts at all times with the scattering centers and emits the radiation. It is necessary to introduce the spectral intensity per unit time by dividing the both parts of Equation (9.71) by the time interval $2T = \int_{-T}^{T} dt$. In order to pass to the losses per unit path length, which were calculated in the preceding sections, we should change the sign and divide the result by the particle velocity. As a result, from Equation (9.71) we obtain

$$\frac{d^2\mathcal{E}_\omega}{dz\,d\omega} = -\frac{e^2k^2}{2\pi^2}\Re\int_0^\infty d\tau\, e^{i(\omega-kv)\tau}\int d^2\vartheta\, d^2\vartheta'(\boldsymbol{\vartheta}'\cdot\boldsymbol{\vartheta})w(\boldsymbol{\vartheta}',\boldsymbol{\vartheta},\tau). \tag{9.72}$$

In order to take the remaining integrations in Equation (9.72), it is convenient to present all expressions in the form

$$\frac{d^2\mathcal{E}_\omega}{dzd\omega} = -\frac{e^2k^2}{2\pi^2}\Re\int_0^\infty d\tau e^{i(\omega-kv)\tau}\int d^2\vartheta\, d^2\vartheta'(\boldsymbol{\vartheta}'\cdot\boldsymbol{\vartheta})[w(\boldsymbol{\vartheta}',\boldsymbol{\vartheta},\tau) - w^{(0)}(\boldsymbol{\vartheta}',\boldsymbol{\vartheta},\tau)], \tag{9.73}$$

where

$$w^{(0)}(\boldsymbol{\vartheta}',\boldsymbol{\vartheta},\tau) = \delta(\boldsymbol{\vartheta}'-\boldsymbol{\vartheta})e^{ikv\vartheta^2/2} \tag{9.74}$$

is the distribution function of a free particle, which does contribute to the radiating energy. The integration with respect to $d^2\vartheta'$ gives the result

$$\int d^2\vartheta'(\boldsymbol{\vartheta}'\cdot\boldsymbol{\vartheta})[w(\boldsymbol{\vartheta}',\boldsymbol{\vartheta},\tau) - w^{(0)}(\boldsymbol{\vartheta}',\boldsymbol{\vartheta},\tau)] = -\frac{2i}{kv}\frac{d}{d\tau}\left[e^{-\vartheta^2 a\tanh b\tau} - e^{ikv\vartheta^2/2}\right]. \tag{9.75}$$

Now we integrate one time by parts over $d\tau$. The infinitely oscillating terms, separated at upper limit, must be dropped. To prove this, it is sufficient to introduce an infinitely small absorption of quanta. Integrating over $d^2\vartheta$, we have

$$\frac{d^2\mathcal{E}_\omega}{dz\,d\omega} = -\frac{e^2k(\omega-kv)}{\pi c}\Re\int_0^\infty d\tau\, e^{i(\omega-kv)\tau}\left(\frac{1}{a}\coth b\tau - \frac{2i}{kv\tau}\right). \tag{9.76}$$

We will transform the remaining integral to a more convenient form and pass to the integration over $d\zeta = bd\tau$. In the plane of complex ζ, it is the integration along the ray 1–2 (Figure 9.1). Consider the closed contour 1–2–3, where 2–3 is an arc of infinite radius. The integrand is analytical in the shaded region; therefore the integral over the closed contour is zero. The integral over the arc of infinite radius also vanishes due to the presence of the cutoff exponent

$$\exp\left\{\frac{i(\omega-kv)}{b}\zeta\right\} = \exp\left\{(-1+i)\frac{\omega-kv}{c\sqrt{kq}}\zeta\right\}.$$

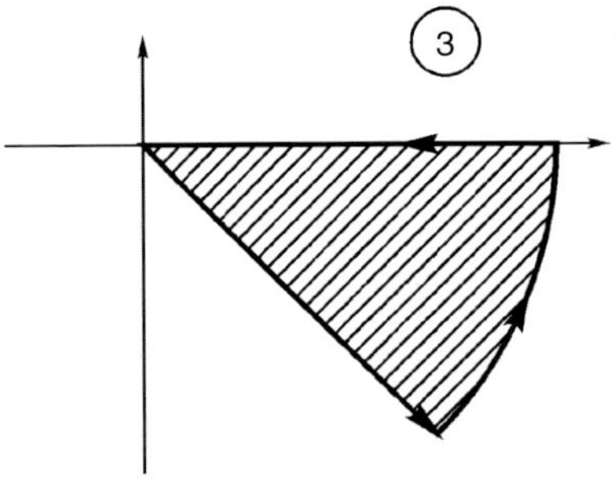

Figure 9.1 Integration contour in a complex plane ζ.

Hence, the integral along the ray 1–2 is equal to the integral over the real semiaxis 1–3:

$$\frac{d^2\mathcal{E}_\omega}{dzd\omega} = \frac{2e^2(\omega - kv)}{\pi c^2}\Re i \int_0^\infty \exp\left\{(-1+i)\frac{\omega - kv}{c\sqrt{kq}}\zeta\right\}\left(\coth\zeta - \frac{1}{\zeta}\right)d\zeta. \quad (9.77)$$

The integral depends on one dimensionless parameter

$$s = \frac{\omega - kv}{2c\sqrt{kq}}. \quad (9.78)$$

Following Migdal (1956), we introduce the function

$$\Phi(s) = 24s^2\int_0^\infty e^{-2sx}\left(\coth x - \frac{1}{x}\right)dx, \quad (9.79)$$

which has asymptotes

$$\Phi(s) \approx 1 \quad \text{at} \quad s \gg 1, \qquad \Phi(s) \approx 6s \quad \text{at} \quad s \ll 1. \quad (9.80)$$

The function $\Phi(s)$ is tabulated in the Migdal's work. The energy loss can be written in the form

$$\frac{d^2\mathcal{E}_\omega}{dzd\omega} = -\frac{e^2kq}{3\pi(\omega - kv)}\Phi(s). \quad (9.81)$$

Analysis of the results. First of all we consider the case where the multiple scattering and polarization of medium are insignificant ($s \gg 1$, $k = \omega/c$, $\omega - kv \approx \omega/2\gamma^2$). From Equations (9.81) and (9.35), we obtain

$$\frac{d^2\mathcal{E}_\omega}{dzd(\hbar\omega)} = -\frac{8}{3}Z^2\alpha r_0^2 n_a\left(\ln\frac{\theta_{max}}{\theta_0} - \frac{1}{2}\right). \quad (9.82)$$

The Bethe–Heitler formula for the same quantity has the form

$$\frac{d^2\mathcal{E}_\omega}{dzd(\hbar\omega)} = -\frac{16}{3}Z^2\alpha r_0^2 n_a\left(\ln 183Z^{-1/3} + \frac{1}{12}\right). \quad (9.83)$$

The logarithmic factors in two last expressions differ because the description of single scattering with the aid of Fokker–Planck equation is inaccurate. By comparing Equations (9.82) and (9.83), we can refine the parameter in the kinetic equation:

$$\ln\frac{\theta_{max}}{\theta_0}-\frac{1}{2}\to 2\ln 183Z^{-1/3},\qquad q=8\pi n_a\left(\frac{Ze^2}{\mathcal{E}}\right)^2\ln 183Z^{-1/3} \tag{9.84}$$

(a small additive term 1/12 is omitted). Upon correction of the logarithmic factor, the final formula (9.81) can be expressed in terms of the Bethe–Heitler spectral density:

$$\frac{d^2\mathcal{E}_\omega}{dzd(\hbar\omega)}=\left(\frac{d^2\mathcal{E}_\omega}{dzd(\hbar\omega)}\right)_{BH}\Phi(s). \tag{9.85}$$

In the general case, using Equation (9.60), the parameter s may be written in the form

$$s=\frac{1}{2\gamma^2}\sqrt{\frac{\omega}{cq}}\left(1+\frac{\omega_{0e}^2\gamma^2}{\omega^2}\right), \tag{9.86}$$

where

$$cq=\frac{\omega_{0e}^2 r_0}{\gamma^2 c}2Z\ln 183Z^{-1/3}. \tag{9.87}$$

The dependence $s(\omega)$ is not monotonic, and we consider separately the case of high frequencies ($\omega\gg\omega_{0e}\gamma$) and the case of low frequencies ($\omega\ll\omega_{0e}\gamma$).

At high frequencies, the medium polarization effect is insignificant, and as the multiple scattering effect, it manifests itself at $s\ll 1$. Using the asymptotic dependence $\Phi(s)\approx 6s$, for this case we find the result, which was first obtained by Landau and Pomeranchuk (1953) (though with inaccurate numerical coefficient):

$$\frac{d^2\mathcal{E}_\omega}{dz\,d(\hbar\omega)}\approx 6s\left(\frac{d^2\mathcal{E}_\omega}{dz\,d(\hbar\omega)}\right)_{BH}=\frac{\alpha}{\pi}\sqrt{\frac{\omega q}{c}}. \tag{9.88}$$

Compared to Bethe–Heitler formula, the radiation is suppressed and the spectrum is different. The multiple scattering effect is enhanced with decreasing ω. However, it affects essentially the cascade length only when the condition $s(\omega)\ll 1$ is satisfied for the most energetic quanta, $\hbar\omega_{max}\approx m_e c^2\gamma$. Let us give an estimate of the parameter s for such frequencies. With the aid of Equation (9.87) we have

$$s(\omega_{max})\approx\frac{1}{\sqrt{\gamma}}\sqrt{\frac{m_e c^3}{\hbar\omega_{0e}^2 r_0}}\frac{1}{\sqrt{32Z\ln 183Z^{-1/3}}}. \tag{9.89}$$

The condition $s(\omega_{max})\ll 1$ is satisfied at $\gamma\gg 4\times 10^6$ for lead, and at $\gamma\gg 10^{10}$ for air. The first value is accessible to modern accelerators, and the second is realized in broad atmospheric showers. An analogous suppression effect appears in the formation of electron–positron pairs by energetic quanta in the field of a nucleus, where it is caused by the multiple scattering of the particles produced.

Now consider the range of low frequencies ($\omega\ll\omega_{0e}\gamma$). With the decrease of frequency in this region, the parameter s increases as $\omega^{-3/2}$, so that we consider the case $s > 1$, but $\omega - kv \approx \omega_{0e}^2/2\omega$. Substituting $\Phi(s) = 1$ in Equation (9.85), we obtain the spectral power emitted by an electron in this range:

$$\frac{d^2\mathcal{E}_\omega}{dz\,d(\hbar\omega)} \approx -\frac{2\alpha q\omega^2}{3\pi\omega_{0e}^2}. \tag{9.90}$$

As compared to the Bethe–Heitler spectrum, here the radiation is strongly suppressed:

$$\frac{d^2\mathcal{E}_\omega}{dz\,d(\hbar\omega)} = \frac{\omega^2}{(\gamma\omega_{0e})^2}\left(\frac{d^2\mathcal{E}_\omega}{dz\,d(\hbar\omega)}\right)_{BH}. \tag{9.91}$$

The effect of medium polarization on bremsstrahlung was discovered by Ter-Mikaelyan (1954) (see his monograph (1972) and review (2003)).

Note, that Equation (9.90) reflects the processes, which accompany the interaction of a relativistic particle with medium in the high frequency range ($\hbar\omega \gtrsim m_e c^2 \gg \hbar\omega_p$) not fully adequately. The permittivity $\varepsilon(\omega) = 1 - \omega_p^2/\omega^2$, which was used in derivation of the dispersion relation (9.60) and of the spectral power (9.90), takes into account the interaction of quanta (or electromagnetic waves) with the electron system only in the low-order approximation with respect to the coupling constant $\alpha = e^2/\hbar c$. In the next orders, there appear real and imaginary corrections to the permittivity. The real corrections will always enter as a sum with ω_p^2/ω^2, and they can be neglected due to their smallness, while the imaginary corrections can give the dominant contribution in some spectral ranges. The imaginary corrections to the permittivity, according to the results of Section 5.3 (see Equation (5.92)), describe the absorption of quanta of corresponding frequency by the medium. The main processes responsible for the absorption of high-energy photons are the Compton scattering (in this case there appear the quanta of lower energy) and the production of electron–positron pairs by photons in the fields of atomic nuclei.

Example 9.6

Express the imaginary part $\varepsilon''(\omega)$ of the permittivity of the medium at high frequencies ($\hbar\omega \gtrsim m_e c^2 \gg \hbar\omega_p$) in terms of the total cross sections $\sigma_C(\omega)$ and $\sigma_p(\omega)$ of the Compton scattering of quanta by electrons of the medium, and of the formation of electron–positron pairs by quanta in the fields of atomic nuclei.

Solution. The processes mentioned above can be taken into account in terms of the permittivity in the following way. Consider the flux of photons of frequency ω with number density of quanta $N(\omega)$. The energy absorbed per unit time per unit volume of the medium is given by $(n_a\sigma_p + n_a Z\sigma_C)c\hbar\omega N(\omega)$. On the other hand, the same quantity can be written in terms of the imaginary part of the permittivity using Equation (5.92):

$$(n_a\sigma_p + n_a Z\sigma_C)c\hbar\omega N(\omega) = \frac{\omega\varepsilon''(\omega)}{4\pi}\overline{E^2}. \tag{1}$$

However, the quantity $\hbar\omega N(\omega)$ in the left-hand side of the equation just gives the electromagnetic energy density:

(2) $$\hbar\omega N(\omega) = \frac{1}{8\pi}(\overline{E^2} + \overline{B^2}) = \frac{1}{4\pi}\overline{E^2}$$

(in the range of high frequencies the real part $\varepsilon'(\omega)$ is close to unity, and the energy density is described by the expression for vacuum). From the comparison of two last expressions, we find

(3) $$\varepsilon''(\omega) = \frac{c(n_a\sigma_p + n_a Z\sigma_C)}{\omega} = \frac{\Gamma_p + \Gamma_C}{\omega},$$

where $\Gamma_p = cn_a\sigma_p$ and $\Gamma_C = cn_a Z\sigma_C$ are the probabilities of absorption of the quantum per unit time due to the formation of the pair by this quantum or due to Compton scattering. As a result, the permittivity takes the form

(4) $$\varepsilon(\omega) = 1 - \frac{\omega_p^2}{\omega^2} + i\frac{\Gamma_p + \Gamma_C}{\omega}.$$

Here the small real contributions from the processes considered above are ignored. The total cross section $\sigma_C(\omega)$ of the Compton scattering was found in Problem I.6.85. In the ultrarelativistic case, we have

(5) $$\sigma_C(\omega) = \pi r_0^2 \frac{m_e c^2}{\hbar\omega} \ln\left(\frac{2\hbar\omega}{m_e c^2}\right).$$

The cross section of the e^+e^--pair formation by the quantum with the energy $\hbar\omega$ in the field of a nucleus is taken from the manuals (Heitler (1954); Berestetskii *et al.* (1982)):

(6)
$$\sigma_p(\omega) = \frac{28}{9}Z^2\alpha r_0^2\left(\ln\frac{2\hbar\omega}{m_e c^2} - \frac{109}{42}\right), \quad \sigma_p(\omega) = \frac{28}{9}Z^2\alpha r_0^2\left(\ln 183Z^{-1/3} - \frac{1}{42}\right).$$

The first expression can be used in the absence of Coulomb screening, and the second in the case of a total screening. However, in both cases the energies of both particles must exceed their rest energy. The screening conditions are determined by Equations (9.42) and (9.44), in which $\mathcal{E}$, $\mathcal{E}'$ are the energies of the electron and positron. These cross sections are not equal zero only at $\hbar\omega > 2m_e c^2$. In the range of moderate relativism, the cross sections of the Compton scattering and of the production of pairs may be comparable in magnitude, while at $\hbar\omega < 2m_e c^2$ the production of pairs is impossible. At $\hbar\omega \gg 137 m_e c^2/Z$ the effect due to pair production prevails over the Compton scattering. ■

Electrons of medium in radiation processes. We have considered above the polarization effects: the coherence effect (changing of the dispersion law of a quantum) and the incoherent one (Compton scattering). However, electrons of medium take a more active part in the radiation of ultrarelativistic particles moving through the matter. Firstly, the relativistic electron can emit a bremsstrahlung quantum, when scattered by a medium electron. In this sense, medium electrons are similar to nuclei with $Z = 1$ (but not to heavy ones). Since each nucleus possesses Z electrons, the contribution of the electrons to bremsstrahlung can approximately be taken into account by replacing the quantity Z^2 by $Z^2 + Z = Z(Z + 1)$ in all formulas obtained above. In particular, such a replacement must be performed in Equation (9.59) for a cascade length. However, the most significant effect is associated with the radiation of quanta by medium electrons excited by a field of a relativistic particle flying by. In the case of Compton scattering, having absorbed the primary quantum, the medium electrons emit photons of lower energies. This radiation is of particular importance in those ranges of the spectrum, where the radiation of the particle itself is suppressed by polarization of the medium (or by a large mass of the particle, see Problems 9.11 and 9.12).

When bremsstrahlung of quanta of frequencies of the order of transition frequencies in atoms occurs, an electron shell participates in this process as a dynamic system capable to be excited. The radiation probability depends on the dynamic polarizability of the atom (i.e., on its polarizability at the radiation frequency). Such processes are referred to as **polarization** (or dynamic) bremsstrahlung. This scope of questions will be touched on in Section 9.2 (see Examples 9.10 and 9.11).

Recommended literature: Rossi (1952); Landau *et al.* (1984); Tamm and Frank (1937); Ginzburg (1987); Ter-Mikhaelyan (1972); Akhieser and Shul'ga (1993); Fermi (1951); Berestetskii *et al.* (1982); Fermi (1940, 1949); Amusya (1990); Basylev and Zhewago (1987)

Problems

9.1 Calculate the differential cross section for the scattering of a relativistic Dirac particle by a Coulomb field of the nucleus, assuming that the nucleus is at rest. Use the first Born approximation. The particles are not spin polarized.

9.2 Calculate the differential cross section for the elastic scattering of a relativistic Dirac particle screened by the Coulomb field of the nucleus $U(r) = \pm(Ze^2/r)\exp(-r/R)$. The screening radius, in the Thomas–Fermi approximation, is $R = a_B Z^{-1/3}$, where $a_B = \hbar^2/m_e e^2$ is the Bohr radius. Use the first Born approximation. The spin polarization of the particles is absent in the original state.

9.3* Calculate the differential cross section for the elastic collision of an unpolarized muon with an electron. Write down the differential cross section as a function of the energy transmitted to the electron in the collision, using the reference frame in which the electron was at rest before the collision.

9.4* Solve Problem 9.3 for the electron–electron collision (Möller scattering).

9.5 Based on the result of Problem 9.4, calculate the losses of a relativistic electron per unit path length in close collisions, when the electrons can be assumed to be free.

9.6 Calculate the intensity dI of the Cherenkov radiation in the particular range $d\omega$ of frequencies in an isotropic dielectric, and also the interval of angles $d\theta$, within which this radiation propagates.

9.7• The collimated monoenergetic stationary beam of fast particles, which has a large cross section, is directed on the boundary of the scattering medium. Neglecting the energy losses and using the Fokker–Planck equation (9.34), calculate the dependence of the distribution function on z and angle ϑ. Find the average scattering angle as a function of z. Evaluate the characteristic length l, at which the isotropization of the beam occurs. Compare it with the characteristic length L of deceleration of particles, which is described by Equation (9.18).

9.8* The collimated monoenergetic stationary beam of fast particles, which has a small ("point-like") cross section, is incident to the boundary of the scattering medium. Find the distribution function of the beam. Evaluate the average square of the beam radius as a function of the penetration depth.

9.9** Using the perturbation theory in the first nonvanishing approximation, in a manner similar to that in Example I.6.19, where the Compton effect was considered, evaluate the cross section of bremsstrahlung of an electron with arbitrary energy at the nonscreened atomic nucleus of charge Ze.

9.10 Show that at γ_0 and $\gamma \gg 1$, the general formula for bremsstrahlung of an electron of arbitrary energy, obtained in Problem 9.9, reduces to the relativistic formula (9.37).

9.11* Find the spectral density of the energy lost by a relativistic particle per unit path length, using the permittivity found in Example 9.6. Use the Fermi method, that is, calculate the flux of the Poynting vector through the cylindrical surface around the trajectory of the particle. Interpret the result in terms of the equivalent photons.

9.12* A relativistic particle moves in the medium with an approximately constant velocity. Calculate the spectral density of the losses per unit path length, due to bremsstrahlung, produced by the particle itself and by electrons of the medium. Compare these two values for the heavy and light particles.

9.2
Macroscopic Mechanisms of Radiation of Fast Particles in Media

In vacuum, in the absence of macroscopic bodies, the electromagnetic waves are emitted only during the accelerated motion of charged particles. In the presence of the matter, the eigenoscillations can be generated in the form of transverse or more complicated waves by particles moving without an acceleration. The examples of such radiation were already considered in Section 9.1 (generation of plasma oscillations, Cherenkov radiation). In this section, we consider in more details the main types of the emission of electromagnetic perturbations by fast particles in media. We do not pretend to give the most general presentation, but consider the typical examples and the solutions of the problems on this subject. The more general statement of a problem about radiation of the given current in substance and its decision can be found in work by Toptygin and Fleishman (2008).

Example 9.7

For the Cherenkov radiation in a transparent dielectric, determine the length of the formation (coherence length) of the radiation, which is similar to the coherence length of bremsstrahlung considered in Section 9.1. Evaluate the threshold energy $\mathcal{E}_0$, beginning from which the Cherenkov radiation becomes possible for a particle moving in a dielectric. Consider the specific case of $\varepsilon(\omega) - 1 \ll 1$.

Solution. The length of radiation formation, l_c, can be defined as the distance at which the particle and the wave emitted by it diverge by a wavelength λ, moving with different velocities:

(1) $$\frac{l_c}{v}\left(v - \frac{c}{\sqrt{\varepsilon(\omega)}}\right) = \lambda.$$

Then we find

(2) $$l_c = \frac{\lambda\beta n}{\beta n - 1}, \qquad \text{where} \qquad n(\omega) = \sqrt{\varepsilon(\omega)}$$

is the refraction index;

(3) $$\mathcal{E}_0 = \frac{mc^2 n}{\sqrt{n^2 - 1}}.$$

At $\varepsilon(\omega) - 1 = 4\pi\alpha(\omega) \ll 1$ we have

(4) $$l_c = \frac{2\lambda}{4\pi\alpha(\omega) - \gamma^{-2}}, \qquad \mathcal{E}_0 = \frac{mc^2}{\sqrt{4\pi\alpha(\omega)}}. \qquad \blacksquare$$

Example 9.8

(Transient radiation). *The (x, y)-plane separates two nonmagnetic media of permittivities ε_1 and ε_2 (Figure 9.2). The particle of a charge e and velocity $v = const$ moves along the positive direction of Oz-axis. Calculate the Fourier harmonics of the electromagnetic field generated by the particle.*

Solution. Expand the Maxwell equations

(1) $$\nabla \times \boldsymbol{\mathcal{E}} = -\frac{1}{c}\frac{\partial \boldsymbol{B}}{\partial t}, \qquad \nabla \times \boldsymbol{B} = \frac{1}{c}\frac{\partial \boldsymbol{D}}{\partial t} + \frac{4\pi e}{c}\boldsymbol{v}\delta(\boldsymbol{r} - \boldsymbol{v}t)$$

into Fourier integral over variables x, y, and t, denoting the wave vector in the plane (x, y) by $\boldsymbol{\kappa}$. Excluding the vector $\boldsymbol{B}$ and projecting the equation obtained on the directions $\boldsymbol{e}_z$ and $\boldsymbol{\kappa}$, we get two equations for the transverse and longitudinal components of the vector $\boldsymbol{E}$:

$$\left(\kappa^2 - \frac{\omega^2\varepsilon}{c^2}\right)E_z + i\frac{\partial}{\partial z}\boldsymbol{\kappa}\cdot\boldsymbol{E} = i\frac{4\pi e\omega}{c^2}e^{i\omega z/v},$$

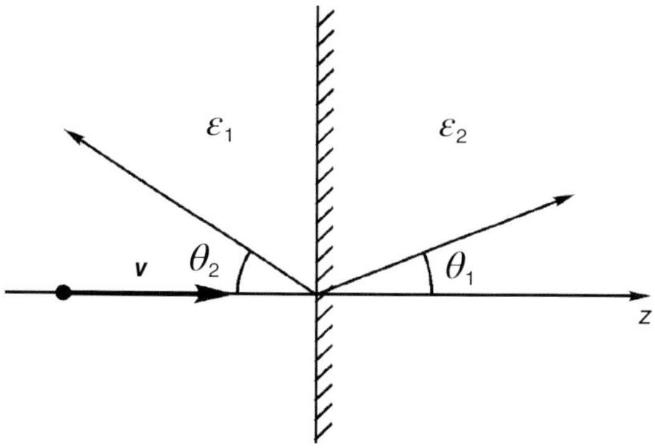

Figure 9.2 Radiation at crossing by a fast particle of sharp border of two dielectrics.

(2) $$\left(\frac{\partial^2}{\partial z^2}+\frac{\omega^2\varepsilon}{c^2}\right)\boldsymbol{\kappa}\cdot\boldsymbol{E}=i\kappa^2\frac{\partial E_z}{\partial z},\qquad z\neq 0,$$

where the Fourier amplitudes of the electric and magnetic fields are denoted by $\boldsymbol{E}$ and $\boldsymbol{B}$. From these equations, we obtain the nonuniform equation for the longitudinal component:

(3) $$\frac{\partial^2 E_z}{\partial z^2}+\left(\frac{\omega^2\varepsilon}{c^2}-\kappa^2\right)E_z=-i\frac{4\pi e\omega}{c^2}\left(1-\frac{c^2}{\varepsilon v^2}\right)e^{i\omega z/v}.$$

The solution of this equation is the sum of the particular solution

(4) $$E_z^{(q)}=-i\frac{4\pi e\omega}{c^2}\left(1-\frac{c^2}{\varepsilon\, v^2}\right)\left(\frac{\omega^2\varepsilon}{c^2}-\kappa^2-\frac{\omega^2}{v^2}\right)^{-1}e^{i\omega z/v},$$

which corresponds to the preset right-hand side, and of the general solution of the uniform equation

(5) $$\frac{\partial^2 E_z}{\partial z^2}+\left(\frac{\omega^2\varepsilon}{c^2}-\kappa^2\right)E_z=0.$$

The particular solution describes the electromagnetic field propagating in a space with a velocity of the particle v. It includes both the quasi-stationary field, produced by the particle, and the field of the Cherenkov radiation, provided the condition $v>c/n(\omega)$ is satisfied (see the solution of Problem 9.14). The field transverse component $\boldsymbol{E}^{(q)}$ can be determined from the second Equation (2):

(6) $$\boldsymbol{\kappa}\cdot\boldsymbol{E}^{(q)}=i\frac{4\pi e\kappa^2}{\varepsilon v}\left(\frac{\omega^2\varepsilon}{c^2}-\kappa^2-\frac{\omega^2}{v^2}\right)^{-1}e^{i\omega z/v}.$$

The second transverse component is absent.

The general solution of the uniform Equation (5) has the form

(7) $$E_z^{(rad)}=Ae^{iqz}+Ce^{-iqz},\qquad q=\sqrt{\frac{\omega^2\varepsilon}{c^2}-\kappa^2},\qquad \Im q>0,$$

where A and C are constants. They must be found from the boundary conditions and the condition restricting the solution. With the given value of q, the restriction at large $|z|$ requires that $A = 0,\ z < 0$ and $C = 0,\ z > 0$:

$$E_{1z}^{(rad)}|_{z<0} = Ce^{-iq_1 z},\ q_1 = \sqrt{\frac{\omega^2\varepsilon_1}{c^2} - \kappa^2} \text{ at } z < 0;$$

(8)
$$E_{2z}^{(rad)}|_{z>0} = Ae^{iq_2 z},\ q_2 = \sqrt{\frac{\omega^2\varepsilon_2}{c^2} - \kappa^2} \text{ at } z > 0.$$

The upper index (rad) shows that this solution can describe the radiation field. For example, in a transparent dielectric the field $E_z^{(rad)}$ has the dispersion relation

(9)
$$k_{1,2}^2 = \kappa^2 + q_{1,2}^2 = \frac{\omega^2\varepsilon_{1,2}}{c^2},$$

which is inherent in the propagating plane monochromatic waves at $\varepsilon_{1,2} > 0$ and $q_{1,2} > 0$. Below we will see that this radiation is due to the sharp change of the permittivity from the value ε_1 up to ε_2. It is referred to as **transient**. The solutions obtained satisfy the condition $\nabla \cdot \boldsymbol{E}^{(rad)} = 0$, which gives the relations

(10)
$$\boldsymbol{\kappa} \cdot \boldsymbol{E}^{(rad)} = q_1 E_z^{(rad)} \quad \text{at} \quad z < 0; \qquad \boldsymbol{\kappa} \cdot \boldsymbol{E}^{(rad)} = -q_2 E_z^{(rad)} \quad \text{at} \quad z > 0.$$

These relations determine the transverse components of the field $\boldsymbol{E}^{(rad)}$ in terms of its longitudinal component.

The conditions at the separation boundary express the continuity of the normal component of the electric induction and of the tangential component of the total electric field:

$$\varepsilon_1(E_{1z}^{(q)} + E_{1z}^{(rad)})|_{z=-0} = \varepsilon_2(E_{2z}^{(q)} + E_{2z}^{(rad)})|_{z=+0},$$

(11)
$$(\boldsymbol{\kappa} \cdot \boldsymbol{E}_1^{(q)} + q_1 E_{1z}^{(rad)})_{z=-0} = (\boldsymbol{\kappa} \cdot \boldsymbol{E}_2^{(q)} - q_2 E_{2z}^{(rad)})_{z=+0}.$$

Note, that the field components $\boldsymbol{E}^{(rad)}$ along the third direction $\boldsymbol{e}_z \times \boldsymbol{\kappa}$ are not connected with the field of the particle by any relations and cannot be produced by the particle. Hence, they are equal to zero in this problem. The constants C and A can be calculated from two algebraic equations (11):

(12)
$$C = i\frac{4\pi e\kappa^2(1 - \varepsilon_2/\varepsilon_1)}{v(\varepsilon_1 q_2 + \varepsilon_2 q_1)(q_1^2 - \omega^2/v^2)}\left\{1 - \frac{\varepsilon_1\beta\omega}{c(q_2 + \omega/v)}\right\},$$

(13)
$$A = -i\frac{4\pi e\kappa^2(1 - \varepsilon_1/\varepsilon_2)}{v(\varepsilon_1 q_2 + \varepsilon_2 q_1)(q_2^2 - \omega^2/v^2)}\left\{1 + \frac{\varepsilon_2\beta\omega}{c(q_1 - \omega/v)}\right\}.$$

They vanish unless the permittivity experience a step-like jump.

Thus, the electric field in the first and second media is determined by the above quantities $\boldsymbol{E}_{1,2} = \boldsymbol{E}_{1,2}^{(q)} + \boldsymbol{E}_{1,2}^{(rad)}$, where $\boldsymbol{E}_{1,2}^{(q)}$ and $\boldsymbol{E}_{1,2}^{(rad)}$ are given by Equations (4), (6), (8), (12), and (13). The magnetic field can be determined from the first of the

Maxwell equations (1). In particular, the part of the magnetic field that is due to the sharp separation boundary, has the form

(14) $$\boldsymbol{B}_1^{(rad)} = \frac{\varepsilon_1\omega}{c\kappa^2}[\boldsymbol{\kappa}\times\boldsymbol{e}_z]Ce^{-iq_1z} \quad \text{at } z<0;$$

$$\boldsymbol{B}_2^{(rad)} = \frac{\varepsilon_2\omega}{c\kappa^2}[\boldsymbol{\kappa}\times\boldsymbol{e}_z]Ae^{iq_2z} \quad \text{at } z>0. \qquad \blacksquare$$

Example 9.9

Assume that in Problem 9.8 both media are transparent dielectrics for the frequencies under consideration. Evaluate the spectrum and the angular distribution of the transient radiation of a particle into the front and rear semi-spheres during the travel time of the particle through the boundary.

Solution. Calculate the total energy Q of the transient radiation through the (x, y)-plane in the regions, $z\to-\infty$ and $z\to\infty$, respectively:

(1) $$Q_1 = \frac{c}{4\pi}\int_{-\infty}^{\infty}dx\int_{-\infty}^{\infty}dy\int_{-\infty}^{\infty}d\omega[\boldsymbol{\mathcal{E}}_1\times\boldsymbol{B}_1]\cdot\boldsymbol{e}_z \quad \text{at } z\to-\infty;$$

(2) $$Q_2 = \frac{c}{4\pi}\int_{-\infty}^{\infty}dx\int_{-\infty}^{\infty}dy\int_{-\infty}^{\infty}d\omega[\boldsymbol{\mathcal{E}}_2\times\boldsymbol{B}_2]\cdot\boldsymbol{e}_z \quad \text{at } z\to+\infty.$$

Here and further on we omit the index (rad). Substitute in Equation (2) the expansion

(3) $$\boldsymbol{\mathcal{E}}_2 = \int \boldsymbol{E}_2(z,\boldsymbol{\kappa},q_2,\omega)\exp(i\boldsymbol{\kappa}\cdot\boldsymbol{r}_\perp - i\omega t)\frac{d^2\kappa}{(2\pi)^2}\frac{d\omega}{2\pi},$$

and the analogous expansion for $\boldsymbol{B}_2$, where q_2 is the longitudinal component of the wave vector. By using the relation

(4) $$\boldsymbol{E}(z,\boldsymbol{\kappa},q,\omega) = \boldsymbol{E}^*(z,-\boldsymbol{\kappa},-q,-\omega),$$

we obtain the radiation energy in the form of the integral over the frequency and the transverse wave vector:

(5) $$Q_2 = \frac{c}{(2\pi)^4}\Re\int_0^{\infty}d\omega\int d^2\kappa[\boldsymbol{E}_2\times\boldsymbol{B}_2^*]\cdot\boldsymbol{e}_z.$$

Now the z-coordinate does not enter in this expression.

With the aid of the formulas from the preceding example, we find

$$[\boldsymbol{E}_2\times\boldsymbol{B}_2^*]\cdot\boldsymbol{e}_z = \frac{q_2}{\kappa}E_{2z}B_2^* = \frac{\varepsilon_2\omega q_2}{c\kappa^2}|A|^2, \qquad \text{where}$$

(6) $$E_{2z} = Ae^{iq_2z}, \qquad B_2^* = \frac{\varepsilon_2\omega}{c\kappa}A^*e^{-iq_2z}.$$

We express all quantities in the integrand (5) in terms of the frequency and the angle θ_2 between the particle velocity $\boldsymbol{v}$ and the wave vector $\boldsymbol{k}_2 = \boldsymbol{\kappa} + q_2\boldsymbol{e}_z$ in medium 2 (see Figure 9.2). We have

(7) $$k_2 = \frac{\omega}{c}\sqrt{\varepsilon_2}, \quad \kappa^2 = k_2^2 - q_2^2 = k_1^2 - q_1^2,$$

$$q_2 = k_2 \cos\theta_2, \quad d^2\kappa = k_2^2 \cos\theta_2 d\Omega_2, \quad \theta_2 \le \frac{\pi}{2}.$$

From these formulas, we also obtain

(8)

$$q_1 = \frac{\omega}{c}\sqrt{\varepsilon_1 - \varepsilon_2 \sin^2\theta_2}, \qquad \varepsilon_1 q_2 + \varepsilon_2 q_1 = k_2[\varepsilon_1 \cos\theta_2 + \sqrt{\varepsilon_1\varepsilon_2 - \varepsilon_2^2 \sin^2\theta_2}\,].$$

By using these results, we take out of the integrand (5) the spectral-angular distribution of the transient radiation forwards:

$$\frac{d^2 I_\omega^{(2)}}{d\omega d\Omega_2} = \frac{e^2\beta^2\sqrt{\varepsilon_2}(\varepsilon_2-\varepsilon_1)^2 \sin^2\theta_2 \cos^2\theta_2 |\beta^2\varepsilon_2 + \beta\sqrt{\varepsilon_1 - \varepsilon_2\sin^2\theta_2} - 1|^2}{\pi^2 c\left(\beta^2\varepsilon_2 \cos^2\theta_2 - 1\right)^2 |\left(\varepsilon_1\cos\theta_2 + \sqrt{\varepsilon_1\varepsilon_2 - \varepsilon_2^2\sin^2\theta_2}\right)\left(\beta\sqrt{\varepsilon_1 - \varepsilon_2\sin^2\theta_2} - 1\right)|^2},$$

$$\cos\theta_2 \ne 1/\beta\sqrt{\varepsilon_2}. \tag{9.92}$$

Here the modulus sign is used since the radicals can be imaginary even for a transparent media. This is because certain waves may experience the total internal reflection at the boundary. The radiation backwards can be calculated in a similar way:

$$\frac{d^2 I_\omega^{(1)}}{d\omega d\Omega_1} = \frac{e^2\beta^2\sqrt{\varepsilon_1}\left(\varepsilon_2-\varepsilon_1\right)^2 \sin^2\theta_1 \cos^2\theta_1 \left|\beta^2\varepsilon_1 - \beta\sqrt{\varepsilon_2 - \varepsilon_1\sin^2\theta_1} - 1\right|^2}{\pi^2 c\left(\beta^2\varepsilon_1 \cos^2\theta_1 - 1\right)^2 \left|\left(\varepsilon_2\cos\theta_1 + \sqrt{\varepsilon_1\varepsilon_2 - \varepsilon_1^2\sin^2\theta_1}\right)\left(\beta\sqrt{\varepsilon_2 - \varepsilon_1\sin^2\theta_1} + 1\right)\right|^2},$$

$$\cos\theta_1 \ne 1/\beta\sqrt{\varepsilon_1}. \tag{9.93}$$

The result is expressed in terms of the angle $\theta_1 \le \pi/2$ between the vector $-\boldsymbol{v}$ and direction of the wave vector $\boldsymbol{k}_1$ in the first medium (Figure 9.2). The particular properties of Equation (9.92) at $\cos\theta_2 = 1/\beta\sqrt{\varepsilon_2}$, and of Equation (9.93) at $\cos\theta_1 = 1/\beta\sqrt{\varepsilon_1}$ are associated with the Cherenkov effect. The radiation in these directions can be determined by going to the limit in a manner similar to that in Section 9.1 for the Cherenkov radiation. ■

The transient radiation was theoretically predicted by Ginzburg and Frank (1946). At present it is a well-developed field of electrodynamics with numerous applications (see Ginzburg and Tsytovich (1990); Platonov and Fleishman (2002)). See also the work by Bolotovskii and Serov (2009) in which transient radiation is calculated without Fourier decomposition.

Transient radiation of surface waves. According to the results of Problem 8.1, the dispersion relation for surface waves at $\mu_1 = \mu_2 = 1$ has the form

$$\kappa^2 = \frac{\omega^2}{c^2}\frac{\varepsilon_1\varepsilon_2}{\varepsilon_1+\varepsilon_2} > 0, \qquad \varepsilon_1\varepsilon_2 < 0, \qquad \varepsilon_1 + \varepsilon_2 < 0. \tag{9.94}$$

The quantities q_1 and q_2, which determine the field $\boldsymbol{E}^{(rad)}$, found in Example 9.8, are given by

$$q_1^2 = \frac{\omega^2}{c^2}\frac{\varepsilon_1^2}{\varepsilon_1+\varepsilon_2} < 0, \qquad q_2^2 = \frac{\omega^2}{c^2}\frac{\varepsilon_2^2}{\varepsilon_1+\varepsilon_2} < 0. \tag{9.95}$$

Hence, it follows that the field of the transient radiation is concentrated in the near-surface region:

$$E_{1z}^{(rad)} = Ce^{|q_1|z} \quad \text{at } z<0; \qquad E_{1z}^{(rad)} = Ae^{-|q_2|z} \quad \text{at } z>0, \tag{9.96}$$

that is, the generation of surface electromagnetic waves by a fast particle takes place. Calculation of the energy emitted in the form of surface waves is not trivial (see Ginzburg and Tsytovich (1990)). The spectral density of the emitted energy has the form

$$\frac{dI_\omega}{d\omega} = \frac{2e^2\beta^2|\varepsilon_2-\varepsilon_1||\varepsilon_1\varepsilon_2|(1+\beta^2|\varepsilon_1+\varepsilon_2|)}{c\sqrt{|\varepsilon_1+\varepsilon_2|}[|\varepsilon_1+\varepsilon_2|+\beta^2\varepsilon_1^2][|\varepsilon_1+\varepsilon_2|+\beta^2\varepsilon_2^2]}. \tag{9.97}$$

Polarization bremsstrahlung.

Example 9.10

A charged particle moves along the straight line in a rarefied atomic gas with a velocity significantly exceeding velocities of atomic electrons. The dynamic polarizability $\alpha(\omega)$ of an individual atom in the uniform alternating field is known. Calculate the spectral distribution of the radiation of electron shells of atoms at frequencies of the order of atomic transition frequencies ω_0, which is produced by the flying-by particle per unit path length. On emission, the atom remains in the ground state.

Solution. The motion of the particle is quasi-classic and weakly disturbed by interaction with atoms under the formulated conditions (emission of soft optical quanta). Hence, one can consider a motion of the particle along a straight line with a uniform velocity. The radiation occurs due to polarization of atomic shells by a field of the traveling particle. At distances $\rho \gg R$, the field near each atom is almost uniform, and it is sufficient to take into account the dipole polarization only. In order to calculate the electric dipole moment of the atom at frequency ω, we use the formula

$$\boldsymbol{p}_\omega = \alpha(\omega)\boldsymbol{E}_\omega, \tag{1}$$

where $\boldsymbol{E}_\omega$ is the Fourier component of the field near the atom.

The energy $\mathcal{E}_\omega^{(at)}$ emitted by nonrelativistic atomic electrons can be calculated by the formula for the electric dipole radiation (5.31):

$$I(t) = \frac{2\ddot{\boldsymbol{p}}^2(t)}{3c^3}, \tag{2}$$

where t is the time which includes retardation. The total energy, emitted by the atomic shell during the flight time of the particle, is given by

$$\mathcal{E}^{(at)} = \int_{-\infty}^{\infty} I(t)dt = \frac{2}{3\pi c^3}\int_0^{\infty} |\ddot{\boldsymbol{p}}_\omega|^2 d\omega = \int_0^{\infty} \mathcal{E}_\omega^{(at)} d\omega, \tag{3}$$

and, hence,

$$\mathcal{E}_\omega^{(at)} = \frac{2}{3\pi c^3}|\ddot{\boldsymbol{p}}_\omega|^2 = \frac{2r_0^2 c}{3\pi}\left|\frac{m\omega^2}{e^2}\alpha(\omega)\right|^2 |\boldsymbol{E}_\omega(\rho)|^2. \tag{4}$$

Here we have used Equation (1), r_0 is the classic radius of an electron.

In order to find the radiation $d\mathcal{E}_\omega/dz$ per unit path length, we must multiply the radiation spectrum of an individual atom (Equation (4)) by the number density of the atoms n_a and integrate over the target parameters:

$$\frac{d\mathcal{E}_\omega}{dz} = \frac{2r_0^2 n_a c}{3\pi}\left|\frac{m\omega^2}{e^2}\alpha(\omega)\right|^2 \int_{\rho_{min}}^{\rho_{max}} |\boldsymbol{E}_\omega(\rho)|^2 2\pi\rho d\rho. \tag{5}$$

The lower limit of the integration must be of the order of several atomic radii R, that is, $\rho_{min} \approx R$, since for smaller target parameters Equation (1) does not hold. The upper limit, in view of the rapid convergence of the integral (see below), can be assumed to be infinite: $\rho_{max} \to \infty$.

The Fourier components of the electromagnetic field of a relativistic particle in vacuum were evaluated in Example I.6.20 (Equations (5) and (6)):

$$\boldsymbol{E}_\omega = \frac{2q\omega}{\gamma v^2}K_1\left(\frac{\omega\rho}{\gamma v}\right)\frac{\boldsymbol{\rho}}{\rho} - i\frac{2q\omega}{\gamma^2 v^2}K_0\left(\frac{\omega\rho}{\gamma v}\right)\frac{\boldsymbol{v}}{v}, \qquad \omega > 0. \tag{6}$$

Upon substitution in Equation (5) and integration over the target parameter, we have

$$\frac{d\mathcal{E}_\omega}{dz} = \frac{8q^2 r_0^2 n_a c u_m^2}{3v^2}\left|\frac{m\omega^2}{e^2}\alpha(\omega)\right|^2\left\{K_0(u_m)K_2(u_m) - \left(1-\frac{1}{\gamma^2}\right)K_1^2(u_m) - \frac{1}{\gamma^2}K_0^2(u_m)\right\},$$

$$u_m = \frac{\omega R}{\gamma v}. \tag{7}$$

Here

$$u_m = \frac{\omega R}{\gamma v} \approx \frac{Rc}{\gamma\lambda v} \ll 1. \tag{8}$$

In the optical range, the condition of smallness is satisfied not only for the relativistic particles, but for the fast ($v \gg v_{at}$), nonrelativistic particles as well, since $R/\lambda \approx 10^{-3}$ at such frequencies. By using Equation (I.1.168), we obtain, with a logarithmic accuracy, the final expression:

$$\frac{d\mathcal{E}_\omega}{dz} = \frac{16q^2 r_0^2 n_a c}{3v^2}\left|\frac{m\omega^2}{e^2}\alpha(\omega)\right|^2 \ln\left(\frac{\gamma v}{\omega R}\right). \tag{9}$$

For relativistic particles, $\gamma \gg 1$, the last formula is valid in the X-ray range where $\lambda < R$. At frequencies $\omega \gg \omega_0$ the atomic electrons behave as free electrons (see

Problem 5.7), and the atomic polarizability takes the form

$$(10) \qquad \alpha(\omega) = \frac{Ze^2}{m\omega^2}, \qquad \omega \gg \omega_0.$$

In this range, the radiation of the relativistic particle per unit path length is given by

$$(11) \qquad \frac{d\mathcal{E}_\omega}{dz} = \frac{16Z^2q^2r_0^2n_a}{3c}\ln\left(\frac{\gamma c}{\omega R}\right). \qquad \blacksquare$$

The radiation considered above is referred to as **polarization (or dynamic)** bremsstrahlung (Amusya *et al.* (1987); Amusya (1990); see recent studies in Korol' *et al.* (2004)). The term "bremsstrahlung" is not exactly adequate in this name, since the radiation is possible without any deceleration of a fast particle when it moves along a straight line with a constant velocity. There are more grounds to call it the transient (in a broad sense) radiation or scattering, since in the given case the radiation is caused by the scattering of the fast-particle field by the microscopic inhomogeneities, that is, by the electron shells of the atoms. The extended presentation of the theory of bremsstrahlung can be found in the monograph (Ginzburg and Tsytovich (1990)). Since this radiation is emitted by a nonrelativistic system, it does not have the sharp forward directivity, even though the radiation can be emitted by a relativistic particle. The anisotropy of directivity is not large like in any electric dipole radiation; however, the intensity can be significant in view of the resonance dependence of the atomic polarizability on a frequency (see Problems 5.13–5.15).

The length $l_\omega = \gamma v/\omega$ under the logarithm in Equation (9) has a meaning of the target parameter leading to the emission at a given frequency. If this length is larger than an interatomic distance of the medium, then in the calculation of the polarization bremsstrahlung, it is necessary to take into account the influence of neighboring atoms, that is, of the permittivity of the medium, on the field of the fast particle and also on the propagation of the emitted photons.

Example 9.11

Consider bremsstrahlung of a fast particle on atoms, taking into account the effect of medium on the electromagnetic field. Analyze, in particular, the effect under conditions when $\varepsilon(\omega) > 1$ and when the Cherenkov radiation takes place. Obtain for this case the radiation spectrum of the particle per unit path length and give a physical interpretation of the result obtained.

Solution. The general approach remains the same as in the preceding example. However, at all stages one should take into account the presence of permittivitiy $\varepsilon(\omega) = \varepsilon'(\omega) + i\varepsilon''(\omega)$ and permeability $\mu = 1$ (the imaginary part ε'' is assumed to be small, except for narrow resonance peaks). As a result, Equation (4) of Example 9.10, which determines the polarization radiation of an atom when a fast particle travels by at the impact parameter ρ, takes the form

$$(1) \qquad \mathcal{E}_\omega^{(at)} = \frac{2\omega^4\sqrt{\varepsilon'}}{3\pi c^3}|\boldsymbol{p}_\omega|^2 = \frac{2r_0^2c\sqrt{\varepsilon'}}{3\pi}\left|\frac{m\omega^2}{e^2}\alpha(\omega)\right|^2 |\boldsymbol{E}_\omega(\boldsymbol{\rho})|^2,$$

where the last factor contains the electric field, whose Fourier harmonics were evaluated in Example 9.1. In order to use them, the quantity $d\mathcal{E}_\omega/dz$ should be expressed by the integral over the wave vector $k_\perp$ (which determines momentum transfer to the medium), instead of the impact parameter:

(2) $$\frac{d\mathcal{E}_\omega}{dz} = \frac{2r_0^2 n_a c\sqrt{\varepsilon'}}{3\pi(2\pi)^2}\left|\frac{m\omega^2}{e^2}\alpha(\omega)\right|^2 \int_0^{1/R} |\boldsymbol{E}_\omega(\boldsymbol{k}_\perp)|^2 2\pi k_\perp \, dk_\perp .$$

Large values of ρ correspond to small $k_\perp$. Therefore, the integration domain over $k_\perp$ is bounded by a circle of radius $1/R$.

Using Equation (9.9), we find the Fourier components of the electric vector, which are transverse and longitudinal relative to the velocity of the particle:

(3) $$\boldsymbol{E}_\omega^\perp(\boldsymbol{k}_\perp) = -i\frac{4\pi q}{v\varepsilon}\frac{\boldsymbol{k}_\perp}{k_\perp^2 + \omega^2/v^2 - \omega^2\varepsilon/c^2},$$

(4) $$E_\omega^\parallel(\boldsymbol{k}_\perp) = -i\frac{4\pi q\omega}{v^2\varepsilon}\frac{1 - v^2\varepsilon/c^2}{k_\perp^2 + \omega^2/v^2 - \omega^2\varepsilon/c^2}.$$

Note, that the denominators in these formulas, when integrated over $dk_\perp^2$, can vanish if the permittivity is real and the condition for the Cherenkov radiation $v > c/\sqrt{\varepsilon'}$ is satisfied. Hence, we have to consider separately two cases.

1. $v < c/\sqrt{\varepsilon'}$. The denominators do not vanish, the small imaginary part ε'' can be ignored, the integral in Equation (2) can be easily evaluated, and the result is close to the one obtained in Example 9.10:

(5) $$\frac{d\mathcal{E}_\omega}{dz} = \frac{16q^2 r_0^2 n_a c}{3v^2\varepsilon'^{3/2}}\left|\frac{m\omega^2}{e^2}\alpha(\omega)\right|^2 \ln\left[\frac{v}{\omega R}(1 - v^2\varepsilon'/c^2)^{1/2}\right],$$

where the logarithm is assumed to be a large number compared to unity. At $\varepsilon' = 1$ Equation (5) reduces to Equation (8) of Example 9.10. The main contribution is due to the transverse field, while the longitudinal component leads to the replacement of the logarithm by the quantity

(6) $$\ln\left[\frac{v}{\omega R}(1 - v^2\varepsilon'/c^2)^{1/2}\right] + \frac{1}{2}\left(1 - \frac{v^2\varepsilon'}{c^2}\right).$$

2. $v > c/\sqrt{\varepsilon'}$. In addition to polarization bremsstrahlung, Cherenkov quanta are generated. While integrating, one should take into account the imaginary part of ε'', which makes the denominator finite at all $k_\perp^2$:

(7) $$k_\perp^2 - \frac{\omega^2}{v^2}\left(\frac{v^2}{\omega^2}\varepsilon' - 1\right) - \frac{i\omega}{c^2}\Gamma(\omega), \qquad \text{where} \qquad \Gamma(\omega) = \omega\varepsilon''(\omega)$$

is the imaginary part of the frequency of the virtual photon, that characterizes the lifetime $\tau = \Gamma^{-1}(\omega)$ of this photon (or the decay time of the field component at the frequency ω). According to the results of Problem 5.15, this quantity can be written in the form

(8) $$\Gamma(\omega) = \omega\varepsilon''(\omega) = \frac{4\pi n_a e^2}{m}\sum_s \frac{f_s\omega^2\gamma_s}{(\omega^2 - \omega_{0s}^2)^2 + \omega^2\gamma_s^2},$$

where f_s are the oscillator strengths, ω_{0s} are the transition frequencies, and γ_s are the damping constants of the excited states of the atom.

Note, that in the more detailed quantum calculation, other quantities would appear in the denominator of Equation (7), which characterize lifetimes of the radiating particle and of the atom. The lifetime of the ground state of atoms at low temperatures is infinite, while the lifetime of a particle in the state with a momentum $\boldsymbol{p}$ is limited. The basic processes changing the state of the particle are the multiple scattering (elastic and nonelastic) by atoms, and the Cherernkov radiation. Omitting the derivation, we give the contribution due to these processes:

(9) $$\tau_p^{-1} = v\sigma n_a + \int \frac{q^2}{\hbar c}\left(1 - \frac{c^2}{v^2\varepsilon'(\omega)}\right) d\omega,$$

where σ is the total cross section of the collision of the particle with the atom, and the integration in the Cherenkov term is over those frequencies at which the integrand is positive. The total width responsible for the broadening of the resonance denominator (7), in view of the processes mentioned above, is given by

(10) $$\Gamma(\omega) = \tau_p^{-1} + \omega\varepsilon''(\omega).$$

The integrals in Equation (2) give

$$\int_0^{1/R^2} \frac{k_\perp^2 dk_\perp^2}{|k_\perp^2 - \frac{\omega^2}{v^2}b(\omega) - \frac{i\omega}{c^2}\Gamma(\omega)|^2} \approx \left(\frac{c}{v}\right)^2 \frac{\pi b\omega}{\Gamma} + 2\ln\left[\frac{v}{\omega R b^{1/2}}\right],$$

(11) $$b(\omega) = \frac{v^2}{c^2}\varepsilon'(\omega) - 1 > 0,$$

(12) $$\int_0^{1/R^2} \frac{k_\perp^2 dk_\perp^2}{|k_\perp^2 - \frac{\omega^2}{v^2}b(\omega) - \frac{i\omega}{c^2}\Gamma(\omega)|^2} \approx \left(\frac{c}{v}\right)^2 \frac{\pi b\omega}{\Gamma},$$

provided $c^2/R^2\omega\Gamma \gg 1$, $c^2 b\omega/v^2\Gamma \gg 1$.

The first term in (11) is due to the pole singularity, and the second term includes all remaining contributions. In the second term, we neglect the terms containing Γ, because of their smallness. As a result, for the spectral density of the radiation per unit path length $d\mathcal{E}_\omega/dz = d\mathcal{E}_\omega/dz|_1 + d\mathcal{E}_\omega/dz|_2$ we have two terms corresponding to two mechanisms of polarization bremsstrahlung:

(13) $$\left.\frac{d\mathcal{E}_\omega}{dz}\right|_1 = \frac{16q^2 r_0^2 n_a c}{3v^2\varepsilon'^{3/2}}\left|\frac{m\omega^2}{e^2}\alpha(\omega)\right|^2 \ln\left[\frac{v}{\omega R}b^{1/2}\right],$$

(14) $$\left.\frac{d\mathcal{E}_\omega}{dz}\right|_2 = \frac{8\pi q^2 r_0^2 n_a cb}{3v^2\varepsilon'^{1/2}}\left|\frac{m\omega^2}{e^2}\alpha(\omega)\right|^2 \frac{\omega}{\Gamma(\omega)}, \qquad v > c/\sqrt{\varepsilon'}.$$

The first term is of the same nature as the radiation at $v < c/\sqrt{\varepsilon'}$ or at $\varepsilon' = 1$: it occurs due to the shaking of an atomic shell by a field of a fast particle. The second term exists only when the Cherenkov effect takes place. Let us rearrange the factors in the second term in such a way, that each of them to have a simple physical meaning:

$$\left.\frac{d\mathcal{E}_{\omega}}{dz}\right|_{2} = \left(n_a \sigma_T \frac{\lambda}{2\pi}\right) \left|\frac{m\omega^2}{e^2}\alpha(\omega)\right|^2 \frac{(q^2/\hbar c)\omega(1 - c^2/v^2\varepsilon')}{\Gamma} \frac{\hbar\omega}{c}. \tag{14a}$$

Here $\sigma_T = 8\pi r_0^2/3$ is the Thomson scattering cross section of a soft quantum by a free electron without frequency change: the first factor $(n_a\sigma_T\lambda/2\pi)$ is the number of atoms in the cylinder with the base σ_T and height of the order of the wavelength of the quantum, λ. The factor $|m\omega^2\alpha(\omega)/e^2|^2$ takes into account that the polarizabilities of an atomic shell and free electrons are different. The product of the first and second factors can be looked upon as the total effective number of free electrons, which scatter Cherenkov's quanta, with account for the binding of the electrons in atoms and for the coherent character of the scattering of the quantum by an atomic shell as a whole. The next dimensionless fraction in (14a) contains the nominator that has a meaning of the partial width associated with the radiation of the Cherenkov quantum. The fraction itself is the ratio of this width to the total width characterizing decay probability of the state of the radiating fast particle. The nominator of the fraction is always less than its denominator. Finally, the last factor has a required dimension of the energy spectral density divided by distance and contains the energy of the quantum. Thus, the mechanism of polarization radiation is the scattering of the Cherenkov quanta by an electron shell of atoms. The Cherenkov mechanism acquires the resonance character, and it will prevail over the usual mechanism, that is, the inequality $d\mathcal{E}_\omega/dz|_2 > d\mathcal{E}_\omega/dz|_1$ will hold at $\omega/\Gamma > 2\ln(v/\omega R)$.

Usual Cherenkov radiation is much more intensive than both components of the polarization radiation. However, the Cherenkov quanta propagate only at the angle $\theta = \arccos(c/v\sqrt{\varepsilon'})$ to the velocity of a fast particle. Polarization bremsstrahlung has the angular distribution that is close to anisotropic. Particularly, it exists beyond the Cherenkov cone as well. In this range of angles, it prevails over other channels of radiation and can be measured. ■

Density effect in magnetic bremsstrahlung. As shown in the previous section, the density effect becomes important in bremsstrahlung of ultrarelativistic electrons in medium at frequencies $\omega < \omega_p\gamma$, where $\omega_p = \sqrt{4\pi n_a Ze^2/m}$ is the plasma frequency. It results in the suppression of radiation and in the decrease of its spectral density. This is due to the effect of medium polarization on the propagation of electromagnetic waves, which changes their dispersion and decreases the coherence length (see Equation (9.60)). The analogous effect takes place in magnetic bremsstrahlung as well. Tsytovich was the first (1951) who pointed out the role of the medium in magnetic bremsstrahlung, though he did not consider the most interesting aspect of the influence of the medium on the high-frequency radiation. Razin (1960) extensively studied the role of plasma for cosmic sources of synchrotron radio emission. He was the first who attributed the important observable fact, that is, the cutoff of spectra of radio sources at low frequencies, to the medium effect. At present, the Razin–Tsytovich effect is universally recognized and often used for interpretation of the radio astronomical data.

Example 9.12

A relativistic electron moves along a circle in the magnetic field $\boldsymbol{B} = const$ in a plasma with the number density n_e of the background electrons. Calculate the spectral-angular and the spectral distribution of the radiation power at frequencies $\omega \gg \omega_p = \sqrt{4\pi n_e e^2/m}$ and $\omega \gg \omega_B = eB/mc$. Analyze the specific cases, compare the results with magnetic bremsstrahlung in vacuum (Problems I.5.83 and I.5.84).

Solution. Since at the given frequencies the permittivity $\varepsilon(\omega) = 1 - \omega_p^2/\omega^2$ is isotropic and close to unity, it is sufficient to take into account only in the exponent of Equation (I.5.77). Align the xy-plane of the reference frame with the plane of the orbit, and place the line of sight in the xz-plane at an angle $\theta \ll 1$ to the Ox-axis (Figure 9.3).

We have

(1) $$\boldsymbol{s}(\tau) = a\boldsymbol{e}_x \sin(v\tau/a) + a\boldsymbol{e}_y[1 - \cos(v\tau/a)], \qquad \boldsymbol{v} = \dot{\boldsymbol{s}},$$

where $a = cp/eB$ is the radius of the orbit. Further, like in Problem I.5.83, we take into account the relativistic effect of the forward radiation and decompose all quantities in small parameters θ, ω_p^2/ω^2, $1/\gamma$ and $(v\tau/a)$. We obtain

$$\boldsymbol{n} \times \boldsymbol{v}(\tau) \approx c\theta\boldsymbol{e}^{(1)} + \frac{c^2\tau}{a}\boldsymbol{e}^{(2)}, \qquad \omega\tau - \boldsymbol{k}\cdot\boldsymbol{s}(\tau) \approx \frac{1}{2}\omega\tau\left(\theta^2 + \frac{1}{\gamma^2} + \frac{\omega_p^2}{\omega^2}\right)$$

(2) $$+\frac{\omega\tau}{3}\left(\frac{c\tau}{a}\right)^2 = \frac{3}{2}\xi\left(x + \frac{1}{3}x^3\right).$$

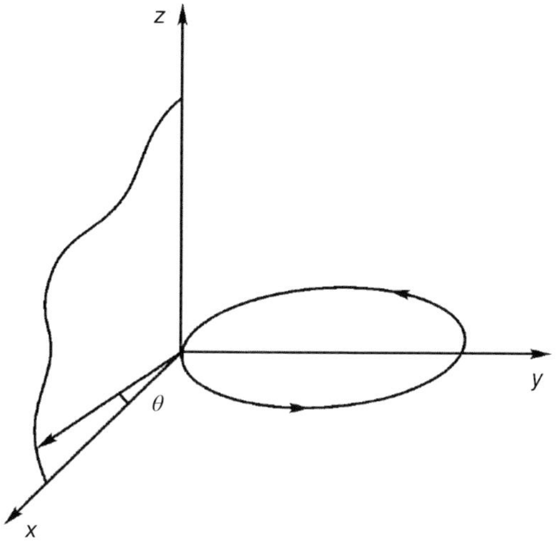

Figure 9.3 Geometry of magnetic bremsstrahlung (see text).

Here the unit vectors $\boldsymbol{e}^{(1)} = \boldsymbol{e}_y$, $\boldsymbol{e}^{(2)} = \boldsymbol{n} \times \boldsymbol{e}_y$, and $\boldsymbol{n}$ are mutually orthogonal,

(3) $$x = \frac{c\tau}{(\theta^2 + \gamma^{-2} + \omega_p^2/\omega^2)^{1/2}}, \qquad \xi = \frac{\omega a}{3c}(\theta^2 + \gamma^{-2} + \omega_p^2/\omega^2)^{3/2}.$$

Afterwards we perform the integration over x using the formulas given in Problem I.5.83. Then we obtain the relation for the spectral-angular distribution, which generalizes Equation (3) of Problem I.5.83:

(4)
$$\frac{d^2 I'_\omega}{d\omega d\Omega} = \frac{e^2}{3\pi^2 c}\left(\frac{a\omega}{c}\right)^2 \left(\theta^2 + \frac{1}{\gamma^2} + \frac{\omega_p^2}{\omega^2}\right)^2 \left[K_{2/3}^2(\xi) + \frac{\theta^2}{\theta^2 + \gamma^{-2} + \omega_p^2/\omega^2} K_{1/3}^2(\xi)\right].$$

After dividing both parts by the orbital period of the particle $T = 2\pi\gamma/\omega_B$, we find the spectral-angular power:

(5)
$$\frac{d^2 I_\omega}{d\omega d\Omega} = \frac{e^2\omega_B}{6\pi^3 c\gamma}\left(\frac{a\omega}{c}\right)^2 \left(\theta^2 + \frac{1}{\gamma^2} + \frac{\omega_p^2}{\omega^2}\right)^2 \left[K_{2/3}^2(\xi) + \frac{\theta^2}{\theta^2 + \gamma^{-2} + \omega_p^2/\omega^2} K_{1/3}^2(\xi)\right].$$

Comparing the last expression with Equation (5) of Problem I.5.83, we see that the presence of plasma essentially changes the radiation distribution over frequencies and angles at $\omega < \gamma\omega_p$. In this range of frequencies, the angular distribution expands, the characteristic angle is given by $\theta_0 \approx \omega_p/\omega > \gamma^{-1}$, and the parameter ξ decreases with the increase of frequency. The spectral shape can be conveniently investigated by integrating Equation (5) over the solid angle. The angular distribution has an axial symmetry, and the angle θ is measured from the xy-plane. Hence, $d\Omega = 2\pi \cos\theta d\theta$, $-\pi/2 \le \theta \le \pi/2$; in the small-angle approximation we have $d\Omega = 2\pi d\theta$ (with $-\infty < \theta < \infty$). The integration over the angle can be performed with the aid of formulas given in Problem I.5.84. The result is

(6) $$\frac{dI_\omega}{d\omega} = \frac{e^2\omega}{\sqrt{3}\pi c}\left(\frac{1}{\gamma^2} + \frac{\omega_p^2}{\omega^2}\right)\int_\eta^\infty K_{5/3}(x)dx, \qquad \eta = \frac{2\omega\gamma}{3\omega_B}\left(\frac{1}{\gamma^2} + \frac{\omega_p^2}{\omega^2}\right)^{3/2}.$$

The parameter $\eta(\omega)$ depends on frequency non-monotonically: it reaches a minimum $\eta_{min} = \sqrt{3}\omega_p/\gamma\omega_B$ at $\omega = \sqrt{2}\gamma\omega_p$. The spectrum shape and the total radiation intensity depend significantly on the minimum value of this parameter. If $\eta_{min} \gg 1$, then, using Equation (I.1.170), we find

(7) $$\frac{dI_\omega}{d\omega} = \frac{e^2\omega}{\sqrt{6\pi}c}\left(\frac{1}{\gamma^2} + \frac{\omega_p^2}{\omega^2}\right)\frac{e^{-\eta(\omega)}}{\sqrt{\eta(\omega)}};$$

the radiation is suppressed by the density effect and is exponentially small. This can occur when a particle is moderately relativistic, and the ratio of the plasma and cyclotron frequencies is large:

(8) $$1 \ll \gamma \ll \frac{\sqrt{3}\omega_p}{\omega_B}.$$

The opposite case, $\eta_{min} \ll 1$, is more interesting. With the aid of Equation (I.1.168) we find

$$\frac{dI_\omega}{d\omega} = 3^{1/6}\Gamma(2/3)\frac{e^2}{\pi c}\left(\frac{\omega_B^2\omega}{\gamma^2}\right)^{1/3}, \qquad \eta \ll 1. \tag{9}$$

This result is the same as that obtained for the magnetic bremsstrahlung in vacuum at $\omega \ll \omega_c = 3\omega_B\gamma^2$ (see the result of Problem I.5.84). At frequencies $\omega \gg \omega_p\gamma/\eta_{min}$ there occurs the exponential decrease of the radiation:

$$\frac{dI_\omega}{d\omega} = \frac{\sqrt{3}e^2\omega_B}{2\sqrt{\pi}c}\left(\frac{\omega}{\omega_c}\right)^{1/2} e^{-2\omega/\omega_c}. \tag{10}$$

In this case, the presence of the plasma does not affect the process. However, the same effect of the exponential suppression of the radiation takes place at low frequencies as well, $\omega \ll \omega_p\gamma\sqrt{\eta_{min}}$:

$$\frac{dI_\omega}{d\omega} = \frac{e^2}{2\sqrt{2\pi}c}\left(\frac{\omega_p\omega_B}{\gamma}\right)^{1/2} \exp\left(-\frac{2\gamma\omega_p^3}{3\omega_B\omega^2}\right). \tag{11}$$

Here the spectrum drastically differs from the spectrum in vacuum. The latter one is described, in the frequency range under consideration, by Equation (9). Further analysis of the medium effect on magnetic bremsstrahlung radiation can be found in Problems 9.32 and 9.33. ■

There are many other radiation processes by charged particles in media. Specifically, in the review of Bolotovskii and Galst'yan (2000) the authors describe a similarity with an elegant theory of diffraction radiation arising when particles travel through apertures in screens.

Recommended literature: Tamm and Frank (1937); Ginzburg and Frank (1946); Frank (1988); Ginzburg (1987, 2001); Basylev and Zhewago (1987); Ginzburg and Tsytovich (1990); Toptygin (2014); Platonov and Fleishman (2002); Ter-Mikhaelyan (1972); Amusya *et al.* (1987); Amusya (1990); Korol' *et al.* (2004)

Problems

9.13* A particle of charge e moves with a velocity $\mathbf{v} = const$ through the homogeneous isotropic medium of permittivity $\varepsilon(\omega)$ and magnetic permeability $\mu = 1$. Determine the components of the electromagnetic field produced by this particle.

9.14* A particle moves with a constant velocity $v = \beta c$ in a nonabsorbing dielectric. Using the results of Problem 9.13, investigate the field created by this particle at large distances from its trajectory. Show that a sufficiently fast particle will emit transverse electromagnetic waves (Cherenkov effect). Find the conditions under which this emission occurs and the total Cherenkov energy loss w per unit path length.

9.15 A particle of charge e moves with a constant velocity through a medium, whose permittivity can be approximately described by formula

$$\varepsilon(\omega) = 1 + \frac{\omega_p^2}{\omega_0^2 - \omega^2}.$$

Determine the Cherenkov radiation energy w per unit path length when the particle velocity satisfy the condition $v^2\varepsilon_0 > c^2$, where ε_0 is the static permittivity. What is the range of angles within which the radiation is concentrated? Obtain the numerical estimate, assuming that $\omega_0 = 6 \times 10^{15}\ \mathrm{s}^{-1}$, $\varepsilon_0 = 2$ and $v = c$.

9.16 Derive the condition $\cos\theta = 1/\beta n$, which determines the direction of propagation of Cherenkov radiation, by considering the interference between separate waves emitted by the particle at various points along its trajectory.

9.17 Cherenkov radiation emitted by a particle can be considered a result of a resonance between proper vibrations of the medium and an induced force associated with the moving particle. Derive the condition for the appearance of the Cherenkov effect comparing the frequencies of the eigenoscillations of the medium and those of the induced force.

9.18 A relativistic particle of velocity v passes through the dielectric plate of thickness l in a direction at right angles to its plane. The refractive index of the plate is n and dispersion can be neglected. Find the duration τ of the burst of Cherenkov radiation, which an observer fixed in relation to the plate will register. Find the energy flux I of Cherenkov radiation through the surface of the plate during the burst. Neglect the edge effect.

9.19 Show that the minimum velocity v_{min} of a particle at which Cherenkov effect becomes possible in a particular direction, satisfies the condition

$$v_{min} \cos\theta = v_g(\omega_m),$$

where v_g is the group velocity of electromagnetic waves in the dielectric, ω_m is the frequency at which the refractive index has maximum, and θ is the angle between the direction of the radiation and the particle velocity. The dielectric can be regarded as nonabsorbing.

9.20* A particle moves with a constant velocity $v = \beta c$ through a nondispersive medium of permittivity ε and magnetic permeability μ. Find the electromagnetic potentials φ and $\mathbf{A}$. Consider two cases, $v < v_{ph}$ and $v > v_{ph}$, where v_{ph} is the phase velocity of electromagnetic waves in the medium.

9.21 A straight-line conductor lying along the Ox-axis is displaced along the Oy-axis with a velocity $v = const$ in a nonabsorbing medium of permittivity $\varepsilon(\omega)$ and magnetic permeability $\mu(\omega)$. The conductor is electrically neutral in the laboratory frame and carries a current I in the direction of the Ox-axis.[3] Find the condition for the emission of Cherenkov radiation.

3) Fast moving, current-carrying beams of particles can occur in accelerators or at discharges of a certain type.

Determine the total energy of this radiation w per unit path length. Calculate the retarding force $\mathbf{f}$ acting per unit length of the conductor under the action of the field produced by it.

Hint. The vector potential has the single component $A_x(y, z, t)$. While performing the inverse Fourier transformation, use the method of going round the poles, that was used in the preceding problem.

9.22 Two point charges e_1 and e_2 move with equal constant velocities v along the same straight line, but at a distance l (measured in the laboratory frame) from each other, in medium of permittivity $\varepsilon(\omega)$ and magnetic permeability $\mu = 1$. Find the energy of Cherenkov radiation w per unit path length. Consider two cases: (a) $e_1 = e_2 = e$ and (b) $e_1 = -e_2 = e$. By passing to the limit, find the Cherenkov energy for a point electric dipole lying along the direction of motion.

9.23* Two point charges $+e$ and $-e$ move with equal constant velocities v at a distance l (measured in the laboratory frame) from each other in the medium of permittivity $\varepsilon(\omega)$ and magnetic permeability $\mu = 1$. A line joining the charges is at an angle α to the direction of motion (l and α are measured in the laboratory frame). Use the method employed in Problem 9.22 to find the Cherenkov energy loss w per unit path length, assuming that l is small as compared with a wavelength.

9.24* A magnetic dipole[4)] moves with a constant velocity $v = \beta c$ in a nonabsorbing medium of permittivity $\varepsilon(\omega)$ and magnetic permeability $\mu(\omega)$. The magnetic moment measured in the laboratory frame is $\boldsymbol{m}$; it is parallel to the direction of velocity. Find the Cherenkov energy losses w per unit path length.

Hint. Use the Fourier transformation to integrate the equations for the potentials. The moving magnetic dipole produces a current $\boldsymbol{j}(\boldsymbol{r}, t) = c \, rot \, \boldsymbol{m}\delta(\boldsymbol{r} - \boldsymbol{v}t)$.

9.25* A fast particle of charge e moves through a nonabsorbing dielectric of permittivity

$$\varepsilon(\omega) = 1 + \frac{\omega_p^2}{\omega_0^2 - \omega^2},$$

where $\omega_p^2 = 4\pi e^2 N/m$, m is a mass of electron. Determine the energy losses $(-d\mathcal{E}/dl)$ per unit path length at distances from the particle trajectory which are greater than the interatomic distance a (the parameter a should be chosen in such a way that for the region $r > a$ the macroscopic consideration will be valid). Elucidate the physical significance of the various terms in the expression for the energy losses.

9.26* A charged particle moves with a velocity $v = \beta c$ through a plasma of permittivity

$$\varepsilon(\omega) = 1 - \frac{\omega_p^2}{\omega^2},$$

4) Neutral system (bunch) of particles, having the magnetic moment, radiates as a magnetic dipole if the wavelength in a medium is much larger than the sizes of the particle bunch.

where $\omega_p = \sqrt{4\pi e^2 N/m}$ is the electron plasma frequency. Find the energy losses $-d\mathcal{E}/dl$ per unit path length due to "distant" collisions, that is, collisions for which $r > a$, where a is the distance at which the macroscopic equations begin to hold.

9.27⋆ A point particle of charge e moves in vacuum in a direction perpendicular to the boundary of a perfect conductor. Determine the spectral and angular distribution of the radiation emitted when the charge enters the conductor. Neglect the acceleration of the charge due to the force of its electrical image induced in the conductor. The velocity of the charge is $v = \beta c$.

Hint. The field in the vacuum is due to the charge and its image moving toward each other with equal constant velocities. When the particle reaches the boundary of the conductor, its charge is instantaneously screened by free electrons, as if the charge and its image suddenly stop at the same point at the boundary.

9.28⋆ A point charge e has a velocity $v = \beta c$ and moves in vacuum in a direction perpendicular to the boundary of a nonabsorbing dielectric of permittivity $\varepsilon(\omega)$ ($\mu = 1$). When the charge enters the dielectric it starts to radiate. Neglecting the acceleration of the charge due to the image force, determine the spectral and angular distribution of the radiation into vacuum (i.e., into the region $z > 0$ in Figure 9.15).

Hint. The charge and current densities produced by the moving particle should be replaced by the equivalent set of harmonic oscillators. In order to determine the field in the wave zone, use the reciprocity theorem (see Example I.5.7): $\boldsymbol{p}_B \cdot \boldsymbol{E}_A(B) = \boldsymbol{p}_A \cdot \boldsymbol{E}_B(A)$. Here $\boldsymbol{E}_B(A)$ is the field produced at the point A by the harmonic dipole oscillator $\boldsymbol{p}_B$ placed at the point B, and $\boldsymbol{E}_A(B)$ is the field produced at the point B by the oscillator $\boldsymbol{p}_A$ at A. Fresnel's formulas can be used to calculate $\boldsymbol{E}_A(B)$, since the point of observation A is at a large distance from the point where the charge enters the dielectric (wave zone).

9.29 Write down the spectral-angular distributions (9.92) and (9.93) of the transient radiation for a nonrelativistic particle, $v/c = \beta \ll 1$. Compare with the results of the two preceding problems.

9.30⋆ Write down the spectral-angular distributions (9.92) and (9.93) of the transient radiation for a nonrelativistic particle, $\gamma = \mathcal{E}/mc^2 \gg 1$. Show that in a certain spectral range the radiation is concentrated in a forward direction. Give the estimate of the cone solid angle of the radiation. Evaluate also the frequency range of the radiation.

9.31⋆ Based on Equations (9.92) and (9.93), find the spectral distribution of the hard transient radiation (at frequencies much greater than the atomic frequencies of the medium) when an ultrarelativistic particle enters the medium from vacuum or when it leaves the medium. Analyze the limiting cases. Calculate the total energy of the transient radiation.

9.32⋆ Generalize the results of Example 9.12 to the case of a relativistic electron spiraling along the homogeneous magnetic field in a plasma. An angle between the electron velocity and the magnetic field (the pitch angle) is Θ. Calculate the spectral-angular distribution of the radiation power.

9.33* Find the spectrum of the synchrotron radiation produced by a cloud of energetic ultrarelativistic electrons in a plasma which have the isotropic distribution function. The number of electrons is N_0. The magnetic field $\boldsymbol{B}$ is homogeneous, and the angle between $\boldsymbol{B}$ and the line of sight $\boldsymbol{n}$ is Θ. Compare the derived spectrum and the spectrum in the absence of the medium (Problem I.5.86).

9.3 Channeling and Radiation Emitted by Fast Particles in Crystals

When fast particles move in regular single crystals, the orientation and interference effects are significant. For example, when moving in particular directions (along atomic chains or in crystalline planes), the paths of particles increase by several times as compared to the paths in media with disordered structure. Due to large length at which bremsstrahlung forms, ultrarelativistic particles emit coherently moving along large segments of their trajectories. We have already dealt with the interference phenomena in crystals, considering scattering of X-rays in Section 6.4. In the section below, we will consider a motion of fast particles and their radiation in crystals.

Interference effects in scattering of fast particles.

Example 9.13

Use the Born approximation to calculate the differential cross section of the elastic scattering of a relativistic electron by a two-atomic system. Express it through the scattering cross section by a single atom. The relative position of the atoms is characterized by vector $\boldsymbol{a}$*, the atoms are identical and each of them produces the Coulomb field, whose radius of the exponential screening is R (see Example 9.2). Consider the various orientations of the initial momentum of the electron relative to the axis of the system* $\boldsymbol{a}$*, and evaluate the coherence length when the scattering occurs at the longitudinal orientation and at the transverse orientation.*

Solution. The scattering potential has the form

(1)
$$U(\boldsymbol{r}) = -Ze^2\left(\frac{\exp(-r/R)}{r} + \frac{\exp(-|\boldsymbol{r}-\boldsymbol{a}|/R)}{|\boldsymbol{r}-\boldsymbol{a}|}\right).$$

The differential scattering cross section is proportional to the Fourier component of the scattering potential

(2)
$$U_{\mathbf{q}} = \int dV\, U(\boldsymbol{r})\exp(-i\boldsymbol{q}\cdot\boldsymbol{r}) = u(\boldsymbol{q})(1+\exp(-i\boldsymbol{q}\cdot\boldsymbol{a})),$$

where $\boldsymbol{q} = \boldsymbol{p}' - \boldsymbol{p}$ is the wave vector transfer in a scattering event, and $u(\boldsymbol{q})$ is the Fourier transform of the potential of an individual atom. Using the last expression, we find the differential cross section

(3) $$\frac{d\sigma}{d\Omega} = \left(\frac{d\sigma}{d\Omega}\right)_C (2 + 2\cos \boldsymbol{q}\cdot\boldsymbol{a}),$$

where the first factor is the cross section of scattering at the single screened Coulomb center, found in Example 9.2. The first term in the brackets gives the doubled cross section of a single atom, that is, the sum of the cross sections of two atoms. The cosine term describes the interference effect.

The interference is important at $|\boldsymbol{q}\cdot\boldsymbol{a}| \lesssim 1$, because at $|\boldsymbol{q}\cdot\boldsymbol{a}| \gg 1$ the cosine will highly oscillate and the interference is washed out. At $|\boldsymbol{q}\cdot\boldsymbol{a}| \ll 1$ the cross section increases four times in comparison with the scattering at a single atom (i.e., the interference enhancement is proportional to the square of the number of atoms). We take into account that the Coulomb scattering is most important at small angles $\theta \leq 1/kR$ (see the cross section derived in Example 9.2). For the longitudinal orientation, we have $\boldsymbol{q}\cdot\boldsymbol{a} \approx ka\theta^2/2$, and the longitudinal correlation length is given by

(4) $$l_{\parallel} \approx \frac{2pR^2}{\hbar}.$$

The transverse correlation length is given by

(5) $$l_{\perp} \approx \frac{1}{k\theta} \approx R. \quad \blacksquare$$

The longitudinal scattering correlation length is proportional to the particle momentum and can reach the macroscopic dimensions for ultrarelativistic particles. The transverse correlation length remains macroscopic quantity for all energies. If there is a chain of atoms with interatomic distances a, then the particles with initial momenta parallel to the chain will interact coherently with $N_c = l_{\parallel}/a$ atoms, and the cross section will increase N_c times as compared to the one-atom scattering cross sections (in the Born approximation provided it is valid). When the orientation is transverse, the cross section is proportional to the number of the atoms.

Example 9.14

A relativistic particle is scattered by a linear chain consisting of N atoms separated by a distance a. Use the Born approximation to calculate the differential and total cross sections of the elastic scattering at $N \gg 1$. Determine the validity criterion of the approximation depending on the orientation of the initial momentum of the particle relative to the alignment of the atoms.

Solution. Calculations are similar to those in the previous example. We obtain the differential cross section

(1) $$\frac{d\sigma}{d\Omega} = \left(\frac{d\sigma}{d\Omega}\right)_C \frac{\sin^2(\boldsymbol{q}\cdot\boldsymbol{a}N/2)}{\sin^2(\boldsymbol{q}\cdot\boldsymbol{a}/2)},$$

where

(2) $$\boldsymbol{q}\cdot\boldsymbol{a} = \frac{1}{2}ak\theta^2,$$

when the scattered particles are incident along the chain. Here we have used the small-angle approximation for scattering of relativistic particles by a screened Coulomb potential. At $ak\theta^2 N/4 \ll 1$ Equation (1) gives:

(3) $$\frac{d\sigma_{\parallel}}{d\Omega} = N^2 \left(\frac{d\sigma}{d\Omega}\right)_C = 4\left(\frac{NZe^2}{pv}\right)^2 \frac{1}{(\theta^2 + \theta_0^2)^2}.$$

All N atoms of the chain act coherently, as a whole. This results in the effective charge of the nucleus NZe and gives the possibility to use the well-known (see Landau and Lifshitz (1977), Quantum Mechanics) criterion of applicability of Born approximation:

(4) $$\frac{NZe^2}{\hbar v} \ll 1.$$

Since $e^2/\hbar c \approx 1/137$, at $N \gg 1$ the Born approximation can be applied only for sufficiently light elements with $Z < 10$. By integrating the relation (3) over the solid angle, we find the total cross section

(5) $$\sigma_{\parallel} = N^2 \sigma_C = 4\left(\frac{NZe^2}{\hbar v}\right)^2 \pi R^2,$$

where R is the screening radius. When the Born approximation can be used, that is, condition (4) is satisfied, the scattering cross section by the chain is small in comparison with the geometric cross section of an atom πR^2. It means that the mutual shadowing of the atoms for the incident wave is small.

When particles are incident normally (Figure 9.4), we have

(6) $$\boldsymbol{q} \cdot \boldsymbol{a} = ak\theta \cos\varphi.$$

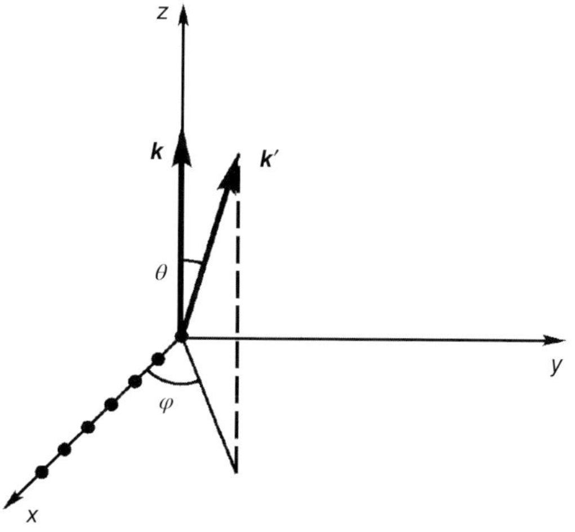

Figure 9.4 The beam of particles is directed perpendicular to an atomic chain.

At $N \gg 1$, we use the representation of the delta-function (I.1.217) and transform the interference factor in Equation (1):

$$\frac{\sin^2(\boldsymbol{q}\cdot\boldsymbol{a}N/2)}{\sin^2(\boldsymbol{q}\cdot\boldsymbol{a}/2)} = \frac{\sin^2(ak\theta\cos\varphi N/2)}{\sin^2(ak\theta\cos\varphi/2)} \underset{N\gg 1}{\rightarrow} N\frac{\pi(ak\theta\cos\varphi/2)^2}{\sin^2(ak\theta\cos\varphi/2)}$$

(7) $$\delta(ak\theta\cos\varphi/2) = \frac{2\pi N}{ak}\delta(\theta\cos\varphi).$$

The interference factor has sharp maxima at $\theta \to 0$ and at $\varphi \to \pi/2$. Their real angular width is of the order of $1/Nak \ll 1$. Integrating Equation (1) over $d\Omega = \theta d\theta d\varphi$ and taking into account (7), we obtain the total cross section

(8) $$\sigma_\perp = N\sigma_C\frac{\pi R}{a}.$$

In this case, the radiation is incoherent, the cross section is proportional to the number of the atoms, and the factor $\pi R/a$ is close to unity, because in the condensed media an interatomic distance is of the order of the atomic size. The validity of Born approximation is the same as for an individual atom, $Ze^2/\hbar v \ll 1$. ■

Axial and planar channeling. When the condition opposite to the Born's criterion (4) is satisfied (4), that is, when $NZe^2/\hbar v \gg 1$, the classical approximation can be applied. Neglecting the diffraction effects, consider the classic motion of a relativistic particle in the field of a long atomic chain. If the initial momentum is parallel to the chain or is directed at a small angle to it, then for a single collision at the impact parameter $\rho \lesssim R$, where R is a radius of a screening, the particle acquires the additional transverse momentum of the order $\Delta p_\perp \approx Ze^2\tau/\rho^2$, where Ze^2/ρ is the force and $\tau \approx \rho/v$ is the collision duration. The particle is deflected by the angle of the order of $\theta \approx \Delta p_\perp/p \approx Ze^2/\rho\beta^2\mathcal{E} \ll 1$, where $\mathcal{E} = mc^2\gamma$ is its relativistic energy. An increment of the impact parameter $\Delta\rho$ will be of the order of $\Delta\rho \approx a\theta$, where a is a distance between atoms of the chain. For sufficiently fast particles (not necessary relativistic) $\Delta\rho \ll \rho$ and a trajectory of the particle is thus a curve close to a straight line at distances of the order of an interatomic distance (Figure 9.5).

The trajectory is just of the same character when a particle is incident at a sufficiently small angle ϑ to the axis of the chain, though not parallel to it. Particles with a negative charge (electrons) are attracted to the axial line and can move largely inside the chain at a distance $\rho \leq R$ from its axis **(axial channeling)**, while the positively charged particles (positrons, protons, positive ions) are concentrated in space between the neighboring parallel closely packed chains. The accurate calculation with the use of the realistic atomic potential is highly time consuming.

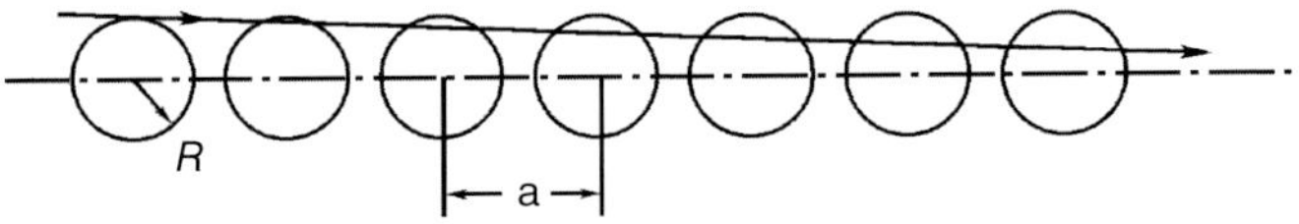

Figure 9.5 Motion of a relativistic particle in a field of a long atomic chain.

However, the character of the particle motion itself suggests the rational simplification of the problem. Namely, we should replace the actual potential of the chain by the axially symmetric potential averaged over the coordinate along the chain:

$$U(\rho) = \frac{1}{L}\int_{-\infty}^{\infty} \sum_{n} u(\rho, z - a_n)\, dz, \tag{9.98}$$

where u is the potential energy of an individual atom, a_n is the atom's coordinate on the chain's axis, and L is the length of the chain.

If closely packed parallel atomic planes can be separated in the crystal, and the initial velocity of the particle is at a small angle to them, we can introduce the potential averaged over positions of the atoms in the plane:

$$U(x) = \frac{1}{S}\int \sum_{n,m} u(x, y + b_m, z + a_n)\, dy\, dz, \tag{9.99}$$

where S is an area of the plane, and (b_m, a_n) are coordinates of the atom in the (y, z)-plane. In this approximation, the problem becomes one-dimensional. Particles can be confined for a long time either near the plane (electrons) or between the neighboring planes (positively charged particles), in the state of the **planar channeling**. A description of a motion of fast particles in crystals on the base of averaged potentials has been introduced by Lindhard (see the review Lindhard (1954)). A more accurate approach suggests, in addition, the averaging over thermal oscillations of atoms in sites crystalline lattice.

The main properties of the channeling can be analyzed from the classical mechanical integrals of motion, without solving equations of motion.

In the field of one-dimensional stationary potential $U(x)$, the total energy of a particle $\mathcal{E}$ is conserved, together with its momentum $p_{\|}$ parallel to a closely packed plane:

$$\frac{mc^2}{\sqrt{1 - \beta_{\|}^2 - \dot{x}^2/c^2}} + U(x) = \mathcal{E} = const, \qquad \frac{mc\beta_{\|}}{\sqrt{1 - \beta_{\|}^2 - \dot{x}^2/c^2}} = p_{\|} = const, \tag{9.100}$$

where $\beta_{\|} = v_{\|}/c$. We express the longitudinal velocity from the second equation:

$$\beta_{\|}^2 = \frac{c^2 p_{\|}^2}{\mathcal{E}_{\|}^2}\left(1 - \frac{\dot{x}^2}{c^2}\right), \tag{9.101}$$

and substitute it in the first equation. In this way, we find the transverse velocity

$$\dot{x}^2 = c^2 \frac{(\mathcal{E} - U(x))^2 - \mathcal{E}_{\|}^2}{(\mathcal{E} - U(x))^2}, \qquad \text{where} \qquad \mathcal{E}_{\|} = \sqrt{m^2c^4 + c^2 p_{\|}^2} = const. \tag{9.102}$$

Further we take into consideration that the method of the averaged potential can be used provided a particle moves at a small angle to the planes, and the energy associated with the transverse motion, $\mathcal{E}_{\perp} = \mathcal{E} - \mathcal{E}_{\|}$, is small in comparison with the total energy $\mathcal{E}$ and with the longitudinal energy $\mathcal{E}_{\|}$. It is comparable to the potential energy U. Thus, we can simplify Equation (9.102) and write down the energy conservation for the transverse motion:

$$\frac{\mathcal{E}_{\parallel}}{c^2}\frac{\dot{x}^2}{2} + U(x) = \mathcal{E}_{\perp} \tag{9.103}$$

Such an expression corresponds to the motion of a nonrelativistic particle with a relativistic effective mass $m_{ef} = \mathcal{E}_{\parallel}/c^2$ in the prescribed potential field $U(x)$.

The character (finite or infinite) of the particle motion in the crystal (along the Ox-axis) depends on the initial conditions of entering the crystal. The electrons are attracted to the closely packed planes, and the potential relief for them has the form shown schematically in Figure 9.6.

If a relativistic electron enters the crystal at the point x_0 and its momentum is inclined at an angle $\theta \ll 1$ to the (yz)-plane, then $\mathcal{E}_{\perp} = \mathcal{E}_{\parallel}\theta^2/2 + U(x_0)$, and $U(x_0) \leq 0$. If the motion is finite (i.e., the particle is trapped in one of the potential wells), the condition $\mathcal{E}_{\perp} < 0$ should be satisfied, that is, $\theta < \sqrt{2|U(x_0)|/\mathcal{E}_{\parallel}} = \theta_c$. Such electrons are referred to as **channeled**. At $\theta > \theta_c$ the motion of the electron is infinite and the electron overcomes the potential barriers. For positrons and positive ions, on the place of potential wells (Figure 9.7) there appear humps, and the wells will be moved in the gaps between the humps. In this case, the channeled particles will be the particles with the transverse energy $\mathcal{E}_{\perp} = \mathcal{E}_{\parallel}\theta^2/2 + U(x_0) < U_{max}$, where $U_{max} > 0$ is a height of the humps, that is, the particles with the flight-in angles $\theta < \sqrt{2(U_{max} - U(x_0))/\mathcal{E}_{\parallel}} = \theta_c$. The particles with $\theta > \theta_c$ will move be over barriers.

The transverse velocity and acceleration of the over-barrier particle are oscillate with the period

$$T = \int_0^b \frac{dx}{|\dot{x}|}, \qquad \dot{x} = \sqrt{\frac{2c^2}{\mathcal{E}_{\parallel}}(\mathcal{E}_{\perp} - U(x))}, \tag{9.104}$$

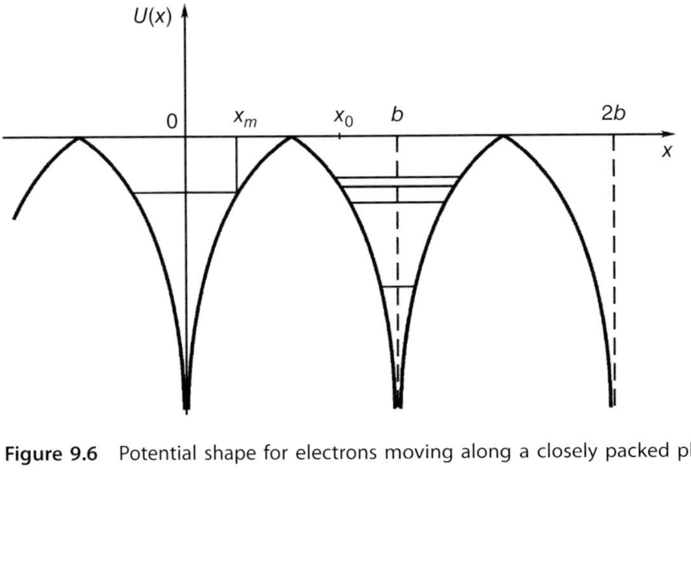

Figure 9.6 Potential shape for electrons moving along a closely packed plane.

where b is a distance between the planes. The period of oscillations of the channeled particle is given by

$$T = 4\int_0^{x_m} \frac{dx}{|\dot{x}|}, \qquad \text{where} \qquad U(x_m) = \mathcal{E}_\perp \tag{9.105}$$

(see Figure 9.6). Here the transverse energy $\mathcal{E}_\perp$ depends on the angle θ and the point x_0 at which the particle enters the crystal. The longitudinal velocity executes periodic oscillations of a small amplitude according to Equation (9.101). Thus, the coordinate of the over-barrier particle increases by b over each period, and the motion parallel to the plane is almost uniform, with the velocity $v_\| \approx v \approx const.$ For the channeled particles, the longitudinal motion is of the same character, and the transverse coordinate x, like the velocity, $\dot{x}$, oscillates with the period (9.105).

The motion of particles will be considered in the model of axial channeling in Examples 9.3 and 9.36.

Here we have presented only a sketchy description of the motion of fast particles in crystals. In particular, we have not discussed the following important phenomena (their analysis can be found in the literature to this section):

(a) The effect of thermal motion of atoms (their oscillations in lattice cites) on the motion of fast particles. This question will be outlined below in connection with radiative processes.

(b) The energy losses of fast particles in crystals and the effect of channeling on this process, and also on the interaction of channeled particles with the nuclei of the crystal. Even the first experiments have shown that for the channeled protons the transparency of the crystal increased by several times (see Thompson (1968)).

(c) Under specific conditions, the oscillations of the channeled and over-barrier particles acquire the stochastic character ("dynamic chaos," see (Akhieser and Shul'ga (1993))). Stochasticity can also be induced by multiple scattering of channeled particles.

(d) The restriction of motion of channeled particles in the transverse direction results in the discrete levels of the transverse energy and requires, in certain cases, quantum mechanical description with the use of Schrödinger equation (see, e.g., Kalashnikov (1981); Basylev and Zhewago (1987)).

Emission of radiation by relativistic particles in crystals.

In crystals, similar to media with a disordered arrangement of atoms, various mechanisms of radiation can occur. We begin with bremsstrahlung due to interaction of a relativistic particle with screened fields of atomic nuclei. The crystal as a whole will be treated as an object producing an external field. The bremsstrahlung cross section will be calculated by the method of equivalent photons, which was used in Examples I.6.20 and 9.4. In the rest frame of the crystal, the electromagnetic potentials are given by

$$\varphi(\boldsymbol{r}) = \int \frac{4\pi Ze}{q^2+\kappa^2} e^{i\mathbf{q}\cdot\mathbf{r}} \sum_n e^{-i\mathbf{q}\cdot\mathbf{r}_n} \frac{d^3q}{(2\pi)^3}, \qquad \kappa = \frac{1}{R}, \qquad \boldsymbol{A} = 0, \tag{9.106}$$

where summation is over all atoms of the crystal, which are equal and placed at the points $\boldsymbol{r}_n$. We will use the model of exponential screening. In the rest frame of the electron, the potentials have the form

$$\varphi'(\mathbf{r}',t') = \frac{Ze\gamma}{2\pi^2}\int \frac{d^3q}{q^2+\kappa^2} e^{i\mathbf{q}_\perp\cdot\mathbf{r}_\perp + iq_\parallel\gamma(z'+vt')} \sum_n e^{-i\mathbf{q}_\perp\cdot\mathbf{r}_{n\perp} - iq_\parallel\gamma z'_n}, \quad \mathbf{A} = -\frac{\mathbf{v}}{c}\varphi'(\mathbf{r}',t'), \tag{9.107}$$

that is, differ from those used in Example 9.4 by the interference factor only.

Calculating the field strengths, we get the relation which determines the spectral density of the equivalent photons (the analog of formula (6) from Example 9.4):

$$\frac{Z^2e^2\gamma}{2\pi^2} 2\int_0^\infty dq'_\parallel \int \frac{q_\perp\, d^2q_\perp}{(q^2+\kappa^2)^2}\left|\sum_n e^{-i\mathbf{q}\cdot\mathbf{r}_n}\right|^2 = \int_0^\infty \hbar\omega' n(\omega')\, d\omega',$$

where $q'_\parallel = \gamma q_\parallel \approx \omega'/c$, $\mathbf{r}_n = (\mathbf{r}_{n\perp}, \gamma z'_n) = (\mathbf{r}_{n\perp}, z_n)$ are the coordinates of the atom in the frame of the crystal. From the obtained relation, we find the spectral density of the photons:

$$n(\omega') = \frac{Z^2e^2\gamma}{\pi^2\hbar c}\frac{1}{\omega'}\int \frac{q_\perp^2\, d^2q_\perp}{(q^2+\kappa^2)^2}\left|\sum_n e^{-i\mathbf{q}\cdot\mathbf{r}_n}\right|^2, \qquad \text{where} \qquad \omega' = c\gamma q_\parallel. \tag{9.108}$$

The bremsstrahlung cross section for the electron in the crystal is found using the evident formula

$$d\sigma_r(\omega) = \int_{\omega'_{min}}^{\omega'_{max}} d\omega' n(\omega') d\sigma_C(\omega', \omega'') \tag{9.109}$$

(see Equation (1) from Problem (6.81)). Here $d\sigma_C(\omega', \omega'')$ is the Compton scattering cross section; ω' and ω'' are the frequencies of the primary and scattered quanta in the frame of the electron. The final frequency and the limits of integration, ω'_{min} and ω'_{max}, should be transformed in the laboratory (crystal) frame. The Compton scattering cross section was found in Problem 6.80:

$$\frac{d\sigma_C}{d(\hbar\omega'')} = \pi r_0^2 \frac{mc^2}{(\hbar\omega')^2}\left[\frac{\omega'}{\omega''} + \frac{\omega''}{\omega'} + \left(\frac{mc^2}{\hbar\omega''} - \frac{mc^2}{\hbar\omega'}\right)^2 - 2\left(\frac{mc^2}{\hbar\omega''} - \frac{mc^2}{\hbar\omega'}\right)\right]. \tag{9.110}$$

Let us now transform this cross section using the relations which were found in Problem I.6.88 from the Lorentz transformation $\omega = \omega''\gamma(1-\cos\theta')$ for the frequency of the scattered quantum in the two reference frames, and from the conservation laws

$$1 - \cos\theta' = \frac{mc^2}{\hbar\omega''} - \frac{mc^2}{\hbar\omega'}$$

in the frame of the electron. We have

$$\frac{\hbar\omega'}{\mathcal{E}}d\omega, \qquad \omega'' = \omega'\frac{\mathcal{E}'}{\mathcal{E}}, \qquad \omega'_{min} = \omega\frac{mc^2}{2\mathcal{E}'}, \qquad \omega'_{max} = \omega\frac{2\gamma\mathcal{E}}{\mathcal{E}'}, \tag{9.111}$$

where $\mathcal{E}$ and $\mathcal{E}'$ are the energies of the electron in the crystal frame before and after the quantum emission, respectively; ω'_{min} and ω'_{max} determine the limits of integration over $d\omega'$ at the particular frequency ω of the emitted quantum in the laboratory frame and at the given $\mathcal{E}$ and $\mathcal{E}' = \mathcal{E} - \hbar\omega$. With the aid of the two first formulas, we write down the Compton scattering cross section in the form

$$d\sigma_C(\omega', \omega) = \pi r_0^2 \frac{mc^2}{\mathcal{E}\omega'} \left[\frac{\mathcal{E}}{\mathcal{E}'} + \frac{\mathcal{E}'}{\mathcal{E}} + \left(\frac{mc^2}{\mathcal{E}'} \right)^2 \left(\frac{\omega}{\omega'} \right)^2 - 2 \frac{mc^2}{\mathcal{E}'} \frac{\omega}{\omega'} \right] d\omega. \quad (9.112)$$

Finally, the cross section of bremsstrahlung in a crystal takes the form of the three-dimensional integral

$$d\sigma_r(\omega) = \frac{1}{\pi} r_0^2 Z^2 \frac{e^2}{\hbar c} \frac{mc^2}{\mathcal{E}} \int_{\omega'_{min}}^{\omega'_{max}} \frac{d\omega'}{\omega'^2} \int \frac{q_\perp^2 \, d^2 q_\perp}{(q^2 + \kappa^2)^2}$$
$$\times \left[\frac{\mathcal{E}}{\mathcal{E}'} + \frac{\mathcal{E}'}{\mathcal{E}} + \left(\frac{mc^2}{\mathcal{E}'} \right)^2 \left(\frac{\omega}{\omega'} \right)^2 - 2 \frac{mc^2}{\mathcal{E}'} \frac{\omega}{\omega'} \right] \left| \sum_n e^{-i\mathbf{q}\cdot\mathbf{r}_n} \right|^2 d\omega. \quad (9.113)$$

Let us remind that $\omega' = c\gamma q_\parallel$. Hence, the integrand in Equation (9.113) can be treated as the differential cross section $d\sigma_r(\boldsymbol{q}, \omega)$ of bremsstrahlung at a frequency ω, which is accompanied by the transfer of the fixed wave vector $\boldsymbol{q} = (\boldsymbol{q}_\perp, q_\parallel)$ to the medium. It differs from the bremsstrahlung cross section for an individual atom $d\sigma_{BH}(\boldsymbol{q}, \omega)$ by the structure factor of the medium:

$$d\sigma_r(\boldsymbol{q}, \omega) = d\sigma_{BH}(\boldsymbol{q}, \omega) \left| \sum_n e^{-i\mathbf{q}\cdot\mathbf{r}_n} \right|^2. \quad (9.114)$$

This relation between the cross sections is natural, since the probability of the radiative transition in Born approximation is proportional to the squared modulus $|U_\mathbf{q}|^2$ of the potential of the external field which can be expressed in terms of the potential $u_\mathbf{q}$ of a single atom:

$$|U_\mathbf{q}|^2 = |u_\mathbf{q}|^2 \left| \sum_n e^{-i\mathbf{q}\cdot\mathbf{r}_n} \right|^2.$$

The method of equivalent photons, as clear from Examples I.6.20 and 9.4 for ultra-relativistic particles, is equivalent to the first Born approximation.

The transition from Equation (9.114) to the bremsstrahlung cross section for a macroscopic body with disordered ensemble of atoms can be performed by representing the structure factor in the form

$$\left| \sum_n e^{-i\mathbf{q}\cdot\mathbf{r}_n} \right|^2 = \sum_{m,n} e^{i\mathbf{q}\cdot(\mathbf{r}_m - \mathbf{r}_n)} = N + \sum_{m \neq n} e^{i\mathbf{q}\cdot(\mathbf{r}_m - \mathbf{r}_n)}, \quad (9.115)$$

where N is the total number of atoms. For the random atomic arrangement, one should average (9.115) over all $\boldsymbol{r}_m$, assuming that all positions of atoms inside the body are equally probable:

$$\langle e^{-i\mathbf{q}\cdot\mathbf{r}_n} \rangle = \frac{1}{V} \int_V e^{-i\mathbf{q}\cdot\mathbf{r}_n} dV_n = 0,$$

provided the body dimension L is large in comparison with q^{-1}. As a result, the last sum in Equation (9.115) vanishes, and we have

$$\left\langle \left| \sum_n e^{-i\mathbf{q}\cdot\mathbf{r}_n} \right|^2 \right\rangle = N, \qquad d\sigma_r(\boldsymbol{q},\omega) = N d\sigma_{BG}(\boldsymbol{q},\omega),$$

which corresponds to the incoherent radiation of N atoms.

However, if the atoms execute even small oscillations in their lattice sites, the structure factor significantly and non-monotonically depends on the wave vector transfer $\boldsymbol{q}$, resulting in the interference maxima and minima, similar to diffusion of X-rays in crystals. In this case, the averaging should be performed by methods of solid-state physics (see Ziman (1964); Ter-Mikhaelyan (1972)), taking into account the quantum character of oscillations of atoms in crystals. The result of averaging is given by

$$\left\langle \left| \sum_n e^{i\mathbf{q}\cdot\mathbf{r}_n} \right|^2 \right\rangle = N(1 - e^{-q^2\overline{u^2}}) + e^{-q^2\overline{u^2}} \left| \sum_n e^{i\mathbf{q}\cdot\mathbf{r}_n^0} \right|^2, \tag{9.116}$$

where

$$\overline{u^2} = \frac{3\hbar^2}{4MT_D}\left[1 + 4\left(\frac{T}{T_D}\right)^2 \int_0^{T_D/T} \frac{t\,dt}{e^t - 1}\right]$$

is the average squared deflection of the atom from the equilibrium position, $\boldsymbol{r}_n^0$ are position vectors of lattice sites (all atoms are assumed to be identical), M is a mass of atom, and T/T_D is the ratio of the crystal temperature to its Debye temperature. At room temperature the quantity $\sqrt{\overline{u^2}}$ is of the order or somewhat less than the screening radius R. Hence, the interference effects in bremsstrahlung appear at sufficiently small wave vector transfers, $qR \lesssim 1$.

In view of Equation (9.116), the bremsstrahlung cross section for a crystal consists of two parts, $d\sigma_r(\boldsymbol{q},\omega) = d\sigma_{nc}(\boldsymbol{q},\omega) + d\sigma_c(\boldsymbol{q},\omega)$, where the incoherent part of the cross section

$$d\sigma_{nc}(\boldsymbol{q},\omega) = N(1 - e^{-q^2\overline{u^2}}) d\sigma_{BH}(\boldsymbol{q},\omega), \tag{9.117}$$

is independent of the direction of the electron propagation in the crystal, relative to its crystallographic axes, while the coherent part

$$d\sigma_c(\boldsymbol{q},\omega) = d\sigma_{BH}(\boldsymbol{q},\omega) e^{-q^2\overline{u^2}} \left| \sum_n e^{i\mathbf{q}\cdot\mathbf{r}_n^0} \right|^2 \tag{9.118}$$

depends on this direction. An analysis of the interference phenomena appearing in Equation (9.118), can be performed similarly to that of the diffraction of X-rays (see Section 6.4 and also Example 9.15).

Example 9.15

A heavy relativistic particle (muon, proton) interacts with a crystalline body of volume V located in air, $\varepsilon(\omega) \approx 1$. Find the intensity of radiation caused by scattering *of the fast-particle field by the electron subsystem of the crystal, at frequencies which are much greater than the atomic frequencies. Use the approach developed in the kinematic theory of the X-ray diffraction (Section 6.4). Investigate the properties of this radiation* ***(diffracted X-ray radiation)****.*

Solution. Heavy particles produce weak bremsstrahlung in the fields of atomic nuclei. Hence, the main contribution to the radiation is due to medium electrons, excited by the field of a fast particle. We consider the range of frequencies for which the permittivity is given by Equation (6.70) and satisfies the condition of smallness (6.71). Equation (6.73) can be used to calculate the field at a large distance from the body where $\boldsymbol{D} = \boldsymbol{E}$. On the right-hand side of Equation (6.73) we substitute the harmonic of the undisturbed field of a particle moving with a constant velocity in vacuum (see Problem I.4.34):

$$\boldsymbol{E}^0_\omega(\boldsymbol{r}) = \int_{-\infty}^{\infty} \boldsymbol{E}(\boldsymbol{r}, t) e^{i\omega t}\, dt. \tag{1}$$

We use the approximation according to which the incident wave is weakly decaying by traveling through the crystal (Born approximation). Writing down the solution of Equation (6.73), we obtain the scattered field

$$\boldsymbol{E}_\omega(\boldsymbol{r}) = -\frac{e^2}{m\omega^2} \int_V G^R_\omega(\boldsymbol{r}, \boldsymbol{r}') \nabla' \times [\nabla' \times \boldsymbol{E}^0_\omega(\boldsymbol{r}') n(\boldsymbol{r}')]\, dV', \tag{2}$$

where $G^R_\omega(\boldsymbol{r}, \boldsymbol{r}')$ is the Green's function (I.5.14) at a large distance from the body ($r \gg r'$):

$$G^R_\omega(\boldsymbol{r}, \boldsymbol{r}') = \frac{1}{r} e^{ikr - i\mathbf{k}\cdot\mathbf{r}'}, \qquad \text{where} \qquad \boldsymbol{k} = \frac{\omega}{c}\frac{\mathbf{r}}{r} \tag{3}$$

is the wave vector of the scattered wave. Here the integral on the right-hand side of Equation (2) is taken over the volume of the body V instead of an infinite space.

Upon integration of Equation (2) by parts, like in the derivation of (6.74), and substitution of the expansion (6.79) of the electron density, we reduce the scattered field to the representation which contains the sum over the vectors $\boldsymbol{b}$ of reciprocal lattice:

$$\boldsymbol{E}_\omega(\boldsymbol{r}) = \frac{e^2}{m\omega^2} \sum_{\mathbf{b}} n_{\mathbf{b}} \boldsymbol{k} \times \left[\boldsymbol{k} \times \int_V \boldsymbol{E}^0_\omega(\boldsymbol{r}') \exp(-i(\boldsymbol{k} - \boldsymbol{b}) \cdot \boldsymbol{r}')\, dV'\right] \frac{e^{ikr}}{r}. \tag{4}$$

The integral, if taken in the infinite limits, would contain the singular factor, namely, the delta-function $\delta(\omega - \boldsymbol{k} \cdot \boldsymbol{v})$ (see the solution of Problem I.5.144 or Example 9.1). However, when integrated over the bounded crystal volume, all integrals are finite:

$$\int_V e^{i(\mathbf{k}-\mathbf{k}')\cdot\mathbf{r}}\, dV = V \delta_{\mathbf{k},\mathbf{k}'}, \tag{5}$$

where the large but finite crystal volume V is separated; $\delta_{\mathbf{k},\mathbf{k}'}$ is the Kronecker symbol with a finite width of the order of $|\boldsymbol{k} - \boldsymbol{k}'| \sim 1/L$ along each direction, and

L is the integration length along the corresponding axis. In a maximum at $\boldsymbol{k} = \boldsymbol{k}'$ the Kronecker symbol is close to unity. Finally, we have

$$\int_V \boldsymbol{E}^0_\omega(\boldsymbol{r}') \exp(-i(\boldsymbol{k} - \boldsymbol{b}) \cdot \boldsymbol{r}')\, dV' = -i \frac{4\pi e' L}{v[(\boldsymbol{k} - \boldsymbol{b})^2 - \omega^2/c^2]} \left(\boldsymbol{k} - \boldsymbol{b} - \frac{\omega \boldsymbol{k}}{c^2}\right), \text{ and}$$

(6)
$$(\boldsymbol{k} - \boldsymbol{b}) \cdot \boldsymbol{v} = \omega.$$

Here L is the crystal length along the particle velocity, L/v is its transient time, and the condition $(\boldsymbol{k} - \boldsymbol{b}) \cdot \boldsymbol{v} = \omega$ appeared from the unwritten Kronecker symbol. Substituting the latter result in equation (4), we have the radiation field (the scattered field of the particle)

$$\boldsymbol{E}_\omega(\boldsymbol{r}) = -i \frac{4\pi e' e^2 L}{m v \omega^2} \sum_{\mathbf{b}} \frac{n_{\mathbf{b}}}{(\boldsymbol{k} - \boldsymbol{b})^2 - \omega^2/c^2} \boldsymbol{k} \times \left[\boldsymbol{k} \times \left(\boldsymbol{b} + \frac{\omega \boldsymbol{v}}{c^2}\right)\right] \frac{e^{ikr}}{r}, \text{ where}$$

(7)
$$(\boldsymbol{k} - \boldsymbol{b}) \cdot \boldsymbol{v} = \omega.$$

The method used for calculating this field is related to the method of equivalent photons. However, in this case, the particle need not be ultrarelativistic, because the electron system reacts to any alternating field of frequency ω through the dielectric function $\varepsilon(\omega)$. The mechanism of radiation under consideration can be referred to as the **polarization mechanism** (see, e.g., Example 9.10); it is due to the polarization of the electron system, produced by the field of the fast particle. In the case of heavy particles, the bremsstrahlung of the particle itself is suppressed by its large mass. In the case of electrons and positrons, coherent bremsstrahlung of the light particle should be added to the radiation under consideration. It seems likely that the interference of both mechanisms can occur.

The relationship between the frequency and the propagation direction of quanta follows from Equation (7):

(8)
$$\cos\theta = \frac{c}{v} - \frac{c(\boldsymbol{b} \cdot \boldsymbol{v})}{v\omega},$$

where θ is an angle between the directions of propagation of the emitted quantum and the fast particle. The spectral-angular distribution of radiation can be found from Equation (I.5.22):

(9)
$$\frac{d^2 I_\omega}{d\omega d\Omega} = \frac{4 e'^2 e^4 L^2}{m^2 v^2 c^3} \left| \sum_{\mathbf{b}} \frac{n_{\mathbf{b}}}{(\boldsymbol{k} - \boldsymbol{b})^2 - \omega^2/c^2} \boldsymbol{n} \times \left(\boldsymbol{b} + \frac{\omega \boldsymbol{v}}{c^2}\right) \right|^2.$$

The direction of the quantum escape $\boldsymbol{n}$ and its fundamental frequency are connected with the vector of the inverse lattice and with each other by Equation (8). At the prescribed $\boldsymbol{b}$, $\boldsymbol{v}$, and $\cos\theta$, the frequency range of the emitted quanta is rather narrow, for example, of the order of several electron volts for the quantum energy 5 – 8 keV. The consideration of experiments on the diffracted X-ray radiation and on its possible practical applications can be found in the review of Ter-Mikhaelyan (2001). ■

Radiation of channeled and over-barrier particles. In the case of channeling, the Born approximation cannot be used, and the character of radiation is essentially changed as compared to the cases considered above. By using the classic theory of radiation, we consider now the radiation energy loss of the channeled particle per unit path length. In both cases of the axial or the planar channeling, the momentum of a particle along a certain (Oz)-axis is conserved, and the averaged potential energy depends only on one or two transverse coordinates. By differentiating the total energy of the particle

$$\mathcal{E} = \sqrt{c^2p^2 + m^2c^4} + U(\rho) \tag{9.119}$$

with respect to time, we have

$$\frac{d\mathcal{E}}{dt} = \boldsymbol{v} \cdot \left(\frac{d\boldsymbol{p}}{dt} + \frac{\partial U}{\partial \rho} \right). \tag{9.120}$$

Afterwards we use the equation of motion of the particle including the radiative friction force $\boldsymbol{F}$, Equation (I.5.118),

$$\frac{d\boldsymbol{p}}{dt} = -\frac{\partial U}{\partial \rho} + \boldsymbol{F}. \tag{9.121}$$

The substitution of this result in Equation (9.120) gives $\dot{\mathcal{E}} = \boldsymbol{v} \cdot \boldsymbol{F}$ and shows that the energy loss of the particle is equal to a work of the radiative friction force. Taking into account that in Equation (I.5.118) $\boldsymbol{H} = 0$, $e\boldsymbol{E} = -\partial U/\partial \boldsymbol{\rho}$ and that $v \approx c$, we find the radiative energy loss of the particle per unit path length:

$$\frac{d\mathcal{E}}{dz} = -\frac{2e^2\gamma^2}{3m^2c^4} \left(\frac{\partial U}{\partial \rho} \right)^2. \tag{9.122}$$

This formula is valid for the over-barrier particles as well, provided they move at a small angle to atomic chains or atomic planes, so that the consideration based on the averaged potential is thus possible. If the potential energy is periodic, the radiation losses (9.122) should be averaged over the period.

This result can be compared with the synchrotron energy losses of a relativistic electron in the uniform magnetic field (see Problem I.5.75):

$$\left. \frac{d\mathcal{E}}{dz} \right|_m = -\frac{2e^4\gamma^2H^2}{3m^2c^4}. \tag{9.123}$$

The quantity e^2H^2 comes from the squared derivative $(\partial U/\partial \rho)^2$, which can be written in terms of the strength of the atomic field $e^2E_{at}^2$. For the atoms with an atomic number Z, we have $E_{at} = Ze/R^2 \approx Z^{5/3}e/a_B^2 \approx 4 \times 10^{10}$ CGSE $\approx 10^{13}$ V/cm, for heavy elements ($Z = 64$). The equivalent magnetic field is equal to 4×10^{10} G.

When an electron moves in medium with randomly placed atoms, the radiation losses are given by Equation (9.58). The ratio of the losses in an amorphous medium to that in a crystalline medium can be expressed in the form

$$\frac{d\mathcal{E}/dz|_{am}}{d\mathcal{E}/dz|_{cr}} \approx \frac{30}{\gamma Z^{4/3}} \left(\frac{n_a \Lambda_C^3}{\alpha^4} \right), \tag{9.124}$$

where n_a is number density of atoms, $\Lambda_C = \hbar/mc$, and $\alpha \approx 1/137$. For condensed media, the factor in the brackets is of the order of unity, and, hence, the radiation of the channeled electrons and positrons significantly exceeds their radiation in amorphous media. The fact that the intensive radiation by relativistic particles is possible in crystals was firstly pointed out by Kumahov (see his monograph Kumakhov, 1986).

Radiation spectrum of channeled particles. We use Equation (I.5.58) for the field of an arbitrarily moving particle in the wave zone:

$$\boldsymbol{E} = \left.\frac{e\boldsymbol{n} \times [(\boldsymbol{n} - \boldsymbol{v}/c) \times \dot{\boldsymbol{v}}]}{c^2 R(1 - \boldsymbol{n}\cdot\boldsymbol{v}/c)^3}\right|_{t'=t-R(t')/c}. \tag{9.125}$$

For the channeled relativistic particle, we assume that $\boldsymbol{v} = const$, and the acceleration $\dot{\boldsymbol{v}}$ is a periodic function of time (see Equations (9.104) and (9.105) for the planar channeling and Problem 9.36 for the axial channeling). Assume, also, that the vectors $\boldsymbol{v}$ and $\dot{\boldsymbol{v}}$ are mutually perpendicular, neglecting thus the terms of the order of γ^{-2}. In order to calculate the spectrum, the Fourier component of the field (9.123) should be found:

$$\boldsymbol{E}_\omega = \int_{-\infty}^{\infty} \boldsymbol{E}(t) e^{i\omega t} dt. \tag{9.126}$$

The field (9.123) is expressed in terms of the retarded time t'; therefore we should pass to the integration over dt', taking into account that the particle velocity is constant:

$$t = t' - \frac{R(t')}{c} \approx \frac{R_0}{c} + \left(1 - \frac{\boldsymbol{n}\cdot\boldsymbol{v}}{c}\right) t'. \tag{9.127}$$

Substituting Equations (9.123) and (9.127) in Equation (9.124), we have

$$\boldsymbol{E}_\omega = \frac{e}{c^2(1 - \boldsymbol{n}\cdot\boldsymbol{v}/c)^2} \frac{e^{i\omega R_0/c}}{R_0} \int_{-\infty}^{\infty} \boldsymbol{n} \times [(\boldsymbol{n} - \boldsymbol{v}/c) \times \dot{\boldsymbol{v}}(t')] \exp\left\{i\omega\left(1 - \frac{\boldsymbol{n}\cdot\boldsymbol{v}}{c}\right) t'\right\} dt'. \tag{9.128}$$

Now the role of frequency belongs to the quantity $\omega' = \omega(1 - \boldsymbol{n}\cdot\boldsymbol{v}/c)$, which, for an ultrarelativistic channeled particle, has the form

$$\omega' \approx \omega\left(1 - \frac{v}{c}\cos\theta\right) \approx \frac{\omega}{2}(\gamma^{-2} + \theta^2). \tag{9.129}$$

Here θ is an angle between the direction of radiation and the channel axis. Thus, the Fourier component of the field is expressed in terms of the Fourier component of the acceleration at the frequency ω':

$$\boldsymbol{E}_\omega = \frac{e}{c^2}\left(\frac{\omega}{\omega'}\right)^2 \frac{e^{i\omega R_0/c}}{R_0} \boldsymbol{n} \times [(\boldsymbol{n} - \boldsymbol{v}/c) \times \dot{\mathbf{v}}_{\omega'}]. \tag{9.130}$$

The spectral-angular distribution of the radiation can be calculated using Equation (I.5.22), taking into account that at high frequencies in the matter $|\boldsymbol{H}_\omega|^2 = |\boldsymbol{E}_\omega|^2$:

$$\frac{d^2 I_\omega}{d\omega\, d\Omega} = \frac{cR_0^2}{4\pi^2} |\boldsymbol{E}_\omega(R_0)|^2. \tag{9.131}$$

Calculating the square of the double vector product in Equation (9.131), we have

$$|\boldsymbol{n}\times[(\boldsymbol{n}-\boldsymbol{v}/c)\times\dot{\boldsymbol{v}}_{\omega'}]|^2 = |\dot{\boldsymbol{v}}_{\omega'}|^2\left(1-\frac{\boldsymbol{n}\cdot\boldsymbol{v}}{c}\right)^2 - |\boldsymbol{n}\cdot\dot{\boldsymbol{v}}_{\omega'}|^2\left(1-\frac{v^2}{c^2}\right).$$

We integrate Equation (9.131) over the solid angle $d\Omega = \theta\, d\theta\, d\varphi = d\varphi\, d\omega'/\omega$, where we have used Equation (9.129) and passed to the integration over ω'. We measure the azimuth angle φ from the direction of the vector $\dot{\mathbf{v}}_{\omega'}$, and write down $|\boldsymbol{n}\cdot\dot{\boldsymbol{v}}_{\omega'}|^2 = |\dot{\boldsymbol{v}}_{\omega'}|^2\theta^2\cos^2\varphi$, where an angle θ should be expressed in terms of ω'. As a result, we have

$$\frac{dI_\omega}{d\omega} = \frac{e^2\omega}{2\pi c^3}\int_{\omega/2\gamma^2}^{\infty}|\dot{\boldsymbol{v}}_{\omega'}|^2\left[1-\frac{\omega}{\gamma^2\omega'}+\frac{\omega^2}{2\gamma^4\omega'^2}\right]\frac{d\omega'}{\omega'^2}. \qquad (9.132)$$

This expression can be used for various cases of motion of channeled and over-barrier particles, for which the approximation of the averaged potential is valid. Let us consider the simplest ("model") case where the planar channeling occurs, and an electron executes harmonic oscillations of frequency Ω and of amplitude x_m, $x(t) = x_m\cos\Omega t$, between the planes. In order to evaluate a circular frequency Ω, we take into account that the channeled relativistic particle has the effective mass $m_{ef}\approx \mathcal{E}/c^2 = \gamma m$ (see Equation (9.103)). Hence, $\Omega = \omega_0/\gamma^{1/2}$, where ω_0 is the frequency of oscillations of the nonrelativistic electron in the potential field U, which for the planar channeling is of the order of $\omega_0 = \sqrt{E_{at}R/mb^2}$, with b being a distance between the crystal planes. In this case,

$$\dot{v}_{\omega'} = \int_{-\infty}^{\infty}\ddot{x}(t)e^{i\omega' t}dt = -\pi\Omega^2 x_m\delta(\omega'-\Omega) \qquad \text{at} \qquad \Omega>0,\ \omega'>0.$$

When this quantity is squared, there appears the divergent expression, that corresponds to an infinite time of electron motion through the crystal:

$$|\dot{v}_{\omega'}|^2 = \frac{1}{2}\pi\Omega^4x_m^2\delta(\omega'-\Omega)\Delta t|_{\Delta t\to\infty}.$$

Therefore, the radiative energy losses should be calculated per unit path length, that is, we have to divide Equation (9.132) by $v\Delta t\approx c\Delta t$ and change the sign:

$$\frac{d^2\mathcal{E}_\omega}{dz\,d\omega} = -\frac{e^2(\Omega x_m)^2\omega}{4c^4}\left[1-\frac{\omega}{\gamma^2\Omega}+\frac{\omega^2}{2\gamma^4\Omega^2}\right]\Theta(2\gamma^2\Omega-\omega). \qquad (9.133)$$

Such a method was used while passing from Equations (9.71) to (9.72)

The presence of the step function means that the spectrum is cut off at the frequency

$$\omega = 2\gamma^2\Omega = 2\gamma^{3/2}\omega_0. \qquad (9.134)$$

The radiation of the channeled particle occurs mainly at these frequencies. Integrating Equation (9.133) over all frequencies, we have

$$\frac{d\mathcal{E}}{dz} = -\frac{e^2\omega_0^4x_m^2}{3c^4}\gamma^2, \qquad (9.135)$$

which is in accordance with the general expression (9.122). In the case of unharmonic oscillations or more complicated rotation motion in the channel, the transverse acceleration of the particle and the radiation spectrum contain a large number of Fourier harmonics. The analysis of a variety of specific cases of the channeling and quasi-channeling (i.e., over-barrier) motions, and also the detailed comparison of the theory with experimental data can be found in recommended literature section.

Recommended literature: Lindhard (1954); Thompson (1968); Ter-Mikhaelyan (1972); Kumakhov and Shirmer (1980); Landau and Lifshitz (1975); Akhieser and Shul'ga (1993); Kumakhov (1986); Kalashnikov (1981); Basylev and Zhewago (1987); Baryshevskii (1982); Ter-Mikhaelyan (2001)

Problems

9.34 Calculate the averaged potential energy of an atomic chain, assuming that for the separate atom it has the form of the Coulomb potential with the exponential screening (see the statement of Problem 9.2), a distance between atoms being b.

9.35 Calculate the averaged potential energy of the atomic plane when for the separate atom it has the form of the Coulomb potential with the exponential screening (see the statement of Problem 9.2). A distance between atoms on the plane is a, and between planes is b.

9.36* Write down the integrals of motion and investigate the character of motion of a relativistic electron in the field of a single atomic chain using the approximation of the averaged potential. Discuss the picture of particle motion in the crystal at small angles to the direction of atomic chains.

9.4
Acceleration of Particles in Turbulent Plasma Media

Properties of cosmic nonequilibrium media. We have seen in Section 9.1 that in the matter under equilibrium the energetic particles quickly lose their energy and finally come to the statistic equilibrium with the medium. However, in nature there exist the processes that accelerate charged particles up to super-high energies when they interact with natural media (i.e., with media existing in nature). This statement is strongly supported by the existence of cosmic rays, which are particles of high energies coming from the Universe and penetrating into the Earth atmosphere. Over dozens of years such particles have been registered both by ground-based instruments and by instruments installed on the interplanetary vehicles beyond the terrestrial magnetosphere. The largest measured energy of such particles is about 3×10^{20} eV, that is, it exceeds the energy reached at modern accelerators of elementary particles by eight orders of magnitude. A macroscopic particle of mass 1 g, moving with sound velocity in air (~330 m/s), acquires the energy of that order.

The majority of accelerated particles of the natural origin with moderate energies ($\mathcal{E} < 10^{12}$ eV) are protons, α-particles (helium nuclei) comprise about 10% and less, and more heavy nuclei less than 1% . Electrons also contribute about 1% to the general flux of cosmic rays.

The question on the origin of particles of so high energies represents one of the most interesting miracle of nature which is not solved. The study of strongly nonequilibrium states of matter suggests the way of solution of this mystery. In Section 7.3, we have considered the examples of nonequilibrium media and elucidated, in particular, that plasma with beams of particles is unstable; electric oscillations in such a plasma can grow up. A significant fraction of matter in stars as well as in interplanetary and interstellar space is in the form of plasma where all kinds of macroscopic motions and magnetic fields can exist. Interstellar matter is extremely rarefied according to terrestrial scales: the number density of particles n is of the order $10 \gtrsim n \gtrsim 0.1\ \text{cm}^{-3}$, and it is in a strongly nonequilibrium state.

Such a state is supported by mechanical motions of celestial bodies and their systems, mainly by internal evolution of stars, which can explode and erupt jets and plasma clusters moving with velocities of tens thousands kilometers per second. Many plasma jets exist beyond our stellar system (Galaxy), moving with relativistic velocities; their origin is far from being clear. However, all observational data obtained justify that in cosmic space there are strongly disturbed regions filled with regularly or randomly moving clouds of plasma with magnetic fields. In such regions, regular or random electric fields can also be present. These fields can accelerate particles up to high energies. The acceleration implies a nonequilibrium state of medium and a large storage of energy which could be converted into the energy of accelerated particles.

A charged particle moving in an electric field from point 1 to point 2 along a certain trajectory acquires the energy $\Delta\mathcal{E} = e\int_{(1)}^{(2)} \boldsymbol{E}\cdot\boldsymbol{v}\,dt$. If averaging over trajectories of various particles in the acceleration region gives $\langle \boldsymbol{E}\rangle \neq 0$, then the acceleration is referred to as **regular.** If $\langle \boldsymbol{E}\rangle = 0$, then the electric field changes its direction along at the trajectory. However, in this case the acceleration is also possible, because the quantity $\langle \int_{(1)}^{(2)} \boldsymbol{E}\cdot\boldsymbol{v}\,dt\rangle \neq 0$ (the particle velocity $\boldsymbol{v}[\boldsymbol{E}]$ is the functional of the electric field and there appear nonzero quantities of type $\langle E^2\rangle$ at averaging). The acceleration in this case is called **stochastic**.

When the medium has high electric conductivity, the electric field may be expressed in terms of the magnetic field and macroscopic velocity of the medium (see Equation (I.4.34))

$$\boldsymbol{E} = -\frac{1}{c}\boldsymbol{u}\times\boldsymbol{B}. \tag{9.136}$$

In this case, the particle acceleration can be interpreted as a result of their collisions with moving "magnetic inhomogeneities," which transfer a fraction of their energy to the particles. In the present section, we consider some the best-studied processes of particle acceleration in nonequilibrium plasma, which can provide sufficiently high energies.

Motion of fast particles in random magnetic field. In Section I.4.2, we have considered various cases of particle motion in a regular magnetic field. Such a field can be treated as a particular function of coordinates and time. However, in interplanetary and interstellar plasmas, besides the large-scale regular magnetic field there exist random magnetic inhomogeneities, caused by the random macroscopic plasma motions, plasma instabilities and various oscillations, which can be developed in nonequilibrium plasmas. A random magnetic field can be realized in many ways. As we have seen in Section 4.3 and Appendix, analytically it can be determined by its average value and correlation tensors of different order.

A trajectory of an individual particle moving in a random electromagnetic field can also be realized in many ways. Generally, it cannot be calculated as a certain functional of a random field. The system of particles moving in random fields should be described by the distribution functions averaged over possible values of these fields. In a rarefied medium, it is often possible to neglect the collisions of fast particles with particles of a background plasma, since the macroscopic field affects their motion much stronger.

Let us derive the kinetic equation for the distribution function of fast particles moving in the magnetic field with random component. The electric field of moving magnetic inhomogeneities (weak magnetohydrodynamic (MHD)-waves with the phase velocity v_{ph}, see Section 4.3) is v_{ph}/c times smaller than their magnetic field; we neglect it in the present problem. Let the total magnetic field consist of two components: the regular field $\boldsymbol{B}_0 = const$ and the random field $\boldsymbol{b}(\boldsymbol{r}, t)$. Averaging over an ensemble of random fields, we obtain $\langle \boldsymbol{b} \rangle = 0$. We assume that the random field is statistically homogeneous and isotropic, and in the plasma-comoving reference frame it is characterized by the correlation tensor

$$T_{\alpha\beta}(\boldsymbol{r}'_1 - \boldsymbol{r}'_2, t_1 - t_2) = \langle b_\alpha(\boldsymbol{r}'_1, t_1) b_\beta(\boldsymbol{r}'_2, t_2) \rangle. \tag{9.137}$$

Under the cosmic conditions, the plasma, as a rule, moves, participating in the large-scale motion, with a certain velocity $\boldsymbol{u}$, which we assume here to be the nonrelativistic constant quantity, $\boldsymbol{u} = const$. In this important case, the position vectors in the laboratory and co-moving reference frames are related by the equality $\boldsymbol{r} = \boldsymbol{r}' + \boldsymbol{u}t$. The velocity v of fast particles under consideration is assumed to be high in comparison with any of the plasma velocities:

$$v \gg u, \qquad v \gg v_{ph} \qquad (v \gg c_s,\ v_A), \tag{9.138}$$

because the velocity v_{ph} of motion of inhomogeneities in the plasma co-moving frame is determined by the sound and Alfven velocities (c_s and v_A, respectively). Under these assumptions the magnetic inhomogeneities in the frame of the medium can be considered static, and we can write down the correlation tensor in the laboratory frame, where the plasma is moving, in the form

$$\langle b_\alpha(\boldsymbol{r}_1, t_1) b_\beta(\boldsymbol{r}_2, t_2) \rangle = T_{\alpha\beta}(\boldsymbol{r}_1 - \boldsymbol{r}_2 - \boldsymbol{u}(t_1 - t_2)). \tag{9.139}$$

Thus, we neglect the phase velocities of MHD waves and their electric fields. However, the plasma velocity $\boldsymbol{u}$ can be significant (solar and stellar winds, $u \approx 300$–1000 km/s; supernova explosions, $u \approx 3000$–30000 km/s); it will be taken into account. The presence of this velocity produces the electric fields which in the laboratory reference frame are given by Equation (9.136).

Let $f(\boldsymbol{r}, \boldsymbol{p}, t)$ be the distribution function of fast particles not averaged over random fields. It satisfies the collisionless kinetic equation

$$\frac{\partial f}{\partial t} + \boldsymbol{v} \cdot \frac{\partial f}{\partial \mathbf{r}} + \boldsymbol{\mathcal{F}} \cdot \frac{\partial f}{\partial \mathbf{p}} = 0. \tag{9.140}$$

The force

$$\boldsymbol{\mathcal{F}} = e\boldsymbol{E} + \frac{e}{c}\boldsymbol{v} \times \boldsymbol{B} = \frac{e}{c}(\boldsymbol{v} - \boldsymbol{u}) \times \boldsymbol{B} \tag{9.141}$$

contains the electric and magnetic components. Let us rewrite Equation (9.140) in the form

$$\frac{\partial f}{\partial t} + \boldsymbol{v} \cdot \frac{\partial f}{\partial \mathbf{r}} + \boldsymbol{B} \cdot \widehat{\boldsymbol{D}} f = 0, \tag{9.142}$$

where

$$\widehat{\boldsymbol{D}} = \frac{e}{c}(\boldsymbol{v} - \boldsymbol{u}) \times \frac{\partial}{\partial \mathbf{p}} \tag{9.143}$$

is the operator changing the magnitude and direction of particle momentum. At this stage, we assume that the plasma velocity $\boldsymbol{u}$ is regular, and, hence, the operator $\widehat{\boldsymbol{D}}$ is also regular.

The case of small-scale inhomogeneities. Experimentally measured are the fluxes produced by many particles moving along their own random trajectories. This results in averaging of the trajectories. Such fluxes can be determined by the distribution function averaged over all possible realizations of random fields:

$$F(\boldsymbol{r}, \boldsymbol{p}, t) = \langle f(\boldsymbol{r}, \boldsymbol{p}, t) \rangle. \tag{9.144}$$

From the accurate collisionless equation (9.140), we will derive the equation for the averaged function F. Consider, first, the simplest case when

$$b \lesssim B_0, \qquad R_0 = \frac{cp}{eB_0} \gg L_c, \tag{9.145}$$

where R_0 is the Larmor radius of the particle in the regular field, and L_c is the correlation length of the random field. This is the case of small-scale inhomogeneities; in the limits of a single random inhomogeneity the trajectory is slightly curved, and the field of this inhomogeneity deflects the particle through a small angle (see Figure 9.7). We will look for the distribution function F of the particles, which already interacted with many inhomogeneities.

Upon separation, the averaged component from the distribution function,

$$f(\boldsymbol{r}, \boldsymbol{p}, t) = F(\boldsymbol{r}, \boldsymbol{p}, t) + \widetilde{f}(\boldsymbol{r}, \boldsymbol{p}, t), \qquad \langle \widetilde{f} \rangle = 0, \tag{9.146}$$

from Equation (9.142) we have the equation

$$\frac{\partial F}{\partial t} + \frac{\partial \widetilde{f}}{\partial t} + \boldsymbol{v} \cdot \frac{\partial F}{\partial \mathbf{r}} + \boldsymbol{v} \cdot \frac{\partial \widetilde{f}}{\partial \mathbf{r}} = \boldsymbol{B}_0 \cdot \widehat{\boldsymbol{D}} F + \boldsymbol{B}_0 \cdot \widehat{\boldsymbol{D}} \widetilde{f} + \boldsymbol{b} \cdot \widehat{\boldsymbol{D}} F + \boldsymbol{b} \cdot \widehat{\boldsymbol{D}} \widetilde{f}. \tag{9.147}$$

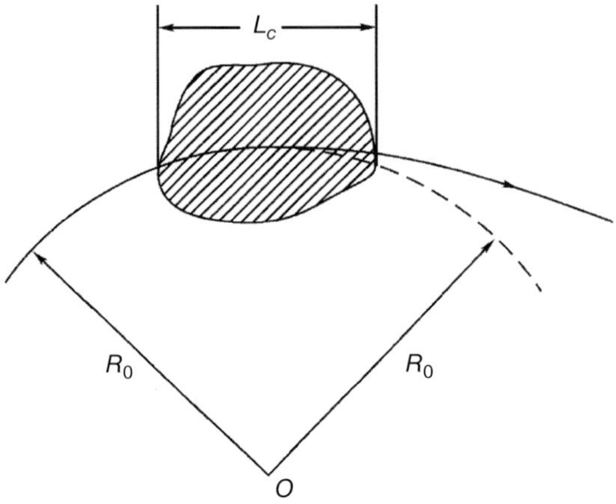

Figure 9.7 Motion of a particle within one small-scale inhomogeneity.

Now the left- and right-hand sides of Equation (9.147) will be averaged over the realizations of the random field. Using the conditions $\langle \boldsymbol{b} \rangle = \langle \tilde{f} \rangle = 0$, we obtain the exact equation

$$\frac{\partial F}{\partial t} + \boldsymbol{v} \cdot \frac{\partial F}{\partial \mathbf{r}} - \boldsymbol{B}_0 \cdot \hat{\boldsymbol{D}} F = \langle \boldsymbol{b}(\boldsymbol{r}) \cdot \hat{\boldsymbol{D}} \tilde{f} \rangle. \tag{9.148}$$

It involves the unknown average of the product of random quantities $\langle \boldsymbol{b}\tilde{f} \rangle$ (the operator $\hat{\boldsymbol{D}}$ is regular). In order to close Equation (9.148), we have to express $\langle \boldsymbol{b} \cdot \hat{\boldsymbol{D}} \tilde{f} \rangle$ in terms of the averaged function F. We perform this procedure approximately, allowing for the inequality

$$|\tilde{f}| \ll F, \tag{9.149}$$

that is, assuming that the additive correction to the distribution function is small compared to its average. We will see that inequality (9.149) follows from Equation (9.145).

Equation (9.148) should be subtracted from Equation (9.147). By doing so, we have

$$\frac{\partial \tilde{f}}{\partial t} + \boldsymbol{v} \cdot \frac{\partial \tilde{f}}{\partial \mathbf{r}} - \boldsymbol{B}_0 \cdot \hat{\boldsymbol{D}} \tilde{f} = \boldsymbol{b} \cdot \hat{\boldsymbol{D}} F + \boldsymbol{b} \cdot \hat{\boldsymbol{D}} \tilde{f} - \langle \boldsymbol{b} \cdot \hat{\boldsymbol{D}} \tilde{f} \rangle.$$

Using the condition (9.149), we drop in the right-hand side of the last equation the two terms which contain the small quantity $\tilde{f}$. Then we obtain the approximate inhomogeneous equation for the random component:

$$\frac{\partial \tilde{f}}{\partial t} + \boldsymbol{v} \cdot \frac{\partial \tilde{f}}{\partial \mathbf{r}} - \boldsymbol{B}_0 \cdot \hat{\boldsymbol{D}} \tilde{f} = \boldsymbol{b} \cdot \hat{\boldsymbol{D}} F \equiv Q(\boldsymbol{r}, \boldsymbol{p}, t). \tag{9.150}$$

Here Q is the right-hand side of the equation, which can be treated as a source for the random quantity $\tilde{f}$. It should be evaluated for the distances of the order of the correlation length L_c and for the times of the order of the time L_c/v, which is the time-of-flight of a particle through the correlation region, since the average $\langle \boldsymbol{b}\tilde{f} \rangle$ is proportional to the correlation tensor of the random field.

The solution of Equation (9.150) can be written in terms of the Green's function of the operator

$$\hat{L} = \frac{\partial}{\partial t} + \boldsymbol{v} \cdot \frac{\partial}{\partial \mathbf{r}} - \boldsymbol{B}_0 \cdot \hat{\boldsymbol{D}} : \qquad \hat{L}G(\boldsymbol{r}, \boldsymbol{p}, t; \boldsymbol{r}', \boldsymbol{p}', t') = \delta(\boldsymbol{r} - \boldsymbol{r}')\,\delta(\boldsymbol{p} - \boldsymbol{p}')\,\delta(t - t'). \tag{9.151}$$

Along the correlation length, the change of velocity u can be neglected because of the condition $u \ll v$. Moreover, if inequalities (9.145) hold, the particle trajectory of the particle within one correlation region is close to a straight line, the change in momentum can be neglected. Hence, the Green's function is approximately equal to the propagator of free motion:

$$G(\boldsymbol{r}, \boldsymbol{p}, t; \boldsymbol{r}', \boldsymbol{p}', t') = \delta(\boldsymbol{r} - \boldsymbol{r}' - \boldsymbol{v}(t - t'))\,\delta(\boldsymbol{p} - \boldsymbol{p}')\,\Theta(t - t'). \tag{9.152}$$

The additive correction to the distribution function takes the form

$$\begin{aligned} \tilde{f} &= \int_{-\infty}^{t} Q(\boldsymbol{r} - \boldsymbol{v}(t - t'), \boldsymbol{p}, t')\,dt' \\ &= \int_{-\infty}^{t} \boldsymbol{b}(\boldsymbol{r} - \boldsymbol{v}(t - t') - \boldsymbol{u}t') \cdot \hat{\boldsymbol{D}}F(\boldsymbol{r} - \boldsymbol{v}(t - t'), \boldsymbol{p}, t')\,dt'. \end{aligned} \tag{9.153}$$

Let us evaluate its order of magnitude: the integration domain over t' is of the order L_c/v, the operator $\hat{\boldsymbol{D}} \sim ev/cp$, $\tilde{f} \sim (ebL_c/cp)F \approx (L_c/r_g)F$, where $r_g \approx cp/eb$ is the Larmor radius of the particle in the random field. Using Equation (9.145), we have $L_c \ll r_g$ and $|\tilde{f}| \ll F$.

The term on the right-hand side of Equation (9.148) is expressed now through the averaged distribution function F and the correlation tensor:

$$\langle \boldsymbol{b}(\boldsymbol{r}) \cdot \hat{\boldsymbol{D}}\tilde{f} \rangle = \hat{D}_\alpha \int_{-\infty}^{t} T_{\alpha\beta}((\boldsymbol{v} - \boldsymbol{u})(t - t'))\hat{D}_\beta F(\boldsymbol{r} - \boldsymbol{v}(t - t'), \boldsymbol{p}, t')dt'. \tag{9.154}$$

Since, the tensor $T_{\alpha\beta}$ is negligible when the argument exceeds L_c, then, by the order of magnitude, $v(t - t') \approx L_c$. The averaged function F changes slightly along the correlation length, if the particle already passed through a few correlation regions. It also slightly changes over the time-of-flight $t - t' \approx L_c/v$ through one correlation region. Therefore, we can replace $\boldsymbol{r}'$ by t in the function F in Equation (9.154) and move this function out of the integral. Then the integral will contain the correlation tensor alone:

$$\langle \boldsymbol{b}(\boldsymbol{r}) \cdot \hat{\boldsymbol{D}}\tilde{f} \rangle = \hat{D}_\alpha \left(\int_0^{\infty} T_{\alpha\beta}((\boldsymbol{v} - \boldsymbol{u})\tau)\,d\tau \right) \hat{D}_\beta F(\boldsymbol{r}, \boldsymbol{p}, t). \tag{9.155}$$

In the case of statistically homogeneous and isotropic random field, the most general form of the correlation tensor is given by

$$T_{\alpha\beta}((\boldsymbol{v} - \boldsymbol{u})\tau) = T((\boldsymbol{v} - \boldsymbol{u})\tau)\,\delta_{\alpha\beta} + P((\boldsymbol{v} - \boldsymbol{u})\tau)\frac{(\boldsymbol{v} - \boldsymbol{u})_\alpha(\boldsymbol{v} - \boldsymbol{u})_\beta}{(\boldsymbol{v} - \boldsymbol{u})^2}, \tag{9.156}$$

where $T(r)$ and $P(r)$ are the scalar function satisfying the conditions

$$T(0) = \frac{1}{3}\langle b^2 \rangle, \qquad P(0) = 0, \tag{9.157}$$

$\langle b^2 \rangle$ is the averaged of squared random field. The second term in Equation (9.156) does not contribute to the right-hand side of Equation (9.155), because $(\boldsymbol{v}-\boldsymbol{u})_\beta \hat{\boldsymbol{D}}_\beta = 0$. Let us define the correlation length by the condition

$$\int_0^\infty T(r)dr = \frac{1}{3}\langle b^2 \rangle L_c. \tag{9.158}$$

Then

$$\int_0^\infty T(|\boldsymbol{v}-\boldsymbol{u}|\tau)d\tau = \frac{\langle b^2 \rangle L_c}{3|\boldsymbol{v}-\boldsymbol{u}|}$$

After substituting Equations (9.155) and (9.158) in the kinetic equation (9.148), it becomes closed:

$$\frac{\partial F}{\partial t} + \boldsymbol{v}\cdot\frac{\partial F}{\partial \mathbf{r}} - \boldsymbol{B}_0\cdot\hat{\boldsymbol{D}}F = \frac{1}{3}\langle b^2 \rangle L_c \hat{\boldsymbol{D}}\frac{1}{|\boldsymbol{v}-\boldsymbol{u}|}\hat{\boldsymbol{D}}F. \tag{9.159}$$

The right-hand side of this equation plays the role of the collision term and describes the interaction of fast particles with moving random inhomogeneities. The quantities $\boldsymbol{B}_0$, $\boldsymbol{u}$, and $\langle b^2 \rangle$ can be the slowly varying functions of coordinates; only then they can be regarded as constant at the correlation length. In addition, $u \ll v$, so that the difference $|\boldsymbol{v}-\boldsymbol{u}|$ does not vanish.

In the medium at rest $\boldsymbol{u} = 0$, and the operator $\hat{\boldsymbol{D}}$ takes the form

$$\hat{\boldsymbol{D}} = \frac{ec}{\mathcal{E}}\hat{\boldsymbol{\mathcal{O}}}, \qquad \text{where} \qquad \hat{\boldsymbol{\mathcal{O}}} = \boldsymbol{p}\times\frac{\partial}{\partial \mathbf{p}} \tag{9.160}$$

is the operator of rotation of the particle momentum. The absolute value of the momentum and the energy of the particles are not changed in this case, and the elastic scattering takes place only. Kinetic equation (9.159) is then simplified:

$$\frac{\partial F}{\partial t} + \boldsymbol{v}\cdot\frac{\partial F}{\partial \mathbf{r}} - \frac{ec}{\mathcal{E}}\boldsymbol{B}_0\cdot\hat{\boldsymbol{\mathcal{O}}}F = \frac{v}{2\Lambda(p)}\hat{\boldsymbol{\mathcal{O}}}^2 F, \tag{9.161}$$

$$\Lambda(p) = \frac{3r_g^2}{2L_c}, \qquad r_g^2 = \frac{c^2p^2}{e^2\langle b^2 \rangle}. \tag{9.162}$$

The quantity $\Lambda(p)$ is the transport path of the particle, that is, a length at which the particle is scattered through the angle of the order of unity. Let r_g^2 denote the squared Larmor radius in the random field and let us consider the case $r_g \gg L_c$, where $\Lambda(p) \gg r_g \gg L_c$ and $\Lambda(p) \propto p^2$. The relation between r_g and Λ can be arbitrary. The operator $\hat{\boldsymbol{\mathcal{O}}}^2$ in the right-hand side of Equation (9.161) is the angular part of the Laplace operator in momentum space. Note that in the present case the transport path depends only on the squared random field and is independent of the spectrum of magnetic inhomogeneities.

Example 9.16

Rearrange and simplify kinetic equation (9.161) for the case in which the momentum distribution of particles is nearly isotropic. Such a situation is realized at *distances $\Delta r \gg \Lambda$ from an arbitrary isotropic source. Represent the distribution function in the form*

$$F(\boldsymbol{r},\boldsymbol{p},t) = \frac{1}{4\pi}\left[N(\boldsymbol{r},p,t) + \frac{3}{v^2}\,\boldsymbol{v}\cdot\boldsymbol{J}(\boldsymbol{r},p,t)\right], \tag{9.163}$$

where

$$N(\boldsymbol{r},p,t) = \int F(\boldsymbol{r},\boldsymbol{p},t)\,d\Omega_p, \qquad \boldsymbol{J}(\boldsymbol{r},p,t) = \int \boldsymbol{v}F(\boldsymbol{r},\boldsymbol{p},t)\,d\Omega_p \tag{9.164}$$

are, respectively, the isotropic part of the distribution function and the flux density of particles with particular energy (under the condition that $J \ll vN$). Derive the closed equation for $N(\boldsymbol{r},p,t)$.

Solution. Substitute Equation (9.163) in Equation (9.161) and separate the terms which are independent of momentum orientation, and the terms proportional to $\boldsymbol{p}/p$. This yields two equations:

$$\frac{\partial N}{\partial t} + \nabla\cdot\boldsymbol{J} = 0, \tag{9.165}$$

$$\frac{\Lambda}{v}\frac{\partial \boldsymbol{J}}{\partial t} + \boldsymbol{J} + \frac{\Lambda}{R_0}\boldsymbol{b}_0\times\boldsymbol{J} = -\frac{v\Lambda}{3}\nabla N, \qquad \boldsymbol{b}_0 = \frac{\boldsymbol{B}_0}{B_0}. \tag{9.166}$$

The first term in Equation (9.166) can be evaluated as $(\Lambda/v)(J/t) \approx (\tau_s/t)J$, where $\Lambda/v = \tau_s$ is the time of isotropization of the particles. The anisotropy of the distribution function will be a weak effect at time intervals $t \gg \tau_s$, when the first term can be dropped. From the remaining Equation, we find the relation between the flux density and the number density gradient:

$$J_\alpha = -\kappa_{\alpha\beta}\nabla_\beta N, \qquad \text{where} \quad \kappa_{11} = \kappa_{22} = \kappa_\perp = \frac{\kappa_\parallel R_0^2}{R_0^2+\Lambda^2}, \quad \kappa_{33} = \frac{v\Lambda}{3} = \kappa_\parallel,$$

$$\kappa_{12} = -\kappa_{21} = \kappa_H = \frac{\kappa_\parallel \Lambda R_0}{R_0^2+\Lambda^2}. \tag{9.167}$$

Here the axis numbered 3 is directed along the field $\boldsymbol{B}_0$.

The diffusion tensor is anisotropic due to the presence of the large-scale magnetic field. It has the same structure as the tensor of electric conductivity in the external magnetic field (see Problem 7.30). Non-diagonal components κ_H describe the effect of twisting of particles by the external magnetic field (the Hall effect). At $B_0 \to 0$ ($R_0 \to \infty$), the diffusion becomes isotropic: $\kappa_{\alpha\beta} = \kappa\delta_{\alpha\beta}$, $\kappa = \kappa_\perp = \kappa_\parallel = v\Lambda/3$, $\kappa_H = 0$. At $R_0 \ll \Lambda$ the tensor becomes strongly anisotropic:

$$\kappa_\perp \approx \kappa_\parallel\left(\frac{R_0}{\Lambda}\right)^2, \qquad \kappa_H \approx \kappa_\parallel\frac{R_0}{L}, \qquad \kappa_\perp \ll \kappa_H \ll \kappa_\parallel. \tag{9.168}$$

Upon substitution, the flux density (9.167) in the continuity equation (9.165), we generally obtain the equation of anisotropic diffusion:

$$\frac{\partial N}{\partial t} = \nabla_\alpha\kappa_{\alpha\beta}\nabla_\beta N. \quad \blacksquare \tag{9.169}$$

Example 9.17

In the case of isotropic diffusion, $\kappa_{\alpha\beta} = \kappa\,\delta_{\alpha\beta}$ *and* $\kappa = const$*, derive the diffusion equation with account for the term* $(\Lambda/v)\,\partial \boldsymbol{J}/\partial t$ *in Equation (9.166).*

Solution. We differentiate Equation (9.165) with respect to time and substitute the derivative

(1) $$\frac{\partial \boldsymbol{J}}{\partial t} = -\frac{v}{\Lambda}(\boldsymbol{J} + \kappa\nabla N).$$

Then, in order to exclude $\nabla \cdot \boldsymbol{J}$, we use Equation (9.165) again and obtain the diffusion equation with the second time derivative:

$$\frac{\partial N}{\partial t} + \frac{\Lambda}{v}\frac{\partial^2 N}{\partial t^2} = \kappa \triangle N. \quad \blacksquare \tag{9.170}$$

The effects, which are described by the more accurate diffusion equation (9.170), are discussed in Problem 9.39.

Diffusion of particles in a strong magnetic field. In nature, the turbulent magnetic field is as a rule characterized by a very wide range of scales. For instance, the scientists have discovered in Galaxy the turbulence with the maximum length scale of the order 3×10^{20} cm $\approx$100 pc, and the minimal length scale by ten orders of magnitude smaller. Therefore, in many cases the above condition $R_0 \gg L_c$ is not satisfied. The following condition is more realistic:

$$L_c \gg R_0 \gg l_m, \tag{9.171}$$

where l_m is the minimal scale of the turbulence, and $R_0 = cp/eB_0$ is the Larmor radius of a particle in a quasi-homogeneous magnetic field. For this case, we will give an estimate of the components of the diffusion tensor. A more accurate consideration based on the averaging of the kinetic Equation can be found in the Toptygin's book (1985).

If there is continuum of scales of turbulent pulsations in a given interval (Figure 9.8), then small-scale ($l < R_0$) and large scale ($l > R_0$) magnetic inhomogeneities will act on the particle in various ways. Small-scale inhomogeneities will cause the scattering of the particle and form its transport path along the regular field $\boldsymbol{B}_0$. This path $\Lambda_\parallel$ can be estimated from Equation (9.162), where L_c has to be replaced by R_0, and $\langle b^2\rangle$ is the turbulent field, which is produced by the small-scale ($R_0 \geq l \geq l_m$) part of the spectrum. We will assume that the turbulent spectrum can be approximated by the power-low with some index ν:

$$\langle b^2\rangle_k = (\nu - 1)\langle B_t^2\rangle \frac{k_c^{\nu-1}dk}{k^\nu}, \qquad \nu > 1, \tag{9.172}$$

where $\langle B_t^2\rangle$ is the total turbulent field. Integrating Equation (9.172) from $k = k_0$ to $k = k_m$ ($k_0 \ll k_m$), we find

$$\langle b^2\rangle = \int_{k_0}^{k_m} \langle b^2\rangle_k dk = \langle B_t^2\rangle \left(\frac{R_0}{L_c}\right)^{\nu-1}.$$

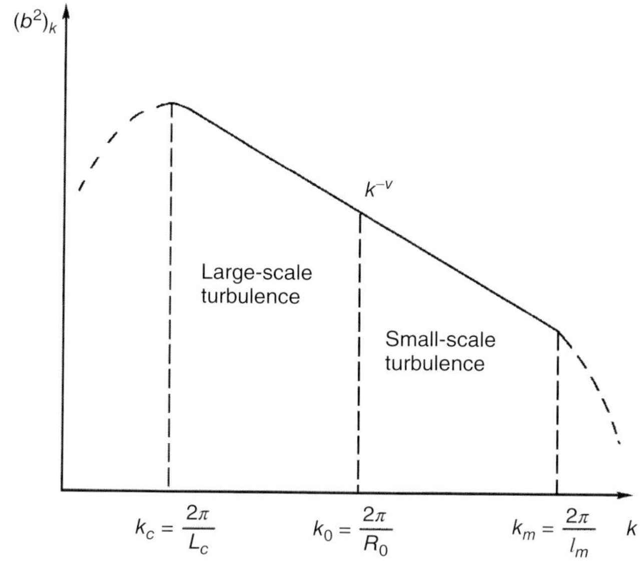

Figure 9.8 Schematic image of a spectrum of a turbulent magnetic field.

Using Equation (9.162), with L_c replaced by R_0, we have

$$\Lambda_{\|} \approx \frac{B_0^2}{\langle B_t^2 \rangle} \left(\frac{R_0}{L_c} \right)^{2-\nu} L_c. \tag{9.173}$$

We have estimated the path up to an unknown dimensionless factor of the order of unity. More accurate expression and other models of turbulence can be found in Toptygin (1985).

The qualitative difference between Equations (9.173) and (9.162) is that in the given case the path of particles depends on the form of turbulent spectrum. The momentum dependence of the form $\Lambda_{\|} \sim p^{2-\nu}$, obtained above, means an increase in the path with increasing energy at $\nu < 2$, the constancy of the path at $\nu = 2$, and a decrease of the path with decreasing energy of the particle at $\nu > 2$. The latter case seems unlikely, because of the absence of a turbulence with such a spectrum. The longitudinal diffusion constant is given by the standard formula $\kappa_{\|} = v\Lambda_{\|}/3$. Let us recall that the Kolmogorov–Obukhov turbulence corresponds to $\nu = 5/3$, and the weak MHD turbulence corresponds to $\nu = 3/2$. The value $\nu = 2$ is realized when the turbulence is produced by shock waves or by other explosions.

Transverse diffusion in random large-scale field.

We now turn to the role of large-scale magnetic inhomogeneities. The fields of scale $l > R_0$ affect the particle adiabatically. The particle moves in such a field conserving the transverse adiabatic invariant $p_{\perp}^2/B = const$ (see Section 4.2). The guiding center of the particle moves mainly along the force line, with small transverse drifts. However, the force line belongs to the resultant large-scale field, which consists of the field $\boldsymbol{B}_0$ and of the large-scale turbulent field (denoted by $\widetilde{\boldsymbol{B}}$). Due to the presence of the random component of the field, the force lines have a rather

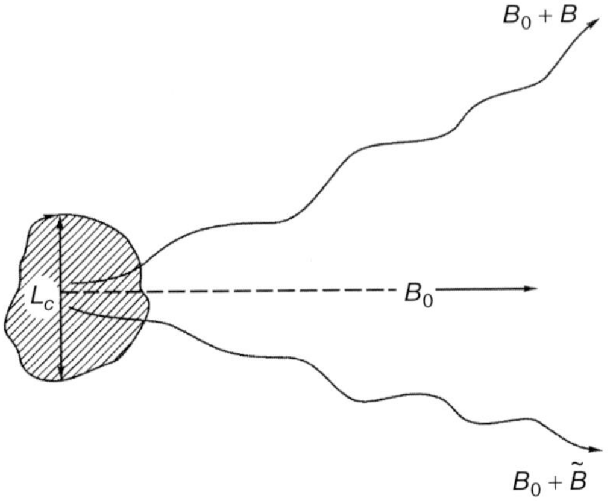

Figure 9.9 Walk of magnetic force lines.

complicated and intricate form. In particular, two close force lines, outgoing from the same correlation region of the field, may become widely separated in several correlation lengths (Figure 9.9). This effect is referred to as a **walk of magnetic force lines**. At $\tilde{B} \ll B_0$ it slightly appears in the longitudinal (relative to $\boldsymbol{B}_0$) motion of particles. However, it can significantly enhance their transverse diffusion.

Let us estimate the transverse diffusion coefficient of particles due to the walk of magnetic force lines. Consider two cases.

(a) Let $\Lambda_\parallel \gg L_c$, the particles travel through the correlation region of the field almost without scattering. The transverse path in the limits of the correlation region (the length of one pitch) is $l_\perp \approx B_t L_c / B_0$, where $B_t \ll B_0$ is the average of the random field (almost all energy of the random field belongs to the large-scale harmonics). In a time t, the particle travels $N = |v_\parallel| t / L_c$ pitches. When a walk is random, the path $\Delta L_\perp$ traveled by particles in the transverse direction during time t is proportional to the square root of the number of pitches, $\Delta L_\perp \approx l_\perp N^{1/2}$. Hence, we have $(\Delta L_\perp)^2 \approx (\langle B_t^2 \rangle / B_0^2) L_c |v_\parallel| t$. Comparing this expression with $\overline{x^2} = 4\kappa_\perp t$ (see Problem 9.38), we find $\kappa_\perp \approx (\langle B_t^2 \rangle / B_0^2) L_c |v_\parallel|$. The last expression should be averaged over the pitch-angle. The more accurate calculation based on kinetic equation gives

$$\kappa_\perp = \frac{1}{3} v \Lambda_\perp, \qquad \Lambda_\perp = \frac{\pi^{1/2} \Gamma(\nu/2) \langle B_t^2 \rangle}{2\Gamma(\nu/2 - 1/2) B_0^2} L_c. \tag{9.174}$$

In the second case, $\Lambda_\parallel \ll L_c$, the particles diffuse through the correlation region. Unfortunately, the attempts to evaluate the transverse diffusion in a simple way, similar to that given above, have not been successful. Here we present only the results (see Bykov and Toptygin (1993)), obtained with the use of more complicated methods. For a weak Alfven turbulence ($\epsilon = \langle \tilde{B}^2 \rangle / B_0^2 \ll 1$) at $v_A L_c \gg v \Lambda_\parallel$, the transverse diffusion coefficient is estimated as

$$\kappa_\perp \approx \epsilon \kappa_\parallel. \tag{9.175}$$

In the interval $v_A L_c \ll \kappa_\| \ll v_A L_c \epsilon^{-1}$ the estimate is changed:

$$\kappa_\perp \approx \epsilon v_A L_c. \tag{9.176}$$

Finally, at $\kappa_\| \gg v_A L_c \epsilon^{-1}$ the fluctuations can be treated as quasi-stationary, and the transverse diffusion coefficient becomes proportional to the forth power of the ratio of the random and regular fields:

$$\kappa_\perp \approx \epsilon^2 \kappa_\|. \tag{9.177}$$

In many cases, the transverse diffusion coefficients caused by a walk of force lines, significantly exceed the coefficient $\kappa_\perp \approx \kappa_\|(R_0/\Lambda_\|)^2$ due to scattering of particles by small inhomogeneities. If the large-scale random field is of the same order of magnitude as the regular field, then the transverse and longitudinal diffusion coefficients become of the same order of magnitude, and the diffusion is close to isotropic.

Kinetic equation in a strong magnetic field. If the regular magnetic field is so strong that the period of Larmor gyration of fast particles, $T = 2\pi\mathcal{E}/ecB_0$, is much less than the duration of the free-flight-time $\tau_s = \Lambda_\|/v$, then the kinetic equation can be written in the approximation of guiding center (see subsection "Approximated methods" of Section I.4.2). Such particles are sometimes called **magnetized.** The distribution function of the particles should be averaged over gyration around force lines of the magnetic field, with the result that the number of arguments decreases to six: $F(p_\perp, p_\|, \boldsymbol{r}, t)$, where $p_\| = p\cos\vartheta$, $p_\perp = p\sin\vartheta$, and ϑ is an angle between the momentum and the magnetic field (pitch-angle). Below we present the kinetic equation for the case under consideration:

$$\frac{\partial F}{\partial t} + v_\| \frac{\partial F}{\partial s} + \dot{p}_\perp \frac{\partial F}{\partial p_\perp} + \dot{p}_\| \frac{\partial F}{\partial p_\|} = \frac{1}{\sin\vartheta}\frac{\partial}{\partial\vartheta} b_s(\vartheta)\sin\vartheta \frac{\partial F}{\partial\vartheta} + \kappa_\perp \triangle_\perp F. \tag{9.178}$$

Here s is the coordinate measured along a magnetic force line; the terms containing $\dot{p}_\perp$ and $\dot{p}_\|$ describe the change of momentum components produced by a slow change of the large-scale field in space and time; the operator with the derivatives over angle accounts for the scattering of particles by small-scale inhomogeneities. It generalizes the operator $\widehat{\boldsymbol{\mathcal{O}}}^2$ entering Equation (9.161); $b_s(\vartheta)$ is the diffusion coefficient over a pitch-angle. Finally, the last term in Equation (9.178) describes the transverse component caused by the walk of magnetic force lines.

If the magnetic field can be regarded as static, then the energy of particles is conserved and the derivatives over the components of the momentum in Equation (9.178) can be transformed with the use of the adiabatic invariant $p_\perp^2/B(s) = const$ (see Equation (4.89)). The equation is slightly simplified and takes the form

$$\frac{\partial F}{\partial t} + v\cos\vartheta \frac{\partial F}{\partial s} + \frac{v}{2B}\frac{\partial B}{\partial s}\sin\vartheta\frac{\partial F}{\partial\vartheta} = \frac{1}{\sin\vartheta}\frac{\partial}{\partial\vartheta} b_s(\vartheta)\sin\vartheta \frac{\partial F}{\partial\vartheta} + \kappa_\perp \triangle_\perp F. \tag{9.179}$$

By going from this equation to the diffusion approximation, we obtain the equation of anisotropic diffusion.

Taking into account the electric field and variation of particle energy. In the preceding considerations, we mostly took into account the magnetic field, assuming that the medium is at rest ($\boldsymbol{u} = 0$). The inclusion of electric fields induced by medium motion leads to energy variation of particles (to their acceleration or deceleration).

Example 9.18

Derive the diffusion type equation from the kinetic equation (9.159) with account for the medium motion. Assume that the velocity $\boldsymbol{u}$ *is the slow function of coordinates,* $\boldsymbol{u} = \boldsymbol{u}(\boldsymbol{r})$*, and slightly varies along the scales* $\sim L_c$*. Use the representation of the distribution function in the form (9.163) and take into account the small parameter* $u/v \ll 1$ *up to terms of the second order.*

Solution. We perform the same calculations as in Example 9.16, but taking account for the velocity $\boldsymbol{u}$ in the second order:

(1) $$\frac{1}{|\boldsymbol{v}-\boldsymbol{u}|} \approx \frac{1}{v}\left[1 + \frac{\boldsymbol{u}\cdot\boldsymbol{v}}{v^2} + \frac{3(\boldsymbol{u}\cdot\boldsymbol{v})^2}{2v^4} - \frac{u^2}{2v^2}\right].$$

We have two equations:

(2) $$\frac{\partial N}{\partial t} + \nabla\cdot\boldsymbol{J} = \frac{u^2}{9\kappa}\left[p^2\frac{\partial^2 N}{\partial p^2} + \left(1+\frac{v^2}{c^2}\right)p\frac{\partial N}{\partial p}\right] + \frac{1}{3\kappa}\boldsymbol{u}\cdot\left(p\frac{\partial \boldsymbol{J}}{\partial p} + \frac{v^2}{c^2}\boldsymbol{J}\right)$$
$$+\frac{1}{R_0 v}[\boldsymbol{u}\times\boldsymbol{b}_0]\cdot\left[p\frac{\partial \boldsymbol{J}}{\partial p} + \left(1+\frac{v^2}{c^2}\right)\boldsymbol{J}\right],$$

(3) $$\frac{\Lambda}{v}\frac{\partial \boldsymbol{J}}{\partial t} + \boldsymbol{J} + \frac{\Lambda}{R_0}\boldsymbol{b}_0\times\boldsymbol{J} = -\kappa\nabla N - \frac{p}{3}\frac{\partial N}{\partial p}\left(\boldsymbol{u} + \frac{\Lambda}{R_0}\boldsymbol{b}_0\times\boldsymbol{u}\right).$$

Compared with Equations (9.165) and (9.166), some new terms have appeared. In particular, the equation of continuity (2) contains now the source (the right-hand side). It is associated with energy variation of particles under the action of the electric field. Hence, now the number of particles with particular energy does not remain constant.

Omitting again the first term in Equation (3), we solve this equation relative to $\boldsymbol{J}$ and obtain the flux density of particles with particular momentum p:

$$J_\alpha = -\kappa_{\alpha\beta}\nabla_\beta N - \frac{p}{3}\frac{\partial N}{\partial p}u_\alpha. \tag{9.180}$$

In addition to the diffusion term, there appear the convection terms in the expression for the current which is proportional to $\boldsymbol{u}$ and is produced by a drag of isotropically distributed particles by moving magnetic inhomogeneities. The convection term has been derived in Problem 9.40 in different manner.

After substituting Equation (9.180) in Equation (2), all terms proportional to u^2, are simplified, leading to a simpler transport equation:

$$\frac{\partial N}{\partial t} = \nabla_\alpha\kappa_{\alpha\beta}\nabla_\beta N - \boldsymbol{u}\cdot\nabla N + \frac{p}{3}\frac{\partial N}{\partial p}\nabla\cdot\boldsymbol{u}. \tag{9.181}$$

The additional terms (as compared to Equation (9.169)) which contain $\boldsymbol{u}$, describe convective transport of particles and variation of their energy.

Acceleration of particles by small-scale electric fields. Before analyzing physical meaning and properties of Equation (9.181), we perform some generalization. Up to now we have considered $\boldsymbol{u}$ as the regular velocity of the medium, which can be a function of time and coordinates and which is weakly changing along the length-scales L_c and times L_c/v. In reality, along with transport of magnetic inhomogeneities by an overall large-scale motion of the medium, their relative motion often occurs. It can be due to fluctuations of regular velocity (solar or stellar winds) or due to evolution of magnetic field itself (MHD waves, rotational discontinuities, solitons, weak shock waves and others). We assume that the spatial scale of velocity fluctuations does not exceed the transport path of particles Λ.

A spread of velocities of the magnetic inhomogeneities is equivalent to the appearance of random small-scale electric fields affecting the energy of the particles (in Equation (9.181) only the large-scale electric field is taken into account). Since the velocities of the magnetic structures $v_{ph} < u \ll v$, then the particle energy is changed by a small value on the correlation length, and the acceleration effect is revealed at large times, when the isotropization of the particles has already occurred. In order to take proper account of the acceleration effect produced by small-scale electric fields, we have to incorporate them into the averaging procedure, which led us previously to Equation (9.159). To avoid cumbersome mathematical operations (they can be found in the Toptygin's book (1983)), we give here a quasi-quantitative (order of magnitude) estimate of this effect. The general form of the equation taking into account the acceleration effect due to electric fields, can be derived from general physical considerations. The random change of the particle momenta under the action of weak electric fields results in the diffusion of the particles in momentum space. This effect can be described by the operator of the same form as in Equation (9.181) for the diffusion in usual space. However, in view of the isotropization of particles it is sufficient to retain only the derivatives of absolute value of the momentum:

$$\nabla_\alpha \kappa_{\alpha\beta} \nabla_\beta N \rightarrow \frac{1}{p^2}\frac{\partial}{\partial p} p^2 D(p) \frac{\partial N}{\partial p}. \tag{9.182}$$

Here $D(p)$ is the diffusion coefficient in momentum space. Its evaluation is described below.

Evaluation of acceleration time. Apart from considering regular motion of the medium, we assume that the velocity $\boldsymbol{u}$ is random: $\langle \boldsymbol{u} \rangle = 0$. The collision frequency with magnetic inhomogeneities (the inverse scattering time) is $\nu_s = 1/\tau_s = v/\Lambda$. This quantity results from random magnetic force $\mathcal{F}_m = (e/c)[\boldsymbol{v} \times \boldsymbol{b}]$ and is proportional to its square $\nu_s \propto \langle \mathcal{F}_m^2 \rangle$ because the linear force vanishes after averaging, $\langle \mathcal{F}_m \rangle = 0$.

The random electric field $\mathcal{F}_e = -(e/c)[\boldsymbol{u} \times (\boldsymbol{B}_0 + \boldsymbol{b})]$ also vanishes after averaging, $\langle \mathcal{F}_e \rangle = 0$. This is because in a plasma at rest (as a whole) the large-scale electric field cannot exist. Therefore, the frequency of acceleration of the particle is proportional to the square of $\mathcal{F}_e$: $\nu_a \propto \langle \mathcal{F}_e^2 \rangle$. Consequently, the acceleration time $\tau_a = 1/\nu_a$ is inverse to the averaged squared force. As a result, we can write down an order of magnitude estimate

$$\nu_a \approx \nu_s \frac{\langle \mathcal{F}_e^2 \rangle}{\langle \mathcal{F}_m^2 \rangle} \approx \frac{v}{\Lambda} \frac{\langle u^2 \rangle}{v^2} \frac{B_0^2 + b^2}{b^2} . \tag{9.183}$$

When the general regular motion occurs, instead of $\langle u^2 \rangle$, the random velocity component will enter the formula, $\langle \Delta u^2 \rangle = \langle u^2 \rangle - \langle \boldsymbol{u} \rangle^2$. The diffusion coefficient and the acceleration frequency ν_a are related by $\nu_a \approx D/p^2$, which results in the estimate

$$D(p) \approx \frac{\langle \Delta u^2 \rangle p^2}{v\Lambda(p)} \frac{B_0^2 + b^2}{b^2} . \tag{9.184}$$

When the medium consists of "plasma clouds" moving in a more rarefied plasma, so that $B_0 < b$, then $D(p) \approx \langle \Delta u^2 \rangle p^2 / v\Lambda(p)$. In the case of MHD waves of small amplitude $b \ll B_0$, the plasma velocity $u \approx v_{ph} b/B_0$, which gives $D(p) \approx v_{ph}^2 p^2 / v\Lambda$. Here the velocity of the disturbance v_{ph} is determined by Alfven velocity v_A and sound velocity c_s. The case in which the disturbance is produced by a strong shock wave requires special consideration (see below).

The transport equation that includes the acceleration effect of particles by the weak random electric fields can be obtained by adding the term (9.182) on the right-hand side of Equation (9.181) :

$$\frac{\partial N}{\partial t} = \nabla_\alpha \kappa_{\alpha\beta} \nabla_\beta N - \boldsymbol{u}_0 \cdot \nabla N + \frac{p}{3} \frac{\partial N}{\partial p} \nabla \cdot \boldsymbol{u}_0 + \frac{1}{p^2} \frac{\partial}{\partial p} p^2 D(p) \frac{\partial N}{\partial p} , \tag{9.185}$$

where $\boldsymbol{u}_0$ is the regular velocity of the medium. Using this equation, it is possible to describe the diffusion of particles in space, their convective transport and their energy variation due to the regular motion of the medium and under the action of the small-scale electric fields.

General properties of the transport equation. Energy source. To analyze the general properties, it is convenient to rewrite Equation (9.185) in a form that would include the divergence of the flux density of particles in the six-dimensional phase space:

$$\frac{\partial N}{\partial t} + \nabla \cdot \boldsymbol{J} + \frac{1}{p^2} \frac{\partial}{\partial p} p^2 S_p = 0. \tag{9.186}$$

Here $\boldsymbol{J}$ is the flux density of the particles of particular energy in coordinate space, determined by Equation (9.180), and

$$S_p = \frac{p}{3} \boldsymbol{u}_0 \cdot \nabla N - D(p) \frac{\partial N}{\partial p} \tag{9.187}$$

is the projection of the flux density of particles on the direction of $\boldsymbol{p}$ in momentum space. Projections on other axes are absent because the particles have been isotropized. The last term on the left-hand side of Equation (9.186) is the divergence of the flux density $\boldsymbol{S}$.

After the integration of Equation (9.186) over $p^2 dp$, we get the continuity equation in coordinate space:

$$\frac{\partial n}{\partial t} + \nabla \cdot \boldsymbol{j} = 0, \tag{9.188}$$

where $n(\mathbf{r},t) = \int_0^\infty N(\mathbf{r},p,t)p^2\,dp$ is the number density of particles of all energies, and $\mathbf{j}(\mathbf{r},t) = \int_0^\infty \mathbf{J}(\mathbf{r},p,t)p^2\,dp$ is the total flux density of the particles.

The integration of Equation (9.186) over $\mathcal{E}p^2 dp$ gives the energy balance equation:

$$\frac{\partial w}{\partial t} + \nabla \cdot \mathbf{q} = Q(\mathbf{r},t). \tag{9.189}$$

Here

$$w(\mathbf{r},t) = \int_0^\infty \mathcal{E}(p)N(\mathbf{r},p,t)p^2\,dp = \overline{\mathcal{E}}n(\mathbf{r},t) \tag{9.190}$$

is the energy density of relativistic particles, and

$$\mathbf{q}(\mathbf{r},t) = \int \mathbf{v}\mathcal{E}(p)F(\mathbf{r},p,t)\,d^3p = \int_0^\infty \mathcal{E}(p)\mathbf{J}(\mathbf{r},p,t)p^2\,dp \tag{9.191}$$

is the energy flux density. The latter quantity consists of the diffusive flux density

$$q_\alpha^{(d)} = -\int_0^\infty \mathcal{E}(p)\kappa_{\alpha\beta}\nabla_\beta N(\mathbf{r},p,t)p^2 dp \tag{9.192}$$

and of the convective flux density

$$\mathbf{q}^{(c)} = (w+P)\mathbf{u}_0. \tag{9.193}$$

The quantity

$$P(\mathbf{r},t) = \frac{1}{3}\int_0^\infty p\,vN(\mathbf{r},p,t)p^2\,dp \tag{9.194}$$

is the gas pressure of fast (possibly relativistic) particles. In the nonrelativistic case $v = p/m$, and Equation (9.194) can be expressed through the average kinetic energy $K = p^2/2m$ of the particles: $P = (2/3)\overline{K}n$. In particular, if the gas is in the state of statistic equilibrium, then $\overline{K} = (3/2)T$, where T is the temperature, and we obtain the well-known expression $P = nT$. In the strongly relativistic case $p\,v \approx p\,c \approx \mathcal{E}$, and the pressure $P = \overline{\mathcal{E}}n/3$, which agrees with the pressure of the gas of photons, provided that $\overline{\mathcal{E}}$ is the energy of a photon (see Problem I.6.27). The density of the convective flux (9.193) can be expressed in the form

$$\mathbf{q}^{(c)} = h(\mathbf{r},t)\mathbf{u}_0, \qquad \text{where} \qquad h = w + P \tag{9.195}$$

is the enthalpy density of the gas of fast particles. This is in accordance with the general expression for the energy flux in the continuum mechanics (see Landau and Lifshitz (1987)). Needless to say that we have assumed that the distribution function of energetic particles sufficiently rapidly decreases at large momenta, so that all integrals converge and all energy averages are finite.

The quantity $Q(\mathbf{r},t)$ on the right-hand side of Equation (9.189) has meaning of the energy source. It consists of two terms, which can be reduced to the following form by integrating by parts:

$$Q(\mathbf{r},t) = \mathbf{u}_0 \cdot \nabla P + \int_0^\infty N(\mathbf{r},p,t)\frac{d}{dp}\left(vp^2 D(p)\right)dp. \tag{9.196}$$

The first term is sign-changing and depends on the angle between the velocity of large-scale motion and the pressure gradient of fast particles. The second term is always nonnegative, because $N \geq 0$ and $d(vp^2D(p))/dp > 0$. The last inequality follows from Equation (9.184): $D(p) \sim p^2/v\Lambda(p)$, and $\Lambda(p)$ increases not rapidly than p^2 (see Equation (9.162)). Therefore, the product $vp^2D(p) \sim p^\alpha$, where $\alpha \geq 2$, and the derivative, which is the quantity of our interest, are positive.

Example 9.19

A cloud of fast particles at the initial instant of time occupied the bounded volume. Obtain and discuss the change dE/dt in the total energy of the particles. There exist the small-scale electric fields in the medium and the large-scale motion with an nonuniform velocity $\boldsymbol{u}(\boldsymbol{r})$.

Solution. We integrate both sides of Equation (9.189) over all space and find the relation

$$\frac{dE}{dt} = -\int P(\boldsymbol{r}, t)\, \nabla \cdot \boldsymbol{u} d^3r + \int d^3r \int_0^\infty N(\boldsymbol{r}, p, t) \frac{d}{dp}(vp^2D(p))\, dp. \quad (9.197)$$

The first term in the right-hand part contains, in the integrand, the divergence of the large-scale velocity, which characterizes the local expansion or compression of the medium: $div\, \boldsymbol{u} > 0$ corresponds to the expansion, $div\, \boldsymbol{u} < 0$ corresponds to the compression. One can easily prove that it is true by tracing a small macroscopic element of mass $M = \rho V$ of the moving medium. In the process of motion $M = const, dM/dt = 0$, and, hence, $(1/V)\, dV/dt = -(1/\rho)\, d\rho/dt$. Now we use the continuity equation for the medium to obtain $d\rho/dt = \partial\rho/\partial t + \boldsymbol{u} \cdot \nabla\rho = -\rho\nabla \cdot \boldsymbol{u}$, $(1/V)\, dV/dt = \nabla \cdot \boldsymbol{u}$.

Thus, the first term in Equation (9.197) describes the energy gain of particles (acceleration) as the medium undergoes the compression, and the particle deceleration as it undergoes the expansion. This effect has the same nature as changes in temperature of an ideal gas with the adiabatic change of the vessel volume: a gas is cooled as the vessel expands and heated as the vessel is compressed. In the case under consideration **the adiabatic energy change** of the particles occurs due to their interaction with macroscopic excitations of the medium (magnetic inhomogeneities).

The second term in Equation (9.197) describes the acceleration of particles due to a random electric field (in other language, due to collisions with randomly moving magnetic inhomogeneities). The latter inhomogeneities, being macroscopic objects, possess the energy greatly exceeding the energy of the fast particle. This happens even at low velocities of the inhomogeneities, because of their large mass. In the process of interaction of the two subsystems (particles and inhomogeneities), the equilibration of their temperatures occurs, that is, the increase in the energy of the particles. The outstanding Italian physicist Fermi (1949) was the first who used the model of acceleration of particles in their collisions with magnetized clouds of cosmic plasma to explain the existence of high energy of cosmic rays. ■

Let us emphasize once again, that all the equations presented above describe the propagation and acceleration of fast particles only, that is, of those particles of the medium whose energy, due to random nonequilibrium processes, turns out to be several times higher than the average energy per one particle. The portion of such particles is rather small. In equilibrium media where particles obey the Maxwell distribution, the fraction of particles with energies significantly exceeding the average energy is exponentially small; in such a media, the accelerated particles do not appear. However, in nonequilibrium media, the fractions of suprathermal particles becomes noticeable, and these particles can be injected in the acceleration process.

It should be noted, that the problem of injection of particles in different accelerated mechanisms is very complicated and needs its solution till now. Therefore, in most cases, it is impossible to determine theoretically which fractions of particles is accelerated in specific situations, and we do not study this question in this chapter. At different ranges of particle energies the various acceleration mechanisms can be involved, so that acceleration processes may consists of many stages.

Acceleration of particles near a shock front. Here we consider the process of particle acceleration near an MHD shock front. The problem has been widely discussed in the literature as the most promising for explaining the origin of relativistic accelerated particles in our stellar system and the Galaxy. The energy gain of particles near the front occurs under two necessary conditions: the presence of turbulent magnetic fields (magnetic inhomogeneities) which induce random walks of particles, and the presence of the flux of scattering centers, whose velocity undergoes a jump (deceleration) at the shock front (Figure 9.10). As a result, the particles near the front can be confined between two walls approaching each other. The walls are realized by the fluxes of magnetic inhomogeneities and have a width

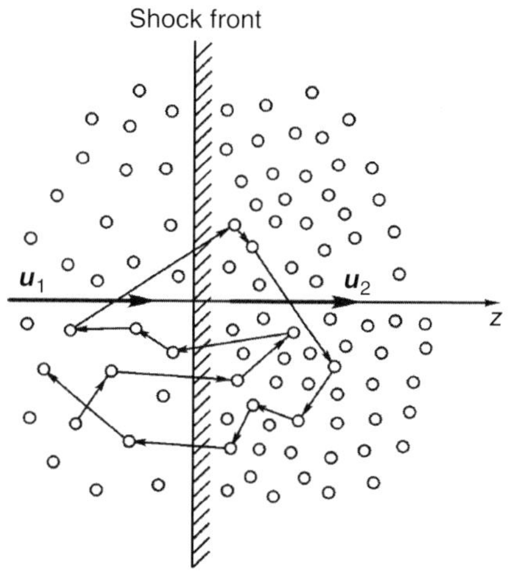

Figure 9.10 Wandering of accelerated particles in the vicinity of a shock front in turbulent medium.

of the order of Λ, that is, of the transport path of the particles. The more times the particle crosses the front, the more energy it gains in collisions with magnetic inhomogeneities. However, sooner or later any particle is drifted beyond the front with flow of matter. The source to increase the particle energy is the gas dynamic energy of the matter in which the shock wave is excited.

Now we give a quantitative estimate and apply Equation (9.197) to the plane shock front, neglecting the turbulent acceleration (i.e., the second term). In the one-dimensional case at the shock front, we have

$$\nabla \cdot \boldsymbol{u} = \frac{\partial u}{\partial z} = -(u_1 - u_2)\delta(z) = -\Delta u \delta(z). \tag{9.198}$$

Therefore, from Equation (9.197) we obtain the energy increase of particles near the front

$$\frac{dE}{dt} = \Delta u \int P(\boldsymbol{r}, t)\, d^2 r > 0, \tag{9.199}$$

where integration is made on a front surface.

However, the situation becomes more complicated in the three-dimensional case, because the shock front, where the compression of matter always occurs, can be followed by the wave of rarefaction in which $\nabla \cdot \boldsymbol{u} > 0$. The energy balance takes the form

$$\frac{dE}{dt} = \Delta u_n \int P(\boldsymbol{r}, t) d^2 r - \int P(\boldsymbol{r}, t) \nabla \cdot \boldsymbol{u}\, d^3 r, \tag{9.200}$$

where first positive term corresponds to the acceleration at the front, and the second term corresponds to the adiabatic deceleration of particles in the rarefaction wave. The result will be determined by the combined action of both processes.

Example 9.20

At the plane stationary shock wave, there appears the source $Q = Q_0 \delta(z)\delta(p - p_0)/p_0^2$, which injects the particles in the regime of acceleration by the shock wave. The medium in the vicinity of the front is homogeneous. Calculate the energy spectrum of the accelerated particles and their distribution in space, neglecting the turbulent acceleration. Consider the cases of relativistic and nonrelativistic particles.

Solution. Write down the unified equation (9.185) for the regions ahead and behind the front:

$$\frac{\partial}{\partial z} \kappa \frac{\partial N}{\partial z} - u \frac{\partial N}{\partial z} + \frac{p}{3} \frac{\partial N}{\partial p} \frac{du}{dz} + \frac{Q_0}{p_0^2} \delta(z)\, \delta(p - p_0) = 0. \tag{9.201}$$

Here the distribution function N should be continuous at the front:

$$N_1 = N_2, \qquad z = 0, \tag{9.202}$$

and the derivatives $\partial N / \partial z$ are finite everywhere. However, the second derivatives have the delta-shaped singularity. To find the second boundary condition, we integrate both sides of the equation over small interval Δz containing the vicinity of the front. We have

$$\kappa_2 \frac{\partial N_2}{\partial z} - \kappa_1 \frac{\partial N_1}{\partial z} - \frac{p}{3}\frac{\partial N}{\partial p}\Delta u + \frac{Q_0}{p_0^2}\delta(p - p_0) = 0, \qquad z = 0, \tag{9.203}$$

where N denotes the general value of the distribution function at the shock front. In the absence of the source, the obtained equation expresses the discontinuity of the particle flux (9.180) at the front. In the presence of the source, the flux exhibits a jump.

Beyond the front, Equation (9.201) takes the form

$$\kappa_{1,2} \frac{\partial^2 N_{1,2}}{\partial z^2} - u_{1,2} \frac{\partial N_{1,2}}{\partial z} = 0 \tag{9.204}$$

and has the solution

$$N_{1,2}(z, p) = A_{1,2}(p) \exp\left(\frac{u_{1,2} z}{\kappa_{1,2}}\right) + D_{1,2}(p). \tag{9.205}$$

To determine the quantities A and D we use the boundary conditions (9.202) and (9.203) at the front and the boundary conditions at infinity: as $z \to \infty$ the distribution function is restricted, and as $z \to -\infty$ it must vanish, because all particles are drifted under the action of magnetic inhomogeneities behind the front. From the boundary conditions at infinity, we find

$$N_1(z, p) = A_1(p) \exp\left(\frac{u_1 z}{\kappa_1}\right), \qquad N_2(p) = D_2(p). \tag{9.206}$$

From the condition at the front, taking into account that $A_1(p) = D_2(p)$ and solving the simple differential first-order equation, we obtain

$$N_2(p) = \frac{3Q_0}{\Delta u p_0^2}\left(\frac{p_0}{p}\right)^{3u_1/\Delta u} \Theta(p - p_0), \qquad N_1(z, p) = N_2(p) e^{u_1 z/\kappa_1}. \tag{9.207}$$

We can pass to the intensity $I(T)$ of the particle flux per unit area, per one steradian and per unit interval of the kinetic energy $T = \mathcal{E} - mc^2$ using the formula

$$I(T)\, dT = vNp^2\, dp = Np^2\, dT. \tag{9.208}$$

In the region behind the front, we have

$$I(T) = \frac{3Q_0}{\Delta u}\left(\frac{T_0^2 + 2mc^2 T_0}{T^2 + 2mc^2 T}\right)^{\gamma/2}, \qquad \text{where} \qquad \gamma = \frac{\sigma + 2}{\sigma - 1}, \qquad \sigma = \frac{u_1}{u_2} = \frac{\rho_2}{\rho_1}. \tag{9.209}$$

The last quantity is the relative compression of the matter by the shock wave.

When the particles are ultrarelativistic ($T \gg mc^2$ and $T_0 \gg mc^2$), the spectrum takes the simple form

$$I(\mathcal{E}) = \frac{3Q_0}{\Delta u}\left(\frac{\mathcal{E}_0}{\mathcal{E}}\right)^{\gamma}, \qquad \mathcal{E} \geq \mathcal{E}_0. \tag{9.210}$$

For nonrelativistic particles (T, $T_0 \ll mc^2$),

$$I(T) = \frac{3Q_0}{\Delta u}\left(\frac{T_0}{T}\right)^{\gamma/2}, \qquad T \geq T_0. \tag{9.211}$$

If the initial energy is nonrelativistic ($T_0 \ll mc^2$), and the final energy is relativistic ($T \gg mc^2$), then the spectrum takes the form

$$I(\mathcal{E}) = \frac{3Q_0}{\Delta u}\frac{(2mc^2 T_0)^{\gamma/2}}{\mathcal{E}^{\gamma}}. \qquad \blacksquare \tag{9.212}$$

Note some characteristic features of particle acceleration by a shock wave.

1. In the one-dimensional case, considered above, for the relativistic energies the spectrum has the power-law form with the index γ, depending only on the force of the shock wave, which is determined by the relative compression σ of the matter at the front. At $\sigma = 4$ (extremely strong adiabatic wave) we have $\gamma = 2$. For the galactic cosmic rays in a wide range of energies, $10 \leq \mathcal{E} \leq 10^6$ GeV, the observed value $\gamma \approx 2.7$ corresponds to $\sigma \approx 2.8$. For the nonrelativistic energies, the power-law index of the spectrum is twice smaller. The index does not depend on turbulence properties in the vicinity of the front.

2. The spatial inhomogeneity in the distribution of fast particles is characterized by the scale

$$L = \frac{\kappa_1}{u_1} = \frac{v\Lambda_1}{3u_1}, \tag{9.213}$$

where Λ is the transport path of particles in the direction at a right angle to the front. This scale for the fast particles, $v \gg u_1$, greatly exceeds the transport path Λ. The scale L sets the direction in which the particles can return upwards along the flux after being at the front. The particles from the layer of width L cross the front repeatedly. The average number of traverses is of the order of v/u_1.

3. The actual shock waves have the finite sizes and exist for a short time. The energy range where the power spectrum has enough time to be established, depends on the age of the shock wave, on the sizes of the front and on the scattering properties of the medium (on the diffusion coefficients κ_1 and κ_2). In particular, the acceleration terminates for particles of sufficiently high energies, for which the transport path Λ_1 approaches the characteristic size of the front. According to the estimates of different authors, a single shock wave produced by a supernova explosion can accelerate protons up to energies $\sim 10^4 - 10^6$ GeV.

4. In the case of the extremely strong shock wave, $\sigma = 4$, the power-law spectral index of accelerated particles $\gamma = 2$, and their energy density $w = \int_{\mathcal{E}_0}^{\mathcal{E}_{max}} \mathcal{E} I(\mathcal{E})\, d\mathcal{E} \sim \ln \mathcal{E}_{max}$ diverges logarithmically at the upper limit. This indicates that the energy transferred to accelerated particles can be a significant fraction of the total mechanic energy of the shock wave. In that case, the problem of particle acceleration has to be solved self-consistently, taking into account the energy removal from the shock wave by the particles and its effect on the structure of the shock wave.

The acceleration in the presence of large-scale stochastic motion of the medium. Let us return to the transport Equation (9.185). The last term describes the acceleration effect of particles by stochastic motions of the medium along the scales not exceeding the transport path Λ. However, random motions can be of much larger scale, $L_0 \gg \Lambda$. As a rule, large-scale motions contain larger energy. Therefore, Equation (9.185) should be generalized for the large-scale stochastic motions. For simplicity, we continue our consideration assuming isotropy ($\kappa_{\alpha\beta} = \kappa\delta_{\alpha\beta}$) and statistic homogeneity of the medium.

We rewrite Equation (9.185) in the form

$$\frac{\partial F}{\partial t} = \kappa \triangle F - \boldsymbol{u} \cdot \nabla F + \frac{p}{3}\frac{\partial F}{\partial p} \nabla \cdot \boldsymbol{u}, \tag{9.214}$$

where F denotes the distribution function. We have neglected the acceleration effect induced by small-scale motions because of its smallness. We assume here that $\boldsymbol{u}$ is the large-scale stochastic quantity:

$$\langle \boldsymbol{u} \rangle = 0, \qquad \text{but} \quad \langle u_\alpha(\boldsymbol{r}_1, t_1) u_\beta(\boldsymbol{r}_2, t_2)\rangle = \int K_{\alpha\beta}(\boldsymbol{k}, \tau) \exp(i\boldsymbol{k} \cdot \boldsymbol{r})\, d^3k \neq 0,$$
$$\boldsymbol{r} = \boldsymbol{r}_1 - \boldsymbol{r}_2, \quad \tau = t_1 - t_2. \tag{9.215}$$

In the statistically homogeneous and isotropic medium, the spectral tensor of the velocities $K_{\alpha\beta}$ has the form

$$K_{\alpha\beta}(\boldsymbol{k}, \tau) = A(k, \tau)\, \delta_{\alpha\beta} + B(k, \tau) k_\alpha k_\beta / k^2. \tag{9.216}$$

If the medium is compressible, then $\nabla \cdot \boldsymbol{u} \neq 0$ and

$$A(k, \tau) + B(k, \tau) \neq 0. \tag{9.217}$$

In the opposite case, the acceleration effect would be absent (in the approximation under consideration).

While averaging Equation (9.214), one should distinguish two cases: $\tau_d \ll \tau_c$ and $\tau_d \gg \tau_c$, where $\tau_d \approx L_0^2/\kappa$ is the time of the diffusive propagation of particles along the distance about the main turbulence scale, and $\tau_c \approx L_0/u$ is the time of their convective transport along the same distance.

First considered the case

$$\tau_d/\tau_c \approx L_0 u / \Lambda v \ll 1. \tag{9.218}$$

The averaging method is the same as when obtaining Equation (9.159). We write down

$$F = N + f, \qquad N = \langle F \rangle, \qquad |f| \ll N, \qquad \langle f \rangle = 0, \tag{9.219}$$

where f is the small randomly nonuniform correction to the averaged distribution function N. From Equation (9.214), we have two equations:

$$\frac{\partial N}{\partial t} = \kappa \triangle N - \langle \boldsymbol{u} \cdot \nabla f \rangle + \frac{p}{3}\frac{\partial}{\partial p}\langle f \nabla \cdot \boldsymbol{u} \rangle, \tag{9.220}$$

$$\frac{\partial f}{\partial t} - \kappa \triangle f = \frac{p}{3}\frac{\partial N}{\partial p} \nabla \cdot \boldsymbol{u}. \tag{9.221}$$

In the last equation, only the linear small terms are kept. Its solution can be written in terms of the Green's function:

$$f = \int G(q, q')\nabla \cdot \boldsymbol{u}(q')\, dq' \frac{p}{3}\frac{\partial N}{\partial p}, \qquad G(q, q') = \frac{1}{(4\pi\kappa\tau)^{3/2}} \exp\left[-\frac{(\boldsymbol{r} - \boldsymbol{r}')^2}{4\kappa\tau}\right], \tag{9.222}$$

where $q = (\boldsymbol{r}, t)$, $\tau = t - t'$. The quantity $\partial N/\partial p$ is written under the integral, because after averaging it slightly depends on the length L_0. By the order of magnitude $f \approx NL_0u/v\Lambda$, and the condition $|f| \ll N$ is satisfied.

The term $\langle \boldsymbol{u} \cdot \nabla f\rangle$ in Equation (9.220) can be rearranged by integrating by parts, taking into account that $G(q, q')$ and $\langle u_\alpha u'_\beta\rangle$ depend on the differences $\boldsymbol{r} - \boldsymbol{r}'$ and $t - t'$:

$$\langle \boldsymbol{u} \cdot \nabla f\rangle = -\int G(q, q')\langle(\nabla \cdot \boldsymbol{u})\,(\nabla' \cdot \boldsymbol{u}')\rangle\, dq' \frac{p}{3}\frac{\partial N}{\partial p}.$$

As a result, the last two terms in Equation (9.220) can be reduced to the form

$$-\langle \boldsymbol{u} \cdot \nabla f\rangle + \frac{p}{3}\frac{\partial}{\partial p}\langle f\nabla \cdot \boldsymbol{u}\rangle = \frac{1}{p^2}\frac{\partial}{\partial p} p^2 D(p)\frac{\partial N}{\partial p},$$

where the diffusion coefficient in momentum space is

$$D(p) = \frac{p^2}{9}\int \langle \nabla \cdot \boldsymbol{u}(q)\, \nabla' \cdot \boldsymbol{u}(q')\rangle G(q, q')\, dq'. \tag{9.223}$$

Equation (9.220) takes the form of the diffusion equation in coordinate space and in momentum space:

$$\frac{\partial N}{\partial t} = \kappa \triangle N + \frac{1}{p^2}\frac{\partial}{\partial p} p^2 D(p)\frac{\partial N}{\partial p}. \tag{9.224}$$

With the aid of Equations (9.216) and (9.217), the diffusion coefficient in momentum space can be expressed in the form of the integral over the turbulent spectrum

$$D = \frac{p^2}{9}\int k^2 d^3k \int_0^\infty [A(k, \tau) + B(k, \tau)]e^{-k^2\kappa\tau}\, d\tau. \tag{9.225}$$

The correlation time of velocities for the scale k^{-1} can be evaluated as $\tau_k \approx (u_k k)^{-1}$, where u_k is the velocity at the scale k^{-1}. For the power-law spectrum $u_k \approx u(k_0/k)^{(v-1)/2}$, where v is the spectral index, and u is the velocity at the main scale $L_0 \sim k_0^{-1}$. If the condition (9.218) is satisfied, the correlation time τ_k is large in comparison with the propagation time of particles $(k^2\kappa)^{-1}$ for all scales up to L_0. Hence, in the functions $A(k, \tau)$ and $B(k, \tau)$ under the integral (9.225) we can set $\tau = 0$. Finally, the integral over $d\tau$ can be easily calculated, and the diffusion coefficient is reduced to the form

$$D(p) = \frac{p^2}{9\kappa}\int [A(k, 0) + B(k, 0)]\, d^3k = \frac{\langle u_{ef}^2\rangle p^2}{9\kappa}, \tag{9.226}$$

where $\langle u_{ef}^2\rangle$ is the integral which is equal to $\langle u^2\rangle$ by the order of magnitude. It is clear that the structure of the diffusion coefficient, as a whole, is close to the structure given by Equation (9.184). However, in this case, the acceleration effect is significantly larger, since the large-scale velocity $\langle u_{ef}^2 \rangle$ greatly exceeds the small-scale quantity $\langle \Delta u^2 \rangle$.

Now consider the case $\tau_d/\tau_c \approx L_0 u/\Lambda v \gg 1$. The main part in spatial diffusion of the particles and in their acceleration is played by the stochastic convection. The calculation becomes more complicated (it can be found in Bykov and Toptygin (1993)). However, the final equation has the same form (9.224) provided the increase of particle energy along the correlation length constitute a small fraction of its kinetic energy. The diffusion coefficients, by the order of magnitude, are

$$\kappa \approx uL_0, \qquad D(p) \approx \frac{\langle u^2 \rangle p^2}{9L_0^2} \tau_c \approx \frac{up^2}{9L_0},$$

where $\tau_c \approx L_0/u$ is the convection time.

There are many other processes of particle acceleration in a plasma. We have discussed only those processes that can accelerate particles to highest energies and could explain the origin of the most energetic particles (cosmic rays) observed in the Galaxy.

Recommended literature: Ginzburg and Syrovatskii (1964); Toptygin (1985); Berezinskii *et al.* (1990) Antonova *et al.* (19883); Berezhko *et al.* (1988); Tverskoy (2004); Mursin (2007); Fleishman and Toptygin (2013)

Problems

9.37 Fast particles are uniformly distributed in space containing the homogeneous magnetic field $\boldsymbol{B}_0$ and random inhomogeneities. Initially (at $t = 0$) their distribution function is known: $F(\theta, \phi, 0) = F_0(\theta, \phi)$, where θ and ϕ are the angles which determine the direction of the momentum in the spherical coordinate system. Using Equation (9.161), calculate the distribution function at $t > 0$. Find the e-folding time of isotropization τ_s (during which the anisotropic part of the distribution function reduces by a factor of e).

9.38 Find the solution of equation of anisotropic diffusion (9.169) for the instantaneous point source of particles $Q(\boldsymbol{r}, t) = \delta(\boldsymbol{r})\,\delta(t)$. Calculate the average sizes of the cloud of particles in three dimensions at $t > 0$.

9.39* Construct the Green's function for the diffusion Equation (9.170) with the second time derivative for an infinite three-dimension space. Analyze the physical meaning of the solution obtained and compare with the solution for the standard diffusion equation which contains only the first time-derivative.

9.40 A medium containing magnetic inhomogeneities moves with the velocity $\boldsymbol{u}$. In the reference frame of the medium at rest, fast particles are described by the isotropic distribution function $F_0 = N(p_0)/4\pi$. Calculate the flux density of particles with given p in the laboratory reference frame.

9.41 Magnetized particles propagate from the stationary point source along the magnetic force lines in the direction of decreasing large-scale field and experience scattering by magnetic inhomogeneities. Using the kinetic Equation (9.179), investigate the dependence of their distribution function on the pitch-angle ϑ and on the distance s along the magnetic line. Restrict yourself by the small-angle approximation. Compare the result with that for scattering of particles in a homogeneous medium (Problem 9.7).

9.42 Relativistic particles ($v \approx c$) are injected into the regime of acceleration by small-scale electric fields. Using equation (9.181) for the spatially uniform case ($\nabla_\beta N = 0,\ u_0 = 0$), calculate the distribution function for arbitrary t. Find the averaged momentum $p(t)$. The initial condition is $N(p, 0) = N_0\, \delta(p - p_0)/p_0^2$, p_0 is the momentum of injection. The transport path is given by Equation (9.162).

9.43★ Perform the same calculations for the case of $\Lambda(p) = const$. The solution should be sought in the form

$$N(p, t) = N_0 (pp_0)^{-3/2} e^{-9\tau/4} f(x, \tau), \qquad \text{where}$$

$$x = \ln \frac{p}{p_0}, \quad \tau = \frac{t}{\tau_a}, \quad \tau_a = \frac{p^2}{D(p)} = const.$$

9.44 In the case considered in the preceding problem, the electric fields are switched off over a finite time due to the damping of turbulence. This process can be described under the assumption that the diffusion coefficient $D(p, t)$ is the function on time, and $D \to 0$ as $t \to \infty$. Find the spectrum of the accelerated particles after the turbulence is damped.

9.45 Relativistic particles are accelerated in a finite region of space containing stationary turbulence which is responsible for the transport path of the particles, $\Lambda = const$. The injection into the acceleration regime occurs uniformly over the whole region, and the source of accelerated particles is given by $Q(p) = Q_0\, \delta(p - p_0)/p_0^2$. Calculate the stationary spectrum of the accelerated particles. Consider the drift of the particles from the acceleration region, by using the relaxation-time approximation and by adding to the equation the term of the form N/τ_d with the required sign. Here τ_d is the average transient time of the particles through the acceleration region (it is of the order of the diffusion time through this region).

9.46 Solve the preceding problem for the spherical region of radius a. Calculate the relaxation time τ_d from the boundary conditions. Find the spherically symmetrical solution that depends on r and p. In the external region, the acceleration is absent, and the diffusion coefficient $\kappa = const$ in coordinate space is the same as in the acceleration region.

9.47★ In the homogeneous medium, the turbulent acceleration of particles occurs with the diffusion coefficient in momentum space of the form $D(p, t) = (p/p_0)^\alpha D_0(t)$, $\alpha \neq 2$, where $D_0(t) \to 0$ as $t \to \infty$. The particles are injected in the acceleration regime with the momentum p_0. Calculate the energy spectrum of accelerated particles after the turbulence is damped.

9.48 At the flat stationary shock front (see Example 9.20) the injection of particles is absent. However, the flux of matter, propagating against the front, contains fast particles with the momentum distribution function $N_0(q)$. Calculate the spectrum of accelerated particles in the vicinity of the front.

9.49* Consider the acceleration of particles by a spherically symmetric shock front at which the supersonic stellar wind turns to the subsonic. Take the following model of the stellar wind:

$$\kappa = \kappa_1 = const, \quad u = u_1 = const \quad \text{at} \quad r < r_0;$$

$$\kappa = \kappa_2(r), \quad u = u_2\left(\frac{r_0}{r}\right)^2 \quad \text{at} \quad r > r_0.$$

The shock wave is located at $r = r_0$, being characterized by the velocity jump $\Delta u = u_1 - u_2 > 0$. In the subsonic region $r > r_0$, the velocity of the medium decreases as r^{-2}, which corresponds to the expansion with a constant density. Specify the sources of particles injected into the acceleration regime by the relation in the form $Q(r,p) = (Q_0/4\pi r_0^2 p_0^2)\,\delta(p - p_0)\,\delta(r - r_0)$. Calculate the spectrum of accelerated particles in momentum space and their distribution in coordinate space. [5)]

9.5
Answers and Solutions

9.1

$$\frac{d\sigma(\theta)}{d\Omega} = \left(\frac{Ze^2}{2pv}\right)^2 \frac{(1 - \beta^2 \sin^2\theta/2)}{\sin^4\theta/2}.$$

This expression without the factor $(1 - \beta^2 \sin^2\theta/2)$ coincides with the Rutherford formula (see Problem I.4.67). The factor gives the relativistic spin correction.

9.2

$$\frac{d\sigma(\theta)}{d\Omega} = \left(\frac{Ze^2}{2pv}\right)^2 \frac{(1 - \beta^2 \sin^2(\theta/2))}{(\sin^2\theta/2 + \hbar^2/4p^2R^2)^2}.$$

In the small-angle approximation,

$$\frac{d\sigma(\theta)}{d\Omega} = \left(\frac{2Ze^2}{pv}\right)^2 \frac{1}{(\theta^2 + \theta_0^2)^2}, \qquad \theta_0 = \hbar/pR.$$

9.3

$$d\sigma = \frac{2\pi e^4}{m_e^2c^2v^2}\frac{d\gamma_e}{(\gamma_e - 1)^2}\left[1 - \beta^2\frac{\gamma_e - 1}{\gamma_m - 1} + \frac{m_e^2c^4}{2\mathcal{E}^2}(\gamma_e - 1)^2\right].$$

Here γ_e is the Lorentz factor of the recoil electron, γ_m is its maximum; v and $\mathcal{E}$ are the velocity and the total relativistic energy of the incident particle.

5) Stellar winds are inherited to many most active stars which have $u_1 \approx (2-3)\times 10^8$ cm/s, $r_0 \approx 3 -$ 10 pc $\approx 10^{19} - 3\times 10^{20}$ cm. These parameters are more moderate for the solar wind: $u_1 \approx 4\times 10^7$ cm/s, $r_0 \approx 10^{15}$ cm (the latter value is not very certain).

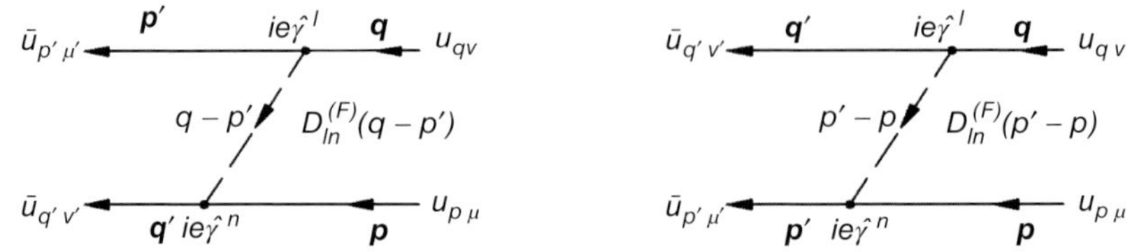

Figure 9.11 Feynman diagrams of $e^- - e^-$ scattering (Möller scattering). Wavy lines represent photon propagator.

9.4

$$d\sigma = \frac{2\pi e^4}{m_e^2 c^4}\frac{d\gamma_e}{\gamma^2 - 1}\left\{\frac{(\gamma-1)^2\gamma^2}{(\gamma_e - 1)^2(\gamma-\gamma_e)^2} - \frac{2\gamma^2+2\gamma-1}{(\gamma-1)(\gamma-\gamma_e)} + 1\right\}.$$

Here γ is the Lorentz factor of the fast electron, γ_e and $\gamma - \gamma_e + 1$ are the Loretz factors of two indistinguishable electrons after collision. The process is described by two Feynman diagrams (Figure 9.11).

9.5

$$\left.\frac{d\mathcal{E}}{dz}\right|_{\gamma_e > \gamma_0} = -n_e \int_{\gamma_0}^{\gamma/2} (\gamma_e - 1) m_e c^2 d\sigma = -\frac{2\pi e^4 n_e}{m_e v^2}\left[\ln\frac{\gamma}{4(\gamma_0 - 1)} + \frac{9}{8}\right].$$

9.6

$$dI(\omega) = \frac{e^2}{c^2 v}\left(1 - \frac{c^2}{n^2(\omega)v^2}\right)\omega\, d\omega, \qquad d\theta = \frac{c}{n^2(\omega) v \sin\theta}\frac{dn}{d\omega}\, d\omega.$$

9.7 The distribution function satisfies the equation of diffusion over angles

(1) $$\frac{\partial f}{\partial z} = \frac{q}{4}\left(\frac{\partial^2 f}{\partial \vartheta_x^2} + \frac{\partial^2 f}{\partial \vartheta_y^2}\right),$$

and the boundary condition $f(\vartheta_x, \vartheta_y, z)|_{z=0} = \delta(\vartheta_x)\,\delta(\vartheta_y)$ (normalization to unity). The diffusion equation was considered in Example 4.6; its solution can be written down by analogy:

(2) $$f(\vartheta_x, \vartheta_y, z) = \frac{1}{\pi q z}\exp\left\{-\frac{\vartheta_x^2 + \vartheta_y^2}{qz}\right\}.$$

Hence,

(3) $$\overline{\vartheta^2}(z) = qz$$

(we integrate within infinite limits because of fast convergence of the integral). From the condition $\overline{\vartheta^2}(l) \approx 1$, the length of isotropization is $l \approx q^{-1}$. The length of deceleration L is evaluated from Equations (9.17) and (9.18):

(4) $$L \approx \frac{T}{dT/dz}, \qquad T = \mathcal{E} - mc^2$$

is the kinetic energy, and $d\mathcal{E}/dz = dT/dz$. The ratio of lengths is given by

(5) $$\frac{l}{L} \approx \frac{dT/dz}{qT} \approx \frac{(pv)^2}{ZTm_e v^2},$$

if we neglect the ratio of the logarithmic factors assuming them to be of the order of unity. In the nonrelativistic case, we have $l/L \approx T/Zm_e v^2 \approx m/Zm_e$, where m is a mass of the fast particle. Heavy particles are decelerated almost without scattering even in the matter with large Z. The length of isotropization for the nonrelativistic electron is Z times smaller than the deceleration length. In the ultrarelativistic case, we have $l/L \approx T/Zm_e c^2$, that is, again heavy particles are decelerated almost without scattering.

9.8

$$f = \frac{12}{\pi^2 q^2 z^4} \exp\left\{-\frac{4}{qz}\left(\vartheta^2 + \frac{3\rho^2}{z^2} - \frac{3\boldsymbol{\rho}\cdot\boldsymbol{\vartheta}}{z}\right)\right\}, \qquad \overline{\rho^2}(z) = \frac{qz^3}{3} = \frac{1}{3}z^2\overline{\vartheta^2}(z).$$

9.9 An ultrarelativistic electron emits a bremsstrahlung quantum in the external stationary field characterized by the potential energy $U(\boldsymbol{r})$. Based on the perturbation theory, consider both the interaction of the electron with the external field and its interaction with the emitted quantum. The probability of the process should be calculated according to the second-order perturbation theory including intermediate states (see Example I.6.19).

The Feynman diagrams for bremsstrahlung differ from the diagrams of Compton effect by replacing the line of the initial quantum by the line of the external field (the dashed line in Figure 9.12; the momentum $\hbar\boldsymbol{q}$ is transferred to the external field). We use the same procedure as in the consideration of Compton effect. The probability of the transition per unit time corresponding to the diagrams in Figure 9.12 is given by the expression (compare with Equation (1) from Example I.6.19)

$$dw_{fi} = \frac{2\pi}{\hbar}\left|\sum_l \left\{\frac{\langle f|\hat{V}|l\rangle\langle l|\hat{U}|i\rangle}{\epsilon_i - \epsilon_l'} + \frac{\langle f|\hat{U}|l\rangle\langle l|\hat{V}|i\rangle}{\epsilon_i - \epsilon_l''}\right\}\right|^2$$

(1) $$\delta(\epsilon_i - \epsilon_f)\frac{\mathcal{V}k^2 dk d\Omega_k}{(2\pi)^3} p^2\, dp\, d\Omega_p.$$

Here $\hat{V}$ is the interaction operator (6.126) of the relativistic electron with quantized electromagnetic field, and $\hat{U} = U(\boldsymbol{r})$ is the potential energy of the electron in the external field. At the vertex with the photon emission, momentum is conserved, because all particles are described by plane waves. At the vertex with the external field the electron momentum is not conserved, the momentum $\hbar\boldsymbol{q}$ is transferred to the external field. The energy of

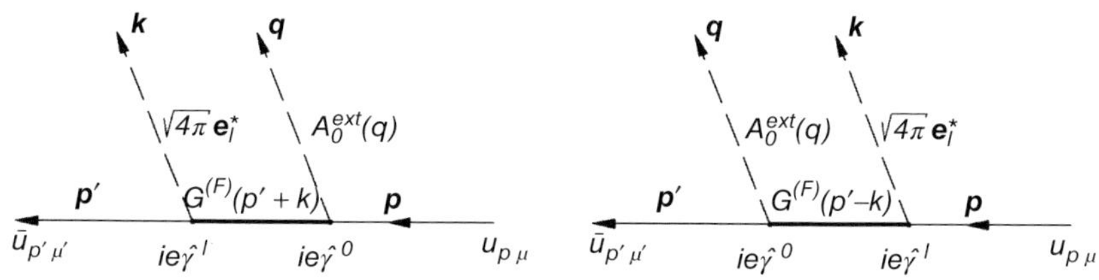

Figure 9.12 The Feynman diagrams of bremsstrahlung.

the initial state is that of electron (electromagnetic quantum is absent):

(2)
$$\epsilon_i = \mathcal{E}(p_0) = \sqrt{c^2 p_0^2 + m_e c^4}.$$

In the final state, there is the electron with positive energy and the emitted photon:

(3)
$$\epsilon_f = \mathcal{E}(p) + \hbar\omega = \sqrt{c^2 p^2 + m_e c^4} + \hbar\omega.$$

The energy ϵ_l' of an intermediate state corresponds to Feynman diagram. In this state, there is only one electron with not determined energy sign:

(4)
$$\epsilon_l' = \mathcal{E}_\lambda(p') = \pm\sqrt{c^2 p'^2 + m_e c^4}, \qquad \boldsymbol{p}' = \boldsymbol{p} + \hbar\boldsymbol{k}, \qquad \epsilon_i - \epsilon_l' = \mathcal{E}(p_0) - \mathcal{E}_\lambda(p'),$$

where the index λ, as in Example I.6.19, enumerates two signs of the energy and two spin states for each bispinors describing intermediate states of the electron.

Feynman diagram corresponds to the energy ϵ_l'' of the intermediate state, containing the electron and the emitted photon:

$$\epsilon_l'' = \hbar\omega + \mathcal{E}_\lambda(p'') = \hbar\omega \pm \sqrt{c^2 p''^2 + m_e c^4}, \qquad \boldsymbol{p}'' = \boldsymbol{p}_0 - \hbar\boldsymbol{k},$$

(5)
$$\epsilon_i - \epsilon_l'' = \mathcal{E}(p) - \mathcal{E}_\lambda(p'') - \hbar\omega.$$

The number of quantum states in Equation (1) is given by the product of quantities that characterize the photon and the electron; the bispinors are normalized to the delta-function in momentum space.

The matrix elements in Equation (1) have the form

$$\langle l|\widehat{U}|i\rangle = \frac{U_{\mathbf{q}}}{(2\pi\hbar)^3}\langle u_\lambda'|u_0\rangle, \qquad \langle f|\widehat{U}|l\rangle = \frac{U_{\mathbf{q}}}{(2\pi\hbar)^3}\langle u|u_\lambda''\rangle, \qquad \text{where}$$

(6)
$$U_{\mathbf{q}} = \int U(\boldsymbol{r}) \exp(i\boldsymbol{q}\cdot\boldsymbol{r})\, d^3r,$$

(7)
$$\hbar\boldsymbol{q} = \boldsymbol{p}_0 - \boldsymbol{p} - \hbar\boldsymbol{k},$$

$$\langle f|\widehat{V}|l\rangle = -e\sqrt{\frac{2\pi\hbar c^2}{\mathcal{V}\omega}}\langle u|\widehat{\alpha}\cdot\boldsymbol{e}^*|u_\lambda'\rangle\,\delta(\boldsymbol{p}' - \boldsymbol{p} - \hbar\boldsymbol{k}),$$

(8)
$$\langle l|\widehat{V}|i\rangle = -e\sqrt{\frac{2\pi\hbar c^2}{\mathcal{V}\omega}}\langle u_\lambda''|\widehat{\alpha}\cdot\boldsymbol{e}^*|u_0\rangle\,\delta(\boldsymbol{p}'' - \boldsymbol{p}_0 + \hbar\boldsymbol{k}).$$

The sum over l in Equation (1) must include summation over four values of λ and integration over intermediate momenta $\boldsymbol{p}'$ and $\boldsymbol{p}''$ of the electron.

Further calculation, including also summation over polarizations of the particles and averaging over the initial states of the electron, should be performed in the same way as in Example I.6.19. The algebraic operations are sufficiently cumbersome. However, the reader can gain useful experience while performing them. The details can be found in the books by Heitler (1956) and Mac-Connel (1962). Below we give the cross section integrated over angles of the emitted quantum and secondary electron for the non-screened Coulomb field of the nucleus $U(r) = -Ze^2/r$:

$$d\sigma_r(\omega) = Z^2 \alpha r_0^2 \frac{d\omega}{\omega}\frac{p}{p_0}\left\{\frac{4}{3} - 2\mathcal{E}_0\mathcal{E}\frac{p^2+p_0^2}{c^2p^2p_0^2} + m_e^2 c\left(\frac{l_0\mathcal{E}}{p_0^3} + \frac{l\mathcal{E}_0}{p^3} - \frac{cl_0 l}{p_0 p}\right)\right.$$

$$+\left[\frac{8\mathcal{E}_0\mathcal{E}}{3c^2p_0p} + \frac{(\hbar\omega)^2(\mathcal{E}_0^2\mathcal{E}^2 + c^4p_0^2p^2)}{c^6p_0^3p^3}\right.$$

(9) $$\left.\left. + \frac{m_e^2c^2\hbar\omega}{2p_0p}\left(\frac{\mathcal{E}_0\mathcal{E} + c^2p_0^2}{c^3p_0^3}l_0 - \frac{\mathcal{E}_0\mathcal{E} + c^2p^2}{c^3p^3}l + \frac{2\hbar\omega\mathcal{E}_0\mathcal{E}}{c^4p_0^2p^2}\right)\right] L\right\},$$

where

(10)

$$L = 2\ln\frac{\mathcal{E}_0\mathcal{E} + c^2p_0p - m_e^2c^4}{m_ec^2\hbar\omega}, \qquad l_0 = 2\ln\frac{\mathcal{E}_0 + cp_0}{m_ec^2}, \qquad l = 2\ln\frac{\mathcal{E} + cp}{m_ec^2}.$$

This cross section is valid for any energies of the electron and quantum which are allowed by conservation law.

9.11 From Maxwell's equations, we determine the monochromatic components of electromagnetic field of a particle moving with a constant velocity $v = const$ in the cylindrical reference frame:

$$E_{\omega z}(\boldsymbol{r}, t) = \frac{iq\omega}{\pi c^2}\left(1 - \frac{1}{\beta^2\varepsilon}\right)K_0(s\rho)e^{i\omega(z/v-t)}, \quad B_{\omega\alpha}(\boldsymbol{r}, t) = \frac{iq\omega}{\pi c^2}K_1(s\rho)e^{i\omega(z/v-t)},$$

(1) $$s^2 = \frac{\omega^2}{v^2} - \frac{\omega^2}{c^2\varepsilon(\omega)}, \quad \Re\, s > 0,$$

where $\varepsilon(\omega)$ is given by Equation (4) of Example 9.6 (details can be found in Example 9.13). With these formulas we calculate the energy lost by the particle per unit path length:

(2) $$-\frac{d\mathcal{E}}{dz} = 2\pi b\int_{-\infty}^{\infty}\frac{c}{4\pi}[\boldsymbol{E}\times\boldsymbol{B}]_\rho\, dt = -2\pi cb\Re\int_0^\infty B^*_{\omega\alpha}E_{\omega z}\, d\omega.$$

From the uncertainty relations, it follows that the radius b must be of the order of the Compton length Λ_C; and $|bs| \ll 1$. Substituting the components (1) in Equation (2), we obtain the spectral power of the losses per unit path length:

(3) $$-\frac{d\mathcal{E}_\omega}{dz\, d(\hbar\omega)} = \hbar\omega\, n(\omega)\frac{\Gamma(\omega)}{c},$$

where

(4) $$n(\omega) = \frac{2}{\pi}\frac{q^2}{\hbar c}\frac{1}{\omega}\ln\frac{m_ec^2}{\hbar\omega[\Gamma^2/\omega^2 + (\omega_p^2/\omega^2 + \gamma^{-2})^2]^{1/4}}$$

is the spectral power of photons, which are equivalent of the self-field of the particle. In derivation of this formula, we assume that the logarithm is much larger than unity; the value under the logarithm is determined up to the factor of the order of unity. In the absence of the medium ($\omega_p = \Gamma = 0$), the quantity (4) changes to the spectral power (I.6.129) of equivalent photons of the relativistic particle in vacuum.

The result (3) has a simple meaning: the product $\Gamma_p n(\omega)\, d\omega$ is the number of quanta absorbed per unit time from the self-field of the particle due to formation of pairs. In other words, it is the number of pairs with the total energy of the particles $\hbar\omega$, which are formed by the field of the particle. This subject is more elucidated by Ternovsky (1960).

The product $\Gamma_p n(\omega)\, d\omega$ gives the number of quanta absorbed due to the Compton effect. Along with this process, other quanta of lower energies are also produced, that is, bremsstrahlung of quanta by medium electrons takes place.

9.12 The radiation of medium electrons can be found by the method used in Problem I.6.87. When $\hbar\omega \gg m_e c^2$, and Compton effect and pair production of pairs operate, for the spectral density of equivalent photons one should use Equation (4) from Problem 9.11. Performing the calculations, we have

$$(1) \quad -\left.\frac{d\mathcal{E}_\omega}{dz\, d(\hbar\omega)}\right|_e = \frac{8}{3} Z n_a \frac{q^2}{\hbar c} \frac{m_e c^2}{\hbar\omega} \ln \frac{m_e c^2}{\hbar\omega[\Gamma^2/\omega^2 + (\omega_p^2/\omega^2 + \gamma^{-2})^2]^{1/4}}.$$

In the spectral region where $\Gamma_C \gg \Gamma_p$, the expression under the logarithm is simplified. Moreover, if $\omega \ll \gamma\omega_p$, the spectral density can be calculated with the use of the Compton cross section from Problem I.6.86. Then

$$(2) \quad -\left.\frac{d^2\mathcal{E}_\omega}{dz\, d(\hbar\omega)}\right|_e = -\frac{16}{3} Z\alpha n_a r_0^2 \left\{1 - \frac{\hbar\omega}{m_e c^2} + \left(\frac{\hbar\omega}{m_e c^2}\right)^2\right\} \ln \frac{m_e c^2}{\hbar\omega_p}, \qquad \hbar\omega < \frac{1}{2} m_e c^2;$$

$$(3) \quad -\left.\frac{d^2\mathcal{E}_\omega}{dz\, d(\hbar\omega)}\right|_e = -\frac{8}{3} Z\alpha n_a r_0^2 \frac{m_e c^2}{\hbar\omega} \left\{1 - \frac{m_e c^2}{4\hbar\omega} + \left(\frac{m_e c^2}{4\hbar\omega}\right)^2\right\} \ln \frac{m_e c^2}{\hbar\omega_p}, \qquad \hbar\omega > \frac{1}{2} m_e c^2.$$

Self-radiation of a heavy particle is strongly suppressed by its large mass, and only the medium electrons will radiate. The radiation of the primary electron or of the positron at $\omega \gg \gamma\omega_{0e}$, in the absence of the Landau–Pomeranchuk effect, is given by the Bethe–Heitler equation (9.83), and at $\omega \ll \gamma\omega_{0e}$ by Ter-Mikaelyan equation (9.91). A comparison of these quantities and Equation (1) shows that at $\omega \gg \gamma\omega_{0e}$ the radiation of the primary electron exceeds the radiation of the medium electrons at least by $\approx Z$ times. However, at $\omega \ll \gamma\omega_{0e}$ the ratio of the primary to the secondary radiation changes and becomes of the order of $Z(\omega/\gamma\omega_{0e})^2$. It can be significantly less than unity, and the electrons of the medium will mainly contribute to the radiation. This is associated with the suppression of primary bremsstrahlung by the medium polarization. These questions and the Landau–Pomeranchuk effect are considered in Toptygin (1964) .

9.13 By decomposition the field vectors into the Fourier integral over space and time coordinates

$$E(R,t) = \int \mathcal{E}(k,\ \omega)\exp(i(k\cdot R-\omega t))\frac{d^3k}{(2\pi)^3}\frac{d\omega}{2\pi}\ \ldots,$$

we obtain, from Maxwell's equations, a system of algebraic equations relative to the Fourier amplitudes:

$$(1)\quad \begin{cases} \kappa\, n\times\mathcal{E}(k,\ \omega) &= \mathcal{H}(k,\ \omega), \\ \kappa\, n\times\mathcal{H}(k,\ \omega) &= -\varepsilon(\omega)\mathcal{E}(k,\ \omega) - i\,\dfrac{8\pi^2 e\mathbf{v}}{\omega^2}\,\delta\Big(\dfrac{\kappa}{c}\,n\cdot v-1\Big), \\ \kappa\,\varepsilon(\omega)n\cdot\mathcal{E}(k,\ \omega) &= -i\,\dfrac{8\pi^2 ec}{\omega^2}\,\delta\Big(\dfrac{\kappa}{c}\,n\cdot v-1\Big), \\ \kappa\, n\cdot\mathcal{H}(k,\ \omega) &= 0. \end{cases}$$

Here $\mathcal{H}(k,\ \omega)$ is the Fourier amplitude of the magnetic field, $k=\omega\kappa n/c$, κ is the parameter which is depends on ω and k, and n is the unit vector. When deriving Equation (1), one should take into account that the amplitude of the Fourier function $\delta(R-vt)$ is equal to $2\pi\,\delta(k\cdot v-\ \omega)$ and $\delta(\alpha x)=(1/|\alpha|)\delta(x)$. From the system of Equation (1), we can determine $\mathcal{E}$ and $\mathcal{H}$:

$$(2)\quad \begin{cases} \mathcal{E}(k,\ \omega) &= -i\,\dfrac{8\pi^2 ec}{\omega^2}\cdot\dfrac{\kappa n-(v/c)\varepsilon}{\varepsilon(\kappa^2-\varepsilon)}\,\delta\Big(\dfrac{\kappa}{c}\,n\cdot v-1\Big), \\ \mathcal{H}(k,\ \omega) &= i\,\dfrac{8\pi^2 e\kappa}{\omega^2}\cdot\dfrac{n\times v}{\kappa^2-\varepsilon}\,\delta\Big(\dfrac{\kappa}{c}n\cdot v-1\Big). \end{cases}$$

In order to determine the fields, we have to inverse the Fourier transformation. First we calculate $E_z(R,t)$. It follows from Equation (2) that

$$\mathcal{E}_z(k,\ \omega) = -i\,\frac{8\pi^2 ec}{\omega^2}\cdot\frac{\kappa\cos\theta-\beta\varepsilon}{\varepsilon(\kappa^2-\varepsilon)}\,\delta(\beta\kappa\cos\theta-1),$$

and, hence,

$$E_z(R,t) =$$

$$= -\frac{ie}{2\pi^2c^2}\int_{-\infty}^{\infty}\omega\, d\,\omega\, e^{-i\omega t}\int_0^{\infty}\kappa^2\,d\kappa\int\frac{\kappa\cos\theta-\beta\varepsilon}{\varepsilon(\kappa^2-\varepsilon)}$$

$$(3)\quad \times\exp\Big\{i\frac{\omega}{c}\kappa[r\sin\theta\cos(\Phi-\varphi)-z\cos\theta]\Big\}\delta(\beta\kappa\cos\theta-1)\sin\theta\, d\theta\, d\Phi.$$

Here r is the component of R in the xy-plane, φ is the angle between r and the Ox-axis, $\beta=v/c$; θ and Φ are the polar angles of the unit vector n.

The integral over Φ is expressed in terms of the Bessel function J_0 of the argument $(\omega/c)\kappa r \sin\theta$. The integral over θ has the form:

$$\text{(4)} \qquad \int_0^{\pi} f(\theta)\delta(\beta\kappa\cos\theta - 1)\sin\theta\, d\theta = \frac{1}{\beta\kappa}\int_{-\beta\kappa}^{\beta\kappa} \varphi(y)\,\delta(y-1)\,dy.$$

It has nonzero value only at $\beta\kappa \geq 1$, and hence the lower limit of κ is equal to $1/\beta$. In Equation (3), this is automatically allowed for because of the presence of the delta-function. However, after integration over dy the delta-function will vanish and it will be necessary to take into account explicitly the lower limit of integration.

The integration of Equation (4) over dy gives

$$\text{(5)} \qquad \frac{1}{\beta\kappa}\varphi(1) = \frac{1}{\beta\kappa} f(\theta)\Big|_{\cos\theta = 1/\beta\kappa}.$$

Substitute Equation (5) in Equation (3) and change the variable κ by $x = \sqrt{\kappa^2 - 1/\beta^2}$; because κ varies in the limits from $1/\beta$ to ∞, x will vary between 0 and ∞. Then $E_z(\boldsymbol{R}, t)$ will be of the form:

$$E_z(\boldsymbol{R}, t) = \frac{ie}{\pi c^2}\int_{-\infty}^{\infty} \omega\, d\,\omega\, \exp\left[i\,\omega\left(\frac{z}{v} - t\right)\right]\left(1 - \frac{1}{\beta^2\varepsilon}\right)\int_0^{\infty} \frac{J_0(\omega r x/c)x\, dx}{x^2 + 1/\beta^2 - \varepsilon}.$$

Integration over x with the aid of formula

$$\text{(6)} \qquad \int_0^{\infty} \frac{J_0(xr)\, x\, dx}{x^2 + k^2} = K_0(kr),$$

gives

$$\text{(7)} \qquad E_z(\boldsymbol{R}, t) = \frac{ie}{\pi c^2}\int_{-\infty}^{\infty}\left(1 - \frac{1}{\beta^2\varepsilon}\right)K_0(sr)\exp\left[i\,\omega\left(\frac{z}{v} - t\right)\right]\,\omega\, d\,\omega,$$

where $s^2 = \omega^2/v^2 - (\omega^2/c^2)\varepsilon(\omega)$. The sign of s should be chosen so that $\Re s > 0$; otherwise the integral over ω will diverge. The integration over ω in Equation (6) can only be completed by specifying the precise form of the function $\varepsilon(\omega)$.

In order to evaluate $E_x(\boldsymbol{R}, t)$, we again begin with integration over Φ. Integration over θ can be carried out with the aid of the delta-function. Subsequent integration over $x = \sqrt{\kappa^2 - 1/\beta^2}$ can be done with the formula

$$\int_0^{\infty} \frac{J_1(xr)x^2\, dx}{x^2 + k^2} = kK_1(kr),$$

which is obtained from Equation (6) by differentiating with respect to r and taking into account that $J_0' = -J_1$ and $K_0' = -K_1$.

The result is

$$E_x(\boldsymbol{R}, t) = \cos\varphi\,\frac{e}{\pi v}\int_{-\infty}^{\infty} \frac{s}{\varepsilon}K_1(sr)\exp\left[i\omega\left(\frac{z}{v} - t\right)\right]\,d\omega.$$

The components $E_y(\boldsymbol{R}, t)$ and $\boldsymbol{H}(\boldsymbol{R}, t)$ can be determined in a similar way. E_y differs from E_x in that $\cos\varphi$ is replaced by $\sin\varphi$; hence, in the cylindrical coordinates we have

$$(8) \qquad E_r(\boldsymbol{R},t) = \frac{e}{\pi v}\int_{-\infty}^{\infty} \frac{s}{\varepsilon} K_1(sr) \exp\left[i\omega\left(\frac{z}{v}-t\right)\right] d\omega, \qquad E_\varphi = 0.$$

For $\boldsymbol{H}$, we obtain

$$(9) \quad H_\varphi(\boldsymbol{R},t) = \frac{e}{\pi c}\int_{-\infty}^{\infty} sK_1(sr) \exp\left[i\omega\left(\frac{z}{v}-t\right)\right] d\omega, \qquad H_z = H_r = 0.$$

According to Equations (7)–(9), the electromagnetic field is axially symmetric.

The above formulas will hold only in the region $r \gg a$, where a is of the order of the interatomic distance. In the region $r \lesssim a$, it is necessary to take into account spatial dispersion of the permittivity.

9.14 As seen from Equations (7)–(9) of Problem 9.13, the monochromatic components of the fields $\boldsymbol{E}_\omega(\boldsymbol{R},t)$ and $\boldsymbol{H}_\omega(\boldsymbol{R},t)$ are

$$(1) \qquad E_{\omega z}(\boldsymbol{R},t) = \frac{ie\,\omega}{\pi c^2}\left(1 - \frac{1}{\beta^2\varepsilon}\right) K_0(sr) \exp\left[i\omega\left(\frac{z}{v}-t\right)\right] \ldots,$$

where

$$(2) \qquad s^2 = \frac{\omega^2}{v^2} - \frac{\omega^2}{c^2}\varepsilon(\omega), \qquad \Re\, s > 0,$$

and K_n is a modified Bessel function.

In the wave zone $|sr| \gg 1$ and, hence, the asymptotic expression (I.1.170) can be used for the functions K_n,

$$(3) \qquad K_n(sr) = \sqrt{\frac{\pi}{2sr}}\, e^{-sr}.$$

It follows from Equation (2) that when $\varepsilon(\omega)$ is real, s is also real, provided $1/\beta^2 > \varepsilon(\omega)$ or $\beta n(\omega) < 1$ ($n(\omega)$ is the refractive index for waves of frequency ω). At $\beta n(\omega) > 1$ the parameter s is purely imaginary.

For real s (in view of Equation (2), $s > 0$), and in the wave zone the field will decay exponentially, and there is no radiation. For a purely imaginary s, the field amplitude in the wave zone will vary as $1/\sqrt{r}$, which corresponds to cylindrical waves. Now we will demonstrate that these waves will diverge, that is, the radiation will, in fact, occur in this case.

Write down s in the form

$$(4) \qquad s = \pm\frac{\omega}{c}\sqrt{\frac{1}{\beta^2} - \varepsilon(\omega)} = \pm i\frac{\omega}{c} n\sqrt{\beta^2 n^2 - 1}$$

and elucidate which sign should be chosen in front of the square root. To do this, we take into account that the dielectric without losses under consideration is the limiting case of a weakly absorbing dielectric with a complex refractive index $n = n' + in''$. In order that the imaginary part of the refractive index, n'', would really describe the absorption of energy (so that the amplitude of the corresponding wave would decay rather than increase), it is necessary to satisfy the condition $n'' > 0$ at $\omega > 0$ and $n'' < 0$ at $\omega < 0$. Assuming that n'' is rather small, we can write

$$\sqrt{\beta^2(n' + in'')^2 - 1} \approx \sqrt{\beta^2 n'^2 - 1}\left(1 + i\frac{\beta n' n''}{\beta^2 n'^2 - 1}\right).$$

Hence, the condition $\Re s > 0$ will be satisfied provided the negative sign is chosen in Equation (4). As n'' tends to zero, we obtain

(5) $$s = -i\frac{\omega n}{c}\sqrt{\beta^2 n^2 - 1}.$$

However, the negative sign just corresponds to diverging waves, since the exponential factor in Equation (1) will take the form

(6) $$\exp i(\boldsymbol{k} \cdot \boldsymbol{R} - \omega t) = \exp i[k(z\cos\theta + r\sin\theta) - \omega t],$$

where $k = \omega n/c$, $\cos\theta = 1/\beta n$, $\sin\theta = \sqrt{1 - 1/\beta^2 n^2}$, with $k\cos\theta = k_z = k_\parallel$ and $k\sin\theta = k_\perp$ being the components of the wave vector.

Thus, when the condition $\beta n(\omega) > 1$ is satisfied the particle moving in the dielectric with a constant velocity $v = \beta c$ emits the electromagnetic waves of frequency ω (Cherenkov radiation). The condition $\beta n > 1$ means that the velocity of the particle should exceed the phase velocity of the wave of frequency ω in the given medium. It follows from the expression for the wave vector $\boldsymbol{k}$, that the radiation occurs at an angleθ to the velocity of the particle, and

(7) $$\cos\theta = \frac{1}{\beta n(\omega)}.$$

This characteristic directional property of the radiation is a consequence of the coherence of waves emitted by the particle at various points along its trajectory (see Example 9.15).

The phase velocity of Cherenkov waves

$$v_{ph} = \frac{\omega}{c} = \frac{c}{n}$$

is the same as for all transverse electromagnetic waves. The polarization of the radiation can be easily determined from Equation (1): the vector $\boldsymbol{H}$ is perpendicular to the plane containing the trajectory of the particle and the wave vector $\boldsymbol{k}$, while the vector $\boldsymbol{E}$ lies in this plane and is perpendicular to $\boldsymbol{k}$ in the wave zone. The fact that $\boldsymbol{k}$ and $\boldsymbol{E}$ are perpendicular may be verified by evaluating the scalar product $\boldsymbol{k} \cdot \boldsymbol{E}_{\omega}$.

The total energy w of Cherenkov radiation per unit path length is equal to the time integral of the flux of the Poynting vector through the infinitely distant cylindrical surface of unit length, surrounding the trajectory of the particle:

(8) $$w = 2\pi r \int_{-\infty}^{\infty} \frac{c}{4\pi}(\boldsymbol{E} \times \boldsymbol{H})_r\, dt = -\frac{cr}{2}\int_{-\infty}^{\infty} H_\varphi E_z\, dt.$$

Going to the Fourier components, Equation (8) can be represented in the form

(9) $$w = -2\pi cr\Re \int_{\beta n(\omega)>1} H^*_{\omega\varphi} E_{\omega z}\, d\omega,$$

where the monochromatic components $H_{\omega\varphi}$ and $E_{\omega z}$ should be taken in the wave zone, and the integration should be carried out over the frequency range in which the radiation condition $\beta n(\omega) > 1$ is satisfied. With the aid of Equations (1)–(3) we have, finally,

(10) $$w = \frac{e^2}{c^2} \int_{\beta(\omega)>1} \left(1 - \frac{c^2}{v^2 n^2}\right) \omega\, d\omega.$$

9.15

$$w = \frac{e^2\omega_0^2}{2v^2}(\beta^2 - 1) + \frac{e^2\omega_0^2}{2v^2}(\varepsilon_0 - 1)\ln\frac{\varepsilon_0}{\varepsilon_0 - 1}.$$

Under the conditions of the problem, $w \approx 5000$ eV/cm. The radiation is concentrated in the cone $\theta_0 \le \theta \le \pi/2$, where $\beta^2\varepsilon_0\cos^2\theta_0 = 1$.

9.16 Each point of the trajectory can be regarded as the source of elementary excitation propagating in the form of a spherical wave with velocity $v_{ph} = c/n$ (Figure 9.13). The front of the resulting wave is the envelope of the elementary spherical waves. The normal to the front is at an angle θ to the trajectory, and $\cos\theta = 1/\beta$ as is clear from the figure.

9.17 The field produced by a charged particle moving with a constant velocity is a result of superposition of plane waves with frequencies $\omega = \boldsymbol{k}\cdot\boldsymbol{v}$, where $\boldsymbol{v}$ is the particle velocity and $\boldsymbol{k}$ is the wave vector. In an infinite dielectric, possible oscillations have frequencies $\omega = kc/n$, where n is the refractive

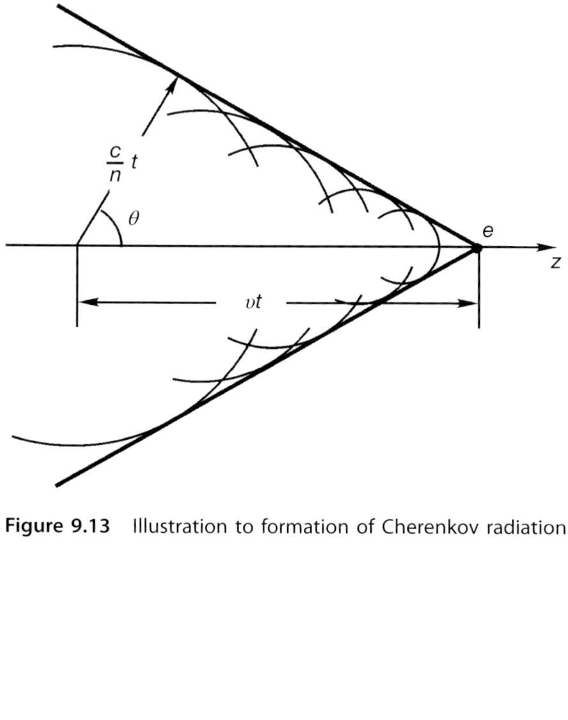

Figure 9.13 Illustration to formation of Cherenkov radiation cone.

index of the medium (eigenoscillations of the medium). From the resonance condition,

$$\frac{kc}{n} = \boldsymbol{k} \cdot \boldsymbol{v} = kv\cos\theta$$

it follows that $\cos\theta = c/vn$. Since $\cos\theta \le 1$, then $vn/c \ge 1$, which is the condition for existence of Cherenkov radiation.

9.18 $\tau = (l/v)\tan^2\theta$, $I = wv\cot^2\theta$, where $\cos\theta = c/nv$, w is the energy of Cherenkov radiation per unit length, calculated in Example 9.14.

9.20 At $\beta n < 1$ (i.e., at $v < v_{ph}$)

(1) $$\varphi = \frac{e}{\varepsilon\sqrt{(z - vt)^2 + r^2(1 - \beta^2 n^2)}}.$$

This solution can be obtained by calculating the Fourier harmonic of the potential with the use of the nonuniform wave equation and by carrying out the inverse Fourier transform. At $\beta n > 1$ the integrand will have a pole at $k^2 = \varepsilon\mu(\boldsymbol{k} \cdot \boldsymbol{v})^2/c^2$. Introduce the cylindrical coordinates in the $\boldsymbol{k}$ -space and write φ in the form

$$\varphi(\boldsymbol{R}, t) = \frac{e}{2\pi^2\varepsilon}\int \frac{\exp[ik_z(z - vt) + ik_\perp r\cos\alpha]}{k_\perp^2 - k_z^2(\beta^2 n^2 - 1)} k_\perp\, dk_\perp\, dk_z\, d\alpha.$$

In order to evaluate, the integral over k_z, we use the theorem on residues. The denominator has zeros at points $k_z = \pm k_\perp/\sqrt{\beta^2 n^2 - 1}$; in order to elucidate the rule of going round these poles, we assume that n has a small imaginary part $n'' > 0$ when $k_z > 0$ (see the analogous analysis in Problem 9.14; here the sign of ω is the same as the sign of k_z, since $\omega = \boldsymbol{k} \cdot \boldsymbol{v}$). Hence, both zeros will be displaced into the lower half-plane of the complex variable k_z. At $z > vt$ the contour of integration should be closed by an arc of infinite radius in the upper half-plane (on this arc the integrand will be zero). Since the denominator has no zeros in the upper half-plane, the integral over k_z will vanish. At $z < vt$ the contour of integration will be closed in the lower half-plane. Both poles contribute to the integral and after integrating we have:

$$\int_{-\infty}^{\infty} \frac{\exp[ik_z(z - vt)]}{k_\perp^2 - k_z^2(\beta^2 n^2 - 1)} dk_z = -\frac{2\pi}{k_\perp\sqrt{\beta^2 n^2 - 1}} \sin\frac{k_\perp(z - vt)}{\sqrt{\beta^2 n^2 - 1}}.$$

The integral over α can be expressed in terms of the Bessel function $J_0(k_\perp r)$ (see Equation (I.1.150)). The last integral over $k_\perp$ is standard; see Gradshtein and Ryzhik (2007). Thus, at $\beta n > 1$ we have:

(2)

$$\varphi(\boldsymbol{R}, t) = \begin{cases} \dfrac{2e}{\varepsilon\sqrt{(z - vt)^2 - r^2(\beta^2 n^2 - 1)}} & \text{at} \quad z < vt - r\sqrt{\beta^2 n^2 - 1}, \\ 0 & \text{otherwise.} \end{cases}$$

The vector potential $\boldsymbol{A}$ can be obtained when multiplying φ by $\varepsilon\mu\boldsymbol{v}/c$.

Equation (2) shows that when the condition for the Cherenkov radiation, $\beta n > 1$, is satisfied, the field is discontinuous. It exists only within the cone whose surface is described by the equation

$$z - vt + r\sqrt{\beta^2 n^2 - 1} = 0. \tag{3}$$

The normal to the surface of the cone is at an angle $\theta = \arccos(1/\beta n)$ to the direction of particle motion. It follows from Equation (3) that the conic wave propagates along the Oz-axis with the velocity of the particle.

Not only electromagnetic waves can have the above structure, but waves of other nature as well. For example, the discontinuous acoustic waves of the above type are excited by a projectile moving in air with the velocity greater than the velocity of sound (ballistic shock wave). The same nature is inherent in waves produced on the surface of water by a fast-moving vessel.

9.21 Cherenkov radiation is emitted provided $\beta n > 1$, where $n(\omega) = \sqrt{\varepsilon(\omega)\mu(\omega)}$; the vector potential is of the form

$$A_x = \frac{iI}{c}\int \frac{\exp[(i\omega/v)(y - vt + \sqrt{\beta^2 n^2 - 1}|z|)}{\sqrt{\beta^2 n^2 - 1}} \cdot \frac{\mu(\omega)\, d\omega}{\omega};$$

$$w = \frac{2I^2}{c^2 v}\int_{\beta n > 1} \frac{\mu(\omega)\, d\omega}{\sqrt{\beta^2 n^2 - 1}}.$$

The retarded force can be calculated from the formula $\boldsymbol{f} = (\boldsymbol{j} \times \boldsymbol{B})/c$, where $\boldsymbol{B}$ should be taken at the point $z = 0,\ y = vt$. The force acts in the direction opposite to the Oy-axis. Its absolute magnitude is equal to the loss of energy per unit path length: $F_y = -w$. This result is a direct consequence of energy conservation.

9.22

$$w = \frac{2e^2}{c^2}\int_{\beta n > 1}\left(1 - \frac{1}{\beta^2 n^2}\right)\left(1 \pm \cos\frac{\omega l}{v}\right)\omega\, d\omega.$$

The positive and negative signs correspond, respectively, to the cases (a) and (b). The spectral density of the radiation emitted by two equal charges differ from the spectral density of the radiation emitted by a single charge by factor $2(1 + \cos(\omega l/v)$. Hence, the intensity of the harmonics with frequencies

$$\omega = \frac{2\pi v}{l} n \qquad (n = 0, 1, 2, \dots)$$

increases by a factor of 4, and the harmonics with frequencies

$$\omega = \frac{\pi v}{l}(2n + 1)$$

will vanish. If the charges differ in sign, the reverse situation will occur. In order to consider the case of a point dipole-oriented along the direction of motion, the quantity $1 - \cos(\omega l/v)$ should be expanded into a series assuming the argument of the cosine to be small. This yields

$$w = \frac{p^2}{c^2 v^2}\int_{\beta n > 1}\left(1 - \frac{1}{\beta^2 n^2}\right)\omega^3\, d\,\omega,$$

where p is the electric dipole moment in the laboratory frame.

9.23

$$w = \frac{p^2}{c^2 v^2} \int_{\beta n > 1} \left(1 - \frac{1}{\beta^2 n^2}\right) \left[\cos^2 \alpha + \frac{1}{2} \sin^2 \alpha (\beta^2 n^2 - 1)\right] \omega^3 \, d\omega,$$

$n = \sqrt{\varepsilon}$, p is the electric dipole moment in the laboratory system.

9.24

$$w = \frac{m^2}{c^2 v^2} \int_{\beta n > 1} \left(1 - \frac{1}{\beta^2 n^2}\right) n^2 \omega^3 \, d\omega.$$

9.25 The energy loss per unit path length is expressed by the integral over time of the flux of energy through the cylindrical surface of unit length and radius a, surrounding the trajectory of the particle. In order to evaluate the losses we can use Equation (9) in the solution of Problem 9.14, provided the fields are taken at $r = a$ and the integration is carried out over all frequencies, from 0 to ∞. Using the expressions for the field components obtained in Problem 9.13 and the specific form of the function $\varepsilon(\omega)$ given in the present problem, we have

$$(1) \qquad -\frac{d\mathcal{E}}{dl} = \frac{2e^2 \omega_0^2}{\pi v^2} \Re i \int_0^\infty \left(\frac{1 - x^2}{\varepsilon_0 - x^2} - \beta^2\right) s^* a K_1(s^* a) K_0(sa) x \, dx,$$

where $x = \omega/\omega_0$, $\varepsilon(0) = \varepsilon_0 = 1 + \omega_p^2/\omega_0^2$ is the static permittivity

$$(2) \qquad s^2 = \frac{\omega_0^2}{c^2}\left(\frac{1}{\beta^2} - 1\right) \frac{b - x^2}{1 - x^2} x^2, \qquad b = \frac{c^2 - \varepsilon_0 v^2}{c^2 - v^2}.$$

It follows from Equation (1) that the imaginary part of the integral is the only one to contribute to the losses. The functions K_0 and K_1 are real for a real argument, and hence the imaginary part of the integral will be determined by the range of x values for which s is complex. As is seen from Equation (2), this range depends on the sign and magnitude of the parameter b. If $b > 0$ (implying $v < c/\sqrt{\varepsilon_0}$), s is purely imaginary for x in the range $(\sqrt{b}, 1)$, and real outside of this range. If $b < 0$ (implying $v > c/\sqrt{\varepsilon_0}$), s is imaginary at $0 \le x \le 1$, and real at $x > 1$.

In addition to the above-mentioned ranges of x, the individual points at which the denominator $\varepsilon_0 - x^2$ of the integrand is zero, $x = \pm\sqrt{\varepsilon_0}$, will also contribute to the imaginary part of the integral. Since the integration is carried out for $x > 0$, it is sufficient to consider the single pole $x = \sqrt{\varepsilon_0} > 1$. If the losses are neglected, this pole is on the real axis. When the losses are taken into account, it is displaced into the lower half-plane of the complex variable ω. This is easily seen from the explicit expression for $\varepsilon(\omega)$ (see Problem 5.13).[6] In order to evaluate the integral correctly, it is necessary either to introduce the damping parameter which should tend to zero after the integration, or to deform slightly the path of integration by-passing the

6) This is consistent with the general theorem according to which $\varepsilon(\omega)$ has no zeros in the upper half-plane (see Section 5.2).

pole by a circle of infinitely small radius in the upper half-plane. We use the second method. Denoting integration over this semicircle by $\frown$, we have

$$\int_{\frown} \frac{1-x^2}{\varepsilon_0 - x^2} s^* a K_1(s^* a) K_0(sa) x\,dx$$

$$= i\frac{1-\varepsilon_0}{2} \cdot \frac{\omega_0 a\sqrt{\varepsilon_0}}{v} K_0\left(\frac{\omega_0 a\sqrt{\varepsilon_0}}{v}\right) K_1\left(\frac{\omega_0 a\sqrt{\varepsilon_0}}{v}\right). \tag{3}$$

Next, evaluate the integral over the region, where s is purely imaginary. Note, that for purely imaginary argument the cylindrical functions K_0 and K_1 are related by

$$s^* a K_1(s^* a) K_0(sa) - sa K_1(sa) K_0(s^* a) = i\frac{\pi}{2},$$

which follows from the properties of the Wronskian of the system of solutions of the Bessel equation. Hence,

(4)

$$\Re\, i \int_{s^2<0} \left(\frac{1-x^2}{\varepsilon_0 - x^2} - \beta^2\right) s^* a K_1(s^* a) K_0(sa) x\,dx = -\frac{\pi}{2} \int_{s^2<0} \left(\frac{1-x^2}{\varepsilon_0 - x^2} - \beta^2\right) x\,dx.$$

The last integral can be evaluated by elementary methods. The limits of integration are chosen as indicated above.

Substituting Equations (3) and (4) into Equation (1), we obtain for $v < c/\sqrt{\varepsilon_0}$

(5)

$$-\frac{d\mathcal{E}}{dl} = \frac{2\pi e^4 N}{m v^2}\left[\frac{2a\omega_0\sqrt{\varepsilon_0}}{v} K_0\left(\frac{a\omega_0\sqrt{\varepsilon_0}}{v}\right) K_1\left(\frac{a\omega_0\sqrt{\varepsilon_0}}{v}\right) - \beta^2 - \ln(1-\beta^2)\right]$$

and for $v > c/\sqrt{\varepsilon_0}$:

(6)

$$-\frac{d\mathcal{E}}{dl} = \frac{2\pi e^4 N}{m v^2}\left[\frac{2a\omega_0\sqrt{\varepsilon_0}}{v} K_0\left(\frac{a\omega_0\sqrt{\varepsilon_0}}{v}\right) K_1\left(\frac{a\omega_0\sqrt{\varepsilon_0}}{v}\right) - \frac{1-\beta^2}{\varepsilon_0 - 1} + \ln\frac{\varepsilon_0}{\varepsilon_0 - 1}\right].$$

The part of the total losses that does not vanish as $a \to \infty$ (terms which do not contain a in Equations (5) and (6)), represent the energy losses due to the emission of transverse waves (Cherenkov effect):

$$-\left(\frac{d\mathcal{E}}{dl}\right)_{V-Ch} \equiv w = \frac{e^2\omega_p^2}{2v^2}[-\beta^2 - \ln(1-\beta^2)] \quad \text{at} \quad v < \frac{c}{\sqrt{\varepsilon_0}}, \tag{7}$$

(8)

$$-\left(\frac{d\mathcal{E}}{dl}\right)_{V-Ch} \equiv w = \frac{e^2\omega_p^2}{2v^2}\left(-\frac{1-\beta^2}{\varepsilon_0 - 1} + \ln\frac{\varepsilon_0}{\varepsilon_0 - 1}\right) \quad \text{at} \quad v > \frac{c}{\sqrt{\varepsilon_0}}.$$

Equation (8) has been obtained in the solution of Example 9.15.

The terms with K_0 and K_1 in Equations (5) and (6), which depend on a, enter as a result of integration round the pole at the point $\omega \equiv \Omega = \sqrt{\omega_0^2 + \omega_p^2}$, at which ε becomes equal to zero. Longitudinal oscillations are excited at these frequencies (see Problems 5.45 and 5.46) and hence the expression

(9) $$-\left(\frac{d\mathcal{E}}{dl}\right)_{pol} = \frac{e^2\omega_p^2\Omega a}{v^3}K_0\left(\frac{\Omega a}{v}\right)K_1\left(\frac{\Omega a}{v}\right)$$

describes the losses on excitation of longitudinal oscillations (polarization losses). At $\Omega a/v \ll 1$ Equation (9) simplifies (see formula (I.1.168)):

(10) $$-\left(\frac{d\mathcal{E}}{dl}\right)_{pol} = \frac{e^2\omega_p^2}{v^2}\ln\frac{v}{\Omega a}.$$

If $\Omega a/v \gg 1$, the quantity$-(d\mathcal{E}/dl)_{pol}$ becomes very small (it is proportional to $\exp(-\Omega a/v)$). Accordingly, the effect of medium polarization is small at low frequencies.

The macroscopic method of calculating the losses presented above was used by Fermi (1940).

9.26

(1) $$-\frac{d\mathcal{E}}{dl} = \frac{e^2\,\omega_p^3 a}{v^3}K_0\left(\frac{\omega_p a}{v}\right)K_1\left(\frac{\omega_p a}{v}\right).$$

If the parameter $\omega_p a/v \ll 1$, which happens at sufficiently large particle velocity, the asymptotic equations (1.168) for the modified Bessel functions can be used. Equation (1) then becomes

(2) $$-\frac{d\mathcal{E}}{dl} = \frac{e^2\omega_p^2}{v^2}\ln\frac{2v}{\gamma\omega_p a}.$$

According to Equations (1) and (2), the energy losses essentially depend on ω_p which is the frequency of longitudinal plasma oscillations.

Cherenkov radiation is absent in a plasma, since at all frequencies $\varepsilon(\omega) < 1$ and, thus, the condition for radiation $\beta^2\varepsilon \geq 1$ is not satisfied (however, Cherenkov radiation may take place if the plasma is placed in a magnetic field).

In a quantum mechanical interpretation, the excitation of plasma oscillations is equivalent to the appearance of certain discrete elementary excitations (these are quasi-particles known as **plasmons**). The energy of each plasmon is equal to $\hbar\omega_p$, where $\hbar = 1.05\times10^{-27}$ erg s is Planck's constant. For metals, the energy $\hbar\omega_p$ lies between ~ 5 and ~ 30 eV. Thus, when plasma oscillations are excited, the particle loses the energy in discrete steps. As was mentioned above, studies of these discrete (or characteristic) losses of the energy provide valuable data on the properties of solids.

9.27 Consider the expansion of the current density (Figure 9.14)

(1) $$j = j_z = \begin{cases} -ev\delta(z - vt)\,\delta(x)\,\delta(y) & \text{at} \quad z \geq 0, \\ -ev\delta(z + vt)\,\delta(x)\,\delta(y) & \text{at} \quad z \leq 0, \ -\infty < t \leq 0, \end{cases}$$

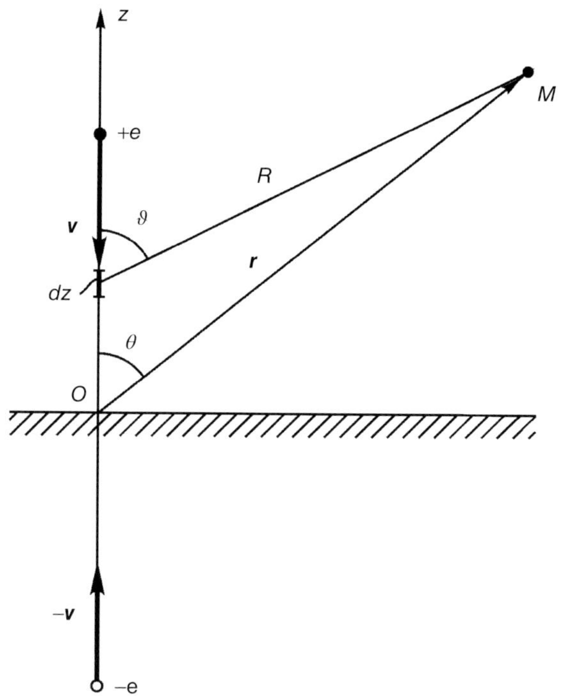

Figure 9.14 Representation of a charge and current of a moving particle as an ensemble of oscillators.

in the Fourier integral with respect to the time:

(2)
$$j = \int j_\omega e^{-i\omega t}\,dt, \qquad j_\omega = \begin{cases} -\dfrac{e}{2\pi}\exp(-i\omega z/v)\delta(x)\delta(y) & \text{at} \quad z \geq 0, \\ -\dfrac{e}{2\pi}\exp(i\omega z/v)\delta(x)\delta(y) & \text{at} \quad z \leq 0. \end{cases}$$

Introduce the polarization vector according to (5.41):

(3)
$$P_\omega = -\frac{j_\omega}{i\omega}.$$

The vector $\boldsymbol{P}_\omega$ lies along the Oz-axis.

Equations (2) and (3) show that the density of the charge and the current of the moving particle are equivalent to a set of harmonic oscillators whose spatial distribution is given by

(4)
$$P_\omega = \begin{cases} -\dfrac{ie}{2\pi\omega}\exp(-i\omega z/v)\,\delta(x)\,\delta(y) & \text{at} \quad z \geq 0, \\ -\dfrac{ie}{2\pi\omega}\exp(i\omega z/v)\,\delta(x)\,\delta(y) & \text{at} \quad z \leq 0. \end{cases}$$

The presence of $\delta(x)\,\delta(y)$ in Equation (4) means that the oscillators are distributed along the line of charge motion.

The oscillators lying along the element dz produce the magnetic field at a point M of the wave zone (see Figure 9.14):

$$(5) \qquad d\boldsymbol{H}_\omega = -\frac{\omega^2 e^{ikR}}{c^2 R^2}(\boldsymbol{P}_\omega \times \boldsymbol{R})\, dz = -\frac{\omega^2 e^{ikR}}{c^2 R} P_\omega \sin\vartheta\, \boldsymbol{e}_\alpha\, dz.$$

After integrating Equation (5) over dz, we obtain the expression for the total field:

$$H_{\omega\alpha} = \frac{ie\omega}{2\pi c^2}\left[\int_{-\infty}^{0} \frac{e^{(i\omega z/v)+ikR}\sin\vartheta}{R}\, dz + \int_{0}^{\infty} \frac{e^{-(i\omega z/v)+ikR}\sin\vartheta}{R}\, dz\right].$$

In the latter expression, the integrands are the products of decreasing and oscillating functions. Therefore, the main contribution to the integral will be that due to the region near $z = 0$. From the view point of physics, this is because the emission takes place at the boundary between the vacuum and the metal. To evaluate the integrals approximately, let $R = r - z\cos\theta$ in the exponents. Expressing $\sin\vartheta$ in terms of R, we have

$$H_{\omega\alpha} = \frac{ie\,\omega e^{ikr} r\sin\theta}{2\pi c^2}\left[\int_{-\infty}^{0} \frac{\exp[i(\omega/v)(1-\beta\cos\theta)z]}{R^2}\, dz + \int_{0}^{\infty} \frac{\exp[-i(\omega/v)(1+\beta\cos\theta)z]}{R^2}\, dz\right].$$

After integrating by parts, these integrals can be expanded into series in powers of $1/R$; retaining only the term linear in this parameter, we obtain

$$(6) \quad H_\alpha = E_\theta = \frac{e\,\omega}{2\pi c^2}\left[\frac{1}{(\omega/v)(1-\beta\cos\theta)} + \frac{1}{(\omega/v)(1+\beta\cos\theta)}\right]\frac{\sin\theta\, e^{ikr}}{r}.$$

The second term in this expression represents the radiation field due to the sudden deceleration of the charge, while the first term represents the radiation due to the image. The intensity of the radiation at frequency ω into the solid angle $d\Omega$ can be determined by

$$(7) \qquad dI(\omega,\theta) = c|E(\omega,\theta)|^2 r^2\, d\Omega = \frac{e^2 v^2}{\pi^2 c^3}\frac{\sin^2\theta\, d\Omega}{(1-\beta^2\cos^2\theta)^2}.$$

In the nonrelativistic limit ($\beta \ll 1$), Equation (7) gives the electric dipole radiation,

$$(8) \qquad dI(\omega,\theta) = \frac{e^2 v^2}{\pi^2 c^3}\sin^2\theta\, d\Omega,$$

whose intensity is proportional to the squared particle velocity. Note, that the intensity of the radiation is independent of the particle mass.

The integrals of Equations (7) and (8) over ω, which give the angular distribution of the total radiation (at all frequencies), will diverge. This is because the metal is assumed to be a perfect conductor. In reality, the metal cannot be regarded as a perfect conductor even in the infrared region, so that at high frequencies Equations (7) and (8) will not hold.

The spectral distribution of the total radiation can be obtained by integrating Equation (7) over the upper hemisphere:

$$I(\omega) = \frac{4e^2v^2}{3\pi c^3}\left(\frac{3(\beta^2+1)}{8\beta^3}\ln\frac{1+\beta}{1-\beta} - \frac{3}{4\beta^2}\right). \tag{9}$$

In the ultrarelativistic limit, when the total energy of the particle $\mathcal{E}$ is much greater than the rest energy mc^2, Equation (9) gives

$$I(\omega) = \frac{2e^2}{\pi c}\ln\frac{\mathcal{E}}{mc^2}. \tag{10}$$

The intensity of the radiation increases logarithmically with the energy.

In the nonrelativistic limit, the expression in brackets becomes equal to unity and

$$I(\omega) = \frac{4e^2v^2}{3\pi c^3}. \tag{11}$$

9.28 The Fourier component of the polarization vector is given by

$$P_\omega = -\frac{ie}{2\pi\,\omega}e^{-i\omega z/v}\,\delta(x)\,\delta(y). \tag{1}$$

To begin with, consider the field at the point A due to oscillators lying in the region $z > 0$ (Figure 9.15). It is sufficient to consider the oscillators lying near the point $z = 0$, since they mainly produce radiation field (see Problem 9.27).

Let us use the reciprocity theorem and choose an oscillator $\mathbf{p}_B$ on the Oz-axis near $z = 0$ (point B), and an oscillator $\boldsymbol{p}_A$ at the point A where the field

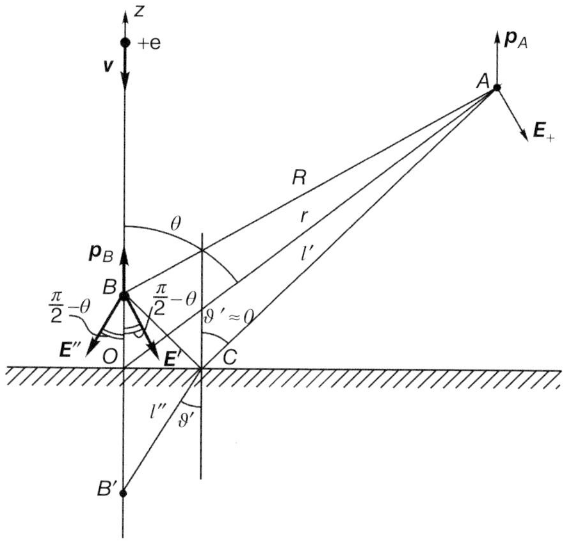

Figure 9.15 Calculation of radiation field at crossing dielectric's border by a moving particle.

is calculated. Suppose that both oscillators are equal in absolute magnitude and parallel to the Oz-axis, and that the distance between them is large compared with the wave length. The oscillator $\boldsymbol{p}_B$ produces the field at the point A whose amplitude $\boldsymbol{E}_+$ is at an angle to the Oz-axis, which is approximately equal to $\pi/2 - \theta$ (see Figure 9.15). The waves arrive from A to B by two routs: directly and after the reflection from the boundary of the dielectric. In the figure, the corresponding amplitudes are denoted by $\boldsymbol{E}'$ and $\boldsymbol{E}''$. They are at the same angle $\pi/2 - \vartheta' \approx \pi/2 - \theta$ to the Oz-axis. Therefore, according to the reciprocity theorem, $\boldsymbol{E}_+ = \boldsymbol{E}' + \boldsymbol{E}''$. Taking into account that in the wave zone of the oscillator $\boldsymbol{H} = \boldsymbol{n} \times \boldsymbol{E}$, we have $\boldsymbol{H}_+ = -\boldsymbol{H}' - \boldsymbol{H}''$ (all three vectors $\boldsymbol{H}_+$, $\boldsymbol{H}'$ and $\boldsymbol{H}''$ are perpendicular to the AOz plane).

The wave emitted by A and arriving at B directly gives rise to the field

$$dH' = \frac{\omega^2 e^{ikR}}{c^2 R} P_\omega \sin\theta \, dz. \tag{2}$$

The amplitude of the reflected wave can be determined from Fresnel formulas, since the distance AC is large and the wave emitted by the point A can be regarded as a plane wave near the point C. Using Equation (9.11) and taking into account the change in phase and that $\vartheta' \approx \theta$, we have

$$dH'' = \frac{\omega^2 f e^{ikR'}}{c^2 R'} P_\omega \sin\theta \, dz, \tag{3}$$

where

$$f = \frac{\varepsilon\cos\theta - \sqrt{\varepsilon - \sin^2\theta}}{\varepsilon\cos\theta + \sqrt{\varepsilon - \sin^2\theta}}, \qquad R' = ACB.$$

The field $\boldsymbol{H}_+$, which is produced at the point A by all oscillators lying in the region $z > 0$, can be obtained by integrating the sum $-(dH' + dH'')$ over dz, from 0 to ∞. The integration is carried out precisely, as in Problem 9.20. The result is

$$H_+ = \frac{ev}{2\pi c^2}\left(\frac{1}{1+\beta\cos\theta} + \frac{f}{1-\beta\cos\theta}\right)\frac{\sin\theta e^{ikr}}{r}. \tag{4}$$

This formula can be easily interpreted by comparison with the analogous formula (6) of Problem 9.27. The first term represents the field of the particle moving in vacuum and suddenly stops at the point $z = 0$; the second term represents the field due to the image $(-ef)$ moving in the dielectric toward the particle and also stopping at the point $z = 0$. In contrast to a perfect conductor, the strength of the image is reduced by a factor f and its magnitude depends on the frequency ω of the given harmonic (through $\varepsilon(\omega)$) and on the position of observation point (through the angle θ).

The field $\boldsymbol{H}_-$ due to the dipoles placing at $z < 0$ can be determined in a similar way. The wave emitted by A will arrive to B after refraction at the separation boundary. Using the Fresnel formulas again, we have

$$dH_- = -\frac{\omega^2}{\varepsilon c^2 R''}(1+f)P_\omega \sin\vartheta' e^{i\varphi} \, dz, \tag{5}$$

where $R'' = l' + l''$ is the length of the broken line ACB' (see Figure 9.15). The phase φ is responsible for retardation, so that

$$\varphi = \frac{\omega}{c} l' + \frac{\omega}{c} \sqrt{\varepsilon}\, l''.$$

At $|z| \ll r$ $(z < 0)$ we have $l' = r + z \tan \vartheta'' \sin \theta$, $l'' = -z/\cos \vartheta''$. According to the refraction law, $\sin \vartheta'' = \sin \vartheta' / \sqrt{\varepsilon}$; replacing ϑ' by θ, we find

$$\varphi = \frac{\omega}{c} r - \frac{\omega}{c} z \sqrt{\varepsilon - \sin^2 \theta}.$$

By integrating Equation (5) over dz from $-\infty$ to 0, we obtain the field due to dipoles which are situated in the region $z < 0$:

(6) $$H_- = -\frac{ev}{2\pi\varepsilon c^2}(1+f)\frac{1}{1 - \beta\sqrt{\varepsilon - \sin^2 \theta}} \cdot \frac{\sin \theta\, e^{ikr}}{r}.$$

The total field at the point A is the sum $H_+ + H_-$. The intensity of radiation of frequency ω within the solid angle $d\Omega$ is

$$dI(\omega, \theta) = \frac{e^2 v^2}{4\pi^2 c^3} A^2(\omega, \theta) \sin^2 \theta\, d\Omega,$$

(7)

$$A(\omega, \theta) = \frac{2\beta \cos \theta}{1 - \beta \cos^2 \theta} + (1+f)\left[\frac{1}{\varepsilon(1 - \beta\sqrt{\varepsilon - \sin^2 \theta})} - \frac{1}{1 - \beta \cos \theta}\right].$$

The quantity A depends on frequency through $\varepsilon(\omega)$.

In the nonrelativistic limit $\beta \ll 1$. Accordingly, we have

(8) $$dI(\omega, \theta) = \frac{e^2 v^2}{\pi^2 c^3} \cdot \frac{(\varepsilon - 1)^2 \sin^2 \theta \cos^2 \theta}{(\varepsilon \cos \theta + \sqrt{\varepsilon - \sin^2 \theta})^2}\, d\Omega.$$

9.29

$$\frac{d^2 I_\omega^{(1)}}{d\omega\, d\Omega_1} = \frac{e^2 \beta^2 \sqrt{\varepsilon_1}(\varepsilon_2 - \varepsilon_1)^2 \sin^2 \theta_1 \cos^2 \theta_1}{\pi^2 c \,|\, \varepsilon_2 \cos \theta_1 + \sqrt{\varepsilon_1 \varepsilon_2 - \varepsilon_1^2 \sin^2 \theta_1}\,|^2},$$

$$\frac{d^2 I_\omega^{(2)}}{d\omega\, d\Omega_2} = \frac{e^2 \beta^2 \sqrt{\varepsilon_2}(\varepsilon_2 - \varepsilon_1)^2 \sin^2 \theta_2 \cos^2 \theta_2}{\pi^2 c \,|\, \varepsilon_1 \cos \theta_2 + \sqrt{\varepsilon_1 \varepsilon_2 - \varepsilon_2^2 \sin^2 \theta_2}\,|^2}.$$

When $\varepsilon_1 = 1$ and $\varepsilon_2 = \varepsilon$, the first formula agrees with the nonrelativistic limit in Problem 9.28. At $\varepsilon_1 = 1$ and $\varepsilon_2 \to \infty$ it transforms to the nonrelativistic case of Problem 9.27.

9.30 It follows from the form of the denominators in Equation (9.92) that the maximum radiation within the small angles $\theta_2 = \theta$ is possible only when the condition $|\varepsilon_{1,2} - 1| \ll 1$ is satisfied. This occurs at $\omega \gg \omega_{p1,2}$, where $\omega_{p1,2}$ are the plasma frequencies of the two media $(\varepsilon_{1,2} = 1 - \omega_{p1,2}^2/\omega^2)$. Let us write Equation (9.92) in the small-angle approximation by using the expansions

$$1 - \beta^2 \varepsilon_2 \cos^2 \theta \approx \gamma^{-2} + \frac{\omega_{p1}^2}{\omega^2} + \frac{\theta^2}{2},$$

(1) $$\beta\sqrt{\varepsilon_1 - \varepsilon_2 \sin^2\theta} - 1 \approx \frac{1}{2}\left(\gamma^{-2} + \frac{\omega_{p1}^2}{\omega^2} + \theta^2\right).$$

We have

$$\frac{d^2 I_\omega^{(2)}}{d\omega\, d\Omega_2} = \frac{e^2}{\pi^2 c}\left(\frac{\Delta\omega_p^2}{\omega^2}\right)^2 \frac{\theta^2}{(\gamma^{-2} + \omega_{p2}^2/\omega^2 + \theta^2/2)^2(\gamma^{-2} + \omega_{p1}^2/\omega^2 + \theta^2)^2},$$

(2) $$\Delta\omega_p^2 = \omega_{p1}^2 - \omega_{p1}^2.$$

Hence, the characteristic angle of the radiation is approximately determined by the condition

(3) $$\theta_0^2 \approx \gamma^{-2} + \frac{\omega_p^2}{\omega^2},$$

where ω_p is the largest of the two plasma frequencies.

The spectrum can be estimated by taking Equation (2) at $\theta \approx \theta_0$ and then multiplying it by θ_0^2:

(4) $$\frac{dI_\omega^{(2)}}{d\omega} \approx \frac{e^2}{\pi^2 c}\left(\frac{\Delta\omega_p^2}{\omega^2}\right)^2 \frac{\theta_0^4}{(\gamma^{-2} + \omega_{p2}^2/\omega^2 + \theta_0^2/2)^2(\gamma^{-2} + \omega_{p1}^2/\omega^2 + \theta_0^2)^2}.$$

At the frequencies satisfying the condition $\omega_p/\omega \geq \gamma^{-1}$, the frequency dependence of the spectrum (4) is weak, since the characteristic angle increases with decreasing frequency; then the nominator and the denominator in Equation (4) increase approximately alike. At $\omega \gg \gamma\omega_p$ the characteristic angle is determined by the factor γ; therefore, the spectrum decreases with frequency as ω^{-4}. Thus, the radiation is mostly generated at frequencies $\sim \gamma\omega_p$ and smaller, although, in principle, the spectrum is restricted by $\omega_{max} = (\gamma - 1)mc^2/\hbar$. The above consideration allows the estimation of the total energy $\Delta\mathcal{E} = \int_0^{\omega_{max}} (dI_\omega^{(2)}/d\omega)\, d\omega$ of the transient radiation. After substituting the characteristic frequency $\gamma\omega_p$ in Equation (4), we multiply its right-hand side by the same frequency and obtain

$$\Delta\mathcal{E} \approx \frac{e^2\omega_p\gamma}{c} = \frac{e^2\omega_p}{c}\frac{\mathcal{E}}{mc^2}.$$

The radiated energy is proportional to the energy of the relativistic particle.

The backwards radiation has no sharp directivity, and the generated frequencies are significantly lower than $\gamma\omega_p$. Hence, the energy is mostly emitted forward.

9.31

$$\frac{dI_\omega^{(2)}}{d\omega} = \frac{2e^2}{\pi c}\left\{\left(\frac{1}{2} + \frac{\omega^2}{\gamma^2\omega_p^2}\right)\ln\left(1 + \frac{\gamma^2\omega_p^2}{\omega^2}\right) - 1\right\},$$

$$\Delta\mathcal{E} = \int_0^{\omega_{max}} \frac{dI_\omega^{(2)}}{d\omega} d\omega = \frac{e^2\omega_p\gamma}{3c} = \frac{e^2\omega_p}{3c}\frac{\mathcal{E}}{mc^2}.$$

9.32 The problem is solved by analogy with Equation (5.85).

$$\frac{d^2 I_\omega}{d\omega\, d\Omega} = \frac{1}{T}\frac{d^2 I'_\omega}{d\omega d\Omega} = \frac{e^2\omega_B}{6\pi^3 c\gamma}\left(\frac{a\,\omega}{c\sin\Theta}\right)^2\left(\theta^2 + \frac{1}{\gamma^2} + \frac{\omega_p^2}{\omega^2}\right)^2$$

(1) $$\times\left[K^2_{2/3}(\xi) + \frac{\theta^2}{\theta^2 + \gamma^{-2} + \omega_p^2/\omega^2}K^2_{1/3}(\xi)\right].$$

where

(2) $$\xi = \frac{\omega\, a}{3c\sin\Theta}(\theta^2 + \gamma^{-2} + \omega_p^2/\omega^2)^{3/2}, \qquad T = \frac{2\pi\gamma}{\omega_B}\sin^2\Theta$$

is the period between the pulses of radiation which will be measured by a distant observer, θ is the angle between the line of sight and the direction of the radiation cone. Note that this result can be used when the pitch-angles $\Theta \gg \theta$.

The modified Bessel functions in Equation (1) decrease exponentially at large values of their argument (see Equation (I.1.170)). Hence, the boundary frequency ω_c of the spectrum can be evaluated by equating the quantity $\xi(\omega, \theta)$ at $\theta = 0$ to unity:

(3) $$\omega_c = \frac{3c\gamma^3\sin\Theta}{a} = 3\omega_B\gamma^2\sin\Theta.$$

9.33 After integrating the above result over the angle θ, as was performed in Problem I.5.86, we find

$$\frac{dP_\omega}{d\omega} = \frac{N_0}{4\sqrt{3}\,\pi^2}\frac{e^2\omega}{c\sin^3\Theta}\left(\frac{1}{\gamma^2} + \frac{\omega_p^2}{\omega^2}\right)\int_\eta^\infty K_{5/3}(x)\,dx,$$

(1) $$\eta = \frac{2\omega\gamma}{3\omega_B\sin\Theta}\left(\frac{1}{\gamma^2} + \frac{\omega_p^2}{\omega^2}\right)^{3/2}.$$

An analysis of this general result is analogous to that presented in Example 9.12. We find the least possible value of the parameter $\eta(\omega)$:

(2) $$\eta_{min} = \frac{\sqrt{3}\omega_p}{\gamma\omega_B\sin\Theta} = \frac{3^{3/2}\gamma\omega_p}{\omega_c} \quad \text{at} \quad \omega = \sqrt{2}\gamma\omega_p.$$

If the parameters are such that $\eta_{min} \gg 1$, the radiation is exponentially weak at all frequencies; it is suppressed by the density effect of the medium. In vacuum, such a suppression is absent (see Problem I.5.86). At $\eta_{min} \ll 1$ we have a wide range of frequencies where the plasma effect is weak. If the condition $\eta < 1$ ($\omega_c > \omega > \gamma\omega_p\eta_{min}^{1/2}$) is satisfied, we use Equation (I.1.168) and find

(3) $$\frac{dP_\omega}{d\omega} = \frac{\sqrt{3}\,\Gamma(2/3)N_0}{4\,\pi^2}\frac{e^2\omega_B}{c\sin^2\Theta}\left(\frac{\omega}{\omega_c}\right)^{1/3}.$$

In vacuum, such a spectrum is formed at all frequencies $\omega < \omega_c$. At higher frequencies, $\omega > \omega_c$, the effect of the medium is also insignificant; however, the spectral density exponentially decreases with frequency:

$$\text{(4)} \qquad \frac{dP_\omega}{d\omega} = \frac{\sqrt{3}\,N_0}{8\,\pi^{3/2}}\frac{e^2\omega_B}{c\sin^2\Theta}\left(\frac{\omega}{\omega_c}\right)^{1/2} e^{-2\omega/\omega_c}.$$

At low frequencies $\omega < \gamma\omega_p\eta_{min}^{1/2}$, the radiation is exponentially suppressed, almost absent:

$$\text{(5)} \qquad \frac{dP_\omega}{d\omega} = \frac{3^{1/4}N_0}{24\pi^{3/2}}\frac{e^2\omega_B}{c\sin^2\Theta}\eta_{min}^{1/2}\exp\left(-\frac{2\gamma^3\omega_p^3}{\omega_c\omega^2}\right).$$

The suppression of radiation at low frequencies take place also in the case in which relativistic electrons are distributed over energies according to a power law. The detailed analysis of spectra of radio sources and the more elaborated theory can be found in the papers of Razin (1960) and Gundarev and Rasin (1995).

9.34 For a single chain

$$U(\rho) = \pm\frac{2Ze^2}{b}K_0\left(\frac{\rho}{R}\right);$$

for many chains

$$U(y,z) = \pm\frac{2Ze^2}{b}\sum_n K_0\left(\frac{|\rho-\rho_n|}{R}\right);$$

where ρ_n are the coordinates of the chains on the $x = const$ plane.

9.35 $U(x) = \pm(2\pi Ze^2R/a^2)e^{-|x|/R}$ for a single plane, $U(x) = \pm(2\pi Ze^2R/a^2)\sum_n e^{-|x-nb|/R}$ for a set of planes.

9.36 If it is possible to separate an individual chain of atoms whose potential is axially symmetric, then in the field of the chain the integrals of motion are the longitudinal momentum of the particle $p_\| = const$, the angular momentum relative to the axis $M = \gamma m\rho^2\dot{\varphi}$ and the total energy of the transverse motion:

$$\text{(1)} \qquad \mathcal{E}_\perp = \frac{\gamma m}{2}(\dot{\rho}^2 + \rho^2\dot{\varphi}^2) + U(\rho) = \frac{\gamma m}{2}\dot{\rho}^2 + \frac{M^2}{2\gamma m\rho^2} + U(\rho),$$

where ρ and φ are the cylindrical coordinates of the particle. The values of all three integrals are determined by the initial conditions. The potential energy $U(\rho)$ of the electron in the field of the chain is negative and decreases exponentially at large ρ. For the motion in the transverse direction to be finite (channeling), the condition

$$\text{(2)} \qquad \frac{\gamma m}{2}\dot{\rho}_0^2 + \frac{M^2}{2\gamma m\rho_0^2} < |U(\rho_0)|,$$

must be satisfied (the values of quantities at the point of entering of the particle into the crystal are denoted by subscript 0). This condition means that the total energy of the electron is negative.

The dependence $\rho(t)$ can be expressed in the implicit form with the aid of energy conservation (1):

(3)
$$\pm\int\frac{d\rho}{\sqrt{(2/\gamma m)[\mathcal{E}_\perp - U_{ef}(\rho)]}} = t, \qquad \text{where} \qquad U_{ef}(\rho) = U(\rho) + \frac{M^2}{2\gamma m\rho^2}$$

is the effective potential energy. At $M \neq 0$, the motion of the particle is confined within the limits $\rho_1 \le \rho \le \rho_2$, which can be determined from the equation $U(\rho_{1,2}) = \mathcal{E}_\perp$. The period of the radial oscillations is

(4)
$$T = 2\int_{\rho_1}^{\rho_2}\frac{d\rho}{\sqrt{(2/\gamma m)[\mathcal{E}_\perp - U_{ef}(\rho)]}}.$$

Over this time interval the azimuthal angle will increase by

(5)
$$\Delta\varphi = \frac{2M}{\gamma m}\int_{\rho_1}^{\rho_2}\frac{d\rho}{\rho^2\sqrt{(2/\gamma m)[\mathcal{E}_\perp - U_{ef}(\rho)]}},$$

and the relativistic particle will travel along the chain by $\Delta z \approx cT$.

If the condition (2) is not satisfied but the angle of entering the crystal is small, the particle will move over-barrier along a complicated trajectory going from one chain to another and falling in the regions where the fields of the chains overlap.

9.37

$$F(\theta, \phi, t) = \sum_{l,m} C^0_{lm} Y_{lm}(\theta, \varphi)\exp\left\{im\Omega t - \frac{v}{2\Lambda}l(l+1)t\right\},$$

$$C^0_{lm} = \int F_0(\theta, \phi)Y^*_{lm}(\theta, \phi)d\Omega_p, \quad \Omega = \frac{ec}{\mathcal{E}}B_0, \quad \tau_s = \frac{\Lambda}{v}.$$

The value of τ_s shows that Λ is the isotropization length.

9.38

$$N(\boldsymbol{r}, t) = \frac{1}{(4\pi t)^{3/2}\kappa_\perp\sqrt{\kappa_\parallel}}\exp\left(-\frac{x^2+y^2}{4\kappa_\perp t} - \frac{z^2}{4\kappa_\parallel t}\right).$$

$$\overline{x^2} = \overline{y^2} = 4\kappa_\perp t, \quad \overline{z^2} = 4\kappa_\parallel t.$$

9.39 The required Green's function satisfies the equation

(1)
$$\frac{\Lambda}{v}\frac{\partial^2 G}{\partial t^2} + \frac{\partial G}{\partial t} - \kappa\triangle G = \delta(\boldsymbol{r} - \boldsymbol{r}')\,\delta(t - t').$$

Let us rewrite Equation (1) in a more transparent form by introducing

(2)
$$u = \frac{v}{\sqrt{3}}, \qquad \varrho = \frac{1}{\kappa} = \frac{3}{v\Lambda}, \qquad u^2\varrho = \frac{v}{\Lambda}, \qquad \boldsymbol{R} = \boldsymbol{r} - \boldsymbol{r}', \qquad \tau = t - t'.$$

Then we obtain

(3)
$$\triangle_R G - \frac{1}{u^2}\frac{\partial^2 G}{\partial\tau^2} - \varrho\frac{\partial G}{\partial\tau} = -\varrho\,\delta(\boldsymbol{R})\,\delta(\tau).$$

This equation can be interpreted as a wave equation in the medium with losses: the first two terms in the left-hand side contain the D'Alembert operator, and the third term, with the first time derivative, describes the dissipative process; u is the velocity of propagation of disturbances.

The solution of Equation (3) can be obtained in the same way as for simpler equation (3) in Example 4.6. The complication arises only when performing the inverse Fourier transform, because it is necessary to bypass the cut in the complex plane. As a result, we obtain the Green's function

$$G(R,\tau) = \frac{u\varrho}{4\pi R}\, e^{-u^2\varrho\tau/2} \left\{ \delta(u\tau - R) + \frac{u\varrho R}{2\sqrt{R^2 - u^2\tau^2}} \right.$$

(4) $$\left. \times J_1\left(\frac{1}{2}\varrho u\sqrt{R^2 - u^2\tau^2}\right)\Theta(u\tau - R)\right\}$$

(the details of the calculations can be found in the book of Mors and Feshbach (1958), Chapter 7).

The first term with the delta-function is similar to the Green's function (5.9) of the wave equation and presents the wave emitted by a point source and propagating with finite velocity u. This wave damps not only due to the geometric factor $1/R$, but to the dissipation in the medium described by $\exp(-u^2\varrho\tau/2)$. The term with the Bessel function presents the purely dissipative effect. However, the presence of the step function provides the finite velocity of propagation of this disturbance. In order to pass to the approximation for the conventional diffusion, one should consider the case when $u\tau \gg R$ and $\varrho u^2\tau \gg 1$, that is, when $v\tau \gg \Lambda$. Using Equation (I.1.162) and the asymptote (I.1.169), we easily reproduce the well-known Green's function of the diffusion equation

(5) $$G(R,\tau) = \frac{1}{(4\pi\kappa\tau)^{3/2}} \exp\left(-\frac{R^2}{4\kappa\tau}\right).$$

It is realized after the particle travels along many typical mean free paths.

9.40 By going to the laboratory reference frame instead of the medium frame, the momentum should be rearranged according to the formulas of the relativistic kinematics (see Chapter I.3):

(1) $$\boldsymbol{p}_0 = \frac{\boldsymbol{p} - \boldsymbol{u}\mathcal{E}/c^2}{\sqrt{1 - u^2/c^2}} \approx \boldsymbol{p} - \frac{\boldsymbol{u}\mathcal{E}}{c^2}.$$

The distribution function in the laboratory frame is given by

(2) $$F(\boldsymbol{p}) = \frac{1}{4\pi} N(|\boldsymbol{p} - \boldsymbol{u}\mathcal{E}/c^2|) \approx \frac{1}{4\pi}\left[N(p) - \frac{\mathcal{E}\boldsymbol{u}\cdot\boldsymbol{p}}{c^2 p}\frac{\partial N}{\partial p}\right].$$

The current density is

(3) $$\boldsymbol{J} = \frac{1}{4\pi}\int \boldsymbol{v}F(\boldsymbol{p})\, d\Omega_p = -\frac{p}{3}\frac{\partial N}{\partial p}\boldsymbol{u}.$$

9.41 We integrate both parts of Equation (9.179) over transverse coordinates taking into account the smallness of transverse diffusion at small pitch-angles. In the stationary case and in the approximation of small pitch-angles, the equation takes the form

$$(1) \qquad \frac{\partial F}{\partial s} + \frac{1}{2B}\frac{\partial B}{\partial s}\vartheta\frac{\partial F}{\partial \vartheta} = \frac{1}{l(s)}\frac{1}{\vartheta}\frac{\partial}{\partial \vartheta}\vartheta\frac{\partial F}{\partial \vartheta} + \frac{1}{2\pi\vartheta}\,\delta(\vartheta)\,\delta(s - s_0).$$

Here the point source is included on the right-hand side, and $l(s) = v/b_s(0)$ is a longitudinal mean free path of the particles. Next we are seeking for the solution in the same way as in Example 9.7 and find the distribution function

$$(2) \quad F(s,\vartheta) = \frac{B(s)}{\pi B(s_0)\overline{\vartheta^2}(s)}\exp\left(-\frac{\vartheta^2}{\overline{\vartheta^2}(s)}\right), \qquad \overline{\vartheta^2}(s) = 4\int_{s_0}^{s}\frac{B(s)\,dz}{B(z)l(z)}.$$

If the quantities B and l are independent of s (homogeneous medium), the square of the scattering angle then increases proportionally to the distance traveled, $s - s_0$. However, in the inhomogeneous medium, the particles can experience not only the isotropization, but can be focused by a regular magnetic field as well. Let, for example, $B \propto s^{-2}$. Such a dependence is approximately inherent in the solar wind at distances not exceeding the radius of the Earth orbit. If the random field varies by the same law, then $l(z) \approx l_0(z/s_0)^{2(2-\nu)}$ (we use the dependence (9.173), $l \propto R_0^{2-\nu}$, for the transport path). In this case, from Equation (2) we find $\overline{\vartheta^2}(s) \propto s^{2\nu-3}$ at $s \gg s_0$. When the turbulence spectral index is $\nu = 3/2$ (weak MHD turbulence), we have $2\nu - 3 = 0$, i.e. the beam of particles propagates without increasing its angular dimensions. The scattering is compensated by the focusing in the decreasing regular magnetic field.

9.42

$$N(p,t) = \frac{N_0}{(4\pi Dt)^{1/2}pp_0}\left\{\exp\left[\frac{(p-p_0)^2}{4Dt}\right] - \exp\left[\frac{(p+p_0)^2}{4Dt}\right]\right\},$$

$$\overline{p}(t) = 4(Dt/\pi)^{1/2} \quad \text{at} \quad \overline{p}(t) \gg p_0.$$

The spectrum decays as the Gaussian exponent at $p \gg \overline{p}(t)$.

9.43

$$f(x,t) = \frac{1}{(4\pi\tau)^{1/2}}\exp\left(-\frac{x^2}{4\tau}\right), \qquad \overline{p}(t) = p_0 e^{4\tau/\tau_a}.$$

In the regime under consideration, the spectrum is very hard (there are many high energy particles), and the energy of an individual particle increases exponentially with time. However, as the energy of the particle increases, its mean free path also begins to increase at a certain stage, slowing down the acceleration rate. Moreover, the region of acceleration has the finite sizes, so that the time which the particle is spent in this region is finite. Hence, the acceleration always occurs up to the finite energy.

9.44 The problem is reduced to the problem considered above by introducing the dimensionless time $\tau = \int_0^t dt'/\tau_a(t')$. The required spectrum is given by the expressions obtained in the preceding problem provided the dimensionless time τ is replaced by the quantity $\tau_0 = \int_0^\infty dt/\tau_a(t)$, where $\tau_a(t) = p^2/D(p,t)$. The diffusion coefficient must decrease faster than t^{-1} as $t \to \infty$.

9.45 We solve the stationary equation

$$\frac{1}{p^2}\frac{d}{dp}\frac{p^4}{\tau_a}\frac{dN}{dp} - \frac{N}{\tau_d} + \frac{Q_0}{p_0^2}\delta(p - p_0) = 0, \tag{1}$$

where $D(p) = p^2/\tau_a$ is the diffusion coefficient in momentum space, $\tau_a = const$. The singular source of particles is replaced by the boundary conditions. At $p = p_0$ the distribution function must be continuous:

$$N|_{p=p_0+0} = N|_{p=p_0-0}. \tag{2}$$

The second boundary condition can be obtained by integrating both sides of Equation (1) over a narrow interval of momenta, which includes $p = p_0$, and by taking into account the restriction upon N:

$$\frac{p_0^4}{\tau_a}\left(\frac{dN}{dp}\bigg|_{p=p_0+0} - \frac{dN}{dp}\bigg|_{p=p_0-0}\right) = -Q_0. \tag{3}$$

Next, we solve the homogeneous Equation (1) ($p \neq p_0$). The result is given by

$N(p) = Ap^{-s_1}$ at $p > p_0$, $\quad N(p) = Bp^{s_2}$ at $p < p_0$, where

$$s_1 = \sqrt{\frac{9}{4} + \frac{\tau_a}{\tau_d}} + \frac{3}{2}, \quad s_2 = \sqrt{\frac{9}{4} + \frac{\tau_a}{\tau_d}} - \frac{3}{2}. \tag{4}$$

From the boundary conditions, we have

$$N(p) = \frac{N_0}{p_0^3}\left(\frac{p_0}{p}\right)^{s_1}, \quad p > p_0, \quad N_0 = \frac{Q_0\tau_a}{s_1 + s_2}. \tag{5}$$

The power law is formed, with the index that depends on the ratio τ_a/τ_d. Let us compare the spectrum obtained with the observed spectrum of cosmic rays. Usually, the flux of particles $I(\mathcal{E})$ per unit area, per steradian and per unit energy range is measured. This quantity is associated with the number density of the particles N in phase space by the relation

$$I(\mathcal{E})\,d\mathcal{E} = vNp^2\,dp = Np^2\,d\mathcal{E}. \tag{6}$$

The observed spectrum is approximated by the power law $I(\mathcal{E}) \propto \mathcal{E}^{-\gamma}$, whose index is $\gamma \approx 2.7$ at photon energies $10 \le \mathcal{E} \le 10^6$ GeV. For ultrarelativistic particles $\mathcal{E} \approx cp$, and, according to Equation (6), the index is $\gamma = s_1 - 2$. Hence, the required parameter is $\tau_a/\tau_d \approx 8$. However, one should take in mind that the above condition is difficult to attain under particular astrophysical conditions in a wide range of energies. Hence, the power law spectrum will be replaced by a steeper function of the type obtained in Example 9.42.

9.46 In Equation (1) of the Problem 9.45, the term $-N/\tau_d$ should be replaced by the quantity

$$\frac{1}{r^2}\frac{\partial}{\partial r}r^2\kappa\frac{\partial N}{\partial r}$$

which represents diffusion in coordinate space. The variables in this equation can be separated, $N(r,p) = R(r)P(p)$; the radial function satisfies the equation

$$\frac{1}{r^2}\frac{\partial}{\partial r}r^2\frac{\partial N}{\partial r} = -\frac{\lambda}{\kappa}R, \tag{1}$$

where λ is the parameter of expansion. In the internal region, the radial function has the form $R(r) = (A/r)\sin kr$, where $k = \sqrt{\lambda/\kappa}$.

In the external region, the radial function is $R(r) = B/r$. Joining the radial functions R and the fluxes of particles $\kappa\, dR/dr$ at $r = a$ allows one to find the values of k: $k_n a = (\pi/2)(2n+1)$, $n = 0, 1, \ldots$ Thus, we have an infinite set of the relaxation times

$$\tau_{dn} = \frac{1}{\lambda_n} = \frac{\tau_{d0}}{(2n+1)^2}, \quad \text{where} \quad \tau_{d0} = \frac{4a^2}{\pi^2\kappa} \tag{2}$$

is the shortest relaxation time, denoted by τ_b in Problem 9.45.

The function $P_n(p)$, which depends on momentum p, can be determined in the same way, as in Problem 9.45, and it has the same form. However, each value n is now characterized by its own power-law index s_{1n}, which increases with n. The distribution function has the form of the sum

$$N(r,t) = \sum_{n=0}^{\infty} A_n R_n(r) P_n(p). \tag{3}$$

If momenta $p \gg p_0$, the term with $n = 0$ plays the main role. The coefficients A_n can be determined by expanding into a series of orthogonal functions $R_n(r)$ (with the weight r^2); the radial dependence of the source, in our case, is given by the step function $\Theta(a - r)$. Looking for an approximate solution with a single relaxation time, we lost the terms with large spectral indices (that is unimportant at high energies) and determined incorrectly spatial distribution of particles in the acceleration region.

9.47 We introduce the dimensionless variables

$$\tau = \frac{1}{p_0^2}\int_0^t D_0(t')dt', \qquad q = \frac{p}{p_0}, \tag{1}$$

and obtain the equation

$$\frac{\partial N}{\partial t} = \frac{1}{q^2}\frac{\partial}{\partial q}q^{2+\alpha}\frac{\partial N}{\partial q}. \tag{2}$$

Equation (2) has the particular solution of the form

$$N_\lambda(q,\tau) = f_\lambda(q)e^{-\lambda^2\tau}. \tag{3}$$

The substitution $f_\lambda(q) = q^{-(1+\alpha)/2}F_\lambda(\xi)$, where $\xi = q^{1-\alpha/2}$ ($\alpha \neq 2$), allows us to come to the Bessel equation for $F_\lambda(\xi)$:

$$\xi^2 F''_\lambda + \xi F'_\lambda + \left[\frac{\lambda^2\xi^2}{(1-\alpha/2)^2} - \frac{(1+\alpha)^2}{(2-\alpha)^2}\right] F_\lambda = 0. \tag{4}$$

The bounded solution of this equation is

$$F_\lambda(\xi) = J_\beta\left(\frac{2\lambda\xi}{|2-\alpha|}\right), \tag{5}$$

where $\beta = (1+\alpha)/|2-\alpha|$ is the index of the Bessel function.

The general solution of the primary Equation (2) can be written in the form of the integral

$$N(q,\tau) = q^{-(1+\alpha)/2}\int_0^t \psi(\lambda)e^{-\lambda^2\tau}J_\beta\left(\frac{2\lambda}{|2-\alpha|}q^{1-\alpha/2}\right)\lambda\, d\lambda. \tag{6}$$

The coefficient of separation $\psi(\lambda)$ is determined according to the initial condition $N(p,0) = N_0\delta(p-p_0)/p_0^2 = N_0 p_0^{-3}\delta(q-1)$ with the aid of the Fourier–Bessel theorem:

$$\psi(\lambda) = \frac{2}{|2-\alpha|}\int_0^\infty N_0 p_0^{-3}\delta(q-1)J_\beta\left(\frac{2\lambda}{|2-\alpha|}q^{1-\alpha/2}\right)$$

$$dq = \frac{2N_0}{|2-\alpha|p_0^3}J_\beta\left(\frac{2\lambda}{|2-\alpha|}\right). \tag{7}$$

Substitution of this expression in Equation (6) and the integration over λ give the solution for any $\alpha \neq 2$:

$$N(q,\tau) = \frac{N_0}{|2-\alpha|p_0^3 q^{(1+\alpha)/2}\tau} I_\beta\left(\frac{2q^{1-\alpha/2}}{(2-\alpha)^2\tau}\right)\exp\left\{-\frac{1+q^{2-\alpha}}{(2-\alpha)^2\tau}\right\}, \tag{8}$$

where I_β is a modified Bessel function. The solution describes the evolution of the spectrum with time. Once the turbulence decays, the spectrum has the form (8), where $\tau = \tau_0 = p_0^{-2}\int_0^\infty D_0(t)\,dt$.

In the case of strong acceleration, such that the typical momenta are much larger than the initial momentum ($q \sim \tau^{1/(2-\alpha)} \gg 1$, $\alpha < 2$), the argument of the Bessel function is small and Equation (8) simplifies,

$$N(q,\tau) = \frac{N_0}{|2-\alpha|^{1+2\beta}\Gamma(1+\beta)\tau_0^{1+\beta}}\exp\left(-\frac{q^{2-\alpha}}{(2-\alpha)^2\tau_0}\right). \tag{9}$$

9.48

$$z < 0, \qquad N_1(z,p) = \alpha\int_{q_{min}}^p N_0(q)\left(\frac{q}{p}\right)^\alpha e^{u_1 z/\kappa_1}\frac{dq}{q} + N_0(p)(1-e^{u_1 z/\kappa_1});$$

$$z > 0, \quad N_2(z,p) = \alpha\int_{q_{min}}^p N_0(q)\left(\frac{q}{p}\right)^\alpha \frac{dq}{q}, \qquad \alpha = \frac{3u_1}{\Delta u},$$

where q_{min} is the lowest momentum of pre-accelerated particles. These particles are injected from the outside (from "infinity"). The second term represents the particles ahead the shock front, which have not interacted with it. The function $N_0(q)$ is cut off at $q > q_{max}$, where q_{max} is the largest momentum of primary particles. Hence, at $p > q_{max}$, the second term becomes zero, and the particle distribution behind the front takes the form $N_2(z, p) \sim p^{-\alpha}$.

9.49 The boundary conditions have the form

$$N_2(r,p) \to 0 \quad \text{as} \quad r \to \infty; \qquad N_1 = N_2, \quad \kappa_2 \frac{\partial N_2}{\partial r} - \kappa_1 \frac{\partial N_1}{\partial r} = \frac{p}{3}\frac{\partial N_1}{\partial p}$$

(1) $$\Delta u - \frac{Q_0}{4\pi r_0^2 p_0^2}\delta(p - p_0) \quad \text{at} \quad r = r_0.$$

In addition, both functions must be bounded everywhere.

The problem can be conveniently solved by the Mellin transformation in variable p:

(2) $$\overline{N}(r,s) = \int_0^\infty N(r,p)p^{s-1}dp.$$

The transport equation for the Mellin transform in the internal region takes the form

(3) $$r\frac{d^2\overline{N}_1}{dr^2} + \left(2 - \frac{u}{\kappa}r\right)\frac{d\overline{N}_1}{dr} - \frac{2us}{3\kappa}\overline{N}_1 = 0.$$

Its solution can be expressed in terms of the degenerate hypergeometric function (see, e.g., Abramowitz and Stegun, 1979):

(4) $$\overline{N}_1(r,s) = A(s)\Phi\left(\frac{2s}{3}, 2, \frac{ur}{\kappa}\right).$$

In the external region, this equation can be written in the form

(5) $$\frac{1}{r^2}\frac{d}{dr}r^2\kappa_2(r)\frac{d\overline{N}_2}{dr} - u_2\frac{r_0^2}{r^2}\frac{d\overline{N}_2}{dr} = 0$$

and has the solution

(6) $$\overline{N}_2(r,s) = B(s)\left\{1 - \exp\left[-\int_r^\infty \frac{u_2 r_0^2}{\kappa_2(r)r^2}\,dr\right]\right\}.$$

By joining of the solutions with the boundary conditions, we come to the relations

$$\overline{N}_1(r,s) = \frac{Q_0 p_0^{s-3}(e^{k_2} - 1)}{4\pi r_0^2 u_2}\frac{\Phi(2s/3, 2, u_1 r/\kappa_1)}{\Psi(s, \Delta u, k_1, k_2)},$$

(7) $$k_1 = \frac{u_1 r_0}{\kappa_1}, \qquad k_2 = \int_{r_0}^\infty \frac{u_2 r_0^2}{\kappa_2(r)r^2}\,dr,$$

(8) $$\Psi = \left[1 - (e^{k_2} - 1)\frac{s\Delta u}{3u_2}\right] \Phi(2s/3, 2, k_1) + \frac{u_1}{u_2}(e^{k_2} - 1)\Phi'(2s/3, 2, k_1),$$

where prime means the derivative of the degenerate hypergeometric function with respect to its last argument. In order to inverse the Mellin transformation, it is necessary to integrate the Mellin transform along the contour which is the straight line parallel to the imaginary axis on the plane of complex s:

(9) $$N(r,p) = \frac{1}{2\pi i}\int_{\beta-i\infty}^{\beta+i\infty} \overline{N}(r,s)p^{-s}ds.$$

To do this, we have to find zeros of Ψ as a function of s. Using recurrent formulas for degenerate hypergeometric functions, we reduce the equation $\Psi = 0$ to

(10) $$s = \frac{3u_1}{\Delta u}\left\{1 - \frac{1}{k_1} + \frac{\Phi(2s/3 - 1, 1, k_1)}{k_1\Phi(2s/3, 2, k_1)}\right\} + \frac{3u_2}{\Delta u(e^{k_2} - 1)}.$$

We are interested in the minimal, by modulus, root of Equation (10), which will be most significant at $p \gg p_0$. As $k_1 \to \infty$ and $k_2 \to \infty$, from Equation (10) we have $s_0 = 3u_1/\Delta u$, which corresponds to acceleration of particles by the plane shock front (see Example 9.20). For the finite but large k_1 and k_2, we use the asymptote of the function Φ and obtain the larger value of the root

(11) $$s_0 = \frac{3u_1}{\Delta u}\left[1 + \frac{2u_2}{k_1\Delta u}\right] + \frac{3u_2}{\Delta u(e^{k_2} - 1)} > \frac{3u_1}{\Delta u}.$$

After performing the inverse Mellin transformation, we have

(12) $$N_1(r,p) = \frac{3Q_0}{4\pi r_0^2\Delta u p_0^3}\frac{\Phi(2s_0/3, 2, u_1 r/\kappa_1)}{\Phi(2s_0/3, 2, k_1)}\left(\frac{p_0}{p}\right)^{s_0},$$

(13)
$$N_2(r,p) = \frac{3Q_0}{4\pi r_0^2\Delta u p_0^3(1 - e^{-k_2})}\left\{1 - \exp\left[-\int_r^{\infty}\frac{u_2 r_0^2}{\kappa_2(r)r^2}dr\right]\right\}\left(\frac{p_0}{p}\right)^{s_0}.$$

In comparison with the plane front, the spectrum turns out to be steeper, i.e. the number of energetic particles is smaller. This is associated both with the fast outflow of particles from the front of a finite width and with the adiabatic deceleration of particles in the supersonic region (at $r < r_0$, $div\,\boldsymbol{u}_1 = 2u_1/r > 0$, the medium expands).

Appendix: Turbulence and Its Description with the Aid of Correlation Tensors

Physical picture of turbulent motion. The motion of a liquid or gas can lose its stability when velocities are sufficiently high. The examples of unstable states of MHD systems are considered in Problems 4.77–4.86. In the unstable medium, random small perturbations of velocity and of other parameters will grow with time until nonlinear and dissipative processes stabilize them at a some equilibrium (for given conditions) level. In the unstable system, the velocity and other parameters acquire the irregular character, their values at a fixed points of space experience random fluctuations (pulsations) in magnitude and direction, the fluctuations of the velocity being not small by comparison with the mean velocity of the flux. Such a motion becomes complicated and irregular and is referred to as **turbulent** as opposed to the regular (laminar) motion at which variations of velocity in space and time are subject to a certain law.

The motion becomes turbulent when the dimensionless parameter $R = ul/\nu$, called the **Reynolds number**, is sufficiently large. Here u is the characteristic velocity of the medium, l is the characteristic length scale of its variations, and ν is the kinematic viscosity. The Reynolds number presents the ratio of the nonlinear inertial term in the equation of motion to the dissipative term. We write the equations of motion for the incompressible neutral liquid in the form (4.35) (without a magnetic force)

$$\frac{\partial \boldsymbol{u}}{\partial t} + (\boldsymbol{u}\cdot\nabla)\boldsymbol{u} = -\frac{1}{\tau}\nabla p + \nu\triangle\boldsymbol{u}, \qquad \nabla\cdot\boldsymbol{u} = 0, \tag{A1}$$

and find

$$R \approx \frac{|(\boldsymbol{u}\cdot\nabla)\boldsymbol{u}|}{|\nu\triangle\boldsymbol{u}|} \approx \frac{ul}{\nu}, \tag{A2}$$

provided spatial derivatives are approximately replaced by the division by the characteristic scale. The critical value R_c of the Reynolds number, at which the stability of the laminar motion terminates and the turbulence appears, is rather high. For instance, for a flow of water in tubes of circular cross section $R_c \approx 10^3 - 10^4$, depending on the quality of surface processing and on the uniformity of the flow at the inlet.

We restrict ourselves by **fully developed turbulence**, when many macroscopic degrees of freedom are excited in the system and there exist motions of different scales in a wide range (vortices, waves, inhomogeneities, etc.). This situation will be realized if the Reynolds number significantly exceeds the threshold value R_c at which the laminar flow passes into the turbulent flow. Rather high Reynolds numbers are characteristic for turbulence in geophysical and astrophysical objects in view of their large characteristic length scales. For instance, in the Earth atmosphere in a town with large-sized buildings ($l \approx 100$ m) at a wind velocity $u \approx 20$ m/s and the kinematic viscosity of air $\nu \approx 0.15$ cm^2/s, we have $R = ul/\nu \approx 10^8$. In gigantic hydrogen molecular clouds of the Galactic disk, according to observations (Ruzmaikin *et al.*, 1988; Vainstein *et al.*, 1989), $u \approx 10^6$ cm/s, $l \approx 10$ pc $\approx 3\times10^{19}$ cm, the temperature attains 100 K and the number density of molecules 100 particles/cm^3. This gives $\nu \approx 3\times10^{15}$ cm^2/c and $R \approx 10^{10}$.

If matter is not ionized, in the subsonic range of velocities ($u \ll c_s$) the vertex motion of various scales are the main elements of the turbulence. At velocities $u \gtrsim c_s$ the acoustic waves and the shock waves become essential phenomena. In a conducting liquid and plasma, turbulent pulsations can be excited by waves of various type, which are possible in corresponding media, for example, by the MHD linear and nonlinear modes.

Averaging turbulent quantities. In order to set a random field of turbulent velocities, it is necessary to indicate certain averaged quantities characterizing such a field. For the statistically equilibrium systems, the averaging is, as a rule, performed over the statistical ensemble of equilibrium states (Gibbs ensemble, see Appendix I.C). However, all turbulent systems are very nonequilibrium, and the probabilities of distributions of macroscopic velocities are unknown for them. Hence, for such systems the method of averaging should be closely connected with the measuring method of corresponding values. Measuring devices performs often the time averaging. The interval of averaging Δt should be chosen in a way to exceed the periods of all basic harmonic components of turbulent pulsations. With such an averaging the pulsations are smoothed up,

$$\langle \boldsymbol{u}(\boldsymbol{r}, t)\rangle \equiv \frac{1}{\Delta t}\int_{\Delta t/2}^{\Delta t/2} \boldsymbol{u}(\boldsymbol{r}, t+t')\,dt' = \boldsymbol{u}_0(\boldsymbol{r}, t), \tag{A3}$$

and the averaged velocity $\boldsymbol{u}_0(\boldsymbol{r}, t)$ becomes a regular function of coordinates and time, slowly varying on time scales Δt. The difference

$$\boldsymbol{u}'(\boldsymbol{r}, t) = \boldsymbol{u}(\boldsymbol{r}, t) - \boldsymbol{u}_0(\boldsymbol{r}, t) \tag{A4}$$

is called **velocity of pulsations**. Its averaged value is equal to zero, $\langle \boldsymbol{u}'\rangle = 0$. However, the two-point correlation tensor is not zero:

$$U_{\alpha\beta}(\boldsymbol{r}_1, t_1; \boldsymbol{r}_2, t_2) = \frac{1}{\Delta t}\int_{-\Delta t/2}^{\Delta t/2} u'_\alpha(\boldsymbol{r}_1, t_1 + t')u'_\beta(\boldsymbol{r}_2, t_2 + t')\,dt'. \tag{A5}$$

For a complete description of the random field of turbulent velocities, one should also have the infinite set of the many-point correlation tensors of higher orders. The method of the description of turbulence with use of correlation tensors for the first time have applied Keller und Friedman (1924).

In terms of these tensors, it is possible to express the average value of any functional $\langle F[\boldsymbol{u}'(\boldsymbol{r}, t)]\rangle$, which is expanded into power series with respect to components of $\boldsymbol{u}'(\boldsymbol{r}, t)$. However, for the approximate solution of many problems it is sufficient to know the average velocity and the second-rank correlation tensor. For example, the density of kinematic energy of the incompressible turbulent medium can be written in the form

$$w_k(\boldsymbol{r}, t) = \frac{1}{2}\tau\{u_0^2 + U_{\alpha\alpha}(\boldsymbol{r}, t; \boldsymbol{r}, t)\}, \tag{A6}$$

where τ is the mass density.

Further we will be interested only in a stationary turbulence, whose average characteristics do not change with time. This means that $U_{\alpha\beta}$ depends only on the time difference $t = t_1 - t_2$ but not on the initial time. If the random velocity field is homogeneous in space as well, that is, its averaged (but instantaneous) characteristics are independent of coordinates, the position vectors $\boldsymbol{r}_1$ and $\boldsymbol{r}_2$ enter $U_{\alpha\beta}$ only through the difference $\boldsymbol{r} = \boldsymbol{r}_1 - \boldsymbol{r}_2$. Correlation tensors of a stationary and homogeneous turbulence depend only on the differences of coordinates and times for each pair of points:

$$U_{\alpha\beta}(\boldsymbol{r}_1, t_1; \boldsymbol{r}_2, t_2) = U_{\alpha\beta}(\boldsymbol{r}_1 - \boldsymbol{r}_2, t_1 - t_2). \tag{A7}$$

At $\boldsymbol{r} = t = 0$ the convolution $U_{\alpha\alpha}(0, 0) = \langle u'^2 \rangle > 0$ gives the averaged square of the pulsation velocity. However, as $t \to \infty$ we have

$$U_{\alpha\beta}(0, t) = \langle u'_\alpha(\boldsymbol{r}, t_1) u'_\beta(\boldsymbol{r}, t_2) \rangle \approx \langle u'_\alpha(\boldsymbol{r}, t_1) \rangle \langle u'_\beta(\boldsymbol{r}, t_2) \rangle = 0, \tag{A8}$$

because after a long time any interrelation of components of the pulsation velocity is weakened, and they can be averaged separately. The time duration τ_c, when the tensor $U_{\alpha\beta}$ is noticeably different from zero, is referred to as **correlation time** or **coherence time** of the velocity. It is possible also to give the formal definition of the correlation time for any pair of random quantities:

$$\tau_c = \frac{1}{U_{\alpha\beta}(0, 0)} \int_0^\infty U_{\alpha\beta}(0, t)\, dt. \tag{A9}$$

The **correlation length**, or **coherence length**, L_c, can be defined in the same fashion:

$$U_{\alpha\beta}(\boldsymbol{r}_1 - \boldsymbol{r}_2, 0) \to 0 \qquad \text{at} \qquad |\boldsymbol{r}_1 - \boldsymbol{r}_2| \gg L_c. \tag{A10}$$

Here we have assumed that the turbulence is statistically isotropic and the correlation lengths are equal for all directions.

Although for nonequilibrium turbulent systems the probability distribution of the velocities and other macroscopic parameters is unknown, the formal ensemble averaging is possible in this case as well. Let us imagine a large (infinite in the limit) number of noninteracting systems which are identical to the primary system. Such a set of macroscopically identical systems is referred to as **ensemble**. Because of random character, velocities u'_α, $u''_\alpha, \ldots$ at equivalent points of space and at the same instant of time will, in general, be different for different systems: $u'_\alpha(\boldsymbol{r}, t) \neq u''_\alpha(\boldsymbol{r}, t) \neq u'''_\alpha(\boldsymbol{r}, t) \neq \ldots$ The ensemble average (or statistical average) is the arithmetical mean

$$\langle u(\boldsymbol{r}_1, t_1) u(\boldsymbol{r}_2, t_2) \rangle = \lim_{N \to \infty} \frac{u'(\boldsymbol{r}_1, t_1) u'(\boldsymbol{r}_2, t_2) + u''(\boldsymbol{r}_1, t_1) u''(\boldsymbol{r}_2, t_2) + \ldots}{N}, \tag{A11}$$

where N is the total number of systems in the ensemble. For many turbulent systems, the ensemble and time averages (when the time of averaging is large enough) are the same. Such systems are referred to as **ergodic**.

The ergodicity condition (e.g., Monin and Yaglom (1965)) for stationary random fields requires sufficiently fast attenuation of correlations with time:

$$\lim_{\Delta t \to \infty} \frac{1}{\Delta t} \int_0^{\Delta t} U_{\alpha\beta}(0, t) dt = 0.$$

In the theoretical analysis of the random fields, the method that is used the most often is the statistical averaging.

Spectral properties of correlation tensors of homogeneous turbulence. One of the main characteristics of a turbulent state is the energy distribution of turbulent pulsations over length scales λ (or over the inverse quantities. i.e. over the wave numbers $k = 2\pi/\lambda$). In the case of electrically neutral medium without a magnetic field, such a distribution is given by the Fourier transform of the second-rank correlation tensor at coincident instants of time:

$$U_{\alpha\beta}(\boldsymbol{r}_1 - \boldsymbol{r}_2, 0) = \int \widetilde{U}_{\alpha\beta} e^{i\mathbf{k}\cdot(\mathbf{r}_1-\mathbf{r}_2)} \frac{d^3k}{(2\pi)^3}. \tag{A12}$$

The turbulent energy E per unit mass is determined by the convolution

$$E = \frac{1}{2} U_{\alpha\alpha}(0, 0) = \frac{1}{16\pi^3} \int \widetilde{U}_{\alpha\beta}(\boldsymbol{k}) d^3k. \tag{A13}$$

If the turbulence is not only homogeneous but isotropic as well, i.e. preferential directions in the medium are absent, the spectral tensor $\widetilde{U}_{\alpha\beta}(\boldsymbol{k})$ may be composed of the vector components k_α and of the invariant unit tensors $\delta_{\alpha\beta}$ and $e_{\alpha\beta\gamma}$. First, we construct the spectral tensor on the base of the symmetric tensors $\delta_{\alpha\beta}$ and $k_\alpha k_\beta / k^2$. It will be determined by two scalar functions $A(k)$ and $B(k)$:

$$\widetilde{U}_{\alpha\beta}(\boldsymbol{k}) = A(k)\, \delta_{\alpha\beta} + B(k)\, \frac{k_\alpha k_\beta}{k^2}. \tag{A14}$$

In the case of incompressible medium, the condition $div\, \boldsymbol{u} = 0$ in the Fourier representation allows one to find that $B(k) = -A(k)$ and to write the expression (A14) in the simple form

$$\widetilde{U}_{\alpha\beta}(\boldsymbol{k}) = A(k) \left(\delta_{\alpha\beta} - \frac{k_\alpha k_\beta}{k^2} \right). \tag{A15}$$

We can easily integrate the spectral tensors (A14) and (A15) over the angles and write down the energy (A13) in the form of the one-dimensional integral over absolute value of the wave vector:

$$E = \int_0^\infty \frac{dE}{dk} dk, \qquad \text{where} \qquad \frac{dE}{dk} = \frac{k^2}{4\pi^2}(3A(k) + B(k)) \tag{A16}$$

is the spectral density of the turbulent energy (per unit mass and per unit interval of wave numbers). When the characteristic velocities of turbulent pulsations, u, are subsonic, the effects, associated with the compressibility of the medium, are small, of the order of $u/c_s \ll 1$, and the spectral density is determined by one scalar function, as seen from Equation (A16) (at $B = -A$):

$$\frac{dE}{dk} = \frac{k^2}{2\pi^2} A(k). \tag{A17}$$

A solid quantitative theory, which would permit to calculate correlation tensors of the turbulence and also the energy distribution over scales, is not yet developed. However, there are some quasi-quantitative models which proved to be adequate for many simplified cases. The most pictorial and widely recognized model of this kind was proposed by Kolmogorov (1941) and Obukhov (1941) for a fully developed homogeneous isotropic and stationary turbulence of an incompressible neutral (nonionized) medium. It is based on the following basic assumptions:

(a) The energy supporting the turbulence produces motions with length scales that are determined by the properties of specific energy sources (we denote the characteristic parameter of these scales by L, which is **the external scale** of turbulence, close to the correlation length L_c mentioned above). In this large-scale region, the turbulence can be inhomogeneous and anisotropic. The averaged energy of the order of ε per unit mass is pumped into the medium per unit time. In the stationary case, this energy must finally be dissipated and converted to heat. The quantity ε also determines the intensity of the turbulent pulsations (see Equation (A20)).

(b) Turbulent energy is mostly contained in vortex motions with length-scales of the order of L. This is the **energy-contained** region of scales (and of corresponding wave numbers $k_0 = 2\pi/L$, see Figure A1). For a fully developed turbulence, we have $L \gg \nu/u$; hence, the dissipation of mechanical energy in the energy-contained region is of minor importance. The energy goes out of this region mainly due to the nonlinear fragmentation of the large-size vortices into smaller-size ones. This is allowed by the nonlinear structure of the equations of continuous medium. The cascade fragmentation occurs in a wide range of scales up to a certain scale l (**internal scale** of the turbulence), at which the energy dissipation of the mechanical motion becomes significant. In the range of scales between L and l, the main process is the nonlinear transfer, over the spectrum, of the all energy obtained from the source, from large-scale to small-scale motions. Since the main nonlinear term $(\boldsymbol{u} \cdot \nabla)\boldsymbol{u}$, which is responsible for this process, is due to the inertial force $d\boldsymbol{u}/dt$, the range of scales $L \gtrsim \lambda \gtrsim l$ is referred to as **inertial**. Nonlinear fragmentation of vortices results in isotropization of the turbulence. Therefore, on scales $\lambda \ll L$ it becomes **locally homogeneous and isotropic**, even if the large-scale motion does not possess this property.

(c) In the inertial interval, the dependence of the velocity of motion u_λ on the scale λ can be found from the following consideration. Assume that the lifetime of the vortex with a length-scale λ is of the order of its revolution $\tau_\lambda \approx \lambda/u_\lambda$ ("strong" turbulence). In a time τ_λ, the energy of this vortex ($u_\lambda^2/2$ per unit mass) will be

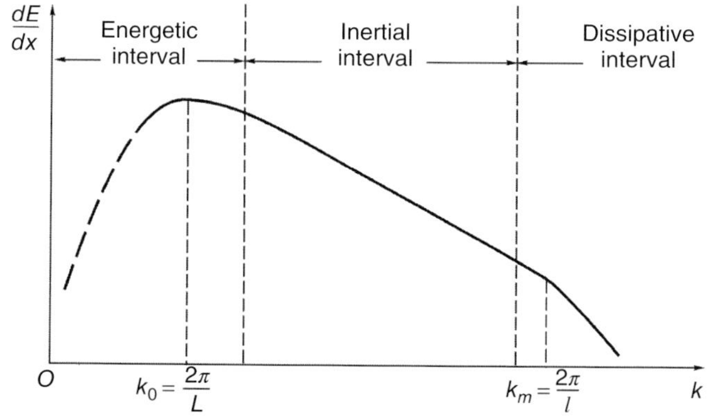

Figure A1 Schematic image of turbulent spectrum in Kolmogorov–Obukhov theory.

transferred to smaller-scale motions. The transferred energy ε per unit time is determined by the sources and given by

$$\varepsilon \approx \frac{u_\lambda^2}{2\tau_\lambda} \approx \frac{u_\lambda^3}{\lambda}. \tag{A18}$$

Hence, we have, by the order of magnitude,

$$u_\lambda \approx (\varepsilon\lambda)^{1/3} \sim \lambda^{1/3} \tag{A19}$$

(which is the**Kolmogorov–Obukhov law** relating velocities to scales). The dependence (A18) also approximately holds on the scale L, and hence,

$$\varepsilon \approx \frac{u^3}{L}, \tag{A20}$$

where u is the average velocity of pulsations, close to their maximum velocity. However, it is clear that the average velocity of the flux should not appear in this estimate (if available). In the above formulas, the quantity u_λ should be read as $u_\lambda = \langle \boldsymbol{u}_\lambda^2 \rangle^{1/2}$, because $\langle \boldsymbol{u}_\lambda \rangle = 0$ for the isotropic turbulence.

(d) The dissipative scale l is determined from the approximate equality of inertial and dissipative terms in this scale, $l^{-1}u_l^2 \approx \nu u_l l^{-2}$, which yields $l \approx \nu/u_l$. With the aid of (A19) we have

$$l \approx \left(\frac{\nu^3}{\varepsilon}\right)^{1/4}. \tag{A21}$$

The scales $\lambda < l$ constitutes the dissipative range in which the turbulence decays.

With the use of (A19) we obtain the spectral energy density (A17). We have

$$\lambda = 2\pi/k, \qquad u_\lambda^2 = \int_k^\infty \langle u_k^2 \rangle 4\pi k^2\, dk \approx 4\pi \langle u_k^2 \rangle k^3,$$

and, hence, for the inertial range we obtain

$$\langle u_k^2 \rangle = \frac{u_\lambda^2}{4\pi k^3} = \left(\frac{2\pi\varepsilon}{k}\right)^{2/3} \frac{1}{4\pi k^3}. \tag{A22}$$

Finally, we can write the energy distribution over the spectrum:

$$\frac{dE}{dk} = \frac{\langle u_k^2 \rangle}{2} 4\pi k^2 = C\varepsilon^{2/3} k^{-5/3} \tag{A23}$$

(**Kolmogorov–Obukhov law** for the spectral energy density in the inertial interval of wave numbers). Certainly, the above derivation looks like an order-of-magnitude estimate. Therefore, the dimensionless constant C must be determined experimentally. In spite of the simplicity of the model, the Kolmogorov–Obukhov law is confirmed by measured characteristics of the atmospheric and sea turbulence, and also of turbulence of molecular clouds of the Galactic disk.

Kinetic helicity. Gyrotropic turbulence. In the more complicated case in which the correlation tensor of velocities contains the unit antisymmetric pseudotensor $e_{\lambda\mu\nu}$, the simplest form of the spectral tensor is

$$\widetilde{U}_{\mu\nu}(\boldsymbol{k}, t) = A(k, t)\left(\delta_{\mu\nu} - \frac{k_\mu k_\nu}{k^2}\right) - iP(k, t) e_{\mu\nu\lambda} k_\lambda. \tag{A24}$$

It is essential here that the function $P(k,t)$ must be pseudoscalar, that is, change the sign with the inversion of the coordinate axes. The turbulence with $P(k,t) \neq 0$ is referred to as **gyrotropic** (or reflectively noninvariant). It can arise, for example, in a rotating body under the action of Coriolis forces (see Vainshtein *et al.* (1980), for detains). Let us show that the gyrotropy can be characterized by the integral parameter

$$\alpha = -\frac{1}{3}\int_0^{\infty} \langle \boldsymbol{u}(\boldsymbol{r}, t+\tau) \cdot rot\, \boldsymbol{u}(\boldsymbol{r}, t)\rangle\, d\tau = -\frac{\tau_c}{3}\langle \boldsymbol{u}(\boldsymbol{r}, 0) \cdot rot\, \boldsymbol{u}(\boldsymbol{r}, 0)\rangle. \tag{A25}$$

The last equation can be considered the definition of the correlation time τ_c. We write

$$\langle \boldsymbol{u}(\boldsymbol{r}, 0) \cdot rot\, \boldsymbol{u}(\boldsymbol{r}, 0)\rangle = e_{\lambda\mu\nu}\frac{\partial}{\partial x_{2\mu}}\langle u_\lambda(\boldsymbol{r}_1, 0) u_\nu(\boldsymbol{r}_2, 0)\rangle \Big|_{\mathbf{r}_1 \to \mathbf{r}_2}$$

$$= -i\, e_{\lambda\mu\nu}\int \frac{d^3k}{(2\pi)^3} k_\mu \widetilde{U}_{\lambda\nu}(\boldsymbol{k}, 0).$$

Substituting (A25) and using (A24), we obtain the relationship of the pseudoscalar α and the pseudoscalar function $P(k, 0)$:

$$\alpha = -\frac{\tau_c}{3\pi^2}\int_0^{\infty} k^4 P(k, 0)\, dk. \tag{A26}$$

Gyrotropic turbulence is very important because it can enhance the Large-scale magnetic field in a conducting medium. This question is considered in Problems 4.81–4.85.

Weak MHD turbulence in incompressible conducting medium with a magnetic field. An external magnetic field B, whose energy density is of the order of

$$\frac{B^2}{8\pi} \approx \frac{\tau u^2}{2}, \tag{A27}$$

can be available in a turbulent medium with the energy-contained scale L and the velocity u in this scale. This means that the Alfvén velocity is of the order of the velocity in the fundamental scale: $u \approx u_A = B/\sqrt{4\pi\tau}$. Even is the magnetic field is originally homogeneous, with time, due to the freezing-in, it acquires random character with the largest scale of inhomogeneity of the order of L. In the incompressible medium, the eigenmodes are the Alfvén waves. In the simple Alfvén wave (of small amplitude), the magnetic and kinetic energies are equal in any scale λ (see formula (19) from Example 4.10):

$$\frac{B_\lambda^2}{8\pi} = \frac{\tau u_\lambda^2}{2}. \tag{A28}$$

The energy flux over the spectrum toward the smaller scales arises due to the nonlinear generation of small-scale modes. The generalization of the Kolmogorov–Obukhov model to the case under consideration can be obtained taking into account that in the scales $l \ll \lambda \ll L$ the energy dissipation is of little importance. Pulsation velocities are small when compared with the velocity in the fundamental scale ($u_\lambda^2 \ll u \approx u_A^2$). We write the energy flux ϵ over the spectrum in the form of expansion in the small ratio $\langle u_\lambda^2\rangle / u_A^2$:

$$\epsilon \approx a\langle u_\lambda^2\rangle / u_A^2 + b(\langle u_\lambda^2\rangle / u_A^2)^2 + \ldots \tag{A29}$$

The first term, proportional to the square of the velocity, can represent only the flux of the dissipated energy. Neglecting the dissipation, we have $a = 0$. The coefficient b in the second term has the dimension of ϵ. It can be composed of the scale λ and the Alfvén velocity u_A, which is the same for motions of all scales: $b = u_A^3/\beta\lambda$, β is the dimensionless coefficient. Finally, from (A28), we obtain the relations

$$\epsilon \approx \frac{\langle u_\lambda^2\rangle^2}{\beta u_A \lambda}, \qquad \langle u_\lambda^2\rangle \approx (\beta\epsilon u_A \lambda)^{1/2}, \tag{A30}$$

which determine the dependence of the turbulent velocity on λ. Going to the distribution of the turbulent energy over wave numbers, as in the derivation of Equation (A22), we find

$$\frac{dE}{dk} \approx C(\varepsilon u_A)^{1/2} k^{-3/2} \tag{A31}$$

Here C is a dimensionless constant. In view of (A28), the energy spectrum of the turbulent magnetic field resembles the spectrum of kinetic energy (A31).

On the dissipative scale l, the conversion of mechanical and magnetic energies to heat begins to play the main role. This process is described by the first term in the expansion (A29). This term, with account for the dimensional factor, can be written in the form $\epsilon \approx u_A\langle u_l^2\rangle / l$. On the other hand, the dissipated energy consists of the absorbed mechanical energy per unit mass $\nu\langle u_l^2\rangle / l^2$ and the analogous quantity related to the magnetic field: $\nu_m\langle Inl^2\rangle / 4\pi\tau l^2 \approx \nu_m\langle u_l^2\rangle / l^2$. Finally, we have $\epsilon \approx (\nu + \nu_m)\langle u_l^2\rangle / l^2$ and, by comparing the two relations for ϵ given above, we have

$$l \approx \frac{\nu + \nu_m}{u_A}. \tag{A32}$$

The theory of turbulence on the basis of correlation tensors has been suggested by Keller und Friedman (1924).[7] A developed turbulence can be considered a result of self-organization in a strongly nonequilibrium and nonlinear open system. Corresponding ideas and methods are described, for instance, by Zelenyi and Milovanov (2004).

Recommended literature: Monin and Yaglom (1965, 1967); Frisch (1998); Frik (2003); Frost and Moulden (1977); Rytov *et al.* (1978); Lichtenberg and Lieberman (1983); Vainshtein *et al.* (1993)

7) Friedman Aleksandr Aleksandrovich (1888–1925), outstanding Soviet physicist and mathematician; theoretically introduced nonstationary (expanding) Universe.

Bibliography

Ablowitz, M.J. and Segur, H. (1981) *Solitons and the Inverse Scattering Transform*, SIAM, Philadelphia.

Abramovitz, M. and Stegun, I.A. (eds) (1964) *Handbook of Mathematical Functions*, U.S. National Bureau of Standards; Dover, New York (1965).

Agranovich, V.M. and Gartstein, Yu.N. (2006) Spatial dispersion and negative refraction of light. *Phys. Usp.*, **49** (10), 1029–1044.

Akhieser, A.I., Akhieser, I.A., Polovin, R.V., Sitenko, A.G., and Stepanov, K.N. (1975) *Plasma Electrodynamics: Linear Theory: Non-Linear Theory and Fluctuations*, vols **1** and **2**, Pergamon Press, Oxford.

Akhieser, A.I. and Shul'ga, N.F. (1993) *Electrodynamics of High Energy in Matter*, Nauka, Moscow.

Akhmanov, S.A., Dyakov, Yu.E., and Chirkin, A.S. (1981) *Introduction in Statistical Radiophysics and Optics*, Nauka, Miscow.

Akhmanov, S.A. and Nikitin, S.Yu. (1998) *Physical Optics*, Publishing House of Moscow University.

Akhmanov, S.A., Vysloukh, V.A., and Chirkin, A.S. (1988) *Optics of Laser Pulses*, Nauka, Miscow.

Akhmediev, N.N. and Ankevich, A. (2003) *Solitons: Nonlinear Pulses and Beams*, Fizmatlit.

Aleksandrov, A.F., Bogdankevich, L.S., and Rukhadze, A.A. (1978) *Foundations of the Plasmas Electrodynamics*, Vyschaya Shkola, Moscow, (in Russian).

Alfvén, H. and Felthammar, C.-G. (1963) *Cosmical Electrodynamics: Fundamental Principles*, Clarendon Press, Oxford.

Amusya, M.Ya. (1990) *Bremsstralung*, Energoatomizdat, Moscow.

Amusya, M.Ya., Buimistrov, V.M., Son, B.A. *et al.* (1987) *Polarization Bremsstralung of Particles and Atoms*, Nauka, Moscow.

Antonova, E.E., Bakhareva, M.F., Lomonosov, V.N., and Tverskoy, B.A. (1988) *Accelerated Mechanisms in Cosmos*, Publishing House of Moscow University.

Barash, Yu.S. and Ginzburg, V.L. (1976) Expressions for the Energy Density and evolved heat in the electrodynamics of a dissipative and absorptive medium. *Sov. Phys. Usp.*, **19**, 263–270.

Bardeen, J. and Schrieffer, J. (1961) *Resent Developments in Superconductivity*, North-Holland Publishing Company, Amsterdam.

Baryshevskii, V.G. (1982) *Channeling, Emission and Reactions in Crystals at High Energies*, Belorusskii Gisudarstvennyi Universitet, Minsk.

Basylev, V.A. and Zhewago, N.K. (1987) *Emission of Fast Particles in Matter and in External Fields*, Nauka, Moscow, (in Russian).

Bass, F.G. and Fuks, I.M. (1972) *Scattering of Waves on Statistical Rough Surface*, Nauka, Moscow.

Berdyshev, A.A. (1992) *Introduction to Quantum Theory of Magnetics*, Publishing House of Ural University, Ekaterinburg, (in Russian).

Berestetskii, V.B., Lifshitz, E.M., and Pitaevskii, L.P. (1982) *Quantum Electrodynamics*, Pergamon Press.

Berezhko, E.G., Yolshin, V.K., Krymskii, G.F., and Petukhov, S.I. (1988) *Generation of Cosmic Rays by Shock Waves*, Nauka, Novosibirsk.

Berezinskii, V.S., Bulanov, S.V., Dogiel, V.A., and Ptuskin, V.S. (1990) in *Astrophysics of Cosmic Rays* (ed. V.L. Ginzburg), North-Holland, Amsterdam.

Blinc, R. and Zeks, B. (1974) *Soft Modes in Ferroelectrics and Antiferroelectrics*, North-Holland Publishing Company. Amsterdam.

Bliokh, K.Yu. and Bliokh, Yu.P. (2004) What are the left-handed media and what is interesting about them? *Phys. Usp.*, **47**, 393–400.

Bloembergen, N. (1965) *Nonlinear Optics*, W.A. Benjamin, Inc., New York, Amsterdam.

Bogolyubov, N.N. (ed.) (1960) *Theory of Superconductivity. Colliction of Articles*, IIL, Moscow, (in Russian).

Boiko, B.B. and Petrov, N.S. (1988) *Reflection of Light from Amplified and Nonlinear Media*, Nauka and Technology, Minsk.

Bolotovskii, B.M. (1985) *Oliver Heaviside*, Nauka, Moscow, (in Russian).

Bolotovskii, B.M. and Galst'yan, E.A. (2000) Diffraction and diffraction radiation. *Phys. Usp.*, **43**, 755–775.

Bolotovskii, B.M. and Serov, A.V. (2009) Features of the transition radiation field. *Phys. Usp.*, **52**, 487–493.

Bolotovskii, B.M. and Stolyarov, S.N. (1974) The up-to-date state of moving bodies electrodynamics. *Sov. Phys. Usp.*, **114** (12), 569, (in Russian).

Bolotovskii, B.M. and Stolyarov, S.N. (1983) *The Fields of Emission Sources in Moving Media. Einstein's collection 1978–1979*, Nauka, Moscow, pp. 173–277, (in Russian).

Born, M. and Huang, Kun. (1954) *Dynamical Theory of Crystal Lattices*, Clarendon Press, Oxford.

Born, M. and Wolf, E. (1964) *Foundations of Optics*, Pergamon Press, Oxford.

Braginskii, S.I. (1965) Transport processes in a plasma. *Rev. Plasma Phys.*, **1**, 205.

Bredov, M.M., Rumiantsev, V.V., and Toptygin, I.N. (2003) *Classical Electrodynamics*, Lan', Saint Petersburg, (in Russian).

Brillouin, L. and Parodi, M. (1956) *Wave Propagation in Periodic Structures*, Malson Dunod, Paris.

Burfoot, J.C. (1967) *Ferroelectrics: An Introduction to the Physical Principles* , D. Van Nostrand Company Ltd., Princeton, NJ, Toronto.

Bykov, V.P. (2006) *Laser's Electrodynamics*, Fismatlit, Moscow, (in Russian).

Bykov, A.M. and Toptygin, I.N. (1993) Particle kinetics in highly turbulent plasmas. Renormalization and self-consistent methods. *Phys. Usp.*, **36** (11), 1020–1052.

Cowly, J.M. (1975) *Diffraction Physics*, North Holland Publishing Company, Amsterdam and New York.

Davydov, A.S. (1976) *The Theory of Solid Body*, Nauka, Moscopw, (in Russian).

de Broglie, L. (1941) *Problemes de Propagations Guidees des Ondes Electromagnetiques*, Gauthier-Villars, Paris.

De Gennes, P.G. (1966) *Superconductivity of Metalls and Alloys*, W.A. Bendjamin, Inc., New York and Amsterdam.

de Groot, S.R. and Mazur, P. (1962) *Non-Equilibrium Thermodynamics*, North-Holland Publishing Company, Amsterdam.

Dodd, R.K., Eilbeck, J.C., Gibbon, J.D., and Morris, H.C. (1984) *Solitons and Nonlinear Wave Equations*, Academic Press, Inc.

Dolgov, O.V., Kirzhnits, D.A., and Losyakov, V.V. (1982) Admissible values of permittivity and magnetic permeability of matter. *JETP*, **56**, 1095.

Dolgov, O.V. and Maksimov, E.G. (1981) Effects of the local field and violation of Kramers-Kronig relations for dielectric permittivity. *Sov. Phys. Usp.*, **135**, 441–477, (in Russian).

Dolin, L.S. (1961) About possibility of comparison of three-dimensional electromagnetic systems with non-uniform anisotropic filling. *Radiofizika (Izvestiya vysshikh uchebnykh savedenii)*, **IV**, 944–967, (in Russian).

Dorofeenko, A.V. *et al.* (2012) Light propagation in composite materials with gain layers. *Phys. Usp.*, **55**, 1080–1097.

Dubinov, A.E. and Mytareva, L.A. (2010) Invisible cloacing of material bodies using the wave flow method. *Phys. Usp.*, **53**, 455–479.

Dubinov, A.E. and Mytareva, L.A. (2012) Is singularity-free cloacing possible? New cloacing ideas (a reply to comments on the previous article). *Phys. Usp.*, **55**, 315–318.

Dykhne, A.M. (1971a) Conductivity of a two-dimensional two-phase system. *JETP*, **32**,63.

Dykhne, A.M. (1971b) Anomalous plasma resistance in a strong magnetic field. *JETP*, **32**, 348.

Fedorov, F.I. (2004) *Optics of Anisotropic Media*, URSS, Moscow, (in Russian).

Feinberg, E.L. (1999) *Propagation of Radiowaves Along a Surface of the Earth*, Fizmatlit, Moscow, (in Russian).

Fermi, E. (1940) Ionization energy losses in gases and condensed media. *Phys. Rev.*, **57**, 485.

Fermi, E. (1949) On the origin of the cosmic radiation. *Phys. Rev.*, **75**, 1169.

Fermi, E. (1950) Nuclear Physics. Notes compiled by J. Orear, A.N. Rosenfeld and N.A. Shluter. Chicago.

Feynman, R.P. (1972) *Statistical Mechanics (A Set of Lectures)*, W.A.Benjamin, Inc.

Fleishman, G.D. (2008) Stochastic Theory of Emission, IKI-RCD, (in Russian).

Fleishman, G.D. and Toptygin, I.N. (2013) *Cosmic Electrodynamics*, Springer, New York.

Fock, V.A. (1962) *Theory of Space, Time and Gravitation*, Addison-Wisley.

Fokin, A.G. (1996) Macroscopic conductivity of Random inhomogeneous media. Calculation methods. *Phys. Usp.*, **39**, 1009–1032.

Frank, I.M. (1988) *Vavilov-Cherenkov's Emission: Theoretical Aspects*, Nauka, Moscow.

(a) Frenkel, Ya.I. (1926) Lehrbuch der Electrodynamik, *Allgemeine Mechanik der Elektrizität*, Bd. 1, Springer, Berlin; (b) Frenkel, Ya.I. (1935) *Electrodynamics: General Theory of Electricity. Collection of selected works*, vol. **I**, Publishing House Academy of Sciences, Leningrad-Moscow, (in Russian).

Frenkel, Ya.I. (1928) Lehrbuch der Electrodynamik, *Makroskopische Elektrodynamik der materiellen Körper*, Bd. 2., Springer, Berlin.

Frenkel, Ya.I. (1935) *Electrodynamics: Macroscopic Electrodynamics of Matter Bodies*, vol. **II**, GONTI, Leningrad-Moscow, (in Russian).

Frik, P.G. (2003) *Turbulence: Approaches and Models*, IKI, Moscow-Izhevsk, (in Russian).

Frisch, U. (1998) Turbulence, Legacy by A.N. Kolmogorov.

Frish, U. (1995) *Turbulence: The Legacy of A.N. Kolmogorov*, Cambridge University Press, Cambridge.

Frohlich, H. (1958) *Theory of Dielactrics*, Oxford at the Clarendon Press, Oxford.

Frost, W. and Moulden, T.H. (1977) *Handbook of Turbulence*, vol. **1**, Plenum Press, New York and London.

Ginzburg, V.L. (1949) Theory of ferroelectric phenomena. *Usp. Fiz. Nauk*, **38**, 390.

Ginzburg, V.L. (1973) The laws of conservation of energy and momentum in emissium of electromagnetic waves (photons) in a medium and the energy-momentum tensor in macroscopic electrodynamics. *Sov. Phys. Usp.*, **16**, 434–439.

Ginzburg, V.L. (1987) *Theoretical Physics and Astrophysics. Additional Chapters*, Nauka, Moscow, (in Russian).

Ginzburg, V.L. (2001) Phase transitions in ferroelectrics: some historical remarks. *Phys. Usp.*, **44**, 1037–1043.

Ginzburg, V.L. (2002) Some remarks on the radiation of charges and multipoles uniformly moving in a medium. *Phys. Usp.*, **45**, 341–344.

Ginzburg, V.L. and Frank, I.L. (1946) Emission in regular interval moving electron, which arises at transition from one medium in another. *Sov. Phys. JETP*, **16**, 15–28, (in Russian).

Ginzburg, V.L. and Landau, L.D. (1950) The theory of superconductivity. *JETP*, **20**, 1064, (in Russian).

Ginzburg, V.L. and Syrovatskii, S.I. (1964) *The Origin of Cosmic Rays*, Macmillan, New York.

Ginzburg, V.L. and Tsytovich, V.N. (1990) *Transition Radiation and Transition Scattering*, Hilger, Bristol.

Ginzburg, V.L. and Ugarov, V.A. (1976) Remarks on forces and the energy-momentum tensor in macroscopic electrodynamics. *Sov. Phys. Usp.*, **19**, 94–101.

Goundyrev, A.G. and Razin, V.A. (1995) On the synchrotron emission of a relativistic electron spiraling in rarefied Plasma.

I. The criteria of the plasma Influence. *Astron. Astrophys. Trans.*, **6**, 229–249.

Gradshtein, I.S. and Ryzhik, I.M. (2007) *Tables of Integrals, Series, and Products*, 4th edn, prepared by Geronimus, Yu.V. and Tseytlin, M.Yu., translation eited by Jeffrey, A., Academic Press, New York.

Grigoriev, I.S. and Meylikhov, E.Z. (eds) (1991) *Physical Quantities. Reference Book*, Energoatomizdat, Moscow.

Grinberg, G.A. (1948) *The Selected Problems of the Mathematical Theory of the Electric and Magnetic Phenomena*, Publishing House AN SSSR.

Guarmat, I. (1970) *Non-Equilibrium Thermodynamics: Field Theory and Variational Principles*, Springer-Verlag, Berlin.

Gurevich, A.G. and Melkov, G.A. (1994) *Magnetic Oscillations and Waves*, Fizmatlit, Moscow, (in Russian).

Gyarmati, I. (1970) *Non-Equilibrium Thermodynamics: Field Theory and Variational Principles*, Springer-verlag, Berlin.

Heer, C.V. (1972) *Statistical Mechanics, Kinetic Theory, and Stochastic Processes*, Academic Press, New York and London.

Heitler, W. (1954) *The Quanrum Theory of Radiation*, the Clarendon Press, Oxford.

Ignatov, A.M. and Rukhadze, A.A. (1981) Ambiguity in the definition of the magnetic permeability of material media. *Sov. Phys. Usp.*, **24**, 795–797.

Ignatovich, F.V. and Ignatovich, V.K. (2012) Optics of anisotropic media. *Phys. Usp.*, **55**, 709–720.

Iorganev, D.V. and Bondarenko, O.V. (2002) *Filament-Optic Cables and Communication Lines*, Eko-trends, Moscow.

Isimaru, A. (1978) *Wave Propagation and Scattering in Random Media: Single Scattering and Transport Theory: Multiple Scattering, Turbulence, Rough Surfaces and Remote Sensing*, vols **1 and 2**, Academic Press, New York.

Isyumov, Yu.A. *et al.* (1994) *Magnetizm of Collectivization Electrons*, Fizmatlit, Moscow, (in Russian).

Isyumov, Yu.A. and Sckryabin, Yu.N. (1994) *Statistical Mechanics of Magnetic Ordered Systems*, Nauka, Moscow.

Jackson, J.D. (1999) *Classical Electrodynamics*, John Wiley & Sons, Inc., New York.

James, R. (1950) *The Optical Principles of the Diffraction of X-Rays*, G. Bell and Sons, London.

Kaganov, M.I., Pustyl'nik, M.B., and Shalaeva, T.I. (1997) Magnons, magnetic polaritons, magnetostatic waves. *Phys. Usp.*, **40**, 181–224.

Kalashnikov, N.P. (1981) *Coherent Interactions of Charged Particles in Monocrystals*, Atomizdat, Moscow.

Kasantsev, V.P. (2002) An example illustrating the potentiality and pecularities of a variation approach to electrostatic problems. *Phys. Usp.*, **45**, 325–330.

Keller, L.V. and Friedman, A.A. (1924) Differentialgleihung für die turbulente Bewegung einer kompressiblen Flüssigkeit. Proceedings of the 1st International Congress of Applied Mechanics, Delft, pp. 395–405.

Kildishev, A.V. and Shalaev, V.M. (2011) Transformation optics and metamaterials. *Phys. Usp.*, **54**, 53–63.

Kirzhnits, D.A. (1976) Are the Kramers-Kronig relations for the dielectric permeability of a material always valid? *Sov. Phys. Usp.*, **19**, 530–537.

Kirzhnits, D.A. (1987) General properties of electromagnetic response functions. *Sov. Phys. Usp.*, **30**, 575–587.

Kittel, C. (1978) *Introduction to Solid State Physics*, John Wiley & Sons, Inc.

Klyshko, D.N. (1960) *Photons and Non-Linear Optics*, Nauka, Moscow.

Klyshko, D.N. (1980) *Physical Foundations of Quantum Electronics*, Nauka, Moscow.

Kolmogorov, A.N. (1941) The local structure of turbulence in incompressible viscous fluid for very large Reynold's numbers. *Akad. Nauk SSSR Dokl.*, **30**, 301–305.

Kolpakov, A.V., Bushuev, V.A., and Kuz'min, R.N. (1978) Dielectric permittivity in the x-ray region. *Sov. Phys. Usp.*, **21**, 959–977.

Korol', A.V., Lyalin, A.G., and Solovyov, A.V. (2004) *Polarization Bremsstrahlung*, Publisher House of SPbGPU.

Krause, F. and Rädler, K.-H. (1980) *Mean-Field Magnetohydrodynamics and Dynamo Therapy*, Academie-Verlag, Berlin.

Kresin, V.Z. (1978) *Superconductivity and Superfluidity*, Nauka, Moscow, (in Russian).

Krivoglaz, M.A. (1967) *The Theory of X-Ray and Heat Neutrons Scattering by Real Crystalls*, Nauka, Moscow, (in Russian).

Kubo, R. (1965a) *Statistical Mechanics*, North-Holland Publishing Company, Amsterdam.

Kubo, R. (1965b) *Thermodynamics*, North-Holland Publishing Company, Amsterdam.

Kubo, R., Toda, M., and Hashitsume, N. (1985) *Statistical Physics II. Nonequilibrium Statistical Mechanics*, Springer-Verlag, Berlin.

Kulikovskii, A.G. and Lyubimov, G.A. (1962) *Magnetic Hydrodynamics*, GIFML, Moscow.

Kumakhov, M.A. and Kumarov, F.F. (1985).

Kumakhov, M.A. (1986) *Emission of Channeling Particles in Crystals*, Atomizdat, Moscow, (in Russian).

Kumakhov, M.A. and Shirmer, G. (1980) *Atomic Scatterings in Crystals*, Atomizdat, Moscow, (in Russian).

Kuz'min, V.L. and Romanov, V.P. (1996) Coherent phenomena in light scattering from disorder systems. *Phys. Usp.*, **39**, 231–260.

Landau, L.D. (1946) Oscillations of the electron plasma. *Sov. Phys. JETP*, **16**, 574, (in Russian).

Landau, L.D. and Lifshitz, E.M. (1975) *The Classical Theory of Fields*, 4th edn, Pergamon Press.

Landau, L.D. and Lifshitz, E.M. (1977) *Quantum Mechanics—Non-Relativistic Theory*, 3rd edn, Pergamon Press, Oxford.

Landau, L.D. and Lifshitz, E.M. (1980) *Statistical Physics. Part 1*, 3rd edn, Butterworth-Heineman.

Landau, L.D., Lifshitz, E.M. and Pitaevskii, L.P. (1984) *Electrodynamics of Continuous Media*, Butterworth-Heinemann.

Landau, L.D. and Lifshitz, E.M. (1987) *Fluid Mechanics*, Butterworth-Heineman.

Landau, L.D. and Pomeranchuk, I.Ya. (1953) *Dokl. AN SSSR*, **92**, 735.

Leontovich, M.A. (1961) Generalization of Kramers-Kronig relations for media with space dispersion. *Sov. Phys. JETP*, **40**, 907.

Leontovich, M.A. (1983) *An Introduction to Thermodynamics. Statistical Physics*, Nauka, Moscov, (in Russian).

Levin, M.L. and Rytov, S.M. (1967) *The Theory of Equilibrium Heat Fluctuations in Electrodynamics*, Nauka, Moscow, (in Russian).

Lichtenberg, A.J. and Lieberman, M.A. (1983) *Regular and Stochastic Motion*, Springer-Verlag, New York and Heidelberg Berlin.

Lifshitz, E.M. and Pitaevskii, L.P. (1980) *Statistical Physics. Part 2*, Butterworth-Heineman.

Lindhard, J. (1954) On the properties of gas of charged particles. *Det. Kong. Danske Vid. Selskab. Dan. Mat. Fys. Med.*, **28** (8), 1–57.

Lines, M. and Glass, A. (1980) *Principles and Application of Ferroelectrics and Related Materials*, Clarendon Press, Oxford.

Loudon, R. (1973) *The Quantum Theory of Light*, Clarendon Press, Oxford.

Mandel, L. and Wolf, E. (1995) *Optical Coherence and Quantum Optics*, Cambridge University Press, Cambridge, London.

Martinson, M.L. and Nedospasov, A.V. (1993) On the charge density inside a conductor carrying a current. *Phys. Usp.*, **36** (1), 23–24.

Maxwell, J.C. (1891) *Treatise on Electrisity and Magnetism*, 3rd edn, vol. **2**, reprint by Dover, New York (1954).

McConnell, J. (1960) *Quantum Particles Dynamics*. North-Holland Publishing Company, Amsterdam.

Migdal, A.B. (1956) Bremsstrahlung and pair production in condensed media at high energies. *Phys. Rev.*, **103**, 1811–1820.

Mikhailovskii, A.B. (1975) *Theory of Plasmas Instability: Nonstabilities of Homogeneous Plasma*, vol. **1**, Atomizdat, Moscow, (in Russian).

Moffatt, H. (1978) *Magnetic Field Generation in Electrically Conducting Fluids*, Cambridge University Press, London.

Monin, A.S. and Yaglom, A.M. (1965) *Statistical Hydromechanics. Part 1*, Nauka, Moscow, (in Russian).

Monin, A.S. and Yaglom, A.M. (1967) *Statistical Hydromechanics. Part 2*, Nauka, Moscow, (in Russian).

Morse, Ph.M. and Feshbach, H. (1953) *Methods of Theoretical Physics, parts 1 and 2.* McGraw Hill Book Company, Inc. New York, Toronto, London.

Mursin, V.S. (2007) *Astrophysics of Cosmic Rays*, Publisher House of Moscow University, Moscow.

Nelson, D.L., Whittingham, M.S., and George, T.F. (eds) (1987) *Chemistry of High-Temperature Superconducters*, American Chemical Society, Washington, DC.

Newell, A.C. (1985) *Solitons in Mathematics and Physics*, Society for Industrial and Applied Mathematics.

Newton, R.G. (1969) *Scattering Theory of Waves and Particles*, McGraw-Hill Book Company, New York.

Nezlin, M.V. (1976) Negative energy waves and the anomalous Doppler effect. *Sov. Phys. Usp.*, **19**, 946–954.

Nikol'skii, V.V. (1973) *Electrodynamics and Propagation of Radio Waves*, Nauka, Moscow, (in Russian).

Niz'ev, V.G. (2002) Dipole-wave theory of electromagnetic diffraction. *Phys. Usp.*, **45**, 553–559.

Novokshenov, V.Yu. (2002) *Introduction in the Theory of Solitons*, IKI, Moscow-izhevsk.

Nussenzveig, H.M. (1972) *Causality and Dispersion Relations*, Academic Press, New York and London.

Obukhov, A.M. (1941) On spectral energy distribution in a turbulent flow. *Izv. AN SSSR: Seriya Geogr. Geophys.*, **5**, 453–456.

Oraevskii, A.N. (1998) Superluminal waves in amplifying media. *Phys. Usp.*, **41**, 1199–1209.

Ostrovskii, L.A. and Potapov, A.I. (1999) *Modulated Waves. Theory and Applications*, The Johns Hopkins University Press, Baltimore, MD and London.

Pake, G. (1962) *Paramagnetic Resonance*, W.A. Benjamin, Inc., New York.

Pamyatnykh, E.A. and Turov, E.A. (2000) *Foundations of Media Electrodynamics in Variable and Non-Homogeneous Fields*, Nauka, Moscow, (in Russian).

Panofsky, W. and Phillips, M. (1963) *Classical Electricity and Magnetism*, Addison-Wesley Publishibg Company, Inc., Cambridge, MA.

Parker, E. (1979) *Cosmical Magnetic Fields*, Clarendon Press, Oxford.

Peierls, R. (1979) *Surprises in Theoretical Physics*, Princeton University Press, Princeton, NJ.

Pendry, J.B. *et al.* (2006) Controlling electromagnetic fields. *Science*, **312**, 1780–1782.

Pikelner, S.B. (1966) *Foundations of Cosmic Electrodynamics*, Nauka, Moscow, (in Russian).

Pinsker, Z.G. (1974) *Dynamic Theory of X-Ray Scattering in crystalls*, Nauka, Moscow, (in Russian).

Pitaevskii, L.P. and Lifshitz, E.M. (1980) *Statistical Physics. Part 2*, Butterworth-Heineman.

Platonov, K.Yu. and Fleishman, G.D. (2002) Transition radiation in media with random inhomogeneities. *Phys. Usp.*, **45**, 235–291.

Polovin, R.V. and Demutskii, V.P. (1987) *Foundations of Magnetic Hydrodynamics*, Energoatomisdat, Moscow, (in Russian).

Potekhin, V.A. and Tatarinov, V.N. (1978) *The Theory of Coherence of Electromagnetic Field*, Svyas, Moscow.

Prokhorov, A.M. (Cheif-ed.) (1984) *Physical Encyclopaedic Dictionary*, "Soviet Encyclopaedia", Moscow.

Razin, V.A. (1960) Theory of radioemission spectra from discret sourses. *Izv. Vysshikh Uchebnykh Zavedenii, Radiofisika*, **3**, 584–594, (in Russian).

Rose-Innes, A.C. and Rhoderick, E.H. (1969) *Introduction to Superconductivity*, Pergamon Press, Oxford.

Rossi, B. (1952) *High Energy Particles*, Prentice-Hall, Englewood Cliffs, NJ and New York.

Rozanov, N.N. (2011) Can even monochromatic radiation ensure ideal invisibility? *Phys. Usp.*, **54**, 763–764.

Ruzmaikin, A.A., Sokolov, D.D., and Shukurov, A.M. (1988) *Magnetic Fields of Galaxies: Astrophysics and Space Physics Library*, vol. **133**, Kluwer Academic Publishers.

Ryskin, N.M. and Trubetskov, D.I. (2000) *Nonlinear Waves*, Fizmatlit, Moscow, (in Russian).

Rytov, S.M. (1976) *An Introduction to Statistical Radiophysics, Part I. Stochastic Processes*, Nauka, Moscow, (in Russian).

Rytov, S.M., Kravtsov, Yu.A., and Tatarskii, V.I. (1978) *An Introduction to Statistical Radiophysics. Part II. Stochastic Fields*, Nauka, Moscow, (in Russian).

Schmidt, V.V. (2000) *The Introduction to Superconductor Physics*, NCNMO, Moscow, (in Russian).

Schrieffer, J.R. (1964) *Theory of Superconductivity*, Publishing House of Pencilvany, Philadelphia.

Schurig, D. *et al.* (2006) Metamaterial electromagnetic cloak at microwave frequencies. *Science*, **314**, 977–980.

Scully, M.O. and Zubairy, M.S. (1997) *Quantum Optics*, Cambridge University Press.

Sedov, L.I. (1972) *Methods of Similarity and Dimensions in Mechanics*, Nauka, Moscow, (in Russian).

Sedov, L.I. (1987) *Similarity and Dimension Methods in Mechanics*, Nauka, Moscow, (in Russian).

Shalashov, A.C. and Gospodchikov, E.D. (2011) Impedance technique for modeling of electromagnetic wave propagation in anisotropic and gyrotropic media. *Phys. Usp.*, **54** (2), 145–165.

Shevchenko, V.V. (2011) Localization of a stationary electromagnetic field by a planar boundary of a metamaterial. *Phys. Usp.*, **54**, 1131–1142.

Shvartsburg, A.B. (1998) Single cycle wave forms and non-periodic waves in dispersive media (exactly solvable models). *Phys. Usp.*, **41**, 77–94.

Shvartsburg, A.B. (2000) Dispersion of electromagnetic waves in stratified and nonstationary media (exactly solvable models). *Phys. Usp.*, **43**, 1201–1228.

Silin, V.P. and Rukhadze, A.A. (1961) *Electromagnetic Properties of Plasma and Plasma-Like Media*, Gosatomisdat, Moscow, (in Russian).

Simonyi, K. (1956) *Theoretische Elektrotechnik*, Veb Deutscher Verlag der Wissenschaften, Berlin.

Sivukhin, D.V. (1977) *Overall Course of Physics: Electricity*, vol. **III**, Nauka, Moscow, (in Russian).

Skobel'tsyn, D.V. (1973) The momentum-energy tensor of the electromagnetic field. *Sov. Phys. Usp.*, **16**, 381–401.

Smythe, W.R. (1950) *Static and Dynamic Electricity*, McGraw-Hill, New York.

Solimeno, S., Crosignani, B., and DiPorto, P. (1986) *Guiding, Diffraction, and Confinement of Optical Radiation*, Academic Press.

Sommerfeld, A. (1952) *Electrodynamics*, Academic Press, New York.

Sommerfeld, A. (1954) *Optics*, Academic Press, New York.

Soroko, L.M. (1971) *The Foundations of the Holography and Coherent Optics*, Nauka, (in Russian).

Stratton, J.A. (1941) *Electromagnetic Theory*, The IEEE Press Series on Electromagnetic Wave Theory, John Wiley & Sons, Inc.

Stroke, G.W. (1966) *An Introduction to Coherent Optics and Holography*, Academic Press.

Strukov, B.A. and Levaniuk, A.P. (1995) *Physical Foundations of Ferroelectric Phenomena in Crystals*, Fismatlit, Moscow.

Svelto, O. (1982) *Principles of Lasers*, Plenum Press, New York and London.

Syrovatskii, S.I. (1957) Magnetic hydrodynamics. *Sov. Phys. Usp.*, **62** (3), 247, (in Russian).

Tamm, I.E. (1976) *Foundations of Electrisity Theory*, Nauka, Moscow, (in Russian).

Tamm, I.E. and Frank, I.M. (1937) Coherent emission of a fast electron in medium. *DAN SSSR*, **14**, 107, (in Russian).

Tatarskii, V.I. (1967) *Propagation of Waves in Turbulent Atmosphere*, Nauka, Moscow, (in Russian).

Ter-Mikhaelyan, M.L. (1972) *The Effect of a Media on Electromagnetic Processes at High Energies*, Publishing House of Armenia, Erevan.

Ter-Mikhaelyan, M.L. (1997) Simple atomic systems in resonant laser fields. *Phys. Usp.*, **40**, 1239–1248.

Ter-Mikhaelyan, M.L. (2001) Electromagnetic radiative processes in periodic media at high energies. *Phys. Usp.*, **44**, 571–596.

Ter-Mikhaelyan, M.L. (2003) High energy electromagnetic processes in amorphous and inhomogeneous media. *Phys. Usp.*, **46**, 1231–1252.

Ternovskii, F.F. (1960) Effect of repeated scattering on own field of the fast charged particle. *Sov. Phys. JETP*, **39**, 491–496, (in Russian).

Thompson, M.W. (1968) The channeling of particles in crystals. *Contemp. Phys.*, **9** (4), 375.

Tilley, D.R. and Tilley, J. (1974) *Superfluidity and Superconductivity*, Van Nostrand Reinhold Company, New York.

Toda, M., Kubo, R., and Saito, N. (1983) *Statistical Physics I. Equilibrium Statistical Mechanics*, Springer-Verlag, Berlin.

Toptygin, I.N. (1964) To the theory of bremsstrahlung and pair production in medium. *Sov. Phys. JETP*, **46**, 851–862.

Toptygin, I.N. (1985) *Cosmic Rays in Interplanetary Magnetic Fields*, D. Reidel Publishing Company, Dordrecht.

Toptygin, I.N. (2014) *Foundations of Classical and Quantum Electrodynamics*, Wiley-VCH Verlag GmbH.

Toptygin, I.N. and Fleishman, G.D. (2008) Eigenmode generation by a given current in anisotropic and gyrotropic media. *Phys. Usp.*, **51**, 363–374.

Tsytovich, V.N. (1951) Synchrotron radiation in medium. *Vestn. Mosk. Univ. Ser. Phys.*, **11**, 27.

Tverskoy, B.A. (2004) *Foundations of Theoretical Cosmophysics. Selected Works*, URSS, Moscow.

Vaganov, R.B. and Katsenelenbaum, B.Z. (1982) *Foundation of the Diffraction Theory*, Nauka, Moscow.

Vainshtein, L.A. (1966a) *Open Resonators and Open Waveguides,–Sovietskoe Radio*, Moscow, (in Russian).

Vainshtein, L.A. (1966b) *Theory of Diffraction and Factorisation Approach*, Sovietskoe Radio, Moscow, (in Russian).

Vainshtein, L.A. (1988) *Electromagnetic Waves*, Radio i Svyas, (in Russian).

Vainshtein, S.I., Zel'dovich, Ya.B., and Rusmaikin, A.A. (1980) *Turbulent Dynamo in Astrophysics*, Nauka, Moscow, (in Russian).

Vainshtein, L.A. and Vakman, D.E. (1983) *Separation of Frequencies in Theory of Oscillations and Waves*, Nauka, Moscow, (in Russian).

Vainshtein, S.I., Bykov, A.M., and Toptygin, I.N. (1993) *Turbulence, Current Sheets and Shocks in Cosmic Plasma*, Gordon and Breach.

Vinogradov, A.I. (2001) *Electrodynamics of Composit Materials*, URSS, Moscow, (in Russian).

Vinogradov, A.I. (2002) On the form of constitutive equations in electrodynamics. *Phys. Usp.*, **45**, 331–338.

Vinogradova, M.B., Rudenko, O.V., and Sukhorukov, A.P. (1979) *Theory of Waves*, Nauka, Moscow, (in Russian).

Vlasov, A.A. (1955) *Macroscopic Electrodynamics*, GITTL, Moscow, (in Russian).

Vuks, V.G. (1973) *Introduction to Microscopic Theory of Ferroelectrics*, Nauka, Moscow, (in Russian).

Walker, G.B. and Lahoz, D.G. (1975) *Nature*, **253**, 339.

White, R. (1983) *Quantum Theory of Magnetism*, Springer-Verlag, Berlin.

Whitham, G.B. (1974) *Linear and Nonlinear Waves*, John Wiley & Sons, Inc.

Zakharov, V.E., Manakov, S.V., Novikov, S.P., and Pitaevskii, L.P. (1980) *The Theory of Solitons: Method of Inverse Problem*, Nauka, Moscow.

Zavadskii, E.A. and Val'kov, V.I. (1980) *Magnetic Phase Transitions*, Naukova Dumka, Kiev.

Zel'dovich, Ya.B. (1956) Magnetic field in conducting turbulent medium in the presence of 2D motion. *Sov. Phys. JETP*, **31**, 154.

Zel'dovich, Ya.B. and Illarionov, A.F. (1972) Scattering of a strong wave by an electron in a magnetic field. *JETP*, **34**, 467.

Ziman, J.M. (1964) *Principles of the Theory of Solids*, University Press, Cambridge.

Index

a
abnormal skin-effect 372
Abraham force 319
acceleration near a shock front 641
acceleration of particles by moving medium 640
adiabatic demagnetization 147
Alfvén waves 214
Alfvén velocity 215
amplification of waves in nonequilibrium media 501
amplitude of scattering 403, 446
analytical properties of the responce function 292
angle of total polarization 378
angle of total reflection 378
anharmonic oscillator 499
anisotropic diffusion 631
antidynamo theorem 224
antiferromagnetism 144
atomic form-factor 407
averaging over the Gibbs ensemble 275
axial and planar channeling 612

b
Babinet principle 399
Bethe-Heitler cross-section 580
Bloch equation 305
Bohr – Van Leeuween theorem 14
Boltzmann distribution 7
Boltzmann equation 10
bound charge density 4
boundary conditions for the field vectors 277
Bragg – Wulf equation 408
Brillouin function 27
Brillouin's formula 302
Brown – Twiss interferometer 474
Buneman instability increment 520
Burgers equation 492

c
capacitance coefficients 40
capacitance of a capacitor 39
cascade length 580
Cauchy-Riemann conditions 50
causality principle 291
cause and consequence 292
charged particle in random magnetic field 626
Cherenkov radiation 572
Clausius – Mossotti formulas 10
cloaking of material bodies 375
"close" and "distant" collisions 566
cnoidal wave 517
coaxial line 527
coherence length, area, and volume 467
coherense (correlation) time and length 465
collision integral in the relaxation time approximation 10
complex degree of coherence 470
complex dielectric permeability 281
complex electric conductivity 281
Complex order parameter 156
complex resistance (impedance) 195
conformal mapping in electrostatics 48
conformity 49
contact potential difference 114
correlation time and correlation length of turbulence 683
Cooper pairs in superconductors 158, 162
correlation field tensors 463
cosmic rays 625
Cotton – Mouton effect 427
Curie law for paramagnetics 26

Electromagnetic Phenomena in Matter: Statistical and Quantum Approaches, First Edition. Igor N. Toptygin.

Curie temperature of ferroelectric 66
Curie–Weiss law 68, 139

d
Debye–Hückel radius 32
Debye potentials 442
density effect in magnetic bremsstrahlung 602
density of energy and Poynting vector in the dispersion medium 302
depolarization coefficients 54
developed turbulence 681
diamagnetic susceptibility 15
diamagnetics 6
dielectric and magnetic susceptibilities 6
dielectric permittivity 7
dielectric permittivity in X-range 405
dielectric susceptibility 6
diffracted X-ray radiation 619
diffraction at a random screen 487
diffraction by single crystall 407
diffusion of particles in momentum space 637
diffusion of particles in random field 626
diffusion of the magnetic field 204
4-dimensional Maxwell equations 310
discontinuities in MHD 218
dispersion equations 284, 363
dispersion measure (DM) 378
dispersion relations 294
dispersion relations in the presence of a spatial dispersion 295
dispersion relations taking into account STR 296
dissipation of electromagnetic energy in medium 297
Drude formula for electric conductivity 12

e
Eddy current 202
eigen oscillations in isotropic media 363
eigenfrequency of the oscillatory circuit 196
eigenmodes in isotropic media 386
Einstein formula for the extinction ratio 484
electric conductivity 6, 10
electric conductivity of random media 128
electric field vector 3
electric induction vector 6
electric polarization 7
electric polarizability tensor 40
electric stress tensor 65
electromagnetic potentials in moving media 311
electromotive force of induction 193
electromotive forces (EMF) 16, 115
electron paramagnetic resonance, EPR 351
electron-photon showers 581
electronic paramagnetic resonance 351
electronic plasma (Langmuir) oscillations 346
electrostatic Green reciprocity theorem 40
electrostriction 65
Elsasser's antidynamo-theorem 224
emission in crystals 615
energy losses in waveguides 524
energy of the field in dispersive media 299
energy spectrum of the accelerated particles 642
energy-momentum tensors of Abraham and Minkowski 317
entropy simple waves 213
equation
– Bragg–Wulf 408
– for electromagnetic potentials in moving media 313
– Helmholtz 442
– Laue 408
– of transfer and acceleration of particles 638
ergodicity 466
evolutionarity of MHD discontinuities 221
example of artificial dielectric 290
external charges and currents 1
external electromotive force 115
external scale of turbulence 685
extinction ratio 484

f
Faraday effect 388, 428
fast and slow simple waves 215
ferroelectrics 66
Ferromagnetic resonance 352
ferromagnetics in alternating fields 305
ferromagnetism 138
Fluctuation-dissipative theorem 482
Fokker–Planck equation 574
force-free magnetic field 223
forces in a magnetic field 137
forces in an alternating field 316
forces in an electric field 63
Fraunhofer diffraction 397
free and bound charges 5
freezing of a magnetic field into conducting medium 210

Fresnel diffraction 395
Fresnel equation for gyrotropic media 385
Fresnel equation of crystaloptics 383
Fresnel's formulas 366

g
generalized electric induction 278
generalized Ohm law 212
generalized Poynting vector 302
generalized susceptibility 479
generation of plasmons 572
geometrical optics 394
Ginzburg–Landau theory 156
Ginzburg–Landau theory of superconductors 156
Green function in moving media 314
Green function for moving media 316
Green function in electrostatics 46
Green's function in moving medium 316
group velocity 323
gyrotropic turbulence 687

h
Hall current 35
Heisenberg ferromagnetic 139
Hertz vector in moving media 313
high-temperature superconductors 149
holography 471
Huygens'–Fresnel principle 391

i
impedance 195
induction EMF 193
inhomogeneous medium 373
inhomogeneous plane waves 364, 370
integral transformation method in electrostatics 45
interference effect in scattering 609
inversion method in electrostatics 45
ion plasma oscillations 348
ion-sound oscillations 348
ionization losses 566
Ising model 142
isotopic effect 152

j
Joule–Lenz law 119
Joule losses 119

k
Kelvin–Helmholtz instability 227, 265
kinematic theory of diffraction 406
kinematic theory of X-ray diffraction 406
kinematic viscosity 209
kinematics of the collision 565
kinetic equation for a stochastic magnetic field 630
kinetic equation in a strong magnetic field 635
kinetic helicity 687
Kirchhoff integral 393
Kirchhoff laws 118
Klausius–Mossotti formulas 10
Kolmogorov–Obukhov model 684
Korteweg–de Vries (KdV) equation 495
Kramers–Kronig dispersion relations 294

l
Landé g-factor 27
Landau damping 287
Landau diamagnetism 30
Landau–Lifshitz equation 306
Landau–Pomeranchuk effect 582
Langevin function 8
law of refraction and reflection 366
"left-handed" media 368
length of formation of radiation 578
length of radiation formation 579
Leontovich boundary condition 371
light pressure and "light attraction" 369
linear responce function 280
local homogeneity and isotropy of turbulent motion 685
longitudinal and transverse waves 284
longitudinal length of the quantum formation 579
longitudinal oscillations in gases and solid bodies 303
Lorenz–Lorentz formulas 10

m
magnetic field amplification 272
magnetic field diffusion 204
magnetic field strength vector 6
magnetic flux quantization 152
magnetic induction vector 3
magnetic permeability 7
magnetic polaritons 410, 425
magnetic polarizability tensor 204
magnetic polarization 10
magnetic pressure 210
magnetic Reynolds number 209
magnetic stress tensor 138
magnetic susceptibility 6
magnetic tensions 210
magnetic viscosity 209
magnetization current 277

magnetoacoustic waves 215
magnetostatics of superconductors 153
magnetostriction 148
magnons (spin waves) 351
material coupling equations 278
Maxwellian distribution 7
Meissner–Ochsenfeld effect 150
metallic waveguides 523
metamaterials 367
method of equivalent photons 576
method of slowly varying amplitude 405
MHD shock wave 219
MHD turbulence 687
Michelson interferometer 466
Michelson stellar interferometer 473
Mie's problem (scattering by conducting sphere) 402
Migdal theory of multiple scattering 582
Minkowski coupling equations 310, 311
momentum flux in dispersive medium 325
multiple scattering 573
mutual coherence function 469

n

negative refraction 367
Nernst principle 147
nonlinear responce 497
nonlinear Schrödinger equation 497
nonreciprocity in ferrodielectrics 525
nuclear magnetic resonance (NMR) 351

o

Ohm law 6
Ohm law with external EMF 115
Onsager principle 120
open resonators 534
optical axes 383
optical theorem 403, 452
ordinary and extraordinary waves 387
overturning of the simple wave 259

p

packet of waves 323
paramagnetic susceptibility 14
Paramagnetics 6
paramagnetics in alternating fields 304
particle acceleration in moving random media 637
Pauli paramagnetism 30
Peltier coefficient 170
Peltier effect 128
phase and group velocities in waveguides 526
photon in waveguide 526
piezoelectric effect 66
plasma dielectric permeabilities 287
plasmons 348
polaritons 410
polaritons in a dielectric 410
polarizability tensor 7, 40
polarization and magnetization currents 276
polarization bremsstrahlung 597
potential coefficients 39
Poynting vector (generalized) 301
pressure of light 379
principle of symmetry of kinetic coefficients 121
propagation velocity of energy and momentum 324
pseudoscalar magnetic potential 130
pyroelectrics 66

q

quality factor of the resonator 531
quantization of the magnetic flux 152
quantum theory of dispersion 289
quasiclassical dispersion model 289

r

radiation (or cascade) length 581
radiation energy losses 578
radiation region 579
radiation spectra of channeled and over-barrier particles 622
Rayleigh formula 512
Rayleigh–Taylor instability 226
Razin–Tsytovich effect 602
reduced equation 399
reflection and refraction on the boundary 364
regions of transparency 298
regular and stochastic acceleration 625
relaxation time approximation 10
resistance coefficients 117
Reynolds number 209
rotation measure 389

s

scattering amplitude 403
scattering by random inhomogeneities 485
scattering in an equilibrium medium 482
self-focusing 500
shock adiabatic equation 219
shock wave
– oscillatory structure 518
– in MHD 219

simple Riemann waves 213
– in MHD 212
skin-effect 201
skin-layer depth 203
slowly varying amplitude method 399
Snell's laws 366
soliton 495, 516
spectral distribution of turbulence 686
spectrum of magnons 351
spin waves 432
spontaneous polarization 66
stability of MHD discontinuities 221
statistical ensemble averaging 276
stress tensor in dispersive medium 325, 326
stress tensor in electrostatics 65
strong MHD discontinuities 216
surface impedance 372
surface waves 521

t
4-tensor of electromagnetic field 309
4-tensor of electromagnetic induction 309
tensor of magnetic polarizability 204
tensor of the complex electric conductivity 281
4-tensors of electromagnetic fields and electromagnetic induction 309
Ter–Mikaelyan effect 582, 588
thermal emf 128
thermodynamic functions of dielectrics 56
thermodynamic functions of magnetics 133
thermodynamics of superconductors 152
theta-pinch 223
Thomson formula 196
Thomson's effect 127
three-halves power law 170
time and space dispersion 280
Tonks effect 226
transient radiation 594
transmission function 487
transparency regions 298
transport phenomena 120
transverse diffusion in a strong magnetic field 633
triangular wave 515
turbulent magnetic viscosity 268
turbulent motion 680
two-fluid plasma model 211
two-liquid magnetic hydrodynamics 211

u
uniqueness of solution of electrostatic problem 38

v
Vavilov–Cherenkov radiation 572
vectors electric field and magnetic induction 4
vectors of electric induction and magnetic field 6
vectors of magnetic and electric polarizations 4
Verdet constant 428
viscous stress tensor 209
visibility 470
visibility (sharpness, contrast range) 470
von Laue equation 408

w
walk of magnetic force lines 634
waves amplification in nonequilibrium media 501
waveguide with filling 525
waveguides 521
work function 76

y
Young experiment 467

z
Zel'dovich antidynamo-theorem 224
z-pinch 223